ATMOSPHERIC CHEMISTRY AND PHYSICS

From Air Pollution to Climate Change

SECOND EDITION

John H. Seinfeld
California Institute of Technology

Spyros N. Pandis
*University of Patras and
Carnegie Mellon University*

A WILEY-INTERSCIENCE PUBLICATION

JOHN WILEY & SONS, INC.

1804945540

Published by John Wiley & Sons, Inc., Hoboken, New Jersey
Published simultaneously in Canada

For general information on our other products and services or for technical support, please contact our
Customer Care Department within the United States at (800) 762-2974, outside the United States at
(317) 572-3993 or fax (317) 572-4002.

Wiley also publishes its books in a variety of electronic formats. Some content that appears in print may
not be available in electronic formats. For more information about Wiley products, visit our web site at
www.wiley.com.

Library of Congress Cataloging-in-Publication Data:

Seinfeld, John H.
    Atmospheric chemistry and physics : from air pollution to climate change / John H. Seinfeld,
    Spyros N. Pandis.– 2nd ed.
        p. cm.
    "A Wiley-Interscience publication."
    Includes bibliographical references and index.
    ISBN-13: 978-0-471-72017-1 (cloth)
    ISBN-10: 0-471-72017-8 (cloth)
    ISBN-13: 978-0-471-72018-8 (paper)
    ISBN-10: 0-471-72018-6 (paper)
    1. Atmospheric chemistry. 2. Air–Pollution–Environmental aspects. 3. Environmental chemistry.
    I. Pandis, Spyros N., 1963- II. Title.

    QC879.6.S45 2006
    660.6'2–dc22   2005056954

Printed in the United States of America
10 9 8 7 6 5 4 3 2 1

To
Benjamin and Elizabeth
and
Angeliki and Nikos

PREFACE TO THE SECOND EDITION

Two considerations motivated us to undertake the Second Edition of this book. First, a number of important developments have occurred in atmospheric science since 1998, the year of the First Edition, and we wanted to update the treatments in several areas of the book to reflect these advances in understanding of atmospheric processes. New chapters have been added on chemical kinetics, atmospheric radiation and photochemistry, global circulation of the atmosphere, and global biogeochemical cycles. The chapters on stratospheric and tropospheric chemistry and organic atmospheric aerosols have been revised to reflect the current state of understanding in this area. The second consideration relates to the style of the book. Our goal in the First Edition was, and continues to be, in the Second Edition, both rigor and thoroughness. The First Edition has been widely used as a course textbook and reference text worldwide; feedback we have received from instructors and students is that additional examples would aid in illustrating the basic theory. The Second Edition contains numerous examples, delineated by vertical bars offsetting the material. In order to prevent an already lengthy book from becoming unwieldy with the new additions, some advanced material from the First Edition, generally of interest to specialists, has been omitted. Problems at the end of the chapters have been thoroughly reconsidered and updated. While many of the problems from the First Edition have been retained, in a number of chapters substantially new problems have been added. These problems have been used in courses at Caltech (California Institute of Technology), Carnegie Mellon, and the University of Patras.

Many colleagues have provided important material, as well as proofreading suggestions. Special appreciation is extended to Wei-Ting Chen, Cliff Davidson, Theodore Dibble, Mark Lawrence, Sally Ng, Tracey Rissman, Ross Salawitch, Charles Stanier, Jason Surratt, Satoshi Takahama, Varuntida Varutbangkul, Paul Wennberg, and Yang Zhang. Finally, Ann Hilgenfeldt and Yvette Grant skillfully prepared the manuscript for the Second Edition.

John H. Seinfeld

Spyros N. Pandis

PREFACE TO THE FIRST EDITION

The study of atmospheric chemistry as a scientific discipline goes back to the eighteenth century, when the principal issue was identifying the major chemical components of the atmosphere, nitrogen, oxygen, water, carbon dioxide, and the noble gases. In the late nineteenth and early twentieth centuries attention turned to the so-called trace gases, species present at less than 1 part per million parts of air by volume (1 μmol per mole). We now know that the atmostphere contains a myriad of trace species, some at levels as low as 1 part per trillion parts of air. The role of trace species is disproportionate to their atmospheric abundance; they are responsible for phenomena ranging from urban photochemical smog, to acid deposition, to stratospheric ozone depletion, to potential climate change. Moreover, the composition of the atmosphere is changing; analysis of air trapped in ice cores reveals a record of striking increases in the long-lived so-called greenhouse gases, carbon dioxide (CO_2), methane (CH_4), and nitrous oxide (N_2O). Within the last century, concentrations of tropospheric ozone (O_3), sulfate (SO_4^{2-}), and carbonaceous aerosols in the Northern Hemisphere have increased significantly. There is evidence that all these changes are altering the basic chemistry of the atmosphere.

Atmospheric chemistry occurs within a fabric of profoundly complicated atmospheric dynamics. The results of this coupling of dynamics and chemistry are often unexpected. Witness the unique combination of dynamical forces that lead to a wintertime polar vortex over Antarctica, with the concomitant formation of polar stratospheric clouds that serve as sites for heterogeneous chemical reactions involving chlorine compounds resulting from anthropogenic chlorofluorocarbons—all leading to the near total depletion of stratospheric ozone over the South Pole each spring; witness the nonlinear, and counterintuitive, dependence of the amount of ozone generated by reactions involving hydrocarbons and oxides of nitrogen (NO_x) at the urban and regional scale—although both hydrocarbons and NO_x are ozone precursors, situations exist where continuous emission of more and more NO_x actually leads to less ozone.

The chemical constituents of the atmosphere do not go through their life cycles independently; the cycles of the various species are linked together in a complex way. Thus a perturbation of one component can lead to significant, and nonlinear, changes to other components and to feedbacks that can amplify or damp the original perturbation.

In many respects, at once both the most important and the most paradoxical trace gas in the atmosphere is ozone (O_3). High in the stratosphere, ozone screens living organisms from biologically harmful solar ultraviolet radiation; ozone at the surface, in the troposphere, can produce adverse effects on human health and plants when present at levels elevated above natural. At the urban and regional scale, significant policy issues concern how to decrease ozone levels by controlling the ozone precursors—hydrocarbons and oxides of nitrogen. At the global scale, understanding both the natural ozone chemistry of the troposphere and the causes of continually increasing background troposheric ozone levels is a major goal.

Aerosols are particles suspended in the atmosphere. They arise directly from emissions of particles and from the conversion of certain gases to particles in the atmosphere. At elevated levels they inhibit visibility and are a human health hazard. There is a growing body of epidemiological data suggesting that increasing levels of aerosols may cause a significant increase in human mortality. For many years it was thought that atmospheric aerosols did not interact in any appreciable way with the cycles of trace gases. We now know that particles in the air affect climate and interact chemically in heretofore unrecognized ways with atmospheric gases. Volcanic aerosols in the stratosphere, for example, participate in the catalytic destruction of ozone by chlorine compounds, not directly, but through the intermediary of NO_x chemistry. Aerosols reflect solar radiation back to space and, in so doing, cool the Earth. Aerosols are also the nuclei around which clouds droplets form—no aerosols, no clouds. Clouds are one of the most important elements or our climate system, so the effect of increasing global aerosol levels on the Earth's cloudiness is a key problem in climate.

Historically the study of urban air pollution and its effects occurred more or less separately from that of the chemistry of the Earth's atmosphere as a whole. Similarly, in its early stages, climate research focused exclusively on CO_2, without reference to effects on the underlying chemistry of the atmosphere and their feedbacks on climate itself. It is now recognized, in quantitative scientific terms, that the Earth's atmosphere is a continuum of spatial scales in which the urban atmosphere, the remote troposphere, the marine boundary layer, and the stratosphere are merely points from the smallest turbulent eddies and the fastest timescales of free-redical chemistry to global circulations and the decadal timescales of the longest-lived trace gases.

The object of this book is to provide a rigorous, comprehensive treatment of the chemistry of the atmosphere, including the formation, growth, dynamics, and propeties of aerosols; the meteorology of air pollution; the transport, diffusion, and removal of species in the atmosphere; the formation and chemistry of clouds; the interaction of atmospheric chemistry and climate; the radiative and climatic effects of gases and particles; and the formulation of mathematical chemical/transport models of the atmosphere. Each of these elements is covered in detail in the present volume. In each area the central results are developed from first principles. In this way, the reader will gain a significant understanding of the science underlying the description of atmospheric processes and will be able to extend theories and results beyond those for which we have space here.

The book assumes that the reader has had introductory courses in thermodynamics, transport phenomena (fluid mechanics and/or heat and mass transfer), and engineering mathematics (differential equations). Thus the treatment is aimed at the senior or first-year graduate level in typical engineering curricula as well as in meterology and atmospheric science programs.

The book is intended to serve as a textbook for a course in atmospheric science that might vary in length from one quarter or semester to a full academic year. Aside from its use as a course textbook, the book will serve as a comprehensive reference book for professionals as well as for those from traditional engineering and scienc disciplines. Two types of appendixes are given: those of a general nature appear at the end of the book and are designated by letters; those of a nature specific to a certain chapter appear with that chapter and are numbered according to the associated chapter.

Numerous problems are provided to enable the reader to evaluate his or her understanding of the material. In many cases the problems have been chosen to extend the results given in the chapter to new situations. The problems are coded with a "degree of difficulty" for the benefit of the student and the instructor. The subscript designation "A" (e.g., 1.1_A in the Problems section of Chapter 1) indicates a problem that involves a straightforward application of material in the text. Those problems denoted "B" require some extension of the ideas in the text. Problems designated "C" encourage the reader to apply concepts from the book to current problems in atmospheric science and go somewhat beyond the level of "B" problems. Finally, those problems denoted "D" are of a degree of difficulty corresponding to "C" but generally require development of a computer program for their solution.

JOHN H. SEINFELD

SPYROS N. PANDIS

CONTENTS

1 The Atmosphere

1.1 HISTORY AND EVOLUTION OF THE EARTH'S ATMOSPHERE

It is generally believed that the solar system condensed out of an interstellar cloud of gas and dust, referred to as the "primordial solar nebula," about 4.6 billion years ago. The atmospheres of the Earth and the other terrestrial planets, Venus and Mars, are thought to have formed as a result of the release of trapped volatile compounds from the planet itself. The early atmosphere of the Earth is believed to have been a mixture of carbon dioxide (CO_2), nitrogen (N_2), and water vapor (H_2O), with trace amounts of hydrogen (H_2), a mixture similar to that emitted by present-day volcanoes.

The composition of the present atmosphere bears little resemblance to the composition of the early atmosphere. Most of the water vapor that outgassed from the Earth's interior condensed out of the atmosphere to form the oceans. The predominance of the CO_2 that outgassed formed sedimentary carbonate rocks after dissolution in the ocean. It is estimated that for each molecule of CO_2 presently in the atmosphere, there are about 10^5 CO_2 molecules incorporated as carbonates in sedimentary rocks. Since N_2 is chemically inert, non-water-soluble, and noncondensable, most of the outgassed N_2 accumulated in the atmosphere over geologic time to become the atmosphere's most abundant constituent.

The early atmosphere of the Earth was a mildly reducing chemical mixture, whereas the present atmosphere is strongly oxidizing. Geochemical evidence points to the fact that atmospheric oxygen underwent a dramatic increase in concentration about 2300 million years ago (Kasting 2001). While the timing of the initial O_2 rise is now well established, what triggered the increase is still in question. There is agreement that O_2 was initially produced by cyanobacteria, the only prokaryotic organisms (*Bacteria* and *Archea*) capable of oxygenic photosynthesis. These bacteria had emerged by 2700 million years ago. The gap of 400 million years between the emergence of cyanobacteria and the rise of atmospheric O_2 is still an issue of debate. The atmosphere from 3000 to 2300 million years ago was rich in reduced gases such as H_2 and CH_4. Hydrogen can escape to space from such an atmosphere. Since the majority of the Earth's hydrogen was in the form of water, H_2 escape would lead to a net accumulation of O_2. One possibility is that the O_2 left behind by the escaping H_2 was largely consumed by oxidation of continental crust. This oxidation might have sequestered enough O_2 to suppress atmospheric levels before 2300 million years ago, the point at which the flux of reduced gases fell below the net photosynthetic production rate of oxygen. The present level of O_2 is maintained by a balance between production from photosynthesis and removal through respiration and decay of organic carbon. If O_2 were not replenished by photosynthesis, the reservoir of surface organic carbon would be completely oxidized in about 20 years, at which time the amount of O_2 in

Atmospheric Chemistry and Physics: From Air Pollution to Climate Change, Second Edition, by John H. Seinfeld and Spyros N. Pandis. Copyright © 2006 John Wiley & Sons, Inc.

the atmosphere would have decreased by less than 1% (Walker 1977). In the absence of surface organic carbon to be oxidized, weathering of sedimentary rocks would consume the remaining O_2 in the atmosphere, but it would take approximately 4 million years to do so.

The Earth's atmosphere is composed primarily of the gases N_2 (78%), O_2 (21%), and Ar (1%), whose abundances are controlled over geologic timescales by the biosphere, uptake and release from crustal material, and degassing of the interior. Water vapor is the next most abundant constituent; it is found mainly in the lower atmosphere and its concentration is highly variable, reaching concentrations as high as 3%. Evaporation and precipitation control its abundance. The remaining gaseous constituents, the *trace gases*, represent less than 1% of the atmosphere. These trace gases play a crucial role in the Earth's radiative balance and in the chemical properties of the atmosphere. The trace gas abundances have changed rapidly and remarkably over the last two centuries.

The study of atmospheric chemistry can be traced back to the eighteenth century when chemists such as Joseph Priestley, Antoine-Laurent Lavoisier, and Henry Cavendish attempted to determine the chemical components of the atmosphere. Largely through their efforts, as well as those of a number of nineteenth-century chemists and physicists, the identity and major components of the atmosphere, N_2, O_2, water vapor, CO_2, and the rare gases, were established. In the late nineteenth–early twentieth century focus shifted from the major atmospheric constituents to trace constituents, that is, those having mole fractions below 10^{-6}, 1 part per million (ppm) by volume. It has become clear that the atmosphere contains a myriad of trace species. The presence of these species can be traced to geologic, biological, chemical, and anthropogenic processes.

Spectacular innovations in instrumentation since 1975 or so have enabled identification of atmospheric trace species down to levels of about 10^{-12} parts per part of air, 1 part per trillion (ppt) by volume. Observations have shown that the composition of the atmosphere is changing on the global scale. Present-day measurements coupled with analyses of ancient air trapped in bubbles in ice cores provide a record of dramatic, global increases in the concentrations of gases such as CO_2, methane (CH_4), nitrous oxide (N_2O), and various halogen-containing compounds. These "greenhouse gases" act as atmospheric thermal insulators. They absorb infrared radiation from the Earth's surface and reradiate a portion of this radiation back to the surface. These gases include CO_2, O_3, CH_4, N_2O, and halogen-containing compounds. The emergence of the Antarctic ozone hole provides striking evidence of the ability of emissions of trace species to perturb large-scale atmospheric chemistry. Observations have documented the essentially complete disappearance of ozone in the Antarctic stratosphere during the austral spring, a phenomenon that has been termed the "Antarctic ozone hole." Observations have also documented less dramatic decreases over the Arctic and over the northern and southern midlatitudes. Whereas stratospheric ozone levels have been eroding, those at ground level in the Northern Hemisphere have, over the past century, been increasing. Paradoxically, whereas ozone in the stratosphere protects living organisms from harmful solar ultraviolet radiation, ozone in the lower atmosphere can have adverse effects on human health and plants.

Quantities of airborne particles in industrialized regions of the Northern Hemisphere have increased markedly since the Industrial Revolution. Atmospheric particles (aerosols) arise both from direct emissions and from gas-to-particle conversion of vapor precursors. Aerosols can affect climate and stratospheric ozone concentrations and have been implicated in human morbidity and mortality in urban areas. The climatic role of atmospheric aerosols arises from their ability to reflect solar radiation back to space and from their role as cloud condensation nuclei. Estimates of the cooling effect resulting from the reflection of solar radiation back to space by aerosols indicate that the cooling effect may be sufficiently large to mask the warming effect of greenhouse gas increases over industrialized regions of the Northern Hemisphere.

The atmosphere is the recipient of many of the products of our technological society. These effluents include products of combustion of fossil fuels and the development of new synthetic chemicals. Historically these emissions can lead to unforeseen consequences in the atmosphere. Classical examples include the realization in the 1950s that motor vehicle emissions could lead to urban smog and the realization in the 1970s that emissions of chlorofluorocarbons from aerosol spray cans and refrigerators could cause the depletion of stratospheric ozone.

The chemical fates of trace atmospheric species are often intertwined. The life cycles of the trace species are inextricably coupled through the complex array of chemical and physical processes in the atmosphere. As a result of these couplings, a perturbation in the concentration of one species can lead to significant changes in the concentrations and lifetimes of other trace species and to feedbacks that can either amplify or damp the original perturbation. An example of this coupling is provided by methane. Methane is the predominant organic molecule in the troposphere and an important greenhouse gas. Methane sources such as rice paddies and cattle can be estimated and are increasing. Methane is removed from the atmosphere by reaction with the hydroxyl (OH) radical, at a rate that depends on the atmospheric concentration of OH. But the OH concentration depends on the amount of carbon monoxide (CO), which itself is a product of CH_4 oxidation as well as a result of fossil fuel combustion and biomass burning. The hydroxyl concentration also depends on the concentration of ozone and oxides of nitrogen. Change in CH_4 can affect the total amount of ozone in the troposphere, so methane itself affects the concentration of the species, OH, that governs its removal.

Depending on their atmospheric lifetime, trace species can exhibit an enormous range of spatial and temporal variability. Relatively long-lived species have a spatial uniformity such that a handful of strategically located sampling sites around the globe are adequate to characterize their spatial distribution and temporal trend. As species lifetimes become shorter, their spatial and temporal distributions become more variable. Urban areas, for example, can require tens of monitoring stations over an area of hundreds of square kilometers in order to characterize the spatial and temporal distribution of their atmospheric components.

The extraordinary pace of the recent increases in atmospheric trace gases can be seen when current levels are compared with those of the distant past. Such comparisons can be made for CO_2 and CH_4, whose histories can be reconstructed from their concentrations in bubbles of air trapped in ice in such perpetually cold places as Antarctica and Greenland. With gases that are long-lived in the atmosphere and therefore distributed rather uniformly over the globe, such as CO_2 and CH_4, polar ice core samples reveal global average concentrations of previous eras. Analyses of bubbles in ice cores show that CO_2 and CH_4 concentrations remained essentially unchanged from the end of the last ice age some 10,000 years ago until roughly 300 years ago, at mixing ratios close to 260 ppm by volume and 0.7 ppm by volume, respectively. (See Section 1.6 for discussion of units.) About 300 years ago methane levels began to climb, and about 100 years ago levels of both gases began to increase markedly. In summary, activities of humans account for most of the rapid changes in the trace gases over the past 200 years—combustion of fossil fuels (coal and oil) for energy and transportation, industrial and agricultural activities, biomass burning (the burning of vegetation), and deforestation.

1.2   CLIMATE

Viewed from space, the Earth is a multicolored marble: clouds and snow-covered regions of white, blue oceans, and brown continents. The white areas make Earth a bright planet; about 30% of the Sun's radiation is reflected immediately back to space. Solar energy that does not reflect off clouds and snow is absorbed by the atmosphere and the Earth's surface. As the surface warms, it sends infrared radiation back to space. The atmosphere, however, absorbs much of the energy radiated by the surface and reemits its own energy, but at much lower temperatures. Aside from gases in the atmosphere, clouds play a major climatic role. Some clouds cool the planet by reflecting solar radiation back to space; others warm the Earth by trapping energy near the surface. On balance, clouds exert a significant cooling effect on Earth, although in some areas, such as the tropics, heavy clouds can markedly warm the regional climate.

The temperature of the Earth adjusts so that solar energy reaching the Earth is balanced by that leaving the planet. Whereas the radiation budget must balance for the entire Earth, it does not balance at each particular point on the globe. Very little solar energy reaches the white, ice-covered polar regions, especially during the winter months. The Earth absorbs most solar radiation near its equator. Over time, though, energy absorbed near the equator spreads to the colder regions of the globe, carried by winds in the atmosphere and by currents in the oceans. This global heat engine, in its attempt to equalize temperatures, generates the climate with which we are all familiar. It pumps energy into storm fronts and powers hurricanes. In the colder seasons, low-pressure and high-pressure cells push each other back and forth every few days. Energy is also transported over the globe by masses of wet and dry air. Through evaporation, air over the warm oceans absorbs water vapor and then travels to colder regions and continental interiors where water vapor condenses as rain or snow, a process that releases heat into the atmosphere. In the oceans, salt helps drive the heat engine. Over some areas, like the arid Mediterranean, water evaporates from the sea faster than rain or river flows can replace it. As seawater becomes increasingly salty, it grows denser. In the North Atlantic, cool air temperatures and excess salt cause the surface water to sink, creating a current of heavy water that spreads throughout the world's oceans. By redistributing energy in this way, the oceans act to smooth out differences in temperature and salinity. Whereas the atmosphere can respond in a few days to a warming or cooling in the ocean, it takes the sea surface months or longer to adjust to changes in energy coming from the atmosphere.

The condition of the atmosphere at a particular location and time is its *weather*; this includes winds, clouds, precipitation, temperature, and relative humidity. In contrast to weather, the *climate* of a region is the condition of the atmosphere over many years, as described by long-term averages of the same properties that determine weather.

Solar radiation, clouds, ocean currents, and the atmospheric circulation weave together in a complex and chaotic way to produce our climate. Until recently, climate was assumed to change on a timescale much much longer than our lifetimes and those of our children. Evidence is indisputable, however, that the release of trace gases to the atmosphere, the "greenhouse gases," has the potential to lead to an increase of the Earth's temperature by several degrees Celsius. The Earth's average temperature rose about 0.6°C over the past century. It has been estimated that a doubling of CO_2 from its pre–Industrial Revolution mixing ratio of 280 ppm by volume could lead to a rise in average global temperature of 1.5–4.5°C. A 2°C warming would produce the warmest climate seen on Earth in 6000 years. A 4.5°C rise would place the world in a temperature regime last experienced in the Mesozoic Era—the age of dinosaurs.

Although an average global warming of a few degrees does not sound like much, it could create dramatic changes in climatic extremes. Observations and global climate model simulations show, for example, that present-day heat waves over Europe and North America coincide with a specific atmospheric circulation pattern that is intensified by ongoing increases in greenhouse gases, indicating that heat waves in these regions will become more intense, more frequent, and longer-lasting (Meehl and Tebaldi 2004). Changes in the timing and amount of precipitation would almost certainly occur with a warmer climate. Soil moisture, critical during planting and early growth periods, will change. Some regions would probably become more productive, others less so. Of all the effects of a global warming, perhaps none has captured more attention than the prospect of rising sea levels [see IPCC (2001), Chapter 11]. This would result from the melting of land-based glaciers and volume expansion of ocean water as it warms. IPCC (2001) estimated the range of global average sea-level rise by 2100 to be 0.11–0.77 m. In the most dramatic scenario, the west Antarctic ice sheet, which rests on land that is below sea level, could slide into the sea if the buttress of floating ice separating it from the ocean were to melt. This would raise the average sea level by 5–6 m (Bentley 1997). Even a 0.3-m rise would have major effects on the erosion of coastlines, saltwater intrusion into the water supply of coastal areas, flooding of marshes, and inland extent of surges from large storms.

To systematically approach the complex subject of climate, the scientific community has divided the problem into two major parts; climate *forcings* and climate *responses*. Climate *forcings* are changes in the energy balance of the Earth that are imposed on it; forcings are measured in units of heat flux—watts per square meter ($W\ m^{-2}$). An example of a forcing is a change in energy output from the Sun. *Responses* are the results of these forcings, reflected in temperatures, rainfall, extremes of weather, sea-level height, and so on.

Much of the variation in the predicted magnitude of potential climate effects resulting from the increase in greenhouse gas levels hinges on estimates of the size and direction of various feedbacks that may occur in response to an initial perturbation of the climate. Negative feedbacks have an effect that damps the warming trend; positive feedbacks reinforce the initial warming. One example of a greenhouse warming feedback mechanism involves water vapor. As air warms, each cubic meter of air can hold more water vapor. Since water vapor is a greenhouse gas, this increased concentration of water vapor further enhances greenhouse warming. In turn, the warmer air can hold more water, and so on. This is an example of a positive feedback, providing a physical mechanism for multiplying the original impetus for change beyond its initial amount.

Some mechanisms provide a negative feedback, which decreases the initial impetus. For example, increasing the amount of water vapor in the air may lead to forming more clouds. Low-level, white clouds reflect sunlight, thereby preventing sunlight from reaching the Earth and warming the surface. Increasing the geographic coverage of low-level clouds would reduce greenhouse warming, whereas increasing the amount of high, convective clouds could enhance greenhouse warming. This is because high, convective clouds absorb energy from below at higher temperatures than they radiate energy into space from their tops, thereby effectively trapping energy. It is not known with certainty whether increased temperatures would lead to more low-level clouds or more high, convective clouds.

1.3   THE LAYERS OF THE ATMOSPHERE

In the most general terms, the atmosphere is divided into lower and upper regions. The lower atmosphere is generally considered to extend to the top of the stratosphere, an altitude of about 50 kilometers (km). Study of the lower atmosphere is known as *meteorology*; study of the upper atmosphere is called *aeronomy*.

The Earth's atmosphere is characterized by variations of temperature and pressure with height. In fact, the variation of the average temperature profile with altitude is the basis for distinguishing the layers of the atmosphere. The regions of the atmosphere are (Figure 1.1):

Troposphere. The lowest layer of the atmosphere, extending from the Earth's surface up to the tropopause, which is at 10–15 km altitude depending on latitude and time of year; characterized by decreasing temperature with height; rapid vertical mixing.

Stratosphere. Extends from the tropopause to the stratopause (From ~ 45 to 55 km altitude); temperature increases with altitude, leading to a layer in which vertical mixing is slow.

Mesosphere. Extends from the stratopause to the mesopause (From ~ 80 to 90 km altitude); temperature decreases with altitude to the mesopause, which is the coldest point in the atmosphere; rapid vertical mixing.

Thermosphere. The region above the mesopause; characterized by high temperatures as a result of absorption of short-wavelength radiation by N_2 and O_2; rapid vertical mixing. The *ionosphere* is a region of the upper mesosphere and lower thermosphere where ions are produced by photoionization.

Exosphere. The outermost region of the atmosphere (>500 km altitude) where gas molecules with sufficient energy can escape from the Earth's gravitational attraction.

Over the equator the average height of the tropopause is about 18 km; over the poles, about 8 km. By convention of the World Meteorological Organization (WMO), the tropopause is defined as the lowest level at which the rate of decrease of temperature with height (the *temperature lapse rate*) decreases to 2 K km^{-1} or less and the lapse rate averaged between this level and any level within the next 2 km does not exceed 2 K km^{-1} (Holton et al. 1995). The tropopause is at a maximum height over the tropics, sloping downward moving toward the poles. The name coined by British meteorologist, Sir Napier Shaw, from the Greek word *tropos*, meaning turning, the troposphere is a region of ceaseless turbulence and mixing. The caldron of all weather, the troposphere contains almost all of the atmosphere's water vapor. Although the troposphere accounts for only a small fraction of the atmosphere's total height, it contains about 80% of its total mass. In the troposphere, the temperature decreases almost linearly with height. For dry air the lapse rate is 9.7 K km^{-1}. The reason for this progressive decline is the increasing distance from the Sun-warmed Earth. At the tropopause, the temperature has fallen to an average of

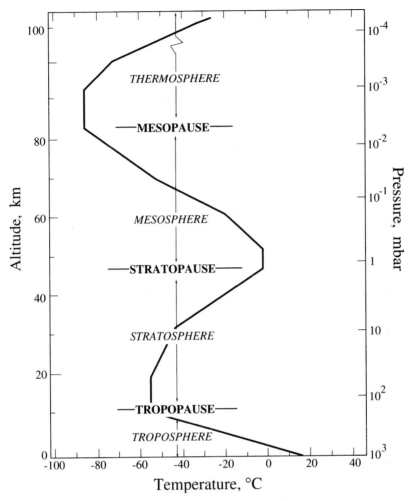

FIGURE 1.1   Layers of the atmosphere.

about 217 K ($-56°$C). The troposphere can be divided into the *planetary boundary layer*, extending from the Earth's surface up to about 1 km, and the *free troposphere*, extending from about 1 km to the tropopause.

As air moves vertically, its temperature changes in response to the local pressure. For dry air, this rate of change is substantial, about 1°C per 100 m (the theory behind this will be developed in Chapter 16). An air parcel that is transported from the surface to 1 km can decrease in temperature from 5 to 10°C depending on its water content. Because of the strong dependence of the saturation vapor pressure on temperature, this decrease of temperature of a rising air parcel can be accompanied by a substantial increase in relative humidity (RH) in the parcel. As a result, upward air motions of a few hundreds of meters can cause the air to reach saturation (RH = 100%) and even supersaturation. The result is the formation of clouds.

Vertical motions in the atmosphere result from (1) convection from solar heating of the Earth's surface, (2) convergence or divergence of horizontal flows, (3) horizontal flow over topographic features at the Earth's surface, and (4) buoyancy caused by the release of latent heat as water condenses. Interestingly, even though an upward moving parcel of air cools, condensation of water vapor can provide sufficient heating of the parcel to maintain the temperature of the air parcel above that of the surrounding air. When this occurs, the parcel is buoyant and accelerates upward even more, leading to more condensation. Cumulus clouds are produced in this fashion, and updraft velocities of meters per second can be reached in such clouds. Vertical convection associated with cumulus clouds is, in fact, a principal mechanism for transporting air from close to the Earth's surface to the mid- and upper-troposphere.

The stratosphere, extending from about 11 km to about 50 km, was discovered at the turn of the twentieth century by the French meteorologist Léon Philippe Teisserenc de Bort. Sending up temperature-measuring devices in balloons, he found that, contrary to the popular belief of the day, the temperature in the atmosphere did not steadily decrease to absolute zero with increasing altitude, but stopped falling and remained constant at 11 km or so. He named the region the stratosphere from the Latin word *stratum* meaning layer. Although an isothermal region does exist from about 11–20 km at midlatitudes, temperature progressively increases from 20 to 50 km, reaching 271 K at the stratopause, a temperature not much lower than the average of 288 K at the Earth's surface. The vertical thermal structure of the stratosphere is a result of absorption of solar ultraviolet radiation by O_3.

1.4 PRESSURE IN THE ATMOSPHERE

1.4.1 Units of Pressure

The unit of pressure in the International System of Units (SI) is newtons per meter squared $(N\ m^{-2})$, which is called the *pascal* (Pa). In terms of pascals, the atmospheric pressure at the surface of the Earth, the so-called standard atmosphere, is 1.01325×10^5 Pa. Another commonly used unit of pressure in atmospheric science is the millibar (mbar), which is equivalent to the hectopascal (hPa) (see Tables A.5 and A.8). The standard atmosphere is 1013.25 mbar.

Because instruments for measuring pressure, such as the manometer, often contain mercury, commonly used units for pressure are based on the height of the mercury column (in millimeters) that the gas pressure can support. The unit mm Hg is often called the *torr* in honor of the scientist, Evangelista Torricelli. A related unit for pressure is the standard atmosphere (abbreviated atm).

We summarize the various pressure units as follows:

$$1\,Pa = 1\,N\,m^{-2} = 1\,kg\,m^{-1}\,s^{-2}$$
$$1\,atm = 1.01325 \times 10^5\,Pa$$
$$1\,bar = 10^5\,Pa$$
$$1\,mbar = 1\,hPa = 100\,Pa$$
$$1\,torr = 1\,mm\,Hg = 134\,Pa$$

Standard atmosphere: $1.01325 \times 10^5\,Pa = 1013.25\,hPa = 1013.25\,mbar = 760\,torr$

The variation of pressure and temperature with altitude in the standard atmosphere is given in Table A.8. Because the millibar (mbar) is the unit most commonly used in the meteorological literature, we will use it when discussing pressure at various altitudes in the in the atmosphere. Mean surface pressure at sea level is 1013 mbar; global mean surface pressure, calculated over both land and ocean, is estimated as 985.5 mbar. The lower value reflects the effect of surface topography; over the highest mountains, which reach an altitude of over 8000 m, the pressure may be as low as 300 mbar. The 850 mbar level, which as we see from Table A.8, is at about 1.5 km altitude, is often used to represent atmospheric quantities, such as temperature, as the first standard meteorological level above much of the topography and a level at which a considerable quantity of heat is located as well as transported.

1.4.2   Variation of Pressure with Height in the Atmosphere

Let us derive the equation governing the pressure in the static atmosphere. Imagine a volume element of the atmosphere of horizontal area dA between two heights, z and $z + dz$. The pressures exerted on the top and bottom faces are $p(z + dz)$ and $p(z)$, respectively. The gravitational force on the mass of air in the volume $= \rho g \, dA \, dz$, with $p(z) > p(z + dz)$ due to the additional weight of air in the volume. The balance of forces on the volume gives

$$(p(z) - p(z + dz)) \, dA = \rho g \, dA \, dz$$

Dividing by dz and letting $dz \to 0$ produce

$$\frac{dp(z)}{dz} = -\rho(z)g \tag{1.1}$$

where $\rho(z)$ is the mass density of air at height z and g is the acceleration due to gravity. From the ideal-gas law, we obtain

$$\rho(z) = \frac{M_{air} \, p(z)}{RT(z)} \tag{1.2}$$

where M_{air} is the average molecular weight of air (28.97 g mol^{-1}). Thus

$$\frac{dp(z)}{dz} = -\frac{M_{air} \, g p(z)}{RT(z)} \tag{1.3}$$

which we can rewrite as

$$\frac{d \ln p(z)}{dz} = -\frac{1}{H(z)} \tag{1.4}$$

where $H(z) = RT(z)/M_{air}g$ is a characteristic length scale for decrease of pressure with height.

The temperature in the atmosphere varies by less than a factor of 2, while the pressure changes by six orders of magnitude (see Table A.8). If the temperature can be taken to be approximately constant, just to obtain a simple approximate expression for $p(z)$, then the pressure decrease with height is approximately exponential

$$\frac{p(z)}{p_0} = e^{-z/H} \tag{1.5}$$

where $H = RT/M_{air} g$ is called the *scale height*.

Since the temperature was assumed to be constant in deriving (1.5), a temperature at which to evaluate H must be selected. A reasonable choice is the mean temperature of the troposphere. Taking a surface temperature of 288 K (Table A.8) and a tropopause temperature of 217 K, the mean tropospheric temperature is 253 K. At 253 K, $H = 7.4$ km.

Number Concentration of Air at Sea Level and as a Function of Altitude The number concentration of air at sea level is

$$n_{air}(0) = \frac{p_0 N_A}{RT}$$

where N_A is Avogadro's number (6.022×10^{23} molecules mol^{-1}) and p_0 is the standard atmospheric pressure (1.013×10^5 Pa). The surface temperature of the U.S. Standard Atmosphere (Table A.8) is 288 K, so

$$n_{air}(0) = \frac{(6.022 \times 10^{23} \text{ molecules mol}^{-1})(1.013 \times 10^5 \text{ N m}^{-2})}{(8.314 \text{ N m mol}^{-1} \text{ K}^{-1})(288 \text{ K})}$$

$$= 2.55 \times 10^{25} \text{ molecules m}^{-3}$$

$$= 2.55 \times 10^{19} \text{ molecules cm}^{-3}$$

Throughout this book we will need to know the number concentration of air molecules as a function of altitude. We can estimate this using the average scale height $H = 7.4$ km and

$$n_{air}(z) = n_{air}(0)e^{-z/H}$$

where $n_{air}(0)$ is the number density at the surface. If we take the mean surface temperature as 288 K, then $n_{air}(0) = 2.55 \times 10^{19}$ molecules cm^{-3}. The table below gives the approximate number concentrations at various altitudes based on the average scale height of 7.4 km and the values from the U.S. Standard Atmosphere:

	n_{air} (molecules cm^{-3})	
z (km)	Approximate	U.S. Standard Atmosphere[a]
0	2.55×10^{19}	2.55×10^{19}
5	1.3×10^{19}	1.36×10^{19}
10	6.6×10^{18}	6.7×10^{18}
15	3.4×10^{18}	3.0×10^{18}
20	1.7×10^{18}	1.4×10^{18}
25	8.7×10^{17}	6.4×10^{17}

[a]See Table A.8.

Total Mass, Moles, and Molecules of the Atmosphere   The total mass of the atmosphere is

$$\int_0^\infty \rho(z)A_e dz$$

where $A_e = 4\pi R_e^2$, the total surface area of the Earth. We can obtain an estimate of the total mass of the atmosphere using (1.5),

$$\text{Total mass} = 4\pi R_e^2 \rho_0 \int_0^\infty e^{-z/H} dz$$
$$= 4\pi R_e^2 \rho_0 H$$

Using $R_e \cong 6400\,\text{km}$, $H \cong 7.4\,\text{km}$, and $\rho_0 \cong 1.23\,\text{kg m}^{-3}$ (Table A.8), we get the rough estimate:

$$\text{Total mass} \cong 4.7 \times 10^{18}\,\text{kg}$$

An estimate for the total number of moles of air in the atmosphere is total mass/M_{air}

$$\text{Total moles} \cong 1.62 \times 10^{20}\,\text{mol}$$

and an estimate of the total number of molecules in the atmosphere is

$$\text{Total molecules} \cong 1.0 \times 10^{44}\,\text{molecules}$$

An accurate estimate of the total mass of the atmosphere can be obtained by considering the global mean surface pressure (985.50 hPa) and the water vapor content of the atmosphere (Trenberth and Smith 2005). The total mean mass of the atmosphere is

$$\text{Total mass (accurate)} = 5.1480 \times 10^{18}\,\text{kg}$$

The mean mass of water vapor in the atmosphere is estimated as $1.27 \times 10^{16}\,\text{kg}$, and the dry air mass of the atmosphere is

$$\text{Total dry mass (accurate)} = 5.1352 \times 10^{18}\,\text{kg}$$

1.5   TEMPERATURE IN THE ATMOSPHERE

Figure 1.2 shows the global average temperature distribution for January over the period 1979–98, as determined from satellite. The heavy dark line denotes the height of the tropopause; the tropopause is highest over the tropics ($\sim$14–15 km) and lowest over the poles ($\sim$8 km). The coldest region of the atmosphere is in the stratosphere just above the tropical tropopause, where temperatures are less than 200 K ($-73°$C). Since Figure 1.2 presents climatology for the month of January, temperatures over the North

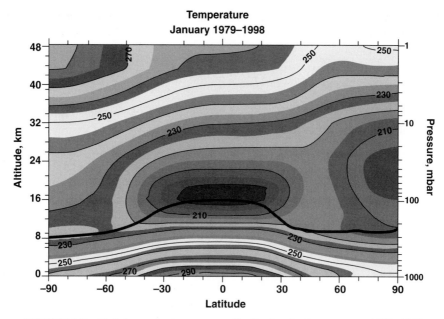

FIGURE 1.2   Global average temperature distribution for January over 1979–1998.

Pole (90° latitude) are colder than those over the South Pole (−90° latitude); the reverse is true in July.

The U.S. Standard Atmosphere (Table A.8) gives mean conditions at 45°N latitude. From Figure 1.2 we note that the change of temperature with altitude varies with latitude. Throughout this book we will need the variation of atmospheric properties as a function of altitude. For this we will generally use the U.S. Standard Atmosphere.

1.6   EXPRESSING THE AMOUNT OF A SUBSTANCE IN THE ATMOSPHERE

The SI unit for the amount of a substance is the mole (mol). The number of atoms or molecules in 1 mol is Avogadro's number, $N_A = 6.022 \times 10^{23} \, \text{mol}^{-1}$. Concentration is the amount (or mass) of a substance in a given volume divided by that volume. *Mixing ratio* in atmospheric chemistry is defined as the ratio of the amount (or mass) of the substance in a given volume to the total amount (or mass) of all constituents in that volume. In this definition for a gaseous substance the sum of all constituents includes all gaseous substances, including water vapor, but *not* including particulate matter or condensed phase water. Thus mixing ratio is just the fraction of the total amount (or mass) contributed by the substance of interest.

The volume mixing ratio for a species i is

$$\xi_i = \frac{c_i}{c_{\text{total}}} \tag{1.6}$$

where c_i is the molar concentration of i and c_{total} is the total molar concentration of air. From the ideal-gas law the total molar concentration at any point in the atmosphere is

$$c_{total} = \frac{N}{V} = \frac{p}{RT} \tag{1.7}$$

Thus the mixing ratio ξ_i and the molar concentration are related by

$$\begin{aligned}
\xi_i &= \frac{c_i}{p/RT} \\
&= \frac{p_i/RT}{p/RT} = \frac{p_i}{p}
\end{aligned} \tag{1.8}$$

where p_i is the partial pressure of i.

Concentration (mol m^{-3}) depends on pressure and temperature through the ideal-gas law. Mixing ratios, which are just mole fractions, are therefore better suited than concentrations to describe abundances of species in air, particularly when spatiotemporal variation is involved. The inclusion of water vapor in the totality of gaseous substances in a volume of air means that mixing ratio will vary with humidity. The variation can amount to several percent. Sometimes, as a result, mixing ratios are defined with respect to dry air.

It has become common use in atmospheric chemistry to describe mixing ratios by the following units:

parts per million (ppm)	10^{-6}	μmol mol^{-1}
parts per billion (ppb)	10^{-9}	nmol mol^{-1}
parts per trillion (ppt)	10^{-12}	pmol mol^{-1}

These quantities are sometimes distinguished by an added v (for volume) and m (for mass), that is,

ppmv	parts per million by volume
ppmm	parts per million by mass

Unless noted otherwise, we will always use mixing ratios by volume and not use the added v. The parts per million, parts per billion, and parts per trillion measures are not SI units; the SI versions are, as given above, μmol mol^{-1}, nmol mol^{-1}, and pmol mol^{-1}.

Water vapor occupies an especially important role in atmospheric science. The water vapor content of the atmosphere is expressed in several ways:

1. Volume mixing ratio, ppm
2. Ratio of mass of water vapor to mass of dry air, g H_2O (kg dry air)$^{-1}$
3. Specific humidity – ratio of mass of water vapor to mass of total air, g H_2O (kg air)$^{-1}$
4. Mass concentration, g H_2O (m^3 air)$^{-1}$
5. Mass mixing ratio, g H_2O (g air)$^{-1}$
6. Relative humidity – ratio of partial pressure of H_2O vapor to the saturation vapor pressure of H_2O at that temperature, $p_{H_2O}/p^0_{H_2O}$.

Relative humidity (RH) is usually expressed in percent:

$$RH = 100 \, \frac{p_{H_2O}}{p_{H_2O}^0} \qquad (1.9)$$

Number Concentration of Water Vapor   The vapor pressure of pure water as a function of temperature can be calculated with the following correlation:

$$p_{H_2O}^0 (T) = 1013.25 \, \exp[13.3185 \, a - 1.97 \, a^2 - 0.6445 \, a^3 - 0.1299 \, a^4] \quad (1.10)$$

where

$$a = 1 - \frac{373.15}{T}$$

(An alternate correlation is given in Table 17.2.)

Let us calculate the number concentration of water vapor n_{H_2O} (molecules cm^{-3}) at RH = 50% and $T = 298$ K:

$$RH \, (\text{in } \%) = 100 \, \frac{p_{H_2O}}{p_{H_2O}^0}$$

$$p_{H_2O} = n_{H_2O} \, RT$$

From the correlation above, at 298 K, we obtain

$$p_{H_2O}^0 = 31.387 \, \text{mbar}$$

Thus

$$n_{H_2O} = \frac{31.387 \times 0.5 \times 100 \times 6.022 \times 10^{23}}{8.314 \times 298 \times 10^6} = 3.81 \times 10^7 \, \text{molecules cm}^{-3}$$

Figure 1.3 shows the U.S. Standard Atmosphere temperature profile at 45°N and that at the equator. Corresponding to each temperature profile is the vertical profile of $p_{H_2O}^0/p$, expressed as mixing ratio. Note that in the equatorial tropopause region, the saturation mixing ratio of water vapor drops to about 4–5 ppm, whereas at the surface it is several percent. Air enters the stratosphere from the troposphere, and this occurs primarily in the tropical tropopause region, where rising air in towering cumulus clouds is injected into the stratosphere. Because this region is so cold, water vapor is frozen out of the air entering the stratosphere. This process is often referred to as "freeze drying" of the atmosphere. As a result, air entering the stratosphere has a water vapor mixing ratio of only 4–5 ppm, and the stratosphere is an extremely dry region of the atmosphere.

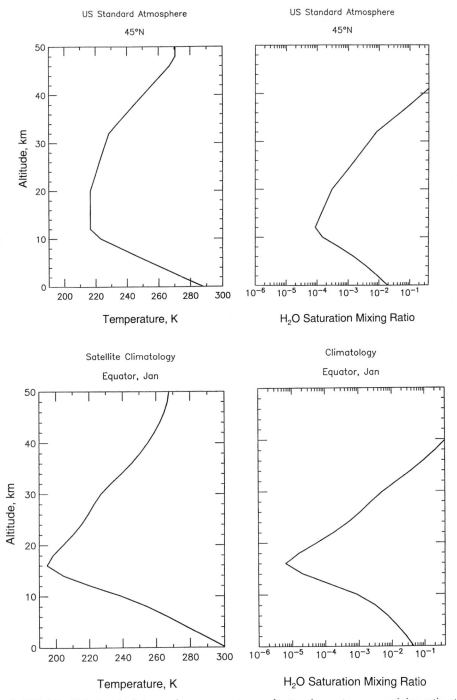

FIGURE 1.3   U.S. Standard Atmosphere temperature and saturation water vapor mixing ratio at 45°N and at the equator.

Conversion from Mixing Ratio to µg m^{-3}   Atmospheric species concentrations are sometimes expressed in terms of mass per volume, most frequently as µg m^{-3}. Given a concentration m_i, in µg m^{-3}, the molar concentration of species i, in mol m^{-3}, is

$$c_i = \frac{10^{-6} m_i}{M_i}$$

where M_i is the molecular weight of species i.

Noting that the total molar concentration of air at pressure p and temperature T is $c = p/RT$, then

$$\text{Mixing ratio of } i \text{ in ppm} = \frac{RT}{p M_i} \times \text{concentration of } i \text{ in µg m}^{-3}$$

If T is in K and p in Pa (see Table A.6 for the value of the molar gas constant, R)

$$\boxed{\text{Mixing ratio of } i \text{ in ppm} = \frac{8.314\,T}{p M_i} \times \text{concentration of } i \text{ in µg m}^{-3}}$$

As an example, let us determine the concentration in µg m^{-3} for O$_3$ corresponding to a mixing ratio of 120 ppb at $p = 1$ atm and $T = 298$ K. The 120 ppb corresponds to 0.12 ppm. Then

$$\text{Concentration in µg m}^{-3} = \frac{p M_i}{8.314\,T} \times \text{Mixing ratio in ppm}$$
$$= \frac{(1.013 \times 10^5)(48)}{8.314(298)} \times 0.12$$
$$= 235.6 \,\text{µg m}^{-3}$$

Now let us convert 365 µg m^{-3} of SO$_2$ to ppm at the same temperature and pressure:

$$\text{Mixing ratio in ppm} = \frac{(8.314)(298)}{(1.013 \times 10^5)(64)} \times 365$$
$$= 0.139 \,\text{ppm}$$

1.7   SPATIAL AND TEMPORAL SCALES OF ATMOSPHERIC PROCESSES

In spite of its apparent unchanging nature, the atmosphere is in reality a dynamic system, with its gaseous constituents continuously being exchanged with vegetation, the oceans, and biological organisms. The so-called cycles of the atmospheric gases involve a number of physical and chemical processes. Gases are produced by chemical processes within the atmosphere itself, by biological activity, volcanic exhalation, radioactive decay, and human industrial activities. Gases are removed from the atmosphere by chemical reactions in the atmosphere, by biological activity, by physical processes in the atmosphere (such as particle formation), and by deposition and uptake by the oceans and land masses. The average lifetime of a gas molecule introduced into the atmosphere can range from seconds to millions of years, depending on the effectiveness of the removal processes. Most of the species considered air pollutants (in a region in which their concentrations exceed substantially the normal background levels) have natural as well as man-made sources. Therefore, in order to assess the effect human-made emissions may have on the atmosphere as a whole, it is essential to understand the atmospheric cycles of the trace gases, including natural and anthropogenic sources as well as predominant removal mechanisms.

While in the air, a substance can be chemically altered in one of two ways. First, the sunlight itself may contain sufficient energy to break the molecule apart, a so-called photochemical reaction. The more frequently occurring chemical alteration, however, takes place when two molecules interact and undergo a chemical reaction to produce new species. Atmospheric chemical transformations can occur homogeneously or heterogeneously. Homogeneous reactions occur entirely in one phase; heterogeneous reactions involve more than one phase, such as a gas interacting with a liquid or with a solid surface.

During transport through the atmosphere, all except the most inert substances are likely to participate in some form of chemical reaction. This process can transform a chemical from its original state, the physical (gas, liquid, or solid) and chemical form in which it first enters the atmosphere, to another state that may have either similar or very different characteristics. Transformation products can differ from their parent substance in their chemical properties, toxicity, and other characteristics. These products may be removed from the atmosphere in a manner very different from that of their precursors. For example, when a substance that was originally emitted as a gas is transformed into a particle, the overall removal is usually hastened since particles often tend to be removed from the air more rapidly than gases.

Once emitted, species are converted at various rates into substances generally characterized by higher chemical oxidation states than their parent substances. Frequently this oxidative transformation is accompanied by an increase in polarity (and hence water solubility) or other physical and chemical changes from the precursor molecule. An example is the conversion of sulfur dioxide (SO_2) into sulfuric acid (H_2SO_4). Sulfur dioxide is moderately water soluble, but its oxidation product, sulfuric acid, is so water soluble that even single molecules of sulfuric acid in air immediately become associated with water molecules. The demise of one substance through a chemical transformation can become another species in situ source. In general, then, a species emitted into the air can be transformed by a chemical process to a product that may have markedly different physico-chemical properties and a unique fate of its own.

The atmosphere can be likened to an enormous chemical reactor in which a myriad of species are continually being introduced and removed over a vast array of spatial and temporal scales. The atmosphere itself presents a range of spatial scales in its motions that spans eight orders of magnitude (Figure 1.4). The scales of motion in the atmosphere vary from tiny eddies of a centimeter or less in size to huge airmass movements of continental

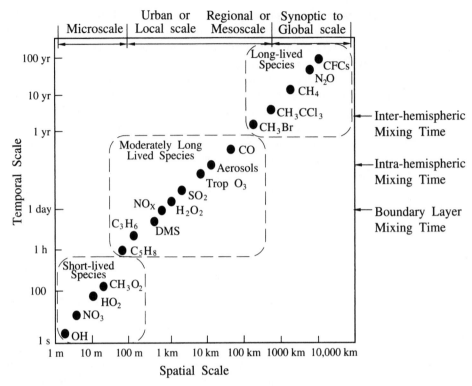

FIGURE 1.4 Spatial and temporal scales of variability for atmospheric constituents.

dimensions. Four rough categories have proved convenient to classify atmospheric scales of motion:

1. *Microscale.* Phenomena occurring on scales of the order of 0–100 m, such as the meandering and dispersion of a chimney plume and the complicated flow regime in the wake of a large building.
2. *Mesoscale.* Phenomena occurring on scales of tens to hundreds of kilometers, such as land–sea breezes, mountain–valley winds, and migratory high- and low-pressure fronts.
3. *Synoptic Scale.* Motions of whole weather systems, on scales of hundreds to thousands of kilometers.
4. *Global Scale.* Phenomena occurring on scales exceeding 5×10^3 km.

Spatial scales characteristic of various atmospheric chemical phenomena are given in Table 1.1. Many of the phenomena in Table 1.1 overlap; for example, there is more or less of a continuum between (1) urban and regional air pollution, (2) the aerosol haze associated with regional air pollution and aerosol–climate interactions, (3) greenhouse gas increases and stratospheric ozone depletion, and (4) tropospheric oxidative capacity and stratospheric ozone depletion. The lifetime of a species is the average time that a molecule of that species resides in the atmosphere before removal (chemical transformation to another species counts as removal). Atmospheric lifetimes vary from less than a second for

TABLE 1.1   Spatial Scales of Atmospheric Chemical Phenomena

Phenomenon	Length scale, km
Urban air pollution	1–100
Regional air pollution	10–1000
Acid rain/deposition	100–2000
Toxic air pollutants	0.1–100
Stratospheric ozone depletion	1000–40,000
Greenhouse gas increases	1000–40,000
Aerosol–climate interactions	100–40,000
Tropospheric transport and oxidation processes	1–40,000
Stratospheric–tropospheric exchange	0.1–100
Stratospheric transport and oxidation processes	1–40,000

the most reactive free radicals to many years for the most stable molecules. Associated with each species is a characteristic spatial transport scale; species with very short lifetimes have comparably small characteristic spatial scales while those with lifetimes of years have a characteristic spatial scale equal to that of the entire atmosphere. With a lifetime of less than 0.01 s, the hydroxyl radical (OH) has a spatial transport scale of only about 1 cm. Methane (CH_4), on the other hand, with its lifetime of about 10 years, can become more or less uniformly mixed over the entire Earth.

The spatial scales of the various atmospheric chemical phenomena shown above result from an intricate coupling between the chemical lifetimes of the principal species and the atmosphere's scales of motion. Much of this book will be devoted to understanding the exquisite interactions between chemical and transport processes in the atmosphere.

> **Units of Atmospheric Emission Rates and Fluxes**   Fluxes of species into the atmosphere are usually expressed on an annual basis, using the prefixes given in Table A.5. A common unit used is teragrams per year, Tg yr^{-1} (1 Tg = 10^{12} g). An alternative is to employ the *metric ton* (1 t = 10^6 g = 10^3 kg). Carbon dioxide fluxes are often expressed as multiples of gigatons, Gt (1 Gt = 10^9 t = 10^{15} g = 1 Pg).

PROBLEMS

1.1$_A$ Calculate the concentration (in molecules cm^{-3}) and the mixing ratio (in ppm) of water vapor at ground level at $T = 298$ K at RH values of 50%, 60%, 70%, 80%, 90%, 95%, and 99%.

1.2$_A$ Determine the concentration (in $\mu g\ m^{-3}$) for N_2O at a mixing ratio of 311 ppb at $p = 1$ atm and $T = 298$ K.

1.3$_A$ A typical global concentration of hydroxyl (OH) radicals is about 10^6 molecules cm^{-3}. What is the mixing ratio corresponding to this concentration at sea level and 298 K?

1.4$_A$ Measurements of dimethyl sulfide (CH_3SCH_3) during the Aerosol Characterization Experiment-1 (ACE-1) conducted November–December 1995 off Tasmania were in the range of 250–500 ng m^{-3}. Convert these values to mixing ratios in ppt at 298 K at sea level.

1.5$_A$ You consume 500 gallons of gasoline per year in your car (1 gal = 3.7879 L). Assume that gasoline can be represented as consisting entirely of C_8H_{18}. Gasoline has a density of 0.85 g cm^{-3}. Assume that combustion of C_8H_{18} leads to CO_2 and H_2O. How many kilograms of CO_2 does your driving contribute to the atmosphere each year? (Problem suggested by T. S. Dibble.)

1.6$_A$ Show that 1 ppm CO_2 in the atmosphere corresponds to 2.1 Pg carbon and therefore that the current atmospheric level of 375 ppm corresponds to 788 Pg C.

REFERENCES

Bentley, C. R. (1997) Rapid sea-level rise soon from West Antarctica Ice Sheet collapse? *Science* **275**, 1077–1078.

Holton, J. R., Haynes, P. H., McIntyre, M. E., Douglass, A. R., Rood, R. B., and Pfister, L. (1995) Stratosphere-troposphere exchange, *Rev. Geophys.* **33**, 403–439.

Intergovernmental Panel on Climate Change (IPCC) (2001) *Climate Change 2001: The Scientific Basis*, Cambridge Univ. Press, Cambridge, UK.

Kasting, J. F. (2001) The rise of atmospheric oxygen, *Science* **293**, 819–820.

Meehl, G. A., and Tebaldi, C. (2004) More intense, more frequent, and longer lasting heat waves in the 21st Century, *Science* **305**, 994–997.

Trenberth, K. E., and Smith, L. (2005) The mass of the atmosphere: A constraint on global analyses, *J. Climate* **18**, 864–875.

Walker, J. C. G. (1977) *Evolution of the Atmosphere*, Macmillan, New York.

2 Atmospheric Trace Constituents

Virtually every element in the periodic table is found in the atmosphere; however, when classifying atmospheric species according to chemical composition, it proves to be convenient to use a small number of major groupings such as

1. Sulfur-containing compounds.
2. Nitrogen-containing compounds.
3. Carbon-containing compounds.
4. Halogen-containing compounds.

Obviously these categories are not exclusive; many sulfur-containing compounds, for example, also include atoms of carbon. And virtually all the atmospheric halogens involve a carbon atom backbone. We do not include in the list above species of the general formula H_xO_y; with the exception of water and hydrogen peroxide (H_2O_2), these are all radical species (e.g., hydroxyl, OH) that play key roles in atmospheric chemistry but do not necessarily require a separate category. Every substance emitted into the atmosphere is eventually removed so that a cycle of the elements in that substance is established. This is called the *biogeochemical cycle* of the element. The biogeochemical cycle of an element or a compound refers to the transport of that substance among atmospheric, oceanic, biospheric, and land compartments, the amounts contained in the different reservoirs, and the rate of exchange among them. The circulation of water among oceans, atmosphere, and continents is a prime example of a biogeochemical cycle. The term *biogeochemical cycle* is often used to describe the global or regional cycles of the "life elements," C, O, N, S, and P, with reservoirs including the atmosphere, the ocean, the sediments, and living organisms.

A condition of "air pollution" may be defined as a situation in which substances that result from anthropogenic activities are present at concentrations sufficiently high above their normal ambient levels to produce a measurable effect on humans, animals, vegetation, or materials. This definition could include any substance, whether noxious or benign; however, the implication is that the effects are undesirable. Traditionally, air pollution has been viewed as a phenomenon characteristic only of large urban centers and industrialized regions. It is now clear that dense urban centers are just "hotspots" in a continuum of trace species concentrations over the entire Earth. Both urban smogs and stratospheric ozone depletion by chlorofluorocarbons are manifestations of what might be termed in the broadest sense as air pollution.

Atmospheric Chemistry and Physics: From Air Pollution to Climate Change, Second Edition, by John H. Seinfeld and Spyros N. Pandis. Copyright © 2006 John Wiley & Sons, Inc.

The main questions that we will seek to answer with respect to classes of atmospheric compounds are as follows:

1. What are the species present in the atmosphere? What are their natural and anthropogenic sources?
2. What chemical reactions do they undergo in the atmosphere? What are the rates of these reactions?
3. What are the products of atmospheric transformations?
4. What effect does the presence of the compound and its chemical transformation products have on the atmosphere?

In this chapter we focus on the first question.

2.1   ATMOSPHERIC LIFETIME

Imagine that we could follow all the individual molecules of a substance emitted into the air. Some might be removed close to their point of emission by contact with airborne droplets or the Earth's surface. Others might get carried high into the atmosphere and be transported a great distance before they ultimately are removed. Averaging the life histories of all molecules of a substance yields an average lifetime or average *residence time* for that substance. This residence time tells us on average how long a representative molecule of the substance will stay in the atmosphere before it is removed. The atmosphere presents two ultimate exits: precipitation and the surface of the Earth itself. Species released into the air must sooner or later leave by one of these two routes.

Atmospheric species removal processes can be conveniently grouped into two categories: dry deposition and wet deposition. *Dry deposition* denotes the direct transfer of species, both gaseous and particulate, to the Earth's surface and proceeds without the aid of precipitation. *Wet deposition*, on the other hand, encompasses all processes by which airborne species are transferred to the Earth's surface in aqueous form (i.e., rain, snow, or fog): (1) dissolution of atmospheric gases in airborne droplets, for example, cloud droplets, rain, or fog; (2) removal of atmospheric particles when they serve as nuclei for the condensation of atmospheric water to form a cloud or fog droplet and are subsequently incorporated in the droplet; and (3) removal of atmospheric particles when the particle collides with a droplet both within and below clouds.

By "particulate matter" we refer to any substance, except pure water, that exists as a liquid or solid in the atmosphere under normal conditions and is of microscopic or submicroscopic size but larger than molecular dimensions. Among atmospheric constituents, particulate matter is unique in its complexity. Airborne particulate matter results not only from direct emissions of particles but also from emissions of certain gases that either condense as particles directly or undergo chemical transformation to a species that condenses as a particle. A full description of atmospheric particles requires specification of not only their concentration but also their size, chemical composition, phase (i.e., liquid or solid), and morphology.

Once particles are in the atmosphere, their size, number, and chemical composition are changed by several mechanisms until ultimately they are removed by natural processes. Some of the physical and chemical processes that affect the "aging" of atmospheric particles are more effective in one regime of particle size than another. In spite of the specific processes that affect particulate aging, the usual residence time of particles in the lower atmosphere does not exceed several weeks. Very close to the ground, the main mechanisms for particle removal are settling and dry deposition on surfaces; whereas at altitudes above about 100 m, precipitation scavenging is the predominant removal mechanism.

As air rises through a cloud and becomes slightly supersaturated with water vapor (i.e., as its relative humidity exceeds 100%), cloud droplets form on condensation nuclei— usually soluble aerosol particles (e.g., microscopic particles of various salts) that exist in the atmosphere at concentrations of $100-3000 \, cm^{-3}$—and grow by condensation of water vapor. As the droplets grow and collide with each other they become raindrops, which grow rapidly as they fall and accrete cloud droplets.

The fundamental physical principle governing the behavior of a chemical in the atmosphere is conservation of mass. In any imaginary volume of air the following balance must hold:

$$\begin{array}{ccccc}
\text{Rate of the} & \text{rate of} & \text{rate of} & \text{rate of} & \text{rate of} \\
\text{species} & - \ \text{species} & + \ \text{introduction} & - \ \text{removal of} \ = & \text{accumulation} \\
\text{flowing in} & \text{flowing out} & \text{(emission) of} & \text{species} & \text{of species in} \\
& & \text{species} & & \text{imaginary volume}
\end{array}$$

This balance must hold from the smallest volume of air all the way up to the entire atmosphere.

If we let Q denote the total mass of the substance in the volume of air; F_{in} and F_{out} the mass flow rates of the substance in and out of the air volume, respectively; P the rate of introduction of the species from sources; and R the rate of removal of the species, then conservation of mass can be expressed mathematically as

$$\frac{dQ}{dt} = (F_{in} - F_{out}) + (P - R) \tag{2.1}$$

If the amount Q of the substance in the volume or reservoir is not changing with time, then Q is a constant and $dQ/dt = 0$. In order for Q to be unchanging, all the sources of the substance to the reservoir must be precisely balanced by the sinks of the substance. This means that

$$F_{in} + P = F_{out} + R \tag{2.2}$$

In such a case *steady-state* conditions are said to hold.

If the volume we are referring to is the entire atmosphere, then $F_{in} = 0$ and $F_{out} = 0$, and for a substance at steady-state conditions, its rate of injection from sources must equal its rate of removal, $P = R$. The average residence time or lifetime τ, in terms of the quantities introduced earlier, is

$$\tau = \frac{Q}{R + F_{out}} \tag{2.3}$$

Since at steady-state conditions $R + F_{out} = P + F_{in}$, the lifetime is also given by

$$\tau = \frac{Q}{P + F_{in}} \tag{2.4}$$

If the entire atmosphere is taken as the reservoir, then under steady-state conditions

$$\tau = \frac{Q}{R} = \frac{Q}{P} \tag{2.5}$$

As an illustration of the concept of mean lifetime, consider all sulfur-containing compounds in the troposphere. If the average mixing ratio of these compounds is 1 part per billion by mass (ppbm) and a steady state is assumed to exist, then with the mass of the troposphere about 4×10^{21} g, the total mass of sulfur-containing compounds in the troposphere is $Q = 4 \times 10^{12}$ g. If natural and anthropogenic sources of sulfur contribute to give a total P of about 200×10^{12} g yr^{-1}, the mean lifetime of sulfur compounds in the troposphere is estimated to be

$$\tau = \frac{4 \times 10^{12} \text{ g}}{200 \times 10^{12} \text{ g yr}^{-1}} = 1 \text{ week}$$

Calculations of lifetimes can be useful in estimating how far from its source a species is likely to remain airborne before it is removed from the atmosphere.

If we consider a particular region of the atmosphere, say, the volume of air over a city or the volume of air in the Northern Hemisphere or the entire stratosphere, we can define a characteristic mixing time for that volume as the time needed to thoroughly mix a chemical in that volume of air. Call the characteristic mixing time τ_M. A reservoir is poorly mixed for a particular species if the characteristic mixing time, τ_M, is not small compared with the species residence time, τ. Note that this means that a particular reservoir can be well mixed for some species and poorly mixed for others, depending on the residence time of each species. Furthermore, as we have seen, the mixing times in the atmosphere are different for different directions. For example, as we have noted, the characteristic vertical mixing time in the troposphere, the time required to mix a species uniformly from the ground up to the tropopause, is about one week; whereas the troposphere's horizontal mixing time, the time required to mix a constituent thoroughly around the globe in the troposphere, is about one year. Thus the troposphere can be considered well mixed for ^{85}Kr, which has a residence time of 10 years; but for sulfur compounds, which are estimated to have a residence time of about one week, the troposphere is not even well mixed vertically.

The stratosphere can be considered well mixed vertically only for atmospheric species with lifetimes substantially exceeding 50 years. In fact, one of the only examples of such a long-lived species is He, which has its source at the Earth's surface and its sink as escape through the very top of the atmosphere into space. Thus the stratosphere is poorly mixed vertically for essentially all atmospheric trace constituents.

Frequently the rate at which a chemical is removed from the atmosphere is proportional to its concentration (first-order loss)—the more that is present, the faster its rate of removal. This is generally the case for both dry deposition at the Earth's surface and scavenging by cloud droplets. Consider a species for which steady-state conditions hold and which is removed at a rate proportional to its concentration with a proportionality constant λ. Such a species is ^{85}Kr, the only significant removal process for which is radioactive decay. For ^{85}Kr, then

$$\tau = \frac{Q}{\lambda Q} = \frac{1}{\lambda} \tag{2.6}$$

Thus to estimate the lifetime of ^{85}Kr does not even require knowledge of its atmospheric abundance Q, but only of the radioactive decay constant. Consequently, it does not matter whether ^{85}Kr is uniformly mixed throughout the entire atmosphere or not to estimate its lifetime. In a case where the removal process is first-order, then even for a poorly mixed species a simple and accurate estimate for its lifetime can be obtained provided that its removal rate constant can be accurately estimated.

Now consider a species with mass Q in the atmosphere that is removed by two independent processes, the first at a rate $k_1 Q$ and the second at a rate $k_2 Q$, where k_1 and k_2 are the first-order removal coefficients. Its overall lifetime is given by

$$\tau = \frac{Q}{(k_1 + k_2)Q} = \frac{1}{k_1 + k_2} \tag{2.7}$$

or

$$\frac{1}{\tau} = k_1 + k_2 \tag{2.8}$$

Process 1, for example, could be dry deposition at the Earth's surface and process 2 cloud scavenging. We can actually associate time constants with the two individual removal processes,

$$\tau_1 = \frac{1}{k_1}, \quad \tau_2 = \frac{1}{k_2} \tag{2.9}$$

where τ_1 can be thought of as the lifetime of the species against removal process 1, and τ_2 the lifetime against process 2. From (2.8) and (2.9) we can express the overall lifetime τ in terms of the two individual removal times τ_1 and τ_2 by

$$\frac{1}{\tau} = \frac{1}{\tau_1} + \frac{1}{\tau_2} \tag{2.10}$$

Equation (2.10) shows that separate removal paths add together to give a total lifetime, like electrical resistances in parallel add to give a total resistance that is even smaller than the smallest resistance. From (2.10)

$$\tau = \frac{\tau_1 \tau_2}{\tau_1 + \tau_2} \tag{2.11}$$

If $\tau_1 \gg \tau_2$, the lifetime associated with removal by process 1 is much longer than that associated with process 2, process 2 is the more effective removal mechanism, and $\tau \simeq \tau_2$. Thus, when there are several competing removal paths, in order to estimate the overall lifetime of a species, focus should always be on improving estimates for the fastest removal rate.

Removal in the troposphere of those compounds that react with the hydroxyl radical occurs according to the chemical reaction

$$OH + A \xrightarrow{k} products$$

The parameter k is the rate constant for the reaction, such that the rate of the reaction is $k[OH][A]$, where the brackets denote the concentration of the species contained therein.

From the analysis of atmospheric residence times, if Q is the total quantity of species A in the troposphere and its rate of removal is $R = k[OH]Q$, then the compound's mean lifetime is, from (2.6)

$$\tau = \frac{Q}{k[OH]Q}$$
$$= \frac{1}{k[OH]} \tag{2.12}$$

where $[OH]$ is an appropriate globally averaged tropospheric concentration of OH radicals.

Let us develop the equations governing the total moles of a species i in the troposphere Q_i. The dynamic material balance can be written as

$$\frac{dQ_i}{dt} = P_i - R_i \tag{2.13}$$

where P_i and R_i represent the source and loss rates. The terms P_i and R_i consist of the following contributions:

$$P_i \begin{cases} P_i^n & \text{natural emissions} \\ P_i^a & \text{anthropogenic emissions} \\ P_i^c & \text{chemical reactions} \end{cases}$$

$$R_i \begin{cases} R_i^d & \text{dry deposition} \\ R_i^w & \text{wet deposition} \\ R_i^c & \text{chemical reactions} \\ R_i^t & \text{transport to the stratosphere} \end{cases}$$

The loss processes are usually represented as first order; for example, $R_i^d = k_i^d Q_i$, where the first-order rate constants, which we will denote by k's, must be specified. Thus (2.13) becomes

$$\frac{dQ_i}{dt} = P_i^n + P_i^a + P_i^c - (k_i^d + k_i^w + k_i^c + k_i^t)Q_i \tag{2.14}$$

If the concentration of the species is not changing, then a steady state may be presumed in which

$$P_i^n + P_i^a + P_i^c - (k_i^d + k_i^w + k_i^c + k_i^t)Q_i = 0 \tag{2.15}$$

The lifetime of species i can be calculated by either

$$\tau_i = \frac{1}{k_i^d + k_i^w + k_i^c + k_i^t} \tag{2.16}$$

or

$$\tau_i = \frac{Q_i}{P_i^n + P_i^a + P_i^c} \tag{2.17}$$

To use (2.16) the individual first-order rate constants for removal must be estimated, whereas in (2.17), estimates for the total number of moles in the troposphere, which can be derived from a concentration measurement, and for the source strength terms are needed. If the k_i values are difficult to specify, mean residence times are often estimated from (2.17).

2.2 SULFUR-CONTAINING COMPOUNDS

Sulfur is present in the Earth's crust at a mixing ratio of less than 500 parts per million by mass and in the Earth's atmosphere at a total volume mixing ratio of less than 1 ppm. Yet, sulfur-containing compounds exert a profound influence on the chemistry of the atmosphere and on climate.

Table 2.1 lists atmospheric sulfur compounds. The principal sulfur compounds in the atmosphere are H_2S, CH_3SCH_3, CS_2, OCS, and SO_2. Sulfur occurs in five oxidation states in the atmosphere. (See Box) Chemical reactivity of atmospheric sulfur compounds is inversely related to their sulfur oxidation state. Reduced sulfur compounds, those with oxidation state -2 or -1, are rapidly oxidized by the hydroxyl radical and, to a lesser extent, by other species, with resulting atmospheric lifetimes of a few days. The water solubility of sulfur species increases with oxidation state; reduced sulfur species occur preferentially in the gas phase, whereas the $S(+6)$ compounds often tend to be found in particles or droplets. Once converted to compounds in the $S(+6)$ state, sulfur species residence times are determined by removal by wet and dry deposition.

Oxidation State The oxidation states of atoms in covalent compounds are obtained by arbitrarily assigning the electrons to particular atoms. For a covalent bond between two identical atoms, the electrons are split equally between the two. When two different atoms are involved, the shared electrons are assigned completely to the atom that has the stronger attraction for the electrons. In the water molecule, for example, oxygen has a greater attraction for electrons than hydrogen, so in assigning the oxidation states of oxygen and hydrogen in H_2O, it is assumed that the oxygen atom possesses all the electrons. This gives the oxygen an excess of two electrons, and its oxidation state is -2. Each hydrogen has no electrons, and the oxidation state of each hydrogen is $+1$.

TABLE 2.1   Atmospheric Sulfur Compounds

Oxidation State	Compound		Chemical Structure	Usual Atmospheric State
	Name	Formula		
-2	Hydrogen sulfide	H_2S	$H-S-H$	Gas
	Dimethyl sulfide (DMS)	CH_3SCH_3	CH_3-S-CH_3	Gas
	Carbon disulfide	CS_2	$S=C=S$	Gas
	Carbonyl sulfide	OCS	$O=C=S$	Gas
	Methyl mercaptan	CH_3SH	CH_3-S-H	Gas
-1	Dimethyl disulfide	CH_3SSCH_3	$CH_3-S-S-CH_3$	Gas
0	Dimethyl sulfoxide	CH_3SOCH_3	$CH_3-\overset{\overset{\displaystyle O}{\|}}{S}-CH_3$	Gas
4	Sulfur dioxide	SO_2	$O=S=O$	Gas
		$SO_2 \cdot H_2O$		Aqueous
	Bisulfite ion	HSO_3^-		Aqueous
	Sulfite ion	SO_3^{2-}		Aqueous
6	Sulfuric acid	H_2SO_4	$HO-\overset{\overset{\displaystyle O}{\|}}{\underset{\underset{\displaystyle O}{\|}}{S}}-OH$	Gas aqueous/aerosol
	Bisulfate ion	HSO_4^-	$HO-\overset{\overset{\displaystyle O}{\|}}{\underset{\underset{\displaystyle O}{\|}}{S}}-O^-$	Aqueous/aerosol
	Sulfate ion	SO_4^{2-}	$^-O-\overset{\overset{\displaystyle O}{\|}}{\underset{\underset{\displaystyle O}{\|}}{S}}-O^-$	
	Methane sulfonic acid (MSA)	CH_3SO_3H	$CH_3-\overset{\overset{\displaystyle O}{\|}}{\underset{\underset{\displaystyle O}{\|}}{S}}-OH$	Gas/aqueous
	Dimethyl sulfone	$CH_3SO_2CH_3$	$CH_3-\overset{\overset{\displaystyle O}{\|}}{\underset{\underset{\displaystyle O}{\|}}{S}}-CH_3$	Gas
	Hydroxymethane sulfonic acid (HMSA)	$HOCH_2SO_3H$	$HOCH_2-\overset{\overset{\displaystyle O}{\|}}{\underset{\underset{\displaystyle O}{\|}}{S}}-OH$	Aqueous

Rules for assigning oxidation states are

1. The oxidation state of an atom in an element is 0.
2. The oxidation state of a monatomic ion is the same as its charge.
3. Oxygen is assigned an oxidation state of -2 in its covalent compounds, such as CO, CO_2, SO_2, and SO_3. An exception to this rule occurs in peroxides, where each oxygen is assigned an oxidation state of -1.
4. In its covalent compounds with nonmetals, hydrogen is assigned an oxidation state of $+1$. Examples include HCl, H_2O, NH_3, and CH_4.
5. In its compounds fluorine is always assigned an oxidation state of -1.
6. The sum of the oxidation states must be zero for an electrically neutral compound. For an ion, the sum must equal the charge of the ion. For example, the sum of oxidation states for the nitrogen and hydrogen atoms in NH_4^+ is $+1$, and the oxidation state of nitrogen is -3. For NO_3^-, the sum of oxidation states is -1. Since oxygen has an oxidation state of -2, nitrogen must have an oxidation state of $+5$. Sometimes, oxidation states are indicated with roman numerals, for example, the sulfur atom in SO_4^{2-} is $+VI$.

Oxidation states of sulfur in various compounds of atmospheric importance are as follows:

$$H_2S = -2$$
$$SO_2 = +4$$
$$SO_3^{2-} = +4$$
$$H_2SO_4 = +6$$
$$SO_4^{2-} = +6$$

Oxidation states of nitrogen in atmospheric species are as follows:

$$NH_3, RNH_2, R_2NH, R_3N = -3$$
$$N_2 = 0$$
$$N_2O = +1$$
$$NO = +2$$
$$HNO_2 = +3$$
$$NO_2 = +4$$
$$HNO_3, NO_3^-, N_2O_5 = +5$$
$$NO_3 = +6$$

Table 2.2 presents estimates of total sulfur emissions to the atmosphere, both anthropogenic and natural, including estimated division between Northern and Southern Hemispheres. Current estimates place total global emissions (excluding seasalt) in the range of 98–120 Tg(S) yr^{-1}. At present, anthropogenic emissions account for about 75% of total sulfur emissions, and 90% of the anthropogenic emissions occur in the Northern

TABLE 2.2 Global Sulfur Emissions Estimates, Tg(S) yr^{-1}

Source	H$_2$S	DMS	CS$_2$	OCS[d]	SO$_2$	SO$_4$	Total[a]
Fossil fuel combustion + industry		Total reduced S: 2.2			70	2.2	71–77 (mid-1980s) (68/6)
Biomass burning	<0.01?	—	<0.01?	0.075	2.8	0.1	2.2–3.0(1.4/1.1)
Oceans	<0.3	15–25	0.08	0.08	—	40–320	15–25(8.4/11.6)[b]
Wetlands	0.006–1.1	0.003–0.68	0.0003–0.06	—	—	—	0.01–2 (0.8/0.2)
Plants + soils	0.17–0.53	0.05–0.16	0.02–0.05	—	—	2–4	0.25–0.78 (0.3/0.2)[c]
Volcanoes	0.5–1.5	—	—	0.01	7–8	2–4	9.3–11.8(7.6/3.0)
Anthropogenic (total)							73–80
Natural (total, without sea salt and soil dust)							25–40
Total							98–120

[a]Numbers in parentheses are fluxes from Northern Hemisphere/Southern Hemisphere.
[b]Excluding seasalt contributions.
[c]Excluding soil dust contributions.
[d]Andreae and Crutzen (1997).
Source: Berresheim et al. (1995).

30

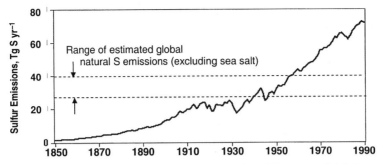

FIGURE 2.1 Estimated global anthropogenic sulfur emissions. The range of global natural sulfur emissions (excluding seasalt) is indicated. [Adapted from Lefohn et al. (1999).]

Hemisphere. Figure 2.1 shows estimates of global anthropogenic sulfur emissions since 1850, and Table 2.3 summarizes observed mixing ratios and atmospheric lifetimes of atmospheric sulfur gases.

2.2.1 Dimethyl Sulfide (CH₃SCH₃)

Dimethyl sulfide (DMS) is the dominant sulfur compound emitted from the world's oceans. DMS was first measured in the surface ocean by Lovelock et al. (1972), who showed that DMS, produced by marine phytoplankton, was ubiquitous in the surface ocean waters. It had been known for a number of years that the global sulfur budget could not be balanced without a substantial flux of a sulfur-containing compound from the oceans to the atmosphere. Lovelock's measurements pointed to DMS as that compound. Since Lovelock's pioneering data, many expeditions have measured DMS concentrations in both ocean waters and the marine atmosphere. Kettle et al. (1999) summarize all the oceanic DMS data available as of 1999, comprising 134 cruises and over 15,000 individual measurements. There is a rough correspondence between areas of high DMS concentrations and blooming of marine phytoplankton that produce DMS. In January, the highest ocean water DMS concentrations occur in the Southern Hemisphere,

TABLE 2.3 Average Lifetimes and Observed Mixing Ratios of Tropospheric Sulfur Compounds

		Mixing Ratio, ppt			
Species	Average Lifetime	Marine air	Clean Continental	Polluted Continental	Free Troposphere
H_2S	2 days	0–110	15–340	0–800	1–13
OCS	7 years	530	510	520	510
CS_2	1 week	30–45	15–45	80–300	≤ 5
CH_3SCH_3	0.5 day	5–400	7–100	2–400	≤ 2
SO_2	2 days	10–200	70–200	100–10,000	30–260
SO_4^{2-}	5 days	5–300[a]	10–120	100–10,000	5–70

[a]Nonseasalt sulfate.

Source: Lelieveld et al. (1997).

especially in Antarctic waters; in July, the highest concentrations are in the Northern Hemisphere oceans.

DMS is thought to originate from the decomposition of dimethyl sulfoniopropionate produced by marine organisms, in particular, phytoplankton (Andreae 1990). Its concentration in the upper layer of the ocean varies between a few nanograms of S per liter to a few micrograms of S per liter. The DMS surface seawater concentration is highly nonuniform; its average concentration is approximately 100 nanograms (ng) of S per liter. It has been observed that the concentration of DMS is dependent on diurnal (Andreae and Barnard 1984) and seasonal variations (Turner and Liss 1985), and on depth and location (Andreae and Raemdonck 1983). On the basis of DMS concentrations in the atmosphere and its Henry's law constant in seawater (see Chapter 7), oceanic DMS concentrations are greatly in excess of those that would be in equilibrium with atmospheric values. The result of this lack of equilibrium is a flux of DMS from the ocean to the atmosphere.

Once the importance of DMS to the global sulfur cycle was established, numerous measurements of DMS concentrations in the marine atmosphere have been conducted. The average DMS mixing ratio in the marine boundary layer (MBL) is in the range of 80–110 ppt but can reach values as high as 1 ppb over entrophic (e.g., coastal, upwelling) waters. DMS mixing ratios fall rapidly with altitude to a few parts per trillion in the free troposphere. After transfer across the air–sea interface into the atmosphere, DMS reacts predominantly with the hydroxyl radical and also with the nitrate (NO_3) radical. Oxidation of DMS is the exclusive source of methane sulfonic acid (MSA) in the atmosphere, and the dominant source of SO_2 in the marine atmosphere. We will return to the atmospheric chemistry of DMS in Chapter 6.

2.2.2 Carbonyl Sulfide (OCS)

Carbonyl sulfide is the most abundant sulfur gas in the global background atmosphere because of its low reactivity in the troposphere and its correspondingly long residence time. It is the only sulfur compound that survives to enter the stratosphere. (An exception is the direct injection of SO_2 into the stratosphere in volcanic eruptions.) In fact, the input of OCS into the stratosphere is considered to be responsible for the maintenance of the normal stratospheric sulfate aerosol layer.

The estimated global budget of OCS is given in Table 2.4. The main atmospheric sources of OCS are OCS emissions at the surface and conversion of CS_2 and dimethyl sulfide (DMS) in the atmosphere. Because of its long tropospheric lifetime, much of the OCS reaches the stratosphere, where photolysis and oxidation lead to SO_2 and eventually to sulfate particles. The OCS mixing ratio in the tropospheric is relatively constant at about 500 ppt (Chin and Davis 1995). Enhanced OCS mixing ratios up to 600 ppt from biomass burning emissions have been found below the tropical tropopause, at altitudes between 10 and 18 km (Notholt, et al. 2003). The other sulfur compounds, CS_2 and DMS, have much shorter tropospheric lifetimes than OCS and therefore do not contribute appreciably to the sulfur budget of the stratosphere under volcanically quiescent periods.

On the basis of atmospheric measurements, Chin and Davis (1995) estimated the total quantity of OCS in the atmosphere to be 5.2 Tg, of which 4.63 Tg is in the troposphere and 0.57 Tg in the stratosphere. Based on the estimated global OCS source strength of 0.86 Tg yr^{-1}, the global atmospheric lifetime of OCS is estimated to be about 6 years. We will return to the global cycle and chemistry of OCS in Chapter 5 in connection with the stratospheric aerosol layer.

TABLE 2.4   Global Budget of Carbonyl Sulfide (OCS)

	Sources, 10^9 g S yr$^{-1\,a}$	
Direct OCS flux from oceans	41	(154)
Indirect OCS flux as CS_2 from oceans	84	(54)
Indirect OCS flux as DMS from oceans	154	(37)
Direct anthropogenic OCS flux	64	(32)
Indirect OCS flux as anthropogenic CS_2	116	(58)
Total sources	459	
	Sinks, 10^9 g S yr$^{-1\,a}$	
OCS uptake by soils	130	(56)
OCS uptake by plants	238	(30)
OCS reaction with OH radical	94	
Total sinks	462	

[a]Estimated uncertainties are shown in parentheses.

Source: Kettle et al. (2002).

2.2.3   Sulfur Dioxide (SO$_2$)

Sulfur dioxide is the predominant anthropogenic sulfur-containing air pollutant. Mixing ratios of SO_2 in continental background air range from 20 ppt to over 1 ppb; in the unpolluted marine boundary layer levels range between 20 and 50 ppt. Urban SO_2 mixing ratios can attain values of several hundred parts per billion. We will consider the atmospheric chemistry of SO_2 in Chapter 6.

2.3   NITROGEN-CONTAINING COMPOUNDS

The strong triple bond of the $N\equiv N$ molecule makes it practically inert; it is extremely stable chemically and is not involved in the chemistry of the troposphere or stratosphere. The important nitrogen-containing trace species in the atmosphere are nitrous oxide (N_2O), nitric oxide (NO), nitrogen dioxide (NO_2), nitric acid (HNO_3), and ammonia (NH_3). The first of these, nitrous oxide (N_2O), is a colorless gas that is emitted almost totally by natural sources, principally by bacterial action in the soil. The gas is employed as an anesthetic and is commonly referred to as "laughing gas." The second, nitric oxide (NO), is emitted by both natural and anthropogenic sources. Nitrogen dioxide (NO_2) is emitted in small quantities from combustion processes along with NO and is also formed in the atmosphere by oxidation of NO. The sum of NO and NO_2 is usually designated as NO_x. Nitric oxide is the major oxide of nitrogen formed during high-temperature combustion, resulting from both the interaction of nitrogen in the fuel with oxygen present in the air and the chemical conversion of atmospheric nitrogen and oxygen at the high temperatures of combustion. Other oxides of nitrogen, such as NO_3 and N_2O_5, exist in the atmosphere in relatively low concentrations but nonetheless participate importantly in atmospheric chemistry. Nitric acid is an oxidation product of NO_2 in the atmosphere. Ammonia (NH_3) is emitted primarily by natural sources. Finally, nitrate and ammonium salts are not emitted in any significant quantities but result from the atmospheric conversion of NO, NO_2, and NH_3.

Nitrogen is an essential nutrient for all living organisms. The primary source of this nitrogen is the atmosphere. However, N_2 is not useful to most organisms until it is "fixed" or converted to a form that can be chemically utilized by the organisms. (Nitrogen fixation refers to the chemical conversion of N_2 to any other nitrogen compound.) The "natural" fixation of N_2 occurs by two types of processes. One is the action of a comparatively few microorganisms that are capable of converting N_2 to ammonia, ammonium ion (NH_4^+), and organic nitrogen compounds. The other natural nitrogen fixation process occurs in the atmosphere by the action of ionizing phenomena, such as cosmic radiation or lightning, on N_2. This process leads to the formation of nitrogen oxides in the atmosphere, which are ultimately deposited on the Earth's surface as biologically useful nitrates.

In addition to natural nitrogen fixation, human activities have led to biological and industrial fixation and fixation by combustion. Humans have increased the cultivation of legumes, which have a symbiotic relationship with certain microorganisms capable of nitrogen fixation. Legumes provide an increase in the soil nitrogen and serve as a valuable food crop. Industrial nitrogen fixation consists primarily of the production of ammonia for fertilizer use. Combustion can also lead to the fixation of nitrogen as NO_x. In the process of nitrification, ammonium is oxidized to NO_2^- and NO_3^- by microbial action. N_2O and NO are byproducts of nitrification; the result is the release of N_2O and NO to the atmosphere. Reduction of NO_3^- to N_2, NO_2, N_2O, or NO is called *denitrification*. Denitrification is accomplished by a number of bacteria and is the process that continually replenishes the atmosphere's N_2. Figure 2.2 depicts the atmospheric nitrogen cycle.

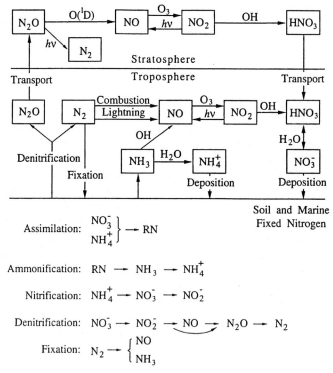

FIGURE 2.2 Processes in the atmospheric cycle of nitrogen compounds. A species written over an arrow signifies reaction with the species from which the arrow originates.

TABLE 2.5   Estimates of the Global N₂O Budget (in TgN/yr) and Values Adopted by IPCC (2001)

Reference:	Mosier et al. (1998b) Kroeze et al. (1999)		Olivier et al. (1998)		IPCC (2001)
Base Year:	1994	Range	1990	Range	1990s
Sources					
Ocean	3.0	1–5	3.6	2.8–5.7	
Atmosphere (NH₃ oxidation)	0.6	0.3–1.2	0.6	0.3–1.2	
Tropical soils					
Wet forest	3.0	2.2–3.7			
Dry savannas	1.0	0.5–2.0			
Temperate soils					
Forests	1.0	0.1–2.0			
Grasslands	1.0	0.5–2.0			
All soils			6.6	3.3–9.9	
Natural subtotal	9.6	4.6–15.9	10.8	6.4–16.8	
Agricultural soils	4.2	0.6–14.8	1.9	0.7–4.3	
Biomass burning	0.5	0.2–1.0	0.5	0.2–0.8	
Industrial sources	1.3	0.7–1.8	0.7	0.2–1.1	
Cattle and feedlots	2.1	0.6–3.1	1.0	0.2–2.0	
Anthropogenic subtotal	8.1	2.1–20.7	4.1	1.3–7.7	6.9
Total sources	17.7	6.7–36.6	14.9	7.7–24.5	
Imbalance (trend)	3.9	3.1–4.7			3.8
Total sinks (stratospheric)	12.3	9–16			12.6
Implied total source	16.2				16.4

Source: IPCC (2001).

2.3.1   Nitrous Oxide (N₂O)

Nitrous oxide (N₂O) is an important atmospheric gas that is emitted predominantly by biological sources in soils and water (Table 2.5). Although by comparison to CO_2 and H_2O, N_2O has a far lower concentration, it is an extremely influential greenhouse gas. This is a result of its long residence time and its relatively large energy absorption capacity per molecule. Per unit mass the global warming potential of N_2O (see Chapter 23) is about 300 times that of CO_2. Tropical soils are the most important individual sources of N_2O to the atmosphere. N_2O is also emitted in smaller quantities by a large number of other sources, such as biomass burning, degassing of irrigation water, agricultural activities, and industrial processes. The oceans are significant N_2O sources. The total preindustrial N_2O source was approximately $10\,Tg(N)\,yr^{-1}$. The current flux of N_2O into the atmosphere that results from anthropogenic activities is estimated by IPCC (2001) to be $6.9\,Tg(N)\,yr^{-1}$.

Nitrous oxide is inert in the troposphere; its major atmospheric sink is photodissociation in the stratosphere (about 90%) and reaction with excited oxygen atoms, $O(^1D)$ (about 10%). Oxidation of N_2O by $O(^1D)$ yields NO, providing the major input of NO to the stratosphere. We will return to this process in Chapter 5. Sources of N_2O exceed estimated sinks by $3.8\,Tg(N)\,yr^{-1}$.

Estimates for the atmospheric lifetime of N_2O come from stratospheric chemical transport models that have been tested against observed N_2O distributions. The best

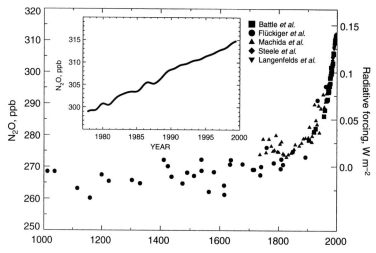

FIGURE 2.3   Atmospheric abundance of N_2O over the last millennium, as determined from ice cores, firn, and whole-air samples (IPCC 2001). Sources of data are indicated, references for which are given in IPCC. The inset contains deseasonalized global averages.

current estimate for the lifetime of N_2O is 120 years. Because of its long lifetime, N_2O exhibits more or less uniform concentrations throughout the troposphere. Ice core records of N_2O show a preindustrial mixing ratio of about 276 ppb. N_2O levels have risen approximately 15% since preindustrial times, reaching 315 ppb in 2000 (Figure 2.3). This observed atmospheric increase is consistent with a difference of 3.8 Tg(N) yr^{-1} excess of sources over sinks, which is in reasonable agreement, given the uncertainties, with the mismatch based on attempting to estimate sources and sinks independently.

2.3.2   Nitrogen Oxides ($NO_x = NO + NO_2$)

The oxides of nitrogen, NO and NO_2, are among the most important molecules in atmospheric chemistry. We will devote in this book considerable attention to their chemistry. Estimated global emissions of NO_x are given in Table 2.6. Aircraft emissions are listed separately in Table 2.6 because they are released predominantly in the free

TABLE 2.6   Estimate of Global Tropospheric NO_x Emissions in Tg N yr^{-1} for Year 2000

Sources	Emissions, Tg N yr^{-1}
Fossil fuel combustion	33.0
Aircraft	0.7
Biomass burning	7.1
Soils	5.6
NH_3 oxidation	—
Lightning	5.0
Stratosphere	<0.5
Total	51.9

Source: IPCC (2001).

troposphere at altitudes of 8–12 km rather than at the surface, and although such emissions are only a small fraction of the total combustion source, they are potentially responsible for a large fraction of the NO_x found at those altitudes at northern midlatitudes.

2.3.3 Reactive Odd Nitrogen (NO_y)

Reactive nitrogen, denoted NO_y, is defined as the sum of the two oxides of nitrogen ($NO_x = NO + NO_2$) and all compounds that are products of the atmospheric oxidation of NO_x. These include nitric acid (HNO_3), nitrous acid (HONO), the nitrate radical (NO_3), dinitrogen pentoxide (N_2O_5), peroxynitric acid (HNO_4), peroxyacetyl nitrate (PAN) ($CH_3C(O)OONO_2$) and its homologs, alkyl nitrates ($RONO_2$), and peroxyalkyl nitrates ($ROONO_2$). Nitric acid (HNO_3) is the major oxidation product of NO_x in the atmosphere. Because of its extreme water solubility, HNO_3 is rapidly deposited on surfaces and in water droplets. Also, in the presence of NH_3, HNO_3 can form an ammonium nitrate (NH_4NO_3) aerosol. The nitrate radical (NO_3) is an important constituent in the chemistry of the troposphere, especially at night. NO_3 is present at night at mixing ratios ranging up to 300 ppt in the boundary layer. Nitrous oxide (N_2O) and ammonia (NH_3) are not considered in this context as reactive nitrogen compounds.

Measurement of total NO_y in the atmosphere provides an important measure of the total oxidized nitrogen content. Concentrations of individual NO_y species relative to the total indicate the extent of interconversion among species. NO_y is indeed closer to a conserved quantity than any of its constituent species (Roberts 1995).

Only during the past two decades have techniques been available with sufficient sensitivity and range of detectability to measure NO_x in nonurban locales (NO_x concentrations below 1 ppb), and as a result the size and reliability of the database needed to define nonurban NO_x concentrations are limited. Measurements taken at isolated rural sites tend to be significantly lower than concentrations measured at less-isolated rural sites and generally range from a few tenths to 1 ppb. Measurements of NO_x in the atmospheric boundary layer and lower free troposphere in remote maritime locations have generally yielded mixing ratios of 0.02–0.04 ppb (20–40 ppt). Although the database is still quite sparse, mixing ratios in remote tropical forests (not under the direct influence of biomass burning) appear to range from 0.02 to 0.08 ppb (from 20 to 80 ppt); the somewhat higher NO_x concentrations found in remote tropical forests, as compared with those observed in remote marine locations, could result from biogenic NO_x emissions from soil.

A summary of the NO_x measurements made in the four regions of the globe mentioned above is presented in Table 2.7. NO_x concentrations decrease sharply as one moves from

TABLE 2.7   Typical Boundary-Layer NO_x Mixing Ratios

Region	NO_x, ppb
Urban–suburban	10–1000
Rural	0.2–10
Remote tropical forest	0.02–0.08
Remote marine	0.02–0.04

Source: National Research Council (1991).

TABLE 2.8   Estimated Annual Global Ammonia Emissions

Source	Amount, Tg N yr^{-1}
Agricultural (domestic animals, synthetic fertilizers, crops)	37.4
Natural (oceans, undisturbed soils, wild animals)	10.7
Biomass burning	6.4
Other (humans and pets, industrial processes, fossil fuels)	3.1
Total	57.6

Source: Bouwman et al. (1997).

urban and suburban to rural sites and then to remote sites over the ocean and tropical forests. The striking difference of three orders of magnitude or more between NO_x concentrations in urban–suburban areas and remote locations is compelling evidence for the dominant role of anthropogenic emissions of NO_x over strong anthropogenic source regions such as North America. Because the ability to measure NO_y was developed only recently, the rural and remote NO_y database is more limited than that for NO_x. Average NO_y concentrations observed at many sites in the United States are quite similar; median mixing ratio values range from 3 to 10 ppb. These are somewhat lower than NO_x mixing ratios typically observed in urban and suburban locations, which range from 10 to 1000 ppb.

2.3.4   Ammonia (NH$_3$)

Ammonia is the primary basic gas in the atmosphere and, after N_2 and N_2O, is the most abundant nitrogen-containing compound in the atmosphere. The significant sources of NH_3 are animal waste, ammonification of humus followed by emission from soils, losses of NH_3-based fertilizers from soils, and industrial emissions (Table 2.8). The ammonium (NH_4^+) ion is an important component of the continental tropospheric aerosol. Because NH_3 is readily absorbed by surfaces such as water and soil, its residence time in the lower atmosphere is estimated to be quite short, about 10 days. Wet and dry deposition of NH_3 are the main atmospheric removal mechanisms for NH_3. In fact, deposition of atmospheric NH_3 and NH_4^+ may represent an important nutrient to the biosphere in some areas. Atmospheric concentrations of NH_3 are quite variable, depending on proximity to a source-rich region. NH_3 mixing ratios over continents range typically between 0.1 and 10 ppb.

2.4   CARBON-CONTAINING COMPOUNDS

2.4.1   Classification of Hydrocarbons

Let us review briefly the classifications of carbon-containing compounds, particularly those of interest in atmospheric chemistry. The carbon atom has four valence electrons and can therefore share bonds with from one to four other atoms. The nature of the carbon–carbon bonding in a hydrocarbon molecule basically governs the properties (as well as the nomenclature) of the molecule.

In some sense the simplest hydrocarbon molecules are those in which all the carbon bonds are shared with hydrogen atoms except for a minimum number required for carbon–carbon bonds. Molecules of this type are referred to as *alkanes* or, equivalently, as *paraffins*. The general chemical formula of alkanes is C_nH_{2n+2}. The first four alkanes having a straight-chain structure are

$$CH_4 \qquad \text{methane}$$
$$CH_3-CH_3 \qquad \text{ethane}$$
$$CH_3-CH_2-CH_3 \qquad \text{propane}$$
$$CH_3-CH_2-CH_2-CH_3 \quad \textit{n-butane}$$

Alkanes need not have a straight-chain structure. If a side carbon chain exists, the name of the longest continuous chain of carbon atoms is taken as the base name, which is then modified to include the type of group. Typical examples of substituted alkanes are (the numbering system is indicated below the carbon atoms):

$$
\begin{array}{ccc}
CH_3 & CH_3 & \\
| & | & \\
CH_3-CH-CH_2-CH-CH_2-CH_3 & & \\
\;\;1\quad\;\;2\quad\;\;3\quad\;\;\;\;4\quad\;\;5\quad\;\;\;6 &
\end{array}
\qquad
\begin{array}{c}
Cl\;\;\;\;Br \\
| \;\;\;\; | \\
CH_3-CH_2-CH-CH-CH_3 \\
\;\;5\quad\;\;\;4\quad\;\;3\quad\;\;2\quad\;\;1
\end{array}
$$

2,4–Dimethylhexane   2-Bromo-3-chloropentane

Alkanes may also be arranged in a ring structure, in which case the molecule is referred to as a *cycloalkane*.

Alkanes generally react by replacement of a hydrogen atom. Once a hydrogen atom is removed from an alkane, the involved carbon atom has an unpaired electron and the molecule becomes a free radical, in this case an alkyl radical. Examples of alkyl radicals are

$$CH_3\cdot \qquad \text{methyl}$$
$$CH_3-CH_2\cdot \qquad \text{ethyl}$$
$$CH_3-CH_2-CH_2\cdot \qquad \textit{n-propyl}$$
$$CH_3-\overset{\cdot}{C}H-CH_3 \qquad \text{isopropyl}$$
$$CH_3-CH_2-CH_2-CH_2\cdot \quad \textit{n-butyl}$$

Alkyl radicals are often simply designated $R\cdot$, where R denotes the chemical formula for any member of the alkyl group. The unpaired electron in a free radical makes the species extremely reactive. Free radicals play an essential role in atmospheric chemistry.

The next class of carbon-containing compounds of interest in atmospheric chemistry is the *alkenes*. In this class two neighboring carbon atoms share a pair of electrons, a so-called double bond. Alkenes are also known as alkylenes or olefins. The location of the carbon atom nearest to the end of the molecule that is the first of the two carbon atoms sharing the double bond is often indicated by the number of the carbon atom. Examples of common alkenes are

$$CH_2{=}CH_2 \qquad \text{ethene (ethylene)}$$
$$CH_3CH{=}CH_2 \qquad \text{propene (propylene)}$$
$$CH_3CH_2CH{=}CH_2 \quad \text{1-butene}$$
$$CH_3CH{=}CHCH_3 \quad \text{2-butene}$$

Molecules with two double bonds are called alkadienes, an example of which is

$$CH_2{=}CH{-}CH{=}CH_2$$

1,3-Butadiene

Molecules with a single triple bond are known as *alkynes*, the first in the series of which is acetylene, $HC{\equiv}CH$.

Double-bonded hydrocarbons may also be arranged in a ring structure. This class of molecules, of which the basic unit is benzene

is called *aromatics*. Other common aromatics are

Toluene   o-Xylene   m-Xylene   p-Xylene

Hydrocarbons may acquire one or more oxygen atoms. Of the oxygenated hydrocarbons, two classes of *carbonyls* that are of considerable importance in the atmosphere are *aldehydes* and *ketones*. In each type of molecule, a carbon atom and an oxygen atom are joined by a double bond. Aldehydes have the general form

$$\begin{matrix} O \\ \| \\ R{-}C{-}H \end{matrix}$$

whereas ketones have the structure

$$\begin{matrix} O \\ \| \\ R{-}C{-}R \end{matrix}$$

Thus the distinction lies in whether the carbon atom is bonded to one or two alkyl groups. Examples of aldehydes and ketones are

$$\begin{matrix} O \\ \| \\ HCH \end{matrix} \qquad \begin{matrix} O \\ \| \\ CH_3CH \end{matrix} \qquad \begin{matrix} O \\ \| \\ CH_3CCH_3 \end{matrix}$$

Formaldehyde   Acetaldehyde   Acetone

$$\begin{matrix} O \\ \| \\ CH_3CCH_2CH_3 \end{matrix} \qquad \qquad \begin{matrix} O \\ \| \\ CH_2{=}CHCH \end{matrix}$$

Methylethylketone   Acrolein

Table 2.9 lists a number of organic species found in the atmosphere.

TABLE 2.9   Atmospheric Organic Species

Type of Compound	General Chemical Formula	Examples
Alkanes	R—H	CH_4, methane
		CH_3CH_3, ethane
Alkenes	$R_1C{=}CR_2$	$CH_2{=}CH_2$, ethene or ethylene
		$CH_3{-}CH{=}CH_2$, propene
Alkynes	$RC{\equiv}CR$	$HC{\equiv}CH$, acetylene
Aromatics	C_6R_6 (cyclic)	C_6H_6, benzene
		$C_6H_5(CH_3)$, toluene
Alcohols	R—OH	CH_3OH, methanol
		CH_3CH_2OH, ethanol
Aldehydes	R—CHO	HCHO, formaldehyde
		CH_3CHO, acetaldehyde
Ketones	RCOR	$CH_3C(O)CH_3$, acetone
Peroxides	R—OOH	CH_3OOH, methylhydroperoxide
Organic acids	R—COOH	HC(O)OH, formic acid
		$CH_3C(O)OH$, acetic acid
Organic nitrates	$R{-}ONO_2$	CH_3ONO_2, methyl nitrate
		$CH_3CH_2ONO_2$, ethyl nitrate
Alkyl peroxy nitrates	RO_2NO_2	$CH_3O_2NO_2$, methyl peroxynitrate
Acylperoxy nitrates	$R{-}C(O)OONO_2$	$CH_3C(O)O_2NO_2$,
		peroxyacetyl nitrate (PAN)
Biogenic compounds	C_5H_8	$CH_2{=}C(CH_3){-}CH{=}CH_2$, isoprene
	$C_{10}H_{16}$	α-pinene, β-pinene
Multifunctional species		$CH_3C(O)CHO$, methylglyoxal
		$CH_2(OH)CHO$, glycolaldehyde

Source: Adapted from Brasseur et al. (1999).

2.4.2   Methane

Methane is the most abundant hydrocarbon in the atmosphere. Table 2.10 summarizes the global sources of CH_4, which are estimated at $598\,Tg(CH_4)\,yr^{-1}$. Methane is removed from the atmosphere through reaction with hydroxyl radicals (OH) in the troposphere, estimated at $506\,Tg(CH_4)\,yr^{-1}$, and by reaction in the stratosphere, estimated at $40\,Tg(CH_4)\,yr^{-1}$. Microbial uptake in soils contributes an estimated $30\,Tg(CH_4)\,yr^{-1}$ removal rate. The estimated imbalance of $+22\,Tg(CH_4)\,yr^{-1}$ between the current sources and sinks of CH_4 in Table 2.10 indicates that methane is accumulating in the atmosphere.

Atmospheric CH_4 concentrations have changed considerably over time. Figure 2.4 shows CH_4 mixing ratios over the past 1000 years. CH_4 has increased from a preindustrial mixing ratio near 700 ppb to a present-day value of 1745 ppb. All the data points are based on CH_4 in air bubbles trapped in ice cores in Antarctica, with the exception of the solid curve, which is based on atmospheric measurements made at Cape Grim, Tasmania.

TABLE 2.10 Estimates of the Global CH$_4$ Budget (in Tg CH$_4$ yr^{-1}) and Values Adopted by IPCC (2001)

Reference: Base Year:	Fung et al. (1991) 1980s	Hein et al. (1997) —	Lelieveld et al. (1998) 1992	Houweling et al. (1999) —	Mosier et al. (1998a) 1994	Olivier et al. (1999) 1990	Cao et al. (1998) —	IPCC (2001) 1998
Natural sources								
Wetlands	115	237	225[b]	145			92	
Termites	20	—	20	20				
Ocean	10	—	15	15				
Hydrates	5	—	10	—				
Anthropogenic sources								
Energy	75	97	110	89	80	109		
Landfills	40	35	40	73		36		
Ruminants	80	90[a]	115	93	80	93[a]		
Waste treatment	—	—[a]	25	—	14	—[a]		
Rice agriculture	100	88	—[b]	—	25–54	60	53	
Biomass burning	55	40	40	40	34	23		
Other	—	—	—	20	15			
Total source	500	587	600					598
			Imbalance (Trend)					+22
Sinks								
Soils	10	—	30	30	44			30
Tropospheric OH	450	489	510					506
Stratospheric loss	—	46	40					40
Total sink	460	535	580					576

[a]Waste treatment included under ruminants.
[b]Rice included under wetlands.

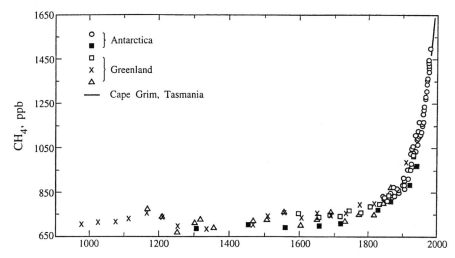

FIGURE 2.4   Methane mixing ratios over the last 1000 years as determined from ice cores from Antarctica and Greenland (IPCC 1995). Different data points indicate different locations. Atmospheric data from Cape Grim, Tasmania, are included to demonstrate the smooth transition from ice core to atmospheric measurements.

2.4.3   Volatile Organic Compounds

The term volatile organic compounds (VOCs) is used to denote the entire set of vapor-phase atmospheric organics excluding CO and CO_2. VOCs are central to atmospheric chemistry from the urban to the global scale.

As an illustration of the variety of organic compounds identified in the atmosphere, Table 2.11 lists the median concentrations of the 25 most abundant nonmethane organic species measured in the 1987 Southern California Air Quality Study.

2.4.4   Biogenic Hydrocarbons

Vegetation naturally releases organic compounds to the atmosphere. In 1960, Went (1960) first proposed that natural foliar emissions of VOCs from trees and other vegetation could have a significant effect on the chemistry of the Earth's atmosphere.

Measurements in wooded and agricultural areas coupled with emission studies from selected individual trees and agricultural crops have demonstrated the ubiquitous nature of biogenic emissions and the variety of organic compounds that can be emitted. Table 2.12 shows the chemical structures of some of the common biogenic hydrocarbons. Each of the compounds shown in Table 2.12 is characterized by an olefinic double bond that renders the molecule highly reactive in the atmosphere, with the result that the lifetimes of these molecules tend to be quite short.

Isoprene (2-methyl-1,3-butadiene, C_5H_8) is unique among the biogenic hydrocarbons in its relationship to photosynthetic activity in a plant. It is emitted from a wide variety of mostly deciduous vegetation in the presence of photosynthetically active radiation, exhibiting a strong increase in emission as temperature increases. Not only do the isoprene and terpenoid emissions vary considerably among plant species, but the biochemical and biophysical processes that control the rate of these emissions also appear to be quite

TABLE 2.11   Median Mixing Ratio of the 25 Most Abundant Nonmethane Organic Compounds Measured in the Summer 1987 Southern California Air Quality Study

	Median Mixing Ratio in Parts per Billion of Carbon[a]
Ethane	27.1
Ethene	22.3
Acetylene	17.3
Propane	56.0
Propene	7.8
i-Butane	19.4
Butane	42.0
i-Pentane	52.4
Pentane	24.0
2-Methylpentane	16.0
3-Methylpentane	11.8
Hexane	10.8
Methylcyclopentane	10.1
Benzene	17.0
3-Methylhexane	7.4
Heptane	6.0
Methylcyclohexane	7.0
Toluene	49.1
Ethylbenzene	7.6
m, p-Xylenes	25.2
o-Xylene	10.0
1,2,4-Trimethylbenzene	8.2
Formaldehyde	9.1
Acetaldehyde	14.8
Acetone	22.4

[a]Parts per billion of carbon (ppbC) is the parts per billion of carbon atoms in the molecule. It is simply the volume mixing ratio of the compound multiplied by the number of carbon atoms in the molecule.

Source: Lurmann and Main (1992).

distinct. Isoprene emissions appear to be a species-dependent byproduct of photosynthesis, photorespiration, or both; there is no evidence that isoprene is stored within or metabolized by plants. As a result, isoprene emissions are temperature- and light-dependent; essentially no isoprene is emitted without illumination. By contrast, terpenoid emissions seem to be triggered by biophysical processes associated with the amount of terpenoid material present in the leaf oils and resins and the vapor pressure of the terpenoid compounds. As a result, terpene emissions do not depend strongly on light (and they typically continue at night), but they do vary with ambient temperature. The dependence of natural isoprene and terpenoid emissions on temperature can result in a large variation in the rate of production of biogenic VOCs over the course of a growing season. An analysis of emissions data by Lamb et al. (1987) indicates that an increase in ambient temperature from 25 to 35°C can result in a factor of 4 increase in the rate of natural VOC emissions from isoprene-emitting deciduous trees and in a factor of 1.5 increase from

TABLE 2.12   Organic Compounds Emitted by Vegetation[a]

Isoprene		α-Pinene	
Camphene		β-Pinene	
2-Carene		Sabinene	
Δ^3-Carene		α-Terpinene	
d-Limonene		γ-Terpinene	
Myrcene		Teripinolene	
Ocimene		β-Phellandrene	
α-Phellandrene		p-Cymene	

[a]In the simplified molecular structures here bonds between carbon atoms are shown. Vertices represent carbon atoms. Hydrogen atoms bonded to the carbons are not explicitly indicated.

terpene-emitting conifers. Thus, all other factors being equal, natural VOC emissions are generally highest on hot summer days.

Biogenic hydrocarbon emission rates from individual plant species have been estimated experimentally by placing small plants or branches in enclosures and measuring the accumulation of emitted compounds. Extensive emission rate measurements have been reported for a relatively limited number of compounds: isoprene and a number of the dominant monoterpenes. Isoprene does appear to be the dominant compound emitted from vegetation.

There have been a number of efforts to compile inventories of biogenic hydrocarbon emissions [e.g., see Lamb et al. (1987, 1993) and Guenther et al. (1995)]. Emission rate measurements from individual plant species are used with empirical algorithms that account for temperature and, for isoprene, light effects to scale up to entire geographic regions, based on land-use data and biomass density factors. Biomass density, expressed in $g\,m^{-2}$ of area, is required to convert individual plant species emission rates, expressed in $mg\,g^{-1}\,min^{-1}$, to emission fluxes, in $mg\,m^{-2}\,min^{-1}$, which are then multiplied by vegetation class land coverage, in m^2, to yield total emissions. Uncertainties in these inventories are large; Lamb et al. (1993) estimated total U.S. biogenic hydrocarbon emissions ranging from 29 to 51 $Tg\,yr^{-1}$.

On a global scale, the largest biogenic hydrocarbon emissions occur in the tropics, with isoprene being the dominant emitted compound. These result from a combination of high temperatures and large biomass densities. During summer months, the maximum flux of biogenic hydrocarbon emissions in the southeastern United States is predicted to be as large as that in the tropics. An estimate of global biogenic VOC emissions appears in Table 2.13. On a global basis, biogenic hydrocarbon emissions far exceed those of anthropogenic hydrocarbons.

2.4.5   Carbon Monoxide

The global sources and sinks of CO are given in Table 2.14. Methane oxidation (by OH) is a major source of CO, as are technological processes (combustion and industrial processes), biomass burning, and the oxidation of nonmethane hydrocarbons. Uncertainties in each of these estimated sources are large. It is estimated that about two-thirds of the CO comes from anthropogenic activities, including oxidation of anthropogenically derived CH_4. The major

TABLE 2.13   Global Biogenic VOC Emission Rate Estimates by Source and Class of Compound, $Tg\,yr^{-1}$

Source	Isoprene	Monoterpenes	ORVOC[a]	Total VOC[b]
Woods	372	95	177	821
Crops	24	6	45	120
Shrub	103	25	33	194
Ocean	0	0	2.5	5
Other	4	1	2	9
Total	503	127	260	1150

[a]Other reactive biogenic VOCs (ORVOC).
[b]These totals include additional nonreactive VOCs not reflected in the columns to the left.

Source: Guenther et al. (1995).

TABLE 2.14   Estimates of Global Tropospheric CO Budget (in Tg(CO) yr^{-1}) and Values Adopted by IPCC (2001)

Reference:	Hauglustaine et al. (1998)	Bergamaschi et al. (2000)	WMO (1998)	IPCC (2001)
Sources				
Oxidation of CH_4		795		800
Oxidation of isoprene		268		270
Oxidation of terpenes		136		~ 0
Oxidation of industrial NMHC		203		110
Oxidation of biomass NMHC		—		30
Oxidation of acetone		—		20
Subtotal in situ oxidation	881	1402		1230
Vegetation		—	100	150
Oceans		49	50	50
Biomass burning		768	500	700
Fossil and domestic fuel		641	500	650
Subtotal direct emissions	1219	1458	1150	1550
Total sources	2100	2860		2780
Sinks				
Surface deposition	190			
OH reaction	1920			

sink for CO is reaction with OH radicals, with soil uptake and diffusion into the stratosphere being minor routes.

Tropospheric CO mixing ratios range from 40 to 200 ppb. Carbon monoxide has a chemical lifetime of 30–90 days on the global scale of the troposphere. Measurements indicate that there is more CO in the Northern Hemisphere than in the Southern Hemisphere; the maximum values are found near the surface at northern midlatitudes. In general, the CO mixing ratio decreases with altitude in the Northern Hemisphere to a free tropospheric average value of about 120 ppb near 45°N. In the Southern Hemisphere CO tends to be more nearly uniformly mixed vertically with a mixing ratio of about 60 ppb near 45°S. Seasonal variations have been established to be about $\pm 40\%$ about the mean in the Northern Hemisphere and $\pm 20\%$ about the mean in the Southern Hemisphere. The maximum concentration is observed to occur during the local spring and the minimum is found during the late summer or early fall.

2.4.6   Carbon Dioxide

Because of its overwhelming importance in global climate, we will delay consideration of CO_2 until Chapter 22, where we address the global cycle of CO_2.

2.5   HALOGEN-CONTAINING COMPOUNDS

Atmospheric halogen-containing compounds are referred to by a variety of names:

Halocarbons—a general term referring to halogen-containing organic compounds

Chlorofluorocarbons(CFCs)—the collective name given to a series of halocarbons containing carbon, chlorine, and fluorine atoms

Hydrochlorofluorocarbons (HCFCs)—halocarbons containing atoms of hydrogen, in addition to carbon, chlorine, and fluorine

Hydrofluorocarbons (HFCs)—halocarbons containing atoms of hydrogen, in addition to carbon and fluorine

Perhalocarbons—halocarbons in which every available carbon bond contains a halogen atom (compounds saturated with halogen atoms)

Halons—bromine-containing halocarbons, especially used as fire extinguishing agents

The earliest interest in halogens in the atmosphere arose from seasalt as a source of gaseous halogens (Eriksson 1959). Synthetic halocarbons have been known for the past century. Chlorofluorocarbons (CFCs) were first synthesized in the late nineteenth century, and their properties as refrigerants were recognized over 60 years ago. Halocarbons achieved widespread industrial use as refrigerants, propellants, and solvents.

As a group of atmospheric chemicals, halogen-containing compounds have a wide variety of anthropogenic and natural sources. They are produced by biological processes in the oceans, from seasalt, from biomass burning, and from industrial synthesis. Their atmospheric lifetimes vary considerably depending on their mechanism of removal, ranging from a few days to several centuries.

Table 2.15 lists atmospheric halogenated organic species with global average concentrations, atmospheric burdens, lifetimes, sources, and sinks. Of the exclusively human-made organic halogenated species, the chlorofluorocarbons are used as refrigerants (CFC-12, HCFC-22), blowing agents (CFC-11, HCFC-22), and cleaning agents (CFC-113). Methyl chloroform (CH_3CCl_3), methylene chloride (CH_2Cl_2), and tetrachloroethene (C_2Cl_4) are used as degreasers and as dry cleaning and industrial solvents. Methyl bromide (CH_3Br) is a widely used agricultural and space fumigant. All the monomethyl halides listed in Table 2.15 have natural sources. Methyl chloride (CH_3Cl) and CH_3Br are also products of biomass burning. Atmospheric levels of $CFCl_3$ and CF_2Cl_2 are shown in Figure 2.5.

Lovelock (1971) first detected SF_6 and $CFCl_3$ in the atmosphere using the electron-capture detector. In landmark work in atmospheric chemistry for which they received the 1995 Nobel Prize in Chemistry, Molina and Rowland (1974) showed that CFCs that are immune to removal in the troposphere could decompose photolytically in the stratosphere to release Cl atoms capable of catalytic destruction of stratospheric ozone. The very lack of chemical reactivity that makes chlorofluorocarbon molecules so intrinsically useful also allows them to survive unchanged in most commercial applications and eventually to be released to the atmosphere in their original gaseous form. The usual tropospheric sinks of oxidation, photodissociation, and wet and dry deposition are ineffective with the chlorofluorocarbons. The only important sink for $CFCl_3$ and CF_2Cl_2 is photodissociation in the midstratosphere (25–40 km). These same CFCs that lead to stratospheric ozone depletion are efficient absorbers of infrared radiation and potentially important greenhouse gases.

There is a sharp demarcation in atmospheric behavior between fully halogenated halocarbons and those containing one or more atoms of hydrogen. Halocarbons containing at least one hydrogen atom, such as CF_2HCl, $CHCl_3$, and CH_3CCl_3, are effectively broken down in the troposphere by reaction with the hydroxyl radical before they can reach the stratosphere. Atmospheric lifetimes of these species range from months to decades.

TABLE 2.15 Atmospheric Halogens

Compound	Generic Name	1998 Mixing Ratio (ppt)	Lifetime (yr)	Sources[a]	Sinks[b]
$CFCl_3$	CFC-11	268	45	A	Strat.hv
CF_2Cl_2	CFC-12	533	100	A	Strat.hv
$CF_2ClCFCl_2$	CFC-113	84	85	A	Strat.hv
CF_2ClCF_2Cl	CFC-114	15	300	A	Strat.hv
CCl_4	Carbon tetrachloride	102	35	A	Strat.hv
CH_3CCl_3	Methyl chloroform	69	4.8	A	Trop. OH
CH_3Cl	Methyl chloride	500	1.5	N(O),BB	Trop. OH
CF_2HCl	HCFC-22	132	11.9	A	Trop. OH
CH_3Br	Methyl bromide	9–10	0.8	N(O)A,BB	Trop. OH
CF_3Br	H-1301	2.5	65	A	Strat.hv
CF_4	Perfluoromethane	80	50,000	A	Meso.hv
SF_6	Sulfur hexafluoride	4.2	3200	A	Meso. electrons
CF_3CHCl_2	HCFC-123		1.4	A	Trop. OH
CF_3CHFCl	HCFC-124		5.9	A	Trop. OH
CH_3CFCl_2	HCFC-141b	10	9.3	A	Trop. OH
CH_3CF_2Cl	HCFC-142b	11	19	A	Trop. OH
$CF_3CF_2CHCl_2$	HCFC-225ca		2.5	A	Trop. OH
$CClF_2CF_2CHClF$	HCFC-225cb		6.6	A	Trop. OH
$CHCl_3$	Chloroform		0.55	A,N(O)	Trop. OH
CH_2Cl_2	Methylene chloride		0.41	A	Trop. OH
CF_3CF_2Cl	CFC-115	7	1700	A	Strat. $O(^1D)$
C_2Cl_4	Tetrachloroethene		0.4	A	Trop. OH

[a]A = anthropogenic; N(O) = natural (oceanic); BB = biomass burning.
[b]Strat. hv = photolysis in stratosphere; Trop. OH = hydroxyl radical reaction in troposphere; Meso. electrons = mesosphere electron impact; Strat. $O(^1D)$ = reactions in stratosphere with excited atomic oxygen.
Sources: IPCC (2001) and Singh (1995).

Chlorofluorocarbons   The term *hydrochlorofluorocarbons* is the collective name given to a series of chemicals with varying number of carbon, hydrogen, chlorine, and fluorine atoms. The somewhat arcane system of numbering these compounds was proposed by the American Society of Heating and Refrigeration Engineers in 1957. For the simpler hydrochlorofluorocarbons, the numbering system may be summarized as follows:

1. The first digit on the right is the number of fluorine (F) atoms in the compound.
2. The second digit from the right is one more than the number of hydrogen (H) atoms in the compound.
3. The third digit from the right, plus one, is the number of carbon (C) atoms in the compound. When this digit is zero (i.e., only one carbon atom in the compound), it is omitted from the number.

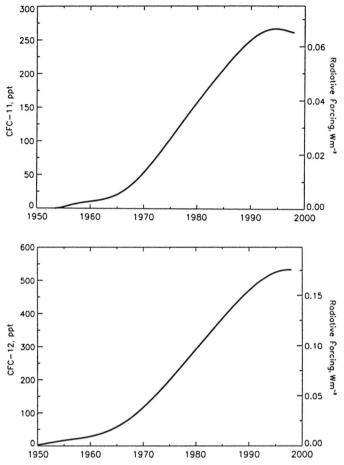

FIGURE 2.5   Global mean CFC-11 ($CFCl_3$) and CFC-12 (CF_2Cl_2) tropospheric abundance from 1950 to 1998 based on smoothed measurements and emission models (IPCC 2001). The radiative forcing of each CFC is shown on the right axis (see Chapter 23).

4. The number of chlorine (Cl) atoms in the compound is found by subtracting the sum of the fluorine and hydrogen atoms from the total number of atoms that can be connected to the carbon atoms.

CCl_2F_2	$C_2Cl_2F_4$	$CHClF_2$	CCl_3F
CFC-12	CFCl-114	CFC-22	CFC-11

2.5.1   Methyl Chloride (CH_3Cl)

Methyl chloride supplies about 0.5 ppb of chlorine to the stratosphere. Prior to 1996, the atmospheric input of CH_3Cl was assumed to be a result of biological processes in the ocean. Tropical terrestrial sources, biomass burning and tropical plants, are now believed to be important. The dominant sink of CH_3Cl is reaction with the OH radical in the

TABLE 2.16   Global Budget of Methyl Chloride (CH_3Cl)

	Sources, $Gg\,yr^{-1}$	
Oceans	600	$(325–1300)^a$
Biomass burning	911	(655–1125)
Tropical plants	910	(820–8200)
Fungi	160	(43–470)
Salt marshes	170	(65–440)
Wetlands	40	(6–270)
Coal combustion	105	(5–205)
Incineration	45	(15–75)
Industrial	10	
Rice	5	
Total sources	2956	(1934–12,085)
	Sinks, $Gg\,yr^{-1}$	
Reaction with OH in troposphere	3180	(2380–3970)
Loss to stratosphere	200	(100–300)
Reaction with Cl in marine boundary layer	370	(180–550)
Microbial degradation in soil	180	(100–1600)
Loss to polar oceans	75	(37–113)
Total sinks	4005	(4599–9288)

^{a}Numbers in parentheses represent the range of estimates.
Source: WMO (2002).

troposphere. Table 2.16 gives the global budget of CH_3Cl. According to the best estimates, the sinks exceed the sources by 1049 $Gg\,yr^{-1}$. It has been suggested that a major, as yet unidentified, source of CH_3Cl exists, probably in the tropics.

2.5.2   Methyl Bromide (CH_3Br)

Methyl bromide (CH_3Br) is the most abundant atmospheric bromocarbon, responsible for over half of the bromine reaching the stratosphere (Schauffler et al. 1998). Because of its toxicity to a broad range of pests, including weeds, insects, and bacteria, CH_3Br is used industrially in fumigation of soils and buildings, preservation of grains in silos, and treatment of fresh fruits and vegetables. Table 2.17 presents an estimate of the global budget of CH_3Br. Sources of CH_3Br include use as a fumigant and as a leaded fuel additive, oceanic production, biomass burning, and plant and marsh emissions. The main anthropogenic sources are biomass burning and soil fumigation. CH_3Br sinks include reaction with OH, photolysis at higher altitudes, loss to soils, chemical and biological degradation in the oceans, and uptake by green plants. The overall mean atmospheric mixing ratio of CH_3Br is between 9 and 10 ppt, with a mean interhemispheric ratio, NH/SH, of 1.3 ± 0.1 (Kurylo et al. 1999). Once in the stratosphere, CH_3Br mixing ratios drop off rapidly with altitude, reflecting the facile release of active bromine.

The ocean acts as both a source and sink for CH_3Br; the net oceanic flux to the atmosphere is negative, estimated as -21 $Gg\,yr^{-1}$ (Table 2.17). Oceanic uptake of CH_3Br is currently estimated at 77 $Gg\,yr^{-1}$. Combined with atmospheric OH reaction and photolysis (86 $Gg\,yr^{-1}$) and soil uptake (46.8 $Gg\,yr^{-1}$), this leads to a total estimated

TABLE 2.17   Global Budget of Methyl Bromide (CH_3Br)

	Sources, $Gg\,yr^{-1}$	
Oceans	56	$(5-130)^{a,b}$
Fumigation		
Soils	26.5	$(16-48)$
Durable	6.6	$(4.8-8.4)$
Perishables	5.7	$(5.4-6.0)$
Structures	2	
Gasoline	5	$(0-10)$
Biomass burning	20	$(10-40)$
Wetlands	4.6	$(?)$
Salt marshes	14	$(7-29)$
Plants (rapeseed)	6.6	$(4.8-8.4)$
Rice fields	1.5	$(0.5-2.5)$
Fungus	1.7	$(0.5-5.2)$
Total sources	151	$(56-290)$
	Sinks, $Gg\,yr^{-1}$	
Reaction with OH and photolysis	86	$(65-107)$
Oceans	77	$(37-133)$
Soils	46.8	$(32-154)$
Plants	?	
Total sinks	210	$(134-394)$

[a]Numbers in parentheses represent the range of estimates.
[b]The ranges in the oceanic source and sink terms must satisfy the accepted range in the net oceanic flux of -3 to $-32\,Gg\,yr^{-1}$. Thus, the lower limit in uptake, $-37\,Gg\,yr^{-1}$, corresponds to the lower limit in emissions of 5 $Gg\,yr^{-1}$, and the upper limit in uptake, $-133\,Gg\,yr^{-1}$, corresponds to the upper limit in emissions of 130 $Gg\,yr^{-1}$.
Source: Kurylo et al. (1999) and Yvon-Lewis (2000).

annual sink of $210\,Gg\,yr^{-1}$. To maintain the net ocean sink of $21\,Gg\,yr^{-1}$, the estimated oceanic emission of CH_3Br must be $56\,Gg\,yr^{-1}$. This results in total estimated emissions of $151\,Gg\,yr^{-1}$, leading to an unbalance (missing sources) in the CH_3Br budget of $59\,Gg\,yr^{-1}$. Research is continuing to identify all natural sources of CH_3Br.

The tropospheric lifetime of CH_3Br can be estimated based on its mixing ratio and its estimated total sink. With a mean tropospheric mixing ratio of 10 ppt and 1.8×10^{20} total moles of air in the atmosphere, the total number of moles of $CH_3Br \cong (10 \times 10^{-12})$ $(1.8 \times 10^{20}) = 18 \times 10^8$ mol CH_3Br. With a molecular weight of 95 g mol^{-1}, this is 1.71×10^{11}g CH_3Br. Given a total sink of 210×10^9g yr^{-1} (Table 2.17), the mean atmospheric residence time of CH_3Br is

$$\tau_{CH_3Br} = \frac{1.71 \times 10^{11}}{2.1 \times 10^{11}} = 0.8\,yr$$

2.6   ATMOSPHERIC OZONE

Ozone (O_3) is a reactive oxidant gas produced naturally in trace amounts in the Earth's atmosphere. Ozone was discovered by C. F. Schönbein in the middle of the nineteenth century; he also was first to detect ozone in air (Schönbein 1840, 1854). Schönbein (1840) suggested the presence of an atmospheric gas having a peculiar odor (the Greek word for "to smell" is *ozein*). Spectroscopic studies in the late nineteenth century showed that ozone is present at a higher mixing ratio in the upper atmospheric layers than close to the ground. Attempts to explain the chemical basis of existence of ozone in the upper atmosphere began nearly 80 years ago. Within the last 40 years, however, while increased understanding of the role of other trace atmospheric species in stratospheric ozone was unfolding, it became apparent that anthropogenically emitted substances have the potential to seriously deplete the natural levels of ozone in the stratosphere. At about the same period, ironically, it was realized that anthropogenic emissions could lead to ozone *increases* in the troposphere. Whereas stratospheric ozone is essential for screening of solar ultraviolet radiation, ozone at ground level can, at elevated concentrations, lead to respiratory effects in humans. This paradoxical dual role of ozone in the atmosphere has, on occasion, led to the dubbing of stratospheric ozone as "good" ozone and tropospheric ozone as "bad" ozone.

Most of the Earth's atmospheric ozone (about 90%) is found in the stratosphere where it plays a critical role in absorbing ultraviolet radiation emitted by the Sun. Figure 2.6 shows the stratospheric ozone at 35°N in September 1996. The peak in ozone molecular number density (concentration) occurs in the region of 20–30 km. The so-called stratospheric ozone layer absorbs virtually all of the solar ultraviolet radiation of wavelengths between 240 and 290 nm. Such radiation is harmful to unicellular organisms and to surface cells of higher plants and animals. In addition, ultraviolet radiation in the wavelength range 290–320 nm, so-called UV-B, is biologically active. A reduction in stratospheric ozone leads to increased levels of UV-B at the ground, which can lead to increased incidence of skin cancer in susceptible individuals. (An approximate rule of thumb is that a 1% decrease in stratospheric ozone leads to a 2% increase in UV-B.) The stratospheric temperature profile

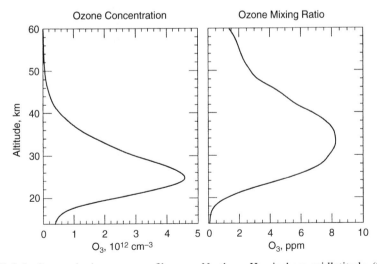

FIGURE 2.6 Stratospheric ozone profile over Northern Hemisphere midlatitude (35°N) in September 1996 as measured by satellite with the Jet Propulsion Laboratory FTIR (Fourier transform infrared) spectrometer. Note that molecular concentration and mixing ratio peak at different altitudes.

(see Figure 1.1) is the result of ozone absorption of radiation. Stratospheric chemistry centers around the chemical processes influencing the abundance of ozone.

As compared with the stratosphere, natural concentrations of tropospheric ozone are small—usually a few tens of parts per billion (ppb) in mixing ratio (molecules of O_3/molecules of air; 10 ppb $= 2.5 \times 10^{11}$ molecules cm^{-3} at sea level and 298 K) versus peak stratospheric mixing ratios of more than 10,000 ppb (10 ppm). Since the atmospheric molecular number density thins out exponentially with altitude, the peak in ozone mixing ratio occurs at a higher altitude than does its peak in concentration. Still, a significant amount of naturally occurring ozone, about 10–15% of the atmospheric total, is found in the troposphere (Fishman et al. 1990). The total amount of O_3 in the atmosphere, stratosphere and troposphere combined, is extremely small. In the pristine, unpolluted troposphere ozone mixing ratios are in the range of 10–40 ppb with somewhat higher mixing ratios in the upper troposphere. Ozone reaches a maximum mixing ratio of about 10 ppm at an altitude of 25–30 km in the stratosphere.

Over the past three decades, the improved ability to monitor atmospheric ozone with ground-sited, aircraft-mounted, and satelliteborne instruments has led to definitive data about changes in the amounts of ozone found in the atmosphere. One of the most significant environmental issues facing the planet is the catalytic destruction of significant portions of the stratospheric ozone layer, most dramatically seen each Antarctic spring, which is the result of stratospheric halogen-induced photochemistry (see Chapter 5). The current and projected loss of stratospheric ozone has stimulated a major, continuing global research program in stratospheric chemistry and an international treaty that restricts the release of chlorofluorocarbons into the atmosphere.

> **The Dobson Unit**   Imagine that all the ozone in a vertical column of air reaching from the Earth's surface to the top of the atmosphere were concentrated in a single layer of pure O_3 at the surface of the Earth, at 273 K and 1.013×10^5 Pa. The thickness of that layer, measured in hundredths of a millimeter, is the column abundance of O_3 expressed in *Dobson units* (DU). The unit is named after G. M. B. Dobson, who, in 1923, produced the first ozone spectrometer, the standard instrument used to measure ozone from the ground. The Dobson spectrometer measures the intensity of solar UV radiation at four wavelengths, two of which are absorbed by ozone and two of which are not.
>
> One DU is equivalent to 2.69×10^{16} molecules O_3 cm^{-2}. A normal value of O_3 column abundance over the globe is about 300 DU, corresponding to a layer of pure O_3 at the surface only 3 mm thick.
>
> If the ozone concentration as a function of altitude is $n_{O_3}(z)$ (molecules cm^{-3}), the O_3 column burden is
>
> $$\bar{n}_{O_3} = \int_0^\infty n_{O_3}(z)\,dz \quad (\text{molecules}\,cm^{-2})$$
>
> To find what this $\bar{n}_{O_3}$ translates to in terms of DU, we need to determine the thickness of the layer of pure O_3 at 273 K and 1 atm corresponding to this burden. Over 1 cm^2 of area, this corresponds to $\bar{n}_{O_3}$ molecules of O_3. The volume (cm^3) occupied by this number of molecules of O_3 can be written as $V = 1(cm^2) \times h(cm)$. One DU corresponds to a thickness, h, of 0.01 mm (0.001 cm).

From the ideal-gas law

$$V = \frac{(\bar{n}_{O_3}/N_A)RT}{p} \times 10^6$$

where $\bar{n}_{O_3}/N_A$ is the number of moles of O_3, and the factor of 10^6 is needed to convert m^3 to cm^3. At 273 K and 1 atmosphere, the thickness of the layer of pure O_3 is

$$h = \frac{(\bar{n}_{O_3})(8.314)(273)}{(1.013 \times 10^5)(6.022 \times 10^{23})} \times 10^6 \quad (cm)$$

$$= 3.72 \times 10^{-20} \bar{n}_{O_3} \quad (cm)$$

For $h = 0.01$ mm $= 0.001$ cm (1 DU), the column burden of O_3 is

$$\bar{n}_{O_3} = 2.69 \times 10^{16} \text{ molecules cm}^{-2}$$

Whereas stratospheric ozone is thinning, tropospheric ozone is increasing (WMO 1986, 1990). Much of the evidence for increased baseline levels of tropospheric ozone comes from Europe, where during the late 1800s there was much interest in atmospheric ozone. Because ozone was known to be a disinfectant, it was believed to promote health (Warneck 1988). Measurements of atmospheric ozone made at Montsouris, near Paris, from 1876 to 1910 have been reanalyzed by Volz and Kley (1988), who recalibrated the original measurement technique. Their analysis showed that surface ozone mixing ratios near Paris over 100 years ago averaged about 10 ppb; current mixing ratios in the most unpolluted parts of Europe average between 20 and 45 ppb (Volz and Kley 1988; Bojkov 1988; Crutzen 1988; Staehelin and Schmid 1991; Oltmans and Levy 1994; Janach 1989). An analysis of ozone measurements made in relatively remote European sites indicates a 1 to 2% annual increase in average concentrations over the past 40 years (Janach 1989).[1] Based on total ozone mapping spectrometer (TOMS) data between the high Andes and the Pacific Ocean, from 1979 to 1992 tropospheric ozone concentrations in the tropical Pacific South America apparently increased by 1.48% per year (Jiang and Yung 1996). The integrated ozone column in the troposphere can be determined from the difference of the measurements of two satellite instruments, TOMS and SAGE, which detect total column ozone and stratospheric ozone, respectively (Fishman et al. 1990, 1992). Tropospheric ozone column densities average about 30 DU, but there is significant variation with season and hemisphere.

2.7 PARTICULATE MATTER (AEROSOLS)

Particles in the atmosphere arise from natural sources, such as windborne dust, seaspray, and volcanoes, and from anthropogenic activities, such as combustion of fuels. Whereas an *aerosol* is technically defined as a suspension of fine solid or liquid particles in a gas,

---

[1] If stratospheric ozone concentrations remained constant, the 10% increase in tropospheric ozone would increase the total column abundance of ozone by about 1%. Thus the additional tropospheric ozone is believed to have counteracted only a small fraction of the stratospheric loss, even if the trends observed over Europe are representative of the entire northern midlatitude region.

TABLE 2.18   Terminology Relating to Atmospheric Particles

Aerosols, aerocolloids, aerodisperse systems	Tiny particles dispersed in gases
Dusts	Suspensions of solid particles produced by mechanical disintegration of material such as crushing, grinding, and blasting; $D_p > 1$ μm
Fog	A term loosely applied to visible aerosols in which the dispersed phase is liquid; usually, a dispersion of water or ice, close to the ground
Fume	The solid particles generated by condensation from the vapor state, generally after volatilization from melted substances, and often accompanied by a chemical reaction such as oxidation; often the material involved is noxious; $D_p < 1$ μm
Hazes	An aerosol that impedes vision and may consist of a combination of water droplets, pollutants, and dust; $D_p < 1$ μm
Mists	Liquid, usually water in the form of particles suspended in the atmosphere at or near the surface of the Earth; small water droplets floating or falling, approaching the form of rain, and sometimes distinguished from fog as being more transparent or as having particles perceptibly moving downward; $D_p > 1$ μm
Particle	An aerosol particle may consist of a single continuous unit of solid or liquid containing many molecules held together by intermolecular forces and primarily larger than molecular dimensions (>0.001 μm); a particle may also consist of two or more such unit structures held together by interparticle adhesive forces such that it behaves as a single unit in suspension or on deposit
Smog	A term derived from smoke and fog, applied to extensive contamination by aerosols; now sometimes used loosely for any contamination of the air
Smoke	Small gasborne particles resulting from incomplete combustion, consisting predominantly of carbon and other combustible materials, and present in sufficient quantity to be observable independently of the presence of other solids. $D_p \geq 0.01$ μm
Soot	Agglomerations of particles of carbon impregnated with "tar," formed in the incomplete combustion of carbonaceous material

common usage refers to the aerosol as the particulate component only (Table 2.18). Emitted directly as particles (primary aerosol) or formed in the atmosphere by gas-to-particle conversion processes (secondary aerosol), atmospheric aerosols are generally considered to be the particles that range in size from a few nanometers (nm) to tens of micrometers (μm) in diameter. Once airborne, particles can change their size and composition by condensation of vapor species or by evaporation, by coagulating with other particles, by chemical reaction, or by activation in the presence of water supersaturation to become fog and cloud droplets. Particles smaller than 1 μm diameter generally have atmospheric concentrations in the range from around ten to several thousand per cm^3; those exceeding 1 μm diameter are usually found at concentrations less than 1 cm^{-3}.

Particles are eventually removed from the atmosphere by two mechanisms: deposition at the Earth's surface (dry deposition) and incorporation into cloud droplets during the formation of precipitation (wet deposition). Because wet and dry deposition lead to relatively short residence times in the troposphere, and because the geographic distribution of particle sources is highly nonuniform, tropospheric aerosols vary widely in concentration and composition over the Earth. Whereas atmospheric trace gases have lifetimes ranging from less than a second to a century or more, residence times of particles in the troposphere vary only from a few days to a few weeks.

2.7.1 Stratospheric Aerosol

The stratospheric aerosol is composed of an aqueous sulfuric acid solution of 60–80% sulfuric acid for temperatures from -80 to $-45°C$, respectively (Shen et al. 1995). The source of the globally distributed, unperturbed background stratospheric aerosol is oxidation of carbonyl sulfide (OCS), which has its sources at the Earth's surface. OCS is chemically inert and water insoluble and has a long tropospheric lifetime. It diffuses into the stratosphere where it dissociates by solar ultraviolet radiation to eventually form sulfuric acid, the primary component of the natural stratospheric aerosol. Other surface-emitted sulfur-containing species, for example, SO_2, DMS, and CS_2, do not persist long enough in the troposphere to be transported to the stratosphere.

A state of unperturbed background stratospheric aerosol may be relatively rare, however, as frequent volcanic eruptions inject significant quantities of SO_2 directly into the lower and midstratosphere. Major eruptions include Agung in 1963, El Chichón in 1982, and Pinatubo in 1991. The subsequent sulfuric acid aerosol clouds can, over a period of months, be distributed globally at optical densities that overwhelm the natural background aerosol. The stratosphere's relaxation to background conditions has a characteristic time on the order of years, so that, given the frequency of volcanic eruptions, the stratospheric aerosol is seldom in a state that is totally unperturbed by volcanic emissions. With an estimated aerosol mass addition of 30 Tg to the stratosphere, the June 1991 eruption of Mt. Pinatubo was the largest in the 20th century and led to enhanced stratospheric aerosol levels for over 2 years.

2.7.2 Chemical Components of Tropospheric Aerosol

A significant fraction of the tropospheric aerosol is anthropogenic in origin. Tropospheric aerosols contain sulfate, ammonium, nitrate, sodium, chloride, trace metals, carbonaceous material, crustal elements, and water. The carbonaceous fraction of the aerosols consists of both elemental and organic carbon. Elemental carbon, also called black carbon, graphitic carbon, or soot, is emitted directly into the atmosphere, predominantly from combustion processes. Particulate organic carbon is emitted directly by sources or can result from atmospheric condensation of low-volatility organic gases. Anthropogenic emissions leading to atmospheric aerosol have increased dramatically over the past century and have been implicated in human health effects (Dockery et al. 1993), in visibility reduction in urban and regional areas (see Chapter 15), in acid deposition (see Chapter 20), and in perturbing the Earth's radiation balance (see Chapter 24).

Table 2.19 presents data summarized by Heintzenberg (1989) and Solomon et al. (1989) on aerosol mass concentrations and composition in different regions of the troposphere. It is interesting to note that average total fine particle mass (that associated with particles of diameter less than about 2 μm) in nonurban continental, (i.e., regional) aerosols is only a factor of 2 lower than urban values. This reflects the relatively long residence time of particles. Correspondingly, the average compositions of nonurban

TABLE 2.19   Mass Concentrations and Composition of Tropospheric Aerosols

Region	Mass ($\mu g\,m^{-3}$)	Percentage Composition				
		C (elem)	C (org)	NH_4^+	NO_3^-	SO_4^{2-}
Remote (11 areas)[a]	4.8	0.3	11	7	3	22
Nonurban continental (14 areas)[a]	15	5	24	11	4	37
Urban (19 areas)[a]	32	9	31	8	6	28
Rubidoux, California[b] (1986 annual average)	87.4	3	18	6	20	6

[a]Heintzenberg(1989).
[b]Solomon et al. (1989).

continental and urban aerosols are roughly the same. The average mass concentration of remote aerosols is a factor of 3 lower than that of nonurban continental aerosols. The elemental carbon component, a direct indicator of anthropogenic combustion sources, drops to 0.3% in the remote aerosols, but sulfate is still a major component. This is attributable to a global average concentration of nonseasalt sulfate of about $0.5\,\mu g\,m^{-3}$. Rubidoux, California, located about 100 km east of downtown Los Angeles, routinely experiences some of the highest particulate matter concentrations in the United States.

2.7.3   Cloud Condensation Nuclei (CCN)

Aerosols are essential to the atmosphere as we know it; if the Earth's atmosphere were totally devoid of particles, clouds could not form. Particles that can become activated to grow to fog or cloud droplets in the presence of a supersaturation of water vapor are termed *cloud condensation nuclei* (CCN). At a given mass of soluble material in the particle there is a critical value of the ambient water vapor supersaturation below which the particle exists in a stable state and above which it spontaneously grows to become a cloud droplet of 10 μm or more diameter. The number of particles from a given aerosol population that can act as CCN is thus a function of the water supersaturation. For marine stratiform clouds, for which supersaturations are in the range of 0.1–0.5%, the minimum CCN particle diameter is 0.05–0.14 μm. CCN number concentrations vary from fewer than $100\,cm^{-3}$ in remote marine regions to many thousand cm^{-3} in polluted urban areas. An air parcel will spend, on average, a few hours in a cloud followed by a few days outside clouds. The average lifetime of a CCN is about 1 week, so that an average CCN will experience 5–10 cloud activation/cloud evaporation cycles before actually being removed from the atmosphere in precipitation.

2.7.4   Sizes of Atmospheric Particles

Atmospheric aerosols consist of particles ranging in size from a few tens of angstroms (Å) to several hundred micrometers. Particles less than 2.5 μm in diameter are generally referred to as "fine" and those greater than 2.5 μm diameter as "coarse." The fine and coarse particle modes, in general, originate separately, are transformed separately, are removed from the atmosphere by different mechanisms, require different techniques for their removal from sources, have different chemical composition, have different optical properties, and differ significantly in their deposition patterns in the respiratory tract. Therefore the distinction between fine and coarse particles is a fundamental one in any discussion of the physics, chemistry, measurement, or health effects of aerosols.

The phenomena that influence particle sizes are shown in an idealized schematic in Figure 2.7, which depicts the typical distribution of surface area of an atmospheric aerosol. Particles can often be divided roughly into modes. The *nucleation* (or *nuclei*) mode comprises particles with diameters up to about 10 nm. The *Aitken* mode spans the size range from about 10 nm to 100 nm (0.1 μm) diameter. These two modes account for the

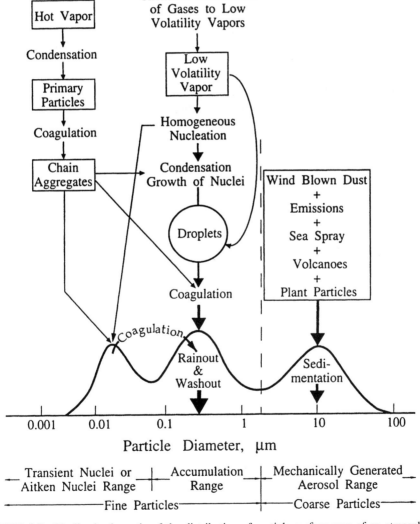

FIGURE 2.7 Idealized schematic of the distribution of particle surface area of an atmospheric aerosol (Whitby and Cantrell 1976). Principal modes, sources, and particle formation and removal mechanisms are indicated.

preponderance of particles by number; because of their small size, these particles rarely account for more than a few percent of the total mass of airborne particles. Particles in the nuclei mode are formed from condensation of hot vapors during combustion processes and from the nucleation of atmospheric species to form fresh particles. They are lost principally by coagulation with larger particles. The *accumulation* mode, extending from 0.1 to about 2.5 μm in diameter, usually accounts for most of the aerosol surface area and a substantial part of the aerosol mass. The source of particles in the accumulation mode is the coagulation of particles in the nuclei mode and from con- densation of vapors onto existing particles, causing them to grow into this size range. The accumulation mode is so named because particle removal mechanisms are least efficient in this regime, causing particles to accumulate there. The *coarse* mode, from > 2.5 μm in diameter, is formed by mechanical processes and usually consists of human-made and natural dust particles. Coarse particles have sufficiently large sedimentation velocities that they settle out of the atmosphere in a reasonably short time. Because removal mechanisms that are efficient at the small and large particle extremes of the size spectrum are inefficient in the accumulation range, particles in the accumulation mode tend to have considerably longer atmospheric residence times than those in either the nuclei or coarse mode.

2.7.5   Sources of Atmospheric Particulate Matter

Significant natural sources of particles include soil and rock debris (terrestrial dust), volcanic action, sea spray, biomass burning, and reactions between natural gaseous emissions. Table 2.20 presents a range of emission estimates of particles generated from natural and anthropogenic sources, on a global basis. Emissions of particulate matter attributable to the activities of humans arise primarily from four source categories: fuel combustion, industrial processes, nonindustrial fugitive sources (roadway dust from paved and unpaved roads, wind erosion of cropland, construction, etc.), and transportation sources (automobiles, etc.).

2.7.6   Carbonaceous Particles

Carbonaceous particles in the atmosphere consist of two major components—graphitic or black carbon (sometimes referred to as "elemental" or "free carbon") and organic material. The latter can be directly emitted from sources or produced from atmospheric reactions involving gaseous organic precursors. Elemental carbon can be produced only in a combustion process and is therefore solely primary. Graphitic carbon particles are the most abundant light-absorbing aerosol species in the atmosphere. Particulate organic matter is a complex mixture of many classes of compounds (Daisey 1980). A major reason for the study of particulate organic matter has been the possibility that such compounds pose a health hazard. Specifically, certain fractions of particulate organic matter, especially those containing polycyclic aromatic hydrocarbons (PAHs), have been shown to be carcinogenic in animals and mutagenic in in vitro bioassays.

As employed in atmospheric chemistry, the term *elemental carbon* refers to carbonaceous material that does not volatilize below a certain temperature, usually about 550°C. Thus, this term is an operational definition based on the volatility properties of the material. The term *soot* is also used to refer to any light-absorbing, combustion-generated carbonaceous material. Perhaps the most widely used term for light-absorbing carbonaceous aerosols is *black carbon*, a term that implies strong absorption across a wide spectrum of visible wavelengths. In their comprehensive review of light absorption by carbonaceous particles, Bond and Bergstrom (2006), following the suggestion of Malm et al., use the term *light-absorbing carbon* to represent this class of material.

2.7.7   Mineral Dust

Mineral dust aerosol arises from wind acting on soil particles. The arid and semiarid regions of the world, which cover about one-third of the global land area, are where the major dust sources are located. The largest global source is the Sahara–Sahel region of northern Africa; central Asia is the second largest dust source. Estimates of global dust emissions are uncertain. Published global dust emission estimates since 2001 range from 1000 to 2150 Tg yr^{-1} (one such estimate appears in Table 2.20), and atmospheric burden estimates range from 8 to 36 Tg (Zender et al. 2004). The large uncertainty range of dust emission estimates is mainly a result of the complexity of the processes that raise dust into the atmosphere. The emission of dust is controlled by both the windspeed and the nature of the surface itself. In addition, the size range of the dust particles is a crucial factor in emissions estimates. It is generally thought that a threshold windspeed as a function of

TABLE 2.20   Global Emission Estimates for Major Aerosol Classes

Source	Estimated Flux, Tg yr^{-1}	Reference
Natural		
Primary		
Mineral dust		Zender et al. (2003)
0.1–1.0 μm	48	
1.0–2.5 μm	260	
2.5–5.0 μm	609	
5.0–10.0 μm	573	
0.1–10.0 μm	1490	
Seasalt	10,100	Gong et al. (2002)
Volcanic dust	30	Kiehl and Rodhe (1995)
Biological debris	50	Kiehl and Rodhe (1995)
Secondary		
Sulfates from DMS	12.4	Liao et al. (2003)
Sulfates from volcanic SO_2	20	Kiehl and Rodhe (1995)
Organic aerosol from biogenic VOC	11.2	Chung and Seinfeld (2002)
Anthropogenic		
Primary		
Industrial dust (except black carbon)	100	Kiehl and Rodhe (1995)
Black carbon	12[a]	Liousse et al. (1996)
Organic aerosol	81[a]	Liousse et al. (1996)
Secondary		
Sulfates from SO_2	48.6[b]	Liao et al. (2003)
Nitrates from NO_x	21.3[c]	Liao et al. (2004)

[a]Tg C.
[b]Tg S.
[c]Tg NO_3^-.

particle size is required to mobilize dust particles into the atmosphere. Although clearly an oversimplification, a single value of the threshold windspeed of 6.5 m s^{-1} at 10 m height has frequently been used in modeling dust emissions in general circulation models (Sokolik 2002). The main species found in soil dust are quartz, clays, calcite, gypsum, and iron oxides, and the properties (especially optical) depend on the relative abundance of the various minerals. Gravitational settling and wet deposition are the major removal processes for mineral dust from the atmosphere. The average lifetime of dust particles in the atmosphere is about 2 weeks, during which dust can be transported thousands of kilometers. Saharan dust plumes frequently reach the Caribbean, and plumes of dust from Asia are detected on the west coast of North America.

Mineral dust is emitted from both natural and anthropogenic activities. Natural emissions arise by wind acting on undisturbed source regions. Anthropogenic emissions result from human activity, including (1) land-use changes that modify soil surface conditions; and (2) climate modifications that, in turn, alter dust emissions. Such modifications include changes in windspeeds, clouds and precipitation, and the amounts of airborne soluble material, such as sulfate, that may become attached to mineral dust particles and render them more susceptible to wet removal.

2.8   EMISSION INVENTORIES

An estimate of emissions of a species from a source is based on a technique that uses "emission factors," which are based on source-specific emission measurements as a function of activity level (e.g., amount of annual production at an industrial facility) with regard to each source. For example, suppose one wants to sample a power plant's emissions of SO_2 or NO_x at the stack. The plant's boiler design and its fuel consumption rate are known. The sulfur and nitrogen content of fuel burned can be used to calculate an emissions factor of kilograms (kg) of SO_2 or NO_x emitted per metric ton (Mg) of fuel consumed.

The U.S. Environmental Protection Agency (EPA) has compiled emission factors for a variety of sources and activity levels (such as production or consumption), reporting the results since 1972 in *AP-42 Compilation of Air Pollutant Emission Factors*, for which supplements are issued regularly. Emission factors currently in use are developed from only a limited sampling of the emissions source population for any given category, and the values reported are an average of those limited samples and might not be statistically representative of the population. For example, 30 source tests of coal-fueled, tangentially fired boilers led to calculations of emission factors that range approximately from 5 to 11 kg NO_x per metric ton of coal burned (Placet et al. 1990). The sample population was averaged and the emission factor for this source type was reported as 7.5 kg NO_x per metric ton of coal. The uncertainties associated with emission factor determinations can be considerable.

The formulation of emission factors for mobile sources, the major sources of VOCs and NO_x, is based on rather complex emission estimation models used in conjunction with data from laboratory testing of representative groups of motor vehicles. Vehicle testing is performed with a chassis dynamometer, which determines the exhaust emission of a vehicle as a function of a specified ambient temperature and humidity, speed, and load cycle. The current specified testing cycle is called the *Federal Test Procedure* (FTP). On the basis of results from this set of vehicle emissions data, a computer model has been developed to simulate for specified speeds, temperatures, and trip profiles, for example, the emission factors to be applied for the national fleet average for all vehicles or any specified distribution of vehicle age and type. These data are then incorporated with activity data on vehicle miles traveled as a function of spatial and temporal allocation factors to estimate emissions.

2.9 BIOMASS BURNING

The intentional burning of land results in a major source of combustion products to the atmosphere. Most of this burning occurs in the tropics. Emissions from burning vegetation are typical of those from any uncontrolled combustion process and include CO_2, CO, NO_x, CH_4 and nonmethane hydrocarbons, and elemental and organic particulate matter. The quantity and type of emissions from a biomass fire depend not only on the type of vegetation but also on its moisture content, ambient temperature, humidity, and local windspeed. Estimates of global emissions of trace gases from biomass burning have already been given in many of the tables in this chapter. For CH_4, for example, it is estimated that $40\,\mathrm{Tg\,yr^{-1}}$ out of a total flux of $598\,\mathrm{Tg\,yr^{-1}}$ (Table 2.10) is a result of biomass burning. For CO, a biomass burning source of $700\,\mathrm{Tg\,yr^{-1}}$ out of a total source of $2780\,\mathrm{Tg\,yr^{-1}}$ is estimated (Table 2.14). For NO_x, biomass burning is estimated to contribute globally $7.1\,\mathrm{Tg\,yr^{-1}}$, as compared to $33\,\mathrm{Tg\,yr^{-1}}$ from fossil-fuel burning (Table 2.6). The effect of emissions from biomass burning on atmospheric chemistry and climate, particularly in the tropics, is critically important. Large-scale savanna burning in the dry season in Africa leads to regional-scale ozone levels that approach those characteristic of urban industrialized regions (60–100 ppb) Fishman et al. 1990, 1992).

APPENDIX 2.1 AIR POLLUTION LEGISLATION

The legislative basis for air pollution abatement in the United States is the 1963 Clean Air Act and its amendments. The Clean Air Act was the first modern environmental law enacted by the U.S. Congress. The original act was signed into law in 1963, and major amendments were made in 1970, 1977, and 1990. The act establishes the federal–state relationship that requires the U.S. Environmental Protection Agency (EPA) to develop National Ambient Air-Quality Standards (NAAQS) and empowers the states to implement and enforce regulations to attain them. The act also requires the EPA to set NAAQS for common and widespread pollutants after preparing criteria documents summarizing scientific knowledge of their detrimental effects. The EPA has established NAAQS for each of six criteria pollutants: sulfur dioxide, particulate matter, nitrogen dioxide, carbon monoxide, ozone, and lead. At certain concentrations and length of exposure these pollutants are anticipated to endanger public health or welfare. The NAAQS are threshold concentrations based on a detailed review of the scientific information related to effects. Concentrations below the NAAQS are expected to have no adverse effects for humans and the environment. Table 2.21 presents the U.S. national primary and secondary ambient air quality standards for ozone, carbon monoxide, nitrogen dioxide, sulfur dioxide, suspended particulate matter, and lead.

The 1990 amendments of the Clean Air Act establish an interstate ozone transport region extending from the Washington, DC metropolitan area to Maine. In this densely

TABLE 2.21   United States National Ambient Air Quality Standards

Pollutant	Primary Standards	Averaging Times	Secondary Standards
Carbon monoxide	9 ppm (10 mg m^{-3})	8-h[a]	None
	35 ppm (40 mg m^{-3})	1-h[a]	None
Lead	1.5 μg m^{-3}	Quarterly average	Same as primary
Nitrogen dioxide	0.053 ppm (100 μg m^{-3})	Annual (arithmetic mean)	Same as primary
Particulate matter (PM$_{10}$)	50 μg m^{-3}	Annual[b] (arithmetic mean)	Same as primary
	150 μg m^{-3}	24-h[a]	Same as primary
Particulate matter (PM$_{2.5}$)	15.0 μg m^{-3}	Annual[c] (arithmetic mean)	Same as primary
	65 μg m^{-3}	24-h[d]	
Ozone	0.08 ppm	8-h[e]	Same as primary
	0.12 ppm	1-h[f]	Same as primary
Sulfur oxides	0.03 ppm	Annual (arithmetic mean)	—
	0.14 ppm	24-h[a]	—
	—	3-h[a]	0.5 ppm (1300 μg m^{-3})

[a]Not to be exceeded more than once per year.

[b]To attain this standard, the expected annual arithmetic mean PM$_{10}$ concentration at each monitor within an area must not exceed 50 μg m^{-3}.

[c]To attain this standard, the 3-year average of the annual arithmetic mean PM$_{2.5}$ concentrations from single or multiple community-oriented monitors must not exceed 15 μg m^{-3}.

[d]To attain this standard, the 3-year average of the 98th percentile of 24-h concentrations at each population-oriented monitor within an area must not exceed 65 μg m^{-3}.

[e]To attain this standard, the 3-year average of the fourth-highest daily maximum 8-h average ozone concentration measured at each monitor within an area over each year must not exceed 0.08 ppm.

[f](a) The standard is attained when the expected number of days per calendar year with maximum hourly average concentration above 0.12 ppm is ≤ 1; (b) the 1-h NAAQS will no longer apply to an area one year after the effective date of the designation of that area for the 8-h ozone NAAQS; the effective designation date for most areas is June 15, 2004 [40 CFR 50.9; see *Federal Register* of April 30, 2004 (69 FR 23996)].

populated region, ozone violations in one area are caused, at least in part, by emissions in upwind areas. A transport commission is authorized to coordinate control measures within the interstate transport region and to recommend to the EPA when additional control measures should be applied in all or part of the region in order to bring any area in the region into attainment. Hence areas within the transport region that are in attainment of the ozone NAAQS might become subject to the controls required for nonattainment areas in that region.

The Clean Air Act requires each state to adopt a plan, a *State Implementation Plan* (SIP), which provides for the implementation, maintenance, and enforcement of the NAAQS. It is, of course, emission reductions that will abate air pollution. Thus the states' plans must contain legally enforceable emission limitations, schedules, and timetables for compliance with such limitations. The control strategy must consist of a combination of measures designed to achieve the total reduction of emissions necessary for the attainment of the air quality standards. The control strategy may include, for example, such measures as emission limitations, emission charges or taxes, closing or relocation of commercial or industrial facilities, periodic inspection and testing of motor vehicle emission control systems, mandatory installation of control devices on motor vehicles, means to reduce motor vehicle traffic, including such measures as parking restrictions and carpool lanes on freeways, and expansion and promotion of the use of mass transportation facilities.

APPENDIX 2.2   HAZARDOUS AIR POLLUTANTS (AIR TOXICS)

Hazardous air pollutants or toxic air contaminants ("air toxics") refer to any substances that may cause or contribute to an increase in mortality or in serious illness, or that may pose a present or potential hazard to human health. Title III of the Clean Air Act Amendments of 1990 completely overhauled the existing hazardous air emission program. Section 112 of the Amendments defines a new process for controlling air toxics that includes the listing of 189 substances, the development and promulgation of Maximum Achievable Control Technology (MACT) standards, and the assessment of residual risk after the implementation of MACT. Any stationary source emitting in excess of $10 \, \text{tons} \, \text{yr}^{-1}$ of any listed hazardous substance, or $25 \, \text{tons} \, \text{yr}^{-1}$ or more of any combination of hazardous air contaminants, is a major source for the purpose of Title III and is subject to regulation. Congress established a list of 189 hazardous air pollutants in the CAA itself. It includes organic chemicals, pesticides, metals, coke-oven emissions, fine mineral fibers, and radionuclides (including radon). This initial list may be revised by the EPA to either add or remove substances. The EPA is required to add pollutants to the list if they are shown to present, through inhalation or other routes of exposure, a threat of adverse human health effects or adverse environmental effects, whether through ambient concentrations, bioaccumulation, deposition, or otherwise.

Congress directed the EPA to list by November 15, 1995, the categories and subcategories of sources that represent 90% of the aggregate emissions of

- Alkylated lead compounds
- Polycylic organic matter
- Hexachlorobenzene
- Mercury
- Polychlorinated biphenyls
- 2,3,7,8-Tetrachlorodibenzofuran
- 2,3,7,8-Tetrachlorodibenzo-*p*-dioxin

Congress further directed the EPA to establish and promulgate emissions standards for such sources by November 15, 2000. The emissions standards must effect the maximum degree of reduction in the listed substance, including the potential for a prohibition on such emissions, taking into consideration costs, any non-air-quality health and environmental impacts, and energy requirements. In establishing these emissions standards, the EPA may also consider health threshold levels, which may be established for particular hazardous air pollutants. Each state may develop and submit to the EPA for approval a program for the implementation and enforcement of emission standards and other requirements for hazardous air pollutants or requirements for the prevention and mitigation of accidental releases of hazardous substances.

TABLE 2.22 Substances Either Confirmed or under Study as Hazardous to Human Health by State of California Air Resources Board (1989)

Substance	Qualitative Health Assessment[a]	Manner of Usage/ Major Sources	Atmospheric Residence Time	Concentrations	
				Ambient Average[b]	Hotspot
Benzene	Human carcinogen	Gasoline	12 days	4.6 ppb (SoCAB)[c]	—
Ethylene dibromide	Probable carcinogen	Gasoline, pesticides	50 days	7.4 ppt (SoCAB)	—
Ethylene dichloride	Probable carcinogen	Gasoline, solvents, pesticides	42 days	19–110 ppt	—
Hexavalent chromium	Human carcinogen	Chrome plating, corrosion inhibitor	—	0.5 ng m^{-3} (SoCAB)	—
Dioxins	Probable carcinogen	Combustion product	1 yr in soil	1.0 pg m^{-3}	—
Asbestos	Human carcinogen	Milling, mining	Unknown; removed by deposition	8–80 fibers m^{-3}	50–500 fibers m^{-3}
Cadmium	Probable carcinogen	Secondary smelters, fuel combustion	7 days; removed by deposition	1–2.5 ng m^{-3}	40 ng m^{-3}
Carbon tetrachloride	Probable carcinogen	CCl$_4$ production, grain fumigant	42 yr	0.13 ppb	0.63 ppb
Ethylene oxide	Probable carcinogen	Sterilization agent, manufacture of surfactants	200 days	50 ppt (SoCAB)	17 ppb
Vinyl chloride	Human carcinogen	Landfill byproduct	2 days	—	0.08–0.34 ppb
Inorganic arsenic	Human carcinogen	Fuel combustion, pesticides	Unknown; removed by deposition	2.4 ng m^{-3} (SoCAB)	200 ng m^{-3}
Methylene chloride	Probable carcinogen	Solvent	0.41 yr	1.1–2.4 ppb	10.7 ppb
Perchloroethylene	Probable carcinogen	Solvent, chemical intermediate	0.4 yr	0.71 ppb	22 ppb
Trichloroethylene	Probable carcinogen	Solvent, chemical	5–8 days	0.22 ppb	—
Nickel	Probable carcinogen	Alloy, plating ceramics, dyes intermediate	Unknown, removed by deposition	7.3 ng m^{-3}	23 ng m^{-3}

Compound	Classification	Use	Lifetime		
Chloroform	Probable carcinogen	Solvent, chemical intermediate	0.55 yr	0.006–0.13 ppb	10 ppb
Formaldehyde	Probable carcinogen	Chemical	3.8–8.6 h	2–39 ppb	—
1,3-Butadiene	Probable carcinogen	Chemical feedstock, resin production	< 1 day	—	0.016 ppb
Acetaldehyde	Probable carcinogen	Motor vehicles	9 h	—	35 ppb
Acrylonitrile	Probable carcinogen	Feedstock, resins, rubber	5.6 days	—	—
Beryllium	Probable carcinogen	Metal alloys, fuel combustion	10 days to be removed by deposition	$0.11–0.22 \ \text{ng m}^{-3}$	—
Dialkylnitrosamines	Probable carcinogen	Chemical feedstock	9.6 h	—	0.3 ppb
p-Dichlorobenzene	Probable carcinogen	Room deodorant, moth repellent	39 days	—	$105–1700 \ \mu\text{g m}^{-3}$
Di-(2-ethylhexyl) phthalate	Probable carcinogen	Plasticizer, resins	1.3–13 h (urban)	$2 \ \mu\text{g m}^{-3}$	—
1,4-Dioxane	Probable carcinogen	Solvent stabilizer, feedstock	3.9 days	—	—
Dimethyl sulfate	Probable carcinogen	Chemical reagent	—	—	—
Ethyl acrylate	Possible carcinogen	Chemical intermediate	12 h	—	—
Hexachlorobenzene	Probable carcinogen	Solvent, pesticide	4 yr in soils	—	—
Lead	Blood system toxic, neurotoxicity	Auto exhaust, fuel additive	7–30 days; removed by deposition	$270–820 \ \text{ng m}^{-3}$	—
Mercury	Neurotoxic	Electronics, paper/pulp manufacture	0.3–2 yr; removed by deposition	0.37–0.49 ppb	1.2 ppb
4,4'-Methylenedianiline	Possible carcinogen	Chemical intermediate	6.4 h	—	—
N-Nitrosomorpholine	Probable carcinogen	Detergents, corrosion inhibitor	<9.6 h	$0.025 \ \text{ng m}^{-3}$	$0.1 \ \text{ng m}^{-3}$

(Continued)

TABLE 2.22 *(Continued)*

Substance	Qualitative Health Assessment[a]	Manner of Usage/ Major Sources	Atmospheric Residence Time	Concentrations	
				Ambient Average[b]	Hotspot
PAHs	Probable carcinogen	Fuel combustion	0.4–40 days; removed by deposition	0.46 ng m^{-3}	—
PCBs	Probable carcinogen	Electronics	3–1700 days;	0.5–14 ng m^{-3}	—
Propylene oxide	Probable carcinogen	Resin manufacture, surfactant	6 days	—	—
Styrene	Probable carcinogen	Chemical feedstock	—	10 ppb	—
Toluene diisocyanates	Possible carcinogen	Raw material polyurethane	26 h	—	—
2,4,6-Trichlorophenol	Probable carcinogen	Herbicide, wood preservative	—	—	—

[a]*Human carcinogen* = sufficient evidence in humans (International Agency for Research on Cancer). *Probable human carcinogen* = limited human or sufficient animal evidence using IARC criteria or EPA guidelines for carcinogen risk assessment. *Possible human carcinogen* = limited animal evidence using IARC criteria or EPA guidelines for carcinogen risk assessment.

[b]Values presented by the California ARB relevant to California.

[c]South Coast Air Basin of California (Los Angeles metropolitan area).

The California Air Resources Board (ARB) (1989) has developed a list of substances of concern in California, called "Status of Toxic Air Contaminant Identification." This list and the organization of substances within it are subject to periodic revision, as needed. The February 1989 Status List grouped substances into three categories. Category I includes identified toxic air contaminants: asbestos, benzene, cadmium, carbon tetrachloride, chlorinated dioxins and dibenzofurans (15 species), chromium (VI), ethylene dibromide and ethylene dichloride, and ethylene oxide. Category IIA contains nine substances that were in the formal review process (1,3-butadiene, chloroform, formaldehyde, inorganic arsenic, methylene chloride, nickel, perchloroethylene, trichloroethylene, and vinyl chloride). Category IIB contains 23 substances not yet reviewed at the time (acetaldehyde, acrylonitrile, beryllium, coke-oven emissions, dialkylnitrosamines, p-dichlorobenzene, di(2-ethylhexyl)-phthalate, 1,4-dioxane, dimethyl sulfate, environmental tobacco smoke, ethyl acrylate, hexachlorobenzene, inorganic lead, mercury, 4,4'-methylenedianiline, N-nitrosomorpholine, PAHs, PCBs, propylene oxide, radionuclides, styrene, toluene diiosocyanates, and 2,4,6-trichlorophenol). Category III includes substances for which additional health information was needed prior to review. These are acrolein, allyl chloride, benzyl chloride, chlorobenzene, chlorophenols/phenol, chloroprene, glycol ethers, maleic anhydride, manganese, methyl bromide, methyl chloroform, nitrobenzene, vinylidene chloride, and xylenes. Available information on these compounds is summarized in Table 2.22.

PROBLEMS

2.1$_A$ Methane (CH_4) has an atmospheric mixing ratio of about 1745 ppb. Carbon monoxide (CO) has a global average mixing ratio in the neighborhood of 100 ppb. Global annual emissions of CH_4 are about 600 Tg (CH_4) yr^{-1} (Table 2.10); global annual CO sources are estimated as 2780 Tg (CO) yr^{-1} (Table 2.14). How can the atmospheric concentration of CH_4 be almost 20 times higher than that of CO given their relative emissions? (Problem suggested by T. S. Dibble.)

2.2$_A$ Figure 2.6 shows ozone molecular number concentration and mixing ratio versus altitude. Why do the molecular number concentration and mixing ratio peak at different altitudes?

2.3$_A$ The total estimated global sink of nitrous oxide (N_2O) is 12.6 Tg N yr^{-1}. The global mean N_2O mixing ratio in 2000 was 315 ppb. On the basis of these two values, estimate the mean lifetime of N_2O in the atmosphere.

2.4$_A$ Calculate the number of DU assuming that the entire atmospheric O_3 column is at a uniform concentration of 3×10^{12} molecules cm^{-3} between 15 km and 30 km and zero elsewhere.

2.5$_A$ Singh et al. (2003) reported airborne measurements of hydrogen cyanide (HCN) and methyl cyanide (CH_3CN, also known as *acetonitrile*) over the Pacific Ocean in 2001. Mean HCN and CH_3CN mixing ratio of 243 ppt and 149 ppt, respectively, were measured. On the basis of these findings and other information, they prepared global budgets for both compounds.

   a. Compute the mean tropospheric column burdens, in molecules cm^{-2}, of each compound according to these values.

b. The authors estimated the following quantities for the two compounds:

	HCN	CH$_3$CN
Annual mean atmospheric burden, Tg (N)	0.44	0.30
Residence time due to OH reaction, months	63	23
Global loss to oceans, Tg (N) yr^{-1}	1.0	0.4

Calculate the residence time of each compound due to the oceanic sink and the overall global mean residence time of each compound, assuming that OH reaction and deposition to the oceans are the only loss processes.

2.6$_A$ Table 2.10 presents the global methane budget, in which sources are estimated to exceed sinks by 22 Tg (CH$_4$) yr^{-1}. Show that this difference corresponds to an increase of 8 ppb yr^{-1} of CH$_4$.

REFERENCES

Andreae, M. O. (1990) Ocean–atmosphere interactions in the global biogeochemical sulfur cycle, *Marine Chem.* **30**, 1–29.

Andreae, M. O., and Barnard, W. R. (1984) The marine chemistry of dimethylsulfide, *Marine Chem.* **14**, 267–279.

Andreae, M. O., and Crutzen, P. J. (1997) Atmospheric aerosols: Biogeochemical sources and role in atmospheric chemistry, *Science* **276**, 1052–1058.

Andreae, M. O., and Raemdonck, H. (1983) Dimethylsulfide in the surface ocean and the marine atmosphere: A global view, *Science* **221**, 744–747.

Bergamaschi, P., Hein, R., Heimann, M., and Crutzen, P. J. (2000) Inverse modeling of the global CO cycle 1. Inversion of CO mixing ratios, *J. Geophys. Res.* **105**, 1909–1927.

Berresheim, H., Wine, P. H., and Davis, D. D. (1995) Sulfur in the atmosphere, in *Composition, Chemistry, and Climate of the Atmosphere*, H. B. Singh, ed., Van Nostrand Reinhold, New York, pp. 251–307.

Bojkov, R. D. (1988) Ozone changes at the surface and in the free troposphere, in *Tropospheric Ozone*, I. S. A. Isaksen, ed., Reidel, Dordrecht, pp. 83–96.

Bond, T. C., Streets, D. G., Yarber, K. F., Nelson, S. M., Woo, J., and Klimant, Z. (2004) A technology-based global inventory of black and organic carbon emissions from combustion, *J. Geophys. Res.* **109**(D14), D14203 (doi: 10.1029/2003JD003697).

Bond, T. C., and Bergstrom, R. W. (2006) Light absorption by carbonaceous particles: An investigative review, *Aerosol Sci. Technol.*, **40**, 27–67.

Bouwman, A. F., Lee, D. S., Asman, W. A. H., Dentener, F. J., Van Der Hoek, K. W., and Oliver, J. G. J. (1997) A global high-resolution emission inventory for ammonia, *Global Biogeochem. Cycles* **11**, 561–587.

Brasseur, G. P., Orlando, J. J., and Tyndall, G. S. (1999) *Atmospheric Chemistry and Global Change*, Oxford Univ. Press, New York.

California Air Resources Board (1989) *Information on Substances for Review as Toxic Air Contaminants*, Report ARB/SSD/89-01, Sacramento, CA.

Cao, M., Gregson, K., and Marshall, S. (1998) Global methane emission from wetlands and its sensitivity to climate change, *Atmos. Environ.* **32**, 3293–3299.

Chin, M., and Davis, D. D. (1995) A reanalysis of carbonyl sulfide as a source of stratospheric background sulfur aerosol, *J. Geophys. Res.* **100**, 8993–9005.

Chung, S. H. and Seinfeld, J. H. (2002) Global distribution and climate forcing of carbonaceous aerosols, *J. Geophys. Res.* **107**(D19), 4407 (doi:10.1029/2001JD001397).

Crutzen, P. J., (1988) Tropospheric ozone: A review, in *Tropospheric Ozone*, I. S. A. Isaksen, ed., Reidel, Dordrecht, pp. 3–32.

Daisey, J. M. (1980) Organic compounds in urban aerosols, *Ann. NY Acad. Sci.* **338**, 50–69.

Dockery, D. W., Pope, C. A. III, Xu, X., Spengler, J. D., Ware, J. H., Fay, M. E., Ferris, B. G. Jr., and Speizer, F. E. (1993) An association between air pollution and mortality in six U.S. cities, *N. Engl. J. Med.* **329**, 1753–1759.

Eriksson, E. (1959) The yearly circulation of chloride and sulfur in nature. Meteorological, geochemical, and pedological implications, *Tellus* **11**, 375–404.

Fishman, J., Watson, C. E., Larson, J. C., and Logan, J. A. (1990) Distribution of tropospheric ozone determined from satellite data, *J. Geophys. Res.* **95**, 3599–3617.

Fishman, J. et al. (1992) Distribution of tropospheric ozone in the tropics from satellite and ozonesonde measurements, *J. Atmos. Terr. Phys.* **54**, 589–597.

Fung, I., John, J., Lerner, J., Matthews, E., Prather, M., Steele, L. P., and Fraser, P. J. (1991) Three-dimensional model synthesis of the global methane cycle, *J. Geophys. Res.* **96**, 13033–13065.

Gong, S. L., Barrie, L. A., and Lazare, M. (2002) Canadian Aerosol Module (CAM): A size-segregated simulation of atmospheric aerosol processes for climate and air quality models 2. Global sea-salt aerosol and its budgets, *J. Geophys. Res.* **107**(D24), 4779 (doi:10.1029/2001JD002004).

Guenther, A. et al. (1995) A global model of natural volatile organic compound emissions, *J. Geophys. Res.* **100**, 8873–8892.

Hauglustaine, D. A., Brasseur, G. P., Walters, S., Rasch, P. J., Müller, J. -F., Emmons, L. K., and Carroll, M. A. (1998) MOZART, a global chemical transport model for ozone and related chemical tracers: 2. Model results and evaluation, *J. Geophys. Res.* **103**, 28291–28335.

Hein, R., Crutzen, P. J., and Heimann, M. (1997) An inverse modeling approach to investigate the global atmospheric methane cycle, *Global Biogeochem. Cycles* **11**, 43–76.

Heintzenberg, J. (1989) Fine particles in the global troposphere—a review, *Tellus* **41B**, 149–160.

Houweling, S., Kaminski, T., Dentener, F., Lelieveld, J., and Heimann, M. (1999) Inverse modeling of methane sources and sinks using the adjoint of a global transport model, *J. Geophys. Res.* **104**, 26137–26160.

Intergovernmental Panel on Climate Change (IPCC) (1995) *Climate Change 1995: The Science of Climate Change*, Cambridge Univ. Press, Cambridge, UK.

Intergovernmental Panel on Climate Change (IPCC) (2001) *Climate Change 2001: The Scientific Basis*, Cambridge Univ. Press, Cambridge, UK.

Janach, W. E. (1989) Surface ozone: Trend details, seasonal variations, and interpretation, *J. Geophys. Res.* **94**, 18289–18295.

Jiang, Y., and Yung, Y. L. (1996) Concentrations of tropospheric ozone for 1979 to 1992 over tropical Pacific South America from TOMS data, *Science* **272**, 714–716.

Kettle, A. J. et al. (1999) A global database of sea surface dimethylsulfide (DMS) measurements and a procedure to predict sea surface DMS as a function of latitude, longitude, and month, *Global Biogeochem. Cycles* **13**, 399–444.

Kettle, A. J., Kuh, U., von Hobe, M., Kesselmeier, J., and Andreae, M. O. (2002) Global budget of atmospheric carbonyl sulfide: Temporal and spatial variations of the dominant sources and sinks, *J. Geophys. Res.* **107**(D22), 4658 (doi:10.1029/2002JD002187).

Kiehl, J. T., and Rodhe, H. (1995) Modeling geographical and seasonal forcing due to aerosols, in *Aerosol Forcing of Climate*, R. J. Charlson and J. Heintzenberg, eds., Wiley, New York, pp. 281–296.

Kroeze, C., Mozier, A., and Bouwman, L. (1999) Closing the N_2O budget: A retrospective analysis, *Global Biogeochem. Cycles* **13**, 1–8.

Kurylo, M. J. et al. (1999) Short-lived ozone-related compounds, in *Scientific Assessment of Ozone Depletion: 1998*, Global Ozone Research and Monitoring Project, Report 44, 2.1-2.56, C. A. Ennis, ed., World Meteorological Organization, Geneva.

Lamb, B., Gay, D., Westberg, H., and Pierce, T. (1993) A biogenic hydrocarbon emission inventory for the U.S.A. using a simple forest canopy model, *Atmos. Environ.* **27A**, 1673–1690.

Lamb, B., Guenther, A., Gay, D., and Westberg, H. (1987) A national inventory of biogenic hydrocarbon emissions, *Atmos. Environ.* **21**, 1695–1705.

Lefohn, A. S., Husar, J. D., and Husar, R. B. (1999) Estimating historical anthropogenic global sulfur emission patterns for the period 1850–1990, *Atmos. Environ.* **33**, 3435–3444.

Lelieveld, J., Roelofs, G. J., Ganzeveld, L., Feichter, J., and Rodhe, H. (1997) Terrestrial sources and distribution of atmospheric sulphur, *Phil. Trans. Roy. Soc. Lond. B* **352**, 149–158.

Lelieveld, J., Crutzen, P., and Dentener, F. J. (1998) Changing concentration, lifetime and climate forcing of atmospheric methane, *Tellus* **50B**, 128–150.

Liao, H., Adams, P. J., Seinfeld, J. H., Mickley, L. J., and Jacob, D. J. (2003) Interactions between tropospheric chemistry and aerosols in a unified GCM simulation, *J. Geophys. Res.* **108**(D1), 4001 (doi: 10.1029/2001JD001260).

Liao, H., Seinfeld, J. H., Adams, P. J., and Mickley, L. J. (2004) Global radiative forcing of coupled tropospheric ozone and aerosols in a unified general circulation model, *J. Geophys. Res.*, **109**, D16207 (doi: 10.1029/2003JD004456).

Liousse, C., Penner, J. E., Chuang, C., Walton, J. J., Eddleman, H., and Cachier, H. (1996) A global three-dimensional model study of carbonaceous aerosols, *J. Geophys. Res.* **101**, 19411–19432.

Lovelock, J. E. (1971) Atmospheric fluorine compounds as indicators of air movements, *Nature* **230**, 379.

Lovelock, J. E., Maggs, R. J., and Rasmussen, R. A. (1972) Atmospheric dimethyl sulfide and the natural sulfur cycle, *Nature* **237**, 452–453.

Lurmann, F. W., and Main, H. H. (1992) *Analysis of the Ambient VOC Data Collected in the Southern California Air Quality Study. Final Report*, ARB Contract A832-130, California Air Resources Board, Sacramento, CA.

Molina, M. J., and Rowland, F. S. (1974) Stratospheric sink for chlorofluoromethanes: Chlorine atom catalyzed destruction of ozone, *Nature* **249**, 810–812.

Mosier, A. R., Duxbury, J. M., Freney, J. R., Heinemeyer, O., Minami, K., and Johnson, D. E. (1998a) Mitigating agricultural emissions of methane, *Climate Change* **40**, 39–80.

Mosier, A., Kroeze, C., Nevison, C., Oenema, O., Seitzinger, S., and van Cleemput, O. (1998b) Closing the global N_2O budget: Nitrous oxide emissions through the agricultural nitrogen cycle— OECD/IPCC/IEA phase II development of IPCC guidelines for national greenhouse gas inventory methodology, *Nutrient Cycling Agroecosyst.* **52**, 225–248.

National Research Council (1991) *Rethinking the Ozone Problem in Urban and Regional Air Pollution*, National Academy Press, Washington, DC.

Notholt, J. et al. (2003) Enhanced upper tropical troposphere OCS: Impact on the stratospheric aerosol layer, *Science* **300**, 307–310.

Olivier, J. G. J., Bouwman, A. F., van der Hoek, K. W., and Berdowski, J. J. M. (1998) Global air emission inventories for anthropogenic sources of NO_x, NH_3 and N_2O in 1990, *Environ. Pollut.* **102**, 135–148.

Olivier, J. G. J., Bouwman, A. F., Berdowski, J. J. M., Veldt, C., Bloos, J. P. J., Visschedijk, A. J. H., van der Maas, C. W. M., and Zasndveld, P. Y. J. (1999) Sectoral emission inventories of greenhouse gases for 1990 on a per country basis as well as on 1×1, *Environ. Sci. Policy* **2**, 241–263.

Oltmans, S. J., and Levy, H. II (1994) Surface ozone measurements from a global network, *Atmos. Environ.* **28**, 9–24.

Placet, M., Battye, R. E., Fehsenfeld, F. C., and Bassett, G. W. (1990) *Emissions Involved in Acidic Deposition Processes. State-of-Science/Technology Report 1. National Acid Precipitation Assessment Program*, U.S. Government Printing Office, Washington, DC.

Prinn, R. G. et al. (2000) A history of chemically and radiatively important gases in air deduced from ALE/GAGE/AGAGE, *J. Geophys. Res.* **105**, 17751–17792.

Roberts, J. M. (1995) Reactive odd-nitrogen (NO_y) in the atmosphere, in *Composition, Chemistry, and Climate of the Atmosphere*, H. B. Singh, ed., Van Nostrand Reinhold, New York, pp. 176–215.

Schauffler, S. M., Atlas, E. L., Flocke, F., Lueb, R. A., Stroud, V., and Travnicek, W. (1998) Measurements of bromine containing organic compounds at the tropical tropopause, *Geophys. Res. Lett.* **25**, 317–320.

Schönbein, C. F. (1840) Beobachtungen über den bei der elektrolysation des wassers und dem ausströmen der gewöhnlichen electrizitat aus spitzen eich entwichelnden geruch, *Ann. Phys. Chem.* **50**, 616.

Schönbein, C. F. (1854) Über verschiedene zustände des sauerstoffs, liebigs, *Ann. Chem.* **89**, 257–300.

Shen, T. L., Wooldrige, P. J., and Molina, M. J. (1995) Stratospheric pollution and ozone depletion, in *Composition, Chemistry, and Climate of the Atmosphere*, H. B. Singh, ed., Van Nostrand Reinhold, New York, pp. 394–442.

Singh, H. B. (1995) Halogens in the atmospheric environment, in *Composition, Chemistry, and Climate of the Atmosphere*, H. B. Singh, ed., Van Nostrand Reinhold, New York, pp. 216–250.

Singh, H. B. et al. (2003) In situ measurements of HCN and CH_3CN over the Pacific Ocean: Sources, sinks, and budgets, *J. Geophys. Res.* **108**(D20), 8795 (doi: 10.1029/2002JD003006).

Sokolik, I. N. (2002) Dust, in *Encyclopedia of Atmospheric Sciences*, J. R. Holton, ed., Elsevier, Amsterdam.

Solomon, P. A., Fall, T., Salmon, L., Cass, G. R., Gray, H. A., and Davidson, A. (1989) Chemical characteristics of PM_{10} aerosols collected in the Los Angeles area, *J. Air Pollut. Control Assoc.* **39**, 154–163.

Staehelin, J., and Schmid, W. (1991) Trend analysis of tropospheric ozone concentrations utilizing the 20-year data set of balloon soundings over Payerne (Switzerland), *Atmos. Environ.* **25A**, 1739–1749.

Turner, S. M., and Liss, P. S. (1985) Measurements of various sulfur gases in a coastal marine environment, *J. Atmos. Chem.* **2**, 223–232.

Volz, A., and Kley, D. (1988) Evaluation of the Montsouris series of ozone measurements made in the nineteenth century, *Nature* **332**, 240–242.

Warneck, P. (1988) *Chemistry of the Natural Atmosphere*, Academic Press, New York.

Went, F. W. (1960) Blue hazes in the atmosphere, *Nature* **187**, 641–643.

Whitby, K. T., and Cantrell, B. (1976) Fine particles, *Proc. Int. Confe. Environmental Sensing and Assessment*, Las Vegas, NV, Institute of Electrical and Electronic Engineers.

World Meteorological Organization (WMO) (1986) *Atmospheric Ozone* 1985, Global Ozone Research and Monitoring Project, Report 16, Geneva.

World Meteorological Organization (WMO) (1990) *Report of the International Ozone Trends Panel: 1988*, Global Ozone Research and Monitoring Project, Report 18, Geneva.

World Meteorological Organization (WMO) (1998) *Scientific Assessment of Ozone Depletion: 1998*. Global Ozone Research and Monitoring Project, Report 44, World Meteorological Organization, Geneva.

World Meteorological Organization (WMO) (2002) *Scientific Assessment of Ozone Depletion: 2002.* Global Ozone Research and Monitoring Project, Report 47, World Meteorological Organization, Geneva.

Yvon-Lewis, S., Methyl bromide in the atmosphere and ocean, *IGACtivities Newsletter*, International Global Atmospheric Chemistry Project, Issue 19, Jan. 9-12, 2000.

Zender, C. S., Bian, H., and Newman, D. (2003) Mineral dust entrainment and deposition (DEAD) model: Description and 1990s dust climatology, *J. Geophys. Res.* **107**(D24), 4416 (doi: 10.1029/2002JD002775).

Zender, C. S., Miller, R. L., and Tegen, I. (2004) Quantifying mineral dust mass budgets: Terminology, constraints, and current estimates, *EOS* **85**(48), 509.

3 Chemical Kinetics

3.1 ORDER OF REACTION

We will consider three types of chemical reaction:

$$
\begin{array}{lll}
\text{First-order (unimolecular)} & \text{A} \rightarrow \text{B} + \text{C} \\
\text{Second-order (bimolecular)} & \text{A} + \text{B} \rightarrow \text{C} + \text{D} \\
\text{Third-order (termolecular)} & \text{A} + \text{B} + \text{M} \rightarrow \text{AB} + \text{M}
\end{array}
$$

The rate (in molecules $cm^{-3}\,s^{-1}$) of a first-order reaction is expressed as

$$
\frac{d[\text{A}]}{dt} = -k_1[\text{A}] \tag{3.1}
$$

where the first-order rate coefficient k_1 has units of s^{-1} (reciprocal seconds).

Few reactions are truly first-order, in that they involve decomposition of a molecule without intervention of a second molecule. The classic example of a true first-order reaction is radioactive decay, such as $^{222}\text{Rn} \rightarrow {}^{218}\text{Po} + \alpha$-particles. In the atmosphere, by far the most important class of first-order reactions is photodissociation reactions in which absorption of a photon of light ($h\nu$) by the molecule induces chemical change. Photodissociation, or photolysis, reactions are written as

$$
\text{A} + h\nu \rightarrow \text{B} + \text{C}
$$

in which $h\nu$ represents a photon of light of frequency ν. In the photolysis of species A, the rate coefficient is customarily denoted by the symbol j_{A}.

Thermal decomposition of a molecule is often represented as first-order, but the energy required for decomposition is usually supplied through collision with another molecule. If the other molecule is an air molecule, it is denoted as M, and the actual reaction is $\text{A} + \text{M} \rightarrow \text{B} + \text{C} + \text{M}$. Since M is at great excess relative to A, its concentration is constant, and the concentration of M can be implicitly included in the reaction rate coefficient; then, the reaction is written simply as $\text{A} \rightarrow \text{B} + \text{C}$.

The rate equation (3.1) can be integrated to give

$$
[\text{A}] = [\text{A}]_0\, e^{-k_1 t} \tag{3.2}
$$

Atmospheric Chemistry and Physics: From Air Pollution to Climate Change, Second Edition, by John H. Seinfeld and Spyros N. Pandis. Copyright © 2006 John Wiley & Sons, Inc.

Thus, species A decays to $1/e$ of its initial concentration in time $\tau = 1/k_1$. This time is referred to as the *e*-folding time of the reaction, or the mean lifetime of A against this reaction.

The rate of a second-order, or bimolecular, reaction is

$$\frac{d[A]}{dt} = -k_2[A][B] \tag{3.3}$$

where the second-order rate coefficient k_2 has units of cm^3 $molecule^{-1}$ s^{-1}.

The termolecular reaction, which is written as $A + B + M \rightarrow AB + M$, actually does not take place as the result of the simultaneous collision of all three molecules A, B, and M. The probability of such an event happening is practically zero. Rather, what actually occurs is that molecules A and B collide to produce an energetic intermediate $AB^{\dagger}$ (the dagger representing vibrational excitation):

$$A + B \rightarrow AB^{\dagger}$$

In order for $AB^{\dagger}$ to proceed to the product AB, its excess energy must be removed through collision with another molecule denoted by M, to which the excess energy is transferred:

$$AB^{\dagger} + M \rightarrow AB + M$$

In the atmosphere, M is the background mixture of N_2 and O_2. Termolecular reactions are usually expressed as

$$A + B + M \rightarrow AB + M$$

or

$$A + B \xrightarrow{M} AB$$

We return to a derivation of the rate of termolecular reactions in Section 3.5.

Lifetime of a Species as a Result of a Chemical Reaction Let us determine the lifetime of a species undergoing a first-order reaction. When a species undergoes a first-order decay, its concentration as a function of time is given by (3.2)

$$[A] = [A]_0 \, e^{-kt}$$

A measure of the relative speed of a chemical reaction is the time required for A to decay to a certain fraction of its initial concentration. For example, the *halflife*, $t_{1/2}$, of a reaction is the time needed for $[A] = [A]_0/2$. A more commonly used measure of the speed of a reaction is derived from (3.2). The *Lifetime*, τ, is the time at which $[A]/[A]_0 = e^{-1} = 0.368$. Thus, $\tau = 1/k$. Halflife and lifetime are related by

$$t_{1/2} = \ln(2)/k = 0.69/k$$

We will use lifetime τ exclusively as a measure of the speed of a chemical reaction. When a substance participates in several chemical reactions, and we are interested in its lifetime as a result of reaction i, that lifetime is referred to as its *lifetime against reaction i*.

The concept of lifetime can be applied to reactions of any order. Determine, for example, the lifetime of each of the reactants in the second-order reaction of nitric oxide and ozone:

$$NO + O_3 \rightarrow NO_2 + O_2 \qquad k(298\,K) = 1.9 \times 10^{-14}\,cm^3\,molecule^{-1}\,s^{-1}$$

The rate of the reaction is $k\,[NO]\,[O_3]$ and depends on the concentrations of both reactants. The lifetime of NO against this reaction is given by

$$\tau_{NO} = \frac{1}{k[O_3]}$$

To calculate τ_{NO}, it is necessary to specify the concentration of O_3. This makes sense because the larger the O_3 concentration, the shorter the lifetime of NO. At the Earth's surface at 298 K, if the O_3 mixing ratio is 50 ppb, then $[O_3] = (50 \times 10^{-9})(2.5 \times 10^{19}) = 1.25 \times 10^{12}$ molecules cm^{-3}. Then

$$\tau_{NO} = \frac{1}{(1.9 \times 10^{-14})(1.25 \times 10^{12})} = 42\,s$$

Conversely, the lifetime of O_3 against this reaction is

$$\tau_{O_3} = \frac{1}{k[NO]}$$

and to calculate τ_{O_3}, a value of [NO] needs to be specified. For example, if the NO mixing ratio is 10 ppb, then the lifetime of O_3 against this reaction is 3.5 min.

3.2 THEORIES OF CHEMICAL KINETICS

A basic goal of the theory of chemical kinetics is to predict the magnitude of the reaction rate coefficient and its temperature dependence. We focus first on bimolecular reactions. The most elementary approach to bimolecular reactions is based on the collision of hard, structureless spheres. This approach is called *collision theory*.

3.2.1 Collision Theory

Imagine that the molecules are like billiard balls, in that there is no interaction between them until they come into contact, and they are impenetrable, so that their centers cannot come any closer than a distance equal to the sum of their radii. When molecules react, there is a rearrangement of their valence electrons, and this usually means that some energy has to be expended in overcoming the energy barrier associated with disturbing the electrons from their original configuration. This is true even if the final configuration of electrons is a more stable arrangement.

Consider a collision between two hard spheres A and B with radii r_A and r_B and velocities v_A and v_B. If we imagine B to be fixed, then the relative velocity of A is $v = v_A - v_B$. If the centers of A and B approach each other at a separation less than or equal to the sum of their radii, then collision occurs. We can therefore define a total cross section S as the effective target area presented to A by B, that is, $S = \pi d^2$, where $d = r_A + r_B$. An A molecule passing with velocity v through n_B stationary B molecules per unit volume will collide with a B molecule whose center lies within an area πd^2 around the path of A. In unit time, an A molecule sweeps out a collision volume $\pi d^2 v$ and undergoes collision with $\pi d^2 v n_B$ B molecules. If there are n_A molecules of A per unit volume, the total number of collisions per unit volume per unit time is

$$Z_{AB} = n_A n_B \, \pi d^2 \, v \qquad (\text{cm}^{-3}\,\text{s}^{-1}) \tag{3.4}$$

The molecules of each species possess a Maxwell distribution of speeds. For molecules of species A, for example, their mean speed from the Maxwell distribution is $\bar{v}_A = (8k_B T/\pi m_A)^{1/2}$, and the relative velocity of the A, B collision partners is

$$\bar{v} = \left(\frac{8k_B T}{\pi\mu}\right)^{1/2} \tag{3.5}$$

where k_B is the Boltzmann constant and $\mu = m_A m_B/(m_A + m_B)$ is the reduced mass; the reduced mass appears because we are interested in the relative speed of approach. Then $\bar{v}$ is the appropriate velocity in (3.4). Thus, we get

$$Z_{AB} = n_A n_B \, \pi d^2 \left(\frac{8k_B T}{\pi\mu}\right)^{1/2} \tag{3.6}$$

Consider the bimolecular reaction

$$A + B \rightarrow C + D$$

If reaction occurred with every collision, then the rate of reaction between A and B would be just

$$R_{AB} = -\frac{dn_A}{dt} = -\frac{dn_B}{dt} = \pi d^2 \left(\frac{8k_B T}{\pi\mu}\right)^{1/2} n_A n_B \tag{3.7}$$

Not every collision will result in reaction; only those collisions that have sufficient kinetic energy to surmount the energy barrier for reaction will lead to reaction. For a Maxwell distribution the fraction of encounters that have energy greater than a barrier E (kJ mol^{-1}) is $\exp(-E/RT)$. The rate of reaction is then

$$R_{AB} = \pi d^2 \left(\frac{8k_B T}{\pi\mu}\right)^{1/2} \exp\left(-\frac{E}{RT}\right) n_A n_B \tag{3.8}$$

and the collision theory bimolecular rate coefficient is

$$k = \underbrace{\pi d^2 \left(\frac{8k_B T}{\pi \mu}\right)^{1/2}}_{A} \exp\left(-\frac{E}{RT}\right) \qquad (3.9)$$

As indicated, the terms multiplying the exponential are customarily denoted by A, the *collision frequency factor*, or simply the *preexponential factor*. Thus, the reaction rate coefficient consists of two components, the frequency with which the reactants collide and the fraction of collisions that have enough energy to overcome the barrier to reaction.

The quantity $(8k_B T/\pi \mu)^{1/2}$ is a molecular speed; at ordinary temperatures its value is about 5×10^4 cm s^{-1}. The value of d for small molecules typical of atmospheric chemistry is around 3×10^{-8} cm. Thus an order of magnitude estimate for A is

$$A \cong 1.5 \times 10^{-10} \quad \text{cm}^3 \text{ molecule}^{-1} \text{ s}^{-1}$$

The collision rate of one molecule with other molecules in a gas at standard temperature and pressure is $\sim 5 \times 10^9$ collisions s^{-1}. A typical molecular vibrational frequency is 3×10^{13} s^{-1}. The mean time between collisions is $\sim 2 \times 10^{-10}$ s. Thus, a typical molecule undergoes on the order of 10^3–10^4 vibrational cycles between collisions.

Actual measured A factors are usually substantially smaller than those based on collision theory. Recall that the molecules have been assumed to be hard spheres, and molecular structure has been assumed not to play a role. In real molecules certain parts of the molecule are more "reactive" than others. As a result, some of the collisions will be ineffective if the colliding molecules are not pointing their reactive "ends" at each other. For example, in the reaction of hydroxyl (OH) radicals with CHBr$_3$

$$\text{OH} + \text{CHBr}_3 \rightarrow \text{H}_2\text{O} + \text{CBr}_3 \qquad k(298\text{ K}) = 1.8 \times 10^{-13} \text{ cm}^3 \text{ molecule}^{-1} \text{ s}^{-1}$$

the small hydrogen atom in CHBr$_3$ is shielded by the three bulky bromine atoms, so only if the OH approaches in just the right direction so as to encounter the H atom will reaction occur. On the other hand, there are a number of reactions whose rate coefficients approach the collision limit. One example is

$$\text{O} + \text{ClO} \rightarrow \text{Cl} + \text{O}_2 \qquad k(298\text{ K}) = 3.8 \times 10^{-11} \text{ cm}^3 \text{ molecule}^{-1} \text{ s}^{-1}$$

In summary, collision theory provides a good physical picture of bimolecular reactions, even though the structure of the molecules is not taken into account. Also, it is assumed that reaction takes place instantaneously; in practice, the reaction itself requires a certain amount of time. The structure of the reaction complex must evolve, and this must be accounted for in a reaction rate theory. For some reactions, the rate coefficient actually decreases with increasing temperature, a phenomenon that collision theory does not describe. Finally, real molecules interact with each other over distances greater than the sum of their hard-sphere radii, and in many cases these interactions can be very important. For example, ions can react via long-range Coulomb forces at a rate that exceeds the collision limit. The next level of complexity is transition state theory.

Evaluation of Collision Theory   The reaction

$$OH + HO_2 \rightarrow H_2O + O_2$$

has a measured rate coefficient

$$k_{OH+HO_2} = 4.8 \times 10^{-11} \exp(250/T) \qquad cm^3 \, molecule^{-1} \, s^{-1}$$

Calculate the collision theory rate coefficient A at $T = 300 \, K$ and compare it to the measured rate coefficient:

$$k_{OH+HO_2}(300 \, K) = 1.1 \times 10^{-10} \qquad cm^3 \, molecule^{-1} \, s^{-1}$$

For the purpose of the collision theory estimate, let us assume that the radii of OH and HO_2 are each 2×10^{-8} cm. Also calculate the fraction of collisions that lead to reaction.

The collision theory rate coefficient A requires the following quantities:

$$\pi d^2 = \pi (2 \times 10^{-10} \, m + 2 \times 10^{-10} \, m)^2 = 5.03 \times 10^{-19} \, m^2$$

$$\mu = \frac{m_{OH} m_{HO_2}}{m_{OH} + m_{HO_2}} = \frac{(2.82 \times 10^{-26})(5.48 \times 10^{-26})}{2.82 \times 10^{-26} + 5.48 \times 10^{-26}}$$

$$= 1.86 \times 10^{-26} \, kg \, molecule^{-1}$$

Then

$$A = \pi d^2 \left(\frac{8 k_B T}{\pi \mu} \right)^{1/2} = (5.03 \times 10^{-19} \, m^2) \left(\frac{8 \times 1.381 \times 10^{-23} \, J \, K^{-1} \times 300 \, K}{3.14 \times 1.86 \times 10^{-26} \, kg} \right)^{1/2}$$

$$= 3.8 \times 10^{-16} \, m^3 \, molecule^{-1} \, s^{-1}$$

$$= 3.8 \times 10^{-10} \, cm^3 \, molecule^{-1} \, s^{-1}$$

Thus, the fraction of collisions that are reactive at 300 K is

$$\frac{1.1 \times 10^{-10}}{3.8 \times 10^{-10}} = 0.29$$

3.2.2   Transition State Theory

Consider the bimolecular reaction

$$A + BC \rightarrow AB + C$$

in which the preexisting chemical bond, B—C, is broken and a new bond, A—B, is formed. Reactions in which both A and BC are molecules are not important at atmospheric temperatures; the amount of electron rearrangement needed produces a large barrier to reaction. On the other hand, reactions in which A is a free radical and BC is a molecule are very important in the atmosphere. In this case, formation of the A—B bond and cleavage of the B—C bond occur virtually simultaneously. Essentially, the electronic energy contained in the first bond is transferred to the second.

The first step in the above reaction is the formation of a transient complex, called the *activated complex* or *transition state*:

$$A + BC \rightarrow ABC^{\ddagger}$$

The transition state can dissociate back to the reactants

$$ABC^{\ddagger} \rightarrow A + BC$$

or to the new products:

$$ABC^{\ddagger} \rightarrow AB + C$$

Transition state theory (also known as *activated-complex theory*) assumes that the transition state is much more likely to decay back to the original reactants than proceed to the stable products; if this is the case, then first two reactions can be assumed to be in equilibrium. The reactive process can then be represented as

$$A + BC \rightleftharpoons ABC^{\ddagger} \rightarrow AB + C$$

The rate of reaction is that at which $ABC^{\ddagger}$ passes to products (as a result of translational or vibrational motions along the reaction coordinate).

We will not go through the full derivation of the transition state theory, for which there are many excellent references [e.g., Laidler (1987), Pilling and Seakins (1995)]. The result is that the rate coefficient is expressed as

$$k = A' \exp\left(\frac{\Delta S}{R}\right) \exp\left(-\frac{E}{RT}\right)$$

where A' is the collision theory preexponential factor and ΔS is the change in entropy, $S(ABC^{\ddagger}) - S(A) - S(BC)$, which is a measure of the molecular rearrangement involved in forming the transition state and can itself be a function of temperature. Complex transition states in which energy can be distributed over many states have large ΔS. In this case the excess energy is less likely to be funneled into the channel that causes the transition state to decompose back to the original reactants.

In many cases the preexponential factor can be considered to be independent of temperature, and the rate coefficient is written as

$$k = A \exp\left(-\frac{E}{RT}\right) \tag{3.10}$$

where A is determined experimentally rather than from theory. Equation (3.10) is called the *Arrhenius form* for k.

3.2.3 Potential Energy Surface for a Bimolecular Reaction

Consider the potential energy surface for the biomolecular reaction (most elementary reactions can be considered as reversible)

$$A + BC \underset{b}{\overset{f}{\rightleftharpoons}} AB + C$$

as shown in Figure 3.1. As the two reactant molecules approach each other, the energy of the reaction system rises. A point is reached, denoted by $ABC^{\ddagger}$, beyond which the energy starts to decrease again. $ABC^{\ddagger}$, the activated complex, is a short-lived intermediate through which the reactants must pass if the encounter is to lead to reaction. By estimating the structure of this transition state the activation energy E may be estimated (Benson, 1976). This point is a saddle point in the potential energy surface of the system. Figure 3.1 shows the relationship between the energies of the process. The activation energy for the forward reaction is E^f; that for the reverse reaction is E^r. The enthalpy of reaction is ΔH_r. Note that

$$E^f - E^r = \Delta H_r$$

The forward reaction (left to right) sketched in Figure 3.1 is exothermic, and $\Delta H_r = H_{products} - H_{reactants}$ is negative. The reverse reaction (right to left) is endothermic and must have an activation energy, E^r, at least as large as ΔH_r. It is customary to identify the activation energy E in (3.10) with E^f (or E^r.)

The height of the energetic barrier E depends on the amount of electronic rearrangement going from reactants to transition state. Molecule–molecule reactions involve a high degree of electronic rearrangement and have large activation energy barriers. Such reactions are usually unimportant at atmospheric temperatures. In radical–molecule reactions the amount of electronic rearrangement is small, often just the transfer of an atom. Activation energy barriers are much lower, and these reactions occur readily at ambient temperatures. In radical–radical reactions typically a single bond is formed from the addition of the two radicals, electronic rearrangement is minimal, and there is

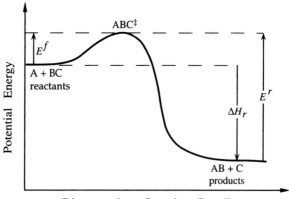

Distance along Reaction Coordinate

FIGURE 3.1   Potential energy surface along the reaction coordinate for a bimolecular reaction.

generally very little electronic energy barrier to reaction. Examples include

$$O + ClO \rightarrow Cl + O_2$$
$$OH + HO_2 \rightarrow H_2O + O_2$$

After initial collision, the adduct will decompose rapidly if an energetically accessible channel exists. For such reactions the activation energy E is often negative, meaning that as temperature increases the rate coefficient decreases. The explanation is that, with virtually no barrier, all collisions have sufficient energy to react. However, less energetic molecules at lower temperatures actually allow the initial adduct to have more time to rearrange itself so as to decompose to the products.

3.3   THE PSEUDO-STEADY-STATE APPROXIMATION

Many chemical reactions involve very reactive intermediate species such as free radicals, which, as a result of their high reactivity, are consumed virtually as rapidly as they are formed and consequently exist at very low concentrations. The pseudo-steady-state approximation[1] (PSSA) is a fundamental way of dealing with such reactive intermediates when deriving the overall rate of a chemical reaction mechanism.

It is perhaps easiest to explain the PSSA by way of an example. Consider the unimolecular reaction $A \rightarrow B + C$, whose elementary steps consist of the activation of A by collision with a background molecule M (a reaction chaperone) to produce an energetic A molecule denoted by A^*, followed by decomposition of A^* to give $B + C$:

$$A + M \underset{1b}{\overset{1f}{\rightleftharpoons}} A^* + M$$

$$A^* \xrightarrow{2} B + C$$

Note that A^* may return to A by collision and transfer of its excess energy to an M. The rate equations for this mechanism are

$$\frac{d[A]}{dt} = -k_{1f}[A][M] + k_{1b}[A^*][M] \tag{3.11}$$

$$\frac{d[A^*]}{dt} = k_{1f}[A][M] - k_{1b}[A^*][M] - k_2[A^*] \tag{3.12}$$

The reactive intermediate in this mechanism is A^*. The PSSA states that the rate of generation of A^* is equal to its rate of disappearance; physically, what this means is that A^* is so reactive that, as soon as an A^* molecule is formed, it reacts by one of its two paths. Thus the PSSA gives

$$k_{1f}[A][M] - k_{1b}[A^*][M] - k_2[A^*] = 0 \tag{3.13}$$

---

[1]The word *pseudos* in Greek means "lie." As we will see shortly, the PSSA is a "lie" for a certain time period, and only after that period is it close to the truth.

From this we find the concentration of A^* in terms of the concentrations of the stable molecules A and M:

$$[A^*] = \frac{k_{1f}[A][M]}{k_{1b}[M] + k_2} \tag{3.14}$$

This expression can be used in (3.11) to give

$$\frac{d[A]}{dt} = -\frac{k_{1f}k_2[M][A]}{k_{1b}[M] + k_2} \tag{3.15}$$

We see that the single overall reaction $A \rightarrow B + C$ with a rate given by (3.15) depends on the concentration of M. If the background species M is in such excess that its concentration is effectively constant, the overall rate can be expressed as $d[A]/dt = -k[A]$, where $k = k_{1f}k_2[M]/(k_{1b}[M] + k_2)$ is a constant. If $k_{1b}[M] \gg k_2$, then $d[A]/dt = -k[A]$, with $k = k_{1f}k_2/k_{1b}$. On the other hand, if $k_{1b}[M] \ll k_2$, then $d[A]/dt = -k_{1f}[M][A]$, and the rate of the reaction depends on the concentration of M.

One comment is in order. The PSSA is based on the presumption that the rates of formation and disappearance of a reactive intermediate are equal. A consequence of this statement is that $d[A^*]/dt = 0$ from (3.12). This should not, however, be interpreted to mean that $[A^*]$ does not change with time. $[A^*]$ is at steady state with respect to $[A]$ and $[M]$. We can, in fact, compute $d[A^*]/dt$. It is

$$\frac{d[A^*]}{dt} = \frac{d}{dt}\frac{k_{1f}[A][M]}{k_{1b}[M] + k_2} \tag{3.16}$$

Which, if [M] is constant, is

$$\frac{d[A^*]}{dt} = -\frac{k_{1f}^2 k_2[M]^2[A]}{(k_{1b}[M] + k_2)^2} \tag{3.17}$$

To reconcile $d[A^*]/dt = 0$ from (3.12) with (3.17), we note that (3.14) is valid only after a short initial time interval needed for the rates of formation and disappearance of A^* to equilibrate so as to establish the steady state. After that time $[A^*]$ adjusts slowly on the timescale associated with changes in $[A]$ so as to maintain that balance. That slow adjustment is given by (3.17).

3.4 REACTIONS OF EXCITED SPECIES

Photolysis and chemical reaction can produce species that are vibrationally or electronically excited. These molecules process more energy than in the ground state. The most important exicted species in atmospheric chemistry is the first electronically excited state of the oxygen atom, $O(^1D)$. The major source of $O(^1D)$ below $\sim 40\,km$ altitude is the photolysis of O_3:

$$O_3 + h\nu \rightarrow O_2 + O(^1D)$$

Most of the $O(^1D)$ is quenched to ground-state atomic oxygen, $O(^3P)$, which we simply denote as O, by collision with an air molecule ($M = N_2$ or O_2),

$$O(^1D) + M \rightarrow O + M \qquad k(298\,K) \cong 3 \times 10^{-11}\,cm^3\,molecule^{-1}\,s^{-1}$$

The $O(^1D)$ atom is important in atmospheric chemistry because it reacts with the very unreactive species, H_2O and N_2O. The reaction with H_2O produces two hydroxyl radicals:

$$O(^1D) + H_2O \rightarrow OH + OH \qquad k(298\,K) = 2.2 \times 10^{-10}\,cm^3\,molecule^{-1}\,s^{-1}$$

This reaction is so fast that, although much of the $O(^1D)$ is just quenched by M, enough of the $O(^1D)$ reacts with H_2O to make this reaction the major source of OH in the atmosphere. We will return to this reaction again and again in Chapters 5 and 6.

Reaction of $O(^1D)$ with N_2O yields two molecules of nitric oxide (NO)

$$O(^1D) + N_2O \rightarrow NO + NO \qquad k(298\,K) = 6.7 \times 10^{-11}\,cm^3\,molecule^{-1}\,s^{-1}$$

and is the principal source of NO in the stratosphere.

Incidentally, the reaction of ground-state atomic oxygen and water vapor

$$O + H_2O \rightarrow OH + OH$$

is endothermic and quite slow, whereas the $O(^1D) + H_2O$ reaction is exothermic and very fast.

3.5 TERMOLECULAR REACTIONS

As noted in Section 3.1, the termolecular reaction $A + B + M \rightarrow AB + M$ does not take place as the result of the simultaneous collision of A, B, and M; rather, A and B react to form an energy-rich intermediate $AB^\dagger$ that subsequently collides with a third molecule M (the reaction chaperone), which removes the excess energy and allows formation of AB. (The dagger denotes vibrational excitation.)

The rate of formation of a product AB in the general system

$$A + B \underset{r}{\overset{a}{\rightleftharpoons}} AB^\dagger$$
$$AB^\dagger + M \overset{s}{\rightarrow} AB + M$$

is given here in (3.18), following the development in Section 3.3:

$$\frac{d[AB]}{dt} = \frac{k_a k_s [A][B][M]}{k_s[M] + k_r} \tag{3.18}$$

This result was first developed by Lindemann and Hinshelwood and often bears their names.

From (3.18), if $k_r \gg k_s[M]$, the reaction is third-order:

$$\frac{d[AB]}{dt} = \frac{k_a k_s}{k_r}[A][B][M] \qquad (3.19)$$

If $k_r \ll k_s[M]$, then the reaction is second-order:

$$\frac{d[AB]}{dt} = k_a[A][B] \qquad (3.20)$$

As the product molecule AB becomes more complex, the value of k_r decreases because the combination energy is distributed among more and more vibrational modes. The concentration of the third body, [M], is usually related directly to the pressure since in the atmosphere M is the sum of N_2 and O_2. The concentration of M at which the reaction rate behavior changes from third-order to second-order is lower the more complex the product molecule. Combination of two hydrogen atoms to form H_2 is third-order all the way up to 10^4 atm. On the other hand, addition of the OH radical to the alkene, 1-butene, C_4H_8, is second-order at all tropospheric pressures.

Consider the combination of two oxygen atoms to form O_2:

$$O + O + M \rightarrow O_2 + M$$

On collision, the newly formed O_2 molecule possesses the combination energy of $O + O$. Unless some energy is removed within the time of one vibrational period, the freshly formed O_2 will decompose back to $O + O$. The excess energy is removed by the third body, M. In this reaction the association complex $O_2^\dagger$ has only a single vibrational mode. Because all the excess energy is directed into that mode, the complex $O_2^\dagger$ rapidly falls back to $O + O$, and consequently the rate coefficient for the formation of O_2 is small. On the other hand, in the combination of two methyl radicals, $CH_3 + CH_3 + M \rightarrow C_2H_6 + M$, the activated complex $C_2H_6^\dagger$ has 18 vibrational modes over which the excess energy may be distributed. The lifetime of $C_2H_6^\dagger$ is relatively long since thousands of vibrations will occur before the excess energy can be concentrated into the particular bond that breaks, returning $C_2H_6^\dagger$ back to two methyl radicals. With a sufficiently large product molecule, the lifetime can be so long that collisional removal of the excess energy by a third molecule M will no longer be rate-determining.

Two reactions of significant atmospheric importance

$$OH + NO_2 + M \rightarrow HNO_3 + M$$

and

$$O + O_2 + M \rightarrow O_3 + M$$

lie just at the point where both second- and third-order kinetics are exhibited in the atmospheric pressure range.

The rate equation (3.18) can be written as pseudo-second-order

$$\frac{d[\text{AB}]}{dt} = k[\text{A}][\text{B}] \tag{3.21}$$

where k can be expressed in terms of the high- and low-pressure limiting values

$$k_\infty = k_a \tag{3.22}$$

$$k_0 = \frac{k_a k_s}{k_r} \tag{3.23}$$

as

$$k = \frac{k_0[\text{M}]k_\infty}{k_0[\text{M}] + k_\infty} \tag{3.24}$$

In between the low- and high-pressure limits is the "falloff region."

Actual experimental data on the pressure variation of the pseudo-second-order rate constant k exhibit a broader transition between the two limits than is predicted by (3.24). Lindemann–Hinshelwood theory makes the assumption that a single collision with a bath gas molecule M is sufficient to deactivate $\text{AB}^\dagger$ to AB. In reality, each collision removes only a fraction of the energy. To account for the fact that not all collisions are fully deactivating, Troe (1983) developed a modification to the Lindemann–Hinshelwood rate expression. In the Troe theory, the right-hand side of (3.24) is multiplied by a broadening factor that is itself a function of k_0/k_∞ (see Table B.2)

$$k(T) = \left\{ \frac{k_0(T)[\text{M}]}{1 + \dfrac{k_0(T)[\text{M}]}{k_\infty(T)}} \right\} F^{\left\{ \left(1 + \left[\log_{10}\left(\frac{k_0(T)[\text{M}]}{k_\infty(T)}\right)\right]^2\right)^{-1} \right\}} \tag{3.25}$$

$$k_0(T) = k_0^{300} \ (T/300)^{-n} \quad \text{cm}^6 \text{ molecule}^{-2} \text{ s}^{-1}$$

$$k_\infty(T) = k_\infty^{300}(T/300)^{-m} \quad \text{cm}^3 \text{ molecule}^{-1} \text{ s}^{-1}$$

and where F often has the value 0.6. Pressure-dependent rate coefficient data are generally fit well by this expression.

Third-order reactions often exhibit decreasing rate with increasing temperature. The higher the temperature, the greater the thermal kinetic energy possessed by the reactants A and B, and the larger the internal vibrational energy stored in the $\text{AB}^\dagger$ molecule. The larger this energy, the higher the chance of dissociation back to reactants and the larger the value of k_r. The rate constants k_a and k_s do not depend strongly on temperature, so in the low-pressure regime since k_r increases as T increases, the overall rate constant decreases. This temperature dependence of k_0 is frequently represented empirically by a factor T^n in the overall rate constant.

Second-Order versus Third-Order Kinetics (suggested by P. O. Wennberg)   The self-reaction of the hydroxyl radical has both a bimolecular channel and a termolecular channel:

$$OH + OH \xrightarrow{1} H_2O + O$$

$$OH + OH + M \xrightarrow{2} H_2O_2 + M$$

The rate coefficient for the bimolecular reaction 1 is $k_1 = 4.2 \times 10^{-12} \exp(-240/T) cm^3$ molecule^{-1}s^{-1}. For the termolecular reaction 2 (see Table B.2), we obtain

$$k_0^{300} = 6.2 \times 10^{-31} \quad cm^6 \, molecule^{-2} \, s^{-1}; \quad n = 1$$
$$k_\infty^{300} = 2.6 \times 10^{-11} \quad cm^3 \, molecule^{-1} \, s^{-1}; \quad m = 0$$

Calculate the rate coefficients of these reactions under two conditions:

Lower stratosphere	$p = 50$ mbar	$T = 220$ K
Surface	$p = 1013$ mbar	$T = 298$ K

For k_1

$$k_1(220 \, K) = 1.4 \times 10^{-12} \, cm^3 \, molecule^{-1} \, s^{-1}$$
$$k_1(298 \, K) = 1.88 \times 10^{-12} \, cm^3 \, molecule^{-1} \, s^{-1}$$

The termolecular rate coefficient is given by (3.25) with $F = 0.6, n = 1$, and $m = 0$. The atmospheric number concentrations are

Lower stratosphere	$[M] = 1.64 \times 10^{18}$ molecules cm^{-3}
Surface	$[M] = 2.46 \times 10^{19}$ molecules cm^{-3}

We obtain the following for the value of the pseudo-second order rate coefficient k_2:

Lower stratosphere	$k_2 = 1.09 \times 10^{-12} \, cm^3 \, molecule^{-1} \, s^{-1}$
Surface	$k_2 = 5.96 \times 10^{-12} \, cm^3 \, molecule^{-1} \, s^{-1}$

Under the lower stratospheric conditions, k_2 is close to its low-pressure limit, whereas at the surface it is in the midrange. The ratio of the rates of reactions 1 and 2 is $R_1/R_2 = k_1/k_2$:

Lower stratosphere	$R_1/R_2 = 1.28$
Surface	$R_1/R_2 = 0.31$

Near the surface, where [M] is at its maximum, the second reaction predominates because a reasonable proportion of the excited H_2O_2 can collisionally deexcite before it falls back to two OH radicals. In the lower stratosphere, [M] is almost an order of magnitude lower than at the surface, and a larger fraction of $H_2O_2^\dagger$ falls apart, so that the first reaction is favored somewhat over the second one.

3.6 CHEMICAL FAMILIES

Consider the following chemical system:

$$A \underset{2}{\overset{1}{\rightleftarrows}} B \overset{3}{\longrightarrow} C$$

Compounds A and B react reversibly, and B reacts irreversibly to C. Assume that the reversible reaction is rapid, compared with the timescale of conversion of B to C. Physically, this means that once a molecule of A reacts to form B, B is much more likely to react back to A than to go on to C; every once in a while, B does react to form C. In this case, A and B achieve an equilibrium on a short timescale, and this equilibrium slowly adjusts as some of the B reacts to form C. We recognize that the A–B equilibrium is just a pseudo-steady state.

Since A and B rapidly interchange, it is useful to view these two species as a *chemical family*. The overall process can then be depicted as shown below, where we enclose the family of A and B within the dashed box and denote the family as A_x, where $A_x = A + B$,

Thus, the A_x family has a sink that occurs at a rate $k_3 [B]$.

It turns out that the concept of chemical families is enormously useful in atmospheric chemistry. We will now analyze the properties of the general A,B,C system above; in subsequent chapters we will see how that analysis applies directly to several key atmospheric components.

Invoking the idea of a pseudo–steady state between A and B, because the rate of the reaction that returns B to A, $k_2 [B]$, far exceeds that converting B to C, $k_3 [B]$, at any instant the concentrations of A and B are related by

$$k_1[A] \cong k_2[B] \tag{3.26}$$

Slow changes in the sink B $\rightarrow$ C only serve to alter this instantaneous steady state between A and B.

What is often of interest with respect to a chemical family is the lifetime of the family. Let's write the full rate equations:

$$\frac{d[A]}{dt} = -k_1[A] + k_2[B] \tag{3.27}$$

$$\frac{d[B]}{dt} = k_1[A] - k_2[B] - k_3[B] \tag{3.28}$$

$$\frac{d([A] + [B])}{dt} = -k_3[B] \tag{3.29}$$

If τ_{A_x} is the lifetime of $A_x = A + B$, and τ_B is the lifetime of B, then, by definition

$$\frac{1}{\tau_{A_x}} = -\frac{1}{[A] + [B]}\frac{d}{dt}([A] + [B]) \tag{3.30}$$

Thus

$$\frac{1}{\tau_{A_x}} = -\frac{1}{[A]+[B]}\frac{d}{dt}([A]+[B]) = \frac{k_3[B]}{[A]+[B]} \tag{3.31}$$

and since

$$\frac{1}{\tau_B} = -\frac{1}{[B]}\frac{d[B]}{dt} = k_3 \tag{3.32}$$

we get

$$\frac{1}{\tau_{A_x}} = \frac{1}{\tau_B}\left(\frac{[B]}{[A]+[B]}\right) \tag{3.33}$$

or

$$\tau_{A_x} = \tau_B\left(1 + \frac{[A]}{[B]}\right) \tag{3.34}$$

where the [A]/[B] ratio is determined from (3.26) as $[A]/[B] = k_2/k_1$. Thus

$$\tau_{A_x} = \tau_B\left(1 + \frac{k_2}{k_1}\right) \tag{3.35}$$

Because of the pseudo–steady state that exists between A and B, the lifetime of the chemical family $A_x = A + B$ is actually longer than that of B alone, by the factor $(1 + k_2/k_1)$.

In an atmospheric mechanism there is a source of A or B or both that sustains the cycle. Let us call these S_A and S_B. In this case, the family can be depicted as follows:

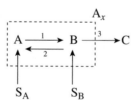

The existence of sources of A and/or B does not alter the instantaneous equilibrium between A and B or the lifetime of A_x, given above. The overall steady state balance on the A_x family is then

$$S_A + S_B = k_3[B]$$

As an example of a chemical family, consider photodissociation of ozone to produce O_2 and ground-state atomic oxygen:

$$O_3 + h\nu \rightarrow O_2 + O$$

The oxygen atoms just combine with O_2, most of the time, to reform ozone:

$$O + O_2 + M \rightarrow O_3 + M$$

In most regions of the atmosphere, the interconversion between O and O_3 is so rapid that the two species are often considered together as the chemical family, $O_x = O + O_3$, which is denoted "odd oxygen."

In summary, a "chemical family" refers to two compounds that interchange with each other sufficiently rapidly such that they tend to behave chemically as a group. As noted, we will encounter a number of such systems in the chapters to come.

3.7 GAS–SURFACE REACTIONS

A number of important chemical reactions in the atmosphere involve a gas molecule striking the surface of an airborne particle. For gas molecules A in three-dimensional random motion, the number of molecules of A striking a unit area per unit time is

$$Z = \frac{1}{4} n_A \bar{v}_A \tag{3.36}$$

where n_A is the gas-phase concentration of A and $\bar{v}_A$ is the mean speed of the A molecules:

$$\bar{v}_A = \left(\frac{8 k_B T}{\pi m_A} \right)^{1/2} \tag{3.37}$$

Then the number of collisions per second with a single spherical particle of radius R_p is $(\frac{1}{4} n_A \bar{v}_A)(4\pi R_p^2)$. Usually one is interested in reaction occurring with an ensemble of particles, all of different sizes. If the particle population has a total surface area per unit volume of air of $A_p (\text{cm}^2 \text{cm}^{-3})$, then the total number of collisions of A molecules with particles is $(\frac{1}{4} n_A \bar{v}_A) A_p$.

When a gas molecule strikes the surface of a particle, usually not every collision leads to reaction. One can define a *reaction efficiency* or *uptake coefficient* γ as the probability of reaction. γ is usually determined experimentally as the ratio of the number of collisions that result in reaction to the theoretical total number of collisions, and it depends in general on temperature and particle type. Therefore, the rate of the heterogeneous reaction is expressed as pseudo-first-order

$$R = \frac{1}{4} \gamma \bar{v}_A A_p \, n_A \tag{3.38}$$

where $\frac{1}{4} \gamma \bar{v}_A A_p$ is the first-order rate coefficient.

An important heterogeneous reaction in the atmosphere is that of gaseous N_2O_5 with water molecules on the surface of atmospheric particles:

$$N_2O_5 + H_2O(s) \rightarrow 2\,HNO_3$$

The ionic state of H_2O on the particle allows this reaction to proceed. For this reaction γ depends on the particle type. This reaction will play an important role in both the stratosphere and troposphere, and we defer a calculation of its rate until these chapters.

Sources of Kinetic Data for Atmospheric Chemistry   Two major sources of evaluated kinetic and photochemical data for atmospheric chemistry are

Sander et al. (2003)

NASA Panel for Data Evaluation

Publication from the Jet Propulsion Laboratory, Pasadena, CA

http://jpldataeval.jpl.nasa.gov/

Atkinson et al. (2004)

IUPAC Subcommittee on Gas Kinetic Data Evaluation for Atmospheric Chemistry

http://www.iupac-kinetic.ch.cam.ac.uk/summary/IUPACsumm_web_latest.pdf

Termolecular Reactions that Lead to Thermally Unstable Products   Several termolecular atmospheric reactions lead to products that can thermally decompose back to the reactants at atmospheric temperatures. Two important reactions of this type are formation of N_2O_5 and $CH_3C(O)O_2NO_2$:

$$NO_2 + NO_3 + M \rightleftarrows N_2O_5 + M$$
$$CH_3C(O)O_2 + NO_2 + M \rightleftarrows CH_3C(O)O_2NO_2 + M$$

N_2O_5 plays an important role in both stratospheric and tropospheric chemistry. The $CH_3C(O)O_2NO_2$ molecule, namely, peroxyacetyl nitrate and abbreviated PAN, is influential in tropospheric chemistry. We will study the chemistry of these two species in Chapters 5 and 6.

Because both forward and reverse reactions involve the participation of air molecules (M), the reaction rate coefficients for the forward and reverse reactions are represented by the general termolecular form (3.25). The reverse (decomposition) reactions in these types of reactions are typically highly temperature-dependent. Reaction rate coefficients for these two reactions are given in both the IUPAC (Atkinson et al. 2004) and JPL (Sander et al. 2003) kinetic data evaluations. In the IUPAC evaluation both forward and reverse rate coefficients are given. The JPL evaluation presents the forward rate coefficients and the equilibrium constants for the reactions. At equilibrium the rates of the forward (f) and reverse (r) reactions are equal, so the equilibrium constant, $K_{f,r}$, is directly related to the forward, k_f, and reverse, k_r, rate coefficients. For example, for N_2O_5 formation,

$$K_{f,r} = \frac{[N_2O_5]}{[NO_2][NO_3]} = \frac{k_f}{k_r} \quad (cm^3\,molecule^{-1})$$

where k_f has second-order units ($cm^3\,molecule^{-1}\,s^{-1}$) and k_r has first-order units (s^{-1}). Given k_f and $K_{f,r}$, one can therefore obtain k_r.

APPENDIX 3   FREE RADICALS

Free radicals are characterized by an odd number of electrons, an unpaired electron in the outer valence shell. These species are exceptionally reactive, as they are always seeking to pair off their lone election. Free radicals play a central role in atmospheric chemistry in both the stratosphere and the troposphere. Important radicals include, for example, OH and HO_2 (both stratosphere and troposphere) and Cl and ClO (stratosphere).

One can represent the bonding in molecules using the *Lewis dot structure*, in which lines represent a pair of bonded electrons and dots represent other electrons. The hydrogen and methane molecules are represented by

$$
\text{H--H} \qquad \overset{\displaystyle \text{H}}{\underset{\displaystyle \text{H}}{\text{H--C--H}}}
$$

Oxygen and nitrogen molecules are

$$
:\ddot{\text{O}}=\ddot{\text{O}}: \qquad :\text{N}\equiv\text{N}:
$$

The triple bond between the two nitrogen atoms makes N_2 an exceptionally stable molecule. The hydroxyl (OH) radical has the structure

$$
\text{H--O·}
$$

NO and NO_2, important throughout the atmosphere, are actually radicals. Their electronic structures are

$$
:\dot{\text{N}}=\ddot{\text{O}}:
$$
$$
:\ddot{\text{O}}-\dot{\text{N}}=\ddot{\text{O}}:
$$

PROBLEMS

3.1$_A$ What are the lifetimes of CHF_2Cl (HCFC-22) and CH_2ClCF_3 (HCFC-133a) by reaction with OH in the troposphere? Assume an average OH concentration of $[OH] = 10^6$ molecules cm^{-3} and an average tropospheric temperature of $T = 250\,\text{K}$. Reaction rate constants are (Sander et al. 2003):

$$
k_{OH+CHF_2Cl} = 1.05 \times 10^{-12} \exp(-1600/T) \qquad \text{cm}^3 \text{ molecule}^{-1}\,\text{s}^{-1}
$$
$$
k_{OH+CH_2ClCF_3} = 5.6 \times 10^{-13} \exp(-1100/T) \qquad \text{cm}^3 \text{ molecule}^{-1}\,\text{s}^{-1}
$$

3.2$_A$ The termolecular reaction

$$OH + NO_2 + M \rightarrow HNO_3 + M$$

is quite important in atmospheric chemistry. Plot the reaction rate coefficient of this reaction as a function of pressure, specifically, [M], at 300 K. Consider the pressure range from 0.1 to 10 atm. At 1 atm, where does the reaction rate constant lie in the transition between third- and second-order kinetics? The expression for the reaction rate coefficient is given in Table B.2.

3.3$_A$ Calculate the lifetime of O atoms against the reaction

$$O + O_2 + M \rightarrow O_3 + M$$

at the Earth's surface at 298 K. The third-order rate constant for this reaction is 4.8×10^{-33} cm^6 molecule^{-2}s^{-1}. How does this lifetime change at 25 km altitude?

3.4$_B$ The reaction rate coefficient at 298 K of the reaction

$$OH + CHF_3 \rightarrow CF_3 + H_2O$$

is 2.8×10^{-16} cm^3 molecule^{-1}s^{-1}. Estimate the rate coefficient with reference to collision theory and on this basis, determine the fraction of collisions that lead to reaction. Assume molecular radii for CHF$_3$ and OH of 2.1×10^{-8} cm and 2.0×10^{-8} cm, respectively.

3.5$_B$ The reaction of OH and ClO has two channels:

$$OH + ClO \rightarrow HO_2 + Cl \qquad k = 7.4 \times 10^{-12} \exp{(270/T)}$$
$$\rightarrow HCl + O_2 \qquad k = 3.2 \times 10^{-13} \exp{(320/T)}$$

Why is the A factor for the second channel so much lower than the first? Note that for both radicals OH and ClO, the unpaired election that reacts is on the O atom.

3.6$_A$ What is the first-order rate coefficient of N$_2$O$_5$ reacting heterogeneously by

$$N_2O_5 + H_2O(s) \rightarrow 2\,HNO_3$$

at 298 K in a population of particles, all of which have 0.2 μm diameter, of overall number concentration 1000 cm^{-3}? The reaction efficiency γ can be taken as 0.1.

3.7$_B$ Alkylperoxynitrates, RO_2NO_2, can be presumed to decompose according to the following mechanism:

$$RO_2NO_2 \underset{2}{\overset{1}{\rightleftharpoons}} RO_2 \cdot + NO_2$$

$$RO_2 \cdot + RO_2 \cdot \overset{3}{\rightarrow} 2\,RO \cdot + O_2$$

$$RO \cdot + NO_2 \overset{4}{\rightarrow} RONO_2$$

Let us assume that a sample of RO_2NO_2 decomposes in a reactor and its decay is observed. We desire to estimate k_1 from that rate of disappearance. To analyze the system we assume that both RO_2 and NO_2 are in pseudo–steady state and that $[RO_2] \simeq [NO_2]$. Show that the observed first-order rate constant for RO_2NO_2 decay is related to the fundamental rate constants of the system by

$$k_{obs} = k_1 \left\{ 1 - \frac{k_2}{k_2 + 2\,k_3} \right\}$$

Thus, given k_{obs} and values for k_2 and k_3, k_1 can be determined.

3.8$_B$ Consider the following reaction system:

$$A \overset{1}{\longrightarrow} B \overset{2}{\longrightarrow} C$$

The concentrations of B and C are zero at $t = 0$.

a. Derive analytical expressions for the exact dynamic behavior of this system over time. Show mathematically under what conditions the pseudo-steady-state approximation (PSSA) can be made for [B].

b. Use the PSSA to derive a simpler set of equations for the concentrations of A, B, and C.

3.9$_B$ The most important oxidizing species for tropospheric compounds is the hydroxyl (OH) radical. A standard way of determining the OH rate constant of a compound is to measure its decay in a reactor in the presence of OH relative to the decay of a second compound, the OH rate constant of which is known. Consider two compounds A and B, A being the one for which the OH rate constant is to be determined and B the reference compound for which its OH rate constant is known. Show that the concentrations of A and B in such a reactor obey the following relation:

$$\ln \frac{[A]_0}{[A]_t} = \frac{k_A}{k_B} \ln \frac{[B]_0}{[B]_t}$$

where $[A]_0$ and $[B]_0$ are the initial concentrations, $[A]_t$ and $[B]_t$ are the concentrations at time t, and k_A and k_B are the OH rate constants. Thus, plotting

$$\ln \frac{[A]_0}{[A]_t} \quad \text{versus} \quad \ln \frac{[B]_0}{[B]_t}$$

yields a straight line with slope k_A/k_B. Knowing k_B allows one to calculate k_A from the slope.

3.10$_C$ Once released at the Earth's surface, a molecule diffuses upward through the troposphere and at any time may be removed by chemical reaction with other species, by absorption into particles and droplets, or by photodissociation. If the removal processes are rapid relative to the rate of diffusion, the species will not get mixed uniformly in the troposphere before it is removed. If, on the other hand, removal is slow relative to the rate of diffusion, the species may have a uniform tropospheric concentration.

Consider a species A whose removal from the atmosphere can be expressed as a first-order reaction, that is, $R_A = -k_A c_A$. If the removal of A is the result of reaction with background species B, then k_A can be a pseudo-first-order rate constant that includes the concentration of B in it. The intrinsic rate constant is given by the Arrhenius expression $k_A = A_0 \exp(-E_a/RT)$.

Let us assume that the vertical concentration distribution of A can be represented generally as $c_A = c_{A_0} \exp(-H_A z)$ by analogy to the exponential decrease of pressure with altitude, $p = p_0 \exp(-Hz)$. Show that the tropospheric lifetime of A over the tropospheric height H_T is given by

$$\tau_T = \frac{1 - e^{-H_A H_T}}{H_A \int_0^{H_T} k_A(T, z) \exp(-H_A z) dz}$$

The tropospheric temperature profile can be approximated by $T(z) = T_0 - \alpha z$, where $T_0 = 293\,\text{K}$ and $\alpha = 5.5\,\text{K km}^{-1}$. Show that the ratio of the lifetime of species A at altitude z to that at the Earth's surface is

$$\frac{\tau_T}{\tau_0} = \frac{[1 - \exp(-H_A H_T)]\exp(-E_a/RT_0)}{H_A \int_0^{H_T} \exp[-E_a/R(T_0 - \alpha z)]\exp(-H_A z)dz}$$

Let us apply the foregoing theory to some trace atmospheric constituents whose principal removal reactions are with the OH radical. Consider CH_3Cl, CHF_2Cl, CH_3SCH_3, and H_2S. For the purpose of the calculation assume that the OH radical concentration is 10^6 molecules cm^{-3}, independent of height. Place the computed values of τ_T/τ_0 for these species on a plot of τ_T/τ_0 versus k_A at surface conditions. Discuss.

REFERENCES

Atkinson, R. et al. (2004) IUPAC Subcommittee on Gas Kinetic Data Evaluation for Atmospheric Chemistry, http://www.iupac-kinetic.ch.cam.ac.uk/summary/IUPACsumm_web_latest.pdf.

Benson, S. W. (1976) *Thermochemical Kinetics*, 2nd ed., Wiley, New York.

Laidler, K. J. (1987) *Chemical Kinetics*, 3rd ed., Harper & Row, London.

Pilling, M. J., and Seakins, P. W. (1995) *Reaction Kinetics*, Oxford Univ. Press, Oxford, UK.

Sander, S. P., Friedl, R. R., Golden, D. M., Kurylo, M. J., Huie, R. E., Orkin, V. L., Moortgat, G. K., Ravishankara, A. R., Kolb, C. E., Molina, M. J., and Finlayson-Pitts, B. J. (2003) *Chemical Kinetics and Photochemical Data for Use in Atmospheric Studies*, Evaluation no. 14, Jet Propulsion Laboratory Publication 02-25 (available at http://jpldataeval.jpl.nasa.gov/download.html).

Troe, J. (1983) Specific rate constants $k(E, J)$ for unimolecular bond fissions, *J. Chem. Phys.* **79**, 6017–6029.

4 Atmospheric Radiation and Photochemistry

4.1 RADIATION

Basically all the energy that reaches the Earth comes from the Sun. The absorption and loss of radiant energy by the Earth and the atmosphere are almost totally responsible for the Earth's weather on both global and local scales. The average temperature on the Earth remains fairly constant, indicating that the Earth and the atmosphere on the whole lose as much energy by reradiation back into space as is received by radiation from the Sun. The accounting for the incoming and outgoing radiant energy constitutes the Earth's energy balance. The atmosphere, although it may appear to be transparent to radiation, plays a very important role in the energy balance of the Earth. In fact, the atmosphere controls the amount of solar radiation that actually reaches the surface of the Earth and, at the same time, controls the amount of outgoing terrestrial radiation that escapes into space.

Radiant energy, arranged in order of its wavelengths λ, is called the *spectrum* of radiation. The electromagnetic spectrum is shown in Figure 4.1. The Sun radiates over the entire electromagnetic spectrum, although, as we will see, most of the energy is concentrated near the visible portion of the spectrum, the narrow band of wavelengths from 400 to 700 nm (0.4–0.7 μm). Our interest will be confined to the so-called optical region, which extends over the near ultraviolet, the visible, and the near infrared, the wavelength range from 200 nm to 100 μm. This range covers most of the solar radiation and that emitted by the Earth's surface and atmosphere. Three interrelated measures are used to specify the location in the electromagnetic spectrum, the wavelength λ, the frequency ν, and the wavenumber $\tilde{\nu} = \lambda^{-1}$. Frequency ν and wavelength λ are related by $\nu = c/\lambda$, where c is the speed of light. In the ultraviolet and visible portion of the spectrum it is common to characterize radiation by its wavelength, expressed either in nanometers (nm) or micrometers (μm). In the infrared part of the spectrum, the wavenumber (cm^{-1}) is frequently used.

Radiation is emitted from matter when an electron drops to a lower level of energy. The difference in energy between the initial and final level, $\Delta\varepsilon$, is related to the frequency of the emitted radiation by

$$\Delta\varepsilon = h\nu = \frac{hc}{\lambda} \tag{4.1}$$

Atmospheric Chemistry and Physics: From Air Pollution to Climate Change, Second Edition, by John H. Seinfeld and Spyros N. Pandis.   Copyright © 2006 John Wiley & Sons, Inc.

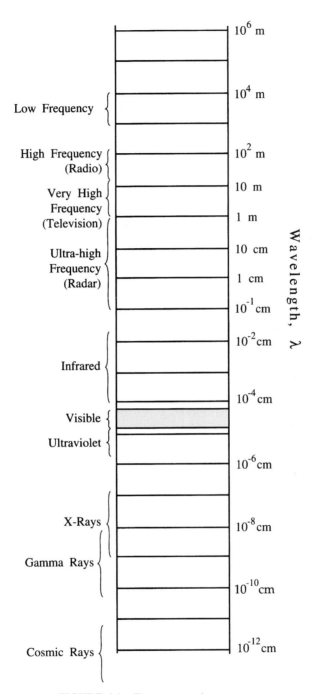

FIGURE 4.1 Electromagnetic spectrum.

where Planck's constant, $h = 6.626 \times 10^{-34}$ J s, and the speed of light in vacuum, $c = 2.9979 \times 10^8$ m s^{-1} (see Table A.6). When the energy difference $\Delta\varepsilon$ is large, the frequency of the excited photon is high (very small wavelength) and the radiation is in the X-ray or gamma-ray region. Equation (4.1) also applies to the absorption of a photon of energy by a molecule. Thus a molecule can absorb radiant energy only if the wavelength of the radiation corresponds to the difference between two of its energy levels. Since the spacing between energy levels is, in general, different for molecules of different composition and shape, the absorption of radiant energy by molecules of differing structure occurs in different regions of the electromagnetic spectrum.

The amount of energy radiated from a body depends largely on the temperature of the body. It has been demonstrated experimentally that at a given temperature there is a maximum amount of radiant energy that can be emitted per unit time per unit area of a body. This maximum amount of radiation for a certain temperature is called the *blackbody radiation*. A body that radiates, for every wavelength, the maximum possible intensity of radiation at a certain temperature is called a *blackbody*. This maximum is identical for every blackbody regardless of its constituency. Thus the intensity of radiation emitted by a blackbody is a function only of the wavelength, absolute temperature, and surface area. The term "blackbody" has no reference to the color of the body. A blackbody can also be characterized by the property that all radiant energy reaching its surface is absorbed.

4.1.1   Solar and Terrestrial Radiation

The Sun is a gaseous sphere of radius about 6.96×10^5 km and of mass about 1.99×10^{30} kg. It is made up of approximately three parts hydrogen and one part helium. In the core of the Sun energy is produced by nuclear reactions (fusion of four H atoms into one He atom, with a small loss of mass). It is believed that energy is transferred to the outer layers mainly by electromagnetic radiation. The outer 500 km of the Sun, called the *photosphere*, emits most of the radiation received on the Earth. Radiation emitted by the photosphere closely approximates that of a blackbody at about 6000 K. The energy spectrum of the Sun as compared with that of a blackbody at 5777 K as received at the top of the Earth's atmosphere is shown in Figure 4.2. The maximum intensity of incident radiation occurs in the visible spectrum at about 500 nm (0.5 µm). In contrast, Figure 4.3 shows the emission of radiant energy from a blackbody at 300 K, approximating the Earth. The peak in radiation intensity occurs at about 10 µm in the invisible infrared.

The monochromatic emissive power of a blackbody $F_B(\lambda)(\text{W m}^{-2}\,\text{m}^{-1})$ is related to temperature and wavelength by

$$F_B(\lambda) = \frac{2\pi c^2 h \lambda^{-5}}{e^{ch/k\lambda T} - 1} \tag{4.2}$$

where k is the Boltzmann constant (Table A.6). (In Chapter 3 we used k_B to represent the Boltzmann constant to avoid confusion with the reaction rate coefficient.)

As can be seen from Figures 4.2 and 4.3, the higher the temperature, the greater is the emissive power (at all wavelengths). We also see that, as temperature increases, the maximum value of $F_B(\lambda)$ moves to shorter wavelengths. The wavelength at which the maximum amount of radiation is emitted by a blackbody is found by differentiating (4.2) with respect to λ, setting the result equal to zero, and solving for λ. The result is approximately $hc/5\,kT$ and with λ expressed in nm and T in kelvin units is

$$\lambda_{\max} = \frac{2.897 \times 10^6}{T} \tag{4.3}$$

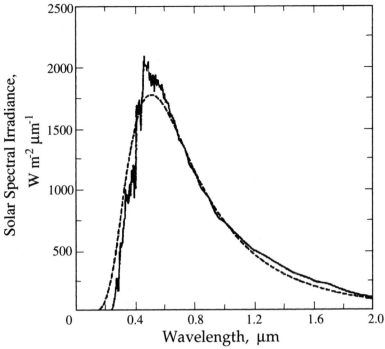

FIGURE 4.2 Solar spectral irradiance $(W\,m^{-2}\,\mu m^{-1})$ at the top of the Earth's atmosphere compared to that of a blackbody at 5777 K (dashed line) (Iqbal 1983). There is a reduction in total intensity of solar radiation from the Sun's surface to the top of the Earth's atmosphere, given by the ratio of the solar constant, $1370\,W\,m^{-2}$, to the integrated intensity of the Sun [see (4.4)]. That ratio is about $1/47,000$. (Reprinted by permission of Academic Press.)

Thus hot bodies not only radiate more energy than cold ones, they do so at shorter wavelengths. The wavelengths for the maxima of solar and terrestrial radiation are 480 nm and $\sim 10,000$ nm, respectively. The Sun, with an effective surface temperature of ~ 6000 K, radiates about 2×10^5 more energy per square meter than the Earth does at 300 K.

If (4.2) is integrated over all wavelengths, the total emissive power F_B $(W\,m^{-2})$ of a blackbody is found to be

$$F_B = \int_0^\infty F_B(\lambda)d\lambda = \sigma T^4 \tag{4.4}$$

where $\sigma = 5.671 \times 10^{-8}\,W\,m^{-2}\,K^{-4}$, the Stefan–Boltzmann constant.

4.1.2 Energy Balance for Earth and Atmosphere

The Earth's climate is controlled by the amount of solar radiation intercepted by the planet and the fraction of that energy that is absorbed. The flux density of solar energy, integrated

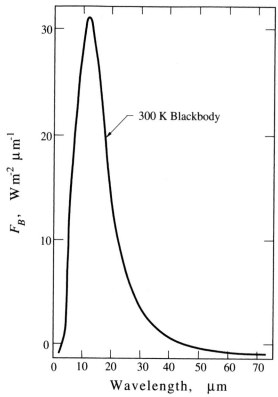

FIGURE 4.3   Spectral irradiance ($W\,m^{-2}\,\mu m^{-1}$) of a blackbody at 300 K.

over all wavelengths, on a surface oriented perpendicular to the solar beam at the Earth's orbit is about $1370\,W\,m^{-2}$. This is called the *solar constant*.[1] Let the solar constant be denoted by $S_0 = 1370\,W\,m^{-2}$. The cross-sectional area of the Earth that intercepts the solar beam is πR^2, where R is the Earth's radius. The surface area of the Earth that receives the radiation is $4\pi R^2$. Thus the fraction of the solar constant received per unit area of the Earth is $(\pi R^2/4\pi R^2) = \frac{1}{4}$ of the solar constant, about $342\,W\,m^{-2}$. Of this incoming solar radiation, a fraction is reflected back to space; that fraction, which we can denote by R_p, is the global mean planetary reflectance or *albedo*. R_p is about 0.3 (Ramanathan 1987; Ramanathan et al. 1989).[2] Contributing to R_p are clouds, scattering by air molecules,

---

[1]Since the late 1970s, regular satellite measurements of the solar constant have been performed (Mecherikunnel et al. 1988). Maximum differences in the value of S_0 among the instruments is about $2\,W\,m^{-2}$, corresponding to a little more than 0.1% of the value of S_0. Over the period 1980–1986 the so-called SMM/ACRIM instrument measured an average value of S_0 of about $1386\,W\,m^{-2}$, whereas that on NIMBUS-7 reported an average S_0 of about $1370\,W\,m^{-2}$.

[2]The average value of the albedo, the incoming radiation that is reflected or scattered back to space without absorption, is usually taken to be somewhere in the range of 30–34%. It is important to note that the albedo varies considerably, depending on the surface of the Earth. For example, in the polar regions, which are covered by ice and snow, the reflectivity of the surface is very high. On the other hand, in the equatorial regions, which are covered largely with oceans, the reflectivity is low, and most of the incoming energy is absorbed by the surface.

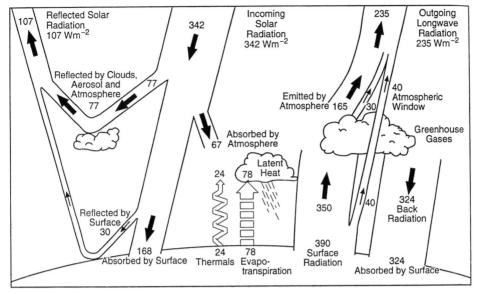

FIGURE 4.4 The Earth's annual and global mean energy balance (Kiehl and Trenberth 1997). Of $342\,\text{W}\,\text{m}^{-2}$ incoming solar radiation, $168\,\text{W}\,\text{m}^{-2}$ is absorbed by the surface. That energy is returned to the atmosphere as sensible heat, latent heat via water vapor, and thermal infrared radiation. Most of this radiation is absorbed by the atmosphere, which, in turn, emits radiation both up and down. (Reprinted by permission of the American Meteorological Society.)

scattering by atmospheric aerosol particles, and reflection from the surface itself, the surface albedo (the surface albedo is denoted as R_s). The fraction $1 - R_p$ represents that fraction of solar shortwave radiation that is absorbed by the Earth–atmosphere system. For $R_p = 0.3$, this corresponds to about $235\,\text{W}\,\text{m}^{-2}$. This amount is matched, on an annual and global average basis, by the longwave infrared radiation emitted from the Earth–atmosphere system to space (Figure 4.4). The infrared radiative flux emitted at the surface of the Earth, about $390\,\text{W}\,\text{m}^{-2}$, substantially exceeds the outgoing infrared flux of $235\,\text{W}\,\text{m}^{-2}$ at the top of the atmosphere. Clouds, water vapor, and the greenhouse gases (GHGs) both absorb and emit infrared radiation. Since these atmospheric constituents are at temperatures lower than that at the Earth's surface, they emit infrared radiation at a lower intensity than if they were at the temperature of the Earth's surface and therefore are net absorbers of energy.

The equilibrium temperature of the Earth can be estimated by a simple model that equates incoming and outgoing energy (Figure 4.5). Incoming solar energy at the surface of the Earth is

$$F_S = \frac{S_0}{4}(1 - R_p) \tag{4.5}$$

For an average blackbody temperature of the Earth–atmosphere system T_e defined on the basis of (4.4), the longwave emitted flux averaged over the globe is

$$F_L = \sigma T_e^4 \tag{4.6}$$

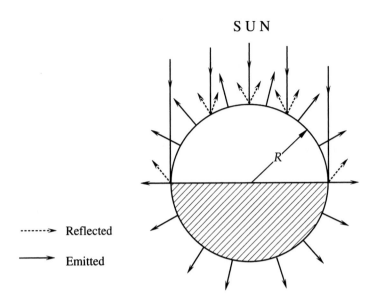

FIGURE 4.5 Global radiative equilibrium.

Equating F_S and F_L yields the following expression for T_e:

$$T_e = \left(\frac{(1 - R_p)S_0}{4\sigma} \right)^{1/4} , \tag{4.7}$$

For $R_p = 0.30$, this equation gives $T_e \simeq 255$ K. If the Earth were totally devoid of clouds, then the global albedo would be about $R_p = 0.15$. With this value of R_p, the equilibrium temperature $T_e = 268$ K. This simple equation predicts that T_e varies about 0.5 K for a $10\,\mathrm{W\,m^{-2}}$ (0.7%) variation in the solar constant, or for a reflectance variation around $R_p = 0.3$ of $\Delta R_p = 0.005$.

In summary, the net outgoing radiative flux of about $235\,\mathrm{W\,m^{-2}}$ corresponds to a blackbody temperature of 255 K ($-18°$C). The surface emission of $390\,\mathrm{W\,m^{-2}}$ corresponds to a blackbody temperature of 288 K. Therefore, the surface temperature is about $33°$C warmer than it would be if the atmosphere were completely transparent (and the planetary albedo were still 0.3).

The net radiative energy input, $F_{\text{net}} = F_S - F_L$, is zero at equilibrium. If a perturbation occurs then the change in net energy input is related to the changes in both solar and longwave components by

$$\Delta F_{\text{net}} = \Delta F_S - \Delta F_L \tag{4.8}$$

To reestablish equilibrium, a temperature change ΔT_e results, which can be related to ΔF_{net} by a parameter λ_0

$$\Delta T_e = \lambda_0 \Delta F_{\text{net}} \tag{4.9}$$

where λ_0, having units $K\,(W\,m^{-2})^{-1}$, is the *climate sensitivity factor*. If we neglect any feedbacks in the climate system, λ_0 can be estimated as $(\partial F_e/\partial T_e)^{-1}$

$$\lambda_0 = \frac{1}{4\sigma T_e^3} = \frac{T_e}{4\,F_L} \tag{4.10}$$

and $\lambda_0 \simeq 0.3\,K\,(W\,m^{-2})^{-1}$. A doubling of the CO_2 abundance from the preindustrial (pre–Industrial Revolution) level is estimated to produce $\Delta F_L = 4.6\,W\,m^{-2}$. With $\lambda_0 = 0.3\,K\,(W\,m^{-2})^{-1}$, this would lead to an increase in the global mean temperature of $\Delta T_e = 1.4\,K$. This temperature increase is less than what climate models predict because of feedbacks that act to enhance warming. As noted earlier, such feedbacks include, for example, the fact that a warmer atmosphere contains more water vapor, and hence an enhanced infrared absorption.

Global climate change is induced by a forcing that disturbs the equilibrium and leads to a nonzero average downward net flux at the top of the atmosphere (TOA):

$$-F_{\text{net}} = \frac{S_0}{4}(1 - R_p) - F_L \tag{4.11}$$

It is customary to write (4.11) in terms of downward flux, $-F_{\text{net}}$, at the TOA; an increase of $-F_{\text{net}}$ corresponds to heating of the planet. Primary forcing can occur as a result of changes of S_0, R_p, or F_L. Changes in incoming solar radiation have resulted from changes in the Earth's orbit and from variations in the Sun's output of energy. Changes in the planetary albedo R_p can result from changes in surface reflectance from human activity (agriculture, deforestation), from changes in the aerosol content of the atmosphere from both natural (volcanoes) and anthropogenic (industrial emissions, biomass burning) causes, and to a lesser extent from changes in levels of gases that absorb solar wavelengths (e.g., ozone). Changes in the emitted longwave flux F_L result primarily from changes (increases) of absorbing gases in the atmosphere and to a lesser extent from changes in aerosols.

We return to this in Chapter 23.

4.1.3 Solar Variability

The amount of solar radiation reaching Earth and Earth's changing orientation to the Sun have been the major causes for climatic change throughout its history. If the Sun's radiation intensity declined by 5–10% and there were no other compensating factors, ice would engulf the planet in less than a century. Although no theory exists to predict future changes in solar output, the effect of changes in Earth's orbit as it travels around the Sun is beginning to be understood. During the past million years, Earth has experienced 10 major and 40 minor episodes of glaciation. All appear to have been controlled by three so-called orbital elements that vary cyclically over time:

1. Earth's tilt changes from $22°$ to $24.5°$ and back again every 41,000 years.

2. The month when Earth is closest to the Sun also varies over cycles of 19,000 and 24,000 years. Currently, Earth is closest to the Sun in January. This month-of-closest-approach factor can make a difference of 10% in the amount of solar radiation reaching a particular location in a given season.

3. The shape of Earth's orbit varies from being nearly circular to being more elliptical with a period of 100,000 years.

The climatic cycles caused by these orbital factors are called *Milankovitch cycles* after the Serbian mathematician Milutin Milankovitch, who first described them in 1920. Superimposed on the Milankovitch cycles are changes in the Sun that occur over days or months or a few years.

Even though studies of ocean cores have shown that these orbital changes are the principal determinant of the times of glaciation, the exact mechanisms by which Earth responds to the orbital changes have not been established. Orbital changes alone appear not to have caused the vast climate shifts associated with glaciation and deglaciation. Feedbacks, such as changes in Earth's reflectivity, amount of particles in the atmosphere, and the carbon dioxide and methane content of the atmosphere, act together with orbital changes to enhance global warming and cooling. The levels of carbon dioxide and methane, as shown in ice core measurements, decrease during times of glaciation and increase during warming periods, although it is not known exactly how or why their concentrations rise and fall. (See Chapter 23.)

The radiative forcing resulting from changing solar output can be obtained by multiplying the change in total solar irradiance by $(1 - R_p)/4$, where R_p is the Earth's albedo. For $R_p = 0.3$, $(1 - R_p)/4 = 0.175$. A 0.1% change in total solar irradiance $(1.4 \, \text{W m}^{-2})$ would be equivalent to a radiative forcing of about $0.2 \, \text{W m}^{-2}$.

4.2   RADIATIVE FLUX IN THE ATMOSPHERE

The essential energy flux in atmospheric chemistry is the flux of solar radiation. The *radiant flux density F* is the radiant energy flux across any surface element, without consideration of the direction; F is measured in watts per square meter (W m^{-2}). The radiant flux density is called the *irradiance E* when the radiation is received on a surface. Thus F and E are often used interchangeably. We will use F in general and E when we are referring specifically to the radiant flux density on a surface. The *radiance L* is the radiant flux as a function of the solid angle $d\omega$ crossing a surface perpendicular to the axis of the radiation beam; L is measured in watts per square meter per steradian $(\text{W m}^{-2} \, \text{sr}^{-1})$. The radiance as a function of direction gives a complete description of the radiative field.

Consider a beam of radiation of radiance L crossing a surface dS with the beam axis making an angle θ to the normal to dS (Figure 4.6). dS projects as $dS \cos \theta$ perpendicular to the beam axis of the radiation, and the radiant flux density dF on dS is

$$dF = L \cos \theta \, d\omega \qquad (4.12)$$

The radiant flux density, or irradiance, on the surface dS is obtained by integrating the radiance over all angles:

$$E = \int L \cos \theta \, d\omega \qquad (4.13)$$

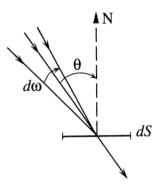

FIGURE 4.6   Relation between radiance and radiant flux density.

When the radiance L is independent of direction, the radiative field is called *isotropic*. In this case, (4.13) can be integrated over the half-space, $\Omega = 2\pi$, and the relation between the irradiance and the radiance is $E = \pi L$.

The irradiance on a horizontal surface is obtained from the incoming radiance by integrating the radiance over the spherical coordinates θ and ϕ

$$E = \int_0^{2\pi} \int_0^{\pi} L(\theta, \phi) \cos \theta \sin \theta \, d\theta \, d\phi \tag{4.14}$$

where the direction of the incoming beam is characterized by the angle θ (see Figure 4.7).

The *spectral radiant flux density, $F(\lambda)$,* is the radiant flux density per unit wavelength interval, expressed in watts per square meter per nanometer ($\text{W m}^{-2} \text{ nm}^{-1}$). Equivalently, when considering radiation incident upon a surface, the *spectral irradiance, $E(\lambda)$,* is expressed as $\text{W m}^{-2} \text{ nm}^{-1}$.

Sometimes the spectral radiant flux density is expressed as a function of frequency ν, that is, $F(\nu)$. Because the frequency ν of radiation is related to its wavelength by (4.1), $\nu = c/\lambda$, $F(\lambda)$ and $F(\nu)$ can be interrelated. Since the flux of energy in a small interval of

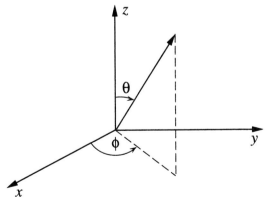

FIGURE 4.7   Coordinates for radiative calculations.

wavelength $d\lambda$ must be equal to that in a small interval of corresponding frequency $d\nu$, then

$$F(\lambda)\, d\lambda = F(\nu)\, d\nu \tag{4.15}$$

Since $d\nu = (c/\lambda^2)|d\lambda|$, we have

$$
\begin{aligned}
F(\lambda) &= \left(\frac{c}{\lambda^2}\right) F(\nu) \\
&= \left(\frac{\nu^2}{c}\right) F(\nu)
\end{aligned}
\tag{4.16}
$$

Generally we will deal with wavelength as the variable rather than frequency, although they can easily be interrelated as indicated in (4.16).

4.3   BEER–LAMBERT LAW AND OPTICAL DEPTH

Consider the propagation of radiation of wavelength λ through a layer of thickness dx perpendicular to a beam of intensity $F(\lambda)$. The extinction of radiation on traversing an infinitesimal pathlength dx is linearly proportional to the amount of matter along the path

$$F(x + dx,\lambda) - F(x,\lambda) = -b(x,\lambda)\, F(x,\lambda)\, dx \tag{4.17}$$

where $b(x,\lambda)$ (units of inverse length) is called the *extinction coefficient* and is proportional to the density of material in the medium. Extinction includes both absorption and scattering, each of which removes photons from the beam. In contrast to absorption, the radiant energy scattered remains in the form of radiation, but its direction is altered from that of the incident radiation. Dividing (4.17) by dx and letting $dx \to 0$ gives

$$\frac{dF(x,\lambda)}{dx} = -b(x,\lambda)\, F(x,\lambda) \tag{4.18}$$

which is known as the *Beer–Lambert law*.

Absorption and scattering occur simultaneously because all molecules (and particles) both absorb and scatter. The extinction coefficient $b(x,\lambda)$ is the sum of absorption and scattering:

$$b = b_a + b_s \tag{4.19}$$

The *optical depth* (dimensionless) at wavelength λ between points x_1 and x_2 is defined as

$$\tau(x_1, x_2; \lambda) = \int_{x_1}^{x_2} b(x, \lambda) dx \tag{4.20}$$

From (4.19), the total optical depth is a sum of the optical depth due to absorption and that due to scattering:

$$\tau = \tau_a + \tau_s \tag{4.21}$$

The most frequently used form of optical depth is that in which the two points are at different altitudes because one is often interested in how solar radiation is attenuated as it traverses the atmosphere. By definition, $\tau = 0$ at the top of the atmosphere (TOA), increasing, like pressure, monotonically from zero at TOA to its value at any altitude z:

$$\tau(z, \lambda) = \int_{z}^{z_{TOA}} b(z', \lambda) dz' \tag{4.22}$$

Let us integrate (4.18) from the TOA to an altitude z. Since z is decreasing in the direction of propagation of radiation, we need to add a minus sign on the left-hand side (LHS) of (4.18):

$$-\frac{dF(z, \lambda)}{dz} = -b(z, \lambda) F(z, \lambda) \tag{4.23}$$

Integrating from z_{TOA} to z and $F_{TOA}(\lambda)$ to $F(z, \lambda)$, we obtain

$$F(z, \lambda) = F_{TOA}(\lambda) \exp(-\tau(z, \lambda)) \tag{4.24}$$

At the altitude at which $\tau(z, \lambda) = 1$, the radiative flux is reduced to $1/e$ of its value at the top of the atmosphere. The *transmittance* of the atmosphere at wavelength λ at height z is defined as the ratio $F(z, \lambda)/F_{TOA}(\lambda)$:

$$\frac{F(z, \lambda)}{F_{TOA}(\lambda)} = \exp(-\tau(z, \lambda)) \tag{4.25}$$

Although, as noted, molecules scatter as well as absorb radiation, our concern at the moment is with molecular absorption. The absorption coefficient for a gas A (b_a^A) is proportional to the number of molecules of A per unit volume, n_A (molecules cm^{-3}). The absorption coefficient divided by the number density is called the molecule's *absorption cross section*:

$$\sigma_A(\lambda) = \frac{b_a^A}{n_A} \quad (\text{cm}^2 \text{ molecule}^{-1}) \tag{4.26}$$

$\sigma_A(\lambda)$ can be regarded as the effective cross-sectional area of a molecule for absorption of photons, and is a measure of the ability of a molecule to absorb photons. Absorption cross sections vary from zero to about 1×10^{-16} cm^2 molecule^{-1}. A value of the absorption cross section of 10^{-18}–10^{-17} cm^2 molecule^{-1} is considered to be large.

Therefore, the optical depth of the atmosphere at the surface as a result of absorption by species A is

$$\tau(0,\lambda) = \int_0^{z_{TOA}} \sigma_A[T(z),\lambda]n_A(z)dz \tag{4.27}$$

The integral in (4.27) presumes that the path of the solar beam is perpendicular to the Earth's surface. This is, of course, the case only when the sun is directly overhead.

The angle measured at the Earth's surface between the Sun and the zenith is called the *solar zenith angle*; it is denoted by the symbol θ_0. When the sun is exactly overhead, $\theta_0 = 0°$; at the horizon $\theta_0 = 90°$. As θ_0 increases, the pathlength of the solar beam through the atmosphere increases. If the relative pathlength of the solar beam from the TOA to the surface when the Sun is directly overhead ($\theta_0 = 0°$) is taken as 1.0, then the pathlength m at θ_0, neglecting the sphericity of the Earth, is

$$m = \frac{1}{\cos\theta_0} = \sec\theta_0 \tag{4.28}$$

The optical depth of the atmosphere increases as the pathlength increases since the solar beam transsects a longer section of the atmosphere, and there is proportionately more opportunity for extinction to occur. Therefore, the more general form of (4.27) is

$$\tau(0,\lambda) = m \int_0^{z_{TOA}} \sigma_A[T(z),\lambda]n_A(z)dz \tag{4.29}$$

Equation (4.28) holds for θ_0 less than about 75°. For larger values of θ_0, m has to be computed, taking into account the path through the spherical atmospheric layers, the

TABLE 4.1   Relation between the Slant Path Optical Depth m and sec θ_0 for a Standard Rayleigh Atmosphere

θ_0	sec θ_0	m
0	1.00	1.00
30	1.15	1.15
60	2.00	1.99
70	2.92	2.90
75	3.86	3.81
80	5.76	5.59
85	11.47	10.32
87	19.11	15.16
89	57.30	26.26
90	∞	38.09

Source: Kasten and Young (1989).

vertical profile of absorbing and scattering species, and the curvature of the optical rays as a result of refraction. Values of m are given in Table 4.1 for the molecular atmosphere, that is, the Rayleigh scattering optical depth.[3]

4.4   ACTINIC FLUX

Sunlight drives the chemistry of the atmosphere by dissociating a number of molecules into fragments that are often highly reactive. Whether a molecule can be dissociated in the atmosphere depends on the probability of an encounter between a photon of appropriate energy and the molecule. The radiative quantity pertinent for photochemical reactions is the photon flux incident on the molecule from all directions, since it does not matter to the molecule from which direction a photon comes. The radiative flux *from all directions* on a volume of air is called the *actinic flux* (*actinic* means "capable of causing photochemical reactions"). We use the symbol I for the actinic flux to distinguish it from the irradiance E; I is usually expressed in units of photons $cm^{-2} s^{-1}$. The *spectral actinic flux*, $I(\lambda)$, expresses the wavelength dependence of the actinic flux; $I(\lambda)$ has units of photons $cm^{-2} s^{-1} nm^{-1}$. Whereas the spectral irradiance is expressed per square meter of area, the actinic flux is usually expressed per square centimeter; the different area units, m^{-2} versus cm^{-2}, arise because molecular absorption cross sections are expressed in terms of cm^2, and the photodissociation rate of a molecule will be seen shortly to be a product of the spectral actinic flux and the molecular absorption cross section.

To compute atmospheric photochemical reaction rates it is necessary to determine the total light intensity incident on a given volume of air, from all directions. The light

---

[3]The TOA radiative flux can be estimated by measuring $E(\lambda)$ on a surface at ground level for various solar zenith angles θ_0 and plotting $\ln E(\lambda)$ versus m and extrapolating to $m = 0$. The slope of the best-fit straight line is $\tau(\lambda)$. This method of calculating $E_\infty(\lambda)$ is called the *Bouguer–Langley method*. Integrating $E_\infty(\lambda)$ over all wavelengths produces the solar constant S_0.

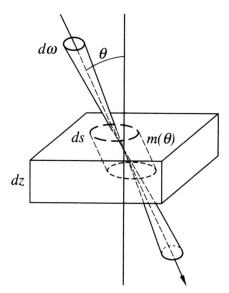

FIGURE 4.8 Interception of radiation emanating from a solid angle $d\omega$ on a surface element of area ds on an atmospheric layer of thickness dz. Pathlength through the layer is $m(\theta)$.

intensity impacting a volume of air includes not only direct solar radiation, but light, either direct from the Sun or reflected from the Earth's surface, that is scattered into the volume by gases and particles, as well as light reflected directly from the Earth's surface.

The actinic flux is related but not equal to the irradiance. The relation between the two quantities has been illustrated carefully by Madronich (1987). Consider an atmospheric layer of infinitesimal vertical thickness dz, illuminated from above (Figure 4.8). The quantity of light incident on the top surface of the layer depends on the direction of incidence of the light, as defined by the spherical coordinates θ, ϕ. This dependence is specified by the spectral radiance $L(\lambda, \theta, \phi)$, expressed in units of photons $\text{cm}^{-2}\,\text{s}^{-1}\,\text{nm}^{-1}\,\text{sr}^{-1}$. The number of photons entering the layer (through ds, in time dt, from solid angle $d\omega$) is

$$L(\lambda, \theta, \phi)\cos\theta\,ds\,d\omega\,dt\,d\lambda$$

To determine the spectral actinic flux in the layer in Figure 4.8 resulting from the spectral radiance $L(\lambda, \theta, \phi)$, we need to jump ahead a bit. The rate of photodissociation of a species A is written as

$$\frac{dn_A}{dt} = -j_A n_A \tag{4.30}$$

where n_A is the molecular number concentration (molecules cm^{-3}) and j_A is the photodissociation rate coefficient (s^{-1}). j_A depends on the available light and the nature of the molecule A, expressed in terms of the product of the probability a photon will be absorbed and the probability of dissociation following absorption of a photon. After entering the layer in Figure 4.8, photons may interact with molecules of A in the layer and be absorbed. For an infinitesimally thin layer, the Beer–Lambert law can be used to compute the number of photons absorbed in time dt in wavelength range $d\lambda$ as

$$\sigma_A(\lambda)n_A m(\theta)L(\lambda, \theta, \phi) \cos \theta \, ds \, d\omega \, dt \, d\lambda$$

where $\sigma_A(\lambda)$ is the absorption cross section (cm^{-2}) of a molecule of A and $m(\theta)$ is the pathlength shown in Figure 4.8. $m(\theta) = dz/\cos \theta$. For each photon absorbed, the probability that the molecule will dissociate is $\phi_A(\lambda)$. (This is called the *quantum yield*.) Thus the number of molecules dissociated in time dt in wavelength range $d\lambda$ is

$$\phi_A(\lambda)\sigma_A(\lambda)n_A L(\lambda, \theta, \phi) ds \, dz \, d\omega \, dt \, d\lambda$$

The total number of dissociations dN_A occurring in the volume in a time interval dt is obtained by integrating over all solid angles, over the upper surface of the layer, and over all wavelengths:

$$dN_A = -\left(dz \int_s ds \right)(n_A \, dt) \int_\lambda \phi_A(\lambda)\sigma_A(\lambda) \left(\int_\omega L(\lambda, \theta, \phi) d\omega \right) d\lambda$$

The first factor on the right-hand side (RHS) is just the total volume of the layer, which can be brought to the left-hand side (LHS) along with dt to produce

$$\frac{dn_A}{dt} = -n_A \underbrace{\int_\lambda \phi_A(\lambda)\sigma_A(\lambda) \left(\int_\omega L(\lambda, \theta, \phi) d\omega \right) d\lambda}_{j_A} \tag{4.31}$$

As indicated, the quantity on the RHS multiplying n_A is j_A. The spectral actinic flux is then the radiative quantity that drives the photodissociation, that is, the quantity that multiplies $\phi_A(\lambda)\sigma_A(\lambda)$ to produce a product that when integrated over all wavelengths produces the photodissociation rate coefficient.

The spectral actinic flux $I(\lambda)$ is then

$$I(\lambda) = \int_\omega L(\lambda, \theta, \phi) d\omega$$

$$= \int_\phi \int_\theta L(\lambda, \theta, \phi) \sin \theta \, d\theta \, d\phi \tag{4.32}$$

It is important to distinguish the actinic flux from the spectral irradiance. The spectral irradiance $E(\lambda)$ is the radiant energy crossing a surface (per unit surface area, time, and wavelength) and is calculated from $L(\lambda, \theta, \phi)$ by (4.14),

$$E(\lambda) = \int_\phi \int_\theta L(\lambda, \theta, \phi) \cos\theta \, \sin\theta \, d\theta \, d\phi \tag{4.33}$$

The factor $\cos\theta$ reflects the change in the projected area of the surface as the angle of incidence is varied. This factor does not appear in the expression for the actinic flux because the projected area and the pathlengths offset exactly. As the angle of incidence is changed from overhead $(\theta = 0°)$ to nearly glancing $(\theta \to 90°)$, the energy (irradiance) incident on the layer decreases, but the actinic flux remains unchanged because the lower intensity is exactly compensated for by the longer pathlength of light through the layer.

4.5   ATMOSPHERIC PHOTOCHEMISTRY

According to Planck's law, the energy of one photon of frequency ν is $h\nu$. In atmospheric photochemistry, the photon that is a reactant in a chemical reaction is written as $h\nu$, for example, for the photolysis of NO_2,

$$NO_2 + h\nu \to NO + O$$

Photon energy can be expressed per mole of a substance by multiplying $h\nu$ by Avogadro's number, 6.022×10^{23} molecules mol^{-1}:

$$\varepsilon = 6.022 \times 10^{23} h\nu$$

$$= 6.022 \times 10^{23} \frac{hc}{\lambda} \tag{4.34}$$

The energy[4] associated with a particular wavelength λ is as follows, with λ in nm:

$$\varepsilon = \frac{1.19625 \times 10^5}{\lambda} \, kJ \, mol^{-1} \tag{4.35}$$

Typical ranges of wavelengths and energies in the portion of the electromagnetic spectrum of interest in atmospheric chemistry are

Name	Typical Wavelength or Range of Wavelengths, nm	Typical Range of Energies, kJ mol^{-1}
Visible		
Red	700	170
Orange	620	190
Yellow	580	210
Green	530	230
Blue	470	250
Violet	420	280
Near ultraviolet	400–200	300–600
Vacuum ultraviolet	200–50	600–2400

[4]A traditional unit used by chemists for expressing energies associated with molecules is kcal mol^{-1}. The conversion factor between kJ mol^{-1} and kcal mol^{-1} is (kJ mol^{-1}) $\times 0.2390 =$ kcal mol^{-1}

Photon energies can be compared with bond energies of molecules. The energy contained in photons of wavelengths near the red end of the visible spectrum is comparable to the bond energies of rather loosely bound chemical species. For example, in the ozone molecule, the $O-O_2$ bond energy is about $105\ kJ\ mol^{-1}$; in NO_2, the $O-NO$ bond energy is about $300\ kJ\ mol^{-1}$ (which corresponds to a wavelength of about $400\ nm$). The lowest energy photons that are capable of promoting chemical reaction lie in the visible region of the electromagnetic spectrum. Wavelengths at which chemical change can occur correspond roughly to the energies at which electronic transitions in molecules take place. Absorption of radiation can occur only if an upper energy level of the molecule exists that is separated from the lower level by an energy equal to that of the incident photon. Small molecules generally exhibit intense electronic absorption at wavelengths shorter than do larger molecules. For example, N_2 and H_2 absorb significantly at wavelengths less than $100\ nm$, while O_2 absorbs strongly for $\lambda < 200\ nm$, H_2O for $\lambda < 180\ nm$, and CO_2 for $\lambda < 165\ nm$.

Photodissociation of O_2   Let us estimate the maximum wavelength of light at which the photodissociation of O_2 into two ground-state oxygen atoms occurs:

$$O_2 + h\nu \rightarrow O + O$$

The enthalpy change for this reaction is $\Delta H = 498.4\ kJ\ mol^{-1}$, which is ε in (4.35):

$$\lambda_{max} = \frac{1.19625 \times 10^5}{498.4}\ nm$$

$$= 240\ nm$$

Thus, O_2 cannot photodissociate at wavelengths longer than about $240\ nm$.

The primary step of a photochemical reaction may be written

$$A + h\nu \rightarrow A^*$$

where A^* is an electronically excited state of the molecule A. The excited molecule A^* may subsequently partake in

Dissociation	$A^* \xrightarrow{1} B_1 + B_2$
Direct reaction	$A^* + B \xrightarrow{2} C_1 + C_2$
Fluorescence	$A^* \xrightarrow{3} A + h\nu$
Collisional deactivation	$A^* + M \xrightarrow{4} A + M$
Ionization	$A^* \xrightarrow{5} A^+ + e$

The *quantum yield* for a specific process involving A^* is defined as the ratio of the number of molecules of A^* undergoing that process to the number of photons absorbed. Since the total number of A^* molecules formed equals the number of photons absorbed, the quantum yield ϕ_i for a specific process i, say, dissociation, is just the fraction of the A^* molecules that participate in path i. The sum of the quantum yields for all possible processes must equal 1.

The rate of formation of A^* is equal to the rate of photon absorption and is written

$$\frac{d[A^*]}{dt} = j_A[A] \tag{4.36}$$

where j_A, having units s^{-1}, is the first-order rate constant for photolysis or the so-called specific absorption rate; j_A is normally taken to be independent of $[A]$. The rate of formation of B_1 in step 1 is

$$\frac{d[B_1]}{dt} = \phi_1\, j_A[A] \tag{4.37}$$

where ϕ_1 is the quantum yield of step 1.

Photodissociation of a molecule can occur when the energy of the incoming photon exceeds the binding energy of the particular chemical bond. Thus the excited species A^* can lie energetically above the dissociation threshold of the molecule. One or more of the products of photodissociation may themselves be electronically excited. Consider the photolysis of ozone:

$$O_3 + h\nu \rightarrow O_2 + O$$

Various combinations of electronic states are possible for the products O and O_2 depending on the wavelength of incident radiation. The lowest energy pair of excited products is $O(^1D) + O_2(^1\Delta_g)$, which form at an expected threshold wavelength of about 305 nm. As noted in Chapter 3, the singlet-D oxygen atom, $O(^1D)$, is the most important electronically excited species in the atmosphere. (Henceforth we will have no need to distinguish electronically excited states of molecular oxygen; we will simply indicate all oxygen molecules emerging from photodissociation of O_3 as O_2.) Recall that the reaction of $O(^1D)$ with water vapor is a source of OH radicals in the entire atmosphere, and $O(^1D)$ reaction with N_2O is the principal source of NO_x in the stratosphere.

To calculate the rate of a photochemical reaction we need to know the number of photons absorbed per unit volume of air containing a given concentration $[A]$ (molecules cm^{-3}) of an absorbing molecule A.

The number of photons absorbed by a molecule A in a wavelength region λ to $\lambda + d\lambda$ is the product of its absorption cross section $\sigma_A(\lambda)$ (cm^2 molecule^{-1}), the spectral actinic flux $I(\lambda)$ (photons $cm^{-2}\,s^{-1}\,nm^{-1}$), and the number concentration of A (molecules cm^{-3}):

$$\sigma_A(\lambda)I(\lambda)d\lambda[A] \quad \text{photons } cm^{-3}\,s^{-1}$$

To calculate the rate of photolysis of A we need to multiply this expression by the quantum yield for photolysis, $\phi_A(\lambda)$. Thus the rate of photolysis in the wavelength region λ to $\lambda + d\lambda$ is

$$\sigma_A(\lambda)\phi_A(\lambda)I(\lambda)d\lambda[A]$$

The total photolysis rate of A is the integral of this expression over all possible wavelengths

$$\left[\int_{\lambda_1}^{\lambda_2} \sigma_A(\lambda)\phi_A(\lambda)I(\lambda)d\lambda \right][A] \quad \text{molecules cm}^{-3}\,\text{s}^{-1} \tag{4.38}$$

where λ_1 and λ_2 are, respectively, the shortest and longest wavelengths at which absorption occurs. For the troposphere, for example, $\lambda_1 = 290\,\text{nm}$.

The quantity in brackets has already been identified as the first-order photolysis rate constant:

$$j_A = \int_{\lambda_1}^{\lambda_2} \sigma_A(\lambda,T)\phi_A(\lambda,T)I(\lambda)\,d\lambda \tag{4.39}$$

The integral in (4.39) is often approximated for computational purposes by a summation over small wavelength intervals

$$j_A = \sum_i \bar{\sigma}_A(\lambda_i, T)\bar{\phi}_A(\lambda_i, T)\bar{I}(\lambda_i)\,\Delta\lambda_i \tag{4.40}$$

where the overbar denotes an average over a wavelength interval $\Delta\lambda_i$ centered at λ_i. The width of the wavelength intervals $\Delta\lambda_i$ is usually dictated by the available resolution for the actinic flux $I(\lambda)$. A typical size of $\Delta\lambda_i$ is 5 nm from 290 nm to over 400 nm, and 10 nm beyond 400 nm. Values of $\sigma(\lambda)$ and $\phi(\lambda)$ may not be available on precisely the same intervals as for $I(\lambda)$, so some interpolation may be necessary.

4.6   ABSORPTION OF RADIATION BY ATMOSPHERIC GASES

Figure 4.9 shows the solar irradiance at the top of the atmosphere and that at sea level. The absorption spectra are quite complex, but they do indicate that absorption is so strong in some spectral regions that no solar energy in those regions reaches the surface of the Earth. As we will see, absorption by O_2 and O_3 is responsible for removal of practically all the incident radiation with wavelengths shorter than 290 nm. On the other hand, atmospheric absorption is not strong from 300 to about 800 nm, forming a "window" in the spectrum. About 40% of the solar energy is concentrated in the region of 400–700 nm. Water vapor absorbs in a complicated way, and mostly in the region where the Sun's and Earth's radiation overlap. From 300 to 800 nm, the atmosphere is essentially transparent. From 800 to 2000 nm, terrestrial longwave radiation is moderately absorbed by water vapor in the atmosphere.

Why molecules absorb in particular regions of the spectrum can be determined only through quantum chemical calculations. In general, the geometry of the molecule explains, for example, why H_2O, CO_2, and O_3 interact strongly with radiation above 400 nm, but N_2 and O_2 do not. In H_2O, for instance, the center of the negative charge is shifted toward the oxygen nucleus and the center of positive charge toward the hydrogen nuclei, leading to a separation between the centers of positive and negative charge, a so-called electric dipole moment. Molecules with dipole moments interact strongly with electromagnetic radiation because the electric field of the wave causes oppositely directed forces and therefore accelerations on electrons and nuclei at one end of the molecule as compared with the other.

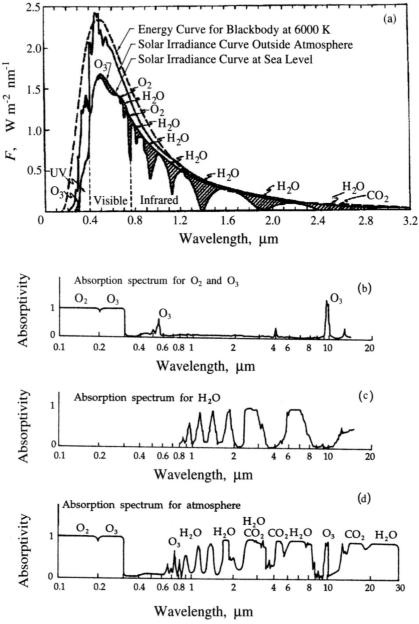

FIGURE 4.9 (a) Solar spectral irradiance at the top of the atmosphere and at sea level. Shaded regions indicate the molecules responsible for absorption. Absorption spectra for (b) molecular oxygen and ozone, (c) water vapor, and (d) the atmosphere, expressed on a scale of 0–1.

Similar arguments hold for ozone; however, nitrogen and oxygen are symmetric and thus are not strongly affected by radiation above 400 nm. The CO_2 molecule is linear but can easily be bent, leading to an induced dipole moment. A transverse vibrational mode exists for CO_2 at 15 μm, just where the Earth emits most of its infrared radiation.

TABLE 4.2   Solar Spectral Irradiance, Normalized to a Solar Constant of 1367 $W\,m^{-2}$

λ, nm	$F(\lambda)$, $W\,m^{-2}\,nm^{-1}$	$\int_0^\lambda F(\lambda')d\lambda'$, $W\,m^{-2}$	λ, nm	$F(\lambda)$, $W\,m^{-2}\,nm^{-1}$	$\int_0^\lambda F(\lambda')d\lambda'$, $W\,m^{-2}$
250.5	0.059	2.092	490.5	2.009	276.7
255.5	0.089	2.387	495.5	1.928	286.4
260.5	0.102	2.967	500.5	1.859	296.1
265.5	0.280	3.921	510.5	1.949	315.3
270.5	0.293	5.257	520.5	1.833	333.5
275.5	0.200	6.245	530.5	1.954	352.3
280.5	0.112	7.111	540.5	1.772	371.1
285.5	0.141	8.364	550.5	1.864	389.8
290.5	0.623	10.44	560.5	1.845	408.4
295.5	0.548	13.19	570.5	1.772	426.7
300.5	0.403	15.51	580.5	1.840	445.2
305.5	0.580	18.26	590.5	1.815	463.3
310.0	0.495	20.54	600.5	1.748	481.1
315.2	0.695	24.03	610.5	1.705	498.7
320.0	0.712	27.46	620.5	1.736	515.7
325.2	0.646	31.15	631.0	1.641	535.0
330.0	1.144	35.86	641.0	1.616	551.4
335.5	0.982	41.69	651.0	1.608	567.5
340.5	0.992	46.17	661.0	1.573	582.8
345.5	0.967	50.77	671.0	1.518	598.2
350.5	1.119	55.52	681.0	1.494	613.3
355.5	1.058	60.69	691.0	1.450	627.9
360.5	0.979	65.26	701.0	1.388	642.2
365.5	1.263	70.56	711.0	1.387	656.1
370.5	1.075	76.52	721.0	1.332	669.7
375.5	1.141	81.75	731.0	1.327	683.0
380.5	1.289	87.77	741.0	1.259	696.0
385.5	0.954	92.27	751.0	1.263	708.8
390.5	1.223	97.67	761.0	1.238	721.3
395.5	1.378	103.0	771.0	1.205	733.4
400.5	1.649	109.5	781.0	1.188	745.3
405.5	1.672	118.0	791.0	1.159	757.1
410.5	1.502	126.3	801.0	1.143	768.5
415.5	1.736	135.1	821.0	1.081	790.6
420.5	1.760	143.8	841.0	1.045	811.8
425.5	1.697	152.3	861.0	0.997	831.6
430.5	1.136	159.8	881.0	0.960	850.9
435.5	1.725	168.3	901.0	0.905	869.7
440.5	1.715	177.1	921.0	0.830	887.1
445.5	1.823	186.7	941.0	0.800	903.5
450.5	2.146	196.8	961.0	0.767	919.1
455.5	2.036	206.9	981.0	0.762	934.3
460.5	2.042	217.1	1002.5	0.745	952.8
465.5	2.044	227.3	1052.5	0.661	987.9
470.5	1.879	237.1	1102.5	0.608	1019
475.5	2.018	247.3	1152.5	0.545	1048
480.5	2.037	257.4	1202.5	0.496	1074
485.5	1.832	267.4	1252.5	0.474	1098

(Continued)

TABLE 4.2   (*Continued*)

λ, nm	$F(\lambda)$, W m^{-2} nm^{-1}	$\int_0^\lambda F(\lambda')d\lambda'$, W m^{-2}	λ, nm	$F(\lambda)$, W m^{-2} nm^{-1}	$\int_0^\lambda F(\lambda')d\lambda'$, W m^{-2}
1302.5	0.438	1120	2302.5	0.066	1313
1352.5	0.387	1140	2402.5	0.054	1319
1402.5	0.353	1159	2517.5	0.047	1325
1452.5	0.323	1176	2617.5	0.041	1329
1502.5	0.296	1191	2702.5	0.036	1332
1552.5	0.273	1205	2832.5	0.031	1336
1602.5	0.247	1218	3025.0	0.024	1342
1652.5	0.234	1230	3235.0	0.019	1346
1702.5	0.217	1241	3425.0	0.015	1349
1752.5	0.187	1251	3665.0	0.012	1353
1802.5	0.169	1260	3855.0	0.010	1355
1852.5	0.148	1267	4085.0	0.008	1357
1902.5	0.133	1274	4575.0	0.005	1360
1952.5	0.126	1281	5085.0	0.003	1362
2002.5	0.116	1287	5925.0	0.002	1364
2107.5	0.093	1298	7785.0	0.001	1366
2212.5	0.075	1307	10075.0	0.000	1367

Source: Fröhlich and London (1986).

Table 4.2 tabulates the solar spectral irradiance, normalized to a solar constant of 1367 W m^{-2}. Solar spectral actinic flux at the surface (0 km), 20, 30, 40, and 50 km is shown in Figure 4.11.

CO$_2$ Absorption in the Atmosphere   The spectral region from about 7 to 13 μm is a window region; nearly 80% of the radiation emitted by the Earth in this region escapes to space. Most of the non-CO$_2$ greenhouse gases, including O$_3$, CH$_4$, N$_2$O, and the chlorofluorocarbons, all have strong absorption bands in this window region. For this reason, relatively small changes in the concentrations of these gases can produce a significant change in the net radiative flux. As the concentration of a greenhouse gas continues to increase, it can absorb more of the radiation in its energy bands. Once an absorption wavelength becomes saturated, further increases in the concentration of the gas have less and less effect on radiative flux. This is called the *band saturation effect*. For CO$_2$, for example, the 15-μm band is already close to saturated. In addition, if a gas absorbs at wavelengths that are also absorbed by other gases, then the effect of increasing concentrations on radiative flux is less than in the absence of band overlap. For example, there is significant overlap between some of the absorption bands of CH$_4$ and N$_2$O; this overlap must be carefully accounted for when calculating the effect of these gases on radiative fluxes.

Even with the band saturation effect, it is incorrect to conclude that because there is already so much CO$_2$ in the atmosphere, more CO$_2$ can have no additional effect on absorption of outgoing radiation. When gases are present in small concentrations, doubling the concentration of the gas will approximately double its absorption. When

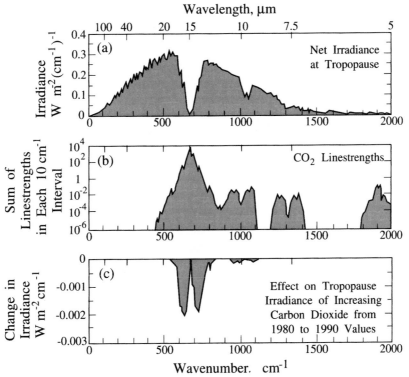

FIGURE 4.10 Effect of CO_2: (a) net infrared irradiance $(W\,m^{-2}\,(cm^{-1})^{-1})$ at the tropopause; (b) representation of the strength of the spectral lines of CO_2 in the thermal infrared (note the logarithmic scale); (c) change in net irradiance at the tropopause as a result of increasing the CO_2 concentration from its 1980–1990 levels, holding all other parameters fixed (IPCC 1995).

an absorbing gas is present in high concentration the effect of further addition is not one-to-one but it is not zero, either. For example, doubling the concentration of CO_2 from its present-day value leads to a 10–20% increase in its total greenhouse effect. Where this increase comes from can be explained as follows. The top frame of Figure 4.10 shows the spectral variation in the infrared radiance at the tropopause, in $W\,m^{-2}\,(cm^{-1})^{-1}$. The Planck function (4.2) determines the shape of the upper envelope of the curve, the maximum energy that can be emitted at a given wavelength and temperature. At typical atmospheric temperatures the maximum lies between 10 and 15 µm. The notches in the spectrum result from the presence of greenhouse gases and clouds. If the atmosphere were transparent to infrared radiation, the irradiance reaching the tropopause would be the same as that leaving the surface.

The middle frame of Figure 4.10 shows the CO_2 absorption spectrum. As noted earlier, a strong absorption band exists near 15 µm. The bottom frame shows the modeled effect of an instantaneous change in CO_2 abundance (change in concentration from 1980 to 1990) with all other factors, such as cloudiness, held fixed. At the center of the 15-µm band, the increase in CO_2 concentration has almost no effect; the CO_2 absorption is indeed saturated in this portion of the spectrum.

Away from this band, however, where CO_2 is less strongly absorbing, the increase in CO_2 does have an effect. As more and more CO_2 is added to the atmosphere, more of its spectrum will become saturated, but there will always be regions of the spectrum that remain unsaturated and thus capable of continuing to absorb infrared radiation. For example, the 10-μm absorption band is about 10^6 times weaker than the peak of the 15 μm band, but its contribution to the irradiance change in the lower frame is important. And as CO_2 concentrations increase, the importance of the 10 μm band will continue to increase relative to the 15-μm band.

4.7   ABSORPTION BY O_2 AND O_3

For wavelengths below about 1000 nm (1 μm), the predominant absorbing species in the atmosphere are O_2 and O_3 (Figure 4.9). The spectral solar actinic flux at various altitudes, as shown in Figure 4.11, can, therefore, be computed fairly accurately by considering only the absorption by O_2 and O_3. The absorption cross sections of O_2 and O_3 are shown in Figures 4.12 and 4.13. All the absorption shown in Figure 4.12 and 4.13 leads to dissociation.

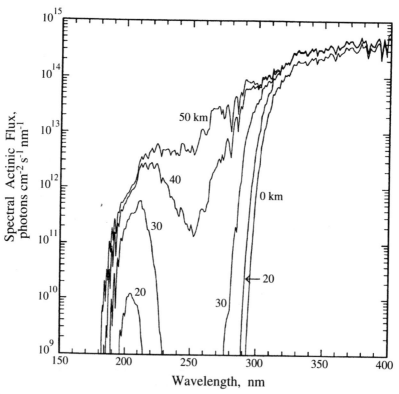

FIGURE 4.11   Solar spectral actinic flux (photons cm^{-2} s^{-1} nm^{-1}) at various altitudes and at the Earth's surface (DeMore et al. 1994).

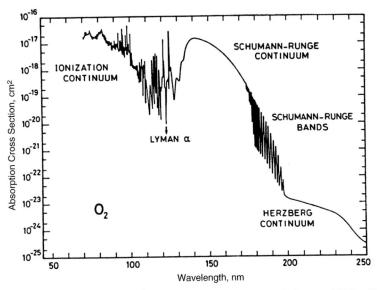

FIGURE 4.12 Absorption cross section of O_2 (Brasseur and Solomon 1984). (With kind permission of Springer Science and Business Media.)

The attenuation of radiation from the top of the atmosphere (TOA) to any altitude z is described by the Beer-Lambert Law (4.24),

$$F(z,\lambda) = F_{TOA}(\lambda)\exp(-\tau(z,\lambda)) \qquad (4.24)$$

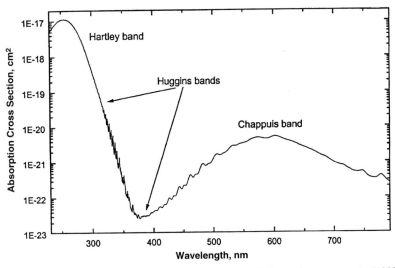

FIGURE 4.13 Absorption cross section of O_3. [Reprinted from Burrows et al. (1999), with permission from Elsevier.]

The optical depth of the atmosphere at the surface as a result of absorption by species A is given by (4.29)

$$\tau_A(0,\lambda) = m \int_0^{z_{TOA}} \sigma_A[T(z),\lambda]n_A(z)dz \qquad (4.29)$$

where $m = \sec\theta_0$ and where $\sigma_A[T(z),\lambda]$ is the absorption cross section of molecule A at temperature T (at altitude z). Here, the species of interest are O_2 and O_3. The total optical depth at the surface is then the sum of those attributable to O_2 and O_3:

$$\tau(0,\lambda) = \tau_{O_2}(0,\lambda) + \tau_{O_3}(0,\lambda) \qquad (4.41)$$

To estimate $\tau(0,\lambda)$ for O_2 and O_3, we neglect the temperature dependence of the absorption cross sections. Then (4.29) becomes for O_2 and O_3

$$\tau_{O_2}(0,\lambda) \cong m\sigma_{O_2}(\lambda) \int_0^{z_{TOA}} n_{O_2}(z)dz \qquad (4.42)$$

$$\tau_{O_3}(0,\lambda) \cong m\sigma_{O_3}(\lambda) \int_0^{z_{TOA}} n_{O_3}(z)dz \qquad (4.43)$$

The integrals in (4.42) and (4.43) are just the total column abundances (molecules cm^{-2}) of O_2 and O_3, respectively.

Let us first estimate the transmittance of the atmosphere at the surface resulting from the absorption of solar radiation by O_2. In (4.42), we need the total column abundance of O_2 (molecules cm^{-2}). From Chapter 1, estimation of the number concentration of air molecules at any altitude can be based on the scale height of the atmosphere

$$n_{air}(z) = n_{air}(0)\exp\left(-\frac{z}{H}\right) \qquad (4.44)$$

where H is the scale height ($\cong 7.4\,km$) and $n_{air}(0) = 2.55 \times 10^{19}$ molecules cm^{-3}. Then

$$\int_0^{\infty} n_{air}(z)dz = n_{air}(0)H \qquad (4.45)$$

The column abundance of O_2 is $(0.21)(2.6 \times 10^{19})(7.4 \times 10^5) \cong 4.0 \times 10^{24}$ molecules cm^{-2}.

In order to obtain an estimate of the atmospheric transmittance as a result of absorption by O_2

$$\frac{F(0,\lambda)}{F_{TOA}(\lambda)} = \exp(-\tau_{O_2}(0,\lambda)) \qquad (4.46)$$

we can characterize each of the three absorption bands for O_2 in Figure 4.12 by its maximum cross section. For this calculation, let us take $m = 1$ (overhead sun). The result is tabulated here:

Absorption Band	σ_{max}, cm^2	τ_{O_2}	$F(0,\lambda)/F_{TOA}(\lambda)$
Schumann–Runge continuum	10^{-17}	4×10^7	~ 0
Schumann–Runge bands (175–205 nm)	10^{-20}	4×10^4	~ 0
Herzberg continuum (185–242 nm)	10^{-23}	40	~ 0

Since $F(0,\lambda)/F_{TOA}(\lambda)$ is effectively zero for each of these bands, which extend up to 242 nm, we see that atmospheric O_2 completely removes all radiation below $\lambda = 242$ nm from reaching the Earth's surface. [In such a case, the species is called *optically thick* in the region; conversely, for a species for which $\exp(-\tau) \cong 1$, the species is *optically thin*.]

To estimate the transmittance resulting from O_3 absorption, the cross section of which is shown in Figure 4.13, we follow the same procedure as for O_2, in which we approximate the cross section of each band by its maximum value. Assuming a column abundance of O_3 of 300 DU (recall 1 DU = 2.69×10^{16} molecules cm^{-2}), the result is

Absorption Band	σ_{max}, cm^2	τ_{O_3}	$F(0,\lambda)/F_{TOA}(\lambda)$
Hartley (220–280 nm)	10^{-17}	80	~ 0
Huggins (310–330 nm)	10^{-19}	0.8	0.15
Chappuis (500–700 nm)	3×10^{-21}	0.024	~ 1.0

At wavelengths $\lambda > 242$ nm, the atmosphere is transparent with respect to O_2. Ozone is the dominant absorber in the range 240–320 nm. Table 4.3 gives estimated surface-level spectral actinic flux at 40°N latitude on January 1 and July 1.

TABLE 4.3 Estimated Ground-Level Actinic Flux $I(\lambda)$ at 40°N Latitude

	$I(\lambda) \times 10^{-14}$ (photons cm^{-2} s^{-1})	
	Noon	Noon
$\Delta\lambda$, nm	January 1	July 1
290–295	0.0	0.0
295–300	0.0	0.031
300–305	0.021	0.335
305–310	0.196	1.25
310–315	0.777	2.87
315–320	1.45	4.02
320–325	2.16	5.08
325–330	3.44	7.34
330–335	3.90	7.79
335–340	4.04	7.72
340–345	4.51	8.33
345–350	4.62	8.33
350–355	5.36	9.45
355–360	5.04	8.71
360–365	5.70	9.65

Source: Finlayson-Pitts and Pitts (1986).

A Somewhat More Accurate Estimate of Absorption of Solar Radiation by O_2 and O_3   We calculate the reduction in solar radiation at the Earth's surface as a result of O_2 and O_3 absorption in the wavelength range from 200 to 320 nm at a solar zenith angle of 45°. The approximate column abundances are

$$O_2 \quad 4 \times 10^{24} \text{ molecules cm}^{-2}$$
$$O_3 \quad 8 \times 10^{18} \text{ molecules cm}^{-2} \ (300 \text{ DU})$$

Absorption cross sections for O_2 and O_3, as well as the solar flux at the top of the atmosphere, in 20 nm increments are

λ, nm	σ_{O_2}, cm^2	σ_{O_3}, cm^2	F_{TOA}, photons cm^{-2} s^{-1}
200	9×10^{-24}	3.2×10^{-19}	1.5×10^{13}
220	4.7×10^{-24}	1.8×10^{-18}	9.6×10^{13}
240	1.2×10^{-24}	8.2×10^{-18}	1.0×10^{14}
260	0	1.1×10^{-17}	2.6×10^{14}
280	0	3.9×10^{-18}	4.2×10^{14}
300	0	3.2×10^{-19}	1.6×10^{15}
320	0	2.2×10^{-19}	2.4×10^{15}

$$F(z,\lambda) = F_{TOA}(\lambda) \exp[-(\tau_{O_2}(z,\lambda) + \tau_{O_3}(z,\lambda))]$$

$$\tau_i(z,\lambda) = m\sigma_i(\lambda) \int_0^\infty n_i(z)dz \quad i = O_2, O_3$$

For solar zenith angle $= 45°$, $m = \sec 45° = 1.414$.

The total transmittance of the atmosphere is defined as the ratio $F(z,\lambda)/F_{TOA} = \exp[-(\tau_{O_2} + \tau_{O_3})] = \exp(-\tau_{O_2})\exp(-\tau_{O_3})$, which is the product of the individual transmittances of O_2 and O_3. We obtain

λ, nm	O_2 Transmittance	O_3 Transmittance	Total Transmittance	$F(0,\lambda)$, photons cm^{-2} s^{-1}
200	7.8×10^{-23}	2.6×10^{-2}	2.0×10^{-24}	3.0×10^{-11}
220	2.8×10^{-12}	1.2×10^{-9}	3.4×10^{-21}	3.3×10^{-7}
240	1.1×10^{-3}	2.3×10^{-41}	2.6×10^{-44}	2.6×10^{-30}
260	1.0	3.0×10^{-55}	3.0×10^{-55}	7.8×10^{-41}
280	1.0	4.7×10^{-20}	4.7×10^{-20}	2.0×10^{-5}
300	1.0	2.6×10^{-2}	2.6×10^{-2}	4.2×10^{13}
320	1.0	8.1×10^{-2}	8.1×10^{-2}	2.0×10^{14}

4.8   PHOTOLYSIS RATE AS A FUNCTION OF ALTITUDE

Photolysis reactions are central to atmospheric chemistry, since the source of energy that drives the entire system of atmospheric reactions is the Sun. The general expression for the first-order rate coefficient j_A for photodissociation of a species A is given by (4.39). Because the rate of a photolysis reaction depends on the spectral actinic flux I and because the spectral actinic flux varies with altitude in the atmosphere, the rate of photolysis reactions depends in general on altitude. The altitude dependence of j_A enters indirectly through the temperature dependence of the absorption cross section and directly through the altitude dependence of the spectral actinic flux, which can be written as $I(z,\lambda)$. We have already seen that the altitude dependence of the solar irradiance in the wavelength range below ~ 1 μm is essentially a result of absorption of solar radiation by O_2 and O_3. From the absorption cross sections for O_2 and O_3, it is possible to calculate $I(z,\lambda)$ at any altitude z, which can then be used in (4.39) to calculate the photolysis rate of any species A as a function of altitude.

The photolysis rate coefficient of a species A can be expressed as a function of altitude from (4.39) as

$$j_A(z) = \int_{\lambda_1}^{\lambda_2} \sigma_A(T,\lambda)\phi_A(T,\lambda)I(z,\lambda)d\lambda \qquad (4.47)$$

Assume that the attenuation of radiation is due solely to absorption by species A. From the Beer–Lambert Law, $I(z,\lambda)$ can be expressed as

$$I(z,\lambda) = I_{TOA}(\lambda)\exp(-\tau_A(z,\lambda)) \qquad (4.48)$$

If we neglect the z dependence of temperature, then

$$\tau_A(z,\lambda) = m\sigma_A(T,\lambda)\int_z^{\infty} n_A(z')dz' \qquad (4.49)$$

To proceed further, consider the case in which the absorbing species A has a uniform mixing ratio, ξ_A (e.g., O_2); then, its concentration can be expressed in terms of the molecular air density as a function of altitude, $n_{air}(z)$:

$$n_A(z) = \xi_A n_{air}(z) \qquad (4.50)$$

The total number concentration of air as a function of altitude falls off approximately with the scale height H of atmospheric pressure:

$$n_{air}(z) = n_{air}(0)\exp\left(-\frac{z}{H}\right) \qquad (4.51)$$

Using (4.51) in (4.49), we get

$$\tau_A(z,\lambda) = m\sigma_A(T,\lambda)\xi_A n_{air}(0)H e^{-z/H} \qquad (4.52)$$

Combing (4.47), (4.48), and (4.52), the photolysis rate coefficient is given by

$$j_A(z) = \int_{\lambda_1}^{\lambda_2} \sigma_A(T,\lambda)\phi_A(T,\lambda)I_{TOA}(\lambda)\exp\{-m\sigma_A(T,\lambda)\xi_A n_{air}(0)e^{-z/H}\}d\lambda \qquad (4.53)$$

The photolysis rate at any altitude z is the product of $j_A(z)$ and the number concentration of A, $n_A(z)$:

$$\frac{dn_A(z)}{dt} = -j_A(z)n_A(z)$$

$$= -j_A(z)\xi_A n_{air}(0)e^{-z/H}$$

$$= -\xi_A n_{air}(0)e^{-z/H} \int_{\lambda_1}^{\lambda_2} \sigma_A(T,\lambda)\phi_A(T,\lambda)I_{TOA}(\lambda)\exp\{-m\sigma_A(T,\lambda)\xi_A n_{air}(0)e^{-z/H}\}d\lambda$$

$$= -\xi_A n_{air}(0) \int_{\lambda_1}^{\lambda_2} \sigma_A(T,\lambda)\phi_A(T,\lambda)I_{TOA}(\lambda)\exp\left\{-\frac{z}{H} - m\sigma_A(T,\lambda)\xi_A n_{air}(0)e^{-z/H}\right\}d\lambda$$

$$(4.54)$$

As one descends toward the Earth's surface, the photolysis rate decreases because the overall actinic flux is decreasing due to increasing absorption in the layers above. As one ascends in the atmosphere, the concentration of the gas $n_A(z)$ simply decreases as the atmosphere's density thins out. These two competing effects combine to produce a maximum in the photolysis rate at a certain altitude. The altitude at which this rate is a maximum is the value of z at which

$$\frac{d}{dz}\left(\frac{dn_A(z)}{dt}\right) = 0 \qquad (4.55)$$

We find that the photolysis rate is a maximum at the altitude z at which

$$m\sigma_A \xi_A n_{air}(0)H\exp(-z/H) = 1 \qquad (4.56)$$

This is just the altitude z where $\tau(z,\lambda) = 1$.

The function in the integral of (4.54) is called a *Chapman function* in honor of Sydney Chapman, pioneer of stratospheric ozone chemistry. We will return to Chapman in the next chapter. The elegant analytical form of the Chapman function results when the mixing ratio of the absorbing gas A is constant throughout the atmosphere, so that $n_A(z)$ scales with altitude according to the scale height H of pressure. Molecular O_2 is the prime example of such an absorbing gas. For a gas with a nonuniform mixing ratio with altitude, one can, of course, numerically evaluate (4.49) for the optical depth at any z on the basis of its actual profile $n_A(z)$.

Figure 4.14 shows the photodissociation rate coefficient for O_2, namely, j_{O_2}, as a function of altitude above 30 km. From 30 to 60 km, the Herzberg continuum provides the dominant contribution to j_{O_2}. At about 60 km, the contribution from the Schumann–Runge bands equals that from the Herzberg continuum; above 60 km, the Schumann–Runge bands predominate until about 80 km, where the Schumann–Runge continuum takes over. In the mid-to upper stratosphere, at solar zenith angle = 0°, an approximate value of j_{O_2} is $j_{O_2} \cong 10^{-9} \text{ s}^{-1}$.

4.9 PHOTODISSOCIATION OF O_3 TO PRODUCE O AND O(^{1}D)

Ozone photodissociates to produce either ground-state atomic oxygen

$$O_3 + h\nu \rightarrow O_2 + O$$

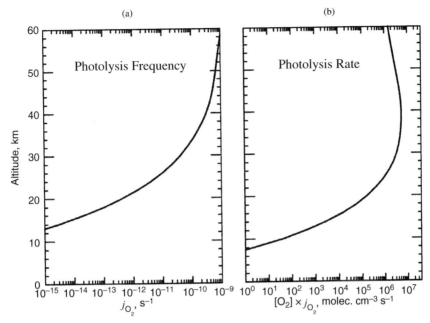

FIGURE 4.14 Photodissociation rate coefficient for O_2, j_{O_2}, and O_2 photolysis rate as a function of altitude at solar zenith angle $\theta_0 = 0°$. (Courtesy Ross J. Salawitch, Jet Propulsion Laboratory.)

or the first electronically excited state of atomic oxygen, $O(^1D)$:

$$O_3 + h\nu \rightarrow O_2 + O(^1D)$$

Ozone always dissociates when it absorbs a visible or UV photon; therefore, the quantum yield for O_3 dissociation is unity. Photodissociation of O_3 in the visible region, the Chappuis band, is the major source of ground state O atoms in the stratosphere. (Since these O atoms just combine with O_2 to reform O_3 with the release of heat, this absorption of visible radiation by O_3 merely converts light into heat.)

Figure 4.15 shows the photodissociation rate coefficients for $O_3 + h\nu \rightarrow O(^1D) + O_2$ [panels (a) and (b)] and for $O_3 + h\nu \rightarrow O + O_2$ [panels (c) and (d)]. Panels (a) and (c) show $j_{O_3 \rightarrow O(^1D)}$ and $j_{O_3 \rightarrow O}$, respectively, as a function of altitude at solar zenith angles $\theta = 0°$ and $85°$. Panels (b) and (d) show the two photolysis rate coefficients at 20 km as a function of solar zenith angle. The photodissociation of O_3 at altitudes below about 30 km is governed mainly by absorption in the Chappuis bands, and this absorption is practically independent of altitude. Above ~ 30 km, absorption in the Hartley bands dominates, the rate of which increases strongly with increasing altitude. In the troposphere, the photodissociation rate coefficients $j_{O_3 \rightarrow O(^1D)}$ and $j_{O_3 \rightarrow O}$ are virtually independent of altitude; at $\theta = 0°$:

$$j_{O_3 \rightarrow O(^1D)} \cong 6 \times 10^{-5}\,\text{s}^{-1}$$

$$j_{O_3 \rightarrow O} \cong 5.5 \times 10^{-4}\,\text{s}^{-1}$$

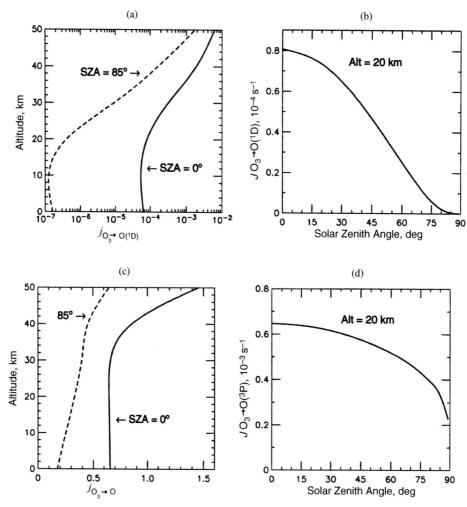

FIGURE 4.15   Photodissociation rate coefficient for O_3 as a function of altitude and solar zenith angle. Panels (a) and (b) are $j_{O_3 \rightarrow O(^1D)}$. Panels (c) and (d) are $j_{O_3 \rightarrow O}$. (Courtesy Ross J. Salawitch, Jet Propulsion Laboratory.)

Because of the importance of $O(^1D)$ to the chemistry of the lower stratosphere and entire troposphere, a great deal of effort has gone into precisely determining its production from O_3 photodissociation. Since solar radiation of wavelengths below ~ 290 nm does not reach the lower stratosphere and troposphere, the wavelength region just above 290 nm is critical for production of $O(^1D)$. Table 4.4 gives the quantum yield for production of $O(^1D)$ as a function of wavelength in this region. On the basis of heat of reaction data, the threshold energy for photodissociation of O_3 to produce $O(^1D)$ is estimated to be $390 \, kJ \, mol^{-1}$, which is equivalent to a wavelength of 305 nm. At wavelengths below 305 nm, the quantum yield for $O(^1D)$ production is indeed close to unity. It is expected that the threshold is shifted to somewhat longer wavelengths (~ 310 nm) because a fraction of the O_3 molecules will have extra vibrational energy that can help overcome the barrier. Michelson et al. (1994) showed that the effect of vibrationally excited O_3

TABLE 4.4   O(^{1}D) Quantum Yield in the Photolysis of O$_3$

Wavelength, nm	298 K	253 K	203 K
306	0.884	0.875	0.872
307	0.862	0.844	0.836
308	0.793	0.760	0.744
309	0.671	0.616	0.585
310	0.523	0.443	0.396
311	0.394	0.298	0.241
312	0.310	0.208	0.152
313	0.265	0.162	0.112
314	0.246	0.143	0.095
315	0.239	0.136	0.090
316	0.233	0.133	0.088
317	0.222	0.129	0.087
318	0.206	0.123	0.086
319	0.187	0.116	0.085
320	0.166	0.109	0.083
321	0.146	0.101	0.082
322	0.128	0.095	0.080
323	0.113	0.089	0.079
324	0.101	0.085	0.078
325	0.092	0.082	0.078
326	0.086	0.080	0.077
327	0.082	0.079	0.077
328	0.080	0.078	0.077

Source: Matsumi et al. (2002).

leads to nonzero quantum yields that vary with temperature up to about 320 nm. Recent data, as reflected in Table 4.4, show that O(^{1}D) is produced well beyond the 310 nm threshold, attributable to so-called spin-forbidden channels that may account for as much as 10% of the overall rate.

The critical role of wavelength in photolysis can be seen by comparing Figure 4.11 and Table 4.4. From $\lambda = 305$ nm to 320 nm, the absorption cross section drops by a factor of ~ 10, and the quantum yield for O(^{1}D) formation drops from 0.9 to about 0.1. Since the $\sigma\phi$ product changes rapidly with λ, the actual rate of production of O(^{1}D) is critically dependent on how $I(\lambda)$ varies with λ. At the surface of the Earth, the spectral actinic flux increases by about an order of magnitude between $\lambda = 300$ nm and $\lambda = 320$ nm (see Table 4.3).

4.10   PHOTODISSOCIATION OF NO$_2$

Since no solar radiation of wavelength shorter than about 290 nm reaches the troposphere, the absorbing species of interest from the point of view of tropospheric chemistry are those that absorb in the portion of the spectrum above 290 nm. Nitrogen dioxide is an extremely important molecule in the troposphere; it absorbs over the entire visible and ultraviolet range of the solar spectrum in the lower atmosphere (Figure 4.16). Between 300 and 370 nm over 90% of the NO$_2$ molecules absorbing will dissociate into NO and O (Figure 4.17). Above 370 nm this percentage drops off rapidly and above

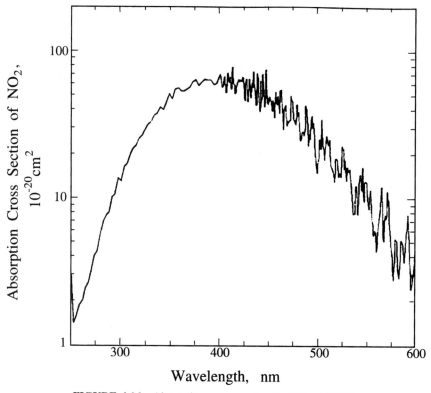

FIGURE 4.16   Absorption cross section for NO_2 at 298 K.

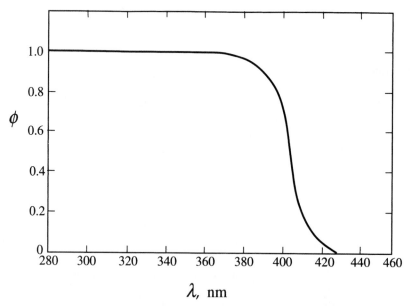

FIGURE 4.17   Primary quantum yield for O formation from NO_2 photolysis, as correlated by Demerjian et al. (1980).

about 420 nm dissociation does not occur. As noted earlier, the bond energy between O and NO, about 300 kJ mol^{-1}, corresponds to the energy contained in wavelengths near 400 nm. At longer wavelengths, there is insufficient energy to promote bond cleavage. The point at which dissociation fails to occur is not sharp because the individual molecules of NO$_2$ do not possess a precise amount of ground-state energy prior to absorption. The gradual transition area (370–420 nm) corresponds to a variation in ground-state energy of about 40 kJ mol^{-1}. This transition curve can be shifted slightly to longer wavelengths by increasing the temperature and therefore increasing the ground-state energy of the system. Table 4.5 gives tabulated values of the NO$_2$ absorption cross section and quantum yield at 273 and 298 K, and Table 4.6 illustrates the calculation of the photolysis rate j_{NO_2} for noon, July 1, at 40°N.

Figure 4.18 shows the NO$_2$ photodissociation rate coefficient, j_{NO_2}, as a function of altitude. The majority of NO$_2$ photolysis occurs for $\lambda > 300$ nm. Since in this region of the

TABLE 4.5 Absorption Cross Section and Quantum Yield for NO$_2$ Photolysis at 273 and 298 K

			Absorption Cross Section		
From: λ, nm	To: λ, nm	$\bar{\lambda}$, nm	At 273 K $10^{20}\sigma$, cm^2	At 298 K $10^{20}\sigma$, cm^2	ϕ
273.97	277.78	275.875	5.03	5.05	1
277.78	281.69	279.735	5.88	5.9	1
281.69	285.71	283.7	7	6.99	1
285.71	289.85	287.78	8.15	8.14	0.999
289.85	294.12	291.985	9.72	9.71	0.9986
294.12	298.51	296.315	11.54	11.5	0.998
298.51	303.03	300.77	13.44	13.4	0.997
303.03	307.69	305.36	15.89	15.8	0.996
307.69	312.5	310.095	18.67	18.5	0.995
312.5	317.5	315	21.53	21.4	0.994
317.5	322.5	320	24.77	24.6	0.993
322.5	327.5	325	28.07	27.8	0.992
327.5	332.5	330	31.33	31	0.991
332.5	337.5	335	34.25	33.9	0.99
337.5	342.5	340	37.98	37.6	0.989
342.5	347.5	345	40.65	40.2	0.988
347.5	352.5	350	43.13	42.8	0.987
352.5	357.5	355	47.17	46.7	0.986
357.5	362.5	360	48.33	48.1	0.984
362.5	367.5	365	51.66	51.3	0.983
367.5	372.5	370	53.15	52.9	0.981
372.5	377.5	375	55.08	54.9	0.979
377.5	382.5	380	56.44	56.2	0.9743
382.5	387.5	385	57.57	57.3	0.969
387.5	392.5	390	59.27	59	0.96
392.5	397.5	395	58.45	58.3	0.92
397.5	402.5	400	60.21	60.1	0.695
402.5	407.5	405	57.81	57.7	0.3575
407.5	412.5	410	59.99	59.7	0.138
412.5	417.5	415	56.51	56.5	0.0603
417.5	422.5	420	58.12	57.8	0.0188

TABLE 4.6   Calculation of the Photolysis Rate of NO_2 at Ground Level, July 1, Noon, 40° N, 298 K

From: λ, nm	To: λ, nm	I^a, photons cm^{-2} s^{-1}	$10^{19}\sigma_{NO_2}$, cm^2	ϕ	j_{NO_2}, s^{-1}
295	300	3.14×10^{12}	1.22	0.9976	3.83×10^{-7}
300	305	3.35×10^{13}	1.43	0.9966	4.78×10^{-6}
305	310	1.24×10^{14}	1.74	0.9954	2.16×10^{-5}
310	315	2.87×10^{14}	2.02	0.9944	5.77×10^{-5}
315	320	4.02×10^{14}	2.33	0.9934	9.31×10^{-5}
320	325	5.08×10^{14}	2.65	0.9924	1.34×10^{-4}
325	330	7.34×10^{14}	2.97	0.9914	2.16×10^{-4}
330	335	7.79×10^{14}	3.28	0.9904	2.53×10^{-4}
335	340	7.72×10^{14}	3.61	0.9894	2.76×10^{-4}
340	345	8.33×10^{14}	3.91	0.9884	3.22×10^{-4}
345	350	8.32×10^{14}	4.18	0.9874	3.43×10^{-4}
350	355	9.45×10^{14}	4.51	0.9864	4.21×10^{-4}
355	360	8.71×10^{14}	4.75	0.9848	4.07×10^{-4}
360	365	9.65×10^{14}	5.00	0.9834	4.75×10^{-4}
365	370	1.19×10^{15}	5.23	0.9818	6.09×10^{-4}
370	375	1.07×10^{15}	5.41	0.9798	5.68×10^{-4}
375	380	1.20×10^{15}	5.57	0.9762	6.51×10^{-4}
380	385	9.91×10^{14}	5.69	0.9711	5.48×10^{-4}
385	390	1.09×10^{15}	5.84	0.9636	6.13×10^{-4}
390	395	1.13×10^{15}	5.86	0.9360	6.18×10^{-4}
395	400	1.36×10^{15}	5.93	0.7850	6.34×10^{-4}
400	405	1.64×10^{15}	5.86	0.4925	4.74×10^{-4}
405	410	1.84×10^{15}	5.89	0.2258	2.45×10^{-4}
410	415	1.94×10^{15}	5.78	0.0914	1.03×10^{-4}
415	420	1.97×10^{15}	5.73	0.0354	4.00×10^{-5}
420	422.5	9.69×10^{14}	5.78	0.0188	1.05×10^{-5}

Total $j_{NO_2} = 8.14 \times 10^{-3}$ s^{-1}

$j_{NO_2} = 0.488$ min^{-1}

^{a}Actinic flux from Finlayson-Pitts and Pitts (1986).

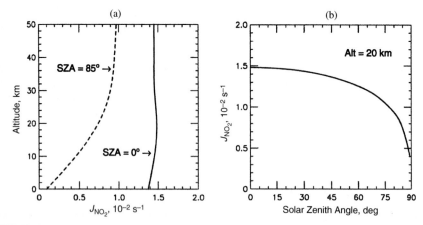

FIGURE 4.18 Photodissociation rate coefficient for NO_2, j_{NO_2}, as a function of altitude. (Courtesy Ross J. Salawitch, Jet Propulsion Laboratory.)

spectrum there is virtually no absorption by O_2 and O_3, the actinic flux is essentially independent of altitude (see Figure 4.11), and as a result, j_{NO_2} is nearly independent of altitude. For the same reason, j_{NO_2} does not drop off with increasing solar zenith angle as rapidly as that for molecules that absorb in the region of $\lambda > 300$ nm.

PROBLEMS

4.1$_A$ For a global mean albedo of $R_p = 0.3$, show that the equilibrium temperature of the Earth (assuming no atmospheric absorption of outgoing infrared radiation) is about 255 K. For $R_p = 0.15$, show that the equilibrium temperature is 268 K. Calculate the variation in solar constant and in global albedo corresponding to a change of 0.5 K in temperature.

4.2$_A$ The enthalpy changes for the following photodissociation reactions are

$$O_2 + h\nu \rightarrow O + O(^1D) \qquad\qquad 164.54 \text{ kcal mol}^{-1}$$
$$O_3 + h\nu \rightarrow O(^1D) + O_2(^1\Delta_g) \qquad 93.33 \text{ kcal mol}^{-1}$$
$$O_3 + h\nu \rightarrow O + O_2 \qquad\qquad\qquad 25.5 \text{ kcal mol}^{-1}$$

Estimate the maximum wavelengths at which these reactions can occur.

4.3$_A$ Convert the values of the surface UV radiation flux obtained for O_2 and O_3 absorption of solar radiation in Section 4.7 to W m^{-2}.

4.4$_A$ The combination of O_2 and O_3 absorption explains the solar spectrum at different altitudes in Figure 4.11. At the altitude where the optical depth $\tau = 1$, the solar irradiance is reduced to $1/e$ of its value at the top of the atmosphere. Considering overhead sun ($\theta = 0°$), estimate the altitude at which $\tau_{O_2} = 1$ for each of the three O_2 absorption bands given in Section 4.7.

4.5$_B$ At $\lambda = 300$ nm, the total transmittance of the atmosphere at a solar zenith angle of $45°$ is ~ 0.026, which is a result of absorption of solar radiation by O_3. At $\lambda = 300$ nm, the absorption cross section of O_3 is $\sigma_{O_3} \cong 3.2 \times 10^{-19}$ cm^2. What column abundance of O_3, expressed in Dobson units, produces this transmittance?

4.6$_B$ Consider the absorption of 240 nm sunlight by O_2. The O_2 cross section at this λ is 1×10^{-24} cm^2 molecule^{-1}. Assuming F_{TOA} photons of this wavelength enter the top of the atmosphere, derive the equation for the attenuation of this light as a result of absorption of these photons as a function of altitude, neglecting scattering, for a solar zenith angle of $0°$. You may assume $n_{air}(0) = 2.6 \times 10^{19}$ molecules cm^{-3} and the atmosphere scale height $H = 7.4$ km. As the Sun moves lower in the sky, how does the absorption change?

4.7$_B$ Show that (4.56) defines the altitude of the maximum photolysis rate of a species.

4.8$_B$ Estimate the altitude at which the photolysis rate of O_2 is maximum corresponding to the three spectral regions: Schumann–Runge continuum, Schumann–Runge bands, and Herzberg continuum. Approximate the O_2 absorption cross section in each of these three regions by its maximum value, 10^{-17}, 10^{-20}, and 10^{-23} cm^2, respectively. Assume a solar zenith angle of $0°$.

4.9$_B$ Plot the photodissociation rate of O_2 as a function of altitude from $z = 20$ to $z = 60$ km in the Herzberg continuum. Approximate the O_2 absorption cross section as 10^{-23} cm^2 and take $\lambda = 200$ nm. Consider solar zenith angles of $0°$ and $60°$.

4.10$_A$ What is the longest wavelength of light, absorption of which by NO_2 leads to dissociation at least 50% of the time? At 20 km, what are the lifetimes of NO_2 by photolysis at solar zenith angles of $0°$ and $85°$?

REFERENCES

Brasseur, G., and Solomon, S. (1984) *Aeronomy of the Middle Atmosphere*. Reidel, Dordrecht, The Netherlands.

Burrows, J. P., Richter, A., Dehn, A., Deters, B., Himmelmann, S., Voigt, S., and Orphal, J. (1999) Atmospheric remote sensing reference data from GOME—2. Temperature-dependent absorption cross sections of O_3 in the 231–794 nm range, *J. Quant. Spectros. Radiative Transf.* **61**, 509–517.

Demerjian, K. L., Schere, K. L., and Peterson, J. T. (1980) Theoretical estimates of actinic (spherically integrated) flux and photolytic rate constants of atmospheric species in the lower troposphere, *Adv. Environ. Sci. Technol.* **10**, 369–459.

DeMore, W. G., Sander, S. P., Golden, D. M., Hampson, R. F., Kurylo, M. J., Howard, C. J., Ravishankara, A. R., Kolb, C. E., and Molina, M. J. (1994) *Chemical Kinetics and Photochemical Data for Use in Stratospheric Modeling*, Evaluation no. 11, JPL Publication 94-26, Jet Propulsion Laboratory, Pasadena, CA.

Finlayson-Pitts, B. J., and Pitts, J. N. Jr. (1986) *Atmospheric Chemistry: Fundamentals and Experimental Techniques*. Wiley, New York.

Fröhlich, C., and London, J., eds. (1986) *Revised Instruction Manual on Radiation Instruments and Measurements*, World Climate Research Program (WCRP) Publication Series 7, World Meteorological Organization/TD No. 149, Geneva.

Intergovernmental Panel on Climate Change (IPCC) (1995) *Climate Change 1994: Radiative Forcing of Climate Change and an Evaluation of the IPCC IS92 Emission Scenarios*, Cambridge Univ. Press, Cambridge, UK.

Iqbal, M. (1983) *An Introduction to Solar Radiation*, Academic Press, Toronto.

Kasten, F., and Young, A. T. (1989) Revised optical air mass tables and approximation formula, *Appl. Opt.* **28**, 4735–4738.

Kiehl, J. T., and Trenberth, K. E. (1997) Earth's annual global mean energy budget, *Bull. Am. Meteorol. Soc.* **78**, 197–208.

Liou, K. N. (1992) *Radiation and Cloud Processes in the Atmosphere*, Oxford Univ. Press, Oxford, UK.

Madronich, S. (1987) Photodissociation in the atmosphere 1. Actinic flux and the effects of ground reflections and clouds, *J. Geophys. Res.* **92**, 9740–9752.

Matsumi, Y., Comes, F. J., Hancock, G., Hofzumahaus, A., Hynes, A. J., Kawasaki, M., and Ravishankara, A. R. (2002) Quantum yields for production of $O(^1D)$ in the ultraviolet photolysis of ozone: Recommendation based on evaluation of laboratory data, *J. Geophys. Res.* **107** (D3), 4024 (doi: 10.1029/2001JD000510).

Mecherikunnel, A. T., Lee, R. B., Kyle, H. L., and Major, E. R. (1988) Intercomparison of solar total irradiance data from recent spacecraft measurements, *J. Geophys. Res.* **93**, 9503–9509.

Michelsen, H. A., Salawitch, R. J., Wennberg, P. O., and Anderson, J. G. (1994) Production of $O(^1D)$ from photolysis of O_3, *Geophys. Res. Lett.* **21**, 2227–2230.

Ramanathan, V. (1987) The role of earth radiation budget studies in climate and general circulation research, *J. Geophys. Res.* **92**, 4075–4095.

Ramanathan, V., Cess, R. D., Harrison, E. F., Minnis, P., Barkstrom, B. R., Ahmad, E., and Hartmann, D. (1989) Cloud-radiative forcing and climate: Results from the earth radiation budget experiment, *Science* **243**, 57–63.

5 Chemistry of the Stratosphere

Ozone is the most important trace constituent of the stratosphere. Discovered in the nineteenth century, its importance as an atmospheric gas became apparent in the early part of the twentieth century when the first quantitative measurements of the ozone column were carried out in Europe. The British scientist Dobson developed a spectrophotometer for measuring the ozone column, and the instrument is still in widespread use. In recognition of his contribution, the standard measure of the ozone column is the Dobson unit (DU) (recall the definition in Chapter 2). Sydney Chapman, a British scientist, proposed in 1930 that ozone is continually produced in the atmosphere by a cycle initiated by photolysis of O_2 in the upper stratosphere. This photochemical mechanism for ozone production in the stratosphere bears Chapman's name. After some time it was realized that the simple Chapman mechanism was not capable of predicting observed ozone profiles in the stratosphere; the cycle predicted too much ozone. Additional chemical reactions that consume ozone were proposed, but it was not until 1970 when the true breakthrough in understanding stratospheric chemistry emerged when Paul Crutzen elucidated the role of nitrogen oxides in stratospheric ozone chemistry, closely followed by investigation of the possible depletion of stratospheric ozone as a result of the catalytic effect of NO_x emitted from a proposed fleet of supersonic aircraft by Harold Johnston.

Shortly thereafter, the effect on stratospheric ozone of chlorine released from human-made (anthropogenic) chlorofluorocarbons was predicted by Mario Molina and F. Sherwood Rowland. For their pioneering studies of atmospheric ozone chemistry, Crutzen, Molina, and Rowland were awarded the 1995 Nobel Prize in Chemistry. It was not until 1985, with the discovery of the Antarctic ozone hole by a team led by the British scientist Joseph Farman, that definitive evidence of the depletion of the stratospheric ozone layer emerged.

Massive annual ozone decreases during the Antarctic spring (September to November) have now been extensively documented since 1985 (Jones and Shanklin 1995). By the early 1990s, total column ozone decreases outside the Antarctic were established. Both ground-based and satellite data have confirmed a downward trend in the total column amount of ozone over midlatitude areas of the Northern Hemisphere in all seasons.

5.1 OVERVIEW OF STRATOSPHERIC CHEMISTRY

About 90% of the atmosphere's ozone is found in the stratosphere, residing in what is commonly referred to as the "ozone layer." At the peak of the ozone layer the O_3 mixing ratio is about 12 ppm. The total amount of O_3 in the Earth's atmosphere is not great; if all the O_3 molecules in the troposphere and stratosphere were brought down to the Earth's surface and distributed uniformly as a layer over the globe, the resulting layer of pure gaseous O_3 would be less than 5 mm thick.

Stratospheric ozone is produced naturally as a result of the photolytic decomposition of O_2. The two oxygen atoms that result each react with another O_2 molecule to produce two molecules of ozone. The overall process, therefore, converts three O_2 molecules to two O_3 molecules. The O_3 molecules produced themselves react with other stratospheric molecules, both natural and anthropogenic; the balance achieved between O_3 produced and destroyed leads to a steady-state abundance of O_3. The concentration of stratospheric O_3 varies with altitude and latitude, depending on sunlight intensity, temperature, and stratospheric air movement.

The greatest O_3 production occurs in the tropical stratosphere because this is the region having the most intense sunlight. The highest amounts of ozone, however, occur at middle and high latitudes. This results because of winds that circulate air in the stratosphere, bringing tropical air rich in ozone toward the poles in fall and winter. Despite the fact that O_3 production is highest in the tropics, total ozone in the tropical stratosphere remains small in all seasons, in part because the thickness of the ozone layer is smallest in the tropics.

Stratospheric O_3 absorbs ultraviolet UV-B solar radiation, in the wavelength region 280–315 nm (see Chapter 4). UV-B radiation is harmful to humans, leading to skin cancer and a suppressed immune system. Excessive exposure to UV-B radiation also can damage plant life and aquatic ecosystems. If stratospheric O_3 is decreased, the amount of UV-B radiation that reaches the Earth's surface increases.

In the 1970s it was discovered that industrial gases containing chlorine and bromine emitted at the Earth's surface, and impervious to chemical destruction or removal by precipitation in the troposphere, eventually reach the stratosphere, where the chlorine and bromine are liberated by photolysis of the molecules. The Cl and Br atoms then attack O_3 molecules, initiating catalytic cycles that remove ozone. These industrial gases include the chlorofluorocarbons (CFCs) and hydrochlorofluorocarbons (HCFCs), previously used in almost all refrigeration and air-conditioning systems, and the "halons," which are used as fire retardants (see Chapter 2). Figure 5.1 shows the distribution of chlorine- and bromine-containing gases in the stratosphere as of 1999. Methyl chloride (CH_3Cl) is virtually the only natural source of chlorine in the stratosphere, accounting for 16% of the chlorine. Natural sources, on the other hand, account for nearly half of the stratospheric bromine. Even though the amount of chlorine is about 170 times that of bromine, on an atom-for-atom basis, Br is about 50 times more effective than Cl in destroying ozone. The fluorine atoms in the molecules end up in chemical forms that do not participate in O_3 destruction.

Because most of these halogen-containing gases have no significant removal processes in the troposphere, their atmospheric lifetimes can be quite long, for example:

$CFCl_3$	(CFC-11)	45 years
CF_2Cl_2	(CFC-12)	100 years
CCl_4	(Carbon tetrachloride)	35 years
CH_3Cl	(Methyl chloride)	1.5 years
$CBrClF_2$	(Halon-1211)	11 years
CF_3Br	(Halon-1301)	65 years
CH_3Br	(Methyl bromide)	0.8 years

The halogen gases enter the stratosphere primarily across the tropical tropopause with air pumped into the stratosphere by deep convection. Air motions in the stratosphere then

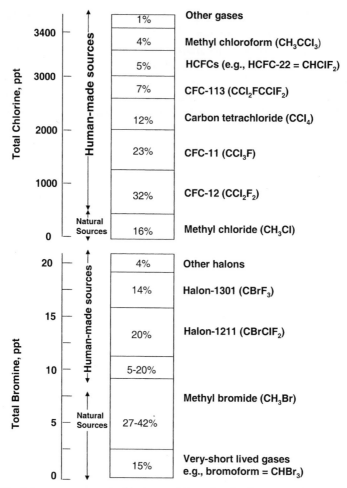

FIGURE 5.1 Primary sources of chlorine and bromine in the stratosphere in 1999 [World Meteorological Organization (WMO) 2002].

transport them upward and toward the poles in both hemispheres. Those gases with long lifetimes are present in comparable amounts throughout the stratosphere in both hemispheres. The atmospheric burden of each gas depends on its lifetime as well as the amount emitted. Ozone-depleting substances are compared in their effectiveness to destroy stratospheric O_3 by an "ozone depletion potential" (ODP). The ODP depends on the amount of each gas present and its chemical efficiency in O_3 depletion.

Much of this chapter will be devoted to carefully working through the chemistry of stratospheric O_3 depletion. Depletion occurs as a result of chemical cycles in which the reactive entity is regenerated over and over. One such cycle that results after the release of a Cl atom from photolysis of one of the molecules in Figure 5.1 is

$$Cl + O_3 \longrightarrow ClO + O_2$$
$$\frac{ClO + O \longrightarrow Cl + O_2}{\text{Net:} \quad O + O_3 \longrightarrow O_2 + O_2}$$

The cycle can be considered to begin with either Cl or ClO. (Recall that atomic oxygen is present from the photolysis of O_2 and O_3.) Because Cl or ClO is reformed each time an O_3 molecule is destroyed, chlorine is actually a catalyst for O_3 destruction. One Cl atom participates in many cycles before it is eventually tied up in a more stable molecule, called a *reservoir species.*

In the early 1980s, based on remote sensing measurements of the total O_3 column over Antarctica, it was discovered that the Antarctic ozone layer was becoming thinner year by year. This depletion of stratospheric ozone became known as the "ozone hole." The stratosphere is generally cloudless because of its small amount of water vapor, but at the extremely cold temperatures of the winter Antarctic stratosphere, ice clouds called *polar stratospheric clouds* (PSCs) form, and these cloud particles serve as sites for ozone depletion chemistry. In short, the reactions that occur on the surfaces of PSCs convert the reservoir molecules, which sequester reactive chlorine, back into ClO, so that virtually all of the reactive chlorine exists in the form of ClO. The spatial extent of the Antarctic ozone hole, at its peak, is almost twice the area of the Antarctic continent. About 3% of the total mass of ozone in the entire global atmosphere is destroyed each winter season. Although temperatures in the winter Arctic stratosphere do not get as low as those over the Antarctic, they do reach levels at which PSCs can form, and ozone depletion occurs over the Arctic as well. Air in the polar stratospheric regions remains relatively isolated from that in the rest of the stratosphere for a long period in winter because strong winds that encircle the poles prevent air movement into or out of the polar vortex. When temperatures warm in spring, PSCs no longer form, and the production of ClO ceases. Also, the polar vortex weakens and air is exchanged with the rest of the stratosphere, bringing ozone-rich air into the region.

The Montreal Protocol on Substances that Deplete the Ozone Layer was signed in 1987 and has since been ratified by over 180 nations. The Protocol establishes legally binding controls on the national production and consumption of ozone-depleting gases. Amendments and Adjustments to the Protocol added new controlled substances, accelerated existing control measures, and scheduled phaseouts of certain gases. The Montreal Protocol provides for the transitional use of HCFCs as substitutes for the major ozone-depleting gases, such as CFC-11 and CFC-12. As a result of the Montreal Protocol, the total amount of ozone-depleting gases in the atmosphere has begun to decrease.

Some Comments on Notation and Quantities of General Use   In this, and succeeding chapters the molecular number concentration of a species is denoted with square brackets ($[O_2]$, $[O_3]$, etc.) in units of molecules cm^{-3}. The rate coefficient of a reaction will carry, as a subscript, either the number of the reaction, for example

$$O + O_3 \overset{4}{\longrightarrow} O_2 + O_2 \qquad k_4$$

or, if the reaction is not explicitly numbered, the reaction itself, for instance

$$O + O_3 \longrightarrow O_2 + O_2 \qquad k_{O+O_3}$$

We will have frequent occasion to require stratospheric temperature and number concentration as a function of altitude. Approximate global values based on the U.S. Standard Atmosphere are given in Table 5.1. The number concentration of air is

TABLE 5.1   Stratospheric Temperature, Pressure, and Atmospheric Number Density

z, km	T, K	p, hPa	p/p_0	[M], molecules cm^{-3}
20	217	55	0.054	1.4×10^{18}
25	222	25	0.025	6.4×10^{17}
30	227	12	0.012	3.1×10^{17}
35	237	5.6	0.0055	1.4×10^{17}
40	251	2.8	0.0028	7.1×10^{16}
45	265	1.4	0.0014	3.6×10^{16}

^{a}[M$_0$] $= 2.55 \times 10^{19}$ molecules cm^{-3} (288 K).

denoted as [M(z)]. A number of atmospheric reactions involve the participation of a molecule of N_2 or O_2 as a third body (recall Chapter 3), and the third body in a chemical reaction is commonly designated as M.

The global mean surface temperature is usually taken as 288 K, at which [M] $= 2.55 \times 10^{19}$ molecules cm^{-3}. (If one chooses 298 K, which is the standard temperature at which chemical reaction rate coefficients are usually reported, then [M] $= 2.46 \times 10^{19}$ molecules cm^{-3}.) In example calculations of chemical reaction rates, in which a value of [M] is needed at the Earth's surface, we will often simply use 2.5×10^{19} molecules cm^{-3} as the approximate value at 298 K.

5.2   CHAPMAN MECHANISM

Ozone formation occurs in the stratosphere above ~ 30 km altitude, where solar UV radiation of wavelengths less than 242 nm slowly dissociates molecular oxygen:

$$O_2 + h\nu \xrightarrow{1} O + O \qquad \text{(reaction 1)}$$

The oxygen atoms react with O_2 in the presence of a third molecule M (N_2 or O_2) to produce O_3:

$$O + O_2 + M \xrightarrow{2} O_3 + M \qquad \text{(reaction 2)}$$

Reaction 2 is, for all practical purposes, the only reaction that produces ozone in the atmosphere. The O_3 molecule formed in that reaction itself strongly absorbs radiation in the wavelength range of 240–320 nm (recall Chapter 4) to decompose back to O_2 and O

$$O_3 + h\nu \xrightarrow{3} O + O_2 \qquad \text{Chappuis bands (400–600 nm)} \qquad \text{(reaction 3)}$$

$$\xrightarrow{3'} O(^1D) + O_2 \qquad \text{Hartley bands ($<$320 nm)} \qquad \text{(reaction 3$'$)}$$

where reaction 3$'$ leads to excited states of both O and O_2 [O_2 ($^1\Delta$)]. As we recall, O(^{1}D) is quenched to ground-state atomic oxygen by collision with N_2 or O_2:

$$O(^1D) + M \longrightarrow O + M$$

The rate coefficient for this reaction is (Table B.2)

$$k_{O(^1D)+M} = 3.2 \times 10^{-11} \exp(70/T) \qquad M = O_2$$
$$= 1.8 \times 10^{-11} \exp(110/T) \qquad M = N_2$$

The mean lifetime of $O(^1D)$ against reaction with M is

$$\tau_{O(^1D)} = \frac{1}{k_{O(^1D)+M}[M]}$$

Choosing 30 km ($T = 227\,K$), and noting that M consists of $0.21\,O_2 + 0.79\,N_2$, $[M] = 3.1 \times 10^{17}$ molecules cm^{-3} (Table 5.1), we obtain

$$\tau_{O(^1D)} \cong 10^{-7}\,s$$

Consequently, $O(^1D)$ is effectively converted instantaeously to ground-state O, and the photodissociation of O_3 by both reactions 3 and 3′ (above) can be considered to produce entirely ground-state O atoms.

Finally, O and O_3 react to reform two O_2 molecules:

$$O + O_3 \xrightarrow{\;4\;} O_2 + O_2 \qquad\qquad \text{(reaction 4)}$$

The mechanism (reactions 1–4 above) for production of ozone in the stratosphere was proposed by Chapman (1930) and bears his name.

The rate equations for [O] and [O_3] in the Chapman mechanism are

$$\frac{d[O]}{dt} = 2j_{O_2}[O_2] - k_2[O][O_2][M] + j_{O_3}[O_3] - k_4[O][O_3] \qquad (5.1)$$

$$\frac{d[O_3]}{dt} = k_2[O][O_2][M] - j_{O_3}[O_3] - k_4[O][O_3] \qquad (5.2)$$

Once an oxygen atom is generated in reaction 1 (above), reactions 2 and 3 proceed rapidly. The characteristic time for reaction of the O atom in reaction 2 is

$$\tau_2 = \frac{1}{k_2[O_2][M]} \qquad (5.3)$$

where k_2 is assumed to be expressed as a termolecular rate coefficient (cm^6 molecule^{-2} s^{-1}). The rate coefficient is (Table B.2)

$$k_2 = 6.0 \times 10^{-34} \quad (T/300)^{-2.4} \qquad \text{cm}^6 \text{ molecule}^{-2}\,\text{s}^{-1}$$

Let us evaluate τ_2 at $z = 30$ and 40 km. Since $[O_2] = 0.21\,[M]$ at any altitude, we obtain

$$\tau_2 = \frac{1}{0.21\,k_2[M]^2} \qquad (5.4)$$

We find

z, km	T, K	[M], molecules cm^{-3}	k_2 (cm^6 molecule^{-2} s^{-1})	τ_2 (s)
30	227	3.1×10^{17}	1.15×10^{-33}	0.04
40	251	7.1×10^{16}	9.1×10^{-34}	1.04

The overall photolysis rate of ozone, j_{O_3}, is the sum of the rates of the reactions 3 and 3$'$ above, and the overall lifetime of O_3 against photolysis is

$$\tau_3 = \frac{1}{j_{O_3 \to O} + j_{O_3 \to O(^1D)}} \tag{5.5}$$

Evaluating τ_3 at 30 km and 40 km at a solar zenith angle of $0°$ (see Figure 4.15) yields

z, km	$j_{O_3 \to O}$, s^{-1}	$j_{O_3 \to O(^1D)}$, s^{-1}	τ_3, s
30	$\sim 7 \times 10^{-4}$	$\sim 5 \times 10^{-4}$	~ 800
40	$\sim 9 \times 10^{-4}$	$\sim 1 \times 10^{-3}$	~ 500

Thus, the photolytic lifetime of an O_3 molecule is on the order of 10 min at these altitudes. However, this lifetime is *not* the overall lifetime of O_3. Once O_3 photodissociates, the O atom formed can rapidly reform O_3 in reaction 2; thus, O_3 photolysis alone does not lead to a loss of O_3. Only reaction 4 truly removes O_3 from the system.

The lifetime of O_3 against reaction 4 is

$$\tau_4 = \frac{1}{k_4 [O]} \tag{5.6}$$

where $k_4 = 8.0 \times 10^{-12} \exp(-2060/T)$ cm^3 molecule^{-1} s^{-1} (Table B.1). To estimate τ_4, we need [O]. Let us assume, for the moment, that once an O atom is produced in reaction 1, reactions 2 and 3 cycle many times before reaction 4 has a chance to take place. If, indeed, there is rapid interconversion between O and O_3, it is useful to view the sum of O and O_3 as a chemical family (see Section 3.6). The sum of O and O_3 is given the designation *odd oxygen* and is denoted O_x. The odd-oxygen family can be depicted as shown in Figure 5.2.

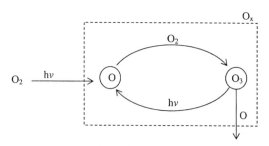

FIGURE 5.2   Odd-oxygen family.

The rapid interconversion between O and O_3 leads to a pseudo-steady-state relationship between the O and O_3 concentrations:

$$\frac{[O]}{[O_3]} = \frac{j_{O_3}}{k_2[O_2][M]} = \frac{j_{O_3}}{0.21\,k_2[M]^2} \qquad (5.7)$$

From the values of k_2, [M], and j_{O_3}, we can estimate the $[O]/[O_3]$ ratio. For example, at 30 and 40 km:

z, km	$[O]/[O_3]$
30	3.0×10^{-5}
40	9.4×10^{-4}

As altitude increases, [M] decreases, so the $[O]/[O_3]$ ratio becomes larger at higher altitudes. Regardless, O_3 is the dominant form of odd oxygen below ~ 50 km.

The lifetime of odd oxygen O_x can be found from the ratio of its abundance to its rate of disappearance:

$$\tau_{O_x} = \frac{[O] + [O_3]}{k_4[O][O_3]}$$

But since $[O_3] \gg [O]$, it follows that

$$\tau_{O_x} \cong \frac{[O_3]}{k_4[O][O_3]} = \frac{1}{k_4[O]}$$
$$= \frac{0.21\,k_2[M]^2}{k_4 j_{O_3}[O_3]} \qquad (5.8)$$

which is just the lifetime of O_3 against reaction 4 (above).

Now, let us obtain an estimate of the lifetime of O_3 against reaction 4, τ_4. Shortly we will see that O_3 concentrations at 30 and 40 km are about 3×10^{12} and 0.5×10^{12} molecules cm^{-3}, respectively. Given these concentrations and the $[O]/[O_3]$ estimates above, we find that

z, km	T, K	k_4, cm^3 molecule^{-1} s^{-1}	$\tau_{O_x} (= \tau_4)$, s
30	227	9.2×10^{-16}	1.2×10^7 (~ 140 days)
40	251	2.2×10^{-15}	10^6 (~ 12 days)

Whereas the photolytic lifetime of O_3 at these altitudes is about 10 min, the overall lifetime of O_3 is on the order of weeks to months, validating the assumption that cycling within the O_x family is rapid relative to loss of O_x. τ_{O_x} varies from about 140 days at 20 km to about 12 days at 40 km. (Even though both j_{O_3} and $[O_3]$ vary with altitude, it is the $[M]^2$ dependence of τ_{O_x} that dominates its behavior as a function of altitude.) At lower altitudes,

O_3 lifetime is sufficiently long for it to be transported intact. At high altitudes, ozone tends to be produced locally rather than imported.

Defining the chemical family $[O_x] = [O] + [O_3]$, and by adding (5.1) and (5.2), we obtain the rate equation governing odd oxygen:

$$\frac{d[O_x]}{dt} = 2j_{O_2}[O_2] - 2k_4[O][O_3] \qquad (5.9)$$

Since O_3 constitutes the vast majority of total O_x, a change in the abundance of O_x is usually just referred to as a change in O_3. Two molecules of odd oxygen are formed on photolysis of O_2, and two molecules of odd oxygen are consumed in reaction 4.

Using (5.7) in (5.9), the reaction rate equation for odd oxygen becomes

$$\frac{d[O_x]}{dt} = 2j_{O_2}[O_2] - 2k_4[O_3]\left(\frac{j_{O_3}[O_3]}{k_2[O_2][M]}\right) \qquad (5.10)$$

Since $[O_x] = [O] + [O_3]$, and $[O] \ll [O_3], d[O_x]/dt \cong d[O_3]/dt$, so (5.10) becomes

$$\frac{d[O_3]}{dt} = 2j_{O_2}[O_2] - \frac{2k_4 j_{O_3}[O_3]^2}{k_2[O_2][M]} \qquad (5.11)$$

After a sufficiently long time, the steady-state O_3 concentration resulting from reactions 1–4 is

$$[O_3]_{ss} = [O_2]\left(\frac{k_2[M]j_{O_2}}{k_4 j_{O_3}}\right)^{1/2} \qquad (5.12)$$

which can be rearranged as

$$[O_3]_{ss} = 0.21\left(\frac{k_2 j_{O_2}}{k_4 j_{O_3}}\right)^{1/2}[M]^{3/2} \qquad (5.13)$$

An equivalent way to express the steady-state O_3 concentration is

$$[O_3]_{ss} = \left(\frac{0.21 k_2}{k_4 j_{O_3}}\right)^{1/2}(j_{O_2}[O_2])^{1/2}[M] \qquad (5.14)$$

The steady-state O_3 concentration is sustained by the continual photolysis of O_2.

The steady-state O_3 concentration should be a maximum at the altitude where the product of the number density of air and the square root of the O_2 photolysis rate is largest. Figure 4.14 shows j_{O_2} as a function of altitude above 30 km. The photolysis rate of O_2, j_{O_2} $[O_2]$ (molecules $cm^{-3}\,s^{-1}$), peaks at about 29 km. This maximum results because even though j_{O_2} continues to increase with z, the number density of O_2, $[O_2] = 0.21\,[M]$, decreases with z. The product $(j_{O_2}[O_2])^{1/2}$ [M] peaks at about 25 km. This is a key result—*the peak in the stratospheric ozone layer occurs at ~ 25 km*. In summary, at high altitudes the concentration of O_3 decreases primarily as a result of a drop in the concentration of O_2, the photolysis of which initiates the formation of O_3. At low altitudes, the O_3 concentration decreases because of a decrease in the flux of photons at the UV wavelengths at which O_2 photodissociates.

An issue that we have not yet addressed is how long it takes to establish the steady-state O_3 concentration at any particular altitude. To determine this we need to return to the rate equation for $[O_3]$, namely, (5.11). If we let $y = [O_3]$, this differential equation is of the form

$$\frac{dy}{dt} = b - ay^2 \tag{5.15}$$

If $y(0) = 0$, the general solution of (5.15) is

$$y(t) = \left(\frac{b}{a}\right)^{1/2} \frac{1 - \exp[-2(ab)^{1/2}t]}{1 + \exp[-2(ab)^{1/2}t]} \tag{5.16}$$

The characteristic approach time of this solution to its steady state is

$$\tau = \frac{1}{2(ab)^{1/2}}$$

Applying this to (5.11), we find the characteristic time needed to establish the O_3 steady state is given by

$$\tau_{O_3}^{ss} = \frac{1}{4}\left(\frac{k_2[M]}{k_4\, j_{O_2}\, j_{O_3}}\right)^{1/2} \tag{5.17}$$

Although j_{O_2} and j_{O_3} decrease as altitude decreases, the exponential increase of [M] with decreasing altitude exerts the dominate influence on $\tau_{O_3}^{ss}$. Thus, we expect this time scale to increase at lower altitudes.

Let us estimate $\tau_{O_3}^{ss}$ as a function of altitude, at a solar zenith angle of $0°$:

z, km	T, K	k_4, cm^3 molecule^{-1} s^{-1}	j_{O_2}, s^{-1}	j_{O_3}, s^{-1}	$\tau_{O_3}^{ss}$, h
20	217	6×10^{-16}	1×10^{-11}	0.7×10^{-3}	~ 1400
25	222	7.5×10^{-16}	2×10^{-11}	0.7×10^{-3}	~ 600
30	227	9.2×10^{-16}	6×10^{-11}	1.2×10^{-3}	~ 160
35	237	1.3×10^{-15}	2×10^{-10}	1.6×10^{-3}	~ 40
40	251	2.2×10^{-15}	5×10^{-10}	1.9×10^{-3}	~ 12
45	265	3.4×10^{-15}	8×10^{-10}	6×10^{-3}	~ 3

Again, we caution that $\tau_{O_3}^{ss}$ is *not* the overall lifetime of an O_3 molecule; rather, this quantity is the characteristic time required for the Chapman mechanism to achieve a steady-state balance after some perturbation. When $\tau_{O_3}^{ss}$ is short relative to other

TABLE 5.2   Chemical Families in Stratospheric Chemistry

Symbol	Name	Components
O_x	Odd oxygen	$O + O_3$
NO_x	Nitrogen oxides	$NO + NO_2$
NO_y	Oxidized nitrogen	$NO + NO_2 + HNO_3 + 2N_2O_5 + ClONO_2 + NO_3 +$ $HOONO_2 + BrONO_2$
HO_x	Hydrogen radicals	$OH + HO_2$
Cl_y	Inorganic chlorine	Sum of all chlorine-containing species that lack a carbon atom ($Cl + 2Cl_2 + ClO + OClO +$ $2Cl_2O_2 + HOCl + ClONO_2 + HCl + BrCl$)
ClO_x	Reactive chlorine	$Cl + ClO$
CCl_y	Organic chlorine	$CF_2Cl_2 + CFCl_3 + CCl_4 + CH_3CCl_3 + CFCl_2CF_2Cl$ $(CFC - 113) + CF_2HCl(CFC - 22)$
Br_y	Inorganic bromine	Sum of all bromine-containing species that lack a carbon atom ($Br + BrO + HOBr + BrONO_2$)

stratospheric processes, it can be assumed that the local O_3 concentration obeys the steady state, (5.12)–(5.14). When $\tau_{O_3}^{ss}$ is relatively long, the Chapman mechanism does not actually attain a steady state. We see that above ~ 30 km, [O_3] can be assumed to be in a local steady state. At these altitudes the O_3 concentration over a year is governed by the seasonal cycle of production and loss terms.

Finally, the O_x chemical family is the first of a number of chemical families that are important in stratospheric chemistry. Table 5.2 summarizes chemical families important in stratospheric chemistry. As we proceed through this chapter, each of these families will emerge.

Figure 5.3 shows the ozone distribution in the stratosphere. Note that the *regions of highest ozone concentration do not coincide with the location of the highest rate of*

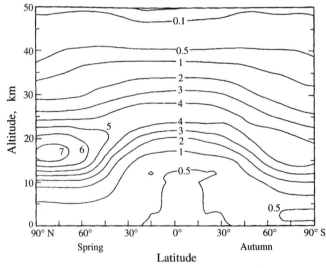

FIGURE 5.3   Zonally averaged ozone concentration (in units of 10^{12} molecules cm^{-3}) as a function of altitude, March 22 (Johnston 1975). Ozone concentration at the equator peaks at an altitude of about 25 km. Over the poles the location of the maximum is between 15 and 20 km. At altitudes above the ozone bulge, O_3 formation is oxygen-limited; below the bulge, O_3 formation is photon-limited.

formation of O_3. Rates of O_3 production are highest at the equator and at about 40 km, altitude, whereas ozone concentrations peak at northern latitudes. Even at the equator, maximum ozone concentrations occur at about 25 km as opposed to 40 km, where the production rate is greatest. At the poles, maximum O_3 production occurs at altitudes even higher than those over the equator, whereas the maximum O_3 concentrations are at lower altitudes. There is even a north-south asymmetry. Figure 5.4 shows the historical total column ozone as a function of latitude and time of year, measured in Dobson units, prior to anthropogenic ozone depletion. Highest ozone column abundances are found at high latitudes in winter, and the lowest values are in the tropics.

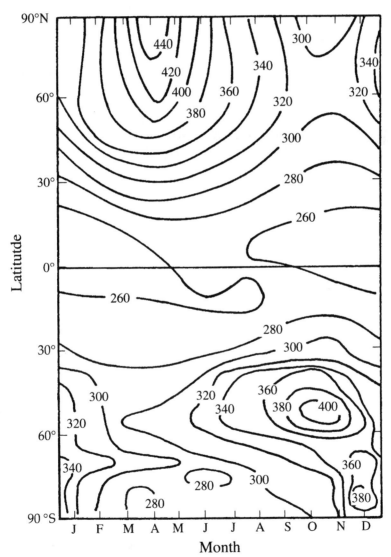

FIGURE 5.4 Zonally averaged total ozone column density (in Dobson units) as a function of latitude and time of year (Dütsch 1974). This distribution is representative of that prior to anthropogenic perturbation of stratospheric ozone.

The steady-state odd oxygen model of (5.10)–(5.12) predicts that the O_3 concentration should be proportional to the square root of the O_2 photolysis rate. We see that, in fact, ozone concentration and O_2 photolysis rate do not peak together. The explanation for the lack of alignment of these two lies in the role of horizontal and vertical transport in redistributing stratospheric airmasses. The ozone bulge in the northern polar regions is a result, for example, of northward and downward air movement in the NH (Northern Hemisphere) stratosphere that transports ozone from high-altitude equatorial regions where ozone production is the largest.

That stratospheric O_3 concentrations are maximum in areas far removed from those where O_3 is being produced suggests that the lifetime of O_3 in the stratosphere is longer than the time needed for the transport to occur. The stratospheric transport timescale from the equator to the poles is of order 3–4 months.

Until about 1964, the Chapman mechanism was thought to be the principal set of reactions governing ozone formation and destruction in the stratosphere. First, improved measurement of the rate constant of reaction 4 (above) indicated that the reaction is considerably slower than previously thought, leading to larger abundances of O_3 as predicted by (5.10)–(5.12). Then, measurements indicated that the actual amount of ozone in the stratosphere is a factor of ~ 2 less than what is predicted by the Chapman mechanism with the more accurate rate constant of reaction 4 (Figure 5.5). It was concluded that significant additional ozone destruction pathways must be present beyond reaction 4.

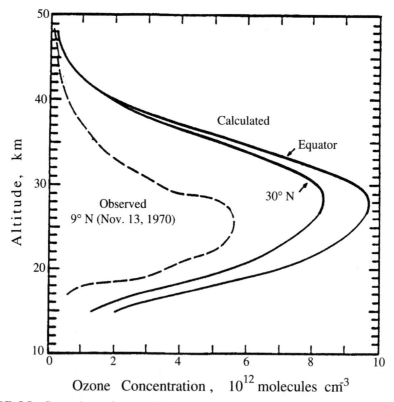

FIGURE 5.5 Comparison of stratospheric ozone concentrations as a function of altitude as predicted by the Chapman mechanism and as observed over Panama (9°N) on November 13, 1970.

For an atmospheric species to contribute to ozone destruction, it would either have to be in great excess or, if a trace species, regenerated in a catalytic cycle. Following the initial studies of the effect of NO_x emissions from supersonic aircraft and chlorofluorocarbons, an intricate and highly interwoven chemistry of catalytic cycles involving nitrogen oxides, hydrogen radicals, chlorine, and bromine emerged in one stunning discovery after another. The result is a tapestry of different catalytic cycles dominating at different altitudes in the stratosphere. The remainder of this chapter develops each of these systems, culminating in a sythesis of the stratospheric ozone depletion cycles not accounted for in the Chapman chemistry.

5.3 NITROGEN OXIDE CYCLES

For a stratospheric species to effectively destroy ozone, it either has to be in great excess or regenerated in a catalytic cycle. As early as 1950, Bates and Nicolet (1950) introduced the idea of a catalytic loss process involving hydrogen radicals, but the true breakthrough in understanding stratospheric ozone chemistry did not occur until the early 1970s when Crutzen (1970) and Johnston (1971) revealed the role of nitrogen oxides in stratospheric chemistry. Reactive nitrogen is produced in the stratosphere from N_2O. N_2O is produced by microbial processes in soils and the ocean and does not undergo reactions in the troposphere (Chapter 2).

5.3.1 Stratospheric Source of NO_x from N_2O

The principal natural source of NO_x ($NO + NO_2$) in the stratosphere is N_2O. Approximately 90% of N_2O in the stratosphere is destroyed by photolysis:

$$N_2O + hv \xrightarrow{\ 1\ } N_2 + O(^1D) \qquad \text{(reaction 1)}$$

The remainder reacts with $O(^1D)$:

$$N_2O + O(^1D) \xrightarrow{\ 2a\ } NO + NO \quad k_{2a} = 6.7 \times 10^{-11} \text{ cm}^3 \text{ molecule}^{-1} \text{ s}^{-1} \quad \text{(reaction 2a)}$$

$$\xrightarrow{\ 2b\ } N_2 + O_2 \quad k_{2b} = 4.9 \times 10^{-11} \text{ cm}^3 \text{ molecule}^{-1} \text{ s}^{-1} \quad \text{(reaction 2b)}$$

Reaction 2a is the main source of NO_x in the stratosphere. About 58% of the $N_2O + O(^1D)$ reaction proceeds via channel 2a, the remaining 42% by channel 2b.

The main influx of tropospheric species into the stratosphere occurs in the tropics, and N_2O enters via this route. Upon crossing the tropopause, N_2O advects slowly upward. During its ascent, N_2O is diluted by N_2O-poor air that mixes in from outside the tropical stratosphere and, at the same time, disappears by reactions 1 and 2. The higher a molecule of N_2O rises in the stratosphere, the more energetic the photons that are encountered, and the more rapid its photodissociation by reaction 1.

Even though the photodissociation of N_2O produces $O(^1D)$, the main source of $O(^1D)$ in the stratosphere is photodissociation of O_3 (reaction $3'$ in Section 5.2), and the concentration of $O(^1D)$ at any altitude is determined by its source from O_3 photolysis and its sink from quenching by $O(^1D) + M$. The concentration of $O(^1D)$ increases with

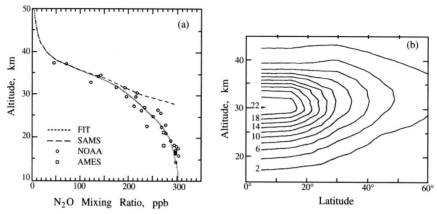

FIGURE 5.6   (a) Vertical profiles of N_2O over the tropics at equinox circa 1980. Circles denote balloonborne measurements at 9°N and 5°S; squares represent aircraft measurements between 1.6°S and 9.9°N. Dashed curve refers to the average of satellite measurements at 5°N, equinox, between 1979 and 1981. This compilation of data was presented by Minschwaner et al. (1993), where the original sources of data can be found. The dotted curve indicates the vertical profile used by Minschwaner et al. to estimate the lifetime of N_2O. (b) Calculated diurnally averaged loss rate for N_2O (in units of 10^{12} molecules cm^{-3} s^{-1}) as a function of altitude and latitude, at equinox. The loss rate includes both photolysis and reaction with $O(^1D)$ (Minschwaner et al. 1993).

altitude, largely because [M] is decreasing. Because the O_3 profile eventually decreases with altitude, the concentration of $O(^1D)$ reaches a maximum at a certain altitude. Also, since the N_2O concentration is continually decreasing with altitude, its rate of destruction, j_{N_2O} [N_2O], reaches a maximum at a certain altitude even though the light intensity is stronger at higher altitudes. Figure 5.6a shows vertical profiles of the N_2O mixing ratio in the tropics. The circles denote balloonborne measurements at 9°N and 5°S; the squares represent aircraft measurements between 1.6°S and 9.9°N. The dashed curve is the average of satellite measurements at 5°N, equinox, between 1979 and 1981. Figure 5.6b shows the diurnally averaged 1980 loss rate for N_2O (molecules cm^{-3} s^{-1}) as a function of altitude and latitude for equinox calculated with a photochemical model (Minschwaner et al. 1993). The loss rate includes N_2O photolysis and reaction with $O(^1D)$. As seen from Figure 5.6, the global loss of N_2O occurs mainly at latitudes between the equator and 30°. Also, N_2O loss rates are largest in the 25–35 km altitude range.

The fractional yield of NO from N_2O at any altitude is the ratio of the number of molecules of NO formed to the total molecules of N_2O reacted:

$$\text{NO yield} = \frac{2k_{2a}[O(^1D)]}{j_{N_2O} + (k_{2a} + k_{2b})[O(^1D)]} \tag{5.18}$$

The stratospheric $O(^1D)$ concentration at any altitude is controlled by the source from O_3 photolysis and the sink by quenching to ground-state atomic oxygen

$$O_3 + h\nu \xrightarrow{\;3\;} O(^1D) + O_2 \tag{reaction 3}$$

$$O(^1D) + M \xrightarrow{\;4\;} O + M \tag{reaction 4}$$

so that the $O(^1D)$ steady-state concentration is given by

$$[O(^1D)] = \frac{j_{O_3 \rightarrow O(^1D)}[O_3]}{k_4[M]} \qquad (5.19)$$

For example, at 30 km and $\theta_o = 45°$, $j_{N_2O} \cong 5 \times 10^{-8}\,s^{-1}$ and $j_{O_3 \rightarrow O(^1D)} \cong 15 \times 10^{-5}\,s^{-1}$. With $k_4 = 3.2 \times 10^{-11}\,cm^3\,molecule^{-1}\,s^{-1}$, the instantaneous steady-state concentration of $O(^1D)$ from (5.19) at 30 km and $\theta_o = 45°$ is ~ 45 molecules cm^{-3}. The NO yield under these conditions from (5.18) is 0.11 molecules of NO formed per molecule of N_2O removed. The NO yield at any altitude would require using 24-h averages of the two j values. The fractional yield of NO from N_2O loss varies with altitude and latitude; yet, the ratio of NO_y ($NO + NO_2 +$ all products of NO_x oxidation) to N_2O is nearly linear with altitudes (until NO_y reaches its peak), suggesting a nearly constant yield with altitude and latitude (Figure 5.7). Transport smoothes out the variations of the fractional yield with

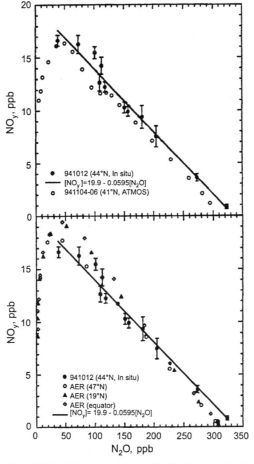

FIGURE 5.7 Stratospheric NO_y mixing ratio versus N_2O mixing ratio observed by balloonborne in situ measurements at 44°N during October and November 1994 (WMO 1998). Solid line is linear fit to the data. Points labeled "AER" are results of a stratospheric two-dimensional (2D) model. [Adapted from Kondo et al. (1996).]

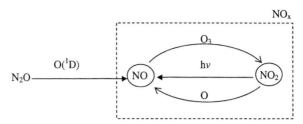

FIGURE 5.8   The NO_x chemical family in relation to stratospheric NO_x cycle 1.

altitudes and latitude. In the uppermost stratosphere, the reaction $N + NO \rightarrow N_2 + O$ converts NO back to N_2; this is the explanation for the lack of adherence of the data to the straight lines in Figure 5.7.

5.3.2 NO_x Cycles

Consider the following cycle involving NO_x that converts odd oxygen ($O_3 + O$) into even oxygen (O_2):

$$NO_x \text{ cycle 1:}\quad NO + O_3 \xrightarrow{1} NO_2 + O_2 \quad k_1 = 3.0 \times 10^{-12} \exp(-1500/T) \text{ (reaction 1)}$$

$$NO_2 + O \xrightarrow{2} NO + O_2 \quad k_2 = 5.6 \times 10^{-12} \exp(180/T) \quad \text{(reaction 2)}$$

$$\text{Net}: \quad \overline{O_3 + O \longrightarrow O_2 + O_2}$$

The characteristic time of reaction 1 with respect to NO removal is $\tau_1 = (k_1[O_3])^{-1}$. At 25 km ($T = 222$ K), $[O_3] \cong 4 \times 10^{12}$ molecules cm^{-3} and $k_1 = 3.6 \times 10^{-15}$ cm^3 molecule^{-1} s^{-1}. Thus, at this altitude, $\tau_1 \cong 70$ s. The characteristic time of reaction 2 relative to reaction of NO_2 is $\tau_2 = (k_2[O])^{-1}$. At 25 km, $k_2 = 12.6 \times 10^{-12}$ cm^3 molecule^{-1} s^{-1}. The O atom concentration is given by (5.7). Using an estimate of $[O] \cong 10^8$ molecules cm^{-3}, we find $\tau_2 \cong 800$ s. Therefore, the rate of the overall cycle of reactions 1 and 2 is controlled by reaction 2.

In competition with reaction 2 is NO_2 photolysis:

$$NO_2 + h\nu \xrightarrow{3} NO + O \tag{reaction 3}$$

which is followed by

$$O + O_2 + M \xrightarrow{4} O_3 + M \tag{reaction 4}$$

If reaction 1 is followed by reaction 3, then no net O_3 destruction takes place because the O atom formed in reaction 3 rapidly combines with O_2 to reform O_3 in reaction 4. If, on the other hand, reaction 2 follows reaction 1, then two molecules of odd oxygen are converted to two O_2 molecules.

These reactions involving the NO_x family can be depicted as in Figure 5.8. The inner cycle interconverting NO and NO_2 is sufficiently rapid that a steady state can be assumed:

$$0 = k_1[NO][O_3] - j_{NO_2}[NO_2] - k_2[NO_2][O] \tag{5.20}$$

The reaction rate equation for odd oxygen is

$$\frac{d[O_x]}{dt} = -k_1[NO][O_3] - k_2[NO_2][O] + j_{NO_2}[NO_2] \qquad (5.21)$$

Adding (5.20) and (5.21), the rate of destruction of odd oxygen by the cycle is

$$\frac{d[O_x]}{dt} = -2k_2[NO_2][O] \qquad (5.22)$$

Let us compare (5.22) to the rate of destruction of odd oxygen by reaction 4 in the Chapman mechanism; so we need to compare the rates of the two reactions:

$$NO_2 + O \longrightarrow NO + O_2 \qquad k_{NO_2+O} = 5.6 \times 10^{-12} \exp(180/T)$$
$$O + O_3 \longrightarrow O_2 + O_2 \qquad k_{O+O_3} = 8.0 \times 10^{-12} \exp(-2060/T)$$

The ratio of the two rates is

$$\frac{k_{NO_2+O}[NO_2]}{k_{O+O_3}[O_3]}$$

Let us evaluate this ratio at 35 km (237 K). From Figure 5.2, $[O_3] \cong 2 \times 10^{12}$ molecules cm^{-3} at this altitude. We will see later that a representative NO_2 concentration is $\sim 10^9$ molecules cm^{-3}. The rate coefficient ratio at this altitude is $k_{NO_2+O}/k_{O+O_3} \cong 9000$. Thus, the ratio of the two rates is

$$\frac{k_{NO_2+O}[NO_2]}{k_{O+O_3}[O_3]} \cong 9000\frac{[NO_2]}{[O_3]} \cong 4.5$$

Even though O_3 is present at roughly 1000-fold greater concentration than NO_2, the large value of the $NO_2 + O$ rate coefficient compensates for this difference to make the $NO_2 + O$ reaction almost 5 times more efficient in odd oxygen destruction than $O + O_3$.

The cycle shown above is most effective in the upper stratosphere, where O atom concentrations are highest. An NO_x cycle that does not require oxygen atoms and therefore is more important in the lower stratosphere is:

$$NO_x \text{ cycle 2}: \qquad NO + O_3 \xrightarrow{1} NO_2 + O_2 \qquad \text{(reaction 1)}$$
$$NO_2 + O_3 \xrightarrow{2} NO_3 + O_2 \qquad \text{(reaction 2)}$$
$$\underline{NO_3 + h\nu \xrightarrow{3} NO + O_2} \qquad \text{(reaction 3)}$$
$$\text{Net}: \quad O_3 + O_3 \longrightarrow O_2 + O_2 + O_2$$

The nitrate radical NO_3 formed in reaction 2, during daytime, rapidly photolyzes (at a rate of about 0.3 s^{-1}). There are two channels for photolysis:

$$NO_3 + h\nu \xrightarrow{a} NO_2 + O \qquad \text{(channel a)}$$
$$\xrightarrow{b} NO + O_2 \qquad \text{(channel b)}$$

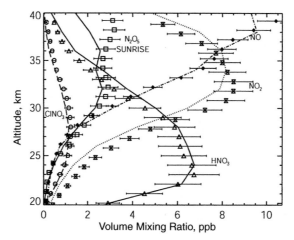

FIGURE 5.9   NO, NO$_2$, HNO$_3$, N$_2$O$_5$, and ClONO$_2$ mixing ratios as a function of altitude (WMO 1998). All species except N$_2$O$_5$ measured at sunset; N$_2$O$_5$ measured at sunrise. Lines are the result of a calculation assuming photochemical steady state over a 24-h period. [Adapted from Sen et al. (1998).]

The photolysis rate by channel a is about 8 times that for channel b. Channel b leads to an overall loss of odd oxygen and is the reaction involved in cycle 2. At night, NO$_3$ formed by reaction 2 does not photodissociate and participates in an important series of reactions that involve stratospheric aerosol particles; we will return to this in Section 5.8.

Figure 5.9 shows mixing ratios of NO, NO$_2$, HNO$_3$, N$_2$O$_5$, and ClONO$_2$ (this compound to be discussed subsequently) measured at 35°N. At 25 km, the predominant NO$_y$ species is HNO$_3$, whereas from 30 to 35 km, NO$_2$ is the major compound. Above 35 km, NO is predominant. We will consider the source of HNO$_3$ shortly.

5.4   HO$_x$ CYCLES

The HO$_x$ family (OH + HO$_2$) plays a key role in stratospheric chemistry. In fact, the first ozone-destroying catalytic cycle identified historically involved hydrogen-containing radicals (Bates and Nicolet 1950).

Production of OH in the stratosphere is initiated by photolysis of O$_3$ to produce O(^{1}D), followed by

$$O(^1D) + H_2O \longrightarrow 2\,OH \qquad k_{O(^1D)+H_2O} = 2.2 \times 10^{-10}\ cm^3\ molecule^{-1}\ s^{-1}$$
$$O(^1D) + CH_4 \longrightarrow OH + CH_3 \qquad k_{O(^1D)+CH_4} = 1.5 \times 10^{-10}\ cm^3\ molecule^{-1}\ s^{-1}$$

Whereas the troposphere contains abundant water vapor, little H$_2$O makes it to the stratosphere; the low temperatures at the tropopause lead to an effective freezing out of water before it can be transported up (a "cold trap" at the tropopause). Mixing ratios of H$_2$O in the stratosphere do not exceed approximately 5–6 ppm. In fact, about half of this water vapor in the stratosphere actually results from the oxidation of methane that has leaked into the stratosphere from the troposphere. Between 20 and 50 km the total rate of

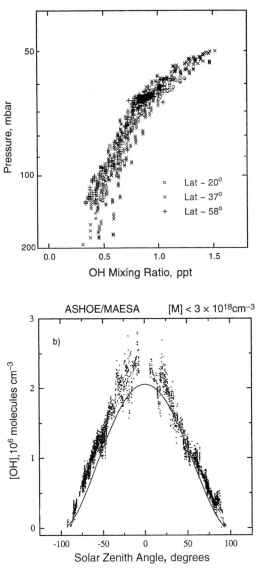

FIGURE 5.10   (a) OH mixing ratio versus altitude (pressure) and (b) OH concentration as a function of solar zenith angle (WMO 1998). Solid curve is prediction by Wennberg et al. (1994).

OH production from the two reactions above is $\sim 2 \times 10^4$ molecules cm^{-3} s^{-1}. About 90% is the result of the O(^{1}D) + H$_2$O reaction; the remaining 10% results from O(^{1}D) + CH$_4$.

Figure 5.10 shows the OH mixing ratio as a function of altitude at three different latitudes and the OH concentration as a function of solar zenith angle. The strong solar zenith angle dependence reflects the photolytic source of OH. Figure 5.11 shows OH, HO$_2$, and total HO$_x$ versus altitude.

OH and HO$_2$ rapidly interconvert so as to establish the HO$_x$ chemical family (Figure 5.12). A reaction that is important in affecting the interconversion between OH

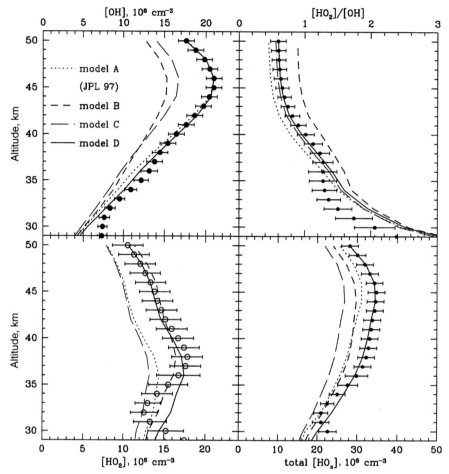

FIGURE 5.11 OH, HO$_2$, and total HO$_x$ versus altitude, as compared with different model predictions (Jucks et al. 1998). Concentration profiles of OH and HO$_2$ measured during midmorning by satellite on April 30, 1997 near Fairbanks, Alaska (65°N). Model predictions: model A—JPL 1997 kinetics; model B—rate of O + HO$_2$ decreased by 50%; model C—rate of O + OH decreased by 20% and rate of OH + HO$_2$ increased by 30%; model D—rates of O + HO$_2$ and OH + HO$_2$ decreased by 25%.

and HO$_2$ in the HO$_x$ cycle is

$$HO_2 + NO \longrightarrow NO_2 + OH \qquad k = 3.5 \times 10^{-12} \exp(250/T)$$

The NO$_2$ formed in this reaction photolyzes

$$NO_2 + h\nu \longrightarrow NO + O$$

followed by

$$O + O_2 + M \longrightarrow O_3 + M$$

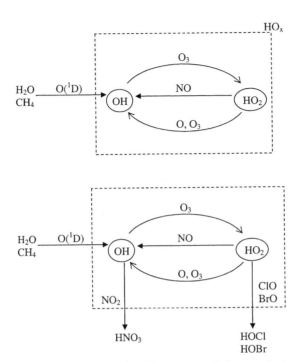

FIGURE 5.12   The stratospheric HO$_x$ family. The upper panel shows only the reactions affecting the inner dynamics of the HO$_x$ system. The lower panel includes additional reactions that affect OH and HO$_2$ levels; these reactions will be discussed subsequently.

As a result, if a molecule of HO$_2$ reacts with NO before it has a chance to react with either O or O$_3$, the result is a do-nothing (null) cycle with respect to removal of O$_3$. Each time a molecule of HO$_2$ follows the bottom path and reacts with O or O$_3$, two molecules of odd oxygen are removed:

$$HO_2 + O \longrightarrow OH + O_2$$
$$HO_2 + O_3 \longrightarrow OH + 2O_2$$

Regeneration of HO$_2$ occurs by

$$OH + O_3 \longrightarrow HO_2 + O_2$$

The resulting catalytic ozone-depletion cycles are

HO$_x$ cycle 1: $OH + O_3 \longrightarrow HO_2 + O_2$
                  $\underline{HO_2 + O \longrightarrow OH + O_2}$
       Net: $O_3 + O \longrightarrow O_2 + O_2$
HO$_x$ cycle 2: $OH + O_3 \longrightarrow HO_2 + O_2$
                  $\underline{HO_2 + O_3 \longrightarrow OH + O_2 + O_2}$
       Net: $O_3 + O_3 \longrightarrow O_2 + O_2 + O_2$

The steady-state relation in the HO_x family is

$$k_{OH+O_3}[OH][O_3] = k_{HO_2+NO}[HO_2][NO] + k_{HO_2+O}[HO_2][O] + k_{HO_2+O_3}[HO_2][O_3] \quad (5.23)$$

The balance on odd oxygen, $O + O_3$, is

$$\frac{d[O_x]}{dt} = \frac{d[O_3]}{dt} + \frac{d[O]}{dt}$$
$$= -k_{OH+O_3}[OH][O_3] - k_{HO_2+O_3}[HO_2][O_3] + j_{NO_2}[NO_2] - k_{HO_2+O}[HO_2][O] \quad (5.24)$$

Using (5.23) in (5.24), we obtain

$$\frac{d[O_x]}{dt} = -2\,k_{HO_2+O_3}[HO_2][O_3] - 2\,k_{HO_2+O}[HO_2][O]$$
$$+ j_{NO_2}[NO_2] - k_{HO_2+NO}[HO_2][NO] \quad (5.25)$$

However, formation of NO_2 by $HO_2 + NO$ is followed immediately by photolysis of NO_2:

$$j_{NO_2}[NO_2] = k_{HO_2+NO}[HO_2][NO] \quad (5.26)$$

Thus (5.25) becomes

$$\frac{d[O_x]}{dt} = -2\,k_{HO_2+O_3}[HO_2][O_3] - 2\,k_{HO_2+O}[HO_2][O] \quad (5.27)$$

and the rates of O_3 depletion by HO_x cycles 1 and 2 are governed by the rates of the $HO_2 + O$ and $HO_2 + O_3$ reactions.

The characteristic time for cycling within the HO_x family is controlled by the rate of which HO_2 is converted back to OH. That interconversion is controlled by the $HO_2 + NO$ reaction at all altitudes. The characteristic time is $\tau_{HO_2} = (k_{HO_2+NO}[NO])^{-1}$. At 30 km, for example, $\tau_{HO_2} \cong 100\,s$. This time is short relative to competing processes, establishing the validity of treating HO_x as a chemical family.

The two HO_x cycles are important at different altitudes. First, we can examine the relative values of the two key rate coefficients, k_{HO_2+O} and $k_{HO_2+O_3}$. Neither is strongly temperature-dependent; roughly, we have

$$k_{HO_2+O} \cong 10^{-10}\,cm^3\,molecule^{-1}\,s^{-1}$$
$$k_{HO_2+O_3} \cong 10^{-15}\,cm^3\,molecule^{-1}\,s^{-1}$$

From Figure 5.11 we note that $[HO_2]$ does not vary strongly with altitude, with a value $\cong 10^7\,molecule\,cm^{-3}$. The major difference arises from the relative concentrations of O and O_3. In the lower stratosphere ($\sim 20\,km$), $[O]/[O_3] \sim 10^{-7}$, so even with a rate coefficient five orders of magnitude larger than that of $HO_2 + O_3$, the $HO_2 + O$ reaction is negligible compared to that of $HO_2 + O_3$. Therefore, HO_x cycle 2 dominates in the lower stratosphere. Conversely, at about 50 km, $[O]/[O_3] \cong 10^{-2}$, so HO_x cycle 1 is predominant.

Let us consider behavior of the HO$_x$ family in the lower stratosphere, where HO$_2$ reacts preferentially with O$_3$ rather than O. The partitioning between HO$_2$ and OH is given by

$$\frac{[HO_2]}{[OH]} = \frac{k_{OH+O_3}[O_3]}{k_{HO_2+NO}[NO]} \tag{5.28}$$

The relative concentrations depend on altitude through the temperature dependence of the reaction rate coefficients and through the concentrations of O$_3$ and NO. At 30 km ($T = 227$ K), the two reaction rate coefficients, $k_{OH+O_3} = 1.7 \times 10^{-12} \exp(-940/T)$ and $k_{HO_2+NO} = 3.5 \times 10^{-12} \exp(250/T)$, have the values (cm^3 molecule^{-1} s^{-1}), $k_{OH+O_3} = 2.7 \times 10^{-14}$ and $k_{HO_2+NO} = 10.5 \times 10^{-12}$. Using $[O_3] \cong 2 \times 10^{12}$ molecules cm^{-3} and an NO mixing ratio of 3 ppb at 30 km ($[NO] \cong 9.3 \times 10^8$ molecules cm^{-3}), we find

$$\frac{[HO_2]}{[OH]} \cong 4.4$$

This estimate is roughly consistent with the ratio at 30 km derived from the vertical profiles of OH and HO$_2$ measured by satellite, as shown in Figure 5.11. At higher altitudes, the concentration of OH becomes roughly comparable to that of HO$_2$. The OH concentration is basically independent of the NO concentration. This can be explained as follows. At constant $[O_3]$, the concentration of HO$_2$ decreases with increasing NO$_x$. The increase of OH that results from the HO$_2$ + NO reaction is offset by a decrease of the rates of the HO$_2$ + O$_3$, HO$_2$ + ClO, and HOCl + $h\nu$ reactions, which themselves generate OH. (We will discuss the latter two reactions shortly.) The net result is that OH becomes more or less independent of the NO concentration. The $[HO_2]/[OH]$ ratio does, however, depend on the NO$_x$ level as in (5.28). Any process that might serve to decrease the level of NO$_x$ would have the effect of shifting this ratio in favor of HO$_2$. And since HO$_2$ is the key species in the rate-determining step of both HO$_x$ cycles, the overall effect of a decrease in NO$_x$ would be an increase in the effectiveness of the HO$_x$ cycles.

As we did for NO$_x$, let us compare the effectiveness of the HO$_x$ cycles with the Chapman mechanism. The appropriate comparison is with HO$_x$ cycle 1, since both involve atomic oxygen. The effectiveness of HO$_x$ cycle 1 relative to reaction 4 in the Chapman mechanism is given by the ratio

$$\frac{k_{HO_2+O}[HO_2][O]}{k_{O+O_3}[O][O_3]} = \frac{k_{HO_2+O}[HO_2]}{k_{O+O_3}[O_3]}$$

At 35 km (237 K), the rate coefficient ratio is

$$\frac{k_{HO_2+O}}{k_{O+O_3}} = \frac{3.0 \times 10^{-11} \exp(200/T)}{8.0 \times 10^{-12} \exp(-2060/T)}$$
$$\cong 5.4 \times 10^4$$

From Figure 5.3, we estimate $[O_3]$ at 35 km as $[O_3] \cong 2 \times 10^{12}$ molecules cm^{-3}. At this altitude, $[HO_2] \cong 1.5 \times 10^7$ molecules cm^{-3}. Thus

$$\frac{k_{HO_2+O}[HO_2]}{k_{O+O_3}[O_3]} \cong (5.4 \times 10^4)(0.75 \times 10^{-5}) \cong 0.4$$

HO_x cycle 1 is therefore about half as effective in destroying O_3 at 35 km as the Chapman cycle.

5.5 HALOGEN CYCLES

At the time of the first tropospheric measurements of anthropogenic halogenated hydrocarbons in the early 1970s, the quantities of the chlorofluorocarbons in the atmosphere were found to be approximately equal to the total amounts ever manufactured. In 1974 Molina and Rowland (Molina and Rowland 1974; Rowland and Molina 1975) realized that chlorofluorocarbons (CFCs), manufactured and used by humans in a variety of technological applications from refrigerants to aerosol spray propellants, have no tropospheric sink and persist in the atmosphere until they diffuse high into the stratosphere where the powerful UV light photolyzes them. The photolysis reactions release a chlorine (Cl) atom, for example, for $CFCl_3$ (CFC-11) and CF_2Cl_2 (CFC-12):

$$CFCl_3 + h\nu \longrightarrow CFCl_2 + Cl$$
$$CF_2Cl_2 + h\nu \longrightarrow CF_2Cl + Cl$$

To photodissociate, CFCs need not rise above most of the atmospheric O_2 and O_3; they are photodissociated at wavelengths in the 185–210 nm spectral window between O_2 absorption of shorter wavelengths and O_3 absorption of longer wavelengths. The maximum loss rate of $CFCl_3$ occurs at about 23 km, whereas that for CF_2Cl_2 takes place in the 25–35 km range. As with N_2O, the bulk of the removal of $CFCl_3$ and CF_2Cl_2 is confined to the tropics, reflecting larger values for photolysis rates in this region.

The only continuous natural source of chlorine in the stratosphere is methyl chloride, CH_3Cl (see Chapter 2). The tropospheric lifetime for CH_3Cl is sufficiently long, 1.5 years (Table 2.15), so some amount of CH_3Cl is transported through the tropopause. In the stratosphere, as in the troposphere, CH_3Cl is removed primarily by reaction with the OH radical. At higher altitudes in the stratosphere a portion of CH_3Cl is photolyzed. Regardless of the path of reaction, the chlorine atom in CH_3Cl is released as active chlorine.

5.5.1 Chlorine Cycles

The chlorine atom is highly reactive toward O_3 and establishes a rapid cycle of O_3 destruction involving the chlorine monoxide (ClO) radical:

ClO_x cycle 1:

$$Cl + O_3 \xrightarrow{1} ClO + O_2 \quad k_1 = 2.3 \times 10^{-11} \exp(-200/T) \quad \text{(reaction 1)}$$
$$\underline{ClO + O \xrightarrow{2} Cl + O_2 \quad k_2 = 3.0 \times 10^{-11} \exp(70/T) \quad \text{(reaction 2)}}$$
$$\text{Net:} \quad O_3 + O \longrightarrow O_2 + O_2$$

Let us estimate the lifetime of the Cl atom against reaction 1 and the lifetime of ClO against reaction 2. To do so we need the concentrations of O_3 and O; these are related by (5.7). At 40 km, for example, where $T = 251$ K, the O_3 concentration can be taken as

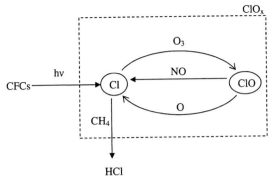

FIGURE 5.13   The stratospheric ClO_x family.

0.5×10^{12} molecules cm^{-3}, and the ratio $[O]/[O_3]$ was shown to be about 9.4×10^{-4}. The two lifetimes at 40 km are

$$\tau_{Cl} = \frac{1}{k_1[O_3]} = 0.2\,s$$

$$\tau_{ClO} = \frac{1}{k_2[O]} = 53\,s$$

Subsequent reactions of the $CFCl_2$ and CF_2Cl radicals lead to rapid release of the remaining chlorine atoms.

Cl and ClO establish the ClO_x chemical family (Figure 5.13). The rapid inner cycle is characterized by ClO formation by reaction 1 and loss by reaction with both O and NO. If ClO reacts with O, the catalytic O_3 depletion cycle above occurs. If ClO reacts with NO, the following cycle takes place:

$$Cl + O_3 \xrightarrow{1} ClO + O_2$$

$$ClO + NO \xrightarrow{3} Cl + NO_2 \qquad k_3 = 6.4 \times 10^{-12} \exp(290/T) \qquad \text{(reaction 3)}$$

$$NO_2 + h\nu \longrightarrow NO + O$$

Net: $\quad \overline{O_3 + h\nu \longrightarrow O + O_2}$

Because the O atom rapidly reforms O_3, this is a null cycle with respect to O_3 removal.

The rate of net O_3 removal by the ClO_x cycle of reactions 1 and 2 is

$$\frac{d[O_x]}{dt} = \frac{d[O_3]}{dt} + \frac{d[O]}{dt} = -2\,k_2[ClO][O] \qquad (5.29)$$

Within the ClO_x chemical family we can compare the relative importance of reactions 2 and 3:

$$\frac{k_{ClO+O}[O]}{k_{ClO+NO}[NO]}$$

At 40 km, $T = 251$ K, $[O]/[O_3] \cong 9.4 \times 10^{-4}$, $[O_3] \cong 0.5 \times 10^{12}$ molecules cm^{-3}, and $[NO] \cong 1 \times 10^9$ molecules cm^{-3}. The rate coefficient values are $k_{ClO+O} \cong 4 \times 10^{-11}$ cm^3 molecule^{-1} s^{-1} and $k_{ClO+O} \cong 2 \times 10^{-11}$ cm^3 molecule^{-1} s^{-1}. Thus

$$\frac{k_{ClO+O}[O]}{k_{ClO+NO}[NO]} \cong 0.9$$

The steady-state ratio [Cl]/[ClO] within the ClO$_x$ family is given by

$$\frac{[Cl]}{[ClO]} = \frac{k_{ClO+O}[O] + k_{ClO+NO}[NO]}{k_{Cl+O_3}[O_3]} \tag{5.30}$$

At 40 km

$$\frac{[Cl]}{[ClO]} \cong 0.008$$

so most of the ClO$_x$ is in the form of ClO; this is a result of the rapidity of the Cl + O$_3$ reaction.

Since the ClO$_x$ cycle of reactions 1 and 2 and the Chapman mechanism (Section 5.2) both involve the O atom, let us compare, as we have done with the NO$_x$ and HO$_x$ cycles, the rate of reaction 2 to that of reaction 4 in the Chapman mechanism:

$$\frac{k_{ClO+O}[ClO]}{k_{O+O_3}[O_3]}$$

At 35 km, the rate coefficient ratio, $k_{ClO+O}/k_{O+O_3} = 3 \times 10^4$. We need the altitude dependence of [ClO].

Figure 5.14 shows the stratospheric chlorine inventory as of November 1994 (35–49°N). At 35 km, the ClO mixing ratio was about 0.5 ppb, which translates to

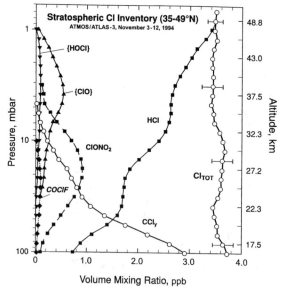

FIGURE 5.14   Stratospheric chlorine inventory (35–49°N), November 3–12, 1994 (WMO 1998). [Adapted from Zander et al. (1996).]

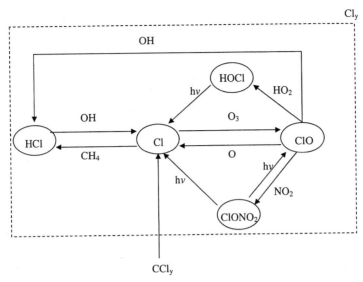

FIGURE 5.15 Cl_y chemical family.

$[ClO] \cong 0.7 \times 10^8$ molecules cm^{-3}. Thus, at this altitude

$$\frac{k_{ClO+O}[ClO]}{k_{O+O_3}[O_3]} \cong 1.1$$

Ozone removal by catalytic chlorine chemistry and the Chapman cycle are, therefore, roughly equal at about 40 km.

The ClO_x cycle of reactions 1 and 2 achieves its maximum effectiveness at about 40 km, roughly the altitude at which ClO achieves its maximum concentration. At higher altitudes, the $Cl + CH_4$ reaction controls Cl/ClO partitioning, and Cl becomes sequestered as HCl. The overall set of reactions in the chlorine cycles involves a few more key reactions. Figure 5.15 shows an expanded view of stratospheric chlorine chemistry. At lower altitudes, the reservoir species, chlorine nitrate, $ClONO_2$, ties up ClO. As altitude increases, [M] decreases, and the rate of the termolecular reaction, $ClO + NO_2 + M \rightarrow ClONO_2 + M$, decreases. The competing formation of the reservoir species HCl and $ClONO_2$ produce the maximum in [ClO] at about 40 km.

An increase of OH produces opposite effects on the NO_x and ClO_x cycles. As OH increases, HCl is converted back to active Cl, enhancing the effectiveness of the ClO_x cycle in O_3 destruction. On the other hand, as OH increases, more NO_2 is converted to the HNO_3 reservoir, decreasing the catalytic effectiveness of the NO_x cycle. (Although HNO_3 itself reacts with OH, $HNO_3 + OH \rightarrow H_2O + NO_3$, releasing NO_x, this reaction is less effective in releasing NO_x than is $OH + NO_2 + M \rightarrow HNO_3 + M$ in sequestering it.)

From Figure 5.15 we can define

$$Cl_y = Cl + ClO + HOCl + ClONO_2 + HCl$$

as the chemical family of reactive chlorine. Note from Figure 5.14 that all the stratospheric chlorine is accounted for, as indicated by the constant value of Cl_{tot} at all altitudes. As

CCl_y, the chlorine contained in CFCs, CH_3Cl, CCl_4, and so on decreases with increasing altitude as a result of photodissociation of the CFCs, the species in Cl_y appear, such that the sum of all chlorine is conserved. Together, HCl and $ClONO_2$ store as much as 99% of the active chlorine. *As a result, only a small change in the abundance of either HCl or $ClONO_2$ can have a profound effect on the catalytic efficiency of the ClO_x cycle.*

At lower altitudes where O atom levels are significantly lower, the following cycle that involves coupling with HO_x is important:

$$
\begin{aligned}
HO_x/ClO_x \text{ cycle 1:} \quad & Cl + O_3 \longrightarrow ClO + O_2 \\
& OH + O_3 \longrightarrow HO_2 + O_2 \\
& ClO + HO_2 \longrightarrow HOCl + O_2 \\
& \underline{HOCl + h\nu \longrightarrow OH + Cl} \\
\text{Net:} \quad & O_3 + O_3 \longrightarrow O_2 + O_2 + O_2
\end{aligned}
$$

In addition, the following HO_x/BrO_x and BrO_x/ClO_x cycles are important in the lower stratosphere ($\sim 20\,\text{km}$):

$$
\begin{aligned}
HO_x/BrO_x \text{ cycle 1:} \quad & Br + O_3 \longrightarrow BrO + O_2 \\
& OH + O_3 \longrightarrow HO_2 + O_2 \\
& BrO + HO_2 \longrightarrow HOBr + O_2 \\
& \underline{HOBr + h\nu \longrightarrow OH + Br} \\
\text{Net:} \quad & O_3 + O_3 \longrightarrow O_2 + O_2 + O_2
\end{aligned}
$$

$$
\begin{aligned}
BrO_x/ClO_x \text{ cycle 1:} \quad & Br + O_3 \longrightarrow BrO + O_2 \\
& Cl + O_3 \longrightarrow ClO + O_2 \\
& BrO + ClO \longrightarrow BrCl + O_2 \\
& \underline{BrCl + h\nu \longrightarrow Br + Cl} \\
\text{Net:} \quad & O_3 + O_3 \longrightarrow O_2 + O_2 + O_2
\end{aligned}
$$

$$
\begin{aligned}
BrO_x/ClO_x \text{ cycle 2:} \quad & Br + O_3 \longrightarrow BrO + O_2 \\
& Cl + O_3 \longrightarrow ClO + O_2 \\
& BrO + ClO \longrightarrow ClOO + Br \\
& \underline{ClOO + M \longrightarrow Cl + O_2} \\
\text{Net:} \quad & O_3 + O_3 \longrightarrow O_2 + O_2 + O_2
\end{aligned}
$$

5.5.2   Bromine Cycles

While there is a total chlorine compound mixing ratio in the stratosphere of approximately 3400 ppt, that for bromine gases is only ~ 20 ppt (Figure 5.1). Remarkably, with 150 times less abundance than chlorine, bromine is approximately as important as chlorine in overall ozone destruction. Methyl bromide (CH_3Br) constitutes about half the source of bromine in the stratosphere (see Section 2.5). The H-atom-containing bromine compounds, CH_3Br, CH_2Br_2, and $CHBr$, release their Br almost immediately on entry into the stratosphere; the

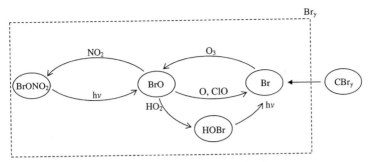

FIGURE 5.16   Br$_y$ chemical family.

halons, $CBrFCl_2$ and $CBrF_3$, release their Br more slowly and therefore at higher altitude. The Br$_y$ system is shown in Figure 5.16. In order to understand why bromine compounds are so much more effective in ozone destruction, we need to consider the role of reservoir species in stratospheric ozone destruction cycles.

5.6   RESERVOIR SPECIES AND COUPLING OF THE CYCLES

Cycles such as HO$_x$ cycles 1 and 2, NO$_x$ cycle 1, and ClO$_x$ cycle 1 would appear to go on destroying O_3 forever. Actually, the cycle can be interrupted when the reactive species, OH, NO$_2$, Cl, and ClO, become tied up in relatively stable species so that they are not available to act as catalysts in the cycles. Knowing the competing reactions that can occur, it is possible to estimate the average number of times each catalytic O_3 destruction cycle will proceed before one of the reactants participates in a competing reaction and terminates the cycle. For example, at current stratospheric concentrations, ClO$_x$ Cycle 1, once initiated, loops, on average, 10^5 times before the Cl atom or the ClO molecule reacts with some other species to terminate the cycle. This means that, on average, one chlorine atom can destroy 100,000 molecules of ozone before it is otherwise removed. The frequency of such cycle termination reactions is thus critical to the overall efficiency of a cycle. The removal of a reactive species can be permanent if the product actually leaves the stratosphere by eventually migrating down to the troposphere, where it is removed from the atmosphere altogether. Examples of cycle-interrupting reactions that can lead to ultimate removal from the atmosphere are

$$OH + NO_2 + M \longrightarrow HNO_3 + M$$
$$Cl + CH_4 \longrightarrow HCl + CH_3$$

Both nitric acid (HNO_3) and hydrogen chloride (HCl) are relatively stable in the stratosphere and some fraction of each migrates back to and is removed from the troposphere by precipitation.

A reactive species can also be temporarily removed from a catalytic cycle and be stored in a reservoir species, which itself is relatively unreactive but is not actually removed from the atmosphere. One of the most important reservoir species in the stratosphere is chlorine nitrate ($ClONO_2$), formed by

$$ClO + NO_2 + M \longrightarrow ClONO_2 + M$$

Chlorine nitrate can photolyze

$$ClONO_2 + h\nu \longrightarrow ClO + NO_2$$
$$\longrightarrow Cl + NO_3$$

thereby releasing Cl or ClO back into the active ClO_x reservoir. Chlorine nitrate is an especially important reservoir species because it stores two catalytic agents, NO_2 and ClO.

Partitioning of chlorine between reactive (e.g., Cl and ClO) and reservoir forms (e.g., HCl and $ClONO_2$) depends on temperature, altitude, and latitude history of an air parcel. For the midlatitude, lower stratosphere, HCl and $ClONO_2$ are the dominant reservoir species for chlorine, constituting over 90% of the total inorganic chlorine (Figure 5.14). From the Cl_y family in Figure 5.15, Cl can be bled off the internal cycle by reacting with methane to produce HCl, and the chlorine atom in HCl can be returned to active Cl if HCl reacts with OH. The abundances of CH_4 and OH will control the amount of Cl sequestered as HCl. ClO can be temporarily removed from active participation in the ClO_x catalytic cycle by reacting with NO_2 to form $ClONO_2$ or with HO_2 to form HOCl. Both $ClONO_2$ and HOCl can photolyze and release active chlorine back into the cycle. The importance of $ClONO_2$ and HOCl as reservoir species will depend on the abundances of NO_2 and HO_2 relative to atomic oxygen.

Lifetime of the Reservoir Species $ClONO_2$ and HCl   Chlorine nitrate ($ClONO_2$) photodissociates back to ClO; its lifetime against photolysis is $\tau_{ClONO_2} = j_{ClONO_2}^{-1}$. In the midstratosphere (15–30 km), $j_{ClONO_2} \cong 5 \times 10^{-5}\,s^{-1}$, so $\tau_{ClONO_2} \sim 5\,h$.

The lifetime of HCl against reaction with OH is $\tau_{HCl} = (k_{OH+HCl}[OH])^{-1}$, where $k_{OH+HCl} = 2.6 \times 10^{-12} \exp(-350/T)$. At 35 km (237 K), $k_{OH+HCl} \cong 6 \times 10^{-13}\,cm^3$ molecule$^{-1}\,s^{-1}$, and $[OH] \cong 10^7$ molecules cm^{-3}, so $\tau_{HCl} \cong 2$ days. At lower altitudes, where OH levels are lower, the HCl lifetime increases to several weeks.

From Figure 5.15 we note the amounts of $ClONO_2$ and HCl relative to those of Cl and ClO. NO_2 and CH_4 are responsible for shifting most of the active chlorine into these reservoir species. Typically

$$\frac{[HCl]}{[Cl_y]} \sim 0.7 \qquad \frac{[ClONO_2]}{[Cl_y]} \sim 0.3 \qquad \frac{[ClO]}{[Cl_y]} \sim 0.02\text{--}0.05$$

Any process that even modestly shifts the balance away from the reservoir species to ClO can have a large impact on ozone depletion.

In the midlatitude lower stratosphere the concentration of ClO is actually controlled by the amount of $ClONO_2$ present:

$$[ClO] \cong \frac{j_{ClONO_2}[ClONO_2]}{k_{ClO+NO_2}[NO_2][M]}$$

(Note that the O atom concentration in the lower stratosphere is too low for its reaction with ClO to compete effectively with the $ClO + NO_2$ reaction.)

The HO_x, NO_x, and ClO_x cycles are all coupled to one another, and their interrelationships strongly govern stratospheric ozone chemistry. The NO_x and ClO_x cycles are coupled by $ClONO_2$. For example, increased emissions of N_2O would lead to increased stratospheric concentrations of NO and hence increased ozone depletion by the NO_x catalytic cycle. Likewise, increasing CFC levels will lead to increased ozone depletion by the ClO_x cycle. However, increased NO_x will lead to an increased level of the $ClONO_2$ reservoir and a mitigation of the chlorine cycle. Thus the net effect on ozone of simultaneous increases in both N_2O and CFCs is less than the sum of their separate effects because of increased formation of $ClONO_2$. Increases in CH_4 or stratospheric H_2O would also act to mitigate the efficiency of NO_x and ClO_x cycles through increased formation of HCl and OH (and hence HNO_3).

Whereas HCl is formed by $Cl + CH_4$, the corresponding reaction of Br atoms with CH_4 is endothermic and extremely slow.[1] (See Br_y reactions in the bottom panel of Figure 5.16.) As a result, the only possible routes for formation of HBr involve reaction of Br with species far less abundant than CH_4, such as HO_2. Even if HBr is formed, the OH + HBr rate coefficient is about 12 times larger than that of OH + HCl, so that the stratospheric lifetime of HBr against OH reaction is a matter of days as compared to about 1 month for HCl. In addition, $BrONO_2$ is considerably less stable than $ClONO_2$ because of rapid photolysis; by comparison, the photolysis lifetimes of $ClONO_2$ and $BrONO_2$ in the lower stratosphere are 6 hours and 10 minutes, respectively.

In summary, the extraordinary effectiveness of the Br catalytic system is a consequence of two factors: (1) Br compounds release their Br rapidly in the stratosphere; and (2) a large fraction of the total Br_y exists in the form of Br + BrO. As a result, Br ozone depletion potentials (see Section 5.11) are about 50 times greater than those of the Cl system.

5.7 OZONE HOLE

In 1985 a team led by British scientist Joseph Farman shocked the scientific community with reports of massive annual decreases of stratospheric ozone over Antarctica in the polar spring (September to October), an observation that the prevailing understanding of stratospheric chlorine chemistry was incapable of explaining (Farman et al. 1985). This phenomenon has been termed the "ozone hole" by the popular press. The British Antarctic Survey has, for many years, been measuring total column ozone levels from ground level at its base at Halley Bay. Data seemed to indicate that column ozone levels had been decreasing since about 1977. When the Farman et al. paper appeared, there was a concern that the instruments aboard the Nimbus 7 satellite, the *total ozone mapping spectrometer* (TOMS) and the *solar backscattered ultraviolet* (SBUV) instrument, had apparently not detected the ozone depletion seen in the ground-based data. It turned out, upon inspection of the satellite data, that the low ozone concentrations were indeed observed, but were being systematically rejected in the database as being outside the reasonable range of data. Once this was discovered, the satellite data confirmed the ground-based measurements of the British Antarctic Survey.

Figure 5.17a shows October mean total column ozone observed over Halley, Antarctica, from 1956 to 1994. The open triangles are the data first reported by Farman

---

[1] At 260 K, the $Br + CH_4$ rate coefficient $\cong 10^{-25}$ cm^3 molecule^{-1} s^{-1}, as compared with that for $Cl + CH_4 \cong$ 10^{-14} cm^3 molecule^{-1} s^{-1}.

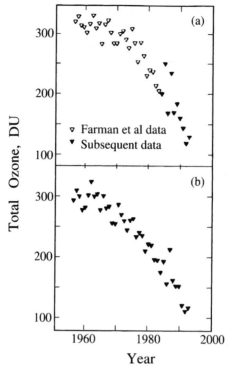

FIGURE 5.17   Column ozone over Halley, Antarctica (Jones and Shanklin 1995). (a) October mean total ozone (in Dobson units) observed over Halley from 1956 to 1994. The open triangles are the data of Farman et al. (1985); solid triangles, subsequent data. (b) The lowest daily value of total ozone observed over Halley in October for the years between 1956 and 1994 (the October minimum). The daily value is the mean of all the readings taken during the day (between 5 and 30). Lowest daily values of total ozone (in DU) for 1996–2001 were as follows:

1996	111
1997	104
1998	90
1999	92
2000	94
2001	99

(Reprinted with permission from *Nature* **376**, Jones, A. E. and Shanklin, J. D. Copyright 1995 Macmillan Magazines Limited.)

et al. (1985); the solid triangles are the subsequent data (Jones and Shanklin, 1995). Figure 5.17b shows the lowest daily value of total column ozone observed over Halley, Antarctica, in October for the years between 1956 and 1994 (the October minimum). The daily value is the mean of all the readings taken during the day. The monthly mean column ozone over Halley Bay, Antarctica (76° S), decreased from a high of 350 Dobson units (DU) in the mid-1970s to values approaching 100 DU. Ozone vertical profiles from balloon measurements showed that the O_3 depletion was occurring at altitudes between 10 and 20 km (Hofmann et al. 1987). The discovery of the ozone hole was surprising not only because of its magnitude but also its location. It was expected that ozone

OZONE VERTICAL PROFILES

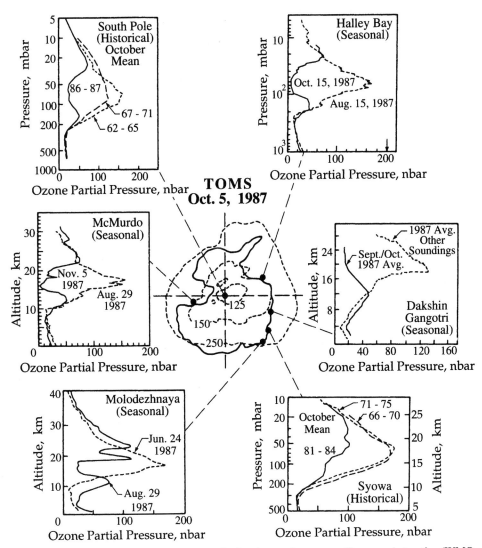

FIGURE 5.18   Observations of the change in October total ozone profiles over Antarctica (WMO 1994). Historical data at South Pole and Syowa show changes in October mean profiles measured in the 1960s and 1970s as compared to more recent observations. Changes in seasonal vertical profiles are shown at the other stations. Isopleths mapped onto Antartica represent TOMS ozone column measurements on Oct. 5, 1987.

depletion resulting from the ClO_x catalytic cycle would manifest itself predominantly at middle and lower latitudes and at altitudes between 35 and 45 km. Figure 5.18 shows historical changes in October ozone profiles over Antarctica.

Although the Antarctic has some of the Earth's highest ozone concentrations during much of the year, most of its ozone is actually made in the tropics and delivered, along with the molecular reservoirs of chlorine, to the Antarctic by large-scale air movement. Arctic ozone is similar. The Antarctic stratosphere is deficient in atomic oxygen because of the absence of the intense UV radiation. As air cools during the Antarctic winter, it descends and develops a westerly circulation. This *polar vortex* develops a core of very cold air. In the winter and early spring the vortex is extremely stable, effectively sealing off air in the vortex from that outside. The polar vortex serves to keep high levels of the imported ozone trapped over Antarctica for the several-month period each year. As the sun returns in September at the end of the long polar night, the temperature rises and the vortex weakens, eventually breaking down in November. Normally the amount of ozone in the polar vortex begins to decrease somewhat as the Antarctic emerges from the months-long austral night in late August and early September. It levels off in October and increases in November. The discovery of the Antarctic ozone hole represented a significant change in historic patterns; the springtime ozone levels decreased to unprecedented levels, with each succeeding year being, more or less, worse than the year before.

It was initially suggested that the Antarctic ozone hole could be explained on the basis of solar cycles or purely atmospheric dynamics. Neither explanation was consistent with observed features of the ozone hole. Chemical explanations based on the gas-phase catalytic cycles described above were advanced. As noted, little ozone is produced in the polar stratosphere as the low Sun elevation (large solar zenith angle) results in essentially no photodissociation of O_2. Thus catalytic cycles that require oxygen atoms were not able to explain the massive ozone depletion. Moreover, CFCs and halons would be most effective in ozone depletion in the Antarctic stratosphere at an altitude of about 40 km, whereas the ozone hole is sharply defined between 12 and 24 km altitude. Also, existing levels of CFCs and halons could lead at most to an O_3 depletion at 40 km of 5–10%, far below that observed.

Molina and Molina (1987) proposed that a mechanism involving the ClO dimer, Cl_2O_2, might be involved:

$$ClO + ClO + M \xrightarrow{1} Cl_2O_2 + M \qquad \text{(reaction 1)}$$

$$Cl_2O_2 + h\nu \xrightarrow{2} Cl + Cl + O_2 \qquad \text{(reaction 2)}$$

$$\underline{2[Cl + O_3 \xrightarrow{3} ClO + O_2]} \qquad \text{(reaction 3)}$$

$$\text{Net:} \quad 2\,O_3 + h\nu \longrightarrow 3\,O_2$$

(Bimolecular reactions of ClO and ClO are slow and can be neglected. The termolecular reaction 1 is facilitated at higher pressures, that is, larger M, and low temperature.) Cl_2O_2 has been shown to have the symmetric structure ClOOCl (McGrath et al. 1990). Photolysis of ClOOCl has two possible channels:

$$ClOOCl + h\nu \xrightarrow{4a} Cl + ClOO \quad (\lambda \sim 350\,\text{nm}) \qquad \text{(reaction 4a)}$$

$$\xrightarrow{4b} ClO + ClO \qquad \text{(reaction 4b)}$$

Only reaction 4a can lead to an ozone-destroying cycle. Indeed, reaction 4a is the main photolysis path (Molina et al., 1990). The ClOO product rapidly decomposes to yield a Cl atom and O_2,

$$ClOO + M \xrightarrow{5} Cl + O_2 + M \qquad \text{(reaction 5)}$$

Thus reaction 2 is seen to be a composite of reactions 4a and 5, leading to the release of both chlorine atoms from Cl_2O_2.

Two similar cycles involving both ClO_x and BrO_x cycles also can take place:

$$ClO + BrO \longrightarrow BrCl + O_2$$
$$BrCl + h\nu \longrightarrow Br + Cl$$
$$Cl + O_3 \longrightarrow ClO + O_2$$
$$\underline{Br + O_3 \longrightarrow BrO + O_2}$$
$$\text{Net:}\quad 2\,O_3 \longrightarrow 3\,O_2$$

$$ClO + BrO \longrightarrow ClOO + Br$$
$$ClOO + M \longrightarrow Cl + O_2 + M$$
$$Cl + O_3 \longrightarrow ClO + O_2$$
$$\underline{Br + O_3 \longrightarrow BrO + O_2}$$
$$\text{Net:}\quad 2O_3 \longrightarrow 3\,O_2$$

Note that atomic oxygen is not required in either cycle.

If sufficient concentrations of ClO and BrO could be generated, then the three cycles shown above could lead to substantial O_3 depletion. However, gas-phase chemistry alone does not produce the necessary ClO and BrO concentrations needed to sustain these cycles.

5.7.1   Polar Stratospheric Clouds

The stratosphere is very dry and generally cloudless. The long polar night, however, produces temperatures as low as 183 K ($-90°C$) at heights of 15–20 km. At these temperatures even the small amount of water vapor condenses to form *polar stratospheric clouds* (PSCs), seen as wispy pink or green clouds in the twilight sky over polar regions.

The conceptual breakthrough in explaining the Antarctic ozone hole occurred when it was realized that PSCs provide the surfaces on which halogen-containing reservoir species are converted to active catalytic species. Then, intense interest ensued: what are PSCs made of—pure ice or ice mixed with other species? Nitrate was detected in PSCs, and both laboratory and theoretical studies showed that nitric acid trihydrate, $HNO_3 \cdot 3H_2O$, denoted NAT, is the thermodynamically stable form of HNO_3 in ice at polar stratospheric temperature (Peter 1996). It was also discovered that some PSCs are liquid particles composed of supercooled ternary solutions of H_2SO_4, HNO_3, and H_2O.

An important implication of the fact that PSCs contain nitrate is that, if the particles are sufficiently large, they can fall out of the stratosphere and thereby permanently remove nitrogen from the stratosphere. The removal of nitrogen from the stratosphere is termed *denitrification*. If nitrate-containing PSCs sediment out of the stratosphere, then that could lead to an appreciably lower supply of nitrate for possible return to NO_x (and conversion of ClO to $ClONO_2$).

Stratospheric NAT particles were first detected in situ in the 1999–2000 SAGE III "ozone loss and validation experiment," carried out in the stratosphere over northern Sweden (Voight et al. 2000; Fahey et al. 2001). The particles identified were large enough (1–$20\,\mu m$ in diameter) to be able to fall a substantial distance before evaporating. The fall of nitrate-containing PSC particles was therefore established as the mechanism by which the stratosphere is denitrified.

5.7.2   PSCs and the Ozone Hole

Gas-phase chemistry associated with the ClO_x and NO_x cycles is not capable of explaining the polar ozone hole phenomenon. The ozone hole is sharply defined between about 12 and 24 km altitude. Polar stratospheric clouds occur in the altitude range 10–25 km. Ordinarily, liberation of active chlorine from the reservoir species HCl and $ClONO_2$ is rather slow, but the PSCs promote the conversion of the major chlorine reservoirs, HCl and $ClONO_2$, to photolytically active chlorine. The first step in the process, absorption of gaseous HCl by PSCs, occurs very efficiently. This step is followed by the heterogeneous reaction of gaseous $ClONO_2$ with the particle,

$$HCl(s) + ClONO_2 \xrightarrow{\ 1\ } Cl_2 + HNO_3(s) \qquad \text{(reaction 1)}$$

where (s) denotes a species on the surface of the ice, with the liberation of molecular chlorine as a gas and the retention of nitric acid in the particles. This is the most important chlorine-activating reaction in the polar stratosphere. (The gas-phase reaction between HCl and $ClONO_2$ is extremely slow.) The solubility of HCl in normal stratospheric aerosol, 50–80 wt% H_2SO_4 solutions, is low. When stratospheric temperatures drop below 200 K, the stratospheric particles absorb water and allow HCl to be absorbed, setting the stage for reaction 1.

Gaseous Cl_2 released from the PSCs in reaction 1 rapidly photolyzes, producing free Cl atoms, while the other product, HNO_3, remains in the ice, leading to the overall removal of nitrogen oxides from the gas phase. This trapping of HNO_3 further facilitates catalytic O_3 destruction by removing NO_x from the system, which might otherwise react with ClO to form $ClONO_2$. The net result of reaction 1 is

$$HCl(s) + ClONO_2 \xrightarrow{\ 1\ } Cl_2 + HNO_3(s) \qquad \text{(reaction 1)}$$

$$Cl_2 + h\nu \xrightarrow{\ 2\ } 2\,Cl \qquad \text{(reaction 2)}$$

$$2[Cl + O_3 \xrightarrow{\ 3\ } ClO + O_2] \qquad \text{(reaction 3)}$$

$$\underline{ClO + NO_2 + M \xrightarrow{\ 4\ } ClONO_2 + M} \qquad \text{(reaction 4)}$$

$$\text{Net:}\quad HCl(s) + NO_2 + 2\,O_3 \longrightarrow ClO + HNO_3(s) + 2\,O_2$$

The reaction between $ClONO_2$ and $H_2O(s)$, which is very slow in the gas phase, also can occur:

$$ClONO_2 + H_2O(s) \xrightarrow{\ 5\ } HOCl + HNO_3(s) \qquad \text{(reaction 5)}$$

The gaseous HOCl rapidly photolyzes to yield a free chlorine atom. HOCl itself can undergo a subsequent heterogeneous reaction (Abbatt and Molina 1992)

$$HOCl + HCl(s) \xrightarrow{\ 6\ } Cl_2 + H_2O \qquad \text{(reaction 6)}$$

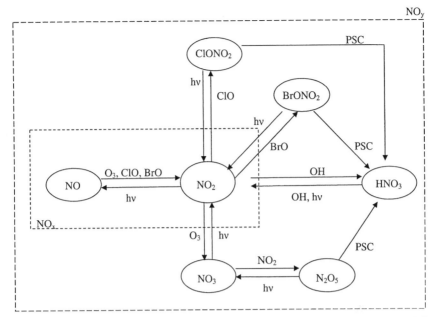

FIGURE 5.19   Catalytic cycles in O_3 depletion involving polar stratospheric clouds (PSCs).

The mechanism of ozone destruction in the polar stratosphere is thus as follows. Two ingredients are necessary: *cold temperatures and sunlight*. The absence of either one of these precludes establishing the destruction mechanism. Cold temperatures are needed to form polar stratospheric clouds to provide the surfaces on which the heterogeneous reactions take place. The reservoir species $ClONO_2$ and N_2O_5 react heterogeneously with PSCs on which HCl has been absorbed to produce gaseous Cl_2, HOCl, and $ClNO_2$. Sunlight is then required to photolyze the gaseous Cl_2, HOCl, and $ClNO_2$ that are produced as a result of the heterogeneous reactions. At sunrise, the Cl_2, $ClNO_2$, and HOCl are photolyzed, releasing free Cl atoms that then react with O_3 by reaction 3. Figure 5.19 depicts the catalytic cycles and the role of PSCs. At first, the ClO just accumulates (recall that the O atoms normally needed to complete the cycle by ClO + O are essentially absent in the polar stratosphere). However, once the ClO concentrations are sufficiently large, the three catalytic cycles presented in the beginning of this section occur. The ClO–ClO cycle accounts for ~ 60% of the Antarctic ozone loss, and the ClO–BrO cycles account for ~ 40% of the removal.

Furthermore, since much of the NO_x is tied up as HNO_3 in PSCs, the normally moderating effect of NO_x, through formation of $ClONO_2$, is absent. In fact, massive ozone depletion requires that the abundance of gaseous HNO_3 be very low. The major process removing HNO_3 from the gas phase at temperatures below about 195 K is the formation of NAT PSCs. Of course, HNO_3 is also removed by HCl(s) + $ClONO_2$, but that is only the NO_x associated with $ClONO_2$, and removal of HNO_3 by this reaction alone would not be sufficient to accomplish the large-scale denitrification that is required; that requires the formation of PSCs and the removal of the nitrogen associated with them. PSCs exhibit a bimodal size distribution in the Antarctic stratosphere, with most of the nitrate concentrated in particles with radii $\geq 1\mu m$. The bimodal size distribution sets the stage for efficient denitrification, with nitrate particles either falling on their own or serving as

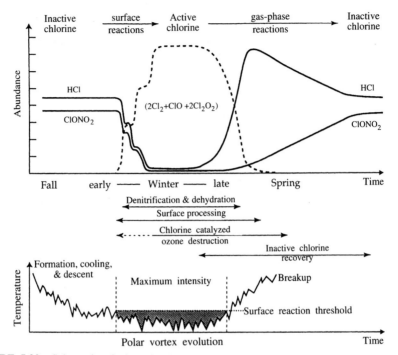

FIGURE 5.20 Schematic of photochemical and dynamical features of polar ozone depletion [WMO (1994), as adapted from Webster et al. (1993)]. Upper panel shows the conversion of chlorine from inactive reservoir forms, ClONO$_2$ and HCl, to active forms, Cl and ClO, in the winter in the lower stratosphere, followed by reestablishment of the inactive forms in spring. Corresponding stages of the polar vortex are indicated in the lower panel, where the temperature scale represents changes in the minimum temperatures in the lower polar stratosphere.

nuclei for condensation of ice (Salawitch et al. 1989). Figure 5.20 is a schematic of the time evolution of the polar stratospheric chlorine chemistry.

In the Antarctic winter vortex, vertical transport within the vortex as well as horizontal transport across the boundaries of the vortex is slower than the characteristic time for the ozone-depleting reactions. The local rate of loss of ozone can be approximated as that of the rate-determining step:

$$\frac{d[O_3]}{dt} \simeq -2k_{\text{ClO+ClO+M}}[\text{ClO}][\text{ClO}][\text{M}]$$

ClO levels are typically elevated by a factor of 100 over their normal levels in the 12 to 24 km altitude range. With ClO mixing ratios in the range of 1 to 1.3 ppb, the above rate predicts complete O$_3$ removal in about 60 days.

Eventually, as the polar air mass warms through breakup of the polar vortex and by absorption of sunlight, the PSCs evaporate, releasing HNO$_3$. The nitric acid vapor photolyzes and reacts with OH to restore gas-phase NO$_x$:

$$\text{HNO}_3 + h\nu \longrightarrow \text{OH} + \text{NO}_2$$
$$\text{HNO}_3 + \text{OH} \longrightarrow \text{H}_2\text{O} + \text{NO}_3$$

Gaseous NO_2 reacts with ClO to again tie up active chlorine as $ClONO_2$. The ClO/HCl ratio is indicative of the course of the ozone destruction process. Most of the atomic chlorine in the stratosphere reacts either with O_3, or with CH_4. The ratio of rate constants, k_{Cl+O_3}/k_{Cl+CH_4}, is about 900 at 200 K. The $Cl + CH_4$ reaction is the principal source of stratospheric HCl, and this reaction governs the recovery rate of HCl following its loss from the PSC-catalyzed reaction $HCl(s) + ClONO_2$. Once PSC chlorine conversion has ceased, HCl recovers to its original amount with a characteristic time of $\Delta[HCl]/k_{Cl+CH_4}$ [Cl][CH_4]. This recovery time is estimated to be about ninety 12-h days, assuming a mean temperature of 200 K, a mean [Cl] of 0.015 ppt, and a mean [CH_4] of 0.8 ppm during the recovery process (Webster et al. 1993). Because an important contribution to stratospheric warming is solar absorption by O_3, and because O_3 levels have been depleted, the usual warmup is delayed, prolonging the duration of the ozone hole.

After discovery of the Antarctic ozone hole a number of field campaigns were mounted to measure concentrations of important species in the ozone depletion cycle. The key active chlorine species in the polar ozone-destroying catalytic cycle is ClO. Simultaneous measurement of ClO and O_3, as shown in Figure 5.21, provided conclusive evidence linking ClO generation to ozone loss. At an altitude near 20 km, ClO mixing ratios reached 1 ppb, several orders of magnitude higher than those in the midlatitude stratosphere, indicating almost total conversion of chlorine to active species within the polar vortex (Anderson et al. 1991). Significant reductions in the column abundances of HCl, $ClONO_2$, and NO_2 are equally important evidence as elevated ClO in verifying the mechanism of catalytic ozone destruction.

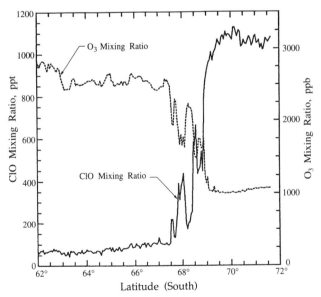

FIGURE 5.21   Simultaneous measurements of O_3 and ClO made aboard the NASA ER-2 aircraft on a flight from Punta Arenas, Chile (53–72°S), on September 16, 1987 (Anderson et al. 1989). As the plane entered the polar vortex, concentrations of ClO increased to about 500 times normal levels while O_3 plummeted.

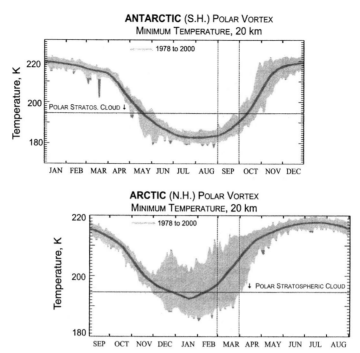

FIGURE 5.22   Minimum air temperatures in the Arctic and Antarctic lower stratosphere. (Courtesy of Paul Newman, NASA Goddard Space Flight Center.)

5.7.3   Arctic Ozone Hole

The ozone hole phenomenon was first discovered in the Antarctic stratosphere. What about the Arctic stratosphere? Recall that two ingredients are necessary to produce the ozone hole: very cold temperatures and sunlight. The Arctic winter stratosphere is generally warmer than the Antarctic by about 10 K. Figure 5.22 shows the distribution of minimum temperatures in the Antarctic and Arctic stratospheres. We see the distribution of generally lower minimum temperatures in the Antarctic versus the Arctic and the overall lower frequency of PSC formation in the Arctic. Thus polar stratospheric clouds are less abundant in the Arctic and, where they do form, they tend to disappear several weeks before solar radiation penetrates the Arctic stratosphere. Also, the Arctic polar vortex is less stable than that over the Antarctic, because Antarctica is a land mass, colder than the water mass over the Arctic. As a result, wintertime transport of ozone toward the north pole from the tropics is stronger than in the Southern Hemisphere. All these factors combine to maintain relatively higher levels of ozone in the Arctic region. Ample evidence for perturbed ozone chemistry does, however, exist over the Arctic including observed ClO levels up to 1.4 ppb (Figure 5.23) (Salawitch et al. 1993; Webster et al. 1993). ClO concentrations are routinely as high in February in the Arctic as in September in the Antarctic but the PSCs disappear sooner in the Arctic. As a result, denitrification is the most important difference between the Antarctic and Arctic; the Arctic experiences only a modest denitrification, whereas denitrification in the Antarctic is massive. The reason for the difference is the sufficient persistence of NAT PSCs in the Antarctic to allow time to settle out of the stratosphere.

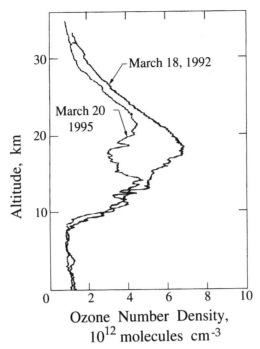

FIGURE 5.23   Vertical O_3 profiles determined by balloonborne sensors over Spitzbergen, Norway (79°N), on March 18, 1992 and March 20, 1995 (Reprinted with permission from *Chemical & Engineering News*, June 12, 1995, p. 21. Copyright 1995 American Chemical Society.)

5.8   HETEROGENEOUS (NONPOLAR) STRATOSPHERIC CHEMISTRY

Evidence now exists for significant loss of O_3 in all seasons in both hemispheres over the decades of the 1980s and 1990s. It is difficult to account for the global average (69°S–69°N) total ozone decrease of 3.5% over the 11 year period, 1979–1989, based on gas-phase chemistry alone; the reduction expected on that basis is only about 1% over that period. The process of unraveling the chemistry responsible for the Antarctic ozone hole led to the realization of the pivotal importance of heterogeneous reactions in the polar stratosphere. When midlatitude ozone losses could not be explained on the basis of gas-phase chemistry alone, it was recognized that midlatitude ozone destruction is also tied to surface reactions. In this case the reactions occur on the sulfuric acid aerosols that are ubiquitous in the stratosphere (Brasseur et al. 1990; Rodriguez et al. 1991).

5.8.1   The Stratospheric Aerosol Layer

The stratosphere contains a natural aerosol layer at altitudes of 12–30 km. This aerosol is composed of small sulfuric acid droplets with size on the order of 0.2 μm diameter and at number concentrations of $1–10\,cm^{-3}$. In the midlatitude lower stratosphere (about 16 km) the temperature is about 220 K, and the particles in equilibrium with 5 ppm H_2O have compositions of 70–75 wt% H_2SO_4. As temperature decreases, these particles absorb water to maintain equilibrium; at 195 K, they are about 40 wt% H_2SO_4.

The role of carbonyl sulfide (OCS) as a source of the natural stratospheric aerosol layer was first pointed out by Crutzen (1976). Because OCS is relatively chemically inert in the troposphere, much of it is transported to the stratosphere where it is eventually photo-dissociated and attacked by O atoms and OH radicals. The gaseous sulfur product of the chemical breakdown of OCS is SO_2, which is subsequently converted to H_2SO_4 aerosol. As noted in Section 2.2.2, an approximate, global average tropospheric OCS mixing ratio is 500 ppt. In the stratosphere the OCS mixing ratio decreases rapidly with altitude, from near 500 ppt at the tropopause to less than 10 ppt at 30 km. Chin and Davis (1995) evaluated existing data on stratospheric OCS concentrations and used the measured vertical profile of OCS to calculate an average OCS stratospheric mixing ratio of 380 ppt.

5.8.2   Heterogeneous Hydrolysis of N_2O_5

In NO_x cycle 2, which is important in the lower stratosphere, the nitrate radical NO_3 is formed from the $NO_2 + O_3$ reaction

$$NO_2 + O_3 \longrightarrow NO_3 + O_2$$

During daytime, NO_3 rapidly photolyzes, but at night, NO_3 reacts with NO_2 to produce dinitrogen pentoxide, N_2O_5:

$$NO_3 + NO_2 + M \longrightarrow N_2O_5 + M$$

N_2O_5 can decompose back to NO_3 and NO_2 either photolytically or thermally. N_2O_5 itself is photolyzed in the 200–400 nm region; since this wavelength region overlaps that of the strongest O_3 absorption, the photolysis lifetime of N_2O_5 depends on the overhead column of ozone. The photolytic lifetime of N_2O_5 is typically on the order of hours at 40 km and many days near 30 km.

The key reaction that N_2O_5 undergoes is with a water molecule to form two molecules of nitric acid. Whereas the gas-phase reaction between N_2O_5 and a H_2O vapor molecule is too slow to be of any importance, the reaction proceeds effectively on the surface of aerosol particles that contain water:

$$N_2O_5 + H_2O(s) \longrightarrow 2\,HNO_3(s)$$

This reaction has been demonstrated in the laboratory to proceed rapidly on the surface of concentrated H_2SO_4 droplets (Mozurkewich and Calvert 1988; Van Doren et al. 1991).

When an N_2O_5 molecule strikes the surface of an aqueous particle, not every collision leads to reaction. In Section 3.7, a *reaction efficiency* or *uptake coefficient* γ was introduced to account for the probability of reaction. Values of γ for this reaction ranging from 0.06 to 0.1 have been reported. The first-order rate efficient for this reaction can be expressed as in (3.38)

$$k_{N_2O_5 + H_2O(s)} = \frac{\gamma}{4} \left(\frac{8\,kT}{\pi m_{N_2O_5}} \right)^{1/2} A_p \qquad (5.31)$$

where $(8kT/\pi m_{N_2O_5})^{1/2}$ is the molecular mean speed of N_2O_5, $m_{N_2O_5}$ is the N_2O_5 molecular mass, and A_p is the aerosol surface area per unit volume ($cm^2\,cm^{-3}$). Thus,

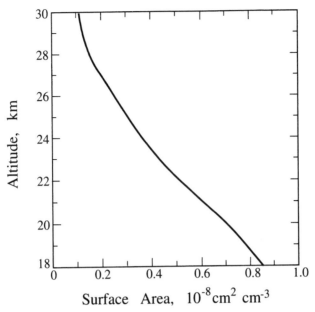

FIGURE 5.24 Variation of stratospheric aerosol surface area with altitude as inferred from satellite measurements. Maximum in surface area concentration occurs at about 18–20 km. [Reprinted from McElroy et al. (1992) with kind permission from Elsevier Science Ltd., The Boulevard, Langford Lane, Kidlington OX5 1GB, UK.]

the reaction occurs at a rate governed by that at which N_2O_5 molecules strike the aerosol surface area times the amount of surface area times the reaction efficiency. The HNO_3 formed in this reaction is assumed to be released to the gas phase immediately since the midlatitude stratosphere is undersaturated with respect to HNO_3.

Figure 5.24 shows a distribution of stratospheric aerosol surface area as a function of altitude from 18 to 30 km inferred from satellite measurements. The surface area units used in Figure 5.24 are $cm^2\ cm^{-3}$, and typical values of the stratospheric surface area at, say, 18 km altitude are about $0.8 \times 10^{-8}\ cm^2\ cm^{-3}$. This is equivalent to $0.8\mu m^2\ cm^{-3}$. As a useful rule of thumb, stratospheric aerosol surface area in the lower stratosphere ranges between 0.5 and $1.0\,\mu m^2\ cm^{-3}$.

Figure 5.25 depicts the complete NO, NO_2, NO_3, N_2O_5, HNO_3 system (top). During daytime, photolysis of NO_3 is so rapid that $[NO_3] \cong 0$, and HNO_3 is formed only by the gas-phase $OH + NO_2 + M \rightarrow HNO_3 + M$ reaction. At night, the system shifts to that shown in the bottom panel, in which HNO_3 forms entirely by the heterogenous path involving stratospheric sulfate aerosol. HNO_3 has a relatively long lifetime (~ 10 days) so that the HNO_3 formed during daylight remains at night, to be augmented by that formed heterogeneously at night.

At night, the NO_3 concentration achieves a steady state given by

$$[NO_3] = \frac{k_{NO_2+O_3}[O_3]}{k_{NO_3+NO_2+M}[M]} \tag{5.32}$$

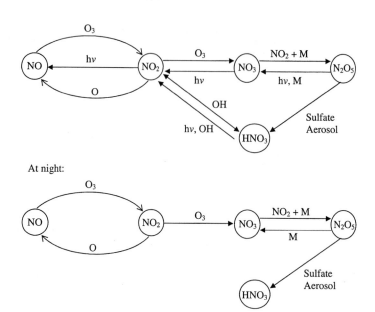

FIGURE 5.25   Stratospheric NO_y system. Lower panel shows only the paths that operate at night.

and N_2O_5 also attains a steady state by

$$[N_2O_5] = \frac{k_{NO_3+NO_2+M}[NO_2][NO_3][M]}{k_{N_2O_5+H_2O(s)}}$$
$$= \frac{k_{NO_2+O_3}[NO_2][O_3]}{k_{N_2O_5+H_2O(s)}} \tag{5.33}$$

Because of its relatively long lifetime, HNO_3 achieves a steady-state concentration over both day and night so that total production and loss are in balance:

$$k_{OH+NO_2+M}[OH][NO_2][M] + 2\,k_{N_2O_5+H_2O(s)}[N_2O_5]$$
$$= k_{OH+HNO_3}[OH][HNO_3] + j_{HNO_3}[HNO_3] \tag{5.34}$$

During daytime when $[N_2O_5] \cong 0$

$$\frac{[NO_2]}{[HNO_3]} = \frac{k_{OH+HNO_3}[OH] + j_{HNO_3}}{k_{OH+NO_2+M}[OH][M]} \tag{5.35}$$

As altitude increases, $[M]$ decreases, and the $[NO_2]/[HNO_3]$ ratio increases. At night, $j_{HNO_3} = 0$ and $[OH] \cong 0$ so HNO_3 builds up. However, over a full daily cycle HNO_3 achieves a steady state given by

$$\frac{[NO_2]}{[HNO_3]} = \frac{k_{OH+HNO_3}[OH] + j_{HNO_3}}{k_{OH+NO_2+M}[OH][M] + 2\,k_{N_2O_5+H_2O(s)}\dfrac{[N_2O_5]}{[NO_2]}}$$
$$= \frac{k_{OH+HNO_3}[OH] + j_{HNO_3}}{k_{OH+NO_2+M}[OH][M] + 2\,k_{NO_2+O_3}[O_3]} \tag{5.36}$$

The "bottleneck" reaction in the heterogeneous formation of HNO_3 is that of NO_2 and O_3. As altitude increases, temperature increases and the rate of this reaction increases; the faster this reaction, the more aerosol surface area is needed to effectively compete for the increased amount of N_2O_5. The conversion of NO_2 to HNO_3 at midlatitude saturates for aerosol loadings close to typical background levels of $0.5\,\mu m^2\,cm^{-3}$ near 20 km; at 30 km, saturation occurs at aerosol loadings of about $3\,\mu m^2\,cm^{-3}$. However, from Figure 5.24 we note that aerosol surface area concentration decreases with increasing altitude, so the amount of NO_x present, as opposed to NO_y, actually increases with altitude because the hydrolysis of N_2O_5 becomes aerosol surface area–limited.

Returning to Figure 5.9, we can now explain the observed altitude variation of HNO_3 and NO_2. The stratospheric aerosol layer resides between about 20 and 24 km, and it is in this region that heterogeneous formation of HNO_3 is most effective. At altitudes above the stratospheric aerosol layer, the daytime formation of HNO_3 by $OH + NO_2 + M$ runs out of [M] as altitude increases; for this reason the HNO_3 concentration drops off with altitude, and the $[NO_2]/[HNO_3]$ ratio increases in accord with (5.36). The conversion of a reservoir species N_2O_5 into a relatively stable species HNO_3 serves to remove NO_2 from the active catalytic NO_x system, reducing the effectiveness of O_3 destruction by the NO_x cycle. The reduction of NO_2, on the other hand, decreases the formation of $ClONO_2$, allowing more ClO to accumulate than in the absence of the $N_2O_5 + H_2O(s)$ reaction. More ClO means that the effectiveness of the ClO_x cycles is increased. Although photolysis of HNO_3, with the H atom derived from aerosol H_2O, provides an additional source of HO_x to the system, this is not predicted to have a major effect on the HO_x cycles. Effectively, the hydrolysis of N_2O_5 on stratospheric aerosol moves NO_2 from $ClONO_2$ with a 1-day lifetime to HNO_3 with a 10-day lifetime, leading to substantially lower NO_x concentrations. As a result, lower stratospheric O_3 becomes more sensitive to halogen and HO_x chemistry and less sensitive to the addition of NO_x.

Effect of N_2O_5 Hydrolysis on the Chemistry of the Lower Stratosphere

$$NO_2 + O_3 \xrightarrow{1} NO_3 + O_2 \qquad \text{(reaction 1)}$$

$$NO_3 + NO_2 + M \xrightarrow{2} N_2O_5 + M \qquad \text{(reaction 2)}$$

$$NO_3 + h\nu \xrightarrow{3a} NO_2 + O \qquad \text{(reaction 3a)}$$

$$\xrightarrow{3b} NO + O_2 \qquad \text{(reaction 3b)}$$

NO_3 forms from reaction 1 and is consumed by reactions 2 and 3. During daytime, photolysis of NO_3 is very efficient; j_{NO_3} from both channels 3a and 3b is about $0.3\,s^{-1}$, giving a mean daytime lifetime of NO_3 of 3 s. At night, reaction 2 is the only removal pathway for NO_3. The reaction rate coefficient of reaction 2 from Table B.2 has $k_0 = 2.2 \times 10^{-30}(T/300)^{-3.9}$ and $k_\infty = 1.5 \times 10^{-12}(T/300)^{-0.7}\,cm^3$ molecule$^{-1}\,s^{-1}$. At 20 km, $T = 220\,K$, $[M] = 2 \times 10^{18}$ molecules cm^{-3}, and the Troe formula gives $k_2 = 1.2 \times 10^{-12}\,cm^3$ molecule$^{-1}\,s^{-1}$. Assuming that NO_2 is present at 1 ppb, the mean lifetime of NO_3 against reaction 2 is 400 s. In summary, the lifetime of a molecule of NO_3 at 20 km increases from 3 s in daytime to about 7 min at night.

Each NO_3 molecule formed at night is converted to N_2O_5 in about 7 min, so over the course of a 12-h night, every time reaction 1 occurs, a molecule of N_2O_5

is formed, with consumption of two NO_2 molecules. NO_x is converted to N_2O_5 at the following rate:

$$\frac{d[NO_x]}{dt} = -2\,k_1[O_3][NO_2] \qquad k_1 = 1.2 \times 10^{-13}\exp(-2450/T)$$

At $T = 220\,K$, $k_1 = 1.7 \times 10^{-18}\,cm^3\,molecule^{-1}\,s^{-1}$. The lifetime of NO_x against reaction 1 is

$$\tau_{NO_x} = \frac{1}{2\,k_1[O_3]}$$

With $[O_3] = 5 \times 10^{12}$ molecules cm^{-3}, $\tau_{NO_x} = 5.9 \times 10^4\,s$. The fractional decay of NO_x is given by $\exp(-t/\tau_{NO_x})$. Over a single 12-h night ($4.3 \times 10^4\,s$), 52% of the NO_x is converted to N_2O_5.

N_2O_5 is hydrolyzed on stratospheric aerosol to produce nitric acid:

$$N_2O_5 + H_2O(s) \xrightarrow{4} 2HNO_3 \qquad\qquad \text{(reaction 4)}$$

The conversion can be represented as a pseudo-first-order reaction with rate coefficient (5.31)

$$k_4 = \frac{\gamma}{4}\left(\frac{8\,kT}{\pi m_{N_2O_5}}\right)^{1/2} A_p$$

which depends on the stratospheric aerosol surface area A_p and the reactive uptake coefficient γ. Previously, we noted that a representative value for γ for reaction 4 is 0.06. Values of k_4 for $A_p = 1 \times 10^{-8}\,cm^2\,cm^{-3}$ and $10 \times 10^{-8}\,cm^2\,cm^{-3}$ at $T = 220\,K$ are

$$k_4 = 3.1 \times 10^{-6}\,s^{-1} \qquad \text{at} \quad A_p = 1 \times 10^{-8}\,cm^2\,cm^{-3}$$
$$= 3.1 \times 10^{-5}\,s^{-1} \qquad \text{at} \quad A_p = 10 \times 10^{-8}\,cm^2\,cm^{-3}$$

The lifetime of N_2O_5 against reaction 4 is just $\tau_{N_2O_5} = 1/k_4$.

What fraction of NO_x is converted to HNO_3 over the nighttime? N_2O_5 is gradually produced throughout the night at a rate determined by reaction 1. As N_2O_5 is produced, N_2O_5 itself is gradually converted to HNO_3 by reaction 4. The system is the classic reactions in series: $A \longrightarrow B \longrightarrow C$. In this case, $A = NO_2$, $B = N_2O_5$, and $C = HNO_3$. (Here, two molecules of C are produced in the $B \rightarrow C$ reaction.) To determine the amount of C formed over the night starting from a given concentration of A at sundown, one must integrate the rate equations:

$$\frac{d[A]}{dt} = -\frac{[A]}{\tau_{NO_x}}$$

$$\frac{d[B]}{dt} = \frac{1}{2}\frac{[A]}{\tau_{NO_x}} - \frac{[B]}{\tau_{N_2O_5}}$$

$$\frac{d[C]}{dt} = \frac{[B]}{\tau_{N_2O_5}}$$

The factor of $\frac{1}{2}$ in the rate equation for [B] reflects the fact that each molecule of B formed requires the consumption of two molecules of A. The fractional conversion of an initial concentration of A, $[A]_0$ into C is $2[C]/[A]_0$, where the factor of 2 accounts for the fact that two molecules of HNO_3 are formed in reaction 4.

The solutions for [A], [B], and [C] are

$$[A(t)] = [A]_0 \exp\left(-\frac{t}{\tau_{NO_x}}\right)$$

$$[B(t)] = \frac{[A]_0/2\tau_{NO_x}}{\frac{1}{\tau_{N_2O_5}} - \frac{1}{\tau_{NO_x}}} \left[\exp\left(-\frac{t}{\tau_{NO_x}}\right) - \exp\left(-\frac{t}{\tau_{N_2O_5}}\right)\right]$$

$$[C(t)] = \frac{[A]_0}{2}\left\{1 + \frac{1}{\tau_{NO_x} - \tau_{N_2O_5}}\left[\tau_{N_2O_5}\exp\left(-\frac{t}{\tau_{N_2O_5}}\right) - \tau_{NO_x}\exp\left(-\frac{t}{\tau_{NO_x}}\right)\right]\right\}$$

The fraction of NO_x converted to HNO_3 over a time period t is thus

$$f_{NO_x \to HNO_3} = 1 + \frac{1}{\tau_{NO_x} - \tau_{N_2O_5}}\left[\tau_{N_2O_5}\exp\left(-\frac{t}{\tau_{N_2O_5}}\right) - \tau_{NO_x}\exp\left(-\frac{t}{\tau_{NO_x}}\right)\right]$$

We are interested in $t = 12\,h = 4.3 \times 10^4\,s$. From before, $\tau_{NO_x} = 5.9 \times 10^4\,s$. For $A_p = 1 \times 10^{-8}\,cm^2\,cm^{-3}$, $\tau_{N_2O_5} = 3.2 \times 10^5\,s$; for $A_p = 10 \times 10^{-8}\,cm^2\,cm^{-3}$, $\tau_{N_2O_5} = 3.2 \times 10^4\,s$. We find

$$f_{NO_x \to HNO_3} = 0.04 \quad \text{at} \quad A_p = 1 \times 10^{-8}\,cm^2\,cm^{-3}$$
$$= 0.26 \quad \text{at} \quad A_p = 10 \times 10^{-8}\,cm^2\,cm^{-3}$$

The conversion rate of NO_x to HNO_3 through N_2O_5 depends on the two lifetimes, $\tau_{NO_x} = (2k_1(T)[O_3])^{-1}$ and $\tau_{N_2O_5} = [\frac{\gamma}{4}(8kT/\pi\,m_{N_2O_5})^{1/2}A_p]^{-1}$. The smaller these two characteristic times, the faster the conversion rate. The greatest sensitivity of the conversion is to T (k_1 is a strong function of temperature) and A_p; the larger T and A_p are, the greater the conversion to nitric acid. The largest values of A_p occur when the stratospheric aerosol is perturbed by a volcano or by formation of PSCs in the polar stratosphere.

5.8.3   Effect of Volcanoes on Stratospheric Ozone

Volcanoes inject gaseous SO_2 and HCl directly into the stratosphere. Because of its large water solubility, HCl is rapidly removed by liquid water. The SO_2 remains and is converted to H_2SO_4 aerosol by reaction with the OH radical. (We will discuss this reaction in Chapter 6 and the nucleation process that leads to fresh particles in Chapter 11.) Thus volcanic eruptions lead to an increase in the amount of the stratospheric aerosol layer (Figure 5.26). As we have just seen, "normal" stratospheric aerosol surface area concentrations lie in the range of $0.5\text{--}1.0\,\mu m^2\,cm^{-3}$. Eruption of Mt. Pinatubo in the Philippines in June 1991, the largest volcanic eruption of the twentieth century, led to average midlatitude stratospheric aerosol surface areas of $20\,\mu m^2\,cm^{-3}$. In the core of the plume, 5 months after the eruption, the aerosol surface area concentration was $35\,cm^2\,cm^{-3}$ (Grainger et al. 1995). The time required for stratospheric aerosol levels to decay back

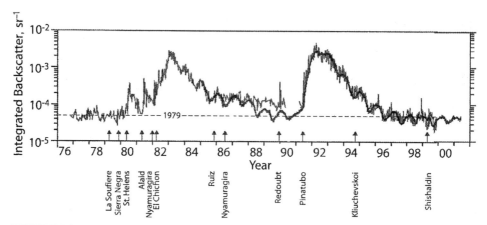

FIGURE 5.26 Time series of the abundance of stratospheric aerosols, as inferred from integrated backscatter measurements (WMO 2002). The data stream beginning in 1976 was acquired by ground-based lidar at Garmisch, Germany (47.5°N). That beginning in 1985 is from the SAGE II satellite over the latitude band 40–50°N. Vertical arrows show major volcanic eruptions. The dashed line indicates the 1979 level. (Data from Garmisch provided courtesy of H. Jäger.)

to "normal" levels following a volcanic eruption is about 2 years. With volcanoes erupting somewhere on Earth every few years or so, the stratosphere is seldom in a state totally uninfluenced by volcanic emissions.

Volcanic injection of large quantities of sulfate aerosol into the stratosphere offers the opportunity to examine the sensitivity of ozone depletion and species concentrations to a major perturbation in aerosol surface area (Hofmann and Solomon 1989; Johnston et al. 1992; Prather 1992; Mills et al. 1993). The increase in stratospheric aerosol surface area resulting from a major volcanic eruption can lead to profound effects on ClO_x-induced ozone depletion chemistry. Because the heterogeneous reaction of N_2O_5 and water on the surface of stratospheric aerosols effectively removes NO_2 from the active reaction system, less NO_2 is available to react with ClO to form the reservoir species $ClONO_2$. As a result, more ClO is present in active ClO_x cycles. Therefore an increase in stratospheric aerosol surface area, as from a volcanic eruption, can serve to make the chlorine present more effective at ozone depletion, *even if no increases in chlorine are occurring.*

Figure 5.27 shows in situ stratospheric data (solid circles) on NO_x/NO_y and ClO/Cl_y ratios plotted as a function of stratospheric aerosol surface area concentration (Fahey et al. 1993). Figure 5.27 shows comparisons of measurements made in September 1991 in a region of the atmosphere that had not been strongly impacted by the Mt. Pinatubo aerosol with those made in March 1992 at a similar latitude and solar zenith angle where the effect of the eruption is clearly present. In the March data the aerosol surface area concentration was $20\,\mu m^2\,cm^{-3}$, a factor of 40 over the "natural" level of $0.5\,\mu m^2\,cm^{-3}$ (vertical dashed line). The solid and dashed curves are model-calculated relationships for these two ratios for the September (solid) and March (dashed) data sets. The open circles on the NO_x/NO_y plot at a surface area of $20\,\mu m^2\,cm^{-3}$ indicate calculations assuming gas-phase chemistry only; the crosses are calculations including heterogeneous hydrolysis of N_2O_5. The measured ratio of NO_x/NO_y decreased as aerosol surface area increased and the ClO/Cl_y ratio increased. Both of these effects are expected as a result of the heterogeneous hydrolysis of N_2O_5 on aerosol surfaces. Increasing aerosol surface area from about 1 to

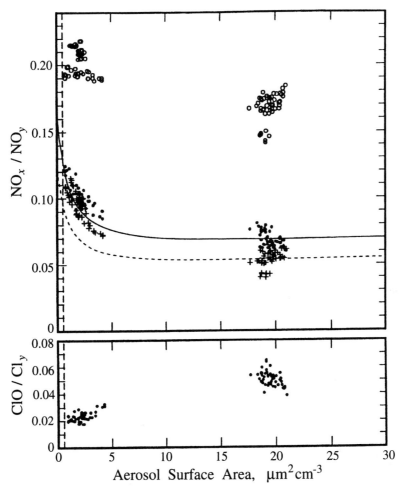

FIGURE 5.27 Measured ratios NO_x/NO_y and ClO/Cl_y (solid circles) shown as a function of stratospheric aerosol surface area (Fahey et al. 1993). The open circles represent simulations of the conditions of the measurements accounting for gas-phase chemistry only; the crosses are the corresponding simulations including the heterogeneous hydrolysis of N_2O_5. The vertical dashed line represents stratospheric background aerosol surface area ($0.5\,\mu m^2\,cm^{-3}$). The curved lines represent the modeled dependence of the two ratios on aerosol surface area for the average conditions in the September (solid) and March (dashed) data sets. (Reprinted with permission from *Nature* **363**, Fahey, D.W., et al. Copyright 1993 Macmillan Magazines Limited.)

$20\,\mu m^2\,cm^{-3}$ led to an increase in the ClO/Cl_y ratio from about 0.025 to about 0.05. Thus this increase in aerosol surface area from a large volcano renders the chlorine already present in the stratosphere *twice as effective* in ozone depletion as in the absence of the volcano. Even at a more modest scale, an increase of aerosol surface area from 1 to $5\,\mu m^2\,cm^{-3}$ is estimated to increase ClO/Cl_y from 0.02 to 0.03 and thereby render the chlorine 50% more effective in ozone depletion. Thus one volcanic eruption, at least during the 2 years or so following the eruption when stratospheric aerosol levels are elevated, can produce the same ozone-depleting effect as a decade of increases in CFC emissions [e.g., see Tie et al. (1994)]. Conversely, in the absence of chlorine in the

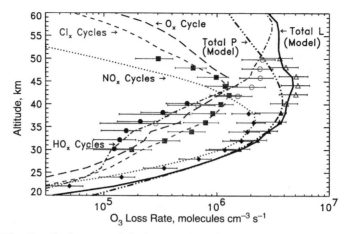

FIGURE 5.28   Contribution to total O_3 loss rate by different catalytic cycles. [After Osterman et al. (1997) updated to current kinetic parameters.]

stratosphere, stratospheric ozone would *increase* after a volcanic eruption as a result of the removal of active NO_x by the heterogeneous $N_2O_5 + H_2O$ reaction.

5.9   SUMMARY OF STRATOSPHERIC OZONE DEPLETION

Figure 5.28 shows the total ozone loss rate as a function of altitude. Chemical destruction of O_3 in the lower stratosphere ($\lesssim 25$ km) is slow. In this region of the stratosphere, O_3 has a lifetime of months. Rates exceeding 10^6 molecules cm^{-3} s^{-1} are achieved only $\gtrsim 28$ km. Figure 5.29 gives the fractional contributions of the O_x, NO_x, HO_x, and halogen cycles to the total ozone loss rate. In many respects, Figures 5.28 and 5.29 represent the culmination of this chapter—the synthesis of how the various ozone depletion cycles interact at

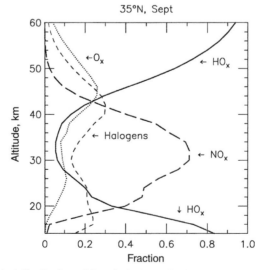

FIGURE 5.29   Vertical distribution of the relative contributions by different catalytic cycles to the O_3 loss rate (WMO 1998). [After Osterman et al. (1997) updated to current kinetic parameters.]

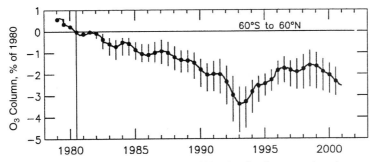

FIGURE 5.30 Change in O_3 column relative to 1980 value for the atmosphere between $60°$S and $60°$N. Measurements based on a variety of satellite and ground-based instruments (WMO 2002).

different altitudes in the stratosphere. Figure 5.30 summarizes global total ozone change relative to the 1964–1980 average.

In order to explain the fractional contributions in Figure 5.29 we can consider both the rate coefficients of the key reactions (Figure 5.31) and the profiles of O, NO_2, HO_2, O_3, BrO, and ClO (Figure 5.32). The rates of the rate-determining reactions in each of the catalytic cycles can then be estimated at different altitudes. For example, at 15 km ($T \sim 215$ K), the rates of key reactions (molecules cm^{-3} s^{-1}) can be estimated as follows:

$$R_{NO_2+O} = k_{NO_2+O}[NO_2][O] \cong (1 \times 10^{-11}) \times (4 \times 10^8) \times (1 \times 10^3) \cong 5.0$$

$$R_{HO_2+O_3} = k_{HO_2+O_3}[HO_2][O_3] \cong (1 \times 10^{-15}) \times (2 \times 10^6) \times (5 \times 10^{11}) \cong 1 \times 10^3$$

$$R_{ClO+BrO \to BrCl} = k_{ClO+BrO \to BrCl}[ClO][BrO] \cong (2 \times 10^{-12}) \times (1 \times 10^3) \times (2 \times 10^6) \cong 3 \times 10^{-3}$$

$$R_{ClO+BrO \to ClOO} = k_{ClO+BrO \to ClOO}[ClO][BrO] \cong (8 \times 10^{-12}) \times (1 \times 10^3) \times (2 \times 10^6) \cong 2 \times 10^{-2}$$

$$R_{BrO+HO_2} = k_{BrO+HO_2}[BrO][HO_2] \cong (4 \times 10^{-11}) \times (2 \times 10^6) \times (2 \times 10^6) \cong 2 \times 10^2$$

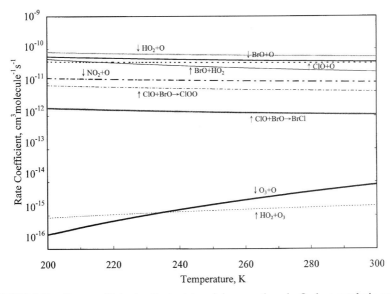

FIGURE 5.31 Rate coefficients of rate-determining reactions in O_3 loss catalytic cycles.

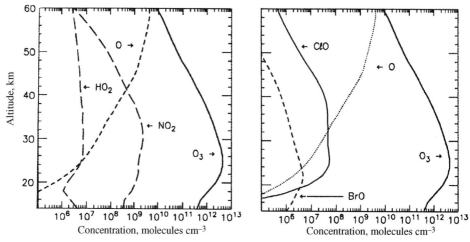

FIGURE 5.32 Stratospheric concentration profiles of key species involved in O_3 loss catalytic cycles (35°N, Sept.).

The total estimated O_3 loss rate is $\sim 1 \times 10^3$ molecules cm^{-3} s^{-1}, approximately 90% of which is due to $HO_2 + O_3$ (HO_x cycle 2) and 10% to $BrO + HO_2$ (HO_x/BrO_x cycle 1).

At 30 km ($T \sim 255$K), a similar analysis leads to

$$R_{O_3+O} = k_{O_3+O}[O_3][O] \cong (8 \times 10^{-16}) \times (3 \times 10^{12}) \times (3 \times 10^7) \cong 8 \times 10^4$$

$$R_{HO_2+O} = k_{HO_2+O}[HO_2][O] \cong (7 \times 10^{-11}) \times (1 \times 10^7) \times (3 \times 10^7) \cong 2 \times 10^4$$

$$R_{NO_2+O} = k_{NO_2+O}[NO_2][O] \cong (1 \times 10^{-11}) \times (2 \times 10^9) \times (3 \times 10^7) \cong 7 \times 10^5$$

$$R_{ClO+O} = k_{ClO+O}[ClO][O] \cong (4 \times 10^{-11}) \times (5 \times 10^7) \times (3 \times 10^7) \cong 6 \times 10^4$$

$$R_{BrO+O} = k_{BrO+O}[BrO][O] \cong (5 \times 10^{-11}) \times (2 \times 10^6) \times (3 \times 10^7) \cong 3 \times 10^3$$

The total estimated O_3 loss rate is $\sim 9 \times 10^5$ molecules cm^{-3} s^{-1}, about 80% of which is attributable to $NO_2 + O$ (NO_x cycle 1) and 10% each to the Chapman cycle and $ClO + O$ (ClO_x cycle 1).

Below ~ 20 km, HO_x cycle 2 is dominant in O_3 removal. Between 20 and 40 km, NO_x cycles dominate O_3 loss; NO_x cycle 2 in the lower portion of this range, and NO_x cycle 1 in the upper portion. Above about 40 km, HO_x cycles are again dominant (HO_x cycle 1). Because of the rapid release of Br in the lower stratosphere, BrO_x cycles are approximately comparable in importance to ClO_x cycles at 20 km. The coupled BrO_x/ClO_x cycle is itself responsible for $\sim 25\%$ of the overall halogen-controlled loss in the lower stratosphere. The halogen cycles achieve two maxima: (1) one in the lowermost stratosphere due to HO_x/ClO_x cycle 1, HO_x/BrO_x cycle 1, and BrO_x/ClO_x cycles 1 and 2; and (2) one in the upper stratosphere due to ClO_x cycle 1.

NO_x is the dominant O_3-reducing catalyst between 25 and 35 km. Because of coupling among the HO_x, ClO_x, and NO_x cycles, and the role that NO_x plays in that coupling, the rate of O_3 consumption is very sensitive to the concentrations of NO and NO_2. Partitioning between OH and HO_2 by $HO_2 + NO \longrightarrow NO_2 + OH$ is controlled by the level of NO. This reaction short-circuits HO_x cycle 4 by reducing the concentration of HO_2.

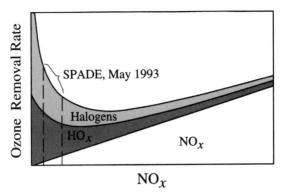

FIGURE 5.33 Qualitative behavior of the ozone removal rate in the lower stratosphere as a function of NO_x level (Wennberg et al. 1994). SPADE stands for *Stratospheric Photochemistry, Aerosols, and Dynamics Expedition*. This is a cumulative plot; the contributions of the various processes add to produce the upper curve.

Figure 5.33 shows schematically the overall O_3 loss rate for the midlatitude lower stratosphere as a function of the NO_x level. At very low NO_x concentrations, the O_3 loss rate increases as NO_x decreases. This is a result of the absence of NO_x to tie up reactive hydrogen and chlorine in reservoir species. As NO_x increases, a critical value is reached where O_3 removal through the reaction, $O + NO_2$, is equal to the sum of O_3 loss by the HO_x and ClO_x cycles. At this point, increases in NO_x result in comparable reductions in the concentrations of HO_2 and ClO. As the chlorine content of the stratosphere has increased over time, this crossover point has been shifted to higher NO_x concentrations. As $[NO_x]$ increases further, the NO_x cycle eventually becomes dominant, and O_3 loss increases linearly with $[NO_x]$.

The dependence of the O_3 loss rate on NO_x at higher altitudes in the stratosphere is different from that in Figure 5.33. At higher altitudes, the increasing amount of atomic oxygen means that the catalytic cycles that rely on O atoms become increasingly important. As a result, substantially higher O_3 loss rates occur in the upper stratosphere than in the lower stratosphere. The fractional contribution to O_3 loss by HO_x becomes dominant above 45 km. As altitude increases, OH and HO_2 concentrations fall off slowly, while those of ClO_x and NO_x fall off rapidly. The ClO/HCl ratio decreases because of a shift in the Cl/ClO ratio favoring Cl and production of HCl from $Cl + CH_4$. The decrease in NO_x at high altitudes results because the $N + NO$ reaction (recall Figure 5.8) begins to become important, even though the NO_x/NO_y ratio shifts in favor of NO_x.

5.10 TRANSPORT AND MIXING IN THE STRATOSPHERE

A latitude–height schematic cross section of the upper troposphere and lower stratosphere is shown in Figure 5.34. The lowermost stratosphere is sometimes referred to as the "middleworld" and that above the 380 K potential temperature level as the "overworld." Tropospheric air enters the stratosphere principally in the tropics, as a result of deep cumulus convection, and then moves poleward in the stratosphere. By overall mass conservation, stratospheric air must return to the troposphere. This return occurs in midlatitudes; the overall circulation is called the *Brewer–Dobson circulation*. The midlatitude

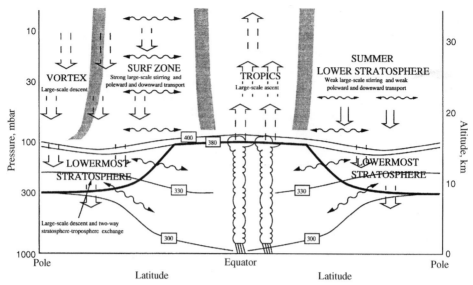

FIGURE 5.34   Stratospheric transport and mixing (UK National Environment Research Council). Temperature isotherms are labeled as potential temperature. Potential temperature θ at any altitude z is the temperature of an air parcel brought from that altitude adiabatically to the surface. It can be calculated from the relation $\theta = T(p/p_0)^{-2/7}$, where T and p are the temperature and pressure at altitude z and p_0 is the surface pressure, usually taken to be 1013 hPa. (Figure available at http://utls.nerc.ac.uk/background/sp/utls_sp_fig1a.pdf.)

stratosphere is disturbed by breaking planetary waves, forming what is referred to as a "surf zone." The rising air in the tropical stratosphere has given rise to the concept of a "tropical pipe," where tropical air is assumed to advect upward with no exchange with midlatitude stratospheric air. Observed vertical profiles of chemical species cannot, however, be reconciled with a strict interpretation of the tropical pipe; some mixing with midlatitude air is required, especially below ~ 22 km. By considering dilution, vertical ascent, and chemical decay, these so-called "leaky pipe" calculations estimate that by 22 km, close to 50% of the air in the tropical upwelling region is of midlatitude origin. Volk et al. (1996) estimated that about 45% of air of extratropical origin accumulates in a tropical air parcel during its 8-month ascent from the tropopause to 21 km. This lateral mixing is depicted by the wavy arrows in Figure 5.34. On analysis of relative vertical profiles of CO_2 and N_2O, Boering et al. (1996) estimated the following vertical ascent rates in the stratosphere:

$$NH\ summer: \quad 0.21\ mm\ s^{-1}\ (17\ m\ day^{-1})$$

$$NH\ winter: \quad 0.31\ mm\ s^{-1}\ (26\ m\ day^{-1})$$

The tropical tropopause, essentially defined by the 380 K potential temperature surface (see Chapter 21), is located at about 16–17 km. Tropical convection occurs up to an altitude of about 11–12 km. Between these two levels lies a transition region between the tropopause and the stratosphere, call the *tropical tropopause layer* (TTL). Together with the lowermost stratosphere, the region is referred to as the *upper troposphere/lower stratosphere* (UT/LS). Convection does penetrate into the TTL and deliver air from the boundary layer, and within the TTL lateral mixing occurs with subtropical air of stratospheric character. Within the TTL, ozone profiles exhibit a continuous transition from low values characteristic of the upper troposphere to the higher values characteristic of the stratosphere, with no special signature at the tropopause itself.

The flux of ozone from the stratosphere to the troposphere is taken as that across the 380 K potential temperature surface at mid and high latitudes. This ozone is of stratospheric origin, because the flux of air from the upper tropical troposphere into the lowermost stratosphere contains very little ozone. The ozone flux from the stratosphere to the troposphere is estimated by general circulation models.

As tropospheric air enters the tropical stratosphere, the cold temperatures encountered freeze out the water vapor and dehydrate the air, on average, to an H_2O mixing ratio of about 4 ppm. Once in the stratosphere, subsequent oxidation of CH_4 and H_2 can lead to H_2O vapor mixing ratios as high as ~ 8 ppm (4 ppm $H_2O + 2 \times 1.75$ ppm $CH_4 +$ 0.5 ppm H_2). Water vapor crossing the extratropical tropopause experiences higher temperatures than air that enters via the tropical tropopause, leading to water vapor mixing ratios in the lowermost stratosphere of tens of ppm (Dessler et al. 1995).

5.11 OZONE DEPLETION POTENTIAL

The stratospheric ozone-depleting potential of a compound emitted at the Earth's surface depends on how much of it is destroyed in the troposphere before it gets to the stratosphere, the altitude at which it is broken down in the stratosphere, and chemistry subsequent to its dissociation. Halocarbons containing hydrogen in place of halogens or containing double bonds are susceptible to attack by OH in the troposphere. (We will consider the mechanisms of such reactions in Chapter 6.) The more effective the tropospheric removal processes, the less of the compound that will survive to reach the stratosphere. Once halocarbons reach the stratosphere their relative importance in ozone depletion depends on the altitude at which they are photolyzed and the distribution of halogen atoms, Cl, Br, and F, contained within the molecule.

Different halocarbons are photolyzed at different wavelengths. Those photolyzed at shorter wavelengths must reach higher altitudes in the stratosphere to be photodissociated. Generally, substitution of fluorine atoms for chlorine atoms shifts absorption to shorter wavelengths. For example, maximum photolysis of $CFCl_3$ occurs at about 25 km, that for CF_2Cl_2 at 32 km, and that for $CClF_2CF_3$ at 40 km (Wayne 1991). Molecule for molecule, more heavily chlorinated halocarbons are more effective in destroying ozone than are those containing more fluorine; they are photolyzed at lower altitudes where the amount of ozone is greater, and the chlorine atom is much more efficient at ozone destruction than fluorine. Atom for atom, however, bromine is, as noted earlier, an even more efficient catalyst for ozone destruction than chlorine. Also, photolysis of bromine-containing compounds occurs lower in the stratosphere, where impact is the greatest. The atmospheric photolytic lifetime is shortest for molecules containing relatively more bromine than chlorine or fluorine, followed next by those with relatively more chlorine.

The effect of a compound on stratospheric ozone depletion is usually assessed by simulating transport and diffusion, chemistry, and photochemistry in a one-dimensional vertical column of air from the Earth's surface to the top of the stratosphere at a given latitude. Two-dimensional models have also been used. One-dimensional models calculate the global averaged vertical variation of the chemical degradation processes. Two-dimensional models allow examination of the calculated effects as a function of latitude and time of year. Ozone changes calculated with two-dimensional models are then averaged with respect to both latitude and season to obtain a single ozone depletion value for the compound of interest. If a constant rate of emission is assumed, then eventually a steady state will be reached such that the concentration of the compound is unchanging in time at all levels of the atmosphere. The *ozone depletion potential* (ODP) of a compound is usually defined as the total steady-state ozone destruction, vertically integrated over the stratosphere, that results per unit mass of species i emitted per year relative to that for a unit mass emission of CFC-11:

$$\text{ODP}_i = \frac{\Delta O_{3_i}}{\Delta O_{3_{\text{CFC}-11}}}$$

Thus ODP_i is a relative measure.

Table 5.3 presents ozone depletion potentials calculated with two-dimensional models, as presented by the World Meteorological Organization (2002). Tropospheric OH reaction is the primary sink for the hydrogen-containing species. As seen in Table 5.3, the ODPs for the hydrogen-containing species are considerably smaller than those of the CFCs, the difference reflecting the more effective chemical removal in the troposphere. Although Table 5.3 gives the ODPs of several brominated halocarbons relative to CFC-11,

TABLE 5.3   Steady-State Ozone Depletion Potentials (ODPs) Derived from Two-Dimensional Models[a]

Trace Gas	ODP
CFC-11 ($CFCl_3$)	1.0
CFC-12 (CF_2Cl_2)	0.82
CFC-113 ($CFCl_2CF_2Cl$)	0.90
CFC-114 (CF_2ClCF_2Cl)	0.85
CFC-115 (CF_2ClCF_3)	0.40
CCl_4	1.20
CH_3CCl_3	0.12
HCFC-22 (CF_2HCl)	0.034
HCFC-123 (CF_3CHCl_2)	0.012
HCFC-124 (CF_3CHFCl)	0.026
HCFC-141b (CH_3CFCl_2)	0.086
HCFC-142b (CH_3CF_2Cl)	0.043
HCFC-225ca ($CF_3CF_2CHCl_2$)	0.017
HCFC-225cb (CF_2ClCF_2CHFCl)	0.017
CH_3Br	0.37
H-1301 (CF_3Br)	12
H-1211 (CF_2ClBr)	5.1
H-1202 (CF_2Br_2)	1.3
H-2402 (CF_2BrCF_2Br)	<8.6
CH_3Cl	0.02

[a]World Meteorological Organization (2002).

the ODPs for bromine-containing compounds should really be compared relative to each other because of the strong coupling of bromine catalytic cycles to chlorine levels in the lower stratosphere. The halon that is generally selected as the reference compound for ODPs of bromine-containing halocarbons is H-1301 (CF_3Br), the halon with the longest atmospheric lifetime and largest ODP.

PROBLEMS

5.1$_A$ The rate coefficients for $O(^1D) + O_3 \longrightarrow 2O_2$ and $O + O_3 \longrightarrow 2O_2$ are given below. In the middle stratosphere, at about 30 km, $T = 230$ K, typical noontime concentrations of $O(^1D)$ and O are 50 molecules cm^{-3} and 7.5×10^7 molecules cm^{-3}, respectively. Evaluate the rates of both O_3 loss processes at this altitude. Which loss process dominates?

$$k_{O(^1D)+O_3} = 1.2 \times 10^{-10} \text{ cm}^3 \text{ molecule}^{-1} \text{ s}^{-1}$$
$$k_{O+O_3} = 8 \times 10^{-12} \exp(-2060/T) \text{ cm}^3 \text{ molecule}^{-1} \text{ s}^{-1}$$

5.2$_A$ Estimate the lifetime of odd oxygen (i.e., ozone) at 20 km and 45 km. It may be assumed that the photolysis rates of O_3 at these two altitudes are (see Figure 4.14)

$$j_{O_3} = 7 \times 10^{-4} \text{ s}^{-1} \quad \text{at} \quad 20 \text{ km}$$
$$= 6 \times 10^{-3} \text{ s}^{-1} \quad \text{at} \quad 45 \text{ km}$$

and that the O_3 concentrations at the two altitudes are (see Figure 5.2)

$$[O_3] = 2 \times 10^{12} \text{ molecules cm}^{-3} \quad \text{at} \quad 20 \text{ km}$$
$$= 0.2 \times 10^{12} \text{ molecules cm}^{-3} \quad \text{at} \quad 45 \text{ km}$$

5.3$_A$ Estimate the stratospheric lifetime of N_2O at 30 km using Figure 5.6.

5.4$_A$ For the catalytic cycle

$$Cl + O_3 \xrightarrow{1} ClO + O_2 \qquad k_1 = 2.9 \times 10^{-11} \exp(-260/T)$$
$$ClO + O \xrightarrow{2} Cl + O_2 \qquad k_2 = 3.0 \times 10^{-11} \exp(70/T)$$

determine the ratio of $[ClO]/[O_3]$ at which the rate of ozone destruction from this cycle equals that from the Chapman cycle at 20 km where $T \cong 220$ K. If the O_3 mixing ratio is 4 ppm, what concentration of ClO does this correspond to?

5.5$_A$ **a.** Write out the catalytic ozone loss cycle that is initiated by the self-reaction of ClO, which is a key cycle for loss of polar ozone.

   b. The ozone abundance at 20 km altitude fell from 2.0 ppm on August 23 to 0.8 ppm on September 22. Assuming that the self-reaction of ClO is the rate-limiting step for ozone loss at 20 km, calculate the mixing ratio of ClO required to account for the observed loss of ozone. Assume $T = 200$ K and $p = 60$ mbar. Since the catalytic cycle involves a photolytic process, it occurs only during periods of daylight. Between August 23 and September 22, air in the polar vortex experiences roughly 8 h of sunlight per day.

5.6$_B$ The dominant reaction producing HCl in the stratosphere is

$$Cl + CH_4 \longrightarrow HCl + CH_3$$

Observations of ClO, HCl, and ClONO$_2$ in the midstratosphere could not, however, be reconciled with this reaction as the sole source of HCl until it was discovered that HCl is formed as a minor channel in the reaction of OH with ClO:

$$OH + ClO \longrightarrow HO_2 + Cl \qquad k = 7.4 \times 10^{-12} \exp(270/T)$$
$$\longrightarrow HCl + O_2 \qquad k = 3.2 \times 10^{-13} \exp(320/T)$$

Compare the strength of this HCl source to that of Cl + CH$_4$ at 20 and 35 km. Assume that $[OH] = 1 \times 10^6$ and 1×10^7 molecules cm^{-3} at 20 and 35 km, respectively. The following conditions can be assumed:

	20 km	35 km
$[NO]$	5×10^8	2×10^9 molecules cm^{-3}
$[O_3]$	3×10^{12}	1×10^{12} molecules cm^{-3}
j_{O_3}	5×10^{-4}	1×10^{-3} s^{-1}
ξ_{CH_4}	1.6	1.2 ppm

5.7$_A$ The steady-state concentration of ClO in the lower stratosphere results from a balance between the ClO + NO$_2$ reaction and photolysis of ClONO$_2$. Given values of j_{ClONO_2} at 27 and 37 km

$$j_{ClONO_2} \cong 10^{-4} \text{ s}^{-1} \qquad \text{at} \quad 27 \text{ km}$$
$$\cong 5 \times 10^{-4} \text{ s}^{-1} \text{ at} \quad 37 \text{ km}$$

evaluate the extent to which the profiles of ClO and ClONO$_2$ in Figure 5.14 agree with the steady-state relation at these two altitudes.

5.8$_A$ Calculate the number of collisions an N$_2$O$_5$ molecule makes with a stratospheric aerosol particle over a 12-h nighttime. Assume a temperature of 217 K and that the aerosol surface area per unit volume, is 0.8×10^{-8} cm^2 cm^{-3}.

5.9$_B$ In June 1991, Mount Pinatubo erupted, injecting $15 - 20 \times 10^6$ metric tons (t) of SO$_2$ into the stratosphere. The subsequent oxidation of SO$_2$ to sulfate aerosol dramatically increased the aerosol surface area in the stratosphere. Ground- and satellite-based instruments recorded a decrease in the stratospheric abundance of NO$_x$ and an increase in the abundance of HNO$_3$.

    a. What is the cause of the decrease in NO_x?

    b. At 28 km, the nonvolcanic background stratospheric aerosol surface area is about $0.2 \, \mu m^2 \, cm^{-3}$. During the 6 months following the eruption, aerosol surface area was 5–10 times higher. O_3 was observed to increase by 15% at this altitude. Why?

    c. After the eruption, large negative changes in O_3 were observed in both the lower stratosphere and in the total column. Explain.

5.10$_B$ In the stratosphere from 15 to 20 km, the cycle

$$OH + O_3 \longrightarrow HO_2 + O_2 \qquad \text{(reaction 1)}$$

$$\underline{HO_2 + O_3 \longrightarrow OH + O_2 + O_2} \qquad \text{(reaction 2)}$$

$$\text{Net:} \quad 2\,O_3 \longrightarrow 3\,O_2$$

is responsible for 30 to 50% of total odd oxygen loss (Cohen et al. 1994; Wennberg et al. 1994). The coupled HO_2/ClO and HO_2/BrO catalytic cycles

$$HO_2 + XO \longrightarrow HOX + O_2 \qquad \text{(reactions 3a, b)}$$

$$HOX + h\nu \longrightarrow OH + X \qquad \text{(reactions 4a, b)}$$

$$OH + O_3 \longrightarrow HO_2 + O_2 \qquad \text{(reaction 1)}$$

$$\underline{X + O_3 \longrightarrow XO + O_2} \qquad \text{(reactions 5a, b)}$$

$$\text{Net:} \quad 2\,O_3 \longrightarrow 3\,O_2$$

where X = Cl(a) or Br(b), are responsible for about 15% of the odd-oxygen loss. The rate-determining reactions are 1 and 3a,b. Interconversion of OH and HO_2 also occurs by

$$HO_2 + NO \longrightarrow NO_2 + OH \qquad \text{(reaction 6)}$$

The interconversion of OH and HO_2 occurs about 10 times faster ($\tau \approx 10 \, s$) than the net processes contributing to increase or decrease of HO_x ($\tau \approx 100 \, s$). Show that the OH to HO_2 ratio owing to these reactions is given by

$$\frac{[OH]}{[HO_2]} = \frac{k_2[O_3] + k_{3a}[ClO] + k_{3b}[BrO] + k_6[NO]}{k_1[O_3]}$$

Using rate constants given in Appendix B, and neglecting reactions 3a and 3b, compute and plot [OH]/[HO$_2$] as a function of [O$_3$] at $T = 215 \, K$ over the range of [O$_3$], $(1–5) \times 10^{12}$ molecules cm^{-3}, and at [NO] $= 7.5 \times 10^8$ molecules cm^{-3}.

5.11$_B$ Le Bras and Platt (1995) have estimated the rate of tropospheric ozone depletion in the Arctic boundary layer as a result of ClO and BrO. Molecular bromine and chlorine would be photolyzed

$$Br_2 + hv \longrightarrow Br + Br \qquad \text{(reaction 1)}$$

$$Cl_2 + hv \longrightarrow Cl + Cl \qquad \text{(reaction 2)}$$

followed by (rate constants at 250 K, mean temperature of the Arctic boundary layer in spring)

$$Br + O_3 \longrightarrow BrO + O_2 \qquad k_3 = 7 \times 10^{-13} \text{ cm}^3 \text{ molecule}^{-1} \text{s}^{-1} \qquad \text{(reaction 3)}$$

$$Cl + O_3 \longrightarrow ClO + O_2 \qquad k_4 = 1 \times 10^{-11} \text{ cm}^3 \text{ molecule}^{-1} \text{s}^{-1} \qquad \text{(reaction 4)}$$

Reconversion of ClO and BrO to halogen atoms occurs particularly through reaction with NO:

$$BrO + NO \longrightarrow Br + NO_2 \qquad k_5 = 2.5 \times 10^{-11} \text{ cm}^3 \text{ molecule}^{-1} \text{s}^{-1} \qquad \text{(reaction 5)}$$

$$ClO + NO \longrightarrow Cl + NO_2 \qquad k_6 = 2 \times 10^{-11} \text{ cm}^3 \text{ molecule}^{-1} \text{s}^{-1} \qquad \text{(reaction 6)}$$

Ozone destruction would then be catalyzed by the following cycle:

$$BrO + BrO \longrightarrow 2\,Br + O_2 \qquad k_7 = 3.2 \times 10^{-12} \text{ cm}^3 \text{ molecule}^{-1} \text{s}^{-1} \qquad \text{(reaction 7)}$$

$$\underline{Br + O_3 \longrightarrow BrO + O_2} \qquad \text{(reaction 3)}$$

$$\text{Net:} \quad O_3 + O_3 \longrightarrow O_2 + O_2 + O_2$$

(The self-reaction of ClO leads to the Cl_2O_2 dimer and not to the direct production of Cl atoms.) A combined BrO–ClO cycle might also occur:

$$ClO + BrO \longrightarrow Br + Cl + O_2 \qquad k_{8a} = 7 \times 10^{-12} \text{ cm}^3 \text{ molecule}^{-1} \text{s}^{-1} \qquad \text{(reaction 8a)}$$

$$\longrightarrow BrCl + O_2 \qquad k_{8b} = 1.2 \times 10^{-12} \text{ cm}^3 \text{ molecule}^{-1} \text{s}^{-1} \qquad \text{(reaction 8b)}$$

$$\longrightarrow OClO + Br \qquad k_{8c} = 9 \times 10^{-12} \text{ cm}^3 \text{ molecule}^{-1} \text{s}^{-1} \qquad \text{(reaction 8c)}$$

The products BrCl and OClO are readily photolyzed

$$BrCl + hv \longrightarrow Br + Cl \qquad \tau_{BrCl} \simeq 53 \text{ s} \qquad (\theta_0 = 70°) \qquad \text{(reaction 9)}$$

$$OClO + hv \longrightarrow O + ClO \qquad \tau_{OClO} \simeq 6 \text{ s} \qquad (\theta_0 = 70°) \qquad \text{(reaction 10)}$$

a. Show that the net reaction of the cycles involving reactions 8a and 8b is

$$2\,O_3 \longrightarrow 3\,O_2$$

b. Show that no net ozone loss occurs as a result of reaction 8c and thus that the total rate of loss of O_3 by reaction 8 is

$$-\frac{d[O_3]}{dt} = 2(k_{8a} + k_{8b})[BrO][ClO]$$

c. Show that the steady-state ratio of [ClO] to [Cl] is

$$\frac{[\text{ClO}]}{[\text{Cl}]} = \frac{k_4[\text{O}_3]}{k_6[\text{NO}] + (k_{8a} + k_{8b})[\text{BrO}]}$$

d. Calculate the [ClO]/[Cl] ratio for [NO] = 0 and 10 ppt, [BrO] = 10 ppt, and [O$_3$] = 40 ppb. Using [Cl] = 8×10^4 molecules cm^{-3}, calculate the concentration of ClO corresponding to NO equal to 0 and 10 ppt.

e. Calculate the lifetime of an individual ClO radical against reaction with NO and BrO at 10 ppt of NO and 15 ppt of BrO.

f. Show that the rate of O$_3$ loss from the combined BrO–ClO cycle with the ClO concentration range calculated in (d) and BrO equal to 15 ppt is (8 to 30) $\times 10^5$ molecules cm^{-3} s^{-1}, or 0.1 to 0.34 ppb h^{-1}. (This rate would consume the initial 40 ppb of O$_3$ within 7–13 days.)

g. Show that the BrO cycle destroys O$_3$ at a rate

$$-\frac{d[\text{O}_3]}{dt} = 2\,k_7[\text{BrO}]^2$$

and that, for the conditions above, is 1.2×10^6 molecules cm^{-3} s^{-1}.

h. Finally, show that the combined rate of O$_3$ loss from both cycles is

$$-\frac{d[\text{O}_3]}{dt} = 2[\text{BrO}]\left\{\frac{(k_{8a} + k_{8b})k_4[\text{Cl}][\text{O}_3]}{k_6[\text{NO}] + (k_{8a} + k_{8b})[\text{BrO}]} + k_7[\text{BrO}]\right\}$$

5.12$_B$ The vertical SO$_2$ column in the plume from Mt. Pinatubo, integrated from the surface to the top of the stratosphere, was 3×10^{16} molecules cm^{-2}. We wish to estimate the amount of aerosol surface area that this SO$_2$ would produce if converted to H$_2$SO$_4$ aerosol. Assume complete conversion of this SO$_2$ to aqueous H$_2$SO$_4$ particles of mean size 0.1 μm diameter and 75% mass concentration H$_2$SO$_4$/25% mass concentration H$_2$O. Calculate the resulting column-integrated aerosol surface area in μm^2 cm^{-2}. To convert this to an aerosol surface area concentration, assume that all the aerosol is confined to a uniform layer 5 km thick.

5.13$_C$ Annual mean global-scale mass exchange between the stratosphere and troposphere can be estimated from the budget of a long-lived trace species, such as CFCl$_3$, that is inert in the troposphere but photodissociates in the stratosphere (Holton et al. 1995). Air parcels passing upward through the tropical tropopause will have CFCl$_3$ mixing ratios characteristic of the well-mixed troposphere; air parcels passing downward from the midstratosphere to the lower stratosphere (see Figure 5.34) will have smaller mixing ratios as a result of photodissociation in the stratosphere. Let

F_c = net flux of CFCl$_3$ into the midstratosphere

F_{au} = upward mass flux of air into the stratosphere in the tropics

F_{ad} = downward mass flux of air out of the midstratosphere into the extra-tropical lower stratosphere

ξ_T = mean volume mixing ratio of $CFCl_3$ in the troposphere

ξ_S = mean volume mixing ratio of $CFCl_3$ transported downward across the control surface

Show that

$$F_c = (F_{au}\xi_T - F_{ad}\xi_S)\frac{M_{CFCl_3}}{M_{air}}$$

where M_{CFCl_3} and M_{air} are the molecular weights of $CFCl_3$ and air. Conservation of mass requires that on average $F_{au} = F_{ad}$. Observed values of ξ_T and ξ_S and an estimate of F_{au} can be used to compute F_c. Applying the preceding equation at the 100 hPa level, using $\xi_S \simeq 0.6\,\xi_T$, $\xi_T = 268$ ppt, at this level and the estimate $F_{au} = 2.7 \times 10^{17}\,kg\,yr^{-1}$, compute F_c.

REFERENCES

Abbatt, J. P. D., and Molina, M. J. (1992) The heterogeneous reaction $HOCl + HCl \longrightarrow Cl_2 + H_2O$ on ice and nitric acid trihydrate: Reaction probabilities and stratospheric implications, *Geophys. Res. Lett.* **19**, 461–464.

Anderson, J. G., Brune, W. H., and Proffitt, M. H. (1989) Ozone destruction by chlorine radicals within the Antarctic vortex—the spatial and temporal evolution of ClO–O anticorrelation based on in situ ER-2 data, *J. Geophys. Res.* **94**, 11465–11479.

Anderson, J. G., Toohey, D. W., and Brune, W. H. (1991) Free radicals within the Antarctic vortex: The role of CFCs in Antarctic ozone loss, *Science* **251**, 39–46.

Bates, D. R., and Nicolet, M. (1950) The photochemistry of atmospheric water vapor, *J. Geophys. Res.* **55**, 301–327.

Boering, K. A., Wofsy, S. C., Daube, B. C., Schneider, H. R., Lowenstein, M., Podolske, J. R., and Conway, J. T. (1996) Stratospheric mean ages and transport rates from observations of carbon dioxide and nitrous oxide, *Science* **274**, 1340–1343.

Brasseur, G., Granier, C., and Walters, S. (1990) Future changes in stratospheric ozone and the role of heterogeneous chemistry, *Nature* **348**, 626–628.

Chapman, S. (1930) A theory of upper atmospheric ozone, *Mem. Roy. Meteorol. Soc.* **3**, 103–125.

Chin, M., and Davis, D. D. (1995) A reanalysis of carbonyl sulfide as a source of stratospheric background sulfur aerosol, *J. Geophys. Res.* **100**, 8993–9005.

Cohen, R. C. et al. (1994) Are models of catalytic removal of O_3 by HO_x accurate? Constraints from in situ measurements of the OH to HO_2 ratio, *Geophys. Res. Lett.* **21**, 2539–2542.

Crutzen, P. J. (1970) The influence of nitrogen oxides on atmospheric ozone content, *Quart. J. Roy. Meteorol. Soc.* **96**, 320–325.

Crutzen, P. J. (1976) The possible importance of OCS for the sulfate layer of the stratosphere, *Geophys. Res. Lett.* **3**, 73–76.

Dessler, A. E., Hintsa, E. J., Weinstock, E. M., Anderson, J. G., and Chan, K. R. (1995) Mechanisms controlling water vapor in the lower stratosphere: "A tale of two stratospheres," *J. Geophys. Res.* **100**, 23167–23172.

Dütsch, H. U. (1974) The ozone distribution in the atmosphere, *Can. J. Chem.* **52**, 1491–1504.

Fahey, D. W. et al. (1993) In situ measurements constraining the role of sulfate aerosols in midlatitude ozone depletion, *Nature*, **363**, 509–514.

Fahey, D. et al. (2001) The detection of large HNO_3-containing particles in the winter Arctic stratosphere, *Science* **291**, 1026–1031.

Farman, J. C., Gardiner, B. G., and Shanklin, J. D. (1985) Large losses of total ozone in Antarctica reveal seasonal ClO_x/NO_x interaction, *Nature* **315**, 207–210.

Grainger, R. G., Lambert, A., Rodgers, C. D., Taylor, F. W., and Deshler, T. (1995) Stratospheric aerosol effective radius, surface area and volume estimated from infrared measurements, *J. Geophys. Res.* **100**, 16507–16518.

Hanson, D. R., and Ravishankara, A. R. (1991) The reaction probabilities of $ClONO_2$ and N_2O_5 on polar stratosphere cloud materials, *J. Geophys. Res.* **96**, 5081–5090.

Hofmann, D., and Solomon, S. (1989) Ozone depletion through heterogeneous chemistry following the eruption of the El Chichón Volcano, *J. Geophys. Res.* **94**, 5029–5041.

Hofmann, D. J., Harder, J. W., Rolf, S. R., and Rosen, J. M. (1987) Balloon borne observations of the development and vertical structure of the Antarctic ozone hole in 1986, *Nature*, **326**, 59–62.

Holton, J. R., Haynes, P. H., McIntyre, M. E., Douglass, A. R., Rood, R. B., and Pfister, L. (1995) Stratosphere–troposphere exchange, *Rev. Geophys.* **33**, 403–439.

Johnston, H. S. (1971) Reduction of stratospheric ozone by nitrogen oxide catalysts from supersonic transport exhaust, *Science* **173**, 517–522.

Johnston, H. S. (1975) Global ozone balance in the natural stratosphere, *Rev. Geophys. Space Phys.* **13**, 637–649.

Johnston, P. V., McKenzie, R. L., Keys, J. G., and Matthews, A. W. (1992) Observations of depleted stratospheric NO_2 following the Pinatubo volcanic eruption, *Geophys. Res. Lett.* **19**, 211–213.

Jones, A. E. and Shanklin, J. D. (1995) Continued decline of total ozone over Halley, Antarctica, since 1985, *Nature* **376**, 409–411.

Jucks, K. W., Johnson, D. G., Chance, K. V., Traub, W. A., Margitan, J. J., Osterman, G. B., Salawitch, R. J., and Sasano, Y. (1998) Observations of OH, HO_2, H_2O, and O_3 in the upper stratosphere: Implications for HO_x photochemistry, *Geophys. Res. Lett.* **25**, 3935–3938.

Kondo, Y., Schmidt, U., Sugita, T., Engel, A., Koike, M., Aimedieu, P., Gunson, M. R., and Rodriguez, J. (1996) NO_2 correlation with N_2O and CH_4 in the midlatitude stratosphere, *Geophys. Res. Lett.* **23**, 2369–2372.

Le Bras, G., and Platt, U. (1995) A possible mechanism for combined chlorine and bromine catalyzed destruction of tropospheric ozone in the Arctic, *Geophys. Res. Lett.* **22**, 599–602.

McElroy, M. B., Salawitch, R. J., and Minschwaner, K. (1992) The changing stratosphere, *Planet. Space Sci.* **40**, 373–401.

McElroy, M. B., Salawitch, R. J., Wofsy, S. C., and Logan, J. A. (1986) Reductions of Antarctic ozone due to synergistic interactions of chlorine and bromine, *Nature* **321**, 759–762.

McGrath, M. P., Clemitshaw, K. C., Rowland, F. S., and Hehre, W. J. (1990) Structures, relative stabilities, and vibrational spectra of isomers of Cl_2O_2: The role of chlorine oxide dimer in Antarctic ozone depleting mechanisms, *J. Phys. Chem.* **94**, 6126–6132.

Mills, M. J., Langford, A. O., O'Leary, T. J., Arpag, K., Miller, H. L., Proffitt, M. H., Sanders, R. W., and Solomon, S. (1993) On the relationship between stratospheric aerosols and nitrogen dioxide, *Geophys. Res. Lett.* **20**, 1187–1190.

Minschwaner, K., Salawitch, R. J., and McElroy, M. B. (1993) Absorption of solar radiation by O_2: implications for O_3 and lifetimes of N_2O, $CFCl_3$, and CF_2Cl_2, *J. Geophys. Res.*, **98**, 10543–10561.

Molina, L. T., and Molina, M. J. (1987) Production of Cl_2O_2 from the self-reaction of the ClO radical, *J. Phys. Chem.* **91**, 433–436.

Molina, M. J. (1991) Heterogeneous chemistry on polar stratospheric clouds, *Atmos. Environ.* **25A**, 2535–2537.

Molina, M. J., and Rowland, F. S. (1974) Stratospheric sink for chlorofluoromethanes: Chlorine atom-catalyzed destruction of ozone, *Nature* **249**, 810–812.

Molina, M. J., Colussi, A. J., Molina, L. T., Schindler, R. N., and Tso, T.-L. (1990) Quantum yield of chlorine-atom formation in the photodissociation of chlorine peroxide (ClOOCl) at 308 nm, *Chem. Phys. Lett.* **173**, 310–315.

Mozurkewich, M., and Calvert, J. G. (1988) Reaction probability of N_2O_5 on aqueous aerosols, *J. Geophys. Res.* **93**, 15889–15896.

Osterman, G. B, Salawitch, R. J., Sen, B., Toon, G. C., Stachnik, R. A., Pickett, H. M., Margitan, J. J., Blavier, J. F., and Peterson, D. B. (1997) Balloon-borne measurements of stratospheric radicals and their precursors: Implications for the production and loss of ozone, *Geophys. Res. Lett.* **24**, 1107–1110.

Peter, T. (1996) Formation mechanisms of polar stratospheric clouds, in *Nucleation and Atmospheric Aerosols 1996*, M. Kulmala and P. E. Wagner, eds., Elsevier, Oxford, UK, pp. 280–291.

Prather, M. J. (1992) Catastrophic loss of stratospheric ozone in dense volcanic clouds, *J. Geophys. Res.* **97**, 10187–10191.

Rodriguez, J. M., Ko, M. K. W., and Sze, N. D. (1991) Role of heterogeneous conversion of N_2O_5 on sulfate aerosols in global ozone loss, *Nature* **352**, 134–137.

Rowland, F. S., and Molina, M. J. (1975) Chlorofluoromethanes in the environment, *Rev. Geophys. Space Phys.* **13**, 1–35.

Salawitch, R. J. et al. (1993) Chemical loss of ozone in the Arctic polar vortex in the winter of 1991–1992, *Science* **261**, 1146–1149.

Salawitch, R. J., Gobbi, G. P., Wofsy, S. C., and McElroy, M. B. (1989) Denitrification in the Antarctic stratosphere, *Nature* **339**, 525–527.

Sen, B., Toon, G. C., Osterman, G. B., Blavier, J. F., Margitan, J. J., Salawitch, R. J., and Yue, G. K. (1998) Measurements of reactive nitrogen in the stratosphere, *J. Geophys. Res.* **103**, 3571–3585.

Solomon, S., Garcia, R. R., Rowland, F. S., and Wuebbles, D. J. (1986) On the depletion of Antarctic ozone, *Nature* **321**, 755–758.

Tie, X. X., Brasseur, G. P., Briegleb, B., and Granier, C. (1994) Two-dimensional simulation of Pinatubo aerosol and its effect on stratospheric ozone, *J. Geophys. Res.* **99**, 20545–20562.

Van Doren, J. M., Watson, L. R., Davidovits, P., Worsnop, D. R., Zahniser, M. S., and Kolb, C. E. (1991) Uptake of N_2O_5 and HNO_3 by aqueous sulfuric acid droplets, *J. Phys. Chem.* **95**, 1684–1689.

Voight, C. et al. (2000) Nitric acid trihydrate (NAT) in polar stratospheric clouds, *Science* **290**, 1756–1758.

Volk, C. M. et al. (1996) Quantifying transport between the tropical and mid-latitude lower stratosphere, *Science* **272**, 1763–1768.

Wayne, R. P. (1991) *Chemistry of Atmospheres*, 2nd ed., Oxford Univ. Press, Oxford, UK.

Webster, C. R., May, R. D., Allen, M., Jaeglé, L., and McCormick, M. P. (1994) Balloon profiles of stratospheric NO_2 and HNO_3 for testing the heterogeneous hydrolysis of N_2O_5 on sulfate aerosols, *Geophys. Res. Lett.* **21**, 53–56.

Webster, C. R., May, R. D., Toohey, D. W., Avallone, L. M., Anderson, J. G., and Solomon, S. (1993) In situ measurements of the ClO/HCl ratio: Heterogeneous processing on sulfate aerosols and polar stratospheric clouds, *Geophys. Res. Lett.* **20**, 2523–2526.

Wennberg, P. O. et al. (1994) Removal of stratospheric O_3 by radicals: In situ measurements of OH, HO_2, NO, NO_2, ClO, and BrO, *Science* **266**, 398–404.

World Meteorological Organization (WMO) (1994) *Scientific Assessment of Ozone Depletion: 1994*. World Meteorological Organization, Geneva, Switzerland.

World Meteorological Organization (WMO) (1998) *Scientific Assessment of Ozone Depletion: 1998*, World Meterological Organization, Geneva.

World Meteorological Organization (WMO) (2002) *Scientific Assessment of Ozone Depletion: 2002*, World Meterological Organization, Geneva, Switzerland.

Zander, R. et al. (1996) The 1994 northern midlatitudes budget of stratospheric chlorine derived from ATMOS/ATLAS-3 observations, *Geophys. Res. Lett.* **23**, 2357–2360.

6 Chemistry of the Troposphere

The troposphere behaves as a chemical reservoir relatively distinct from the stratosphere. Transport of species from the troposphere into the stratosphere is much slower than mixing within the troposphere itself. A myriad of species are emitted at the Earth's surface, and those with chemical lifetimes less than about a year or so are destroyed in the troposphere. Even though solar radiation of the most energetic wavelengths is removed in the stratosphere, light of sufficiently energetic wavelengths penetrates into the troposphere to promote significant photochemical reactions in the troposphere. A factor important in tropospheric chemistry is the relatively high concentration of water vapor. The chemistry of the stratosphere involves the reactions that form and destroy ozone; ozone formation and removal is, likewise, central to the chemistry of the troposphere.

In the troposphere, for every 1,000,000,000 molecules, about 35 of them are, on average, O_3. Ground-level O_3 mixing ratios in the troposphere range from 20 to 60 ppb. Values can exceed 100 ppb in urban areas and broad regions downwind of urban complexes, especially under conditions in which a stationary high-pressure system produces several days of high temperatures and stagnant conditions. One of the most notable events in the eastern United States was one that occurred during the second week of June 1998; ozone mixing ratios exceeding 90 ppb were observed at every surface monitoring site in an area extending from Maine to Virginia to Ohio. Even higher values were found in the cities of the northeastern corridor. Ozone levels exceeding 200 ppb are considered a severe air pollution episode. The highest recorded hourly average O_3 mixing ratio in North America was apparently 680 ppb, in downtown Los Angeles in 1955. The current U.S. National Ambient Air Quality Standard for O_3 is an 8-h average of 80 ppb (see Chapter 2).

With the emergence of Los Angeles photochemical smog in the late 1940s and early 1950s and the identification of O_3 as its principal ingredient, those involved in the effort began to understand the nature of this new form of air pollution. In two pioneering articles published in *Industrial and Engineering Chemistry*, Arie J. Haagen-Smit identified ozone as the principal component of photochemical smog and postulated that a free-radical chain reaction mechanism involving organics and oxides of nitrogen was responsible for O_3 buildup (Haagen-Smit 1952; Haagen-Smit et al. 1953). Although the species oxidizing the hydrocarbons and other organic compounds was unknown, Haagen-Smit did deduce the critical fact that nitrogen oxides play the role of a catalyst in ozone formation. It was long known that hydroxyl (OH) radicals react rapidly with hydrocarbons, but whether OH radicals played any significant role in the chemistry of the troposphere was unknown.

The hint came in 1969, and then the breakthrough came in 1971. With information on the ^{14}C content of atmospheric CO, Bernard Weinstock deduced that the atmospheric lifetime of CO was 0.1 year, a value far smaller than what had been believed up to that time (Weinstock 1969). The short residence time implied that CO is being removed from the atmosphere far faster than any known mechanism could explain. Noting that a likely process for conversion of CO to CO_2 in the stratosphere is

$$CO + OH \longrightarrow CO_2 + H$$

Weinstock suggested that the same reaction could be an effective removal mechanism for tropospheric CO and noted that to maintain the needed OH levels would require some kind of regenerative chain mechanism for tropospheric OH, although he did not know what that mechanism was. It was known that photodissociation of O_3 in the wavelength range of 290–320 nm produces excited atomic oxygen, $O(^1D)$, a small fraction of which reacts with atmospheric water vapor to produce OH radicals. Hiram ("Chip") Levy showed that even that small fraction is sufficient to produce OH radicals in the atmosphere at a rate exceeding 10^5 molecules $cm^{-3}\,s^{-1}$, making OH the dominant oxidizing free radical in the troposphere (Levy 1971).

Ozone in the troposphere is generated from two major classes of precursors: volatile organic compounds (VOCs) and oxides of nitrogen (NO_x, which denotes the sum of NO and NO_2). The process of ozone formation is initiated by the reaction of the OH radical with organic molecules. The subsequent reaction sequence is catalyzed by NO_x, in an interwoven network of free-radical reactions. Most of the direct emission of NO_x to the atmosphere is in the form of NO. NO_2 is formed in the atmosphere by conversion of NO. As in the stratosphere, the two oxides of nitrogen are generally grouped together as NO_x because interconversion between NO and NO_2 is rapid (with a tropospheric timescale of ~ 5 min) compared to the timescale for organic oxidation (one to several hours). NO_x mixing ratios range from 5 to 20 ppb in urban areas, about 1 ppb in rural areas during regionwide episodes, and from 10 to 100 ppt in the remote troposphere.

In the remote troposphere, ozone formation is sustained by the oxidation of carbon monoxide (CO) and methane (CH_4), each through reaction with OH. Both CO and CH_4 are long-lived species, with atmospheric lifetimes against OH reaction of about 2 months and 9 years, respectively. Ozone formation in the urban and regional atmosphere is driven by much shorter-lived volatile organic compounds emitted from anthropogenic and biogenic sources. These include alkenes, aromatics, and oxygenated organic species. Because two classes of precursor compounds are involved in tropospheric ozone formation, volatile organic compounds (VOCs) and oxides of nitrogen (NO_x), a key question is how changing VOC and NO_x levels affect the amount of ozone formed. We will see that ozone formation depends critically on the level of NO_x.

We begin this chapter with hydroxyl radical generation by O_3 photolysis and then proceed first to the chemistry of NO_x as we develop the overall organic/NO_x chemistry of the troposphere.

6.1 PRODUCTION OF HYDROXYL RADICALS IN THE TROPOSPHERE

The key to understanding tropospheric chemistry begins with the hydroxyl (OH) radical. Because the OH radical is unreactive toward O_2, once produced, it survives to react with virtually all atmospheric trace species. The most abundant oxidants in the atmosphere are O_2 and O_3, but these molecules have large bond energies and are generally unreactive, except with certain free radicals; this leaves the OH radical as the primary oxidizing species in the troposphere.

We learned in Section 3.6 and Chapter 5 that O_3 photolysis (at wavelengths $< 319\,nm$) to produce both ground-state (O) and excited singlet ($O(^1D)$) oxygen atoms is important in both the stratosphere and troposphere:

$$O_3 + hv \xrightarrow{1a} O_2 + O \qquad \text{(reaction 1a)}$$

$$\xrightarrow{1b} O_2 + O(^1D) \qquad \text{(reaction 1b)}$$

The ground-state O atom combines rapidly with O_2 to reform O_3:

$$O + O_2 + M \xrightarrow{2} O_3 + M \qquad \text{(reaction 2)}$$

so reaction 1a followed by reaction 2 has no net chemical effect (a null cycle). However, when $O(^1D)$ is produced, since the spontaneous $O(^1D) \longrightarrow O$ transition is forbidden, it must react with another atmospheric species. Most often, $O(^1D)$ collides with N_2 or O_2, removing its excess energy and quenching $O(^1D)$ to its ground state:

$$O(^1D) + M \xrightarrow{3} O + M \qquad \text{(reaction 3)}$$

Since the ground-state O atom then just reacts with O_2 by reaction 2 to replenish O_3, this path, consisting of reactions 1b, 3, and 2, is just another null cycle. Occasionally, however, $O(^1D)$ collides with a water molecule and produces two OH radicals:

$$O(^1D) + H_2O \xrightarrow{4} 2\,OH \qquad \text{(reaction 4)}$$

Reaction 4 is, in fact, the only gas-phase reaction in the troposphere able to break the H—O bond in H_2O.

Let us compute the rate of formation of OH radicals from reactions 1–4. $O(^1D)$ is sufficiently reactive that the pseudo-steady-state approximation can be invoked for its concentration. Thus

$$j_{O_3 \to O(^1D)}[O_3] = (k_3[M] + k_4[H_2O])[O(^1D)]$$

and

$$[O(^1D)] = \frac{j_{O_3 \to O(^1D)}[O_3]}{k_3[M] + k_4[H_2O]} \qquad (6.1)$$

At 298 K, the rate coefficient values are (Table B.1)

$$k_3 \; (M = O_2) \quad 4.0 \times 10^{-11}\,cm^3\,molecule^{-1}\,s^{-1}$$
$$k_3 \; (M = N_2) \quad 2.6 \times 10^{-11}\,cm^3\,molecule^{-1}\,s^{-1}$$
$$k_4 \qquad\qquad\;\; 2.2 \times 10^{-10}\,cm^3\,molecule^{-1}\,s^{-1}$$

For the atmospheric N_2/O_2 mixture, the value of k_3 is 2.9×10^{-11} cm^3 molecule^{-1} s^{-1}. The first-order photolysis rate coefficient of $O_3 \longrightarrow O(^1D)$ at the Earth's surface at solar zenith angle $0°$ is 6×10^{-5} s^{-1} (Figure 4.15). Equation (6.1) can be written as

$$[O(^1D)] = \frac{j_{O_3 \to O(^1D)} \xi_{O_3}[M]}{k_3[M] + k_4 \xi_{H_2O}[M]}$$

$$= \frac{j_{O_3 \to O(^1D)} \xi_{O_3}}{k_3 + k_4 \xi_{H_2O}} \tag{6.2}$$

where ξ_{O_3} and ξ_{H_2O} are the mixing ratios of O_3 and H_2O. Under typical conditions at midlatitudes at the surface at noon, we can assume $\xi_{O_3} \cong 50\,\text{ppb} = 50 \times 10^{-9}$. To determine the mixing ratio of H_2O vapor we need the relative humidity. $\xi_{H_2O} = p_{H_2O}/p_s$ and $p_{H_2O} = \text{RH}$ (as a fraction) $\times p_{H_2O}^{\circ}$. At 288 K, for example, $p_{H_2O}^{\circ} = 0.0167 p_s$, and $\xi_{H_2O} = 0.0167\,\text{RH}$. At 50% RH, $\xi_{H_2O} = 0.0083$; at 90% RH, $\xi_{H_2O} = 0.015$. Even at 90% RH, the second term in the denominator of (6.2) is about one order of magnitude smaller than the first term. Thus, to estimate $[O(^1D)]$, the second term can be neglected:

$$[O(^1D)] \cong \frac{j_{O_3 \to O(^1D)} \xi_{O_3}}{k_3}$$

The rate of production of OH from O_3 photolysis is $P_{OH} = 2k_4[O(^1D)][H_2O]$, which gives

$$P_{OH} = \frac{2 j_{O_3 \to O(^1D)} k_4}{k_3} \frac{[H_2O]}{[M]} [O_3] \tag{6.3}$$

Once an $O(^1D)$ is formed, it is either quenched back to ground-state O or reacts with H_2O to yield 2 OH radicals. The number of OH radicals produced per $O(^1D)$ formed is the ratio of twice the rate of reaction 4 to the rate of reaction 1b:

$$\epsilon_{OH} = \frac{2 k_4[O(^1D)][H_2O]}{j_{O_3 \to O(^1D)}[O_3]}$$

$$= \frac{2k_4[O(^1D)][H_2O]}{(k_3[M] + k_4[H_2O])[O(^1D)]}$$

$$= \frac{2 k_4[H_2O]}{k_3[M] + k_4[H_2O]}$$

$$\cong \frac{2 k_4[H_2O]}{k_3[M]} = \frac{2k_4 \xi_{H_2O}}{k_3} \tag{6.4}$$

At 298 K, $k_4/k_3 = 7.6$. We can evaluate ϵ_{OH} as a function of RH at the surface at 298 K:

RH(%)	10	25	50	80
ϵ_{OH}	0.047	0.12	0.23	0.38

At 80% RH, close to 40% of the $O(^1D)$ formed leads to OH radicals.

The hydroxyl radical does not react with any of the major constituents of the atmosphere, such as N_2, O_2, CO_2, or H_2O, yet it is the most important reactive species in the troposphere. Indeed, OH reacts with most trace species in the atmosphere, and its importance derives from both its high reactivity toward other molecules and its relatively high concentration. Were OH simply to react with other species and not in some manner be regenerated, its concentration would be far too low, in spite of its high reactivity, to be an important player in tropospheric chemistry. The key is that, when reacting with atmospheric trace gases, OH is generated in catalytic cycles, leading to sustained concentrations on the order of 10^6 molecules cm^{-3} during daylight hours.

By using a chemical mechanism to simulate tropospheric chemistry, it is possible to estimate the atmospheric concentration of OH. Highest OH levels are predicted in the tropics, where high humidities and strong actinic fluxes lead to a high rate of OH production from O_3 photolysis to $O(^1D)$. In addition, OH levels are predicted to be about 20% higher in the Southern Hemisphere as a result of the large amounts of CO produced by human activities in the Northern Hemisphere that act to reduce OH through reaction with it. Hydroxyl radical levels are a factor of ~ 5 higher over the continents than over the oceans. Confirmation of the levels predicted by the chemical mechanisms is usually based on balancing budgets of species that are known to be consumed only by OH. For example, methyl chloroform (CH_3CCl_3) is removed from the atmosphere almost solely by reaction with OH, so the global average concentration of OH determines the mean residence time of CH_3CCl_3. From the history of methyl chloroform emissions, which are entirely anthropogenic and are known to reasonable accuracy, and its present atmospheric level, it is possible to infer its residence time and then to compare the OH level corresponding to that residence time to the level predicted theoretically from tropospheric chemical mechanisms (Prinn et al. 1992). A global mean tropospheric concentration of OH of 1.0×10^6 molecules cm^{-3} is a good value to use for estimations.

Can Tropospheric Ozone Levels Be Explained by Downward Transport from the Stratosphere?   Before we begin the study of tropospheric chemistry, we pose the question—can tropospheric O_3 levels be explained solely on the basis of the O_3 flux from the stratosphere to the troposphere? We can address this question indirectly through the OH radical (Jacob 1999). The principal source of the OH radical in the troposphere is O_3 photolysis (Section 6.1). The principal sink of tropospheric OH is reaction with CO and CH_4. Given estimated global budgets for CO and CH_4, we can first determine the quantity of OH in the atmosphere needed to oxidize the emitted CO and CH_4.

An estimate of the global sink of CH_4 by OH reaction was given in Table 2.10 as $506\,Tg\ CH_4\ yr^{-1}$. The corresponding estimate for the global sink of CO, given in Table 2.14, is $1920\,Tg\ CO\ yr^{-1}$. These translate to

$$CH_4:\ 506\,Tg\ CH_4\ yr^{-1} \cong 3.2 \times 10^{13}\ mol\ CH_4\ yr^{-1}$$

$$CO:\ 1920\,Tg\ CO\ yr^{-1} \cong \underline{6.9 \times 10^{13}\ mol\ CO\ yr^{-1}}$$

$$\text{Total OH needed} \qquad \sim 10^{14}\ mol\ OH\ yr^{-1}$$

Can this amount of OH can be supplied by photolysis of the O_3 that is transported down from the stratosphere? The global stratosphere to troposphere flux of O_3 is estimated as $1-2 \times 10^{13}$ mol O_3 yr^{-1}. Each O_3 molecule can produce, at most, two OH molecules, so the maximum OH production rate is $2 - 4 \times 10^{13}$ mol OH yr^{-1}. This OH production rate, solely from stratospheric O_3, is anywhere from a factor of 2.5 to 5 too low to balance the global CH_4 and CO budgets. We conclude on the basis of this simple calculation that there must be a major in situ source of O_3 in the troposphere.

6.2 BASIC PHOTOCHEMICAL CYCLE OF NO₂, NO, AND O₃

When NO and NO_2 are present in sunlight, ozone formation occurs as a result of the photolysis of NO_2 at wavelengths <424 nm:

$$NO_2 + h\nu \xrightarrow{1} NO + O \qquad \text{(reaction 1)}$$

$$O + O_2 + M \xrightarrow{2} O_3 + M \qquad \text{(reaction 2)}$$

There are no significant sources of ozone in the atmosphere other than reaction 2. Once formed, O_3 reacts with NO to regenerate NO_2:

$$O_3 + NO \xrightarrow{3} NO_2 + O_2 \qquad \text{(reaction 3)}$$

Let us consider for a moment the dynamics of a system in which only these three reactions are taking place. Let us assume that known initial concentrations of NO and NO_2, $[NO]_0$ and $[NO_2]_0$, in air are placed in a reactor of constant volume at constant temperature and irradiated. The rate of change of the concentration of NO_2 after the irradiation begins is given by

$$\frac{d[NO_2]}{dt} = -j_{NO_2}[NO_2] + k_3[O_3][NO]$$

Treating $[O_2]$ as constant, there are four species in the system: NO_2, NO, O, and O_3. We could write the dynamic equations for NO, O, and O_3 just as we have done for NO_2. For example, the equation for [O] is

$$\frac{d[O]}{dt} = j_{NO_2}[NO_2] - k_2[O][O_2][M]$$

Since the oxygen atom is so reactive that it disappears by reaction 2 virtually as fast as it is formed by reaction 1, one can invoke the pseudo-steady-state approximation and thereby assume that the rate of formation is equal to the rate of disappearance:

$$j_{NO_2}[NO_2] = k_2[O][O_2][M]$$

The steady-state oxygen atom concentration in this system is then given by

$$[O] = \frac{j_{NO_2}[NO_2]}{k_2[O_2][M]} \qquad (6.5)$$

Note that [O] is not constant; rather it varies with $[NO_2]$ in such a way that at any instant a balance is achieved between its rate of production and loss. Thus, the oxygen atom concentration adjusts to changes in the NO_2 concentration many orders of magnitude faster than the NO_2 concentration changes, so that on a timescale of the NO_2 dynamics, [O] always appears to satisfy (6.5).

These three reactions will reach a point where NO_2 is destroyed and reformed so fast that a steady-state cycle is maintained. Let us compute the steady-state concentrations of NO, NO_2, and O_3 achieved in this cycle. (The steady-state concentration of oxygen atoms is already given by (6.5).) The steady-state ozone concentration is given by

$$[O_3] = \frac{j_{NO_2}[NO_2]}{k_3[NO]} \tag{6.6}$$

This expression, resulting from the steady-state analysis of reactions 1–3, has been named the *photostationary state relation*. We note that the steady-state ozone concentration is proportional to the $[NO_2]/[NO]$ ratio. We now need to compute $[NO_2]$ and $[NO]$. These are obtained from conservation of nitrogen

$$[NO] + [NO_2] = [NO]_0 + [NO_2]_0$$

and the stoichiometric reaction of O_3 with NO:

$$[O_3]_0 - [O_3] = [NO]_0 - [NO]$$

Solving for $[O_3]$, we obtain the relation for the ozone concentration formed at steady state by irradiating any mixture of NO, NO_2, O_3, and excess O_2 (in which only reactions 1–3 are important):

$$[O_3] = -\frac{1}{2}\left([NO]_0 - [O_3]_0 + \frac{j_{NO_2}}{k_3}\right)$$
$$+ \frac{1}{2}\left\{\left([NO]_0 - [O_3]_0 + \frac{j_{NO_2}}{k_3}\right)^2 + \frac{4j_{NO_2}}{k_3}([NO_2]_0 + [O_3]_0)\right\}^{1/2} \tag{6.7}$$

If $[O_3]_0 = [NO]_0 = 0$, (6.7) reduces to

$$[O_3] = \frac{1}{2}\left\{\left[\left(\frac{j_{NO_2}}{k_3}\right)^2 + \frac{4j_{NO_2}}{k_3}[NO_2]_0\right]^{1/2} - \frac{j_{NO_2}}{k_3}\right\} \tag{6.8}$$

We will see later that a typical value of j_{NO_2}/k_3 expressed in mixing ratio units is 10 ppb, so we can compute the ozone mixing ratio attained as a function of the initial mixing ratio of NO_2 with $[O_3]_0 = [NO]_0 = 0$:

$[NO_2]_0$, ppb	$[O_3]$, ppb
100	27
1000	95

If, on the other hand, $[NO_2]_0 = [O_3]_0 = 0$, then $[O_3] = 0$. This is clear since in the absence of NO_2 there is no means to produce atomic oxygen and therefore ozone. Thus the maximum steady-state ozone concentration would be achieved with an initial charge of pure NO_2. The mixing ratios of ozone attained in urban and regional atmospheres are often greater than those in the sample calculation. Since most of the NO_x emitted is in the form of NO and not NO_2, the concentration of ozone reached, if governed solely by reactions 1–3, cannot account for the actual observed concentrations. It must be concluded that reactions other than 1–3 are important in tropospheric air in which relatively high ozone concentrations occur.

Characteristic Time for the Photochemical NO_x Cycle to Achieve Steady State   The photochemical NO_x cycle can be written concisely as the reversible reactions

$$NO_2 + h\nu \underset{\underset{3}{\longleftarrow}}{\overset{1+2}{\underset{O_2}{\longrightarrow}}} NO + O_3$$

where reaction 1 is the rate-determining step of the forward reaction. The rate equation for $[O_3]$ is

$$\frac{d[O_3]}{dt} = j_{NO_2}[NO_2] - k_3[NO][O_3]$$

The characteristic relaxation time to steady state is based on the reverse reaction:

$$\tau = \frac{1}{k_3[NO]}$$

At 298 K, $k_3 = 1.9 \times 10^{-14}\,cm^3$ molecule^{-1} s^{-1}. Let us evaluate τ at an NO mixing ratio of 1 ppb at 1 atm. Thus, $[NO] = 10^{-9}[M] = (10^{-9})(2.5 \times 10^{19}) = 2.5 \times 10^{10}$ molecules cm^{-3}, and

$$\tau = \frac{1}{(1.9 \times 10^{-14})(2.5 \times 10^{10})} = 0.22 \times 10^4\,s = 0.6\,h$$

At an NO mixing ratio of 10 ppb, $\tau \cong 3.6$ min.

6.3 ATMOSPHERIC CHEMISTRY OF CARBON MONOXIDE

We now turn to the tropospheric chemistry of carbon-containing compounds, the simplest of which, in many respects, is CO. The atmospheric oxidation of CO exhibits many of the key features of that of much more complex organic molecules, and so it is the ideal place to begin a study of the chemistry of the troposphere.

Carbon monoxide reacts with the hydroxyl radical

$$CO + OH \overset{1}{\longrightarrow} CO_2 + H \qquad\qquad \text{(reaction 1)}$$

and the hydrogen atom formed in reaction 1 combines so quickly with O_2 to form the hydroperoxyl radical HO_2

$$H + O_2 + M \longrightarrow HO_2 + M$$

that, for all intents and purposes, we can simply write reaction 1 as

$$CO + OH \xrightarrow[O_2]{1} CO_2 + HO_2 \qquad \text{(reaction 1)}$$

The addition of an H atom to O_2 weakens the O—O bond in O_2, and the resulting HO_2 radical reacts much more freely than O_2 itself. When NO is present, the most important atmospheric reaction that the HO_2 radical undergoes is with NO:

$$HO_2 + NO \xrightarrow{2} NO_2 + OH \qquad \text{(reaction 2)}$$

We have already encountered this important reaction in the stratosphere.

The hydroperoxyl radical also reacts with itself to produce hydrogen peroxide (H_2O_2):[1]

$$HO_2 + HO_2 \xrightarrow{3} H_2O_2 + O_2 \qquad \text{(reaction 3)}$$

Hydrogen peroxide is a temporary reservoir for HO_x ($OH + HO_2$):

$$H_2O_2 + h\nu \xrightarrow{4} OH + OH \qquad \text{(reaction 4)}$$

$$H_2O_2 + OH \xrightarrow{5} HO_2 + H_2O \qquad \text{(reaction 5)}$$

Reaction 4 returns two HO_x species, whereas in reaction 5 one HO_x is lost to H_2O. The NO_2 formed in reaction 2 participates in the photochemical NO_x ($NO + NO_2$) cycle:

$$NO_2 + h\nu \xrightarrow{6} NO + O \qquad \text{(reaction 6)}$$

$$O + O_2 + M \xrightarrow{7} O_3 + M \qquad \text{(reaction 7)}$$

$$NO + O_3 \xrightarrow{8} NO_2 + O_2 \qquad \text{(reaction 8)}$$

Finally, termination of the chain occurs when OH and NO_2 react to form nitric acid:

$$OH + NO_2 + M \xrightarrow{9} HNO_3 + M \qquad \text{(reaction 9)}$$

This reaction removes both HO_x and NO_x from the system.

The CO oxidation is represented in terms of the HO_x family in Figure 6.1. P_{HO_x} denotes the rate of OH generation from O_3 photolysis. For simplicity, we do not show a flux from H_2O_2 back into HO_x. Within the HO_x family, OH and HO_2 rapidly cycle between

---

[1]Reaction 3 has both bimolecular and termolecular channels (Table B.1). At 298 K at the surface ($[M] = 2.46 \times 10^{19}$ molecules cm^{-3}) the second-order rate coefficients for the two channels are 1.7×10^{-12} and 1.2×10^{-12} cm^3 molecule^{-1} s^{-1}, respectively.

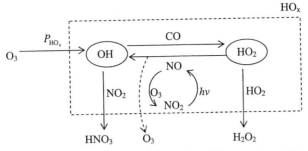

FIGURE 6.1   Reactions involving the HO_x ($OH + HO_2$) family in CO oxidation.

themselves, so that a steady-state OH/HO_2 partitioning is established, which depends on the NO_x level. Figure 6.2 shows CO oxidation from the perspective of the NO_x chemical family. The heavy arrows interconnecting NO and NO_2 in Figure 6.2, which represent the photochemical NO_x cycle, indicate that this cycle occurs more frequently than either the $NO + HO_2$ or $OH + NO_2$ reactions. As a result, the partitioning between NO and NO_2 within the NO_x family is approximately controlled by the photostationary state relation (6.6):

$$[O_3] = \frac{j_{NO_2}[NO_2]}{k_{NO+O_3}[NO]}$$

If the ratio of $[NO_2]$ to $[NO]$ increases, then the steady-state O_3 concentration increases.

Each time an $HO_2 + NO$ reaction occurs, an additional O_3 molecule is produced as the resulting NO_2 molecule photolyzes. In Figure 6.1 and 6.2, this is indicated by the dashed lines leading to O_3. This is a "new" O_3 molecule because the NO_2 molecule formed from $HO_2 + NO$ did not require an O_3 molecule in its formation. Thus, the rate of production of O_3 is simply equal to the rate of the $HO_2 + NO$ reaction:

$$P_{O_3} = k_{HO_2+NO}[HO_2][NO] \tag{6.9}$$

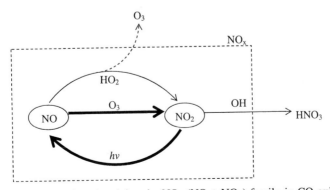

FIGURE 6.2   Reactions involving the NO_x ($NO + NO_2$) family in CO oxidation.

It turns out that the overall behavior of the CO oxidation system depends critically on the overall level of NO_x. (Recall that there is also an effect of NO_x level on stratospheric chemistry.) This can be seen most clearly by examining the nature of the system at the limits of low and high NO_x.

6.3.1 Low NO_x Limit

Let us obtain the expression for the rate of O_3 formation in the CO system in the low NO_x limit. The overall steady state HO_x balance is (see Figure 6.1)

$$P_{HO_x} = 2\,k_{HO_2+HO_2}[HO_2]^2 + k_{OH+NO_2}[OH][NO_2] \tag{6.10}$$

The rate of generation of HO_x is balanced by loss through both $HO_x + HO_x$ and $HO_x + NO_x$ reactions; the two loss terms above can be denoted as

$$HHL = 2\,k_{HO_2+HO_2}[HO_2]^2 \tag{6.11}$$

$$NHL = k_{OH+NO_2}[OH][NO_2] \tag{6.12}$$

At low NO_x, the principal sink of HO_x is the $HO_2 + HO_2$ reaction. Neglecting the $OH + NO_2$ reaction, we obtain the steady-state concentration of HO_2 under low NO_x conditions:

$$[HO_2] \cong \left(\frac{P_{HO_x}}{2\,k_{HO_2+HO_2}} \right)^{1/2} \tag{6.13}$$

Substituting (6.13) into (6.9), we obtain the rate of O_3 generation in the low NO_x limit as

$$P_{O_3} = k_{HO_2+NO} \left(\frac{P_{HO_x}}{2\,k_{HO_2+HO_2}} \right)^{1/2} [NO] \tag{6.14}$$

Thus, the rate of O_3 production increases linearly with the NO concentration and proportionally to the square root of the HO_x generation rate. Note that P_{O_3} is independent of the CO level; this is because NO_x is the limiting reactant at the low NO_x limit.

6.3.2 High NO_x Limit

At high NO_x, the steady-state HO_x balance is

$$P_{HO_x} \cong k_{OH+NO_2}[OH][NO_2] \tag{6.15}$$

so the steady-state concentration of OH is

$$[OH] = \frac{P_{HO_x}}{k_{OH+NO_2}[NO_2]} \tag{6.16}$$

In order to compute P_{O_3} from (6.9), we need to know $[HO_2]$. Within the HO_x family, OH and HO_2 rapidly interconvert (Figure 6.1), so we can neglect the effect of the $OH + NO_2$ and $HO_2 + HO_2$ reactions on their steady-state partitioning

$$k_{CO+OH}[CO][OH] = k_{HO_2+NO}[HO_2][NO] \tag{6.17}$$

so

$$[HO_2] = \frac{k_{CO+OH}[CO][OH]}{k_{HO_2+NO}[NO]} \tag{6.18}$$

Using (6.16) for [OH], we obtain

$$[HO_2] = \frac{k_{CO+OH}P_{HO_x}[CO]}{k_{HO_2+NO}k_{OH+NO_2}[NO][NO_2]} \tag{6.19}$$

Finally, substituting (6.19) into (6.9), the rate of O_3 production in the high NO_x limit is

$$P_{O_3} = \frac{k_{CO+OH}P_{HO_x}[CO]}{k_{OH+NO_2}[NO_2]} \tag{6.20}$$

Therefore, in the high NO_x limit, the rate of O_3 production increases linearly with both the CO concentration and the rate of HO_x generation but decreases with increasing NO_x. Since ample NO_x exists, the rate of the overall photochemical cycle is controlled by the $CO + OH$ reaction; in fact, there is so much NO_x available that as NO_x increases, the $OH + NO_2$ termination reaction increases in importance, effectively limiting the OH/HO_2 cycling and thereby decreasing the amount of O_3 that can be formed.

6.3.3 Ozone Production Efficiency

An important measure of atmospheric ozone-forming oxidation cycles is the *ozone production efficiency* (OPE). Because NO_x gets cycled back and forth between NO and NO_2 in O_3 generation, NO_x can be viewed as the catalyst in O_3 formation. One can define the OPE as the number of molecules of O_3 formed per each NO_x removed from the system. The rate of production of O_3 is given by (6.9):

$$P_{O_3} = k_{HO_2+NO}[HO_2][NO] \tag{6.9}$$

The rate at which NO_x is removed from the system is

$$L_{NO_x} = k_{OH+NO_2}[OH][NO_2] \tag{6.21}$$

Thus

$$OPE = \frac{P_{O_3}}{L_{NO_x}} = \frac{k_{HO_2+NO}[HO_2][NO]}{k_{OH+NO_2}[OH][NO_2]} \tag{6.22}$$

Let us evaluate OPE as a function of the NO_x level for the CO system. OPE depends on two ratios: $[HO_2]/[OH]$ and $[NO]/[NO_2]$. We need to obtain expressions for these two ratios. Two equations are needed to determine $[OH]$ and $[HO_2]$; we choose the overall HO_x balance (6.10)

$$P_{HO_x} = 2 k_{HO_2+HO_2}[HO_2]^2 + k_{OH+NO_2}[OH][NO_2] \qquad (6.10)$$

and the steady-state balance for $[HO_2]$:

$$k_{OH+CO}[OH][CO] = k_{HO_2+NO}[HO_2][NO] + 2 k_{HO_2+HO_2}[HO_2]^2 \qquad (6.23)$$

Equation (6.23) can be combined with (6.10) to eliminate $[OH]$ and yield a quadratic equation for $[HO_2]$

$$a[HO_2]^2 + b[HO_2] + c = 0$$

where

$$a = 2 k_{HO_2+HO_2}\left(1 + \frac{k_{OH+NO_2}[NO_2]}{k_{OH+CO}[CO]}\right)$$

$$b = \frac{k_{HO_2+NO}\, k_{OH+NO_2}[NO_2][NO]}{k_{OH+CO}[CO]}$$

$$c = -P_{HO_x}$$

The quadratic equation can be solved to yield $[HO_2]$ as a function of $[CO]$, $[NO]$, and $[NO_2]$. The $[HO_2]/[OH]$ ratio can be written as

$$\frac{[HO_2]}{[OH]} = \frac{1}{\dfrac{k_{HO_2+NO}[NO]}{k_{OH+CO}[CO]} + \dfrac{2 k_{HO_2+HO_2}[HO_2]}{k_{OH+CO}[CO]}} \qquad (6.24)$$

where the second term in the denominator is the deviation from the rapid OH/HO_2 cycling approximation (6.18). This term reflects the effect of the $HO_2 + HO_2$ reaction in perturbing the OH/HO_2 ratio from that in which all HO_2 is assumed to react with NO. The $[NO]/[NO_2]$ ratio is determined from the photochemical NO_x cycle:

$$\frac{[NO]}{[NO_2]} = \frac{j_{NO_2}}{k_{NO+O_3}[O_3]} \qquad (6.6)$$

Figure 6.3 shows the ozone production efficiency for CO oxidation at 298 K at the Earth's surface as a function of $[NO_x] = [NO] + [NO_2]$ for NO_x mixing ratios from 1 ppt to 100 ppb. At the ground-level conditions of Figure 6.3, we assume that $P_{HO_x} = 1$ ppt s^{-1}, that $[NO]/[NO_2] = 0.1$, and a CO mixing ratio of 200 ppb. The OPE is largest at the lowest concentration of NO_x; at these low levels, NO_x termination by $OH + NO_2$ is suppressed and each NO_x participates in more O_3 production cycles. At 100 ppb NO_x, OPE approaches zero, as the concentration of NO_2 is so large that the $OH + NO_2$ reaction occurs preferentially relative to propagation of the cycle.

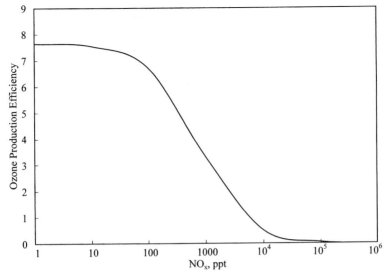

FIGURE 6.3   Ozone production efficiency (OPE) for atmospheric CO oxidation as a function of the NO_x (NO + NO_2) level. Conditions are 298 K at the Earth's surface, $P_{HO_x} = 1 \, ppt \, s^{-1}$, [NO]/[$NO_2$] = 0.1, and CO mixing ratio of 200 ppb.

Dependence of P_{O_3} on NO_x Abundance in CO Oxidation   One of the key aspects of tropospheric chemistry is the dependence of ozone production on the NO_x abundance. We have derived relationships for P_{O_3} for the CO system in the limits of low and high NO_x. Here we examine how P_{O_3} depends on the NO_x abundance over the complete range of NO_x levels. To do this, we will fix the rate of HO_x production, P_{HO_x}, and vary the NO concentration at a fixed NO_2/NO ratio. Under conditions of high HO_2 radical abundance relative to NO_x, the primary chain-terminating reaction is the $HO_x + HO_x$ reaction, $HO_2 + HO_2$. This condition is referred to as NO_x-limited. At sufficiently high NO_x levels, chain termination results from the $HO_x + NO_x$ reaction, $OH + NO_2$. This condition is called NO_x-saturated. By varying the NO_x concentration, we can explore the point at which the system crosses over from NO_x-limited to NO_x-saturated conditions. The crossover point occurs at the NO concentration where $\partial P_{O_3}/\partial[NO] = 0$. The actual value of the NO concentration at this crossover point depends on the values of P_{HO_x} and the NO_2/NO ratio.

The calculation described above is presented by Thornton et al. (2002), in modeling ozone concentrations downwind of Nashville, Tennessee. For numerical values, assume 298 K at the Earth's surface, [NO_2]/[NO] = 7, and $P_{HO_x} = 0.1, 0.6$, or $1.2 \, ppt \, s^{-1}$. The NO_2/NO ratio of 7 approximates that at midday in a regional continental atmosphere. In order to avoid having to represent the detailed chemistry of the many hydrocarbon species present in such an atmosphere, Thornton et al. (2002) assumed that the ozone-forming characteristics of the airmass could be simulated as if all the hydrocarbons were replaced by CO at a mixing ratio of 4500 ppb. This is feasible because the atmospheric oxidation of CO exhibits most of the major chemical features of that of more complex organic molecules.

Figure 6.4 shows [HO_2], P_{O_3}, and HHL and NHL as a function of [NO] from 10 to 2500 ppt. The behavior exhibited in Figure 6.4 is very basic to tropospheric

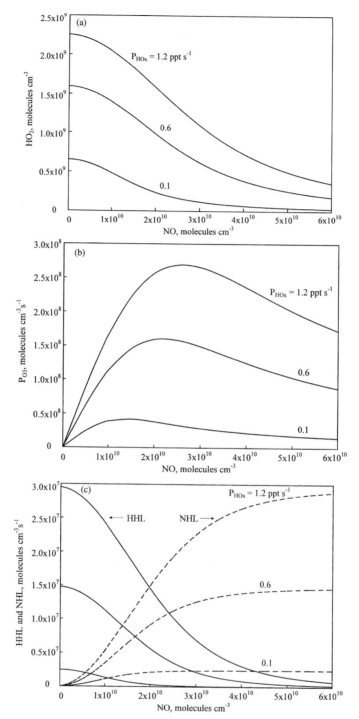

FIGURE 6.4   Characteristics of the atmospheric oxidation of CO as a function of NO_x level. Conditions are 298 K at the Earth's surface, $[NO_2]/[NO] = 7$; three values of P_{HO_x} (0.1, 0.6, and 1.2 ppt s^{-1}) are considered. A similar calculation has been presented by Thornton et al. (2002). (a) HO_2 concentration; (b) ozone production rate, P_{O_3}; (c) HHL and NHL [see (6.11) and (6.12)].

chemistry, and its ramifications will appear again and again. $[HO_2]$ decreases as $[NO]$ increases; this behavior is a result of the fact that as the overall NO_x level increases, the $OH + NO_2$ termination reaction becomes increasingly important, more effectively removing HO_x from the system. The O_3 production rate P_{O_3} achieves a maximum at a particular value of $[NO]$; the larger the value of P_{HO_x}, the larger the value of both P_{O_3} and $[NO]$ at the maximum. Panel (c) shows the HO_x and NO_x loss rates, HHL and NHL. HHL is at its maximum at low $[NO]$, where $[HO_2]$ is at its maximum. Likewise, NHL is largest at high $[NO]$, where $[NO_2]$ is at its maximum. As P_{HO_x} increases, $[HO_2]$ is larger at any given NO level. P_{O_3} attains a maximum reflecting the competition between decreasing HO_2 as NO increases and the NO increase itself. At higher P_{HO_x}, $[HO_2]$ will remain high for larger values of $[NO]$, thus shifting the maximum in P_{O_3}. The maximum in P_{O_3} occurs at a larger value of NO than that at which HHL $=$ NHL. HHL depends on $[HO_2]^2$ versus $P_{O_3} \sim [HO_2] [NO]$. At any value of $[NO]$, HHL $+$ NHL $= P_{HO_x}$. As $[NO]$ increases, HHL decreases with the square of $[HO_2]$, while the decrease of $[HO_2]$ is somewhat compensated for by the increase of $[NO]$ in P_{O_3}. As a result, the HHL/NHL crossover occurs at a smaller value of $[NO]$ than that at which P_{O_3} reaches its maximum.

6.3.4   Theoretical Maximum Yield of Ozone from CO Oxidation

The theoretical maximum yield of O_3 per $CO + OH$ reaction would occur if NO_x concentrations were sufficiently high that every HO_2 radical reacts with NO rather than with itself and termination of the chain by the $OH + NO_2$ reaction were neglected. The resulting mechanism would be

$$CO + OH \xrightarrow{O_2} CO_2 + HO_2$$
$$HO_2 + NO \longrightarrow NO_2 + OH$$
$$NO_2 + h\nu \xrightarrow{O_2} NO + O_3$$

$$\text{Net: } CO + 2O_2 + h\nu \longrightarrow CO_2 + O_3$$

While it is informative to see that one O_3 molecule could theoretically result from each $CO + OH$ reaction, this condition can never be achieved. If NO_x levels are sufficiently high to keep HO_2 from reacting with itself, they are also sufficiently high so that some NO_2 must react with OH to form HNO_3, thereby terminating the chain reaction.

6.4   ATMOSPHERIC CHEMISTRY OF METHANE

The principal oxidation reaction of methane, CH_4, is with the hydroxyl radical:

$$CH_4 + OH \xrightarrow{1} CH_3 + H_2O \qquad \text{(reaction 1)}$$

As in the case of the hydrogen atom, the methyl radical, CH_3, reacts virtually instantaneously with O_2 to yield the methyl peroxy radical, CH_3O_2

$$CH_3 + O_2 + M \longrightarrow CH_3O_2 + M$$

so that the CH_4–OH reaction may be written concisely as

$$CH_4 + OH \xrightarrow[O_2]{1} CH_3O_2 + H_2O$$

The rate coefficient for reaction 1 is $k_1 = 2.45 \times 10^{-12}$ exp $(-1775/T)$ cm^3 molecule^{-1} s^{-1}. At $T = 273$ K and $[OH] = 10^6$ molecules cm^{-3}, the lifetime of CH_4 against OH reaction is about 9 years. Despite its long lifetime, because of its high concentration (about 1750 ppb), CH_4 exerts a dominant effect on the chemistry of the background troposphere.

Under tropospheric conditions, the methyl peroxy radical can react with NO, NO_2, HO_2 radicals, and itself; the reactions with NO and HO_2 are the most important. The reaction with NO leads to the methoxy radical, CH_3O, and NO_2:

$$CH_3O_2 + NO \xrightarrow{2} CH_3O + NO_2 \qquad \text{(reaction 2)}$$

The only important reaction for the methoxy radical under tropospheric conditions is with O_2 to form formaldehyde (HCHO) and the HO_2 radical:

$$CH_3O + O_2 \longrightarrow HCHO + HO_2$$

The $CH_3O + O_2$ reaction is so rapid that reaction 2 can be written as

$$CH_3O_2 + NO \xrightarrow[O_2]{2} HCHO + HO_2 + NO_2$$

The reaction of CH_3O_2 with HO_2 leads to the formation of methyl hydroperoxide, CH_3OOH

$$CH_3O_2 + HO_2 \xrightarrow{3} CH_3OOH + O_2 \qquad \text{(reaction 3)}$$

which can photolyze or react with OH,

$$CH_3OOH + h\nu \xrightarrow{4} CH_3O + OH \qquad \text{(reaction 4)}$$

$$CH_3OOH + OH \xrightarrow{5a} H_2O + CH_3O_2 \qquad \text{(reaction 5a)}$$

$$\xrightarrow{5b} H_2O + CH_2OOH \qquad \text{(reaction 5b)}$$

$$\downarrow \text{fast}$$

$$HCHO + OH$$

The lifetime of CH_3OOH in the troposphere resulting from photolysis and reaction with OH is ~ 2 days.

The fate of CH_3OOH has a significant impact on the methane oxidation chain. Its formation by reaction 3 removes two radicals, CH_3O_2 and HO_2. The photolysis of CH_3OOH returns two radicals, CH_3O and OH. The reaction of CH_3OOH with OH leads to an overall loss of radicals, however. In reaction 5a, OH abstracts the hydrogen atom from the —OOH group; the net result is that one OH radical is destroyed and one CH_3O_2 radical is regenerated. Some fraction of the CH_3O_2 radicals generated in reaction 5a can react with HO_2 by reaction 3 to reform CH_3OOH; in this manner, reactions 5a and 3 constitute a catalytic cycle for the removal of OH and HO_2. Reaction 5b proceeds by abstraction of an H atom from the —CH_3 group by OH, followed by breaking of the weak O—O bond in CH_2OOH. The OH is regenerated, and a molecule of formaldehyde is formed.

Formaldehyde is a first-generation oxidation product of CH_4 and, it turns out, of many other hydrocarbons. Indeed, the chemistry of formaldehyde is common to virtually all mechanisms of tropospheric chemistry. Formaldehyde undergoes two main reactions in the atmosphere, photolysis

$$HCHO + h\nu \xrightarrow{6a} H + HCO \qquad \text{(reaction 6a)}$$

$$\xrightarrow{6b} H_2 + CO \qquad \text{(reaction 6b)}$$

and reaction with OH,

$$HCHO + OH \xrightarrow{7} HCO + H_2O \qquad \text{(reaction 7)}$$

As we have already noted, the hydrogen atom combines immediately with O_2 to yield HO_2. The formyl radical, HCO, also reacts very rapidly with O_2 to yield the hydroperoxyl radical and CO:

$$HCO + O_2 \longrightarrow HO_2 + CO$$

Because of the rapidity of this reaction, the formaldehyde reactions may be written concisely as

$$HCHO + h\nu \xrightarrow[O_2]{6a} 2\,HO_2 + CO$$

$$\xrightarrow{6b} H_2 + CO$$

$$HCHO + OH \xrightarrow[O_2]{7} HO_2 + CO + H_2O$$

Approximate atmospheric photolysis rates of HCHO are

$$HCHO + h\nu \xrightarrow{6a} 2\,HO_2 + CO \qquad j_{HCHO_a} \sim 3 \times 10^{-5}\,s^{-1}$$

$$\xrightarrow{6b} H_2 + CO \qquad j_{HCHO_b} \sim 4 \times 10^{-5}\,s^{-1}$$

Thus, the lifetimes against these two photolysis pathways are $\tau_{HCHO_a} \sim 9\,h$ and $\tau_{HCHO_b} \sim 7\,h$. The rate coefficient for reaction 7, $k_7 = 9 \times 10^{-12}\,cm^3$ molecule^{-1} s^{-1}, gives an HCHO lifetime against OH reaction ($[OH] = 10^6$ molecules cm^{-3}) of $\tau_7 \sim 31\,h$. While the rates of the two photolysis channels, 6a and 6b, are roughly comparable, their effect on the HO_x radical pool is quite different; reaction 6a is a source of two HO_x radicals, whereas reaction 6b leads to permanent loss of radicals.

Virtually every CH_4 molecule is converted to formaldehyde, and the HCHO lifetime is relatively short.[2] Both photolysis and OH reaction of HCHO lead to formation of CO. Thus, CO is the principal product of methane oxidation. As a result, the CO oxidation chemistry of the previous section automatically becomes part of the CH_4 oxidation chain. Rewritten here, this includes reaction of HO_2 with NO to regenerate the OH radical

$$HO_2 + NO \xrightarrow{8} NO_2 + OH \qquad \text{(reaction 8)}$$

HO_x-NO_x termination by nitric acid formation

$$OH + NO_2 + M \xrightarrow{9} HNO_3 + M \qquad \text{(reaction 9)}$$

and HO_x-HO_x termination to form hydrogen peroxide:

$$HO_2 + HO_2 \xrightarrow{10} H_2O_2 + O_2 \qquad \text{(reaction 10)}$$

Eventually CO is converted to CO_2 on a several-month timescale to complete the CH_4 oxidation chain. Table 6.1 summarizes the tropospheric methane oxidation chain. The overall reaction sequence leading eventually to CO_2, through the HCHO and CO intermediate "stable" products, is shown in Figure 6.5.

The theoretical maximum yield of O_3 from CH_4 oxidation would occur when NO_x levels are sufficiently high that the peroxy radicals HO_2 and CH_3O_2 react exclusively with NO and all the formaldehyde formed photolyzes by the radical path. The theoretical maximum yield is 4 O_3 molecules per each CH_4 molecule oxidized:

$$CH_4 + OH \xrightarrow{O_2} CH_3O_2 + H_2O$$
$$CH_3O_2 + NO \xrightarrow{O_2} HCHO + HO_2 + NO_2$$
$$HCHO + h\nu \xrightarrow{2O_2} CO + 2HO_2$$
$$3(HO_2 + NO \longrightarrow NO_2 + OH)$$
$$\underline{4(NO_2 + h\nu \xrightarrow{O_2} NO + O_3)}$$
$$\text{Net:}\quad CH_4 + 8\,O_2 \longrightarrow CO + H_2O + 2\,OH + 4\,O_3$$

Since the theoretical maximum yield of O_3 from oxidation of CO is one more molecule of O_3, the theoretical maximum yield of O_3 from one CH_4 molecule, all the way to CO_2 and H_2O, is

$$CH_4 + 10\,O_2 \longrightarrow CO_2 + H_2O + 2\,OH + 5\,O_3$$

Such a maximum yield is, of course, not realized in the actual atmosphere because of all the competing reactions that have been neglected.

[2]Since

$$\frac{1}{\tau_{HCHO}} = \frac{1}{\tau_{6a}} + \frac{1}{\tau_{6b}} + \frac{1}{\tau_7}$$

the overall lifetime of HCHO for the conditions described above is 3.6 h.

TABLE 6.1   Methane Oxidation Mechanism

	Reaction	Rate Coefficient, $(cm^3 \; molecule^{-1} \; s^{-1})^a$
1.	$CH_4 + OH \xrightarrow{O_2} CH_3O_2 + H_2O$	$2.45 \times 10^{-12} \exp(-1775/T)$
2.	$CH_3O_2 + NO \xrightarrow{O_2} HCHO + HO_2 + NO_2$	$2.8 \times 10^{-12} \exp(300/T)$
3.	$CH_3O_2 + HO_2 \longrightarrow CH_3OOH + O_2$	$4.1 \times 10^{-13} \exp(750/T)$
4.	$CH_3OOH + h\nu \xrightarrow{O_2} HCHO + HO_2 + OH$	Depends on light intensity
5a.	$CH_3OOH + OH \longrightarrow H_2O + CH_3O_2$	$3.6 \times 10^{-12} \exp(200/T)^b$
5b.	$\longrightarrow H_2O + HCHO + OH$	$1.9 \times 10^{-12\,b}$
6a.	$HCHO + h\nu \xrightarrow{O_2} 2HO_2 + CO$	Depends on light intensity
6b.	$\longrightarrow H_2 + CO$	—
7.	$HCHO + OH \xrightarrow{O_2} HO_2 + CO + H_2O$	9.0×10^{-12}
8.	$HO_2 + NO \longrightarrow NO_2 + OH$	$3.5 \times 10^{-12} \exp(250/T)$
9.	$OH + NO_2 \xrightarrow{M} HNO_3$	See Table B.2
10.	$HO_2 + HO_2 \longrightarrow H_2O_2 + O_2$	$2.3 \times 10^{-13} \exp(600/T)$ $+ 4.3 \times 10^{-14} \exp(1000/T)$
11.	$CO + OH \xrightarrow{O_2} CO_2 + HO_2$	$1.5 \times 10^{-13}(1 + 0.6 p_{atm})$
12.	$H_2O_2 + h\nu \longrightarrow OH + OH$	Depends on light intensity
13.	$H_2O_2 + OH \longrightarrow HO_2 + H_2O$	$2.9 \times 10^{-12} \exp(-160/T)$
14.	$NO_2 + h\nu \xrightarrow{O_2} NO + O_3$	Depends on light intensity
15.	$NO + O_3 \longrightarrow NO_2 + O_2$	$3.0 \times 10^{-12} \exp(-1500/T)$
16.	$HO_2 + O_3 \longrightarrow HO + 2O_2$	$1.0 \times 10^{-14} \exp(-490/T)$
17.	$OH + O_3 \longrightarrow HO_2 + O_2$	$1.7 \times 10^{-12} \exp(-940/T)$

[a] Values from Sander et al. (2003) unless noted otherwise.
[b] Atkinson et al. (2004)

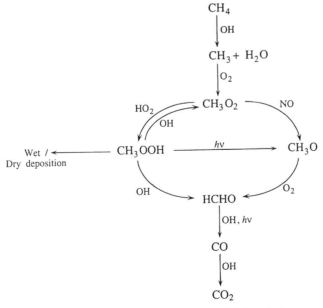

FIGURE 6.5   Atmospheric methane oxidation chain.

The chemistry of the background troposphere is dominated by that of CO and CH_4. In continental regions where human emissions significantly influence the volatile organic compound composition of the atmosphere, the chemistry is more complex than that of CO and CH_4 alone. Despite the complexities that arise as a result of the larger molecules, the basic elements of the chemistry are similar to those of CO and CH_4: initiation by OH attack, formation of O_3 as a result of peroxy radical–NO reactions, and termination by $HO_x + HO_x$ and $HO_x + NO_x$ reactions.

6.5   THE NO_x AND NO_y FAMILIES

6.5.1   Daytime Behavior

The NO_x family comprises NO and NO_2: $NO_x = NO + NO_2$. During daytime, NO and NO_2 interconvert by the photochemical NO_x cycle, which is shown in Figure 6.2 by the heavy lines. The steady-state NO/NO_2 ratio from this cycle is given by (6.6)

$$\frac{[NO]}{[NO_2]} = \frac{j_{NO_2}}{k_{NO+O_3}[O_3]}$$

Let us estimate this ratio at 298 K under typical urban conditions and noontime Sun: $NO_x = 100\,ppb$, $O_3 = 100\,ppb$, $j_{NO_2} = 0.015\,s^{-1}$, $[M] = 2.5 \times 10^{19}$ molecules cm^{-3}, and $k_{NO+O_3} = 1.9 \times 10^{-14}\,cm^3$ molecule^{-1} s^{-1}. One finds that $[NO]/[NO_2] \cong 0.32$, so $NO \cong 24\,ppb$ and $NO_2 \cong 76\,ppb$. At noon, j_{NO_2} is at a maximum, and NO constitutes the largest possible fraction of NO_x. A typical 12-h daytime average ratio is $[NO]/[NO_2] \cong 0.1$.

The principal daytime removal path for NO_x is

$$OH + NO_2 + M \longrightarrow HNO_3 + M$$

At surface (300 K, 1 atm) conditions, $k_{OH+NO_2} \sim 1 \times 10^{-11}\,cm^3$ molecule^{-1} s^{-1}. At $[OH] \cong 10^6$ molecules cm^{-3}, the lifetime of NO_2 during daytime is about 1 day. The lifetime of the NO_x chemical family under daytime conditions is given by applying (3.34):

$$\tau_{NO_x} = \tau_{NO_2}\left[1 + \frac{[NO]}{[NO_2]}\right]$$

$$= \tau_{NO_2}\left[1 + \frac{j_{NO_2}}{k_{NO+O_3}[O_3]}\right] \tag{6.25}$$

The table below shows $[NO]/[NO_2]$ and τ_{NO_x} as a function of altitude.

z, km	T, K	$[NO]/[NO_2]^a$	τ_{NO_x}, days^a
0	288	0.72	1.8
5	256	2.6	4.2
10	223	12.6	18.6

^{a}Assumes $j_{NO_2} = 0.015\,s^{-1}$ (independent of altitude), O_3 mixing ratio $= 50\,ppb$, and $[OH] = 10^6$ molecules cm^{-3}.

The value j_{NO_2} is essentially constant with altitude in the troposphere. For the calculation, the ozone mixing ratio can be taken as essentially uniform vertically, but $[O_3]$ decreases with altitude because of the decrease of the number concentration of air. The ratio $[NO]/[NO_2] < 1$ at the surface, increasing to about 12 at 10 km. Two factors contribute to this increase. First, k_{NO+O_3} decreases as temperature decreases, slowing down the return of NO to NO_2. The second factor is the decrease of $[O_3]$ with altitude, also serving to slow down the rate of the $NO + O_3$ reaction. The lifetime of NO_x increases from between 1 and 2 days at the surface to about 2 weeks in the upper troposphere. The relatively short lifetime at the surface is a result of the fact that most of the NO_x is in the form of NO_2 at the surface, and the $OH + NO_2$ reaction dominates the lifetime of NO_x. In the upper troposphere, the opposite condition holds; with most of the NO_x in the form of NO, the net removal of NO_x by $OH + NO_2$ is slowed considerably.

6.5.2 Nighttime Behavior

At night, NO_2 does not photolyze, and, as a result, the chemistry of the NO_x family is entirely different from that during daytime. Any NO present at night reacts rapidly with O_3 $(k_{NO+O_3} = 1.9 \times 10^{-14} \text{ cm}^3 \text{ molecule}^{-1} \text{ s}^{-1}$ at 298 K)[3]; as a result, almost all NO_x at night is converted to NO_2. We recall from Chapter 5 that NO_2 reacts with O_3 to produce the nitrate (NO_3) radical:

$$NO_2 + O_3 \xrightarrow{1} NO_3 + O_2 \qquad k_1 = 1.2 \times 10^{-13} \exp(-2450/T) \text{ cm}^3 \text{ molecule}^{-1} \text{ s}^{-1}$$

(reaction 1)

Reaction 1 is the only direct source of the NO_3 radical in the atmosphere.

During daytime, NO_3 radicals photolyze rapidly via two paths

$$NO_3 + h\nu \ (\lambda < 700 \text{ nm}) \longrightarrow NO + O_2$$
$$NO_3 + h\nu \ (\lambda < 580 \text{ nm}) \longrightarrow NO_2 + O$$

with a noontime lifetime of ~ 5 s, and react with NO

$$NO_3 + NO \longrightarrow 2 NO_2$$

sufficiently rapidly that NO and NO_3 cannot coexist at mixing ratios of a few parts per trillion (ppt) or higher. For typical daytime conditions of $[NO_2] = 40$ ppb, $[O_3] = 50$ ppb, and $[NO] = 40$ ppb, the maximum NO_3 mixing ratio will be less than 1 ppt. At nighttime, however, when NO concentrations drop near zero, the NO_3 mixing ratio can reach 100 ppt or more.

At night, the nitrate radical reacts with NO_2 to produce N_2O_5

$$NO_2 + NO_3 + M \xrightarrow{2} N_2O_5 + M \qquad \text{(reaction 2)}$$

and N_2O_5 itself can thermally decompose back to NO_2 and NO_3:

$$N_2O_5 + M \xrightarrow{3} NO_2 + NO_3 + M \qquad \text{(reaction 3)}$$

---

[3] In Section 3.1 the lifetime of NO at 298 K in the presence of 50 ppb O_3 was estimated to be 42 s.

Reactions 2 and 3 establish an equilibrium on a timescale of only a few minutes:

$$NO_2 + NO_3 + M \underset{3}{\overset{2}{\rightleftharpoons}} N_2O_5 + M \qquad K_{2,3} = \frac{[N_2O_5]}{[NO_2][NO_3]}$$

The equilibrium constant $K_{2,3} = 3.0 \times 10^{-27} \exp(10990/T)$ cm^3 molecule^{-1} (Sander et al. 2003). The value of $K_{2,3}$ at different altitudes in the troposphere is

z, km	T, K	$K_{2,3}$, cm^3 molecule^{-1}
0	288	1.1×10^{-10}
5	256	1.3×10^{-8}
10	223	7.6×10^{-6}

As temperature decreases and NO_2 levels increase, the equilibrium is shifted more and more to the right.

NO_3 mixing ratios at night in urban plumes have been observed to reach values of a few hundred ppt, and values up to 40 ppt are common in more remote regimes. N_2O_5 mixing ratios of up to 3 ppb have been observed near Boulder, Colorado (Brown et al. 2003a) and up to 200 ppt in the area of San Francisco Bay (Wood et al. 2004).

Whereas gas-phase reactions of N_2O_5 are quite slow, we have already seen the importance in the stratosphere of the heterogeneous (particle-phase) hydrolysis of N_2O_5:

$$N_2O_5 + H_2O(s) \overset{4}{\longrightarrow} 2\,HNO_3 \qquad \text{(reaction 4)}$$

Together with the $OH + NO_2$ reaction, reaction 4 is one of the major paths for removal of NO_x from the atmosphere.

Because NO_3 and N_2O_5 are related through reactions 2 and 3, it is useful to introduce a chemical family $NO_3^* = NO_3 + N_2O_5$. The dynamics of the NO_3^* family are depicted in Figure 6.6. Following the same analysis as in the NO_x lifetime, we find that the lifetime of the NO_3^* family at night is

$$\tau_{NO_3^*} = \tau_{N_2O_5}\left[1 + \frac{[NO_3]}{[N_2O_5]}\right] \qquad (6.26)$$

The lifetime of an N_2O_5 molecule is that against reaction with $H_2O(s)$:

$$\tau_{N_2O_5} = \frac{1}{k_4} \qquad (6.27)$$

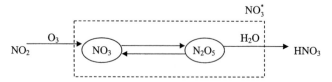

FIGURE 6.6   Chemical family $NO_3^* = NO_3 + N_2O_5$.

At typical boundary-layer conditions, $k_4 \cong 3 \times 10^{-4}\,\text{s}^{-1}$, so $\tau_{N_2O_5} \cong 0.9\,\text{h}$. The NO$_3$/N$_2O_5$ ratio is obtained from the equilibrium constant $K_{2,3}$:

$$\frac{[NO_3]}{[N_2O_5]} = \frac{1}{K_{2,3}[NO_2]} \qquad (6.28)$$

Combining the last three relations, we obtain

$$\tau_{NO_3^*} = \frac{1}{k_4}\left[1 + \frac{1}{K_{2,3}[NO_2]}\right] \qquad (6.29)$$

At mixing ratios of NO$_2 \geq 1$ ppb, most of the NO$_3^*$ is in the form of N$_2$O$_5$, and $\tau_{NO_3^*} \cong \tau_{N_2O_5}$. For example, at an NO$_2$ mixing ratio of 2 ppb, we obtain

z, km	T, K	$\tau_{NO_3^*}$, h
0	288	1.1
5	256	0.93
10	228	0.93

For a more complete analysis of timescales in the NO$_3$/N$_2$O$_5$ system, we refer the reader to Brown et al. (2003b).

6.6   OZONE BUDGET OF THE TROPOSPHERE AND ROLE OF NO$_x$

6.6.1   Ozone Budget of the Troposphere

The tropospheric ozone budget can be calculated by global chemical transport models. Table 6.2 presents the tropospheric O$_3$ budget of Wang et al. (1998). The budget is for the

TABLE 6.2   Global Budget for Tropospheric Ozone

	Global	Northern Hemisphere	Southern Hemisphere
Sources, Tg O$_3$ yr^{-1}			
In situ chemical production	4100	2620	1480
Transport from stratosphere	400	240	160
Total	4500	2860	1640
Sinks, Tg O$_3$ yr^{-1}			
In situ chemical loss	3680	2290	1390
Dry deposition	820	530	290
Total	4500	2820	1680
Interhemispheric transport Tg O$_3$ yr^{-1}	0	−40	40
Burden, Tg O$_3$	310	180	130
Residence time, days	25	23	28

Source: Wang et al. (1998).

air column extending up to 150 mbar. The annual budget amounts were computed for the entire odd-oxygen family O_x to account for reservoir species that, on dissociation or reaction, effectively release O_3 to the atmosphere. Since O_3 accounts for over 95% of O_x, the budgets of O_x and O_3 are essentially equivalent. Chemical production dominates the source of tropospheric O_3 (4100 Tg yr^{-1}), as compared with 400 Tg yr^{-1} estimated for transport down from the stratosphere. The O_3 sink is also dominated by chemical loss (3700 Tg yr^{-1}). Dry deposition accounts for about 800 Tg yr^{-1} of loss. In situ chemical production of O_3 results primarily from reactions of peroxy radicals with NO; in situ chemical loss of O_3 results mainly from $O(^1D) + H_2O$, $O_3 + HO_2$, and $O_3 + OH$. In O_3 production, about 70% of the peroxy radical + NO reactions are $HO_2 + NO$, about 20%, $CH_3O_2 + NO$, and the remainder larger peroxy radical + NO reactions. Ozone loss occurs roughly as $O(^1D) + H_2O \sim 40\%$; $O_3 + HO_2 \sim 40\%$; $O_3 + OH \sim 10\%$. The interhemispheric difference in the O_3 burden is not as great as the difference in NO_x emissions; the explanation is that the ozone production efficiency in the Southern Hemisphere is higher than that in the Northern Hemisphere, 46 mol O_3/mol NO_x in the SH versus 23 mol O_3/mol NO_x in the NH, as computed by Wang et al. (1998). The global mean residence time of a molecule of O_3 is computed to be about 25 days.

6.6.2   Role of NO_x

In situ chemical formation and consumption dominate the balance of tropospheric ozone. The local concentration of NO is critical in determining whether the atmosphere in a particular region is either a source or a sink of O_3. An approximate assessment of this can be made by considering the fate of the HO_2 radical. In the CO oxidation system it was seen that the rate of O_3 production is equal to the rate of the $HO_2 + NO$ reaction, (6.9). The competing reaction, $HO_2 + O_3 \longrightarrow OH + 2O_2$, leads to ozone loss. Thus, the ratio of the rates of these two reactions is indicative of whether a particular region of the atmosphere is one in which O_3 is being produced or consumed:

$$\frac{R_{HO_2+O_3}}{R_{HO_2+NO}} = \frac{k_{HO_2+O_3}[O_3]}{k_{HO_2+NO}[NO]}$$

For a given level of O_3, the concentration of NO at which this ratio is unity can be called the *breakeven concentration* of NO; below the breakeven concentration, O_3 is consumed and above it, O_3 is produced. At 298 K, the ratio of the two rate coefficients is

$$\frac{k_{HO_2+O_3}}{k_{HO_2+NO}} = 2.3 \times 10^{-4}$$

In remote regions the O_3 mixing ratio is about 20 ppb. Thus, the breakeven NO mixing ratio in such regions is about 5 ppt. This amount of NO is roughly equivalent to 15–20 ppt NO_x.

In CH_4 oxidation, O_3 production occurs as a result of both $HO_2 + NO$ and $CH_3O_2 + NO$ reactions. A competition exists between NO and HO_2 for reaction with the CH_3O_2 radical; the former reaction leads to O_3 production and the latter reaction to formation of CH_3OOH. The $CH_3O_2 + NO$ and $CH_3O_2 + HO_2$ rate coefficients are of roughly comparable magnitude (Table B.1). On the basis of approximate HO_2 radical concentrations, Logan et al. (1981) calculated that the reaction of CH_3O_2 with NO exceeds that with HO_2 for NO mixing ratios > 30 ppt. (Note that CH_3OOH may serve as only a temporary sink for HO$_x$.)

Because of O_3 consumption by photolysis, the NO breakeven concentration at which net O_3 production occurs is somewhat larger than the value based solely on the ratio of the $HO_2 + O_3$ and $HO_2 + NO$ reactions. The approximate crossover NO$_x$ concentration between O_3 destruction and production is considered to be about 30 ppt. Ozone mixing ratios in the boundary layer over the remote Pacific Ocean are only about 5 to 6 ppb; NO$_x$ levels are about 10 ppt. Thus, this region of the atmosphere tends to be below the crossover point.

Local production and loss of O_3 in the background troposphere can be estimated from

$$P_{O_3} = \{k_{HO_2+NO}[HO_2] + k_{CH_3O_2+NO}[CH_3O_2]\}[NO]$$
$$L_{O_3} = k_{O(^1D)+H_2O}[O(^1D)][H_2O] + k_{HO_2+O_3}[HO_2][O_3]$$

Figure 6.7 shows calculated, 24-hour-average production and loss rates for the free troposphere above Hawaii during the MLOPEX (Mauna Loa Photochemistry Experiment) as a function of NO$_x$ mixing ratio. The O_3 loss rate is seen to be almost independent of NO$_x$, at about 5×10^5 molecules cm^{-3} s^{-1}. For an O_3 mixing ratio of 40 ppb, this loss rate gives an O_3 lifetime of 17 days. Data for upper tropospheric concentrations over Hawaii indicate that [NO$_x$] is typically ~ 30 ppt, with midday [NO] at ~ 10 ppt (Ridley et al. 1992). From Figure 6.7 we see that at these levels O_3 production and loss are just about in balance, with loss predicted to be slightly greater.

Ozone lifetimes in the troposphere vary significantly depending on altitude, latitude, and season. Lifetimes are shorter in the summer than in the winter as a result of the higher

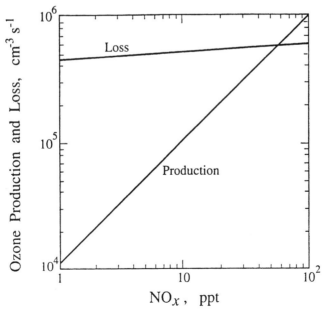

FIGURE 6.7   Calculated 24-h average O_3 production and loss rates for the free troposphere above Hawaii during the MLOPEX as a function of the NO$_x$ mixing ratio (Liu et al. 1992).

solar fluxes in the summer. Lifetimes are also shorter at the surface because of the higher water vapor concentration near the surface. At higher latitudes, lifetimes increase because of the reduced solar intensity. At 20°N, for example, it is estimated that O_3 lifetimes at the surface are about 5 days in summer and 17 days in winter, whereas at 40°N these increase to 8 days in summer and 100 days in winter. At 20°N, at 10 km altitude, estimated summer and winter O_3 lifetimes are 30 days and 90 days, respectively, increasing by a factor of ~ 6 from those at the surface.

The Troposphere–Stratosphere Transition from a Chemical Perspective   The transition from troposphere to stratosphere is traditionally defined on the basis of the reversal of the atmospheric temperature profile. That transition is also dramatically reflected in how the concentrations of trace species vary with altitude both below and above the tropopause. Of these, HO_2 and OH exhibit perhaps the most profound differences across the tropopause (Wennberg et al. 1995). The major difference between the stratosphere and the troposphere is what controls the interconversion between OH and HO_2 within the HO_x family. In the lower stratosphere the cycle within the HO_x family is

$$OH + O_3 \longrightarrow HO_2 + O_2$$
$$HO_2 + O_3 \longrightarrow OH + 2O_2$$

In the lower stratosphere the HO_2/OH ratio is described by the extension of (5.28):

$$\frac{[HO_2]}{[OH]} = \frac{k_{OH+O_3}[O_3]}{k_{HO_2+NO}[NO] + k_{HO_2+O_3}[O_3]}$$

This ratio varies from about 4 to 7 and decreases as [NO] increases. [OH] itself is essentially independent of [NO] and depends almost entirely on solar zenith angle. This independence of OH on NO is a result of the fact that the increase of OH that results from the $HO_2 + NO$ reaction is offset almost exactly by a decrease of the rates of reactions that generate OH. This occurs because the HO_2 that participates in the $HO_2 + NO$ reaction is not otherwise available for other reactions.

Whereas OH to HO_2 conversion in the lower stratosphere occurs mainly by $OH + O_3$, that in the upper troposphere occurs mainly by $OH + CO$ (Lanzendorf et al. 2001),

$$OH + CO \xrightarrow{O_2} HO_2 + CO_2$$
$$HO_2 + NO \longrightarrow OH + NO_2$$

The HO_2/OH ratio in the upper troposphere can be approximated as

$$\frac{[HO_2]}{[OH]} = \frac{k_{OH+CO}[CO]}{k_{HO_2+NO}[NO]}$$

As one proceeds up in the troposphere, the NO_2/NO_x ratio decreases (recall Section 6.6.1), achieving its lowest value at the tropopause, and then increases on moving into the stratosphere. The increase of NO_2 relative to NO in the lower stratosphere is the result of the $HO_2 + NO$ reaction. (The NO_x/NO_y ratio is more or less constant in the upper troposphere, falling off as one goes into the stratosphere. This falloff reflects the influence of the stratospheric aerosol layer in promoting the heterogeneous formation of HNO_3.) Hydroxyl radical levels in the upper troposphere vary from ~ 0.01 to 0.1 ppt. In the lower stratosphere OH depends on solar zenith angle and ranges up to ~ 1 ppt. As noted above, OH is essentially independent of NO in the lower stratosphere, whereas in the upper troposphere OH decreases as NO decreases. This fundamentally different behavior of OH with respect to changes in NO also characterizes the troposphere–stratosphere transition.

6.7 TROPOSPHERIC RESERVOIR MOLECULES

6.7.1 H_2O_2, CH_3OOH, and HONO

In the CO oxidation cycle, hydrogen peroxide, H_2O_2, is a reservoir molecule for HO_x:

$$HO_2 + HO_2 \longrightarrow H_2O_2 + O_2$$
$$H_2O_2 + h\nu \longrightarrow OH + OH$$
$$H_2O_2 + OH \longrightarrow H_2O + HO_2$$

In the CH_4 oxidation cycle, methyl hydroperoxide, CH_3OOH, is also a HO_x reservoir:

$$CH_3O_2 + HO_2 \longrightarrow CH_3OOH + O_2$$
$$CH_3OOH + h\nu \longrightarrow CH_3O + OH$$
$$CH_3OOH + OH \longrightarrow HCHO + OH + H_2O$$

Nitrous acid, HONO, which is formed by a heterogeneous reaction involving NO_2 and H_2O (Calvert et al. 1994), is a reservoir for both HO_x and NO_x. HONO dissociates by photolysis to regenerate OH and NO:

$$HONO + h\nu \longrightarrow OH + NO$$

Formed overnight, HONO photodissociates upon sunrise to inject a pulse of OH into the early-morning atmosphere.

6.7.2 Peroxyacyl Nitrates (PANs)

The class of compounds of general formula $RC(O)OONO_2$ called *peroxyacyl nitrates* (PANs) was first discovered in the early 1950s as components of photochemical smog. The first two compounds in the series are

$$\overset{\displaystyle O}{\overset{\displaystyle \|}{CH_3C}}OONO_2 \qquad \text{peroxyacetyl nitrate}$$

$$\overset{\displaystyle O}{\overset{\displaystyle \|}{CH_3CH_2C}}OONO_2 \qquad \text{peroxypropionyl nitrate}$$

Peroxyacetyl nitrate is the first compound in the series of PANs, which itself is usually called PAN. One route of formation of PAN is the OH reaction with acetaldehyde:

$$CH_3CHO + OH \xrightarrow{1} CH_3CO + H_2O$$
$$CH_3CO + O_2 \xrightarrow[fast]{} CH_3C(O)O_2$$

Since the second reaction is very fast, it can be combined with reaction 1:

$$CH_3CHO \xrightarrow[O_2]{1} CH_3C(O)O_2 + H_2O$$

As do other peroxy radicals, the peroxyacetyl radical, $CH_3C(O)O_2$, reacts with NO:

$$CH_3C(O)O_2 + NO \xrightarrow{2} NO_2 + CH_3C(O)O$$
$$CH_3C(O)O + O_2 \xrightarrow[fast]{} CH_3O_2 + CO_2$$

Again, we can incorporate the fast reaction into its precursor to give

$$CH_3C(O)O_2 + NO \xrightarrow[O_2]{2} NO_2 + CH_3O_2 + CO_2$$

The peroxyacetyl radical also reacts with NO_2 to form PAN

$$CH_3C(O)O_2 + NO_2 + M \underset{4}{\overset{3}{\rightleftharpoons}} CH_3C(O)O_2NO_2 + M$$

and PAN thermally decomposes back to its reactants by reaction 4.

Initially thought to be of importance only in polluted urban atmospheres, PAN has been identified as one of the most abundant reactive nitrogen-containing species in the troposphere (Roberts 1990). PAN mixing ratios ~ 100 ppt are present in the Northern Hemisphere free troposphere, although its abundance is highly variable (Singh et al. 1995). Near the tropics, mixing ratios around 10 ppt are often prevalent.

PAN acts as a reservoir species for both $CH_3C(O)O_2$ radicals and NO_x. Because of this, the atmospheric lifetime of PAN is important; if its lifetime is relatively long, PAN can act as an effective reservoir for NO_x. Potential atmospheric removal processes for PAN include thermal decomposition (reaction 4 above), UV photolysis, and OH reaction. PAN is not highly water-soluble; it is more soluble than NO or NO_2 but considerably less soluble than HNO_3. Thus, wet deposition is a minor removal process. Dry deposition is also unimportant. The PAN–OH rate constant is $<3 \times 10^{-14}$ cm^3 molecule^{-1} s^{-1}, and OH reaction is not an effective removal process. PAN absorbs UV radiation up to 350 nm (Libuda and Zabel 1995; Talukdar et al. 1995). Thus, thermal decomposition and photolysis are the principal removal processes for PAN.

Figure 6.8 shows the first-order loss rate of PAN as a function of altitude by thermal decomposition and photodissociation. The thermal decomposition rate coefficient for PAN can be obtained from the forward rate coefficient for reaction 3, k_3, and the equilibrium constant for reactions 3 and 4, $K_{3,4}$ (Sander et al. 2003),

$$k_4 = k_3/K_{3,4}$$

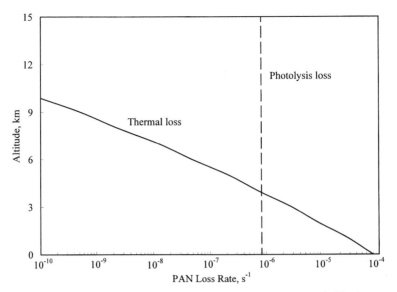

FIGURE 6.8   Atmospheric loss rate of PAN as a function of altitude.

where

$$k_3([\mathrm{M}], T) = \left[\frac{k_0[\mathrm{M}]}{1 + \frac{k_0[\mathrm{M}]}{k_\infty}}\right] F^{\left\{\left(1 + \left[\log_{10}\left(\frac{k_0[\mathrm{M}]}{k_\infty}\right)\right]^2\right)^{-1}\right\}}$$

$$k_0 = 9.7 \times 10^{-29}\,(T/300)^{-5.6}$$
$$k_\infty = 9.3 \times 10^{-12}\,(T/300)^{1.5}$$
$$F = 0.6$$

and

$$K_{3,4} = 9.0 \times 10^{-29} \exp(14000/T) \qquad \mathrm{cm}^3\,\mathrm{molecule}^{-1}$$

The PAN absorption cross section has been determined empirically to be

$$\sigma_{\mathrm{PAN}} = 4 \times 10^{-8}\,\exp(-0.102\,\lambda) \qquad \mathrm{cm}^2\,\mathrm{molecule}^{-1}$$

where λ is in nm. Typical actinic fluxes in summer are given in Table 4.3.

From Figure 6.8 we see that at the Earth's surface (288 K) the lifetime of PAN against thermal decomposition is about 3 h, whereas that against photodissociation is about 13 days. Because the photolytic loss of PAN is approximately independent of altitude and the rate of thermal decomposition is strongly temperature dependent, a point is reached, at about 7 km, where the two rates become equal; above that altitude, photolysis is the more important loss process. At the temperature of the upper troposphere, PAN is an effective reservoir for NO_x; because PAN is transported in the upper troposphere, this amounts to a mechanism for long-range transport of NO_x.

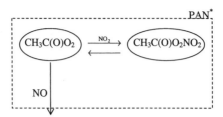

FIGURE 6.9   PAN* chemical family.

Temporary Storage of NO$_x$ by PAN   PAN acts as a reservoir species for both CH$_3$C(O)O$_2$ radicals and NO$_2$. In order to calculate the lifetime for storage of NO$_x$ by PAN, one can consider the sum of CH$_3$C(O)O$_2$ and CH$_3$C(O)O$_2$NO$_2$ as a chemical family PAN* (Figure 6.9). Let us calculate the lifetime of this chemical family, as compared to that for the thermal decomposition of PAN:

$$\tau_{PAN^*} = \frac{1}{k_{CH_3C(O)O_2+NO}[NO]}\left(1 + \frac{[CH_3C(O)O_2NO_2]}{[CH_3C(O)O_2]}\right) \qquad (6.30)$$

When dealing with chemical families, we generally assume that equilibrium is established within the family. We may not be certain that this is the case here because of the significant perturbation to this equilibrium caused by the CH$_3$C(O)O$_2$ + NO reaction. Since CH$_3$C(O)O$_2$ itself is a free radical, its concentration should obey a steady-state relation. Thus we obtain

$$\tau_{PAN^*} = \frac{1}{k_{CH_3C(O)O_2+NO}[NO]}\left(1 + \frac{k_{CH_3C(O)O_2+NO_2}[NO_2] + k_{CH_3C(O)O_2+NO}[NO]}{k_{CH_3C(O)O_2NO_2+M}}\right)$$

$$(6.31)$$

The lifetime of PAN* can be compared with the thermal decomposition lifetime of PAN itself in Figure 6.8. The lifetime of the PAN* family is about 5 times as long as that of PAN itself for these conditions.

Acetone   The first measurement of HO$_x$ radicals in the upper troposphere in 1994–1996 revealed a more photochemically active region than had been expected (Brune et al. 1998; Wennberg et al. 1998). Observed HO$_x$ levels tended to be 2–4 times higher than that based simply on O$_3$ photolysis to O(^{1}D) followed by reaction of O(^{1}D) with H$_2$O. One explanation for these observations is transport of photochemically active species from the lower to the upper troposphere. In addition, injection of air from the surface, carrying high levels of HO$_x$ reservoir species through deep convection is another source of HO$_x$. Such species can include CH$_3$OOH, H$_2$O$_2$, and aldehydes (principally HCHO). Along with these reservoir species, acetone (CH$_3$C(O)CH$_3$) is a source of HO$_x$ in the upper troposphere. The 1997 SONEX aircraft campaign over the North Atlantic provided the first simultaneous measurements of HO$_x$, H$_2$O$_2$, CH$_3$OOH, HCHO, O$_3$, H$_2$O, and acetone and allowed an assessment of the extent to which HO$_x$ precursors influence the HO$_x$ chemistry of the upper troposphere (Brune et al. 1999; Faloona et al. 2000). The primary source of HO$_x$, besides O(^{1}D) + H$_2$O, was found to be acetone photolysis.

Acetone, $CH_3C(O)CH_3$, is an ubiquitous atmospheric species having a mixing ratio of about 1 ppb in rural sites in a variety of locations (Singh et al. 1994, 1995). Under extremely clean conditions, ground-level background mixing ratios of 550 ppt have been found throughout the NH troposphere. In the free troposphere, acetone mixing ratios on the order of 500 ppt are present at northern midlatitudes, declining to about 200 ppt at southern latitudes (Singh et al. 1995). From atmospheric data and three-dimensional photochemical models, a global acetone source of 40–60 Tg yr^{-1} has been estimated, composed of 51% secondary formation from the atmospheric oxidation of precursor hydrocarbons (principally propane, isobutane, and isobutene), 26% direct emission from biomass burning, 21% direct biogenic emissions, and 3% primary anthropogenic emissions (Singh et al. 1994). Atmospheric removal of acetone is estimated to result from photolysis (64%), OH reaction (24%), and deposition (12%). The average lifetime of acetone in the atmosphere is estimated to be 16 days (Singh et al. 1995).

By virtue of its photooxidation chemistry (see Section 6.11), acetone is a source of HO$_x$ radicals in the upper troposphere. Under the dry conditions of the upper troposphere, where $O(^1D) + H_2O$ is relatively slow, acetone makes an important additional contribution to HO$_x$. Photolysis of acetone ($\lambda < 360$ nm) yields two HO$_2$ and two HCHO molecules (when $[NO] \gg [NO_2]$), and 30% of the HCHO molecules photolyze via the radical-forming branch to yield two more HO$_2$ molecules. Thus the HO$_x$ yield from the photolysis of acetone is ~ 3.2 (the result of $2 + 4 \times 0.3$), as compared to a yield of 2 from the reaction of $O(^1D) + H_2O$. A simple photochemical model calculation for the upper troposphere at the equinox (40°N, 11 km, 50 ppb O$_3$, 90 ppm H$_2$O, and 0.5 ppb acetone) produced 24-h average HO$_x$ production rates of 7×10^3 molecules cm^{-3} s^{-1} from $O(^1D) + H_2O$ and 9×10^3 molecules cm^{-3} s^{-1} from photolysis of acetone (Singh et al. 1995).

6.8   RELATIVE ROLES OF VOC AND NO$_x$ IN OZONE FORMATION

6.8.1   Importance of the VOC/NO$_x$ Ratio

The hydroxyl radical is the key reactive species in the chemistry of ozone formation. The VOC–OH reaction initiates the oxidation sequence. There is a competition between VOCs and NO$_x$ for the OH radical. At a high ratio of VOC to NO$_x$ concentration, OH will react mainly with VOCs; at a low ratio, the NO$_x$ reaction can predominate. Hydroxyl reacts with VOC and NO$_2$ at an equal rate when the VOC : NO$_2$ concentration ratio is a certain value; this value depends on the particular VOC or mix of VOCs present, as the OH rate constants of VOCs differ for each VOC species.

At ambient conditions the second-order rate constant for the OH + NO$_2$ reaction is, in mixing ratio units, approximately 1.7×10^4 ppm^{-1} min^{-1}. Considering an average urban mix of VOCs, an average VOC–OH rate constant, expressed on a per carbon atom basis, is about 3.1×10^3 ppmC^{-1} min^{-1}. Using this value for an average VOC–OH rate constant, the ratio of the OH–NO$_2$ to OH–VOC rate constants is about 5.5. Thus, when the VOC : NO$_2$ concentration ratio is approximately 5.5 : 1, with the VOC concentration expressed on a carbon atom basis, the rates of reaction of VOC and NO$_2$ with OH are equal. If the VOC : NO$_2$ ratio is less than 5.5 : 1, reaction of OH with NO$_2$ predominates over reaction of OH with VOCs. The OH–NO$_2$ reaction removes OH radicals from the active VOC oxidation cycle, retarding the further production of O_3. On the other hand, when the ratio exceeds 5.5 : 1, OH reacts preferentially with VOCs. At a minimum, no new radicals are produced or destroyed; however, in actuality, photolysis of intermediate products generated by the OH–VOC reactions generates new radicals, accelerating O_3 production.

Imagine starting with a given mixture of VOCs and NO_x. Because OH reacts about 5.5 times more rapidly with NO_2 than with VOCs, NO_x tends to be removed from the system faster than VOCs.[4] In the absence of fresh NO_x emissions, as the system reacts, NO_x is depleted more rapidly than VOCs, and the instantaneous VOC : NO_2 ratio will increase with time. Eventually the concentration of NO_x becomes sufficiently low as a result of the continual removal of NO_x by the OH–NO_2 reaction that OH reacts preferentially with VOCs to keep the ozone-forming cycle going. At very low NO_x concentrations, peroxy radical–peroxy radical reactions begin to become important.

The essential role of NO_x in ozone formation is evident in the CO oxidation mechanism (Section 6.4). For example, in the low NO_x limit (NO_x-limited), the rate of O_3 formation increases linearly as [NO] increases and the rate is independent of [CO]. In the high NO_x limit (NO_x-saturated), the rate of O_3 formation increases with [CO], but decreases as [NO_x] increases. The explanation for the behavior in the high NO_x limit is that, with ample NO_x available, as NO_x increases, the rate of the OH + NO_2 termination reaction increases, removing both HO_x and NO_x from the system, limiting OH – HO_2 cycling, and thereby decreasing the rate of O_3 formation.

At a given level of VOC, there exists a NO_x concentration at which a maximum amount of ozone is produced, an optimum VOC : NO_x ratio. For ratios less than this optimum ratio, NO_x increases lead to ozone decreases; conversely, for ratios larger than this optimum ratio, NO_x increases lead to ozone increases.

6.8.2   Ozone Isopleth Plot

The dependence of O_3 production on the initial amounts of VOC and NO_x is frequently represented by means of an *ozone isopleth diagram*. Such a diagram is a contour plot of maximum O_3 concentrations achieved as a function of initial VOC and NO_x concentrations. The diagram is generated by contour plotting the predicted ozone maxima obtained from a large number of simulations with an atmospheric VOC/NO_x chemical mechanism with varying initial concentrations of VOC and NO_x while all other variables are constant.

Figure 6.10 is an ozone isopleth plot for Atlanta. To generate this plot, ozone formation was simulated in a hypothetical well-mixed box of air from the ground to the mixing height that is transported over an emissions grid from the region of most intense source emissions, the center city, to the downwind location of maximum ozone concentration. The mixing height rise throughout the day increases the volume of the cell, leading to dilution of the cell's contents. VOC and O_3 in the air above the cell are entrained into the cell as the mixing height rises. Chemical transformations in the cell are described by a chemical reaction mechanism appropriate for the VOC mixture. The cell for Atlanta initially contained 600 ppbC of anthropogenic controllable VOCs, 38 ppbC of background, uncontrollable VOCs, and 100 ppb of NO_x. The air above the cell was assumed to

---

[4] The crossover VOC : NO_2 ratio of 5.5 that we have been using in our discussion applies, more or less, to an average urban VOC mix. Because individual VOC–OH rate constants vary significantly, this ratio will also vary significantly if only a single VOC is present.

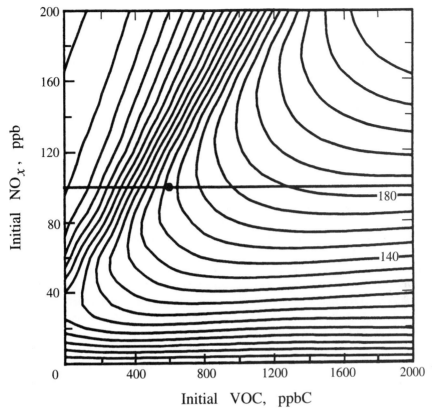

FIGURE 6.10   Ozone isopleth plot based on simulations of chemistry along air trajectories in Atlanta (Jeffries and Crouse 1990). Each isopleth is 10 ppb higher in O$_3$ as one moves upward and to the right.

contain 20 ppbC VOC and 40 ppb of O$_3$. The cell begins moving from center city at 0800 LDT and moves to the suburbs over a 14-h period. The initial mixing height was 250 m and the maximum mixing height was 1500 m. The chemical mechanism used for the simulation was the *carbon bond IV mechanism* (Gery et al. 1989). The peak ozone concentration predicted for these conditions was 114.6 ppb. We describe moving box models of this type in Chapter 25.

The base case described above corresponds to the dot on the line at initial NO$_x$ equal to 100 ppb in Figure 6.10. The full plot is generated by systematically varying the initial VOC and NO$_x$ concentrations and running the same scenario. The ozone ridge in Figure 6.10 separates the regions of low VOC : NO$_x$ ratio, above the ridge line, and high VOC : NO$_x$ ratio, below the ridge line. In Figure 6.10, the maximum O$_3$ concentration is not zero at 0.0 ppmC initial VOC because the scenario included 38 ppbC surface background VOC and 20 ppbC VOC aloft.

The ozone ridge line identifies the maximum O$_3$ concentration that can be achieved at a given VOC level, allowing the NO$_x$ level to vary. If the NO$_x$ level is either increased or decreased relative to that at the ridge line, the maximum amount of O$_3$ produced at that VOC level will decrease. The region of the ozone isopleth plot below the ridge line is what we have denoted as "NO_x-limited"; that above the ridge line as "NO_x saturated." It is apparent that above the ridge line a reduction in NO_x can lead to an increase in maximum O_3. Also, below the ridge line at low NO_x levels there is a region where large reductions in organics have almost no effect on maximum O_3. The ridge line defines the combination of initial VOC and NO_x at which all the NO_x is converted into nitrogen-containing products by the end of the simulation so that there is no NO left to participate in NO-to-NO_2 conversions nor any NO_2 left to photolyze. From a point on the ridge line if more VOC is added to the initial mixture to move into the NO_x-limited regime the consumption of NO_x occurs sooner than at the ridge line. Because of the increasing predominance of peroxy radicals relative to NO_x, termination reactions are not favored relative to propagation and the amount of O_3 increases.

To explain the NO_x-saturated region imagine a vertical line at constant initial VOC of 600 ppbC. The maximum O_3 at this initial VOC is about 142 ppb at 80 ppb NO_x. At low initial NO_x, the average number of cycles an OH makes before termination is low as there is insufficient NO to convert all the HO_2 into OH and radical–radical reactions lead to termination. As NO_x increases to the ridge line and beyond, the average number of OH cycles first increases and then begins to decrease because of increased termination from OH + NO_2 rather than OH propagation occurring by OH + VOC. Peak OH cycling occurs just about at the ridge line. The amount of O_3 produced per OH–VOC reaction increases as NO_x increases, since the increased NO is more competitive for RO_2 radicals and therefore produces more NO_2. Above the ridge line, as initial NO_x is increased, the absence of sufficient radicals, however, makes the system increasingly unable to recycle NO. Also, fresh OH production from O_3 photolysis is reduced because O_3 appears later in the process and at a lower concentration. Thus increased OH termination as initial NO_x is increased above the ridge line and reduction in fresh OH production together result in less production of O_3.

Relation of O_3 to NO_y   From the definition of the *ozone production efficiency*, the signature of an NO_x molecule lost is the appearance of a number of O_3 molecules, the specific number depending on atmospheric conditions and the HC and NO_x levels. Thus, the O_3 concentration attained in an airmass should be correlated with the quantity $[NO_y]-[NO_x]$, which is the total concentration of products of NO_x oxidation (HNO_3, PAN, etc.). That this correlation should exist in airmasses was first pointed out by Trainer (1991, 1993), and it has been subsequently pursued in numerous studies [e.g., Kleinman (1994, 1997), Carpenter et al. (2000)]. To obtain a good correlation between $[O_3]$ and $[NO_y]-[NO_x]$, O_3 production must have occurred within a day or so in the airmass, before significant removal of NO_y can take place, for example by wet and dry deposition of HNO_3.

Kleinman et al. (1994) suggested the linear least-squares correlation

$$[O_3] = 27 + 11.4 \, ([NO_y] - [NO_x]) \tag{6.32}$$

(all concentrations in ppb) for the southeastern United States. The intercept of 27 ppb can be interpreted as an eastern North American background O_3 level so that the NO_y associated with that O_3 has since been removed, leaving behind the longer-lived O_3. Equation (6.32) suggests that the OPE for air in the southeastern United States was, at the time period on which the correlation is based, about 11. In general, the more remote the airmass, the lower the NO_x level, and the less

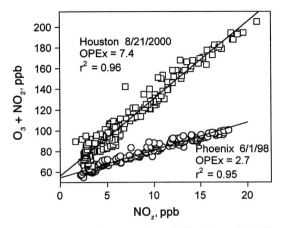

FIGURE 6.11 Comparison of ozone production efficiencies on high O$_3$ days in Phoenix, Arizona and Houston, Texas. (Courtesy Larry J. Kleinman, Brookhaven National Laboratory.)

important the NO$_x$ removal reactions. In this case, each NO$_x$ molecule is a more effective producer of O$_3$, giving a larger slope of the correlation between [O$_3$] and [NO$_z$] = [NO$_y$]−[NO$_x$].

The overall linear relation between [O$_3$] and [NO$_y$] − [NO$_x$] has proved to be an effective way to compare the ozone-forming potential of different airmasses. Figure 6.11 shows [O$_3$] + [NO$_2$] versus [NO$_y$]−[NO$_x$] for Houston, Texas and Phoenix, Arizona on specific dates in 2000 and 1998, respectively, when high O$_3$ concentrations were achieved in both cities. The observed maximum in O$_3$ in Houston (194 ppb) was more than double that observed in Phoenix (93 ppb). The slopes of the straight-line fits to the data indicate that the number of O$_3$ molecules formed per molecule of NO$_x$ converted to oxidation products in Houston is more than twice that in Phoenix. (In each city about 20 ppb of NO$_x$ was consumed in forming O$_3$.) The higher O$_3$ production efficiency in Houston is a result of both high humidity [high rate of O(^{1}D) + H$_2$O] and a high overall radical production rate, arising in part from HCHO produced from alkene oxidation.

6.9 SIMPLIFIED ORGANIC/NO$_x$ CHEMISTRY

The chemistry of the background troposphere is fueled largely by the oxidation of CH$_4$ and CO. Still, even the most pristine regions of the troposphere contain a variety of other volatile organic compounds (VOCs), arising from both biogenic and human emissions. After this section we will consider the chemistry of individual classes of organics. The OH radical is the dominant species oxidizing VOCs, although for alkenes O$_3$ is also an important oxidant.

By analogy to CH$_4$, reaction of OH with many hydrocarbons (RH) leads to alkyl peroxy radicals (RO$_2$):

$$\left.\begin{array}{l} RH + OH \longrightarrow R + H_2O \\ R + O_2 + M \xrightarrow[\text{fast}]{} RO_2 + M \end{array}\right\} \quad RH + OH \xrightarrow{O_2} RO_2 + H_2O$$

The alkyl peroxy radical reacts with NO:

$$RO_2 + NO \longrightarrow RO + NO_2$$
$$\xrightarrow{M} RONO_2$$

The second reaction increases in importance monotonically as the size of R increases. The alkoxy radical (RO) reacts rapidly with O_2:

$$RO + O_2 \xrightarrow[fast]{} R'CHO + HO_2$$

This reaction is the sole fate of small alkoxy radicals. Larger alkoxy radicals can undergo other reactions; these need not be considered at this point. The correspondence of the $RO + O_2$ reaction to the $CH_3O + O_2$ reaction is evident, where HCHO is the carbonyl product of the latter reaction. The higher carbonyl $R'CHO$ can, in general, continue to be oxidized.

The HO_2 radical reacts with NO to regenerate OH

$$HO_2 + NO \longrightarrow NO_2 + OH$$

and the main HO_x–HO_x and HO_x–NO_x termination reactions are

$$HO_2 + HO_2 \longrightarrow H_2O_2 + O_2$$
$$RO_2 + HO_2 \longrightarrow ROOH + O_2$$
$$OH + NO_2 \xrightarrow{M} HNO_3$$

The resulting generalized mechanism is summarized in Table 6.3, where we have given representative rate coefficients.

TABLE 6.3   Generalized VOC/NO_x Mechanism

Reaction	Rate Constant (298 K)
1. $RH + OH \xrightarrow{O_2} RO_2 + H_2O$	$26.3 \times 10^{-12\,a}$
2. $RO_2 + NO \xrightarrow{O_2} NO_2 + R'CHO + HO_2$	$7.7 \times 10^{-12\,b}$
3. $HO_2 + NO \longrightarrow NO_2 + OH$	8.1×10^{-12}
4. $OH + NO_2 \xrightarrow{M} HNO_3$	1.1×10^{-11} (at 1 atm)
5. $HO_2 + HO_2 \longrightarrow H_2O_2 + O_2$	2.9×10^{-12}
6. $RO_2 + HO_2 \longrightarrow ROOH + O_2$	$5.2 \times 10^{-12\,c}$
7. $NO_2 + h\nu \xrightarrow{O_2} NO + O_3$	Depends on light intensity[d]
8. $O_3 + NO \longrightarrow NO_2 + O_2$	1.9×10^{-14}

[a]Rate coefficient for propene (Table B.4). Other reactions consider R equal to CH_3. Propene is selected because it is a relatively important constituent of the urban atmosphere. Even though OH-propene reaction proceeds by OH addition to the double bond of propene (Section 6.10.2), the net result after O_2 attack on the initial radical formed is a peroxy radical.
[b]Rate coefficient for $CH_3O_2 + NO$.
[c]Rate coefficient for $CH_3O_2 + HO_2$.
[d]Typical photolysis rate coefficient for NO_2 is $j_{NO_2} = 0.015\,s^{-1}$.

Let us integrate the rate equations for the mechanism in Table 6.3 for a variety of initial RH and NO$_x$ concentrations for a 10-h period to see how the amount of O$_3$ formed and its rate of formation depend on the amount of NO$_x$ and the initial ratio of RH to NO$_x$. We assume that the source of OH is constant at $P_{HO_x} = 0.1 \, \text{ppt s}^{-1}$. If HO$_x$ = OH + HO$_2$ + RO$_2$, we can assume that HO$_x$ is in steady state, in which P_{HO_x} is balanced by reactions 4, 5, and 6. It can also be assumed that the RO$_2$ steady state is the result of a balance between reactions 1 and 2, and that HO$_2$ steady state is the result of a balance between reactions 2 and 3. We will assume that initially [NO]/[NO$_2$] = 2 and that the initial concentration of O$_3$ is that from the photostationary state corresponding to the initial concentrations of NO and NO$_2$ and the value of j_{NO_2} given in Table 6.3. With the aid of the steady-state expressions for [OH], [HO$_2$], and [RO$_2$], the concentrations of these radical species can be expressed as functions of [RH], [NO], and [NO$_2$]. We assume that over the timescale of the simulation, reactions of the products, R$'$CHO, H$_2$O$_2$, and ROOH, can be neglected. Therefore, one needs to solve the four differential equations for [RH], [NO], [NO$_2$], and [O$_3$]. (Ordinarily, one would use the photostationary state relation (6.6) to determine the O$_3$ concentration. However, in a rapidly reacting system like this one where the HO$_2$ + NO and RO$_2$ + NO reactions are not necessarily small relative to O$_3$ + NO, (6.6) does not hold exactly. As a result, one needs to integrate the O$_3$ rate equation explicitly.)

Figure 6.12 shows isopleths of the maximum O$_3$ mixing ratio achieved over a 10-h period generated from the mechanism in Table 6.3. We see that the simple mechanism in Table 6.3 is able to produce the characteristic features of the ozone isopleth plot in Figure 6.10, based on a much more complex mechanism.

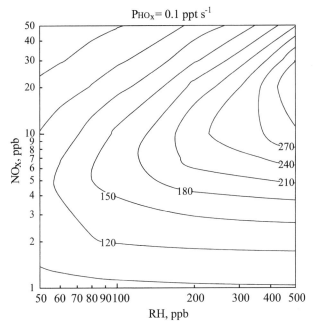

FIGURE 6.12 Isopleths of maximum O$_3$ mixing ratio achieved over a 10-h period by integrating the rate equations arising from the mechanism in Table 6.3.

6.10   CHEMISTRY OF NONMETHANE ORGANIC COMPOUNDS IN THE TROPOSPHERE

The essential elements of tropospheric chemistry have been presented in Sections 6.1–6.9. The remainder of this chapter is devoted to an in-depth treatment of the tropospheric chemistry of individual classes of organic species. For many readers, for example, those in a first course in atmospheric chemistry, it may not be necessary to proceed beyond this point in this chapter. All problems at the end of the chapter can be answered based on the material in Sections 6.1–6.9.

There are a number of excellent reviews of tropospheric chemistry that contain more details on mechanisms than can be included here.

General tropospheric chemistry:
    Atkinson (1989, 1990, 1994)
    Atkinson and Arey (2003b)
Alkenes
    Atkinson et al. (2000)
Aromatics
    Calvert et al. (2002)
Oxygenated organic compounds
    Mellouki et al. (2003)
Biogenic organics
    Atkinson and Arey (2003a)

6.10.1   Alkanes

Under tropospheric conditions, alkanes react with OH and NO_3 radicals; the latter process generally is of minor ($\leq 10\%$) importance as an atmospheric loss process under daytime conditions. Both reactions proceed via H-atom abstraction from C—H bonds[5]

$$RH + OH\cdot \longrightarrow R\cdot + H_2O$$
$$RH + NO_3 \longrightarrow R\cdot + HNO_3$$

to produce the alkyl radical, R. Any H atom in the alkane is susceptible to OH attack. Generally, the OH radical will tend to abstract the most weakly bound hydrogen atom in the molecule. The overall rate constant reflects the number of available hydrogen atoms and the strengths of the C—H bonds for each of these. Correlations have been developed for calculating the OH rate constant of alkanes that account for the number of primary (—CH_3), secondary, (—CH_2—), and tertiary ($>CH$) hydrogen atoms in the molecule (Atkinson 1987, 1994). Hydroxyl attack on a tertiary hydrogen atom is generally faster than that on a secondary H atom and is the slowest for primary H atoms. For propane, $CH_3CH_2CH_3$, for example, these structure–activity correlations predict that 70% of the OH reaction occurs by H-atom abstraction from the secondary carbon atom (—CH_2—) and 30% from the —CH_3 groups.

---

[5]In the remaining sections of this chapter we will explicitly denote free radicals with a dot ($\cdot$) in order to identify more clearly radicals in the mechanisms.

As with the methyl radical, the resulting alkyl (R) radical reacts rapidly, and exclusively, with O_2 under atmospheric conditions to yield an alkyl peroxy radical (RO_2) [see the comprehensive reviews of the chemistry of RO_2 radicals by Lightfoot et al. (1992) and Wallington et al. (1992)]:

$$R \cdot + O_2 + M \xrightarrow[\text{fast}]{} RO_2 \cdot + M$$

These alkyl peroxy radicals can be classed as primary, secondary, or tertiary depending on the availability of H atoms: $RCH_2OO \cdot$ (primary); $RR'CHOO \cdot$ (secondary); $RR'R''COO \cdot$ (tertiary). The alkyl radical–O_2 addition occurs with a room-temperature rate constant of $\geq 10^{-12}\,cm^3\,molecule^{-1}\,s^{-1}$ at atmospheric pressure. Given the high concentration of O_2, the $R + O_2$ reaction can be considered as instantaneous relative to other reactions occurring such as those that form R in the first place. Henceforth, the formation of an alkyl radical can be considered to be equivalent to the formation of an alkyl peroxy radical.

Under tropospheric conditions, these alkyl peroxy (RO_2) radicals react with NO, via two pathways:

$$RO_2 \cdot + NO \xrightarrow{a} RO \cdot + NO_2$$
$$\xrightarrow[M]{b} RONO_2$$

For alkyl peroxy radicals, reaction a can form the corresponding alkoxy ($RO \cdot$) radical together with NO_2, or the corresponding alkyl nitrate, reaction b, with the yield of the alkyl nitrate increasing with increasing pressure and with decreasing temperature. For secondary alkyl peroxy radicals at 298 K and 760 torr total pressure, the alkyl nitrate yields increase monotonically from <0.014 for a C_2 alkane up to ~ 0.33 for a C_8 alkane (Atkinson 1990). It has been shown that CH_3ONO_2 and $C_2H_5ONO_2$ may have a substantial natural source in the oceans (Chuck et al. 2002).

Alkyl peroxy radicals react with NO_2 by combination to yield the peroxynitrates,

$$RO_2 \cdot + NO_2 + M \longrightarrow ROONO_2 + M$$

Limiting high pressure rate constants for $\geq C_2$ alkyl peroxy radicals are identical to that for the $C_2H_5O_2 \cdot$ radical: $k = 9 \times 10^{-12}\,cm^3\,molecule^{-1}\,s^{-1}$, independent of temperature over the range 250 to 350 K.

Alkyl peroxy radicals also react with HO_2 radicals

$$RO_2 \cdot + HO_2 \cdot \longrightarrow ROOH + O_2$$

or with other RO_2 radicals. The self-reaction of $RO_2 \cdot$ and $RO_2 \cdot$ proceeds by the three pathways

$$R_1R_2CHO_2 \cdot + R_1R_2CHO_2 \cdot \xrightarrow{a} 2\,R_1R_2CHO \cdot + O_2$$
$$\xrightarrow{b} R_1R_2CHOH + R_1R_2CO + O_2$$
$$\xrightarrow{c} R_1R_2CHOOCHR_1R_2 + O_2$$

Pathway b is not accessible for tertiary RO_2 radicals, and pathway c is expected to be of negligible importance.

Alkoxy (RO·) radicals are formed in the reaction of alkyl peroxy (RO_2·) radicals with NO. Subsequent reactions of alkoxy radicals determine to a large extent the products resulting from the atmospheric oxidation of VOCs (Orlando et al. 2003). Alkoxy radicals react under tropospheric conditions via a variety of processes: unimolecular decomposition, unimolecular isomerization, or reaction with O_2. Alkoxy radicals with fewer than five carbon atoms are too short to undergo isomerization; for these the competitive processes are unimolecular decomposition versus reaction with O_2. The general alkoxy radical–O_2 reaction involves abstraction of a hydrogen atom by O_2 to produce an HO_2 radical and a carbonyl species:

$$RO\cdot + O_2 \longrightarrow R'CHO + HO_2\cdot$$

Rate constants for the $CH_3O\cdot + O_2$ and $C_2H_5O\cdot + O$ reactions are given in Table B.1. For primary ($RCH_2O\cdot$) and secondary ($R_1R_2CHO\cdot$) alkoxy radicals formed from the alkanes (Atkinson 1994),[6]

$$k(RCH_2O\cdot + O_2) = 6.0 \times 10^{-14}\exp(-550/T)\ \text{cm}^3\ \text{molecule}^{-1}\ \text{s}^{-1}$$
$$= 9.5 \times 10^{-15}\ \text{at } 298\ \text{K}$$
$$k(R_1R_2CHO\cdot + O_2) = 1.5 \times 10^{-14}\exp(-200/T)\ \text{cm}^3\ \text{molecule}^{-1}\ \text{s}^{-1}$$
$$= 8 \times 10^{-15}\ \text{at } 298\ \text{K}$$

Tertiary alkoxy radicals are not expected to react with O_2 because of the absence of a readily available hydrogen atom.

Unimolecular decomposition, on the other hand, produces an alkyl radical and a carbonyl:

$$RCH_2O\cdot \longrightarrow R\cdot + HCHO$$
$$RR_1CHO\cdot \longrightarrow R\cdot + R_1CHO$$
$$RR_1R_2CO\cdot \longrightarrow R\cdot + R_1C(O)R_2$$

Atkinson (1994) presents a correlation that allows one to determine the relative importance of O_2 reaction and decomposition for a particular alkoxy radical. Generally, reaction with O_2 is the preferred path for primary alkoxy radicals that have C-atom chains of two or fewer C atoms in length attached to the carbonyl group.

To illustrate alkoxy radical isomerization, let us consider the OH reaction of n-pentane. The n-pentane–OH reaction proceeds as follows to produce the 2-pentoxy radical:

(2-Pentoxy)

[6]Note a slight difference in the value of the preexponential factor between that in Table B.1 and that recommended by Atkinson (1994).

The 2-pentoxy radical can then react with O_2, decompose, or isomerize:

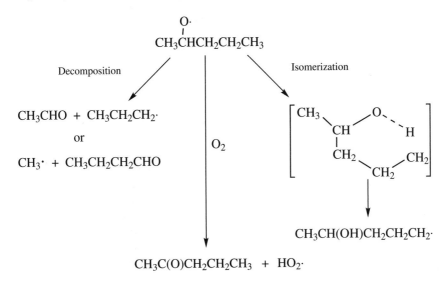

Rate constants for alkoxy radical isomerizations can be combined with rate constants for alkoxy radical decomposition and reaction with O_2 to predict the relative importance of the three pathways (Atkinson 1994). Alkoxy radicals can also react with NO and NO_2, but under ambient tropospheric conditions these reactions are generally of negligible importance.

Atmospheric Photooxidation of Propane   To illustrate alkane chemistry, consider the atmospheric photooxidation of propane:

$$CH_3CH_2CH_3 + OH\cdot \longrightarrow CH_3\dot{C}HCH_3 + H_2O$$
$$CH_3\dot{C}HCH_3 + O_2 \xrightarrow{fast} CH_3CH(O_2\cdot)CH_3$$
$$CH_3CH(O_2\cdot)CH_3 + NO \longrightarrow NO_2 + CH_3CH(O\cdot)CH_3$$
$$CH_3CH(O\cdot)CH_3 + O_2 \xrightarrow[\text{(Acetone)}]{fast} CH_3C(O)CH_3 + HO_2\cdot$$

Internal H-atom abstraction predominates in the initial OH attack. As we have been doing, we can eliminate the fast reactions from the mechanism by combining them with the foregoing rate-determining step. Thus the four reactions above may be expressed more concisely as

$$CH_3CH_2CH_3 + OH\cdot \xrightarrow{O_2} CH_3CH(O_2\cdot)CH_3$$
$$CH_3CH(O_2\cdot)CH_3 + NO \xrightarrow{O_2} NO_2 + CH_3C(O)CH_3 + HO_2\cdot$$

If we further take the liberty of assuming that the $CH_3CH(O_2\cdot)CH_3$ radical is produced and consumed only in these two reactions, a good assumption in this case, these two reactions can be written as a single overall reaction:

$$CH_3CH_2CH_3 + OH\cdot + NO \longrightarrow NO_2 + CH_3C(O)CH_3 + HO_2\cdot$$

where the reaction converts one molecule of NO to one molecule of NO_2. If we assume further that the sole fate of the HO_2 radical is reaction with NO, we can eliminate HO_2 from the right-hand side:

$$CH_3CH_2CH_3 + OH\cdot + 2\,NO \longrightarrow 2\,NO_2 + CH_3C(O)CH_3 + OH\cdot$$

In writing the reaction this way, we can clearly see that the net effect of the hydroxyl radical attack on propane is conversion of two molecules of NO to NO_2, the production of one molecule of acetone, $CH_3C(O)CH_3$ and the regeneration of the hydroxyl radical. Thus the photooxidation of propane can be viewed as a chain reaction mechanism in which the active species, the hydroxyl radical, is regenerated. For larger alkanes, such as n-butane, the atmospheric photooxidation mechanisms become more complex, although they continue to exhibit the same essential features of the propane degradation path. Two important issues arise in the reaction mechanisms of the higher alkanes. The first is that some fraction of the peroxyalkyl–NO reactions lead to alkyl nitrates rather than NO_2 and an alkoxy radical. The second is that the larger alkoxy radicals may isomerize as well as react with O_2. Figure 6.13 shows the mechanism of the n-butane–OH reaction (Jungkamp et al. 1997).

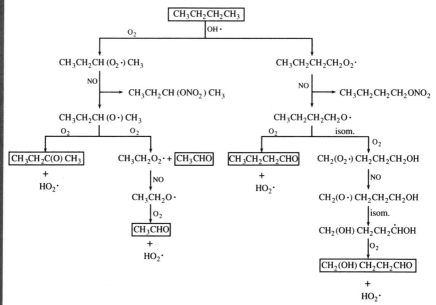

FIGURE 6.13 Atmospheric photooxidation mechanism for n-butane. The only significant reaction of n-butane is with the hydroxyl radical. Approximately 85% of that reaction involves H-atom abstraction from an internal carbon atom and 15% from a terminal carbon atom. In the terminal H-atom abstraction path, the $CH_3CH_2CH_2CH_2O\cdot$ alkoxy radical is estimated to react with O_2 25% of the time and isomerize 75% of the time. The second isomerization is estimated to be a factor of 5 faster than the first isomerization of the $CH_3CH_2CH_2CH_2O\cdot$ radical, so that competition with O_2 reaction is not considered at this step. The predominant fate of α-hydroxy radicals is reaction with O_2. For example, $\cdot CH_2OH + O_2 \longrightarrow HCHO + HO_2\cdot$, and $CH_3\dot{C}HOH + O_2 \longrightarrow CH_3CHO + HO_2\cdot$. In the n-butane mechanism, the α-hydroxy radical, $CH_2(OH)CH_2CH_2\dot{C}HOH$ reacts rapidly with O_2 to form 4-hydroxy-1-butanal, $CH_2(OH)CH_2CH_2CHO$. In the internal H-atom abstraction path, the alkoxy radical $CH_3CH_2CH(O\cdot)CH_3$ reacts with O_2 to yield methyl ethyl ketone (MEK), $CH_3CH_2C(O)CH_3$, and decomposes to form CH_3CHO and $CH_3CH_2\cdot$, which, after reaction with O_2 and NO and O_2 again, yields another molecule of CH_3CHO and HO_2.

6.10.2   Alkenes

We now proceed to the atmospheric chemistry of alkenes (or olefins). Alkenes are constituents of gasoline fuels and motor vehicle exhaust emissions. This class of organic compounds accounts for about 10% of the nonmethane organic compound concentration in the Los Angeles air basin (Lurmann and Main 1992) and other U.S. cities (Chameides et al. 1992). Because of their high reactivity with respect to ozone formation, alkenes are important contributors to overall ozone formation in urban areas. By now we fully expect that alkenes will react with the hydroxyl radical, and that is indeed the case. Because of the double bonded carbon atoms in alkene molecules, they will also react with ozone, the NO_3 radical, and atomic oxygen. The reaction with ozone can be an important alkene oxidation path, whereas that with oxygen atoms is generally not competitive with the other paths because of the extremely low concentration of O atoms. Let us begin with the hydroxyl radical reaction mechanism and focus on the simplest alkene, ethene (C_2H_4).

a. OH Reaction   We just saw that the initial step in OH attack on an alkane molecule is abstraction of a hydrogen atom to form a water molecule and an alkyl radical. In the case of alkenes, OH adds to the double bond rather than abstracting a hydrogen atom.[7] The ethene–OH reaction mechanism is

$$C_2H_4 + OH\cdot \longrightarrow HOCH_2CH_2\cdot$$

$$HOCH_2CH_2\cdot + O_2 \xrightarrow[\text{fast}]{} HOCH_2CH_2O_2\cdot$$

$$HOCH_2CH_2O_2\cdot + NO \longrightarrow NO_2 + HOCH_2CH_2O\cdot$$

The $HOCH_2CH_2O\cdot$ radical then decomposes and reacts with O_2:

$$HOCH_2CH_2O\cdot \xrightarrow{0.72} HCHO + \cdot CH_2OH$$

$$HOCH_2CH_2O\cdot + O_2 \xrightarrow{0.28} HOCH_2CHO + HO_2\cdot$$

The numbers over the arrows indicate the fraction of the reactions that lead to the indicated products at 298 K. Finally, the $\cdot CH_2OH$ radical reacts with O_2 to give formaldehyde and a hydroperoxyl radical:

$$\cdot CH_2OH + O_2 \xrightarrow[\text{fast}]{} HCHO + HO_2\cdot$$

---

[7]Hydrogen atom abstraction from —CH_3 groups accounts generally for <5% of the overall OH reaction of ethene and the methyl-substituted ethenes (propene, 2-methyl propene, the 2-butenes, 2-methyl-2-butene, and 2,3-dimethyl-2-butene). For alkenes with alkyl side chains, perhaps up to 10% of the OH reaction proceeds by H-atom abstraction, but we will neglect that path here.

Following our procedure of condensing the mechanism by eliminating the fast reactions, we have

$$C_2H_4 + OH\cdot \xrightarrow{O_2} HOCH_2CH_2O_2\cdot$$

$$HOCH_2CH_2O_2\cdot + NO \xrightarrow{2\,O_2} NO_2 + 0.72(HCHO + HCHO + HO_2\cdot)$$
$$+ 0.28(HOCH_2CHO + HO_2\cdot)$$

Finally, assuming that $HOCH_2CH_2O_2$ is produced and consumed only in these two reactions, we can write an overall reaction as

$$C_2H_4 + OH\cdot + NO \longrightarrow NO_2 + 1.44\,HCHO + 0.28\,HOCH_2CHO + HO_2\cdot$$

As we did with propane, we can assume that HO_2 reacts solely with NO, to give

$$C_2H_4 + OH\cdot + 2\,NO \longrightarrow 2\,NO_2 + 1.44\,HCHO + 0.28\,HOCH_2CHO + OH\cdot$$

We see that the overall result of hydroxyl radical attack on ethene is conversion of two molecules of NO to NO_2, formation of 1.44 molecules of formaldehyde and 0.28 molecules of glycol aldehyde ($HOCH_2CHO$), and regeneration of a hydroxyl radical. By comparing this mechanism to that of propane, we see the similarities in the NO to NO_2 conversion and the formation of oxygenated products.

For monoalkenes, dienes, or trienes with nonconjugated $C{=}C$ bonds, the OH radical can add to either end of the $C{=}C$ bond. For propene, for example, we obtain

$$CH_3CH = CH_2 + OH\cdot \xrightarrow{0.65} CH_3\dot{C}HCH_2OH$$
$$\xrightarrow{0.35} CH_3CH(OH)CH_2\cdot$$

For dienes with conjugated double bonds, such as 1,3-butadiene and isoprene (2-methyl-1,3-butadiene), OH radical addition to the $C{=}C{-}C{=}C$ system is expected to occur at positions 1 and/or 4:

$$> C{=}C{-}C{=}C <$$
$$\;\;1\;\;\;2\;\;\;3\;\;\;4$$

Alkene–OH reactions proceed via OH radical addition to the double bond to form a β-hydroxyalkyl radical (Atkinson, et al. 2000)

TABLE 6.4   Carbonyl Yields from 1-Alkene–OH Reactions

1-Alkene	Yield	
	HCHO	RCHO
Propene	0.86	0.98 (acetaldehyde)
1-Butene	—	0.94 (propanal)
1-Pentene	0.88	0.73 (butanal)
1-Hexene	0.57	0.46 (pentanal)
I-Heptene	0.49	0.30 (hexanal)
1-Octene	0.39	0.21 (heptanal)

Source: Atkinson et al. (1995b).

followed by rapid addition of O_2 to yield the corresponding β-hydroxyalkyl peroxy radicals:

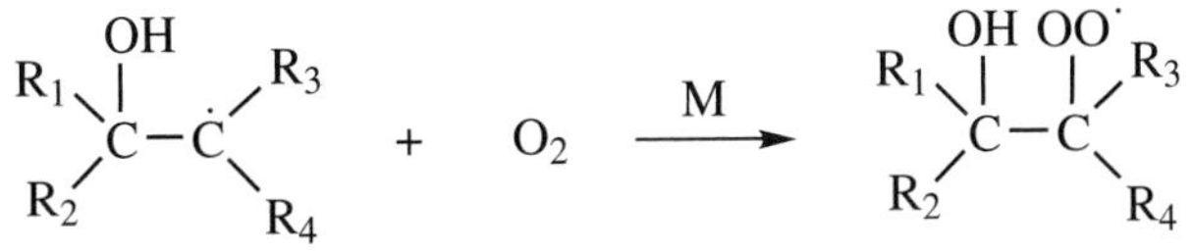

In the presence of NO, the β-hydroxyalkyl peroxy radical reacts with NO to form either the β-hydroxylalkoxy radical plus NO_2 or the β-hydroxynitrate:[8]

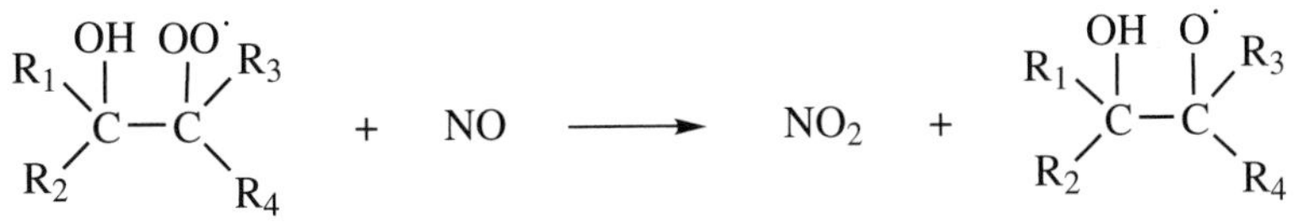

Rate constants for the reactions of β-hydroxyalkyl peroxy radicals with NO are essentially identical to those for the reaction of NO with $> C_2$ alkyl peroxy radicals formed from alkanes.

The β-hydroxyalkoxy radicals can then decompose, react with O_2, or isomerize. Available data show that, apart from ethene, for which reaction of the $HOCH_2CH_2O \cdot$ radical with O_2 and decomposition are competitive, the β-hydroxyalkoxy radicals formed subsequent to OH addition to $\geq C_3$ alkenes undergo decomposition and the reaction with O_2 is negligible.

The decomposition reaction is

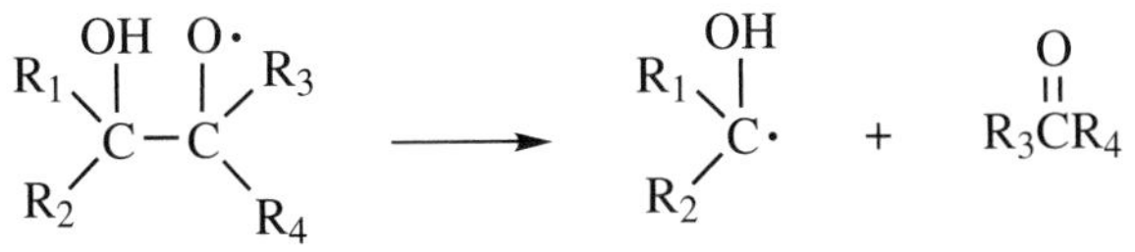

Carbonyl yields from alkene–OH reactions are summarized in Table 6.4. The yields of HCHO and RCHO arising from cleavage of the —C≡C— bond of 1-alkenes

---

[8]The β-hydroxynitrate formation pathway accounts for only ~ 1–1.5% of the overall NO reaction pathway at 298 K for propene (Shepson et al. 1985). The yields of β-hydroxynitrates from the propene–OH and 1-butene–OH reactions are about a factor of 2 lower than those of alkyl nitrates from the propane–OH and n-butane–OH reactions. These observations suggest that the formation yields of β-hydroxynitrates from the OH reaction with higher 1-alkenes could also be a factor of 2 lower than those from the reactions with the corresponding n-alkanes.

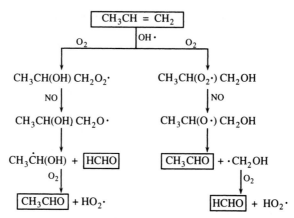

FIGURE 6.14   Propene–OH reaction mechanism.

$RCH=CH_2$ decrease monotonically from ≥ 0.90 for propene and 1-butene to 0.21 to 0.39 for 1-octene. H-atom abstraction from the CH_2 groups in the 1-alkenes is expected to account for an increasing fraction of the overall OH radical reaction as the carbon number of the 1-alkenes increases, with about 15% of the 1-heptene reaction being estimated to proceed by H-atom abstraction from the secondary CH_2 groups. The propene–OH reaction mechanism is shown in Figure 6.14.

b. NO₃ Reaction   Because of its strong oxidizing capacity and its relatively high nighttime concentrations, the NO_3 radical can play an important role in the nighttime removal of atmospheric organic species. Although the reaction of NO_3 with trace gases is significantly slower than that of OH, NO_3 can be present in much higher concentrations than OH, so that the overall reaction with many species is comparable for OH and NO_3 radicals. Since the reaction of OH with organic species is 10 to 1000 times faster than that of NO_3, the oxidation potential of the two radicals is in the same ballpark.

Alkenes react with the nitrate radical (Atkinson et al. 2000). As in OH–alkene reactions, NO_3 adds to the double bond and H-atom abstraction is relatively insignificant:

$$
\begin{array}{c}
R_1 \diagdown \\
\diagup C = C \diagdown \\
R_2 R_4
\end{array}
\;\diagup R_3 \; + \; NO_3 \longrightarrow
\begin{array}{c}
\overset{ONO_2}{\underset{}{|}} \\
R_1 \diagdown \;\; \diagup R_3 \\
C - C \\
R_2 \diagup \diagdown R_4 \;\cdot
\end{array}
$$

This is followed by rapid O_2 addition

$$
\begin{array}{c}
\overset{ONO_2}{\underset{}{|}} \\
R_1 \diagdown \;\; \diagup R_3 \\
C - C \\
R_2 \diagup \diagdown R_4 \;\cdot
\end{array}
\; + \; O_2 \longrightarrow
\begin{array}{c}
\overset{ONO_2}{\underset{}{|}} \\
R_1 \diagdown \;\; \diagup R_3 \\
C - C \\
R_2 \diagup \underset{OO\cdot}{\overset{|}{\diagdown}} R_4
\end{array}
$$

to produce the β-nitratoalkyl peroxy radical, subsequent reactions of which are

$$R_1R_2C(ONO_2)\text{-}CR_3R_4(OO\cdot) + NO \longrightarrow R_1R_2C(ONO_2)\text{-}CR_3R_4(O\cdot) + NO_2$$

$$\longrightarrow R_1R_2C(ONO_2)\text{-}CR_3R_4(ONO_2)$$

The latter reaction is expected to be minor.

Further reactions of β-nitratoalkoxy radicals include

$$R_1R_2C(ONO_2)\text{-}CR_3R_4(O\cdot) + O_2 \longrightarrow R_1R_2C(ONO_2)\text{-}C(R_4)O + HO_2\cdot$$

if $R_3 = H$

$$R_1R_2C(ONO_2)\text{-}CR_3R_4(O\cdot) \longrightarrow R_1CR_2(O) + R_3CR_4(O) + NO_2$$

$$R_1R_2C(ONO_2)\text{-}CR_3R_4(O\cdot) + NO_2 \longrightarrow R_1R_2C(ONO_2)\text{-}CR_3R_4(ONO_2)$$

of which the last reaction is expeced to be minor. Figure 6.15 shows the mechanism of the propene–NO_3 reaction.

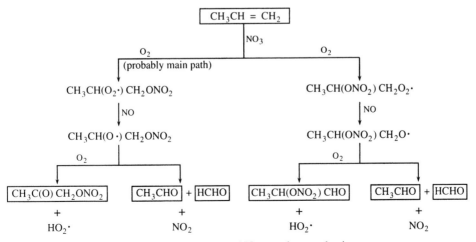

FIGURE 6.15 Propene–NO_3 reaction mechanism.

c. Ozone Reaction   The presence of the double bond renders alkenes susceptible to reaction with ozone. Reactions with ozone are, in fact, competitive with the daytime OH radical reactions and the nighttime NO_3 radical reaction as a tropospheric loss process for the alkenes. The ozone–alkene reaction proceeds via initial O_3 addition to the olefinic double bond, followed by rapid decomposition of the resulting molozonide

with the relative importance of the reaction pathways (a) and (b) being generally assumed to be approximately equal.

Carbonyl product yields for the reaction of alkenes with O_3 are generally consistent with the initial alkene–O_3 reaction given above:

$$\text{Alkene} + O_3 \longrightarrow 1.0\,\text{primary carbonyl} + 1.0\,\text{biradical}$$

The kinetics and products of the gas-phase alkene–O_3 reaction have been studied extensively (Atkinson et al. 2000) and are reasonably well understood for a large number of the smaller alkenes. The major mechanistic issue concerns the fate, under atmospheric conditions, of the initially energy-rich Criegee biradical, which can be collisionally stabilized or can undergo unimolecular decomposition:

$$[R_1CH_2\dot{C}(R_2)O\dot{O}]^{\ddagger} \longrightarrow R_1CH_2\dot{C}(R_2)O\dot{O} \quad \text{(stabilization)}$$
$$\longrightarrow R_1CH_2C(O)R_2 + O$$
$$\longrightarrow [R_1CH_2C(O)OR_2]^{\ddagger} \longrightarrow \text{decomposition}$$
$$\longrightarrow [R_1CH = C(OOH)R_2]^{\ddagger} \longrightarrow R_1\dot{C}HC(O)R_2 + OH\cdot$$

At atmospheric pressure, O atoms are not formed in any appreciable amount, so the second path can generally be neglected.

Hydroxyl radicals have been observed to be formed from alkene–O_3 reactions, sometimes with close to a unit yield (1 molecule of OH per 1 molecule of alkene reacted). Atkinson et al. (1995a) reported OH radical yields from a series of alkene–O_3 reactions:

Alkene	OH Yield
1-Pentene	0.37
1-Hexene	0.32
1-Heptene	0.27
1-Octene	0.18
2,3-Dimethyl-1-butene	0.50
Cyclopentene	0.61
1-Methylcyclohexene	0.90

(Estimated uncertainties in these yields are a factor of ~ 1.5.)

The reaction pathways of Criegee biradicals are generally well established for the first two compounds in the series although the exact fractions that proceed via each individual path are still open to question (Atkinson et al. 2000):

$$[\dot{C}H_2OO\cdot]^{\ddagger} \xrightarrow{M} \dot{C}H_2OO\cdot \quad \text{(stabilization)}$$
$$\longrightarrow CO_2 + H_2$$
$$\longrightarrow CO + H_2O$$
$$\xrightarrow{O_2} 2HO_2\cdot + CO_2$$
$$\longrightarrow H\dot{C}O + OH\cdot$$
$$[CH_3\dot{C}HOO\cdot]^{\ddagger} \xrightarrow{M} CH_3\dot{C}HOO\cdot$$
$$\longrightarrow CH_3\cdot + CO + OH\cdot$$
$$\xrightarrow{O_2} CH_3\cdot + CO_2 + HO_2\cdot$$
$$\longrightarrow H\dot{C}O + CH_3O\cdot$$
$$\longrightarrow CH_4 + CO_2$$
$$\longrightarrow CH_3OH + CO$$

The stabilized biradicals can react with a number of species:

$$R\dot{C}HO\dot{O} + H_2O \longrightarrow RC(O)OH + H_2O$$
$$+ NO \longrightarrow RCHO + NO_2$$
$$+ NO_2 \longrightarrow RCHO + NO_3$$
$$+ SO_2 \xrightarrow{H_2O} RCHO + H_2SO_4$$
$$+ CO \longrightarrow \text{products}$$

In addition, biradicals such as $(CH_3)_2\dot{C}OO\cdot$ may undergo unimolecular isomerization

Rate constants for reactions of $\dot{C}H_2OO\cdot$ radicals with the following species, relative to the reaction with SO_2, are

HCHO	0.25
CO	0.0175
H_2O	0.00023
NO_2	0.014

It appears that the reaction of stabilized biradicals with H_2O will predominate under atmospheric conditions (Atkinson 1994).

6.10.3 Aromatics

Aromatic compounds are of great interest in the chemistry of the urban atmosphere because of their abundance in motor vehicle emissions and because of their reactivity with respect to ozone and organic aerosol formation. The major atmospheric sink for aromatics is reaction with the hydroxyl radical. Whereas rate constants for the OH reaction with aromatics have been well characterized (Calvert et al. 2002), mechanisms of aromatic oxidation following the initial OH attack have been highly uncertain. Aromatic compounds of concern in urban atmospheric chemistry are given in Figure 6.16.

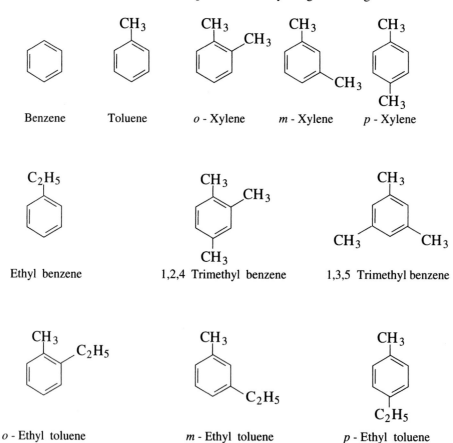

FIGURE 6.16 Aromatic compounds of interest in tropospheric chemistry.

The aromatic–OH radical reaction proceeds via two pathways: (a) a minor one (of order 10%) involving H-atom abstraction from C—H bonds of, for benzene, the aromatic ring, or for alkyl-substituted aromatic hydrocarbons, the alkyl-substituent groups; and (b) a major reaction pathway (of order 90%) involving OH radical addition to the aromatic ring. For example, for toluene these reaction pathways are:

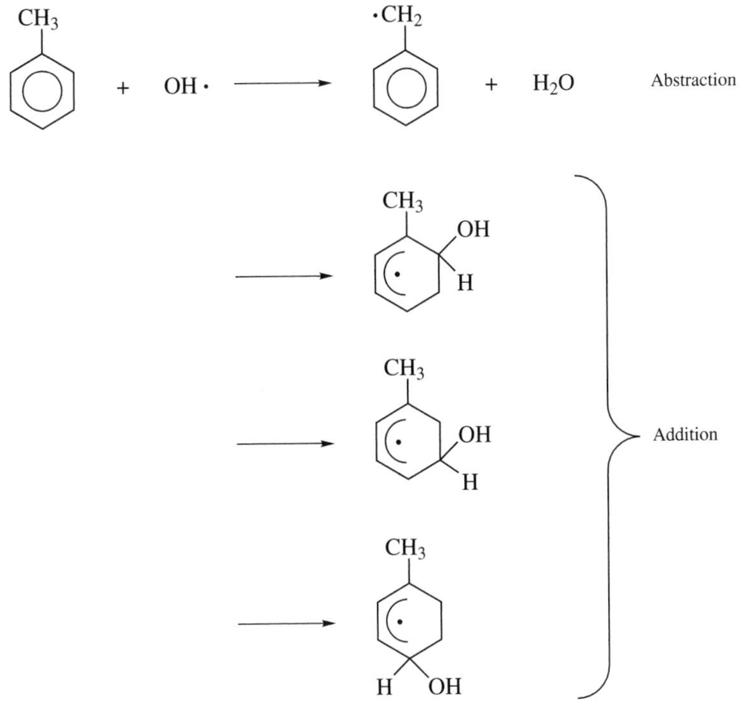

For the first addition product the structure above denotes the radicals:

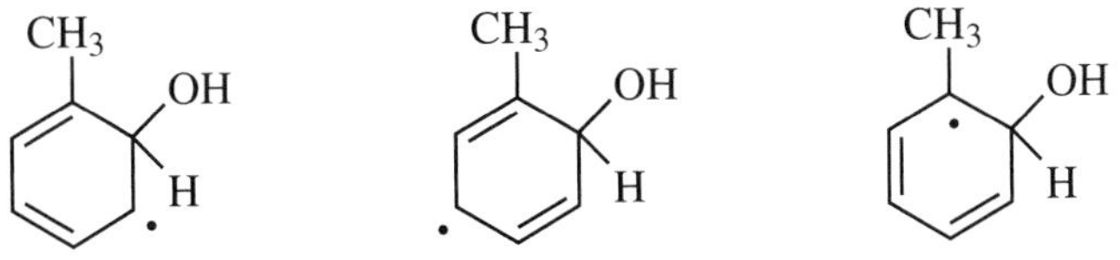

The H-atom abstraction pathway leads mainly to the formation of aromatic aldehydes

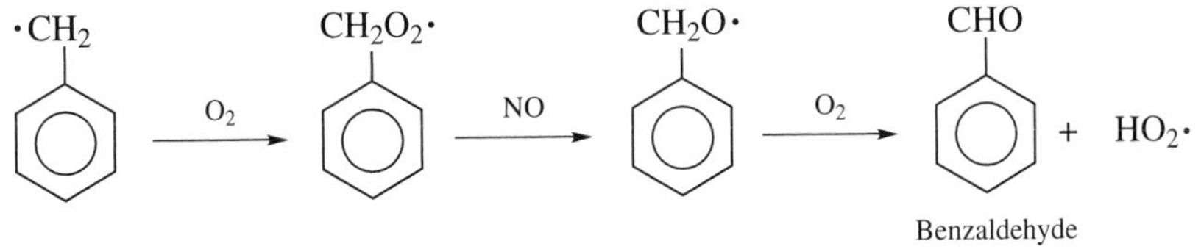

As noted above, this H-atom abstraction pathway is minor, accounting for <10% of the overall OH radical reaction for benzene and the alkyl-substituted aromatic hydrocarbons.

The radicals resulting from OH addition to the aromatic ring are named as follows:

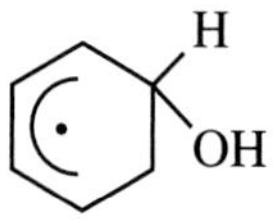

Hydroxycyclohexadienyl
radical
(from benzene)

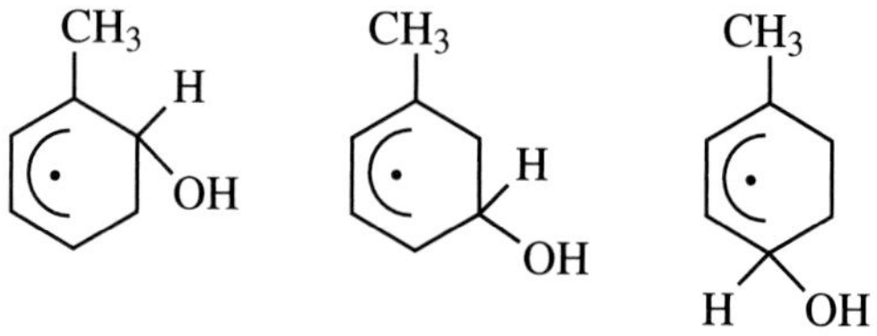

Methyl hydroxycyclohexadienyl
radicals
(from toluene)

For toluene, and other aromatics, there are several possible sites of attack for the OH radical. Some sites are less sterically hindered than others or are favored because of stabilizations resulting from group interactions. Andino et al. (1996) have performed ab initio calculations to determine the most energetically favored structures resulting from OH addition to aromatic compounds. For toluene the most favored structure is that resulting from OH addition at the ortho position:[9]

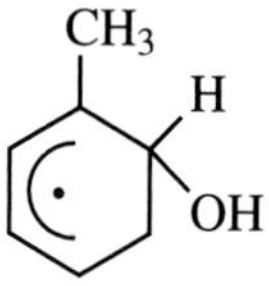

[In general, the preferred place of OH addition to an aromatic is a position ortho to a substituent methyl group (Andino et al. 1996).]

Following formation of the OH adduct, the adduct can react with O_2 or NO_2. The O_2 reaction path is

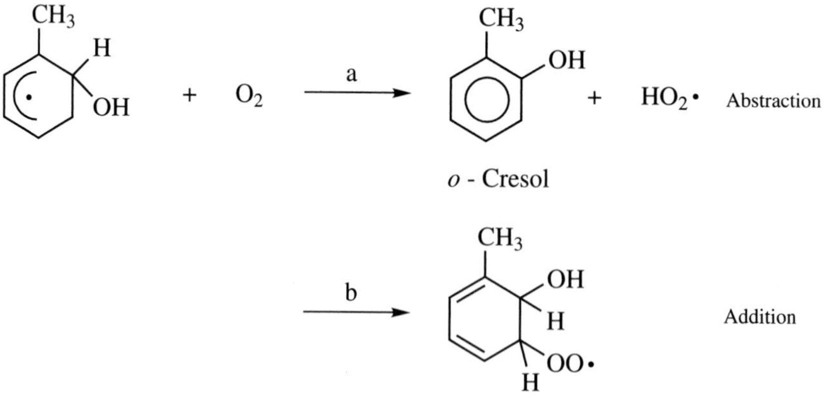

o - Cresol

---

[9]OH addition to the meta and para positions of toluene yield structures that are only 1 to 2 kcal mol^{-1} less favorable than addition at the ortho site and thus cannot be ruled out categorically. For our purposes, we will consider OH addition at the ortho site only.

The location of O_2 addition in the product shown above is that most favored energetically (Andino et al. 1996). The H-atom abstraction reaction to yield phenolic compounds, such as *o*-cresol, has been shown to be relatively minor, accounting for $\sim 16\%$ of the overall OH radical mechanism for toluene (Calvert et al. 2002). The NO_2 reaction of the OH adduct leads to nitroaromatics:

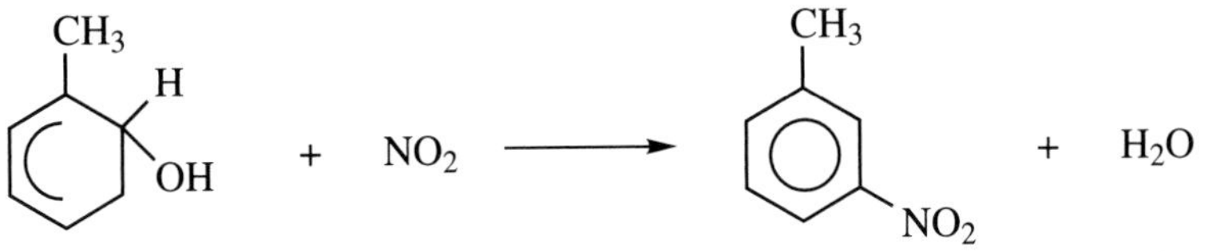

m-Nitrotoluene

Rate constants for the methyl hydroxycyclohexadienyl radical with O_2 and NO_2 are $\sim 5 \times 10^{-16}$ cm^3 molecule^{-1} s^{-1} and $\sim 3 \times 10^{-11}$ cm^3 molecule^{-1} s^{-1}, respectively (Knispel et al. 1990; Zetzsch et al. 1990; Goumri et al. 1992; Atkinson 1994). Based on these rate constants, the NO_2 reaction with the toluene-OH adduct will be of significance for NO_2 concentrations exceeding about 9×10^{12} molecules cm^{-3} (300 ppb).

Alkyl peroxy radicals generally react with NO to form alkoxy radicals (assuming sufficient NO is present). Aromatic peroxy radicals, in contrast, are believed to cyclize, forming bicyclic radicals. For the product of reaction b above, the energetically favored bicyclic radical is

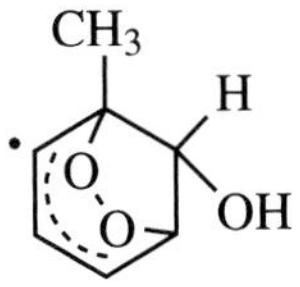

After bicyclic radical formation, O_2 rapidly adds to the radical, forming a bicyclic peroxy radical, for example

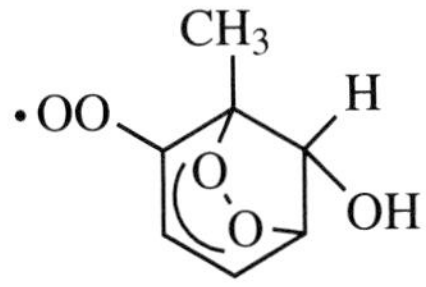

This radical is then expected to react with NO to form a bicyclic oxy radical and NO_2, such as

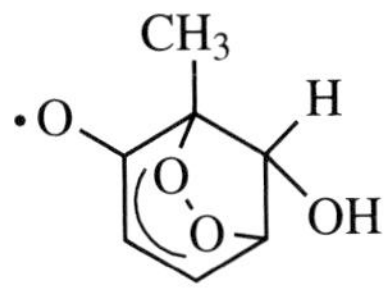

The only path for this bicyclic oxy radical is fragmentation via favorable β-scission reactions. For the above radical, such scission would produce

Observed ring fragmentation products of the toluene–OH reaction include the following:

| Glyoxal | Methylglyoxal | Methyl butenedial | 1,4-Butenedial |

6.10.4   Aldehydes

Aldehydes are important constituents of atmospheric chemistry. We have already seen the role played by formaldehyde in the chemistry of the background troposphere. Aldehydes are formed in the atmosphere from the photochemical degradation of other organic compounds. Aldehydes undergo photolysis, reaction with OH radicals, and reaction with NO$_3$ radicals. Reaction with NO$_3$ radicals is of relatively minor importance as a consumption process for aldehydes, thus the major loss processes involve photolysis and reaction with OH radicals.

Formaldehyde photolyzes and reacts with OH, as seen earlier. Acetaldehyde photolyzes by

$$CH_3CHO + h\nu \longrightarrow CH_4 + CO$$
$$\longrightarrow CH_3\cdot + H\dot{C}O$$

Data on absorption cross sections and quantum yields for higher aldehydes are summarized by Atkinson (1994).

Hydroxyl radical reaction with aldehydes involves H-atom abstraction to produce the corresponding acyl (RCO) radical

$$RCHO + OH \cdot \longrightarrow R\dot{C}O + H_2O$$

which rapidly adds O_2 to yield an acyl peroxy radical:

$$R\dot{C}O + O_2 + M \xrightarrow[\text{fast}]{} RC(O)OO \cdot + M$$

These acyl peroxy radicals then react with NO or NO_2, the latter leading to peroxyacyl nitrates, $RC(O)OONO_2$:

$$RC(O)OO \cdot + NO \longrightarrow RC(O)O \cdot + NO_2$$
$$RC(O)O \cdot \longrightarrow R \cdot + CO_2$$
$$RC(O)OO \cdot + NO_2 + M \rightleftharpoons RC(O)OONO_2 + M$$

6.10.5   Ketones

This class of organic compounds is exemplified by acetone and its higher homologues. As for the aldehydes, photolysis and reaction with the OH radical are the major atmospheric loss processes (Atkinson 1989; Mellouki et al. 2003). The limited experimental data available indicate that, with the exception of acetone (see Figure 6.17), photolysis is probably of minor importance. Reaction with the OH radical is then the major

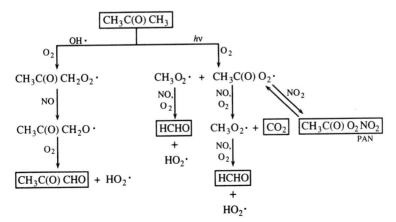

FIGURE 6.17   Atmospheric photooxidation mechanism for acetone.

tropospheric loss process. For example, for methyl ethyl ketone the OH radical can attack any of the three carbon atoms that contain hydrogen atoms

$$OH\cdot + CH_3CH_2C(O)CH_3 \xrightarrow[O_2]{a} H_2O + CH_3\overset{\overset{\displaystyle OO\cdot}{|}}{C}HC(O)CH_3$$

$$\xrightarrow[O_2]{b} H_2O + CH_3CH_2C(O)CH_2OO\cdot$$

$$\xrightarrow[O_2]{c} H_2O + \cdot OOCH_2CH_2C(O)CH_3$$

with a being the major reaction pathway. Subsequent reaction of this particular radical with NO leads to

$$CH_3\overset{\overset{\displaystyle OO\cdot}{|}}{C}HC(O)CH_3 + NO \longrightarrow NO_2 + CH_3\overset{\overset{\displaystyle O\cdot}{|}}{C}HC(O)CH_3$$

$$CH_3C(O\cdot)HC(O)CH_3 \longrightarrow CH_3CHO + CH_3\dot{C}O$$

The major reaction products from the atmospheric reactions of the ketones are aldehydes and PAN precursors.

6.10.6 α, β-Unsaturated Carbonyls

These compounds, exemplified by acrolein (CH_2=CHCHO), crotonaldehyde (CH_3CH=CHCHO), and methyl vinyl ketone (CH_2=CHC(O)CH$_3$), are known to react with ozone and with OH radicals. Photolysis and NO$_3$ radical reaction are of minor importance. Under atmospheric conditions the O$_3$ reactions are also of minor significance, leaving the OH radical reaction as the major loss process. For the aldehydes, OH radical reaction can proceed via two reaction pathways: OH radical addition to the double bond and H-atom abstraction from the—CHO group (Atkinson, 1989). These α,β-unsaturated aldehydes are expected to ultimately give rise to α-dicarbonyls such as glyoxal and methylglyoxal. For the α,β-unsaturated ketones such as methyl vinyl ketone the major atmospheric reaction with the OH radical occurs only by OH radical addition to the double bond. Again, α-dicarbonyls, together with aldehydes and hydroxyaldehydes, are formed as products.

6.10.7 Ethers

The aliphatic ethers, such as dimethyl ether and diethyl ether, react under atmospheric conditions essentially solely with the OH radical, via H-atom abstraction from C—H bonds (Wallington et al. 1988, 1989; Atkinson 1989; Japar et al. 1990, 1991; Wallington and Japar 1991; Mellouki et al. 2003). The reaction mechanism for dimethyl ether is

$$CH_3OCH_3 + OH\cdot \xrightarrow{O_2} CH_3OCH_2O_2\cdot$$

$$CH_3OCH_2O_2\cdot + NO \longrightarrow NO_2 + CH_3OCH_2O\cdot$$

$$CH_3OCH_2O\cdot + O_2 \longrightarrow HC(O)OCH_3 + HO_2\cdot$$

where the carbon-containing product in the last reaction is methyl formate.

6.10.8   Alcohols

The reaction sequences for the simpler aliphatic alcohols under atmospheric conditions are known (Atkinson 1989; Mellouki et al. 2003); these involve H-atom abstraction, mainly from the α C—H bonds. For example, the methanol–OH reaction is

$$CH_3OH + OH\cdot \longrightarrow H_2O + \cdot CH_2OH$$
$$\longrightarrow H_2O + CH_3O\cdot$$

with the first reaction pathway accounting for $\sim 85\%$ of the overall reaction at 298 K. Since, as shown earlier, both the $\cdot CH_2OH$ and $CH_3O\cdot$ radicals react with O_2 to yield formaldehyde and HO_2, the overall methanol–OH reaction can be written as

$$CH_3OH + OH\cdot \xrightarrow{O_2} H_2O + HO_2\cdot + HCHO$$

The ethanol–OH reaction proceeds as follows

$$OH\cdot + CH_3CH_2OH \longrightarrow H_2O + \dot{C}H_2CH_2OH \quad (\sim 5\%)$$
$$\longrightarrow H_2O + CH_3CHOH \quad (\sim 90\%)$$
$$\longrightarrow H_2O + CH_3CH_2O\cdot \quad (\sim 5\%)$$

where the branching ratios are those at 298 K. The second two channels result in identical products under atmospheric conditions, $HO_2 + CH_3CHO$. The first channel forms the intermediate CH_2CH_2OH, which, under atmospheric conditions, leads to the same products as the OH + ethene reaction. Using the ethene–OH mechanism given earlier, the overall ethanol–OH reaction mechanism can be written as

$$C_2H_5OH + OH\cdot + 0.05\,NO \longrightarrow 0.05\,NO_2 + 0.014\,HOCH_2CHO + 0.072\,HCHO$$
$$+ 0.95\,CH_3CHO + HO_2\cdot + H_2O$$

where the principal products are acetaldehyde and the HO_2 radical.

Free tropospheric concentrations of methanol range from about 700 ppt at northern midlatitudes to about 400 ppt at southern latitudes (Singh et al. 1995). In general, ethanol abundance in the free troposphere is an order of magnitude lower than that of methanol. Average lifetimes of CH_3OH and C_2H_5OH in the atmosphere are on the order of 16 days and 4 days, respectively.

6.11   ATMOSPHERIC CHEMISTRY OF BIOGENIC HYDROCARBONS

A great variety of organic compounds are emitted by vegetation. These biogenic compounds are highly reactive in the atmosphere. They are basically alkenes or cycloalkenes, and their atmospheric chemistry is generally analogous to that of alkenes. Because of the presence of C=C double bonds these molecules are susceptible to attack by O_3 and NO_3, in addition to the customary reaction with OH radicals.

TABLE 6.5 Estimated Tropospheric Lifetimes of Organic Compounds

	Lifetime Against Reaction with			
	OH[a]	O_3^b	NO_3^c	$h\nu$
n-Butane	5.7 days	—	1.7 months	
Propene	6.6 h	1.6 days	5.9 h	
Benzene	12 days	—	—	
Toluene	2.4 days	—	1.1 month	
m-Xylene	7.4 h	—	10 days	
Formaldehyde	1.5 days	—	4 days	4 h
Acetaldeyde	11 h	—	20 h	5 days
Acetone	66 days	—	—	38 days
Isoprene	1.7 h	1.3 days	0.8 h	
α-Pinene	3.4 h	4.6 h	6 min	
β-Pinene	2.3 h	1.1 days	15 min	
Camphene	3.5 h	18 days	1.8 h	
2-Carene	2.3 h	1.7 h	1.8 min	
3-Carene	2.1 h	10 h	3.3 min	
d-Limonene	1.1 h	1.9 h	2.7 min	
Terpinolene	49 min	17 min	0.4 min	

[a]12-h daytime OH concentration of 1.5×10^6 molecules cm^{-3} (0.06 ppt).
[b]24-h average O_3 concentration of 7×10^{11} molecules cm^{-3} (30 ppb).
[c]12-h average NO_3 concentration of 4.8×10^8 molecules cm^{-3} (20 ppt).

Tropospheric lifetimes of organic species due to reaction with OH, NO_3, and O_3 can be estimated by combining the rate constant data (Appendix B) with estimated ambient tropospheric concentrations of OH, NO_3, and O_3. Resulting tropospheric lifetimes of a number of organic species, including several biogenic hydrocarbons, with respect to these gas-phase reactions are given in Table 6.5. An important point to note is that the atmospheric lifetimes of the biogenic hydrocarbons are relatively short compared to those of other organic species. The OH radical and ozone reactions are estimated to be of generally comparable importance during the daytime, and the NO_3 radical reaction is important at night.

If one had to single out the most important biogenic hydrocarbon in atmospheric chemistry, it would be isoprene. Isoprene has the chemical formula

$$CH_2{=}\overset{\displaystyle \overset{CH_3}{|}}{C}{-}CH{=}CH_2$$

Another name for isoprene is 2-methyl-1,3-butadiene. When illustrating the reactions of isoprene it is convenient to use the shorthand chemical structure (see Table 2.12),

where the double bonds are indicated by double lines, and each vertex and the end of each line indicates a carbon atom with the requisite number of hydrogen atoms. Isoprene reacts with OH radicals, NO_3 radicals, and O_3 (Paulson et al. 1992a,b; Paulson and Seinfeld

1992; Atkinson and Arey 2003) and rate constants for these reactions are known (see Tables B.9 and B.10).

The OH–isoprene reaction proceeds almost entirely by addition of the OH radical to the C=C double bonds. Formaldehyde, methacrolein (CH_2=C(CH_3)CHO), and methyl vinyl ketone (CH_2=CHC(O)CH_3) have been identified in the laboratory as major products of the OH–isoprene reaction. Figure 6.18 shows the initial steps of OH attack on

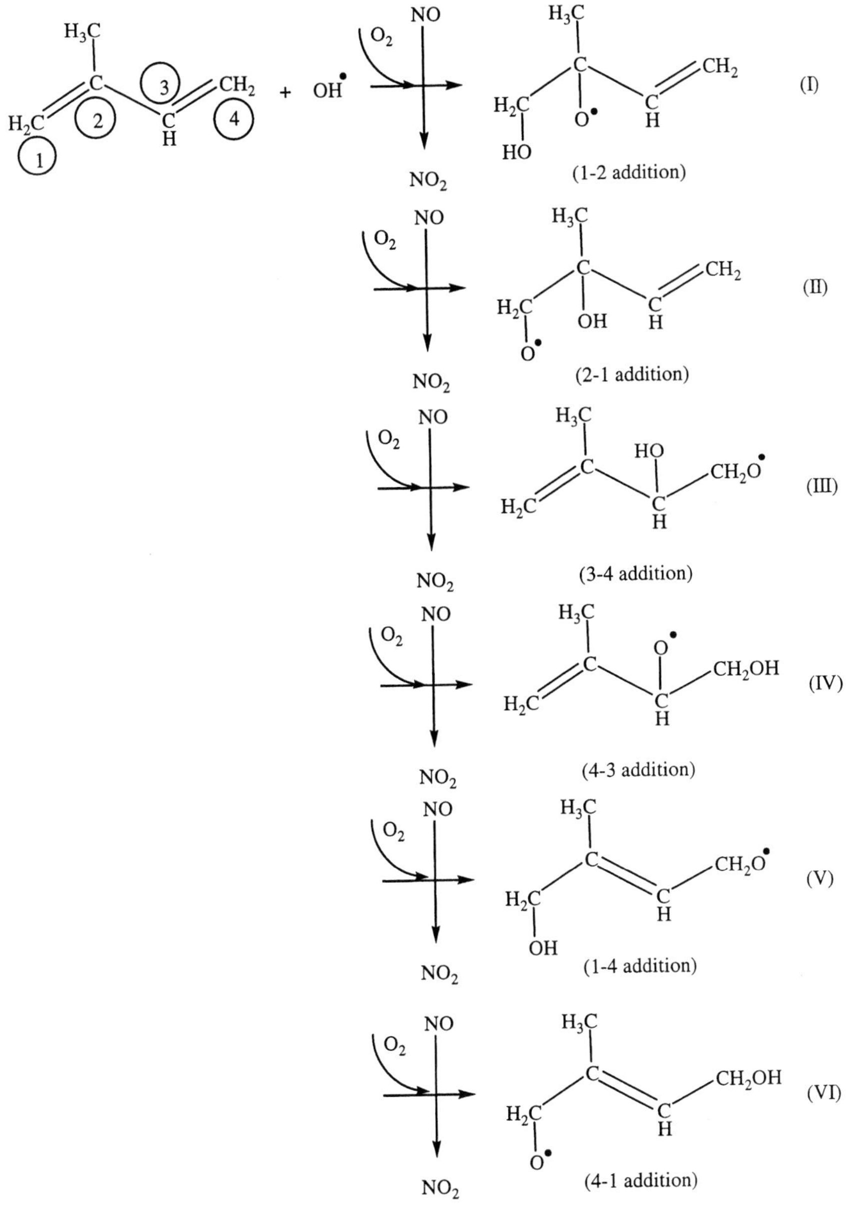

FIGURE 6.18 Isoprene–OH reaction mechanism. OH can add to four different positions in the isoprene molecule, denoted 1–4. We show the pathways from OH addition, followed by O_2 and NO reaction, leading to six alkoxy radicals (I–VI). Nitrate formation in the NO reaction step is not shown.

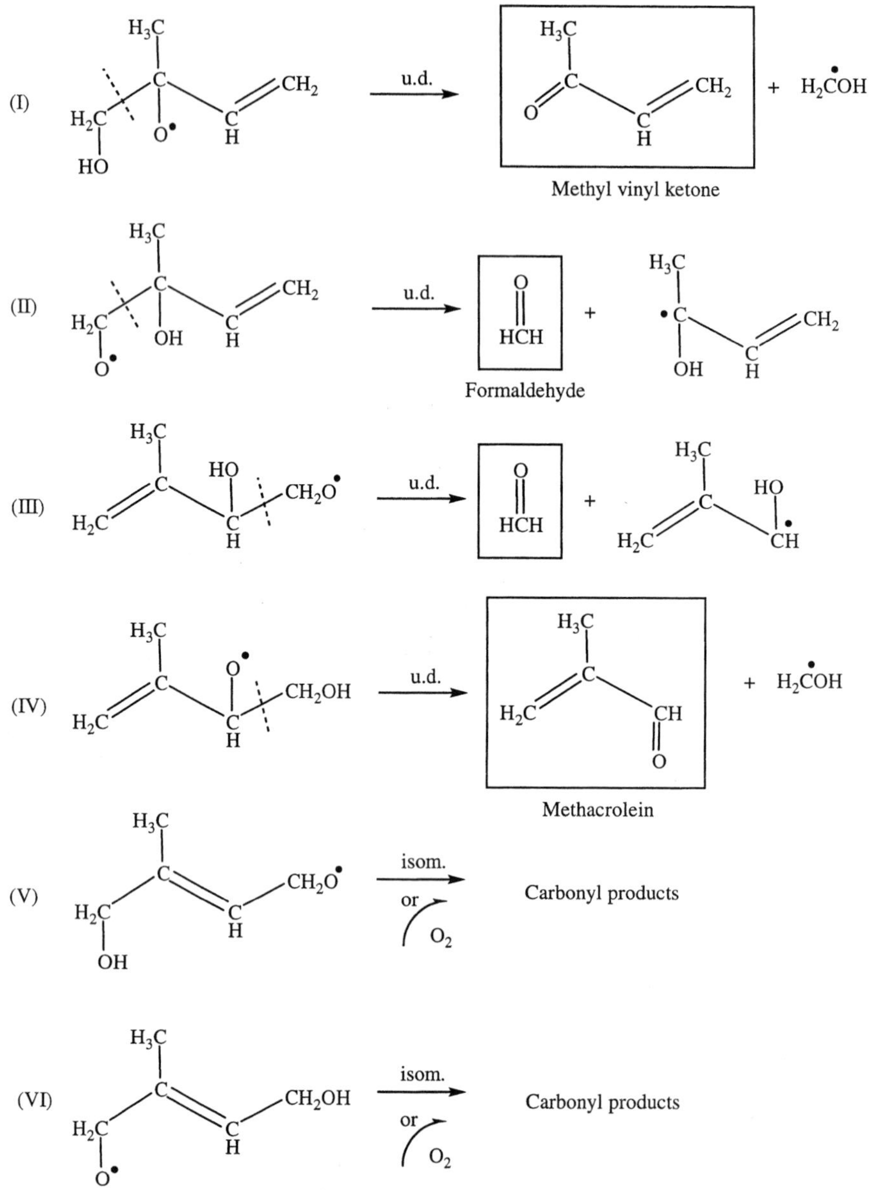

FIGURE 6.18   (*Continued*)

isoprene. The result, after O_2 addition and reaction with NO, are the six alkoxy radicals denoted I–VI. Addition of OH to the terminal carbon atoms (1 or 4) produces allylic radicals, to which subsequent addition of O_2 may occur at carbon atoms either β- or δ- to the OH group. β-Addition leads to radicals I–IV; δ- addition leads to radicals V and VI. Figure 6.18 shows the atmospheric fates of the six alkoxy radicals. Radicals I–IV undergo unimolecular decomposition (UD); V and VI isomerize or react with O_2. The major

products, methacrolein

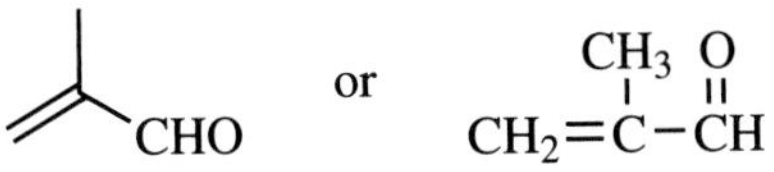

or

$$CH_2=\overset{\overset{\displaystyle CH_3}{|}}{C}-\overset{\overset{\displaystyle O}{||}}{CH}$$

and methyl vinyl ketone

or

$$CH_3-\overset{\overset{\displaystyle O}{||}}{C}-CH=CH_2$$

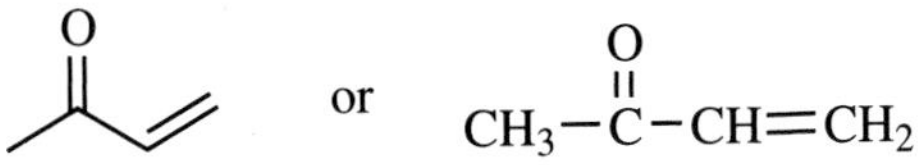

are indicated.

The O_3–isoprene reaction proceeds by initial addition of O_3 to the C=C double bonds to form two primary ozonides, each of which decomposes to two sets of carbonyl plus biradical products (a minor channel is apparently formation of a 1,2-epoxymethyl butene). This mechanism is consistent with the formation of formaldehyde, methacrolein, and methyl vinyl ketone. As in other O_3–alkene reactions, OH radicals are observed in significant yield, about 0.27 molecule of OH per O_3–isoprene reaction.

The NO_3–isoprene reaction proceeds by NO_3 addition to the C=C double bonds, with addition at position 1 dominating over that at position 4. Consistent with laboratory product data, NO_3 + isoprene involves the formation of $\cdot OOCH_2CH=C(CH_3)$ CH_2ONO_2, which, in the presence of NO, forms the corresponding alkoxy radical. This alkoxy radical can react with O_2 forming HO_2 and $CH_3C(CH_2ONO_2)=CHCHO$, isomerize by H-atom abstraction from the —CH_2ONO_2 group, or isomerize by H-atom abstraction from the —CH_3 group, giving rise to C_5-hydroxynitrato carbonyls or formaldehyde plus a C_4-nitrato carbonyl.

6.12 ATMOSPHERIC CHEMISTRY OF REDUCED NITROGEN COMPOUNDS

Reduced nitrogen compounds emitted into the atmosphere include ammonia (NH_3), hydrogen cyanide (HCN), and possibly their higher homologs such as the aliphatic and aromatic amines RNH_2, $RR'NH$, and $RR'R''N$ and the nitriles RCN, where R, R', R'' = alkyl or aryl group.

6.12.1 Amines

The major gas-phase atmospheric reactions of the amines involve the OH radical:

$$NH_3 + OH\cdot \longrightarrow H_2O + NH_2\cdot$$
$$CH_3NH_2 + OH\cdot \longrightarrow H_2O + CH_3\dot{N}H$$
$$(CH_3)_2NH + OH\cdot \longrightarrow H_2O + (CH_3)_2N\cdot$$
$$\longrightarrow H_2O + \dot{C}H_2NHCH_3$$

Amines also react with gaseous nitric acid to form the corresponding nitrate salts.

The reaction of ammonia and nitric acid vapor to give ammonium nitrate is a major route to form particulate nitrate (see Chapter 10),

$$NH_3 + HNO_3 \rightleftharpoons NH_4NO_3(s)$$

Ammonia, being highly soluble and reactive with atmospheric acids, is removed preferentially by those routes rather than through reaction with the OH radical, which is relatively slow.

6.12.2 Nitriles

The available experimental data suggest that this class of reduced nitrogen compounds will react mainly with OH radicals under tropospheric conditions. Hydrogen cyanide (HCN) reacts only slowly with OH radicals, with a room-temperature rate constant at atmospheric pressure of $3 \times 10^{-14}\, cm^3$ molecule^{-1} s^{-1} (Atkinson 1989). For the organic nitriles, the available evidence shows that the OH reaction occurs via H-atom abstraction from the alkyl groups; for example

$$CH_3CN + OH\cdot \longrightarrow H_2O + \dot{C}H_2CN$$

probably leading ultimately to the formation of CN radicals.

6.12.3 Nitrites

Methyl nitrite is the first member of this class of organic compounds, and photolysis is its only important atmospheric loss process

$$CH_3ONO + h\nu \longrightarrow CH_3O\cdot + NO$$

with a photolysis lifetime of ~ 10–15 min at solar noon.

6.13 ATMOSPHERIC CHEMISTRY (GAS PHASE) OF SULFUR COMPOUNDS

6.13.1 Sulfur Oxides

From a thermodynamic perspective, sulfur dioxide (SO_2) has a strong tendency to react with oxygen in air, by the stoichiometric (nonelementary) reaction:

$$2\,SO_2 + O_2 \longrightarrow 2\,SO_3$$

The rate of this reaction is so slow under catalyst-free conditions in the gas phase that it can be totally neglected as a source of atmospheric SO_3.

Sulfur dioxide reacts under tropospheric conditions via both gas- and aqueous-phase processes (see Chapter 7) and is also removed physically via dry and wet deposition. With respect to gas-phase reaction, reaction with the OH radical is dominant (Stockwell and Calvert 1983)

$$SO_2 + OH \cdot + M \longrightarrow HOSO_2 \cdot + M$$

followed by the regeneration of the chain-carrying HO_2 radical

$$HOSO_2 \cdot + O_2 \longrightarrow HO_2 \cdot + SO_3$$

Sulfur trioxide, in the presence of water vapor, is converted rapidly to sulfuric acid:

$$SO_3 + H_2O + M \longrightarrow H_2SO_4 + M$$

The lifetime of SO_2 based on reaction with the OH radical, at typical atmospheric levels of OH, is about one week. SO_2 is one of the gases that is reasonably efficiently removed from the atmosphere by dry deposition (see Chapter 19). (At a dry deposition velocity of about $1\ cm\ s^{-1}$, the lifetime of SO_2 by dry deposition in a 1 km deep boundary layer is about 1 day.) When clouds are present, the removal of SO_2 can be enhanced even beyond that attributable to dry deposition.

6.13.2 Reduced Sulfur Compounds (Dimethyl Sulfide)

Reduced sulfur-containing species RSR$'$ react with OH and NO_3 radicals. For H_2S, the dominant tropospheric removal process involves OH radical reaction:

$$H_2S + OH \cdot \longrightarrow SH \cdot + H_2O$$

The atmospheric lifetime of H_2S by this reaction is about 70 h. It appears that the SH radical undergoes a series of reactions resulting in the ultimate formation of SO_2 (Friedl et al. 1985). For CH_3SH both OH and NO_3 reactions lead to the group, $CH_3S(OH)H$, which is thought to proceed, through CH_3S, to HCHO, SO_2, and CH_3SO_3H.

Dimethyl sulfide (DMS), CH_3SCH_3, is the largest natural contributor to the global sulfur flux (see Section 2.2.1). The DMS–OH rate constant, approximately $5 \times 10^{-12}\ cm^3$ molecule^{-1} s^{-1} at 298 K, exceeds that of DMS–NO_3 by a factor of 4. (In contrast, the reactions of H_2S and CH_3SH with NO_3 are, respectively, 6000 and 40 times slower than that with OH.) The DMS lifetime in the marine atmosphere as a result of both OH and NO_3 reactions is on the order of one to several days, with the majority of the path occurring by OH at low latitudes and by NO_3 in colder, darker regions. Because of the photochemical source of OH, DMS removal by OH occurs only during daytime, leading to a pronounced diel cycle in DMS concentration.

The DMS–OH reaction proceeds via H-atom abstraction or OH addition to the sulfur atom in the DMS molecule (Yin et al. 1990a,b; Barone et al. 1995, 1996; Turnipseed et al. 1996; Ravishankara et al. 1997):

$$CH_3SCH_3 + OH \cdot \longrightarrow CH_3SCH_2 \cdot + H_2O$$

$$\underset{M}{\overset{M}{\rightleftharpoons}} CH_3\dot{S}(OH)CH_3$$

The abstraction path is favored at higher temperatures, the addition path at lower temperature. At 298 K, about 80% of the reaction proceeds by abstraction; at 285 K the two channels are approximately equal. The $CH_3SCH_2\cdot$ radical behaves as an alkyl radical (Stickel et al. 1993; Wallington et al. 1993):

$$CH_3SCH_2\cdot + O_2 + M \longrightarrow CH_3SCH_2O_2\cdot + M$$
$$CH_3SCH_2O_2\cdot + NO \longrightarrow CH_3SCH_2O\cdot + NO_2$$
$$CH_3SCH_2O\cdot \longrightarrow CH_3S\cdot + HCHO$$

The last reaction occurs rapidly, so that, once formed, the $CH_3SCH_2O\cdot$ radical can be assumed to decompose immediately (Tyndall and Ravishankara 1991). In the marine boundary layer, where NO_x levels are relatively low, $CH_3SCH_2O_2\cdot$ radicals react with HO_2 radicals as well as NO. There are two channels for this reaction:

$$CH_3SCH_2O_2\cdot + HO_2\cdot \longrightarrow CH_3SCH_2OOH + O_2$$
$$\longrightarrow CH_3SCHO + H_2O + O_2$$

The $CH_3S\cdot$ radical reacts with O_2 (Turnipseed et al. 1992):

$$CH_3S\cdot + O_2 \xrightarrow{M} CH_3SO_2\cdot$$
$$\longrightarrow CH_2S + HO_2\cdot$$

$CH_3S\cdot$ radicals may also react with O_3 and NO_2.

The DMS–OH adduct, $CH_3\dot{S}(OH)CH_3$, reacts with O_2 to form products. The rate constant for this reaction is $1.8 \times 10^{-12} \exp(-150/T)$ cm^3 molecule^{-1} s^{-1}. This reaction has been postulated to yield dimethyl sulfoxide (DMSO), $CH_3S(O)CH_3$, as the main product (Barnes et al. 1994),

$$CH_3\dot{S}(OH)CH_3 + O_2 \longrightarrow CH_3S(O)CH_3 + HO_2\cdot$$

An overall mechanism for the DMS–OH reaction is shown in Figure 6.19. Many of the details of this mechanism are still uncertain. The principal stable products of the oxidation are DMSO, $DMSO_2$, MSA, SO_2, and H_2SO_4. The ratio of MSA to SO_4^{2-} is indicative of the path that is followed subsequent to formation of CH_3S in the abstraction branch. This ratio is measured in the marine atmosphere as the ratio of MSA to nonseasalt sulfate (nss-SO_4). (Nonseasalt sulfate, the amount of sulfate present in particles in excess of that expected from seasalt particles, is a direct measure of the sulfuric acid formed.) Measurements as a function of latitude indicate that this ratio is quite temperature-dependent, with the ratio varying from about 0.1 near the equator to close to 0.4 in Antarctic waters. Thus the colder the temperature, the more favored the path to MSA formation as opposed to that to SO_2 and eventually to sulfuric acid. This behavior is consistent with a competition between a radical decomposition step, with a fairly large activation energy, namely

$$CH_3SO_2\cdot \longrightarrow CH_3\cdot + SO_2$$

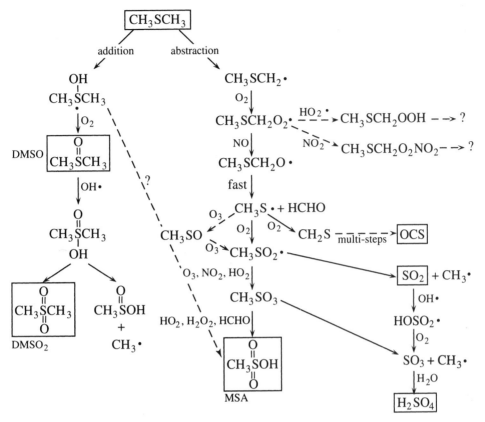

FIGURE 6.19 Mechanism for OH reaction of dimethyl sulfide (DMS).

and a bimolecular reaction, with a fairly small activation energy:

$$CH_3SO_2\cdot + O_3 \longrightarrow CH_3SO_3 + O_2$$

Higher temperatures favor the decomposition step relative to the bimolecular reaction. This behavior is also consistent with a path from the DMS-OH adduct to MSA, as indicated by the dashed line in Figure 6.19.

An interesting finding with implications for the global sulfur cycle is that OCS is a minor product of the DMS–OH reaction, under NO_x-free conditions, with a yield of 0.7% S (Barnes et al. 1996). Even at an OCS yield of only 0.7% of DMS oxidized, this path would represent a significant input to the global cycle of OCS, corresponding to an OCS source strength in the range of 0.10–0.28 Tg yr^{-1} of OCS.

The DMS–NO_3 reaction proceeds initially by an H-atom abstraction (Jensen et al. 1992)

$$CH_3SCH_3 + NO_3 \longrightarrow CH_3SCH_2\cdot + HNO_3$$

and therefore leads directly to CH_3S.

Because all the steps of the DMS oxidation chemistry by OH and NO_3 are not completely understood, some investigators have used measured marine DMS and SO_2 concentrations as a means of inferring important aspects of the mechanism such as the yield of SO_2 from DMS oxidation [e.g., see Yvon et al. (1996) and Yvon and Saltzman (1996)]. The latter investigators estimated SO_2 yields from DMS oxidation to range from 27 to 54% for conditions in the tropical Pacific marine boundary layer.

6.14   TROPOSPHERIC CHEMISTRY OF HALOGEN COMPOUNDS

Halogen compounds arise in the troposphere from the chemical degradation of (partially) halogenated organic compounds that originate from a variety of natural and anthropogenic sources and by the liberation of halogen compounds from seasalt aerosol. Natural sources of gaseous halocarbons include the oceans, which release methyl halides (CH_3Cl, CH_3Br, and CH_3I) and polyhalogenated species ($CHBr_3$, CH_2Br_2). Methyl halides, notably CH_3Br, are also produced by biomass burning. Industrial replacements for fully halogenated CFCs that can be chemically degraded in the troposphere are also a source of tropospheric halogen compounds.

Sea salt contains (by weight) 55.7% Cl, 0.19% Br, and 0.00002% I. Depletion of the Cl and Br content of marine aerosol relative to bulk seawater, as measured by Cl/Na and Br/Na ratios, indicates that there is some net flux of these two halogens to the gas phase. Interestingly, the ratio I/Na in marine aerosol is typically much *greater* than that in seawater, often by a factor of 1000. The large enrichment for iodine in seasalt aerosols relative to seawater has been attributed, in part, to the enhanced level of organic I compounds in the surface organic layer on the ocean that become incorporated in the aerosol formation mechanism.

6.14.1   Chemical Cycles of Halogen Species

Once in the atmosphere, organic halogen molecules are broken down by direct photolysis or by hydroxyl radical attack. Either path leads to the release of atomic halogen. For example, OH reaction of methyl chloride leads to

$$CH_3Cl + OH\cdot \longrightarrow \cdot CH_2Cl + H_2O$$

$$\cdot CH_2Cl + O_2 + M \longrightarrow CH_2ClO_2\cdot + M$$

$$CH_2ClO_2\cdot + NO \longrightarrow NO_2 + HCHO + Cl$$

Halogen atoms are highly reactive toward hydrocarbons, leading to the formation of hydrogen halides through hydrogen abstraction. For Cl atoms, for example

$$Cl + RH \longrightarrow R\cdot + HCl$$

F and Cl atoms react readily in this manner. Br atoms can abstract hydrogen atoms only from HO_2 or aldehydes; I atoms are even less reactive. The alternative to the $Cl + RH$ reaction is oxidation of the halogen atom ($X = Cl$, Br, I) by ozone:

$$X + O_3 \longrightarrow XO + O_2$$

Because of the decreasing reactivity of the halogens, as they go from F to I, toward H-containing compounds to form HX, the fraction of free halogen atoms that react by path $X + O_3$, as opposed to $X + RH$, is: F, $\sim 0\%$; Cl, $\sim 50\%$; Br, $\sim 99\%$; I, $\sim 100\%$ (Platt, 1995).

The halogen halide, HX, can itself react with OH

$$HX + OH\cdot \longrightarrow X + H_2O$$

which returns the halogen atom X to the halogen reservoir. The halogen oxide radicals can undergo a number of reactions. These include photolysis (important for $X = I$, Br, and, to a minor extent, Cl)

$$XO + h\nu \longrightarrow X + O$$

reaction with NO,

$$XO + NO \longrightarrow X + NO_2$$

and reaction with HO_2:

$$XO + HO_2 \longrightarrow HOX + O_2$$

In airmasses influenced by anthropogenic emissions (NO mixing ratio on the order of 1 ppb), $[XO]/[X]$ ratios are on the order of 10–100 ($X = Cl$) and unity ($X = Br$, I).

Reactions of the nitrogen oxides NO_2 or N_2O_5 with NaX contained in seasalt aerosol can lead to the formation of XNO or XNO_2, respectively, for example

$$N_2O_5 + NaX(s) \longrightarrow XNO_2 + NaNO_3(s)$$

(Rossi 2003).

Hydrogen halides can be liberated from seasalt aerosol by the action of strong acids, such as H_2SO_4 and HNO_3:

$$H_2SO_4 + 2\,NaX(s) \longrightarrow 2\,HX + Na_2SO_4(s)$$

A consequence of the presence of reactive halogen species in the troposphere is that because the rate constants for halogen atom, particularly Cl, reactions with hydrocarbons are significantly larger than those for the corresponding OH + HC reactions, hydrocarbons can be effectively removed by reaction with halogen atoms.

6.14.2   Tropospheric Chemistry of CFC Replacements: Hydrofluorocarbons (HFCs) and Hydrochlorofluorocarbons (HCFCs)

Hydrofluorocarbons (HFCs) and hydrochlorofluorocarbons (HCFCs) are replacement compounds for chlorofluorocarbons (CFCs). HCFCs and HFCs contain at least one hydrogen atom and therefore are susceptible to reaction with OH radicals in the troposphere. Since HFCs contain no chlorine they do not give rise to catalytic O_3 destruction. Although HCFCs do contain chlorine, scavenging by OH radicals in the troposphere largely prevents these compounds from having a sufficiently long tropospheric lifetime to survive to be transported into the stratosphere. Figure 6.20 depicts the atmospheric degradation of HFCs, HCFCs, and other CFC substitutes.

   The dominant loss process for HFCs and HCFCs in the atmosphere is reaction with the OH radical. Rate constants for the OH reaction with a wide variety of HFCs and HCFCs are available (Table B.1). A mechanism for the OH reaction of the generalized HFC or HCFC species

$$CX_3CYZH$$
$$(X = H, Cl, \text{ or } F)$$
$$(Y = Cl)$$
$$(Z = H, Cl, \text{ or } F)$$

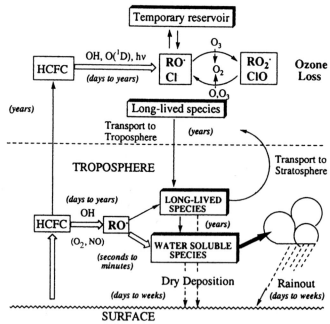

FIGURE 6.20   Atmospheric degradation of HFCs, HCFCs, and other CFC substitutes. Timescales for different processes are given in italics.

is given in Figure 6.21. The first step is H-atom abstraction to yield a haloalkyl radical, which reacts with O_2 to form the corresponding peroxy radical. The peroxy radicals may, under tropospheric conditions, react with NO, NO_2, and HO_2 radicals. Background tropospheric abundances of NO, NO_2, and HO_2 are quite similar. Since rate constants for reaction of halogenated peroxy radicals with NO are about 3 times larger than for reaction with NO_2 and HO_2, the dominant loss process for these radicals is by reaction with NO to give the corresponding alkoxy radicals.

There are several possible reaction paths for the haloalkoxy radicals:

Carbon–halogen bond cleavage:

$$CX_3CYZO\cdot \longrightarrow CX_3CYO + Z \qquad (Z = Cl)$$

Carbon–carbon bond cleavage:

$$CX_3CYZO\cdot \longrightarrow CX_3\cdot + CYZO$$

O_2 reaction:

$$CX_3CYZO\cdot + O_2 \longrightarrow CX_3CYO + HO_2\cdot \qquad (Z = H)$$

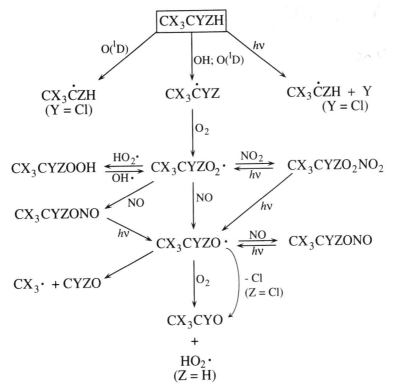

FIGURE 6.21 Generalized atmospheric degradation mechanism for CX_3CYZH. X, Y, and Z may be F, Cl, Br, or H.

Sidebottom (1995) has summarized experimental data pertaining to the fate of the haloalkoxy radicals:

1. CX_2ClO radicals (X = H, Cl, or F) eliminate a Cl atom except for CH_2ClO, where reaction with O_2 is the dominant reaction.
2. CH_2FO and CHF_2O radicals react with O_2 to give the corresponding carbonyl fluorides and HO_2. Fluorine atom elimination or reaction with O_2 is unimportant for CF_3O radicals. Loss of CF_3O is largely determined by reaction with hydrocarbons and nitrogen oxides.
3. CX_3CH_2O radicals (X = H, Cl, or F) react predominantly with O_2 to form the aldehyde and HO_2 radicals.
4. CX_3CCl_2O and CX_3CFClO radicals decompose by Cl atom elimination rather than carbon–carbon bond fission.
5. CX_3CF_2O radicals undergo carbon–carbon bond breaking.
6. CX_3CHYO radicals (Y = Cl or F) have two important reaction channels. The relative importance of the carbon–carbon bond-breaking process and reaction with O_2 is a function of temperature, O_2 pressure, and the total pressure and hence varies considerably with altitude. (Laboratory results for the CF_3CHFO radical, when used in tropospheric model calculations, indicate that 35–40% of HFC-134a released into the atmosphere will be converted into CF_3CFO.)

Halogenated aldehydes, CX_3CHO, may further degrade by photolysis or by reaction with OH, with typical lifetimes on the order of several days. Photolysis to form CX_3 radicals is the dominant removal process if the quantum yield is close to unity.

Trifluoromethyl radicals, CF_3, are produced from HFCs and HCFCs that contain the CF_3 group. As a haloalkyl radical, CF_3 will rapidly add O_2, then react with NO, to give CF_3O radicals. Once present in the stratosphere, the effectiveness of CF_3O_2 and CF_3O radicals in catalytic O_3 destruction will depend on their reactivity with O_3 relative to that with other species. It appears that reactions of these radicals with O_3 is slow compared to that with CH_4 and NO, leading to a catalytic chain length of less than unity (as compared to catalytic chain lengths of 1000–10,000 for ClO_x).

Ozone Destruction in the Polar Tropospheric Boundary Layer   A sudden disappearance of O_3 in ground-level air at Alert (82.5°N, 62.3°W) at polar sunrise was first noted in 1985. Ozone mixing ratios dropped from 30 to 40 ppb to almost zero in the time span of a few hours to a day (Barrie et al. 1988). Continuous, ground-level O_3 records between February and June show that, while proceeding from dark winter to sunlit spring, sudden O_3 depletion events occur, paralleling the increase of sunlight at polar sunrise. A strong observed anticorrelation between O_3 and bromine suggested that bromine is chemically implicated in the ozone removal (Bottenheim et al. 1990; Sturges et al. 1993). The sum of particulate Br and gaseous HBr has been observed to peak in the spring months of March and April throughout the Arctic. It has been suggested that the active component of gas-phase bromine might be bromine oxide (BrO) radicals, possible sources of which (Le Bras and Platt 1995) include the following:

1. Formation of BrNO or $BrNO_2$ by reaction of NO_2 or N_2O_5, respectively, with NaBr contained in seasalt particles. BrNO and $BrNO_2$ are rapidly photolyzed to yield Br atoms.
2. Photochemical reaction of sea salt Br^- on the surface of aerosol particles to form Br_2, followed by

$$Br_2 + hv \longrightarrow 2\,Br$$
$$Br + O_3 \longrightarrow BrO + O_2$$

3. Cycling of inorganic bromine compounds, whereby HBr, HOBr, and $BrONO_2$ are converted back to Br_2 by aqueous-phase chemistry on sulfuric acid aerosol.
4. Free-radical-induced release of Br_2 from sea salt aerosols by aqueous-phase oxidation of Br^-.

The O_3-depletion cycle might involve photolysis of Br_2, followed by the $Br + O_3$ reaction and

$$BrO + BrO \longrightarrow 2\,Br + O_2$$

PROBLEMS

6.1$_A$ **a.** Compute the number of OH radicals produced per $O(^1D)$ generated from O_3 photolysis for lower stratospheric conditions (240 K; $[H_2O]/[M] = 6 \times 10^{-6}$) and for planetary boundary layer conditions (298 K; RH = 0.5).

       b. Calculate the concentration of $O(^1D)$ and the production rate of OH in the lower stratosphere ($[O_3] = 3 \times 10^{12}$ molecules cm^{-3}) and in the boundary layer ($[O_3] = 8 \times 10^{11}$ molecules cm^{-3}).

6.2$_A$ Compute the ozone production efficiency for CO oxidation at the Earth's surface (298 K) assuming a CO level of 200 ppb, an O_3 level of 50 ppb, and a NO level of 40 ppb. Assume $j_{NO_2} = 0.015\,s^{-1}$. You may use the rapid cycling approximation for $[HO_2]/[OH]$.

6.3$_A$ Compute the daytime average $[NO]/[NO_2]$ ratio (12 h) at the Earth's surface for a fixed O_3 mixing ratio of 100 ppb. Assume that the peak noontime value of $j_{NO_2} = 0.015\,s^{-1}$ and that j_{NO_2} is a parabolic function of time, rising from zero at 6:00 a.m. and returning to zero at 6:00 p.m.

6.4$_B$ We wish to estimate the effect of jet aircraft emissions of NO_x on ozone production in the mid- and upper troposphere. Consider the following conditions: $[M] = 7 \times 10^{18}$ molecules cm^{-3}, $T = 230$ K, $[O_3]/[M] = 70$ ppb, and $[CO]/[M] = 100$ ppb.

   a. Assuming that the ratio of $[NO]/[NO_2]$ is determined by the photostationary state, with $j_{NO_2} = 0.01$ s^{-1}, calculate the ratio under these conditions.

   b. The oxidation of CO will dominate the behavior of OH and HO_2. Using $k_{OH+CO} = 1.8 \times 10^{-13}$ cm^3 molecule^{-1} s^{-1} and $k_{HO_2+NO} = 1.0 \times 10^{-11}$ cm^3 molecule^{-1} s^{-1}, determine the $[HO_2]/[OH]$ ratio for

$$NO_x = 10,\ 100,\ 1000,\ \text{and } 10^4 \text{ ppt}$$

   c. Show that for these conditions the production rate of O_3 equals the loss rate of CO.

   d. In this airmass, HO_x is being produced at a rate of 10^4 molecules cm^{-3} s^{-1}. Assuming that the only significant loss processes for HO_x are

$$OH + HO_2 \longrightarrow H_2O + O_2 \qquad k(230\,K) = 1.4 \times 10^{-10} \text{ cm}^3 \text{ molecule}^{-1} \text{ s}^{-1}$$
$$OH + NO_2 + M \longrightarrow HNO_3 + M \quad k(230\,K, [M] = 7 \times 10^{18})$$
$$= 4.8 \times 10^{-13} \text{ cm}^3 \text{ molecule}^{-1} \text{ s}^{-1}$$

   calculate for the different NO_x levels above the number of O_3 molecules produced for each HO_x produced.

   e. Assuming this chemistry occurs for 12 h each day, what is the production rate of O_3 per day for the different NO_x levels above?

6.5$_A$ Determine the lifetime of NO_2 at night against the reaction $NO_2 + O_3 \longrightarrow NO_3 + O_2$ at $z = 0$, 5, and 10 km. Assume an O_3 mixing ratio of 50 ppb.

6.6$_B$ We consider here the formation of N_2O_5 at night. Assume at the beginning of nighttime, the levels of NO_2 and O_3 are 20 ppb and 50 ppb, respectively. As NO_3 forms from the reaction

$$NO_2 + O_3 \longrightarrow NO_3 + O_2 \quad k = 1.2 \times 10^{-13} \exp(-2450/T)$$

NO_2 is converted to NO_3. The NO_3 that forms reacts with NO_2 to form N_2O_5 and the following equilibrium can be assumed to be established:

$$NO_2 + NO_3 + M \rightleftarrows N_2O_5 + M \quad K = 3.0 \times 10^{-27} \exp(10990/T) \text{ cm}^3 \text{ molecule}^{-1}$$

Assume that $[O_3]$ remains constant at its initial value throughout the night (This is an approximation because O_3 reacts with NO and is converted to NO_2.)

Compute $[NO_2]$, $[NO_3]$, and $[N_2O_5]$ over a 12-h nighttime. Consider conditions at the Earth's surface at temperatures of 298 and 273 K. To obtain an upper-limit estimate of the amount of N_2O_5 that forms, assume that the concentration of NO_2 is determined only from the first reaction above, that is, do not account for the loss of NO_2 by conversion to N_2O_5 via the equilibrium. (A more accurate estimate of the amount of N_2O_5 formed is obtained by solving the coupled reaction rate equations for NO_2, NO_3, and N_2O_5 using the forward and reverse rate coefficients of the N_2O_5-forming reaction.)

6.7$_B$ Formaldehyde (HCHO) and acetaldehyde (CH_3CHO) are important carbonyl species in the atmosphere. Both photodissociate and react with OH:

$$HCHO + h\nu \xrightarrow{2O_2} 2HO_2 + CO$$
$$\longrightarrow H_2 + CO$$
$$HCHO + OH \xrightarrow{O_2} HO_2 + CO + H_2O$$
$$CH_3CHO + h\nu \xrightarrow{2O_2} CH_3O_2 + HO_2 + CO$$
$$CH_3CHO + OH \xrightarrow{O_2} CH_3C(O) + H_2O$$

a. Assume that initial concentrations of HCHO and CH_3CHO are added to a mixture of NO_x and air in a laboratory reactor and that at $t = 0$ photolysis begins. Write out a mechanism for the major chemical reactions that take place in the system.

b. Assuming that the NO_x concentration is sufficiently high that HO_x–HO_x termination reactions can be neglected, reduce the mechanism in (a). (On the timescale of the laboratory reactor, oxidation of CO can be neglected.)

c. Write an expression for the rate of O_3 formation in this system, P_{O_3}, for the mechanism of (b). Making any reasonable assumptions, express P_{O_3}, in terms of the concentrations of stable compounds in the system.

d. Give an expression for the ozone production efficiency in this system.

e. If this experiment were repeated at a temperature of, say, 20°C higher, how would you expect P_{O_3} to change? Why?

6.8$_B$ The effectiveness of PAN formation depends on the rate constant ratio, $\alpha = k_{CH_3C(O)O_2+NO_2}/k_{CH_3C(O)O_2+NO}$. This ratio can be measured in the laboratory by producing peroxyacetyl radicals, $CH_3C(O)O_2\cdot$, by photolyzing biacetyl, $CH_3C(O)C(O)CH_3$ in the presence of O_2. Show that in such a system, with NO and NO_2 both initially present at concentrations $[NO]_0$ and $[NO_2]_0$, if $[PAN(t)]$ and $[CO_2(t)]$ are measured as a function of time, and if temperatures are used at which PAN decomposition is slow, α can be determined from

$$\alpha = \frac{[PAN(t)][NO]_0}{[CO_2(t)][NO_2]_0}$$

If it is possible to measure PAN only as a function of time, then show that

$$\alpha = \left(\frac{[\text{PAN}(t)]_{\text{NO}=0}}{[\text{PAN}(t)]} - 1\right)^{-1} \frac{[\text{NO}]_0}{[\text{NO}_2]_0}$$

where $[\text{PAN}(t)]_{\text{NO}=0}$ is the PAN concentration at time t in the absence of initial NO.

6.9$_B$ Regions of the troposphere can be ozone-producing or ozone-depleting depending on the local level of NO_x. The principal chemical sink of O_3 is O_3 photolysis followed by $\text{O}(^1\text{D}) + \text{H}_2\text{O}$. For example, at $10°\text{S}$ at the surface, $j_{\text{O}_3 \to \text{O}(^1\text{D})} \approx 7 \times 10^{-6}\,\text{s}^{-1}$, leading to a photolysis lifetime of O_3 of about 11 days. The chemical sink of O_3 next in importance to photolysis is the reaction,

$$\text{HO}_2 + \text{O}_3 \longrightarrow \text{OH} + 2\,\text{O}_2$$

The principal chemical source of O_3 in the troposphere is production through the methane oxidation chain. The level of NO is critical in this chain in dictating the fate of the HO_2 and CH_3O_2 radicals. The reactions $\text{HO}_2 + \text{NO}$ and $\text{CH}_3\text{O}_2 + \text{NO}$ lead to O_3 production, whereas $\text{HO}_2 + \text{O}_3$ leads to O_3 removal. Consider conditions at the surface at 298 K.

a. Determine the mixing ratio of NO at which the rate of the $\text{HO}_2 + \text{NO}$ reaction just equals that of the $\text{HO}_2 + \text{O}_3$ reaction if $\text{O}_3 = 20\,\text{ppb}$. Assume 298 K.

b. To what level of NO_x does the NO level determined in (a) correspond under noontime conditions? Assume $j_{\text{NO}_2} = 0.015\,\text{s}^{-1}$.

c. A competition also exists between NO and HO_2 for the CH_3O_2 radical, and the path depends on the concentration of HO_2 and NO. Assume that the local OH concentration is 10^6 molecules cm^{-3}. Compute the local HO_2 concentration assuming that reaction with CO is the main sink of OH and HO_2–HO_2 self-reaction is the main sink of HO_2. Assume a CO mixing ratio of 100 ppb.

d. Determine the NO mixing ratio for the conditions of (c) at which the reaction of CH_3O_2 with NO just equals that with HO_2. To what NO_x level does this NO value correspond?

e. It is desired to prepare a plot of the rates of both O_3 production, P_{O_3}, and loss, L_{O_3}, (molecules $\text{cm}^{-3}\,\text{s}^{-1}$) as a function of NO. Consider NO mixing ratios from 1 to 100 ppt. Ozone production occurs as a result of HO_2 and CH_3O_2 reacting with NO, whereas O_3 loss occurs by photolysis and reaction with HO_2. Plot P_{O_3} and L_{O_3} as a function of NO. To estimate the photolysis term in L_{O_3}, assume 50% RH and a local O_3 mixing ratio of 20 ppb. Note any assumptions you make in estimating $[\text{CH}_3\text{O}_2]$.

f. Based on the NO mixing at which $P_{\text{O}_3} = L_{\text{O}_3}$, would you judge the following regions of the troposphere to be O_3-producing or O_3-destroying?

> Remote Pacific Ocean marine boundary layer
>
> Downtown Los Angeles
>
> Rural Southeastern United States

REFERENCES

Andino, J. M., Smith, J. N., Flagan, R. C., Goddard, W. A. III, and Seinfeld, J. H. (1996) Mechanism of atmospheric photooxidation of aromatics: A theoretical study, *J. Phys. Chem.* **100**, 10967–10980.

Atkinson, R. (1987) A structure–activity relationship for the estimation of rate constants for the gas-phase reactions of OH radicals with organic compounds, *Int. J. Chem. Kinet.* **19**, 799–828.

Atkinson, R. (1989) Kinetics and mechanisms of the gas-phase reactions of the hydroxyl radical with organic compounds, *J. Phys. Chem. Ref. Data, Monograph 1*, 1–246.

Atkinson, R. (1990) Gas-phase tropospheric chemistry of organic compounds: A review, *Atmos. Environ.* **24A**, 1–41.

Atkinson, R. (1994) Gas-phase tropospheric chemistry of organic compounds, *J. Phys. Chem. Ref. Data, Monograph 2*, 1–216.

Atkinson, R., and Arey, J. (2003a) Gas-phase tropospheric chemistry of biogenic volatile organic compounds: A review, *Atmos. Environ.* **37**, 5197–5219.

Atkinson, R., and Arey, J. (2003b) Atmospheric degradation of volatile organic compounds, *Chem. Rev.* **103**, 4605–4638.

Atkinson, R., Baulch, D. L., Cox, R. A., Crowley, J. N., Hampson Jr., R. F., Jenkin, M. E., Kerr, J. A., Rossi, M. J., and Troe, J. (2004) Summary of evaluated kinetic and photochemical data for atmospheric chemistry (available at http://www.iupac-kinetic.ch.cam.ac.uk/summary/IUPAC-summ_web_latest.pdf).

Atkinson, R., Calvert, J. G., Kerr, J. A., Madronich, S., Moortgat, G. K., Wallington, T. J., and Yarwood, G. (2000) *The Mechanism of Atmospheric Oxidation of the Alkenes,* Oxford Univ. Press, Oxford, UK.

Atkinson, R., Tuazon, E. C., and Aschmann, S. M. (1995a) Products of the gas-phase reactions of O_3 with alkenes, *Environ. Sci. Technol.* **29**, 1860–1866.

Atkinson, R., Tuazon, E. C., and Aschmann, S. M. (1995b) Products of the gas-phase reactions of a series of 1-alkenes and 1-methylcyclohexene with the OH radical in the presence of NO, *Environ. Sci. Technol.* **29**, 1674–1680.

Barnes, I., Becker, K. H., and Patroescu, I. (1994) The tropospheric oxidation of dimethyl sulfide: A new source of carbonyl sulfide, *Geophys. Res. Lett.* **21**, 2389–2392.

Barnes, I., Becker, K. H., and Patroescu, I. (1996) FT-IR product study of the OH-initiated oxidation of dimethyl sulphide: Observation of carbonyl sulphide and dimethyl sulphoxide, *Atmos. Environ.* **30**, 1805–1814.

Barone, S. B., Turnipseed, A. A., and Ravishankara, A. R. (1995) Role of adducts in the atmospheric oxidation of dimethyl sulfide, *Faraday Discuss.* **100**, 39–54.

Barone, S. B., Turnipseed, A. A., and Ravishankara, A. R. (1996) Reaction of OH with dimethyl sulfide (DMS). 1. Equilibrium constant for OH + DMS reaction and the kinetics of the OH·DMS + O_2 reaction, *J. Phys. Chem.* **100**, 14694–14702.

Barrie, L. A., Bottenheim, J. W., Schnell, R. C., Crutzen, P. J., and Rasmussen, R. A. (1988) Ozone destruction and photochemical reactions at polar sunrise in the lower Arctic atmosphere, *Nature* **334**, 138–141.

Bottenheim, J. W., Barrie, L. W., Atlas, E., Heidt, L. E., Niki, H., Rasmussen, R. A., and Shepson, P. B. (1990) Depletion of lower tropospheric ozone during Arctic spring: The Polar Sunrise Experiment 1988, *J. Geophys. Res.* **95**, 18555–18568.

Brown, S. B., Stark, H., Ryerson, T. B., Williams, E. J., Nicks, D. K., Trainer, M., Fehsenfeld, F. C., and Ravishankara, A. R. (2003a) Nitrogen oxides in the nocturnal boundary layer: Simultaneous in situ measurements of NO_3, N_2O_5, NO_2, NO, and O_3, *J. Geophys. Res.* **108**(D9), 4299 (doi:10.1029/2002JD002917).

Brown, S. B., Stark, H., and Ravishankara, A. R. (2003b) Applicability of the steady state approximation to the interpretation of atmospheric observations of NO_3 and N_2O_5, *J. Geophys. Res.* **108**(D17), 4539 (doi:10.1029/2003JD003407).

Brune, W. H., Faloona, I. C., Tan, D., Weinheimer, A. J., Campos, T., Ridley, B. A., Vay, S. A., Collins, J. E., Sachse, G. W., Jaegle, L., and Jacob, D. J. (1998) Airborne in-situ OH and HO_2 observations in the cloud-free troposphere and lower stratosphere during SUCCESS, *Geophys. Res. Lett.* **25**, 1701–1704.

Brune, W. H. et al. (1999) OH and HO_2 chemistry in the North Atlantic free troposphere, *Geophys. Res. Lett.* **26**, 3077–3080.

Calvert, J. G., Atkinson, R., Becker, K. H., Kamens, R. M., Seinfeld, J. H., Wallington, T. J., and Yarwood, G. (2002) *Mechanisms of Atmospheric Oxidation of Aromatic Hydrocarbons*, Oxford Univ. Press, Oxford, UK.

Calvert, J. G., Yarwood, G., and Dunker, A. (1994) An evaluation of the mechanism of nitrous acid formation in the urban atmosphere, *Res. Chem. Intermediates* **20**, 463–502.

Carpenter, L. J., Green, T. J., Mills, G. P., Bauguitte, S., Penkett, S. A., Zanis, P., Schuepbach, E., Schmidbauer, N., Monks, P. S., and Zellweger, C. (2000) Oxidized nitrogen and ozone production efficiencies in the springtime free troposphere over the Alps, *J. Geophys. Res.* **105**, 14547–14559.

Chameides, W. L. et al. (1992) Ozone precursor relationships in the ambient atmosphere, *J. Geophys. Res.* **97**, 6037–6055.

Chuck, A. L., Turner, S. M., and Liss, P. S. (2002) Direct evidence for a marine source of C_1 and C_2 alkyl nitrates, *Science* **297**, 1151–1154.

Faloona, I. et al. (2000) Observations of HO_x in the upper troposphere during SONEX, *J. Geophys. Res.* **105**, 3771–3783.

Friedl, R. R., Brune, W. H., and Anderson, J. G. (1985) Kinetics of SH with NO_2, O_3, O_2, and H_2O_2, *J. Phys. Chem.* **89**, 5505–5510.

Gery, M. W., Whitten, G. Z., Killus, J. P., and Dodge, M. C. (1989) A photochemical kinetics mechanism for urban and regional scale computer modeling, *J. Geophys. Res.* **94**, 12925–12956.

Goumri, A., Elmaimouni, L., Sawerysyn, J.-P., and Devolder, P. (1992) Reaction rates at $(297 \pm 3)K$ of four benzyl-type radicals with O_2, NO, and NO_2 by discharge flow/laser induced fluorescence, *J. Phys. Chem.* **96**, 5395–5400.

Haagen-Smit, A. J. (1952) Chemistry and physiology of Los Angeles smog, *Ind. Eng. Chem.* **44**, 1342–1346.

Haagen-Smit, A. J., Bradley, C. E., and Fox, M. M. (1953) Ozone formation in photochemical oxidation of organic substances, *Ind. Eng. Chem.* **45**, 2086–2089.

Jacob, D. J. (1999) *Introduction to Atmospheric Chemistry*, Princeton Univ. Press, Princeton, NJ.

Japar, S. M., Wallington, T. J., Richert, J. F. O., and Ball, J. C. (1990) The atmospheric chemistry of oxygenated fuel additives: *t*-Butyl alcohol, dimethyl ether, and methyl *t*-butyl ether, *Int. J. Chem. Kinet.* **22**, 1257–1269.

Japar, S. M., Wallington, T. J., Rudy, S. J., and Chang, T. Y. (1991) Ozone-forming potential of a series of oxygenated organic compounds, *Environ. Sci. Technol.* **25**, 415–420.

Jeffries, H. E., and Crouse, R. (1990) *Scientific and Technical Issues Related to the Application of Incremental Reactivity*, Dept. Environmental Sciences and Engineering, Univ. North Carolina, Chapel Hill, NC.

Jensen, N. R., Hjorth, J., Skov, H., and Restelli, G. (1992) Products and mechanism of the gas-phase reaction of NO_3 with CH_3SCH_3, CD_3SCD_3, CH_3SH, and CH_3SSCH_3, *J. Atmos. Chem.* **14**, 95–108.

Jungkamp, T. P. W., Smith, J. N., and Seinfeld, J. H. (1997) Atmospheric oxidation mechanism of *n*-butane: The fate of alkoxy radicals, *J. Phys. Chem.* **101**, 4392–4401.

Kleinman, L. I., Daum, P. H., Lee, J. H., Lee, Y. -N., Nunnermacker, L. J., Springston, S. R., Newman, L., Weinstein-Lloyd, J., and Sillman, S. (1997) Dependence of ozone production on NO and hydrocarbons in the troposphere, *Geophys. Res. Lett.* **24**, 2299–2302.

Kleinman, L., Lee Y.-N., Springston, S. R., Nunnermacker, L., Zhou, X., Brown, R., Hallock, K., Klotz, P., Leaky, D., Lee, J. H., and Newman, L. (1994) Ozone formation at a rural site in the southeastern United States, *J. Geophys. Res.* **99**, 3469–3482.

Knispel, R., Koch, R., Siese, M., and Zetzsch, C. (1990) Adduct formation of OH radicals with benzene, toluene, and phenol and consecutive reactions of the adducts with NO_x and O_2, *Ber. Bunsenges. Phys. Chem.* **94**, 1375–1379.

Lanzendorf, E. J., Hanisco, T. F., Wennberg, P. O., Cohen, R. C., Stimpfle, R. M., Anderson, J. G., Gao, R. S., Margitan, J. J., and Bui, T. P. (2001) Establishing the dependence of [HO_2]/[OH] on temperature, halogen loading, O_3, and NO_x based on in situ measurements from the NASA ER-2, *J. Phys. Chem. A* **105**, 1535–1542.

Le Bras, G., and Platt, U. (1995) A possible mechanism for combined chlorine and bromine catalyzed destruction of tropospheric ozone in the Arctic, *Geophys. Res. Lett.* **22**, 599–602.

Levy, H. (1971) Normal atmosphere: Large radical and formaldehyde concentrations predicted, *Science* **173**, 141–143.

Libuda, H. G. and Zabel, F. (1995) UV absorption cross sections of acetyl peroxynitrate and trifluoroacetyl peroxynitrate at 298 K, *Ber. Bunsenges. Phys. Chem.* **99**, 1205–1213.

Lightfoot, P. D., Cox, R. A., Crowley, J. N., Destriau, M., Hayman, G. D., Jenkin, M. E., Moortgat, G. K., and Zabel, F. (1992) Organic peroxy radicals: kinetics, spectroscopy and tropospheric chemistry, *Atmos. Environ.* **26A**, 1805–1961.

Liu, S. C. et al. (1992) A study of the photochemistry and ozone budget during the Mauna Loa Observatory photochemistry experiment, *J. Geophys. Res.* **97**, 10463–10471.

Logan, J. A., Prather, M. J., Wofsy, S. C., and McElroy, M. B. (1981) Tropospheric chemistry: a global perspective, *J. Geophys. Res. C: Oceans Atmos.* **86**, 7210–7254.

Lurmann, F. W., and Main, H. H. (1992) *Analysis of the Ambient VOC Data Collected in the Southern California Air Quality Study*, Final Report, ARB Contract A832-130, California Air Resources Board, Sacramento, CA.

Mellouki, A., Le Bras, G., and Sidebottom, H. (2003) Kinetics and mechanisms of the oxidation of oxygenated organic compounds in the gas phase, *Chem. Rev.* **103**, 5077–5096.

Orlando, J. J., Tyndall, G. S., and Wallington, T. J. (2003) The atmospheric chemistry of alkoxy radicals, *Chem. Rev.* **103**, 4657–4689.

Paulson, S. E., and Seinfeld, J. H. (1992) Development and evaluation of a photooxidation mechanism for isoprene, *J. Geophys. Res.* **97**, 20703–20715.

Paulson, S. E., Flagan, R. C., and Seinfeld, J. H. (1992a) Atmospheric photooxidation of isoprene Part I: The hydroxyl radical and ground state atomic oxygen reactions, *Int. J. Chem. Kinet.* **24**, 79–101.

Paulson, S. E., Flagan, R. C., and Seinfeld, J. H. (1992b) Atmospheric photooxidation of isoprene Part II: The ozone-isoprene reaction, *Int. J. Chem. Kinet.* **24**, 103–125.

Platt, U. (1995) The chemistry of halogen compounds in the Arctic troposphere, in *Tropospheric Oxidation Mechanisms*, K. H. Becker, ed., European Commission, Report EUR 16171 EN, Luxbourg, pp. 9–20.

Prinn, R. et al. (1992) Global average concentration and trend for hydroxyl radicals deduced from ALE/GAGE trichloroethane (methyl chloroform) data for 1978–1990, *J. Geophys. Res.* **97**, 2445–2461.

Ravishankara, A. R., Rudich, Y., Talukdar, R., and Barone, S. (1997) Oxidation of atmospheric reduced sulphur compounds: Perspective from laboratory studies, *Phil. Trans. Roy. Soc. Lond. B* **352**, 171–182.

Ridley, B. A., Madronich, S., Chatfield, R. B., Walega, J. G., Shetter, R. E., Carroll, M. A., and Montzka, D. D. (1992) Measurements and model simulations of the photostationary state during the Mauna Loa Observatory photochemistry experiment: Implications for radical concentrations and ozone production and loss rates, *J. Geophys. Res.* **97**, 10375–10388.

Roberts, J. M. (1990) The atmospheric chemistry of organic nitrates, *Atmos. Environ.* **24A**, 243–287.

Rossi, M. J. (2003) Heterogeneous reactions on salts, *Chem. Rev.* **103**, 4823–4882.

Sander, S. P., Friedl, R. R., Golden, D. M., Kurylo, M. J., Huie, R. E., Orkin, V. L., Moortgat, G. K., Ravishankara, A. R., Kolb, C. E., Molina, M. J., and Finlayson-Pitts, B. J. (2003) *Chemical Kinetics and Photochemical Data for Use in Atmospheric Studies*, Evaluation no. 14, Jet Propulsion Laboratory Publication 02-25 (available at http://jpldataeval.jpl.nasa.gov/download.html).

Shepson, P. B., Kleindienst, T. E., Edney, E. O., Namie, G. R., Pittman, J. H., Cupitt, L. T., and Claxton, L. D. (1985) The mutagenic activity of irradiated toluene/NO_x/H_2O/air mixtures, *Environ. Sci. Technol.* **19**, 249–255.

Sidebottom, H. (1995) Degradation of HFCs and HCFCs in the atmosphere, in *Tropospheric Oxidation Mechanisms*, K. H. Becker, ed., European Commission, Report EUR 16171 EN, Luxembourg, pp. 153–162.

Singh, H. B., Kanakidou, M., Crutzen, P. J., and Jacob, D. J. (1995) High concentrations and photochemical fate of oxygenated hydrocarbons in the global troposphere, *Nature* **378**, 50–54.

Singh, H. B., O'Hara, D., Hereth, D., Sachse, W., Blake, D. R., Bradshaw, J. D., Kanakidou, M., and Crutzen, P. J. (1994) Acetone in the atmosphere: distribution, sources, and sinks, *J. Geophys. Res.* **99**, 1805–1819.

Stickel, R. E., Zhao, Z., and Wine, P. H. (1993) Branching ratios for hydrogen transfer in the reaction of OD radicals with CH_3SCH_3 and $CH_3SC_2H_5$, *Chem. Phys. Lett.* **212**, 312–318.

Stockwell, W. R., and Calvert, J. G. (1983) The mechanism of the $HO–SO_2$ reaction, *Atmos. Environ.* **17**, 2231–2235.

Sturges, W. T., Schnell, R. C., Dutton, G. S., Garcia, S. R., and Lind, J. A. (1993) Spring measurements of tropospheric bromine at Barrow, Alaska, *Geophys. Res. Lett.* **20**, 201–204.

Talukdar, R. K., Burkholder, J. B., Schmoltner, A. M., Roberts, J. M., Wilson, R. R., and Ravishankara, A. R. (1995) Investigation of the loss processes for peroxyacetyl nitrate in the atmosphere: UV photolysis and reaction with OH, *J. Geophys. Res.* **100**, 14163–14173.

Thornton, J. A. et al. (2002) Ozone production rates as a function of NO_x abundances and HO_x production rates in the Nashville urban plume; *J. Geophys. Res.* **107**(D12), 4146 (doi: 10.1029/2001JD000932).

Trainer, M. et al. (1993) Correlations of ozone with NO_x in photochemically aged air, *J. Geophys. Res.* **98**, 2917–2925.

Trainer, M. et al. (1991) Observations and modeling of the reactive nitrogen photochemistry at a rural site, *J. Geophys. Res.* **96**, 3045–3063.

Turnipseed, A. A., Barone, S. B., and Ravishankara, A. R. (1996) Reaction of OH with dimethyl sulfide. 2. Products and mechanisms, *J. Phys. Chem.* **100**, 14703–14713.

Turnipseed, A. A., Barone, S. B., and Ravishankara, A. R. (1992) Observation of CH_3S addition to O_2 in the gas phase, *J. Phys. Chem.* **96**, 7502–7505.

Tyndall, G. S., and Ravishankara, A. R. (1991) Atmospheric oxidation of reduced sulfur species, *J. Phys. Chem.* **23**, 483–527.

Wallington, T. J., and Japar, S. M. (1991) Atmospheric chemistry of diethyl ether and ethyl *tert*-butyl ether, *Environ. Sci. Technol.* **25**, 410–415.

Wallington, T. J., Andino, J. M., Skewes, L. M., Siegl, W. O., and Japar, S. M. (1989) Kinetics of the reaction of OH radicals with a series of ethers under simulated atmospheric conditions at 295 K, *Int. J. Chem. Kinet.* **21**, 993–1001.

Wallington, T. J., Dagaut, P., and Kurylo, M. J. (1992) Ultraviolet absorption cross-sections and reaction kinetics and mechanisms for peroxy radicals in the gas phase, *Chem. Rev.* **92**, 667–710.

Wallington, T. J., Ellermann, T., and Nielsen, O. J. (1993) Atmospheric chemistry of dimethyl sulfide-UV spectra and self-reaction kinetics of CH_3SCH_2 and $CH_3SCH_2O_2$ radicals and kinetics of the reactions $CH_3SCH_2 + O_2 \longrightarrow CH_3SCH_2O_2$ and $CH_3SCH_2O_2 + NO \longrightarrow CH_3SCH_2O + NO_2$, *J. Phys. Chem.* **97**, 8442–8449.

Wallington, T. J., Liu, R. Z., Dagaut, P., and Kurylo, M. J. (1988) The gas-phase reactions of hydroxyl radicals with a series of aliphatic ethers over the temperature range 240–440 K, *Int. J. Chem. Kinet.* **20**, 41–49.

Wang, Y., Logan, J. A. and Jacob, D. J. (1998) Global simulation of tropospheric O_3-NO_x-hydrocarbon chemistry. 2. Model evaluation and global ozone budget, *J. Geophys. Res.* **103**(D9), 10727–10755.

Weinstock, B. (1969) Carbon monoxide: Residence time in the atmosphere, *Science* **166**, 224–225.

Wennberg, P. O., Hanisco, T. F., Cohen, R. C., Stimpfle, R. M., Lapson, L. B., and Anderson, J. G. (1995) In situ measurements of OH and HO_2 in the upper troposphere and stratosphere, *J. Atmos. Sci.* **52**, 3413–3420.

Wennberg, P. O. et al. (1998) Hydrogen radicals, nitrogen radicals, and the production of O_3 in the upper troposphere, *Science* **279**, 49–53.

Wood, E. C., Bertram, T. H., Wooldridge, P. J., and Cohen, R. C. (2004) Measurements of N_2O_5, NO_2, and O_3 east of the San Francisco Bay, *Atmos. Chem. Phys. Discuss.* **4**, 6645–6665.

Yin, F., Grosjean, D., and Seinfeld, J. H. (1990a) Photooxidation of dimethyl sulfide and dimethyl disulfide, I: Mechanism development, *J. Atmos. Chem.* **11**, 309–364.

Yin, F., Grosjean, D., Flagan, R. C., and Seinfeld, J. H. (1990b) Photooxidation of dimethyl sulfide and dimethyl disulfide, II: Mechanism evaluation, *J. Atmos. Chem.* **11**, 365–399.

Yvon, S. A., Saltzman, E. S., Cooper, D. J., Bates, T. S., and Thompson, A. M. (1996) Atmospheric sulfur cycling in the tropical Pacific marine boundary layer (12°S, 135°W)—a comparison of field data and model results. I. Dimethylsulfide, *J. Geophys. Res.* **101**, 6899–6909.

Yvon, S. A., and Saltzman, E. S. (1996) Atmospheric sulfur cycling in the tropical Pacific marine boundary layer (12°S, 135°W)—a comparison of field data and model results. 2. Sulfur dioxide, *J. Geophys. Res.* **101**, 6911–6918.

Zetzsch, C., Koch, R., Siese, M., Witte, F., and Devolder, P. (1990) *Proc. 5th European Symp. Physico-Chemical Behavior of Atmospheric Pollutants*, Reidel, Dordrecht, The Netherlands, p. 320.

7 Chemistry of the Atmospheric Aqueous Phase

7.1 LIQUID WATER IN THE ATMOSPHERE

Water is abundant on our planet, distinguishing Earth from all other planets in the solar system. More than 97% of Earth's water is in the oceans, with 2.1% in the polar ice caps and 0.6% in aquifers. The atmosphere contains only about one part in a hundred thousand (0.001%) of Earth's available water. However, the transport and phase distribution of this relatively small amount of water (estimated total liquid equivalent volume of 13,000 km^3) are some of the most important features of Earth's climate. The existence of varying pressures and temperatures in the atmosphere and at the Earth's surface causes water to constantly transfer among its gaseous, liquid, and solid states. Clouds, fogs, rain, dew, and wet aerosol particles represent different forms of that water. Aqueous atmospheric particles play a major role in atmospheric chemistry, atmospheric radiation, and atmospheric dynamics.

The mass concentration of water vapor as a function of temperature and relative humidity in the atmosphere is presented in Figure 7.1. The maximum water vapor capacity of the atmosphere at a given temperature is given by the saturation line (100% RH). The ability of the atmosphere to hold water vapor decreases exponentially with temperature; that is, at 30°C it can contain as much as 30.3 g(water) m^{-3} but at 0°C it becomes saturated at a concentration of 4.8 g(water) m^{-3}. This basic thermodynamic property of the atmosphere is the reason for the creation of liquid water. Consider an air parcel that has initially a temperature of 20°C and contains a water vapor concentration of 8.4 g m^{-3}. The water vapor saturation concentration of the atmosphere at 20°C is 17.3 g(water) m^{-3} and the RH of the air parcel is equal to $100 \times (8.4/17.3) = 48.5\%$ (point A in Figure 7.1). Let us assume that the air parcel cools down to 5°C by rising and expanding in the atmosphere without any interaction with its surroundings, that is, maintaining its water vapor concentration. At the new state (point B in Figure 7.1) the water vapor saturation concentration of the atmosphere is 6.8 g(water) m^{-3} and the air parcel is supersaturated by 1.6 g(water) m^{-3}. This extra water vapor then condenses on available aerosol particles forming a cloud with a liquid water content of 1.6 g(water) m^{-3}, and the air parcel's RH returns to 100% with a water vapor concentration of 6.8 g(water) m^{-3} (point C). This description of the creation of a cloud is highly simplified, but still conveys the main thermodynamic processes. A more detailed picture will be presented later.

Clouds cover approximately 60% of the Earth's surface. Average global cloud coverage over the oceans is estimated at 65% and over land at 52% (Warren et al. 1986, 1988).

Atmospheric Chemistry and Physics: From Air Pollution to Climate Change, Second Edition, by John H. Seinfeld and Spyros N. Pandis. Copyright © 2006 John Wiley & Sons, Inc.

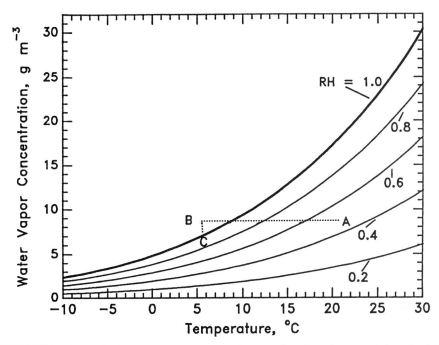

FIGURE 7.1   Atmospheric water vapor concentration as a function of temperature and relative humidity.

The occurrence of clouds shows dramatic geographical variation and is generally restricted to the lowest 4–6 km of the troposphere. Clouds form and evaporate repeatedly. Only a small fraction (around 10%) of the clouds that form actually generate precipitation. Thus nine out of ten clouds evaporate without ever generating rain droplets. Even in cases where precipitation develops, the raindrops often evaporate on their way falling through the cloud-free air, so the drops never reach the surface.

Clouds come in many different forms, and their characteristics reveal the meteorological properties of the atmosphere. In the troposphere four groups of clouds are recognized depending on the altitude of their bases and their vertical development: low-level, midlevel, high-level, and so-called clouds with vertical development. These groups are also subdivided depending on their structure. The three most common types are stratus, cumulus, and cirrus. *Stratus* clouds (St) are layers of low, often gray clouds that have little definition and rarely produce precipitation. Their average thickness varies from 200 to 1500 m. *Cumulus* (Cu) clouds are detached and dense, rising in mounds from a level base. *Cirrus* (Ci) clouds are high clouds usually consisting of ice and having a silken appearance. These primary three categories are often combined to provide additional cloud types. For example, a low-level cloud that is broken up in a wave pattern is referred to as *stratocumulus* (Sc), and a midlevel cloud displaying the same broken structure is called an *altocumulus* (Ac). A precipitating cloud is denoted as *nimbus*. Cloud types and the frequency of their occurrence are summarized in Table 7.1. The layered tropospheric clouds (stratus and cirrus) generally are created during the passage of a cold or warm front. Vertically structured clouds are unrelated to frontal passages, forming as a result of the intense heating of low-lying air, rapid ascent of the heated airmass, and water condensation as the air becomes supersaturated.

TABLE 7.1   Types and Properties of Clouds

Type	Height of Base, km	Frequency (%) of Occurrence over Oceans[a]	Areal Coverage over Oceans (%)[a]	Frequency (%) of Occurrence over Land[a]	Areal Coverage over Land (%)[a]
Low-level					
Stratocumulus (Sc)	0–2 ⎫	45	34	27	18
Stratus (St)	0–2 ⎭				
Nimbostratus (Ns)	0–4	6	6	6	5
Midlevel					
Altocumulus (Ac)	2–7 ⎫	46	22	35	21
Altostratus (As)	2–7 ⎭				
High-level					
Cirrus (Ci)					
Cirrostratus (Cs)	7–18	37	13	47	23
Cirrocumulus (Cc)					
Clouds with vertical development					
Cumulus (Cu)	0–3	33	12	14	5
Cumulonimbus (Cb)	0–3	10	6	7	4

[a]Overlapping clouds often coexist over the same area.

Sources: Warren et al. (1986, 1988).

The liquid water content of typical clouds, given the symbol L and often abbreviated LWC, varies from approximately 0.05 to 3 g(water) m^{-3}, with most of the observed values in the 0.1 to 0.3 g(water) m^{-3} region. A frequency distribution for LWC average values for stratus, stratocumulus, altostratus, and altocumulus clouds is given in Figure 7.2 (Heymsfield 1993).

The liquid water content is often expressed for convenience as liquid water mixing ratio w_L in (volume water/volume air) and is related to L by

$$w_L \text{ (vol water/vol air)} = 10^{-6} L \text{ (g m}^{-3}) \qquad (7.1)$$

The cloudwater mixing ratio w_L varies from 5×10^{-8} to 3×10^{-6}.

Cloud droplet sizes vary from a few micrometers to 50 μm with average diameters usually in the 10–20 μm range. The physics of clouds is discussed in detail in Chapter 17.

The microphysical structure of fogs is similar to that of clouds. Typical fog liquid water contents vary from 0.02 to 0.5 g m^{-3} and fog droplets have sizes from a few micrometers to 40 μm.

7.2   ABSORPTION EQUILIBRIA AND HENRY'S LAW

The equilibrium of a species A between the gas and aqueous phase can be represented by

$$A(g) \rightleftharpoons A(aq) \qquad (7.2)$$

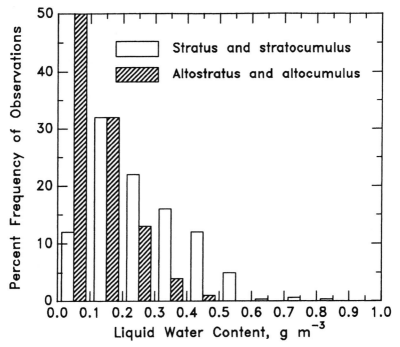

FIGURE 7.2 Frequency distribution for liquid water content average values for various cloud types over Europe and Asia.

The equilibrium between gaseous and dissolved A is usually expressed by the so-called Henry's law coefficient H_A

$$[A(aq)] = H_A p_A \tag{7.3}$$

where p_A is the partial pressure of A in the gas phase (atm) and $[A(aq)]$ is the aqueous-phase concentration of A $(mol\,L^{-1})$ in equilibrium with p_A. The customary units of the Henry's law coefficient H_A are $mol\,L^{-1}\,atm^{-1}$. The unit $mol\,L^{-1}$ is usually written as M, a notation that we will use henceforth. Henry's law constant values are reported in the literature in several different unit systems that may give rise to some confusion. For example, if the gas-phase concentration is expressed in moles per liter of air, and the aqueous-phase concentration in moles per liter of water, the Henry's law constant is dimensionless (actually it is in liters of air per liters of water). Some investigators define the Henry's law constant H'_A by the reverse of (7.3), $p_A = H'_A[A]$; then $H'_A = 1/H_A$. We are going to use the definition of (7.3), but when making use of published Henry's law data, one should always check the definition of the Henry's law constant. By our definition, soluble gases have large Henry's law coefficients. Finally, note that the aqueous-phase concentration of A given by (7.3) does not depend on the amount of liquid water available or on the size of the droplet.

We should emphasize at this point that Henry's law is strictly applicable only to dilute solutions. If the solution is not sufficiently dilute, the concentration of the solute, $[A(aq)]$, in equilibrium with p_A deviates from Henry's law. These deviations from ideal behavior will be examined in Chapter 10. When considering the behavior of atmospheric gases at typical concentrations in equilibrium with cloud or fog droplets or large natural bodies such as lakes, Henry's law is generally accepted as a good approximation.

Table 7.2 gives the Henry's law coefficients for some atmospheric gases in liquid water at 298 K. The values given reflect only the physical solubility of the gas, that is, only the equilibrium (7.2) regardless of the subsequent fate of A once dissolved. Several of the species given in Table 7.2, once dissolved, either undergo acid–base equilibria or react with water. We will consider the effect of these further processes shortly. A detailed compilation of Henry's law coefficients for species important in environmental chemistry can be found in Sander (1999).

TABLE 7.2   Henry's Law Coefficients of Some Atmospheric Gases

Species[a]	$H(\mathrm{M\,atm^{-1}})$ at 298 K
O_2	1.3×10^{-3}
NO	1.9×10^{-3}
C_2H_4	4.8×10^{-3}
NO_2	1.0×10^{-2}
O_3	1.1×10^{-2}
N_2O	2.5×10^{-2}
CO_2	3.4×10^{-2}
H_2S	0.1
DMS	0.56
HCl	1.1
SO_2	1.23
NO_3	1.8
CH_3ONO_2	2.0
CH_3O_2	6
OH	25
HNO_2	49
NH_3	62
CH_3OH	220
CH_3OOH	310
$CH_3C(O)OOH$	473
HO_2	5.7×10^3
HCOOH	3.6×10^3
$HCHO^b$	2.5
CH_3COOH	8.8×10^3
H_2O_2	1×10^5
HNO_3	2.1×10^5

[a] The values given reflect only the physical solubility of the gas regardless of the subsequent fate of the dissolved species. These constants do not account for dissociation or other aqueous-phase transformations.

[b] The value is 6.3×10^3 if the diol formation is included.

The temperature dependence of an equilibrium constant such as Henry's law coefficient is given by the van't Hoff equation (Denbigh 1981)

$$\frac{d \ln H_A}{dT} = \frac{\Delta H_A}{RT^2} \tag{7.4}$$

where ΔH_A is the reaction enthalpy at constant temperature and pressure. ΔH_A is a function of temperature, but over small temperature ranges it is approximately constant, and therefore by integrating (7.4) with respect to temperature,

$$H_A(T_2) = H_A(T_1) \exp\left(\frac{\Delta H_A}{R}\left(\frac{1}{T_1} - \frac{1}{T_2}\right)\right) \tag{7.5}$$

Table 7.3 gives the values of $\Delta H_A(298\ \text{K})$ for several species of atmospheric interest. Henry's law coefficients generally increase in value as the temperature decreases, reflecting a greater solubility of the gas at lower temperatures. For example, the Henry's law constant for O_3 increases from 1.1×10^{-2} to $2.35 \times 10^{-2}\ \text{M atm}^{-1}$ as T decreases from 298 to 273 K, and that for SO_2 from 1.23 to 3.28 M atm^{-1} over the same temperature range.

The definition of how soluble a gas is in water is relative, as a gas may be regarded as soluble in one context but insoluble in another. For example, acetone with $H_A = 25.6\ \text{M atm}^{-1}$ is very soluble compared to aliphatic hydrocarbons but is insoluble compared to formaldehyde with $H_A = 6300\ \text{M atm}^{-1}$. A useful solubility benchmark for

TABLE 7.3   Heat of Dissolution for Henry's Law Coefficients

Species	ΔH_A (kcal mol^{-1}) at 298 K
CO_2	-4.85
NH_3	-8.17
SO_2	-6.25
H_2O_2	-14.5
HNO_2	-9.5
NO_2	-5.0
NO	-2.9
CH_3O_2	-11.1
CH_3OH	-9.7
PAN	-11.7
HCHO	-12.8
HCOOH	-11.4
HCl	-4.0
CH_3OOH	-11.1
$CH_3C(O)OOH$	-12.2
O_3	-5.04

Source: Pandis and Seinfeld (1989a).

atmospheric applications is the distribution of a species between the gas and the aqueous phases in a typical cloud; species that reside mainly in the gas phase are considered insoluble, and species that are almost exclusively in the aqueous phase are considered very soluble. Intermediate species, with significant fractions in both phases, are considered moderately soluble. Let us formulate the above statements mathematically by defining the aqueous/gas-phase distribution factor.

The distribution factor, f_A, of a species A is defined as the ratio of its aqueous-phase mass concentration c_{aq}[g(L of air)$^{-1}$] to its gas-phase mass concentration c_g[g(L of air)$^{-1}$],

$$f_A = \frac{c_{aq}}{c_g} \tag{7.6}$$

Note that the aqueous-phase concentration is expressed per volume of air and not per volume of solution, so f_A is dimensionless. The distribution factor goes to infinity for very soluble species ($c_g \ll c_{aq}$) and approaches zero as the solubility of A approaches zero ($c_g \gg c_{aq}$). Assuming Henry's law equilibrium one can show that

$$f_A = 10^{-6} H_A RT L = H_A RT w_L \tag{7.7}$$

where H_A is in M atm^{-1}, R is the ideal-gas constant equal to 0.08205 atm L mol^{-1} K^{-1}, T is the temperature in K, and L is the cloud/fog liquid water content in g m^{-3}. The factor 10^{-6} is a result of the units used in (7.7).

The fractions of A that exist in the gas X_g^A and aqueous phases X_{aq}^A are given by

$$X_g^A = \frac{1}{1+f_A} \quad X_{aq}^A = \frac{f_A}{1+f_A} \tag{7.8}$$

Combining (7.7) and (7.8), we obtain

$$X_{aq}^A = \frac{10^{-6} H_A RTL}{1 + 10^{-6} H_A RTL} = \frac{H_A RT w_L}{1 + H_A RT w_L} \tag{7.9}$$

The aqueous-phase fraction of A is plotted as function of the Henry's law constant and the cloud liquid water content in Figure 7.3 for a range of expected liquid water content values. For species with Henry's law constants smaller than 400 M atm^{-1} less than 1% of their mass is dissolved in the aqueous phase inside a cloud. Such species include O_3, NO, NO_2, and all the atmospheric hydrocarbons. A significant fraction (more than 10%) of a species resides in the aqueous phase in the atmosphere only if its Henry's law constant exceeds 5000 M atm^{-1}.

The above analysis suggests that species with a Henry's law coefficient lower than about 1000 M atm^{-1} will partition strongly toward the gas phase and are considered relatively insoluble for atmospheric applications. Species with Henry's law coefficients between 1000 and 10,000 M atm^{-1} are moderately soluble, and species with even higher Henry's law coefficients are considered very soluble. As can be seen from Table 7.2, remarkably few gases fall into the very soluble category. This does not imply, however, that only very soluble gases are important in atmospheric aqueous-phase chemistry.

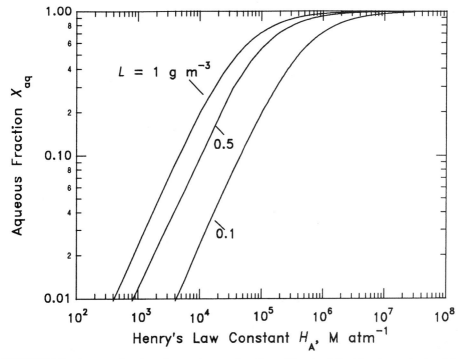

FIGURE 7.3   Aqueous fraction of a species as a function of the cloud liquid water content and the species' Henry's law constant.

7.3   AQUEOUS-PHASE CHEMICAL EQUILIBRIA

On dissolution in water a number of species dissociate into ions. This dissociation is a reversible reaction that reaches equilibrium rapidly. In Chapter 12 we will discuss the timescale needed to achieve such equilibrium.

7.3.1   Water

Water itself ionizes to form a hydrogen ion, H^+, and a hydroxide ion, OH^-:

$$H_2O \rightleftharpoons H^+ + OH^- \tag{7.10}$$

At equilibrium, we obtain

$$K'_w = \frac{[H^+][OH^-]}{[H_2O]} \tag{7.11}$$

where $K'_w = 1.82 \times 10^{-16}$ M at 298 K. However, the concentration of H_2O molecules is so large (around 55.5 M), and so few ions are formed, that $[H_2O]$ is virtually constant. Therefore the molar concentration of pure water is incorporated into the equilibrium constant to give

$$K_w = [H^+][OH^-] \tag{7.12}$$

TABLE 7.4   Thermodynamic Data for Aqueous Equilibrium Constants

Equilibrium	K at 298 K (M)	ΔH_A at 298 K, kcal mol^{-1}
$H_2O \rightleftharpoons H^+ + OH^-$	1.0×10^{-14}	13.35
$CO_2 \cdot H_2O \rightleftharpoons H^+ + HCO_3^-$	4.3×10^{-7}	1.83
$HCO_3^- \rightleftharpoons H^+ + CO_3^{2-}$	4.7×10^{-11}	3.55
$SO_2 \cdot H_2O \rightleftharpoons H^+ + HSO_3^-$	1.3×10^{-2}	-4.16
$HSO_3^- \rightleftharpoons H^+ + SO_3^{2-}$	6.6×10^{-8}	-2.23
$NH_3 \cdot H_2O \rightleftharpoons NH_4^+ + OH^-$	1.7×10^{-5}	8.65

where $K_w = K'_w[H_2O] = 1.0 \times 10^{-14}\,M^2$ at 298 K (Table 7.4). For pure water, each water molecule dissociates, producing one hydrogen and one hydroxide ion and therefore $[H^+] = [OH^-]$. Thus at 298 K, $[H^+] = [OH^-] = 1.0 \times 10^{-7}\,M$. Note that the total concentration of ions in pure water is only 0.2 μM compared to 55.5 M for pure water itself. Therefore there are roughly 300 million water molecules for each ion in pure water. As a result of this small fraction of dissociated water molecules, pure water is considered a very weak electrolyte and has a very small electrical conductivity.

Defining the pH of water as

$$pH = -\log_{10}[H^+] \tag{7.13}$$

we see that pH = 7.0 for pure water at 298 K.

7.3.2   Carbon Dioxide–Water Equilibrium

On dissolution, CO_2 hydrolyzes; that is, a molecule of carbon dioxide combines with a water molecule to form one molecule of $CO_2 \cdot H_2O$. CO_2 (aq) and $CO_2 \cdot H_2O$ are alternative notations for the same chemical species corresponding to A(aq) in (7.2). Dissolved carbon dioxide, $CO_2 \cdot H_2O$, dissociates twice to form the carbonate and bicarbonate ions:

$$CO_2(g) + H_2O \rightleftharpoons CO_2 \cdot H_2O \tag{7.14}$$

$$CO_2 \cdot H_2O \rightleftharpoons H^+ + HCO_3^- \tag{7.15}$$

$$HCO_3^- \rightleftharpoons H^+ + CO_3^{2-} \tag{7.16}$$

The equilibrium constants for these reactions are

$$K_{hc} = H_{CO_2} = \frac{[CO_2 \cdot H_2O]}{p_{CO_2}} \tag{7.17}$$

$$K_{c1} = \frac{[H^+][HCO_3^-]}{[CO_2 \cdot H_2O]} \tag{7.18}$$

$$K_{c2} = \frac{[H^+][CO_3^{2-}]}{[HCO_3^-]} \tag{7.19}$$

where K_{hc} is the equilibrium constant for the hydrolysis of CO_2, and K_{c1} and K_{c2} denote the first and second dissociation equilibrium constants for dissolved CO_2.

Note that liquid water concentration has already been incorporated into the hydrolysis constant and that K_{hc} is identical to the Henry's law coefficient for carbon dioxide, H_{CO_2}.

The concentrations of the species in solution are given by

$$[CO_2 \cdot H_2O] = H_{CO_2} p_{CO_2} \qquad (7.20)$$

$$[HCO_3^-] = \frac{K_{c1}[CO_2 \cdot H_2O]}{[H^+]} = \frac{H_{CO_2} K_{c1} p_{CO_2}}{[H^+]} \qquad (7.21)$$

$$[CO_3^{2-}] = \frac{K_{c2}[HCO_3^-]}{[H^+]} = \frac{H_{CO_2} K_{c1} K_{c2} p_{CO_2}}{[H^+]^2} \qquad (7.22)$$

The total dissolved carbon dioxide $[CO_2^T]$ is then

$$[CO_2^T] = [CO_2 \cdot H_2O] + [HCO_3^-] + [CO_3^{2-}] = H_{CO_2} p_{CO_2} \left(1 + \frac{K_{c1}}{[H^+]} + \frac{K_{c1} K_{c2}}{[H^+]^2}\right) \quad (7.23)$$

Noting the similarity of this expression to Henry's law, we can define the *effective Henry's law constant* for CO_2, $H_{CO_2}^*$, as

$$H_{CO_2}^* = H_{CO_2} \left(1 + \frac{K_{c1}}{[H^+]} + \frac{K_{c1} K_{c2}}{[H^+]^2}\right) \qquad (7.24)$$

and the total dissolved carbon dioxide is given by

$$[CO_2^T] = H_{CO_2}^* p_{CO_2} \qquad (7.25)$$

The effective Henry's law constant always exceeds the Henry's law constant

$$H_{CO_2}^* > H_{CO_2} \qquad (7.26)$$

and therefore the total amount of CO_2 dissolved always exceeds that predicted by Henry's law for CO_2 alone. Note that while the Henry's law constant H_{CO_2} depends only on temperature, the effective Henry's law constant $H_{CO_2}^*$ depends on both temperature and the solution pH.

For pH < 5, the dissolved carbon dioxide does not dissociate appreciably and its effective Henry's law constant is, for all practical purposes, equal to its Henry's law constant. For a gas-phase CO_2 mixing ratio equal to 330 ppm, the equilibrium aqueous-phase concentration is 11.2 μM (Figure 7.4). As the pH increases to values higher than 5, $CO_2 \cdot H_2O$ starts dissociating and the dissolved total carbon dioxide increases exponentially. However, even at pH 8, $H_{CO_2}^*$ is only 1.5 M atm^{-1}, and practically all the available carbon dioxide is still in the gas phase. The aqueous-phase concentration of total carbon dioxide increases to hundreds of μM for alkaline water.

Let us assume momentarily that the atmosphere contains only CO_2 and water. What is the pH of cloud and rainwater in this system? Use an ambient mixing ratio of 350 ppm and a temperature of 298 K. The concentrations of the ions in solution will satisfy the electroneutrality equation

$$[H^+] = [OH^-] + [HCO_3^-] + 2[CO_3^{2-}] \qquad (7.27)$$

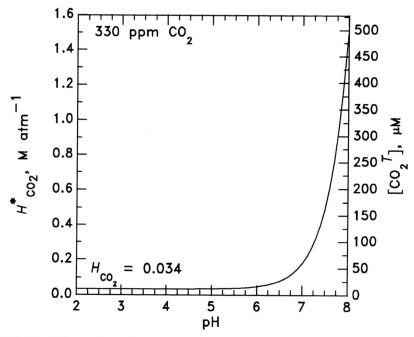

FIGURE 7.4   Effective Henry's law constant for CO_2 as a function of the solution pH. Also shown is the corresponding equilibrium total dissolved CO_2 concentration $[CO_2^T]$ for a CO_2 mixing ratio of 330 ppm.

which can be used to calculate $[H^+]$. Using (7.12), (7.21), and (7.22) for $[OH^-]$, $[HCO_3^-]$, and $[CO_3^{2-}]$, respectively, we can place (7.27) in the form of an equation with a single unknown, the hydrogen ion concentration

$$[H^+] = \frac{K_w}{[H^+]} + \frac{H_{CO_2}K_{c1}p_{CO_2}}{[H^+]} + \frac{2\,H_{CO_2}K_{c1}K_{c2}p_{CO_2}}{[H^+]^2} \tag{7.28}$$

or after rearranging:

$$[H^+]^3 - (K_w + H_{CO_2}K_{c1}p_{CO_2})[H^+] - 2\,H_{CO_2}K_{c1}K_{c2}p_{CO_2} = 0 \tag{7.29}$$

Even if some CO_2 dissolves in the water, this amount as we saw will be small and will not change the gas-phase concentration of CO_2. Given the temperature, which determines the values of K_w, H_{CO_2}, K_{c1}, and K_{c2}, $[H^+]$ can be computed from (7.29), from which all other ion concentrations can be obtained. At an ambient CO_2 mixing ratio of 350 ppm at 298 K the solution pH is 5.6. This value is often cited as the pH of "pure" rainwater.

7.3.3   Sulfur Dioxide–Water Equilibrium

The behavior of SO_2 in aqueous solution is qualitatively similar to that of carbon dioxide. Absorption of SO_2 in water results in

$$SO_2(g) + H_2O \rightleftharpoons SO_2 \cdot H_2O \tag{7.30}$$
$$SO_2 \cdot H_2O \rightleftharpoons H^+ + HSO_3^- \tag{7.31}$$
$$HSO_3^- \rightleftharpoons H^+ + SO_3^{2-} \tag{7.32}$$

with

$$H_{SO_2} = \frac{[SO_2 \cdot H_2O]}{p_{SO_2}} \tag{7.33}$$

$$K_{s1} = \frac{[H^+][HSO_3^-]}{[SO_2 \cdot H_2O]} \tag{7.34}$$

$$K_{s2} = \frac{[H^+][SO_3^{2-}]}{[HSO_3^-]} \tag{7.35}$$

where $K_{s1} = 1.3 \times 10^{-2}$ M and $K_{s2} = 6.6 \times 10^{-8}$ M at 298 K. The concentrations of the dissolved species are given by

$$[SO_2 \cdot H_2O] = H_{SO_2} p_{SO_2} \tag{7.36}$$

$$[HSO_3^-] = \frac{K_{s1}[SO_2 \cdot H_2O]}{[H^+]} = \frac{H_{SO_2} K_{s1} p_{SO_2}}{[H^+]} \tag{7.37}$$

$$[SO_3^{2-}] = \frac{K_{s2}[HSO_3^-]}{[H^+]} = \frac{H_{SO_2} K_{s1} K_{s2} p_{SO_2}}{[H^+]^2} \tag{7.38}$$

The corresponding ionic concentrations in equilibrium with 1 ppb of SO_2 are shown as functions of the pH in Figure 7.5.

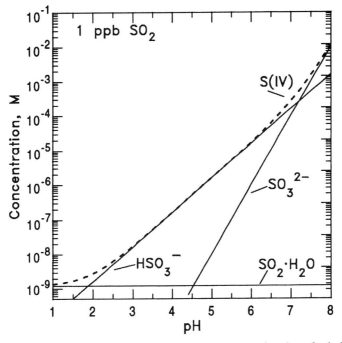

FIGURE 7.5   Concentrations of S(IV) species and total S(IV) as a function of solution pH for an SO_2 mixing ratio of 1 ppb at 298 K.

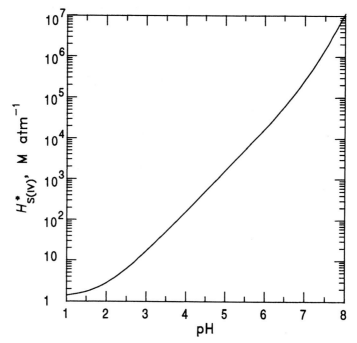

FIGURE 7.6   Effective Henry's law constant for SO_2 as a function of solution pH at 298 K.

The total dissolved sulfur in solution in oxidation state 4, referred to as S(IV) (see Section 2.2), is equal to

$$[S(IV)] = [SO_2 \cdot H_2O] + [HSO_3^-] + [SO_3^{2-}] \qquad (7.39)$$

The total dissolved sulfur, S(IV), can be related to the partial pressure of SO_2 over the solution by

$$[S(IV)] = H_{SO_2} \, p_{SO_2} \left[1 + \frac{K_{s1}}{[H^+]} + \frac{K_{s1}K_{s2}}{[H^+]^2} \right] \qquad (7.40)$$

or, if we define the effective Henry's law coefficient for SO_2, $H_{S(IV)}^*$, as

$$H_{S(IV)}^* = H_{SO_2} \left[1 + \frac{K_{s1}}{[H^+]} + \frac{K_{s1}K_{s2}}{[H^+]^2} \right] \qquad (7.41)$$

the total dissolved sulfur dioxide is given by

$$[S(IV)] = H_{S(IV)}^* \, p_{SO_2} \qquad (7.42)$$

The effective Henry's law constant for SO_2 increases by almost seven orders of magnitude as the pH increases from 1 to 8 (Figure 7.6). The effect of the acid–base equilibria is to "pull" more material into solution than predicted on the basis of Henry's law alone. The Henry's law coefficient for SO_2 alone, H_{SO_2}, is 1.23 M atm^{-1} at 298 K, while for the same

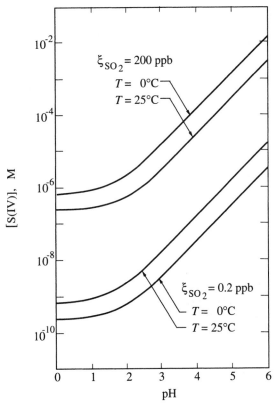

FIGURE 7.7 Equilibrium dissolved S(IV) as a function of pH, gas-phase mixing ratio of SO_2, and temperature.

temperature, the effective Henry's law coefficient for S(IV), $H^*_{S(IV)}$ is $16.4 \, M \, atm^{-1}$ for pH = 3, $152 \, M \, atm^{-1}$ for pH = 4, and $1524 \, M \, atm^{-1}$ for pH = 5. Equilibrium S(IV) concentrations for SO_2 gas-phase mixing ratios of 0.2 to 200 ppb, and over a pH range of 1–6, vary approximately from 0.001 to 1000 μm (Figure 7.7).

S(IV) Concentrations in Different Environments   Calculate [S(IV)] as a function of pH for SO_2 mixing ratios of 0.2 and 200 ppb over the range from pH = 0–6 and for temperatures of 0°C and 25°C.

   The concentration of S(IV) is given by (7.40). As we have seen, at 298 K, $H_{SO_2} = 1.23 \, M \, atm^{-1}$, $K_{s1} = 1.3 \times 10^{-2} \, M$, and $K_{s2} = 6.6 \times 10^{-8} \, M$. We can calculate the values of these constants for $T = 273 \, K$ using (7.5). The required parameters are given in Table 7. A.1 in Appendix 7 in the end of this chapter. We find that at 273 K, $H_{SO_2} = 3.2 \, M \, atm^{-1}$, $K_{s1} = 2.55 \times 10^{-2} \, M$, and $K_{s2} = 10^{-7} \, M$. Figure 7.7 shows [S(IV)] as a function of pH for these two SO_2 concentrations at 0°C and 25°C. The concentration of [S(IV)] increases dramatically as pH increases. The concentration of $SO_2 \cdot H_2O$ does not depend on the pH, and the abovementioned increase (for constant T and ξ_{SO_2}) is due exclusively to the increased concentrations of HSO_3^- and SO_3^{2-}. Temperature has a significant effect on the [S(IV)] concentration. For example for pH = 5 and 0.2 ppb SO_2, the equilibrium concentration increases from 0.16 to 0.82 μM, that is by a factor of ∼5 as the temperature decreases from 298 to 273 K.

S(IV) Composition and pH   Let us compute the mole fractions of the three S(IV) species as a function of solution pH. The mole fractions are found by combining (7.36)–(7.38) with (7.40):[1]

$$x_{SO_2 \cdot H_2O} = \frac{[SO_2 \cdot H_2O]}{[S(IV)]} = \left(1 + \frac{K_{s1}}{[H^+]} + \frac{K_{s1}K_{s2}}{[H^+]^2}\right)^{-1} \tag{7.43}$$

$$x_{HSO_3^-} = \frac{[HSO_3^-]}{[S(IV)]} = \left(1 + \frac{[H^+]}{K_{s1}} + \frac{K_{s2}}{[H^+]}\right)^{-1} \tag{7.44}$$

$$x_{SO_3^{2-}} = \frac{[SO_3^{2-}]}{[S(IV)]} = \left(1 + \frac{[H^+]}{K_{s2}} + \frac{[H^+]^2}{K_{s1}K_{s2}}\right)^{-1} \tag{7.45}$$

Figure 7.8 shows these three mole fractions as a function of pH at 298 K. At pH values lower than 2, S(IV) is mainly in the form of $SO_2 \cdot H_2O$. At higher pH values the HSO_3^- fraction increases, and in the pH range from 3 to 6 practically all S(IV) occurs as HSO_3^-. At pH values higher than 7, SO_3^{2-} dominates. Since these different S(IV) species can be expected to have different chemical reactivities, if a chemical reaction occurs in solution involving either HSO_3^- or SO_3^{2-}, we expect that the rate of the reaction will depend on pH since the concentration of these species depends on pH.

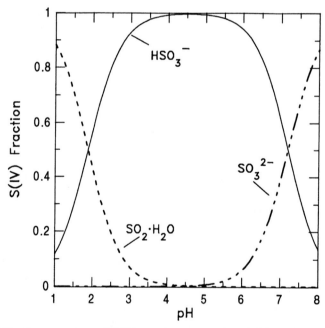

FIGURE 7.8   Concentrations of S(IV) species expressed as S(IV) mole fractions. These fractions are independent of the gas-phase SO_2 concentration.

---

[1]In aquatic chemistry a common notation for these mole fractions is α_0, α_1, and α_2, respectively (Stumm and Morgan 1996).

7.3.4 Ammonia–Water Equilibrium

Ammonia is the principal basic gas in the atmosphere. Absorption of NH_3 in H_2O leads to

$$NH_3 + H_2O \rightleftharpoons NH_3 \cdot H_2O \tag{7.46}$$

$$NH_3 \cdot H_2O \rightleftharpoons NH_4^+ + OH^- \tag{7.47}$$

with

$$H_{NH_3} = \frac{[NH_3 \cdot H_2O]}{p_{NH_3}} \tag{7.48}$$

$$K_{a1} = \frac{[NH_4^+][OH^-]}{[NH_3 \cdot H_2O]} \tag{7.49}$$

where at 298 K, $H_{NH_3} = 62\,M\,atm^{-1}$ and $K_{a1} = 1.7 \times 10^{-5}\,M$. NH_4OH is an alternative notation often used instead of $NH_3 \cdot H_2O$. The concentration of NH_4^+ is given by

$$[NH_4^+] = \frac{K_{a1}[NH_3 \cdot H_2O]}{[OH^-]} = \frac{H_{NH_3}K_{a1}}{K_w}p_{NH_3}[H^+] \tag{7.50}$$

The total dissolved ammonia $[NH_3^T]$ is simply

$$[NH_3^T] = [NH_3 \cdot H_2O] + [NH_4^+] = H_{NH_3}p_{NH_3}\left(1 + \frac{K_{a1}[H^+]}{K_w}\right) \tag{7.51}$$

and the ammonium fraction is given by

$$\frac{[NH_4^+]}{[NH_4^T]} = \frac{K_{a1}[H^+]}{K_w + K_{a1}[H^+]} \tag{7.52}$$

Note that for pH values lower than 8, $K_{a1}[H^+] \gg K_w$ and

$$[NH_3^T] \simeq [NH_4^+]$$

So under atmospheric conditions practically all dissolved ammonia in clouds is in the form of ammonium ion. The aqueous-phase concentrations of $[NH_4^+]$ in equilibrium with 1 ppb of $NH_3(g)$ are shown in Figure 7.9. The partitioning of ammonia between the gas and aqueous phases inside a cloud can be calculated using (7.9) and the effective Henry's law coefficient for ammonia, $H_{NH_3}^* = H_{NH_3}K_{a1}[H^+]/K_w$. If the cloud pH is less than 5 practically all the available ammonia will be dissolved in cloudwater (Figure 7.10).

7.3.5 Nitric Acid–Water Equilibrium

Nitric acid is one of the most water-soluble atmospheric gases with a Henry's law constant (at 298 K) of $H_{HNO_3} = 2.1 \times 10^5\,M\,atm^{-1}$. After dissolution

$$HNO_3(g) \rightleftharpoons HNO_3(aq) \tag{7.53}$$

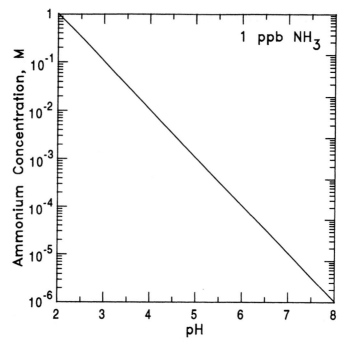

FIGURE 7.9   Ammonium concentration as a function of pH for a gas-phase ammonia mixing ratio of 1 ppb at 298 K.

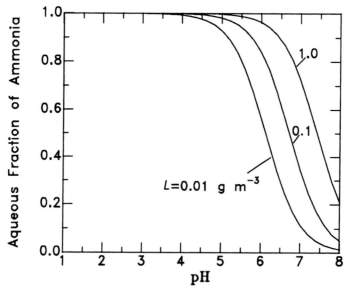

FIGURE 7.10   Equilibrium fraction of total ammonia in the aqueous phase as a function of pH and cloud liquid water content at 298 K.

the nitric acid dissociates readily to nitrate

$$HNO_3(aq) \rightleftharpoons NO_3^- + H^+ \qquad (7.54)$$

increasing further its solubility. The dissociation constant for nitrate is $K_{n1} = 15.4\,M$ at 298 K, where by definition

$$K_{n1} = \frac{[NO_3^-][H^+]}{[HNO_3(aq)]} \qquad (7.55)$$

The total dissolved nitric acid $[HNO_3^T]$ will then be

$$[HNO_3^T] = [HNO_3(aq)] + [NO_3^-] \qquad (7.56)$$

and using Henry's law, we obtain

$$[HNO_3(aq)] = H_{HNO_3}\, p_{HNO_3} \qquad (7.57)$$

Substituting this into the dissociation equilibrium equation, we have

$$[NO_3^-] = \frac{H_{HNO_3} K_{n1}}{[H^+]} p_{HNO_3} \qquad (7.58)$$

The last three equations can be combined to provide an expression for the total dissolved nitric acid in equilibrium with a given nitric acid vapor concentration,

$$[HNO_3^T] = H_{HNO_3}^* p_{HNO_3} = H_{HNO_3}\left(1 + \frac{K_{n1}}{[H^+]}\right) p_{HNO_3} \qquad (7.59)$$

where $H_{HNO_3}^* = H_{HNO_3}(1 + (K_{n1}/[H^+]))$ is the effective Henry's law coefficient for nitric acid.

Note that because the dissociation constant K_{n1} has such a high value, it follows that

$$(K_{n1}/[H^+]) \gg 1 \qquad (7.60)$$

for any cloud pH of atmospheric interest. Therefore $[NO_3^-] \gg [HNO_3(aq)]$ for all atmospheric clouds and one can safely assume that dissolved nitric acid exists in clouds exclusively as nitrate. In other words, nitric acid is a strong acid and upon dissolution in water it dissociates completely to nitrate ions.

The effective Henry's law constant for nitrate is then

$$H_{HNO_3}^* \simeq H_{HNO_3} \frac{K_{n1}}{[H^+]} = \frac{3.2 \times 10^6}{[H^+]} \qquad (7.61)$$

where $H_{HNO_3}^*$ is in $M\,atm^{-1}$ and $[H^+]$ in M.

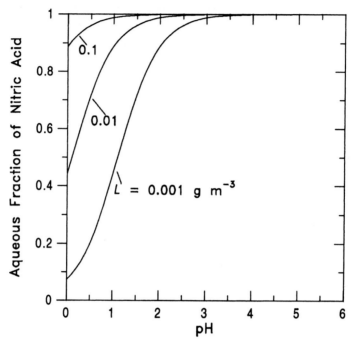

FIGURE 7.11   Equilibrium fraction of total nitric acid in the aqueous phase as a function of pH and cloud liquid water content at 298 K assuming ideal solution.

Partitioning of Nitric Acid inside a Cloud   Calculate the fraction of nitric acid that will exist in the aqueous phase inside a cloud as a function of the cloud liquid water content ($L = 0.001$–1 g m^{-3}) and pH. What does one expect in a typical cloud with liquid water content in the 0.1–1 g m^{-3} and pH in the 2–7 ranges?

Using (7.9) for the aqueous fraction and substituting $H_A = H^*_{HNO_3} = 3.2 \times 10^6/[H^+]$ (the Henry's law coefficient in that equation is the effective Henry's law coefficient for all the dissociating species), we can calculate the aqueous fraction of nitric acid as a function of pH for different cloud liquid water contents. The results are shown in Figure 7.11. For $L = 1$ g m^{-3} the aqueous fraction is practically 1 for all pH values of interest. For all clouds of atmospheric interest ($L > 0.1$ g m^{-3}) and for all pH values (pH>2), nitric acid at equilibrium is completely dissolved in cloudwater and its gas-phase concentration inside a cloud is expected to be practically zero. For example, for pH = 3, $H^*_{HNO_3} = 3.2 \times 10^9$ M atm^{-1}, and for a nitrate concentration of 100 μM, the corresponding equilibrium gas-phase mixing ratio is only 0.03 ppt(3×10^{-14} atm).

7.3.6   Equilibria of Other Important Atmospheric Gases

Hydrogen Peroxide   Hydrogen peroxide is soluble in water with a Henry's law constant $H_{H_2O_2} = 1 \times 10^5$ M atm^{-1} at 298 K. It dissociates to produce HO$_2^-$

$$H_2O_2(aq) \rightleftharpoons HO_2^- + H^+ \tag{7.62}$$

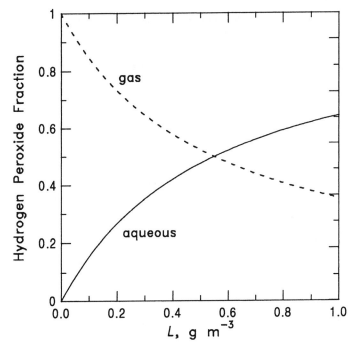

FIGURE 7.12   Equilibrium fraction of total hydrogen peroxide in the aqueous phase as a function of cloud liquid water content at 298 K.

but is a rather weak electrolyte with a dissociation constant of $K_{h1} = 2.2 \times 10^{-12}$ M at 298 K. Using the dissociation equilibrium

$$\frac{[HO_2^-]}{[H_2O_2(aq)]} = \frac{K_{h1}}{[H^+]} \tag{7.63}$$

and this concentration ratio remains less than 10^{-4} for pH values less than 7.5. Therefore, for most atmospheric applications, the dissociation of dissolved hydrogen peroxide can be neglected. The equilibrium partitioning of H_2O_2 between the gas and aqueous phases can be calculated from (7.9) using the Henry's law coefficient $H_{H_2O_2}$ and is shown in Figure 7.12. H_2O_2 exists in appreciable amounts in both the gas and aqueous phases inside a typical cloud. For example, for a cloud liquid water content of 0.2 g m^{-3}, roughly 30% of the H_2O_2 will be dissolved in the cloudwater whereas the remaining 70% will remain in the cloud interstitial air.

Ozone   Ozone is slightly soluble in water with a Henry's law constant of only 0.011 M atm^{-1} at 298 K. For a cloud gas-phase mixing ratio of 100 ppb O_3, the equilibrium ozone aqueous-phase concentration is 1.1 nM and practically all ozone remains in the gas phase.

Oxides of Nitrogen   Nitric oxide (NO) and nitrogen dioxide (NO_2) are also characterized by small solubility in water (Henry's law coefficients 0.002 and 0.01 M atm^{-1} at 298 K). A negligible fraction of these species is dissolved in cloudwater, and their aqueous-phase concentrations are estimated to be on the order of 1 nM or even smaller.

Formaldehyde   Formaldehyde, upon dissolution in water, undergoes hydrolysis to yield the *gem-diol*, methylene glycol

$$HCHO(aq) + H_2O \rightleftharpoons H_2C(OH)_2 \tag{7.64}$$

with a hydration constant K_{HCHO} defined by

$$K_{HCHO} = \frac{[H_2C(OH)_2]}{[HCHO(aq)]} \tag{7.65}$$

that has a value of 2530 at 298 K (Le Henaff 1968). This rather large value of K_{HCHO} suggests that the hydration is essentially complete ($>99.9\%$) and that practically all dissolved formaldehyde will exist in its gem-diol form.

Formaldehyde has a Henry's law constant $H_{HCHO} = 2.5 \, M \, atm^{-1}$ at 298 K (Betterton and Hoffmann 1988), but its water solubility is enhanced by several orders of magnitude as a result of the diol formation. Combining the Henry's law equilibrium and the hydrolysis relationships, we can calculate the effective Henry's law coefficient for the total dissolved formaldehyde H_{HCHO}^* as

$$H_{HCHO}^* = H_{HCHO}(1 + K_{HCHO}) \simeq H_{HCHO} K_{HCHO} \tag{7.66}$$

with a value of $6.3 \times 10^3 \, M \, atm^{-1}$. In the literature, the effective Henry's law coefficient for formaldehyde is often quoted instead of the intrinsic constant. Therefore, for thermodynamic equilibrium assuming that all dissolved formaldehyde exists as $H_2C(OH)_2$:

$$[H_2C(OH)_2] = H_{HCHO}^* \, p_{HCHO} \tag{7.67}$$

Most of the available formaldehyde remains in the gas phase inside a typical cloud (Figure 7.13).

Formic and Other Atmospheric Acids   The most abundant carboxylic acids in the atmosphere are formic and acetic acid, although more than 100 aliphatic, olefinic, and aromatic acids have been detected in the atmosphere [Graedel et al. (1986); see also Chapter 14]. These acids are weak electrolytes, and their partial dissociation enhances their solubility.

Formic acid has a Henry's law coefficient of $3600 \, M \, atm^{-1}$ at 298 K and a dissociation constant $K_f = 1.8 \times 10^{-4} \, M$. From the dissociation equilibrium

$$HCOOH(aq) \rightleftharpoons HCOO^- + H^+$$

one can calculate the dissociated fraction of formic acid as

$$\frac{[HCOO^-]}{[HCOOH(aq)]} = \frac{K_f}{[H^+]} \tag{7.68}$$

At pH $= 3.74$, 50% of the dissolved formic acid has dissociated. Above pH 4, most of the dissolved formic acid exists in its ionic form, while for pH values below 3 most of it is undissociated. The concentrations of HCOOH(aq) and HCOO$^-$ in equilibrium with 1 ppb

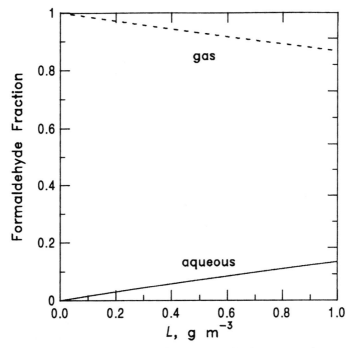

FIGURE 7.13 Equilibrium fraction of total formaldehyde in the aqueous phase as a function of cloud liquid water content at 298 K.

of gas-phase formic acid vary from $3.6\,\mu M$ at pH $= 2$ to $6500\,\mu M$ at pH $= 7$. These results illustrate that significant formic acid concentrations may be found in cloudwater or rainwater that has not been acidified.

For pH values below 4 most of the formic acid remains in the cloud interstitial air and only a small fraction (less than 10%) dissolves in the cloud water. For pH values above 7 practically all the available formic acid is transferred into the aqueous phase and only traces remain in the interstitial air. In the intermediate pH regime from 4 to 7 the formic acid equilibrium partitioning varies considerably depending on the cloud liquid water content and pH.

Acetic acid has a higher Henry's law coefficient ($8800\,M\,atm^{-1}$ at 298 K) than formic acid but dissociates less (dissociation constant $1.7 \times 10^{-5}\,M$ at 298 K). The overall result is that for pH values lower than 4, acetic acid is more soluble than formic acid. However, above pH 4 formic acid is more soluble due to its more efficient dissociation.

OH and HO$_2$ Radicals   The hydroperoxyl radical, HO_2, is a weak acid and on dissolution in the aqueous phase dissociates,

$$HO_2(aq) \rightleftharpoons O_2^- + H^+ \tag{7.69}$$

with an equilibrium constant $K_{HO_2} = 3.5 \times 10^{-5}\,M$ at 298 K (Perrin 1982). For HO_2 one defines then the total dissolved concentration as

$$[HO_2^T] = [HO_2(aq)] + [O_2^-] \tag{7.70}$$

and then its effective Henry's law constant is given by

$$H^*_{HO_2} = H_{HO_2}\left(1 + \frac{K_{HO_2}}{[H^+]}\right) \tag{7.71}$$

where $H^*_{HO_2} = 5.7 \times 10^3 \, M \, atm^{-1}$ is the Henry's law constant for HO_2 at 298 K. The dissociation of HO_2 enhances its dissolution in the aqueous phases for pH values above 4.5 and $H^*_{HO_2} = 2 \times 10^5 \, M \, atm^{-1}$ for pH = 6.

The Henry's law constant of OH at 298 K is $25 \, M \, atm^{-1}$, so at equilibrium and for pH values lower than 6 most OH and HO_2 will be present in the gas phase inside a cloud (see Figure 7.3). For gas-phase mixing ratios of OH and HO_2 of 0.3 and 40 ppt, respectively, the corresponding equilibrium concentrations in the aqueous phase are, for pH = 4, $[OH] = 7.5 \times 10^{-12} \, M$ and $[HO_2] = 0.08 \, \mu M$. The Henry's law equilibrium concentration of HO_2 increases to $2.9 \, \mu M$ at pH = 6.

7.4 AQUEOUS-PHASE REACTION RATES

Rates of reaction of aqueous-phase species are generally expressed in terms of moles per liter (M) of solution per second. It is often useful to express aqueous-phase reaction rates on the basis of the gas-phase properties, especially when comparing gas-phase and aqueous-phase reaction rates. In this way both rates are expressed on the same basis.

To place our discussion on a concrete basis, let us say we have a reaction of S(IV) with a dissolved species A

$$S(IV) + A(aq) \rightarrow products \tag{7.72}$$

the rate of which is given by

$$R_a = k[A(aq)][S(IV)] \quad (M \, s^{-1}) \tag{7.73}$$

where R_a is in $M \, s^{-1}$, the aqueous-phase concentrations [S(IV)] and [A(aq)] are in M, and the reaction constant k is in $M^{-1} \, s^{-1}$. The reaction rate can be expressed in moles of SO_2 per *liter of air* per second by multiplying R_a by the liquid water mixing ratio $w_L = 10^{-6} \, L$:

$$R'_a = k w_L[A(aq)][S(IV)] = 10^{-6} \, k \, L[A(aq)][S(IV)] \quad (mol(L \, of \, air)^{-1} \, s^{-1}) \tag{7.74}$$

The moles per liter of air can be then converted to equivalent SO_2 partial pressure for 1 atmosphere total pressure by applying the ideal-gas law to obtain

$$R''_a = 3.6 \times 10^6 \, L R T \, R_a \quad (ppb \, h^{-1}) \tag{7.75}$$

where L is in $g \, m^{-3}$, $R = 0.082 \, atm \, L \, K^{-1} \, mol^{-1}$, and T is in K. For example, an aqueous-phase reaction rate of $1 \, \mu M \, s^{-1}$ in a cloud with a liquid water content of $0.1 \, g \, m^{-3}$ at 288 K is equivalent to a gas-phase oxidation rate of $8.5 \, ppb \, h^{-1}$. A nomogram relating aqueous-phase reaction rates in $\mu M \, s^{-1}$ to equivalent gas-phase reaction rates in $ppb \, h^{-1}$ as a function of the cloud liquid water content L at 288 K is given in Figure 7.14.

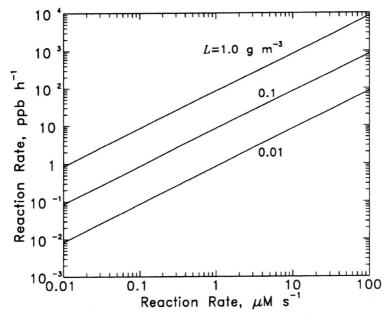

FIGURE 7.14 Nomogram relating aqueous reaction rates, in $\mu M\,s^{-1}$, to equivalent gas-phase reaction rates in $ppb\,h^{-1}$, at $T = 288\,K$, $p = 1$ atm, and given liquid water content.

Reaction rates are sometimes also expressed as a fractional conversion rate in $\%\,h^{-1}$. The rate R_a'' given above can be converted to a fractional conversion rate by dividing by the mixing ratio, ξ_{SO_2}, of SO_2 in ppb and multiplying by 100

$$\bar{R}_a = 3.6 \times 10^8 \frac{R_a LRT}{\xi_{SO_2}} \quad (\%\,h^{-1}) \tag{7.76}$$

where ξ_{SO_2} is in ppb. Let us assume that both S(IV) and A are in thermodynamic equilibrium with Henry's law constants $H^*_{S(IV)}$ and H_A, respectively. Combining (7.73), (7.76), and (7.3), one obtains

$$\bar{R}_a = 3.6 \times 10^{-10}\, k\, H_A H^*_{S(IV)}\, \xi_A\, LRT \quad (\%\,h^{-1}) \tag{7.77}$$

where k is in $M\,s^{-1}$, H_A and $H^*_{S(IV)}$ are in $M\,atm^{-1}$, ξ_A is in ppb, L is in $g\,m^{-3}$, $R = 0.082\,atm\,L\,K^{-1}\,mol^{-1}$, and T is in K. Note that the SO_2 oxidation rate given by (7.77) is independent of the SO_2 concentration and depends only on the mixing ratio of A, the cloud liquid water content, and the temperature. Equation (7.77) should be used only if A exists in both gas and aqueous phases and Henry's law equilibrium is satisfied by both S(IV) and A. If the two species do not satisfy Henry's law, (7.76) can still be used.

The characteristic time for SO_2 oxidation τ_{SO_2} (s) can be calculated from (7.76) as

$$\tau_{SO_2} = \frac{100}{\bar{R}_a} = \frac{\xi_{SO_2}}{10^3\, R_a\, LRT} \tag{7.78}$$

7.5 S(IV)–S(VI) TRANSFORMATION AND SULFUR CHEMISTRY

The aqueous-phase conversion of dissolved SO_2 to sulfate, also referred to as S(VI) since sulfur is in oxidation state 6, is considered the most important chemical transformation in cloudwater. As we have seen dissolution of SO_2 in water results in the formation of three chemical species: hydrated SO_2 ($SO_2 \cdot H_2O$), the bisulfite ion (HSO_3^-), and the sulfite ion (SO_3^{2-}). At the pH range of atmospheric interest (pH = 2–7) most of the S(IV) is in the form of HSO_3^-, whereas at low pH (pH < 2), all of the S(IV) occurs as $SO_2 \cdot H_2O$. At higher pH values (pH > 7), SO_3^{2-} is the preferred S(IV) state (Figure 7.8). Since the individual dissociations are fast, occurring on timescales of milliseconds or less (see Chapter 12), during a reaction consuming one of the three species, $SO_2 \cdot H_2O$, HSO_3^-, or SO_3^{2-}, the corresponding aqueous-phase equilibria are reestablished instantaneously. As we saw earlier, the dissociation of dissolved SO_2 enhances its aqueous solubility so that the total amount of dissolved S(IV) always exceeds that predicted by Henry's law for SO_2 alone and is quite pH dependent.

Several pathways for S(IV) transformation to S(VI) have been identified involving reactions of S(IV) with O_3, H_2O_2, O_2 (catalyzed by Mn(II) and Fe(III)), OH, SO_5^-, HSO_5^-, SO_4^-, PAN, CH_3OOH, $CH_3C(O)OOH$, HO_2, NO_3, NO_2, N(III), HCHO, and Cl_2^- (Pandis and Seinfeld 1989a). There is a large literature on the reaction kinetics of aqueous sulfur chemistry. We present here only a few of the most important rate expressions available.

7.5.1 Oxidation of S(IV) by Dissolved O_3

Although ozone reacts very slowly with SO_2 in the gas phase, the aqueous-phase reaction

$$S(IV) + O_3 \rightarrow S(VI) + O_2 \tag{7.79}$$

is rapid. This reaction has been studied by many investigators (Erickson et al. 1977; Larson et al. 1978; Penkett et al. 1979; Harisson et al. 1982; Maahs 1983; Martin 1984; Hoigné et al. 1985; Lagrange et al. 1994; Botha et al. 1994). A detailed evaluation of existing experimental kinetic and mechanistic data suggested the following expression for the rate of the reaction of S(IV) with dissolved ozone for a dilute solution (subscript 0) (Hoffmann and Calvert 1985):

$$R_0 = -\frac{d[S(IV)]}{dt} = (k_0[SO_2 \cdot H_2O] + k_1[HSO_3^-] + k_2[SO_3^{2-}])[O_3] \tag{7.80}$$

with $k_0 = 2.4 \pm 1.1 \times 10^4 \, M^{-1} \, s^{-1}$, $k_1 = 3.7 \pm 0.7 \times 10^5 \, M^{-1} \, s^{-1}$ and, $k_2 = 1.5 \pm 0.6 \times 10^9 \, M^{-1} \, s^{-1}$. The activation energies recommended by Hoffmann and Calvert (1985) are based on the work of Erickson et al. (1977) and are $46.0 \, kJ \, mol^{-1}$ for k_1 and $43.9 \, kJ \, mol^{-1}$ for k_2.

The complex pH dependency of the effective rate constant of the S(IV)–O_3 reaction is shown in Figure 7.15. Over the pH range of 0 to 2 the slope is close to 0.5 (corresponding to an $[H^+]^{-0.5}$ dependence), while over the pH range of 5 to 7 the slope of the plot is close to 1 (corresponding to an $[H^+]^{-1}$ dependence). In the transition regime between pH 2 and 4 the slope is 0.34. Some studies focusing on pH values lower than 4 have produced rate laws that yield erroneous rates when extrapolated to higher pH values.

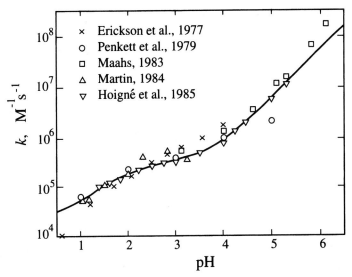

FIGURE 7.15 Second-order reaction rate constant for the S(IV)–O_3 reaction defined according to $d[S(VI)]/dt = k[O_3(aq)]$ [S(IV)], as a function of solution pH at 298 K. The curve shown is the three-component expression of (7.80) and the symbols are the corresponding measurements by a number of investigators.

Hoffmann and Calvert (1985) proposed that the S(IV)–O_3 reaction proceeds by nucleophilic attack on ozone by $SO_2 \cdot H_2O$, HSO_3^-, and SO_3^{2-}. Considerations of nucleophilic reactivity indicate that SO_3^{2-} should react more rapidly with ozone than HSO_3^-, and HSO_3^- should react more rapidly in turn than $SO_2 \cdot H_2O$. This order is reflected in the relative numerical values of k_0, k_1, and k_2. An increase in the aqueous-phase pH results in an increase of $[HSO_3^-]$ and $[SO_3^{2-}]$ equilibrium concentrations, and therefore in an increase of the overall reaction rate. For an ozone gas-phase mixing ratio of 30 ppb, the reaction rate varies from less than 0.001 $\mu M\,h^{-1}$ (ppb SO_2)$^{-1}$ at pH 2 (or less than 0.01% $SO_2(g)\,h^{-1}$ (g water/m^3 air)$^{-1}$) to 3000 $\mu M\,h^{-1}$ (ppb SO_2)$^{-1}$ at pH 6 (7000% $SO_2(g)\,h^{-1}$ (g water/m^3 air)$^{-1}$) (Figure 7.16). A typical gas-phase SO_2 oxidation rate by OH is on the order of 1% h^{-1} and therefore S(IV) heterogeneous oxidation by ozone is significant for pH values greater than 4. The strong increase of the reaction rate with pH often renders this reaction self-limiting; production of sulfate by this reaction lowers the pH and slows down further reaction. The ubiquitousness of atmospheric ozone guarantees that this reaction will play an important role both as a sink of gas-phase SO_2 and as a source of cloudwater acidification as long as the pH of the atmospheric aqueous phase exceeds about 4.

The S(IV)–O_3 reaction rate is not affected by the presence of metallic ion traces: Cu^{2+}, Mn^{2+}, Fe^{2+}, Fe^{3+}, and Cr^{2+} (Maahs 1983; Martin 1984; Lagrange et al. 1994).

Effect of Ionic Strength Lagrange et al. (1994) suggested that the rate of the S(IV)–O_3 reaction increases linearly with the ionic concentration of the solution. The *ionic strength* of a solution I is defined as

$$I = \frac{1}{2}\sum_{i=1}^{n} m_i z_i^2 \qquad (7.81)$$

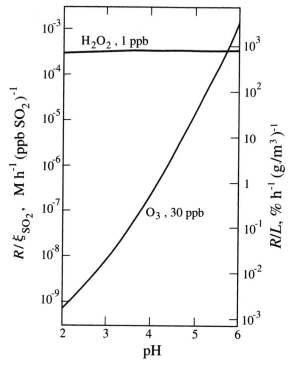

FIGURE 7.16   Rate of aqueous-phase oxidation of S(IV) by ozone (30 ppb) and hydrogen peroxide (1 ppb), as a function of solution pH at 298 K. Gas-aqueous equilibria are assumed for all reagents. R/ξ_{SO_2} represents the aqueous phase reaction rate per ppb of gas-phase SO_2. R/L represents rate of reaction referred to gas-phase SO_2 pressure per $(g\,m^{-3})$ of cloud liquid water content.

where n is the number of different ions in solution, m_i and z_i are the molality (solute concentration in mol kg^{-1} of solvent) and number of ion charges of ion i, respectively. Experiments at 18°C supported the following ionic strength correction

$$R = (1 + FI)R_0 \tag{7.82}$$

where F is a parameter characteristic of the ions of the supporting electrolyte, I is the ionic strength of the solution in units of mol L^{-1}, and R and R_0 are the reaction rates at ionic strengths I and zero, respectively. The measured values of F were

$$F = 1.59 \pm 0.3 \quad \text{for NaCl}$$
$$F = 3.71 \pm 0.7 \quad \text{for Na}_2\text{SO}_4 \tag{7.83}$$

Values for $NaClO_4$ and NH_4ClO_4 were smaller than 1.1. These results indicate that the ozone reaction can be 2.6 times faster in a seasalt particle with an ionic strength of 1 M than in a solution with zero ionic strength.

7.5.2 Oxidation of S(IV) by Hydrogen Peroxide

Hydrogen peroxide, H_2O_2, is one of the most effective oxidants of S(IV) in clouds and fogs (Pandis and Seinfeld 1989a). H_2O_2 is very soluble in water and under typical ambient conditions its aqueous-phase concentration is approximately six orders of magnitude higher than that of dissolved ozone. The S(IV)–H_2O_2 reaction has been studied in detail by several investigators (Hoffmann and Edwards 1975; Penkett et al. 1979; Cocks et al. 1982; Martin and Damschen 1981; Kunen et al. 1983; McArdle and Hoffmann 1983) and the published rates all agree within experimental error over a wide range of pH (0 to 8) (Figure 7.17). The reproducibility of the measurements suggests a lack of susceptibility of this reaction to the influence of trace constituents. The rate expression is (Hoffmann and Calvert 1985)

$$-\frac{d[\mathrm{S(IV)}]}{dt} = \frac{k[\mathrm{H}^+][\mathrm{H_2O_2}][\mathrm{HSO_3^-}]}{1 + K[\mathrm{H}^+]} \tag{7.84}$$

with $k = 7.5 \pm 1.16 \times 10^7 \, \mathrm{M^{-2} \, s^{-1}}$ and $K = 13 \, \mathrm{M^{-1}}$ at 298 K.

Noting that H_2O_2 is a very weak electrolyte, that $[\mathrm{H}^+][\mathrm{HSO_3^-}] = H_{\mathrm{SO_2}} K_{s1} p_{\mathrm{SO_2}}$, and that, for pH > 2, $1 + K[\mathrm{H}^+] \simeq 1$, one concludes that the rate of this reaction is practically pH independent over the pH range of atmospheric interest. For a $H_2O_2(g)$ mixing ratio of 1 ppb the rate is roughly $300 \, \mu\mathrm{M} \, \mathrm{h^{-1}} \, (\mathrm{ppb \, SO_2})^{-1}$ (700% $SO_2(g) \, \mathrm{h^{-1}}$ (g water/$\mathrm{m^3}$ air)$^{-1}$). The near pH independence can also be viewed that the pH dependences of the S(IV) solubility and of the reaction rate constant cancel each other. The reaction is very fast and indeed both field measurements (Daum et al. 1984) and theoretical studies (Pandis and Seinfeld 1989b) have suggested that, as a result, $H_2O_2(g)$ and $SO_2(g)$ rarely coexist in

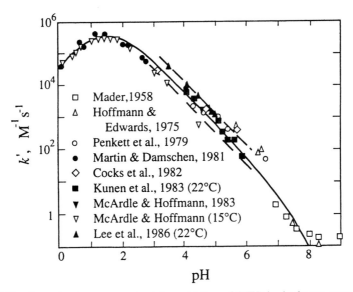

FIGURE 7.17 Second-order rate constant for oxidation of S(IV) by hydrogen peroxide defined according to $d[\mathrm{S(VI)}]/dt = k[\mathrm{H_2O_2(aq)}] \, [\mathrm{S(IV)}]$ as a function of solution pH at 298 K. The solid curve corresponds to the rate expression (7.84). Dashed lines are arbitrarily placed to encompass most of the experimental data shown.

clouds and fogs. The species with the lowest concentration before cloud or fog formation is the limiting reactant and is rapidly depleted inside the cloud or fog layer.

The reaction proceeds via a nucleophilic displacement by hydrogen peroxide on bisulfite as the principal reactive S(IV) species (McArdle and Hoffmann 1983)

$$HSO_3^- + H_2O_2 \rightleftharpoons SO_2OOH^- + H_2O \tag{7.85}$$

and then the peroxymonosulfurous acid, SO_2OOH^-, reacts with a proton to produce sulfuric acid:

$$SO_2OOH^- + H^+ \rightarrow H_2SO_4 \tag{7.86}$$

The latter reaction becomes faster as the medium becomes more acidic.

7.5.3   Oxidation of S(IV) by Organic Peroxides

Organic peroxides have also been proposed as potential aqueous-phase S(IV) oxidants (Graedel and Goldberg 1983; Lind and Lazrus 1983; Hoffmann and Calvert 1985; Lind et al. 1987).

Methylhydroxyperoxide, CH_3OOH, reacts with HSO_3^- according to

$$HSO_3^- + CH_3OOH + H^+ \rightarrow SO_4^{2-} + 2H^+ + CH_3OH \tag{7.87}$$

with a reaction rate given in the atmospheric relevant pH range 2.9–5.8 by

$$R7_{7.87} = k_{7.87}[H^+][CH_3OOH][HSO_3^-] \tag{7.88}$$

where $k_{7.87} = 1.7 \pm 0.3 \times 10^7 \, M^{-2} \, s^{-1}$ at 18°C and the activation energy is 31.6 kJ mol^{-1} (Hoffmann and Calvert 1985; Lind et al. 1987). Because of the inverse dependence of $[HSO_3^-]$ on $[H^+]$ the overall rate is pH independent (Figure 7.18).

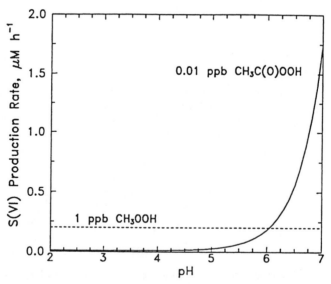

FIGURE 7.18   Reaction rates for the oxidation of S(IV) by CH_3OOH and $CH_3C(O)OOH$ as a function of pH for mixing ratios $[SO_2(g)] = 1$ ppb, $[CH_3OOH(g)] = 1$ ppb, and $[CH_3C(O)OOH(g)] = 0.01$ ppb.

The corresponding reaction involving peroxyacetic acid is

$$HSO_3^- + CH_3C(O)OOH \rightarrow SO_4^{2-} + H^+ + CH_3COOH \qquad (7.89)$$

with a rate law for pH 2.9–5.8 given by

$$R_{7.89} = (k_a + k_b[H^+])[HSO_3^-][CH_3C(O)OOH] \qquad (7.90)$$

where $k_a = 601\,M^{-1}\,s^{-1}$ and $k_b = 3.64 \pm 0.4 \times 10^7\,M^{-2}\,s^{-1}$ at 18°C (Hoffmann and Calvert 1985; Lind et al. 1987). The overall rate of the reaction increases with increasing pH (Figure 7.18).

The Henry's law constants of methylhydroperoxide and peroxyacetic acid are more than two orders of magnitude lower than that of hydrogen peroxide (Table 7.1). Typical mixing ratios are on the order of 1 ppb for CH_3OOH and 0.01 ppb for $CH_3C(O)OOH$. Applying Henry's law, the corresponding equilibrium aqueous-phase concentrations are $0.2\,\mu M$ and 5 nM, respectively. These rather low concentrations result in relatively low S(IV) oxidation rates on the order of $1\,\mu M\,h^{-1}$ (for a SO_2 mixing ratio of 1 ppb), or equivalently in rates of less than 1% $SO_2\,h^{-1}$ for a typical cloud liquid water content of $0.2\,g\,m^{-3}$ (Figure 7.18). As a result these reactions are of minor importance for S(IV) oxidation under typical atmospheric conditions and represent only small sinks for the gas-phase methyl hydroperoxide (0.2% $CH_3OOH\,h^{-1}$) and peroxyacetic acid (0.7% $CH_3C(O)OOH\,h^{-1}$) (Pandis and Seinfeld 1989a).

7.5.4 Uncatalyzed Oxidation of S(IV) by O_2

The importance of the reaction of S(IV) with dissolved oxygen in the absence of any metal catalysts (iron, manganese) has been a controversial issue. Solutions of sodium sulfite in the laboratory oxidize slowly in the presence of oxygen (Fuller and Crist 1941; Martin 1984). However, observations of Tsunogai (1971) and Huss et al. (1978) showed that the rate of the uncatalyzed reaction is negligible. The observed rates can be explained by the existence of very small amounts of catalyst such as iron (concentrations lower than $0.01\,\mu M$) that are extremely difficult to exclude. It is interesting to note that for real cloud droplets there will always be traces of catalyst present (Table 7.5), so the rate of an "uncatalyzed" reaction is irrelevant (Martin 1984).

TABLE 7.5 Manganese and Iron Concentrations in Aqueous Particles and Drops

Medium	Manganese (μM)	Iron (μM)
Aerosol (haze)	0.1–100	100–1000
Clouds	0.01–10	0.1–100
Rain	0.01–1	0.01–10
Fog	0.1–10	1–100

Source: Martin (1984).

7.5.5   Oxidation of S(IV) by O_2 Catalyzed by Iron and Manganese

Iron Catalysis   S(IV) oxidation by O_2 is known to be catalyzed by Fe(III) and Mn(II):

$$S(IV) + \tfrac{1}{2}O_2 \xrightarrow{\text{Mn}^{2+},\text{Fe}^{3+}} S(VI) \tag{7.91}$$

This reaction has been the subject of considerable interest (Hoffmann and Boyce 1983; Hoffmann and Jacob 1984; Martin 1984; Hoffmann and Calvert 1985; Clarke and Radojevic 1987), but significantly different measured reaction rates, rate laws, and pH dependences have been reported (Hoffmann and Jacob 1984). Martin and Hill (1987a,b) showed that this reaction is inhibited by increasing ionic strength, the sulfate ion, and various organics and is even self-inhibited. They explained most of the literature discrepancies by differences in these factors in various laboratory studies.

In the presence of oxygen, iron in the ferric state, Fe(III), catalyzes the oxidation of S(IV) in aqueous solutions. Iron in cloudwater exists both in the Fe(II) and Fe(III) states and there are a series of oxidation–reduction reactions cycling iron between these two forms (Stumm and Morgan 1996). Fe(II) appears not to directly catalyze the reaction and is first oxidized to Fe(III) before S(IV) oxidation can begin (Huss et al. 1982a,b). The equilibria involving Fe(III) in aqueous solution are

$$Fe^{3+} + H_2O \rightleftharpoons FeOH^{2+} + H^+$$
$$FeOH^{2+} + H_2O \rightleftharpoons Fe(OH)_2^+ + H^+$$
$$Fe(OH)_2^+ + H_2O \rightleftharpoons Fe(OH)_3(s) + H^+$$
$$2\,FeOH^{2+} \rightleftharpoons Fe_2(OH)_2^{4+}$$

with Fe^{3+}, $FeOH^{2+}$ $Fe(OH)_2^+$, and $Fe_2(OH)_2^{4+}$ soluble and $Fe(OH)_3$ insoluble. The concentration of Fe^{3+} can be calculated from the equilibrium with solid $Fe(OH)_3$ (Stumm and Morgan 1996)

$$Fe(OH)_3(s) + 3\,H^+ \rightleftharpoons Fe^{3+} + 3\,H_2O$$

as

$$[Fe^{3+}] \simeq 10^3 [H^+]^3 \quad \text{in M at 298 K}$$

and for a pH of 4.5, $[Fe^{3+}] = 3 \times 10^{-11}$ M.

For pH values from 0 to 3.6 the iron-catalyzed S(IV) oxidation rate is first order in iron, is first order in S(IV), and is inversely proportional to $[H^+]$ (Martin and Hill 1987a):

$$R_{7.91} = -\frac{d[S(IV)]}{dt} = k_{7.91} \frac{[Fe(III)][S(IV)]}{[H^+]} \tag{7.92}$$

This reaction is inhibited by ionic strength and sulfate. Accounting for these effects the reaction rate constant is given by

$$k_{7.91} = k_{7.91}^* \frac{10^{-2\sqrt{I}/(1+\sqrt{I})}}{1 + 150[S(VI)]^{2/3}} \tag{7.93}$$

where [S(VI)] is in M. A rate constant $k_{7.91}^* = 6\,s^{-1}$ was recommended by Martin and Hill (1987a). Sulfite appears to be almost equally inhibiting as sulfate. This does not pose a problem for regular atmospheric conditions ([S(IV)] < 0.001 M), but the preceding rate expressions should not be applied to laboratory studies where the S(IV) concentrations exceed 0.001 M. This reaction, according to the rate expressions presented above is very slow under typical atmospheric conditions in this pH regime (0–3.6).

The rate expression for the same reaction changes completely above pH 3.6. This suggests that the mechanism of the reaction differs in the two pH regimes and is probably a free radical chain at high pH and a nonradical mechanism at low pH (Martin et al. 1991). The low solubility of Fe(III) above pH 3.6 poses special experimental problems. At high pH the reaction rate depends on the actual amount of iron dissolved in solution, rather than on the total amount of iron present in the droplet. In this range the reaction is second-order in dissolved iron (zero-order above the solution iron saturation point) and first order in S(IV). The reaction is still not very well understood and Martin et al. (1991) proposed the following phenomenological expressions (in $M\,s^{-1}$)

$$pH\ 4.0: \quad -\frac{d[S(IV)]}{dt} = 1 \times 10^9 [S(IV)][Fe(III)]^2 \qquad (7.94)$$

$$pH\ 5.0\ to\ 6.0: \quad -\frac{d[S(IV)]}{dt} = 1 \times 10^{-3}[S(IV)] \qquad (7.95)$$

$$pH\ 7.0: \quad -\frac{d[S(IV)]}{dt} = 1 \times 10^{-4}[S(IV)] \qquad (7.96)$$

for the following conditions:

$$[S(IV)] \simeq 10\,\mu M, \quad [Fe(III)] > 0.1\,\mu M, \quad I < 0.01\,M,$$
$$[S(VI)] < 100\,\mu M, \quad and \quad T = 298\,K. \qquad (7.97)$$

Note that iron does not appear in the pH 5 to 7 rates because it is assumed that a trace of iron will be present under normal atmospheric conditions. This reaction is important in this high pH regime (Pandis and Seinfeld 1989b; Pandis et al. 1992).

Martin et al. (1991) also found that noncomplexing organic molecules (e.g., acetate, trichloroacetate, ethyl alcohol, isopropyl alcohol, formate, allyl alcohol) are highly inhibiting at pH values of 5 and above and are not inhibiting at pH values of 3 and below. They calculated that for remote clouds formate would be the main inhibiting organic, but by less than 10%. On the contrary, near urban areas formate could reduce the rate of the catalyzed oxidation by a factor of 10–20 in the high-pH regime.

Manganese Catalysis   The manganese-catalyzed S(IV) oxidation rate was initially thought to be inversely proportional to the [H$^+$] concentration. Martin and Hill (1987b) suggested that ionic strength, not hydrogen ion, accounts for the pH dependence of the rate. The manganese-catalyzed reaction obeys zero-order kinetics in S(IV) in the concentration regime above 100 μM S(IV),

$$-\frac{d[S(IV)]}{dt} = k_0[Mn]^2 \qquad (7.98)$$

$$k_0 = k_0^*\, 10^{-4.07\sqrt{I}/(1+\sqrt{I})} \qquad (7.99)$$

with $k_0^* = 680\,\mathrm{M}^{-1}\,\mathrm{s}^{-1}$ (Martin and Hill 1987b). For S(IV) concentrations below $1\,\mu\mathrm{M}$ the reaction is first order in S(IV)

$$-\frac{d[\mathrm{S(IV)}]}{dt} = k_0[\mathrm{Mn}][\mathrm{S(IV)}] \qquad (7.100)$$

$$k_0 = k_0^* \, 10^{-4.07\sqrt{I}/(1+\sqrt{I})} \qquad (7.101)$$

with $k_0^* = 1000\,\mathrm{M}^{-1}\,\mathrm{s}^{-1}$ (Martin and Hill 1987b). It is still not clear which rate law is appropriate for use in atmospheric calculations, although Martin and Hill (1987b) suggested the provisional use of the first-order, low S(IV) rate.

Iron/Manganese Synergism When both Fe^{3+} and Mn^{2+} are present in atmospheric droplets, the overall rate of the S(IV) reaction is enhanced over the sum of the two individual rates. Martin (1984) reported that the rates measured were 3 to 10 times higher than expected from the sum of the independent rates. Martin and Good (1991) obtained at pH 3.0 and for $[\mathrm{S(IV)}] < 10\,\mu\mathrm{M}$ the following rate law:

$$
\begin{aligned}
-\frac{d[\mathrm{S(IV)}]}{dt} = {}& 750[\mathrm{Mn(II)}][\mathrm{S(IV)}] + 2600[\mathrm{Fe(III)}][\mathrm{S(IV)}] \\
& + 1.0 \times 10^{10}[\mathrm{Mn(II)}][\mathrm{Fe(III)}][\mathrm{S(IV)}]
\end{aligned}
\qquad (7.102)
$$

and a similar expression for pH 5.0 in agreement with the work of Ibusuki and Takeuchi (1987).

7.5.6 Comparison of Aqueous-Phase S(IV) Oxidation Paths

We will now compare the different routes for SO_2 oxidation in aqueous solution as a function of pH and temperature. In doing so, we will set the pH at a given value and calculate the instantaneous rate of S(IV) oxidation at that pH. The rate expressions used and the parameters in the rate expressions are given in Table 7.6. Figure 7.19 shows the

TABLE 7.6   Rate Expressions for Sulfate Formation in Aqueous Solution Used in Computing Figure 7.19

Oxidant	Rate Expression, $-d[\mathrm{S(IV)}]/dt$	Reference
O_3	$(k_0[\mathrm{SO_2 \cdot H_2O}] + k_1[\mathrm{HSO_3^-}] + k_2[\mathrm{SO_3^{2-}}])[\mathrm{O_3(aq)}]$ $k_0 = 2.4 \times 10^4\,\mathrm{M}^{-1}\,\mathrm{s}^{-1}$ $k_1 = 3.7 \times 10^5\,\mathrm{M}^{-1}\,\mathrm{s}^{-1}$ $k_2 = 1.5 \times 10^9\,\mathrm{M}^{-1}\,\mathrm{s}^{-1}$	Hoffmann and Calvert (1985)
H_2O_2	$k_4[\mathrm{H^+}][\mathrm{HSO_3^-}][\mathrm{H_2O_2(aq)}]/(1 + K[\mathrm{H^+}])$ $k_4 = 7.45 \times 10^7\,\mathrm{M}^{-1}\,\mathrm{s}^{-1}$ $K = 13\,\mathrm{M}^{-1}$	Hoffmann and Calvert (1985)
Fe(III)	$k_5[\mathrm{Fe(III)}][\mathrm{SO_3^{2-}}]^a$ $k_5 = 1.2 \times 10^6\,\mathrm{M}^{-1}\,\mathrm{s}^{-1}$ for pH ≤ 5	Hoffmann and Calvert (1985)
Mn(II)	$k_6[\mathrm{Mn(II)}][\mathrm{S(IV)}]$ $k_6 = 1000\,\mathrm{M}^{-1}\,\mathrm{s}^{-1}$ (for low S(IV))	Martin and Hill (1987b)
NO_2	$k_7[\mathrm{NO_2(aq)}][\mathrm{S(IV)}]$ $k_7 = 2 \times 10^6\,\mathrm{M}^{-1}\,\mathrm{s}^{-1}$	Lee and Schwartz (1983)

[a]This is an alternative reaction rate expression for the low-pH region. Compare with (7.92) and (7.95).

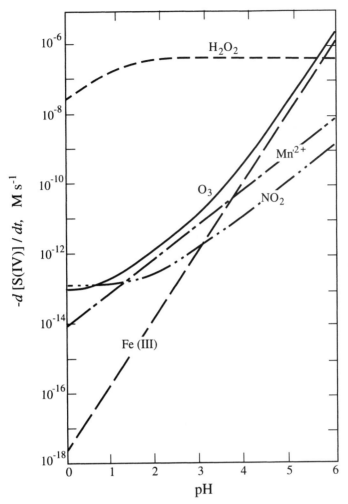

FIGURE 7.19 Comparison of aqueous-phase oxidation paths. The rate of conversion of S(IV) to S(VI) as a function of pH. Conditions assumed are $[SO_2(g)] = 5$ ppb; $[NO_2(g)] = 1$ ppb; $[H_2O_2(g)] = 1$ ppb; $[O_3(g)] = 50$ ppb; $[Fe(III)] = 0.3\,\mu M$; $[Mn(II)] = 0.03\,\mu M$.

oxidation rates in $\mu M\,h^{-1}$ for the different paths at 298 K for the conditions

$$[SO_2(g)] = 5\,\text{ppb} \qquad [H_2O_2(g)] = 1\,\text{ppb}$$
$$[NO_2(g)] = 1\,\text{ppb} \qquad [O_3(g)] = 50\,\text{ppb}$$
$$[Fe(III)(aq)] = 0.3\,\mu M \qquad [Mn(II)(aq)] = 0.03\,\mu M$$

We see that under these conditions oxidation by dissolved H_2O_2 is the predominant pathway for sulfate formation at pH values less than roughly 4–5. At pH ≥ 5 oxidation by O_3 starts dominating and at pH 6 it is 10 times faster than that by H_2O_2. Also, oxidation of S(IV) by O_2 catalyzed by Fe and Mn may be important at high pH, but uncertainties in the rate expressions at high pH preclude a definite conclusion. Oxidation of S(IV) by NO_2 is unimportant at all pH for the concentration levels above.

The oxidation rate of S(IV) by OH cannot be calculated using the simple approach outlined above. Since the overall rate depends on the propagation and termination rates of the radical chain, it depends, in addition to the S(IV) and OH concentrations, on those of HO_2, HCOOH, HCHO, and so on, and its determination requires a dynamic chemical model.

The inhibition of most oxidation mechanisms at low pH results mainly from the lower overall solubility of SO_2 with increasing acidity. H_2O_2 is the only identified oxidant for which the rate is virtually independent of pH.

The effect of temperature on oxidation rates is a result of two competing factors. First, at lower temperatures, higher concentrations of gases are dissolved in equilibrium (see Figure 7.7 for SO_2), which lead to higher reaction rates. On the other hand, rate constants in the rate expressions generally decrease as temperature decreases. The two effects therefore act in opposite directions. Except for Fe- and Mn-catalyzed oxidation, the increased solubility effect dominates and the rate increases with decreasing temperature. In the transition-metal-catalyzed reaction, the consequence of the large activation energy is that as temperature decreases the overall rate of sulfate formation for a given SO_2 concentration decreases.

As we have noted, it is often useful to express aqueous-phase oxidation rates in terms of a fractional rate of conversion of SO_2. Assuming cloud conditions with $1\,g\,m^{-3}$ of liquid water content, we find that the rate of oxidation by H_2O_2 can exceed $500\%\,h^{-1}$ (Figure 7.16).

7.6   DYNAMIC BEHAVIOR OF SOLUTIONS WITH AQUEOUS-PHASE CHEMICAL REACTIONS

To compare the rates of various aqueous-phase chemical reactions we have been calculating instantaneous rates of conversion as a function of solution pH. In the atmosphere a droplet is formed, usually by nucleation around a particle, and is subsequently exposed to an environment containing reactive gases. Gases then dissolve in the droplet, establishing an initial pH and composition. Aqueous-phase reactions ensue and the pH and composition of the droplet start changing accordingly. In this section we calculate that time evolution. We neglect any changes in droplet size that might result; for dilute solutions such as those considered here, this is a good assumption. We focus on sulfur chemistry because of its atmospheric importance.

In our calculations up to this point we have simply assumed values for the gas-phase partial pressures. When the process occurs over time, one must consider also what is occurring in the gas phase. Two assumptions can be made concerning the gas phase:

1. *Open System.* Gas-phase partial pressures are maintained at constant values, presumably by continous infusion of new air. This is a convenient assumption for order of magnitude calculations, but it is often not realistic. Up to now we have implicitly treated the cloud environment as an open system. Note that mass balances (e.g., for sulfur) are not satisfied in such a system, because there is continuous addition of material.

2. *Closed System.* Gas-phase partial pressures decrease with time as material is depleted from the gas phase. One needs to describe mathematically both the gas- and aqueous-phase concentrations of the species in this case. Calculations are more

involved, but the resulting scenarios are more representative of actual atmospheric behavior. The masses of the various elements (sulfur, nitrogen, etc.) are conserved in a closed system.

7.6.1 Closed System

The basic assumption in the closed system is that the total quantity of each species is fixed. Consider, as an example, the cloud formation (liquid water mixing ratio w_L) in an air parcel that has initially a H_2O_2 partial pressure $p_{H_2O_2}^0$ and assume that no reactions take place. If we treat the system as open, then at equilibrium the aqueous-phase concentration of H_2O_2 will be given by

$$[H_2O_2(aq)]_{open} = H_{H_2O_2} p_{H_2O_2}^0 \qquad (7.103)$$

Note that for an open system, the aqueous phase concentration of H_2O_2 is independent of the cloud liquid water content.

For a closed system, the total concentration of H_2O_2 per liter (physical volume) of air, $[H_2O_2]_{total}$, will be equal to the initial amount of H_2O_2 or

$$[H_2O_2]_{total} = \frac{p_{H_2O_2}^0}{RT} \qquad (7.104)$$

After the cloud is formed, this total H_2O_2 is distributed between gas and aqueous phases and satisfies the mass balance

$$[H_2O_2]_{total} = \frac{p_{H_2O_2}}{RT} + [H_2O_2(aq)]_{closed} \, w_L \qquad (7.105)$$

where $p_{H_2O_2}$ is the vapor pressure of H_2O_2 after the dissolution of H_2O_2 in cloudwater. If, in addition, Henry's law is assumed to hold, then

$$[H_2O_2(aq)]_{closed} = H_{H_2O_2} p_{H_2O_2} \qquad (7.106)$$

Combining the last three equations,

$$[H_2O_2(aq)]_{closed} = \frac{H_{H_2O_2} p_{H_2O_2}^0}{1 + H_{H_2O_2} w_L RT} \qquad (7.107)$$

The aqueous-phase concentration of H_2O_2, assuming a closed system, decreases as the cloud liquid water content increases, reflecting the increase in the amount of liquid water content available to accommodate H_2O_2. Figure 7.20 shows $[H_2O_2(aq)]$ as a function of the liquid water content $L = 10^{-6} w_L$, for initial H_2O_2 mixing ratios of 0.5, 1, and 2 ppb. For soluble species, such as H_2O_2, the open and closed system assumptions lead to significantly different concentration estimates, with the open system resulting in the higher concentration.

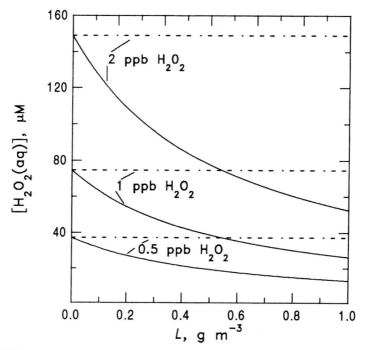

FIGURE 7.20 Aqueous-phase H_2O_2 as a function of liquid water content for 0.5, 1, and 2 ppb H_2O_2. Dashed lines correspond to an open system and solid lines to a closed one.

The fraction of the total quantity of H_2O_2 that resides in the liquid phase is given by (7.9) and is shown in Figure 7.12. A species like H_2O_2 with a large Henry's law coefficient $(1 \times 10^5 \text{ M atm}^{-1})$ will have a significant fraction of the fixed total quantity in the aqueous phase. As w_L increases, more liquid is available to accommodate the gas and the aqueous fraction increases, while its aqueous-phase concentration, as we saw above, decreases.

For species with low solubility, such as ozone, $1 + H w_L R T \simeq 1$, and the two approaches result in essentially identical concentration estimates (recall that $w_L \leq 10^{-6}$ for most atmospheric clouds).

Aqueous-Phase Concentration of Nitrate inside a Cloud (Closed System) A cloud with liquid water content w_L forms in an air parcel with HNO_3 partial pressure equal to $p^0_{HNO_3}$. Assuming that the air parcel behaves as a closed system, calculate the aqueous-phase concentration of $[NO_3^-]$.

Using the ideal-gas law, we obtain

$$[HNO_3]_{total} = \frac{p^0_{HNO_3}}{RT} \tag{7.108}$$

The available HNO_3 will be distributed between gas and aqueous phases

$$[HNO_3]_{total} = \frac{p_{HNO_3}}{RT} + ([HNO_3(aq)] + [NO_3^-])w_L \tag{7.109}$$

where Henry's law gives

$$[HNO_3(aq)] = H_{HNO_3}p_{HNO_3} \tag{7.110}$$

and dissociation equilibrium is

$$K_{n1} = \frac{[NO_3^-][H^+]}{[HNO_3(aq)]} \tag{7.111}$$

Solving these equations simultaneously yields

$$[HNO_3(aq)] = \frac{H_{HNO_3}}{1 + w_L H_{HNO_3}^* RT} p_{HNO_3}^0 \tag{7.112}$$

$$[NO_3^-] = \frac{K_{n1}}{[H^+]} \frac{H_{HNO_3}}{1 + w_L H_{HNO_3}^* RT} p_{HNO_3}^0 \tag{7.113}$$

The last equation can be simplified by using our insights from Section 7.3.5, that is: $w_L H_{HNO_3}^* RT \gg 1$ for typical clouds and that $H_{HNO_3} \approx H_{HNO_3} K_{n1}/[H^+]$. Using these two simplifications, we find that $[NO_3^-] = p_{HNO_3}^0/(w_L RT))$. This is the expected result for a highly water soluble strong electrolyte. All the nitric acid will be dissolved in the cloud water and all of it will be present in its dissociated form, NO_3^-.

7.6.2   Calculation of Concentration Changes in a Droplet with Aqueous-Phase Reactions

Let us consider a droplet that at $t = 0$ is immersed in air containing SO_2, NH_3, H_2O_2, O_3, and HNO_3. Equilibrium is immediately established between the gas and aqueous phases. As the aqueous-phase oxidation of S(IV) to S(VI) proceeds, the concentrations of all the ions adjust so as to satisfy electroneutrality at all times

$$\begin{aligned}[H^+] + [NH_4^+] = {}&[OH^-] + [HSO_3^-] + 2[SO_3^{2-}] \\ &+ 2[SO_4^{2-}] + [HSO_4^-] + [NO_3^-]\end{aligned} \tag{7.114}$$

where the weak dissociation of H_2O_2 has been neglected. The concentrations of each ion except sulfate and bisulfate can be expressed in terms of $[H^+]$ using the equilibrium constant expressions:

$$[NH_4^+] = \frac{H_{NH_3}K_{a1}}{K_w} p_{NH_3}[H^+] \tag{7.115}$$

$$[OH^-] = \frac{K_w}{[H^+]} \tag{7.116}$$

$$[HSO_3^-] = \frac{H_{SO_2}K_{s1}}{[H^+]} p_{SO_2} \tag{7.117}$$

$$[SO_3^{2-}] = \frac{H_{SO_2}K_{s1}K_{s2}}{[H^+]^2} p_{SO_2} \tag{7.118}$$

$$[NO_3^-] = \frac{H_{HNO_3}K_{n1}}{[H^+]} p_{HNO_3} \tag{7.119}$$

If we define

$$[S(VI)] = [SO_4^{2-}] + [HSO_4^-] + [H_2SO_4(aq)] \tag{7.120}$$

then

$$[HSO_4^-] = \frac{K_{7.123}[H^+][S(VI)]}{[H^+]^2 + K_{7.123}[H^+] + K_{7.123}K_{7.124}} \tag{7.121}$$

$$[SO_4^{2-}] = \frac{K_{7.123}K_{7.124}[S(VI)]}{[H^+]^2 + K_{7.123}[H^+] + K_{7.123}K_{7.124}} \tag{7.122}$$

where $K_{7.123}$ and $K_{7.124}$ are the equilibrium constants for the reactions

$$H_2SO_4(aq) \rightleftharpoons H^+ + HSO_4^- \tag{7.123}$$

$$HSO_4^- \rightleftharpoons H^+ + SO_4^{2-} \tag{7.124}$$

respectively. For all practical purposes $K_{7.123}$ may be considered to be infinite since virtually no undissociated sulfuric acid will exist in solution, and $K_{7.124} = 1.2 \times 10^{-2}$ M at 298 K. Therefore the equations can be simplified to

$$[HSO_4^-] = \frac{[H^+][S(VI)]}{[H^+] + K_{7.124}} \tag{7.125}$$

$$[SO_4^{2-}] = \frac{K_{7.124}[S(VI)]}{[H^+] + K_{7.124}} \tag{7.126}$$

The concentration changes are computed as follows assuming that all compounds in the system remain always in thermodynamic equilibrium. At $t = 0$, assuming that there is an initial sulfate concentration $[S(VI)] = [S(VI)]_0$, the electroneutrality equation is solved to determine $[H^+]$ and, thereby, the concentrations of all dissolved species. If an open system is assumed, then the gas-phase partial pressures will remain constant with time. For a closed system, the partial pressures change with time as outlined above. The concentration changes are computed over small time increments of length Δt. The sulfate present at any time t is equal to that of time $t - \Delta t$ plus that formed in the interval Δt:

$$[S(VI)]_t = [S(VI)]_{t-\Delta t} + \left(\frac{d[S(VI)]}{dt}\right)_{t-\Delta t} \Delta t \tag{7.127}$$

The sulfate formation rate $d[S(VI)]/dt$ is simply the sum of the reaction rates that produce sulfate in this system, namely, that of the H_2O_2–$S(IV)$ and O_3–$S(IV)$ reactions. The value of $S(VI)$ at time t is then used to calculate $[HSO_4^-]$ and $[SO_4^{2-}]$ at time t. These concentrations are then substituted into the electroneutrality equation to obtain the new $[H^+]$ and the concentrations of other dissolved species. This process is then just repeated over the total time of interest.

Let us compare the evolution of open and closed systems for an identical set of starting conditions. We choose the following conditions at $t = 0$:

$$[S(IV)]_{total} = 5\,ppb, \quad [HNO_3]_{total} = 1\,ppb, \quad [NH_3]_{total} = 5\,ppb$$
$$[O_3]_{total} = 5\,ppb, \quad [H_2O_2]_{total} = 1\,ppb, \quad w_L = 10^{-6}$$
$$[S(VI)]_0 = 0$$

For the closed system, there is no replenishment, whereas in the open system the partial pressures of all species in the gas phase are maintained constant at their initial values and aqueous-phase concentrations are determined by equilibrium.

Solving the equilibrium problem at $t = 0$ we find that the initial pH $= 6.17$, and that the initial gas-phase mixing ratios are

$$\xi_{SO_2} = 3.03\,\text{ppb}, \quad \xi_{NH_3} = 1.87\,\text{ppb}, \quad \xi_{HNO_3} = 8.54 \times 10^{-9}\,\text{ppb}$$
$$\xi_{O_3} = 5\,\text{ppb}, \quad \xi_{H_2O_2} = 0.465\,\text{ppb}$$

We will use the initial conditions presented above for both the open and closed systems. In the open system these partial pressures remain constant throughout the simulation, whereas in the closed system they change. A comment concerning the choice of O_3 concentration is in order. We selected the uncharacteristically low value of 5 ppb so as to be able to show the interplay possible between both the H_2O_2 and O_3 oxidation rates and in the closed system, the role of depletion as the reaction proceeds. Figures 7.21 and 7.22 show the sulfate concentration and pH, respectively, in the open and closed systems over a 60-min period.

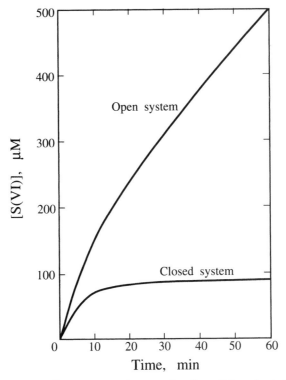

FIGURE 7.21 Aqueous sulfate concentration as a function of time for both open and closed systems. The conditions for the simulation are $[S(IV)]_{total} = 5$ ppb; $[NH_3]_{total} = 5$ ppb; $[HNO_3]_{total} = 1$ ppb; $[O_3]_{total} = 5$ ppb; $[H_2O_2]_{total} = 1$ ppb; $w_L = 10^{-6}$; $pH_0 = 6.17$.

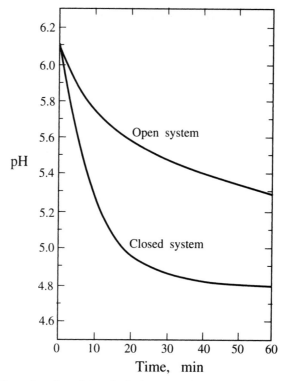

FIGURE 7.22   pH as a function of time for both open and closed systems. Same conditions as in Figure 7.21.

More sulfate is produced in the open system than in the closed system, but the pH decrease is less in the open system. This behavior is a result of several factors:

1. Sulfate production due to H_2O_2 and O_3 is less in the closed system because of depletion of H_2O_2 and O_3.
2. Continued replenishment of NH_3 provides more neutralization in the open system.
3. The low pH in the closed system drives S(IV) from solution.
4. The lower pH in the closed system depresses the rate of sulfate formation by O_3.
5. S(IV) is continually depleted in the closed system.
6. Total HNO_3 decreases in the open system due to the decrease in pH.

In the open system the importance of H_2O_2 in sulfate formation grows significantly, from 20% initially to over 80%, whereas in the closed system the increasing importance of H_2O_2 is enhanced at lower pH but suppressed by the depletion of H_2O_2. In the closed system, after about 30 min, the H_2O_2 has been depleted, the pH has decreased to 4.8 slowing down the ozone reaction, and sulfate formation ceases. The fractions of SO_2, O_3, and H_2O_2 reacted in the closed system at 60 min are 0.43, 0.23, and 0.997, respectively. The ozone reaction produces 44.4% and 53.9% of the sulfate in the open and closed systems, respectively.

APPENDIX 7.1   THERMODYNAMIC AND KINETIC DATA

Refer to Tables 7.A.1–7.A.7.

TABLE 7.A.1   Equilibrium Reactions

Equilibrium Reaction	K_{298} (M or M atm^{-1})^a	$-\Delta H/R$ (K)	Reference
$SO_2 \cdot H_2O \rightleftharpoons HSO_3^- + H^+$	1.3×10^{-2}	1960	Smith and Martell (1976)
$HSO_3^- \rightleftharpoons SO_3^{2-} + H^+$	6.6×10^{-8}	1500	Smith and Martell (1976)
$H_2SO_4(aq) \rightleftharpoons HSO_4^- + H^+$	1000		Perrin (1982)
$HSO_4^- \rightleftharpoons SO_4^{2-} + H^+$	1.02×10^{-2}	2720	Smith and Martell (1976)
$H_2O_2(aq) \rightleftharpoons HO_2^- + H^+$	2.2×10^{-12}	−3730	Smith and Martell (1976)
$HNO_3(aq) \rightleftharpoons NO_3^- + H^+$	15.4	8700^b	Schwartz (1984)
$HNO_2(aq) \rightleftharpoons NO_2^- + H^+$	5.1×10^{-4}	−1260	Schwartz and White (1981)
$CO_2 \cdot H_2O \rightleftharpoons HCO_3^- + H^+$	4.3×10^{-7}	−1000	Smith and Martell (1976)
$HCO_3^- \rightleftharpoons CO_3^{2-} + H^+$	4.68×10^{-11}	−1760	Smith and Martell (1976)
$NH_4OH \rightleftharpoons NH_4^+ + OH^-$	1.7×10^{-5}	−450	Smith and Martell (1976)
$H_2O \rightleftharpoons H^+ + OH^-$	1.0×10^{-14}	−6710	Smith and Martell (1976)
$HCHO(aq) \overset{H_2O}{\rightleftharpoons} H_2C(OH)_2(aq)$	2.53×10^3	4020	Le Hanaf (1968)
$HCOOH(aq) \rightleftharpoons HCOO^- + H^+$	1.8×10^{-4}	−20	Martell and Smith (1977)
$HCl(aq) \rightleftharpoons H^+ + Cl^-$	1.74×10^6	6900	Marsh and McElroy (1985)
$Cl_2^- \rightleftharpoons Cl + Cl^-$	5.26×10^{-6}		Jayson et al. (1973)
$NO_3(g) \rightleftharpoons NO_3(aq)$	2.1×10^5	8700	Jacob (1986)
$HO_2(aq) \rightleftharpoons H^+ + O_2^-$	3.50×10^{-5}		Perrin (1982)
$HOCH_2SO_3^- \rightleftharpoons {}^-OCH_2SO_3^- + H^+$	2.00×10^{-12}		Sorensen and Andersen (1970)

^{a}The temperature dependence is represented by

$$K = K_{298} \exp\left[-\frac{\Delta H}{R}\left(\frac{1}{T} - \frac{1}{298}\right)\right]$$

where K is the equilibrium constant at temperature T (in K).
^{b}Value for equilibrium: $HNO_3(g) \rightleftharpoons NO_3^- + H^+$.

TABLE 7.A.2 Oxygen-Hydrogen Chemistry

Reaction	$k_{298}{}^a$	$-E/R$ (K)	Reference
1. $H_2O_2 \xrightarrow{h\nu} 2\,OH$			Graedel and Weschler (1981)
2. $O_3 \xrightarrow{h\nu,H_2O} H_2O_2 + O_2$			Graedel and Weschler (1981)
3. $OH + HO_2 \longrightarrow H_2O + O_2$	7.0×10^9	-1500	Sehested et al. (1968)
4. $OH + O_2^- \longrightarrow OH^- + O_2$	1.0×10^{10}	-1500	Sehested et al. (1968)
5. $OH + H_2O_2 \longrightarrow H_2O + HO_2$	2.7×10^7	-1700	Christensen et al. (1982)
6. $HO_2 + HO_2 \longrightarrow H_2O_2 + O_2$	8.6×10^5	-2365	Bielski (1978)
7. $HO_2 + O_2^- \xrightarrow{H_2O} H_2O_2 + O_2 + OH^-$	1.0×10^8	-1500	Bielski (1978)
8. $O_2^- + O_2^- \xrightarrow{2H_2O} H_2O_2 + O_2 + 2OH^-$	<0.3		Bielski (1978)
9. $HO_2 + H_2O_2 \longrightarrow OH + O_2 + H_2O$	0.5		Weinstein and Bielski (1979)
10. $O_2^- + H_2O_2 \longrightarrow OH + O_2 + OH^-$	0.13		Weinstein and Bielski (1979)
11. $OH + O_3 \longrightarrow HO_2 + O_2$	2×10^9		Staehelin et al. (1984)
12. $HO_2 + O_3 \longrightarrow OH + 2O_2$	$<1 \times 10^4$		Sehested et al. (1984)
13. $O_2^- + O_3 \xrightarrow{H_2O} OH + 2O_2 + OH^-$	1.5×10^9	-1500	Sehested et al. (1983)
14. $OH^- + O_3 \xrightarrow{H_2O} H_2O_2 + O_2 + OH^-$	70		Staehelin and Hoigné (1982)
15. $HO_2^- + O_3 \longrightarrow OH + O_2^- + O_2$	2.8×10^6	-2500	Staehelin and Hoigné (1982)
16. $H_2O_2 + O_3 \longrightarrow H_2O + 2O_2$	$7.8 \times 10^{-3}[O_3]^{-0.5}$		Martin et al. (1981)

^{a}In appropriate units of M and s^{-1}.

TABLE 7.A.3 Carbonate Chemistry

Reaction	k_{298}	$-E/R$ (K)	Reference
17. $HCO_3^- + OH \longrightarrow H_2O + CO_3^-$	1.5×10^7	-1910	Weeks and Rabani (1966)
18. $HCO_3^- + O_2^- \longrightarrow HO_2^- + CO_3^-$	1.5×10^6	0	Schmidt (1972)
19. $CO_3^- + O_2^- \xrightarrow{H_2O} HCO_3^- + O_2 + OH^-$	4.0×10^8	-1500	Behar et al. (1970)
20. $CO_3^- + H_2O_2 \longrightarrow HO_2 + HCO_3^-$	8.0×10^5	-2820	Behar et al. (1970)

TABLE 7.A.4 Chlorine Chemistry

Reaction	k_{298}	$-E/R$ (K)	Reference
21. $Cl^- + OH \longrightarrow ClOH^-$	4.3×10^9	-1500	Jayson et al. (1973)
22. $ClOH^- \longrightarrow Cl^- + OH$	6.1×10^9	0	Jayson et al. (1973)
23. $ClOH^- \xrightarrow{H^+} Cl + H_2O$	$2.1 \times 10^{10}[H^+]$	0	Jayson et al. (1973)
24. $Cl \xrightarrow{H_2O} ClOH^- + H^+$	1.3×10^3	0	Jayson et al. (1973)
25. $HO_2 + Cl_2^- \longrightarrow 2\,Cl^- + O_2 + H^+$	4.5×10^9	-1500	Ross and Neta (1979)
26. $O_2^- + Cl_2^- \longrightarrow 2\,Cl^- + O_2$	1.0×10^9	-1500	Ross and Neta (1979)
27. $HO_2 + Cl \longrightarrow Cl^- + O_2 + H^+$	3.1×10^9	-1500	Graedel and Goldberg (1983)
28. $H_2O_2 + Cl_2^- \longrightarrow 2\,Cl^- + HO_2 + H^+$	1.4×10^5	-3370	Hagesawa and Neta (1978)
29. $H_2O_2 + Cl \longrightarrow Cl^- + HO_2 + H^+$	4.5×10^7	0	Graedel and Goldberg (1983)
30. $OH^- + Cl_2^- \longrightarrow 2Cl^- + OH$	7.3×10^6	-2160	Hagesawa and Neta (1978)

TABLE 7.A.5   Nitrite and Nitrate Chemistry

Reaction	k_{298}	$-E/R$ (K)	Reference
31. $NO + NO_2 \xrightarrow{H_2O} 2NO_2^- + 2H^+$	2.0×10^8	-1500	Lee (1984a)
32. $NO_2 + NO_2 \xrightarrow{H_2O}$ $NO_2^- + NO_3^- + 2H^+$	1.0×10^8	-1500	Lee (1984a)
33. $NO + OH \longrightarrow OH_2^- + H^+$	2.0×10^{10}	-1500	Strehlow and Wagner (1982)
34. $NO_2 + OH \longrightarrow NO_3^- + H^+$	1.3×10^9	-1500	Gratzel et al. (1970)
35. $HNO_2 \xrightarrow{h\nu} NO + OH$			Rettich (1978)
36. $NO_2^- \xrightarrow{h\nu, H_2O} NO + OH + OH^-$			Graedel and Weschler (1981)
37. $HNO_2 + OH \longrightarrow NO_2 + H_2O$	1.0×10^9	-1500	Rettich (1978)
38. $NO_2^- + OH \longrightarrow NO_2 + OH^-$	1.0×10^{10}	-1500	Treinin and Hayon (1970)
39. $HNO_2 + H_2O_2 \xrightarrow{H^+}$ $NO_3^- + 2H^+ + H_2O$	$6.3 \times 10^3 [H^+]$	-6693	Lee and Lind (1986)
40. $NO_2^- + O_3 \longrightarrow NO_3^- + O_2$	5.0×10^5	-6950	Damschen and Martin (1983)
41. $NO_2^- + CO_3^- \longrightarrow NO_2 + CO_3^{2-}$	4.0×10^5	0	Lilie et al. (1978)
42. $NO_2^- + Cl_2^- \longrightarrow NO_2 + 2Cl^-$	2.5×10^8	-1500	Hagesawa and Neta (1978)
43. $NO_2^- + NO_3 \longrightarrow NO_2 + NO_3^-$	1.2×10^9	-1500	Ross and Neta (1979)
44. $NO_3^- \xrightarrow{h\nu, H_2O} NO_2 + OH + OH^-$			Graedel and Weschler (1981)
45. $NO_3 \xrightarrow{h\nu} NO + O_2$			Graedel and Weschler (1981)
46. $NO_3 + HO_2 \longrightarrow NO_3^- + H^+ + O_2$	4.5×10^9	-1500	Jacob (1986)
47. $NO_3 + O_2^- \longrightarrow NO_3^- + O_2$	1.0×10^9	-1500	Jacob (1986)
48. $NO_3 + H_2O_2 \longrightarrow NO_3^- + H^+ + HO_2$	1.0×10^6	-2800	Chameides (1984)
49. $NO_3 + Cl^- \longrightarrow NO_3^- + Cl$	1.0×10^8	-1500	Ross and Neta (1979)

TABLE 7.A.6   Organic Chemistry

Reaction	k_{298}	$-E/R$ (K)	Reference
50. $H_2C(OH)_2 + OH \xrightarrow{O_2}$ $HCOOH + HO_2 + H_2O$	2.0×10^9	-1500	Bothe and Schulte-Frohlinde (1980)
51. $H_2C(OH)_2 + O_3 \longrightarrow Products$	0.1	0	Hoigné and Bader (1983a)
52. $HCOOH + OH \xrightarrow{O_2} CO_2 + HO_2 + H_2O$	2.0×10^8	-1500	Scholes and Willson (1967)
53. $HCOOH + H_2O_2 \longrightarrow Product + H_2O$	4.6×10^{-6}	-5180	Shapilov and Kostyukovskii (1974)
54. $HCOOH + NO_3 \xrightarrow{O_2}$ $NO_3^- + H^+ + CO_2 + HO_2$	2.1×10^5	-3200	Dogliotti and Hayon (1967)
55. $HCOOH + O_3 \longrightarrow CO_2 + HO_2 + OH$	5.0	0	Hoigné and Bader (1983b)

(Continued)

TABLE 7.A.6   (*Continued*)

Reaction	k_{298}	$-E/R$ (K)	Reference
56. $HCOOH + Cl_2^- \xrightarrow{O_2,}$ $\qquad CO_2 + HO_2 + 2\,Cl^- + H^+$	6.7×10^3	-4300	Hagesawa and Neta (1978)
57. $HCOO^- + OH \xrightarrow{O_2,} CO_2 + HO_2 + OH^-$	2.5×10^9	-1500	Anbar and Neta (1967)
58. $HCOO^- + O_3 \longrightarrow CO_2 + OH + O_2^-$	100.0	0	Hoigné and Bader (1983b)
59. $HCOO^- + NO_3 \xrightarrow{O_2,} NO_3^- + CO_2 + HO_2$	6.0×10^7	-1500	Jacob (1986)
60. $HCOO^- + CO_3^- \xrightarrow{O_2,H_2O}$ $\qquad CO_2 + HCO_3^- + HO_2 + OH^-$	1.1×10^5	-3400	Chen et al. (1973)
61. $HCOO^- + Cl_2^- \xrightarrow{O_2} CO_2 + HO_2 + 2\,Cl^-$	1.9×10^6	-2600	Hagasawa and Neta (1978)
62. $CH_3C(O)O_2NO_2 \longrightarrow NO_3^- + \text{Products}$	4.0×10^{-4}	0	Lee (1984b)
63. $CH_3O_2 + HO_2 \longrightarrow CH_3OOH + O_2$	4.3×10^5	-3000	Jacob (1986)
64. $CH_3O_2 + O_2^- \xrightarrow{H_2O} CH_3OOH + O_2 + OH^-$	5.0×10^7	-1600	Jacob (1986)
65. $CH_3OOH + h\nu \xrightarrow{O_2,} HCHO + OH + HO_2$			Graedel and Wechsler(1981)
66. $CH_3OOH + OH \longrightarrow CH_3O_2 + H_2O$	2.7×10^7	-1700	Jacob (1986)
67. $CH_3OH + OH \longrightarrow HCHO + HO_2 + H_2O$	4.5×10^8	-1500	Anbar and Neta (1967)
68. $CH_3OH + CO_3^- \xrightarrow{O_2,} HCHO + HO_2 + HCO_3^-$	2.6×10^3	-4500	Chen et al. (1973)
69. $CH_3OH + Cl_2^- \xrightarrow{O_2,}$ $\qquad HCHO + HO_2 + H^+ + 2\,Cl^-$	3.5×10^3	-4400	Hagesawa and Neta (1978)
70. $CH_3OOH + OH \longrightarrow HCHO + OH + H_2O$	1.9×10^7	-1800	Jacob (1986)
71. $CH_3OH + NO_3 \xrightarrow{O_2,}$ $\qquad NO_3^- + H^+ + HCHO + HO_2$	1.0×10^6	-2800	Dogliotti and Hayon (1967)

APPENDIX 7.2   ADDITIONAL AQUEOUS-PHASE SULFUR CHEMISTRY

7.A.1   S(IV) Oxidation by the OH Radical

Free radicals, such as OH and HO_2, can be scavenged heterogeneously from the gas phase by cloud droplets or produced in the aqueous phase. More than 30 aqueous-phase reactions involving OH and HO_2 have been proposed (Graedel and Weschler 1981; Chameides and Davis 1982; Graedel and Goldberg 1983; Schwartz 1984; Jacob 1986; Pandis and Seinfeld 1989a).

Chameides and Davis (1982) first proposed that the reaction of aqueous hydroxyl radicals with HSO_3^- and SO_3^{2-} may represent a significant pathway for the conversion of S(IV) to sulfate in cloudwater. The reaction chain is initiated by the attack of OH on HSO_3^- and SO_3^{2-} to form the persulfite radical anion, SO_3^- (Huie and Neta 1987):

$$HSO_3^- + OH \rightarrow SO_3^- + H_2O \qquad (7.A.1)$$

$$SO_3^{2-} + OH \rightarrow SO_3^- + OH^- \qquad (7.A.2)$$

These reactions are elementary and have rate constants, at 298 K, $k_{7.A.1} = 4.5 \times 10^9\,M^{-1}\,s^{-1}$ and $k_{7.A.2} = 5.2 \times 10^9\,M^{-1}\,s^{-1}$, respectively (Huie and Neta 1987). The existence of SO_3^- has been well established (Hayon et al. 1972; Buxton et al. 1996).

TABLE 7.A.7 Sulfur Chemistry

Reaction	k_{298}	$-E/R$ (K)	Reference
72.[a] $S(IV) + O_3 \longrightarrow S(VI) + O_2$	2.4×10^4	-5530	Hoffmann and Calvert (1985)
	3.7×10^5	-5280	
	1.5×10^9		
73.[a] $S(IV) + H_2O_2 \longrightarrow S(VI) + H_2O$	7.5×10^7	-4430	McArdle and Hoffman (1983)
74.[a] $S(IV) + \frac{1}{2}O_2 \xrightarrow{Mn^{2+},Fe^{3+}} S(IV)$	See text		Martin et al. (1991)
75. $SO_3^- + OH \xrightarrow{O_2} SO_5^- + OH^-$	5.2×10^9	-1500	Huie and Neta (1987)
76. $HSO_3^- + OH \xrightarrow{O_2} SO_5^- + H_2O$	4.5×10^9	-1500	Huie and Neta (1987)
77. $SO_5^- + HSO_3^- \xrightarrow{O_2} HSO_5^- + SO_5^-$	2.5×10^4	-3100	Huie and Neta (1987)
$SO_5^- + SO_3^{2-} \xrightarrow{O_2,H_2O} HSO_5^- + SO_5^- + OH^-$	2.5×10^4	-2000	Huie and Neta (1987)
78. $SO_5^- + O_2^- \xrightarrow{H_2O} HSO_5^- + OH^- + O_2$	1.0×10^8	-1500	Jacob (1986)
79. $SO_5^- + HCOOH \xrightarrow{O_2} HSO_5^- + CO_2 + HO_2$	200	-5300	Jacob (1986)
80. $SO_5^- + HCOO^- \xrightarrow{O_2} HSO_5^- + CO_2 + O_2^-$	1.4×10^4	-4000	Jacob (1986)
81. $SO_5^- + SO_5^- \longrightarrow 2\,SO_4^- + O_2$	6.0×10^8	-1500	Huie and Neta (1987)
82. $HSO_5^- + HSO_3^- + H^+ \longrightarrow 2\,SO_4^{2-} + 3H^+$	7.1×10^6	-3100	Betterton and Hoffmann (1988)
83. $HSO_5^- + OH \longrightarrow SO_5^- + H_2O$	1.7×10^7	-1900	Jacob (1986)
84. $HSO_5^- + SO_4^- \longrightarrow SO_5^- + SO_4^{2-} + H^+$	$<1.0 \times 10^5$	0	Jacob (1986)
85. $HSO_5^- + NO_2^- \longrightarrow HSO_4^- + NO_3^-$	0.31	-6650	Jacob (1986)
86. $HSO_5^- + Cl^- \longrightarrow SO_4^{2-} + Product$	1.8×10^{-3}	-7050	Jacob (1986)
87. $SO_4^- + HSO_3^- \xrightarrow{O_2} SO_4^{2-} + H^+ + SO_5^-$	1.3×10^9	-1500	Jacob (1986)
88. $SO_4^- + SO_3^{2-} \xrightarrow{O_2} SO_4^{2-} + SO_5^-$	5.3×10^8	-1500	Jacob (1986)
89. $SO_4^- + HO_2 \longrightarrow SO_4^{2-} + H^+ + O_2$	5.0×10^9	-1500	Jacob (1986)
90. $SO_4^- + O_2^- \longrightarrow SO_4^{2-} + O_2$	5.0×10^9	-1500	Jacob (1986)
91. $SO_4^- + OH^- \longrightarrow SO_4^{2-} + OH$	8.0×10^7	-1500	Jacob (1986)
92. $SO_4^- + H_2O_2 \longrightarrow SO_4^{2-} + H^+ + HO_2$	1.2×10^7	-2000	Ross and Neta (1979)
93. $SO_4^- + NO_2^- \longrightarrow SO_4^{2-} + NO_2$	8.8×10^8	-1500	Jacob (1986)
94. $SO_4^- + HCO_3^- \xrightarrow{O_2} SO_4^{2-} + H^+ + CO_3^-$	9.1×10^6	-2100	Ross and Neta (1979)
95. $SO_4^- + HCOO^- \xrightarrow{O_2} SO_4^{2-} + CO_2 + HO_2$	1.7×10^8	-1500	Jacob (1986)

(Continued)

TABLE 7.A.7 (Continued)

	Reaction	k_{298}	$-E/R$ (K)	Reference
96.	$SO_4^- + Cl^- \longrightarrow SO_4^{2-} + Cl$	2.0×10^8	-1500	Ross and Neta (1979)
97.	$SO_4^- + HCOOH \xrightarrow{O_2} SO_4^{2-} + H^+ + CO_2 + HO_2$	1.4×10^6	-2700	Jacob (1986)
98.[a]	$S(IV) + CH_3C(O)O_2NO_2 \longrightarrow S(VI)$	6.7×10^{-3}	0	Lee (1984a)
99.	$HSO_3^- + CH_3OOH \xrightarrow{H^+} SO_4^{2-} + 2H^+ + CH_3OH$	2.3×10^7	-3800	Lind et al. (1987)
100.[a]	$HSO_3^- + CH_3C(O)OOH \longrightarrow SO_4^{2-} + H^+ + CH_3COOH$	5.0×10^7	-4000	Lind et al. (1987)
101.	$S(IV) + HO_2 \longrightarrow S(VI) + OH$	6.0×10^2	0	Lind et al. (1987)
		1.0×10^6	0	Hoffmann and Calvert (1985)
	$S(IV) + O_2^- \xrightarrow{H_2O} S(VI) + OH + OH^-$	1.0×10^5	0	Hoffmann and Calvert (1985)
102.	$SO_4^- + CH_3OH \xrightarrow{O_2} SO_4^{2-} + HCHO + H^+ + HO_2$	2.5×10^7	-1800	Dogliotti and Hayon (1967)
103.	$HSO_3^- + NO_3 \longrightarrow NO_3^- + H^+ + SO_3^- + SO_3^-$	1.0×10^8	0	Chameides (1984)
104.	$2\,NO_2 + HSO_3^- \xrightarrow{H_2O} SO_4^{2-} + 3H^+ + 2NO_2^-$	2.0×10^6	0	Lee and Schwartz (1983)
105a.[b]	$S(IV) + N(III) \longrightarrow S(VI) + Product$	1.4×10^2	0	Martin (1984)
105b.[c]	$2\,HSO_3^- + NO_2^- \longrightarrow OH^- + Product$	4.8×10^3	-6100	Oblath et al. (1981)
106.	$HCHO + HSO_3^- \longrightarrow HOCH_2SO_3^-$	7.9×10^2	-4900	Boyce and Hoffmann (1984)
	$HCHO+SO_3^{2-} \xrightarrow{H_2O} HOCH_2SO_3^- + OH^-$	2.5×10^7	-1800	Boyce and Hoffmann (1984)
107.	$HOCH_2SO_3^- + OH^- \longrightarrow SO_3^{2-} + HCHO + H_2O$	3.6×10^3	-4500	Munger et al. (1986)
108.	$HOCH_2SO_3^- + OH \xrightarrow{O_2} SO_5^- + HCHO + H_2O$	2.6×10^8	-1500	Olson and Fessenden (1992)
109.	$HSO_3^- + Cl_2^- \xrightarrow{O_2} SO_5^- + 2Cl^- + H^+$	3.4×10^8	-1500	Huie and Neta (1987)
	$SO_3^{2-} + Cl_2^- \xrightarrow{O_2} SO_5^- + 2Cl^-$	3.4×10^8	-1500	Huie and Neta (1987)

[a] Reaction with "nonelementary" rate expression. See text.
[b] For pH $\leq$ 3.
[c] For pH $>$ 3.

SO_3^- then reacts rapidly with dissolved oxygen to produce the peroxymonosulfate radical, SO_5^-

$$SO_3^- + O_2 \longrightarrow SO_5^- \tag{7.A.3}$$

with a rate constant $k_{7.A.3} = 1.5 \times 10^9 \, \text{M}^{-1} \, \text{s}^{-1}$ at 298 K (Huie and Neta 1984).

The fate of SO_5^- is reaction via a series of pathways to produce HSO_5^-, SO_4^-, and S(VI), creating a relatively complicated reaction mechanism (Figure 7.A.1). The reactions of SO_5^- with S(IV)

$$SO_5^- + HSO_3^- \longrightarrow HSO_5^- + SO_3^- \tag{7.A.4}$$

$$SO_5^- + SO_3^{2-} \xrightarrow{H_2O} HSO_5^- + SO_3^- + OH^- \tag{7.A.5}$$

are slow (Huie and Neta 1987), and the main SO_5^- sink in this mechanism is the self-reaction

$$SO_5^- + SO_5^- \longrightarrow 2\,SO_4^- + O_2 \tag{7.A.6}$$

This reaction can also produce peroxydisulfate, $S_2O_8^{2-}$

$$SO_5^- + SO_5^- \longrightarrow S_2O_8^{2-} + O_2 \tag{7.A.7}$$

but with a rate four times less than reaction (7.A.6) (Table 7.A.8). The sulfate radical, SO_4^-, produced by reaction (7.A.6) reacts rapidly with HSO_3^- producing sulfate,

$$SO_4^- + HSO_3^- \longrightarrow SO_3^- + H^+ + SO_4^{2-} \tag{7.A.8}$$

and the propagation cycle from S(IV) to S(IV) is completed.

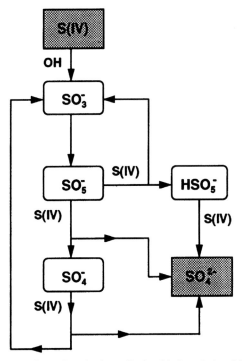

FIGURE 7.A.1   Schematic of reactions in the radical oxidation chain of S(IV) by the OH radical.

TABLE 7.A.8   Reaction Mechanism for the S(IV)–OH Aqueous Phase Reaction

	Reaction	$k(\mathrm{M}^{-1}\,\mathrm{s}^{-1})$ (at 298 K)	Reference
1.	$OH + HSO_3^- \rightarrow SO_3^- + H_2O$	4.5×10^9	Huie and Neta (1987)
2.	$OH + SO_3^{2-} \rightarrow SO_3^- + OH^-$	5.2×10^9	Huie and Neta (1987)
3.	$SO_3^- + O_2 \rightarrow SO_5^-$	1.5×10^9	Huie and Neta (1987)
4.	$SO_5^- + HSO_3^- \rightarrow HSO_5^- + SO_3^-$	2.5×10^4	Huie and Neta (1987)
5.	$SO_5^- + SO_3^{2-} \xrightarrow{H_2O} HSO_5^- + SO_3^- + OH^-$	2.5×10^4	Huie and Neta (1987)
6.	$SO_5^- + HSO_3^- \rightarrow SO_4^- + SO_4^{2-} + H^+$	7.5×10^4	Huie and Neta (1987)
7.	$SO_5^- + SO_3^{2-} \rightarrow SO_4^- + SO_4^{2-}$	7.5×10^4	Huie and Neta (1987)
8.	$SO_4^- + HSO_3^- \rightarrow SO_4^{2-} + SO_3^- + H^+$	7.5×10^8	Wine et al. (1989)
9.	$SO_4^- + SO_3^{2-} \rightarrow SO_4^{2-} + SO_3^-$	5.5×10^8	Deister and Warneck (1990)
10.	$SO_5^- + SO_5^- \rightarrow 2\,SO_4^- + O_2$	6×10^8	Huie and Neta (1987)
11.	$SO_3^- + SO_3^- \rightarrow S_2O_6^{2-}$	7×10^8	Huie and Neta (1987)
12.	$SO_4^- + SO_4^- \rightarrow S_2O_8^{2-}$	4.5×10^8	Buxton et al. (1996)
13.	$SO_5^- + SO_5^- \rightarrow S_2O_8^{2-} + O_2$	1.4×10^8	Huie and Neta (1987)
14.	$HSO_5^- + HSO_3^- + H^+ \rightarrow 2\,SO_4^{2-} + 3\,H^+$	7.1×10^{6a}	Betterton and Hoffmann (1988)

[a]The rate constant is in $\mathrm{M}^{-2}\,\mathrm{s}^{-1}$.

During the propagation of the reaction chain (Figure 7.A.1) multiple sulfate ions are created for each attack of OH on S(IV), and as a result the sulfate production rate exceeds the rate of the reactions (7.A.1) and (7.A.2). This effect can be studied by simplifying the mechanism of Table 7.A.8. For pH values lower than 6, $[HSO_3^-] \gg [SO_3^{2-}]$ and as their reactions in the mechanism have similar rate constants (Table 7.A.8), we can neglect as a first approximation the SO_3^{2-} reactions (reactions 2, 5, 7, and 9 in Table 7.A.8). Assuming that the rates of reactions 11 and 12 are also small, the steady-state concentrations of SO_3^-, SO_4^-, SO_5^-, and HSO_5^- can be calculated by setting

$$\frac{d[SO_3^-]}{dt} = \frac{d[SO_4^-]}{dt} = \frac{d[SO_5^-]}{dt} = \frac{d[HSO_5^-]}{dt} = 0 \tag{7.A.9}$$

resulting in the following steady-state expressions:

$$[HSO_5^-]_{ss} = \frac{k_4}{k_{14}[H^+]}[SO_5^-]_{ss} \tag{7.A.10}$$

$$[SO_4^-]_{ss} = \frac{k_6}{k_8}[SO_5^-]_{ss} + \frac{2k_{10}}{k_8[HSO_3^-]}[SO_5^-]_{ss}^2 \tag{7.A.11}$$

$$[SO_3^-]_{ss} = \frac{k_4 + k_6}{k_3[O_2]}[SO_5^-]_{ss}[HSO_3^-] + \frac{2(k_{10} + k_{13})}{k_3[O_2]}[SO_5^-]_{ss}^2 \tag{7.A.12}$$

$$[SO_5^-]_{ss} = \sqrt{\frac{k_1}{2k_{13}}}[OH]^{1/2}[HSO_3^-]^{1/2} \tag{7.A.13}$$

The steady-state concentration of SO_5^- can be calculated directly from (7.A.13) and then one can calculate the concentrations of the other radicals using (7.A.10) to (7.A.12).

Finally, the sulfate production rate by the above mechanism is given by

$$\frac{d[S(VI)]}{dt} = k_6[SO_5^-][HSO_3^-] + k_8[SO_4^-][HSO_3^-] + 2k_{14}[HSO_5^-][HSO_3^-][H^+] \tag{7.A.14}$$

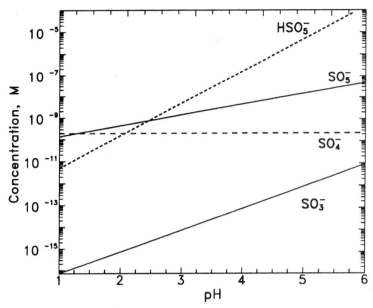

FIGURE 7.A.2 Estimated steady-state radical concentrations as a function of pH for a mixing ratio of $SO_2(g) = 1$ ppb and $[OH(aq)] = 1 \times 10^{-12}$ M at 298 K.

The steady-state concentrations of the various radicals and the sulfate production rate, for $[OH(aq)] = 7.5 \times 10^{-12}$ M and $\xi_{SO_2} = 1$ ppb, as a function of pH are shown in Figures 7.A.2 and 7.A.3. Note that the rate increases with pH, due mainly to the increasing S(IV) solubility, but remains less than $0.2\ \mu M\ s^{-1}$ for pH less than 5.

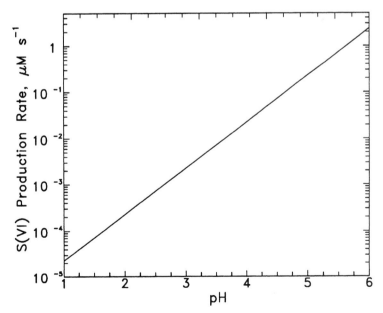

FIGURE 7.A.3 Sulfate production rate from the S(IV)–OH reaction as a function of pH for a mixing ratio of $SO_2(g) = 1$ ppb and $[OH(aq)] = 1 \times 10^{-12}$ M at 298 K.

The sulfate radicals SO_4^- and SO_5^- participate in a series of additional reactions in ambient clouds that complicate even further the above chain and modify significantly the overall reaction rate. These reactions are discussed subsequently.

7.A.2   Oxidation of S(IV) by Oxides of Nitrogen

Nitrogen dioxide has limited water solubility and its resulting low aqueous-phase concentration suggests that the reaction

$$2\,NO_2 + HSO_3^- \xrightarrow{\;H_2O\;} 3\,H^+ + 2\,NO_2^- + SO_4^{2-} \qquad (7.A.15)$$

should be of minor importance in most cases. This reaction has been studied by Lee and Schwartz (1983) and was described as one that is first-order in both NO_2 and S(IV)

$$R_{7.A.15} = k_{6.117}[S(IV)][NO_2] \qquad (7.A.16)$$

with a pH-dependent rate constant $k_{7.A.15}$. At pH 5.0, $k_{7.A.15} = 1.4 \times 10^5\,M^{-1}\,s^{-1}$ but at pH 5.8 and 6.4 only a lower limit, $k_{7.A.15} = 2 \times 10^6\,M^{-1}\,s^{-1}$, could be determined.[2] Whereas this reaction is of secondary importance at the concentrations and pH values representative of clouds, for fogs occurring in urban areas with high NO_2 concentrations this reaction could be a significant pathway for S(IV) oxidation, if the atmosphere has sufficient neutralizing capacity, for example, high $NH_3(g)$ concentration (Pandis and Seinfeld 1989b).

7.A.3   Reaction of Dissolved SO₂ with HCHO

HSO_3^- and SO_3^{2-} in clouds and fogs react with dissolved formaldehyde to produce hydroxymethanesulfonate, $HOCH_2SO_3H$ (HMS) (Boyce and Hoffmann 1984),[3]

$$HCHO(aq) + HSO_3^- \rightarrow HOCH_2SO_3^- \qquad (7.A.17)$$
$$HCHO(aq) + SO_3^{2-} \rightarrow {}^-OCH_2SO_3^- \qquad (7.A.18)$$

The elementary formaldehyde–S(IV) reactions have rate constants $k_{7.A.17} = 7.9 \times 10^2$ and $k_{7.A.18} = 2.5 \times 10^7\,M^{-1}\,s^{-1}$, respectively (Boyce and Hoffmann 1984). Reactions (7.A.17) and (7.A.18) involve HCHO and not its diol form, $H_2C(OH)_2$. HMS is a strong acid, dissociating completely in clouds to the hydroxymethanesulfonate ion (HMSA), $HOCH_2SO_3^-$:

$$HOCH_2SO_3H \rightleftharpoons HOCH_2SO_3^- + H^+ \qquad (7.A.19)$$

$HOCH_2SO_3^-$ can dissociate once more to $^-OCH_2SO_3^-$

$$HOCH_2SO_3^- \rightleftharpoons {}^-OCH_2SO_3^- + H^+ \qquad (7.A.20)$$

---

[2]The evaluation of this rate expression was considered tentative by Lee and Schwartz, in view of evidence for the formation of a long-lived intermediate species.

[3]Sulfur in HMS is also in oxidation state IV and measurements of S(IV) in clouds include the HMS contribution. In this book we have not included HMS in the definition of S(IV).

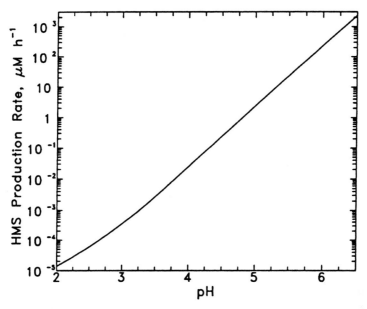

FIGURE 7.A.4 HMSA aqueous-phase production rate as a function of pH for mixing ratios of $SO_2(g) = 1$ ppb and $HCHO(g) = 1$ ppb.

but this second dissociation is weak, with $k_{7.A.20} = 2 \times 10^{-12}$ M, and, for all practical purposes HMS exists as $HOCH_2SO_3^-$ in the atmospheric aqueous phase (Sorensen and Anderson 1970). Therefore the HMSA formation rate R_f is given by

$$R_f = (k_{7.A.17}[HSO_3^-] + k_{7.A.18}[SO_3^{2-}])[HCHO] \qquad (7.A.21)$$

and is shown as a function of pH in Figure 7.A.4. The reaction rate increases exponentially with pH, because of the increasing concentrations of HSO_3^- and SO_3^{2-}, and becomes appreciable above pH 5. The overall rate is more or less equal to that of the SO_3^{2-} reaction for pH values higher than 3.

HMSA reacts with OH^- to reform SO_3^{2-} and formaldehyde

$$HOCH_2SO_3^- + OH^- \rightarrow HCHO(aq) + SO_3^{2-} + H_2O \qquad (7.A.22)$$

with a rate given by

$$R_{7.A.22} = k_{7.A.22}[HOCH_2SO_3^-][OH^-] \qquad (7.A.23)$$

and a second-order rate constant $k_{7.A.22} = 3.6 \times 10^3$ M^{-1} s^{-1} (Kok et al. 1986). The characteristic time for dissociation, $1/(k_{7.A.22}[OH^-])$, is 770 h at pH 4, 7 h at pH 6, and 45 minutes at pH 7. The decomposition of HMSA can be neglected for acidic solutions but becomes appreciable as the solution approaches neutrality. Because several hours are generally necessary to achieve equilibrium between its formation and decomposition reactions, the HMSA concentration is usually not in equilibrium with HSO_3^- and HCHO in atmospheric clouds (see Problem 7.8).

An interesting feature of HMSA chemistry is that the high-pH conditions that are most conducive to its production (Figure 7.A.4) are not suitable for its preservation. However, if the cloud or fog pH is initially high and then decreases as a result of S(IV) oxidation, then high concentrations of HMSA can be attained and maintained in the cloud (Munger et al. 1986).

Potential pathways for the destruction of HMSA in cloudwater or fogwater include reactions with $O_3(aq)$, $H_2O_2(aq)$, and the hydroxyl radical OH(aq). Hoigné et al. (1985) observed no direct reaction between ozone and HMSA. HMSA is also resistant to oxidation by H_2O_2 (Kok et al. 1986). The reaction between HMSA and OH results in the production of the SO_5^- radical

$$HOCH_2SO_3^- + OH \xrightarrow{O_2} SO_5^- + HCHO + H_2O \tag{7.A.24}$$

and links HMSA with the S(IV) radical oxidation chain discussed in the previous section. The rate of this reaction is pH independent and has a second-order rate constant $k_{7.A.24} = 2.6 \times 10^8 \, M^{-1} \, s^{-1}$ (Olson and Fessenden 1992). The lifetime of HMSA due to the attack of OH, $1/(k_{7.A.24}[OH])$, varies from approximately 1 h for $[OH(aq)] = 10^{-12} \, M$, to 10 hours for $[OH] = 10^{-13} \, M$. Oxidation by OH is expected to be the main sink of HMSA during the daytime in typical clouds or fogs (Jacob 1986; Pandis and Seinfeld 1989a).

HMSA has been observed in different environments at concentrations as high as 300 μM near sources of SO_2 and HCHO (Munger et al. 1984, 1986). Its formation explains the relatively high S(IV) concentrations that have been reported in high-pH environments because, we recall, HMSA is a member of the S(IV) family. In such cases the lifetime of dissolved $SO_2(HSO_3^-$ and $SO_3^{2-})$ should be extremely short (Munger et al. 1986), so the measured S(IV) was mostly HMSA and not HSO_3^- or SO_3^{2-}.

APPENDIX 7.3   AQUEOUS-PHASE NITRITE AND NITRATE CHEMISTRY

7.A.4 NO_x Oxidation

Aqueous-phase oxidation of oxides of nitrogen is used by the chemical industry for the production of nitric acid. However, NO_2 aqueous-phase oxidation in water (Lee 1984a)

$$NO_2(aq) + NO_2(aq) \xrightarrow{H_2O} NO_2^- + NO_3^- + 2H^+ \tag{7.A.25}$$

with a rate constant of $1 \times 10^8 \, M^{-1} \, s^{-1}$, by NO (Lee, 1984a)

$$NO(aq) + NO_2(aq) \xrightarrow{H_2O} 2NO_2^- + 2H^+ \tag{7.A.26}$$

with a rate constant of $2 \times 10^8 \, M^{-1} \, s^{-1}$, and by OH

$$NO(aq) + OH(aq) \rightarrow NO_2^- + H^+ \tag{7.A.27}$$

$$NO_2(aq) + OH(aq) \rightarrow NO_3^- + H^+ \tag{7.A.28}$$

(rate constants $2 \times 10^{10} \, M^{-1} \, s^{-1}$ and $1.3 \times 10^9 \, M^{-1} \, s^{-1}$, respectively) all proceed far too slowly under ambient conditions to contribute either to the removal of these nitrogen oxides or to cloudwater acidification. For example, the aqueous-phase concentrations of NO and NO_2 are below 1 nM (Section 7.3.6) and therefore the production of nitrate from reaction 7.A.25 proceeds with rates less than $0.3 \, \mu M \, h^{-1}$, contributing a negligible amount to the aqueous-phase nitrate concentration.

7.A.5   Nitrogen Radicals

The NO_3 radical (either directly or as N_2O_5) is probably the most reactive nitrogen species in the aqueous phase during nighttime (it is negligible during daytime because of the rapid photolysis of $NO_3(g)$). NO_3 and N_2O_5 are both very soluble in water and are a potential source of nitrate. Jacob (1986) estimated a Henry's law coefficient of the order of $2.1 \times 10^5 \, M \, atm^{-1}$ for NO_3 and assumed that N_2O_5 is completely transferred to the aqueous phase at equilibrium. N_2O_5 reacts rapidly with water to produce nitrate[4]

$$N_2O_5(aq) + H_2O \rightarrow 2 \, H^+ + 2 \, NO_3^- \qquad (7.A.29)$$

while NO_3 is converted to nitrate by chloride ion

$$NO_3 + Cl^- \rightarrow NO_3^- + Cl(aq) \qquad (7.A.30)$$

with a rate constant $1 \times 10^8 \, M^{-1} \, s^{-1}$ and produces the chlorine radical, Cl (Ross and Neta 1979; Chameides 1986). For a chloride concentration of $10 \, \mu M$, the lifetime of NO_3 in cloudwater as a result of reaction (7.A.30) is 1 ms, a value indicative of the highly reactive nature of NO_3. In environments with low chloride concentrations, NO_3 reacts with HSO_3^-

$$NO_3(aq) + HSO_3^- \rightarrow NO_3^- + H^+ + SO_3^- \qquad (7.A.31)$$

and produces sulfate radicals, SO_3^-, and nitrate with $k_{7.A.31} = 1 \times 10^8 \, M^{-1} \, s^{-1}$. Chameides (1986) calculated a typical nighttime continental cloud concentration of $[NO_3] = 10^{-12} \, M$. Using this concentration and assuming $\xi_{SO_2} = 1 \, ppb$, one estimates that the reaction proceeds with a rate of $0.06 \, \mu M \, h^{-1}$ at pH 4 and $6 \, \mu M \, h^{-1}$ at pH 6. Radicals produced during the NO_3 reactions participate in a number of additional reactions propagating the S(IV) oxidation radical chain (Pandis and Seinfeld 1989a).

---

[4]In Chapters 5 and 6 we considered

$$N_2O_5(g) + H_2O(aq) \rightarrow 2 \, HNO_3$$

where $H_2O(aq)$ is understood to be water in an atmospheric particle or droplet. Here we indicate the subsequent dissociation of HNO_3.

APPENDIX 7.4 AQUEOUS-PHASE ORGANIC CHEMISTRY

One of the most important aqueous-phase reactions involving organic species is attack of the OH radical on hydrated formaldehyde

$$H_2C(OH)_2 + OH(aq) \xrightarrow{O_2} HCOOH(aq) + HO_2(aq) + H_2O \qquad (7.A.32)$$

with a rate constant $k_{7.A.32} = 2 \times 10^9 \, M^{-1} \, s^{-1}$ at 298 K (Bothe and Schulte-Frohlinde, 1980). Assuming $[OH(aq)] = 10^{-12}$ M and that aqueous-phase formaldehyde is in Henry's law equilibrium with $\xi_{HCHO} = 1$ ppb, $[H_2C(OH)_2] = 6.3 \, \mu M$, and the reaction proceeds with a rate of $36 \, \mu M \, h^{-1}$. Using Figure 7.14 and a typical cloud liquid water content of $0.1 \, g \, m^{-3}$, the equivalent rate in gas-phase concentrations is $0.1 \, ppb \, h^{-1}$. Therefore for typical continental clouds this reaction has the potential to consume formaldehyde at a rate of $10\% \, h^{-1}$ and also to produce formic acid.

The produced formic acid reacts rapidly with dissolved OH

$$HCOO^- + OH \xrightarrow{O_2} CO_2 + HO_2 + OH^- \qquad (7.A.33)$$

$$HCOOH + OH \xrightarrow{O_2} CO_2 + HO_2 + H_2O \qquad (7.A.34)$$

with reaction rate constants of $k_{7.A.33} = 2.5 \times 10^9$ and $k_{7.A.34} = 2 \times 10^8 \, M^{-1} \, s^{-1}$, respectively (Anbar and Neta 1967). The formic acid produced is therefore converted to carbon dioxide by the same radical that led to its production.

The ratio of the rates of formic acid production to destruction r_f is given by

$$r_f = \frac{k_{7.A.32}[H_2C(OH)_2]}{k_{7.A.33}[HCOO^-] + k_{7.A.34}[HCOOH(aq)]} \qquad (7.A.35)$$

Assuming Henry's law equilibrium for formic acid and formaldehyde, we obtain

$$[H_2C(OH)_2] = H^*_{HCHO}p_{HCHO} \qquad (7.A.36)$$

$$[HCOOH(aq)] = H_f p_{HCOOH} \qquad (7.A.37)$$

$$[HCOO^-] = \frac{K_f H_f}{[H^+]} p_{HCOOH} \qquad (7.A.38)$$

the ratio r_f is given by

$$r_f = \frac{k_{7.A.32} H^*_{HCHO}[H^+]}{H_f(k_{7.A.34}[H^+] + k_{7.A.33}K_f)} \frac{p_{HCHO}}{p_{HCOOH}} \qquad (7.A.39)$$

which depends on both the pH and the relative availability of formaldehyde and formic acid (Figure 7.A.5). The ratio of the gas-phase partial pressures of HCHO and HCOOH

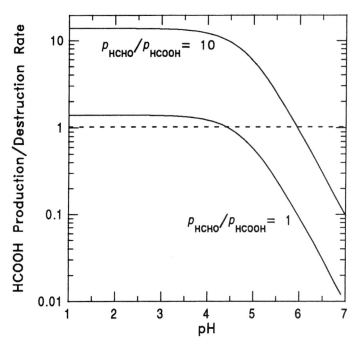

FIGURE 7.A.5 Ratio of the instantaneous formic acid production to destruction rates in a cloud as a function of pH and the ratio of the partial pressures of HCHO and HCOOH.

almost always exceeds unity and usually is around 10; therefore for the pH range of atmospheric interest a net production of HCOOH is expected. The dynamics of these processes during the lifetime of a cloud have been investigated by Chameides (1984) and Jacob (1986).

Additional relatively minor organic aqueous-phase reactions of formaldehyde, formic acid, methanol, CH_3OOH, CH_3O_2, and $CH_3C(O)OOH$ are discussed by Pandis and Seinfeld (1989a).

APPENDIX 7.5   OXYGEN AND HYDROGEN CHEMISTRY

Ozone is not produced at all in the aqueous phase, but at least 12 different chemical pathways consuming ozone have been identified. Because of the limited aqueous-phase solubility of ozone, none of these reactions is rapid enough to influence directly the gas-phase ozone concentration (Pandis and Seinfeld 1989a). The fastest of these reactions is that with the O_2^- radical resulting from the dissociation of HO_2 [reaction (7.69)]:

$$O_3(aq) + O_2^- \xrightarrow{H_2O} OH + 2O_2 + OH^- \tag{7.A.40}$$

Although the rate of this reaction is estimated to be around 0.1% $O_3(g)$ h^{-1}, because of the low ozone solubility this rate results in a lifetime of $O_3(aq)$ on the order of 1 s.

The OH radical in aqueous solution participates in a series of reactions both as a reactant and as a product. Its main sink for remote continental clouds is its reaction with hydrated formaldehyde discussed above. Other important sinks are reactions with hydrogen peroxide, formic acid, and S(IV). The main aqueous-phase sources of OH are reaction of O_2^- with O_3 and photolysis of hydrogen peroxide. Secondary sources are photolysis of NO_3^- and oxidation of S(IV) with HO_2.

HO$_2$ is another highly reactive radical, reacting with a series of species and produced in a number of reactions. The main aqueous-phase sources of HO_2 are usually the main sinks of OH and vice versa. HO_2 is a major aqueous-phase source of hydrogen peroxide

$$HO_2(aq) + O_2^- \xrightarrow{H_2O} H_2O_2(aq) + O_2 + OH^- \qquad (7.A.41)$$

and therefore it accelerates indirectly the oxidation of S(IV) to S(VI). The reaction of O_2^- with SO_5^- to produce HSO_5^-,

$$SO_5^- + O_2^- \xrightarrow{H_2O} HSO_5^- + OH^- + O_2 \qquad (7.A.42)$$

is an additional pathway for the conversion of S(IV) to sulfate by the OH reaction, shown in Figure 7.A.1. S(IV) also results directly with HO_2^- and O_2^- (Hoffmann and Calvert 1985)

$$S(IV) + HO_2(aq) \rightarrow S(VI) + OH \qquad (7.A.43)$$

$$S(IV) + O_2^- \xrightarrow{H_2O} S(VI) + OH + OH^- \qquad (7.A.44)$$

but since these reactions are significantly slower than the S(IV) oxidation by OH, they are of minor importance.

Another significant HO_2(aq) sink is the reaction of the superoxide ion, O_2^-, with bicarbonate ion to form HO_2^-:

$$HCO_3^- + O_2^- \rightarrow HO_2^- + CO_3^- \qquad (7.A.45)$$

This reaction was reported by Schmidt (1972), although subsequently Schwartz (1984) concluded that there is insufficient evidence for its occurrence in cloudwater. If the rate constant of $1.5 \times 10^6 \, M^{-1} \, s^{-1}$ proposed by Schmidt is correct, this reaction represents another relatively important aqueous-phase source of H_2O_2.

PROBLEMS

7.1$_A$ Calculate the pH of pure water for an ambient CO_2 mixing ratio of 350 ppm as a function of temperature from 0 to 30°C. What will be the value of the pH at 25°C if the CO_2 mixing ratio doubles?

7.2$_A$ We wish to compute the total dissolved S(IV) and the pH of a solution exposed to gaseous SO_2 when an initial source of ions is present. Assume that an aqueous solution contains an initial concentration [Ex] of univalent ions from a strong acid or base (e.g., HCl, NaOH). Thus, before exposure to SO_2,

$$[H^+]_0 = [Ex] + [OH^-]_0$$

where

$$[Ex] > 0 \quad \text{for a strong acid}$$
$$[Ex] < 0 \quad \text{for a strong base}$$

Thus

$$[Ex] = [H^+]_0 - K_w/[H^+]_0$$

For a strong acid, $[H^+]_0 \gg K_w/[H^+]_0$, so $[Ex] \simeq [H^+]_0$. Calculate [S(IV)] and $[H^+]$ as a function of the SO_2 mixing ratio with pH_0 as a parameter. Consider ξ_{SO_2} values from 1 to 100 ppb and pH_0 values from 4 to 8.

7.3$_B$ Given an aqueous-phase reaction rate R_a ($M s^{-1}$), show that with a liquid water mixing ratio w_L, the comparable gas-phase rate in ppb h^{-1} is

$$R_g \text{ (ppb } h^{-1}) = 2.95 \times 10^{11} \, w_L \, TR_a/p$$

and expressed in % h^{-1} is

$$R_g \text{ (% } h^{-1}) = 2.95 \times 10^{13} \, w_L \, TR_a/pX$$

where X is the total (gas + aqueous) equivalent gas-phase mixing ratio of the reactant species.

a. We would like to know the rate of sulfate formation through an aqueous S(IV)–O_3 reaction in a haze aerosol of $w_L = 10^{-10}$. Assume $T = 298 \, K$, pH = 5.6, and $p_{O_3} = 5 \times 10^{-8}$ atm (50 ppb). Given the total SO_2 (gas + aqueous) as 5 ppb, calculate the rate of sulfate formation in ppb h^{-1} and in % h^{-1}.

b. If a rate of SO_2 to sulfate conversion inside a cloud is observed to be 10% h^{-1}, then if $p_{SO_2} = 5 \times 10^{-9}$ atm (5 ppb), and $T = 298 \, K$, pH = 5.6, and $p_{O_3} = 5 \times 10^{-8}$, and if $w_L = 10^{-6}$, is the reaction with O_3 capable of generating this rate?

c. What minimum gas-phase H_2O_2 concentration is needed to account for a 1% h^{-1} conversion of SO_2 to sulfate when pH = 3, $w_L = 10^{-10}$, and $T = 298 \, K$? Discuss the result.

7.4$_B$ Consider the surface-catalyzed oxidation of SO_2 on solid particles. Let us assume that the process consists of absorption of SO_2 gas on the surface followed by reaction with O_2 and H_2O_2 to produce a molecule of H_2SO_4

adsorbed on the surface. The rate of the process is controlled by the rate of diffusion of SO_2 to the particle but stops when the surface of the particle is covered with sulfuric acid molecules. We assume that one molecule of H_2SO_4 occupies 30 Å^2 of area. With these assumptions determine the total concentration of sulfate (as $\mu g\,m^{-3}$ of sulfur) that is possible under the following sets of conditions:

	A	B
Total particles (cm^{-3})	10^6	10^3
Mean particle diameter (μm)	0.1	1.0

7.5$_A$ The importance of dissolved iron catalyzing the oxidation of S(IV) to sulfate within aqueous droplets in the atmosphere is uncertain. Assuming, for the purposes of a conservative estimate, that 0.1% of the total iron in the atmosphere is soluble and that iron is present at a level of $1\,\mu g\,m^{-3}$ and that a phase volume ratio of $w_L = 10^{-10}$ exists, at a pH of 4 and an ambient SO_2 level of 5 ppb, calculate the rate of SO_2 oxidation that would result, in % h^{-1}.

7.6$_A$ We want to estimate the rate of S(VI) formation in cloud droplets under two sets of conditions. Condition A represents a polluted cloud, whereas B is for a cloud in continental background conditions:

	A	B
ξ_{SO_2} (ppb)	50	5
T(K)	298	298
L(g m^{-3})	0.1	1
pH	3.5–4.5	4.5–6.0
ξ_{O_3} (ppb)	100	40
[H_2O_2(aq)] (μM)	47	5.9
[Mn^{2+}] (μM)	1	0.002
[Fe^{3+}] (μM)	1	0.033

7.7$_C$ Repeat the open- and closed-system calculations of Section 7.6 using $p_{O_3} = 5 \times 10^{-8}$ atm (50 ppb). Discuss your results.

7.8$_B$ It is often assumed that the HMSA concentration can be calculated assuming thermodynamic equilibrium between

$$HCHO + SO_3^{2-} \rightarrow {}^-OCH_2SO_3^-$$

and

$$HOCH_2SO_3^- + OH^- \rightarrow HCHO + HSO_3^-$$

a. Derive the equilibrium expression and estimate the equilibrium constant as a function of the pH.

b. What is the equilibrium composition of an open system with pH = 6, $\xi_{SO_2}(g) = 1$ ppb and $\xi_{HCHO}(g) = 1$ ppb neglecting HMSA formation.

c. What is the equilibrium composition of (b) incorporating the equilibrium of (a)?

d. Write differential equations describing the rate of change of S(IV) and HMSA. Assume that in the system initially all dissolved sulfur [given by adding your solutions in (c)] is in the HMSA form. When will the system arrive at the equilibrium state calculated in (c)?

7.9$_D$ We consider here a chemical model for the acidification of fog droplets (Hoffmann and Jacob 1984). Fog droplets are assumed to form around cloud condensation nuclei (CCN) such as NaCl. At time $t = 0$ droplets are assumed to condense suddenly in an atmospheric "closed box" containing CCN and the trace gases CO_2, SO_2, NH_3, HNO_3, H_2O_2, and O_3. As soon as the droplets form, they receive an initial chemical loading from the water-soluble fraction of the CCN and immediately establish Henry's law equilibria. The concentrations of the dissolved species at equilibrium are obtained by solving the electroneutrality equation

$$[H^+] + [NH_4^+] + [Na^+] + 3[Fe^{3+}] + 2[Mn^{2+}]$$
$$= [OH^-] + [HCO_3^-] + 2[CO_3^{2-}] + [HSO_3^-] + 2[SO_3^{2-}]$$
$$+ 2[SO_4^{2-}] + [NO_3^-] + [Cl^-]$$

where we have assumed the CCN to be NaCl, $FeCl_3$, and $MnCl_3$. Calculate the formation of S(VI) in the liquid phase over a period of 4 h by numerically integrating the appropriate rate equations. In your numerical integration use $\Delta t = 0.001$ min during the first 10 min and $\Delta t = 0.01$ min thereafter. The following initial concentrations are assumed:

$$\xi_{CO_2}(g) = 330\,\text{ppm} \qquad \xi_{O_3}(g) = 10\,\text{ppb}$$
$$\xi_{NH_3}(g) = 5\,\text{ppb} \qquad \xi_{H_2O_2} = 1\,\text{ppb}$$
$$\xi_{HNO_3}(g) = 3\,\text{ppb} \qquad [Fe^{3+}] = 33.3\,\mu\text{M}$$
$$\xi_{SO_2}(g) = 20\,\text{ppb} \qquad [Mn^{2+}] = 2.5\,\mu\text{M}$$

Assuming that the fog has a liquid water content of $0.1\,\text{g m}^{-3}$ and the temperature is 283 K, plot the sulfate formed, the pH, and the total aqueous-phase S(IV) and S(VI) as a function of time. Discuss your results in terms of the mechanisms contributing to S(IV) oxidation over the course of the 4-h period.

REFERENCES

Anbar, M., and Neta, P. (1967) A compilation of specific bimolecular rate constants for the reactions of hydrated electrons, hydrogen atoms and hydroxyl radicals with inorganic and organic compounds in aqueous solutions, *Int. J. Appl. Radiat. Isotopes* **18**, 493–523.

Behar, D., Czapski, G., and Duchovny, I. (1970) Carbonate radical in flash photolysis and pulse radiolysis of aqueous carbonate solutions, *J. Phys. Chem.* **74**, 2206–2210.

Betterton, E. A., and Hoffmann, M. R. (1988) Oxidation of aqueous SO_2 by peroxymonosulfate, *J. Phys. Chem.* **92**, 5962–5965.

Bielski, B. H. J. (1978) Reevaluation of the spectral and kinetic properties of HO_2 and O_2 free radicals, *Photochem. Photobiol.* **28**, 645–649.

Botha, C. F., Hahn, J., Pienaar, J. J., and Vaneldik, R. (1994) Kinetics and mechanism of the oxidation of sulfur (IV) by ozone in aqueous solutions, *Atmos. Environ.* **28**, 3207–3212.

Bothe, E., and Schulte-Frohlinde, D. (1980) Reaction of dihydroxymethyl radical with molecular oxygen in aqueous solution, *Anorg. Chem. Org. Chem.* **35**, 1035–1039.

Boyce, S. D., and Hoffmann, M. R. (1984) Kinetics and mechanism of the formation of hydroxymethanesulfonic acid at low pH, *J. Phys. Chem.* **88**, 4740–4746.

Buxton, G. V., McGowan, S., Salmon, G. A., Williams, J. E., and Wood, N. D. (1996) A study of the spectra and reactivity of oxysulfur-radical anions involved in the chain oxidation of S(IV): A pulse and γ-radiolysis study, *Atmos. Environ.* **30**, 2483–2493.

Chameides, W. L. (1984) The photochemistry of a marine stratiform cloud, *J. Geophys. Res.* **89**, 4739–4755.

Chameides, W. L. (1986) Possible role of NO_3 in the nighttime chemistry of a cloud, *J. Geophys. Res.* **91**, 5331–5337.

Chameides, W. L., and Davis, D. D. (1982) The free radical chemistry of cloud droplets and its impact upon the composition of rain, *J. Geophys. Res.* **87**, 4863–4877.

Chen S., Cope V. W., and Hoffman M. Z. (1973) Behavior of CO_3 radicals generated in the flash photolysis of carbonatoamines complexes of cobalt(III) in aqueous solutions, *J. Phys. Chem.* **77**, 1111–1116.

Christensen, H., Sehested, K., and Corfitzen, H. (1982) Reactions of hydroxyl radicals with hydrogen peroxide at ambient and elevated temperatures, *J. Phys. Chem.* **86**, 1588–1590.

Clarke, A. G., and Radojevic, M. (1987) Oxidation of SO_2 in rainwater and its role in acid rain chemistry, *Atmos. Environ.* **21**, 1115–1123.

Cocks, A. T., McElroy, W. L., and Wallis, P. G. (1982) *The Oxidation of Sodium Sulphite Solutions by Hydrogen Peroxide*, Central Electricity Research Laboratories, Report RD/L2215N81.

Damschen, D. E., and Martin, L. R. (1983) Aqueous aerosol oxidation of nitrous acid by O_2, O_3, and H_2O_2, *Atmos. Environ.* **17**, 2005–2011.

Daum, P. H., Kelly, T. J., Schwartz, S. E., and Newman, L. (1984) Measurements of the chemical composition of stratiform clouds, *Atmos. Environ.* **18**, 2671–2684.

Deister, U., and Warneck, P. (1990) Photooxidation of SO_3^{2-} in aqueous solution, *J. Phys. Chem.* **94**, 2191–2198.

Denbigh, K. (1981) *The Principles of Chemical Equilibrium*, 4th ed., Cambridge Univ. Press, Cambridge, UK.

Dogliotti, L., and Hayon, E. (1967) Flash photolysis of persulfate ions in aqueous solutions. Study of sulfate and ozonide radical ions, *J. Phys. Chem.* **71**, 2511–2516.

Erickson, R. E., Yates, L. M., Clark, R. L., and McEwen, D. (1977) The reaction of sulfur dioxide with ozone in water and its possible atmospheric significance, *Atmos. Environ.* **11**, 813–817.

Fuller, E. C., and Crist, R. H. (1941) Rate of oxidation of sulfite ions by oxygen, *J. Am. Chem. Soc.* **63**, 1644–1650.

Graedel, T. E., and Goldberg, K. I. (1983) Kinetic studies of raindrop chemistry, 1. Inorganic and organic processes, *J. Geophys. Res.* **88**, 10865–10882.

Graedel, T. E., Hawkins, D. T., and Claxton, L. D. (1986) *Atmospheric Chemical Compounds: Sources, Occurrence, and Bioassay*, Academic Press, Orlando.

Graedel, T. E., and Weschler, C. J. (1981) Chemistry within aqueous atmospheric aerosols and rain-drops, *Rev. Geophys.* **19**, 505–539.

Gratzel, M., Henglein A., and Taniguchi S. (1970) Pulsradiolytische beobachtungen uber die reduktion des NO_3^--ions unt uber bildungund zerfall des persalpetrigen saure in wassriger losung, *Ber. Bundsenges. Phys. Chem.* **74**, 292–298.

Hagesawa, K., and Neta, P. (1978) Rate constants and mechanisms of reaction for Cl_2^- radicals, *J. Phys. Chem.* **82**, 854–857.

Hales, J. M., and Drewes, D. R. (1979) Solubility of ammonia in water at low concentrations, *Atmos. Environ.* **13**, 1133–1147.

Harrison, H., Larson, T. V., and Monkman, C. S. (1982) Aqueous phase oxidation of sulfites by ozone in the presence of iron and manganese, *Atmos. Environ.* **16**, 1039–1041.

Hayon, R., Treinin, A., and Wilf, J. (1972) Electronic spectra, photochemistry and auto-oxidation mechanism of the sulfite–bisulfite–pyrosulfite systems, *J. Am. Chem. Soc.* **94**, 47–57.

Heymsfield, A. J. (1993) Microphysical structures of stratiform and cirrus clouds, in *Aerosol-Cloud-Climate Interactions*, P. V. Hobbs, ed., Academic Press, San Diego.

Hoffmann, M. R., and Boyce, S. D. (1983) Catalytic autooxidation of aqueous sulfur dioxide in relationship to atmospheric systems, *Adv. Environ. Sci. Technol.* **12**, 148–189.

Hoffmann, M. R., and Calvert, J. G. (1985) *Chemical Transformation Modules for Eulerian Acid Deposition Models*, Vol. 2, *The Aqueous-Phase Chemistry*, EPA/600/3-85/017, U.S. Environmental Protection Agency, Research Triangle Park, NC.

Hoffmann, M. R., and Edwards, J. O. (1975) Kinetics of oxidation of sulfite by hydrogen peroxide in acidic solution, *J. Phys. Chem.* **79**, 2096–2098.

Hoffmann, M. R., and Jacob, D. J. (1984) Kinetics and mechanisms of the catalytic oxidation of dissolved sulfur dioxide in aqueous solution: An application to nighttime fog water chemistry, in *SO_2, NO, and NO_2 Oxidation Mechanisms: Atmospheric Considerations*, J. G. Calvert, ed., Butterworth, Stoneham, MA, pp. 63–100.

Hoigné, J., and Bader, H. (1983a) Rate constants of reactions of ozone with organic and inorganic compounds in water, 1, Non-dissociating organic compounds, *Water Res.* **17**, 173–183.

Hoigné, J., and Bader, H. (1983b) Rate constants of reactions of ozone with organic and inorganic compounds in water. 2. Dissociating organic compounds, *Water Res.* **17**, 185–194.

Hoigné, J., Ader, H., Haag, W. R., and Staehelin, J. (1985) Rate constants of reactions of ozone with organic and inorganic compounds in water, III, *Water Res.* **19**, 993–1004.

Huie, R. E., and Neta, P. (1984) Chemical behavior of SO_3^- and SO_5^- radicals in solution, *J. Phys. Chem.* **88**, 5665–5669.

Huie, R. E., and Neta, P. (1987) Rate constants for some oxidations of S(IV) by radicals in aqueous solutions, *Atmos. Environ.* **21**, 1743–1747.

Huss, A. Jr., Lim, P. J., and Eckert, C. A. (1978) On the "uncatalyzed" oxidation of sulfur(IV) in aqueous solutions, *J. Am. Chem. Soc.* **100**, 6252–6253.

Huss, A. Jr., Lim, P. K., and Eckert, C. A. (1982a) Oxidation of aqueous SO_2. 1. Homogeneous manganese(II) and iron(II) catalysis at low pH, *J. Phys. Chem.* **86**, 4224–4228.

Huss, A. Jr., Lim, P. K., and Eckert, C. A. (1982b) Oxidation of aqueous SO_2. 2. High pressure studies and proposed reaction mechanisms, *J. Phys. Chem.* **86**, 4229–4233.

Ibusuki, T., and Takeuchi, K. (1987) Sulfur dioxide oxidation by oxygen catalyzed by mixtures of manganese(II) and iron(III) in aqueous solutions at environmental reaction conditions, *Atmos. Environ.* **21**, 1555–1560.

Jacob, D. J. (1986) Chemistry of OH in remote clouds and its role in the production of formic acid and peroxymonosulfate, *J. Geophys. Res.* **91**, 9807–9826.

Jayson, G. G., Parsons, B. J., and Swallow, A. J. (1973) Some simple, highly reactive, inorganic chlorine derivatives in aqueous-solution, *Trans. Faraday Soc.* **69**, 1597–1607.

Kok, G. L., Gitlin, S. N., and Lazrus, A. L. (1986) Kinetics of the formation and decomposition of hydroxymethanesulfonate, *J. Geophys. Res.* **91**, 2801–2804.

Kozac-Channing, L. F., and Heltz, G. R. (1983) Solubility of ozone in aqueous solutions of 0–0.6 M ionic strength at 5–30 °C, *Environ. Sci. Technol.* **17**, 145–149.

Kunen, S. M., Lazrus, A. L., Kok, G. L., and Heikes, B. G. (1983) Aqueous oxidation of SO_2 by hydrogen peroxide, *J. Geophys. Res.* **88**, 3671–3674.

Lagrange, J., Pallares, C., and Lagrange, P. (1994) Electrolyte effects on aqueous atmospheric oxidation of sulfur dioxide by ozone, *J. Geophys. Res.* **99**, 14595–14600.

Larson, T. V., Horike, N. R., and Harrison, H. (1978) Oxidation of sulfur dioxide by oxygen and ozone in aqueous solution: A kinetic study with significance to atmospheric processes, *Atmos. Environ.* **12**, 1597–1611.

Latimer, W. M. (1952) *The Oxidation States of the Elements and Their Potentials in Aqueous Solutions.* Prentice-Hall, New York, pp. 70–89.

Le Henaff, P. (1968) Methodes d'etude et proprietes des hydrates, hemiacetals et hemiacetals derives des aldehydes et des cetones, *Bull. Soc. Chim. Fr.* 4687–4700.

Lee, Y. N. (1984a) Atmospheric aqueous-phase reactions of nitrogen species, in *Gas–Liquid Chemistry of Natural Waters*, Vol. 1, BNL 51757, Brookhaven National Laboratory, Brookhaven, NY, pp. 20/1–20/10.

Lee, Y. N. (1984b) Kinetics of some aqueous-phase reactions of peroxyacetyl nitrate, in *Gas–Liquid Chemistry of Natural Waters*, Vol. 1, BNL 51757, Brookhaven National Laboratory, Brookhaven, NY, pp. 21/1–21/7.

Lee Y., N., and Lind, J. A. (1986) Kinetics of aqueous-phase oxidation of nitrogen(III) by hydrogen peroxide, *J. Geophys. Res.* **91**, 2793–2800.

Lee, Y. N., Senum, G. I., and Gaffney, J. S. (1983) Peroxyacetyl nitrate (PAN) stability, solubility, and reactivity—implications for tropospheric nitrogen cycles and precipitation chemistry, *5th Int. Conf. Commission on Atmospheric Chemistry and Global Pollution, Symp. Tropospheric Chemistry*, Oxford, UK.

Lee, Y. N., and Schwartz, S. E. (1983) Kinetics of oxidation of aqueous sulfur(IV) by nitrogen dioxide, in *Precipitation Scavenging, Dry Deposition and Resuspension*, Vol. 1, H. R. Pruppacher, R. G. Semonin, and W. G. N. Slinn, eds., Elsevier, New York.

Lee, Y. N., Shen, J., Klotz, P. J., Schwartz, S. E., and Newman, L. (1986) Kinetics of the hydrogen perodixe sulfur(IV) reaction in rainwater collected at northeastern U.S. site, *J. Geophys. Res.* **91**, 13264–13274.

Lilie, J., Henglein, A., and Hanrahan, R. J. (1978) Reactions of the carbonate radical anion with organic and inorganic solutes in aqueous solution, paper presented at 176th meeting of American Chemical Society, Miami Beach, Florida.

Lind, J. A., and Lazrus, A. L. (1983) Aqueous-phase oxidation of sulfur(IV) by some organic peroxides, *EOS Trans.* **64**, 670.

Lind, J. A., Lazrus, A. L., and Kok, G. L. (1987) Aqueous phase oxidation of sulfur(IV) by hydrogen peroxide, methylhydroperoxide, and peroxyacetic acid, *J. Geophys. Res.* **92**, 4171–4177.

Maahs, H. G. (1983) Kinetics and mechanism of the oxidation of S(IV) by ozone in aqueous solution with particular reference to SO_2 conversion in nonurban tropospheric clouds, *J. Geophys. Res.* **88**, 10721–10723.

Mader, P. M. (1958) Kinetics of the hydrogen peroxide-sulfite reaction in alkaline solution, *J. Am. Chem. Soc.* **80**, 2634–2639.

Marsh, A. R. W., and McElroy, W. J. (1985) The dissociation constant and Henry's law constant of HCl in aqueous solution, *Atmos. Environ.* **19**, 1075–1080.

Martell, A. E., and Smith, R. M. (1977) *Critical Stability Constants,* Vol. 3, *Other Organic Ligands*, Plenum, New York.

Martin, L. R. (1984) Kinetic studies of sulfite oxidation in aqueous solution, in *SO₂, NO, and NO₂ Oxidation Mechanisms: Atmospheric Considerations*, J. G. Calvert, ed., Butterworth, Stoneham, MA, pp. 63–100.

Martin, L. R., and Damschen, D. E. (1981) Aqueous oxidation of sulfur dioxide by hydrogen peroxide at low pH, *Atmos. Environ.* **15**, 1615–1621.

Martin, L. R., Damschen, D. E., and Judeikis, H. S. (1981) *Sulfur Dioxide Oxidation Reactions in Aqueous Solution*, EPA 600/7-81-085. U.S. Environmental Protection Agency, Research Triangle Park, NC.

Martin, L. R., and Good, T. W. (1991) Catalyzed oxidation of sulfur dioxide in solution: The iron–manganese synergism, *Atmos. Environ.* **25A**, 2395–2399.

Martin, L. R., and Hill, M. W. (1987a) The iron catalyzed oxidation of sulfur: Reconciliation of the literature rates, *Atmos. Environ.* **21**, 1487–1490.

Martin, L. R., and Hill, M. W. (1987b) The effect of ionic strength on the manganese catalyzed oxidation of sulfur(IV), *Atmos. Environ.* **21**, 2267–2270.

Martin, L. R., Hill, M. W., Tai, A. F., and Good, T. W. (1991) The iron catalyzed oxidation of sulfur(IV) in aqueous solution: Differing effects of organics at high and low pH, *J. Geophys. Res.* **96**, 3085–3097.

McArdle, J. V., and Hoffmann, M. R. (1983) Kinetics and mechanism of the oxidation of aquated sulfur dioxide by hydrogen peroxide at low pH, *J. Phys. Chem.* **87**, 5425–5429.

Munger, J. W., Jacob, D. J., and Hoffmann, M. R. (1984) The occurence of bisulfite–aldehyde addition products in fog- and cloudwater, *J. Atmos. Chem.* **1**, 335–350.

Munger, J. W., Tiller, C., and Hoffmann, M. R. (1986) Identification of hydroxymethanesulfonate in fog water, *Science* **231**, 247–249.

Oblath, S. B., Markowitz, S. S., Novakov, T., and Chang, S. G. (1981) Kinetics of the formation of hydroxylamine disulfonate by reaction of nitrite with sulfites, *J. Phys. Chem.* **85**, 1017–1021.

Olson, T. M., and Fessenden, R. W. (1992) Pulse radiolysis study of the reaction of OH radicals with methanesulfonate and hydroxymethanesulfonate, *J. Phys. Chem.* **96**, 3317–3320.

Pandis, S. N., and Seinfeld, J. H. (1989a) Sensitivity analysis of a chemical mechanism for aqueous-phase atmospheric chemistry, *J. Geophys. Res.* **94**, 1105–1126.

Pandis, S. N., and Seinfeld, J. H. (1989b) Mathematical modeling of acid deposition due to radiation fog, *J. Geophys. Res.* **94**, 12911–12923.

Pandis, S. N., Seinfeld, J. H., and Pilinis, C. (1992) Heterogeneous sulphate production in an urban fog, *Atmos. Environ.* **26**, 2509–2522.

Penkett, S. A. (1972) Oxidation of SO₂ and the other atmospheric gases by ozone in aqueous solution, *Nature* **240**, 105–106.

Penkett, S. A., Jones, B. M. R., Brice, K. A., and Eggleton, A. E. J. (1979) The importance of atmospheric ozone and hydrogen peroxide in oxidizing sulfur dioxide in cloud and rainwater, *Atmos. Environ.* **13**, 123–137.

Perrin, D. D. (1982) *Ionization Constants of Inorganic Acids and Bases in Aqueous Solution,* 2nd ed., Pergamon Press, New York.

Rettich, T. R. (1978) Some photochemical reactions of aqueous nitric acid, *Diss. Abstr. Int. B* **38**, 5968.

Ross, A. B., and Neta, P. (1979) *Rate Constants for Reactions of Inorganic Radicals in Aqueous Solutions*, NSRDS-NBS 65, National Bureau of Standards, U.S. Department of Commerce, Washington, DC.

Sander, R. (1999) Compilation of Henry's law constants for inorganic and organic species of potential importance in environmental chemistry (version 3) (available at http://www.mainz.mpg.de/~ sander/res/henry.html).

Schmidt, K. H. (1972) Electrical conductivity techniques for studying the kinetics of radiation induced chemical reactions in aqueous solutions, *Int. J. Radiat. Phys. Chem.* **4**, 439–468.

Scholes, G., and Willson, R. L. (1967) γ-Radiolysis of aqueous thymine solutions. Determination of relative reaction rates of OH radicals, *Trans. Faraday Soc.* **63**, 2982–2993.

Schwartz, S. E. (1984) Gas- and aqueous-phase chemistry of HO_2 in liquid water clouds, *J. Geophys. Res.* **89**, 11589–11598.

Schwartz, S. E., and Freiberg, J. E. (1981) Mass transport limitation to the rate of reaction of gases in liquid droplets: Application in oxidation of SO_2 in aqueous solution, *Atmos. Environ.* **15**, 1129–1144.

Schwartz, S. E., and White, W. H. (1981) Solubility equilibrium of the nitrogen oxides and oxyacids in dilute aqueous solution, *Adv. Environ. Sci. Eng.* **4**, 1–45.

Sehested, K., Holcman, J., and Hart, E. J. (1983) Rate constants and products of the reactions of e_{aq}^-, O_2^-, and H with ozone in aqueous solutions, *J. Phys. Chem.* **87**, 1951–1954.

Sehested, K., Holcman, J., Bjergbakke, E., and Hart, E. J. (1984) A pulse radiolytic study of the reaction of OH + O_3 in aqueous medium, *J. Phys. Chem.* **88**, 4144–4147.

Sehested, K., Rasmussen, O. L., and Fricke, H. (1968) Rate constants for OH with HO_2, O_2^-, and $H_2O_2^+$ from hydrogen peroxide formation in pulse-irradiated oxygenated water, *J. Phys. Chem.* **72**, 626–631.

Shapilov, O. D., and Kostyukovskii, Y. L. (1974) Reaction kinetics of hydrogen peroxide with formic acid in aqueous solutions, *Kinet. Katal.* **15**, 1065–1067.

Smith, R. M., and Martell, A. E. (1976) *Critical Stability Constants*, Vol. 4, *Inorganic Complexes*, Plenum Press, New York.

Snider, J. R., and Dawson, G. A. (1985) Tropospheric light alcohols, carbonyls, and acetonitrile: concentrations in the southwestern United States and Henry's law data, *J. Geophys. Res.* **90**, 3797–3805.

Sorensen, P. E., and Andersen, V. S. (1970) The formaldehyde–hydrogen sulphite system in alkaline aqueous solution: kinetics, mechanism, and equilibria, *Acta Chem. Scand.* **24**, 1301–1306.

Staehelin, J., and Hoigné, J. (1982) Decomposition of ozone in water: Rate of initiation by hydroxide ions and hydrogen peroxide, *Environ. Sci. Technol.* **16**, 676–681.

Staehelin, J., Buhler, R. E., and Hoigné, J. (1984) Ozone decomposition in water studied by pulse radiolysis. 2. OH and HO_4 as chain intermediates, *J. Phys. Chem.* **88**, 5999–6004.

Strehlow, H., and Wagner, I. (1982) Flash photolysis in aqueous nitrite solutions, *Z. Phys. Chem. Wiesbaden* **132**, 151–160.

Stumm, W., and Morgan, J. J. (1996) *Aquatic Chemistry*, 3rd ed., Wiley, New York.

Treinin, A., and Hayon, E. (1970) Absorption spectra and reaction kinetics of NO_2, N_2O_3, and N_2O_4 in aqueous solutions, *J. Am. Chem. Soc.* **92**, 5821–5828.

Tsunogai, S. (1971) Oxidation rate of sulfite in water and its bearing on the origin of sulfate in meteoric precipitation, *Geochem. J.* **5**, 175–185.

Wagman, D. D., Evans, W. H., Parker, V. B., Schumm, R. H., Halow, I., Bailey, S. M., Churney, K. L., and Nuttall, R. L. (1982) The NBS tables of chemical thermodynamic properties, *J. Phys. Chem. Ref. Data* **11**, 2.1–2.392.

Warren S. G., Hahn C. J., London J., Chervin R. M., and Jenne R. L. (1986) *Global Distribution of Total Cloud Cover and Cloud Type Amounts over Land*, NCAR Technical note TN-273 + STR, National Center for Atmospheric Research, Boulder, CO.

Warren S. G., Hahn C. J., London J., Chervin R. M., and Jenne R. L. (1988) *Global Distribution of Total Cloud Cover and Cloud Type Amounts over Land*, NCAR Technical note TN-317 + STR, National Center for Atmospheric Research, Boulder, CO.

Weeks, J. L., and Rabani, J. (1966) The pulse radiolysis of deaerated aqueous carbonate solutions, *J. Phys. Chem.* **70**, 2100–2106.

Weinstein, J., and Bielski, B. H. J. (1979) Kinetics of the interaction of HO_2 and O_2^- radicals with hydrogen peroxide; the Haber-Weiss reaction, *J. Am. Chem. Soc.* **101**, 58–62.

Wine, P. H., Tang, Y., Thorn, R. P., Wells, J. R., and Davis D. D. (1989) Kinetics of aqueous phase reactions of the SO_4^- radical with potential importance in cloud chemistry, *J. Geophys. Res.* **94**, 1085–1094.

8 Properties of the Atmospheric Aerosol

8.1 THE SIZE DISTRIBUTION FUNCTION

The atmosphere, whether in urban or remote areas, contains significant concentrations of aerosol particles sometimes as high as 10^7–10^8 cm^{-3}. The diameters of these particles span over four orders of magnitude, from a few nanometers to around 100 μm. To appreciate this wide size range one just needs to consider that the mass of a 10-μm-diameter particle is equivalent to the mass of one billion 10-nm particles. Combustion-generated particles, such as those from automobiles, power generation, and woodburning, can be as small as a few nanometers and as large as 1 μm. Windblown dust, pollens, plant fragments, and seasalt are generally larger than 1 μm. Material produced in the atmosphere by photochemical processes is found mainly in particles smaller than 1 μm. The size of these particles affects both their lifetime in the atmosphere and their physical and chemical properties. It is therefore necessary to develop methods of mathematically characterizing aerosol size distributions. For the purposes of this chapter we neglect the effect of particle shape and consider only spherical particles.

An aerosol particle can be considered to consist of an integer number k of molecules or monomers. The smallest aerosol particle could be defined in principle as that containing two molecules. The aerosol distribution could then be characterized by the number concentration of each cluster, that is, by N_k, the concentration (per cm^3 of air) of particles containing k molecules. Although rigorously correct, this discrete method of characterizing the aerosol distribution cannot be used in practice because of the large number of molecules that make up even the smallest aerosol particles. For example, a particle with a diameter of 0.01 μm contains approximately 10^4 molecules and one with a diameter of 1 μm, around 10^{10}.

A complete description of the aerosol size distribution can also include an accounting of the size of each particle. Even if such information were available, a list of the diameters of thousands of particles, which would vary as a function of time and space, would be cumbersome. A first step in simplifying the necessary accounting is division of the particle size range into discrete intervals and calculation of the number of particles in each size bin. Information for an aerosol size distribution using 12 size intervals is shown in Table 8.1. Such a summary of the aerosol size distribution requires only 25 numbers (the boundaries of the size sections and the corresponding concentrations) instead of the diameters of all the particles. This distribution is presented in the form of a histogram in Figure 8.1. Note that the enormous range of the aerosol particle sizes makes the presentation of the full size

Atmospheric Chemistry and Physics: From Air Pollution to Climate Change, Second Edition, by John H. Seinfeld and Spyros N. Pandis. Copyright © 2006 John Wiley & Sons, Inc.

TABLE 8.1   Example of Segregated Aerosol Size Information

Size Range, μm	Concentration, cm^{-3}	Cumulative, cm^{-3}	Concentration, $\mu m^{-1} cm^{-3}$
0.001–0.01	100	100	11,111
0.01–0.02	200	300	20,000
0.02–0.03	30	330	3,000
0.03–0.04	20	350	2,000
0.04–0.08	40	390	1,000
0.08–0.16	60	450	750
0.16–0.32	200	650	1,250
0.32–0.64	180	830	563
0.64–1.25	60	890	98
1.25–2.5	20	910	16
2.5–5.0	5	915	2
5.0–10.0	1	916	0.2

distribution difficult. The details of the size distribution lost by showing the whole range of diameters are illustrated in the inset of Figure 8.1.

The size distribution of a particle population can also be described by using its cumulative distribution. The cumulative distribution value for a size section is defined as the concentration of particles that are smaller than or equal to this size range. For example, for the distribution of Table 8.1, the value of the cumulative distribution for the 0.03–0.04 μm size range indicates that there are 350 particles cm^{-3} that are smaller than

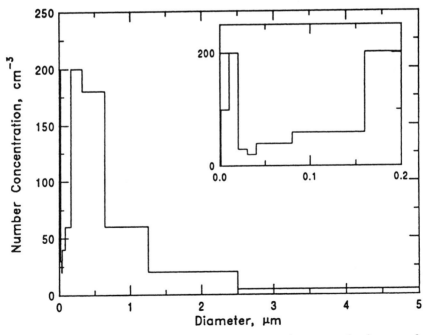

FIGURE 8.1 Histogram of aerosol particle number concentrations versus the size range for the distribution of Table 8.1. The diameter range 0–0.2 μm for the same distribution is shown in the inset.

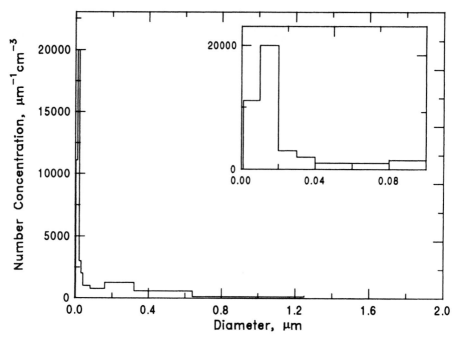

FIGURE 8.2 Aerosol number concentration normalized by the width of the size range versus size for the distribution of Table 8.1. The diameter range 0–0.1 μm for the same distribution is shown in the inset.

0.04 μm. The last value of the cumulative distribution indicates the total particle number concentration.

Use of size bins with different widths makes the interpretation of absolute concentrations difficult. For example, one may want to find out in which size range there are a lot of particles. The number concentrations in Table 8.1 indicate that there are 200 particle cm^{-3} in the range from 0.01 to 0.02 μm and another 200 particle cm^{-3} from 0.16 to 0.32 μm. However, this comparison of the concentration of particles covering a size range of 10 nm with that over a 160 nm range favors the latter. To avoid such biases, one often normalizes the distribution by dividing the concentration with the corresponding size range. The result is a concentration expressed in μm^{-1} cm^{-3} (Table 8.1) and is illustrated in Figure 8.2. The distribution changes shape, but now the area below the curve is proportional to the number concentration. Figure 8.2 indicates that roughly half of the particles are smaller than 0.1 μm. A plot like Figure 8.1 may be misleading, as it indicates that almost all particles are larger than 0.1 μm. If a logarithmic scale is used for the diameter (Figure 8.3) both the large- and small-particle regions are depicted, but it now erroneously appears that the distribution consists almost exclusively of particles smaller than 0.1 μm.

Using a number of size bins to describe an aerosol size distribution generally results in loss of information about the distribution structure inside each bin. While this may be acceptable for some applications, our goal in this chapter is to develop a rigorous mathematical framework for the description of the aerosol size distribution. The issues discussed in the preceding example provide valuable insights into how we should express and present ambient aerosol size distributions.

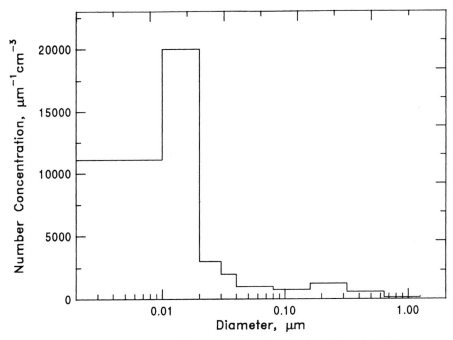

FIGURE 8.3   Same as Figure 8.2 but plotted against the logarithm of the diameter.

8.1.1   The Number Distribution $n_N(D_p)$

In the previous section, the value of the aerosol distribution n_i for a size interval i was expressed as the ratio of the absolute aerosol concentration N_i of this interval and the size range ΔD_p. The aerosol concentration can then be calculated by

$$N_i = n_i \, \Delta D_p$$

The use of arbitrary intervals ΔD_p can be confusing and makes the intercomparison of size distributions difficult. To avoid these complications and to maintain all the information regarding the aerosol distribution, one can use smaller and smaller size bins, effectively taking the limit $\Delta D_p \to 0$. At this limit, ΔD_p becomes infinitesimally small and equal to dD_p. Then one can define the size distribution function $n_N(D_p)$, as follows:

$$n_N(D_p) \, dD_p = \text{number of particles per cm}^3 \text{ of air having diameters}$$
$$\text{in the range } D_p \text{ to } (D_p + dD_p)$$

The units of $n_N(D_p)$ are $\mu\text{m}^{-1}\,\text{cm}^{-3}$, and the total number of particles per cm^3, N_t is then just

$$N_t = \int_0^\infty n_N(D_p) dD_p \qquad (8.1)$$

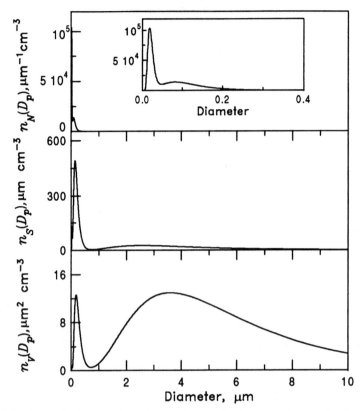

FIGURE 8.4   Atmospheric aerosol number, surface, and volume continuous distributions versus particle size. The diameter range 0–0.4 μm for the number distribution is shown as an inset.

By using the function $n_N(D_p)$ we implicitly assume that the number distribution is no longer a discrete function of the number of molecules but a continuous function of the diameter D_p. This assumption of a continuous size distribution is valid beyond a certain number of molecules, say, around 100. In the atmosphere most of the particles have diameters smaller than 0.1 μm and the number distribution function $n_N(D_p)$ usually exhibits a narrow spike near the origin (Figure 8.4).

The cumulative size distribution function $N(D_p)$ is defined as

$$N(D_p) = \text{number of particles per cm}^3 \text{ having diameter smaller than } D_p$$

The function $N(D_p)$, in contrast to $n_N(D_p)$, represents the actual particle concentration in the size range 0–D_p and has units of cm^{-3}. By definition it is related to $n_N(D_p)$ by

$$N(D_p) = \int_0^{D_p} n_N\left(D_p^*\right) dD_p^* \tag{8.2}$$

D_p^* is used as the integrating dummy variable in (8.2) to avoid confusion with the upper limit of the integration, D_p. Differentiating (8.2), the size distribution function can be written as

$$n_N(D_p) = dN/dD_p \qquad (8.3)$$

and $n_N(D_p)$ can be also viewed as the derivative of the cumulative aerosol size distribution function $N(D_p)$. Both sides of (8.3) represent the same aerosol size distribution, and the notation dN/dD_p is often used instead of $n_N(D_p)$. To conform to the common notation, we will also express the distributions in this manner.

> **Example 8.1**   For the distribution of Figure 8.4, how many particles of diameter 0.1 μm exist?
>
> According to the inset of Figure 8.4, $n_N(0.1\ \mu m) = 13{,}000\ \mu m^{-1}\ cm^{-3}$. However, this is not the number of particles of diameter 0.1 μm (it even has the wrong units). To calculate the number of particles we need to multiply n_N by the width of the size range ΔD_p. But if we are interested only in particles with $D_p = 0.1$ μm this size range is zero and therefore there are zero particles of diameter exactly equal to 0.1 μm.
> Let us try to rephrase the question posed above.

> **Example 8.2.**   For the distribution of Figure 8.4, how many particles with diameter in the range 0.1–0.11 μm exist?
>
> The size distribution is practically constant over this narrow range with $n_N(0.1\ \mu m) = 13{,}000\ \mu m^{-1}\ cm^{-3}$. The width of the region is $0.11 - 0.1 = 0.01$ μm and there are $0.01 \times 13{,}000 = 130$ particles cm^{-3} with diameters between 0.1 and 0.11 μm for this size distribution.

The above examples indicate that while n_N is a unique description of the aerosol size distribution (it does not depend on definitions of size bins, etc.), one should be careful with its physical interpretation.

8.1.2   The Surface Area, Volume, and Mass Distributions

Several aerosol properties depend on the particle surface area and volume distributions with respect to particle size. Let us define the aerosol surface area distribution $n_S(D_p)$ as

$$n_S(D_p)dD_p = \text{surface area of particles per cm}^3 \text{ of air having}$$
$$\text{diameters in range } D_p \text{ to } (D_p + dD_p)$$

and let us consider all particles as spheres. All the particles in this infinitesimally narrow size range have effectively the same diameter D_p, and each of them has surface area πD_p^2. There are $n_N(D_p)\,dD_p$ particles in this size range and therefore their surface area is $\pi D_p^2 n_N(D_p)d D_p$. But then by definition

$$n_S(D_p) = \pi D_p^2 n_N(D_p) \quad (\mu m\ cm^{-3}) \qquad (8.4)$$

The total surface area S_t of the aerosol per cm^3 of air is then

$$S_t = \pi \int_0^\infty D_p^2 n_N(D_p)\,dD_p = \int_0^\infty n_S(D_p)\,dD_p \quad (\mu m^2\,cm^{-3}) \qquad (8.5)$$

and is equal to the area below the $n_S(D_p)$ curve in Figure 8.4. The aerosol volume distribution $n_V(D_p)$ can be defined as

$$n_V(D_p)\,dD_p = \text{volume of particles per cm}^3 \text{ of air having diameters}$$
$$\text{in the range } D_p \text{ to } (D_p + dD_p)$$

and therefore

$$n_V(D_p) = \frac{\pi}{6} D_p^3 n_N(D_p) \quad (\mu m^2\,cm^{-3}) \qquad (8.6)$$

The total aerosol volume per cm^3 of air V_t is

$$V_t = \frac{\pi}{6} \int_0^\infty D_p^3 n_N(D_p)\,dD_p = \int_0^\infty n_V(D_p)\,dD_p \quad (\mu m^3\,cm^{-3}) \qquad (8.7)$$

and is equal to the area below the $n_V(D_p)$ curve in Figure 8.4.

If the particles all have density ρ_p (g cm^{-3}) then the distribution of particle mass with respect to particle size, $n_M(D_p)$, is

$$n_M(D_p) = \left(\frac{\rho_p}{10^6}\right) n_V(D_p) = \left(\frac{\rho_p}{10^6}\right)\left(\frac{\pi}{6}\right) D_p^3 n_N(D_p) \quad (\mu g\,\mu m^{-1}\,cm^{-3}) \qquad (8.8)$$

where the factor 10^6 is needed to convert the units of density ρ_p from g cm^{-3} to μg μm^{-3}, and to maintain the units for $n_M(D_p)$ as μg μm^{-1} cm^{-3}.

Because particle diameters in an aerosol population typically vary over several orders of magnitude, use of the distribution functions, $n_N(D_p)$, $n_S(D_p)$, $n_V(D_p)$, and $n_M(D_p)$, is often inconvenient. For example, all the structure of the number distribution depicted in Figure 8.4 occurs in the region from a few nanometers to 0.3 μm diameter, a small part of the 0–10 μm range of interest. To circumvent this scale problem, the horizontal axis (abscissa) can be scaled in logarithmic intervals so that several orders of magnitude in D_p can be clearly seen (Figure 8.5). Plotting $n_N(D_p)$ on semilog axes gives, however, a somewhat distorted picture of the aerosol distribution. The area below the curve no longer corresponds to the aerosol number concentration. For example, Figure 8.5 appears to suggest that more than 90% of the particles are in the smaller mode centered around 0.02 μm. In reality, the numbers of particles in the two modes are almost equal; this can be seen when the distributions are expressed as a function of the logarithm of diameter, which we discuss in the next section (Figure 8.6).

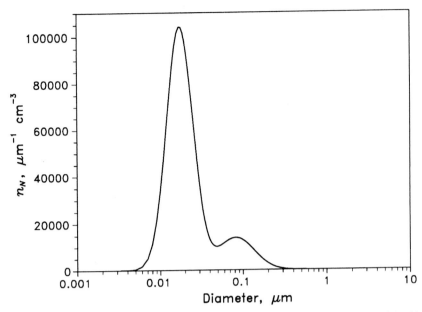

FIGURE 8.5   The same aerosol distribution as in Figure 8.4, plotted against the logarithm of the diameter.

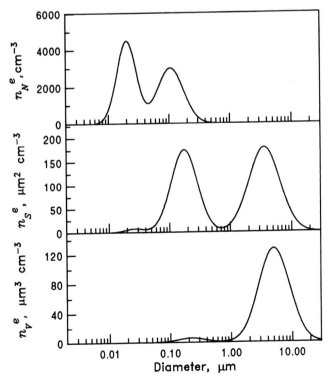

FIGURE 8.6   The same aerosol distribution as in Figures 8.4 and 8.5 expressed as a function of $\log D_p$ and plotted against $\log D_p$. Also shown are the surface and volume distributions. The areas below the three curves correspond to the total aerosol number, surface, and volume, respectively.

8.1.3   Distributions Based on ln D_p and log D_p

Expressing the aerosol distributions as functions of ln D_p or log D_p instead of D_p is often the most convenient way to represent the aerosol size distribution. Formally, we cannot take the logarithm of a dimensional quantity. Thus, when we write ln D_p, we really mean ln $(D_p/1)$, where the "reference" particle diameter is 1 μm and is not explicitly indicated. We can therefore define the number distribution function $n_N^e(\ln D_p)$ as

$$n_N^e(\ln D_p)\,d \ln D_p = \text{number of particles per cm}^3 \text{ of air in the size range}$$
$$\ln D_p \text{ to } (\ln D_p + d \ln D_p)$$

The units of $n_N^e(\ln D_p)$ are cm^{-3} since ln D_p is dimensionless. The total number concentration of particles N_t is

$$N_t = \int_{-\infty}^{\infty} n_N^e(\ln D_p)\,d \ln D_p \quad (\text{cm}^{-3}) \tag{8.9}$$

The limits of integration in (8.9) are from $-\infty$ to ∞ as the independent variable is ln D_p.

The surface area and volume distributions as functions of ln D_p can be defined similarly to those with respect to D_p

$$n_S^e(\ln D_p) = \pi D_p^2 n_N^e(\ln D_p) \quad (\mu\text{m}^2\text{cm}^{-3})$$
$$n_V^e(\ln D_p) = \frac{\pi}{6} D_p^3 n_N^e(\ln D_p) \quad (\mu\text{m}^3\text{cm}^{-3}) \tag{8.10}$$

with

$$S_t = \pi \int_{-\infty}^{\infty} D_p^2 n_N^e(\ln D_p)\,d \ln D_p$$
$$= \int_{-\infty}^{\infty} n_S^e(\ln D_p)\,d \ln D_p \quad (\mu\text{m}^2\,\text{cm}^{-3}) \tag{8.11}$$
$$V_t = \frac{\pi}{6} \int_{-\infty}^{\infty} D_p^3 n_N^e(\ln D_p)\,d \ln D_p$$
$$= \int_{-\infty}^{\infty} n_V^e(\ln D_p)\,d \ln D_p \quad (\mu\text{m}^3\,\text{cm}^{-3}) \tag{8.12}$$

These aerosol distributions can also be expressed as functions of the base 10 logarithm log D_p, defining $n_N^\circ(\log D_p)$, $n_S^\circ(\log D_p)$, and $n_V^\circ(\log D_p)$. Note that n_N, n_N^e, and n_N° are different mathematical functions, and, for the same diameter D_p, they have different arguments, namely, D_p, ln D_p, and log D_p. The expressions relating these functions will be derived in the next section.

Using the notation $dN/dS/dV$ = the differential number/surface/volume of particles in the size range D_p to $D_p + dD_p$ we have

$$dN = n_N(D_p)dD_p = n_N^e(\ln D_p)d\ln D_p = n_N^\circ(\log D_p)d\log D_p \qquad (8.13)$$

$$dS = n_S(D_p)dD_p = n_S^e(\ln D_p)d\ln D_p = n_S^\circ(\log D_p)d\log D_p \qquad (8.14)$$

$$dV = n_V(D_p)dD_p = n_V^e(\ln D_p)d\ln D_p = n_V^\circ(\log D_p)d\log D_p \qquad (8.15)$$

On the basis of that notation, the various size distributions are

$$
\begin{aligned}
n_N(D_p) &= \frac{dN}{dD_p} & n_N^e(\ln D_p) &= \frac{dN}{d\ln D_p} & n_N^\circ(\log D_p) &= \frac{dN}{d\log D_p} \\[2mm]
n_S(D_p) &= \frac{dS}{dD_p} & n_S^e(\ln D_p) &= \frac{dS}{d\ln D_p} & n_S^\circ(\log D_p) &= \frac{dS}{d\log D_p} \qquad (8.16)\\[2mm]
n_V(D_p) &= \frac{dV}{dD_p} & n_V^e(\ln D_p) &= \frac{dV}{d\ln D_p} & n_V^\circ(\log D_p) &= \frac{dV}{d\log D_p}
\end{aligned}
$$

and they represent the derivatives of the cumulative number/surface area/volume distributions $N(D_p)/S(D_p)/V(D_p)$ with respect to D_p, $\ln D_p$, and $\log D_p$, respectively.

8.1.4 Relating Size Distributions Based on Different Independent Variables

It is often necessary to relate a size distribution based on one independent variable, say, D_p, to one based on another independent variable, say, $\log D_p$. Such a relation can be derived based on (8.13). The number of particles dN in an infinitesimal size range D_p to $D_p + dD_p$ is the same regardless of the expression used for the description of the size distribution function. Thus in the particular case of $n_N(D_p)$ and $n_N^\circ(\log D_p)$

$$n_N(D_p)dD_p = n_N^\circ(\log D_p)d\log D_p \qquad (8.17)$$

Since $d\log D_p = d\ln D_p/2.303 = dD_p/2.303D_p$, (8.17) becomes

$$n_N^\circ(\log D_p) = 2.303D_p n_N(D_p) \qquad (8.18)$$

Similarly

$$n_S^\circ(\log D_p) = 2.303D_p n_S(D_p) \qquad (8.19)$$

$$n_V^\circ(\log D_p) = 2.303D_p n_V(D_p) \qquad (8.20)$$

The distributions with respect to D_p are related to those with respect to $\ln D_p$ by

$$n_N^e(\ln D_p) = D_p n_N(D_p) \tag{8.21}$$

$$n_S^e(\ln D_p) = D_p n_S(D_p) \tag{8.22}$$

$$n_V^e(\ln D_p) = D_p n_V(D_p) \tag{8.23}$$

This procedure can be generalized to relate any two size distribution functions $n(u)$ and $n(v)$ where both u and v are related to D_p. The generalization of (8.17) is

$$n(u)\,du = n(v)\,dv \tag{8.24}$$

and dividing both sides by dD_p

$$n(u) = n(v)\frac{(dv/dD_p)}{(du/dD_p)} \tag{8.25}$$

8.1.5 Properties of Size Distributions

It is often convenient to summarize the features of an aerosol distribution using one or two of its properties (mean particle size, spread of distribution) than by using the full function $n_N(D_p)$. Growth of particles corresponds to a shifting of parts of the distribution to larger sizes or simply an increase of the mean particle size. These properties are called the *moments* of the distribution, and the two most often used are the mean and the variance.

Let us assume that we have a discrete distribution consisting of M groups of particles, with diameters D_k and number concentrations N_k, $k = 1, 2, \ldots, M$. The number concentration of particles is therefore

$$N_t = \sum_{k=1}^{M} N_k \tag{8.26}$$

The mean particle diameter, $\bar{D}_p$, of the population is

$$\bar{D}_p = \frac{\sum_{k=1}^{M} N_k D_k}{\sum_{k=1}^{M} N_k} = \frac{1}{N_t}\sum_{k=1}^{M} N_k D_k \tag{8.27}$$

The variance σ^2, a measure of the spread of the distribution around the mean diameter $\bar{D}_p$, is defined by

$$\sigma^2 = \frac{\sum_{k=1}^{M} N_k(D_k - \bar{D}_p)^2}{\sum_{k=1}^{M} N_k} = \frac{1}{N_t}\sum_{k=1}^{M} N_k(D_k - \bar{D}_p)^2 \tag{8.28}$$

A value of σ^2 equal to zero would mean that every one of the particles in the distribution has precisely diameter $\bar{D}_p$. An increasing σ^2 indicates that the spread of the distribution around the mean diameter D_p is increasing.

We will usually deal with aerosol distributions in continuous form. Given the number distribution $n_N(D_p)$, (8.27) and (8.28) can be written in continuous form to define the mean particle diameter of the distribution by

$$\bar{D}_p = \frac{\int_0^\infty D_p n_N(D_p)\,dD_p}{\int_0^\infty n_N(D_p)\,dD_p} = \frac{1}{N_t}\int_0^\infty D_p n_N(D_p)\,dD_p \qquad (8.29)$$

and the variance of the distribution by

$$\sigma^2 = \frac{\int_0^\infty (D_p - \bar{D}_p)^2 n_N(D_p)\,dD_p}{\int_0^\infty n_N(D_p)\,dD_p} = \frac{1}{N_t}\int_0^\infty (D_p - \bar{D}_p)^2 n_N(D_p)\,dD_p \qquad (8.30)$$

Table 8.2 presents a number of other mean values that are often used in characterizing an aerosol size distribution.

TABLE 8.2 Mean Values Often Used in Characterizing an Aerosol Size Distribution

Property	Defining Relation	Description
Number mean diameter $\bar{D}_p$	$\bar{D}_p = \frac{1}{N_t}\int_0^\infty D_p n_N(D_p)\,dD_p$	Average diameter of the population
Median diameter D_{med}	$\int_0^{D_{\mathrm{med}}} n_N(D_p)\,dD_p = \frac{1}{2}N_t$	Diameter below which one-half the particles lie and above which one-half the particles lie
Mean surface area $\bar{S}$	$\bar{S} = \frac{1}{N_t}\int_0^\infty n_S(D_p)\,dD_p$	Average surface area of the population
Mean volume $\bar{V}$	$\bar{V} = \frac{1}{N_t}\int_0^\infty n_V(D_p)\,dD_p$	Average volume of the population
Surface area mean diameter D_S	$N_t \pi D_S^2 = \int_0^\infty n_S(D_p)\,dD_p$	Diameter of the particle whose surface area equals the mean surface area of the population
Volume mean diameter D_V	$N_t \frac{\pi}{6} D_V^3 = \int_0^\infty n_V(D_p)\,dD_p$	Diameter of the particle whose volume equals the mean volume of the population
Surface area median diameter D_{S_m}	$\int_0^{D_{S_m}} n_S(D_p)\,dD_p = \frac{1}{2}\int_0^\infty n_S(D_p)\,dD_p$	Diameter below which one-half the particle surface area lies and above which one-half the particle surface area lies
Volume median diameter D_{V_m}	$\int_0^{D_{V_m}} n_V(D_p)\,dD_p = \frac{1}{2}\int_0^\infty n_V(D_p)\,dD_p$	Diameter below which one-half the particle volume lies and above which one-half the particle volume lies
Mode diameter D_{mode}	$\left(\dfrac{dn_N(D_p)}{dD_p}\right)_{D_{\mathrm{mode}}} = 0$	Local maximum of the number distribution

8.1.6   The Lognormal Distribution

A measured aerosol size distribution can be reported as a table of the distribution values for dozens of diameters. For many applications carrying around hundreds or thousands of aerosol distribution values is awkward. In these cases it is often convenient to use a relatively simple mathematical function to describe the atmospheric aerosol distribution. These functions are semiempirical in nature and have been chosen because they match well observed shapes of ambient distributions. Of the various mathematical functions that have been proposed, the lognormal distribution (Aitchison and Brown 1957) often provides a good fit and is regularly used in atmospheric applications. A series of other distributions are discussed in the next section.

The normal distribution for a quantity u defined from $-\infty < u < \infty$ is given by

$$n(u) = \frac{N}{(2\pi)^{1/2}\sigma_u}\exp\left(-\frac{(u-\bar{u})^2}{2\sigma_u^2}\right) \tag{8.31}$$

where $\bar{u}$ is the mean of the distribution, σ_u^2 is the variance and

$$N = \int_{-\infty}^{\infty} n(u)\,du \tag{8.32}$$

The normal distribution has the characteristic bell shape, with a maximum at $\bar{u}$. The standard deviation, σ_u, quantifies the width of the distribution, and 68% of the area below the curve is in the range $\bar{u} \pm \sigma_u$.

A quantity u is *lognormally distributed* if its logarithm is normally distributed. Either the natural ($\ln u$) or the base 10 logarithm ($\log u$) can be used, but since the former is more common, we will express our results in terms of $\ln D_p$. An aerosol population is therefore log-normally distributed if $u = \ln D_p$ satisfies (8.31), or

$$n_N^e(\ln D_p) = \frac{dN}{d\ln D_p} = \frac{N_t}{(2\pi)^{1/2}\ln \sigma_g}\exp\left(-\frac{\left(\ln D_p - \ln \bar{D}_{pg}\right)^2}{2\ln^2 \sigma_g}\right) \tag{8.33}$$

where N_t is the total aerosol number concentration, and $\bar{D}_{pg}$ and σ_g are for the time being the two parameters of the distribution. Shortly we will discuss the physical significance of these parameters. The distribution $n_N(D_p)$ is often used instead of $n_N^e(\ln D_p)$. Combining (8.21) with (8.33)

$$n_N(D_p) = \frac{dN}{dD_p} = \frac{N_t}{(2\pi)^{1/2}D_p\ln \sigma_g}\exp\left(-\frac{\left(\ln D_p - \ln \bar{D}_{pg}\right)^2}{2\ln^2 \sigma_g}\right) \tag{8.34}$$

A lognormal aerosol distribution with $\bar{D}_{pg} = 0.8$ μm and $\sigma_g = 1.5$ is depicted in Figure 8.7.

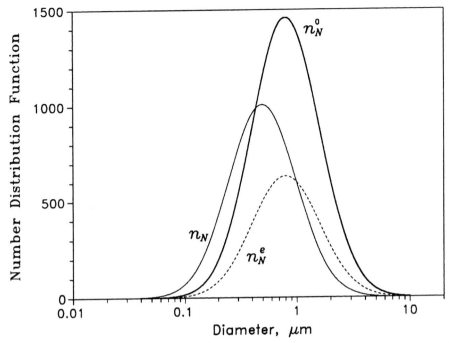

FIGURE 8.7   Aerosol distribution functions, $n_N(D_p)$, $n_N^\circ(\log D_p)$ and $n_N^e(\ln D_p)$ for a lognormally distributed aerosol distribution $\bar{D}_{pg} = 0.8$ μm and $\sigma_g = 1.5$ versus $\log D_p$. Even if all three functions describe the same aerosol population, they differ from each other because they use a different independent variable. The aerosol number is the area below the $n_N^\circ(\log D_p)$ curve.

We now wish to examine the physical significance of the two parameters $\bar{D}_{pg}$ and σ_g. To do so we will use the cumulative size distribution $N(D_p)$.

If the aerosol distribution is lognormal, $n_N(D_p)$ is given by (8.34) and therefore

$$N(D_p) = \frac{N_t}{(2\pi)^{1/2}\ln \sigma_g} \int_0^{D_p} \frac{1}{D_p^*} \exp\left[-\frac{\left(\ln D_p^* - \ln \bar{D}_{pg}\right)^2}{2\ln^2 \sigma_g}\right] dD_p^* \qquad (8.35)$$

To evaluate this integral we let

$$\eta = \left(\ln D_p^* - \ln \bar{D}_{pg}\right)/\sqrt{2}\ln \sigma_g \qquad (8.36)$$

and we obtain

$$N(D_p) = \frac{N_t}{\sqrt{\pi}} \int_{-\infty}^{\left(\ln D_p - \ln \bar{D}_{pg}\right)/\sqrt{2}\ln \sigma_g} e^{-\eta^2} d\eta \qquad (8.37)$$

The *error function* erf z is defined as

$$\mathrm{erf}\, z = \frac{2}{\sqrt{\pi}} \int_0^z e^{-\eta^2} d\eta \qquad (8.38)$$

and erf$(0) = 0$, erf$(\infty) = 1$. If we divide the integral in (8.37) into one from $-\infty$ to 0 and the second from 0 to $(\ln D_p - \ln \bar{D}_{pg})/\sqrt{2} \ln \sigma_g$, then the first integral is seen to be equal to $\sqrt{\pi}/2$ and the second to $(\sqrt{\pi}/2)$erf$[(\ln D_p - \ln \bar{D}_{pg})/\sqrt{2} \ln \sigma_g]$. Thus for the lognormal distribution

$$N(D_p) = \frac{N_t}{2} + \frac{N_t}{2}\,\text{erf}\left(\frac{\ln(D_p/\bar{D}_{pg})}{\sqrt{2}\ln \sigma_g}\right) \tag{8.39}$$

For $D_p = \bar{D}_{pg}$, since erf$(0) = 0$

$$N(\bar{D}_{pg}) = \frac{N_t}{2} \tag{8.40}$$

and we see that $\bar{D}_{pg} = D_{\text{med}}$ is the *median diameter*, that is, the diameter for which exactly one-half of the particles are smaller and one-half are larger. To understand the role of σ_g let us consider the diameter $D_{p\sigma}$ for which $\sigma_g = D_{p\sigma}/\bar{D}_{pg}$. At that diameter, using (8.39)

$$N(D_{p\sigma}) = N_t\left[\frac{1}{2} + \frac{1}{2}\,\text{erf}\left(\frac{1}{\sqrt{2}}\right)\right] = 0.841\,N_t \tag{8.41}$$

Thus σ_g is the ratio of the diameter below which 84.1% of the particles lie to the median diameter and is termed the *geometric standard deviation*. A monodisperse aerosol population has $\sigma_g = 1$. For any distribution, 67% of all particles lie in the range from $\bar{D}_{pg}/\sigma_g$ to $\bar{D}_{pg}\sigma_g$ and 95% of all particles lie in the range from $\bar{D}_{pg}/2\sigma_g$ to $2\bar{D}_{pg}\sigma_g$.

Let us calculate the mean diameter $\bar{D}_p$ of a lognormally distributed aerosol. By definition, the mean diameter is found from

$$\bar{D}_p = \frac{1}{N_t}\int_0^\infty D_p n_N(D_p)\,dD_p \tag{8.42}$$

which we wish to evaluate in the case of $n_N(D_p)$ given by (8.34). Therefore

$$\bar{D}_p = \frac{1}{\sqrt{2\pi}\ln \sigma_g}\int_0^\infty \exp\left(-\frac{(\ln D_p - \ln \bar{D}_{pg})^2}{2\ln^2 \sigma_g}\right)dD_p \tag{8.43}$$

After evaluating the integral one finds that

$$\bar{D}_p = \bar{D}_{pg}\exp\left(\frac{\ln^2 \sigma_g}{2}\right) \tag{8.44}$$

We see that the mean diameter of a lognormal distribution depends on both $\bar{D}_{pg}$ and σ_g and increases as σ_g increases.

8.1.7 Plotting the Lognormal Distribution

The cumulative distribution function $N(D_p)$ for a lognormally distributed aerosol population is given by (8.39). Defining the normalized cumulative distribution

$$\bar{N}(D_p) = \frac{N(D_p)}{N_t} \tag{8.45}$$

one obtains

$$\bar{N}(D_p) = \frac{1}{2} + \frac{1}{2}\, \mathrm{erf}\left(\frac{\ln D_p - \ln \bar{D}_{pg}}{\sqrt{2} \ln \sigma_g}\right) \tag{8.46}$$

The cumulative distribution function can be plotted on a log–probability graph using one of the scientific computer graphics programs. In these diagrams the x axis is logarithmic, $\log(x)$, and the y axis is scaled like the error function, $\mathrm{erf}(y)$. This scaling compresses the scale near the median (50% point) and expands the scale near the ends. In these graphs the cumulative distribution function of a log-normal distribution is a straight line (Figure 8.8). The point at $\bar{N}(D_p) = 0.5$ occurs when $D_p = \bar{D}_{pg}$. Therefore the geometric mean, or median, of the distribution is the value of D_p where the straight line plot of $\bar{N}(D_p)$ crosses the 50th percentile.

The point at which $\bar{N}(D_p) = 0.84$ occurs for $\ln D_{p+\sigma} = \ln \bar{D}_{pg} + \ln \sigma_g$ or $D_{p+\sigma} = \bar{D}_{pg}\sigma_g$. The slope of the line is related to the geometric standard deviation of the distribution. Lognormal distributions with the same standard deviation when plotted on probability

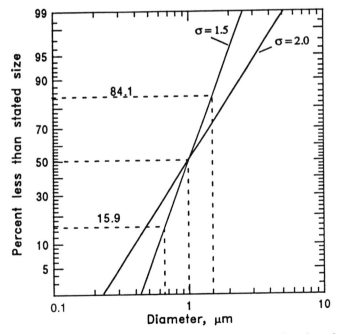

FIGURE 8.8 Cumulative lognormal aerosol number distributions plotted on log–probability paper. The distributions have mean diameter of 1 μm and $\sigma_g = 2$ and 1.5, respectively.

coordinates are parallel to each other. A small standard deviation corresponds to a narrow distribution and to a steep line on the log–probability graph (Figure 8.8). The geometric standard deviation can be calculated as the ratio of the diameter $D_{p+\sigma}$ for which $\bar{N}(D_{p+\sigma}) = 0.84$ to the median diameter

$$\sigma_g = D_{p+\sigma}/\bar{D}_{pg} \tag{8.47}$$

8.1.8 Properties of the Lognormal Distribution

We have discussed the properties of the lognormal distribution for the number concentration. The next step is examination of the surface and volume distributions corresponding to a lognormal number distribution given by (8.34). Since $n_S(D_p) = \pi D_p^2 n_N(D_p)$ and $n_V(D_p) = (\pi/6)D_p^3 n_N(D_p)$, let us determine the forms of $n_S(D_p)$ and $n_V(D_p)$ when $n(D_p)$ is lognormal. From (8.34) one gets

$$n_S(D_p) = \frac{\pi D_p^2 N_t}{(2\pi)^{1/2}D_p \ln \sigma_g} \exp\left(-\frac{(\ln D_p - \ln \bar{D}_{pg})^2}{2\ln^2 \sigma_g}\right) \tag{8.48}$$

By letting $D_p^2 = \exp(2 \ln D_p)$, expanding the exponential, and completing the square in the exponent, (8.48) becomes

$$n_S(D_p) = \frac{\pi N_t}{(2\pi)^{1/2}D_p \ln \sigma_g} \exp\left(2 \ln \bar{D}_{pg} + 2\ln^2 \sigma_g\right)$$

$$\times \exp\left(-\frac{\left[\ln D_p - \left(\ln \bar{D}_{pg} + 2\ln^2 \sigma_g\right)\right]^2}{2\ln^2 \sigma_g}\right) \tag{8.49}$$

Thus we see that if the number distribution $n_N(D_p)$ is lognormal, the surface distribution $n_S(D_p)$ is also lognormal with the same geometric standard deviation σ_g as the parent distribution and with the surface median diameter given by

$$\ln \bar{D}_{pgS} = \ln \bar{D}_{pg} + 2\ln^2 \sigma_g \tag{8.50}$$

The calculations above can be repeated for the volume distribution, and one can show that

$$n_V(D_p) = \frac{\pi D_p^3 N_t}{6(2\pi)^{1/2}D_p \ln \sigma_g} \exp\left(-\frac{(\ln D_p - \ln \bar{D}_{pg})^2}{2\ln^2 \sigma_g}\right) \tag{8.51}$$

or by letting $D_p^3 = \exp(3 \ln D_p)$, expanding the exponential, and completing the square in the exponent, (8.51) becomes

$$n_V(D_p) = \frac{(\pi/6)N_t}{(2\pi)^{1/2}D_p \ln \sigma_g} \exp\left(3 \ln \bar{D}_{pg} + \tfrac{9}{2}\ln^2 \sigma_g\right)$$

$$\times \exp\left(-\frac{\left[\ln D_p - \left(\ln \bar{D}_{pg} + 3\ln^2 \sigma_g\right)\right]^2}{2\ln^2 \sigma_g}\right) \tag{8.52}$$

Therefore if the number distribution $n_N(D_p)$ is lognormal, the volume distribution $n_V(D_p)$ is also lognormal with the same geometric standard deviation σ_g as the parent distribution and with the volume median diameter given by

$$\ln \bar{D}_{pgV} = \ln \bar{D}_{pg} + 3\ln^2 \sigma_g \qquad (8.53)$$

The constant standard deviation for the number, surface, and volume distributions for any lognormal distribution is one of the great advantages of this mathematical representation.

Plotting the surface and volume distributions of a log-normal aerosol distribution on a log–probability graph would also result in straight lines parallel to each other (same standard deviation). For the distribution shown in Figure 8.8 with $\bar{D}_{pg} = 1.0$ μm and $\sigma_g = 2.0$, the resulting surface area and volume median diameters are approximately 2.6 μm and 4.2 μm, respectively.

The Power-Law Distribution A variety of other mathematical functions have been proposed for the description of atmospheric aerosol distributions. The power law, or Junge, distribution was one of the first used in atmospheric science (Pruppacher and Klett 1980),

$$n_N^\circ(\log D_p) = \frac{C}{(D_p)^\alpha}$$

where C and α are positive constants. Values of α from 2 to 5 are used for ambient aerosol distributions.

1. What is the shape of the power law distribution when plotted in log–log coordinates? What is the meaning of α and C based on this plot?
2. Calculate the volume distribution function n_V° for the power-law distribution.
3. Figure 8.9 shows a typical urban aerosol size distribution and its fit by a power-law distribution using $C = 91.658$ and $\alpha = 3.746$. Using these graphs and also Figure 8.6, discuss the advantages and disadvantages of the use of the power law distribution for atmospheric aerosols.

1. Taking the logarithm of both sides of the power-law equation, we find that $\log n_N^\circ = \log C - \alpha \log D_p$, and therefore this distribution will be a straight line on a log–log plot with slope equal to $-\alpha$. For $D_p = 1$ μm we find that $C = n_N^\circ$ and, as a result, the parameter C is the value of the distribution function at 1 μm.

2. Using (8.10), the volume distribution function will be

$$n_V^\circ(\log D_p) = (\pi/6)D_p^3 n_N^\circ(\log D_p) = (\pi C/6)D_p^{3-\alpha}$$

This is once more a straight line on a log–log plot.

3. The main advantage of the power-law distribution is its mathematical simplicity compared to other distribution functions. It is much easier to perform calculations with the simple power-law expression compared to the lognormal distribution given by (8.33). Figure 8.9 indicates that the power-law distribution can provide a reasonable approximation to atmospheric aerosol number distributions

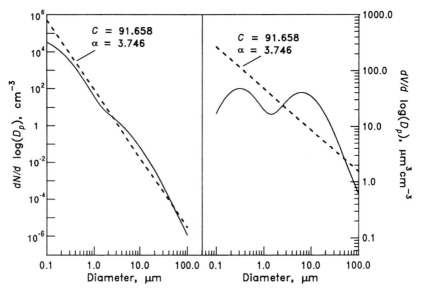

FIGURE 8.9 Fitting of an urban aerosol number distribution with a power-law distribution (left) and comparison of the corresponding volume distributions (right). Even if the power-law distribution appears to match the number distribution, it fails to reproduce the volume distribution.

(Leaitch and Isaac 1991). However, the power-law function is a monotonically decreasing function. We have already seen (e.g., in Figure 8.6) that this is not the case for ambient aerosol size distributions that have considerably more structure. The power-law distribution cannot fit the maxima and minima in a distribution like that of Figure 8.6. It can only fit the continuously decreasing part of the distribution for diameters greater than 0.1 μm (Figure 8.9). The derived power-law distributions are accurate only over a limited size range, and extrapolation to smaller or larger sizes may introduce significant errors.

Another weakness of the power-law distribution is that, even if it is a reasonable approximation of the number distribution for $D_p > 0.1$ μm, it cannot provide a reasonable approximation of the typical volume distributions (Figure 8.9). Atmospheric aerosol volume distributions always have multiple modes for $D_p > 0.1$ μm, and the power-law volume distribution that we calculated above is also a monotonic function (continuously decreasing for $\alpha > 3$ and increasing for $\alpha < 3$). For the distribution of Figure 8.9, the power-law distribution grossly overpredicts the volume of the submicrometer particles and seriously underpredicts the volume of the coarse aerosol. Therefore use of a fitted power-law distribution for the calculation of aerosol properties that depend on powers of the diameter (e.g., optical properties, condensation rates) should be avoided.

8.2 AMBIENT AEROSOL SIZE DISTRIBUTIONS

As a result of particle emission, in situ formation, and the variety of subsequent processes, the atmospheric aerosol distribution is characterized by a number of modes. The volume or

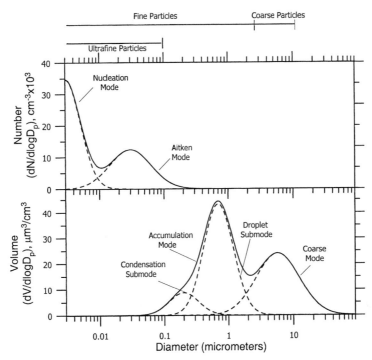

FIGURE 8.10 Typical number and volume distributions of atmospheric particles with the different modes.

mass distribution is dominated in most areas by two modes (Figure 8.10, lower panel): the accumulation (from ~ 0.1 to ~ 2 μm) and the coarse mode (from ~ 2 to ~ 50 μm). Accumulation-mode particles are the result of primary emissions; condensation of secondary sulfates, nitrates, and organics from the gas phase; and coagulation of smaller particles. In a number of cases the accumulation mode consists of two overlapping submodes; the condensation mode and the droplet mode (Figure 8.10, upper panel) (John et al. 1990). The condensation submode is the result of primary particle emissions, and growth of smaller particles by coagulation and vapor condensation. The droplet submode is created during the cloud processing of some of the accumulation-mode particles. Particles in the coarse mode are usually produced by mechanical processes such as wind or erosion (dust, seasalt, pollens, etc.). Most of the material in the coarse mode is primary, but there are some secondary sulfates and nitrates.

A different picture of the ambient aerosol distribution is obtained if one focuses on the number of particles instead of their mass (Figure 8.10, upper panel). The particles with diameters larger than 0.1 μm, which contribute practically all the aerosol mass, are negligible in number compared to the particles smaller than 0.1 μm. Two modes usually dominate the aerosol number distribution in urban and rural areas: the nucleation mode (particles smaller than 10 nm or so) and the Aitken nuclei (particles with diameters between 10 and 100 nm or so). The nucleation mode particles are usually fresh aerosols created in situ from the gas phase by nucleation. The nucleation mode may or may not be present depending on the atmospheric conditions. Most of the Aitken nuclei start their atmospheric life as primary particles, and secondary material condenses on them as they are transported through the atmosphere. The nucleation-mode particles have negligible mass (e.g., 100,000 particles cm^{-3} with a diameter equal to 10 nm have a mass concentration of <0.05 μg m^{-3}) while the larger Aitken nuclei form the accumulation mode in the mass distribution.

Particles with diameters larger than 2.5 μm are identified as *coarse* particles, while those with diameters less than 2.5 μm are called *fine* particles. The fine particles include most of the total number of particles and a large fraction of the mass. The fine particles with diameters smaller than 0.1 μm are often called *ultrafine* particles.

Atmospheric aerosol size distributions are often described as the sum of n lognormal distributions

$$n_N^{\circ}\left(\log D_p\right) = \sum_{i=1}^{n} \frac{N_i}{(2\pi)^{1/2}\log \sigma_i} \exp\left(-\frac{\left(\log D_p - \log \bar{D}_{pi}\right)^2}{2 \log^2 \sigma_i}\right) \tag{8.54}$$

where N_i is the number concentration, $\bar{D}_{pi}$ is the median diameter, and σ_i is the standard deviation of the ith lognormal mode. In this case $3n$ parameters are necessary for the description of the full aerosol distribution. Characteristics of model aerosol distributions are presented in Table 8.3 following the suggestions of Jaenicke (1993).

8.2.1 Urban Aerosols

Urban aerosols are mixtures of primary particulate emissions from industries, transportation, power generation, and natural sources and secondary material formed by gas-to-particle conversion mechanisms. The number distribution is dominated by particles smaller than 0.1 μm, while most of the surface area is in the 0.1–0.5 μm size range. On the contrary, the aerosol mass distribution usually has two distinct modes, one in the submicrometer regime (referred to as the "accumulation mode") and the other in the coarse-particle regime (Figure 8.11).

The aerosol size distribution is quite variable in an urban area. Extremely high concentrations of fine particles (less than 0.1 μm in diameter) are found close to sources (e.g., highways), but their concentration decreases rapidly with distance from the source (Figure 8.12).

There is roughly an order of magnitude more particles close to a major road compared to the average urban concentration. However, the concentration of these particles decays rapidly because of dilution in a characteristic distance of roughly 100 m from the road (Zhu et al. 2002). The increase in mass concentration next to major roads is usually smaller, roughly 10–20% of the urban background. Part of this mass increase is in the Aitken and accumulation modes because of the fresh combustion particles, but a significant part of it can be in the coarse mode due to resuspension of dust particles from traffic (Figure 8.13).

The mass concentrations of the accumulation and coarse particle modes are comparable for most urban areas. The Aitken and nucleation modes, with the exception of areas close to combustion sources, contain negligible volume (Figures 8.11 and 8.13). Most of the aerosol surface area is in particles of diameters 0.1–0.5 μm in the accumulation mode (Figure 8.11). Because of this availability of area, transfer of material from the gas phase during gas-to-particle conversion occurs preferentially on them.

TABLE 8.3 Parameters for Model Aerosol Distributions Expressed as the Sum of Three Lognormal Modes

Type	Mode I			Mode II			Mode III		
	N (cm^{-3})	D_p (μm)	$\log \sigma$	N (cm^{-3})	D_p (μm)	$\log \sigma$	N (cm^{-3})	D_p (μm)	$\log \sigma$
Urban	9.93×10^4	0.013	0.245	1.11×10^3	0.014	0.666	3.64×10^4	0.05	0.337
Marine	133	0.008	0.657	66.6	0.266	0.210	3.1	0.58	0.396
Rural	6650	0.015	0.225	147	0.054	0.557	1990	0.084	0.266
Remote continental	3200	0.02	0.161	2900	0.116	0.217	0.3	1.8	0.380
Free troposphere	129	0.007	0.645	59.7	0.250	0.253	63.5	0.52	0.425
Polar	21.7	0.138	0.245	0.186	0.75	0.300	3×10^{-4}	8.6	0.291
Desert	726	0.002	0.247	114	0.038	0.770	0.178	21.6	0.438

Source: Jaenicke (1993).

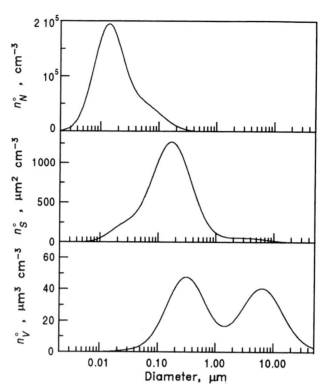

FIGURE 8.11 Typical urban aerosol number, surface, and volume distributions.

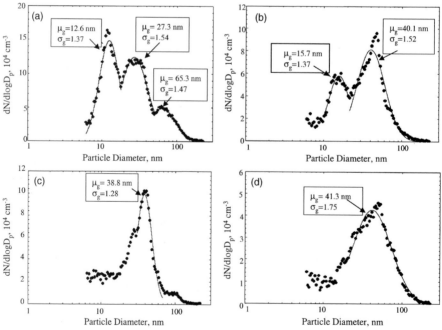

FIGURE 8.12 Measured and fitted multimodal number distributions at different distances downwind from a major road in Los Angeles (a) 30 m downwind, (b) 60 m downwind, (c) 90 m downwind, and (d) 150 m downwind. Please note the different scale for the *y* axis. Modal parameters given are the geometric mean diameter and geometric standard deviation (Zhu et al. 2002).

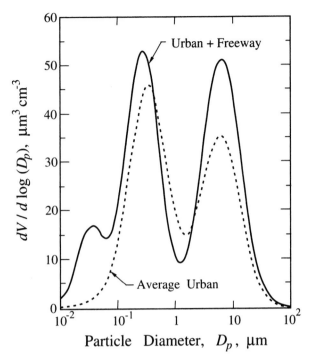

FIGURE 8.13 Aerosol volume distributions next to a source (freeway) and for average urban conditions.

The sources and chemical compositions of the fine and coarse urban particles are different. Coarse particles are generated by mechanical processes and consist of soil dust, seasalt, fly ash, tire wear particles, and so on. Aitken and accumulation mode particles contain primary particles from combustion sources and secondary aerosol material (sulfate, nitrate, ammonium, secondary organics) formed by chemical reactions resulting in gas-to-particle conversion (see Chapters 10 and 14).

The main mechanisms of transfer of particles from the Aitken to accumulation mode is coagulation (Chapter 13) and growth by condensation of vapors formed by chemical reactions (Chapter 12) onto existing particles. Coagulation among accumulation mode particles is a slow process and does not transfer particles to the coarse mode.

Processing of accumulation and coarse mode aerosols by clouds (Chapter 17) can also modify the concentration and composition of these modes. Aqueous-phase chemical reactions take place in cloud and fog droplets, and in aerosol particles at relative humidities approaching 100%. These reactions can lead to production of sulfate (Chapter 7) and after evaporation of water, a larger aerosol particle is left in the atmosphere. This transformation can lead to the formation of the *condensation mode* and the *droplet mode* (Hering and Friedlander 1982; John et al. 1990; Meng and Seinfeld 1994).

Terms often used to describe the aerosol mass concentration include total suspended particulate matter (TSP) and PM_x (particulate matter with diameter smaller than x μm). TSP refers to the mass concentration of atmospheric particles smaller than 40–50 μm, while $PM_{2.5}$ and PM_{10} are routinely monitored. Typical $PM_{2.5}$ and PM_{10} concentrations in large cities are shown in Figure 8.14.

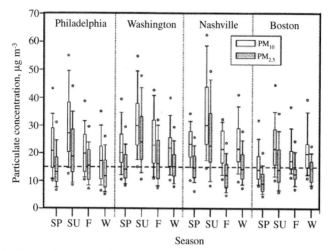

FIGURE 8.14 Seasonal concentrations of $PM_{2.5}$ and PM_{10} in four cities. The data show the lowest, lowest tenth percentile, lowest quartile, median highest quartile, highest tenth percentile, and highest daily average value. The dashed line shows the level of the annual $PM_{2.5}$ standard in the United States.

8.2.2 Marine Aerosols

In the absence of significant transport of continental aerosols, particles over the remote oceans are largely of marine origin (Savoie and Prospero 1989). Marine atmospheric particle concentrations are normally in the range of 100–300 cm^{-3}. Their size distribution is usually characterized by three modes (Figure 8.15): the Aitken ($D_p < 0.1$ µm) the

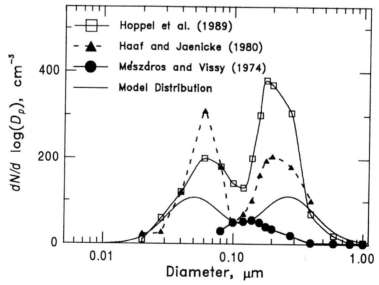

FIGURE 8.15 Measured marine aerosol number distributions and a model distribution used to represent average conditions.

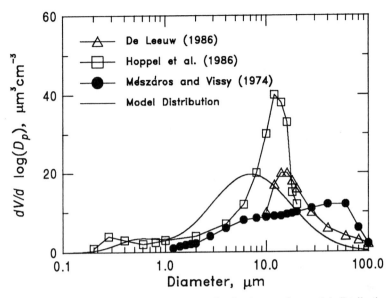

FIGURE 8.16 Measured marine aerosol volume distributions and a model distribution used to represent average conditions.

accumulation ($0.1 < D_p < 0.6$ μm), and the coarse ($D_p > 0.6$ μm) (Fitzgerald 1991). Typically, the coarse-particle mode, representing 95% of the total mass but only 5–10% of the particle number (Figure 8.16), results from the evaporation of seaspray produced by bursting bubbles or wind-induced wave breaking (Blanchard and Woodcock 1957; Monahan et al. 1983). Typical seasalt aerosol concentrations in the marine boundary layer (MBL) are around 5–30 cm^{-3} (Blanchard and Cipriano 1987; O'Dowd and Smith 1993). For a comprehensive treatment of seasalt aerosols, we refer the reader to Lewis and Schwartz (2005).

Figures 8.15 and 8.16 show number and volume aerosol distributions in clean maritime air measured by several investigators (Mészáros and Vissy 1974; Hoppel et al. 1989; Haaf and Jaenicke 1980; De Leeuw 1986) and a model marine aerosol size distribution. The distributions of Hoppel et al. (1989) and De Leeuw (1986) were obtained at windspeeds of less than 5 m s^{-1} in the subtropical and North Atlantic, respectively. The distribution of Mészáros and Vissy (1974) is an average of spectra obtained in the South Atlantic and Indian Oceans during periods when the average windspeed was 12 m s^{-1}. It is difficult to determine the extent to which the differences in these size distributions are the result of differences in sampling location and meteorological conditions such as windspeed (which affects the concentrations of the larger particles), or to uncertainties inherent in the different measurement methods.

8.2.3 Rural Continental Aerosols

Aerosols in rural areas are mainly of natural origin but with a moderate influence of anthropogenic sources (Hobbs et al. 1985). The number distribution is characterized by two modes at diameters about 0.02 and 0.08 μm, respectively (Jaenicke 1993), while the mass distribution is dominated by the coarse mode centered at around 7 μm (Figure 8.17).

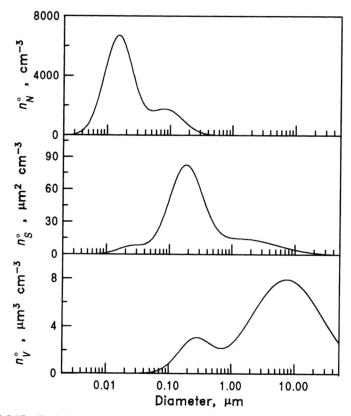

FIGURE 8.17   Typical rural continental aerosol number, surface, and volume distributions.

The mass distribution of continental aerosol not influenced by local sources has a small accumulation mode and no nuclei mode. The PM_{10} concentration of rural aerosols is around 20 μg m^{-3}.

8.2.4   Remote Continental Aerosols

Primary particles (e.g., dust, pollens, plant waxes) and secondary oxidation products are the main components of remote continental aerosol (Deepak and Gali 1991). Aerosol number concentrations average around 1000–10,000 cm^{-3}, and PM_{10} concentrations are around 10 μg m^{-3} (Bashurova et al. 1992; Koutsenogii et al. 1993; Koutsenogii and Jaenicke 1994). For the continental United States PM_{10} concentrations in remote areas vary from 5 to 25 μg m^{-3} and $PM_{2.5}$ from 3 to 17 μg m^{-3} (U.S. EPA 1996). Particles smaller than 2.5 μm in diameter represent 40–80% of the PM_{10} mass and consist mainly of sulfate, ammonium, and organics. The aerosol number distribution may be characterized by three modes at diameters 0.02, 0.1, and 2 μm (Jaenicke 1993) (Figure 8.18).

8.2.5   Free Tropospheric Aerosols

Background free tropospheric aerosol is found in the mid- and upper troposphere above the clouds. The modes in the number distribution correspond to mean diameters of 0.01 and 0.25 (Jaenicke 1993) (Figure 8.19). The middle troposphere spectra typically indicate

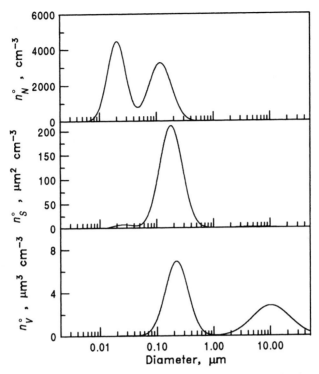

FIGURE 8.18   Typical remote continental aerosol number, surface, and volume distributions.

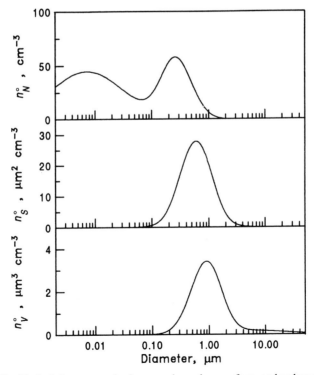

FIGURE 8.19   Typical free tropospheric aerosol number, surface, and volume distributions.

more particles in the accumulation mode relative to lower tropospheric spectra, suggesting precipitation scavenging and deposition of smaller and larger particles (Leaitch and Isaac 1991). The low temperature and low aerosol surface area make the upper troposphere suitable for new particle formation, and a nucleation mode is often present in the number distribution (Figure 8.19).

8.2.6 Polar Aerosols

Polar aerosols, found close to the surface in the Arctic and Antarctica, reflect their aged character; their concentrations are very low. Collections of data from aerosol measurements in the Arctic have been presented by a number of investigators (Rahn 1981; Shaw 1985; Heintzenberg 1989; Ottar 1989). The number distribution appears practically monodisperse (Ito and Iwai 1981) with a mean diameter of approximately 0.15 μm; two more modes at 0.75 and 8 μm (Shaw 1986; Jaenicke et al. 1992) (Figure 8.20) dominate the mass distribution.

During the winter and early spring (February to April) the Arctic aerosol has been found to be influenced significantly by anthropogenic sources, and the phenomenon is commonly referred to as "Arctic haze" (Barrie 1986). During this period the aerosol number concentration increases to over 200 cm^{-3}. The nucleation mode mean diameter is at 0.05 μm and the accumulation mode at 0.2 μm (Covert and Heintzenberg 1993)

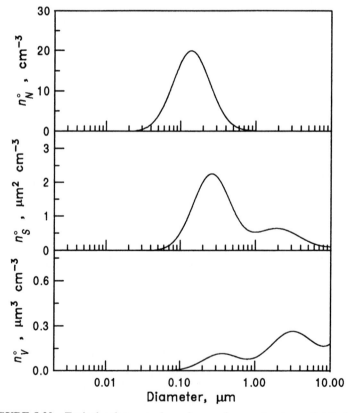

FIGURE 8.20 Typical polar aerosol number, surface, and volume distributions.

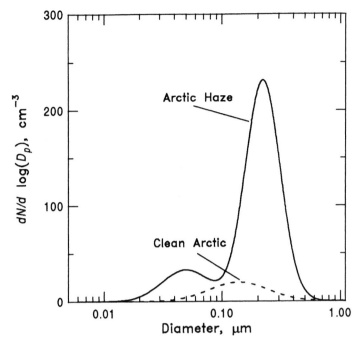

FIGURE 8.21 Comparison of the aerosol distribution during Arctic haze with the typical polar distribution.

(Figure 8.21). Similar measurements have been reported by Heintzenberg (1980), Radke et al. (1984), and Shaw (1984).

The polar aerosol contains carbonaceous material from midlatitude pollution sources, sulfate, seasalt from the surrounding ocean, and mineral dust from arid regions of the corresponding hemisphere. Aerosol PM_{10} concentrations in the polar regions are less than 5 μg m^{-3} with sulfate representing roughly 40% of the mass.

8.2.7 Desert Aerosols

Desert aerosol, of course present over deserts, actually extends considerably over adjacent regions such as oceans (Jaenicke and Schutz 1978; d'Almeida and Schutz 1983; Li et al. 1996). The shape of its size distribution is similar to that of remote continental aerosol but depends strongly on the wind velocity. Its number distribution tends to exhibit three overlapping modes at diameters of 0.01 μm or less, 0.05 μm, and 10 μm, respectively (Jaenicke 1993) (Figure 8.22). An average composition of soils and crustal material is shown in Table 8.4. The soil composition is similar to that of the crustal rock, with the exception of the soluble elements such as Ca, Mg, and Na, which have lower relative concentrations in the soil.

Individual dust storms from the Sahara desert have been shown to transfer material from the northwest coast of Africa, across the Atlantic, to the east coast of the United States (Ott et al. 1991). For example, Prospero et al. (1987) suggested that enough Saharan dust is carried into the Miami area to significantly reduce visibility during the summer months. Similar dust transport occurs from the deserts of Asia across the Pacific Ocean

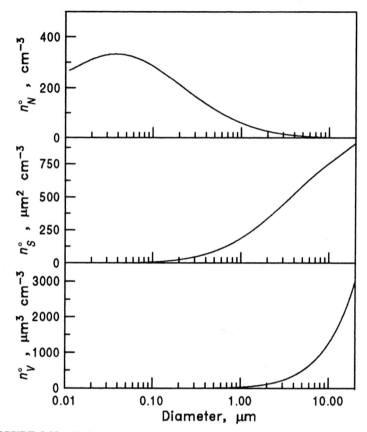

FIGURE 8.22   Typical desert aerosol number, surface, and volume distributions.

TABLE 8.4   Average Abundances of Major Elements in Soil and Crustal Rock

	Elemental Abundance (ppm by mass)	
Element	Soil	Crustal Rock
Si	330,000	311,000
Al	71,300	77,400
Fe	38,000	34,300
Ca	13,700	25,700
Mg	6,300	33,000
Na	6,300	31,900
K	13,600	29,500
Ti	4,600	4,400
Mn	850	670
Cr	200	48
V	100	98
Co	8	12

Source: Warneck (1988).

TABLE 8.5 Properties of Atmospheric Aerosol Types

Type	Number (cm^{-3})	PM$_1$(μg m^{-3})	PM$_{10}$(μg m^{-3})
Urban (polluted)	$10^5 - 4 \times 10^6$	30–150	100–300
Marine	100–400	1–4	10
Rural	2000–10,000	2.5–8	10–40
Remote continental	50–10,000	0.5–2.5	2–10

(Prospero 1995). While particles as large as 100 μm in diameter are found in the source regions, only particles smaller than 10 μm are transported over long distances, often farther than 5000 km.

The average number and volume concentration of the major aerosol types are summarized in Table 8.5.

8.3 AEROSOL CHEMICAL COMPOSITION

Atmospheric aerosol particles contain sulfates, nitrates, ammonium, organic material, crustal species, seasalt, metal oxides, hydrogen ions, and water. From these species sulfate, ammonium, organic and elemental carbon, and certain transition metals are found predominantly in the fine particles. Crustal materials, including silicon, calcium, magnesium, aluminum, and iron, and biogenic organic particles (pollen, spores, plant fragments) are usually in the coarse aerosol fraction. Nitrate can be found in both the fine and coarse modes. Fine nitrate is usually the result of the nitric acid/ammonia reaction for the formation of ammonium nitrate, while coarse nitrate is the product of coarse particle/nitric acid reactions.

A typical urban aerosol size/composition distribution is shown in Figure 8.23 (Wall et al. 1988). These results indicate that sulfate, nitrate, and ammonium have two modes in the 0.1–1.0 μm size range (the condensation and droplet modes), and a third one over 1 μm (coarse mode). The condensation mode has a peak around 0.2 μm and is the result of condensation of secondary aerosol components from the gas phase. The droplet mode peaks around 0.7 μm in diameter and its existence is attributed to heterogeneous, aqueous-phase reactions discussed in Chapter 7 (Meng and Seinfeld 1994). More than half of the nitrate is found in the coarse mode together with most of the sodium and chloride. This coarse nitrate is the result of reactions of nitric acid with sodium chloride or aerosol crustal material (see Chapter 10). This is an interesting case where secondary aerosol matter (nitrate) is formed through the reaction of a naturally produced material (seasalt or dust) and an anthropogenic pollutant (nitric acid).

More than 40 trace elements are routinely found in atmospheric particulate matter samples. These elements arise from dozens of different sources including combustion of coal, oil, woodburning, steel furnaces, boilers, smelters, dust, waste incineration, and brake wear. Depending on their sources, these elements can be found in either the fine or the coarse mode. Concentrations of selected elements together with the size mode where these elements are usually found are shown in Table 8.6. The concentrations of these elements even for similar pollution levels vary over almost three orders of magnitude, indicating the strong effect of local sources. In general, elements such as lead, iron, and copper have the highest concentrations, while elements such as cobalt, mercury, and

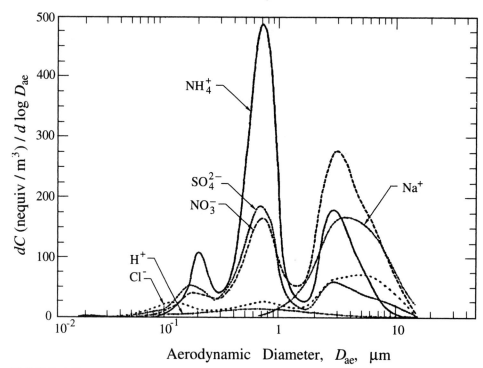

FIGURE 8.23 Measured size distributions of aerosol sulfate, nitrate, ammonium, chloride, sodium, and hydrogen ion in Claremont, CA (Wall et al. 1988).

TABLE 8.6 Concentrations (ng m^{-3}) and Size Distribution of Various Elements Found in Atmospheric Particles

		Concentration (ng m^{-3})		
Element	Mode[a]	Remote	Rural	Urban
Fe	F and C	0.6–4,200	55–14,500	130–13,800
Pb	F	0.01–65	2–1,700	30–90,000
Zn	F	0.03–450	10–400	15–8,000
Cd	F	0.01–1	0.4–1,000	0.2–7,000
As	F	0.01–2	1–28	2–2,500
V	F and C	0.01–15	3–100	1–1,500
Cu	F and C	0.03–15	3–300	3–5,000
Mn	F and C	0.01–15	4–100	4–500
Hg	—	0.01–1	0.05–160	1–500
Ni	F and C	0.01–60	1–80	1–300
Sb	F	0–1	0.5–7	0.5–150
Cr	F and C	0.01–10	1–50	2–150
Co	F and C	0–1	0.1–10	0.2–100
Se	F and C	0.01–0.2	0.01–30	0.2–30

[a]F = fine mode; C = coarse mode.

Source: Schroeder et al. (1987).

TABLE 8.7 Comparison of Ambient Fine and Coarse Particles

	Fine Particles	Coarse Particles
Formation pathways	Chemical reactions Nucleation Condensation Coagulation Cloud/fog processing	Mechanical disruption Suspension of dusts

Composition	Sulfate
	Nitrate
	Ammonium
	Hydrogen ion
	Elemental carbon (EC)
	Organic compounds
	Water
	Metals (Pb, Cd, V, Ni,
	  Cu, Zn, Mn, Fe, etc.)
		Resuspended dust
		Coal and oil fly ash
		Crustal element (Si, Al, Ti,
		  Fe) oxides
		$CaCO_3$, NaCl
		Pollen, mold, spores
		Plant, animal debris
		Tire wear debris

Solubility	Largely soluble, hygroscopic	Largely insoluble and non-hygroscopic

Sources	Combustion (coal, oil,
	  gasoline, diesel, wood)
	Gas-to-particle conversion
	  of NO_x, SO_3, and VOCs
	Smelters, mills, etc.
		Resuspension of industrial
		  dust and soil
		Suspension of soil (farming,
		  mining, unpaved roads)
		Biological sources
		Construction/demolition
		Ocean spray

Atmospheric lifetime	Days to weeks	Minutes to days
Travel distance	100s to 1000s of km	< to 10s of km

antimony are characterized by low concentrations. Elements produced during combustion usually exist in the form of oxides (e.g., Fe_2O_3, Fe_3O_4, Al_2O_3), but their chemical form is in general uncertain.

A summary of chemical information regarding the coarse and fine modes is presented in Table 8.7.

The composition of seasalt reflects the composition of seawater enriched in organic material (marine-derived sterols, fatty alcohols, and fatty acids) that exists in the surface layer of the oceans (Schneider and Gagosian 1985). Seawater contains 3.5% by weight seasalt, and when first emitted, the seasalt composition is the same as that of seawater (Table 8.8). Reactions on seasalt particles modify its chemical composition; for example, sodium chloride reacts with sulfuric acid vapor to produce sodium sulfate and hydrochloric acid vapor

$$H_2SO_4(g) + 2NaCl \rightarrow Na_2SO_4 + 2HCl(g)$$

leading to an apparent "chloride deficit" in the marine aerosol.

TABLE 8.8   Composition of Seasalt [a]

Species	Percent by Weight
Cl	55.04
Na	30.61
SO_4^{2-}	7.68
Mg	3.69
Ca	1.16
K	1.1
Br	0.19
C (noncarbonate)	$3.5 \times 10^{-3} - 8.7 \times 10^{-3}$
Al	$4.6 \times 10^{-4} - 5.5 \times 10^{-3}$
Ba	1.4×10^{-4}
I	1.4×10^{-4}
Si	$1.4 \times 10^{-4} - 9.4 \times 10^{-3}$
NO_3^-	$3 \times 10^{-6} - 2 \times 10^{-3}$
Fe	$5 \times 10^{-5} - 5 \times 10^{-4}$
Zn	$1.4 \times 10^{-5} - 4 \times 10^{-5}$
Pb	$1.2 \times 10^{-5} - 1.4 \times 10^{-5}$
NH_4^+	$1.4 \times 10^{-6} - 1.4 \times 10^{-5}$
Mn	$2.5 \times 10^{-6} - 2.5 \times 10^{-5}$
V	9×10^{-7}

[a]Based on the composition of seawater and ignoring atmospheric transformations.

8.4   SPATIAL AND TEMPORAL VARIATION

The concentration of atmospheric aerosols varies considerably in space and time. This variability of the aerosol concentration field is determined by meteorology and the emissions of aerosols and their precursors. For example, the annual average concentration of $PM_{2.5}$ in North America varies by more than an order of magnitude as one moves from the clean remote to the polluted urban areas of Mexico City and southern California (Figure 8.24). Sulfate dominates the fine aerosol composition in the eastern United States, while organics are major contributors to the aerosol mass everywhere. Nitrates are major components of the $PM_{2.5}$ in the western United States. The EC makes a relatively small contribution to the particle mass in many areas, but because of its ability to absorb light and its toxicity, it is an important component of atmospheric particulate matter.

The composition of the particles in a given area often changes significantly during the year. Species that are produced photochemically usually have higher concentrations during the summer. The higher summertime concentrations of sulfate in the northeastern United States (Figure 8.25) lead to high fine aerosol concentrations and low visibility. On the other hand, aerosol nitrate concentrations often peak in the winter, even if nitric acid is a secondary species produced photochemically. Indeed, there is usually more nitric acid available during the summer in most locations. However, owing to the higher summertime temperatures it remains mostly in the gas phase as nitric acid vapor. During the winter, almost all the nitric acid that is available is transferred to the particulate phase after reaction with ammonia to form ammonium nitrate (see Chapter 10), leading to the higher aerosol nitrate concentrations (Figure 8.25). The seasonal behavior of the organic

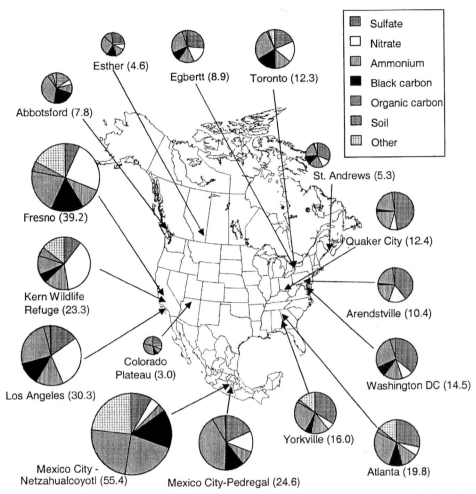

FIGURE 8.24 PM$_{2.5}$ concentration and chemical composition at various sites in North America (NARSTO 2003). Chemical species for each site are presented clockwise in the pie charts in the same order as in the legend.

particulate matter varies for each location. During the winter there are usually additional sources of organic particles because of the additional heating needs. The mixing heights are also lower, causing an increase in aerosol concentrations. On the other hand, the photochemical production of secondary organic aerosol contributes significant organic aerosol matter during the summer. The result of these competing effects is often a relatively constant organic PM concentration (Figure 8.25). Ammonia is transferred to the particulate phase in an effort to neutralize the acidic components forming ammonium sulfates and ammonium nitrate. The ammonium concentration pattern usually follows that of the sum of sulfate and nitrate.

Variation in atmospheric aerosol concentrations over shorter timescales (e.g., hours) is often quite significant as well. Figure 8.26 shows a rather extreme example of the sulfate concentration in an area increasing from 5 to 60 μg m^{-3} over a period of hours because of the transport of pollution to the area. Then the concentration decreases rapidly with rain

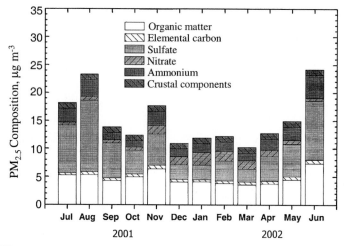

FIGURE 8.25   Annual variation of the $PM_{2.5}$ concentration and composition in Pittsburgh, PA (Wittig et al. 2004a).

and returns to around 20 μg m^{-3}. The concentration of aerosol nitrate is quite sensitive to temperature and often reaches a maximum when the temperature is at its minimum, shortly before sunrise (Figure 8.27).

For species with strong local sources the daily concentration pattern is often determined by the strength of atmospheric mixing. The atmospheric mixing height tends to be lower during the nighttime, so these species often reach high concentrations very early in the morning. The OC and EC concentrations in urban areas sometimes exhibit this behavior (Figure 8.28), peaking during the early stages of rush hour traffic at around 6 am. During the afternoon rush-hour, even if the traffic emissions are similar to those in the morning, the concentrations are considerably lower because they are diluted over a much larger boundary-layer volume. These diurnal variations can be used to determine the importance of local emissions. Areas where most of the organic aerosol is a result of local primary

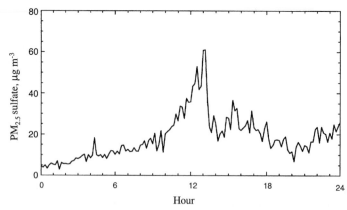

FIGURE 8.26 $PM_{2.5}$ sulfate concentrations measured on July 21, 2001 in Pittsburgh, PA (Wittig et al. 2004b).

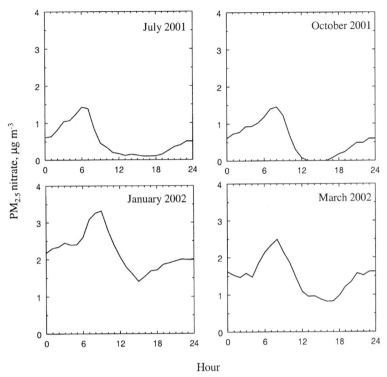

FIGURE 8.27 Measured average $PM_{2.5}$ nitrate concentration as a function of time during the four seasons in Pittsburgh, PA (Wittig et al. 2004b). The concentration pattern is dominated by the temperature variation in this area with the highest concentrations observed during the periods with the lowest temperature (nighttime, winter).

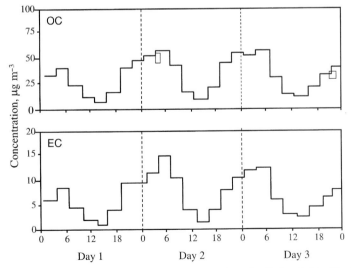

FIGURE 8.28 Measured PM_{10} OC and EC concentrations in Bakersfield, CA during 1995 Integrated Monitoring Study.

traffic emissions exhibit the behavior seen in Figure 8.28. Cities where local traffic is a much smaller source of particles are characterized by a relatively small difference between early morning and afternoon organic aerosol concentrations.

8.5 VERTICAL VARIATION

The vertical distribution of aerosol mass concentration typically shows an exponential decrease with altitude up to a height H_p and a rather constant profile above that altitude (Gras 1991). The aerosol mass concentration as a function of height can then be expressed as

$$M(z) = M(0) \exp\left(-\frac{z}{H_p}\right) \tag{8.55}$$

where $M(0)$ is the surface concentration and H_p the scale height. Jaenicke (1993) proposed values of H_p equal to 900 m for the marine, 730 m for the remote continental, 2000 m for the desert, and 30,000 m for the polar aerosol types. The corresponding vertical aerosol mass concentration profiles are shown in Figure 8.29.

The aerosol number concentration may increase or decrease exponentially with altitude and one suggestion of a form of the profile is (Jaenicke 1993)

$$N(z) = N(0)\left[\exp\left(\frac{-z}{|H_p'|}\right) + \left(\frac{N_B}{N(0)}\right)^n\right]^n \tag{8.56}$$

where

$$n = \frac{H_p'}{|H_p'|} \tag{8.57}$$

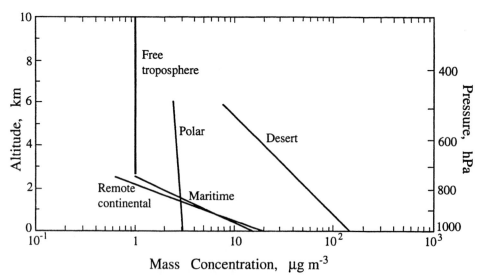

FIGURE 8.29 Representative vertical distribution of aerosol mass concentration (Jaenicke 1993).

and N_B is the number concentration of the background aerosol aloft. For marine aerosol H'_p varies from -290 to 440 m. Note that if H'_p is negative $n = -1$, and (8.56) can be rewritten as

$$N(z) = N(0) \left[\exp\left(\frac{-z}{|H'_p|}\right) + \left(\frac{N(0)}{N_B}\right) \right]^{-1} \tag{8.58}$$

Because in this case $N(0) \ll N_B$, the equation has the correct limiting behavior both for $z \to 0$ and $z \to \infty$.

These vertical profiles are rough representations of long-term averages. Significant variability is observed in aerosol concentrations in anthropogenic plumes, areas influenced by local sources, or during nucleation events in the free troposphere.

PROBLEMS

8.1$_A$ Given the following data on the number of aerosol particles in the size ranges listed, tabulate and plot the normalized size distributions $\bar{n}_N(D_p) = n_N(D_p)/N_t$ and $\bar{n}_N^\circ(\log D_p) = n_N^\circ(\log D_p)/N_t$ as discrete histograms.

Size Interval, μm	Mean of Size Interval, μm	Number of Particles in Interval
0–0.2	0.1	10
0.2–0.4	0.3	80
0.4–0.6	0.5	132
0.6–0.8	0.7	142
0.8–1.0	0.9	138
1.0–1.2	1.1	112
1.2–1.4	1.3	75
1.4–1.6	1.5	65
1.6–1.8	1.7	52
1.8–2.1	1.95	65
2.1–2.7	2.4	62
2.7–3.6	3.15	32
3.6–5.1	4.35	35

8.2$_A$ For the data given in Problem 8.1, plot the surface area and volume distributions $n_S(D_p)$, $n_S^\circ(\log D_p)$, $n_V(D_p)$ and $n_V^\circ(\log D_p)$ in both nonnormalized and normalized form as discrete histograms.

8.3$_A$ You are given an aerosol size distribution function $n_M(m)$ such that $n_M(m)\,dm$ = aerosol mass per cm^3 of air contained in particles having masses in the range m to $m + dm$. It is desired to convert that distribution function to a mass distribution based on $\log D_p$. Show that

$$n_M^o \left(\log D_p \right) = 6.9 \, m \, n_M(m)$$

8.4$_B$ Show that the variance of the size distribution of a lognormally distributed aerosol is

$$\bar{D}_p^2 \left[\exp\left(\ln^2 \sigma_g \right) - 1 \right]$$

8.5$_A$ Starting with semilogarithmic graph paper, construct a log–probability coordinate axis and show that a lognormal distribution plots as a straight line on these coordinates.

8.6$_A$ The data given below were obtained for a lognormally distributed aerosol size distribution:

Size Interval, μm	Geometric Mean of Size Interval, μm	Number of Particles in Interval[a]
0.1–0.2	0.1414	50
0.2–0.4	0.2828	460
0.4–0.7	0.5292	1055
0.7–1.0	0.8367	980
1.0–2.0	1.414	1705
2.0–4.0	2.828	680
4.0–7.0	5.292	102
7.0–10	8.367	10
10–20	14.14	2

[a]Assume that the particles are spheres with density $\rho_p = 1.5 \text{ g cm}^{-3}$.

a. Complete this table by computing the following quantities: $\Delta N_i / \Delta D_{pi}$, $\Delta N_i / N \Delta D_{pi}$, $\Delta S_i / \Delta D_{pi}$, $\Delta S_i / S \Delta D_{pi}$, $\Delta M_i / \Delta D_{pi}$, $\Delta M_i / M \Delta D_{pi}$, $\Delta N_i / \Delta \log D_{pi}$, $\Delta N_i / N \Delta \log D_{pi}$, $\Delta S_i / \Delta \log D_{pi}$, $\Delta S_i / S \Delta \log D_{pi}$, and $\Delta M_i / \Delta \log D_{pi}$, $\Delta M_i / M \Delta \log D_{pi}$, where M = particle mass.

b. Plot $\Delta N_i / \Delta \log D_{pi}$, $\Delta S_i / \Delta \log D_{pi}$, and $\Delta M_i / \Delta \log D_{pi}$ as histograms.

c. Determine the geometric mean diameter and geometric standard deviation of the lognormal distribution to which these data adhere and plot the continuous distributions on the three plots from part (b).

8.7$_B$ For a lognormally distributed aerosol different mean diameters can be defined by

$$\bar{D}_{pv} = \bar{D}_{pg} \exp\left(v \ln^2 \sigma_g \right)$$

where ν is a parameter that defines the particular mean diameter of interest. Show that

Diameter	ν
Mode (most frequent value)	-1
Geometric mean or median	0
Number (arithmetic) mean	0.5
Surface area mean	1
Mass mean	1.5
Surface area median	2
Volume median	3

Plot a normalized lognormal particle size distribution over a range of D_p from 0 to 7 μm with $\bar{D}_{pg} = 1.0$ μm and $\sigma_g = 2.0$ and identify each of the above diameters on the plot. *Hint:* You may find this integral of use:

$$\int_{L_1}^{L_2} e^{ru} \exp\left(-\frac{(u-\bar{u})^2}{2\sigma_u^2}\right) du$$

$$= (\pi/2)^{1/2} \sigma_u e^{r\bar{u}} e^{r^2 \sigma_u^2/2} \left[\text{erf}\left(\frac{L_2 - (\bar{u} + r\sigma_u^2)}{\sqrt{2}\sigma_u}\right) - \text{erf}\left(\frac{L_1 - (\bar{u} + r\sigma_u^2)}{\sqrt{2}\sigma_u}\right) \right]$$

8.8$_B$ The modified gamma distribution (Deirmendjian 1969) has been proposed as another function that approximates ambient aerosol size distributions, $n_N(D_p) = A D_p^b \exp(-B D_p^c)$, where A, b, B, and c are all positive constants.

    a. Plot this size distribution for the following combinations of its parameters:

        $A = 100$, $b = 3$, $B = 2$ and $c = 1, 2, 3$

        $A = 100$, $b = 2$, $c = 2$ and $B = 1.5, 2, 2.5$

        $A = 100$, $c = 2$, $B = 2$ and $b = 3, 4, 5$

    b. Calculate the diameter D_m at which the distribution function reaches a maximum as a function of the distribution parameters.

    c. Calculate the total aerosol number concentration as a function of the distribution parameters.

    d. Using the results of parts (a)–(c), discuss the effect of the four parameters on the shape of the distribution function.

    e. Discuss the strengths and weaknesses of this function for the fitting of atmospheric aerosol size distributions.

8.9$_B$ Assume that an aerosol has a lognormal distribution with $\bar{D}_{pg} = 5.5$ μm and $\sigma_g = 1.36$.

a. Plot the number and volume distributions of this aerosol on log–probability paper.

b. It is desired to represent this aerosol by a distribution of the form

$$F_V(D_p) = 1 - \exp\left(-cD_p^b\right)$$

where $F_V(D_p)$ is the fraction of the total aerosol volume in particles of diameter less than D_p. Determine the values of the constants c and b needed to match this distribution to the given aerosol.

8.10$_B$ Given the following size frequency for a dust:

Size Interval (μm)	% by Number
7–17.5	10
17.5–21	10
21–25	10
25–28	10
28–30	10
30–33	10
33–36	10
36–41	10
41–49	10
49–70	10

a. Plot the cumulative frequency distributions (in %) of the number, surface area, and mass on linear graph paper assuming all particles are spheres with $\rho_p = 1.6$ g cm^{-3}.

b. Is this a lognormally distributed dust?

8.11$_B$ The following particle size distribution data are available for an aerosol:

D_p	% by Volume Less than
9.8	3.2
13.8	10.0
19.6	26.7
27.7	46.8
39.1	72.0
55.3	87.5

a. What are the volume median diameter and geometric standard deviation of the volume distribution of this aerosol?

b. What is the surface area median diameter?

REFERENCES

Aitchison, J., and Brown, J. A. C. (1957) *The Lognormal Distribution Function*, Cambridge Univ. Press, Cambridge, UK.

Barrie, L. A. (1986) Arctic air pollution: An overview of current knowledge, *Atmos. Environ.* **20**, 643–663.

Bashurova, V. S. et al. (1992) Measurements of atmospheric condensation nuclei size distributions in Siberia, *J. Aerosol Sci.* **23**, 191–199.

Blanchard, D. C., and Cipriano, R. J. (1987) Biological regulation of climate, *Nature* **330**, 526.

Blanchard, D. C., and Woodcock, A. H. (1957) Bubble formation and modification in the sea and its meteorological significance, *Tellus* **9**, 145–152.

Covert, D. S., and Heintzenberg, J. (1993) Size distributions and chemical properties of aerosol at NY Ålesund, Svalbard, *Atmos. Environ.* **27A**, 2989–2997.

d'Almeida, G. A., and Schutz, L. (1983) Number, mass and volume distributions of mineral aerosol and soils of the Sahara, *J. Climate Appl. Meteorol.* **22**, 233–243.

Deepak, A., and Gali, G. (1991) *The International Global Aerosol Program (IGAP) Plan*, Deepak Publishing, Hampton, VA.

Deirmendjian, D. (1969) *Electromagnetic Scattering on Spherical Polydispersions*, Elsevier, New York.

De Leeuw, G. (1986) Vertical profiles of giant particles close above the sea surface, *Tellus* **38B**, 51–61.

Fitzgerald, J. W. (1991) Marine aerosols: A review, *Atmos. Environ.* **25A**, 533–545.

Gras, J. L. (1991) Southern hemisphere tropospheric aerosol microphysics, *J. Geophys. Res.* **96**, 5345–5356.

Haaf, W., and Jaenicke, R. (1980) Results of improved size distribution measurements in the Aitken range of atmospheric aerosols, *J. Aerosol Sci.* **11**, 321–330.

Heintzenberg, J. (1980) Particle size distribution and optical properties of Arctic haze, *Tellus* **32**, 251–260.

Heintzenberg, J. (1989) Arctic haze: air pollution in polar regions, *Ambio* **18**, 50–55.

Hering, S. V., and Friedlander, S. K. (1982) Origins of aerosol sulfur size distributions in the Los Angeles basin, *Atmos. Environ.* **16**, 2647–2656.

Hobbs, P. V., Bowdle, D. A., and Radke, L. F. (1985) Particles in the lower troposphere over the high plains of the United States. 1. Size distributions, elemental compositions, and morphologies, *J. Climate Appl. Meteorol.* **24**, 1344–1356.

Hoppel, W. A., Fitzgerald, J. W., Frick, G. M., Larson, R. E., and Mack, E. J. (1989) *Atmospheric Aerosol Size Distributions and Optical Properties in the Marine Boundary Layer over the Atlantic Ocean*, NRL Report 9188, Washington, DC.

Ito, T., and Iwai, K. (1981) On the sudden increase in the concentration of aitken particles in the Antarctic atmosphere, *J. Meteorol. Soc. Jpn.* **59**, 262–271.

Jaenicke, R. (1993) Tropospheric aerosols, in *Aerosol–Cloud–Climate Interactions*, P. V. Hobbs, ed., Academic Press, San Diego, CA, pp. 1–31.

Jaenicke, R., and Schutz, L. (1978) Comprehensive study of physical and chemical properties of the surface aerosol in the Cape Verde Islands region, *J. Geophys. Res.* **83**, 3583–3599.

Jaenicke R., Dreiling V., Lehmann E., Koutsenogii, P. K., and Stingl, J. (1992) Condensation nuclei at the German Antarctic Station Vonneymayer, *Tellus* **44B**, 311–317.

John, W., Wall, S. M., Ondo, J. L., and Winklmayr, W. (1990) Modes in the size distributions of atmospheric inorganic aerosol, *Atmos. Environ.* **24A**, 2349–2359.

Koutsenogii, P. K., and Jaenicke, R. (1994) Number concentration and size distribution of atmospheric aerosol in Siberia, *J. Aerosol Sci.* **25**, 377–383.

Koutsenogii, P. K., Bufetov, N. S., and Drosdova, V. I. (1993) Ion composition of atmospheric aerosol near Lake Baikal, *Atmos. Environ.* **27A**, 1629–1633.

Leaitch, W. R., and Isaac, G. A. (1991) Tropospheric aerosol size distributions from 1982 to 1988 over Eastern North America, *Atmos. Environ.* **25A**, 601–619.

Lewis, E. R., and Schwartz S. E. (2005) *Sea Salt Aerosol Production: Mechanisms, Methods, Measurements, and Models*, American Geophysical Union, Washington, DC.

Li, X., Maring, H., Savoie, D., Voss, K., and Prospero, J. M. (1996) Dominance of mineral dust in aerosol light scattering in the North Atlantic trade winds, *Nature* **380**, 416–419.

Meng, Z., and Seinfeld, J. H. (1994) On the source of the submicrometer droplet mode of urban and regional aerosols. *Aerosol Sci. Technol.* **20**, 253–265.

Mészáros, A., and Vissy, K. (1974) Concentration, size distribution and chemical nature of atmospheric aerosol particles in remote ocean areas, *J. Aerosol Sci.* **5**, 101–109.

Monahan, E. C., Fairall, C. W., Davidson, K. L., and Jones-Boyle, P. (1983) Observed interrelationships amongst 10-m-elevation winds, oceanic whitecaps, and marine aerosols, *Quart. J. Roy. Meteorol. Soc.* **109**, 379–392.

NARSTO (2003) *Particulate Matter Science for Policy Makers*, Electric Power Research Institute, Palo Alto, CA.

O'Dowd, C. D., and Smith, M. H. (1993) Physicochemical properties of aerosols over the Northeast Atlantic: Evidence for wind-speed related submicron sea-salt aerosol production, *J. Geophys. Res.* **98**, 1137–1149.

Ott, S. T., Ott, A., Martin, D. W., and Young, J. A. (1991) Analysis of trans-Atlantic saharan dust outbreak based on satellite and GATE data, *Mon. Weather Rev.* **119**, 1832–1850.

Ottar, B. (1989) Arctic air pollution: A Norwegian perspective, *Atmos. Environ.* **23**, 2349–2356.

Prospero, J. M. (1995) The atmospheric transport of particles to the ocean, in *SCOPE Report: Particle Flux in the Ocean*, V. Ittekkot, S. Honjo, and P. J. Depetris, eds., Wiley, New York.

Prospero, J. M., Nees, R. T., and Uematsu, M. (1987) Deposition rate of particulate and dissolved aluminum derived from sahara dust in precipitation in Miami, Florida, *J. Geophys. Res.* **92**, 14723–14731.

Pruppacher, H. R., and Klett, J. D. (1980) *Microphysics of Cloud and Precipitation*, Reidel, Dordrecht, The Netherlands.

Radke, L. F., Lyons, J. H., Hegg, D. A., and Hobbs, P. V. (1984) Airborne observations of Arctic aerosols. I: Characteristics of Arctic haze, *Geophys. Res. Lett.* **11**, 369–372.

Rahn, K. (1981) Relative importance of North America and Eurasia as sources of Arctic aerosol, *Atmos. Environ.* **15**, 1447–1456.

Savoie, D. L., and Prospero, J. M. (1989) Comparison of oceanic and continental sources of non-seasalt sulfate over the Pacific ocean, *Nature* **339**, 685–687.

Schneider, J. K., and Gagosian, R. B. (1985) Particle size distribution of lipids in aerosols off the coast of Peru, *J. Geophys. Res.* **90**, 7889–7898.

Schroeder, W. H., Dobson, M., Kane, D. M., and Johnson, N. D. (1987) Toxic trace elements associated with airborne particulate matter: A review, *J. Air Pollut. Cont. Assoc.* **37**, 1267–1285.

Shaw, G. E. (1984) Microparticle size spectrum of Arctic haze, *Geophys. Res. Lett.* **11**, 409–412.

Shaw, G. E. (1985) Aerosol measurements in Central Alaska 1982–1984, *Atmos. Environ.* **19**, 2025–2031.

Shaw, G. E. (1986) On physical properties of aerosols at Ross Island, Antarctica, *J. Aerosol Sci.* **17**, 937–945.

United States Environmental Protection Agency (U.S. EPA) (1996) *Air Quality Criteria for Particulate Matter*, EPA/600/P–95/001, Research Triangle Park, NC.

Wall, S. M., John, W., and Ondo, J. L. (1988) Measurement of aerosol size distributions for nitrate and major ionic species, *Atmos. Environ.* **22**, 1649–1656.

Warneck, P. (1988) *Chemistry of the Natural Atmosphere*, Academic Press, San Diego.

Wittig, A. E., Anderson, N., Khlystov, A. Y., Pandis, S. N., Davidson, C., and Robinson, A. L. (2004a) Pittsburgh Air Quality Study overview, *Atmos. Environ.* **38**, 3107–3125.

Wittig, A. E., Takahama, S., Khlystov, A. Y., Pandis, S. N., Hering, S., Kirby, B., and Davidson, C. (2004b) Semi-continuous PM$_{2.5}$ inorganic composition measurements during the Pittsburgh Air Quality Study, *Atmos. Environ.* **38**, 3201–3213.

Zhu, Y., Hinds, W. C., Kim, S., Shen, S., and Sioutas, C. (2002) Study of ultrafine particles near a major highway with heavy-duty diesel traffic, *Atmos. Environ.* **36**, 4323–4335.

9 Dynamics of Single Aerosol Particles

In this chapter, we focus on the processes involving a single aerosol particle in a suspending fluid and the interaction of the particle with the suspending fluid itself. We begin by considering how to characterize the size of the particle in an appropriate way in order to describe transport processes involving momentum, mass, and energy. We then treat the drag force exerted by the fluid on the particle, the motion of a particle through a fluid due to an imposed external force and resulting from the bombardment of the particle by the molecules of the fluid. Because of its importance in atmospheric aerosol processes and aqueousphase chemistry, mass transfer to single particles will be treated separately in Chapter 12.

9.1 CONTINUUM AND NONCONTINUUM DYNAMICS: THE MEAN FREE PATH

As we begin our study of the dynamics of aerosols in a fluid (e.g., air), we would like to determine, from the perspective of transport processes, how the fluid "views" the particle or equivalently how the particle "views" the fluid that surrounds it. On the microscopic scale fluid molecules move in a straight line until they collide with another molecule. After collision, the molecule changes direction, moves for a while until it collides with another molecule, and so on. The average distance traveled by a molecule between collisions with other molecules is defined as its *mean free path.* Depending on the relative size of a particle suspended in a gas and the mean free path of the gas molecules around it, we can distinguish two cases. If the particle size is much larger than the mean free path of the surrounding gas molecules, the gas behaves, as far as the particle is concerned, as a continuous fluid. The particle is so large and the characteristic lengthscale of the motion of the gas molecules so small that an observer of the system sees a particle in a continuous fluid. At the other extreme, if the particle is much smaller than the mean free path of the surrounding fluid, an outside observer of the system (Figure 9.1) sees a small particle and gas molecule moving discretely around it. The particle is small enough that it resembles another gas molecule.

As usual in transport phenomena, one seeks an appropriate dimensionless group that reflects the relative lengthscales outlined above. The key dimensionless group that defines the nature of the suspending fluid relative to the particle is the *Knudsen number* (Kn)

$$Kn = \frac{2\lambda}{D_p} = \frac{\lambda}{R_p} \qquad (9.1)$$

Atmospheric Chemistry and Physics: From Air Pollution to Climate Change, Second Edition, by John H. Seinfeld and Spyros N. Pandis. Copyright © 2006 John Wiley & Sons, Inc.

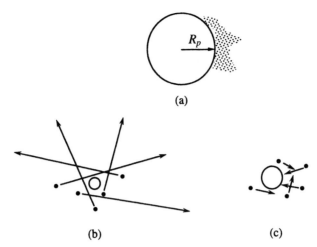

FIGURE 9.1 Schematic of the three regimes of suspending fluid–particle interactions: (a) continuum regime (Kn → 0), (b) free molecule (kinetic) regime (Kn → ∞), and (c) transition regime (Kn ∼ 1).

where λ is the mean free path of the fluid, D_p the particle diameter, and R_p its radius. Thus the Knudsen number is the ratio of two lengthscales, a length characterizing the "graininess" of the fluid with respect to the transport of momentum, mass, or heat, and a length scale characterizing the particle size, its radius.

Before we discuss the role of the Knudsen number, we need to consider the calculation of the mean free path for a vapor. It will soon be necessary to calculate the mean free path both for a pure gas and for gases composed of mixtures of several components. Note that even though air consists of molecules of N_2 and O_2, it is customary to talk about the mean free path of air, λ_{air}, as if air were a single chemical species.

Mean Free Path of a Pure Gas   Let us start with the simplest case, a particle suspended in a pure gas B. If we are interested in characterizing the nature of the suspending gas relative to the particle, the mean free path that appears in the definition of the Knudsen number is λ_{BB}. The subscript denotes that we are interested in collisions of molecules of B with other molecules of B. Ordinarily, air will be the predominant vapor species in such a situation. The mean free path λ_{BB} has been defined as the average distance traveled by a B molecule between collisions with other B molecules. The mean speed of gas molecules of B, $\bar{c}_B$ is (Moore 1962, p. 238)

$$\bar{c}_B = \left(\frac{8RT}{\pi M_B}\right)^{1/2} \tag{9.2}$$

where M_B is the molecular weight of B. Note that larger molecules move more slowly, while the overall mean speed of a gas increases with temperature. The mean speed of N_2 at 298 K is, according to (9.2), $\bar{c}_{N_2} = 474\,\mathrm{m\,s^{-1}}$ and for oxygen $\bar{c}_{O_2} = 444\,\mathrm{m\,s^{-1}}$. Molecular velocities of other atmospheric gases at 298 K are shown in Table 9.1.

Let us estimate what happens to a B molecule during a unit of time, say, a second. During this second the molecule travels on average $(\bar{c}_B \times 1\,\mathrm{s})\,\mathrm{m}$. If during the same

TABLE 9.1 Molecular Velocities of Some Atmospheric Gases at 298 K

Gas	Molecular Weight	Mean Velocity, m s^{-1}
NH_3	17	609
Air	28.8	468
HCl	36.5	416
HNO_3	63	316
H_2SO_4	98	254
$(CH_2)_3(COOH)_2$	132	219

second it undergoes Z_{BB} collisions, then its mean free path will be by definition

$$\lambda_{BB} = \frac{\bar{c}_B}{Z_{BB}} \tag{9.3}$$

Thus to calculate λ_{BB} we need to first calculate the collision rate of B molecules, Z_{BB}. Let σ_B be the diameter of a B molecule. In 1 s a molecule travels a distance $\bar{c}_B$ and collides with all molecules whose centers are in the cylinder of radius σ_B and height $\bar{c}_B$. Note that two molecules of diameter σ_B collide when the distance between their centers is σ_B. If N_B is the number of B molecules per unit volume, then the number of molecules in the cylinder is $\pi \sigma_B^2 \bar{c}_B N_B$. Above we have calculated the number of collisions assuming that one molecule of B is moving while the rest are immobile and in the process we have underestimated the frequency of collisions. In general, all particles are moving in random directions and we need to account for this motion by estimating their relative speed. If two particles move in opposite directions, their relative speed is $2\,\bar{c}_B$ (Figure 9.2). If they move in the same direction, their relative speed is zero, while for a $90°$ angle their relative

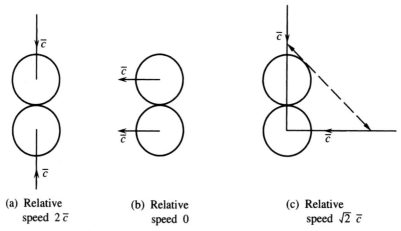

(a) Relative speed $2\,\bar{c}$ (b) Relative speed 0 (c) Relative speed $\sqrt{2}\,\bar{c}$

FIGURE 9.2 Relative speeds (RSs) of molecules for (a) head-on collision (RS $= 2\bar{c}$), (b) grazing collision (RS $= 0$), and (c) right-angle collision (RS $= \sqrt{2}\,\bar{c}$). For molecules moving in the same direction with the same velocity, the relative velocity of approach is zero. If they approach head-on, the relative velocity of approach is $2\bar{c}$. If they approach at $90°$, the relative velocity of approach is the sum of the velocity components along the line.

velocity of approach is $\sqrt{2}\,\bar{c}_B$ (Figure 9.2). One can prove that the latter situation represents the average, so we can write

$$Z_{BB} = \sqrt{2}\pi\sigma_B^2 \bar{c}_B N_B \tag{9.4}$$

and the mean free path λ_{BB} is given by

$$\lambda_{BB} = \frac{1}{\sqrt{2}\pi\sigma_B^2 N_B} \tag{9.5}$$

Note that the larger the molecule size, σ_B, and the higher the gas concentration, the smaller the mean free path.

Unfortunately, even though (9.5) provides valuable insights into the dependence of λ_{BB} on the gas concentration and molecular size, it is not convenient for the estimation of the mean free path of a pure gas, because one needs to know the diameter of the molecule σ_B, a rather ill-defined quantity as most molecules are not spherical. To make things even worse, the mean free path of a gas cannot be measured directly. However, the mean free path can be theoretically related to measurable gas microscopic properties, such as viscosity, thermal conductivity, or molecular diffusivity. One therefore can use measurements of the above gas properties to estimate theoretically the gas mean free path. For example, the mean free path of a pure gas can be related to the gas viscosity using the kinetic theory of gases

$$\lambda_{BB} = \frac{2\mu_B}{p(8M_B/\pi RT)^{1/2}} \tag{9.6}$$

where μ_B is the gas viscosity (in $kg\ m^{-1}\ s^{-1}$), p is the gas pressure (in Pa), and M_B is the molecular weight of B.

Calculation of the Air Mean Free Path   The air viscosity at $T = 298$ K and $p = 1$ atm is $\mu_{air} = 1.8 \times 10^{-5}\ kg\ m^{-1}s^{-1}$. The air mean free path at $T = 298$ K and $p = 1$ atm is then found using (9.6) to be

$$\lambda_{air}(298\ K, 1\ atm) = 0.0651\ \mu m \tag{9.7}$$

Thus for standard atmospheric conditions, if the particle diameter exceeds 0.2 μm or so, Kn < 1, and with respect to atmospheric properties, the particle is in the *continuum regime*. In that case, the equations of continuum mechanics are applicable. When the particle diameter is smaller than 0.01 μm, the particle exists in more or less a rarified medium and its transport properties must be obtained from the kinetic theory of gases. This Kn ≫ 1 limit is called the *free molecule* or *kinetic regime*. The particle size range intermediate between these two extremes (0.01–0.2 μm) is called the *transition regime*, and there the particle transport properties result from combination of the two other regimes.

The mean free path of air varies with height above the Earth's surface as a result of pressure and temperature changes (Chapter 1). This change for standard atmospheric

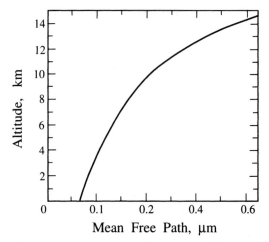

FIGURE 9.3   Mean free path of air as a function of altitude for the standard U.S. atmosphere (Hinds 1999).

conditions (see Table A.7) is shown in Figure 9.3. The net result is an increase of the air mean free path with altitude, to roughly $0.2\,\mu m$ at $10\,km$.

Mean Free Path of a Gas in a Binary Mixture   If we are interested in the diffusion of a vapor molecule A toward a particle, both of which are contained in a background gas B (e.g., air), then the description of the diffusion process depends on the value of the Knudsen number defined on the basis of the mean free path λ_{AB}. The mean free path λ_{AB} is defined as the average distance traveled by a molecule of A before it encounters another molecule of A or B. Note that because ordinarily the concentration of A molecules is several orders of magnitude lower than that of the background gas B (air), collisions between A molecules can be neglected, and the collisions between A and B are practically equal to the total number of collisions for an A molecule. The Knudsen number in the case of interest is given by

$$Kn = \frac{2\lambda_{AB}}{D_p} \tag{9.8}$$

and we need to estimate λ_{AB}. Jeans showed that the effective mean free path of molecules of A, λ_{AB}, in a binary mixture of A and B is (Davis 1983)

$$\lambda_{AB} = \frac{1}{\sqrt{2}\pi N_A \sigma_A^2 + \pi(1+z)^{1/2} N_B \sigma_{AB}^2} \tag{9.9}$$

where N_A and N_B are the molecular number concentrations of A and B, σ_A and σ_{AB} are the collision diameters for binary collisions between molecules of A and molecules of A and B, respectively, where

$$\sigma_{AB} = \frac{\sigma_A + \sigma_B}{2} \tag{9.10}$$

and $z = m_A/m_B = M_A/M_B$ is the ratio of molecular masses (or molecular weights) of A and B. The first term in the denominator of (9.9) accounts for the collisions between A molecules, while the second for the collisions between A and B molecules. If the concentration of species A is very low (a good assumption for almost all atmospheric situations), $N_A \ll N_B$ and (9.9) can be simplified by neglecting the collisions between A molecules as

$$\lambda_{AB} = \frac{1}{\pi(1+z)^{1/2} N_B \sigma_{AB}^2} \tag{9.11}$$

Note that the molecular concentration N_B can be calculated from the ideal-gas law $N_B = p/kT$, where p is the pressure of the system. The mean free path of the trace gas A in the background gas does not depend on the concentration of A itself. This is not a surprise, as we have assumed that the concentration of A is so low that A molecules never get to interact with each other. However, the mean free path of A depends on the sizes of the A and B molecules, and on the temperature and pressure of the mixture.

The mean free path once more is usually calculated not by (9.11) because of the difficulty of directly measuring σ_{AB}, but from the binary diffusivity of A in B, D_{AB}. This diffusivity can be either measured directly or calculated theoretically from the Chapman–Enskog theory for binary diffusivity (Chapman and Cowling 1970) by

$$D_{AB} = \frac{3}{8\pi} \frac{[\pi k^3 T^3 (1+z)/(2m_A)]^{1/2}}{\rho \sigma_{AB}^2 \Omega_{AB}^{(1,1)}} \tag{9.12}$$

where $\Omega_{AB}^{(1,1)}$ is the collision integral, which has been tabulated by Hirschfelder et al. (1954) as a function of the reduced temperature $T^* = kT/\varepsilon_{AB}$, where ε_{AB} is the Lennard-Jones molecular interaction parameter. For hard spheres $\Omega_{AB}^{(1,1)} = 1$, and for this case the following relationship connects the mean free path λ_{AB}, and the binary diffusivity D_{AB}

$$\lambda_{AB} = \frac{32}{3\pi(1+z)} \frac{D_{AB}}{\bar{c}_A} \tag{9.13}$$

Note the appearance of the molecular mass ratio $z = M_A/M_B$ in (9.13). Many investigators have assumed $z \ll 1$ either explicitly or implicitly and this has been the source of some confusion. We can identify certain limiting cases for (9.13):

$$
\begin{aligned}
\lambda_{AB} &= 3.397 \frac{D_{AB}}{\bar{c}_A} & z \ll 1 \\
&= 1.7 \frac{D_{AB}}{\bar{c}_A} & z = 1 \\
&= \frac{3.397}{z} \frac{D_{AB}}{\bar{c}_A} & z \gg 1
\end{aligned}
\tag{9.14}
$$

Additional relationships have been proposed to determine the mean free path in terms of D_{AB}. From zero-order kinetic theory, Fuchs and Sutugin (1971) showed that

$$\lambda_{AB} = 3 \frac{D_{AB}}{\bar{c}_A} \tag{9.15}$$

while Loyalka et al. (1989) used

$$\lambda_{AB} = \frac{4}{\sqrt{\pi}} \frac{D_{AB}}{\bar{c}_A} = 2.257 \frac{D_{AB}}{\bar{c}_A} \tag{9.16}$$

An additional relationship between the mean free path and the binary diffusivity can be derived using the kinetic theory of gases. The derivation relies on a simple argument involving the flux of gas molecules across planes separated by a distance λ. Consider the simplest case, only a single gas, some of the molecules of which are painted red. Assume that the number N' of red molecules is greater in one direction along the x axis, and consequently, if the total pressure is uniform throughout the gas, the number N'' of unpainted molecules must also vary along the x direction. Let us define the "mean free path" for diffusion as λ, so that λ is the distance both left and right of the plane at x where the molecules (both painted and unpainted) experienced their last collisions. We are purposely not defining λ precisely at this point. Figure 9.4 depicts planes at $x^* + \lambda$ and $x^* - \lambda$.

For molecules in three-dimensional random motion, the number of molecules striking a unit area per unit time is $\frac{1}{4}N\bar{c}$. If λ is the average distance from the control surface at which the molecules crossing the x^* surface originated, then the left-to-right flux of painted molecules is $\frac{1}{4}\bar{c}N'(x^* - \lambda)$, while the right-to-left is $\frac{1}{4}\bar{c}N'(x^* + \lambda)$.

The net left-to-right flux of painted molecules through the plane of x^* is (in molecules $cm^{-2} s^{-1}$)

$$J = \frac{1}{4}\bar{c}[N'(x^* - \lambda) - N'(x^* + \lambda)] \tag{9.17}$$

Expanding both $N'(x^* - \lambda)$ and $N'(x^* + \lambda)$ in Taylor series about x^*, we obtain

$$J = -\frac{1}{2}\bar{c}\lambda \left(\frac{\partial N'}{\partial x}\right)_{x=x^*} \tag{9.18}$$

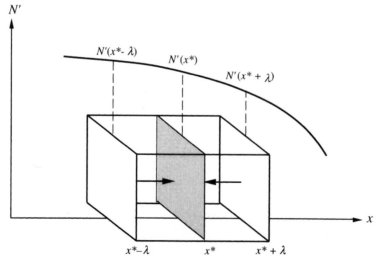

FIGURE 9.4 Control surfaces for molecular diffusion as envisioned in the elementary kinetic theory of gases.

Comparing (9.18) with the continuum expression $J = -D(\partial N'/\partial x)$ results in $D = 0.5\bar{c}\lambda$ or, equivalently

$$\lambda = 2\frac{D}{\bar{c}} \tag{9.19}$$

Since the red molecules differ from the others only by a coat of paint, λ and D apply to all molecules of the gas. Thus the diffusional mean free path λ is defined as a function of the molecular diffusivity of the vapor and its mean speed by (9.19).

Expressions (9.13), (9.15), (9.16), and (9.19) have different numerical constants and their use leads to mean free paths λ_{AB} that differ by as much as a factor of 2 for typical atmospheric gases. The consequences of using these different expressions are discussed in Chapter 12. In the remaining sections of this chapter we focus on the interactions of particles with a single gas, air, with a mean free path given by (9.6) and (9.7).

9.2 THE DRAG ON A SINGLE PARTICLE: STOKES' LAW

We start our discussion of the dynamical behavior of aerosol particles by considering the motion of a particle in a viscous fluid. As the particle is moving with a velocity u_∞, there is a drag force exerted by the fluid on the particle equal to F_{drag}. This drag force will always be present as long as the particle is not moving in a vacuum. To calculate F_{drag}, one must solve the equations of fluid motion to determine the velocity and pressure fields around the particle.

The velocity and pressure in an incompressible Newtonian fluid are governed by the equation of continuity (a mass balance)

$$\frac{\partial u_x}{\partial x} + \frac{\partial u_y}{\partial y} + \frac{\partial u_z}{\partial z} = 0 \tag{9.20}$$

and the Navier–Stokes equations (a momentum balance) (Bird et al. 1960), the x component of which is

$$\rho\left(\frac{\partial u_x}{\partial t} + u_x\frac{\partial u_x}{\partial x} + u_y\frac{\partial u_x}{\partial y} + u_z\frac{\partial u_x}{\partial z}\right) = -\frac{\partial p}{\partial x} + \mu\left(\frac{\partial^2 u_x}{\partial x^2} + \frac{\partial^2 u_x}{\partial y^2} + \frac{\partial^2 u_x}{\partial z^2}\right) + \rho g_x \tag{9.21}$$

where $\mathbf{u} = (u_x, u_y, u_z)$ is the velocity field, $p(x, y, z)$ is the pressure field, μ is the viscosity of the fluid, and g_x is the component of the gravity force in the x direction. To simplify our discussion let us assume without loss of generality that $g_x = 0$. The y and z components of the Navier–Stokes equations are similar to (9.21).

Let us nondimensionalize the Navier–Stokes equations by introducing a characteristic velocity u_0 and characteristic length L and defining the dimensionless variables

$$U_x = \frac{u_x}{u_0}, \quad U_y = \frac{u_y}{u_0}, \quad U_z = \frac{u_z}{u_0}, \quad \xi_x = \frac{x}{L}, \quad \xi_y = \frac{y}{L}, \quad \xi_z = \frac{z}{L} \tag{9.22}$$

and the dimensionless time and pressure:

$$\tau = \frac{t\,u_0}{L} \quad \text{and} \quad \phi = \frac{pL}{\rho U_0^2}$$

Then (9.20) and (9.21) can be rewritten using the definitions presented above

$$\frac{\partial U_x}{\partial \xi_x} + \frac{\partial U_y}{\partial \xi_y} + \frac{\partial U_z}{\partial \xi_z} = 0 \tag{9.23}$$

and

$$\frac{\partial U_x}{\partial \tau} + U_x \frac{\partial U_x}{\partial \xi_x} + U_y \frac{\partial U_x}{\partial \xi_y} + U_z \frac{\partial U_x}{\partial \xi_z} = -\frac{\partial \phi}{\partial \xi_x} + \frac{1}{\text{Re}} \left(\frac{\partial^2 U_x}{\partial \xi_x^2} + \frac{\partial^2 U_x}{\partial \xi_y^2} + \frac{\partial^2 U_x}{\partial \xi_z^2} \right) \tag{9.24}$$

where $\text{Re} = u_0 L \rho / \mu$ is the *Reynolds number*, representing the ratio of inertial to viscous forces in the flow. Note that all the parameters of the problem have been neatly combined into one dimensionless number, Re. The above nondimensionalization provides us with considerable insight, namely, that the nature of the flow will depend exclusively on the Reynolds number.

For flow around a particle submerged in a fluid, the characteristic lengthscale L is the diameter of the particle D_p, and u_0 can be chosen as the speed of the undisturbed fluid upstream of the body, u_∞. Therefore

$$\text{Re} = \frac{\rho u_\infty D_p}{\mu}$$

One could also use the radius R_p of the particle as L and then define Re as $\rho u_\infty R_p / \mu$. Clearly, these differ only by a factor of 2. We will use the Reynolds number Re defined on the basis of the particle diameter in our subsequent discussion.

When viscous forces dominate inertial forces, $\text{Re} \ll 1$, and the type of flow that results is called a low-Reynolds-number flow or creeping flow. In this case the Navier–Stokes equations can be simplified as one can neglect the left-hand-side (LHS) terms of (9.24) (note that $1/\text{Re}$ then is a large number) to obtain at steady state:

$$\frac{\partial \phi}{\partial \xi_x} = \frac{1}{\text{Re}} \left(\frac{\partial^2 U_x}{\partial \xi_x^2} + \frac{\partial^2 U_x}{\partial \xi_y^2} + \frac{\partial^2 U_x}{\partial \xi_z^2} \right) \tag{9.25}$$

The solution of (9.23) and (9.25) to obtain the velocity and pressure distribution around a sphere was first obtained by Stokes. The assumptions invoked to obtain the solution are (1) an infinite medium, (2) a rigid sphere, and (3) no slip at the surface of the sphere. For the solution details, we refer the reader to Bird et al. (1960, p. 132).

Using the spherical coordinate system defined in Figure 9.5, the pressure field around the particle is given by (Bird et al. 1960)

$$p = p_0 - \frac{3}{2} \frac{\mu u_\infty R_p}{r^2} \cos \theta \tag{9.26}$$

where R_p is the particle radius, p_0 is the pressure in the plane $z = 0$ far from the sphere, u_∞ is the approach velocity far from the sphere, and gravity has been neglected.

Our objective is to calculate the net force exerted by the fluid on the sphere in the direction of the flow. This force consists of two contributions. At each point on the surface of the sphere there is a force acting perpendicular to the surface. This is the normal force.

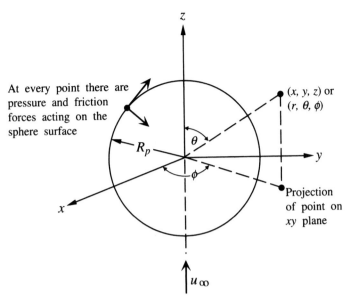

FIGURE 9.5   Coordinate system used in describing the flow of a fluid about a rigid sphere.

At each point there is also a tangential force exerted by the fluid due to the shear stress caused by the velocity gradients in the vicinity of the surface.

To obtain the normal force on the sphere, one integrates the component of the pressure acting perpendicularly to the surface. Then the normal force F_n is found to be

$$F_n = 2\pi\, \mu R_p u_\infty \tag{9.27}$$

The calculation of the tangential force requires the calculation of the shear stress $\tau_{r\theta}$ and then its integration over the particle surface to find the tangential force F_t

$$F_t = 4\pi\, \mu R_p u_\infty \tag{9.28}$$

Both forces act in the z direction (Figure 9.5) and the total drag exerted by the fluid on the sphere is

$$F_{\text{drag}} = F_n + F_t = 6\pi\, \mu R_p u_\infty \tag{9.29}$$

which is known as *Stokes' law*. Note that the case of a stationary sphere in a fluid moving with velocity u_∞ is entirely equivalent to that of a sphere moving with a velocity u_∞ through a stagnant fluid. In both cases the force exerted by the fluid on the particle is given by (9.29).

9.2.1   Corrections to Stokes' Law: The Drag Coefficient

Stokes' law has been derived for Re $\ll$ 1, neglecting the inertial terms in the equation of motion. If Re $= 1$, the drag predicted by Stokes' law is 13% low, due to the errors introduced by the assumption that inertial terms are negligible. To account for these terms, the drag force is usually expressed in terms of an empirical *drag coefficient* C_D as

$$F_{\text{drag}} = \frac{1}{2} C_D A_p \rho u_\infty^2 \tag{9.30}$$

where A_p is the projected area of the body normal to the flow. Thus for a spherical particle of diameter D_p

$$F_{\text{drag}} = \frac{1}{8} \pi C_D \rho D_p^2 u_\infty^2 \tag{9.31}$$

where the following correlations are available for the drag coefficient as a function of the Reynolds number:

$$
\begin{aligned}
C_D &= \frac{24}{\text{Re}} & \text{Re} \lesssim 1 \quad \text{(Stokes' law)} \\
C_D &= 18.5\,\text{Re}^{-0.6} & \text{Re} \gtrsim 1
\end{aligned}
\tag{9.32}
$$

Note for $C_D = 24/\text{Re}$, the drag force calculated by (9.31) is $F_{\text{drag}} = 3\pi\,\mu D_p u_\infty$, equivalent to Stokes' law.

To gain a feeling for the order of magnitude of Re for typical aerosol particles, the Reynolds numbers of spherical particles falling at their terminal velocities in air at 298 K and 1 atm are shown in Table 9.2. Thus for particles smaller than 20 μm (virtually all atmospheric aerosols) Stokes' law is an accurate formula for the drag exerted by the air. For larger particles (rain and large cloud droplets) or for particles in rapid motion one needs to use the drag coefficient correlations presented above.

9.2.2   Stokes' Law and Noncontinuum Effects: Slip Correction Factor

Stokes' law is based on the solution of equations of continuum fluid mechanics and therefore is applicable to the limit Kn → 0. The nonslip condition used as a boundary condition is not applicable for high Kn values. When the particle diameter D_p approaches the same magnitude as the mean free path λ of the suspending fluid (e.g., air), the drag force exerted by the fluid is smaller than that predicted by Stokes' law. To account for

TABLE 9.2   Reynolds Number for Particles in Air Falling at Their Terminal Velocities at 298 K

Diameter, μm	Re
0.1	7×10^{-9}
1	2.8×10^{-6}
10	2.5×10^{-3}
20	0.02
60	0.4
100	2
300	20

TABLE 9.3   Slip Correction Factor C_c for Spherical Particles in Air at 298 K and 1 atm

D_p, μm	C_c
0.001	216
0.002	108
0.005	43.6
0.01	22.2
0.02	11.4
0.05	4.95
0.1	2.85
0.2	1.865
0.5	1.326
1.0	1.164
2.0	1.082
5.0	1.032
10.0	1.016
20.0	1.008
50.0	1.003
100.0	1.0016

noncontinuum effects that become important as D_p becomes smaller and smaller, the *slip correction factor* C_c is introduced into Stokes' law, written now in terms of particle diameter D_p as

$$F_{\text{drag}} = \frac{3\pi \, \mu u_\infty D_p}{C_c} \tag{9.33}$$

where

$$C_c = 1 + \frac{2\lambda}{D_p}\left[1.257 + 0.4\exp\left(-\frac{1.1D_p}{2\lambda}\right)\right] \tag{9.34}$$

A number of investigators over the years have determined the values for the numerical coefficients used in the expression above. Allen and Raabe (1982) have reanalyzed Millikan's data (based on experiments performed between 1909 and 1923) to produce the updated set of parameters shown above.

Values of C_c as a function of the particle diameter D_p in air at 25°C are given in Table 9.3. The slip correction factor is generally neglected for particles exceeding 10 μm in diameter, as the correction is less than 2%. On the other hand, the drag force for a 0.1 μm in diameter particle is reduced by almost a factor of 3 as a result of this slip correction.

9.3   GRAVITATIONAL SETTLING OF AN AEROSOL PARTICLE

Up to this point, we have considered the drag force on a particle moving at a steady velocity u_∞ through a quiescent fluid. Recall that this case is equivalent to the flow of a fluid at velocity u_∞ past the stationary particle. The motion of the particle, however, arises in the first place because of the action of some external force on the particle such as gravity. The drag force arises as soon as there is a difference between the velocity of the particle and that of the fluid. The basis of the description of the behavior of a particle in a fluid is an equation of motion. To derive the equation of motion for a particle of mass m_p, let us begin with a force balance on the particle, which we write in vector form as

$$m_p \frac{d\mathbf{v}}{dt} = \sum_i \mathbf{F}_i \qquad (9.35)$$

where $\mathbf{v}$ is the velocity of the particle and $\mathbf{F}_i$ is the ith force acting on the particle.

For a particle falling in a fluid there are two forces acting on it, the gravitational force $m_p\mathbf{g}$ and the drag force $\mathbf{F}_{\text{drag}}$. Therefore, for Re $<$ 0.1, the equation of motion becomes

$$m_p \frac{d\mathbf{v}}{dt} = m_p\mathbf{g} + \frac{3\pi\mu D_p}{C_c}(\mathbf{u} - \mathbf{v}) \qquad (9.36)$$

where the second term of (9.36) is the corrected Stokes drag force on a particle moving with velocity $\mathbf{v}$ in a fluid having velocity $\mathbf{u}$. Equation (9.36) implicitly assumes that even though the particle motion is unsteady, this acceleration is slow enough that Stokes' law applies at any instant. This equation can be rewritten as

$$\tau \frac{d\mathbf{v}}{dt} = \tau\mathbf{g} + \mathbf{u} - \mathbf{v} \qquad (9.37)$$

where

$$\tau = \frac{m_p C_c}{3\pi\mu D_p} \qquad (9.38)$$

is the characteristic *relaxation time* of the particle.

Let us consider the case of a particle in a quiescent fluid ($\mathbf{u} = \mathbf{0}$) starting with zero velocity and let us take the z axis as positive downward. Then the equation of motion becomes

$$\tau \frac{dv_z}{dt} = \tau g - v_z \quad v_z(0) = 0 \qquad (9.39)$$

and its solution is

$$v_z(t) = \tau g[1 - \exp(-t/\tau)] \qquad (9.40)$$

For $t \gg \tau$, the particle attains a characterstic velocity, called its *terminal settling velocity* $v_t = \tau g$ or

$$v_t = \frac{m_p C_c g}{3\pi\mu D_p} \qquad (9.41)$$

**TABLE 9.4   Characteristic Time Required
for Reaching Terminal Settling Velocity**

D_p, μm	τ, s
0.05	4×10^{-8}
0.1	9.2×10^{-8}
0.5	1×10^{-6}
1.0	3.6×10^{-6}
5.0	7.9×10^{-5}
10.0	3.14×10^{-4}
50.0	7.7×10^{-3}

For a spherical particle of density ρ_p in a fluid of density ρ, $m_p = (\pi/6)D_p^3(\rho_p - \rho)$, where the factor $(\rho_p - \rho)$ is needed to account for both gravity and buoyancy. However, since generally $\rho_p \gg \rho$, $m_p = (\pi/6)D_p^3\rho_p$ and (9.41) can be rewritten in the more convenient form:

$$v_t = \frac{1}{18}\frac{D_p^2\rho_p g C_c}{\mu} \tag{9.42}$$

The timescale τ indicates the time required by the particle to reach this terminal settling velocity and is given in Table 9.4. The relaxation time τ also describes the time required by a particle entering a fluid stream, to approach the velocity of the stream. Thus the characteristic time of most particles of interest to achieve steady motion in air is extremely short.

Settling velocities of unit density spheres in air at 1 atm and 298 K as computed from (9.42) are given in Figure 9.6. Submicrometer particles settle extremely slowly, only a few

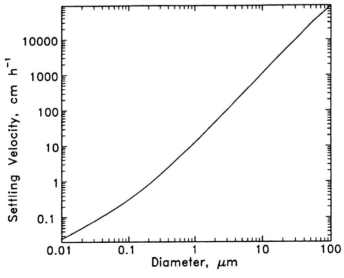

FIGURE 9.6   Settling velocity of particles in air at 298 K as a function of their diameter.

centimeters per hour. Particles larger than $10\,\mu\mathrm{m}$ settle with speeds exceeding $10\,\mathrm{m\,h}^{-1}$ and therefore are expected to have short atmospheric lifetimes.

Our analysis so far is applicable to $\mathrm{Re} < 0.1$ or particles smaller than about $20\,\mu\mathrm{m}$ (Table 9.2). For larger particles, one needs to use the drag coefficient as an empirical means of representing the drag force for higher Reynolds numbers. The equation along the direction of motion of the particle in scalar form, assuming no gas velocity, is then

$$m_p \frac{dv_z}{dt} = m_p g - \frac{1}{8}\pi\frac{C_D}{C_c}\rho D_p^2 v_z^2 \tag{9.43}$$

At steady-state $v_z = v_t$, the particle reaches its terminal velocity given by

$$v_t = \left(\frac{4 g D_p C_c \rho_p}{3 C_D \rho}\right)^{1/2} \tag{9.44}$$

However, as C_D is a function of Re and therefore v_t, we have only an implicit expression for v_t in (9.44). One needs then to solve (9.44) numerically with C_D calculated by (9.32) or one can use the following technique (Flagan and Seinfeld 1988).

If we form the product

$$C_D \mathrm{Re}^2 = \frac{C_D v_t^2 D_p^2 \rho^2}{\mu^2} \tag{9.45}$$

and substitute into this the v_t given by (9.44), we obtain

$$C_D \mathrm{Re}^2 = \frac{4 D_p^3 \rho\rho_p g C_c}{3\mu^2} \tag{9.46}$$

$C_D\mathrm{Re}^2$ can be calculated from (9.32) and one can prepare the plot of $C_D\mathrm{Re}^2$ versus Re shown in Figure 9.7. The terminal velocity can now be calculated as follows. First, we calculate $C_D\mathrm{Re}^2$ using (9.46). Then using Figure 9.7, we calculate Re. Then

$$v_t = \frac{\mu\mathrm{Re}}{\rho D_p}$$

and there is no need to solve the system of nonlinear algebraic equations.

> **Settling Velocity**   Calculate the settling velocity of a 200-μm-diameter droplet with density $\rho_p = 1\,\mathrm{g\,cm}^{-3}$. What would be the value if one uses Stokes' law?
>
> For a drop with $D_p = 200\,\mu\mathrm{m}$ using (9.34), $C_c = 1$ and therefore from (9.46) $C_D\,\mathrm{Re}^2 = 385$. Using Figure 9.7, we find that the corresponding Reynolds number is roughly 10. Now the terminal velocity can be calculated from the definition of Re and it is approximately $75\,\mathrm{cm\,s}^{-1}$.
>
> Using Stokes' law given by (9.42) we calculate $v_t = 120\,\mathrm{cm\,s}^{-1}$. Stokes' law overestimates the settling speed of such a droplet by 60%.

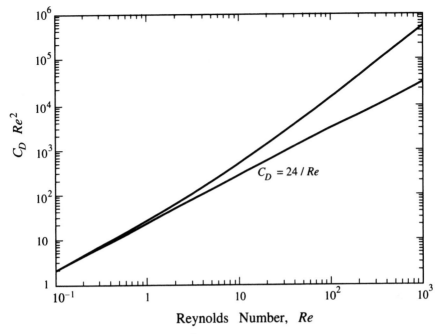

FIGURE 9.7 $C_D\text{Re}^2$ as a function of Re for a sphere.

9.4   MOTION OF AN AEROSOL PARTICLE IN AN EXTERNAL FORCE FIELD

The force balance presented in (9.35) describes the motion of a particle in a force field. As long as the particle is not moving in a vacuum, the drag force will always be present, so let us remove the drag force from the summation of forces

$$m_p \frac{d\mathbf{v}}{dt} = \frac{3\pi\mu D_p}{C_c}(\mathbf{u} - \mathbf{v}) + \sum_i \mathbf{F}_{ei} \qquad (9.47)$$

where $\mathbf{F}_{ei}$ denotes external force i (those forces arising from external potential fields, such as gravity and electrical forces).

Situations in which a charged particle moves in an electric field are important in several gas-cleaning devices and aerosol measurements. If a particle has an electric charge q in an electric field of strength $\mathbf{E}$, an electrostatic force

$$\mathbf{F}_{ee} = q\mathbf{E}$$

acts on the particle. The equation of motion for a particle of charge q moving at velocity $\mathbf{v}$ in a fluid with velocity $\mathbf{u}$ in the presence of an electric field of strength $\mathbf{E}$ is

$$m_p \frac{d\mathbf{v}}{dt} = \frac{3\pi\mu D_p}{C_c}(\mathbf{u} - \mathbf{v}) + q\mathbf{E} \qquad (9.48)$$

At steady state in the absence of a background fluid velocity, the particle velocity is such that the electrical force is balanced by the drag force and

$$\mathbf{v}_e = \frac{qC_c}{3\pi \mu D_p}\mathbf{E} \tag{9.49}$$

where $\mathbf{v}_e$ is termed the *electrical migration velocity*. Note that the characteristic time for relaxation of the particle velocity to its steady-state value is still given by $\tau = m_p C_c / 3\pi\mu D_p$ regardless of the external force influencing the particle. Thus, as long as τ is small compared to the characteristic time of change in the electric force, the particle velocity is given by (9.49). Defining the *electrical mobility* of a charged particle B_e as

$$B_e = \frac{qC_c}{3\pi \mu D_p} \tag{9.50}$$

then the electrical migration velocity is given by

$$\mathbf{v}_e = B_e\mathbf{E} \tag{9.51}$$

9.5 BROWNIAN MOTION OF AEROSOL PARTICLES

Particles suspended in a fluid are continuously bombarded by the surrounding fluid molecules. This constant bombardment results in a random motion of the particles known as *Brownian motion*. A satisfactory description of this irregular motion ("random walk") can be obtained ignoring the detailed structure of the particle-fluid molecule interaction if we assume that what happens to the aerosol fluid system at a given time t depends only on the system state at time t. Stochastic processes with this property are known as *Markov processes*.

In an effort to understand quantitatively Brownian motion, let us consider a particle that is settling in air owing to the action of gravity. As we have seen, the particle eventually reaches a terminal velocity that depends on the size of the particle and the viscosity of the air. A drag force is generated, depending on the velocity of the particle, that acts in a direction opposite to the direction of motion of the particle. If our particle is sufficiently large, say, 1 µm or larger, then the individual bombardment by microscopic gas molecules will have little effect on its motion that will be determined more or less solely by the continuum fluid drag force and gravity. However, if we consider a particle that is only a few nanometers, a size comparable to that of the gas molecules, then its motion will exhibit fluctuations from the random collisions that it experiences with gas molecules.

Let us consider a particle that is initially at the origin of our coordinate system. Assuming that the only force acting on the particles is that resulting from molecular bombardment by fluid molecules, the particle will start moving randomly from its original position and after time t will be at location $\mathbf{r}_1 = (x_1, y_1, z_1)$. If we repeat the same experiment with a second particle, we will find it at $\mathbf{r}_2 = (x_2, y_2, z_2)$ after the same period. Let us continue this experiment with an entire population or an ensemble of particles. If we average the displacements $\langle \mathbf{r} \rangle$ of all these particles, we expect the average $\langle \mathbf{r} \rangle$ to be zero since there is no preferred direction in Brownian motion. Can we then say anything quantitative about Brownian motion? We know that the average mean displacements $\langle x \rangle$, $\langle y \rangle$, $\langle z \rangle$ of a particle ensemble will be zero, but this is not enough. We need a measure of the intensity of Brownian motion, something that will allow us to distinguish between particles moving slowly and randomly and particles moving rapidly and randomly. The traditional measure of such intensity is the mean square displacement of all particles $\langle r^2 \rangle$, or for the three directions $\langle x^2 \rangle$, $\langle y^2 \rangle$, and $\langle z^2 \rangle$. Note that these means cannot be zero, as averages of positive quantities. We expect that the higher the intensity of the motion, the larger the mean square displacements. Since the mean square displacement is an important descriptor of the Brownian motion process, let us see what we can learn about this quantity.

Equations (9.35) and (9.47) provide a convenient framework for the analysis of forces acting on particles. These equations simply state that the acceleration experienced by the particle is proportional to the sum of forces acting on the particle. We have used this equation so far for "deterministic" forces, namely, the gravity, drag, and electrical forces. We now need to use the stochastic Brownian force, which is simply the product of the particle mass m_p and the random acceleration $\mathbf{a}$ caused by the bombardment by the fluid molecules. Then the equation of motion is

$$m_p \frac{d\mathbf{v}}{dt} = -\frac{3\pi \mu D_p}{C_c} \mathbf{v} + m_p \mathbf{a} \tag{9.52}$$

Dividing by m_p, (9.52) becomes

$$\frac{d\mathbf{v}}{dt} = -\frac{1}{\tau} \mathbf{v} + \mathbf{a} \tag{9.53}$$

where τ is the relaxation time of the particle. The random acceleration $\mathbf{a}$ is a discontinuous term, since it represents the random force exerted by the suspending fluid molecules that imparts an irregular, jerky motion to the particle. The equation of motion written to include the Brownian motion has its roots in two worlds: the macroscopic world represented by the drag force and the microscopic world presented by the Brownian force. The decomposition of the equation of motion into continuous and discontinuous pieces in (9.53) is an ad hoc assumption that is intuitively appealing and more important leads to successful predictions of particle behavior. Equation (9.53) was first formulated by the French physicist, Paul Langevin, in 1908 and is referred to as the *Langevin equation*. This equation will be the starting point in our effort to calculate the mean square displacement $\langle r^2 \rangle$.

Let us begin by taking the dot product of $\mathbf{r}$ and (9.53):

$$\mathbf{r} \cdot \frac{d\mathbf{v}}{dt} = -\frac{1}{\tau} \mathbf{r} \cdot \mathbf{v} + \mathbf{r} \cdot \mathbf{a} \tag{9.54}$$

Then ensemble averaging this equation (over many particles) gives

$$\left\langle \mathbf{r} \cdot \frac{d\mathbf{v}}{dt} \right\rangle = -\frac{1}{\tau} \langle \mathbf{r} \cdot \mathbf{v} \rangle + \langle \mathbf{r} \cdot \mathbf{a} \rangle \tag{9.55}$$

Since we assume that there is no preferred direction in $\mathbf{a}$ (directional isotropy of collisions), $\langle \mathbf{r} \cdot \mathbf{a} \rangle$ will be equal to zero, giving

$$\left\langle \mathbf{r} \cdot \frac{d\mathbf{v}}{dt} \right\rangle = -\frac{1}{\tau} \langle \mathbf{r} \cdot \mathbf{v} \rangle \tag{9.56}$$

Now since

$$\frac{d}{dt} \langle \mathbf{r} \cdot \mathbf{v} \rangle = \left\langle \mathbf{r} \cdot \frac{d\mathbf{v}}{dt} \right\rangle + \left\langle \frac{d\mathbf{r}}{dt} \cdot \mathbf{v} \right\rangle \tag{9.57}$$

or, equivalently

$$\left\langle \mathbf{r} \cdot \frac{d\mathbf{v}}{dt} \right\rangle = \frac{d}{dt} \langle \mathbf{r} \cdot \mathbf{v} \rangle - \langle v^2 \rangle \tag{9.58}$$

(9.56) becomes

$$\frac{d}{dt} \langle \mathbf{r} \cdot \mathbf{v} \rangle = -\frac{1}{\tau} \langle \mathbf{r} \cdot \mathbf{v} \rangle + \langle v^2 \rangle \tag{9.59}$$

The term $\frac{1}{2} m_p \langle v^2 \rangle$ is the kinetic energy of the system and as energy is partitioned equally in all three directions, each with an energy of $\frac{1}{2} kT$ for a total of $\frac{3}{2} kT$, we obtain that $\langle v^2 \rangle = 3kT/m_p$. Thus (9.59) becomes

$$\frac{d}{dt} \langle \mathbf{r} \cdot \mathbf{v} \rangle = -\frac{1}{\tau} \langle \mathbf{r} \cdot \mathbf{v} \rangle + \frac{3kT}{m_p} \tag{9.60}$$

Integrating this ordinary differential equation for $\langle \mathbf{r} \cdot \mathbf{v} \rangle$ we find

$$\langle \mathbf{r} \cdot \mathbf{v} \rangle = \frac{3kT\tau}{m_p} + c \exp(-t/\tau) \tag{9.61}$$

Now we note that

$$\langle \mathbf{r} \cdot \mathbf{v} \rangle = \left\langle \mathbf{r} \cdot \frac{d\mathbf{r}}{dt} \right\rangle = \frac{1}{2} \frac{d}{dt} \langle r^2 \rangle \tag{9.62}$$

so that (9.61) becomes

$$\frac{1}{2} \frac{d}{dt} \langle r^2 \rangle = \frac{3kT\tau}{m_p} + c \exp(-t/\tau) \tag{9.63}$$

We saw in Section 9.3 that for $t \gg \tau$, the particle velocity relaxes to a pseudo-steady-state value. We assume that to be the case here, namely, that the Brownian motion of the particle is sufficiently slow, that the particle has time to "relax" after each fluctuating impulse. Under this assumption, we drop the exponential in (9.63) to obtain

$$\frac{1}{2}\frac{d}{dt}\langle r^2 \rangle = \frac{3kT\tau}{m_p} \tag{9.64}$$

which, on integration, becomes

$$\langle r^2 \rangle = \frac{6kT\tau}{m_p}t = \frac{2kTC_c t}{\pi\mu D_p} \tag{9.65}$$

The Brownian motion can be assumed to be isotropic so $\langle x^2 \rangle = \langle y^2 \rangle = \langle z^2 \rangle = \frac{1}{3}\langle r^2 \rangle$. Thus

$$\langle x^2 \rangle = \langle y^2 \rangle = \langle z^2 \rangle = \frac{2kTC_c}{3\pi\mu D_p}t \tag{9.66}$$

This result, first derived by Einstein by a different route, has been confirmed experimentally. It indicates that the mean square distance traversed by a Brownian particle is proportional to the length of time it has experienced such motion.

We should note that we have obtained the foregoing results in a more or less formal manner without attempting to justify from a rigorous mathematical point of view the validity of the Langevin equation as the basic description of the particle motion. The theoretical results we have presented can be rigorously justified. A good starting point for the reader wishing to go deeply into its theory is the classic article by Chandrasekhar (1943), which is reprinted in Wax (1954).

9.5.1 Particle Diffusion

The movement of particles due to Brownian motion can also be viewed as a macroscopic diffusion process. Let us discuss the connection between these two different perspectives on the same process. If $N(x, y, z, t)$ is the number concentration of particles undergoing Brownian motion, then we can define a Brownian diffusivity D, such that

$$\frac{\partial N(x, y, z, t)}{\partial t} = D\nabla^2 N(x, y, z, t) \tag{9.67}$$

If we relate the Brownian diffusivity D to the mean square displacements given by (9.66), then (9.67) can provide a convenient framework for describing aerosol diffusion. To do so, let us repeat the experiment above, namely, let us follow the Brownian diffusion of N_0 particles placed at $t = 0$ at the $y - z$ plane. To simplify our discussion we assume that N does not depend on y or z. Multiplying (9.67) by x^2 and integrating the resulting equation over x from $-\infty$ to ∞, we get

$$\int_{-\infty}^{+\infty} x^2 \frac{\partial N}{\partial t} dx = \int_{-\infty}^{+\infty} x^2 D \frac{\partial^2 N}{\partial x^2} dx \qquad (9.68)$$

The LHS can also be written as

$$\int_{-\infty}^{+\infty} x^2 \frac{\partial N}{\partial t} dx = N_0 \frac{\partial \langle x^2 \rangle}{\partial t} \qquad (9.69)$$

and the RHS of (9.68) as

$$\int_{-\infty}^{+\infty} x^2 D \frac{\partial^2 N}{\partial x^2} dx = 2DN_0 \qquad (9.70)$$

Combining (9.68), (9.69), and (9.70) results in

$$\frac{\partial \langle x^2 \rangle}{\partial t} = 2D \qquad (9.71)$$

or after integration

$$\langle x^2 \rangle = 2Dt \qquad (9.72)$$

We can now equate this result for $\langle x^2 \rangle$ with that of (9.66) to obtain an explicit relation for D

$$D = \frac{kTC_c}{3\pi \mu D_p} \qquad (9.73)$$

which, without the correction factor C_c, is the *Stokes–Einstein–Sutherland relation*. Note that for particles that are larger than the mean free path of air, $C_c \simeq 1$ and their diffusivity varies as D_p^{-1}. As expected, larger particles diffuse more slowly. In the other extreme, when $D_p \ll \lambda$, $C_c = 1 + 1.657(2\lambda/D_p)$ and D can be approximated by $2(1.657)\lambda kT / 3\pi\mu D_p^2$. Therefore, in the free molecule regime, D varies as D_p^{-2}.

Diffusion coefficients for particles ranging from 0.001 to 10.0 μm diameter in air at 20°C are shown in Figure 9.8. The change from D_p^{-2} dependence is indicated by the change of slope of the line of D versus D_p.

The importance of Brownian diffusion as compared to gravitational settling can be judged by comparing the distances a particle travels as a result of each process (Twomey 1977). Over a time of 1 s a 1-μm-radius particle diffuses a distance of about 4 μm, while it falls about 200 μm under gravity. A 0.1 μm radius particle, on the other hand, in 1 s, diffuses a distance of about 20 μm compared to a fall distance of 4 μm. Even though a 1 μm particle's motion is dominated by inertia and gravity, it still diffuses several times its own radius in 1 s. The motion of the 0.1 μm particle is dominated by Brownian

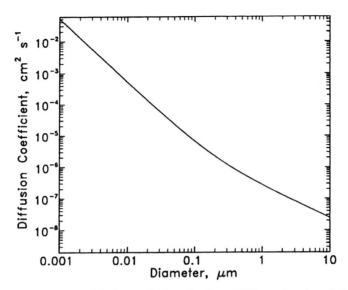

FIGURE 9.8   Aerosol diffusion coefficients in air at 20°C as a function of diameter.

diffusion. For a 0.01-μm-radius particle, Brownian diffusion further outweighs gravity; its diffusive displacement in 1 second is almost 1000 times its displacement because of gravity.

Note that the magnitude of diffusion coefficients of gases is on the order of $0.1 \, \text{cm}^2 \, \text{s}^{-1}$. Therefore a 0.1 μm particle diffuses in a quiescent gas roughly 10,000 times more slowly than a gas molecule, and Brownian diffusion is not expected to be an efficient transport mechanism for aerosols in the atmosphere.

9.5.2   Aerosol Mobility and Drift Velocity

In the development of Brownian motion up to this point, we have assumed that the only external force acting on the particle is the fluctuating Brownian force $m_p \mathbf{a}$. If we generalize (9.52) to include an external force $\mathbf{F}_{ext}$, we get

$$m_p \frac{d\mathbf{v}}{dt} = \mathbf{F}_{ext} - \frac{m_p}{\tau} \mathbf{v} + m_p \mathbf{a} \qquad (9.74)$$

As before, assuming that we are interested in times for which $t \gg \tau$, and taking mean values, the approximate force balance is at steady state:

$$0 = \mathbf{F}_{ext} - \frac{m_p}{\tau} \langle \mathbf{v} \rangle \qquad (9.75)$$

The ensemble mean velocity $\langle \mathbf{v} \rangle$ is identified as the *drift velocity* $\mathbf{v}_{drift}$, where

$$\mathbf{v}_{drift} = \frac{\mathbf{F}_{ext} \tau}{m_p} \qquad (9.76)$$

The drift velocity is the mean velocity experienced by the particle population due to the presence of the external force $\mathbf{F}_{ext}$. For example, in the case where the external force is simply gravity, $\mathbf{F}_{ext} = m_p \mathbf{g}$, and the drift velocity (or settling velocity) will simply be $\mathbf{v}_{drift} = \mathbf{g}\tau$ [see also (9.41)]. When the external force is electrical, the drift velocity is the electrical migration velocity [see also (9.49)]. Therefore our analysis presented in the previous sections is still valid even after the introduction of Brownian motion.

It is customary to define the generalized particle *mobility B* by

$$\mathbf{v}_{drift} = B\mathbf{F}_{ext} \tag{9.77}$$

Therefore the particle mobility is given by

$$B = \frac{\tau}{m_p} = \frac{C_c}{3\pi \mu D_p} \tag{9.78}$$

The mobility can also be viewed as the drift velocity that would be attained by the particles under unit external force. Recall (9.50), which is the mobility in the special case of an electrical force. By definition, the electrical mobility is related to the particle mobility by $B_e = qB$, where q is the particle charge. A particle with zero charge, has a mobility given by (9.78) and zero electrical mobility.

Finally, the Brownian diffusivity can be written in terms of the mobility [see also (9.73)] by

$$D = BkT \tag{9.79}$$

a result known as the *Einstein relation*.

Gravitational Settling and the Vertical Distribution of Aerosol Particles   Let us consider the simultaneous Brownian diffusion and gravitational settling of particles above a surface at $z = 0$. At $t = 0$, a uniform concentration $N_0 = 1000 \, \text{cm}^{-3}$ of particles is assumed to exist for $z > 0$ and at all times the concentration of particles right at the surface is zero as a result of their removal at the surface.

1. What is the particle concentration as a function of height and time, $N(z, t)$?
2. What is the removal rate of particles at the surface?

The concentration distribution of aerosol particles in a stagnant fluid in which the particles are subject to Brownian motion and in which there is a velocity v_t in the $-z$ direction is described by

$$\frac{\partial N}{\partial t} - v_t \frac{\partial N}{\partial z} = D \frac{\partial^2 N}{\partial z^2} \tag{9.80}$$

subject to the conditions

$$
\begin{aligned}
N(z, 0) &= N_0 \\
N(0, t) &= 0 \\
N(z, t) &= N_0 \quad z \to \infty
\end{aligned} \tag{9.81}
$$

where the z coordinate is taken as vertically upward.

The solution of (9.80) and (9.81) for the vertical profile of the number distribution $N(z, t)$ is

$$N(z, t) = \frac{N_0}{2} \left[1 + \text{erf}\left(\frac{z + v_t t}{2\sqrt{Dt}} \right) - \exp\left(-\frac{v_t z}{D} \right) \text{erfc}\left(\frac{z - v_t t}{2\sqrt{Dt}} \right) \right] \qquad (9.82)$$

We can calculate the deposition rate of particles on the $z = 0$ surface from the expression for the flux of particles at $z = 0$,

$$J = D \left(\frac{\partial N}{\partial z} \right)_{z=0} + v_t N(0, t) \qquad (9.83)$$

Recall that $N(0, t) = 0$ in (9.83). Combining (9.82) and (9.83) we obtain

$$J = N_0 \left\{ \frac{v_t}{2} \left[1 - \text{erf}\left(-\frac{v_t t}{2\sqrt{Dt}} \right) \right] + \left(\frac{D}{\pi t} \right)^{1/2} \exp\left(-\frac{v_t^2 t}{4D} \right) \right\} \qquad (9.84)$$

According to (9.84), there is an infinite removal flux at $t = 0$, because of our artificial specification of an infinite concentration gradient at $z = t = 0$. We can identify a characteristic time τ_{ds} for the system

$$\tau_{ds} = \frac{4D}{v_t^2} \qquad (9.85)$$

and observe the following limiting behavior for the particle flux at short and long times:

$$J(t) = N_0 \left[\left(\frac{D}{\pi t} \right)^{1/2} + \frac{v_t}{2} \right] \qquad t \ll \tau_{ds}$$

$$J(t) = N_0 v_t \qquad t \gg \tau_{ds} \qquad (9.86)$$

Thus, at very short times, the deposition flux is that resulting from diffusion plus one-half that due to settling, whereas for long times the deposition flux becomes solely the settling flux. For particles of radii 0.1 μm and 1 μm in air (at 1 atm, 298 K), τ_{ds} is about 80 s and 0.008 s, respectively, assuming a density of 1 g cm^{-3}. For times longer than that, Brownian motion does not have any effect on the particle motion. The aerosol number concentration and removal flux are shown in Figures 9.9 and 9.10. The system reaches a steady state after roughly 100 s and at this state 0.23 particles are deposited per second on each cm^2 of the surface (Figure 9.10). Note that the concentration profile changes only over a shallow layer of approximately 1 mm above the surface (Figure 9.9). The depth of this layer is proportional to D/v_t. Note that the example above is not representative of the ambient atmosphere, where there is turbulence and possibly also sources of particles at the ground.

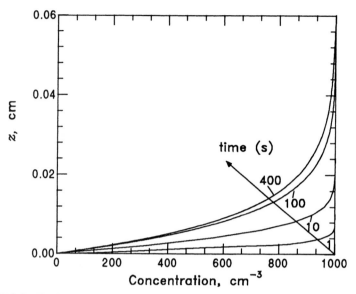

FIGURE 9.9   Evolution of the concentration profile (times 1, 10, 100, and 400 s) of an aerosol population settling and diffusing in stationary air over a flat perfectly absorbing surface. The particles are assumed to be monodisperse with $D_p = 0.2\,\mu m$ and have an initial concentration of $1000\,cm^{-3}$.

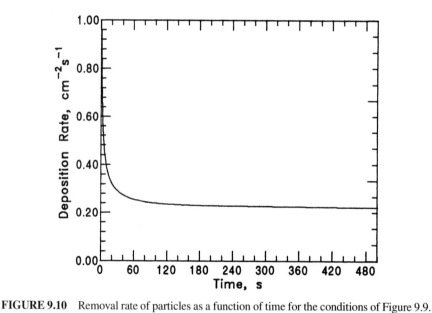

FIGURE 9.10   Removal rate of particles as a function of time for the conditions of Figure 9.9.

9.5.3   Mean Free Path of an Aerosol Particle

The concept of mean free path is an obvious one for gas molecules. In the Brownian motion of an aerosol particle there is not an obvious length that can be identified as a mean free path. This is depicted in Figure 9.11 showing plane projections of the paths followed by an air molecule and an aerosol particle of radius roughly equal to 1 μm. The trajectories

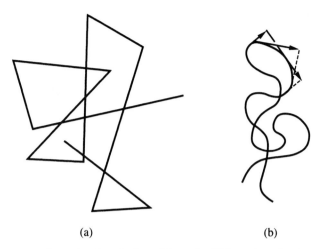

(a)   (b)

FIGURE 9.11   A two-dimensional projection of the path of (a) an air molecule and (b) the center of a 1-μm particle. Also shown is the apparent mean free path of the particle.

of the gas molecules consist of straight segments, each of which represents the path of the molecule between collisions. At each collision the direction and speed of the molecule are changed abruptly. With aerosol particles the mass of the particle is so much greater than that of the gas molecules with which it collides that the velocity of the particle changes negligibly in a single collision. Appreciable changes in speed and direction occur only after a large number of collisions with molecules, resulting in an almost smooth particle trajectory.

The particle motion can be characterized by a mean thermal speed $\bar{c}_p$:

$$\bar{c}_p = \left(\frac{8kT}{\pi m_p}\right)^{1/2} \tag{9.87}$$

To obtain the mean free path λ_p, we recall that in Section 9.1, using kinetic theory, we connected the mean free path of a gas to measured macroscopic transport properties of the gas such as its binary diffusivity. A similar procedure can be used to obtain a particle mean free path λ_p from the Brownian diffusion coefficient and an appropriate kinetic theory expression for the diffusion flux. Following an argument identical to that in Section 9.1, diffusion of aerosol particles can be viewed as a mean free path phenomenon so that

$$D = \frac{1}{2}\bar{c}_p\lambda_p \tag{9.88}$$

and the mean free path λ_p combining (9.73), (9.87), and (9.88) is then

$$\lambda_p = \frac{C_c}{6\mu}\sqrt{\frac{\rho kTD_p}{3}} \tag{9.89}$$

TABLE 9.5   Characteristic Quantities in Aerosol Brownian Motion

D_p, µm	D, cm^2 s^{-1}	$\bar{c}_p$, cm s^{-1}	τ, s	λ_p (µm)
0.002	1.28×10^{-2}	4965	1.33×10^{-9}	6.59×10^{-2}
0.004	3.23×10^{-3}	1760	2.67×10^{-9}	4.68×10^{-2}
0.01	5.24×10^{-4}	444	6.76×10^{-9}	3.00×10^{-2}
0.02	1.30×10^{-4}	157	1.40×10^{-8}	2.20×10^{-2}
0.04	3.59×10^{-5}	55.5	2.98×10^{-8}	1.64×10^{-2}
0.1	6.82×10^{-6}	14.0	9.20×10^{-8}	1.24×10^{-2}
0.2	2.21×10^{-6}	4.96	2.28×10^{-7}	1.13×10^{-2}
0.4	8.32×10^{-7}	1.76	6.87×10^{-7}	1.21×10^{-2}
1.0	2.74×10^{-7}	0.444	3.60×10^{-6}	1.53×10^{-2}
2.0	1.27×10^{-7}	0.157	1.31×10^{-5}	2.06×10^{-2}
4.0	6.1×10^{-8}	5.55×10^{-2}	5.03×10^{-5}	2.8×10^{-2}
10.0	2.38×10^{-8}	1.40×10^{-2}	3.14×10^{-4}	4.32×10^{-2}
20.0	1.38×10^{-8}	4.96×10^{-3}	1.23×10^{-3}	6.08×10^{-2}

Certain quantities associated with the Brownian motion and the dynamics of single aerosol particles are shown as a function of particle size in Table 9.5. All tabulated quantities in Table 9.5 depend strongly on particle size with the exception of the apparent mean free path λ_p, which is of the same order of magnitude right down to molecular sizes, with atmospheric values $\lambda_p \simeq 10-60$ nm.

9.6   AEROSOL AND FLUID MOTION

In our discussion so far, we have assumed that the aerosol particles are suspended in a stagnant fluid. In most atmospheric applications, the air is in motion and one needs to describe simultaneously the air and particle motion. Equation (9.36) will be the starting point of our analysis.

Actually (9.36) is a simplified form of the full equation of motion, which is (Hinze 1959)

$$
m_p \frac{dv}{dt} = \frac{3\pi\mu D_p}{C_c}(\mathbf{u} - \mathbf{v}) + V_p\rho\frac{d\mathbf{u}}{dt} + \frac{V_p}{2}\rho\left(\frac{d\mathbf{u}}{dt} - \frac{d\mathbf{v}}{dt}\right)
$$
$$
+ \frac{3D_p^2}{2}(\pi\rho\mu)^{1/2}\int_0^t \frac{(d\mathbf{u}/dt') - (d\mathbf{v}/dt')}{(t - t')^{1/2}}dt' + \sum_i \mathbf{F}_{ei}
\tag{9.90}
$$

where V_p is the particle volume. The second term on the RHS is due to the pressure gradient in the fluid surrounding the particle, caused by acceleration of the gas by the particle. The third term is the force required to accelerate the apparent mass of the particle relative to the fluid. Finally, the fourth term, the Basset history integral, accounts for the force arising as a result of the deviation of fluid velocity from steady state. In most situations of interest for aerosol particle in air, the second, third, and fourth terms on the RHS of (9.90) are neglected. Assuming that gravity is the only external force exerted on the particle, we again obtain (9.36).

Neglecting the gravitational force and particle inertia leads to the zero-order approximation that $\mathbf{v} \simeq \mathbf{u}$; that is, the particle follows the streamlines of the airflow. This approximation is often sufficient for most atmospheric applications, such as turbulent

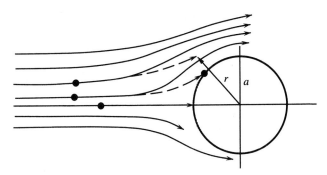

FIGURE 9.12 Schematic diagram of particles and fluid motion around a cylinder. Streamlines are shown as solid lines, while the dashed lines are aerosol paths.

dispersion. However, it is often necessary to quantify the deviation of the particle trajectories from the fluid streamlines (Figure 9.12).

A detailed treatment of particle flow around objects, in channels of various geometries, and so on, is beyond the scope of this book. Treatments are provided by Fuchs (1964), Hinds (1999), and Flagan and Seinfeld (1988). We will focus our analysis on a few simple examples demonstrating the important concepts.

9.6.1 Motion of a Particle in an Idealized Flow (90° Corner)

Let us consider an idealized flow, shown in Figure 9.13, in which an airflow makes an abrupt 90° turn in a corner maintaining the same velocity (Crawford 1976). We would like to determine the trajectory of an aerosol particle originally on the streamline $y = 0$, which turns at the origin $x = 0$.

The trajectory of the particle is governed by (9.37). Neglecting gravity, the x and y components of the equation of motion are after the turning point

$$\tau \frac{dv_x}{dt} + v_x = 0$$

$$\tau \frac{dv_y}{dt} + v_y = U$$

(9.91)

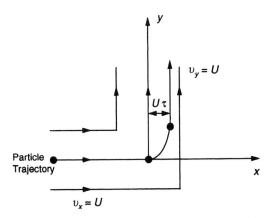

FIGURE 9.13 Motion of an air particle in a flow making a 90° turn with no change in velocity.

and as $v_x = dx/dt$ and $v_y = dy/dt$, we get

$$\tau \frac{d^2x}{dt^2} + \frac{dx}{dt} = 0$$

$$\tau \frac{d^2y}{dt^2} + \frac{dy}{dt} = U$$

(9.92)

subject to

$$x(0) = 0, \quad y(0) = 0, \quad \left(\frac{dx}{dt}\right)_{t=0} = U, \quad \left(\frac{dy}{dt}\right)_{t=0} = 0$$

(9.93)

Solving (9.92) subject to (9.93) gives the particle coordinates as a function of time:

$$x(t) = U\tau[1 - \exp(-t/\tau)]$$

(9.94)

$$y(t) = -U\tau[1 - \exp(-t/\tau)] + Ut$$

(9.95)

We see that for $t \gg \tau$, the particle trajectory is described by $x(t) = U\tau$ and $y(t) = Ut$. Thus the particle eventually ends up at a distance $U\tau$ to the right of its original fluid streamline. Larger particles with high relaxation times will move, because of their inertia, significantly to the right, while small particles, with $\tau \to 0$, will follow closely their original streamline. For example, for a 2-μm-diameter particle moving with a speed $U = 20\,\mathrm{m\,s^{-1}}$ and having density $p_p = 2\,\mathrm{g\,cm^{-3}}$, we find that the displacement is 0.48 mm, while for a 20-μm-diameter particle $U\tau = 4.83$ cm.

The flow depicted in Figure 9.13 is the most idealized one representing the stagnation flow of a fluid toward a flat plane (see also Figure 9.14). If we imagine that in Figure 9.14 there is a flat plate at $x = 0$ and that $y = 0$ is the line of symmetry, then all the particles that initially are a distance smaller than $x_0 = U\tau$ from the line of symmetry will collide with

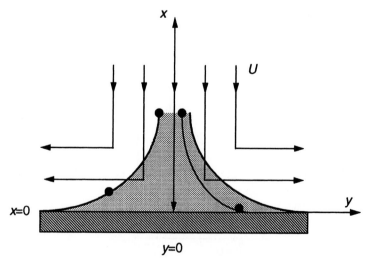

FIGURE 9.14   Idealized flow toward a plate.

the flat plate. On the other hand, particles outside this area will be able to turn and, avoiding the plane, continue flowing parallel to it.

A more detailed treatment of the flow in such situations is presented by Flagan and Seinfeld (1988) considering more realistic fluid streamlines.

9.6.2 Stop Distance and Stokes Number

Let us consider a particle moving with a speed U in a stagnant fluid with no forces acting on it. The particle will slow down because of the drag force exerted on it by the fluid and eventually stop after moving a distance s_p. We can calculate this *stop distance* of the particle employing (9.39), neglecting gravity, and noting that $v = ds/dt$. The motion of the particle is described by

$$\tau \frac{ds^2}{dt^2} + \frac{ds}{dt} = 0, \quad s(0) = 0, \quad \left(\frac{ds}{dt}\right)_{t=0} = U \tag{9.96}$$

with solution

$$s(t) = \tau U \left(1 - \exp\left(-\frac{t}{\tau}\right)\right) \tag{9.97}$$

Note that as $t \gg \tau, s(t) \rightarrow s_p$, where

$$s_p = \tau U \tag{9.98}$$

is the particle stop distance. For a 1-μm-diameter particle, with an initial speed of $10 \, \text{m s}^{-1}$, for example, the stop distance is $36 \, \mu\text{m}$.

Let us write the equation of motion (9.37) without the gravitational term, in dimensionless form. To do so we introduce a characteristic fluid velocity u_0 and a characteristic length L both associated with the flow of interest. We define dimensionless time t^*, distance x^*, and velocity v^* by

$$t^* = \frac{tu_0}{L}, \quad x^* = \frac{x}{L}, \quad u_x^* = \frac{u_x}{u_0} \tag{9.99}$$

Placing (9.37) in dimensionless form gives

$$\frac{\tau u_0}{L} \frac{d^2 x^*}{dt^{*2}} + \frac{dx^*}{dt^*} = u_x^* \tag{9.100}$$

Note that all the variables of the system have been combined in the dimensionless group $\tau u_0/L$, which is called the *Stokes number* (St):

$$\text{St} = \frac{\tau u_0}{L} = \frac{D_p^2 \rho_p C_c u_0}{18 \mu L} \tag{9.101}$$

Note that the Stokes number is the ratio of the particle stop distance s_p to the characteristic length of the flow L. As particle mass decreases, the Stokes number also decreases. A small Stokes number implies that the particle is able to adopt the fluid velocity very quickly. Since the dimensionless equation of motion depends only on the Stokes number, equality between two geometrically similar flows indicates similarity of the particle trajectories.

9.7   EQUIVALENT PARTICLE DIAMETERS

Up to this point we have considered spherical particles of a known diameter D_p and density ρ_p. Atmospheric particles are sometimes nonspherical and we seldom have information about their density. Also a number of techniques used for atomospheric aerosol size measurement actually measure the particle's terminal velocity or its electrical mobility. In these cases we need to define an equivalent diameter for the nonspherical particles or even for the spherical particles of unknown density or charge. These equivalent diameters are defined as the diameter of a sphere, which, for a given instrument, would yield the same size measurement as the particle under consideration. A series of diameters have been defined and are used for such particles.

9.7.1   Volume Equivalent Diameter

The volume equivalent diameter D_{ve} is the diameter of a sphere having the same volume as the given nonspherical particle. If the volume V_p of the nonspherical particle is know then:

$$D_{ve} = \frac{6}{\pi} V_p^{1/3} \qquad (9.102)$$

For a spherical particle the volume equivalent diameter is equal to its physical diameter, $D_{ve} = D_p$.

To account for the shape effects during the flow of nonspherical particles, Fuchs (1964) defined the shape factor χ as the ratio of the actual drag force on the particle F_D to the drag force F_D^{ve} on a sphere with diameter equal to the volume equivalent diameter of the particle:

$$\chi = \frac{F_D}{F_D^{ve}} \qquad (9.103)$$

The dynamic shape factor is almost always greater than 1.0 for irregular particles and flows at small Reynolds numbers and is equal to 1.0 for spheres. For a nonspherical particle of a given shape χ is not a constant but changes with pressure, particle size, and as a result of particle orientation in electric or aerodynamic flow fields.

The dynamic shape factor for flow in the continuum regime is equal to 1.08 for a cube, 1.12 for a 2-sphere cluster, 1.15 for a compact 3-sphere cluster, and 1.17 for a

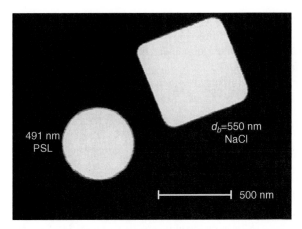

FIGURE 9.15 A micrograph of a single NaCl particle with electrical mobility equivalent diameter of 550 nm. The dry NaCl is almost cubic with rounded edges. Also shown is a polystyrene latex (PSL) particle with diameter of 491 nm (Zelenyuk et al. 2006).

compact 4-sphere cluster (Hinds 1999). These values are averaged over all orientations of the particle, which is the usual situation for atmospheric aerosol flows (Re < 0.1) because of the Brownian motion of the particles. Liquid and organic atmospheric particles are spherical for all practical purposes. Dry NaCl crystals have cubic shape (Figure 9.15), while dry $(NH_4)_2SO_4$ is approximately but not exactly spherical (Figure 9.16). Dynamic shape factors ranging from 1.03 to 1.07 have been measured in the laboratory (Zelenyuk et al. 2005) with the higher values observed for larger particles with diameters of ~ 500 nm.

Nonspherical particles are subjected to a larger drag force compared to their volume equivalent spheres because $\chi > 1$ and therefore settle more slowly. The terminal settling velocity of a nonspherical particle is then [following the same approach as in the derivation of (9.42)]

$$v_t = \frac{1}{18} \frac{D_{ve}^2 \rho_p g C_c(D_{ve})}{\chi \mu} \qquad (9.104)$$

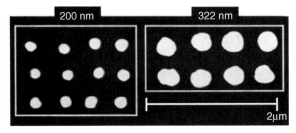

FIGURE 9.16 Ammonium sulfate particles with electrical mobility diameters of 200 and 322 nm (Zelenyuk et al. 2006).

Volume Equivalent Diameter   An approximately cubic NaCl particle with density $2.2\,\text{g cm}^{-3}$ has a terminal settling velocity of $1\,\text{mm s}^{-1}$ in air at ambient conditions. Calculate its volume equivalent diameter and its physical size using the continuum regime shape factor.

The terminal settling velocity is given by (9.104), so after some rearrangement; we obtain

$$D_{ve}^2\, C_c(D_{ve}) = (18\, v_t\, \mu\chi)/(\rho_p\, g) \tag{9.105}$$

and substituting $v_t = 10^{-3}\,\text{m s}^{-1}, \mu = 1.8 \times 10^{-5}\,\text{kg m}^{-1}\,\text{s}^{-1}, \chi = 1.08$ for a cube, $\rho_p = 2200\,\text{kg m}^{-3}$ for NaCl we find that

$$D_{ve}^2\, C_c(D_{ve}) = 1.62 \times 10^{-11}\,\text{m}^2 = 0.162\,\mu\text{m}^2$$

This equation needs to be solved numerically using (9.34) for calculation of the slip correction, $C_c(D_{ve})$, to obtain $D_{ve} = 0.328\,\mu\text{m}$. One can also estimate the value iteratively assuming as a first guess that $C_c = 1$ and then $D_{ve} = 0.402\,\mu\text{m}$. This suggests that for this particle $D_{ve} \gg 2\lambda$ and the exponential term in (9.34) will be much smaller than the 1.257 term. With this simplification we are left with the following quadratic equation for D_{ve}

$$D_{ve}^2 + 0.163\, D_{ve} - 0.162 = 0$$

with $D_{ve} = 0.329\,\mu\text{m}$ as the positive solution. Our simplification of the slip correction expression resulted in an error of only 1 nm.

To calculate the physical size L of the particle, we only need to equate the volume of the cube to the volume of the sphere:

$$L^3 = (\pi/6)D_{ve}^3 \text{ and } L = 0.264\,\mu\text{m}$$

This calculation requires knowledge about the shape of the particle and its density.

9.7.2 Stokes Diameter

The diameter of a sphere having the same terminal settling velocity and density as the particle is defined as its *Stokes diameter* D_{St}. For irregularly shaped particles D_{St} is the diameter of a sphere that would have the same terminal velocity. The Stokes diameter for $Re < 0.1$ can then be calculated using (9.42) as

$$D_{St} = \left(\frac{18 v_t \mu}{\rho_p g C_c(D_{St})} \right)^{1/2} \tag{9.106}$$

where v_t is the terminal velocity of the particle, μ is viscosity of air, and ρ_p is the density of the particle that needs to be known. Evaluation of (9.106) because of the dependence of C_c on D_{St} requires in general the solution of a nonlinear algebraic equation with one unknown. For spherical particles by its definition $D_{St} = D_p$ and the Stokes diameter is equal to the physical diameter.

Stokes Diameter   What is the relationship connecting the volume equivalent diameter and the Stokes diameter of a nonspherical particle with dynamic shape factor χ for Re < 0.1?

Calculate the Stokes diameter of the NaCl particle of the previous example. The two approaches (dynamic shape factor combined with the volume equivalent diameter and the Stokes diameter) are different ways to describe the drag force and terminal settling velocity of a nonspherical particle. The terminal velocity of a nonspherical particle with a volume equivalent diameter D_{ve} is given by (9.104),

$$v_t = \frac{1}{18} \frac{D_{ve}^2 \rho_p g C_c(D_{ve})}{\chi \mu}$$

By definition this settling velocity can be also written as a function of its Stokes diameter as

$$v_t = \frac{1}{18} \frac{D_{St}^2 \rho_p g C_c(D_{St})}{\mu}$$

Combining these two and simplifying we find that

$$D_{St} = D_{ve} \left(\frac{C_c(D_{ve})}{\chi C_c(D_{St})} \right)^{1/2} \tag{9.107}$$

The Stokes diameter of the NaCl particle can be calculated from (9.106):

$$D_{St}^2 C_c(D_{St}) = 0.150 \, \mu m^2$$

and solving numerically $D_{St} = 0.313 \, \mu m$. This value corresponds to the volume equivalent diameter that we would calculate based on the terminal velocity of the particle if we did not know that the particle was nonspherical.

For most atmospheric aerosol measurements the density of the particles is not known and the Stokes diameter cannot be calculated from the measured particle terminal velocity. This makes the use of the Stokes diameter more difficult and requires the introduction of other equivalent diameters that do not require knowledge of the particle density.

9.7.3   Classical Aerodynamic Diameter

The diameter of a unit density sphere, $\rho_p^\circ = 1 \, g \, cm^{-3}$, having the same terminal velocity as the particle is defined as its classical aerodynamic diameter, D_{ca}. The classical aerodynamic diameter is then given by

$$D_{ca} = \left(\frac{18 v_t \mu}{\rho_p^\circ g C_c(D_{ca})} \right)^{1/2} \tag{9.108}$$

Dividing (9.108) by (9.106), one can then find the relationship between the classical aerodynamic and the Stokes diameter as

$$D_{ca} = D_{St}(\rho_p/\rho_p^\circ)^{1/2} [C_c(D_{St})/C_c(D_{ca})]^{1/2} \qquad (9.109)$$

For spherical particles we replace D_{St} with D_p in this equation to find that

$$D_{ca} = D_p \left(\frac{\rho_p}{\rho_p^\circ}\right)^{1/2} \left(\frac{C_c(D_p)}{C_c(D_{ca})}\right)^{1/2} \qquad (9.110)$$

For a spherical particle of nonunit density the classical aerodynamic diameter is different from its physical diameter and it depends on its density. Aerosol instruments like the cascade impactor and aerodynamic particle sizer measure the classical aerodynamic diameter of atmospheric particles, which is in general different from the physical diameter of the particles even if they are spherical.

Aerodynamic Diameter   Calculate the aerodynamic diameter of spherical particles of diameters equal to 0.01, 0.1, and 1 μm. Assume that their density is $1.5 \, \text{g cm}^{-3}$, which is a typical average density for multicomponent atmospheric particles.
    Using (9.110), we obtain

$$D_{ca}^2 C_c(D_{ca}) = 1.5 \, D_p^2 C_c(D_p)$$

for $D_p = 0.01$ μm, $C_c = 22.2$, and $D_{ca}^2 C_c(D_{ca}) = 3.33 \times 10^{-3}$ μm^2 with solution $D_{ca} = 0.015$ μm. Repeating the same calculation we find that for $D_p = 0.1$ μm, $C_c = 2.85$, and $D_{ca}^2 C_c(D_{ca}) = 0.043$ μm^2, resulting in $D_{ca} = 0.135$ μm. Finally, for $D_p = 1$ μm we find that $D_{ca} = 1.242$ μm. For typical atmospheric particles the aerodynamic and physical diameters are quite different (more than 20% in this case) with the discrepancy increasing for smaller particles.

Vacuum Aerodynamic Diameter   Calculate the ratio of the aerodynamic to the physical diameter of a spherical particle of density ρ_p in the continuum and the free molecular regimes.
    Using equation (9.110) yields

$$D_{ca}/D_p = (\rho_p/\rho_p^\circ)^{1/2} [C_c(D_p)/C_c(D_{ca})]^{1/2} \qquad (9.111)$$

For conditions in the continuum regime the slip correction factor is practically unity and

$$D_{ca}/D_p = (\rho_p/\rho_p^\circ)^{1/2} \qquad (9.112)$$

If the particle is in the free molecular regime, then $2\lambda \gg D_p$ and the second term dominates the RHS of (9.34), which simplifies to

$$C_c(D_p) = 2.514 \, \lambda/D_p$$

Combining this and simplifying, we find that in the free molecular regime:

$$D_{ca}/D_p = (\rho_p/\rho_p^\circ) \tag{9.113}$$

The aerodynamic diameter of a spherical particle with diameter D_p and nonunit density will therefore depend on the mean free path of the air molecules around it and thus also on pressure [see (9.6)]. For low pressures resulting in high Knudsen numbers, the particle will be in the free molecular regime and the aerodynamic diameter will be proportional to the density of the particle and given by (9.113). This diameter is often called the *vacuum aerodynamic diameter* or the *free molecular regime aerodynamic diameter* of the particle. A number of aerosol instruments that operate at low pressures such as the aerosol mass spectrometer measure the vacuum aerodynamic diameter of particles. In the other extreme, for high pressures resulting in low Knudsen numbers the particle will be in the continuum regime and its aerodynamic diameter will be proportional to the square root of its density (9.112). This is known as the *continuum regime aerodynamic diameter.*

The aerodynamic diameter of the particle changes smoothly from its vacuum to its continuum value as the Knudsen number decreases.

9.7.4   Electrical Mobility Equivalent Diameter

Electrical mobility analyzers, like the differential mobility analyzer, classify particles according to their electrical mobility B_e given by (9.50). The electrical mobility equivalent diameter D_{em} is defined as the diameter of a particle of unit density having the same electrical mobility as the given particle. Particles with the same D_{em} have the same migration velocity in an electric field. Particles with equal Stokes diameters that carry the same electrical charge will have the same electrical mobility.

For spherical particles assuming that the particle and its mobility equivalent sphere have the same charge then $D_{em} = D_p = D_{ve}$. For nonspherical particles one can show that

$$D_{em} = D_{ve}\chi \frac{C_c(D_{em})}{C_c(D_{ve})} \tag{9.114}$$

Instruments such as the differential mobility analyzer (DMA) (Liu et al. 1979) size particles according to their electrical mobility equivalent diameter.

PROBLEMS

9.1$_A$ (a) Knowing a particle's density ρ_p and its settling velocity v_t, show how to determine its diameter. Consider both the non-Stokes and the Stokes law regions. (b) Determine the size of water droplet that has $v_t = 1\,\text{cm s}^{-1}$ at $T = 20°\text{C}$, 1 atm.

9.2$_A$ (a) A unit density sphere of diameter $100\,\mu\text{m}$ moves through air with a velocity of $25\,\text{cm s}^{-1}$. Compute the drag force offered by the air. (b) A unit density sphere of diameter $1\,\mu\text{m}$ moves through air with a velocity of $25\,\text{cm s}^{-1}$. Compute the drag force offered by the air.

9.3$_A$ Calculate the terminal settling velocities of silica particles $(\rho_p = 2.65\,\mathrm{g\,cm}^{-3})$ of $0.05\,\mu m$, $0.1\,\mu m$, and $0.5\,\mu m$, and $1.0\,\mu m$ diameters.

9.4$_A$ Calculate the terminal settling velocities of 0.001, 0.1, 1, 10, and 100 μm diameter water droplets in air at a pressure of 0.1 atm.

9.5$_A$ Develop a table of terminal settling velocities of water drops in still air at $T = 20°C$, 1 atm. Consider drop diameters ranging from 1.0 to 1000 μm. (Note that for drop diameters exceeding about 1 mm the drops can no longer be considered spherical as they fall. In this case one must resort to empirical correlations. We do not consider that complication here.)

9.6$_A$ What is the stop distance of a spherical particle of 1 μm diameter and density $1.5\,\mathrm{g\,cm}^{-3}$ moving in still air at 298 K with a velocity of $1\,\mathrm{m\,s}^{-1}$?

9.7$_B$ A 0.2-μm-diameter particle of density $1\,\mathrm{g\,cm}^{-3}$ is being carried by an airstream at 1 atm and 298 K in the y direction with a velocity of $100\,\mathrm{cm\,s}^{-1}$. The particle enters a charging device and aquires a charge of two electrons (the charge of a single electron is 1.6×10^{-19} C) and moves into an electric field of constant potential gradient $E_x = 1000\,\mathrm{V\,cm}^{-1}$ perpendicular to the direction of flow.

   a. Determine the characteristic relaxation time of the particle.

   b. Determine the particle trajectory assuming that it starts at the origin at time zero.

   c. Repeat the calculation for a 50-nm-diameter particle.

9.8$_C$ At $t = 0$ a uniform concentration N_0 of monodisperse particles exists between two horizontal plates separated by a distance h. Assuming that both plates are perfect absorbers of particles and the particles settle with a settling velocity v_t, determine the number concentration of particles as a function of time and position. The Brownian diffusivity of the particles is D.

9.9$_B$ The dynamic shape factor of a chain that consists of 4 spheres is 1.32. The diameter of each sphere is 0.1 μm. Calculate the terminal settling velocity of the particle in air at 298 K and 1 atm. What is the error if the shape factor is neglected? Assume the density of the spheres is $2\,\mathrm{g\,cm}^{-3}$.

9.10$_B$ Derive (9.110) using the appropriate force balance for the motion of a charged particle in an electric field.

9.11$_B$ Spherical particles with different diameters can have the same electrical mobility if they have a different number of elementary charges. Calculate the diameters of particles that have an electrical mobility equal to that of a singly charged particle with $D_p = 100\,\mathrm{nm}$ assuming that they have 2, 3, or 4 charges. Assume $T = 298$ K and 1 atm.

REFERENCES

Allen, M. D., and Raabe, O. G. (1982) Reevaluation of Millikan's oil drop data for the motion of small particles in air, *J. Aerosol Sci.* **13**, 537–547.

Bird, R. B., Stewart, W. E., and Lightfoot, E. N. (1960) *Transport Phenomena*, Wiley, New York.

Chandrasekhar, S. (1943) Stochastic problems in physics and astronomy, *Rev. Modern Phys.* **15**, 1–89.

Chapman, S., and Cowling, T. G. (1970) *The Mathematical Theory of Non-uniform Gases*, Cambridge Univ. Press, Cambridge, UK.

Crawford, M. (1976) *Air Pollution Control Theory*, McGraw-Hill, New York.

Davis, E. J. (1983) Transport phenomena with single aerosol particles, *Aerosol Sci. Technol.* **2**, 121–144.

Flagan, R. C., and Seinfeld, J. H. (1988) *Fundamentals of Air Pollution Engineering*, Prentice-Hall, Englewood Cliffs, NJ.

Fuchs, N. A. (1964) *The Mechanics of Aerosols*, Pergamon, New York.

Fuchs, N. A., and Sutugin, A. G. (1971) High dispersed aerosols, in *Topics in Current Aerosol Research*, G. M. Hidy and J. R. Brock, eds., Pergamon, New York, pp. 1–60.

Hinds, W. C. (1999) *Aerosol Technology, Properties, Behavior, and Measurement of Airborne Particles* Wiley, New York.

Hinze, J. O. (1959) *Turbulence*, McGraw-Hill, New York.

Hirshfelder, J. O., Curtiss, C. O., and Bird, R. B. (1954) *Molecular Theory of Gases and Liquids*, Wiley, New York.

Liu, B. Y. H., Pui, D. Y. H., and Kapadia, H. (1979) Electrical aerosol analyzer: history, principle, and data reduction, in *Aerosol Measurement*, D. A. Lundgren, ed., Univ. Presses of Florida, Gainesville.

Loyalka, S. K., Hamood, S. A., and Tompson, R. V. (1989) Isothermal condensation on a spherical particle, *Phys. Fluids A* **1**, 358–362.

Moore, W. J. (1962) *Physical Chemistry*, 3rd ed., Prentice-Hall, Englewood Cliffs, NJ.

Twomey, S. (1977) *Atmospheric Aerosols*, Elsevier, New York.

Wax, N. (1954) *Selected Papers on Noise and Stochastic Processes*, Dover, New York.

Zelenyuk, A., Cai, Y., and Imre, D. (2006) From agglomerates of spheres to irregularly shaped particles: Determination of dynamic shape factors from measurements of mobility and vacuum aerodynamic diameters, *Aerosol Sci. Technol.* **40**, 197–217.

10 Thermodynamics of Aerosols

Several chemical compounds (water, ammonia, nitric acid, organics, etc.) can exist in both the gas and aerosol phases in the atmosphere. Understanding the partitioning of these species between the vapor and particulate phases requires an analysis of the thermodynamic properties of aerosols. Since the most important "solvent" for constituents of atmospheric particles and drops is water, we will pay particular attention to the thermodynamic properties of aqueous solutions.

10.1 THERMODYNAMIC PRINCIPLES

An atmospheric air parcel can be viewed thermodynamically as a homogeneous system that may exchange energy, work, and mass with its surroundings. Let us assume that an air parcel contains k chemical species and has a temperature T, pressure p, and volume V. There are n_i moles of species i in the parcel.

The first section of this chapter is a review of fundamental chemical thermodynamic principles focusing on the chemical potential of species in the gas, aqueous, and solid phases. Further discussion of fundamentals of chemical thermodynamics can be found in Denbigh (1981). Chemical potentials form the basis for the development of a rigorous mathematical framework for the derivation of the equilibrium conditions between different phases. This framework is then applied to the partitioning of inorganic aerosol components (sulfate, nitrate, chloride, ammonium, and water) between the gas and particulate phases. The behavior of organic aerosol components will be discussed in Chapter 14.

10.1.1 Internal Energy and Chemical Potential

In addition to the macroscopic kinetic and potential energy that the air parcel may have, it has internal energy U, arising from the kinetic and potential energy of the atoms and molecules in the system. Let us assume that the state of the air parcel changes infinitesimally (e.g., it rises slightly) but there is no mass exchange between the air parcel and its surroundings, namely, that the air parcel is a closed system. Then, according to the first law of thermodynamics, the infinitesimal change of internal energy dU is given by

$$dU = dQ + dW \tag{10.1}$$

where dQ is the infinitesimal amount of heat that is absorbed by the system and dW is the infinitesimal amount of work that is done on the system. Equation (10.1) can also be viewed as a definition of the internal energy of the system U.

Atmospheric Chemistry and Physics: From Air Pollution to Climate Change, Second Edition, by John H. Seinfeld and Spyros N. Pandis.   Copyright © 2006 John Wiley & Sons, Inc.

The infinitesimal work done on the system by its surroundings is equal to

$$dW = -p\,dV \tag{10.2}$$

where p is the pressure of the system and dV its infinitesimal volume change. Note that, if the parcel expands, dV is positive and the work done on the system is negative (or alternatively the work done by the system is positive). Because of this expansion, if there is no heat exchange, $dU < 0$ and the internal energy of the system will decrease. On the contrary, if the parcel volume decreases, $dV < 0$, $dW > 0$ and if $dQ = 0$ the internal energy of the system increases.

A thermodynamically reversible process is defined as one in which the system changes infinitesimally slowly from one equilibrium state to the next. According to the second law of thermodynamics, the heat added to a system during a reversible process dQ_{rev} is given by

$$dQ_{rev} = T\,dS \tag{10.3}$$

where S is the entropy of the system. The entropy is another property of the system (like the temperature, volume, and pressure) measuring the degree of disorder of the elements of the system; the more disorder the greater the entropy. Combining (10.1), (10.2), and (10.3)

$$dU = T\,dS - p\,dV \tag{10.4}$$

for a closed system. This equation contains the whole of knowledge obtained from the basic thermodynamic laws for a closed system undergoing a reversible change.

In our discussion so far we have assumed that the system is closed. According to (10.4), if the number of moles of all system species $n_1, n_2, \ldots, n_k$ remains constant, the change in internal energy U of the system depends only on changes in S and V. However, for variable composition we must have

$$U = U(S, V, n_1, n_2, \ldots, n_k)$$

and thus the total differential of U is

$$dU = \left(\frac{\partial U}{\partial S}\right)_{V,n_i} dS + \left(\frac{\partial U}{\partial V}\right)_{S,n_i} dV + \sum_{i=1}^{k} \left(\frac{\partial U}{\partial n_i}\right)_{S,V,n_j} dn_i \tag{10.5}$$

In this expression, the subscript n_i in the first two partial derivatives implies that the amounts of all species are constant during the variation in question. On the other hand, the last partial derivative assumes that all but the ith substance are constant. Note that for a closed system $dn_i = 0$ and from (10.5)

$$dU = \left(\frac{\partial U}{\partial S}\right)_{V,n_i} dS + \left(\frac{\partial U}{\partial V}\right)_{S,n_i} dV \tag{10.6}$$

Comparing (10.4) and (10.6) that are both valid for a closed system, we obtain

$$T = \left(\frac{\partial U}{\partial S}\right)_{V,n_i} \qquad \text{and} \qquad -p = \left(\frac{\partial U}{\partial V}\right)_{S,n_i} \tag{10.7}$$

Finally, (10.5) can be rewritten as

$$dU = T\,dS - p\,dV + \sum_{i=1}^{k}\left(\frac{\partial U}{\partial n_i}\right)_{S,V,n_j} dn_i \qquad (10.8)$$

Let us define the chemical potential of species i, specifically, μ_i, as

$$\mu_i = \left(\frac{\partial U}{\partial n_i}\right)_{S,V,n_j} \qquad (10.9)$$

so that (10.8) can be written as

$$dU = T\,dS - p\,dV + \sum_{i=1}^{k}\mu_i\,dn_i \qquad (10.10)$$

The chemical potential μ_i has an important function in the system's thermodynamic behavior analogous to pressure or temperature. A temperature difference between two bodies determines the tendency of heat to pass from one body to another while a pressure difference determines the tendency for bodily movement. We will show that a difference in chemical potential can be viewed as the cause for chemical reaction or for mass transfer from one phase to another. The chemical potential μ_i greatly facilitates the discussion of open systems, or of closed systems that undergo chemical composition changes.

10.1.2 The Gibbs Free Energy, G

Calculation of changes dU of the internal energy of a system U requires the estimation of changes of its entropy S, volume V, and number of moles n_i. For chemical applications, including atmospheric chemistry, it is inconvenient to work with entropy and volume as independent variables. Temperature and pressure are much more useful. The study of atmospheric processes can therefore be facilitated by introducing other thermodynamic variables in addition to the internal energy U. One of the most useful is the Gibbs free energy G, defined as

$$G = U + pV - TS \qquad (10.11)$$

Differentiating (10.11)

$$dG = dU + p\,dV + V\,dp - T\,dS - S\,dT \qquad (10.12)$$

and combining (10.12) with (10.10), one obtains

$$dG = -S\,dT + V\,dp + \sum_{i=1}^{k}\mu_i\,dn_i \qquad (10.13)$$

Note that one can propose using (10.13) as an alternative definition of the chemical potential:

$$\mu_i = \left(\frac{\partial G}{\partial n_i}\right)_{T,p,n_j} \tag{10.14}$$

Both definitions (10.9) and (10.14) are equivalent. Equation (10.13) is the basis for chemical thermodynamics.

For a system at constant temperature $(dT = 0)$ and pressure $(dp = 0)$

$$dG = \sum_{i=1}^{k} \mu_i \, dn_i \tag{10.15}$$

For a system under constant temperature, pressure, and with constant chemical composition $(dn_i = 0)$, $dG = 0$, or the system has a constant Gibbs free energy. Equation (10.13) provides the means of calculating infinitesimal changes in the Gibbs free energy of the system. Let us assume that the system under discussion is enlarged m times in size, its temperature, pressure, and the relative proportions of each component remaining unchanged. Under such conditions the chemical potentials, which do not depend on the overall size of the system, remain unchanged. Let the original value of the Gibbs free energy of the system be G and the number of moles of species i, n_i. After the system is enlarged m times, these quantities are now mG and mn_i. The change in Gibbs free energy of the system is

$$\Delta G = mG - G = (m-1)G \tag{10.16}$$

and the changes in the number of moles are

$$\Delta n_i = mn_i - n_i = (m-1)n_i \tag{10.17}$$

and as T, P, and μ_i are constant using (10.15)

$$\Delta G = \sum_{i=1}^{k} \mu_i \, \Delta n_i$$

or

$$G = \sum_{i=1}^{k} \mu_i \, n_i \tag{10.18}$$

Equation (10.18) applies in general and provides additional significance to the concept of chemical potential. The Gibbs free energy of a system containing k chemical compounds can be calculated by

$$G = \mu_1 n_1 + \mu_2 n_2 + \cdots + \mu_k n_k$$

that is, by summation of the products of the chemical potentials and the number of moles of each species. Note that for a pure substance

$$\mu_i = \frac{G}{n_i} \qquad (10.19)$$

and thus the chemical potential is the value of the Gibbs free energy per mole of the substance. One should note that both (10.13) and (10.18) are applicable in general. It may appear surprising that T and p do not enter explicitly in (10.18). To explore this point a little further, differentiating (10.18)

$$dG = \sum_{i=1}^{k} n_i \, d\mu_i + \sum_{i=1}^{k} \mu_i \, dn_i \qquad (10.20)$$

and combining with (10.13), we obtain

$$-S dT + V \, dp = \sum_{i=1}^{k} n_i \, d\mu_i \qquad (10.21)$$

This relation, known as the *Gibbs–Duhem equation*, shows that when the temperature and pressure of a system change there is a corresponding change of the chemical potentials of the various compounds.

10.1.3 Conditions for Chemical Equilibrium

The second law of thermodynamics states that the entropy of a system in an adiabatic $(dQ = 0)$ enclosure increases for an irreversible process and remains constant in a reversible one. This law can be expressed as

$$dS \geq 0 \qquad (10.22)$$

Therefore a system will try to increase its entropy and when the entropy reaches its maximum value the system will be at equilibrium. One can show that for a system at constant temperature and pressure the criterion corresponding to (10.22) is

$$dG \leq 0 \qquad (10.23)$$

or that a system will tend to decrease its Gibbs free energy. For a proof the reader is referred to Denbigh (1981).

Consider the reaction $A \rightleftharpoons B$, and let us assume that initially there are n_A moles of A and n_B moles of B. The Gibbs free energy of the system is, using (10.18)

$$G = n_A \mu_A + n_B \mu_B$$

If the system is closed

$$n_T = n_A + n_B = \text{constant}$$

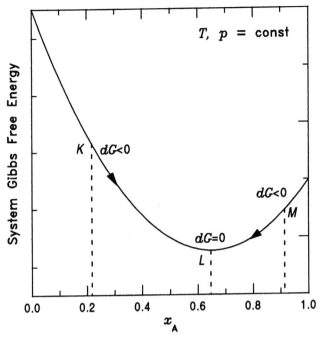

FIGURE 10.1 Sketch of the Gibbs free energy for a closed system where the reaction $A \rightleftharpoons B$ takes places versus the mole fraction of A.

and this equation can be rewritten as

$$G = n_T(x_A \mu_A + (1 - x_A)\mu_B)$$

where

$$x_A = \frac{n_A}{n_A + n_B} = \frac{n_A}{n_T}$$

the mole fraction of A in the system. Let us assume that the Gibbs free energy of the system is that shown in Figure 10.1. If at a given moment the system is at point K, (10.23) suggests that $dG \leq 0$ and G will tend to decrease, so x_A will increase, B will be converted to A, and the system will move to the right.

If the system at a given moment is at point M, once more $dG \leq 0$, so the system will move to the left (A will be converted to B). At point L, the Gibbs free energy is at a minimum. The system cannot spontaneously move to the left or right because then the Gibbs free energy would increase, violating (10.23). If the system is forced to move, then it will return to this equilibrium state. Therefore, for a constant T and p, the point L and the corresponding composition is the equilibrium state of the system and $(x_A)_L$ the corresponding mole fraction of A. At this point $dG = 0$.

Let us consider a general chemical reaction

$$aA + bB \rightleftharpoons cC + dD$$

which can be rewritten mathematically as

$$aA + bB - cC - dD = 0 \qquad (10.24)$$

The Gibbs free energy of the system is given by

$$G = n_A \mu_A + n_B \mu_B + n_C \mu_C + n_D \mu_D$$

If dn_A moles of A react, then, according to the stoichiometry of the reaction, they will also consume $(b/a)dn_A$ moles of B and produce $(c/a)dn_A$ moles of C and $(d/a)dn_A$ moles of D. The corresponding change of the Gibbs free energy of the system at constant T and p is, according to (10.15)

$$
\begin{aligned}
dG &= \mu_A \, dn_A + \mu_B \, dn_B + \mu_C \, dn_C + \mu_D \, dn_D \\
&= \mu_A \, dn_A + \frac{b}{a}\mu_B \, dn_A - \frac{c}{a}\mu_C \, dn_A - \frac{d}{a}\mu_D \, dn_A \qquad (10.25)\\
&= \left(\mu_A + \frac{b}{a}\mu_B - \frac{c}{a}\mu_C - \frac{d}{a}\mu_D \right) dn_A
\end{aligned}
$$

At equilibrium $dG = 0$ and therefore the condition for equilibrium is

$$\mu_A + \frac{b}{a}\mu_B - \frac{c}{a}\mu_C - \frac{d}{a}\mu_D = 0$$

or

$$a\mu_A + b\mu_B - c\mu_C - d\mu_D = 0 \qquad (10.26)$$

Let us try to generalize our conclusions so far. The most general reaction can be written as

$$\sum_{i=1}^{k} v_i A_i = 0 \qquad (10.27)$$

where k is the number of species, A_i, participating in the reaction, and v_i the corresponding stoichiometric coefficients (positive for reactants, negative for products). One can easily extend our arguments for the single reaction (10.24) to show that the general condition for equilibrium is

$$\sum_{i=1}^{k} v_i \mu_i = 0 \qquad (10.28)$$

This is the most general condition of equilibrium *of a single reaction* and is applicable whether the reactants and products are solids, liquids, or gases.

If there are multiple reactions taking place in a system with k species

$$\begin{aligned}\sum_{i=1}^{k} \nu_{i1} A_i &= 0 \\ \sum_{i=1}^{k} \nu_{i2} A_i &= 0 \\ &\vdots \\ \sum_{i=1}^{k} \nu_{in} A_i &= 0\end{aligned} \tag{10.29}$$

the equilibrium condition applies to each one of these reactions and therefore at equilibrium

$$\sum_{i=1}^{k} \nu_{ij}\, \mu_i = 0, \qquad j = 1, \ldots, n \tag{10.30}$$

where ν_{ij} is the stoichiometric coefficient of species i in reaction j (there are n reactions and k species).

Reactions in the H_2SO_4–NH_3–HNO_3 System Let us assume that the following reactions take place:

$$2\,NH_3(g) + H_2SO_4(g) \rightleftharpoons (NH_4)_2SO_4(s)$$
$$NH_3(g) + HNO_3(g) \rightleftharpoons NH_4NO_3(s)$$

Calculate the chemical potential of sulfuric acid as a function of the chemical potentials of nitric acid and the two solids.

At equilibrium the chemical potentials of gas-phase NH_3, and H_2SO_4 and solids $(NH_4)_2SO_4$ and NH_4NO_3 satisfy

$$\begin{aligned}2\,\mu_{NH_3} + \mu_{H_2SO_4} - \mu_{(NH_4)_2SO_4} &= 0 \\ \mu_{NH_3} + \mu_{HNO_3} - \mu_{NH_4NO_3} &= 0\end{aligned} \tag{10.31}$$

From the second equation $\mu_{NH_3} = \mu_{NH_4NO_3} - \mu_{HNO_3}$. Substituting this into the first, we find that

$$\mu_{H_2SO_4} = 2\,\mu_{HNO_3} - \mu_{NH_4NO_3} + \mu_{(NH_4)_2SO_4} \tag{10.32}$$

Determination of the equilibrium composition of this multiphase system therefore requires determination of the chemical potentials of all species as a function of the corresponding concentrations, temperature, and pressure.

10.1.4 Chemical Potentials of Ideal Gases and Ideal-Gas Mixtures

In this section we will discuss the chemical potentials of species in the gas, aqueous, and aerosol phases. In thermodynamics it is convenient to set up model systems to which the behavior of ideal systems approximates under limiting conditions. The important models for atmospheric chemistry are the ideal gas and the ideal solution. We will define these ideal systems using the chemical potentials and then discuss other definitions.

The Single Ideal Gas   We define the ideal gas as a gas whose chemical potential $\mu(T,p)$ at temperature T and pressure p is given by

$$\mu(T,p) = \mu^\circ(T, 1\,\text{atm}) + RT \ln p \qquad (10.33)$$

where μ° is the standard chemical potential defined at a pressure of 1 atm and therefore is a function of temperature only. R is the ideal-gas constant. Pressure p actually stands for the ratio $(p/1\,\text{atm})$ and is dimensionless. This definition suggests that the chemical potential of an ideal gas at constant temperature increases logarithmically with its pressure.

Differentiating (10.33) with respect to pressure at constant temperature, we obtain

$$\left(\frac{\partial \mu}{\partial p}\right)_T = \left(\frac{\partial \mu^\circ}{\partial p}\right)_T + RT \frac{d \ln p}{dp} \qquad (10.34)$$

But μ° is not a function of pressure and therefore its derivative with respect to pressure is zero, so for an ideal gas

$$\left(\frac{\partial \mu}{\partial p}\right)_T = \frac{RT}{p} \qquad (10.35)$$

Using (10.13), one can calculate the derivative of G with respect to pressure

$$\left(\frac{\partial G}{\partial p}\right)_T = V \qquad (10.36)$$

but for a system consisting of n moles of a single gas [see (10.18)]

$$G = n\mu \qquad (10.37)$$

and therefore from (10.36) and (10.37)

$$\left(\frac{\partial \mu}{\partial p}\right)_T = \frac{V}{n} \qquad (10.38)$$

Combining (10.35) and (10.38), we see that our definition of an ideal gas entails

$$pV = nRT \qquad (10.39)$$

the traditional ideal-gas law.

Deviations from ideal gas behavior are customarily expressed in terms of the compressibility factor $C = pV/(nRT)$. Both dry air and water vapor have compressibility factors C, in the range $0.998 < C < 1$ for the pressure and temperature ranges of atmospheric interest (Harrison 1965). Hence both dry air and water vapor can be treated as ideal gases with an error of less than 0.2% for all the conditions of atmospheric interest.

The Ideal-Gas Mixture   A gaseous mixture is defined as ideal if the chemical potential of its ith component satisfies

$$\mu_i = \mu_i^\circ(T) + RT \ln p + RT \ln y_i \qquad (10.40)$$

where $\mu_i^\circ(T)$ is the standard chemical potential of species i, p is the total pressure of the mixture, and y_i is the gas mole fraction of compound i. For $y_i = 1$ (pure component), (10.40) is simplified to (10.33) and therefore μ_i° is precisely the same for both equations. This standard chemical potential is the Gibbs free energy per mole (recall for pure compounds $G = \mu n$) of the gas in the pure state and pressure of 1 atm.

By defining the partial pressure of compound i as

$$p_i = y_i p \qquad (10.41)$$

a more compact form of (10.40) is

$$\mu_i = \mu_i^\circ(T) + RT \ln p_i \qquad (10.42)$$

One can also prove that (10.42) is equivalent to the more traditional ideal-gas mixture definition:

$$p_i V = n_i RT$$

The atmosphere can be treated as an ideal gas mixture with negligible error.[1]

10.1.5   Chemical Potential of Solutions

Atmospheric aerosols at high relative humidities are aqueous solutions of species such as ammonium, nitrate, sulfate, chloride, and sodium. Cloud droplets, rain, and so on are also aqueous solutions of a variety of chemical compounds.

---

[1] For a discussion of non-ideal-gas mixtures, the reader is referred to Denbigh (1981). Discussion of these mixtures is not necessary here, as the behavior of all gases in the atmosphere can be considered ideal for all practical purposes.

Ideal Solutions   A solution is defined as ideal if the chemical potential of *every* component is a linear function of the logarithm of its aqueous mole fraction x_i, according to the relation

$$\mu_i = \mu_i^*(T,p) + RT \ln x_i \tag{10.43}$$

A multicomponent solution is ideal only if (10.43) is satisfied by every component. A solution, in general, approaches ideality as it becomes more and more dilute in all but one component (the solvent). The standard chemical potential μ_i^* is the chemical potential of pure species $i(x_i = 1)$ at the same temperature and pressure as the solution under discussion. Note that in general μ_i^* is a function of both T and p but does not depend on the chemical composition of the solution.

Let us discuss the relationship of the preceding definition with Henry's and Raoult's laws, which are often used to define ideal solutions. Assuming that an ideal solution of i is in equilibrium with an ideal gas mixture, we have

$$I(g) \rightleftharpoons I(aq)$$

and at equilibrium

$$\mu_i(g) = \mu_i(aq)$$

according to (10.28). Using (10.42) and (10.43)

$$\mu_i^\circ(T) + RT \ln p_i = \mu_i^*(T,p) + RT \ln x_i$$

or

$$p_i = \exp\left(\frac{\mu_i^* - \mu_i^\circ}{RT}\right) x_i = K_i(T,p)x_i \tag{10.44}$$

The standard chemical potentials μ_i^* and μ_i° are functions only of temperature and pressure, and therefore the constant K_i is independent of the solution's composition.

If $x_i = 1$ in (10.44), then $K_i(T,p)$ is equal to the vapor pressure of the pure component i, p_i°, and the equation can be rewritten as

$$p_i = p_i^\circ x_i \tag{10.45}$$

Equation (10.45) states that the vapor pressure of a gas over a solution is equal to the product of the pure component vapor pressure and its mole fraction in the solution. The lower the mole fraction in the solution, the more the vapor pressure of the gas over the solution drops. Thus (10.45) is the same as *Raoult's law*.

Most solutions of practical interest satisfy (10.43) only in certain chemical composition ranges and not in others. Let us focus on a binary solution of A and B. If the solution is

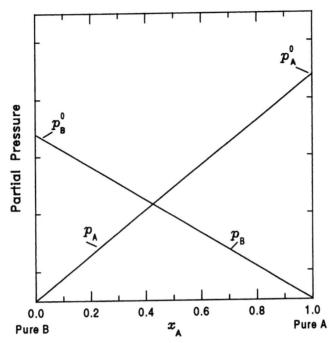

FIGURE 10.2 Equilibrium partial pressures of the components of an ideal binary mixture as a function of the mole fraction of A, x_A.

ideal for every composition, then the partial pressures of A and B will vary linearly with the mole fraction of B (Figure 10.2). When $x_A = 0$, the mixture consists of pure B and the equilibrium partial pressure of B over the solution is p_B° and of A zero. The opposite is true at the other end, where $x_A = 1$ and $p_A = p_A^{\circ}$.

The partial pressures of A and B in a realistic mixture are shown in Figure 10.3. Note that the relationships between p_A, p_B, and x_A are nonlinear with the exceptions of the limits of $x_A \rightarrow 0$ and $x_A \rightarrow 1$. When $x_A \rightarrow 1$, we have a dilute solution of B in A. In this regime

$$p_A \simeq p_A^{\circ} x_A \tag{10.46}$$

and Raoult's law applies to A. In the same regime

$$p_B = H_B' x_B \tag{10.47}$$

where H_B' is a constant calculated from the slope of the p_B line as $x_A \rightarrow 1$. This relationship corresponds to *Henry's law*, and H_B' is the Henry's law constant[2] (based on

---

[2] For a dilute aqueous solution (10.47) is equivalent to

$$[B(aq)] = (0.018 \, H_B')^{-1} p_B$$

The Henry's law constant H_B defined in Chapter 7 is then related to H_B' by

$$H_B = (0.018 \, H_B')^{-1}$$

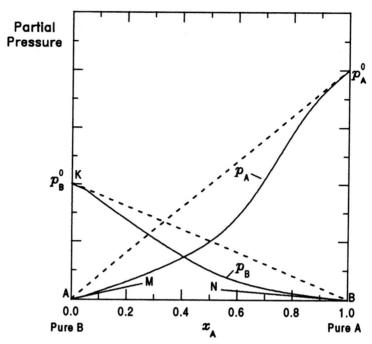

FIGURE 10.3 Equilibrium partial pressures of the components of a nonideal mixture of A and B. Dashed lines correspond to ideal behavior.

mole fraction) for B in A (equal to the slope of the line BN). At the other end $(x_A \to 0)$ we have

$$p_B \simeq p_B^\circ x_B$$
$$p_A = H'_A x_A \tag{10.48}$$

and B obeys Raoult's law while A obeys Henry's law.

Summarizing, if a solution is ideal over the whole composition range (often called "perfect" solution), (10.44) is satisfied for every x_i. In this case K_i is equal to the vapor pressure of pure i, which is also equal in this case to the Henry's law constant of i. Nonideal solutions approach ideality when the concentrations of all components but one approach zero. In that case the solutes satisfy Henry's law $(p_i = H'_i x_i)$, where the solvent satisfies Raoult's law $(p_j = p_j^\circ x_j)$.

Nonideal Solutions   Atmospheric aerosols are usually concentrated aqueous solutions that deviate significantly from ideality. This deviation from ideality is usually described by introducing the activity coefficient, γ_i, and the chemical potential is given by

$$\mu_i = \mu_i^*(T,p) + RT \ln(\gamma_i x_i) \tag{10.49}$$

The activity coefficient γ_i is in general a function of pressure and temperature together with the mole fractions of *all* substances in solution. For an ideal solution $\gamma_i = 1$. The standard chemical potential μ_i^* is defined as the chemical potential at the hypothetical state for which $\gamma_i \to 1$ and $x_i \to 1$ (Denbigh 1981). The product of the mole fraction x_i of a solution component and its activity coefficient γ_i is defined as the *activity*, α_i, of the component

$$\alpha_i = \gamma_i x_i \tag{10.50}$$

and the chemical potential of a species i is then given by

$$\mu_i = \mu_i^*(T,p) + RT \ln \alpha_i \tag{10.51}$$

For convenience, the amount of a species in solution is often expressed as a *molality* rather than as a mole fraction. The molality of a solute is its amount in moles per kilogram of solvent. For an aqueous solution containing n_i moles of solute and n_w moles of water (molecular weight $0.018 \, \text{kg mol}^{-1}$) the molality, m_i, of the solute is

$$m_i = \frac{n_i}{0.018 \, n_w} \tag{10.52}$$

Another measure of solution concentration is the molarity, expressed in moles of solute per liter of solvent (denoted by M). For water solutions at ambient conditions, because 1 liter weighs 1 kilogram, the molality and molarity of a solution are practically equal.

Traditionally, the activity of the solvent is almost always defined on the mole fraction scale ((10.40) and (10.50)), but the activity coefficient of the solute is often expressed on the molality scale:

$$\mu_i = \mu_i^\circ + RT \ln (\gamma_i m_i) \tag{10.53}$$

In this case μ_i° is the value of the chemical potential as $m_i \to 1$ and $\gamma_i \to 1$.

The activity coefficients γ_i are determined experimentally by a series of methods including vapor pressure, freezing-point depression, osmotic pressure, and solubility measurements (Denbigh 1981).

Pure Solid Compounds   The chemical potential of a pure solid compound i can be easily derived from (10.43) by setting $x_i = 1$, so that

$$\mu_i(\text{solid}) = \mu_i^*(T,p) \tag{10.54}$$

The chemical potential of the solid is therefore equal to its standard potential and is a function only of temperature and pressure.

Solutions of Electrolytes Most of the inorganic aerosol components dissociate on dissolution; for example, NH_4NO_3 dissociates forming NH_4^+ and NO_3^-. The concentration of each ion in the aqueous solution is traditionally expressed on the molality scale and the chemical potential of each ion in a NH_4NO_3 solution is

$$\mu_{NH_4^+} = \mu_{NH_4^+}^{\circ} + RT \ln(\gamma_{NH_4^+} m_{NH_4^+})$$

$$\mu_{NO_3^-} = \mu_{NO_3^-}^{\circ} + RT \ln(\gamma_{NO_3^-} m_{NO_3^-})$$

where $m_{NH_4^+}$ and $m_{NO_3^-}$ are the ion molalities and $\gamma_{NH_4^+}$ and $\gamma_{NO_3^-}$ the corresponding activity coefficients. The dissociation reaction

$$NH_4NO_3 \rightleftharpoons NH_4^+ + NO_3^-$$

is at equilibrium and therefore the chemical potential of NH_4NO_3 satisfies

$$\mu_{NH_4NO_3} = \mu_{NH_4^+} + \mu_{NO_3^-}$$

or

$$\mu_{NH_4NO_3} = \mu_{NH_4^+}^{\circ} + \mu_{NO_3^-}^{\circ} + RT \ln(\gamma_{NH_4^+} \gamma_{NO_3^-} m_{NH_4^+} m_{NO_3^-}) \tag{10.55}$$

The binary activity coefficient for NH_4NO_3 can be defined as

$$\gamma_{NH_4NO_3}^2 = \gamma_{NH_4^+} \gamma_{NO_3^-} \tag{10.56}$$

and (10.55) can be rewritten as

$$\mu_{NH_4NO_3}^* = \mu_{NH_4^+}^{\circ} + \mu_{NO_3^-}^{\circ} + RT \ln(\gamma_{NH_4NO_3}^2 m_{NH_4^+} m_{NO_3^-})$$

If the electrolyte dissociates completely and the initial molality of NH_4NO_3 is $m_{NH_4NO_3}$, then

$$m_{NO_3^-} = m_{NH_4^+} = m_{NH_4NO_3}$$

and

$$\mu_{NH_4NO_3}^* = \mu_{NH_4^+}^{\circ} + \mu_{NO_3^-}^{\circ} + RT \ln(\gamma_{NH_4NO_3}^2 m_{NH_4NO_3}^2) \tag{10.57}$$

10.1.6 The Equilibrium Constant

The equilibrium expression (10.28) can be used to obtain a useful expression for aerosol equilibrium calculations. Let us consider the general reaction

$$\sum_{i=1}^{k} \nu_i \mu_i = 0$$

Then substituting (10.51) into (10.28)

$$\sum_{i=1}^{k} v_i(\mu_i^{\circ} + RT \ln \alpha_i) = 0$$

or

$$\prod_{i=1}^{k} \alpha_i^{v_i} = K \tag{10.58}$$

$$K = \exp\left(-\frac{1}{RT}\sum_{i=1}^{k} v_i\mu_i^{\circ}\right) \tag{10.59}$$

If, for example, the species participating in the reaction are all gases, then the activity can be replaced with the partial pressures and

$$\prod_{i=1}^{k} p_i^{v_i} = K \tag{10.60}$$

Equilibrium Constant for Ammonium Sulfate Formation   Let us determine the equilibrium constant for the following reaction:

$$H_2SO_4(g) + 2\,NH_3(g) \rightleftharpoons (NH_4)_2SO_4(s)$$

For this reaction $v_{H_2SO_4} = 1$, $v_{NH_3} = 2$, and $v_{(NH_4)_2SO_4} = -1$. The condition for equilibrium is

$$2\,\mu_{NH_3} + \mu_{H_2SO_4} - \mu_{(NH_4)_2SO_4} = 0$$

or following (10.59) and noting that $\alpha = 1$ for solids

$$\exp\left(\frac{\mu_{(NH_4)_2SO_4}^{*} - 2\,\mu_{NH_3}^{\circ} - \mu_{H_2SO_4}^{\circ}}{RT}\right) = K(T) = p_{NH_3}^{2}\, p_{H_2SO_4}$$

and therefore at equilibrium the product of the square of ammonia partial pressure and of sulfuric acid partial pressure should equal a constant. The value of the constant is a strong function of temperature.

10.2 AEROSOL LIQUID WATER CONTENT

Water is an important component of atmospheric aerosols. Most of the water associated with atmospheric particles is chemically unbound (Pilinis et al. 1989). At very low relative

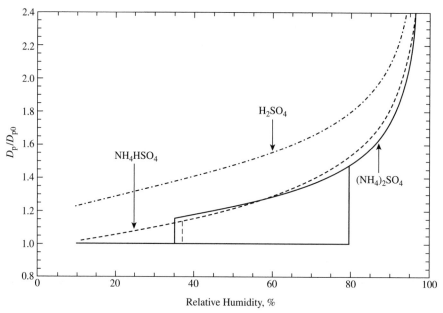

FIGURE 10.4   Diameter change of $(NH_4)_2SO_4$, NH_4HSO_4, and H_2SO_4 particles as a function of relative humidity. D_{p0} is the diameter of the particle at 0% RH.

humidities, atmospheric aerosol particles containing inorganic salts are solid. As the ambient relative humidity increases, the particles remain solid until the relative humidity reaches a threshold value characteristic of the particle composition (Figure 10.4). At this RH, the solid particle spontaneously absorbs water, producing a saturated aqueous solution. The relative humidity at which this phase transition occurs is known as the *deliquescence relative humidity* (DRH). Further increase of the ambient RH leads to additional water condensation onto the salt solution to maintain thermodynamic equilibrium (Figure 10.4). On the other hand, as the RH over the wet particle is decreased, evaporation of water occurs. However, the solution generally does not crystallize at the DRH, but remains supersaturated until a much lower RH at which crystallization occurs (Junge 1952; Richardson and Spann 1984; Cohen et al. 1987). This hysteresis phenomenon with different deliquescence and crystallization points is illustrated in Figure 10.4 for $(NH_4)_2SO_4$. The relative humidities of deliquescence for some inorganic salts, which are common constituents of ambient aerosols, are given in Table 10.1.

One should also note that some aerosol species do not exhibit deliquescent behavior. Species like H_2SO_4 are highly hygroscopic, and therefore the water content associated with them changes smoothly as the RH increases or decreases (Figure 10.4).

For each relative humidity a single salt can exist in either of two states: as a solid or as an aqueous solution. For relative humidities lower than the deliquescence relative humidity the Gibbs free energy of the solid salt is lower than the energy of the corresponding solution and the salt remains in the solid state (Figure 10.5). As the relative humidity increases the Gibbs free energy of the corresponding solution state decreases, and at the DRH it becomes equal to the energy of the solid. When the RH increases further the solution represents the lower energy state and the particle spontaneously absorbs water to form a saturated salt solution. This deliquescence transition is accompanied by a

**TABLE 10.1   Deliquescence Relative Humidities
of Electrolyte Solutions at 298 K**

Salt	DRH (%)
KCl	84.2 ± 0.3
Na_2SO_4	84.2 ± 0.4
NH_4Cl	80.0
$(NH_4)_2SO_4$	79.9 ± 0.5
NaCl	75.3 ± 0.1
$NaNO_3$	74.3 ± 0.4
$(NH_4)_3H(SO_4)_2$	69.0
NH_4NO_3	61.8
$NaHSO_4$	52.0
NH_4HSO_4	40.0

Sources: Tang (1980) and Tang and Munkelwitz (1993).

significant increase in the mass of the particle (Figure 10.4). For even higher RHs the solution state is the preferable one. When the RH decreases reaching the DRH the energies of the two states become once more equal. However, as the RH decreases further, for the particle to attain the lower energy state (solid), all the water in the particle needs to evaporate. This is physically difficult, as salt nuclei need to be formed and salt crystals to grow around them. In the atmosphere, where these salts are suspended in air, this transition does not occur at this point and the particle remains liquid. As the RH keeps decreasing the

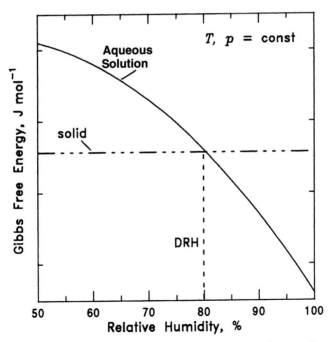

FIGURE 10.5   Gibbs free energy of a solid salt and its aqueous solution as a function of RH. At the DRH these energies become equal.

water in the particle keeps evaporating and the particle becomes a supersaturated solution. The solution eventually reaches a critical supersaturation, and nucleation (crystallization) takes place, forming at last a solid particle at RH significantly lower than the DRH (Figure 10.4).

10.2.1 Chemical Potential of Water in Atmospheric Particles

Water vapor exists in the atmosphere in concentrations on the order of grams per m^3 of air while its concentration in the aerosol phase is less than $1\,mg\,m^{-3}$ of air. As a result, transport of water to and from the aerosol phase does not affect the ambient vapor pressure of water in the atmosphere. This is in contrast to the cloud phase, where a significant amount of water exists in the form of cloud droplets (see Chapter 17). Thus the ambient RH can be treated as a known constant in aerosol thermodynamic calculations. Considering the equilibrium

$$H_2O(g) \rightleftharpoons H_2O(aq)$$

and using the criterion for thermodynamic equilibrium and the corresponding chemical potentials

$$\mu_{H_2O(g)} = \mu_{H_2O(aq)}$$

or

$$\mu_{H_2O}^{\circ} + RT \ln p_w = \mu_{H_2O}^{*} + RT \ln \alpha_w \tag{10.61}$$

where p_w is the water vapor pressure (in atm) and α_w is the water activity in solution. For pure water in equilibrium with its vapor, $\alpha_w = 1$ and $p_w = p_w^{\circ}$ (the saturation vapor pressure of water at this temperature); therefore

$$\mu_{H_2O}^{*} - \mu_{H_2O}^{\circ} = RT \ln p_w^{\circ} \tag{10.62}$$

Using (10.62) in (10.61) yields

$$\alpha_w = \frac{p_w}{p_w^{\circ}} = \frac{RH}{100} \tag{10.63}$$

because the ratio p_w/p_w° is by definition equal to the relative humidity expressed in the 0 to 1 scale. Thus the water activity in an atmospheric aerosol solution is equal to the RH (in the 0.0–1.0 scale). This result simplifies significantly equilibrium calculations for atmospheric aerosol, because for each RH the water activity for any liquid aerosol solution is fixed.

Water equilibrium between the gas and aerosol phases at the point of deliquescence requires that the deliquescence relative humidity of a salt will then satisfy

$$\frac{DRH}{100} = \alpha_{ws} \tag{10.64}$$

TABLE 10.2   Solubility of Common Aerosol Salts in Water as a Function of Temperature ($n = A + BT + CT^2$, $n =$ mol of solute per mol of water)

Salt	n (at 298 K)	A	B	C
$(NH_4)_2SO_4$	0.104	0.1149	-4.489×10^{-4}	1.385×10^{-6}
Na_2SO_4	0.065	0.3754	-1.763×10^{-3}	2.424×10^{-6}
$NaNO_3$	0.194	0.1868	-1.677×10^{-3}	5.714×10^{-6}
NH_4NO_3	0.475	4.298	-3.623×10^{-2}	7.853×10^{-5}
KCl	0.086	-0.2368	1.453×10^{-3}	-1.238×10^{-6}
NaCl	0.111	0.1805	-5.310×10^{-4}	9.965×10^{-7}

where α_{ws} is the water activity of the saturated solution of the salt at that temperature. The water activity values can be calculated from thermodynamic arguments using aqueous salt solubility data (Cohen et al. 1987; Pilinis and Seinfeld 1987; Pilinis et al. 1989).

10.2.2   Temperature Dependence of the DRH

The DRH for a single salt varies with temperature. The vapor–liquid equilibrium of a salt S can be expressed by the following reactions

$$H_2O(g) \rightleftharpoons H_2O(aq) \tag{a}$$

$$H_2O(aq) + n\,S(s) \rightleftharpoons n\,S(aq) \tag{b}$$

where n is the solubility of S in water in moles of solute per mole of water (Table 10.2). The energy that is released in reaction (a) is the heat of condensation of water vapor, which is equal to the negative value of its heat of vaporization, $-\Delta H_v$. The heat that is absorbed in reaction (b) is the enthalpy of solution of the salt ΔH_s. This enthalpy can be readily calculated from the heats of formation tabulated in standard thermodynamic tables. Values of ΔH_s are shown in Table 10.3 (Wagman et al. 1966). The overall enthalpy change ΔH for the two reactions is

$$\Delta H = n\,\Delta H_s - \Delta H_v \tag{10.65}$$

The change of the vapor pressure of water over a solution with temperature is given by the Clausius–Clapeyron equation (Denbigh 1981)

$$\frac{d\ln p_w}{dT} = -\frac{\Delta H}{RT^2} \tag{10.66}$$

TABLE 10.3   Enthalpy of Solution for Common Aerosol Salts at 298 K

Salt	ΔH_s, kJ mol^{-1}
$(NH_4)_2SO_4$	6.32
Na_2SO_4	-9.76
$NaNO_3$	13.24
NH_4NO_3	16.27
KCl	15.34
NaCl	1.88

which for this case becomes

$$\frac{d \ln p_w}{dT} = \frac{\Delta H_v}{RT^2} - n\frac{\Delta H_s}{RT^2} \tag{10.67}$$

Applying the Clausius–Clapeyron equation to pure water, we also obtain

$$\frac{d \ln p_w^\circ}{dT} = \frac{\Delta H_v}{RT^2} \tag{10.68}$$

where p_w° is the saturation vapor pressure of water at temperature T. Combining (10.67) and (10.68)

$$\frac{d \ln(p_w/p_w^\circ)}{dT} = -n\frac{\Delta H_s}{RT^2} \tag{10.69}$$

and substituting (10.63) into (10.69) and applying it to the DRH, we obtain

$$\frac{d \ln(\mathrm{DRH}/100)}{dT} = -n\frac{\Delta H_s}{RT^2} \tag{10.70}$$

The solubility n can be written as a polynomial in T (see Table 10.2), and the equation can be integrated from $T_0 = 298\,\mathrm{K}$ to T to give

$$\ln\frac{\mathrm{DRH}(T)}{\mathrm{DRH}(T_0)} = \frac{\Delta H_s}{R}\left[A\left(\frac{1}{T} - \frac{1}{T_0}\right) - B\ln\frac{T}{T_0} - C(T - T_0)\right] \tag{10.71}$$

or

$$\mathrm{DRH}(T) = \mathrm{DRH}(298)\exp\left\{\frac{\Delta H_s}{R}\left[A\left(\frac{1}{T} - \frac{1}{298}\right) - B\ln\frac{T}{298} - C(T - 298)\right]\right\}$$
$$\tag{10.72}$$

The only assumption used in (10.72) is that the heat of solution is almost constant from 298 K to T. Wexler and Seinfeld (1991) proposed a similar expression assuming constant solubility, namely, $B = C = 0$ in (10.72), while expression (10.72) was derived by Tang and Munkelwitz (1993).

Dependence of the $(NH_4)_2SO_4$, NH_4NO_3, and $NaNO_3$ DRH on Temperature
Calculate the DRH of $(NH_4)_2SO_4$, NH_4NO_3, and $NaNO_3$ at 0°C, 15°C, and 30°C.
    The DRH at 25°C for the three salts is given in Table 10.1. Using (10.72) and the corresponding parameter values from Tables 10.3 and 10.2, we find that

Salt	Deliquescence Relative Humidity		
	0°C	15°C	30°C
$(NH_4)_2SO_4$	81.8	80.6	79.5
NH_4NO_3	76.6	68.1	58.5
$NaNO_3$	80.9	76.9	73.0

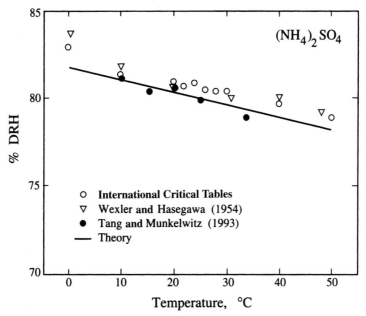

FIGURE 10.6   Deliquescence RH as a function of temperature for $(NH_4)_2SO_4$. (Reprinted from *Atmos. Environ.* **27A**, Tang, I. N., and Munkelwitz, H. R., Composition and temperature dependence of the deliquescence properties of hygroscopic aerosols, 467–473. Copyright 1993, with kind permission from Elsevier Science Ltd., The Boulevard, Langford Lane, Kidlington OX5 1GB, UK.)

These results indicate that the DRH for $(NH_4)_2SO_4$ is practically constant within the temperature range of atmospheric interest while for the other two salts it varies significantly.

The predictions of (10.72) for ammonium sulfate are compared with measurements from a series of investigators in Figure 10.6.

10.2.3   Deliquescence of Multicomponent Aerosols

Multicomponent aerosol particles exhibit behavior similar to that of single-component salts. As the ambient RH increases the salt mixture is solid, until the ambient RH reaches the deliquescence point of the mixture, at which the aerosol absorbs atmospheric moisture and produces a saturated solution. A typical set of data of multicomponent particle deliquescence, growth, evaporation, and then crystallization is shown in Figure 10.7 for a KCl–NaCl particle. Note that the DRH for the mixed-salt particle occurs at 72.7% RH, which is lower than the DRH of either NaCl (75.3%) or KCl (84.2%).

Following Wexler and Seinfeld (1991), let us consider two electrolytes in a solution exposed to the atmosphere. The change of the DRH of a single-solute aqueous solution when another electrolyte is added can be calculated using the Gibbs–Duhem equation, (10.21). For constant T and p and for a solution containing two electrolytes (1 and 2) and water (w):

$$n_1 d\mu_1 + n_2 d\mu_2 + n_w d\mu_w = 0 \qquad (10.73)$$

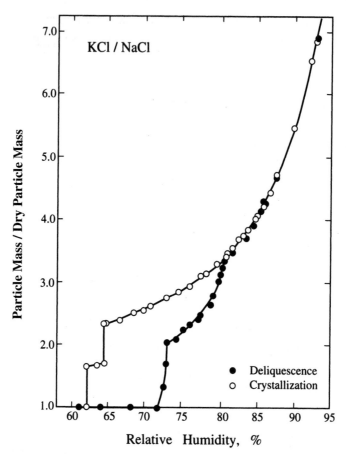

FIGURE 10.7 Hygroscopic growth and evaporation of a mixed-salt particle composed initially of 66% mass KCl and 34% mass NaCl. (Reprinted from *Atmos. Environ.*, **27A**, Tang I. N., and Munkelwitz, H. R., Composition and temperature dependence of the deliquescence properties of hygroscopic aerosols, 467–473. Copyright 1993, with kind permission from Elsevier Science Ltd., The Boulevard, Langford Lane, Kidlington OX5 1GB, UK.)

where n_1, n_2, and n_w are the numbers of moles of electrolytes 1, 2, and water, respectively, while μ_1, μ_2, and μ_w are the corresponding chemical potentials. Let us assume that initially electrolyte 1 is in equilibrium with solid salt 1 and the solution does not yet contain electrolyte 2. As electrolyte 2 is added to the solution, the chemical potential of electrolyte 1 does not change, because it remains in equilibrium with its solid phase. Thus $d\mu_1 = 0$ in (10.73). The chemical potentials of electrolyte 2 and water can be expressed using (10.51) to get

$$n_2\, d \ln \alpha_2 + n_w\, d \ln \alpha_w = 0 \qquad (10.74)$$

Accounting for the fact that $n_2/n_w = M_w m_2/1000$, where m_2 is the molality of electrolyte 2 and M_w is the molecular weight of water, we obtain

$$m_2\, d \ln \alpha_2 + \frac{1000}{M_w} d \ln \alpha_w = 0 \qquad (10.75)$$

Integration of the last equation from $m'_2 = 0$ to $m'_2 = m_2$ gives

$$\ln \frac{\alpha_w(m_2)}{\alpha_w(0)} = -\frac{M_w}{1000} \int_0^{m_2} \frac{m'_2}{\alpha_2(m'_2)} \frac{d\alpha_2(m'_2)}{dm'_2} dm'_2 \tag{10.76}$$

Wexler and Seinfeld (1991) have argued that $d\alpha_2/dm_2 \geq 0$ and therefore the integral is positive, and then

$$\alpha_w(m_2) \leq \alpha_w(m_2 = 0) \tag{10.77}$$

Hence the activity of water decreases as electrolyte 2 is added to the system, until the solution becomes saturated in that electrolyte, too. The aerosol is exposed to the atmosphere and therefore its DRH also decreases.

The preceding analysis can be extended to aerosols containing more than two salts. Thus one can prove that water activity reaches a minimum at the deliquescence point of the aerosol. Another consequence of this analysis is that the DRH of a mixed salt is always lower than the DRH of the individual salts in the particle. Wexler and Seinfeld (1991) solved (10.76) for the case of the system containing NH_4NO_3 and NH_4Cl. Their calculations at 303 K are depicted in Figure 10.8.

When there is only NH_4Cl in the particle ($x_{NH_4NO_3} = 0$) the DRH of the particle is 77.4%. If the RH is below 77.4%, the particle is solid; if it is above 77.4% it is

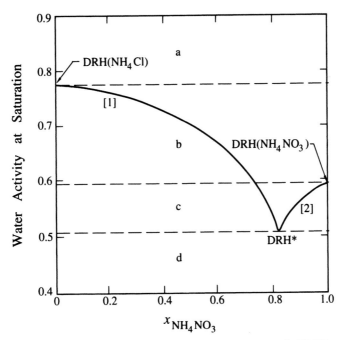

FIGURE 10.8 Water activity at saturation for an aqueous solution of NH_4NO_3 and NH_4Cl at 303 K. (Reprinted from *Atmos. Environ.*, **25A**, Wexler, A. S., and Seinfeld, J. H., Second-generation inorganic aerosol model, 2731–2748. Copyright 1991, with kind permission from Elsevier Science Ltd., The Boulevard, Langford Lane, Kidlington OX5 1GB, UK.)

an aqueous solution of NH_4^+ and Cl^-. (Note that the calculations in Figure 10.8 are at 303 K.)

There are seven different RH composition regimes in the figure:

(a) RH > DRH(NH_4Cl) > DRH(NH_4NO_3). When the RH exceeds the DRH of both compounds, the aerosol is an aqueous solution of NH_4^+, NO_3^-, and Cl^-.

(b_1) DRH(NH_4Cl) > RH > DRH(NH_4NO_3)—left of [1]. In this regime the aerosol consists of solid NH_4Cl in equilibrium with an aqueous solution of NH_4^+, NO_3^-, and Cl^-.

(b_2) DRH(NH_4Cl) > RH > DRH(NH_4NO_3)—right of [1]. If the aerosol contains enough NH_4NO_3 so that its composition is to the right of line [1], the aerosol is an aqueous solution of NH_4^+, NO_3^-, and Cl^-. In this regime addition of NH_4NO_3 results in complete dissolution of NH_4Cl even if the RH is higher than its DRH.

(c_1) DRH(NH_4Cl) > DRH(NH_4NO_3) > RH > DRH*—left of [1]. The aerosol consists of solid NH_4Cl in equilibrium with a solution of NH_4^+, NO_3^-, and Cl^-.

(c_2) Right of [1], left of [2]. In this regime there is no solid phase and the aerosol consists exclusively of an aqueous solution of NH_4^+, NO_3^-, and Cl^-.

(c_3) Right of [2]. Here the aerosol consists of solid NH_4NO_3 in equilibrium with an aqueous solution of NH_4^+, NO_3^-, and Cl^-.

(d) RH < DRH*. If the relative humidity is below DRH* = 51%, the aerosol consists of solid NH_4NO_3 and solid NH_4Cl.

In conclusion, for this mixture of salts, if both are present, the aerosol will contain some liquid water for RH > 51%. However, if we are below the solid line in Figure 10.8, the aerosol will also contain a solid phase in equilibrium with the solution. Only above the solid line is the particle completely liquid.

It is instructive to follow the changes in a particle of a given composition as RH increases. For example, consider a particle consisting of 40% NH_4NO_3 and 60% NH_4Cl as RH increases from 40 to 90%, assuming that there is no evaporation or condensation of these salts. At the beginning (RH = 40%) the particle is solid and it remains solid until RH reaches 51.4% (= DRH*). At this point, the particle deliquesces and consists of two phases—a solid phase consisting of solid NH_4Cl and an aqueous solution with the composition corresponding to the *eutonic point* ($x_{NH_4NO_3} = 0.811, x_{NH_4Cl} = 0.189$). As the RH increases further, more NH_4Cl dissolves in the solution and the composition of the aqueous solution follows line [1]. For example, at RH = 60% the aerosol consists of solid NH_4Cl and a solution of composition $x_{NH_4NO_3} = 0.73$, $x_{NH_4Cl} = 0.27$. At 70% RH, most of the NH_4Cl has dissolved and the composition of the solution is close to the net particle composition $x_{NH_4NO_3} = 0.42$, $x_{NH_4Cl} = 0.58$. Finally, when the RH reaches 71%, all the NH_4Cl dissolves and the particle consists of only one phase (the aqueous one) with composition equal to the particle composition. Note that there is only one step change in the mass of the particle corresponding to the mutual DRH* (e.g., see Figure 10.7). Further increases in the RH cause continuous changes of the aerosol mass and not step changes. On the contrary, step changes are observed during particle evaporation and will be discussed subsequently.

This discussion of phase transitions has been extended to mixtures of more than three salts by Potukuchi and Wexler (1995a,b). The phase diagrams become now three

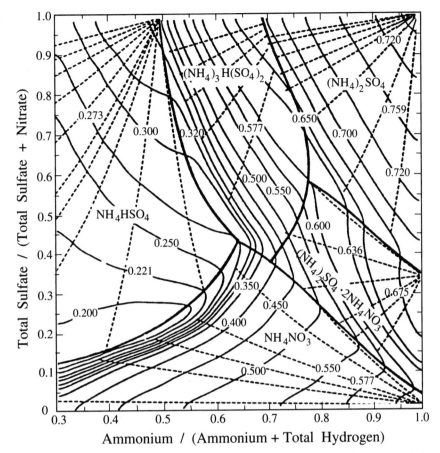

FIGURE 10.9 Deliquescence relative humidity contours (solid lines) for aqueous solutions of $H^+ - NH_4^+ - HSO_4^- - SO_4^{2-} - NO_3^-$ shown together with the lines (dashed) describing aqueous-phase composition with relative humidity. Labels on the contours represent the deliquescence relative humidity values. Total hydrogen is total moles of protons and bisulfate ions. Total sulfate is number of moles of sulfate and bisulfate ions. X = ammonium (ammonium + total hydrogen); Y = total sulfate (total sulfate + nitrate). (Reprinted from *Atmos. Environ.* **29**, Potukuchi, S., and Wexler, A. S., Identifying solid-aqueous phase transitions in atmospheric aerosols. II Acidic solutions, 3357–3364. Copyright 1995, with kind permission from Elsevier Science Ltd., The Boulevard, Langford Lane, Kidlington OX5 1GB, UK.)

dimensional (Figure 10.9). Note that the solid phases that appear are $(NH_4)_2SO_4$, NH_4HSO_4, letovicite $((NH_4)_3H(SO_4)_2)$, NH_4NO_3, and the double salt $(NH_4)_2SO_4 \cdot 2NH_4NO_3$. The labels on the contours show relative humidities at deliquescence. The bold lines in this figure are the phase boundaries separating the various solid phases. On these lines both phases coexist. For example, let us assume that there is enough ammonia present so $X = 1$ in Figure 10.9. If sulfate dominates then $Y = 1$ and the system is in the upper right-hand corner of the diagram with a DRH of 80% corresponding to $(NH_4)_2SO_4$. If there is enough nitrate available so that $Y = 0.3$ then the DRH is 68% corresponding to $(NH_4)_2SO_4 \cdot 2 NH_4NO_3$.

The mutual deliquescence points of a series of pairs are given in Table 10.4.

TABLE 10.4 Deliquescence RH (DRH*) at Mutual Solubility Point at 303 K

Compound 1	Compound 2	DRH*	DRH$_1$	DRH$_2$
NH$_4$NO$_3$	NaCl	42.2	59.4	75.2
NH$_4$NO$_3$	NaNO$_3$	46.3	59.4	72.4
NH$_4$NO$_3$	NH$_4$Cl	51.4	59.4	77.2
NaNO$_3$	NH$_4$Cl	51.9	72.4	77.2
NH$_4$NO$_3$	(NH$_4$)$_2$SO$_4$	52.3	59.4	79.2
NaNO$_3$	NaCl	67.6	72.4	75.2
NaCl	NH$_4$Cl	68.8	75.2	77.2
NH$_4$Cl	(NH$_4$)$_2$SO$_4$	71.3	77.2	79.2

Source: Wexler and Seinfeld (1991).

10.2.4 Crystallization of Single and Multicomponent Salts

The behavior of inorganic salts when RH is decreased is different from that discussed in Sections 10.2.2 and 10.2.3. For example, for (NH$_4$)$_2$SO$_4$, as RH decreases below 80% (the DRH of (NH$_4$)$_2$SO$_4$), the particle water evaporates, but not completely. The particle remains liquid until a RH of approximately 35%, where crystallization finally occurs (Figure 10.4). The RH at which the particle becomes dry is often called the *efflorescence RH* (ERH). This hysteresis phenomenon is characteristic of most salts. For such salts, knowledge of the RH alone is insufficient for determining the state of the aerosol in the RH between the efflorescence and deliquescence RH. One needs to know the RH history of the particle.

The ERH of a salt cannot be calculated from thermodynamic principles such as the DRH; rather, it must be measured in the laboratory. These measurements are challenging because of the potential effects of impurities, observation time, and size of the particle. Measured ERH values at ambient temperatures are shown in Table 10.5 (Martin 2000). Pure NH$_4$HSO$_4$, NH$_4$NO$_3$, AND NaNO$_3$ aqueous particles do not readily crystallize even at RH values close to zero.

Particles consisting of multiple salts exhibit a more complicated behavior (Figure 10.7). For example, the evaporation of a KCl–NaCl particle is characterized by two step changes: the first at 65% with the formation of KCl crystals and the second at 62% with the

TABLE 10.5 Efflorescence Relative Humidity at 298 K

Salt	ERH (%)
(NH$_4$)$_2$SO$_4$	35 $\pm$ 2
NH$_4$HSO$_4$	Not observed
(NH$_4$)H(SO$_4$)$_2$	35
NH$_4$NO$_3$	Not observed
Na$_2$SO$_4$	56 $\pm$ 1
NaCl	43 $\pm$ 3
NaNO$_3$	Not observed
NH$_4$Cl	45
KCl	59

Source: Martin (2000).

complete drying of the particles and the crystallization of the remaining NaCl. The ERH of a particle consisting of multiple salts depends on the composition of the particle. For example, as H_2SO_4 is added to an $(NH_4)_2SO_4$ particle, its ERH decreases from approximately 35% to around 20% for a molar ratio of $NH_4^+ : SO_4^{2-} = 1.5$ to approximately zero for $NH_4^+ : SO_4^{2-} = 1$ (ammonium bisulfate) (Martin et al. 2003). For a particle consisting of approximately 1:1 $(NH_4)_2SO_4:NH_4NO_3$, the ERH is around 20%, while for a 1:2 molar ratio it decreases to around 10%. As a result, particles that are either very acidic (containing ammonium bisulfate) or contain significant amounts of ammonium nitrate tend to remain liquid during their atmospheric lifetime even when the ambient RH is quite low (Shaw and Rood 1990).

Inclusions of insoluble dust minerals ($CaCO_3$, Fe_2O_3, etc.) can increase the efflorescence RH of salts (Martin 2000). For example, for $(NH_4)_2SO_4$, the ERH can increase from 35% to 49% when $CaCO_3$ inclusions are present. These mineral inclusions provide well ordered atomic arrays and thus assist in the formation of crystals at higher RH and lower solution supersaturations. Soot, on the other hand, appears not to be an effective nucleus for crystallization of salts because it does not contain a regular array of atoms that can induce order at least locally in the aqueous medium (Martin 2000).

10.3 EQUILIBRIUM VAPOR PRESSURE OVER A CURVED SURFACE: THE KELVIN EFFECT

In our discussion so far of aerosol thermodynamics, we have assumed that the aqueous aerosol solution has a flat surface. However, the key aspect that distinguishes the thermodynamics of atmospheric particles and drops is their curved interface. In this section we investigate the effect of curvature on the vapor pressure of a species A over the particle surface. Our approach will be to relate this vapor pressure to that over a flat surface. To do so we begin by considering the change of Gibbs free energy accompanying the formation of a single drop of pure A of radius R_p containing n molecules of the substance:

$$\Delta G = G_{droplet} - G_{pure\ vapor}$$

Let us assume that the total number of molecules of vapor initially is N_T; after the drop forms the number of vapor molecules remaining is $N_1 = N_T - n$. Then, if g_v and g_l are the Gibbs free energies of a molecule in the vapor and liquid phases, respectively

$$\Delta G = N_1 g_v + n g_l + 4\pi R_p^2 \sigma - N_T g_v$$

where $4\pi R_p^2 \sigma$ is the free energy associated with an interface with radius of curvature R_p and surface tension σ. Surface tension is the amount of energy required to increase the area of a surface by 1 unit. This equation can be rewritten as

$$\Delta G = n(g_l - g_v) + 4\pi R_p^2 \sigma \tag{10.78}$$

Note that the number of molecules in the drop, n, and the drop radius R_p are related by

$$n = \frac{4\pi R_p^3}{3 v_l} \tag{10.79}$$

where v_l is the volume occupied by a molecule in the liquid phase. Thus, combining (10.78) and (10.79),

$$\Delta G = \frac{4\pi R_p^3}{3v_l}(g_l - g_v) + 4\pi R_p^2 \sigma \tag{10.80}$$

We now need to evaluate $g_l - g_v$, the difference in the Gibbs free energy per molecule of the liquid and vapor states. Using (10.13) at constant temperature and because $dn_i = 0$, $dg = v\,dp$ or $g_l - g_v = (v_l - v_v)dp$. Since $v_v \gg v_l$ for all conditions of interest to us, we can neglect v_l relative to v_v in this equation, giving $g_l - g_v \cong -v_v\,dp$. The vapor phase is assumed to be ideal so $v_v = kT/p$. Thus, integrating, we obtain

$$g_l - g_v = -kT \int_{p_A^\circ}^{p_A} \frac{dp}{p} \tag{10.81}$$

where p_A° is the vapor pressure of pure A over a flat surface and p_A is the actual equilibrium partial pressure over the liquid. Then

$$g_l - g_v = -kT \ln\frac{p_A}{p_A^\circ} \tag{10.82}$$

We can define the ratio p_A/p_A° as the saturation ratio S. Substituting (10.82) into (10.80), we obtain the following expression for the Gibbs free energy change:

$$\Delta G = -\frac{4}{3}\pi R_p^3 \frac{kT}{v_l} \ln S + 4\pi R_p^2 \sigma \tag{10.83}$$

Figure 10.10 shows a sketch of the behavior of ΔG as a function of R_p. We see that if $S < 1$, both terms in (10.83) are positive and ΔG increases monotonically with R_p. On the

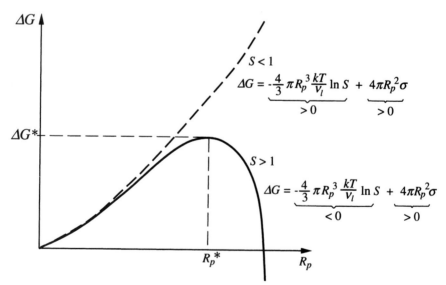

FIGURE 10.10   Gibbs free energy change for formation of a droplet of radius R_p from a vapor with saturation ratio S.

other hand, if $S > 1$, ΔG contains both positive and negative contributions. At small values of R_p the surface tension term dominates and the behavior of ΔG is similar to the $S < 1$ case. As R_p increases, the first term on the RHS of (10.83) becomes more important so that ΔG reaches a maximum value ΔG^* at $R_p = R_p^*$ and then decreases. The radius corresponding to the maximum can be calculated by setting $\partial \Delta G / \partial R_p = 0$ in (10.83) to get

$$R_p^* = \frac{2\sigma v_l}{kT \ln S} \tag{10.84}$$

Since ΔG is a maximum, as opposed to a minimum, at $R_p = R_p^*$, the equilibrium at this point is metastable. Equation (10.84) relates the equilibrium radius of a droplet of a pure substance to the physical properties of the substance, σ and v_l, and to the saturation ratio S of its environment. Equation (10.84) can be rearranged recalling the definition of the saturation ratio as

$$p_A = p_A^{\circ} \exp\left(\frac{2\sigma v_l}{kT R_p}\right) \tag{10.85}$$

Expressed in this form, the equation is frequently referred to as the *Kelvin equation*. The Kelvin equation can also be expressed in terms of molar units as

$$p_A = p_A^{\circ} \exp\left(\frac{2\sigma M}{RT \rho_l R_p}\right) \tag{10.86}$$

where M is the molecular weight of the substance, and ρ_l is the liquid-phase density.

The Kelvin equation tells us that *the vapor pressure over a curved interface always exceeds that of the same substance over a flat surface*. A rough physical interpretation of this so-called Kelvin effect is depicted in Figure 10.11. The vapor pressure of a liquid is determined by the energy necessary to separate a molecule from the attractive forces exerted by its neighbors and bring it to the gas phase. When a curved interface exists, as in a small droplet, there are fewer molecules immediately adjacent to a molecule on the surface than when the surface is flat. Consequently, it is easier for the molecules on the surface of a small drop to escape into the vapor phase and the vapor pressure over a curved interface is always greater than that over a plane surface.

Table 10.6 contains surface tensions for water and a series of organic molecules. At 298 K the organic surface tensions are about one-third that of water and their molecular volumes (M/ρ_l) range from three to six times that of water.

FIGURE 10.11   Effect of radius of curvature of a drop on its vapor pressure.

TABLE 10.6   Surface Tensions and Densities of Selected Compounds at 298 K[a]

Compound	Molecular Weight	Density, $g\,cm^{-3}$	Surface Tension, $dyn\,cm^{-1}$
Water	18	1.0	72
Benzene	78.11	0.879	28.21
Acetone	58.08	0.787	23.04
Ethanol	46.07	0.789	22.14
Styrene	104.2	0.906	31.49
Carbon tetrachloride	153.82	1.594	26.34

[a] σ has units in the SI system of $N\,m^{-1}$ ($=kg\,s^{-2}$). In the cgs system σ has units $dyn\,cm^{-1}$. $1\,dyn\,cm^{-1} = 10^{-3}\,kg\,s^{-2}$.

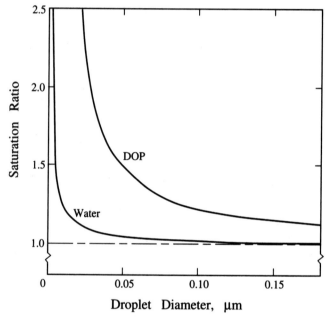

FIGURE 10.12   Saturation ratio versus droplet size for pure water and an organic liquid (dioctyl phthalate DOP) at 20°C.

Figure 10.12 gives the magnitude of the Kelvin effect p_A/p_A° for a pure water droplet and a typical organic compound (DOP) as a function of droplet diameter. We see that for water the vapor pressure is increased by 2.1% for a 0.1-μm and 23% for a 0.01-μm drop over that for a flat interface. Roughly, we may consider 50 nm diameter as the point at which the Kelvin effect begins to become important for aqueous particles. For higher-molecular-weight organics, the Kelvin effect is significant for even larger particles and should be included in calculations for particles with diameters smaller than about 200 nm.

10.4   THERMODYNAMICS OF ATMOSPHERIC AEROSOL SYSTEMS

10.4.1   The H_2SO_4–H_2O System

The importance of sulfuric acid in aerosol formation in a humid environment has been emphasized in a number of studies (Kiang et al. 1973; Mirabel and Katz 1974;

Jaecker-Voirol and Mirabel 1989; Kulmala and Laaksonen 1990; Laaksonen et al. 1995).
It is instructive to study first this simplified binary system before starting the discussion of
more complicated atmospheric aerosol systems.

Sulfuric acid is very hygroscopic, absorbing significant amounts of water even at
extremely low relative humidities. In Figure 10.13, sulfuric acid aqueous solution
properties are shown as a function of the mass fraction of H_2SO_4 in solution. Included in
Figure 10.13 are the equilibrium relative humidity over a flat solution surface, RH, the
solution density, ρ, the boiling point, B.P., and the surface tension, σ. In addition, the
solution normality (in N or equivalents per liter), the mass concentration, $x_{H_2SO_4}\rho$,
the particle growth factor (the ratio of the actual diameter of the droplet D_p to that if the
water associated with that were completely evaporated, D_{p0}), D_p/D_{p0}, and the Kelvin
parameter $D_{p0} \ln \left(p_{H_2SO_4}/p_{H_2SO_4}^{\circ}\right)$ are shown in the same figure.

> **Calculation of the Properties of a H_2SO_4 Droplet (Liu and Levi 1980)**   Consider a
> 1-μm H_2SO_4-H_2O droplet in equilibrium at 50% RH. Using Figure 10.13, calculate
> the following:
>
> 1. The H_2SO_4 concentration in the solution
> 2. The density of the solution
> 3. The boiling point of the solution
> 4. The droplet surface tension
> 5. The solution normality
> 6. The mass concentration of H_2SO_4 in the solution
> 7. The size of the droplet if all the water were removed
> 8. The Kelvin effect parameter
> 9. The size of this droplet at 90% RH.

Using the curve labeled RH and the corresponding scale at the right of the graph,
we find that the solution concentration is 42.5% H_2SO_4 or that $x_{H_2SO_4} = 0.425$ g
H_2SO_4 per g of solution. The rest of the solution properties can be calculated using
this solution concentration, the appropriate curve, and the appropriate scale. At this
concentration the solution has a density, ρ, of 1.32 g cm^{-3}, a boiling point of 115°C,
a surface tension, σ, of 76 dyn cm^{-1}, a normality of 11 N, and a mass concentration,
$x_{H_2SO_4}\rho$, of 0.55 g H_2SO_4 cm^{-3} of solution. The particle size growth factor,
D_p/D_{p0}, is 1.48. Thus the H_2SO_4–H_2O solution droplet would become a droplet of
pure H_2SO_4 of $1/1.48 = 0.68$ μm if all the water were removed. The Kelvin effect
parameter $D_{p0} \ln \left(p_{H_2SO_4}/p_{H_2SO_4}^{\circ}\right)$ is 11.3×10^{-4} μm, indicating that

$$\ln \left(\frac{p_{H_2SO_4}}{p_{H_2SO_4}^{\circ}}\right) = \frac{11.3 \times 10^{-4}}{0.68} = 16.62 \times 10^{-4}$$

Thus $\left(p_{H_2SO_4}/p_{H_2SO_4}^{\circ}\right) = 1.0017$, and in this case the Kelvin effect is negligible;
the increase in water vapor pressure due to the droplet curvature is only 0.166%
above that of a flat surface. The particle size growth factor at 90% relative humidity
is 2.12. Thus the 1 μm diameter drop at 50% RH will grow to become a drop of
$(2.12/1.48)(1) = 1.43$ μm at 90% RH.

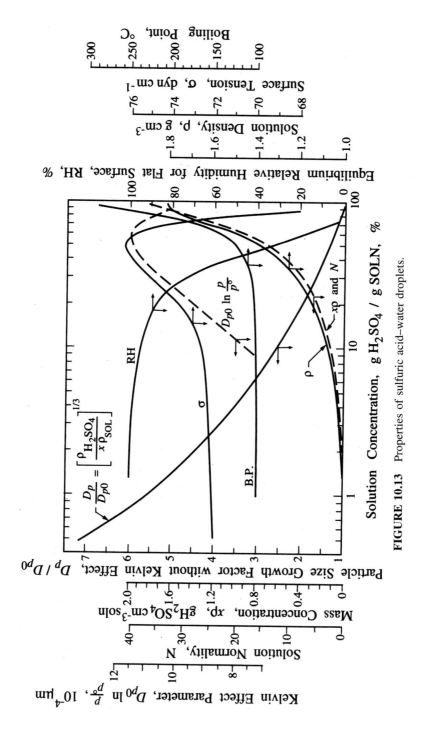

FIGURE 10.13 Properties of sulfuric acid–water droplets.

466

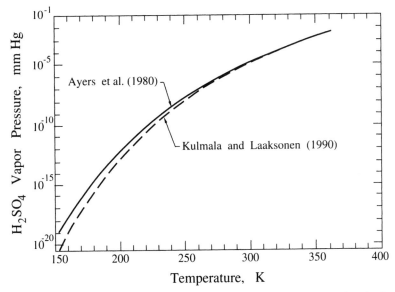

FIGURE 10.14 Vapor pressure of sulfuric acid as a function of temperature using the Ayers et al. (1980) and Kulmala and Laaksonen (1990) estimates. (Reprinted with permission from Kulmala, M. and Laaksonen, A. Binary nucleation of water sulfuric acid system. Comparison of classical theories with different H_2SO_4 saturation vapor pressure, *J. Chem. Phys.* **93**, 696–701. Copyright 1990 American Institute of Physics.)

The saturation vapor pressure of pure sulfuric acid, $p^\circ_{H_2SO_4}$, has been an elusive quantity. Roedel (1979), Ayers et al. (1980), and Kulmala and Laaksonen (1990) have argued that $p^\circ_{H_2SO_4} = 1.3 \pm 1.0 \times 10^{-8}$ atm $(13 \pm 10\,ppb)$ at 296 K. The temperature dependence of the saturation vapor pressure as a function of temperature is shown in Figure 10.14.

The variation of the vapor pressure of H_2SO_4 in a mixture with water over a flat surface as a function of composition at ambient temperatures is shown in Figure 10.15. Information on Figures 10.13, 10.14, and 10.15 can be combined to obtain most of the required information about sulfuric acid properties in the atmosphere. For example, let us calculate the expected H_2SO_4 concentrations, focusing only on the $H_2SO_4 - H_2O$ system. If the relative humidity exceeds 50%, the H_2SO_4 concentration in solution will be less than 40% by mass, and the H_2SO_4 mole fraction is less than 0.1 (Figure 10.13). For a temperature equal to 20°C, this corresponds to equilibrium vapor pressures less than 10^{-12} mm Hg. Under all conditions the concentration of H_2SO_4 in the gas phase is much less than the aerosol sulfate concentration.

The role of the Kelvin effect on the composition of atmospheric $H_2SO_4 - H_2O$ droplets is illustrated in Figure 10.16. If the effect were negligible the equilibrium aerosol composition would not be a function of its size. This is the case for particles larger than 0.1 μm in diameter. For smaller particles the H_2SO_4 mole fraction in the droplet is highly dependent on particle size. Also notice that for a fixed droplet size, the water concentration increases as the relative humidity increases.

The sulfuric acid–water system reactions are shown in Table 10.7. Note that the reaction $H_2SO_4(g) \rightleftharpoons H_2SO_4(aq)$ is not included as $H_2SO_4(aq)$ can be assumed to completely dissociate to HSO_4^- in the atmosphere for all practical purposes.

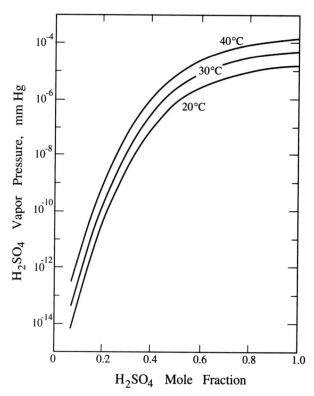

FIGURE 10.15 Equilibrium vapor pressure of H_2SO_4 in a mixture with water over a flat surface as a function of composition and temperature.

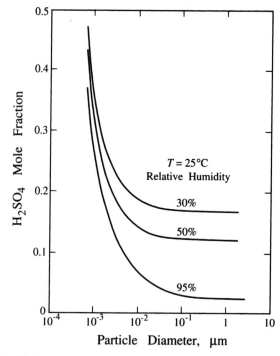

FIGURE 10.16 Equilibrium concentration of H_2SO_4 in a spherical droplet of H_2SO_4 and H_2O as a function of relative humidity and droplet diameter.

TABLE 10.7 CHEMICAL REACTIONS OCCURRING IN ATMOSPHERIC AEROSOLS

Reaction	Equilibrium Constant Value		
	$K\,(298)^a$	a	b
$NaCl(s) + HNO_3(g) \rightleftharpoons NaNO_3(s) + HCl(g)$	3.96	5.50	-2.18
$HSO_4^- \rightleftharpoons H^+ + SO_4^{2-}$	1.01×10^{-2} $(mol\,kg^{-1})$	8.85	25.14
$NH_3(g) + HNO_3(g) \rightleftharpoons NH_4^+ + NO_3^-$	4.0×10^{17} $(mol^2\,kg^{-2}\,atm^{-2})$	64.7	11.51
$HCl(g) \rightleftharpoons H^+ + Cl^-$	2.03×10^6 $(mol^2\,kg^{-2}\,atm^{-1})$	30.21	19.91
$NH_3(g) + HCl(g) \rightleftharpoons NH_4^+ + Cl^-$	2.12×10^{17} $(mol^2\,kg^{-2}\,atm^{-2})$	65.08	14.51
$Na_2SO_4(s) \rightleftharpoons 2\,Na^+ + SO_4^{2-}$	0.48 $(mol^3\,kg^{-3})$	0.98	39.57
$(NH_4)_2SO_4(s) \rightleftharpoons 2\,NH_4^+ + SO_4^{2-}$	1.425 $(mol^3\,kg^{-3})$	-2.65	38.55
$HNO_3(g) \rightleftharpoons H^+ + NO_3^-$	3.638×10^6 $(mol^2\,kg^{-2}\,atm^{-1})$	29.47	16.84
$NH_4Cl(s) \rightleftharpoons NH_3(g) + HCl(g)$	1.039×10^{-16} (atm^2)	-71.04	2.40
$NH_3(g) + HNO_3(g) \rightleftharpoons NH_4NO_3(s)$	3.35×10^{16} (atm^{-2})	75.11	-13.5
$NaCl(s) \rightleftharpoons Na^+ + Cl^-$	37.74 $(mol^2\,kg^{-2})$	-1.57	16.89
$NaHSO_4(s) \rightleftharpoons Na^+ + HSO_4^-$	2.44×10^4 $(mol^2\,kg^{-2})$	0.79	4.53
$NaNO_3(s) \rightleftharpoons Na^+ + HNO_3^-$	11.97 $(mol^2\,kg^{-2})$	-8.22	16.0

where

$$K(T) = K(298)\exp\left\{ a\left(\frac{298}{T} - 1\right) + b\left[1 + \ln\left(\frac{298}{T}\right) - \frac{298}{T}\right]\right\}$$

^{a}Equilibrium constant values in this table are based on products divided by reactants.

The reaction

$$H_2SO_4(g) \rightleftharpoons H^+ + HSO_4^-$$

can often be neglected in atmospheric aerosol calculations as the vapor pressure of $H_2SO_4(g)$ is practically zero over atmospheric particles. The whole system is then described by the bisulfate dissociation reaction,

$$HSO_4^- \rightleftharpoons H^+ + SO_4^{2-}$$

with an equilibrium constant at 298 K (Table 10.7) given by

$$1.01 \times 10^{-2}\,(mol\,kg^{-1}) = \frac{\gamma^2_{H^+,SO_4^{2-}}\,m_{H^+}m_{SO_4^{2-}}}{\gamma_{HSO_4^-}\,m_{HSO_4^-}}$$

The molar ratio of HSO_4^- to SO_4^{2-} is obtained by

$$\frac{m_{HSO_4^-}}{m_{SO_4^{2-}}} = 99 \frac{\gamma_{H^+,SO_4^{2-}}^2}{\gamma_{HSO_4^-}} m_{H^+}$$

As the ratio is proportional to the hydrogen ion concentration, it is expected that HSO_4^- will be present in acidic particles.

10.4.2   The Sulfuric Acid–Ammonia–Water System

Consider a simplified system containing only H_2SO_4, NH_3, and water. The problem we are interested in is the following: given the temperature (T), RH, and available ammonia and sulfuric acid, determine the aerosol composition.

A series of compounds may exist in the aerosol phase including solids like letovicite $((NH_4)_3H(SO_4)_2)$, $(NH_4)_2SO_4$, and NH_4HSO_4, but also aqueous solutions of NH_4^+, SO_4^{2-}, HSO_4^-, and $NH_3(aq)$. The possible reactions in the system are shown in Table 10.7.

For environments with low NH_3 availability, sulfuric acid exists in the aerosol phase in the form of H_2SO_4. As the NH_3 available increases, H_2SO_4 is converted to HSO_4^- and its salts, and finally, if there is an abundance of NH_3, to SO_4^{2-} and its salts. Figure 10.17 illustrates this behavior for an environment with low relative humidity (30%) and a fixed amount of total H_2SO_4 available equal to 10 $\mu g\,m^{-3}$. For total (gas and aerosol) ammonia concentrations less than 0.8 $\mu g\,m^{-3}$ the aerosol phase consist mainly of $H_2SO_4(aq)$ while some $NH_4HSO_4(s)$ is also present. A significant amount of water accompanies the $H_2SO_4(aq)$ even at this low relative humidity. This regime is characterized by an ammonia/sulfuric acid molar ratio of approximately less than 0.5. When the ammonia/sulfuric acid molar ratio is between 0.5 and 1.25, $NH_4HSO_4(s)$ is the dominant aerosol component for this system. When the ratio reaches unity (ammonia concentration 1.8 $\mu g\,m^{-3}$) the salt $(NH_4)_3H(SO_4)_2(s)$ (letovicite) is formed in the aerosol phase, gradually replacing $NH_4HSO_4(s)$. For a molar ratio of 1.25, $(NH_4)_3H(SO_4)_2(s)$ is the dominant aerosol component and for a value equal to 1.5 the aerosol phase consists almost exclusively of letovicite. For even higher ammonia concentrations $(NH_4)_2SO_4$ starts forming. At an ammonia concentration equal to 3.6 $\mu g\,m^{-3}$ (molar ratio 2), the aerosol consists of only $(NH_4)_2SO_4(s)$. If the total ammonia concentration is lower than 3.6 $\mu g\,m^{-3}$, practically all exists in the aerosol phase in the form of sulfate salts. Further increases of the available ammonia do not change the aerosol composition, but rather the excess ammonia remains in the gas phase as $NH_3(g)$.

The total aerosol mass during the variation of total ammonia in the system is also shown in Figure 10.17. One would expect that an increase of the availability of NH_3, an aerosol precursor, would result in a monotonic increase of the total aerosol mass. This is not the case for at least the ammonia-poor conditions (ammonia/sulfuric acid molar ratio less than 1). The increase of NH_3 in this range results in a reduction of the $H_2SO_4(aq)$ and the accompanying water. The overall aerosol mass decreases mainly because of the loss of water, reaching a minimum for an ammonia concentration of 1.8 $\mu g\,m^{-3}$. Further increases of ammonia result in increases of the overall aerosol mass. This nonlinear response of the aerosol mass to changes in the concentration of an aerosol precursor is encountered often in atmospheric aerosol thermodynamics.

The behavior of the same system at a higher relative humidity, 75%, is illustrated in Figure 10.18. Recall that the deliquescence relative humidities for the NH_4HSO_4 and

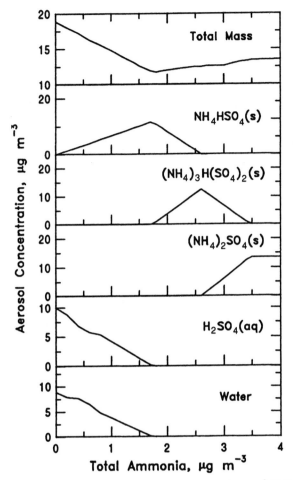

FIGURE 10.17 Aerosol composition for a system containing $10 \, \mu g \, m^{-3}$ H_2SO_4 at 30% RH and $T = 298 \, K$ as a function of the total (gas plus aerosol) concentration. The composition was calculated using the approach of Pilinis and Seinfeld (1987).

$(NH_4)_3H(SO_4)_2$ salts at this temperature are 40% and 69%, respectively (Table 10.1). As a result, the aerosol is a liquid solution of H_2SO_4, HSO_4^-, SO_4^{2-}, and NH_4^+ for the ammonia concentration range where the formation of these salts is favored. Once more, for ammonia/sulfuric acid molar ratios less than 0.5, H_2SO_4 dominates; for ratios between 0.5 and 1.5, HSO_4^- is the main aerosol component; while for higher ammonia concentrations, sulfate is formed. When there is enough ammonia to completely neutralize the available sulfate, forming $(NH_4)_2SO_4$, the liquid aerosol loses its water, becoming solid (Figure 10.18). The deliquescence relative humidity of $(NH_4)_2SO_4$ is 80% at this temperature, and assuming that we are on the deliquescence branch of the hysteresis curve, the aerosol will exist as solid. These changes in the aerosol chemical composition and the accompanying phase transformation change significantly the hygroscopic character of the aerosol, resulting in a monotonic decrease of its water mass. The net result is a significant reduction of the overall aerosol mass with increasing ammonia for this system with an abrupt change accompanying the phase transition from liquid to solid.

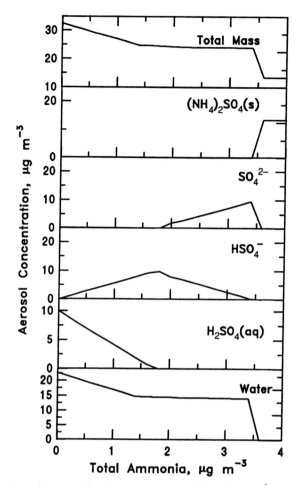

FIGURE 10.18   Aerosol composition for a system containing $10\,\mu g\,m^{-3}$ H_2SO_4 at 75% RH and $T = 298\,K$ as a function of the total (gas plus aerosol) concentration. The composition was calculated using the approach of Pilinis and Seinfeld (1987).

Summarizing, in very acidic atmospheres (NH_3/H_2SO_4 molar ratio less than 0.5) the aerosol particles exist primarily as H_2SO_4 solutions. For acidic atmospheres (NH_3/H_2SO_4 molar ratio in the 0.5–1.5 range) the particles consist mainly as bisulfate. If there is sufficient ammonia to neutralize the available sulfuric acid, $(NH_4)_2SO_4$ (or an NH_4^+ and SO_4^{2-} solution) is the prefered composition of the aerosol phase. For acidic atmospheres all the available ammonia is taken up by the aerosol phase and only for NH_3/H_2SO_4 molar ratios above 2 can ammonia also exist in the gas phase.

10.4.3   The Ammonia–Nitric Acid–Water System

Ammonia and nitric acid can react in the atmosphere to form ammonium nitrate, NH_4NO_3:

$$NH_3(g) + HNO_3(g) \rightleftharpoons NH_4NO_3(s) \qquad (10.87)$$

Ammonium nitrate is formed in areas characterized by high ammonia and nitric acid concentrations and, as we will see in the next section, low sulfate concentrations.

Depending on the ambient RH, ammonium nitrate may exist as a solid or as an aqueous solution of NH_4^+ and NO_3^-. Equilibrium concentrations of gaseous NH_3 and HNO_3, and the resulting concentration of solid or aqueous NH_4NO_3, can be calculated from fundamental thermodynamic principles using the method presented by Stelson and Seinfeld (1982a). The procedure is composed of several steps, requiring as input the ambient temperature and relative humidity. First, the equilibrium state of NH_4NO_3 is defined. If the ambient relative humidity is less than the deliquescence relative humidity (DRH), given by

$$\ln(DRH) = \frac{723.7}{T} + 1.6954 \qquad (10.88)$$

then the equilibrium state of NH_4NO_3 is a solid. For example, at 298 K the corresponding DRH is 61.8%, while at 288 K it increases to 67%.

At low RH, when NH_4NO_3 is a solid, the equilibrium condition for (10.87) is

$$\mu_{NH_3} + \mu_{HNO_3} = \mu_{NH_4NO_3} \qquad (10.89)$$

and using the definitions of the chemical potentials of ideal gases and solids:

$$\exp\left(\frac{\mu_{NH_4NO_3}^* - \mu_{NH_3}^\circ - \mu_{HNO_3}^\circ}{RT}\right) = K_p(T) = p_{NH_3}p_{HNO_3} \qquad (10.90)$$

The dissociation constant, $K_p(T)$, is therefore equal to the product of the partial pressures of NH_3 and HNO_3. K_p can be estimated by integrating the van't Hoff equation (Denbigh 1981). The resulting equation for K_p in units of ppb^2 (assuming 1 atm of total pressure) is

$$\ln K_p = 84.6 - \frac{24220}{T} - 6.1 \ln\left(\frac{T}{298}\right) \qquad (10.91)$$

This dissociation constant is shown as a function of temperature in Figure 10.19. Note that the constant is quite sensitive to temperature changes, varying over more than two

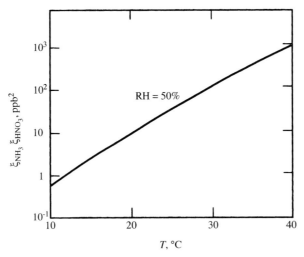

FIGURE 10.19 NH_4NO_3 equilibrium dissociation constant as a function of temperature at RH = 50%.

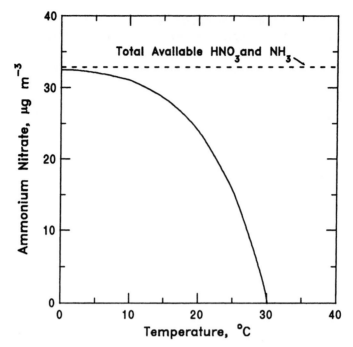

FIGURE 10.20 $NH_4NO_4(s)$ concentration as a function of temperature for a system containing $7\,\mu g\,m^{-3}$ NH_3 and $26.5\,\mu g\,m^{-3}$ HNO_3 at 30% RH. The difference between the total available mass (dashed line) and the NH_4NO_3 mass remains in the gas phase as $NH_3(g)$ and $HNO_3(g)$.

orders of magnitude for typical ambient conditions. Lower temperatures correspond to lower values of the dissociation constant, and therefore lower equilibrium values of the NH_3 and HNO_3 gas-phase concentrations. Thus lower temperatures shift the equilibrium of the system toward the aerosol phase, increasing the aerosol mass of NH_4NO_3 (Figure 10.20).

Solid Ammonium Nitrate Formation   Calculate the $NH_4NO_3(s)$ mass for a system containing $17\,\mu g\,m^{-3}$ of NH_3 and $63\,\mu g\,m^{-3}$ of HNO_3 at 308 K, 30% RH, and 1 atm.

At 308 K, according to (10.88), the DRH of NH_4NO_3 is 57.1% and therefore at 30% RH the NH_4NO_3, if it exists in stable equilibrium, will be solid. Let us first determine whether the atmosphere is saturated with NH_3 and HNO_3 by calculating the product of the corresponding mixing ratios. For 308 K and 1 atm,

$$\text{Mixing ratio of species } i \text{ in ppb} = \frac{25.6}{M_i} \times (\text{concentration of } i \text{ in } \mu g\,m^{-3})$$

and $\xi_{NH_3} = \xi_{HNO_3} = 25.6\,\text{ppb}$. Thus $\xi_{NH_3}\xi_{HNO_3} = 655\,\text{ppb}^2$. Using (10.91), we find that $K_p(308\,\text{K}) = 318\,\text{ppb}^2$ (in mixing ratio units) and

$$\xi_{NH_3}\xi_{HNO_3} > K_p$$

The system is therefore supersaturated with ammonia and nitric acid and a fraction of them will be transferred to the aerosol phase to establish equilibrium, given mathematically by (10.90). Let us assume that x ppb of NH_3 is transferred to the aerosol phase. Because reaction (10.87) requires 1 mole of HNO_3 for each mole of NH_3 reacting, x ppb of HNO_3 will also be transferred to the aerosol phase. The gas phase will contain $(25.6 - x)$ ppb of HNO_3 and $(25.6 - x)$ ppb of NH_3. The system will be in equilibrium and therefore the new gas-phase mixing ratios will satisfy (10.90), or

$$(25.6 - x)^2 = K_p = 318$$

resulting in $x = 7.8$ ppb. Therefore 7.8 ppb of NH_3 and 7.8 ppb of HNO_3 will form solid ammonium nitrate. These correspond to $5.2\ \mu g\ m^{-3}$ of NH_3 and $16.1\ \mu g\ m^{-3}$ of HNO_3 for a total of $21.3\ \mu g\ m^{-3}$ of $NH_4NO_3(s)$.

Ammonium Nitrate Solutions   At relative humidities above that of deliquescence, NH_4NO_3 will be found in the aqueous state. The corresponding dissociation reaction is then

$$NH_3(g) + HNO_3(g) \rightleftharpoons NH_4^+ + NO_3^- \qquad (10.92)$$

Stelson and Seinfeld (1981) have shown that solution concentrations of 8–26 M can be expected in wetted atmospheric aerosol. At such concentrations the solutions are strongly nonideal, and appropriate thermodynamic activity coefficients are necessary for thermodynamic calculations. Tang (1980), Stelson and Seinfeld (1982a–c), and Stelson et al. (1984) have developed activity coefficient expressions for aqueous systems of nitrate, sulfate, ammonium, nitric acid, and sulfuric acid at concentrations exceeding 1 M.

Following the approach illustrated in Sections 10.1.5 and 10.1.6, one can show that the condition for equilibrium of reaction (10.92) is

$$\mu_{HNO_3} + \mu_{NH_3} = \mu_{NH_4^+} + \mu_{NO_3^-} \qquad (10.93)$$

or after using the definitions of the corresponding chemical potentials

$$\exp\left(\frac{\mu_{HNO_3}^\circ + \mu_{NH_3}^\circ - \mu_{NO_3^-}^\diamond - \mu_{NH_4^+}^\diamond}{RT}\right) = \frac{\gamma_{NH_4^+}\gamma_{NO_3^-}m_{NH_4^+}m_{NO_3^-}}{p_{HNO_3}p_{NH_3}} \qquad (10.94)$$

The right-hand side is the equilibrium constant, and defining

$$\gamma_{NH_4NO_3}^2 = \gamma_{NH_4^+}\gamma_{NO_3^-} \qquad (10.95)$$

the equilibrium relationship becomes

$$K_{AN} = \frac{\gamma_{NH_4NO_3}^2 m_{NH_4^+}m_{NO_3^-}}{p_{HNO_3}p_{NH_3}} \qquad (10.96)$$

The equilibrium constant can be calculated by

$$K_{AN} = 4 \times 10^{17} \exp\left\{64.7\left(\frac{298}{T} - 1\right) + 11.51\left[1 + \ln\left(\frac{298}{T}\right) - \frac{298}{T}\right]\right\} \qquad (10.97)$$

where K_{AN} is in $mol^2\,kg^{-2}\,atm^{-2}$.

Solution of (10.96) for a given temperature requires calculation of the corresponding molalities. These concentrations depend not only on the aerosol nitrate and ammonium but also on the amount of water in the aerosol phase. Therefore calculation of the aerosol solution composition requires estimation of the aerosol water content. As we have seen in Section 10.2.1, the water activity will be equal to the relative humidity (expressed in the 0–1 scale). While this is very useful information, it is not sufficient for the water calculation. One needs to relate the tendency of the aerosol components to absorb moisture with their availability and the availability of water given by the relative humidity. In atmospheric aerosol models (Hanel and Zankl 1979; Cohen et al. 1987; Pilinis and Seinfeld 1987; Wexler and Seinfeld 1991) the water content of aerosols is usually predicted using the ZSR relationship (Zdanovskii 1948; Stokes and Robinson 1966)

$$W = \sum_i \frac{C_i}{m_{i,0}(a_w)} \qquad (10.98)$$

where W is the mass of aerosol water in kg of water per m^3 of air, C_i is the aqueous-phase concentration of electrolyte i in moles per m^3 of air, and $m_{i,0}(a_w)$ is the molality ($mol\,kg^{-1}$) of a single-component aqueous solution of electrolyte i that has a water activity $a_w = RH/100$. Equations (10.96) and (10.98) form a nonlinear system that has as a solution the equilibrium composition of the system. The calculation requiring the activity coefficient functions for (10.96) is quite involved and is usually carried out by appropriate thermodynamic codes.

Figure 10.21 depicts the results of such a computation, namely, the product of the mixing ratios of ammonia and nitric acid over a solution as a function of relative humidity. Assuming that there is no $(NH_4)_2SO_4$ present, this corresponds to the line $Y = 1.00$. The product of the mixing ratios decreases rapidly with relative humidity, indicating the shifting of the equilibrium toward the aerosol phase. The presence of water allows NH_4NO_3 to dissolve in the liquid aerosol particles and increases its aerosol concentration.

Defining the equilibrium dissociation constant as $\xi_{HNO_3}\xi_{NH_3}$ shown in Figure 10.21, the equilibrium concentrations of $NH_3(g)$, $HNO_3(g)$, NH_4^+, and NO_3^- (all in $mol\,m^{-3}$ of air) can be calculated for a specific relative humidity. Note that by using (10.96) K_p is related to K_{AN} by

$$K_p = \frac{\gamma_{NH_4NO_3}^2\, m_{NH_4^+}\, m_{NO_3^-}}{K_{AN}} \qquad (10.99)$$

> **Aqueous Ammonium Nitrate Formation**   Calculate the ammonium nitrate concentration for a system containing 5 ppb of total (gas and aerosol) NH_3 and 6 ppb of total HNO_3 at 25°C at 80% RH.
>
>   Using Figure 10.21, we find that the equilibrium product is approximately 15 ppb². The actual product of the total concentrations is $5 \times 6 = 30$ ppb², and

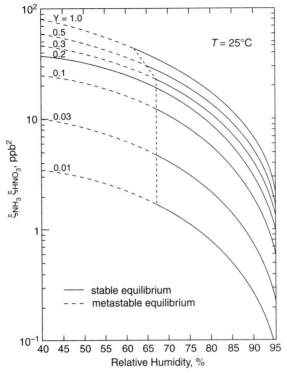

FIGURE 10.21 NH_4NO_3 equilibrium dissociation constant for an ammonium sulfate–nitrate solution as a function of relative humidity and ammonium nitrate ionic strength fraction defined by

$$Y = \frac{[NH_4NO_3]}{[NH_4NO_3] + 3[(NH_4)_2SO_4]}$$

at 298 K (Stelson and Seinfeld 1982c). Pure NH_4NO_3 corresponds to $Y = 1$.

therefore the system is supersaturated with ammonia and nitric acid and some ammonium nitrate will form. If x ppb of ammonium nitrate is formed, then there will be $5 - x$ ppb of NH_3 and $6 - x$ ppb of HNO_3 left. At equilibrium, the product of these remaining concentrations will be the equilibrium product and therefore:

$$(5 - x)(6 - x) = 15 \text{ or } x = 1.6 \text{ ppb}$$

There is one more mathematical solution of the problem ($x = 9.4$ ppb), but obviously it is not acceptable because it exceeds the total available concentrations of NH_3 and HNO_3 in the system. Therefore the amount of aqueous ammonium nitrate will be 1.6 ppb or approximately $5.2 \, \mu g \, m^{-3}$.

Figure 10.22 shows the calculated ammonium and nitrate for a closed system as a function of its RH. The ammonium nitrate concentration increases by a factor of 8 in this example as the RH increases from 65% to 95%. Further increases of the RH will lead to the complete dissolution of practically all the available nitric acid to the particulate phase (see also Chapter 7).

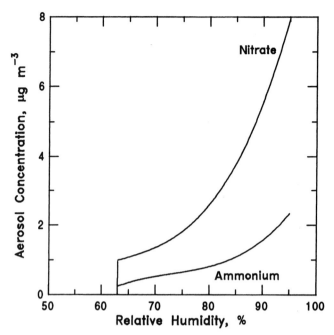

FIGURE 10.22 Calculated aerosol ammonium and nitrate concentrations as a function of relative humidity for an atmosphere with $[TA] = 5\,\mu g\,m^{-3}$, $[TN] = 10\,\mu g\,m^{-3}$, and $T = 300\,K$.

It is important to note the variation of the product of $p_{NH_3} p_{HNO_3}$ and the corresponding aerosol composition over several orders of magnitude under typical atmospheric conditions. This variation may introduce significant difficulties in the measurement of aerosol NH_4NO_3 as small changes in RH and T will shift the equilibrium and cause evaporation/condensation of the aerosol sample.

10.4.4 The Ammonia–Nitric Acid–Sulfuric Acid–Water System

The system of interest consists of the following possible components:

Gas phase: NH_3, HNO_3, H_2SO_4, H_2O.
Solid phase: NH_4HSO_4, $(NH_4)_2SO_4$, NH_4NO_3, $(NH_4)_2SO_4 \cdot 2\,NH_4NO_3$,
$\quad (NH_4)_2SO_4 \cdot 3\,NH_4NO_3$, $(NH_4)_3H(SO_4)_2$.
Aqueous phase: NH_4^+, H^+, HSO_4^-, SO_4^{2-}, NO_3^-, H_2O.

Two observations are useful in determining a priori the composition of the aerosol that exists in such a system:

1. Sulfuric acid possesses an extremely low vapor pressure.
2. $(NH_4)_2SO_4$ solid or aqueous is the preferred form of sulfate.

The second observation means that, if possible, each mole of sulfate will remove 2 moles of ammonia from the gas phase, and the first observation implies that the amount of sulfuric acid in the gas phase will be negligible.

Based on these observations, we can define two regimes of interest, ammonia-rich and ammonia-poor. If [TA], [TS], and [TN] are the total (gas + aqueous + solid) molar concentrations of ammonia, sulfate, and nitrate, respectively, then the two cases are

1. Ammonia-poor, [TA] < 2[TS].
2. Ammonia-rich, [TA] > 2[TS].

Case 1: Ammonia-Poor. In this case there is insufficient NH_3 to neutralize the available sulfate. Thus the aerosol phase will be acidic. The vapor pressure of NH_3 will be low, and the sulfate will tend to drive the nitrate to the gas phase. Since the NH_3 partial pressure will be low, the NH_3–HNO_3 partial pressure product will also be low so ammonium nitrate levels will be low or zero. Sulfate may exist as bisulfate.

Case 2: Ammonia-Rich. In this case there is excess ammonia, so that the aerosol phase will be neutralized to a large extent. The ammonia that does not react with sulfate will be available to react with nitrate to produce NH_4NO_3.

The behavior in these two different regimes is depicted in Figure 10.23. The transition occurs at approximately $3.5 \, \mu g \, m^{-3}$. At very low ammonia concentrations, sulfuric acid and bisulfate constitute the aerosol composition. As ammonia increases ammonium nitrate becomes a significant aerosol constituent. The aerosol liquid water content responds nonlinearly to these changes, reaching a minimum close to the transition between the two regimes.

The accurate calculation of the ammonium nitrate concentrations for this rather complicated system is usually performed using computational thermodynamic models (see Section 10.4.6). However, one can often estimate the ammonium nitrate concentration in an air parcel with a rather simple approach taking advantage of appropriate diagrams that summarize the results of the computational models. For ammonia-poor systems one can assume as a first-order approximation that the ammonium nitrate concentration is very small. For ammonia-rich systems when the particles are solid, the approach is the same as that of Section 10.4.3 with one important difference. In these calculations one uses the *free ammonia* instead of the total ammonia. The free ammonia in the system is defined as the total ammonia minus the ammonia required to neutralize the available sulfate. So in molar units, we have

$$[Free \, ammonia] = [TA] - 2[TS]$$

NH_4NO_3 Calculation for Solid Particles in the NH_3–HNO_3–H_2SO_4 System   Let us repeat the first example of Section 10.4.3 (for a system with $17 \, \mu g \, m^{-3}$ of NH_3, $63 \, \mu g \, m^{-3}$ HNO_3, $T = 308 \, K$, $RH = 30\%$) assuming now that there are also $10 \, \mu g \, m^{-3}$ of sulfate.

We first need to estimate the NH_3 necessary for the full neutralization of the available sulfate, which is equal to $10 \times 2 \times 17/98 = 3.5 \, \mu g \, m^{-3}$ NH_3. Therefore, from the available ammonia only $17 - 3.5 = 13.5 \, \mu g \, m^{-3}$ will be free for the potential formation of NH_4NO_3. Converting the concentrations to ppb, the nitric acid mixing ratio will be 25.6 ppb, while the free ammonia mixing ratio will now be

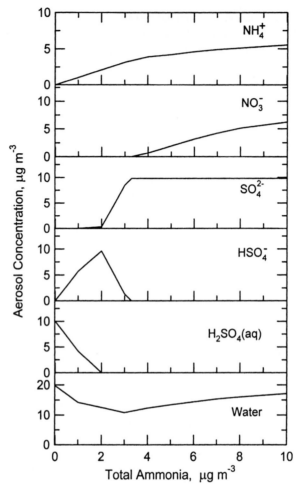

FIGURE 10.23   Calculated aerosol composition as a function of total ammonia for an atmosphere with $[\text{TS}] = 10\,\mu\text{g m}^{-3}$, $[\text{TN}] = 10\,\mu\text{g m}^{-3}$, $T = 298\,\text{K}$, and RH = 70%.

20.3 ppb. The mixing ratio product is then $520\,\text{ppb}^2$ and exceeds the equilibrium mixing ratio product $K_p(308\,\text{K}) = 318\,\text{ppb}^2$. The system is therefore supersaturated with nitric acid and free ammonia and x ppb of ammonium nitrate will be formed. The ammonium nitrate concentration can be found from the equilibrium condition for the gas phase:

$$(25.6 - x)(20.3 - x) = K_p = 318$$

resulting in $x = 4.9$ ppb or $13.4\,\mu\text{g m}^{-3}$ of NH_4NO_3. For zero sulfate we found in the previous example that $21.3\,\mu\text{g m}^{-3}$ of NH_4NO_3 will be formed. Therefore for this system the removal of $10\,\mu\text{g m}^{-3}$ of sulfate results in the removal of $10 \times (96 + 36)/96 = 13.7\,\mu\text{g m}^{-3}$ of $(NH_4)_2SO_4$ and the increase of the NH_4NO_3 by $21.3 - 13.4 = 6.9\,\mu\text{g m}^{-3}$. Roughly 50% of the concentration of the reduced ammonium sulfate will be replaced by ammonium nitrate.

For ammonia-rich aqueous systems the ammonium nitrate calculation is complicated by the dependence of K_p not only on T and RH but also on the sulfate concentration. This dependence is usually expressed by the parameter Y (Figure 10.21). The parameter Y is the ionic strength fraction of ammonium sulfate and is calculated as

$$Y = \frac{[NH_4NO_3]}{[NH_4NO_3] + 3[(NH_4)_2SO_4]} \qquad (10.100)$$

As the $(NH_4)_2SO_4$ in the particles increases compared to the ammonium nitrate, the parameter Y decreases and the equilibrium product of ammonia and nitric acid decreases. The additional ammonium and sulfate ions make the aqueous solution a more favorable environment for ammonium nitrate, shifting its partitioning toward the particulate phase. Therefore, addition of ammonium sulfate to aqueous aerosol particles will tend to increase the concentration of ammonium nitrate in the particulate phase, and, vice versa, reductions of ammonium sulfate can lead to decreases of ammonium nitrate for aqueous aerosol.

The parameter Y depends on the unknown ammonium nitrate concentration, making an iterative solution necessary. For this approach, one assumes a value of the ammonium nitrate concentration, calculates the corresponding Y, and finds from Figure 10.21 the corresponding K_p. Then one can follow the same approach as that for solid ammonium nitrate and calculate the actual ammonium nitrate concentration. If the assumed value and the calculated one are the same, then this value is the solution of the problem. If not, these values are used to improve the initial guess and the process is repeated.

Calculation of Aqueous Ammonium Nitrate Concentration Estimate the ammonium nitrate concentration in aqueous atmospheric particles in an area characterized by the following conditions:

Total (gas and aerosol) ammonia: $10\,\mu g\,m^{-3}$

Total (gas and aerosol) nitric acid: $10\,\mu g\,m^{-3}$

Sulfate: $10\,\mu g\,m^{-3}$

$T = 298\,K$ and $RH = 75\%$

We first need to estimate the NH_3 necessary for the full neutralization of the available sulfate, which is equal to $10 \times 2 \times 17/98 = 3.5\,\mu g\,m^{-3}$ NH_3. Therefore from the available ammonia only $10 - 3.5 = 6.5\,\mu g\,m^{-3}$ NH_3 will be available for the potential formation of ammonium nitrate. Converting all the concentrations to ppb:

$$Free\ NH_3 = 8.314 \times 298 \times 6.5/(100 \times 17) = 9.5\,ppb$$
$$Total\ HNO_3 = 8.314 \times 298 \times 10/(100 \times 53) = 4.7\,ppb$$
$$Sulfate = 8.314 \times 298 \times 10/(100 \times 96) = 2.6\,ppb$$

The product of the free ammonia and total nitric acid concentrations is $\xi_{NH_3}\xi_{HNO_3} = 9.5 \times 4.7 = 44.6\,ppb^2$. This value exceeds the corresponding equilibrium concentration products at 298 K and 70% RH for all values of Y (Figure 10.21), and therefore some ammonium nitrate will be formed.

As a first guess, we assume that 2 ppb of NH_4NO_3 will be formed from 2 ppb of NH_3 and 2 ppb of HNO_3 leaving in the gas phase 7.5 ppb NH_3 and 2.7 ppb HNO_3. The corresponding gas-phase concentration product $\xi_{NH_3} \xi_{HNO_3}$ will then be 20.3 ppb^2. We need to estimate now the equilibrium concentration production for the aqueous aerosol and compare it to this value. The Y for this composition is $Y = 2/(2 + 3 \times 2.6) = 0.2$, and using Figure 10.21, we find that K_p is approximately 8 ppb^2. Obviously, the use of approximate values from a graph introduces some uncertainty in these calculations.

For this trial, we were left with a gas phase concentration product of 20.3 ppb^2 and an aqueous aerosol phase with an equilibrium product of 8 ppb^2. These two are not equal, and therefore we need to increase the assumed ammonium nitrate concentration.

Let us now improve our first guess by assuming that 3 ppb of NH_4NO_3 will be formed from 3 ppb NH_3 and 3 ppb HNO_3. There will be 6.5 ppb NH_3 and 1.7 ppb HNO_3 remaining in the gas phase, resulting in a concentration product of 11 ppb^2. For the particulate phase the corresponding $Y = 3/(3 + 3 \times 2.6) = 0.3$, and using Figure 10.21 K_p is approximately 10 ppb^2. The assumed composition brings the system quite close to equilibrium, and therefore approximately 3 ppb of ammonium nitrate is dissolved in the aqueous particles. This corresponds to roughly $2.2\,\mu g\,m^{-3}$ of NH_4^+ and $6\,\mu g\,m^{-3}$ of NO_3^-. The total concentration of inorganic particles will be $10 + 3.5 + 2.2 + 6 = 21.7\,\mu g\,m^{-3}$. For more accurate results, these calculations are routinely performed today by suitable thermodynamic codes (see Section 10.4.6).

The formation of ammonium nitrate is often limited by the availability of one of the reactants. Figure 10.24 shows the ammonium concentration as a function of

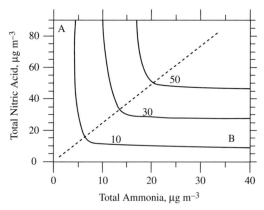

FIGURE 10.24   Isopleths of predicted particulate nitrate concentration ($\mu g\,m^{-3}$) as a function of total (gaseous + particulate) ammonia and nitric acid at 293 K and 80% RH for a system containing $25\,\mu g\,m^{-3}$ of sulfate.

the total available ammonia and the total available nitric acid for a polluted area. The upper left part of the figure (area A) is characterized by relatively high total nitric acid concentrations and relatively low ammonia. Large urban areas are often in this regime. The isopleths are almost parallel to the y- axis in this area, so decreases in nitric acid availability do not affect significantly the NH_4NO_3 concentration in the area. On the other hand, decreases in ammonia result in significant decreases in ammonium nitrate. NH_4NO_3 formation is limited by ammonia in area A. The opposite behavior is observed in area B. Here, NH_4NO_3 is not sensitive to changes in ammonia concentrations but responds readily to changes in nitric acid availability. Rural areas with significant ammonia emissions are often in this regime, where NH_4NO_3 formation is nitric acid-limited. The boundary between the nitric acid-limited and ammonia-limited regimes depends on temperature, relative humidity, sulfate concentrations and also the concentrations of other major inorganic aerosol components.

Summarizing, reductions of sulfate concentrations affect the inorganic particulate matter concentrations in two different ways. Part of the sulfate may be replaced by nitric acid, leading to an increase of the ammonium nitrate content of the aerosol. The sulfate decrease frees up ammonia to react with nitric acid and transfers it to the aerosol phase. If the particles are aqueous, there is also a second effect. The reduction of the sulfate ions in solution increases the equilibrium vapor pressure product of ammonia and nitric acid of the particles (see Figure 10.21), shifting the ammonium nitrate partitioning toward the gas phase. So for aqueous particles the substitution of sulfate by nitrate is in some cases relatively small. These interactions are illustrated in Figure 10.25. Sulfate reductions are accompanied by an increase of the aerosol nitrate and a reduction of the aerosol ammonium, water, and total mass. However, the reduction of the mass is nonlinear. For example, a reduction of total sulfate of $20\,\mu g\,m^{-3}$ (from 30 to $10\,\mu g\,m^{-3}$) results in a net reduction of the dry aerosol mass (excluding water) of only $12.9\,\mu g\,m^{-3}$, because of the increase of the aerosol nitrate by $10\,\mu g\,m^{-3}$. Ammonium decreases by only $2.9\,\mu g\,m^{-3}$.

10.4.5   Other Inorganic Aerosol Species

Aerosol Na^+ and Cl^- are present in substantial concentrations in regions close to seawater. Sodium and chloride interact with several aerosol components (Table 10.7). A variety of solids can be formed during these reactions including ammonium chloride, sodium nitrate, sodium sulfate, and sodium bisulfate, while $HCl(g)$ may be released to the gas phase.

Addition of NaCl to an urban aerosol can have a series of interesting effects, including the reaction of NaCl with HNO_3:

$$NaCl(s) + HNO_3(g) \rightleftharpoons NaNO_3(s) + HCl(g)$$

As a result of this reaction more nitrate is transferred to the aerosol phase and is associated with the large seasalt particles. At the same time, hydrochloric acid is liberated and the aerosol particles appear to be "chloride"-deficient. This deficiency may also be a result of

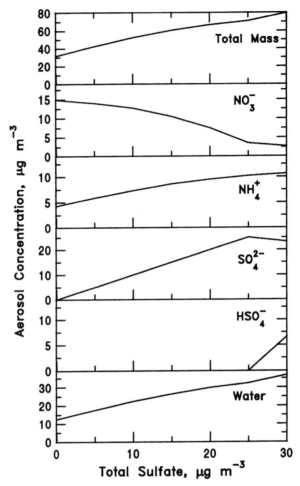

FIGURE 10.25 Calculated aerosol composition as a function of total sulfate for an atmosphere with $[\text{TA}] = 10 \, \mu g \, m^{-3}$, $[\text{TN}] = 20 \, \mu g \, m^{-3}$, $T = 298 \, K$, and $RH = 70\%$.

the reactions

$$2 \, \text{NaCl(s)} + \text{H}_2\text{SO}_4(\text{g}) \rightleftharpoons \text{Na}_2\text{SO}_4(\text{s}) + 2 \, \text{HCl(g)}$$
$$\text{NaCl(s)} + \text{H}_2\text{SO}_4(\text{g}) \rightleftharpoons \text{NaHSO}_4(\text{s}) + \text{HCl(g)}$$

10.4.6 Inorganic Aerosol Thermodynamic Models

Thermodynamic calculations for multicomponent atmospheric particulate matter involve either the minimization of the Gibbs free energy of the system (see Section 10.1.3) or solution of a system of nonlinear algebraic equations [see equation (10.30)] corresponding to the equilibrium reactions taking place in the system. However, there are two major complications: (1) the composition of the system (e.g., existence of a given salt) is not known a priori; and (2) if the particles are aqueous, the activity coefficient of each ion depends on the concentrations of all other ions. A number of mathematical models have been developed since the mid-1980s to solve these challenging problems numerically, starting with EQUIL (Bassett and Seinfeld 1983), MARS (Saxena et al. 1986), and SEQUILIB (Pilinis and Seinfeld 1987). These original models have evolved through the use of more accurate coefficient models and better numerical algorithms (Kim et al. 1993; Nenes et al. 1998). Some of them are readily available through the Web, including AIM (http://mae.ucdavis.edu/~wexler/aim/project.htm) and ISORROPIA (http://nenes .eas.gatech.edu/ISORROPIA). AIM describes the thermodynamics of the system most accurately (Clegg et al. 1998a,b), while ISORROPIA (Nenes et al. 1998) has been optimized for computational speed and use in three-dimensional chemical transport models (see Chapter 25).

PROBLEMS

10.1$_A$ Show that for a particle containing both solid and aqueous ammonium sulfate

$$\mu^*_{(NH_4)_2SO_4} = 2\,\mu^\diamond_{NH_4^+} + \mu^\diamond_{SO_4^{2-}} + RT\ \ln(4\,\gamma^3_{(NH_4)_2SO_4}m^3_{(NH_4)_2SO_4})$$

10.2$_A$ Consider a 0.01-μm-diameter sulfuric acid–water droplet at 50% relative humidity. What is the increase in the equilibrium vapor pressure over the curved droplet surface over that for the corresponding flat surface?

10.3$_B$ The total (gas plus aerosol) ammonia and nitric acid mixing ratios in an urban area are 10 ppb and 15 ppb, respectively. There is little sulfate in the area.

  a. Assuming a relative humidity equal to 30%, calculate the equilibrium aerosol ammonium nitrate concentration in $\mu g\ m^{-3}$ as a function of temperature (from 0 to 35°C).

  b. For a temperature equal to 25°C calculate the equilibrium aerosol ammonium nitrate concentration as a function of the relative humidity.

  c. For a temperature equal to 15°C and relative humidity equal to 30%, calculate the equilibrium aerosol ammonium nitrate concentration in $\mu g\ m^{-3}$ for an ammonia mixing ratio in the range of 0–20 ppb. Is the relationship between the aerosol ammonium nitrate concentration and the ammonia concentration linear? Why?

  d. If the aerosol ammonium nitrate concentration is measured to be $2\,\mu g\ m^{-3}$ at 15°C and 30% RH, can you calculate the corresponding equilibrium concentrations of ammonia and nitric acid in the gas phase? Why?

  e. Using the results above, discuss the factors that favor the formation of aerosol ammonium nitrate.

10.4$_B$ Repeat the calculations of Problem 10.3 parts (a)–(c) for an area characterized by a sulfate concentration equal to $4\,\mu g\ m^{-3}$. In part (c) use 2 ppb of ammonia as the lower limit of the mixing ratio range. Note that a fraction of the available ammonia will be used to neutralize the available sulfate.

10.5$_B$ Calculate the ratio of the wet particle mass to the dry particle mass for aerosol consisting of (a) ammonium sulfate, (b) sodium sulfate, (c) sodium nitrate, and (d) sodium chloride at their deliquescence points for temperature equal to 5 and 25°C.

10.6$_B$ A coastal area is characterized by seasalt (NaCl) concentrations of 10 μg m^{-3}.

    a. Calculate the aerosol concentration and composition at 25°C given a relative humidity of 50% for a total nitric acid concentration from 0 to 100 μg m^{-3}, zero ammonia and sulfate.

    b. Repeat the calculation assuming that the total ammonia concentration is 10 μg m^{-3} and there is also 5 μg m^{-3} of sulfate.

REFERENCES

Ayers, G. P., Gillet, R. W., and Gras, J. L. (1980) On the vapor pressure of sulfuric acid, *Geophys. Res. Lett.* **7**, 433–436.

Bassett, M. E., and Seinfeld, J. H. (1983) Atmospheric equilibrium model of sulfate and nitrate aerosols, *Atmos. Environ.* **17**, 2237–2252.

Clegg, S. L., Brimblecombe, P., and Wexler, A. S. (1998a) A thermodynamic model of the system H$^+$–NH$_4^+$–SO$_4^{2-}$–NO$_3^-$–H$_2$O at tropospheric temperatures, *J. Phys. Chem.* **102A**, 2137–2154.

Clegg, S. L., Brimblecombe, P., and Wexler, A. S. (1998b) A thermodynamic model of the system H$^+$–NH$_4^+$–Na$^+$–SO$_4^{2-}$–NO$_3^-$–Cl$^-$–H$_2$O at 298.15 K, *J. Phys. Chem.* **102A**, 2155–2171.

Cohen, M. D., Flagan, R. C., and Seinfeld, J. H. (1987) Studies of concentrated electrolyte solutions using the electrodynamic balance. 2. Water activities for mixed-electrolyte solutions, *J. Phys. Chem.* **91**, 4575–4582.

Denbigh, K. (1981) *The Principles of Chemical Equilibrium*, 4th ed., Cambridge Univ. Press, Cambridge, UK.

Hanel, G., and Zankl, B. (1979) Aerosol size and relative humidity: Water uptake by mixtures of salts, *Tellus* **31**, 478–486.

Harrison, L. P. (1965) *Humidity and Moisture*, Vol. 3A, Prentice-Hall, Englewood Cliffs, NJ.

Jaecker-Voirol, A., and Mirabel, P. (1989) Heteromolecular nucleation in the sulfuric acid-water system, *Atmos. Environ.* **23**, 2053–2057.

Junge, C. (1952) Die konstitution des atmospharischen aerosols, *Ann. Meteorol.* **5**, 1–55.

Kiang, C. S., Stauffer, D., Mohnen, V. A., Bricard, J., and Vigla, D. (1973) Heteromolecular nucleation theory applied to gas-to-particle conversion, *Atmos. Environ.* **7**, 1279–1283.

Kim, Y. P., Seinfeld, J. H., and Saxena, P. (1993) Atmospheric gas-aerosol equilibrium: I. Thermodynamic model, *Aerosol Sci. Technol.* **19**, 157–181.

Kulmala, M., and Laaksonen, A. (1990) Binary nucleation of water sulfuric acid system. Comparison of classical theories with different H$_2$SO$_4$ saturation vapor pressures, *J. Chem. Phys.* **93**, 696–701.

Laaksonen, A., Talanquer, V., and Oxtoby, D. W. (1995) Nucleation-measurements, theory, and atmospheric applications, *Ann. Rev. Phys. Chem.* **46**, 489–524.

Liu, B. Y. H., and Levi, J. (1980) Generation of submicron sulfuric acid aerosol by vaporization and condensation, in *Generation of Aerosols and Facilities for Exposure Experiments*, K. Willeke, ed., Ann Arbor Science, Ann Arbor, MI, pp. 317–336.

Martin, S. T. (2000) Phase transitions of aqueous atmospheric particles, *Chem. Rev.* **100**, 3403–3453.

Martin, S. T., Schlenker, J. C., Malinowski, A., Hung H. M., and Rudich, Y. (2003) Crystallization of atmospheric sulfate-nitrate-ammonium particles, *Geophys. Res. Lett.* **30**, 2102 (doi: 10.1029/2003GL017930).

Mirabel, P. and Katz, J. L. (1974) Binary homogeneous nucleation as a mechanism for the formation of aerosols, *J. Phys. Chem.* **60**, 1138–1144.

Nenes, A., Pilinis, C., and Pandis, S. N. (1998) ISORROPIA: A new thermodynamic model for multiphase multicomponent inorganic aerosols, *Aquatic Geochem.* **4**, 123–152.

Pilinis, C., and Seinfeld, J. H. (1987) Continued development of a general equilibrium model for inorganic multicomponent atmospheric aerosols, *Atmos. Environ.* **21**, 2453–2466.

Pilinis, C., Seinfeld, J. H., and Grosjean, D. (1989) Water content of atmospheric aerosols, *Atmos. Environ.* **23**, 1601–1606.

Potukuchi, S. and Wexler, A. S. (1995a) Identifying solid-aqueous phase transitions in atmospheric aerosols. I. Neutral acidity solutions, *Atmos. Environ.* **29**, 1663–1676.

Potukuchi, S. and Wexler, A. S. (1995b) Identifying solid-aqueous phase transitions in atmospheric aerosols. II. Acidic solutions, *Atmos. Environ.* **29**, 3357–3364.

Richardson, C. B., and Spann, J. F. (1984) Measurement of the water cycle in a levitated ammonium sulfate particle, *J. Aerosol Sci.* **15**, 563–571.

Roedel, W. (1979) Measurement of sulfuric acid saturation vapor pressure: Implications for aerosols formation by heteromolocular nucleation, *J. Aerosol Sci.* **10**, 175–186.

Saxena, P., Seigneur, C., Hudischewskyj, A. B., and Seinfeld J. H. (1986) A comparative study of equilibrium approaches to the chemical characterization of secondary aerosols, *Atmos. Environ.* **20**, 1471–1484.

Shaw, M. A., and Rood, M. J. (1990) Measurement of the crystallization humidities of ambient aerosol particles, *Atmos. Environ.* **24A**, 1837–1841.

Stelson, A. W., and Seinfeld, J. H. (1981) Chemical mass accounting of urban aerosol, *Environ. Sci. Technol.* **15**, 671–679.

Stelson, A. W., and Seinfeld, J. H. (1982a) Relative humidity and temperature dependence of the ammonium nitrate dissociation constant, *Atmos. Environ.* **16**, 983–993.

Stelson, A. W., and Seinfeld, J. H. (1982b) Relative humidity and pH dependence of the vapor pressure of ammonium nitrate-nitric acid solutions at 25°C, *Atmos. Environ.* **16**, 993–1000.

Stelson, A. W., and Seinfeld, J. H. (1982c) Thermodynamic prediction of the water activity, NH_4NO_3 dissociation constant, density and refractive index for the NH_4NO_3-$(NH_4)_2SO_4$-H_2O system at 25°C, *Atmos. Environ.* **16**, 2507–2514.

Stelson, A. W., Bassett, M. E., and Seinfeld, J. H. (1984) Thermodynamic equilibrium properties of aqueous solutions of nitrate, sulfate, and ammonium, in *Chemistry of Particles, Fogs, and Rain*, J. L. Durham, ed., Butterworth, Boston, pp. 1–52.

Stokes, R. H. and Robinson, R. A. (1966) Interactions in aqueous nonelectrolyte solutions. I. Solute-solvent equilibria, *J. Phys. Chem.* **70**, 2126–2130.

Tang, I. N. (1980) On the equilibrium partial pressures of nitric acid and ammonia in the atmosphere, *Atmos. Environ.* **14**, 819–828.

Tang, I. N., and Munkelwitz, H. R. (1993) Composition and temperature dependence of the deliquescence properties of hygroscopic aerosols, *Atmos. Environ.* **27A**, 467–473.

Wagman, D. D., Evans, W. H., Halow, I., Parker, V. B., Bailey, S. M., and Schumm, R. H. (1966) *Selected Values of Chemical Thermodynamic Properties*, National Bureau of Standards Technical Note 270, U.S. Department of Commerce, Washington, DC.

Wexler, A., and Hasegawa, S. (1954) Relative humidity-temperature relationships of some unsaturated salt solutions in the temperature range 0 to 50°C, *J. Res. Nat. Bur. Std.* **53**, 19–26.

Wexler, A. S., and Seinfeld, J. H. (1991) Second-generation inorganic aerosol model, *Atmos. Environ.* **25A**, 2731–2748.

Zdanovskii, A. B. (1948) New methods for calculating solubilities of electrolytes in multicomponent systems, *Zhur. Fiz. Kim.* **22**, 1475–1485.

11 Nucleation

Nucleation plays a fundamental role whenever condensation, precipitation, crystallization, sublimation, boiling, or freezing occur. A transformation of a phase α, say, a vapor, to a phase β, say, a liquid, does not occur the instant the free energy of β is lower than that of α. Rather, small nuclei of β must form initially in the α phase. This first step in the phase transformation, the nucleation of clusters of the new phase, can actually be very slow. For example, at a relative humidity of 200% at 20°C (293 K), far above any relative humidity achieved in the ambient atmosphere, the rate at which water droplets nucleate homogeneously is about 10^{-54} droplets per cm^3 per second. Stated differently, it would take about 10^{54} s (1 year is $\sim 3 \times 10^7$ s) for one droplet to appear in 1 cm^3 of air. Yet, we know that droplets are readily formed in air at relative humidities only slightly over 100%. This is a result of the fact that water nucleates on foreign particles much more readily than it does on its own. Once the initial nucleation step has occurred, the nuclei of the new phase tend to grow rather rapidly. Nucleation theory attempts to describe the rate at which the first step in the phase transformation process occurs—the rate at which the initial very small nuclei appear. Whereas nucleation can occur from a liquid phase to a solid phase (crystallization) or from a liquid phase to a vapor phase (bubble formation), our interest will be in nucleation of trace substances and water from the vapor phase (air) to the liquid (droplet) or solid phase.

Nucleation can occur in the absence or presence of foreign material. *Homogeneous nucleation* is the nucleation of vapor on embryos comprised of vapor molecules only, in the absence of foreign substances. *Heterogeneous nucleation* is the nucleation on a foreign substance or surface, such as an ion or a solid particle. In addition, nucleation processes can be *homomolecular* (involving a single species) or *heteromolecular* (involving two or more species). Consequently, four types of nucleation processes can be identified:

1. Homogeneous–homomolecular: self-nucleation of a single species. No foreign nuclei or surfaces involved.

2. Homogeneous–heteromolecular: self-nucleation of two or more species. No foreign nuclei or surfaces involved.

3. Heterogeneous–homomolecular: nucleation of a single species on a foreign substance.

4. Heterogeneous–heteromolecular: nucleation of two or more species on a foreign substance.

Atmospheric Chemistry and Physics: From Air Pollution to Climate Change, Second Edition, by John H. Seinfeld and Spyros N. Pandis. Copyright © 2006 John Wiley & Sons, Inc.

Homogeneous nucleation occurs in a supersaturated vapor phase. The degree of supersaturation of a solute A in air at temperature T is defined by the *saturation ratio*

$$S = p_A/p_A^s(T) \tag{11.1}$$

where p_A is the partial pressure of A and $p_A^s(T)$ is the saturation vapor pressure[1] of A in equilibrium with its liquid phase at temperature T. $S < 1$ for subsaturated vapor, $S = 1$ for saturated vapor, and $S > 1$ for supersaturated vapor. The saturation ratio can also be defined in terms of the molecular number concentration of species A as $S = N_A/N_A^s(T)$, where N_A is the molecular number concentration of A in the gas phase, and N_A^s is the molecular number concentration in a saturated vapor at equilibrium. The superscript s will be used to denote conditions at saturation $(S = 1)$.

An unsaturated or saturated vapor may become supersaturated by undergoing various thermodynamic processes, such as isothermal compression, isobaric cooling, and adiabatic expansion. In the first of these processes the vapor temperature remains constant, whereas it decreases in the latter two processes.

In a dilute mixture of a vapor A in air at saturation $(S = 1)$, an instantaneous snapshot would show that nearly all molecules of A exist independently or in small clusters containing two, three, or possibly four molecules; larger clusters of solute molecules would be extremely rare. If the larger clusters, as rare as they are, could be followed in time, most would be found to be very short-lived; they grow rapidly and then disappear rapidly. Moreover, the concentration of independent solute molecules, so-called monomers, will generally be very much larger than the concentration of all other clusters combined. At saturation equilibrium the average concentrations of all clusters are constant. Thus, over time, any forward growth process, that is, the addition of monomers to a cluster, is matched by the corresponding reverse process, the loss of single molecules from a cluster.

At saturation, $S = 1$, nucleation of a stable phase of A from the gas phase does not occur; for nucleation to take place it is necessary that $S > 1$. (We will see later that when two vapor species A and B are nucleating together, their individual saturation ratios need not exceed 1.0 for nucleation to occur.) When the saturation ratio is raised above unity, there is an excess of monomer molecules over that at $S = 1$. These excess monomers bombard clusters and produce a greater number of clusters of larger size than exist at $S = 1$. If the value of S is sufficiently large, then sufficiently large clusters can be formed so that some clusters exceed a critical size, enabling them to grow rapidly to form a new phase. The so-called critical cluster will be seen to be of a size such that its rate of growth is equal to its rate of decay. Clusters that fluctuate to a size larger than the critical size will likely continue to grow to macroscopic size, whereas those smaller than the critical size most likely will shrink. The nucleation rate is the net number of clusters per unit time that grow past the critical size.

The goal of this chapter is to present a comprehensive treatment of nucleation processes in the atmosphere. There is ample evidence for the occurrence of heterogeneous nucleation (the formation of water droplets is the most obvious example); the presence of

---

[1]In this chapter it will prove to be advantageous to use p_A^s as the saturation vapor pressure rather than p_A° as used up to this point.

homogeneous nucleation is less easy to identify. Identifying homogeneous nucleation processes in the atmosphere continues to stand as an important research issue in atmospheric science. We begin the chapter with the classical theories of single-component and binary homogeneous nucleation. This discussion includes a brief summary of laboratory techniques for measuring nucleation rates and of more sophisticated approaches than the classical nucleation theories. The binary system that is perhaps the most important to atmospheric science is sulfuric acid and water, and we therefore explore this system in some detail. Heterogeneous nucleation can occur on ions, insoluble particles, or soluble particles. Little is known about the importance of ion-induced nucleation in the atmosphere; in fact, only very recently have certain key theoretical aspects of ion-induced nucleation been elucidated. We present a short summary of the physics and current status of the theory of ion-induced nucleation but do not attempt to extrapolate these to the atmosphere. This area remains as one to be elucidated by further research. By far the most important heterogeneous nucleation processes in the atmosphere involve nucleation of vapor molecules on aerosol particles, both insoluble and soluble. Nucleation of water vapor on soluble particles is the process by which cloud and fog droplets form in the atmosphere. Without such nucleation the Earth would be a very different place than it is; it is an enormously important process in atmospheric physics. We delay the treatment of the nucleation of water on soluble particles until we discuss cloud physics in Chapter 17.

We have endeavored to present a self-contained development of the classical theory of homogeneous nucleation. Those readers who do not need to deal with the step-by-step derivations may wish to skip directly to the useful expressions for critical cluster size and nucleation rate. The critical cluster size for nucleation of a single substance is given by (11.35), and the corresponding nucleation rate is given by (11.47). Predicting the rate of binary nucleation of two components is not as straightforward as that for a single species. The classical theory of binary homogeneous nucleation is addressed in Section 11.5, with particular focus on the atmospherically important system of H_2SO_4–H_2O.

11.1 CLASSICAL THEORY OF HOMOGENEOUS NUCLEATION: KINETIC APPROACH

The theory of nucleation is based on a set of rate equations for the change of the concentrations of clusters of different sizes as the result of the gain and loss of molecules (monomers). It is usually assumed that the temperatures of all clusters are equal to that of the background gas. This is not strictly correct; phase change involves the evolution of heat. What this assumption implies is that clusters undergo sufficient collisions with molecules of the background gas so that they become thermally equilibrated on a timescale that is short compared to that associated with the gain and loss of monomers. Wyslouzil and Seinfeld (1992) extended the dynamic cluster balance to include an energy balance, but we will not consider this aspect here. Generally the assumption of an isothermal process is a reasonable one for atmospheric nucleation.

To develop the rate equation for clusters it is assumed that clusters grow and shrink via the acquisition or loss of single molecules. Cluster–cluster collision events are so rare that they can be ignored, as can those in which a cluster fissions into two or more clusters. Moreover, from the principle of microscopic reversibility, that is, that at equilibrium every forward process has to be matched by its corresponding reverse process, it follows that if

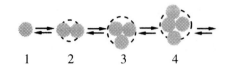

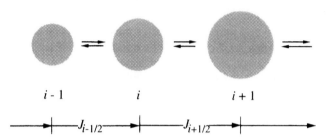

FIGURE 11.1 Cluster growth and evaporation processes.

clusters grow only by the addition of single molecules, evaporation also occurs one molecule at a time.

Let $N_i(t)$ be the number concentration of clusters containing i molecules (monomers) at time t. By reference to Figure 11.1, $N_i(t)$ is governed by the following rate equation

$$\frac{dN_i}{dt} = \beta_{i-1} N_{i-1}(t) - \gamma_i N_i(t) - \beta_i N_i(t) + \gamma_{i+1} N_{i+1}(t) \qquad (11.2)$$

where β_i is the forward rate constant for the collision of monomers with a cluster of size i (a so-called i-mer) and γ_i is the reverse rate constant for the evaporation of monomers from an i-mer.

Equation (11.2) provides the basis for studying transient nucleation. For example, if the monomer concentration is abruptly increased at $t = 0$, what is the time-dependent development of the cluster distribution? Physically, in such a case there is a transient period over which the cluster concentrations adjust to the perturbation in monomer concentration, followed eventually by the establishment of a pseudo-steady-state cluster distribution. Since the characteristic time needed to establish the steady-state cluster distribution is generally short compared to the timescale over which typical monomer concentrations might be changing in the atmosphere, we can assume that the distribution of clusters is always at a steady state corresponding to the instantaneous monomer concentration. There are instances, generally in liquid-to-solid phase transitions, where transient nucleation can be quite important (Shi et al. 1990), although we do not pursue this aspect here.

We define $J_{i+1/2}$ as the net rate (cm^{-3} s^{-1}) at which clusters of size i become clusters of size $i + 1$. This net rate is given by

$$J_{i+1/2} = \beta_i N_i - \gamma_{i+1} N_{i+1} \qquad (11.3)$$

If, for a given monomer concentration (or saturation ratio S), the cluster concentrations can be assumed to be in a steady state, then the left-hand side (LHS) equals zero in (11.2). At steady state from (11.2) we see that all the fluxes must be equal to a single, constant flux J:

$$J_{i+1/2} = J, \quad \text{all } i \qquad (11.4)$$

Let us define a quantity f_i by

$$\beta_i f_i = \gamma_{i+1} f_{i+1} \qquad (11.5)$$

and then, after setting $J_{i+1/2} = J$, divide (11.3) by (11.5)

$$\frac{J}{\beta_i f_i} = \frac{N_i}{f_i} - \frac{N_{i+1}}{f_{i+1}} \qquad (11.6)$$

where we have used

$$f_{i+1} = \frac{\beta_i}{\gamma_{i+1}} f_i \qquad (11.7)$$

Then, setting $f_1 = 1$ and summing (11.6) from $i = 1$ to some maximum value i_{max} gives

$$J \sum_{i=1}^{i_{max}} \frac{1}{\beta_i f_i} = N_1 - \frac{N_{i_{max}}}{f_{i_{max}}} \qquad (11.8)$$

Let us examine the behavior of f_i with i:

$$f_i = \frac{\beta_1}{\gamma_2} \frac{\beta_2}{\gamma_3} \cdots \frac{\beta_{i-1}}{\gamma_i}$$
$$= \prod_{j=1}^{i-1} \frac{\beta_j}{\gamma_{j+1}} \qquad (11.9)$$

The ratio β_{i-1}/γ_i is the ratio of the rate at which a cluster of size $i-1$ gains a monomer to that at which a cluster of size i loses a monomer. Under nucleation conditions ($S > 1$) this ratio must be larger than 1; that is, the forward rate must exceed the reverse rate. For large i, f_i must increase with i by a factor larger than 1 for each increment of i. Also, based on the relative cluster concentrations, $N_{i_{max}} < N_1$, since even under nucleation conditions monomers are still by far the most populous entities. Thus, by choosing i_{max} sufficiently large, the second term on the right-hand side (RHS) of (11.8) will be negligible compared to the first. Then

$$J = N_1 \left(\sum_{i=1}^{i_{max}} \frac{1}{\beta_i f_i} \right)^{-1} \qquad (11.10)$$

Since f_i is an increasing function of i, the summation in (11.10) can be extended to infinity, with convergence guaranteed by the rapid falloff of $1/f_i$. Thus

$$J = N_1 \left(\sum_{i=1}^{\infty} \frac{1}{\beta_i \, f_i} \right)^{-1} \tag{11.11}$$

The net flux J is the nucleation rate, and (11.11) is a direct expression for the nucleation rate as a function of the forward and reverse rate constants for the clusters.

11.1.1   The Forward Rate Constant β_i

From the kinetic theory of gases the number of collisions occurring between molecular entities A and B in a gas per unit time and unit volume of the gas is given by (3.6), with slightly different notation

$$Z_{AB} = \left(\frac{8\pi \, kT}{m_{AB}} \right)^{1/2} \sigma_{AB}^2 N_A N_B \tag{11.12}$$

where m_{AB}, the reduced mass, equals $m_A m_B / (m_A + m_B)$, and σ_{AB}, the collision diameter, equals $r_A + r_B$, the sum of the molecular radii. To determine the collision rate between a monomer and an i-mer, let A represent the monomer and B the i-mer:

$$Z_{1i} = (8\pi \, kT)^{1/2} \left(\frac{1}{m_1} + \frac{1}{m_i} \right)^{1/2} (r_1 + r_i)^2 N_1 N_i \tag{11.13}$$

The i-mer mass is equal to i times the monomer mass ($m_i = im_1$). Assuming the monomer and i-mer densities are equal, $v_i = iv_1$. Thus

$$Z_{1i} = \left(\frac{8\pi \, kT}{m_1} \right)^{1/2} \left(\frac{3v_1}{4\pi} \right)^{2/3} \left(1 + \frac{1}{i} \right)^{1/2} (1 + i^{1/3})^2 N_1 N_i \tag{11.14}$$

Since the surface area of a monomer

$$a_1 = 4\pi \left(\frac{3v_1}{4\pi} \right)^{2/3} \tag{11.15}$$

then

$$Z_{1i} = \left(\frac{8\pi \, kT}{m_1} \right)^{1/2} \frac{a_1}{4\pi} \left(1 + \frac{1}{i} \right)^{1/2} (1 + i^{1/3})^2 \, N_1 N_i \tag{11.16}$$

or

$$Z_{1i} = \left(\frac{kT}{2\pi \, m_1} \right)^{1/2} a_1 \left(1 + \frac{1}{i} \right)^{1/2} (1 + i^{1/3})^2 N_1 N_i \tag{11.17}$$

The number of collisions occurring between monomer and a single i-mer per unit time (the forward rate constant) is

$$\beta_i = \left(\frac{kT}{2\pi m_1}\right)^{1/2}\left(1+\frac{1}{i}\right)^{1/2}(1+i^{1/3})^2\, a_1 N_1 \tag{11.18}$$

which can be written in terms of the monomer partial pressure, $p_1 = N_1 kT$, as

$$\beta_i = \frac{p_1}{(2\pi m_1 kT)^{1/2}}\left(1+\frac{1}{i}\right)^{1/2}(1+i^{1/3})^2 a_1 \tag{11.19}$$

The expression for β_i is sometimes multiplied by an *accommodation coefficient* that is the fraction of monomer molecules impinging on a cluster that stick. The accommodation coefficient is generally an unknown function of cluster size. Its effect on the nucleation rate is relatively small since, as will be shown later, the nucleation rate depends on the ratio of forward to reverse rates and in this ratio the accommodation coefficient cancels. Henceforth, we will assume the accommodation coefficient is unity.

The saturation ratio, $S = p_1/p_1^s$, where p_1^s is the same as p_A^s in (11.1). Replacing p_1 by p_1^s in (11.19) gives the forward rate constant at saturation, β_1^s. Then, the forward rate constant under nucleation conditions $(S > 1)$ can be written as

$$\beta_i = S\beta_i^s \tag{11.20}$$

11.1.2 The Reverse Rate Constant γ_i

The reverse rate constant γ_i governs the rate at which molecules leave a cluster. This quantity is more difficult to evaluate theoretically than the forward rate constant β_i. What we do know is that, as long as the number concentration of monomer molecules is much smaller than that of the carrier gas, the rate at which a cluster loses monomers should depend only on the cluster size and temperature and be independent of monomer partial pressure. Thus the evaporation rate constant under nucleation conditions $(S > 1)$ should be identical to that in the saturated $(S = 1)$ vapor:

$$\gamma_i = \gamma_i^s \tag{11.21}$$

Two different approaches are commonly employed to obtain γ_i^s. One is based on the knowledge of the distribution of clusters in the saturated vapor, and the second is based on the Kelvin equation. These two approaches lead to similar results. In what immediately follows we will employ the cluster distribution at saturation to infer γ_i^s; the approach based on the Kelvin equation is given in Section 11.2.

It is important at this point to define three different cluster distributions that arise in nucleation theory. First is the saturated $(S = 1)$ equilibrium cluster distribution, N_i^s, which we will always denote with a superscript s. Second is the steady-state cluster distribution at

$S > 1$ and a constant net growth rate of J, N_i. At saturation $(S = 1)$ all $J_{i+1/2} = 0$, whereas at steady-state nucleation conditions all $J_{i+1/2} = J$. There is a third distribution that we will not explicitly introduce until the next section. It is the hypothetical, *equilibrium* distribution of clusters corresponding to $S > 1$. Thus it corresponds to all $J_{i+1/2} = 0$, but $S > 1$. Because of the constraint of zero flux, this third distribution is called the *constrained equilibrium* distribution, N_i^e. We will distinguish this distribution by a superscript e.

11.1.3   Derivation of the Nucleation Rate

From (11.9), f_i can be written as

$$f_i = \prod_{j=1}^{i-1} \frac{\beta_j}{\gamma_{j+1}} \tag{11.22}$$

Now, using (11.20) and (11.21), this becomes

$$f_i = S^{i-1} \prod_{j=1}^{i-1} \frac{\beta_j^s}{\gamma_{j+1}^s} \tag{11.23}$$

We now have a direct expression for the nucleation rate, (11.11), with f_i given by (11.23). The difficulty in using this equation lies, as noted, in properly evaluating the evaporation coefficients γ_i^s.

To approach this problem consider the formation of a dimer from two monomers at saturation $(S = 1)$ as a reversible chemical reaction:

$$A_1 + A_1 \rightleftharpoons A_2$$

This is not a true chemical reaction but rather a physical agglomeration of two monomers. The forward rate constant is β_1^s, and the reverse rate constant is γ_2^s. The ratio β_1^s/γ_2^s is just the equilibrium constant for the reaction. Similarly, the formation of a trimer can be written as a reversible reaction between a monomer and a dimer

$$A_1 + A_2 \rightleftharpoons A_3$$

with an equilibrium constant β_2^s/γ_3^s. The product

$$\left(\frac{\beta_1^s}{\gamma_2^s} \right) \left(\frac{\beta_2^s}{\gamma_3^s} \right)$$

is just the equilibrium constant for the overall reaction

$$3A_1 \rightleftharpoons A_3$$

Thus the product of rate constant ratios in (11.23) is just the equilibrium constant for the formation of an i-mer from i individual molecules under saturated equilibrium conditions:

$$iA_1 \rightleftharpoons A_i$$

If ΔG_i is the Gibbs free energy change for the "reaction" above, then (11.23) can be written as

$$f_i = S^{i-1} \exp(-\Delta G_i / kT) \tag{11.24}$$

We have thus expressed the quantity f_i in terms of the Gibbs free energy change for formation of a cluster of size i at saturation ($S = 1$). Now we need to determine ΔG_i.

It is at this point that the major assumption of classical nucleation theory is invoked. It is supposed that the i molecules are converted to a cluster by the following pathway. First, the molecules are transferred from the gas phase (at partial pressure p_1^s) to the liquid phase at the same pressure. The free energy change for this step is zero because the two phases are at saturation equilibrium at the same pressure. Next, a droplet of i molecules is "carved out" of the liquid and separated from it. This step involves the creation of an interface between the gas and liquid and does entail a free energy change. It is assumed that the free-energy change associated with this step is given by[2]

$$\Delta G_i = \sigma a_i \tag{11.25}$$

where σ is the surface tension of the solute A, and, as before, a_i is the surface area of a cluster of size i. If the cluster is taken to be spherical and to have the same volume per molecule v_1 as the bulk liquid, then

$$a_i = (36\pi)^{1/3} \, v_1^{2/3} i^{2/3} \tag{11.26}$$

The surface tension σ in (11.25) is taken as that of the bulk liquid monomer; thus it is assumed that clusters of a small number of molecules exhibit the same surface tension as the bulk liquid. This is the major assumption underlying classical nucleation theory and it has been given the name of the *capillarity approximation.*

If we define the dimensionless surface tension,

$$\theta = (36\pi)^{1/3} \, v_1^{2/3} \sigma / kT \tag{11.27}$$

then the Gibbs free energy change at saturation is just

$$\Delta G_i = \theta k T i^{2/3} \tag{11.28}$$

Combining (11.11), (11.20), (11.24), and (11.28) produces the following direct expression for the nucleation rate:

$$J = N_1 \left(\sum_{i=1}^{\infty} \frac{1}{\beta_i^s S^i \exp(-\theta i^{2/3})} \right)^{-1} \tag{11.29}$$

If we examine the terms in the denominator of the summation, some mathematical approximations are possible. The first term, β_i^s, increases as $i^{2/3}$ for large i [see (11.19)].

---

[2]The free-energy change associated with the formation of an i-mer is also referred to as the *reversible work* and is given the symbol W_i.

The second term, S^i, grows exponentially with i since $S > 1$. The third term, $\exp(-\theta i^{2/3})$, decreases exponentially as i increases, as $i^{2/3}$. The product, $S^i \exp(-\theta i^{2/3})$, initially decreases rapidly as i increases, reaches a minimum, and then begins to increase as S^i begins to dominate. The terms in the summation are largest near the minimum in the denominator. The minimum point of the denominator, i^*, can be found from solving

$$\frac{d}{di}\left(\beta_i^s S^i \exp(-\theta i^{2/3})\right) = 0 \tag{11.30}$$

In practice, because β_i^s varies slowly with i relative to the other two terms, with little loss of accuracy it is removed from the summation and replaced with its value at the minimum, $\beta_{i^*}^s$,

$$J = N_1 \left(\frac{1}{\beta_{i^*}^s} \sum_{i=1}^{\infty} \frac{1}{S^i \exp(-\theta i^{2/3})}\right)^{-1} \tag{11.31}$$

If we let

$$g_i = \theta i^{2/3} - i \ln S \tag{11.32}$$

then (11.31) becomes

$$J = N_1 \left(\frac{1}{\beta_{i^*}^s} \sum_{i=1}^{\infty} \frac{1}{\exp(-g_i)}\right)^{-1} \tag{11.33}$$

and (11.30) can be replaced by

$$\frac{d}{di} \exp(-g_i) = 0 \tag{11.34}$$

Setting the derivative equal to zero yields

$$
\begin{aligned}
i^* &= \left(\frac{2\theta}{3 \ln S}\right)^3 \\
&= \frac{32\pi}{3} \frac{v_1^2 \sigma^3}{(kT)^3 (\ln S)^3}
\end{aligned}
\tag{11.35}
$$

This is the critical cluster size for nucleation.

If i^* is sufficiently large, then the summation in (11.33) can be replaced with an integral

$$J = N_1 \beta_{i^*}^s \left(\int_0^{\infty} \exp(g(i))\, di\right)^{-1} \tag{11.36}$$

where the lower limit can be changed to zero. The function $g(i)$ can be expanded in a Taylor series around its minimum:

$$g(i) \simeq g(i^*) + \frac{dg}{di}\bigg|_{i^*} (i - i^*) + \frac{1}{2}\frac{d^2g}{di^2}\bigg|_{i^*} (i - i^*)^2 + \cdots \tag{11.37}$$

By definition of i^*, the second term on the RHS is zero, so

$$g(i) \simeq g(i^*) + \frac{1}{2}\frac{d^2g}{di^2}\bigg|_{i^*} (i - i^*)^2 \tag{11.38}$$

Then

$$\exp(g(i)) \simeq \exp(g(i^*)) \exp\left(\frac{1}{2}\frac{d^2g}{di^2}\bigg|_{i^*} (i - i^*)^2\right) \tag{11.39}$$

From (11.32)

$$\frac{d^2g}{di^2}\bigg|_{i^*} = -\frac{2}{9}\theta i^{*-4/3} \tag{11.40}$$

Then (11.36), with the aid of (11.39) and (11.40), becomes

$$J = N_1 \beta_{i^*}^s \exp(-g(i^*)) \left(\int_0^\infty \exp\left[-\frac{b}{2}(i - i^*)^2\right] di\right)^{-1} \tag{11.41}$$

where $b = \frac{2}{9}\theta i^{*-4/3}$. The integral in (11.41) has the form of the error function (see (8.38)). To convert the integral into the standard form of the error function, we define a new integration variable, $y = i - i^*$. In so doing, the lower limit of integration becomes $-i^*$. We again invoke the approximation that i^* is large and replace the lower limit by $-\infty$. We obtain

$$J = N_1 \beta_{i^*}^s \exp\left(-g(i^*)\right) \left(\frac{b}{2\pi}\right)^{1/2} \tag{11.42}$$

or

$$J = N_1 \beta_{i^*}^s \frac{1}{3}\left(\frac{\theta}{\pi}\right)^{1/2} i^{*-2/3} \exp\left[-\frac{4}{27}\frac{\theta^3}{(\ln S)^2}\right] \tag{11.43}$$

At saturation $(S = 1)$, the monomer partial pressure is p_1^s. At nucleation conditions $(S > 1)$, $p_1 = Sp_1^s$. Thus

$$\beta_{i^*}^s = \frac{1}{S}\beta_{i^*} \tag{11.44}$$

so

$$J = \frac{N_1}{S} \beta_{i*} \frac{1}{3} \left(\frac{\theta}{\pi}\right)^{1/2} i^{*-2/3} \exp\left[-\frac{4}{27} \frac{\theta^3}{(\ln S)^2}\right] \tag{11.45}$$

is the classical homogeneous nucleation theory expression for the nucleation rate.

The nucleation rate (11.45) can be written in terms of the original variables by using the definitions of β_i, θ, and i^*. In doing so, it is customary to approximate β_{i*} as follows:

$$\begin{aligned}
\beta_{i*} &= \frac{p_1}{(2\pi m_1 kT)^{1/2}} \left(1 + \frac{1}{i^*}\right)^{1/2} (1 + i^{*1/3})^2 a_1 \\
&\simeq \frac{p_1}{(2\pi m_1 kT)^{1/2}} i^{*2/3} a_1
\end{aligned} \tag{11.46}$$

With (11.46) and the definitions of θ and i^*, (11.45) becomes

$$J = \left(\frac{2\sigma}{\pi m_1}\right)^{1/2} \frac{v_1 N_1^2}{S} \exp\left[-\frac{16\pi}{3} \frac{v_1^2 \sigma^3}{(kT)^3 (\ln S)^2}\right] \tag{11.47}$$

Equation (11.47) is the classical homogeneous nucleation theory expression for the nucleation rate. It is the expression that we will use to predict nucleation rates of different pure substances as a function of saturation ratio.

The next section presents an alternate derivation of a classical homogeneous nucleation rate expression based on a somewhat different approach. The development in the next section provides a fuller appreciation of classical theory and its underlying assumptions, but the section may be skipped by those interested primarily in application of the theory. We note that several useful tables of physical properties appear in Section 11.2

11.2   CLASSICAL HOMOGENEOUS NUCLEATION THEORY: CONSTRAINED EQUILIBRIUM APPROACH

11.2.1   Free Energy of i-mer Formation

In the previous chapter we derived the Kelvin, or Gibbs–Thomson, equation [see (10.82) and (10.85)]

$$g_v - g_l = \mu_v - \mu_l = kT \ln\frac{p_A}{p_A^s} = \frac{2\sigma v_1}{R} \tag{11.48}$$

where $\mu_v - \mu_l$ is the difference in chemical potential between a molecule in the vapor and liquid phases. Thus the i-mer's vapor pressure exceeds that of the same substance at the same temperature over a flat surface. Letting $S = p_A/p_A^s$ (since the i-mer must exist in a vapor of partial pressure p_A) yields

$$\mu_v - \mu_l = kT \ln S \tag{11.49}$$

A partial pressure less than $p_A^s (S < 1)$ indicates $\mu_v < \mu_l$. In this case the vapor is stable and will not condense. If the partial pressure exceeds $p_A^s (S > 1)$, $\mu_v > \mu_l$, the i-mer will tend to grow at the vapor's expense since the system gravitates toward its lowest chemical potential.

A transfer of i molecules from the vapor phase forms an i-mer of radius r. The accompanying free energy change is

$$\Delta G_i = (\mu_l - \mu_v)i + 4\pi\sigma r^2 \tag{11.50}$$

Thus, from (11.49), and (11.50), we obtain

$$\Delta G_i = 4\pi\sigma r^2 - \frac{4\pi}{3}\frac{kT \ln S}{v_1} r^3 \tag{11.51}$$

The free energy change of i-mer formation contains two terms. The first is the free-energy increase as a result of the formation of the i-mer surface area. The second term is the free energy decrease from the change in chemical potential on going from the vapor to the condensed phase.

For small r, the increase in free energy resulting from formation of surface area $(\sim r^2)$ mathematically dominates the free-energy decrease from formation of the condensed phase $(\sim r^3)$. ΔG_i is sketched in Figure 10.10. Free energy increases for a subsaturated vapor $(S < 1)$ only as the i-mer radius is increased. On the other hand, the free energy of a supersaturated vapor $(S > 1)$ initially increases until the bulk free energy decrease overtakes the surface free-energy increase. ΔG_i achieves a maximum at a certain critical i-mer radius r^* (and corresponding i^*),

$$r^* = \frac{2\sigma v_1}{kT \ln S} \tag{11.52}$$

The corresponding value of the number of molecules at the critical size is given by (11.35).

The effective vapor pressure on an i-mer with critical radius r^* is the prevailing partial pressure p_A. This critical size i-mer is therefore in equilibrium with the vapor. It is, however, only a metastable equilibrium; the i^*-mer sits precariously on the top of the ΔG curve. If another vapor molecule attaches to it, it will plunge down the curve's right-hand side and grow. If a molecule evaporates from the i^*-mer, it will slide back down the left-hand side and evaporate.

The free energy barrier height is obtained by substituting (11.52) into (11.51):

$$\Delta G^* = \frac{4\pi}{3}\sigma r^{*2}$$
$$= \frac{16\pi}{3}\frac{v_1^2 \sigma^3}{(kT \ln S)^2} \tag{11.53}$$

Increasing the saturation ratio S decreases both the free-energy barrier and the critical i-mer radius. From (11.35) we already knew that it decreases the value of i^*.

Equation (11.52) relates the equilibrium radius of a droplet of pure substance to the physical properties of the substance, σ and v_1, and to the saturation ratio S of its

TABLE 11.1   Critical Number and Radius for Water Droplets

	$T = 273$ K[a]		$T = 298$ K[b]	
S	r^*, Å	i^*	r^*, Å	i^*
1	∞	∞	∞	∞
2	17.3	726	15.1	482
3	10.9	182	9.5	121
4	8.7	87	7.6	60
5	7.5	58	6.5	39

[a] $\sigma = 75.6$ dyn cm^{-1}; $v_1 = 2.99 \times 10^{-23}$ cm^3 molecule^{-1}.
[b] $\sigma = 72$ dyn cm^{-1}; $v_1 = 2.99 \times 10^{-23}$ cm^3 molecule^{-1}.

environment. Equation (11.52) can be rearranged so that the equilibrium saturation ratio is given as a function of the radius of the drop:

$$\ln S = \frac{2\sigma v_1}{kTr^*} \tag{11.54}$$

Expressed in this form, the equation is frequently referred to as the Kelvin equation.

Table 11.1 gives i^* and r^* for water at $T = 273$ K and 298 K as a function of S. As expected, we see that as S increases both i^* and r^* decrease. Table 11.2 contains surface tension and density data for five organic molecules, and values of i^* and r^* for these five substances at $T = 298$ K are given in Table 11.3. The critical radius r^* depends on the product σv_1. For the five organic liquids $\sigma < \sigma_{\text{water}}$ but $v_1 > v_{1\text{water}}$. The organic surface tensions are about one-third that of water, and their molecular volumes range from 3 to 6 times that of water. The product σv_1 for ethanol is approximately the same as that of water, and consequently we see that the r^* values for the two species are virtually identical. Since i^* involves an additional factor of v_1, even though the r^* values coincide for ethanol and water, the critical numbers i^* differ appreciably because of the large size of the ethanol molecule.

11.2.2   Constrained Equilibrium Cluster Distribution

Equation (11.51) can be written in terms of the dimensionless surface tension θ as

$$\Delta G_i = kT\theta i^{2/3} - ikT \ln S \tag{11.55}$$

TABLE 11.2   Surface Tensions and Densities of Five Organic Species at 298 K

Species	M	ρ_ℓ, g cm^{-3}	$v_1 \times 10^{23}$, cm^3 molecule^{-1}	σ, dyn cm^{-1} [a]
Acetone (C_3H_6O)	58.08	0.79	12.25	23.04
Benzene (C_6H_6)	78.11	0.88	14.75	28.21
Carbon tetrachloride (CCl_4)	153.82	1.60	16.02	26.34
Ethanol (C_2H_6O)	46.07	0.79	9.694	22.14
Styrene (C_8H_8)	104.2	0.91	19.10	31.49

[a] The traditional unit for surface tension is dyn cm^{-1}. This unit is equivalent to g s^{-2}.
Source: Handbook of Chemistry and Physics, 56th ed.

TABLE 11.3   Critical Number and Radius for Five Organic Species at 298 K

Species		2	3	4	5
			S		
Acetone	i^*	265	67	33	21
	r^* (Å)	19.8	12.5	9.9	8.5
Benzene	i^*	706	177	88	56
	r^* (Å)	29.2	18.4	14.6	12.6
Carbon tetrachloride	i^*	678	170	85	54
	r^* (Å)	29.6	18.7	14.8	12.7
Ethanol	i^*	147	37	18	12
	r^* (Å)	15.1	9.5	7.5	6.5
Styrene	i^*	1646	413	206	132
	r^* (Å)	42.2	26.6	21.1	18.2

Recall that the constrained equilibrium cluster distribution is the hypothetical equilibrium cluster distribution at a saturation ratio $S > 1$. The constrained equilibrium cluster distribution obeys the usual Boltzmann distribution

$$N_i^e = N_1 \exp(-\Delta G_i/kT) \tag{11.56}$$

where N_1 is the total number of solute molecules in the system. Using (11.55) in (11.56)

$$N_i^e = N_1 \exp(-\theta i^{2/3} + i \ln S) \tag{11.57}$$

For small i and very large i, (11.57) departs from reality. When $i = 1$, N_i^e should be identical to N_1, but (11.57) does not produce this identity. The failure to approach the proper limit as $i \to 1$ is a consequence of the capillarity approximation and the specific form, $\theta i^{2/3}$, chosen for the surface free energy of an i-mer. It is not a fundamental problem with the theory. There have been several attempts to modify the formula for N_i^e so that N_i^e equals N_1 when $i = 1$ (Girshick and Chiu 1990; Wilemski 1995). Some of these take the form of replacing $i^{2/3}$ in (11.57) by $(i^{2/3} - 1)$ or $(i - 1)^{2/3}$. Such modifications can be regarded as making the surface tension σ an explicit function of i. We will not deal further with these corrections here. The mismatch at $i = 1$ turns out not to be a problem because we are generally interested in N_i^e for values of i quite a bit larger than 1.

At the other extreme, N_i^e cannot exceed N_1, but (11.57) predicts that at sufficiently large i, N_i^e can be greater than N_1. N_i^e and the actual cluster distribution N_i are sketched as a function of i in Figure 11.2. Equation (11.57) is no longer valid when the accumulated value of the product $i N_i^e$ begins to approach N_1 in magnitude.

Another consistency issue arises with (11.57). The law of mass action (Appendix 11) requires that the equilibrium constant for formation of an i-mer, $iA_1 \rightleftharpoons A_i$

$$K_i(T) = \frac{N_i^e}{(N_1)^i} \tag{11.58}$$

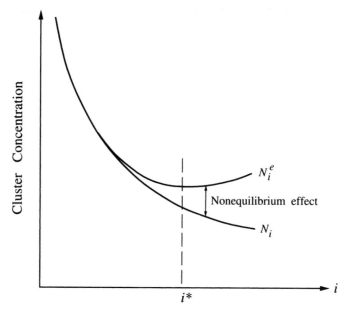

FIGURE 11.2 Constrained equilibrium and steady-state cluster distributions, N_i^e and N_i, respectively.

depend on i and T but not on N_1 or the total pressure if the gas behaves ideally. The distribution

$$N_i^e = N_1^s \exp(-\theta i^{2/3} + i \ln S) \tag{11.59}$$

where N_1^s is the monomer number density at saturation, satisfies mass action, in that

$$K_i(T) = (N_1^s)^{1-i} \exp(\theta(i - i^{2/3})) \tag{11.60}$$

is a function only of i and T. This correction (11.59), originally due to Courtney (1961), is equivalent to introducing the factor $(1/S)$ in (11.44).

Since i^* is at the maximum of ΔG_i, the N_i^e distribution should be a minimum at i^*. Physically, we do not expect that for $i > i^*$, the cluster concentrations should increase with increasing i. Thus N_i^e given by (11.57) is taken to be valid only up to i^*. Even though N_i^e as given by (11.57) is not accurate for $i \to 1$ and $i \to \infty$, it can be considered as accurate in the region close to and below i^*. Fortunately, it is precisely this region in which the nucleation flux is desired.

11.2.3   The Evaporation Coefficient γ_i

To determine the monomer escape frequency from an i-mer, assume that the i-mer is in equilibrium with the surrounding vapor. Then the i-mer vapor pressure equals the system vapor pressure. By (11.48), we obtain

$$p_1 = p_1^s \exp\left(\frac{2\sigma v_1}{kTr}\right) \tag{11.61}$$

At equilibrium the escape frequency equals the collision frequency. Thus using (11.19)

$$\gamma_i = \beta_i$$

$$= \frac{p_1}{(2\pi m_1 kT)^{1/2}} \left(1 + \frac{1}{i}\right)^{1/2} (1 + i^{1/3})^2 a_i \tag{11.62}$$

$$= \frac{p_1^s}{(2\pi m_1 kT)^{1/2}} \left(1 + \frac{1}{i}\right)^{1/2} (1 + i^{1/3})^2 a_i \exp\left(\frac{2\sigma v_1}{kTr}\right)$$

γ_i is thus a function of p_1^s, m_1, σ, and v_1 and is independent of the actual species partial pressure. Thus (11.62) holds at any value of the saturation ratio.

By combining (11.62) with the definition of i^*, γ_i can be written as

$$\gamma_i = \frac{p_1^s}{(2\pi m_1 kT)^{1/2}} \left(1 + \frac{1}{i}\right)^{1/2} (1 + i^{1/3})^2 a_1 S^{\{(i^*/i)^{1/3}\}} \tag{11.63}$$

Taking γ_i from (11.63), based on equilibrium, and using (11.19) with $p_1 = Sp_1^s$, we can form the nonequilibrium ratio of γ_i to β_i as

$$\frac{\gamma_i}{\beta_i} = S^{\{(i^*/i)^{1/3} - 1\}} \tag{11.64}$$

We see from (11.64) that subcritical i-mers $(i < i^*)$ tend to evaporate since their evaporation frequency exceeds their collision frequency. Critical size i-mers $(i = i^*)$ are in equilibrium with the surrounding vapor $(\gamma_i = \beta_i)$. Supercritical clusters $(i > i^*)$ grow since monomers tend to condense on them faster than monomers evaporate.

Figure 11.3 shows γ_i and β_i as a function of i for various values of S. The evaporation and growth curves intersect at the i^* values for the particular value of S. As expected, i^* decreases as S increases.

11.2.4 Nucleation Rate

At constrained equilibrium

$$\beta_i N_i^e = \gamma_{i+1} N_{i+1}^e \tag{11.65}$$

The ratio of equilibrium cluster concentrations between neighboring clusters is

$$\frac{N_{i+1}^e}{N_i^e} = \frac{\beta_i}{\gamma_{i+1}} \tag{11.66}$$

Using (11.19) and (11.63), we obtain

$$\frac{N_{i+1}^e}{N_i^e} = \frac{\left(1 + \frac{1}{i}\right)^{1/2} (1 + i^{1/3})^2}{\left(1 + \frac{1}{i+1}\right)^{1/2} (1 + (i+1)^{1/3})^2} S^{-\{[i^*/(i+1)]^{1/3} - 1\}} \tag{11.67}$$

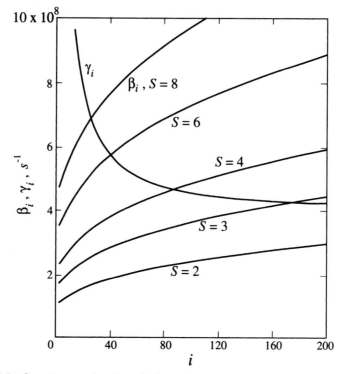

FIGURE 11.3 β_i and γ_i as a function of i for various values of the saturation ratio S. The two quantities are equal at the critical i values.

When (11.67) is successively multiplied from $i = i$ to $i = 1$, it yields

$$\frac{N_1^e}{N_i^e} = \frac{2^{5/2}}{\left(1 + \frac{1}{i}\right)^{1/2}\left(1 + i^{1/3}\right)^2} \prod_{j=2}^{i} S^{-\left\{(i^*/j)^{1/3} - 1\right\}} \tag{11.68}$$

This equation is essentially an alternate way of writing (11.57) in terms of i^*. It differs slightly from (11.57) in that we have retained the $(1 + 1/i)^{1/2}$ and $(1 + i^{1/3})^2$ terms instead of approximating them simply by $i^{2/3}$. Figure 11.4 shows N_i^e/N_1 as a function of i for $S = 6$ and $i^* = 40$.

The constrained equilibrium cluster distribution, N_i^e, is based on a supposed equilibrium existing for $S > 1$. In actuality, when $S > 1$, a nonzero cluster current J exists, which is the nucleation rate. Let the actual steady-state cluster distribution be denoted by N_i. When i_{max} is sufficiently larger than i^*, the actual cluster distribution approaches zero,

$$\lim_{i \to i_{max}} N_i = 0 \tag{11.69}$$

The traditional condition used at $i = 1$ is that the constrained equilibrium distribution and the actual distribution start from the same value:

$$\lim_{i \to 1} \frac{N_i}{N_i^e} = 1 \tag{11.70}$$

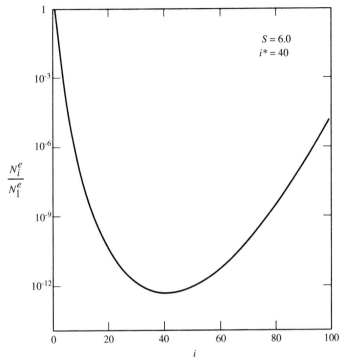

FIGURE 11.4 N_i^e/N_1 as a function of i for $S = 6$ and $i^* = 40$. The increase for $i > i^*$ is physically unrealistic.

From (11.3) and (11.65), we have

$$J = \beta_i N_i - \gamma_{i+1} N_{i+1}$$
$$= \beta_i N_i^e \left[\frac{N_i}{N_i^e} - \frac{N_{i+1}}{N_{i+1}^e} \right] \tag{11.71}$$

Summing successive applications of (11.71) from $i = i_{\max} - 1$ to $i = 1$ yields

$$\sum_{j=1}^{i_{\max}-1} \frac{J}{\beta_j N_j^e} = \frac{N_1}{N_1^e} - \frac{N_{i_{\max}}}{N_{i_{\max}}^e} \tag{11.72}$$

Using (11.69) and (11.70), we obtain

$$J = \left(\sum_{j=1}^{i_{\max}-1} \frac{1}{\beta_j N_j^e} \right)^{-1} \tag{11.73}$$

Thus the unknown steady-state cluster distribution N_i, disappears, allowing the nucleation rate to be expressed in terms of the constrained equilibrium distribution N_i^e. We see that (11.73) is equivalent to (11.11), in that (11.65), which defines the constrained equilibrium distribution N_i^e, is identical to (11.5) that defines f_i.

We have demonstrated the essential equivalence of the two approaches to deriving the nucleation rate. We can obtain a corresponding expression to (11.47) for the nucleation rate from (11.73) and the expressions for β_i, (11.19), and N_i^e, (11.68). When expressions for β_i and N_i^e are substituted into (11.73), the following equation is obtained for the nucleation rate:

$$J = a_1 \frac{N_1^2}{S} \left(\frac{kT}{2\pi m_1} \right)^{1/2} 2^{5/2} \left(1 + \sum_{j=2}^{i_{max}-1} \prod_{k=2}^{j} S^{\{(i^*/k)^{1/3}-1\}} \right)^{-1} \tag{11.74}$$

From the point of view of computationally implementing (11.74), J varies little with the value of i_{max} as long as $i_{max} \gg i^*$.

11.3 RECAPITULATION OF CLASSICAL THEORY

The classical theory of homogeneous nucleation dates back to pioneering work by Volmer and Weber (1926), Farkas (1927), Becker and Döring (1935), Frenkel (1955), and Zeldovich (1942). The expression for the constrained equilibrium concentration of clusters (11.57) dates back to Frenkel. The classical theory is based on a blend of statistical and thermodynamic arguments and can be approached from a kinetic viewpoint (Section 11.1) or that of constrained equilibrium cluster distributions (Section 11.2). In either case, the defining crux of the classical thoery is reliance on the capillarity approximation wherein bulk thermodynamic properties are used for clusters of all sizes.

Let us summarize what we have obtained in our development of homogeneous nucleation theory. In order for homogeneous nucleation to occur, it is necessary that clusters surmount the free-energy barrier at $i = i^*$. The higher the saturation ratio S, the smaller the critical cluster size i^*. At saturation conditions ($S = 1$) clusters containing more than a very few molecules are exceedingly rare and nucleation does not occur. At supersaturated conditions ($S > 1$) the saturated equilibrium cluster distribution N_i^s is perturbed to produce a steady-state cluster distribution N_i. The nucleation flux J results from a chain of bombardments of vapor molecules on clusters that produces a net current through the cluster distribution. In deriving the nucleation rate expression from the thermodynamic approach, it is useful to define a hypothetical equilibrium cluster distribution, called the "constrained equilibrium distribution."

To maintain the nucleation rate at a constant value J it is necessary that the saturation ratio of vapor be maintained at a constant value.[3] Without outside reinforcement of the vapor concentration, the saturation ratio will eventually fall as a result of depletion of vapor molecules to form stable nuclei. The case in which the vapor concentration is augmented by a source of fresh vapor is referred to as nucleation with a continuously reinforced vapor.

Let us evaluate the homogeneous nucleation rate for water as a function of the saturation ratio S. To do so, we use (11.47). Table 11.4 gives the nucleation rate at 293 K for saturation ratios ranging from 2 to 10. By comparing Tables 11.1 and 11.4, the effect of temperature on i^* can be seen also. We see that the nucleation rate of water varies over 70

---

[3]It is sometimes supposed that to maintain a constant S, i_{max} monomers reenter the system at the same rate that i_{max}-mers leave. The i_{max}-mers are said to be broken up by Maxwell demons.

TABLE 11.4 Homogeneous Nucleation Rate and Critical Cluster Size for Water at $T = 293$ K[a]

S	i^*	$J(cm^{-3}s^{-1})$
2	523	5.02×10^{-54}
3	131	1.76×10^{-6}
4	65	1.05×10^{6}
5	42	1.57×10^{11}
6	30	1.24×10^{14}
7	24	8.99×10^{15}
8	19	1.79×10^{17}
9	16	1.65×10^{18}
10	14	9.17×10^{18}

[a] $\sigma = 72.75 \, dyn \, cm^{-1}$
$v_1 = 2.99 \times 10^{-23} \, cm^3 \, molecule^{-1}$
$m_1 = 2.99 \times 10^{-23} \, g \, molecule^{-1}$
$p^s_{H_2O} = 2.3365 \times 10^4 \, g \, cm^{-1} \, s^{-2}$

orders of magnitude from $S = 2$ to $S = 10$. As S increases, the critical cluster size i^* at 293 K decreases from 523 at $S = 2$ to 14 at $S = 10$. From Tables 11.1 and 11.4 we see that at $S = 5$, i^* varies from 58 to 42 to 39 as T increases from 273 K to 293 K to 298 K.

The question arises as to how closely the predictions from (11.47) and (11.74) agree. Derivations of these two expressions were based on similar assumptions, but these expressions do not agree exactly because of approximations that are made in each derivation. It is useful to compare nucleation rates of water predicted by (11.74) with those in Table 11.4 determined from (11.47):

S	(11.47)	(11.74)
3	1.76×10^{-6}	4.52×10^{-3}
5	1.57×10^{11}	4.03×10^{14}
7	8.99×10^{15}	2.06×10^{19}
9	1.65×10^{18}	3.26×10^{21}

We see that (11.74) predicts a nucleation rate about 3 orders of magnitude larger than that predicted by (11.47). In most areas of science a discrepancy of prediction of 3 orders of magnitude between two equations representing the same phenomenon would be a cause for serious concern. As we noted, nucleation rates of water vary over 70 orders of magnitude for the range of saturation ratios considered; in addition, the rate expressions are extremely sensitive to small changes in parameters. Thus, a difference of 3 orders of magnitude, while certainly not trivial, is well within the sensitivity of the rate expressions. As we noted at the end of Section 11.1, we will use (11.47) henceforth as the basic classical theory expression for the homogeneous nucleation rate of a single substance.

11.4 EXPERIMENTAL MEASUREMENT OF NUCLEATION RATES

The traditional method of studying gas–liquid nucleation involves the use of a cloud chamber. In such a chamber the saturation ratio S is changed until, at a given temperature, droplet formation is observable. Because once clusters reach the critical size for nucleation, subsequent droplet growth is rapid, the rate of formation of macroscopically observable droplets is assumed to be that of formation of critical nuclei. In such a device it is difficult to measure the actual rate of nucleation because the nucleation rate changes so rapidly with S. J is very small for S values below a critical saturation ratio S_c and very large for $S > S_c$. Thus what is actually measured is the value of S_c, defined rather arbitrarily by the point where the rate of appearance of droplets is $1\,\mathrm{cm}^{-3}\,\mathrm{s}^{-1}$.

11.4.1   Upward Thermal Diffusion Cloud Chamber

An improvement on the cloud chamber is the upward thermal diffusion cloud chamber (Katz 1970; Heist 1986; Hung et al. 1989). In this device a liquid pool at the bottom of a container is heated from below to vaporize it partially. The upper surface of the container is held at a lower temperature, establishing a temperature gradient in the vapor between bottom and top. The total pressure of gas in the container (background carrier gas plus nucleating vapor) is approximately uniform. However, the partial pressure of the nucleating vapor p_A decreases linearly with height in the chamber because of the diffusion gradient between the pool of liquid at the bottom and the top of the chamber. The saturation vapor pressure of the vapor, p_A^s, depends exponentially on temperature, and therefore as the temperature decreases with height, the saturation vapor pressure decreases rapidly. The result is that the saturation ratio, $S = p_A/p_A^s$, has a rather sharp peak at a height about three-fourths of the way up in the chamber, and nucleation occurs only in the narrow region where S is at its maximum (Figure 11.5). The temperatures at the bottom and top of the chamber are then adjusted so that the nucleation rate at the point of maximum S is about $1\,\mathrm{cm}^{-3}\,\mathrm{s}^{-1}$. As droplets form and grow they fall under the influence of gravity; continued evaporation of the liquid pool serves to replenish the vapor molecules of the nucleating substance so that the experiment can be run in steady state. Nucleation rates that can be measured in the upward thermal diffusion cloud chamber vary roughly from 10^{-4} to $10^3\,\mathrm{cm}^{-3}\,\mathrm{s}^{-1}$.

11.4.2   Fast Expansion Chamber

Another effective technique to study vapor-to-liquid nucleation is an expansion cloud chamber (Schmitt 1981, 1992; Wagner and Strey 1981; Strey and Wagner 1982;

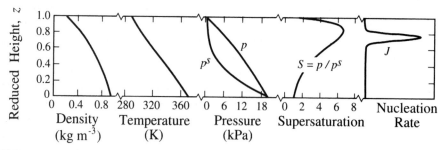

FIGURE 11.5   Upward thermal diffusion cloud chamber. Typical cloud chamber profiles of total gas-phase density, temperature, partial pressure p (*n*-nonane in this example), equilibrium vapor pressure p^s, saturation ratio S, and nucleation rate J, as a function of dimensionless chamber height at $T = 308.4$ K, $S = 6.3$, and total $p = 108.5$ kPa. (Reprinted with permission from Katz, J. L., Fisk, J. A. and Chakarov, V. M. Condensation of a Supersaturated vapor IX. Nucleation of ions, *J. Chem. Phys.* **101**. Copyright 1994 American Institute of Physics.)

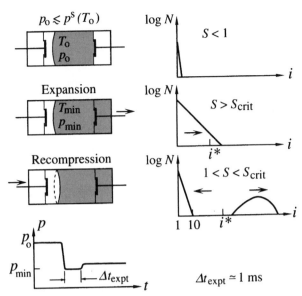

FIGURE 11.6 Fast expansion chamber. Initially undersaturated $(S<1)$ at p_0, T_0; for a time interval of typically 1 ms supercritically supersaturated $(S > S_{crit})$, where S_{crit} is the saturation ratio corresponding to a threshold nucleation rate, at p_{min}, T_{min}; finally still supersaturated, but subcritical $(1 < S < S_{crit})$ to permit droplet growth only. Sketch at right indicates the distribution of clusters with size at the three stages of operation.

Strey et al. 1994). Initially an enclosed volume saturated with the vapor under consideration is allowed to come to thermal equilibrium. The volume is then abruptly expanded, and the carrier gas and vapor decrease in temperature (Figure 11.6). While the expansion slightly decreases the pressure (and density) of the vapor, the partial pressure is still much greater than the saturation vapor pressure at the lower temperature. The vapor is thus supersaturated, and quite large supersaturations can be achieved with the expansion technique. After expansion, the chamber is immediately recompressed to an intermediate pressure (and saturation ratio). Typical durations of the expansion pulse are the order of milliseconds. In order for a condition of steady-state nucleation to be achieved, the time needed to establish the steady-state cluster distribution corresponding to the value of S at the peak must be significantly shorter than the time the system is held at its highest supersaturation. Since J is such a strong function of S, the immediate recompression quenches the nucleation that occurs during the brief expansion pulse. The temperature at the peak supersaturation is calculated from the properties of the carrier gas and the vapor based on the initial temperature, the initial pressure, and the pressure after expansion. The expansion is adiabatic, because of the short duration of the expansion. After recompression the system remains supersaturated and particles nucleated during the supercritical nucleation pulse grow by condensation. The growing drops are illuminated by a laser beam and their number concentration is measured by an appropriate optical technique. From the measured droplet concentration the nucleation rate during the supercritical nucleation pulse can be determined. The duration of the nucleation pulse (about 10^{-3} s) is short compared to characteristic times for growth of the nucleated droplets; thus nucleation is decoupled from droplet growth. The fast expansion chamber is capable of allowing measurement of nucleation rates up to $10^{10} \, \text{cm}^{-3} \, \text{s}^{-1}$.

A variant of the fast expansion method is the shock tube. This device consists of a high-pressure section containing a mixture of the vapor and the carrier gas and a low-pressure section separated from the high-pressure section by a thin diaphragm. An adiabatic expansion is induced by breaking the diaphragm. The expansion results in vapor supersaturation and nucleation. Simultaneously a shock wave starts to travel through the low-pressure gas. The shock wave reflects back from a constriction, causing a recompression and quenching nucleation. By proper design the nucleation pulse can be restricted to a duration of 0.1 ms (Peters and Paikert 1989). Nucleation rates are then determined from the resulting droplet number concentration using light scattering techniques.

11.4.3 Turbulent Mixing Chambers

Another method of nucleation measurement that differs from both diffusion and expansion chambers involves the rapid turbulent mixing of two gas streams (Wyslouzil et al. 1991a,b). This method is particularly suited to studies of binary nucleation. Two carrier gas-vapor streams are led to a device where rapid turbulent mixing takes place. The two-component vapor mixture is supersaturated and begins to nucleate immediately. The stream passes to a tube where nucleation may continue and the nucleated particles grow. Residence time in the flow tube is the order of seconds. When the nucleated particle concentration is sufficiently low, droplet growth does not deplete the vapor appreciably, and constant nucleation conditions can be assumed.

11.4.4 Experimental Evaluation of Classical Homogeneous Nucleation Theory

The experimental techniques outlined above have been used to study the nucleation of a wide range of substances (Heist and He 1994). The agreement between different methods is generally good for most substances, although there exist notable exceptions.

We present here only one example of experimentally measured nucleation rates, those for n-butanol. Viisanen and Strey (1994) measured homogeneous nucleation rates of n-butanol in argon in an expansion chamber as a function of saturation ratio in the temperature range 225–265 K. In this temperature range the equilibrium vapor pressure and surface tension of n-butanol are known with sufficient accuracy that a quantitative comparison of the observed nucleation rates with those predicted by classical theory could be performed. Figure 11.7 shows the measured nucleation rates (the data points) and the predictions of the classical theory (solid lines). Nucleation rates ranging from about 10^5 to $10^9 \, \text{cm}^{-3} \, \text{s}^{-1}$ were measured. To check for possible influence of the carrier gas, measurements were also carried out with helium and xenon as carrier gases. The authors did not observe a significant dependence of the nucleation rate on the carrier gas. From Figure 11.7 it is seen that the classical theory consistently underestimates nucleation rates by a maximum of about two orders of magnitude, exhibiting a quite similar temperature trend over the range of temperatures considered. The experimental nucleation rates in Figure 11.7 are reported accurate to within a factor of 2. Because of the steepness of the J–S curves, this error is equivalent to an uncertainty of 3% in saturation ratio.

Two effects are observed in homogeneous nucleation experiments for all substances. First, the nucleation rate is always a steep function of saturation ratio S. The second feature common to all systems is that the critical saturation ratio S_c decreases as T increases, and J increases as T increases at constant S. Also, critical nuclei become smaller as S increases and as T increases.

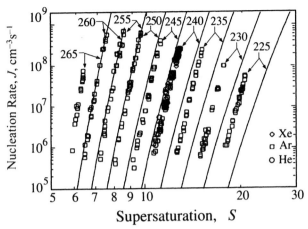

FIGURE 11.7 Nucleation rates measured in supersaturated n-butanol vapor as a function of saturation ratio for various temperatures ranging from 225 to 265 K (Viisanen and Strey 1994). Different symbols indicate different carrier gases at 240 K. Predictions of classical nucleation theory are shown by the lines. (Reprinted with permission from Viisanen, Y., and Strey, R., V. M. Homogeneous Nucleation Rates for n-Butanol, *J. Chem. Phys.* **101**. Copyright 1994 American Institute of Physics.)

Critical supersaturations (those for which $J = 1\,\mathrm{cm}^{-3}\,\mathrm{s}^{-1}$) are predicted reasonably accurately by the classical theory. When compared with actual nucleation rate data, the classical theory, however, often does not give the observed temperature dependence of the nucleation rate. It generally ranges from being several orders of magnitude too low at low temperatures to being several orders of magnitude too high at high temperatures (the data in Figure 11.7 are closer to the theory than might have been expected). Classical theory also predicts significantly lower critical supersaturations than experimentally observed in associated vapors and highly polar fluids (Laaksonen et al. 1995).

11.5 MODIFICATIONS OF THE CLASSICAL THEORY AND MORE RIGOROUS APPROACHES

The principal limitation to the classical theory is seen as the capillarity approximation, the attribution of bulk properties to the critical cluster (see Figure 11.8). Most modifications to the classical theory retain the basic capillarity approximation but introduce correction factors to the model [e.g., see Hale (1986), Dillmann and Meier (1989, 1991), and Delale and Meier (1993)].

Computer simulation offers a way to avoid many of the limitations inherent in classical and related theories. In principle, nucleation can be simulated by starting with a supersaturated gas-like configuration and solving the equations of motion of each molecule. Eventually a liquid drop would emerge from the simulation. Because typical simulation volumes are the order of $10^{-20}\,\mathrm{cm}^3$ and simulation times are the order of $10^{-10}\,\mathrm{s}$, for any particle formation to result, nucleation rates must be of order $10^{30}\,\mathrm{cm}^{-3}\,\mathrm{s}^{-1}$, far higher than anything amenable to laboratory study. This means that higher supersaturations than normal need to be simulated. Under realistic experimental conditions, critical or near-critical clusters, even under relatively high rates of nucleation,

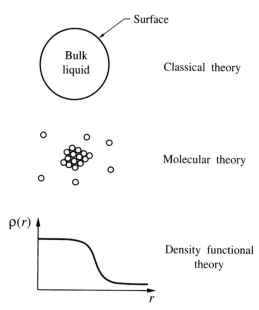

FIGURE 11.8 Three approaches to nucleation theory and the nature of the cluster that results.

are very rare entities surrounded by a nearly ideal gas of monomer and carrier gas. Thus research has focused on simulating an isolated cluster of a given size to determine its free energy. The variation of that free energy with number of monomers can then be used in nucleation theory to relate forward and backward rate constants or to estimate the height of the free energy barrier to nucleation (Reiss et al. 1990; Weakleim and Reiss 1993).

In classical nucleation theory the density of a cluster is assumed to be equal to the bulk liquid density, and the free energy of the surface (surface tension) is taken as that of a bulk liquid with a flat interface. A cluster is then completely defined by its number of molecules or its radius. In general, there is no reason why the density at the center of a cluster should equal the bulk liquid density, nor is there any reason why the density profile should match that at a planar interface (Figure 11.8). The density $\rho(r)$ should not be constrained other than to require that it approach the bulk vapor density at large distances from the cluster. Whereas the condition for the critical nucleus in classical theory is that the derivative of the free energy with respect to radius r be equal to zero, when ρ is allowed to vary with position r, the free energy is now a function of that density. The free energy has a minimum at the uniform vapor density and a second, lower minimum at the uniform liquid density. Between these two minima lies a saddle point at which the functional derivative of the free energy with respect to the density profile is zero. The *density functional thoery* has been applied to nucleation by Oxtoby and colleagues (Oxtoby 1992a,b). Application of the theory requires specification of an intermolecular potential.

11.6 BINARY HOMOGENEOUS NUCLEATION

We have seen that in homogeneous–homomolecular nucleation, nucleation does not occur unless the vapor phase is supersaturated with respect to the species. When two or more vapor species are present, neither of which is supersaturated, nucleation can still take place as long as the participating vapor species are supersaturated with respect to a liquid solution droplet. Thus heteromolecular nucleation can occur when a mixture of vapors is subsaturated with respect to the pure substances as long as there is supersaturation with respect to a solution of these substances. The theory of homogeneous–heteromolecular nucleation parallels that of homogeneous–homomolecular nucleation extended to include two or more nucleating vapor species. We consider here only two such species, binary homogeneous nucleation, the extension to more than two species being straightforward in principle if not in fact. Figure 11.9 depicts clusters in binary nucleation of acid and water.

In classical homogeneous-homomolecular nucleation theory the rate of nucleation can be written in the form

$$J = C \exp(-\Delta G^*/kT) \tag{11.75}$$

where ΔG^* is the free energy required to form a critical nucleus. This same form will apply in binary homogeneous nucleation theory. The first theoretical treatment of binary

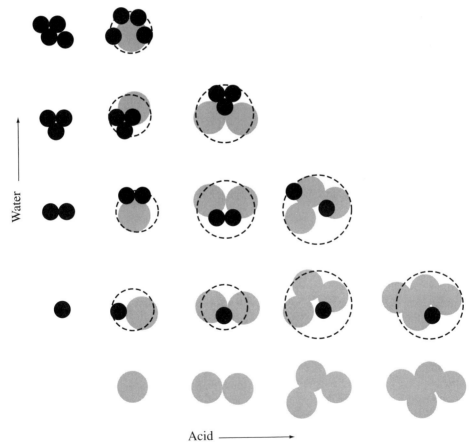

FIGURE 11.9 Clusters in binary nucleation. Proceeding vertically up the left side of the figure are clusters of pure water; proceeding horizontally along the bottom of the figure are clusters of pure acid molecules. The darker circles denote water molecules; the dashed lines indicate the boundary of the cluster.

nucleation dates back to Flood (1934), but it was not until 1950 that Reiss (1950) published a complete treatment of binary nucleation.

The free energy of formation of a nucleus containing n_A molecules of species A and n_B molecules of species B is given by

$$\Delta G = n_A(\mu_{A\ell} - \mu_{Ag}) + n_B(\mu_{B\ell} - \mu_{Bg}) + 4\pi r^2 \sigma \qquad (11.76)$$

where r is the radius of the droplet and $\mu_{A\ell}, \mu_{B\ell}$ and μ_{Ag}, μ_{Bg} are the chemical potentials of components A and B in the mixed droplet (ℓ) and in the gas phase (g), respectively. From before we can express the difference in chemical potential between the liquid and vapor phases as

$$\mu_{j\ell} - \mu_{jg} = -kT \ln \frac{p_j}{p_{j_{sol}}}, \qquad j = A, B \qquad (11.77)$$

where p_j is the partial pressure of component j in the gas phase and $p_{j_{sol}}$ is the vapor pressure of component j over a *flat solution* of the same composition as the droplet. The following quantities can be defined:

$$a_{Ag} = p_A/p_A^s = \text{activity of A in gas phase}$$
$$a_{Bg} = p_B/p_B^s = \text{activity of B in gas phase}$$
$$a_{A\ell} = p_{A_{sol}}/p_A^s = \text{activity of A in the liquid phase}$$
$$a_{B\ell} = p_{B_{sol}}/p_B^s = \text{activity of B in the liquid phase}$$

Here p_A^s and p_B^s are the saturation vapor pressures of A and B over a flat surface of pure A and B, respectively. Thus ΔG can be written as

$$\Delta G(n_A, n_B, T) = -n_A kT \ln \frac{a_{Ag}}{a_{A\ell}} - n_B kT \ln \frac{a_{Bg}}{a_{B\ell}} + 4\pi r^2 \sigma \qquad (11.78)$$

where a_{Ag} and a_{Bg} correspond to the saturation ratios of A and B, respectively, in the homomolecular system. Again, $p_{A_{sol}}$ and $p_{B_{sol}}$ are the partial vapor pressures of A and B over a flat surface of the solution. The radius r is given by

$$\frac{4}{3}\pi r^3 \rho = n_A m_A + n_B m_B \qquad (11.79)$$

Whereas in the case of a single vapor species, ΔG is a function of the number of molecules, i, here ΔG depends on both n_A and n_B. Nucleation will occur only if $\mu_{j\ell} - \mu_{jg}$ is negative, that is, if both vapors A and B are supersaturated with respect to their *vapor pressures over the solution* (*not* with respect to their pure component vapor pressure as in the case of homomolecular nucleation). In such a case, the first two terms on the RHS of (11.76) are negative. The last term on the RHS is always positive. For very small values of n_A and n_B, ΔG is dominated by the surface energy term, and ΔG increases as n_A and n_B (and size) increase. Eventually a point is reached where ΔG achieves a maximum. Instead of considering ΔG as a function only of i, it is now necessary to view ΔG as a surface in the $n_A - n_B$ plane. Reiss (1950) showed that the three-dimensional surface $\Delta G(n_A, n_B)$

has a saddle point that represents the minimum height of the free-energy barrier. The saddle point is one at which the curvature of the surface is negative in one direction (that of the path) and positive in the direction at right angles to it. The saddle point represents a "mountain pass" that clusters have to surmount in order to grow and become stable (Figure 11.10).

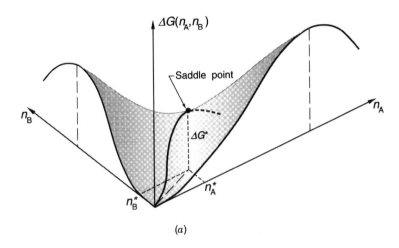

(a)

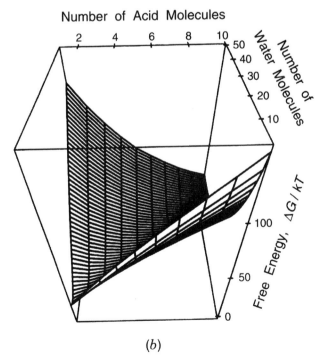

(b)

FIGURE 11.10 Free energy of cluster formation, $\Delta G(n_A, n_B)$, for binary nucleation: (a) schematic diagram of saddle point in the ΔG surface; (b) ΔG surface for $H_2SO_4-H_2O$ system at 298 K.

The height of the free-energy barrier at the saddle point is denoted by ΔG^*. This point may be found by solving the two equations:

$$\left(\frac{\partial \Delta G}{\partial n_A}\right)_{n_B} = \left(\frac{\partial \Delta G}{\partial n_B}\right)_{n_A} = 0 \tag{11.80}$$

Differentiating (11.78) and expressing n_A and n_B in terms of partial molar volumes $\bar{v}_A$ and $\bar{v}_B$, we obtain (Mirabel and Katz 1974)

$$RT \ln \frac{p_A}{p_{A_{sol}}} = \frac{2\sigma \bar{v}_A}{r} - \frac{3 x_B \bar{v}}{r} \frac{d\sigma}{dx_B} \tag{11.81}$$

$$RT \ln \frac{p_B}{p_{B_{sol}}} = \frac{2\sigma \bar{v}_B}{r} + \frac{3(1 - x_B)\bar{v}}{r} \frac{d\sigma}{dx_B} \tag{11.82}$$

where $x_B = n_B/(n_A + n_B)$ and the molar volume of solution $\bar{v} = (1 - x_B)\bar{v}_A + x_B \bar{v}_B$. (Note that this definition of $\bar{v}$ is an approximation since in mixing two species in general $1 \, cm^3$ of A mixed with $1 \, cm^3$ of B does not produce precisely $2 \, cm^3$ of solution.)

It has been shown that (11.81) and (11.82) need to be revised somewhat to locate the saddle point [e.g., see Mirabel and Reiss (1987)]. In short, the revised theory distinguishes between the composition of the surface layer and that of the interior of the cluster. The "revised" theory leads to the following equations to locate the saddle point

$$\Delta \mu_A + \frac{2\sigma \bar{v}_A}{r^*} = 0 \tag{11.83}$$

$$\Delta \mu_B + \frac{2\sigma \bar{v}_B}{r^*} = 0 \tag{11.84}$$

or

$$r^* = \frac{-2\sigma \bar{v}}{(1 - x_B^*) \Delta \mu_A + x_B^* \Delta \mu_B} \tag{11.85}$$

and

$$\bar{v}_B \Delta \mu_A = \bar{v}_A \Delta \mu_B \tag{11.86}$$

Both the classical and the revised classical theory lead to

$$\Delta G^* = \frac{4}{3} \pi \sigma r^{*2} \tag{11.87}$$

where

$$\Delta G^* = \frac{16}{3} \pi \sigma^3 \frac{\bar{v}^2}{(kT \ln S^*)^2} \tag{11.88}$$

and

$$S^* = S_A^{(1 - x_B^*)} S_B^{x_B^*} \tag{11.89}$$

Under conditions for which $d\sigma/dx_B$ is small, which is the case for many of the binary systems of interest, the results given by (11.81), (11.82) and (11.83), (11.84) are only slightly different.

The expression for the preexponential factor C in (11.75) was first derived by Reiss (1950)

$$C = \frac{\beta_A \beta_B}{\beta_A \sin^2 \phi + \beta_B \cos^2 \phi}(N_A + N_B)a^*Z \qquad (11.90)$$

where N_A and N_B are the gas-phase number concentrations of A and B, respectively, and

$$\beta_j = \frac{p_j}{(2\pi m_j kT)^{1/2}}, \qquad j = A, B \qquad (11.91)$$

where a^* is the surface area of the critical sized cluster, ϕ is the angle between the direction of growth at the saddle point and the n_A axis, and Z is the so-called Zeldovich nonequilibrium factor. In Reiss' (1950) theory ϕ is given by the direction of steepest descent on the free-energy surface. Stauffer (1976) subsequently showed that the direction of the nucleation path is not simply given by that of the steepest descent of the free-energy surface alone, but rather by a combination of both the impinging rate of the condensing species and the free-energy surface. In particular, he showed that if the arrival rates of the two species at the cluster are different, the direction of the nucleation path is shifted somewhat from the steepest-descent direction. For practical purposes, ϕ can be determined from

$$\tan \phi = \frac{n_B^*}{n_A^* + n_B^*} \qquad (11.92)$$

The Zeldovich factor generally has a value ranging from 0.1 to 1 and thus has only a slight effect on the rate of nucleation.

Mirabel and Clavelin (1978) have derived the limiting behavior of binary homogeneous nucleation theory when the concentration of one of the vapor species becomes very small. For a low concentration of one species, say, B, the preexponential factor C simplifies from (11.90) but one has to distinguish two cases:

1. If the new phase (critical cluster) is *dilute* with respect to component B (e.g., the critical cluster contains 1 or 2 molecules of B), then $\phi = 0$, and the flow of critical clusters through the saddle point is parallel to the n_A axis. In this case, C reduces to

$$C = \beta_A N_A a^* \qquad (11.93)$$

   which is similar to the preexponential factor for one-component nucleation of A.

2. If the critical cluster is not dilute with respect to component B, but B is still present at a very small concentration in the vapor phase, then in general $\beta_A \sin^2 \phi \gg \beta_B \cos^2 \phi$ and C reduces to

$$C = \frac{\beta_B N_A a^*}{\sin^2 \phi} \qquad (11.94)$$

The rate of nucleation is then controlled by species B. After one molecule of B strikes a cluster, many A molecules impinge such that an equilibrium with respect to A is achieved until another B molecule arrives and the process is repeated all over again.

11.7   BINARY NUCLEATION IN THE H_2SO_4–H_2O SYSTEM

The most important binary nucleation system in the atmosphere is that of sulfuric acid and water. Doyle (1961) was the first to publish predicted nucleation rates for the H_2SO_4–H_2O system. His calculations showed that, even in air of relative humidity less than 100%, extremely small amounts of H_2SO_4 vapor are able to induce nucleation. Mirabel and Katz (1974) reexamined Doyle's predictions using a more appropriate equilibrium vapor pressure for pure H_2SO_4. Further refinements were made by Heist and Reiss (1974) who examined the influence of hydrate formation in the H_2SO_4–H_2O system. [More recent summaries of much of the work on H_2SO_4–H_2O are given by Jaecker-Voirol and Mirabel (1989), Kulmala and Laaksonen (1990), Laaksonen et al. (1995), and Noppel et al. (2002).] For this binary system, A refers to H_2O and B to H_2SO_4.

In a gas containing water vapor, H_2SO_4 molecules exist in various states of hydration: $H_2SO_4 \cdot H_2O$, $H_2SO_4 \cdot 2H_2O$, ..., $H_2SO_4 \cdot h\,H_2O$. Nucleation of H_2SO_4 and H_2O is then not just the nucleation of individual molecules of H_2SO_4 and H_2O, but involves cluster formation between individual H_2O molecules and already-hydrated molecules of H_2SO_4 (Jaecker-Voirol et al. 1987; Jaecker-Voirol and Mirabel 1988). The Gibbs free energy of cluster formation, including hydrate formation, is

$$\Delta G_{\mathrm{hyd}} = \Delta G_{\mathrm{unhyd}} - kT \ln C_h \tag{11.95}$$

where the correction factor C_h due to hydration is

$$C_h = \left[\frac{1 + K_1 p_{\mathrm{A_{sol}}} + \cdots + K_1 K_2 \cdots K_h p_{\mathrm{A_{sol}}}^h}{1 + K_1 p_A + \cdots + K_1 K_2 \cdots K_h p_A^h} \right]^{n_B} \tag{11.96}$$

where $p_{\mathrm{A_{sol}}}$ is the partial vapor pressure of water (species A) over a flat surface of the solution, p_A is the partial pressure of water vapor in the gas, n_B is the number of H_2SO_4 molecules (species B) in the cluster, h is the number of water molecules per hydrate, and K_i are the equilibrium constants for hydrate formation. The equilibrium constants are those for the reactions

$$H_2O + H_2SO_4 \cdot (h-1)H_2O \rightleftharpoons H_2SO_4 \cdot h\,H_2O$$

and can be evaluated using the following expression

$$K_h = \exp\left(-\Delta G_h^\circ / kT \right) \tag{11.97}$$

and

$$\Delta G_h^\circ = KT \ln p_{\mathrm{A_{sol}}} + \frac{2\sigma m_A}{r} \left(\frac{1}{\rho} + \frac{X_B}{\rho^2} \frac{\partial \rho}{\partial X_B} \right) \tag{11.98}$$

TABLE 11.5   Equilibrium Constants K_h at 298 K for the Successive Addition of a Water Molecule to a Hydrated Sulfuric Acid Molecule

			$H_2O + H_2SO_4 \cdot (h-1)H_2O \rightleftharpoons H_2SO_4 \cdot h\ H_2O$							
h	1	2	3	4	5	6	7	8	9	10
K_h	1430	54.72	14.52	8.12	5.98	5.04	4.62	4.41	4.28	4.27

Source: Jaecker-Voirol and Mirabel (1989).

where ρ is the density of the solution and X_B is the mass fraction of B in the cluster,

$$X_B = \frac{n_B m_B}{n_A m_A + n_B m_B} \tag{11.99}$$

The $\partial\rho/\partial X_B$ term in (11.98) can often be neglected as small, so (11.98) becomes

$$\Delta G_h^\circ = kT \ln p_{A_{sol}} + \frac{2\sigma v_A}{r} \tag{11.100}$$

where v_A is the partial molecular volume of water. Hydrate formation can have a relatively large effect on the nucleation rate by changing the shape of the free-energy surface and therefore the location and value of ΔG at the saddle point. The first 10 hydration constants for H$_2$SO$_4$ and H$_2$O at 298 K are given in Table 11.5. [Constants at $-50°$C, $-25°$C, $0°$C, $50°$C, $75°$C, and $100°$C, in addition, are given by Jaecker-Voirol and Mirabel (1989).]

Sulfuric acid–water nucleation falls into the category where component B (H$_2$SO$_4$) is present at a very small concentration relative to that of A (H$_2$O), yet the critical cluster is not dilute with respect to H$_2$SO$_4$. Thus the preexponential factor (11.94) applies. The rate of binary nucleation in the H$_2$SO$_4$–H$_2$O system, accounting for the effect of hydrates and using (11.94), is given by

$$J = \frac{(8\pi kT)^{1/2} C_h N_A Z}{\sin^2 \phi} \sum_{h=0}^{h_{max}} (\delta^2 \mu^{-1/2} N_h) \exp(-\Delta G^*/kT) \tag{11.101}$$

where δ is the sum of the radii of the critical nucleus and the hydrate (or a free acid molecule), μ is the reduced mass of the critical nucleus and the hydrate (or free acid molecule), and N_h is the number concentration of hydrates. As before, ϕ is determined from (11.92). Note that in (11.101), ΔG^* refers to ΔG_{unhyd}.

To calculate the nucleation rate of H$_2$SO$_4$–H$_2$O, first $\Delta G(n_A, n_B, T)$ has to be evaluated to locate the saddle point. One can construct a table of ΔG as a function of n_A and n_B and the saddle point can be found by inspection of the values. To calculate ΔG one needs values for activities, densities, and surface tensions as functions of temperature. The value of the saturation vapor pressure of H$_2$SO$_4$ has a major effect on the nucleation rate. Kulmala and Laaksonen (1990) have evaluated available expressions for the H$_2$SO$_4$ saturation vapor pressure and derived a recommended expression (Table 11.6). Table 11.7 illustrates a particular calculation of ΔG as a function of n_A and n_B (neglecting hydration).

Predicting the nucleation rate of H$_2$SO$_4$–H$_2$O on the basis of the theory presented above is somewhat involved; appropriate thermodynamic data need to be assembled, including the equilibrium constants for H$_2$SO$_4$ hydration. Figures 11.11a–d show predicted nucleation rates (cm^{-3} s^{-1}) as a function of $N_{H_2SO_4}$, the total number concentration of

TABLE 11.6 Saturation Vapor Pressure of H_2SO_4

$$\ln p_{H_2SO_4}^s = \ln p_{H_2SO_4,0}^s + \frac{\Delta H_v(T_0)}{R}\left[-\frac{1}{T} + \frac{1}{T_0} + \frac{0.38}{T_c - T_0}\left(1 + \ln\frac{T_0}{T} - \frac{T_0}{T}\right)\right]$$

$\Delta H_v(T_0)/R = 10156$

$T_c = 905$ K

$T_0 = 360$ K $\qquad \ln p_{H_2SO_4,0}^s = -(10156/T_0) + 16.259$ (atm)

Source: Kulmala and Laaksonen (1990).

H_2SO_4 molecules in the gas phase at relative humidities of 20%, 50%, 80%, and 100% at temperatures ranging from $-50°C$ to $25°C$. Figure 11.12 gives the composition of the critical nucleus, for four different temperatures, in terms of the mole fraction of H_2SO_4 in the droplet, at a rate of nucleation of $J = 1\ cm^{-3}\ s^{-1}$. These curves give a good indication of the composition for other rates since the composition of the critical cluster is quite insensitive to the nucleation rate.

The implications of hydrate formation for the binary nucleation of H_2SO_4–H_2O are the following. In the gas phase, most of the H_2SO_4 molecules (over, say, 95%) are hydrated. Hydrates consisting of one H_2SO_4 molecule greatly exceed those containing two or more H_2SO_4 molecules. The presence of hydrates reduces the rate of binary homogeneous nucleation of H_2SO_4–H_2O as compared to that in the absence of hydrates by a factor of $10^5 - 10^6$. Although quantitative predictions of the nucleation rate differ between the older theory (without hydrates) and the newer theory (with hydrates), the conclusion that extremely small amounts of H_2SO_4 are able to nucleate subsaturated H_2O vapor still holds.

The extraordinary dependence of the H_2SO_4–H_2O nucleation rate on temperature, relative humidity, and H_2SO_4 concentration means that the inevitable uncertainties in atmospheric measurements produce as much uncertainty in nucleation rate predictions as do potential uncertainties in the theory. This steep variation, although a liability for predicting the nucleation rate, is an asset for estimating atmospheric conditions when

TABLE 11.7 Values of ΔG ($\times 10^{12}$ erg) as a Function of n_A and n_B (A = H_2O; B = H_2SO_4) Pressures Used Are $p_A = 2 \times 10^{-4}$ and $p_B = 1.15 \times 10^{-11}$ mm Hg

$n_A\backslash n_B$	75	76	77	78	79	80	81	82
128	5.64	5.82	5.94	6.0	6.02	5.99	5.94	5.86
129	5.51[a]	5.73	5.88	5.97	6.01	6.02	5.98	5.91
130	5.63	5.61	5.80	5.93	6.00	6.02	6.01	5.96
131	5.73	5.54	5.70	5.86	5.96	6.01	6.02	6.00
132	5.91	5.65	5.56	5.77	5.91	5.99	6.02	6.02
133	5.89	5.75	5.57	5.66	5.84	5.95	6.01	6.08
134	5.95	5.83	5.68	5.51	5.74	5.89	5.98	6.02
135	6.00	5.91	5.77	5.60	5.62	5.81	5.93	6.00
136	6.04	5.97	5.85	5.70	5.51	5.71	5.87	5.97
137	6.06	6.01	5.92	5.79	5.63	5.57	5.78	5.91
138	6.08	6.05	5.98	5.87	5.72	5.54	5.67	5.84
139	6.07	6.07	6.02	5.94	5.81	5.64	5.52	5.74

[a]Single underline represents the path of minimum free energy; double underlined value is the saddle-point free energy.

Source: Hamill et al. (1977).

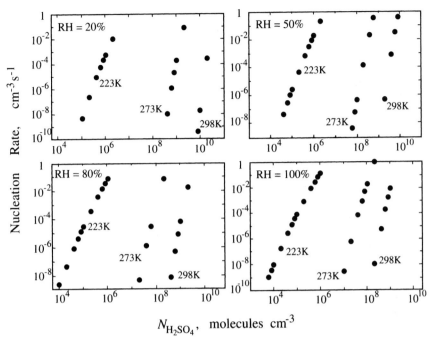

FIGURE 11.11 H$_2$SO$_4$–H$_2$O binary nucleation rate (cm^{-3} s^{-1}) as a function of $N_{H_2SO_4}$, calculated from classical binary nucleation theory (Kulmala and Laaksonen 1990): (a) relative humidity = 20%; (b) RH = 50%; (c) RH = 80%; (d) RH = 100%.

nucleation is likely to occur. Using $J = 1\,\text{cm}^{-3}\,\text{s}^{-1}$ as the critical rate above which nucleation is significant, the critical gas-phase sulfuric acid concentration that produces such a rate can be evaluated from the following empirical fit to nucleation rate calculations (Jaecker-Voirol and Mirabel 1989; Wexler et al. 1994):

$$C_{\text{crit}} = 0.16\,\exp(0.1\,T - 3.5\,\text{RH} - 27.7) \tag{11.102}$$

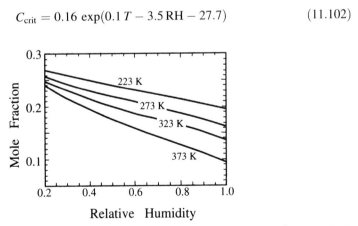

FIGURE 11.12 Composition of the critical H$_2$SO$_4$–H$_2$O nucleus, calculated for $J = 1\,\text{cm}^{-3}\,\text{s}^{-1}$, as a function of RH. The different curves correspond to the temperatures shown. (Reprinted from *Atmos. Environ.* **23**, Jaecker-Voirol, A. and Mirabel, P., Heteromolecular nucleation in the sulfuric acid-water system, p. 2053, 1989, with kind permission from Elsevier Science Ltd, The Boulevard, Langford Lane, Kidlington OX5 1 GB, UK.)

where T is in K, RH is the relative humidity on a scale of 0 to 1, and C_{crit} is in $\mu g \ m^{-3}$. Thus, when the gas-phase concentration of H_2SO_4 exceeds C_{crit}, one can assume that H_2SO_4–H_2O nucleation commences.

Noppel et al. (2002) have continued the development of the hydrate model using ab initio calculations of the structures of small water–sulfuric acid clusters. The predicted nucleation rates by the Noppel et al. (2002) model are generally higher than those predicted by previous models. The nucleation rates have been parameterized for use in chemical transport models by Vehkamäki et al. (2002).

11.8   HETEROGENEOUS NUCLEATION

Nucleation of vapor substances in the atmosphere can occur both homogeneously and heterogeneously. The large amount of prevailing aerosol surface area in the atmosphere under all but the most pristine conditions provides a substrate for heterogeneous nucleation. Heterogeneous nucleation may take place at significantly lower super-saturations than homogeneous nucleation.

11.8.1   Nucleation on an Insoluble Foreign Surface

Let us consider nucleation of a liquid droplet on an insoluble foreign surface (Figure 11.13). The heterogeneous nucleation rate is defined as the number of critical nuclei appearing on a unit surface area per unit time. The classical expression for the free energy of critical cluster formation on a flat solid (insoluble) surface is

$$\Delta G^* = \Delta G^*_{hom} f(m) \tag{11.103}$$

where ΔG^*_{hom} is the free energy for homogeneous critical nucleus formation and the contact parameter m is the cosine of the contact angle θ between the nucleus and the substrate (Figure 11.13)

$$m = \cos \theta = \frac{\sigma_{SV} - \sigma_{SL}}{\sigma_{LV}} \tag{11.104}$$

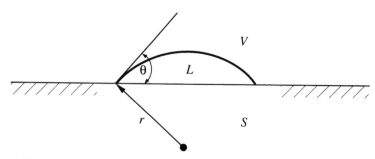

FIGURE 11.13   Cluster formation on an insoluble flat surface. θ is the contact angle between the cluster and the flat substrate.

where the σ terms are the interfacial tensions between the different phases, V denoting the gas phase, L the nucleus, and S the substrate. The function $f(m)$, given by

$$f(m) = \frac{1}{4}(2 + m)(1 - m)^2 \qquad (11.105)$$

assumes values between 0 and 1. When the contact angle θ is zero, $m = 1$, and $f(m) = 0$, and nucleation begins as soon as the vapor is supersaturated; that is, the free-energy barrier to critical cluster formation is zero. If $\theta = 180°$, $f(m) = 1$, and heterogeneous nucleation is not favored over homogeneous nucleation.

The discussion above applies to heterogeneous nucleation on flat surfaces. For atmospheric particles the curvature of the surface complicates the situation. Fletcher (1958) showed that the free energy of formation ΔG^* of an embryo of critical radius r^* on a spherical nucleus of radius R is given by

$$\Delta G^* = \Delta G_{\text{hom}} f(m, x) = \frac{16\pi}{3} \frac{v_1 \sigma_{\text{LV}}}{(kT \ln S)^2} f(m, x) \qquad (11.106)$$

where σ_{LV} is the surface tension between the embryo phase and the surrounding phase, v_1 represents the volume per molecule, S is the supersaturation, and $f(m,x)$ accounts for the geometry of the embryo and is now given by

$$f(m, x) = 1 + \left(\frac{1 - mx}{g}\right) + x^3 \left[2 - 3\left(\frac{x - m}{g}\right) + \left(\frac{x - m}{g}\right)^3\right] + 3mx^2\left(\frac{x - m}{g} - 1\right) \qquad (11.107)$$

with

$$g = (1 + x^2 - 2mx)^{1/2} \qquad (11.108)$$

where m is given by (11.104) and

$$x = \frac{R}{r^*} \qquad (11.109)$$

The critical radius for the embryo is given by

$$r^* = \frac{2v_1 \sigma_{\text{LV}}}{kT \ln S} \qquad (11.110)$$

which corresponds to (11.52). One can show that for $x = 0$ (11.106) corresponds to homogeneous nucleation and for x approaching infinity it corresponds to (11.103). The contact angle with the nucleating liquid is an important parameter for heterogeneous nucleation. Small contact angles make the corresponding insoluble particles good sites for nucleation while larger contact angles inhibit nucleation. The direct relationship between the contact angle and the critical supersaturation for small contact angles and supersaturations was estimated by McDonald (1964) as

$$\cos\theta = 1 - \left(\frac{x-1}{x}\right)(0.662 + 0.022 \ln R)^{1/2} \ln S_c \qquad (11.111)$$

where S_c is the critical supersaturation corresponding to a nucleation rate of 1 embryo per particle per second. For formation of water droplets on insoluble particles in clouds in ambient supersaturations (usually less than 3%), equation (11.111) suggests that the contact angle should be less than 12°.

Describing the interaction between a nucleus of a condensing solute and a solid substrate by a single macroscopic parameter is clearly an approximation. Moreover, the assumption that the nucleus grows exclusively by impingement of vapor molecules neglects the surface diffusion of adsorbed solute molecules, which is an effective mechanism for delivery of monomers to the cluster. Furthermore, the surface is assumed to be perfectly planar and devoid of imperfections that are likely to accelerate nucleation. Whereas these limitations of the classical approach to heterogeneous nucleation on an insoluble substrate are well recognized, experimental data to evaluate the predictions of the classical theory are scarce. One problem is the need to produce sufficiently smooth surfaces to conform to the assumptions of the theory.

11.8.2   Ion-Induced Nucleation

Phase transition from supersaturated vapor to a more stable liquid phase necessarily involves the formation of clusters and their growth within the vapor phase. In the case of homogeneous nucleation, a cluster is formed among the molecules of condensing substance themselves, while in the case of ion-induced nucleation, it will be formed preferentially around the ion, for the electrostatic interaction between the ion and the condensing molecules always lowers the free energy of formation of the cluster.

The presence of ions has been shown experimentally to enhance the rate of nucleation of liquid drops in a supersaturated vapor. Katz et al. (1994) showed, for example, that the nucleation rate of n-nonane, measured in an upward thermal diffusion cloud chamber, at an ion density of 16×10^6 ions cm^{-3}, increased by a factor of 2500 over that in the absence of ions. These investigators also confirmed experimentally that the nucleation rate is directly proportional to the ion density. The phenomenon of ion-induced nucleation plays an important role in atmospheric condensation, particularly in the ionosphere. While both positive and negative ions increase the nucleation rate, many substances exhibit a preference for ions of one sign over the other.

Understanding the effect of ions on nucleation would be very difficult if the ions' specific chemical characteristic had a significant effect on their nucleating efficiency. Since frequently the number of molecules of the nucleating species in the critical-sized cluster is on the order of tens to hundreds of molecules, it is reasonable to assume that an ion buried near its center has little effect on the cluster's surface properties other than those due to its charge. Thus the ion creates a central force field that makes it more difficult for a molecule to evaporate than it would be from an uncharged but otherwise identical cluster. Ion self-repulsion and the typical low density of ions ensure that almost all such clusters contain only one ion. Molecules arrive at and evaporate from the ion–vapor molecule clusters. The critical cluster size for nucleation is that for which the evaporation rate of single vapor molecules is equal to their arrival rate. The effect of the ion imbedded in the cluster on the arrival rate of vapor molecules is considerably weaker than that on the evaporation rate. This difference leads to a significant decrease in the critical cluster size in the presence of ions relative to that for homogeneous nucleation because the reduced evaporation rate means that the two rates become equal at a smaller monomer arrival rate. The smaller monomer arrival rate translates into a lower saturation ratio of vapor molecules needed to achieve the same rate of nucleation as that in the absence of ions.

From a kinetic point of view, a cluster is formed and changes its size by the condensation and evaporation of molecules. At a given temperature, the evaporation rate of molecules (per unit time and unit area) is independent of S, but determined primarily by i, while the condensation rate (per unit time and unit area) is approximately proportional to S and independent of i. Now, the intermolecular interaction energy per molecule in the cluster increases with i. Thus, in the case of homogeneous nucleation, the evaporation rate decreases monotonically with increasing i, approaching the bulk liquid value equal to the condensation rate at saturation ($S = 1$). When $S > 1$, there is only one cluster size i^* at which the rate of evaporation and that of condensation balance to yield a critical nucleus at unstable equilibrium with the vapor phase. On the other hand, when an ion is introduced into the cluster, it attracts molecules, thereby decreasing the evaporation rate. Since the electrostatic field decays as r^{-2}, the evaporation rate decrease should be most significant for small i and become negligible as i approaches infinity. When the ion–molecule interaction energy is sufficiently large, the condensation rate balances the evaporation rate even if $S \leq 1$. This implies that the evaporation rate first increases with i, achieves a maximum at a certain value of i, and then decreases while approaching the bulk value. Thus, for $S > 1$, there are two values of i at which the evaporation rate and the condensation rate balance, the smaller of which is a stable subcritical cluster and the larger is the unstable critical nucleus (Figure 11.14). For $S \geq S'_{max}$, the condensation rate is always larger than the evaporation rate for any cluster size i; hence the cluster formation and its growth are spontaneous.

The cluster free energy, as a function of number of molecules i within the cluster, is shown for the typical case of ion-induced nucleation in Figure 11.14. The free-energy curves are in fact consistent with the kinetic point of view given above. Thus, for an appropriate value of the supersaturation, the free-energy curve shows a local minimum and a maximum for the case of ion-induced nucleation, corresponding, respectively, to the stable subcritical cluster and the unstable critical cluster. The local minimum disappears for the case of homogeneous nucleation.

The earliest attempt to calculate the Gibbs free-energy change of cluster formation in ion-induced nucleation is due to Thomson (1906), based on the theory of capillarity, where a nucleus is represented as a bulk liquid enclosed by an interface of zero thickness with an ion placed at the center. The free-energy change of cluster formation is then given in terms of the thermodynamic quantities such as the surface tension, dielectric constants of the bulk phases, and so on. The free energy change from the Thomson theory depends on q^2, where q is the ion charge, and therefore is incapable of expressing a dependence on the sign of q.

Physically, the dependence of the ion-induced nucleation rate of a substance on the sign of the ion charge must arise from some asymmetry in the molecular interactions. Such asymmetry should, in principle, manifest itself in a sign dependence of the relevant thermodynamic quantities such as the surface tension. To account for the ion charge preference in ion-induced nucleation requires a statistical mechanical theory, which assumes an intermolecular potential as the fundamental information required to evaluate the relevant thermodynamic properties.

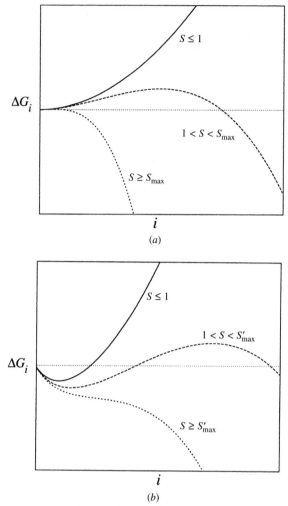

FIGURE 11.14 Ion-induced nucleation: (a) free energy of cluster formation in homogeneous nucleation; (b) free energy of cluster formation in ion-induced nucleation. S_{max} and S'_{max} are the maximum values of the saturation ratio for which no barrier to nucleation exists in homogeneous and ion-indued nucleation, respectively.

Kusaka et al. (1995a) present a density functional theory for ion-induced nucleation of dipolar molecules. Asymmetry is introduced into a molecule by placing a dipole moment at some fixed distance from its center. As a result of the asymmetric nature of the molecules and their interactions with the ion, the free energy change acquires a dependence on the ion charge. The predicted free-energy change shows a sign preference, resulting in a difference in the nucleation rate by a factor of $10–10^2$, for realistic values of model parameters. The sign effect is found to decrease systematically as the supersaturation is increased. The asymmetry of a molecule is shown to be directly responsible for the sign preference in ion-induced nucleation.

Kusaka et al. (1995b) also present a density functional theory for ion-induced nucleation of polarizable multipolar molecules. For a fixed orientation of a molecule, the ion–molecule interaction through the molecular polarizability is independent of the sign of the ion charge, while that through the permanent multipole moments is not. As a result of this asymmetry, the reversible work acquires a dependence on the sign of the ion charge.

11.9 NUCLEATION IN THE ATMOSPHERE

The first evidence of in situ particle formation in the atmosphere was provided by John Aitken at the end of the nineteenth century (Aitken 1897). He built the first apparatus to measure the number of dust and fog particles in the atmosphere. However, little progress was made in understanding what causes new particle formation or how widespread it might be for almost a century. In the 1990s the development of instruments capable of measuring the size distribution of particles as small as 3 nm led to the discovery that nucleation and growth of new particles is a rather common event in many areas around the world (Kulmala et al. 2004). Areas where frequent nucleation bursts have been observed include

1. The free troposphere, especially near the outflows of convective clouds and close to the tropopause
2. The boreal forest
3. Eastern United States, Germany, and other European countries
4. Coastal environments in Europe and the United States

One surprising finding is that nucleation events take place even in relatively polluted urban areas. For example, Stanier et al. (2004) reported observations of nucleation events in Pittsburgh, PA during 30% of the days of a full year. The events take place on sunny, relatively clean days. A comprehensive review of atmospheric nucleation observations can be found in Kulmala et al. (2004).

A typical nucleation event is shown in Figure 11.15. The relatively smooth growth of the particles, even as different air parcels are passing by the measurement site, suggests

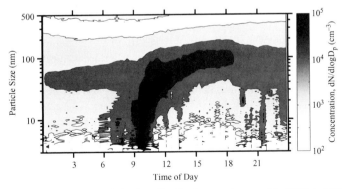

FIGURE 11.15 Evolution of particle size distribution on August 11, 2001 in Pittsburgh, PA, a day with new particle formation and growth. Particle number concentration (z axis) is plotted against time of day (x axis) and particle diameter (y axis). Measurements by Stanier et al. (2004).

that these events take place over distances of hundreds of kilometers. These smooth growth curves are characteristic of regional nucleation events. During such events, the growth of nucleated particles continues throughout the day, even as the wind changes direction. The widespread occurrence of these events has been confirmed by measurements at multiple stations (Kulmala et al. 2001; Stanier et al. 2004).

Mechanisms that have been proposed to explain these nucleation events include

- Sulfuric acid–water binary nucleation (Kulmala and Laaksonen 1990)
- Sulfuric acid–ammonia-water ternary nucleation (Kulmala et al. 2000)
- Nucleation of organic vapors (Hoffmann et al. 1997)
- Ion-induced nucleation (Yu and Turco 2000)
- Halogen oxide nucleation (Hoffmann et al. 2001).

The smallest measurable particle size using current instruments is around 3 nm. The diameter of critical clusters is significantly smaller than this, so the nucleation rate itself cannot be directly measured at the present time. The formation rate of 3 nm particles is usually in the range of 0.01–$10\,cm^{-3}\,s^{-1}$ in the boundary layer and up to $100\,cm^{-3}\,s^{-1}$ in urban areas. Rates as high as $10^5\,cm^{-3}\,s^{-1}$ have been observed in coastal areas and industrial plumes.

The new particles formed during nucleation events have short lifetimes owing to their coagulation with larger particles. As a result, the growth of these particles to larger sizes is a crucial process for the formation of observable particles. Newly formed particles grow with rates in the 1–20 nm h^{-1} range, depending on the availability of condensable gases. For polar regions the observed growth rates are as low as 0.1 nm h^{-1}. The growth rates during the summer are usually several times greater than those during the winter. Condensation of sulfuric acid is often the dominant process for the growth of these particles, especially in polluted areas with high SO_2 concentrations (Gaydos et al. 2005). However, field measurements and model simulations have indicated that the condensation of sulfuric acid alone is often not sufficient to grow these nuclei. To aid in the growth of these particles, the condensation of organic species (Kerminen et al. 2000) and heterogeneous reactions (Zhang and Wexler 2002) have been suggested. While organic compounds having a very low saturation vapor pressure are the most likely candidates for assisting the growth of fresh nuclei, the identity of these compounds remains unknown.

The few observations of nucleation in the free troposphere are consistent with binary sulfuric acid–water nucleation. In the boundary layer a third nucleating component or a totally different nucleation mechanism is clearly needed. Gaydos et al. (2005) showed that ternary sulfuric acid–ammonia–water nucleation can explain the new particle formation events in the northeastern United States through the year. These authors were able to reproduce the presence or lack of nucleation in practically all the days both during summer and winter that they examined (Figure 11.16). Ion-induced nucleation is expected to make a small contribution to the major nucleation events in the boundary layer because it is probably limited by the availability of ions (Laakso et al. 2002). Homogeneous nucleation of iodine oxide is the most likely explanation for the rapid formation of particles in coastal areas (Hoffmann et al. 2001). It appears that different nucleation processes are responsible for new particle formation in different parts of the atmosphere. Sulfuric acid is a major component of the nucleation growth process in most cases.

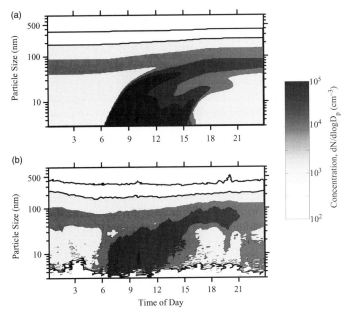

FIGURE 11.16 Comparison of (a) predicted and (b) measured size distributions as a function of time for July 27, 2001 in Pittsburgh, PA. Particle number concentration (z axis) is plotted against time of day (x axis) and particle diameter (y axis) (Gaydos et al. 2005).

APPENDIX 11 THE LAW OF MASS ACTION

The rate of an irreversible chemical reaction can generally be written in the form $r = kF(c_i)$, where k is the rate constant of the reaction and $F(c_i)$ is a function that depends on the composition of the system as expressed by concentrations c_i. The rate constant k does not depend on the composition of the system—hence the term *rate constant*. Frequently, the function $F(c_i)$ can be written as

$$F(c_i) = \prod_i c_i^{\alpha_i} \qquad (11.A.1)$$

where the product is taken over all components of the system. The exponent α_i is called the *order of the reaction* with respect to species i. The algebraic sum of the exponents is the total order of the reaction. If the α_i are equal to the stoichiometric coefficients $|v_i|$ for reactants and equal to zero for all other components, then

$$F(c_i) = \prod_i c_i^{|v_i|} \qquad \text{for reactants only} \qquad (11.A.2)$$

This form is termed the *law of mass action*. When a reaction is reversible, its rate can generally be expressed as the difference between the rates in the forward and reverse directions. The net rate can be written as

$$r = k_f \prod_i c_i^{\alpha_i^f} - k_r \prod_i c_i^{\alpha_i^r} \qquad (11.A.3)$$

At equilibrium, $r = 0$ and

$$\frac{k_f}{k_r} = \prod_i \frac{c_i^{\alpha_i^r}}{c_i^{\alpha_i^f}} \tag{11.A.4}$$

Now the ratio k_f/k_r depends only on temperature. The equilibrium constant K_c, defined by

$$K_c = \prod_i c_i^{\nu_i} \tag{11.A.5}$$

also depends only on temperature. Consequently

$$\prod_i \frac{c_i^{\alpha_i^r}}{c_i^{\alpha_i^f}} \tag{11.A.6}$$

must be a certain function of

$$\prod_i c_i^{\nu_i}$$

that is

$$\prod_i \frac{c_i^{\alpha_i^r}}{c_i^{\alpha_i^f}} = f\left(\prod_i c_i^{\nu_i}\right)$$

This relation must hold regardless of the value of the concentrations. This is so if

$$\prod_i \frac{c_i^{\alpha_i^r}}{c_i^{\alpha_i^f}} = \left(\prod_i c_i^{\nu_i}\right)^n \tag{11.A.7}$$

with $\alpha_i^r - \alpha_i^f = \nu_i n$ for all i. In particular, then

$$\frac{k_f}{k_r} = K_c^n \tag{11.A.8}$$

Ordinarily $n = 1$, so

$$\frac{k_f}{k_r} = K_c \tag{11.A.9}$$

PROBLEMS

11.1$_A$ Calculate the homogeneous nucleation rate of ethanol at 298 K for saturation
ratios from 2 to 7 using (10.47) and (10.74). Comment on any differences.

11.2$_B$ For the homogeneous nucleation of water at 20°C at a saturation ratio $S = 3.5$, calculate the sensitivity of the nucleation rate to small changes in saturation ratio and surface tension; that is, find x and y in

$$\frac{\Delta J}{J} = x\left(\frac{\Delta S}{S}\right) + y\left(\frac{\Delta \sigma}{\sigma}\right)$$

11.3$_B$ The nucleation rate can be expressed in the form

$$J = C\exp(-\Delta G^*/kT)$$

If the preexponential term C is not a strong function of S, show that

$$\frac{d \ln J}{d \ln S} \simeq i^*$$

This result is called the *nucleation theorem.*

11.4$_B$ Calculate the H_2SO_4 gas-phase concentration (in molecules cm^{-3}) that produces a binary homogeneous nucleation rate of 1 cm^{-3} s^{-1} at RHs of 20%, 40%, 60%, 80%, and 100% at 273 K, 298 K, and 323 K. Comment on the results.

REFERENCES

Aitken, J. (1897) On some nuclei of cloudy condensation, *Trans. Royal Soc.* XXXIX.

Becker, R., and Döring, W. (1935) Kinetische behandlung der keimbildung in übersättigten dämpfen, *Ann. Phys. (Leipzig)* **24**, 719–752.

Courtney, W. G. (1961) Remarks on homogeneous nucleation, *J. Chem. Phys.* **35**, 2249–2250.

Delale, C. F., and Meier, G. E. A. (1993) A semiphenomenological droplet model of homogeneous nucleation from the vapor phase, *J. Chem. Phys.* **98**, 9850–9858.

Dillmann, A., and Meier, G. E. A. (1989) Homogeneous nucleation of supersaturated vapors, *Chem. Phys. Lett.* **160**, 71–74.

Dillmann, A., and Meier, G. E. A. (1991) A refined droplet approach to the problem of homogeneous nucleation from the vapor phase, *J. Chem. Phys.* **94**, 3872–3884.

Doyle, G. J. (1961) Self-nucleation in the sulfuric acid–water system, *J. Chem. Phys.* **35**, 795–799.

Farkas, L. (1927) Keimbildungsgeschwindigkeit in übersättigen dämpfen, *Z. Phys. Chem.* **125**, 236–242.

Fletcher N. H. (1958) Size effect in heterogeneous nucleation, *J. Chem. Phys.* **29**, 572–576.

Flood, H. (1934), Tröpfenbildung in übersättigten äthylalkohol-wasserdampfgemischen, *Z. Phys. Chem.* **A170**, 286–294.

Frenkel, J. (1955) *Kinetic Theory of Liquids.* Dover, New York (first published in 1946).

Gaydos T. M., Stanier, C. O., and Pandis S. N. (2005) Modeling of in situ ultrafine atmospheric particle formation in the eastern United States, *J. Geophys. Res.* **110** (doi: 10.1029/2004JD004683).

Girshick, S. L., and Chiu, C.-P. (1990) Kinetic nucleation theory: A new expression for the rate of homogeneous nucleation from an ideal supersaturated vapor, *J. Chem. Phys.* **93**, 1273–1277.

Hale, B. N. (1986) Application of a scaled homogeneous nucleation rate formalism to experimental data at $T < T_C$, *Phys. Rev. A* **33**, 4156–4163.

Hamill, P., Kiang, C. S., and Cadle, R. D. (1977) The nucleation of H_2SO_4–H_2O solution aerosol particles in the stratosphere, *J. Atmos. Sci.* **34**, 150–162.

Heist, R. H. (1986) Nucleation and growth in the diffusion cloud chamber, in *Handbook of Heat and Mass Transfer*, Gulf Publishing, Houston, TX, pp. 487–521.

Heist, R. H., and He, H. (1994) Review of vapor to liquid homogeneous nucleation experiments from 1968 to 1992, *J. Phys. Chem. Ref. Data* **23**, 781–805.

Heist, R. H., and Reiss, H. (1974) Hydrates in supersaturated sulfuric acid–water vapor, *J. Chem. Phys.* **61**, 573–581.

Hoffmann, T., Odum, J. R., Bowman, F., Collins, D., Klockow, D., Flagan, R. C., and Seinfeld, J. H. (1997) Formation of organic aerosols from the oxidation of biogenic hydrocarbons, *J. Atmos. Chem.* **26**, 189–222.

Hoffmann, T., O'Dowd, C. D., and Seinfeld, J. H. (2001) Iodine oxide homogeneous nucleation: An explanation for coastal new particle production, *Geophys. Res. Lett.* **28**, 1949–1952.

Hung, C. H., Krasnopoler, M., and Katz, J. L. (1989) Condensation of a supersaturated vapor. 8. The homogeneous nucleation of *n*-nonane, *J. Chem. Phys.* **90**, 1856–1865.

Jaecker-Voirol, A., and Mirabel, P. (1988) Nucleation rate in a binary mixture of sulfuric acid and water vapor, *J. Phys. Chem.* **92**, 3518–3521.

Jaecker-Voirol, A., and Mirabel, P. (1989) Heteromolecular nucleation in the sulfuric acid–water system, *Atmos. Environ.* **23**, 2053–2057.

Jaecker-Voirol, A., Mirabel, P., and Reiss, H. (1987) Hydrates in supersaturated binary sulfuric acid–water vapor: A reexamination, *J. Chem. Phys.* **87**, 4849–4852.

Katz, J. L. (1970) Condensation of a supersaturated vapor. I. The homogeneous nucleation of the *n*-alkanes, *J. Chem. Phys.* **52**, 4733–4748.

Katz, J. L., Fisk, J. A., and Chakarov, V. M. (1994) Condensation of a supersaturated vapor IX. Nucleation on ions, *J. Chem. Phys.* **101**, 2309–2318.

Kerminen, V. M., Virkkula, A., Hillamo, R., Wexler, A. S., and Kulmala, M. (2000) Secondary organics and atmospheric cloud condensation nuclei production, *J. Geophys. Res.* **105**, 9255–9264.

Kulmala, M., and Laaksonen, A. (1990) Binary nucleation of water–sulfuric acid system: Comparison of classical theories with different H_2SO_4 saturation vapor pressures, *J. Chem. Phys.* **93**, 696–701.

Kulmala, M., Pirjola, L., and Mäkelä, J. M. (2000) Stable sulphate clusters as a source of new atmospheric particles, *Nature* **404**, 66–69.

Kulmala, M., Dal Maso, M., Mäkelä, J. M., Pirjola, L., Väkevä, M., Aalto, P., Miikkulainen, P., Hämeri, K., and O'Dowd, C. D. (2001) On the formation, growth and composition of nucleation mode particles, *Tellus B* **53**, 479–490.

Kulmala, M., Vehkamäki, H., Petäjä, T., Dal Maso, M., Lauri, A., Kerminen, V. M., Birmili, W., and McMurry, P. H. (2004) Formation and growth rates of ultrafine particles: A review of observations, *J. Aerosol Sci.* **35**, 143–176.

Kusaka, I., Wang, Z.-G., and Seinfeld, J. H. (1995a) Ion-induced nucleation: A density functional approach, *J. Chem. Phys.* **102**, 913–924.

Kusaka, I., Wang, Z.-G., and Seinfeld, J. H. (1995b) Ion-induced nucleation. II. Polarizable multipolar molecules, *J. Chem. Phys.* **103**, 8993–9009.

Laakso, L., Mäkelä, J. M., Pirjola, L., and Kulmala, M. (2002) Model studies of ion-induced nucleation in the atmosphere, *J. Geophys. Res.* **107** (doi: 10.1029/2002JD002140).

Laaksonen, A., Talanquer, V., and Oxtoby, D. W. (1995) Nucleation: Measurements, theory, and atmospheric applications, *Annu. Rev. Phys. Chem.* **46**, 489–524.

McDonald, J. E. (1964) Cloud nucleation on insoluble particles, *J. Atmos. Sci.* **21**, 109.

Mirabel, P., and Clavelin, J. L. (1978) On the limiting behavior of binary homogeneous nucleation theory, *J. Aerosol Sci.* **9**, 219–225.

Mirabel, P., and Katz, J. L. (1974) Binary homogeneous nucleation as a mechanism for the formation of aerosols, *J. Chem. Phys.* **60**, 1138–1144.

Mirabel, P., and Reiss, H. (1987) Resolution of the "Renninger–Wilemski problem" concerning the identification of heteromolecular nuclei, *Langmuir* **3**, 228–234.

Noppel, M., Vehkamäki, H., and Kulmala, M. (2002) An improved model for hydrate formation in sulfuric acid-water nucleation, *J. Chem. Phys.* **116**, 218–228.

Oxtoby, D. W. (1992a) Homogeneous nucleation: Theory and experiment, *J. Phys. Condensed Matter.* **4**, 7627–7650.

Oxtoby, D. W. (1992b) Nucleation, in *Fundamentals of Inhomogeneous Fluids*, D. Henderson, ed., Marcel Dekker, New York, pp. 407–442.

Peters, F., and Paikert, B. (1989) Experimental results on the rate of nucleation in supersaturated *n*-propanol, ethanol, and methanol vapors, *J. Chem. Phys.* **91**, 5672–5678.

Reiss, H. (1950) The kinetics of phase transition in binary systems, *J. Chem. Phys.* **18**, 840–848.

Reiss, H., Tabazadeh, A., and Talbot, J. (1990) Molecular theory of vapor phase nucleation: The physically consistent cluster, *J. Chem. Phys.* **92**, 1266–1274.

Schmitt, J. L. (1981) Precision expansion cloud chamber for homogeneous nucleation studies, *Rev. Sci. Instrum.* **52**, 1749–1754.

Schmitt, J. L. (1992) Homogeneous nucleation of liquid from the vapor phase in an expansion cloud chamber, *Metall. Trans. A* **23A**, 1957–1961.

Shi, G., Seinfeld, J. H., and Okuyama, K. (1990) Transient kinetics of nucleation, *Phys. Rev. A* **41**, 2101–2108.

Stanier, C. O., Khlystov, A. Y., and Pandis, S. N. (2004) Nucleation events during the Pittsburgh Air Quality Study: Description and relation to key meteorological, gas phase, and aerosol parameters, *Aerosol Sci. Technol.* **38**, 1–12.

Stauffer, D. (1976) Kinetic theory of two-component ("heteromolecular") nucleation and condensation, *J. Aerosol Sci.* **7**, 319–333.

Strey, R., and Wagner, P. E. (1982) Homogeneous nucleation of 1-pentanol in a two-piston expansion chamber for different carrier gases, *J. Phys. Chem.* **86**, 1013–1015.

Strey, R., Wagner, P. E., and Viisanen, Y. (1994) The problem of measuring homogeneous nucleation rates and molecular contents of nuclei: Progress in the form of nucleation pulse measurements, *J. Phys. Chem.* **98**, 7748–7758.

Thomson, J. J. (1906) *Conduction of Electricity through Gases*, Cambridge Univ. Press, Cambridge, UK.

Vehkamäki, H., Kulmala, M., Napari, I., Lehtinen, K. E. J., Timmreck, C., Noppel, M., and Laaksonen, A. (2002) An improved parameterization for sulfuric acid-water nucleation rates for tropospheric and stratospheric conditions, *J. Geophys. Res.* **107** (doi: 10.1029/2002JD002184).

Viisanen, Y., and Strey, R. (1994) Homogeneous nucleation rates for *n*-butanol, *J. Chem. Phys.* **101**, 7835–7843.

Volmer, M., and Weber, A. (1926) Keimbildung in übersättigten gebilden, *Z. Phys. Chem.* **119**, 277–301.

Wagner, P. E., and Strey, R. (1981) Homogeneous nucleation rates of water vapor measured in a two-piston expansion chamber, *J. Phys. Chem.* **85**, 2694–2698.

Weakleim, C. L., and Reiss, H. (1994) The factor $1/S$ in the classical theory of nucleation, *J. Phys. Chem.* **98**, 6408–6412.

Weakleim, C. L., and Reiss, H. (1993) Toward a molecular theory of vapor-phase nucleation. III. Thermodynamic properties of argon clusters from Monte Carlo simulations and a modified liquid drop theory, *J. Chem. Phys.* **99**, 5374–5383.

Wexler, A. S., Lurmann, F. W., and Seinfeld, J. H. (1994) Modelling urban and regional aerosols: I. Model development, *Atmos. Environ.* **28**, 531–546.

Wilemski, G. (1995) The Kelvin equation and self-consistent nucleation theory, *J. Chem. Phys.* **103**, 1119–1126.

Wyslouzil, B. E., and Seinfeld, J. H. (1992) Nonisothermal homogeneous nucleation, *J. Chem. Phys.* **97**, 2661–2670.

Wyslouzil, B. E., Seinfeld, J. H., Flagan, R. C., and Okuyama, K. (1991a) Binary nucleation in acid–water systems. I. Methanesulfonic acid–water, *J. Chem. Phys.* **94**, 6827–6841.

Wyslouzil, B. E., Seinfeld, J. H., Flagan, R. C., and Okuyama, K. (1991b) Binary nucleation in acid–water systems. II. Sulfuric acid–water and a comparison with methanesulfonic acid–water, *J. Chem. Phys.* **94**, 6842–6850.

Yu, F., and Turco, R. P. (2000) Ultrafine aerosol formation via ion-mediated nucleation, *Geophys. Res. Lett.* **27**, 883–886.

Zeldovich, Y. B. (1942) Theory of new-phase formation: Cavitation, *J. Exp. Theor. Phys. (USSR)* **12**, 525–538.

Zhang, K. M., and Wexler, A. S. (2002) A hypothesis for growth of fresh atmospheric nuclei, *J. Geophys. Res.* **107** (doi: 10.1029/2002JD002180).

12 Mass Transfer Aspects of Atmospheric Chemistry

12.1 MASS AND HEAT TRANSFER TO ATMOSPHERIC PARTICLES

Mass and energy transport to or from atmospheric particles accompanies their growth or evaporation. We would like to develop mathematical expressions describing the mass transfer rates between condensed and gas phases. The desired expressions for the vapor concentrations and temperature profiles around a growing or evaporating particle can be obtained by solving the appropriate mass and energy conservation equations.

Let us consider a particle of pure species A in air that also contains vapor molecules of A. Particle growth or evaporation depends on the direction of the net flux of vapor molecules relative to the particle. As we saw in Chapter 8, the mass transfer process will depend on the particle size relative to the mean free path of A in the surrounding environment. We will therefore start our discussion from the simpler case of a relatively large particle (mass transfer in the continuum regime) and then move to the other extreme (mass transfer in the kinetic regime).

12.1.1 The Continuum Regime

The unsteady-state diffusion of species A to the surface of a stationary particle of radius R_p is described by

$$\frac{\partial c}{\partial t} = -\frac{1}{r^2}\frac{\partial}{\partial r}(r^2 \tilde{J}_{A,r}) \tag{12.1}$$

where $c(r,t)$ is the concentration of A, and $\tilde{J}_{A,r}(r,t)$ is the molar flux of A (moles area^{-1} time^{-1}) at any radial position r. This equation is simply an expression of the mass balance in an infinitesimal spherical cell around the particle. The molar flux of species A through stagnant air is given by Fick's law (Bird et al. 1960)

$$\tilde{J}_{A,r} = x_A(\tilde{J}_{A,r} + \tilde{J}_{air,r}) - D_g\frac{\partial c}{\partial r} \tag{12.2}$$

Atmospheric Chemistry and Physics: From Air Pollution to Climate Change, Second Edition, by John H. Seinfeld and Spyros N. Pandis. Copyright © 2006 John Wiley & Sons, Inc.

where x_A is the mole fraction of A, $\tilde{J}_{air,r}$ the radial flux of air at position r, and D_g the diffusivity of A in air. Since air is not transferred to or from the particle, $\tilde{J}_{air,r} = 0$ at all r. Assuming dilute conditions, an assumption applicable under almost all atmospheric conditions, $x_A \simeq 0$ and (12.2) can be rewritten as

$$\tilde{J}_{A,r} = -D_g \frac{\partial c}{\partial r} \tag{12.3}$$

Combining (12.1) and (12.3), we obtain

$$\frac{\partial c}{\partial t} = D_g \left(\frac{\partial^2 c}{\partial r^2} + \frac{2}{r} \frac{\partial c}{\partial r} \right) \tag{12.4}$$

which is valid for transfer of A to a particle under dilute conditions. If c_∞ is the concentration of A far from the particle, c_s is its vapor-phase concentration at the particle surface, and the particle is initially in an atmosphere of uniform A with a concentration equal to c_∞, the corresponding initial and boundary conditions for (12.4) are

$$c(r,0) = c_\infty, \qquad r > R_p \tag{12.5}$$
$$c(\infty,t) = c_\infty \tag{12.6}$$
$$c(R_p,t) = c_s \tag{12.7}$$

The solution of (12.4) subject to (12.5) to (12.7) is (Appendix 12)

$$c(r,t) = c_\infty - \frac{R_p}{r}(c_\infty - c_s) + \frac{2R_p}{r\sqrt{\pi}}(c_\infty - c_s) \int_0^{(r-R_p)/2\sqrt{D_g t}} e^{-\xi^2} d\xi \tag{12.8}$$

The time dependence of the concentration at any radial position r is given by the third term on the right-hand side (RHS) of (12.8). Note that for large values of t, the upper limit of integration approaches zero and the concentration profile approaches its steady state given by

$$c(r) = c_\infty - \frac{R_p}{r}(c_\infty - c_s) \tag{12.9}$$

In Section 12.2.1 we will show that the characteristic time for relaxation to the steady-state value is on the order of 10^{-3} s or smaller for all particles of atmospheric interest. Rearranging (12.9), at steady state, we obtain

$$\frac{c(r) - c_\infty}{c_s - c_\infty} = \frac{R_p}{r} \tag{12.10}$$

The total flow of A (moles time^{-1}) toward the particle is denoted by J_c, the subscript c referring to the continuum regime, and is given by

$$J_c = -4\pi R_p^2 (\tilde{J}_A)_{r=R_p} \tag{12.11}$$

or using (12.9) and (12.3):

$$J_c = 4\pi R_p D_g (c_\infty - c_s) \qquad (12.12)$$

If $c_\infty > c_s$, the flow of molecules of A is toward the particle and if $c_\infty < c_s$ vice versa. The result given above was first obtained by Maxwell (1877) and (12.12) is often called the *Maxwellian flux*. Note that as c is the molar concentration of A, the units of J_c are moles per time. On the contrary, the units of $\tilde{J}_A$ are moles per area per time.

A mass balance on the growing or evaporating particle is

$$\frac{\rho_p}{M_A} \frac{d}{dt} \left(\frac{4}{3} \pi R_p^3 \right) = J_c \qquad (12.13)$$

where ρ_p is the particle density and M_A the molecular weight of A. Combining (12.12) with (12.13) gives

$$\frac{dR_p}{dt} = \frac{D_g M_A}{\rho_p R_p} (c_\infty - c_s) \qquad (12.14)$$

When c_∞ and c_s are constant, (12.14) can be integrated to give

$$R_p^2 = R_{p0}^2 + \frac{2 D_g M_A}{\rho_p} (c_\infty - c_s) t \qquad (12.15)$$

The use of the time-independent steady-state profile given by (12.9) to calculate the change of the particle size with time in (12.15) may seem inconsistent. Use of the steady-state diffusional flux to calculate the particle growth rate implies that the vapor concentration profile near the particle achieves steady state before appreciable growth occurs. Since growth does proceed hundreds of times more slowly than diffusion, the profile near the particle in fact remains at its steady-state value at all times. Growth of atmospheric particles for a constant gradient of $M_A(c_\infty - c_s) = 1\,\mu g\ m^{-3}$ between the bulk and surface concentrations of A is depicted in Figure 12.1.

Temperature Effects   During the condensation/evaporation of a particle latent heat is released/absorbed at the particle surface. This heat can be released either toward the particle or toward the exterior gas phase. As mass transfer continues, the particle surface temperature changes until the rate of heat transfer balances the rate of heat generation/consumption. The formation of the external temperature and vapor concentration profiles must be related by a steady-state energy balance to determine the steady-state surface temperature at all times during the particle growth.

The steady-state temperature distribution around a particle is governed by

$$u_r \frac{dT}{dr} = \alpha \frac{1}{r^2} \frac{d}{dr} \left(r^2 \frac{dT}{dr} \right) \qquad (12.16)$$

where $\alpha = k/\rho c_p$ is the thermal diffusivity of air and u_r is the mass average velocity at radial position r. The convective velocity u_r is the net result of the fluid motion due

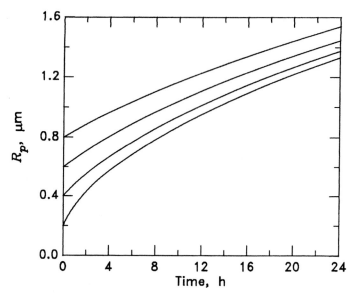

FIGURE 12.1   Growth of aerosol particles of different initial radii as a function of time for a constant concentration gradient of $1\,\mu g\,m^{-3}$ between the aerosol and gas phases ($D_g = 0.1\,cm^2\,s^{-1}$, $\rho_p = 1\,g\,cm^{-3}$).

to the concentration gradients (Pesthy et al. 1981). Equation (12.16) should be solved subject to

$$T(R_p) = T_s$$
$$T(\infty) = T_\infty$$

For a dilute system the first term in (12.16) can be neglected and (12.16) is simplified to the pure conduction equation

$$\frac{d^2T}{dr^2} + \frac{2}{r}\frac{dT}{dr} = 0 \tag{12.17}$$

with solution

$$T = T_\infty + \frac{R_p}{r}(T_s - T_\infty) \tag{12.18}$$

The criterion for neglecting the convective term in (12.16) is

$$\frac{D_g}{\alpha}\left(\frac{M_A}{M_{air}}\right)\ln\left(\frac{1 - x_{As}}{1 - x_{A\infty}}\right) \ll 1 \tag{12.19}$$

where M_{air} is the molecular weight of air and x_{As} and $x_{A\infty}$ are the mole fractions of A at the particle surface and far away from it. This convective flow is often referred to as *Stefan flow* and (12.19) provides a quantitative criterion for determining when it can be neglected. In most applications involving mass and heat transfer to atmospheric particles it can be neglected (Davis 1983).

Up to this point we have been avoiding the complications of the coupled mass and energy balances by treating c_s and T_s as known. The surface temperature is in general unknown and c_s depends on it. To determine T_s we need to write an energy balance on the particle

$$\tilde{J}_{A,r=R_p}\Delta H_v(4\pi R_p^2) = k\left(\frac{dT}{dr}\right)_{r=R_p}(4\pi R_p^2) + k_p\left(\frac{dT_p}{dr}\right)_{r=R_p}(4\pi R_p^2) \tag{12.20}$$

where k and k_p are the thermal conductivities of air and the particle, respectively; T and T_p are the air and particle temperatures; and ΔH_v is the molar heat released. The left-hand side (LHS) is the latent heat contribution to the energy balance, while the RHS includes the rates of heat conduction outward into the gas and inward from the particle surface. Chang and Davis (1974) solved the coupled mass and energy balances numerically. Their numerical solution shows that the last term in (12.20) can be neglected, indicating that the energy ΔH_v is transferred entirely to the gas phase.

Combining (12.18) and (12.9) with (12.20), we obtain

$$k(T_s - T_\infty) = \Delta H_v D_g(c_\infty - c_s) \tag{12.21}$$

where c_s is in general a function of T_s. For convenience this equation is often written as

$$\frac{T_s - T_\infty}{T_\infty} = \frac{\Delta H_v D_g}{kT_\infty}(c_\infty - c_s) \tag{12.22}$$

If $c_\infty \gg c_s$, then the temperature difference between the particle and the ambient gas is $\Delta H_v D_g c_\infty / k$. For slowly evaporating species and small heat of vaporization this temperature change is sufficiently small so that isothermal conditions are approached and (12.9) can be used.

12.1.2 The Kinetic Regime

For molecules in three-dimensional random motion the number of molecules Z_N striking a unit area per unit time is (Moore 1962)

$$Z_N = \frac{1}{4}N\bar{c}_A \tag{12.23}$$

where $\bar{c}_A$ is the mean speed of the molecules:

$$\bar{c}_A = \left(\frac{8kT}{\pi m_A}\right)^{1/2} \tag{12.24}$$

Under these conditions the molar flow J_k (moles time^{-1}) to a particle of radius R_p is

$$J_k = \pi R_p^2 \bar{c}_A \alpha (c_\infty - c_s) \qquad (12.25)$$

where α is the molecular accommodation coefficient [not to be confused with the thermal diffusivity in (12.16)]. The ratio of this kinetic regime flow to the continuum regime flow J_c is

$$\frac{J_k}{J_c} = \frac{\alpha \bar{c}_A}{4 D_g} R_p \qquad (12.26)$$

The accommodation coefficient will be assumed equal to unity in the next section, and the implications of this assumption will be discussed in Section 12.1.4.

12.1.3   The Transition Regime

The steady-state flow of vapor molecules to a sphere, when the particle is sufficiently large compared to the mean free path of the diffusing vapor molecules, is given by Maxwell's equation (12.12). Since this equation is based on the solution of the continuum transport equation, it is no longer valid when the mean free path of the diffusing vapor molecules becomes comparable to the particle diameter. At the other extreme, the expression based on the kinetic theory of gases (12.25) is also not valid in this intermediate regime where $\lambda \approx D_p$. When Kn ≈ 1, the phenomena are said to lie in the *transition regime*.

The concentration distributions of the diffusing species and background gas in the transition regime are governed rigorously by the Boltzmann equation. Unfortunately, there does not exist a general solution to the Boltzmann equation valid over the full range of Knudsen numbers for arbitrary masses of the diffusing species and the background gas. Consequently, most investigations of transport phenomena avoid solving directly the Boltzmann equation and restrict themselves to an approach based on so-called *flux matching*. Flux matching assumes that the noncontinuum effects are limited to a region $R_p \leq r \leq \Delta + R_p$ beyond the particle surface and that continuum theory applies for $r \geq \Delta + R_p$. The distance Δ is then of the order of the mean free path λ and within this inner region the basic kinetic theory of gases is assumed to apply.

Fuchs Theory   The matching of continuum and free molecule fluxes dates back to Fuchs (1964), who suggested that by matching the two fluxes at $r = \Delta + R_p$, one may obtain a boundary condition on the continuum diffusion equation. This condition is, assuming unity accommodation coefficient

$$4\pi R_p^2 \left(\frac{1}{4} \bar{c}_A \right) [c(R_p + \Delta) - c_s] = D \left(\frac{dc}{dr} \right)_{r = R_p + \Delta} 4\pi (R_p + \Delta)^2 \qquad (12.27)$$

Then, the steady-state continuum transport equation for a dilute system is

$$\frac{d^2 c}{dr^2} + \frac{2}{r} \frac{dc}{dr} = 0 \qquad (12.28)$$

Using as boundary conditions (12.27) and $c(\infty) = c_\infty$ one obtains the solution

$$c(r) = c_\infty - \frac{R_p}{r}(c_\infty - c_s)\beta_F \qquad (12.29)$$

where the correction factor β_F is given by

$$\beta_F = \frac{[1 + (\Delta/R_p)]\bar{c}_A R_p}{\bar{c}_A R_p + 4D[1 + (\Delta/R_p)]} \qquad (12.30)$$

Relating the binary diffusivity and the mean free path using $D_{AB}/\lambda_{AB}\bar{c}_A = \frac{1}{3}$ and letting $Kn = \lambda_{AB}/R_p$, one obtains

$$\frac{J}{J_c} = 0.75\frac{1 + Kn\Delta/\lambda_{AB}}{0.75 + Kn + (\Delta/\lambda_{AB})Kn^2} \qquad (12.31)$$

Note that the definition of the mean free path by $D_{AB}/\lambda_{AB}\bar{c}_A = \frac{1}{3}$ implies, using (12.26), that, for $\alpha = 1$

$$\frac{J_k}{J_c} = \frac{3}{4\,Kn} \qquad (12.32)$$

and the Fuchs relation (12.31) also implies, using (12.32), that

$$\frac{J}{J_k} = \frac{1 + Kn\Delta/\lambda_{AB}}{1 + Kn\Delta/\lambda_{AB} + 0.75\,Kn^{-1}} \qquad (12.33)$$

The value of Δ used in the expressions above was not specified in the original theory and must be adjusted empirically or estimated by independent theory. Several choices for Δ have been proposed; the simplest, due to Fuchs, is $\Delta = 0$. Other suggestions include $\Delta = \lambda_{AB}$ and $\Delta = 2D_{AB}/\bar{c}_A$ (Davis 1983).

Fuchs–Sutugin Approach Fuchs and Sutugin (1971) fitted Sahni's (1966) solution to the Boltzmann equation for $z \ll 1$, where $z = M_A/M_{air}$ is the molecular weight ratio of the diffusing species and air, to produce the following transition regime interpolation formula:

$$\frac{J}{J_c} = \frac{1 + Kn}{1 + 1.71\,Kn + 1.33\,Kn^2} \qquad (12.34)$$

Equation (12.34) is based on results for $z \ll 1$ and therefore is directly applicable to light molecules in a heavier background gas. The mean free path included in the definition of the Knudsen number in (12.34) is given by

$$\lambda_{AB} = \frac{3D_{AB}}{\bar{c}_A}$$

For $Kn \to 0$ both (12.31) and (12.34) reduce to the correct limit $J/J_c = 1$. For the kinetic limit $Kn \to \infty$ both (12.31) and (12.34) give $J_k/J_c = 3/(4\,Kn)$.

Dahneke Approach   Dahneke (1983) used the flux matching approach of Fuchs but, assuming that $\Delta = \lambda_{AB}$ and defining $D_{AB}/(\lambda_{AB}\bar{c}_A) = \frac{1}{2}$, obtained

$$\frac{J}{J_c} = \frac{1 + Kn}{1 + 2\,Kn(1 + Kn)} \tag{12.35}$$

where $Kn = \lambda_{AB}/R_p$. The mean free path included in the definition of the Knudsen number in (12.35) is given by

$$\lambda_{AB} = \frac{2\,D_{AB}}{\bar{c}_A}$$

Note here that for $Kn \to 0$, $J/J_c \to 1$ as expected. On the other hand, for $Kn \to \infty$, $J/J_c \to 1/(2\,Kn)$. This limit is in agreement with (12.26) because $D_{AB}/(\lambda_{AB}\bar{c}_A) = \frac{1}{2}$ and therefore the expressions are consistent.

Loyalka Approach   Loyalka (1983) constructed improved interpolation formulas for mass transfer in the transition regime by solving the BGK model [Bhatnagar, Gross, and Krook (Bhatnagar et al. 1954)] of the Boltzmann equation to obtain

$$\frac{J}{J_k} = \frac{\sqrt{\pi}\,Kn(1 + 1.333\,Kn)}{1 + 1.333\,Kn + (1.333\sqrt{\pi}Kn + \zeta_c)Kn} \tag{12.36}$$

The mean free path used by Loyalka was defined by

$$\lambda_{AB} = \frac{4}{\sqrt{\pi}}\frac{D_{AB}}{\bar{c}_A} \tag{12.37}$$

and the mass transfer jump coefficient had a value $\zeta_c = 1.0161$. Williams and Loyalka (1991) pointed out that (12.36) does not have the correct shape near the free-molecule limit.

Sitarski–Nowakowski Approach   None of the approaches given above describes the dependence of the transition regime mass flux on the molecular mass ratio z of the condensing/evaporating species and the surrounding gas. Sitarski and Nowakowski (1979) applied the 13-moment method of Grad (Hirschfelder et al. 1954) to solve the Boltzmann equation to obtain

$$\frac{J}{J_k} = \frac{Kn(1 + a\,Kn)}{b + c\,Kn + d\,Kn^2} \tag{12.38}$$

$$a = \frac{3\,\beta(1 + z)^2}{4(3 + 5z)}, \qquad b = \frac{4(9 + 10z)}{15\pi(1 + z)^2}$$

$$c = \frac{\beta(1 + 2z)}{\pi(3 + 5z)} + \frac{1}{2\beta}, \qquad d = \frac{9(1 + z)^2}{8(3 + 5z)} \tag{12.39}$$

where $\beta = 1$ for unity accommodation coefficient and $z = M_A/M_{air}$ is the molecular weight ratio. This result is obviously incorrect near the free-molecule regime, because in

TABLE 12.1   Transition Regime Formulas for Diffusion of Species A in a Background Gas B to an Aerosol

J/J_c	Mean Free Path Definition	Reference
$\dfrac{0.75\,\alpha(1 + \mathrm{Kn}\Delta/\lambda_{AB})}{0.75\,\alpha + \mathrm{Kn} + (\Delta/\lambda_{AB})\mathrm{Kn}^2}$	$\dfrac{3D_{AB}}{\bar{c}_A}$	Fuchs (1964)
$\dfrac{0.75\,\alpha(1 + \mathrm{Kn})}{\mathrm{Kn}^2 + \mathrm{Kn} + 0.283\,\mathrm{Kn}\alpha + 0.75\,\alpha}$	$\dfrac{3D_{AB}}{\bar{c}_A}$	Fuchs and Sutugin (1971)
$\dfrac{1 + \mathrm{Kn}}{1 + 2\,\mathrm{Kn}(1 + \mathrm{Kn})/\alpha}$	$\dfrac{2D_{AB}}{\bar{c}_A}$	Dahneke (1983)
$\dfrac{1 + 1.333\,\mathrm{Kn}}{1 + 1.333\,\mathrm{Kn} + (1.333\sqrt{\pi}\mathrm{Kn} + 1)\mathrm{Kn}}$	$\dfrac{4}{\sqrt{\pi}}\dfrac{D_{AB}}{\bar{c}_A}$	Loyalka (1983)
$\dfrac{b(1 + a\mathrm{Kn})}{b + c\mathrm{Kn} + d\,\mathrm{Kn}^2}$	$4b\dfrac{D_{AB}}{\bar{c}_A}$ [see (12.39) for a, b, c, and d]	Sitarski and Nowakowski (1979)

the limit $\mathrm{Kn} \to \infty$, (12.38) yields $J \to 0.666\,J_k$. Therefore we expect (12.38) to be in error for relatively high values of Kn.

Table 12.1 summarizes the transition regime expressions that we have presented in this section. Predictions of mass transfer rates of the preceding four theories are shown as a function of the particle diameter in Figure 12.2. All approaches give comparable results

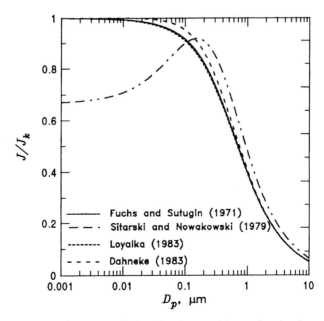

FIGURE 12.2 Mass transfer rate predictions for the transition regime by the approaches of (a) Fuchs and Sutugin (1971), (b) Dahneke (1983), (c) Loyalka (1983), and (d) Sitarski and Nowakowski (1979) ($z = 15$) as a function of particle diameter. Accommodation coefficient $\alpha = 1$.

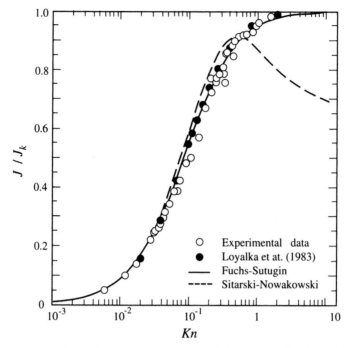

FIGURE 12.3   Comparison of experimental dibutyl phthalate evaporation data with the theories of Loyalka et al. (1989), Sitarski and Nowakowski (1979) (for $z = 15$), and the equation of Fuchs and Sutugin (1970). (Reprinted from *Aerosol Science and Technology*, **25**, Li and Davis, 11–21. Copyright 1995, with kind permission from Elsevier Science Ltd., The Boulevard, Langford Lane, Kidlington OX5 1GB, UK.)

for particle diameters larger than 0.2 µm, even if they employ different definitions of the Knudsen number and different functional dependencies of the mass transfer rate on the Knudsen number. This agreement indicates that as long as one uses a mean free path consistent with the mass transfer theory the final result will differ little from theory to theory. The theory of Sitarski and Nowakowski (1979), although it is the only one that includes an explicit dependence of the mass transfer rate on z, gives erroneous results for particles smaller than 0.2 µm in this case (Figure 12.2). The dependence of the rate itself on z is rather weak and for $z = 5$–15, the Fuchs, Dahneke, and Loyalka formulas are in agreement with the Sitarski–Nowakowski results. Li and Davis (1995) compared the results of the above mentioned theories with measurements of the evaporation rates of dibutyl phthalate (DBP) in air (Figure 12.3). All theories are in agreement with the data with the exception of the theory of Sitarski and Nowakowski (1979), which exhibits deviations for Kn > 0.2.

12.1.4   The Accommodation Coefficient

Up to this point we have assumed that once a vapor molecule encounters the surface of a particle its probability of sticking is unity. This assumption can be relaxed by introducing an accommodation coefficient α, where $0 \leq \alpha \leq 1$. The flux of a gas A to a spherical particle in the kinetic regime is then given by (12.25).

The transition regime formulas can then be extended to account for imperfect accommodation by multiplying the LHS of (12.27) by α. The Fuchs expression in (12.31) becomes

$$\frac{J}{J_c} = 0.75\,\alpha\,\frac{1 + Kn\Delta/\lambda_{AB}}{0.75\,\alpha + Kn + (\Delta/\lambda_{AB})Kn^2} \qquad (12.40)$$

and

$$\frac{J_k}{J_c} = \frac{3\,\alpha}{4\,Kn} \qquad (12.41)$$

The expression (12.35) by Dahneke (1983) becomes

$$\frac{J}{J_c} = \frac{1 + Kn}{1 + 2\,Kn(1 + Kn)/\alpha} \qquad (12.42)$$

where as the Fuchs–Sutugin (1971) approach gives

$$\frac{J}{J_c} = \frac{0.75\,\alpha(1 + Kn)}{Kn^2 + Kn + 0.283\,Kn\alpha + 0.75\,\alpha} \qquad (12.43)$$

The formula of Loyalka is applicable only for $\alpha = 1$, but the theory of Sitarski and Nowakowski (1979) can be used for any accommodation coefficient setting

$$\beta = \frac{\alpha}{2 - \alpha} \qquad (12.44)$$

Figure 12.4 shows mass transfer rates as a function of particle diameter for the three approaches for accommodation coefficient values of 1, 0.1, and 0.01.

12.2 MASS TRANSPORT LIMITATIONS IN AQUEOUS-PHASE CHEMISTRY

Dissolution of atmospheric species into cloud droplets followed by aqueous-phase reactions involves the following series of steps:

1. Diffusion of the reactants from the gas phase to the air–water interface
2. Transfer of the species across the interface
3. Possible hydrolysis/ionization of the species in the aqueous phase
4. Aqueous-phase diffusion of the ionic and nonionic species inside the cloud drop
5. Chemical reaction inside the droplet

These steps must occur during the production of sulfate or for other aqueous-phase chemical reactions within cloud drops. In Chapter 7 we studied the aqueous-phase kinetics of a number of reactions, corresponding to step 5 above. Chapter 7 provided the tools for the

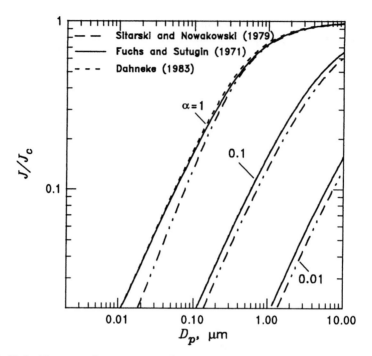

FIGURE 12.4 Mass transfer rates as a function of particle diameter for accommodation coefficient values 1.0, 0.1, and 0.01 for the approaches of Sitarski and Nowakowski (1979), Fuchs and Sutugin (1970), and Dahneke (1983).

calculation of the sulfate production rate at a given point inside a cloud drop, provided that we know the reactant concentrations at this point. For some of the Chapter 7 calculations we assumed that the cloud droplets were saturated with reactants, or equivalently that the concentration of the species inside the droplet is uniform and satisfies at any moment Henry's law equilibrium with the bulk gas phase. In mathematical terms we assumed that the aqueous-phase concentration of a species A at location r, at time t, satisfies

$$[A(r,t)] = H_A^* p_\infty(t) \tag{12.45}$$

where H_A^* is the effective Henry's law coefficient for A in M atm^{-1} and p_∞ is the A partial pressure far from the droplet in the bulk gas phase. Use of (12.45) simplified significantly our order of magnitude estimates in Chapter 7, but its validity relies on the following assumptions:

1. The time to establish a gas-phase steady-state concentration profile of A around the cloud droplet is short.
2. Gas-phase diffusion is rapid enough that the concentration of A around the droplet is approximately constant. Mathematically, this implies that $p_\infty(t) = p(R_p,t)$, where $p(R_p,t)$ is the partial pressure of A at the droplet surface.
3. The transfer of A across the interface is sufficiently rapid, so that local Henry's law equilibrium is always satisfied. Mathematically, this is equivalent to $[A(R_p,t)] = H_A^* p(R_p,t)$.

4. The time to establish a steady-state concentration profile inside the droplet is very short.

5. The time to establish the ionization/hydrolysis equilibria is very short.

6. Aqueous-phase diffusion is rapid enough, so that the concentration of A inside the droplet is approximately constant.

If all six of these assumptions are satisfied, then (12.45) is valid and the calculations are simplified considerably. Our goal now will be to first quantify the rates of the five necessary process steps (gas-phase diffusion, interfacial transport, ionization, aqueous-phase diffusion, reaction) calculating appropriate timescales. Then, we will compare these rates to those of aqueous-phase chemical reactions. Finally, we will integrate our conclusions, developing overall reaction rate expressions that take into consideration, when necessary, the effects of the mass transport limitations.

Figure 12.5 depicts schematically the gas- and aqueous-phase concentrations of A in and around a droplet. The aqueous-phase concentrations have been scaled by $H_A^* RT$, to remove the difference in the units of the two concentrations. This scaling implies that the two concentration profiles should meet at the interface if the system satisfies at that point Henry's law. In the ideal case, described by (12.45), the concentration profile after the scaling should be constant for any r. However, in the general case the gas-phase mass transfer resistance results in a drop of the concentration from $c_A(\infty)$ to $c_A(R_p)$ at the air–droplet interface. The interface resistance to mass transfer may also cause deviations from Henry's law equilibrium indicated in Figure 12.5 by a discontinuity. Finally, aqueous-phase transport limitations may result in a profile of the concentration of A in the aqueous phase from $[A(R_p)]$ at the droplet surface to $[A(0)]$ at the center. All these mass transfer limitations, even if the system can reach a pseudo–steady state, result in reductions of the concentration of A inside the droplet, and slow down the aqueous-phase chemical reactions.

Our discussion has indicated that the existence of several mass transfer processes occurring simultaneously can be significantly simplified by our understanding of the corresponding timescales. We start by analyzing the various processes independently.

12.2.1 Characteristic Time for Gas-Phase Diffusion to a Particle

Let us assume that we introduce a particle of radius R_p consisting of a species A in an atmosphere with a uniform gas-phase concentration of A equal to c_∞. Initially, the concentration profile of A around the particle will be flat and eventually after time τ_{dg} it will relax to its steady state. This timescale, τ_{dg}, corresponds to the time required by gas-phase diffusion to establish a steady-state profile around a particle. It should not be confused with the timescale of equilibration of the particle with the surrounding atmosphere. We assume that c_∞ remains constant and that the concentration of A at the particle surface (equilibrium) concentration is c_s and also remains constant.

This problem was studied in Section 12.1.1, and the change of the concentration profile with time is given by (12.8). The characteristic timescale of the problem can be derived by nondimensionalizing the differential equation describing the problem, namely, (12.4). The characteristic lengthscale of the problem is the particle radius R_p, the characteristic concentration c_∞, and the characteristic timescale, our unknown τ_{dg}. We define dimensionless variables by dividing the problem variables by their characteristic values,

$$\tau = \frac{t}{\tau_{dg}} \qquad x = \frac{r}{R_p} \qquad \phi = \frac{c - c_\infty}{c_s - c_\infty} \qquad (12.46)$$

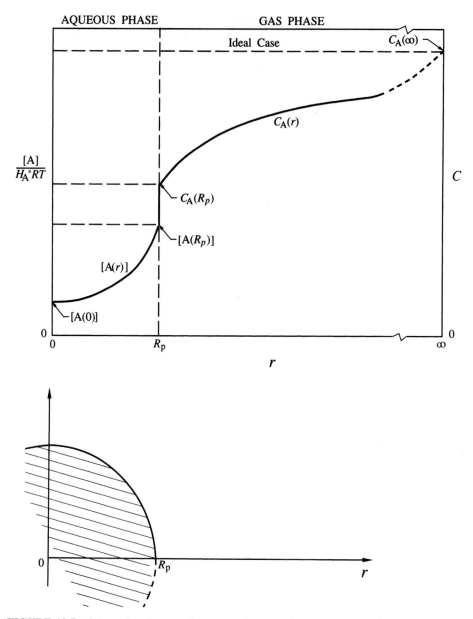

FIGURE 12.5 Schematic of gas- and aqueous-phase steady-state concentration profiles for the case where there are gas-phase, interfacial, and aqueous-phase mass transport limitations. Also shown is the ideal case where there are no mass transport limitations.

Equation (12.4) can be rewritten as

$$\left(\frac{R_p^2}{\tau_{dg}D_g}\right)\frac{\partial\phi}{\partial\tau} = \frac{\partial^2\phi}{\partial x^2} + \frac{2}{x}\frac{\partial\phi}{\partial x} \tag{12.47}$$

$$\phi(x,0) = 0 \qquad \phi(\infty,\tau) = 0 \qquad \phi(1,\tau) = 1$$

Note that all the scaling in the problem is included in the dimensionless term $(R_p^2/\tau_{dg}D_g)$ and, as the rest of the terms in the differential equation are of order one, this term should also be of order unity. Therefore

$$\tau_{dg} \simeq \frac{R_p^2}{D_g} \qquad (12.48)$$

This timescale can also be obtained from the complete solution of the problem in (12.8). Note that the time-dependent term approaches zero as the upper limit of the integration approaches zero or when $2\sqrt{D_g t} \gg r - R_p$ or equivalently when $t \gg (r - R_p)^2/4\,D_g$. For a point close to the particle surface $(r - R_p)^2 \leq R_p^2$ and

$$\tau_{dg} = \frac{R_p^2}{4\,D_g} \qquad (12.49)$$

Note that for a point where $r \gg R_p$, the time-dependent term is practically zero because of the existence of the term (R_p/r) in front of the integral. One should not be bothered by the different numerical factor in the two timescales in (12.48) and (12.49). These timescales are, by definition, order-of-magnitude estimates and as such either value is sufficient for estimation purposes.

The characteristic timescale of (12.49) can be evaluated for a typical molecular gas-phase diffusivity of $D_g = 0.1\ \text{cm}^2\text{s}^{-1}$ as a function of the particle radius R_p:

$R_p(\mu m)$	0.1	10	100
$\tau_{dg}(s)$	2.5×10^{-10}	2.5×10^{-6}	2.5×10^{-4}

We conclude that the characteristic time for attaining a steady-state concentration profile around particles of atmospheric size is smaller than 1 ms. Since we are interested in changes that occur in atmospheric particles and droplets over timescales of several minutes, we can safely neglect this millisecond transition and assume that the concentration is always at steady state and the concentration profile is given in the continuum regime by (12.9) and the flux by (12.12).

12.2.2 Characteristic Time to Achieve Equilibrium in the Gas–Particle Interface

The gas–particle interface offers a resistance for mass transfer. We next want to obtain expressions for the characteristic timescales associated with establishing phase equilibrium at this interface. Different definitions and estimates of this timescale exist in the literature (Schwartz and Freiberg 1981; Kumar 1989a), leading to considerable debate (Schwartz 1989; Kumar 1989b; Shi and Seinfeld 1991). We seek to determine the characteristic time for saturation of the droplet as governed by interfacial transport. For this calculation we assume that at $t = 0$, a droplet of pure water is immersed in air containing species A. We assume that the gas-phase partial pressure of A at the droplet surface remains constant at p_0 throughout the process. As A is absorbed into the liquid, its concentration will eventually reach its equilibrium value, C_{eq}. We assume that this equilibrium is described by Henry's law $C_{eq} = H_A p_0$. The timescale for the increase of the droplet surface concentration from the initial zero value to C_{eq} is τ_p, the aqueous-phase interfacial timescale.

At the liquid surface, molecules are arriving from the gas, molecules are leaving the surface back to the gas, and molecules are diffusing into the liquid phase. Let us term these fluxes R_g^-, R_g^+, and R_l, respectively. Since the surface is presumed to have no thickness $R_g^- = R_g^+ + R_l$. The flux of molecules from the gas to the interface is given by the kinetic theory of gases

$$R_g^- = \frac{1}{4}\alpha \bar{c}_A N_A \tag{12.50}$$

where α is the accommodation coefficient, $\bar{c}_A = (8RT/\pi M_A)^{1/2}$ the mean speed of molecules, R the gas constant, M_A the molecular weight of A, and $N_A = (p_0/RT)$ the molecular number concentration of A. This flux can be then rewritten as

$$R_g^- = \frac{\alpha p_0}{(2\pi M_A RT)^{1/2}} \tag{12.51}$$

The rate of evaporation R_g^+ of A from the liquid depends on the surface concentration of A, $[A(R_p, t)]$. The evaporation process does not "know" whether equilibrium has been reached; it simply expels molecules at a rate dependent on $[A(R_p, t)]$. If the two phases were at equilibrium, then $R_g^+ = R_g^-$, where R_g^- is given by (12.51). If the two phases are not in equilibrium, we replace p_0 by the corresponding $p_s(t)$, where $p_s(t)$ is the gas-phase partial pressure of A just above the surface of the droplet at any time t. In so doing

$$R_g^+ = \frac{\alpha p_s(t)}{(2\pi M_A RT)^{1/2}} \tag{12.52}$$

This amounts to saying that the rate of evaporation from the droplet at any time can be calculated from the *equilibrium rate* corresponding to the instantaneous partial pressure of A just above the surface of the drop. From Henry's law, $[A(R_p, t)] = H_A^* p_s(t)$, and therefore we can rewrite (12.52) using $[A(R_p, t)]$, the concentration of A at the droplet surface as

$$R_g^+ = \frac{\alpha[A(R_p, t)]}{H_A^* (2\pi M_A RT)^{1/2}} \tag{12.53}$$

The net flux across the interface toward the liquid phase is

$$R_l = R_g^- - R_g^+ = \frac{\alpha}{(2\pi M_A RT)^{1/2}}\left(p_0 - \frac{[A(R_p, t)]}{H_A^*}\right) \tag{12.54}$$

Note that in this case when interfacial equilibrium is reached $p_0 = [A(R_p)]/H_A^*$ and the net flux into the droplet is zero (the droplet is filled with A). The evolution of the aqueous-phase concentration of A within the drop $[A(r, t)]$ is governed by the diffusion equation,

$$\frac{\partial[A]}{\partial t} = D_{aq}\left[\frac{\partial^2[A]}{\partial r^2} + \frac{2}{r}\frac{\partial[A]}{\partial r}\right] \tag{12.55}$$

in which D_{aq} is the diffusion coefficient of A in water. Using (12.54) to describe the flux into the drop, we have the following initial and boundary conditions:

$$[A(r, 0)] = 0$$

$$\left(\frac{\partial [A]}{\partial r}\right)_{r=0} = 0$$

$$-D_{aq}\left(\frac{\partial [A]}{\partial r}\right)_{r=R_p} = \frac{\alpha}{(2\pi M_A RT)^{1/2}}\left(p_0 - \frac{[A(R_p, t)]}{H_A^*}\right) \qquad (12.56)$$

Equation (12.55) subject to (12.56) can be solved by either separation of variables or Laplace transform (Kumar 1989a) to obtain

$$[A(r, t)] = p_0 H_A^*\left[1 - \sum_{n=1}^{\infty} \frac{2K R_p^2 \sin(\beta_n r/R_p)}{r D_{aq}(N + N^2 + \beta_n^2)\sin\beta_n}\exp\left(\frac{-\beta_n^2 D_{aq}t}{R_p^2}\right)\right] \qquad (12.57)$$

where

$$K = \frac{\alpha}{H_A^*\sqrt{2\pi M_A RT}} \qquad N = \frac{\alpha R_p}{H_A^* D_{aq}\sqrt{2\pi M_A RT}} - 1 \qquad (12.58)$$

and β_n is the nth positive root of

$$\beta \cot\beta + N = 0 \qquad (12.59)$$

Kumar (1989a) showed that the infinite series in (12.57) converges rapidly and that for the timescale calculation use of only the first term is sufficient; neglecting the subsequent terms introduces an error less than a few percent in the solution. The desired timescale is obtained from the first exponential term as

$$\tau_p = \frac{R_p^2}{\beta_1^2 D_{aq}} \qquad (12.60)$$

where β_1 is the first positive root of (12.59). For very soluble gases Kumar (1989a) showed that the timescale can be approximated by

$$\tau_p \simeq \frac{R_p H_A^*\sqrt{2\pi M_A RT}}{3\alpha} \qquad (12.61)$$

while for gases with very low solubility (low H_A^*)

$$\tau_p \simeq \frac{R_p^2}{\pi^2 D_{aq}} \qquad (12.62)$$

Let us estimate these timescales for a 10-μm-radius droplet at 298 K. For O_3, a gas with low water solubility, the calculated timescale using (12.62) is 5×10^{-3} s. For SO_2 at pH 5, and an accommodation coefficient of 0.1, the timescale using (12.61) is approximately 0.05 s. Since the timescale depends linearly on the effective Henry's law constant, it will be even smaller for lower pH values, while it will increase to approximately 1 s at pH 6. For H_2O_2 with an accommodation coefficient on water over 0.2, the timescale is less than 0.1 s. For relatively soluble species like NH_3 the characteristic time is roughly 1 s for $\alpha = 1$ and 18 s for $\alpha = 0.1$. For extremely soluble species like nitric acid the timescale calculated by (12.61) is on the order of days. However, such a calculation is flawed, because in the derivation of the timescale we assumed that there is a constant vapor pressure p_0 at the surface of the droplet. Such an assumption is not applicable to extremely soluble species, like HNO_3, that dissolve practically completely in the aqueous phase and whose gas-phase concentration is reduced by several orders of magnitude. If one accounts for the reduction in p_0 during the filling of the droplet with nitric acid, then a timescale of less than 1 s is calculated (Jacob 1985; Kumar 1989a).

These estimates indicate that establishment of the Henry's law equilibrium at the interface is a rapid process both for insoluble species and for soluble species with accommodation coefficients above 0.1. For the most important species in atmospheric aqueous-phase chemistry (SO_2, O_3, HNO_3, NH_3) the timescale is less than 1 s for typical atmospheric conditions.

12.2.3 Characteristic Time of Aqueous Dissociation Reactions

The next process in the chain is ionization. For SO_2, for example, we know that subsequent to absorption we have

$$SO_2 \cdot H_2O \rightleftharpoons H^+ + HSO_3^-$$

We are interested in determining the characteristic time to establish this equilibrium. To do this, we wish to derive an expression for the characteristic time to reach equilibrium of the general reversible reaction

$$A \rightleftharpoons B + C$$

with k_f the reaction constant for the first-order forward reaction and k_r for the second-order reverse reaction. At equilibrium

$$k_f[A]_e = k_r[B]_e[C]_e$$

Let us define the extent of reaction η by $[A] = [A]_0 - \eta$, $[B] = [B]_0 + \eta$, and $[C] = [C]_0 + \eta$, and the equilibrium extent η_e by $[A]_e = [A]_0 - \eta_e$, $[B]_e = [B]_0 + \eta_e$, and $[C]_e = [C]_0 + \eta_e$. Then we can define $\Delta\eta = \eta - \eta_e$ to get

$$[A] = [A]_e - \Delta\eta$$
$$[B] = [B]_e + \Delta\eta$$
$$[C] = [C]_e + \Delta\eta \tag{12.63}$$

The rate of disappearance of A is given by

$$\frac{d[A]}{dt} = -k_f[A] + k_r[B][C] \tag{12.64}$$

Using (12.63) together with the equilibrium relation, we obtain from (12.64)

$$\frac{d\Delta\eta}{dt} = -a\,\Delta\eta - b\,\Delta\eta^2 \tag{12.65}$$

where $a = k_f + k_r[B]_e + k_r[C]_e$ and $b = k_r$. Integrating this equation from $\Delta\eta = \Delta\eta_0$ at $t = 0$, we obtain

$$\frac{\Delta\eta}{\Delta\eta_0}\frac{a + b\,\Delta\eta_0}{a + b\,\Delta\eta} = e^{-at} \tag{12.66}$$

Combining this result together with (12.63) gives

$$\frac{[A] - [A]_e}{[A]_0 - [A]_e}\left[\frac{1 + (b/a)([A]_e - [A]_0)}{1 + (b/a)([A]_e - [A])}\right] = e^{-at} \tag{12.67}$$

where

$$\frac{b}{a} = \frac{1}{(k_f/k_r) + [B]_e + [C]_e} \tag{12.68}$$

where k_f/k_r is the equilibrium constant K.

Let us apply the foregoing analysis to the case of

$$SO_2 \cdot H_2O \rightleftharpoons H^+ + HSO_3^-$$

We note from (12.67) and (12.68) that for all atmospheric conditions of interest the terms $[SO_2 \cdot H_2O]_e - [SO_2 \cdot H_2O]_0$ and $[SO_2 \cdot H_2O]_e - [SO_2 \cdot H_2O]$ have the same order of magnitude as $[SO_2 \cdot H_2O]$ and also that

$$\frac{[SO_2 \cdot H_2O]}{K_{s1} + [H^+]_e + [HSO_3^-]_e} \ll 1 \tag{12.69}$$

For example, this term for pH = 4 and $p_{SO_2} = 10^{-9}$ atm (1 ppb) is approximately equal to 0.01. Therefore, setting the expression in brackets in (12.67) equal to unity, this expression for the approach to equilibrium simplifies to

$$\frac{[SO_2 \cdot H_2O] - [SO_2 \cdot H_2O]_e}{[SO_2 \cdot H_2O]_0 - [SO_2 \cdot H_2O]_e} = e^{-at} \tag{12.70}$$

and the characteristic time that we are seeking is just a^{-1}. Thus the characteristic time to achieve equilibrium is

$$\tau_i = \frac{1}{k_f + k_r[\text{H}^+]_e + k_r[\text{HSO}_3^-]_e} \qquad (12.71)$$

For pH $= 4$, $k_f = 3.4 \times 10^6$ s^{-1}, and $k_r = 2 \times 10^8$ M^{-1} s^{-1}, we find that

$$\tau_i \simeq 2 \times 10^{-7} \text{ s}$$

Repeating the calculation at other pH values, we find that this timescale depends only weakly on pH, remaining below 1 μs. Schwartz and Freiberg (1981) estimated the timescale for the second dissociation of S(IV). Using an approach similar to the one outlined above, they found that this timescale increases with increasing pH and remains below 10^{-3} s for pH values below 8. Thus, in the case of the sulfur equilibria, it can be assumed that ionization equilibrium is achieved virtually instantaneously upon absorption of SO_2.

12.2.4   Characteristic Time of Aqueous-Phase Diffusion in a Droplet

Consider unsteady-state diffusion of a dissolved species in a droplet, initially free of solute, when the surface concentration is raised to C_s at $t = 0$. The problem can then be stated mathematically as

$$\frac{\partial C}{\partial t} = D_{aq}\left[\frac{\partial^2 C}{\partial r^2} + \frac{2}{r}\frac{\partial C}{\partial r}\right] \qquad (12.72)$$

$$C(r,0) = 0$$

$$\left(\frac{\partial C}{\partial r}\right)_{r=0} = 0$$

$$C(R_p, t) = C_s \qquad (12.73)$$

Note that the problem statement is very similar to that solved for interfacial transport with a different boundary condition at the droplet surface. A standard separation of variables solution of (12.72) and (12.73) gives

$$\frac{C(r,t)}{C_s} = 1 + \frac{R_p}{r}\sum_{n=1}^{\infty}(-1)^n\frac{2}{n\pi}\sin\frac{n\pi r}{R_p}\exp\left(-\frac{n^2\pi^2 D_{aq}t}{R_p^2}\right) \qquad (12.74)$$

The characteristic time for aqueous-phase diffusion τ_{da} is obtained from the first term in the exponential in the solution:

$$\tau_{da} = \frac{R_p^2}{\pi^2 D_{aq}} \qquad (12.75)$$

Note that the timescale is proportional to the square of the droplet radius. A typical value of D_{aq} is $10^{-5}\,cm^2\,s^{-1}$, so we can evaluate τ_{da} as function of R_p. For a typical cloud droplet of $10\,\mu m$ radius, $\tau_{da} = 0.01\,s$. For a rather large cloud droplet of $100\,\mu m$ radius the timescale is $1\,s$. For raindrops of size $1\,mm$ the timescale becomes appreciable as it is equal to $100\,s$.

12.2.5 Characteristic Time for Aqueous-Phase Chemical Reactions

To develop an expression for the characteristic time for aqueous-phase chemical reaction, let us consider as an example the oxidation of S(IV). The characteristic time can be calculated by dividing the S(IV) aqueous-phase concentration by its oxidation rate:

$$\tau_{ra} = -\frac{[S(IV)]}{d[S(IV)]/dt} \qquad (12.76)$$

It is also possible to define a characteristic time for chemical reaction relative to the gas-phase SO_2 concentration:

$$\tau_{rg} = -\frac{[SO_2(g)]}{d[S(IV)]/dt} \qquad (12.77)$$

If the aqueous-phase concentration is uniform and satisfies Henry's law equilibrium, then $[S(IV)] = H^*_{S(IV)}RT[SO_2(g)]$, and the two timescales are related by

$$\tau_{rg} = \frac{\tau_{ra}}{H^*_{S(IV)}RT} \qquad (12.78)$$

12.3 MASS TRANSPORT AND AQUEOUS-PHASE CHEMISTRY

The reactants for aqueous-phase atmospheric reactions are transferred to the interior of cloud droplets from the gas phase by a series of mass transport processes. We would like to compare the rates of mass transport in the gas phase, at the gas–water interface, and in the aqueous phase in an effort to quantify the mass transport effects on the rates of aqueous-phase reactions. If there are no mass transport limitations, the gas and aqueous phases will remain at Henry's law equilibrium at all times. Our objective will be to identify cases where mass transport limits the aqueous-phase reaction rates and then to develop approaches to quantify these effects.

First, we note that the characteristic time of the aqueous-phase dissociation reactions are short compared with all timescales of interest. Thus the aqueous ionic equilibria can be assumed to hold at all points in the droplet, and these characteristic times no longer need be considered.

We now compare the rate of each mass transport step to the aqueous-phase reaction rate. If the mass transport rate exceeds the aqueous-phase reaction rate, then mass transport does not limit the aqueous-phase kinetics.

12.3.1 Gas-Phase Diffusion and Aqueous-Phase Reactions

In the atmosphere under standard conditions the mean free path of most reacting molecules is on the order of 0.1 μm, much less than the radius of cloud droplets (≥ 1 μm). Hence the mass transport of gas molecules to (or from) cloud droplets may be treated as a continuum process. The characteristic time τ_{dg} for the establishment of a steady-state concentration profile around the droplets in a cloud is, as we saw, less than a millisecond. We can therefore assume that the concentration profile of a reactant gas in the air surrounding a droplet will be given by (12.9) and the flux to the droplet by (12.12).

The flux to the droplet per unit volume of the droplet J_v can be calculated by dividing the molar flux J_c (with units mol s^{-1}) by the droplet volume, as

$$J_v = \frac{3 D_g}{R_p^2}(c_\infty - c_s) \tag{12.79}$$

where J_v is in M s^{-1}. For a given bulk gas-phase concentration c_∞, this flux to a droplet reaches a maximum value when $c_s = 0$; that is, the surface concentration becomes zero. In this case the maximum flux $J_v^{\max}$ is given by

$$J_v^{\max} = \frac{3 D_g}{R_p^2} c_\infty \tag{12.80}$$

Equation (12.80) provides an important insight. For a given droplet size, gas-phase diffusion can provide a reactant to the aqueous phase at a rate that cannot exceed $J_v^{\max}$. This maximum uptake rate is shown in Figure 12.6 as a function of the drop diameter. For a droplet of diameter equal to 10 μm, assuming a gas-phase diffusivity $D_g = 0.1 \text{cm}^2\text{s}^{-1}$, gas-phase diffusion can provide as much as 50 μM s^{-1} ppb^{-1} of reagent. This rate is quite substantial, indicating that gas-phase diffusion is a rather efficient process for the average cloud droplet. The larger the droplet, the smaller the maximum molar uptake rate for any reagent.

Let us return to our reacting system, where a species A diffuses around a droplet of radius R_p and then reacts with an aqueous-phase reaction rate $R_{aq}(\text{M s}^{-1})$. We assume here that interfacial and aqueous-phase transport are so rapid that we can neglect them. At steady-state, the diffusion rate to the particle will be equal to the reaction rate, or

$$\frac{3 D_g}{R_p^2}(c_\infty - c_s) = R_{aq} \tag{12.81}$$

and in terms of partial pressure of A, using the ideal-gas law, we obtain

$$p_A(\infty) - p_A(R_p) = \frac{R_p^2 RT}{3 D_g} R_{aq} \tag{12.82}$$

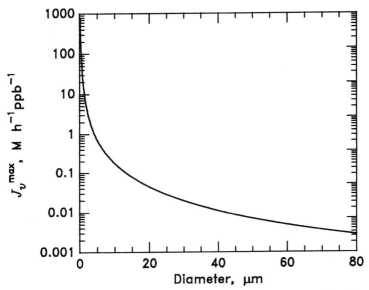

FIGURE 12.6 Maximum molar uptake rate per ppb of gas-phase reagent allowed by gas-phase diffusion at $T = 298\,\text{K}$ and for $D_g = 0.1\,\text{cm}^2\,\text{s}^{-1}$.

where $p_A(\infty)$ and $p_A(R_p)$ are the partial pressures of A far from the droplet and at the droplet surface, respectively. Equation (12.82) permits us to formulate a criterion for the absence of gas-phase mass transport limitation (Schwartz 1986)

$$\frac{p_A(\infty) - p_A(R_p)}{p_A(\infty)} \le \varepsilon_g \tag{12.83}$$

where ε_g is an arbitrarily selected small number close to zero. This criterion states that there is no gas-phase mass transport limitation when the partial pressure of the reactant is approximately constant around the drop (see also Figure 12.5). Combining (12.82) and (12.83), this criterion sets an upper bound for the aqueous-phase reaction rate that can be maintained by gas-phase diffusion:

$$R_{aq} \le \varepsilon_g \frac{3 D_g p_A(\infty)}{R_p^2 RT} \tag{12.84}$$

As long as the aqueous-phase reaction rate is less than the value specified by (12.84), the reaction will not be limited by gas-phase diffusion.

If there is no deviation from equilibrium, Henry's law will be satisfied for A. For first-order kinetics, we obtain

$$R_{aq} = k_1[A] = k_1 H_A^* p_A(\infty)$$

where $k_1(\text{s}^{-1})$ is the first-order rate constant. We obtain the bound as

$$k_1 H_A^* \le \varepsilon_g \frac{3 D_g}{R_p^2 RT} \tag{12.85}$$

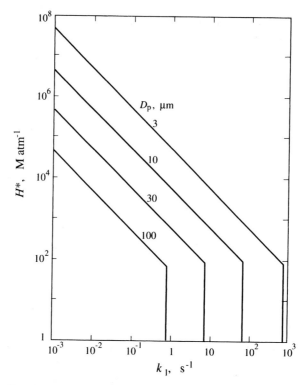

FIGURE 12.7 Gas-phase and aqueous-phase mass transport limitation for a species with $D_g = 0.1\,\text{cm}^2\,\text{s}^{-1}$, $D_{aq} = 10^{-5}\,\text{cm}^2\,\text{s}^{-1}$. The lines represent onset (10%) of mass transport limitation for the indicated values of drop diameter. Diagonal sections represent mass transport limitation; vertical sections represent aqueous-phase limitation (Schwartz 1986).

This criterion for gas-phase diffusion limitation is illustrated in Figure 12.7 for a series of droplet diameters and an arbitrarily chosen $\varepsilon_g = 0.1$. In Figure 12.7 the inequality (12.85) corresponds to the area below and to the left of the lines. For a given situation if the point (k_1, H_A^*) is to the left of the corresponding line in Figure 12.7, then gas-phase mass transport limitation does not exceed 10%. Similar plots, introduced by Schwartz (1984), provide an easy way to ascertain whether there is a mass transport limitation for a given condition of interest.

12.3.2 Aqueous-Phase Diffusion and Reaction

In Section 12.2.4 we showed that after time $\tau_{da} = R_p^2/\pi^2 D_{aq}$ the concentration profile inside a cloud droplet becomes uniform. The characteristic time for aqueous-phase reaction was found to be equal to $\tau_{ra} = [A]/R_{aq}$. If the characteristic aqueous-phase diffusion time is much less than the characteristic reaction time, then aqueous-phase diffusion will be able to maintain a uniform concentration profile inside the droplet. In this case, there will be no concentration gradients inside the drop and therefore no aqueous-phase mass transport limitations on the aqueous-phase kinetics. This criterion can be written as

$$\tau_{da} \leq \varepsilon \tau_{ra}$$

where ε is another arbitrarily chosen small number (i.e., $\varepsilon = 0.1$). Assuming first-order kinetics, $R_{aq} = k_1[A]$, and the criterion immediately above becomes

$$k_1 \leq \varepsilon \frac{\pi^2 D_{aq}}{R_p^2} \tag{12.86}$$

For example, for $D_{aq} = 10^{-5}\,\mathrm{cm}^2\,\mathrm{s}^{-1}, \varepsilon = 0.1$, and $R_p = 5\,\mu\mathrm{m}$, the criterion corresponds to

$$k_1 \leq 39.5\,\mathrm{s}^{-1}$$

and does not depend on the value of the Henry's law constant for A. This criterion corresponds to vertical lines in Figure 12.7 at values of k_1 calculated by (12.86). Points to the left of these lines are limited to an extent less than 10% by aqueous-phase mass transport.

12.3.3 Interfacial Mass Transport and Aqueous-Phase Reactions

Despite the fact that for typical cloud droplets gas-phase mass transport is in the continuum regime, mass transport across the air–water interface is, ultimately, a process involving individual molecules. Therefore the kinetic theory of gases sets an upper limit to the flux of a gas to the air–water interface. This rate is given by (12.25) and depends on the value of the accommodation coefficient.

Following our approach in Section 12.3.1 the molar flux per droplet volume through the interface, J_{kv}, will be

$$J_{kv} = \frac{J_k}{\frac{4}{3}\pi R_p^3}$$

or using (12.25)

$$J_{kv} = \frac{3\bar{c}_A \alpha}{4\,R_p}(c_\infty - c_s) \tag{12.87}$$

This rate reaches a maximum value for $c_s = 0$. Converting to partial pressure, $J_{kv}^{\max}$, the maximum molar uptake rate through the interface will be

$$J_{kv}^{\max} = \frac{3\,\bar{c}_A \alpha p_A(\infty)}{4\,R_p RT} \tag{12.88}$$

It is instructive to compare this rate with the maximum gas-phase diffusion rate $J_v^{\max}$ given by (12.80). As shown in Figure 12.8, for $\alpha \geq 0.1$ the maximum interfacial mass transport rate exceeds the maximum gas-phase diffusion rate. Therefore for $\alpha \geq 0.1$, gas-phase diffusion is more restrictive for atmospheric cloud droplets, while for $\alpha \leq 10^{-3}$, interfacial mass transport is the rate-limiting step.

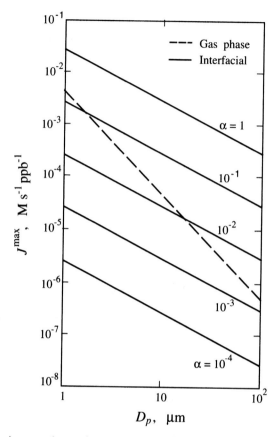

FIGURE 12.8   Maximum molar uptake rate per ppb of gas-phase reagent as a function of cloud drop diameter, as controlled by gas-phase diffusion, or interfacial transport for various accommodation coefficient values at $T = 298\,\mathrm{K}$ and for $D_g = 0.1\,\mathrm{cm^2\,s^{-1}}$ and $M_\mathrm{A} = 30\,\mathrm{g\,mol^{-1}}$ (Schwartz 1986).

The net rate of material transfer across the interface into solution per droplet volume R_l is given, modifying (12.54), as

$$R_l = \frac{3\alpha}{H_\mathrm{A}^* R_p (2\pi M_\mathrm{A} RT)^{1/2}} \left(H_\mathrm{A}^* p_0 - [\mathrm{A}]_s \right) \tag{12.89}$$

where p_0 is the partial pressure of A at the interface, and $[\mathrm{A}]_s$ is the aqueous-phase concentration of A at the droplet surface. The difference between $H_\mathrm{A}^* p_0$ and $[\mathrm{A}]_s$ is the step change at the droplet surface shown in Figure 12.5. Equation (12.89) can be used to define a criterion for the absence of interfacial mass transport limitation (Schwartz 1986) by

$$\frac{H_\mathrm{A}^* p_0 - [\mathrm{A}]_s}{H_\mathrm{A}^* p_0} \le \varepsilon \tag{12.90}$$

where ε is once more an arbitrary small number (i.e., $\varepsilon = 0.1$). The physical interpretation of this criterion is that as long as the deviation from Henry's law at the interface is less than, say, 10%, there is no interfacial mass transport limitation.

Considering a system at steady state with only interfacial mass transport and aqueous-phase reactions taking place (the rest of the mass transport steps are assumed to be fast), then $R_{aq} = R_l$, or

$$H_A^* p_0 - [A]_s = \frac{H_A^* R_p (2\pi M_A RT)^{1/2}}{3\alpha} R_{aq} \tag{12.91}$$

Combining (11.90) and (11.91) the criterion for absence of interfacial mass transport limitation can be rewritten as

$$R_{aq} \leq \varepsilon \frac{3\alpha p_0}{R_p (2\pi M_A RT)^{1/2}} \tag{12.92}$$

For first-order kinetics and no interfacial transport limitation, $R_{aq} = k_1 [A] \simeq k_1 H_A^* p_0$ and therefore, using (12.92), we obtain

$$k_1 H_A^* \leq \varepsilon \frac{3\alpha}{R_p (2\pi M_A RT)^{1/2}} \tag{12.93}$$

This criterion is illustrated in Figure 12.9 for $\varepsilon = 0.1$. A range of bounds is indicated for droplet diameters varying from 3 to 100 μm. Comparing Figures 12.9 and 12.7, one

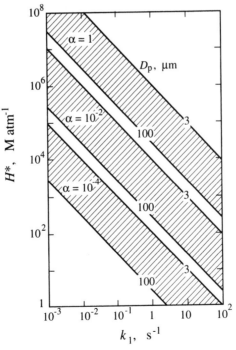

FIGURE 12.9 Interfacial mass transport limitation. The several bands represent the onset (10%) of mass transport limitation for the indicated values of the accommodation coefficient α. The width of each band corresponds to the drop diameter range 3–100 μm for $M_A = 30$ g mol^{-1} (Schwartz 1986).

concludes that interfacial mass transport may be more or less controlling than gas-phase diffusion, depending on the value of the accommodation coefficient α.

In the criteria (12.85), (12.86), and (12.93), ε is the fractional reduction in aqueous-phase reaction rate due to mass transport limitations that triggers our concern. Satisfaction of these inequalities for, say, $\varepsilon = 0.1$, indicates that the decrease of the aqueous-phase reaction rate by the corresponding mass transfer process is less than 10%.

12.3.4   Application to the S(IV)–Ozone Reaction

We would like to use the tools developed above to determine whether aqueous-phase reaction between ozone and S(IV) is limited by mass transport in a typical cloud. The subsequent analysis was first presented by Schwartz (1988). Typical ambient conditions that need to be examined include droplet diameters in the 10–30 μm range, cloud pH values in the 2–6 range, and cloud temperatures varying from 0°C to 25°C. The reaction rate for the S(IV)–O_3 reaction is given by

$$R = k_{O_3}[O_3][S(IV)] \tag{12.94}$$

where k_{O_3} is the temperature- and pH-dependent effective rate constant for the reaction (see Chapter 7). This rate constant is plotted as a function of the pH in Figure 7.15.

We need now to evaluate the pseudo-first-order coefficient for ozone and S(IV), which is included in the criteria of (12.85), (12.86), and (12.93). The effective first-order rate coefficient of ozone, k_{1,O_3} is

$$k_{1,O_3} = k_{O_3}[S(IV)] = k_{O_3}H^*_{S(IV)}p_{SO_2} \tag{12.95}$$

and similarly that for S(IV) is

$$k_{1,S(IV)} = k_{O_3}[O_3] = k_{O_3}H_{O_3}p_{O_3} \tag{12.96}$$

assuming Henry's law equilibrium. The partial pressures used in (12.95) and (12.96) are those after establishing phase equilibrium in the cloud.

Using the graphical analysis technique of Schwartz (1984, 1986, 1987), we can now examine the mass transport limitation in the S(IV)–O_3 reaction for the above range of ambient conditions assuming $\xi_{O_3} = 30$ ppb and $\xi_{SO_2} = 1$ ppb (Figure 12.10).

Each of the criteria (12.85), (12.86), and (12.93) appears in Figure 12.10 as a straight line in the $\log k_1 - \log H^*$ space. The gas-phase and interfacial mass transport lines have slope -1, while the aqueous-phase transport line is vertical. Mass transport limitation is less than 10% for points (k_1, H^*) to the left of the inequality lines. Lines are plotted for two temperatures and two droplet diameters. The lines given for interfacial mass transport limitations reflect the accommodation coefficient measurements of $\alpha_{O_3} = 5 \times 10^{-4}$ (Tang and Lee 1987) and $\alpha_{SO_2} = 0.08$.

We turn our attention first to ozone (Figure 12.10). Since ozone solubility does not depend on pH, all points (k_{1,O_3}, H_{O_3}) at a given temperature for different pH values fall on a horizontal line shown on the lower part of the graph. The increase with pH of both the S(IV) solubility and the effective reaction constant k_{O_3} leads to an increase of k_{1,O_3} with increasing pH. The points (k_{1,O_3}, H_{O_3}) lie well below the bounds of gas-phase mass

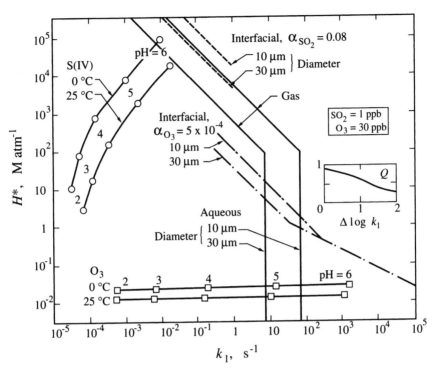

FIGURE 12.10 Examination of mass transport limitation in the S(IV)–O_3 reaction. Mass transport limitation is absent for points below and to the left of the indicated bounds (Schwartz 1988).

transport limitation, and hence such limitation is unimportant. Similarly, no interfacial mass transport limitation is indicated, despite the low ozone accommodation coefficient. Interfacial mass transport limitation may occur only under rather unusual circumstances, corresponding to pH values higher than 7 and/or high SO_2 concentrations. In contrast, significant aqueous-phase mass transport limitation is predicted for 10 μm drops at pH $\geq$ 5.3 and for 30 μm at pH $\geq$ 4.8 (Schwartz 1988). Thus one can assume cloud droplet saturation with O_3 at Henry's law equilibrium with gaseous O_3 at pH values lower than 4.8, but aqueous-phase mass transport limitations should be considered for calculations for higher pH values. For higher SO_2 concentrations the points (k_{1,O_3}, H_{O_3}) are shifted to the right. For example, for $\xi_{SO_2} = 10$ ppb all points for ozone shift to the right by one log unit on the abscissa scale. For a 10 ppb SO_2 concentration the onset of aqueous-phase mass transport limitation occurs at pH 4.8 for 10-μm droplets and at 4.4 for 30-μm droplets.

The values of $(k_{1,S(IV)}, H_{S(IV)}^*)$ are also shown in Figure 12.10. As the pH increases, both the SO_2 solubility $H_{S(IV)}^*$ and the effective reaction constant k_{O_3} increase and the corresponding points start approaching the lines reflecting mass transport limitations. However, the location of the points indicates that there is essentially no such limitation over the entire range of conditions examined; the only exception is gas-phase limitation for 30-μm drops at 0°C and pH $\geq$ 5.9. Once more, for higher gas-phase ozone concentrations, the effective reaction constant k_{1,O_3} increases and the points shift to the right. For an increase of O_3 to 300 ppb and a shifting of the corresponding S(IV) points by a full log unit, there would be no appreciable mass transport limitation for 30-μm drops for pH $\leq$ 5.4.

12.3.5   Application to the S(IV)–Hydrogen Peroxide Reaction

An analogous examination of mass transport limitations can be carried out for the aqueous-phase reaction of S(IV) with H_2O_2 with a reaction rate given by

$$R = k_{H_2O_2}[H_2O_2][S(IV)] \qquad (12.97)$$

where $k_{H_2O_2}$ is the pH- and temperature-dependent effective rate constant for the reaction (see Chapter 7). The effective first-order rate coefficients are then defined as

$$k_{1,H_2O_2} = k_{H_2O_2}[S(IV)] = k_{H_2O_2}H^*_{S(IV)}p_{SO_2}$$
$$k_{1,S(IV)} = k_{H_2O_2}[H_2O_2] = k_{H_2O_2}H_{H_2O_2}p_{H_2O_2} \qquad (12.98)$$

The points $(k_{1,S(IV)}, H^*_{S(IV)})$ and $(k_{1,H_2O_2}, H_{H_2O_2})$ are plotted in Figure 12.11 following the analysis of Schwartz (1988). The bound for H_2O_2 interfacial limitation was drawn using a value of $\alpha_{H_2O_2} = 0.2$ (Gardner et al. 1987).

Let us first examine the points for H_2O_2. Both the H_2O_2 solubility and the effective first-order rate constant, k_{1,H_2O_2}, depend only weakly on pH, resulting in tightly clustered points. The reason for the pH independence of k_{1,H_2O_2} is, as we saw in Chapter 7, the nearly opposite pH dependences of $k_{H_2O_2}$ and $H^*_{S(IV)}$ that cancel each other in (12.97).

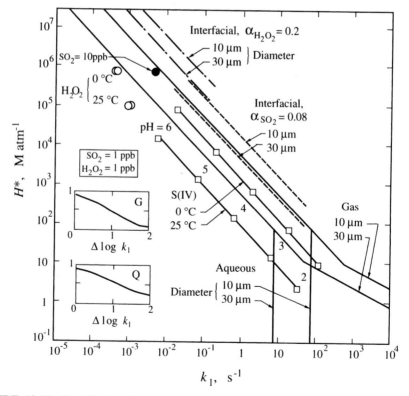

FIGURE 12.11   Examination of mass transport limitation in the S(IV)–H_2O_2 reaction. Mass transport limitation is absent for points below and to the left of the indicated bounds (Schwartz 1988).

Comparison of the location of these points plotted for $\xi_{SO_2} = 1$ ppb shows that gas-phase, aqueous-phase, and interfacial mass transport limitations are absent for H_2O_2 even for 30-μm drops for all cases. For higher partial pressures of SO_2, the points $(k_{1,H_2O_2}, H_{H_2O_2})$ should be displaced appropriately to the right. The point denoted by the solid circle in Figure 12.11 corresponds to $\xi_{SO_2} = 10$ ppb at 0°C and is therefore one log unit to the right of the corresponding $\xi_{SO_2} = 1$ ppb point. In this case there is appreciable gas-phase mass transport limitation for 30-μm droplets and no limitation for 10-μm droplets.

For S(IV) both $k_{1,S(IV)}$ and $H^*_{S(IV)}$ are pH-dependent (Figure 12.11). At 25°C there is no gas-phase or interfacial mass transport limitation for $\xi_{H_2O_2} = 1$ ppb.[1]

Aqueous-phase limitations occur for pH < 2.8 according to Figure 12.11. For temperature 0°C the points $(k_{1,S(IV)}, H^*_{S(IV)})$ are displaced upward and to the right of the 25°C, indicating an increase with decreasing temperature not only of solubility but also of the first-order rate constant. In this low-temperature case, there are significant mass transport limitations for all pH values for 30-μm drops. For 10-μm drops, even at 0°C, there are no mass transport limitations.

12.3.6 Calculation of Aqueous-Phase Reaction Rates

The discussion above indicates that appreciable mass transport limitation is absent under most atmospheric conditions of interest, although instances of substantial limitations under certain conditions of pH, temperature, and reagent concentrations exist. In the latter cases we would like to estimate the aqueous-phase reaction rate taking into account reductions in these rates due to the mass transport limitations. The graphical method presented in the previous sections is a useful method to estimate if there are mass transport limitations present. Our goal in this section is to derive appropriate mathematical expressions to quantify these rates.

No Mass Transport Limitations   In this case there are no appreciable concentration gradients outside or inside the droplet, or across the interface. As a result, Henry's law is applicable and for the first-order reaction

$$A(aq) \rightarrow B(aq)$$

the rate R_{aq} is given by

$$R_{aq} = k[A(aq)] = k H_A p_A \qquad (12.99)$$

where k is the corresponding rate constant, H_A the Henry's law coefficient of A, and p_A its gas-phase partial pressure.

Aqueous-Phase Mass Transport Limitation   The concentration of a species A, specifically, $C(r, t)$, undergoing aqueous-phase diffusion and irreversible reaction inside a cloud droplet, is governed by

$$\frac{\partial C}{\partial t} = D_{aq}\left(\frac{\partial^2 C}{\partial r^2} + \frac{2}{r}\frac{\partial C}{\partial r}\right) + R_{aq}(r, t) \qquad (12.100)$$

---

[1]Schwartz (1988) noted that because the above calculation was made for an open system, with a fixed $\xi_{H_2O_2} = 1$ ppb, because of the solubility of H_2O_2 it actually corresponds to a *total* amount of approximately 9.3 ppb.

where D_{aq} is its aqueous-phase diffusivity and $R_{aq}(r, t)$ the aqueous-phase reaction rate. This equation was solved in Section 12.2.4 neglecting the reaction term. Let us focus here on the case of a first-order reaction

$$R_{aq}(r, t) = -kC(r, t) \tag{12.101}$$

and assume that the system is at steady state. Equation (12.100) then becomes

$$\frac{d^2C}{dr^2} + \frac{2}{r}\frac{dC}{dr} = \frac{k}{D_{aq}}C \tag{12.102}$$

with boundary conditions

$$C(R_p) = C_s$$
$$\left(\frac{dC}{dr}\right)_{r=0} = 0 \tag{12.103}$$

where C_s is the concentration at the droplet surface. The solution of (12.102) under these conditions is

$$C(r) = C_s \frac{R_p}{r} \frac{\sinh(qr/R_p)}{\sinh q} \tag{12.104}$$

where we have introduced the dimensionless parameter

$$q = R_p\sqrt{\frac{k}{D_{aq}}} \tag{12.105}$$

Steady-state concentration profiles calculated from (12.104) are shown in Figure 12.12. Note that as $q \to 0$, the profiles become flat, while for larger q values significant concentration gradients develop inside the drop.

Because of the linearity of the system (first-order reaction), the average reaction rate over the drop $\langle R \rangle$, will be proportional to the average concentration of A, namely, $\langle C \rangle$

$$\langle R \rangle = k\langle C \rangle \tag{12.106}$$

where

$$\langle C \rangle = \frac{3}{4\pi R_p^3} \int_0^{R_p} 4\pi r^2 C(r)dr \tag{12.107}$$

Integration of $C(r)$ given in (12.104) yields

$$\langle C \rangle = 3C_s\left(\frac{\coth q}{q} - \frac{1}{q^2}\right) \tag{12.108}$$

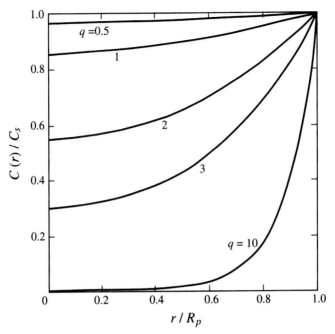

FIGURE 12.12 Steady-state radial concentration profiles of concentration of reactant species $C(r)$ relative to the concentration at the surface of the drop C_s for a drop radius R_p as a function of the dimensionless parameter q (Schwartz and Freiberg 1981).

The average concentration $\langle C \rangle$ for $q \geq 1$ is significantly less than the surface concentration C_s, taking the value $0.34\,C_s$ for $q = 1$ and $0.1\,C_s$ for $q \simeq 20$.

The overall rate of aqueous-phase reaction of the droplet R_{aq} can now be calculated as function of the droplet concentration by

$$R_{aq} = QkC_s \tag{12.109}$$

where

$$Q = 3\left(\frac{\coth q}{q} - \frac{1}{q^2}\right), \qquad q = R_p\sqrt{\frac{k}{D_{aq}}} \tag{12.110}$$

Note that when there are no other mass transfer limitations the droplet surface concentration will be in equilibrium with the bulk gas phase and

$$C_s = H_A^* p_A \tag{12.111}$$

so that

$$R_{aq} = QkH_A^* p_A \tag{12.112}$$

In conclusion, when aqueous-phase mass transport limitations are present, one needs to calculate the correction factor Q and then multiply the overall reaction rate by it. This correction factor is shown in the inset of Figure 12.10 as a function of the distance of the point of interest from the line indicating the onset of mass transport limitation, $\Delta \log k_1$. For example, the O_3 point corresponding to pH 6, 25°C falls 1.2 log units to the right of the bound for 10 μm drops. Using $\Delta \log k_1 = 1.2$ in the inset, we find that $Q = 0.4$, and the rate will be 40% of the rate corresponding to a uniform ozone concentration in the drop.

For further discussion of the correction factor Q and applications to other species, the reader is referred to Jacob (1986). Our analysis is based on a steady-state solution of the problem. The solution to the time-dependent problem has been presented by Schwartz and Freiberg (1981). These authors concluded that the rate of uptake exceeds the steady-state rate by less than 7% after $t = 1/k$ and 1% after $t = 2/k$. This approach to steady state is quite rapid and therefore the conclusions reached above are robust.

Gas-Phase Limitation   The problem of coupled gas-phase mass transport and aqueous-phase chemistry was solved in Section 12.3.1 resulting in (12.82). Solving for the aqueous-phase reaction term R_{aq} and noting that in this case Henry's law will be satisfied at the interface $(p_A(R_p) = C_{aq}/H_A^*)$:

$$R_{aq} = \frac{3 D_g}{R_p^2 RT} \left(p_A(\infty) - \frac{C_{aq}}{H_A^*} \right) \tag{12.113}$$

This equation indicates that when the aqueous-phase reaction rate is limited by gas-phase mass transport, at steady-state the aqueous-phase reaction rate is only as fast as the mass transport rate.

Interfacial Limitation   The solution of the steady-state problem for only interfacial limitation is given by (12.91). The aqueous-phase reaction rate is then

$$R_{aq} = \frac{3 \alpha}{H_A^* R_p (2\pi M_A RT)^{1/2}} [H_A^* p_A(\infty) - C_{aq}] \tag{12.114}$$

Gas-Phase Plus Interfacial Limitation   The expressions developed above can be combined to develop a single expression that includes both gas-phase and interfacial mass transport effects. Schwartz (1986) showed that in this case

$$R_{aq} = k_{mt} \left(p_A(\infty) - \frac{C_{aq}}{H_A^*} \right) \tag{12.115}$$

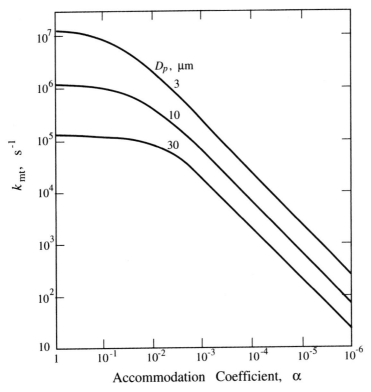

FIGURE 12.13 Mass transfer constant k_{mt} accounting for gas-phase diffusion and interfacial transport as a function of the droplet diameter and the accommodation coefficient (Schwartz 1986).

where

$$k_{mt} = \left[\frac{R_p^2}{3 D_g} + \frac{R_p}{3\alpha} \sqrt{\frac{2\pi M_A}{RT}} \right]^{-1} \qquad (12.116)$$

The mass transfer coefficient k_{mt} for gas-phase plus interfacial mass transport has units s^{-1}. The rate R_{aq} in (12.115) is in equivalent gas-phase concentration units, but the conversion to aqueous-phase units is straightforward multiplying by H_A^*. The mass transfer coefficient k_{mt} as a function of the accommodation coefficient α and the droplet radius is shown in Figure 12.13. For values of $\alpha > 0.1$ the mass transfer rate is not sensitive to the exact value of α. However, for $\alpha < 0.01$, surface accommodation starts limiting the mass transfer rate to the drop, and k_{mt} decreases with decreasing α for all droplet sizes.

Uptake Coefficient γ   We have seen that a fundamental parameter that determines the transfer rate of trace gases into droplets is the mass accommodation coefficient α, which is defined as

$$\alpha = \frac{\text{number of molecules entering the liquid phase}}{\text{number of molecular collisions with the surface}}$$

In most circumstances other processes, such as gas-phase diffusion to the droplet surface and liquid-phase solubility, also limit gas uptake. In a laboratory experiment α can generally not be measured directly. Rather, what is accessible is the measured overall flux $\tilde{J}_A$ of gas A to the droplet (mol A per area per time). That flux can be expressed in terms of a measured *uptake coefficient*, $\gamma \leq 1$, that multiplies the kinetic collision rate per unit of droplet surface area as

$$\tilde{J}_A = \frac{1}{4} c_A(\infty) \bar{c}_A \gamma \tag{12.117}$$

where $c_A(\infty)$ is the concentration of A in the bulk gas. Since $\tilde{J}_A$ is measured and $c_A(\infty)$ and $\bar{c}_A$ are known, γ can be determined from the experimental uptake data. The uptake coefficient γ inherently contains all the processes that affect the rate of gas uptake, including the mass accommodation coefficient at the surface. The experimental task is to separate these effects and determine α from γ.

To analyze laboratory uptake data it is necessary to determine how γ depends on the other parameters of the system. Let us assume that potentially any or all of gas-phase diffusion, interfacial transport, and aqueous-phase diffusion may be influential. We assume steady-state conditions and that species A is consumed by a first-order aqueous phase reaction (12.101). At steady state the rate of transfer of species A across the gas–liquid interface, given by (12.115), must be equal to that as a result of simultaneous aqueous-phase diffusion and reaction, (12.112):

$$QkC_{aq} = k_{mt} \left[p_A(\infty) - \frac{C_{aq}}{H_A^*} \right] \tag{12.118}$$

This equation can be solved to obtain an expression for C_{aq}, the aqueous-phase concentration of A just at the droplet surface. The correction factor Q depends on the dimensionless parameter q as given in (12.110). In the regime of interest where $q \gg 1$, $Q \simeq 3/q$, and this is the case we will consider. Thus (12.118) yields

$$C_{aq} = \frac{k_{mt} p_A(\infty)}{\dfrac{k_{mt}}{H_A^*} + \dfrac{3\sqrt{k}\sqrt{D_{aq}}}{R_p}} \tag{12.119}$$

Then the rate of transfer of A across the gas–liquid interface, in mol s^{-1}, $(3 k C_{aq}/q)(\frac{4}{3}\pi R_p^3)$, is equated to the defining equation for γ:

$$\frac{3\sqrt{k}\sqrt{D_{aq}}}{R_p} C_{aq} \left(\frac{4}{3}\pi R_p^3 \right) = \frac{1}{4} \frac{p_A(\infty)}{RT} \bar{c}_A \gamma (4\pi R_p^2) \tag{12.120}$$

Using (12.116) and (12.119) in (12.120), we can obtain the basic equation that relates γ to the other parameters of the system:

$$\frac{1}{\gamma} = \frac{R_p \bar{c}_A}{4 D_g} + \frac{1}{\alpha} + \frac{\bar{c}_A}{4 RT H_A^* \sqrt{k D_{aq}}} \tag{12.121}$$

The terms on the RHS of (12.121) show the three contributions to overall resistance to uptake: gas-phase diffusion, mass accommodation at the surface, and interfacial

TABLE 12.2   Measured Mass Accommodation (α) and Uptake (γ) Coefficients on Aqueous Surfaces

Species	α	γ	T, K	Reference
HNO_3		0.07	268	Van Doren et al. (1990, 1991)
		0.19	293	
HCl	0.15		283	Watson et al. (1990)
N_2O_5		0.04	283	Van Doren et al. (1991)
SO_2	0.11		273–290	Worsnop et al. (1989)
H_2O_2	0.23		273	Worsnop et al. (1989)
CH_3COOH	0.17		255	Jayne et al. (1991)
	0.03		295	
HCHO	0.02		267	Jayne et al. (1992)
CH_3CHO	>0.03		267	
$CH_3C(O)CH_3$	0.066		260	Duan et al. (1993)
	0.013		285	
MSA	0.17		260	DeBruyn et al. (1994)
	0.1		285	
O_3	0.0005			Tang and Lee (1987)
HO_2	$>0.2^a$			Mozurkewich et al. (1987)
NO_3		0.0002		Rudich et al. (1996)
$NH_4 NO_3$	0.8		293	Dassios and Pandis (1998)
	0.5		300	
$H_2SO_4{}^b$	0.7		298	Jefferson et al. (1997)
	0.2		298	

^{a}On 20M $(NH_4)_2SO_4$.
^{b}On NaCl aerosol with high stearic acid coverage.

transport/aqueous-phase diffusion, respectively. Thus we see that the uptake coefficient γ is equal to the mass accommodation coefficient α only if the other resistances are negligible. In experimental evaluation, γ is measured and all other quantities in (12.121) except for α are known or can be measured separately; this provides a means to determine α.

Table 12.2 presents values of measured mass accommodation and uptake coefficients for a variety of atmospheric gases. For in-depth treatment of uptake of gas molecules by liquids, we refer the reader to Davidovits et al. (1995) and Nathanson et al. (1996).

A general kinetic model framework for the description of mass transport and chemical reactions at the gas–particle interface has been developed by Pöschl et al. (2005).

12.3.7   An Aqueous-Phase Chemistry/Mass Transport Model

The following equations can be used to describe the evolution of the aqueous-phase concentration of a species A, C_{aq}, and each partial pressure, incorporating mass transfer effects

$$\frac{dp}{dt} = -k_{mt} w_L p + \frac{1}{H_A^*} k_{mt} C_{aq} w_L \qquad (12.122)$$

$$\frac{dC_{aq}}{dt} = \frac{k_{mt}}{RT} p - \frac{k_{mt}}{H_A^* RT} C_{aq} - Q R_{aq} \qquad (12.123)$$

where w_L is the cloud liquid water volume fraction, p (atm) is the bulk partial pressure of A in the cloud, C_{aq} is the corresponding aqueous-phase concentration at the surface of the drop, H_A^* is the effective Henry's law coefficient, and k_{mt} and Q are the coefficients given by (12.116) and (12.110), respectively. Note that the concentrations described by these equations are the surface aqueous-phase concentrations, and in cases where there are aqueous-phase mass transport limitations R_{aq} should be expressed as a function of the surface concentrations.

The equations above have been the basis of most atmospheric aqueous-phase chemistry models that include mass transport limitations [e.g., Pandis and Seinfeld (1989)]. These equations simply state that the partial pressure of a species in the cloud interstitial air changes due to mass transport to and from the cloud droplets (incorporating both gas and interfacial mass transport limitations). The aqueous-phase concentrations are changing also due to aqueous-phase reactions that may be limited by aqueous-phase diffusion included in the factor Q.

12.4   MASS TRANSFER TO FALLING DROPS

For a stationary drop of radius R_p for continuum regime transport, we have seen that steady-state transport is rapidly achieved and the flux to the drop is given by (12.12). The flux per unit droplet surface, $\tilde{J}_A = J_c/(4\pi R_p^2)$, is then

$$\tilde{J}_A = \frac{D_g}{R_p}(c_\infty - c_s) \tag{12.124}$$

When the drop is in motion, calculation of the flux of gas molecules to the droplet surface is considerably more involved. The flux is usually defined in terms of a mass transfer coefficient k_c as

$$\tilde{J}_A = k_c(c_\infty - c_s) \tag{12.125}$$

For a stationary drop comparing (12.124) and (12.125) $k_c = D_g/R_p$. In an effort to estimate k_c for a moving drop one defines the dimensionless Sherwood number in terms of k_c as

$$\mathrm{Sh} = \frac{k_c D_p}{D_g} \tag{12.126}$$

For diffusion to a stationary drop $\mathrm{Sh} = 2$. When the drop is falling, one usually resorts to empirical correlations for Sh as a function of the other dimensionless groups of the problem, for example (Bird et al. 1960)

$$\mathrm{Sh} = 2 + 0.6\,\mathrm{Re}^{1/2}\mathrm{Sc}^{1/3} \tag{12.127}$$

where the Reynolds number, $\mathrm{Re} = v_t D_p/v_{air}$, and the Schmidt number $\mathrm{Sc} = v_{air}/D_g$, v_t being the terminal velocity of the droplet and v_{air} the kinematic viscosity of air.

We have seen that for a 1 mm drop the characteristic time for aqueous-phase diffusion, τ_{da}, is on the order of 100 s. Thus, if the only mechanism for mixing is molecular diffusion, for larger drops, aqueous-phase diffusion may create significant concentration gradients inside the drop. However, for falling drops of radii $R_p > 0.1$ mm internal circulations develop (Pruppacher and Klett 1997), the timescale for which is on the order of $R_p \mu_l / v_t \mu_{air}$, with μ_l the viscosity of the drop. For a 1 mm radius drop, this time scale is on the order of 10^{-2} s, very short compared with that taken by the drop falling to the ground. Therefore it can be assumed that for drops larger than about 0.1 mm internal circulations will produce a well-mixed interior.

12.5 CHARACTERISTIC TIME FOR ATMOSPHERIC AEROSOL EQUILIBRIUM

Aerosol particles in the atmosphere contain a variety of volatile compounds (ammonium, nitrate, chloride, volatile organic compounds) that can exist either in the particulate or in the gas phase. We estimate in this section the timescales for achieving thermodynamic equilibrium between these two phases and apply them to typical atmospheric conditions. The problem is rather different compared to the equilibration between the gas and aqueous phases in a cloud discussed in the previous section. Aerosol particles are solid or concentrated aqueous solutions (cloud droplets are dilute aqueous solutions), they are relatively small, and aqueous-phase reactions in the aerosol phase can be neglected to a first approximation because of the small liquid water content.

If the air surrounding an aerosol population is supersaturated with a compound A, the compound will start condensing on the surface of the particles in an effort to establish thermodynamic equilibrium. The gas-phase concentration of A will decrease, and its particulate-phase concentration will increase until equilibrium is achieved. The characteristic time for the two phases to equilibrate due to the depletion of A in the gas phase will be inversely proportional to the total flux of A to the aerosol phase.

For solid-phase aerosol particles the equilibrium concentration of A is constant and does not change as A is transferred to the aerosol phase. However, if the aerosol contains water and A is water-soluble, the equilibrium concentration of A will increase as condensation proceeds, and equilibration between the two phases will be accelerated. This change in equilibrium concentration is a result of changes in the chemical composition of the particle as A is transferred to the particulate phase. The cases of solid and aqueous phases will be examined separately in subsequent sections, based on the analysis of Wexler and Seinfeld (1990).

12.5.1 Solid Aerosol Particles

If a species A is transferred to the solid aerosol phase, as

$$A(g) \rightleftharpoons A(s)$$

the system will achieve thermodynamic equilibrium when the gas-phase concentration of A becomes equal to the equilibrium concentration c_{eq}. Recall that c_{eq} will be equal to the saturation concentration of A at this temperature, which is also equal to the concentration of A at the particle surface. The equilibrium concentration c_{eq} depends only on temperature and not on the particle composition. If the concentration of A far from the aerosol surface is c_∞, then the flux of A to a single particle is given by

$$J_1 = 4\pi R_p D_A f(\text{Kn}, \alpha)(c_\infty - c_{eq}) \tag{12.128}$$

where R_p is the particle radius, D_A is the gas-phase diffusivity of A, and $f(\text{Kn}, \alpha)$ is the correction to the mass transfer flux owing to noncontinuum effects and imperfect accommodation. For example, if the Fuchs–Sutugin (1971) approach is used, then $f(\text{Kn}, \alpha)$ is given by the RHS of (12.43). Note, that, if $c_\infty = c_{eq}$, the flux between the two phases is zero, and the aerosol is in equilibrium with the surrounding gas phase.

If the aerosol population is monodisperse and consists of N particles per cm^3, then the total flux J from the gas phase to the aerosol phase will be

$$J = NJ_1 = 4\pi N R_p D_A f(\text{Kn}, \alpha)(c_\infty - c_{eq}) \tag{12.129}$$

This flux will be equal to the rate of change of the concentration c_∞:

$$\frac{dc_\infty}{dt} = -4\pi N R_p D_A f(\text{Kn}, \alpha)(c_\infty - c_{eq}) \tag{12.130}$$

The characteristic time for gas-phase concentration change and equilibrium establishment τ_s can be estimated by nondimensionalizing (12.130). The characteristic concentration is c_{eq} and the characteristic timescale is τ_s, so we define

$$\phi = \frac{c_\infty}{c_{eq}}, \qquad \tau = \frac{t}{\tau_s}$$

and (12.130) becomes

$$\frac{d\phi}{d\tau} = -u(\phi - 1) \tag{12.131}$$

with

$$u = 4\pi N R_p D_A f(\text{Kn}, \alpha)\tau_s \tag{12.132}$$

Once more, all the physical information about the system has been included in the dimensionless parameter u. Setting $u \simeq 1$, we find that

$$\tau_s \simeq \frac{1}{4\pi N R_p D_A f(\text{Kn}, \alpha)} \tag{12.133}$$

We can express this timescale in terms of the aerosol mass concentration m_p given by

$$m_p = \frac{4}{3}\pi R_p^3 \rho_p N \tag{12.134}$$

where ρ_p is the aerosol density, to get

$$\tau_s = \frac{\rho_p R_p^2}{3 D_A m_p f(\mathrm{Kn}, \alpha)} \tag{12.135}$$

Equation (12.135) suggests that the equilibration timescale will increase for larger aerosol particles and cleaner atmospheric conditions (lower m_p). The timescale does not depend on the thermodynamic properties of A, as it is connected solely to the gas-phase diffusion of A molecules to a particle. The timescale in (12.135) varies from seconds to several hours as the particle radius increases from a few nanometers to several micrometers. We can extend the analysis from a monodisperse aerosol population to a population with a size distribution $n(R_p)$. The rate of change for the bulk gas-phase concentration of A in that case is

$$\frac{dc_\infty}{dt} = -4\pi D_A (c_\infty - c_{eq}) \int_0^\infty n(R_p) R_p f(\mathrm{Kn}, \alpha) dR_p \tag{12.136}$$

and the equilibration timescale will be given by

$$\tau_s = \frac{1}{4\pi D_A \int_0^\infty n(R_p) R_p f(\mathrm{Kn}, \alpha) dR_p} \tag{12.137}$$

Wexler and Seinfeld (1992) estimated this timescale for measured aerosol size distributions in southern California and found values lower than 10 min for most cases. The largest values of τ_s was 15 min. We can approximate (12.137) by assuming that

$$\int_0^\infty n(R_p) R_p f(\mathrm{Kn}, \alpha) dR_p \simeq N \bar{R}_p \bar{f} \tag{12.138}$$

where $\bar{R}_p$ is the number mean radius of the distribution and $\bar{f}$ is the value of $f(\mathrm{Kn}, \alpha)$ corresponding to this radius. Using this rough approximation the timescale τ_s for a polydisperse aerosol population can be approximated by

$$\tau_s \simeq \frac{1}{4\pi N \bar{R}_p D_A \bar{f}} \tag{12.139}$$

For a polluted environment, $N \geq 10^5 \, \mathrm{cm}^{-3}$, $\bar{R}_p \geq 0.01 \, \mu\mathrm{m}$, and assuming $\alpha = 0.1$, $\tau_s \leq 800 \, \mathrm{s}$ in agreement with the detailed calculations of Wexler and Seinfeld (1992). Therefore, for typical polluted conditions, this timescale is expected to be a few minutes or less. For aerosol populations characterized by low number concentrations, however, this timescale can be on the order of several hours (Wexler and Seinfeld 1990).

12.5.2 Aqueous Aerosol Particles

Equation (12.130) is also applicable to aqueous aerosol particles, but with one significant difference. During the condensation of A and the reduction of c_∞, the concentration c_{eq} at the particle surface changes. The timescale τ_s calculated in the previous section remains applicable, but there is an additional timescale τ_a characterizing the change of c_{eq}.

The molality of A in the aqueous phase, m_A, is given (see Chapter 10) by $m_A = n_A/(0.018\, n_w)$, where n_A and n_w are the molar aerosol concentrations per m^3 of air of A and water, respectively. In general, the water concentration n_w changes during transport of soluble species between the two phases. For the purposes of this calculation, let us assume that this change is small (i.e., only a small amount of A is transferred) and assume that n_w is approximately constant. The rate of change of the molality of A is given in this case by

$$\frac{dm_A}{dt} \simeq \frac{1}{0.018\, n_w} \frac{dn_A}{dt} \tag{12.140}$$

The rate of change of moles of A in the aerosol phase dn_A/dt will be equal to the flux J given by (12.129), so

$$\frac{dm_A}{dt} \simeq \frac{4\pi N R_p D_A f(\text{Kn}, \alpha)}{0.018\, n_w} (c_\infty - c_{eq}) \tag{12.141}$$

In this equation, c_∞ and c_{eq} are the gas-phase concentrations of A far from the particle and at the particle surface expressed in mol m^{-3}. The gas-phase equilibrium concentration c_{eq} is related to the liquid-phase molality by an equilibrium constant K_A (kg m^{-3}), which is a function of the composition of the particle and the ambient temperature, such that

$$c_{eq} = K_A \gamma_A m_A \tag{12.142}$$

where γ_A is the activity coefficient of A. For activity coefficients of order unity, we obtain

$$c_{eq} = K_A m_A \tag{12.143}$$

and after differentiation

$$\frac{dc_{eq}}{dt} \simeq K_A \frac{dm_A}{dt} \tag{12.144}$$

Combining (12.144) with (12.141), we find that

$$\frac{dc_{eq}}{dt} = \frac{4\pi N R_p D_A f(\text{Kn}, \alpha) K_A}{0.018\, n_w} (c_\infty - c_{eq}) \tag{12.145}$$

Assuming that the system is open and c_∞ remains constant during the condensation of A, then we can calculate the timescale τ_a from (12.145). This timescale corresponds to the establishment of equilibrium between the gas and liquid aerosol phases as a result of changes in the aqueous-phase concentration of A. Following the nondimensionalization procedure of the previous section, we find

$$\tau_a = \frac{0.018\, n_w}{4\pi N R_p D_A f(\text{Kn}, \alpha) K_A} \tag{12.146}$$

Noting that $m_w = 0.018\,n_w$ is the mass concentration of aerosol water in $kg\,m^{-3}$, (12.146) can be rewritten using (12.133) as

$$\tau_a = \frac{m_w}{K_A}\tau_s \qquad (12.147)$$

The timescale τ_a increases with increasing aerosol water content. The higher the aerosol water concentration, the slower the change of the aqueous-phase concentration of A resulting from the condensation of a given mass of A, and the slower the adjustment of the equilibrium concentration c_{eq} to the background concentration c_∞. The timescale τ_a will be larger for high RH conditions that result in high aerosol liquid water content. This timescale also depends on the thermodynamic properties of A, namely, the equilibrium constant K_A. Equation (12.142) suggests that the more soluble a species is the lower its K_A and the larger the timescale calculated from (12.147).

Let us consider, for example, dissolution of NH_3 and HNO_3

$$NH_4^+ + NO_3^- \rightleftharpoons NH_3(g) + HNO_3(g)$$

with an equilibrium constant K of $4 \times 10^{-15}\,kg^2\,m^{-6}$ at 298 K (Stelson and Seinfeld 1982). The equilibrium constant K_A used in (12.147) can be calculated as $K_A = \sqrt{K}$ (Wexler and Seinfeld 1990), and therefore $K_A = 63\,\mu g\,m^{-3}$ (and is a strong function of temperature). The liquid water content of urban aerosol masses at high RH is of the same magnitude as this K_A and therefore for NH_4NO_3, $\tau_a \approx \tau_p$. For lower temperatures K_A decreases and the timescale τ_a increases.

Summarizing, diffusive transport between the gas and aerosol phases eventually leads to their equilibration, with two timescales governing the approach to equilibrium. One timescale, τ_s, characterizes the approach due to changes in the gas-phase concentration field, and the other timescale, τ_a, due to changes in the aqueous-phase concentrations (if the aerosol is not solid). These equilibration timescales are not necessarily the same, because during condensation the partial pressures at the particle surface (equilibrium concentrations) may change more slowly or rapidly than the ambient concentrations. For solid particles, the surface concentrations do not change, τ_a is infinite, and the approach to equilibrium is governed by τ_s. For aqueous aerosols both timescales are applicable as the system approaches equilibrium through both pathways, with the shorter timescale governing the equilibration process (Wexler and Seinfeld 1992).

The preceding analysis suggests that for polluted airmasses, high temperatures, and small aerosol sizes, the equilibrium timescale is on the order of a few seconds. On the other hand, under conditions of low aerosol mass concentrations, low temperatures, and large particle sizes, the timescale can be on the order of several hours and the aerosol phase may not be in equilibrium with the gas phase. For these conditions, the submicrometer particles are in equilibrium with the gas phase while the coarse particles may not be in equilibrium (Capaldo et al. 2000).

The existence of equilibrium between the atmospheric gas and aerosol phases is generally supported by the available field measurements (Doyle et al. 1979; Stelson et al. 1979; Hildemann et al. 1984; Pilinis and Seinfeld 1988; Russell et al. 1988). For example, the comparison of field measurements with theory by Takahama et al. (2004) is shown in Figure 12.14.

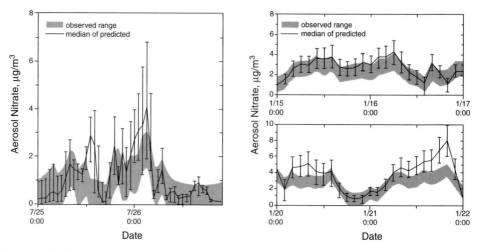

FIGURE 12.14   Time series of observed and predicted $PM_{2.5}$ nitrate concentrations for selected periods in July 2001 and January 2002 in Pittsburgh, PA. Error bars extend to the fifth and 95th percentiles of the distribution function associated with each prediction. The shaded areas bound the interval between the fifth and 95th percentiles of the observations. [Reprinted from Takahama et al. (2004). Reproduced by permission of American Geophysical Union.]

Our analysis so far has assumed that all particles in the aerosol population have similar chemical composition. Meng and Seinfeld (1996) numerically investigated cases where the aerosol population consists of two groups of particles with different compositions. They concluded that while the timescale of equilibration of the fine aerosol particles is indeed on the order of minutes or less, the coarse particles may require several hours or even days to achieve thermodynamic equilibrium with the surrounding atmosphere.

APPENDIX 12   SOLUTION OF THE TRANSIENT GAS-PHASE DIFFUSION PROBLEM EQUATIONS (12.4)–(12.7)

To solve (12.4)–(12.7), let us define a new dependent variable $u(r, t)$ by

$$u(r,t) = \frac{r}{R_p} \frac{c_\infty - c(r,t)}{c_\infty}$$

Therefore

$$\frac{c(r,t)}{c_\infty} = 1 - u(r,t)\frac{R_p}{r}$$

and the problem is transformed to

$$\frac{\partial u}{\partial t} = D_g \frac{\partial^2 u}{\partial r^2}$$
$$u(r,0) = 0, \qquad r > R_p$$
$$u(\infty, t) = 0$$
$$u(R_p, t) = 1$$

We can solve this problem by Laplace or similarity transforms. Let us use the latter. Let

$$u(\eta) = u(r, t)$$

and

$$\eta = \frac{r - R_p}{2\sqrt{D_g t}}$$

Then the equation reduces to

$$\frac{d^2 u}{d\eta^2} + 2\eta \frac{du}{d\eta} = 0$$

the solution of which is

$$u(\eta) = A + B \int_0^{\eta} e^{-\xi^2} d\xi$$

to be evaluated subject to

$$u(0) = 1$$
$$u(\infty) = 0$$

Therefore, determining A and B, we have

$$u(\eta) = 1 - \frac{2}{\sqrt{\pi}} \int_0^{\eta} e^{-\xi^2} d\xi$$

or equivalently

$$u(r, t) = 1 - \frac{2}{\sqrt{\pi}} \int_0^{(r - R_p)/2\sqrt{D_g t}} e^{-\xi^2} d\xi$$

By definition

$$c(r, t) = c_\infty \left[1 - u(r, t) \frac{R_p}{r} \right]$$

resulting in (12.8). The molar flow of species A into the droplet at any time t is

$$4\pi R_p^2 D_g \left(\frac{\partial c}{\partial r} \right)_{r = R_p} = 4\pi R_p^2 c_\infty D_g \left[\frac{R_p}{r^2} - \frac{2 R_p}{r^2 \sqrt{\pi}} \int_0^{(r - R_p)/2\sqrt{D_g t}} e^{-\xi^2} d\xi \right.$$

$$\left. + \frac{2 R_p}{r \sqrt{\pi}} \frac{1}{2\sqrt{D_g t}} \exp\left(-\frac{(r - R_p)^2}{4 D_g t} \right) \right]_{r = R_p}$$

$$= 4\pi R_p^2 c_\infty D_g \left(\frac{1}{R_p} + \frac{1}{\sqrt{\pi D_g t}} \right)$$

Then the total quantity of A that has been transferred into the particle from $t = 0$ to t is

$$M(t) = \int_0^t 4\pi R_p^2 c_\infty D_g \left(\frac{1}{R_p} + \frac{1}{\sqrt{\pi D_g t'}} \right) dt'$$

$$= 4\pi R_p D_g c_\infty \left(t + \frac{2R_p \sqrt{t}}{\sqrt{\pi D_g}} \right)$$

PROBLEMS

12.1$_A$ Consider the growth of a particle of dibutyl phthalate (DBP) in air at 298 K containing DBP at a background mole fraction of 0.10. The vapor pressure of DBP is sufficiently low that it may be assumed to be negligible.

  a. Compute the steady-state mole fraction and temperature profiles around a DBP particle of 1 µm diameter. Assume continuum regime conditions to hold.

  b. Compute the flux of DBP (mol cm^{-2} s^{-1}) at the particle surface.

  c. Can you neglect the Stefan flow in this calculation?

  d. Evaluate the temperature rise resulting from the condensation of the vapor on the particle.

  e. Plot the particle radius as a function of time starting at 0.5 µm and assuming that the background mole fraction of DBP remains constant at 0.10.

The following parameters may be used in the calculation

$$\Delta H_v / R = 8930 \, \text{K}$$
$$D = 0.0282 \, \text{cm}^2 \, \text{s}^{-1}$$
$$R / M_{\text{air}} \hat{c}_p = 0.274$$
$$M_{\text{DBP}} / M_{\text{air}} = 9.61$$

12.2$_B$ In Problem 12.1 for the purpose of computing the mass and energy transport rates to the particle it was assumed that continuum conditions held.

  a. Evaluate the validity of using continuum transport theory for the conditions of Problem 12.1.

  b. Plot the particle radius as a function of time under the conditions of Problem 12.1 for a DBP particle initially of 0.5 µm radius assuming that the accommodation coefficient is $\alpha = 0.1$, 0.01, and 0.001.

  c. Plot the particle radius as a function of time under the conditions of Problem 12.1 for a DBP particle initially of 0.01 µm radius given that the accommodation coefficient is $\alpha = 0.1$, 0.01, and 0.001.

12.3$_C$ Consider the transient absorption of SO_2 by a droplet initially free of any dissolved SO_2. Calculate the time necessary for a water droplet of radius $10\,\mu m$ to fill up with SO_2 in an open system (constant $\xi_{SO_2} = 1\,ppb$), for constant droplet pH 4 and 6.

    a. Assuming that the absorption is irreversible,

    b. Assuming that the absorption is reversible.

    c. Discuss your results.

    d. Discuss qualitatively how your results would change if the system were closed and the cloud liquid water content were $0.1\,g\,m^{-3}$.

12.4$_B$ Consider the transient dissolution of HNO_3 ($\xi_{HNO_3} = 1\,ppb$) by a monodisperse cloud droplet population ($R_p = 10\,\mu m$, $L = 0.2\,g\,m^{-3}$) at 298 K, assuming that the cloudwater has initially zero nitrate concentration. As a simplification, assume that the cloud has constant pH equal to 4.5.

    a. How long will it take for the cloud to reach equilibrium, if the cloud is described as an open system?

    b. How long will it take if the system is described as closed?

    c. Discuss the implications of your results for the mathematical modeling of clouds.

12.5$_B$ Solve (12.72) and (12.73) to obtain (12.74).

12.6$_B$ A $(NH_4)_2SO_4$ aerosol particle of initial diameter equal to 10 nm grows as a result of condensation of H_2SO_4 and NH_3 at 50% RH and 298 K.

    a. Assuming that the H_2SO_4 vapor has a constant mixing ratio of 1 ppt and that the vapor pressure of ammonium sulfate can be neglected, calculate the time required for the particle to grow to 0.05, 0.1, and $1\,\mu m$ diameter. Assume that there is always an excess of ammonia available to neutralize the condensing sulfuric acid.

    b. What would be the corresponding times if the atmospheric relative humidity were 90%?

12.7$_B$ Estimate the time required for the complete evaporation of an organic aerosol particle as a function of its initial diameter $D_{p0} = 0.05$–$10\,\mu m$ and its equilibrium vapor pressure $p_0 = 10^{-11}$–$10^{-8}\,atm$. Assume that the organic has a surface tension of $30\,dyn\,cm^{-1}$, density $0.8\,g\,cm^{-3}$, $D_g = 0.1\,cm^2\,s^{-1}$, $\alpha = 1$, and the background gas-phase concentration is zero. Assume $M_A = 200\,g\,mol^{-1}$.

12.8$_C$ Fresh seasalt particles in the marine atmosphere are alkaline and can serve as sites for the heterogeneous oxidation of SO_2 by ozone. Assuming that the diameter of these particles varies from 0.5 to $20\,\mu m$, their pH varies from 5 to 8, their liquid water content is $100\,\mu g\,m^{-3}$, and typical mixing ratios are 50 ppt for SO_2 and 30 ppb for ozone:

    a. Calculate the sulfate production rate assuming that both reactants are in Henry's law equilibrium between the gas and aerosol phases.

    b. Is this rate limited by gas-phase diffusion, or interfacial transport? Quantify the effects of these limitations on the sulfate production rates estimated in (a).

12.9$_A$ Plot the mass transfer coefficient k_c for raindrops falling through air as a function of their diameter. Assume $T = 298\,K$ and the diffusing species to be SO_2. Use the theory of terminal velocities of drops developed in Chapter 9. Consider drop sizes of $R_p = 0.1$ and 1 mm.

12.10$_B$ Table 12.P.1 gives the forward rate constants and parameters A, B, and C in the equilibrium constant expression

$$\log K = A + B\frac{1000}{T} + C\left(\frac{1000}{T}\right)^2$$

for a number of aqueous-phase equilibria. For each of the reactions given in the table evaluate the characteristic time to reach equilibrium at $T = 298\,K$ and $T = 273\,K$. Discuss your results.

TABLE 12.P.1   Forward Rate Constants k_f and Parameters A, B, and C in the Equilibrium Constant Expression $\log K = A + B(1000/T) + C(1000/T)^2$ for Aqueous Equilibria

Reaction	$k_f\ (s^{-1})$	A	B	C
$HSO_4^- \rightleftharpoons SO_4^{2-} + H^+$	1.8×10^7	-5.95	1.18	0
$SO_2 \cdot H_2O \rightleftharpoons HSO_3^- + H^+$	6.1×10^4	-4.84	0.87	0
$HSO_3^- \rightleftharpoons SO_3^{2-} + H^+$	5.4×10^0	-8.86	0.49	0
$CO_2 \cdot H_2O \rightleftharpoons HCO_3^- + H^+$	2.1×10^4	-14.25	5.19	-0.85
$HCO_3^- \rightleftharpoons CO_3^{2-} + H^+$	4.3×10^{-2}	-13.80	2.87	-0.58
$HO_2 \rightleftharpoons O_2^- + H^+$	2.3×10^3	-4.9	0	0

REFERENCES

Bhatnagar, P. L., Gross, E. P., and Krook, M. (1954) A model for collision processes in gases. I. Small amplitude processes in charged and neutral one-component systems, *Phys. Rev.* **94**, 511–525.

Bird, R. B., Stewart, W. E., and Lightfoot, E. N. (1960) *Transport Phenomena*, Wiley, New York.

Capaldo, K. P., Pilinis, C., and Pandis, S. N. (2000) A computationally efficient hybrid approach for dynamic gas/aerosol transfer in air quality models, *Atmos. Environ.* **34**, 3617–3627.

Chang X., and Davis, E. J. (1974) Interfacial conditions and evaporation rates of a liquid droplet, *J. Colloid Interface Sci.* **47**, 65–76.

Dahneke, B. (1983) Simple kinetic theory of Brownian diffusion in vapors and aerosols, in *Theory of Dispersed Multiphase Flow*, R. E. Meyer, ed., Academic Press, New York, pp. 97–133.

Dassios, K. G., and Pandis, S. N. (1999) The mass accommodation coefficient of ammonium nitrate aerosol, *Atmos. Environ.* **33**, 2993–3003.

Davidovits, P., Hu, J. H., Worsnop, D. R., Zahniser, M. S., and Kolb, C. E. (1995) Entry of gas molecules into liquids, *Faraday Discuss.* **100**, 65–82.

Davis, E. J. (1983) Transport phenomena with single aerosol particles, *Aerosol Sci. Technol.* **2**, 121–144.

DeBruyn, W. J., Shorter, J. A., Davidovits, P., Worsnop, D. R., Zahniser, M. S., and Kolb, C. E. (1994) Uptake of gas phase sulfur species methanesulfonic acid, dimethylsulfoxide, and dimethyl sulfone by aqueous surfaces, *J. Geophys. Res.* **99**, 16927–16932.

Doyle, G. J., Tuazon, E. C., Graham, R. A., Mischke, T. M., Winer, A. M., and Pitts, J. N. Jr. (1979) Simultaneous concentrations of ammonia and nitric acid in a polluted atmosphere and their equilibrium relationship to particulate ammonium nitrate, *Environ. Sci. Technol.* **13**, 1416–1419.

Duan, S. X., Jayne, J. T., Davidovits, P., Worsnop, D. R., Zahniser, M. S., and Kolb, C. E. (1993) Uptake of gas-phase acetone by water surfaces, *J. Phys. Chem.* **97**, 2284–2288.

Fuchs, N. A. (1964) *Mechanics of Aerosols*, Pergamon, New York.

Fuchs, N. A., and Sutugin, A. G. (1971) High dispersed aerosols, in *Topics in Current Aerosol Research (Part 2)*, G. M. Hidy and J. R. Brock, eds., Pergamon, New York, pp. 1–200.

Gardner, J. A., Watson, L. R., Adewuyi, Y. G., Davidovits, P., Zahniser, M. S., Worsnop, D. R., and Kolb, C. E. (1987) Measurement of the mass accommodation coefficient of $SO_2(g)$ on water droplets, *J. Geophys. Res.* **92**, 10887–10895.

Hildemann, L. M., Russell, A. G., and Cass, G. R. (1984) Ammonia and nitric acid concentrations in equilibrium with atmospheric aerosols: Experiment vs theory, *Atmos. Environ.* **18**, 1737–1750.

Hirschfelder, J. O., Curtiss, C. F., and Bird, R. B. (1954) *Molecular Theory of Gases and Liquids,* Wiley, New York.

Jacob, D. J. (1985) Comment on "The photochemistry of a remote stratiform cloud," *J. Geophys. Res.* **90**, 5864.

Jacob, D. J. (1986) Chemistry of OH in remote clouds and its role in the production of formic acid and peroxymonosulfate, *J. Geophys. Res.* **91**, 9807–9826.

Jayne, J. T., Duan, S.X., Davidovits, P., Worsnop, D. R., Zahniser, M. S., and Kolb, C. E. (1991) Uptake of gas-phase alcohol and organic acid molecules by water surfaces, *J. Phys. Chem.* **95**, 6329–6336.

Jayne, J. T. Duan, S. X., Davidovits, P., Worsnop, D. R., Zahniser, M. S., and Kolb, C. E. (1992) Uptake of gas-phase aldehydes by water surfaces, *J. Phys. Chem.* **96**, 5452–5460.

Jefferson, A., Eisele, F. L., Ziemann, P. J., Weber, R. J., Marti, J. J., and McMurray, P. H. (1997) Measurements of the H_2SO_4 mass accommodation coefficient onto polydisperse aerosol, *J. Geophys. Res.* **102**, 19021–19028.

Kumar, S. (1989a) The characteristic time to achieve interfacial phase equilibrium in cloud drops, *Atmos. Environ.* **23**, 2299–2304.

Kumar, S. (1989b) Discussion of "The characteristic time to achieve interfacial phase equilibrium in cloud drops," *Atmos. Environ.* **23**, 2893.

Li, W., and Davis E. J. (1995) Aerosol evaporation in the transition regime, *Aerosol Sci. Technol.* **25**, 11–21.

Loyalka, S. K. (1983) Modeling of condensation on aerosols, *Prog. Nucl. Energy* **12**, 1–8.

Maxwell, J. C. (1877) in *Encyclopedia Britannica*, Vol. 2, p. 82.

Meng, Z., and Seinfeld, J. H. (1996) Timescales to achieve atmospheric gas-aerosol equilibrium for volatile species, *Atmos. Environ.* **30**, 2889–2900.

Moore, W. J. (1962) *Physical Chemistry*, 3rd ed., Prentice-Hall, Englewood Cliffs, NJ.

Mozurkewich, M., McMurry, P. H., Gupta, A., and Calvert, J. G. (1987) Mass accommodation coefficient for HO_2 radicals on aqueous particles, *J. Geophys. Res.* **192**, 4163–4170.

Nathanson, G. M., Davidovits, P., Worsnop, D. R., and Kolb, C. E. (1996) Dynamics and kinetics at the gas-liquid interface, *J. Phys. Chem.* **100**, 13007–13020.

Pandis, S. N., and Seinfeld, J. H. (1989) Sensitivity analysis of a chemical mechanism for aqueous-phase atmospheric chemistry, *J. Geophys. Res.* **94**, 1105–1126.

Pesthy, A. J., Flagan, R. C., and, Seinfeld, J. H. (1981) The effect of a growing aerosol on the rate of homogeneous nucleation of a vapor, *J. Colloid Interface Sci.* **82**, 465–479.

Pilinis, C., and Seinfeld, J. H. (1988) Development and evaluation of an Eulerian photochemical gas–aerosol model, *Atmos. Environ.* **22**, 1985–2001.

Pöschl, U., Rudich, Y., and Ammann, M. (2005) Kinetic model framework for aerosol and cloud surface chemistry and gas-particle interactions: Part 1—general equations, parameters, and terminology, *Atmos. Chem. Phys. Discuss.* **5**, 2111–2191.

Pruppacher, H. R., and Klett, J. D. (1997) *Microphysics of Clouds and Precipitation*, 2nd ed., Kluwer, Boston.

Rudich, Y., Talukdar, R. K., and Ravishankara, A. R. (1996) Reactive uptake of NO_3 on pure water and ionic solutions, *J. Geophys, Res.* **101**, 21023–21031.

Russell, A. G., McCue, K. F., and Cass, G. R. (1988) Mathematical modeling of the formation of nitrogen-containing air pollutants. 1. Evaluation of an Eulerian photochemical model, *Environ. Sci. Technol.* **22**, 263–271.

Sahni, D. C. (1966) The effect of a black sphere on the flux distribution of an infinite moderator, *J. Nucl. Energy* **20**, 915–920.

Schwartz, S. E. (1984) Gas aqueous reactions of sulfur and nitrogen oxides in liquid water clouds, in *SO_2, NO, and NO_2 Oxidation Mechanisms: Atmospheric Considerations*, J. G. Calvert, ed., Butterworth, Boston, pp. 173–208.

Schwartz, S. E. (1986) Mass transport considerations pertinent to aqueous-phase reactions of gases in liquid-water clouds, in *Chemistry of Multiphase Atmospheric Systems*, W. Jaeschke, ed., Springer-Verlag, Berlin, pp. 415–471.

Schwartz, S. E. (1987) Both sides now: The chemistry of clouds, *Ann. NY Acad. Sci.* **502**, 83–114.

Schwartz, S. E. (1988) Mass transport limitation to the rate of in-cloud oxidation of SO_2: Re-examination in the light of new data, *Atmos. Environ.* **22**, 2491–2499.

Schwartz, S. E. (1989) Discussion of "The characteristic time to achieve interfacial phase equilibrium in cloud drops," *Atmos. Environ.* **23**, 2892–2893.

Schwartz, S. E., and Freiberg, J. E. (1981) Mass transport limitation to the rate of reaction of gases and liquid droplets: application to oxidation of SO_2 and aqueous solution, *Atmos. Environ.* **15**, 1129–1144.

Shi, B., and Seinfeld, J. H. (1991) On mass transport limitation to the rate of reaction of gases in liquid droplets, *Atmos. Environ.* **25A**, 2371–2383.

Sitarski, M., and Nowakowski, B. (1979) Condensation rate of trace vapor on Kundsen aerosols from solution of the Boltzmann equation, *J. Colloid Interface Sci.* **72**, 113–122.

Stelson, A. W., Friedlander, S. K., and Seinfeld J. H. (1979) A note on the equilibrium relationship between ammonia and nitric acid and particulate ammonium nitrate, *Atmos. Environ.* **13**, 369–371.

Stelson, A. W., and Seinfeld, J. H. (1982) Relative humidity and temperature dependence of the ammonium nitrate dissociation constant, *Atmos. Environ.* **16**, 983–992.

Takahama, S., Wittig, A. E., Vayenas, D. V., Davidson, C. I., Pandis, S. N. (2004) Modeling the diurnal variation of nitrate during the Pittsburgh Air Quality Study, *J. Geophys. Res.* **109**, D16S06 (doi: 10.1029/2003JD004149).

Tang, I. N., and Lee, J. H. (1987) Accommodation coefficients for ozone and SO_2: Implications for SO_2 oxidation in cloudwater, *ACS Symp. Ser.* **349**, 109–117.

Van Doren, J. M., Watson, L. R., Davidovits, P., Worsnop, D. R., Zahniser, M. S., and Kolb, C. E. (1990) Temperature dependence of the uptake coefficients of HNO_3, HCl, and N_2O_5 by water droplets, *J. Phys. Chem.* **94**, 3265–3272.

Van Doren, J. M., Watson, L. R., Davidovits, P., Worsnop, D. R., Zahniser, M. S., and Kolb, C. E. (1991) uptake of N_2O_5 and HNO_3 by aqueous sulfuric acid droplets, *J. Phys. Chem.* **95**, 1684–1689.

Watson, L. R., Van Doren, J. M., Davidovits, P., Worsnop, D. R., Zahniser, M. S., and Kolb, C. E. (1990) Uptake of HCl molecules by aqueous sulfuric acid droplets as a function of acid concentration, *J. Geophys. Res.* **95**, 5631–5638.

Wexler, A. S., and Seinfeld, J. H. (1990) The distribution of ammonium salts among a size and composition dispersed aerosol, *Atmos. Environ.* **24A**, 1231–1246.

Wexler, A. S., and Seinfeld, J. H. (1992) Analysis of aerosol ammonium nitrate: Departures from equilibrium during SCAQS, *Atmos. Environ.* **26A**, 579–591.

Wiliams, M. M. R. and Loyalka, S. K. (1991) *Aerosol Science: Theory and Practice,* Pergamon, New York.

Worsnop, D. R., Zahniser, M. S., Kolb, C. E., Gardner, J. A., Watson, L. R., Van Doren, J. M., Jayne, J.T., and Davidovits, P. (1989) Temperature dependence of mass accommodation of SO_2 and H_2O_2 on aqueous surfaces, *J. Phys. Chem.* **93**, 1159–1172.

13 Dynamics of Aerosol Populations

Up to this point we have considered the physics and chemistry of atmospheric aerosols in terms of the behavior of a single particle. In this chapter we focus our attention on a population of particles that interact with each other. We will treat changes that occur in the population when a vapor compound condenses on the particles and when the particles collide and adhere. We begin by discussing a series of mathematical approaches for the description of the particle distribution. Next, we calculate the rate of change of a particle distribution due to mass transfer of material from or to the gas phase. We continue by calculating the rate at which two spherical particles in Brownian motion will collide and then, in an Appendix, consider coagulation induced by velocity gradients in the fluid, by differential settling of particles, and compute the enhancement or retardation to the coagulation rate due to interparticle forces. The remainder of the chapter is devoted to solution of the equations governing the size distribution of an aerosol and examination of the physical regimes of behavior of an aerosol population.

13.1 MATHEMATICAL REPRESENTATIONS OF AEROSOL SIZE DISTRIBUTIONS

Observed aerosol number distributions are usually expressed as $n°(\log D_p)$, that is, using $\log D_p$ as an independent variable (see Chapter 8). However, it is often convenient to use alternative definitions of the particle size distribution to facilitate the calculations of its rate of change. The distribution can be expressed in several forms assuming a spatially homogeneous aerosol of uniform chemical composition.

13.1.1 Discrete Distribution

An aerosol distribution can be described by the number concentrations of particles of various sizes as a function of time. Let us define $N_k(t)$ as the number concentration (cm^{-3}) of particles containing k monomers, where a monomer can be considered as a single molecule of the species representing the particle. Physically, the discrete distribution is appealing since it is based on the fundamental nature of the particles. However, a particle of size 1 μm contains on the order of 10^{10} monomers, and description of the submicrometer aerosol distribution requires a vector $(N_2, N_3, \ldots, N_{10^{10}})$ containing 10^{10} numbers. This makes the use of the discrete distribution impractical for most atmospheric aerosol applications. We will use it in the subsequent sections for instructional purposes and as an intermediate step toward development of the continuous general dynamic equation.

Atmospheric Chemistry and Physics: From Air Pollution to Climate Change, Second Edition, by John H. Seinfeld and Spyros N. Pandis. Copyright © 2006 John Wiley & Sons, Inc.

13.1.2 Continuous Distribution

The continuous distribution, introduced in Chapter 8, is usually a more useful concept in practice. The volume of the particle is often chosen for convenience as the independent variable for the continuous size distribution function $n(v,t)(\mu m^{-3}\, cm^{-3})$, where $v = \frac{1}{6}\pi D_p^3$ is the volume of a particle with diameter D_p. Thus $n(v,t)dv$ is defined as the number of particles per cubic centimeter having volumes in the range from v to $v + dv$. Note that the total aerosol number concentration $N_t(t)$ (cm^{-3}) is then given by

$$N_t(t) = \int_0^\infty n(v,t)dv \tag{13.1}$$

The mass of a particle, m, or the diameter, D_p, can also be used as the independent variable for the mathematical description of the aerosol distribution. We saw in Chapter 8 that these functions are not equal to each other but can be easily related. For example, if the radius R_p is used as an independent variable for the distribution function $n_R(R_p,t)$, then the concentration of particles dN in the size range R_p to $(R_p + dR_p)$ is given by $dN = n_R(R_p,t)\, dR_p$. But the same number of particles is equal to $n(v,t)\, dv$, where $v = \frac{4}{3}\pi R_p^3$ or $dv = 4\pi R_p^2\, dR_p$, so

$$n_R(R_p,t) = 4\pi R_p^2 n(v,t) \tag{13.2}$$

Similar expressions can easily be developed for other functions describing size distributions using (8.25).

13.2 CONDENSATION

When a vapor condenses on a particle population or when material evaporates from the aerosol to the gas phase, the particle diameters change and the size distribution of the population $n(v,t)$ changes shape. Assuming that the particles are not in equilibrium with the gas phase—that is, the vapor pressure of the particles is not equal to the partial pressure of this compound in the gas phase—we want to calculate the rate of change $(\partial n(v,t)/\partial t)_{growth/evap}$.

13.2.1 The Condensation Equation

We saw in Chapter 12 that the rate of change of the mass of a particle of diameter D_p as a result of transport of species i between the gas and aerosol phases is

$$\frac{dm}{dt} = \frac{2\pi D_p D_i M_i}{RT} f(Kn,\alpha)(p_i - p_{eq,i}) \tag{13.3}$$

where D_i is the diffusion coefficient for species i in air, M_i is its molecular weight, and $f(Kn,\alpha)$ is the correction due to noncontinuum effects and imperfect surface accommodation. The difference between the vapor pressure of i far from the particle p_i and the equilibrium vapor pressure $p_{eq,i}$ constitutes the driving force for the transport of species i.

Let us define the condensation growth rate $I_v(v)$ as the rate of change of the *volume* of a particle of volume v. Assuming that the aerosol particle contains only one species, it will

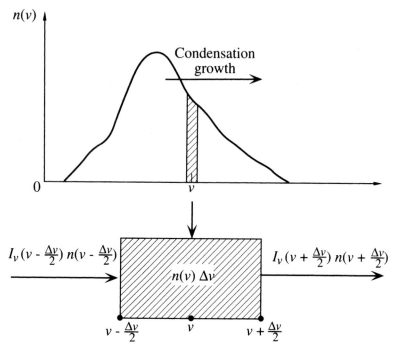

FIGURE 13.1   Schematic of differential mass balance for the derivation of growth equation for a particle population.

have constant density ρ_p; therefore from (13.3) and since $v = \frac{1}{6}\pi D_p^3$

$$I_{(v)} = \frac{dv}{dt} = \frac{2\pi^{2/3}(6v)^{1/3} D_i M_i}{\rho_p RT} f(\text{Kn}, \alpha)(p_i - p_{\text{eq},i}) \tag{13.4}$$

Let us assume that a particle population is growing due to condensation, so the size distribution is moving to the right (Figure 13.1). We focus our attention on an infinitesimal slice Δv of the distribution centered at v that extends from $v - (\Delta v/2)$ to $v + (\Delta v/2)$. The number of particles in this slice at time t is $n(v, t)\Delta v$. As a result of condensation and subsequent growth, particles enter the slice from the left and exit from the right. The instantaneous rate of entry from the left is proportional to the number concentration of particles at the left boundary of the slice, $n(v - (\Delta v/2), t)$, and their volume growth rate, $I_v(v - (\Delta v/2), t)$. So particles enter the slice from the left with a rate equal to $I_v(v - (\Delta v/2), t)n(v - (\Delta v/2), t)$. Note that the units of the product are indeed number of particles per time. Applying the same arguments, particles grow out of this distribution slice from the right boundary with a rate $I_v(v + (\Delta v/2), t)\, n(v + (\Delta v/2), t)$. After a period Δt the number of particles inside this fixed slice will change from $n(v, t)\Delta v$ to $n(v, t + \Delta t)\, \Delta v$. This number change will equal the net flux into the slice during this period, or mathematically

$$n(v, t + \Delta t)\Delta v - n(v, t)\Delta v$$

$$= I_v\left(v - \frac{\Delta v}{2}, t\right)n\left(v - \frac{\Delta v}{2}, t\right)\Delta t - I_v\left(v + \frac{\Delta v}{2}, t\right)n\left(v + \frac{\Delta v}{2}, t\right)\Delta t \tag{13.5}$$

Rearranging the terms and considering the limits $\Delta t \to 0$ and $\Delta v \to 0$ yield

$$\lim_{\Delta t \to 0} \frac{n(v, t + \Delta t) - n(v, t)}{\Delta t}$$
$$= - \lim_{\Delta v \to 0} \frac{I_v(v + (\Delta v/2), t)\, n(v + (\Delta v/2), t) - I_v(v - (\Delta v/2), t)n(v - (\Delta v/2), t)}{\Delta v} \quad (13.6)$$

and noting that the limits are by definition equal to partial derivatives, one finally gets

$$\frac{\partial n(v, t)}{\partial t} = -\frac{\partial}{\partial v}[I_v(v, t)n(v, t)] \quad (13.7)$$

where $I_v(v, t)$ is given by (13.4).

Equation (13.7) is called the *condensation equation* and describes mathematically the rate of change of a particle size distribution $n(v, t)$ due to the condensation or evaporation flux $I_v(v, t)$, neglecting other processes that may influence the distribution shape (sources, removal, coagulation, nucleation, etc). A series of alternative forms of the condensation equation can be written depending on the form of the size distribution used or the expression for the condensation flux. For example, if the mass of a particle is used as the independent variable, one can show that the condensation equation takes the form

$$\frac{\partial n(m, t)}{\partial t} + \frac{\partial}{\partial m}[I_m(m, t)\, n(m, t)] = 0 \quad (13.8)$$

where $I_m(m, t)$, the rate of mass change of a particle due to condensation, based on (13.3), is

$$I_m(m, t) = \frac{2\pi D_i M_i}{RT}\left(\frac{6m}{\rho_p \pi}\right)^{1/3} f(\text{Kn}, \alpha)(p_i - p_{\text{eq},i}) \quad (13.9)$$

Finally, if the diameter is the independent variable of choice, the number distribution is given by $n_D(D_p, t)$ and the condensation equation is

$$\frac{\partial n_D(D_p, t)}{\partial t} + \frac{\partial}{\partial D_p}[I_D(D_p, t)n_D(D_p, t)] = 0 \quad (13.10)$$

where $I_D(D_p)$ is the rate of diameter change of a particle as a result of condensation or evaporation:

$$I_D(D_p, t) = \frac{dD_p}{dt} = \frac{4D_i M_i}{RT D_p \rho_p}f(\text{Kn}, \alpha)(p_i - p_{\text{eq},i}) \quad (13.11)$$

Equations (13.7), (13.8), and (13.10) are equivalent descriptions of the same process using different independent variables for the description of the size distribution.

For sufficiently large particles (the so-called continuum regime) and assuming unity accommodation coefficient, $f(\mathrm{Kn}, \alpha) = 1$, the continuum diffusion growth law based on particle diameter is

$$I_D(D_p, t) = \frac{1}{D_p}\left[\frac{4D_i M_i}{RT\rho_p}(p_i - p_{\mathrm{eq},i})\right] \tag{13.12}$$

Assuming that the condensation driving force is maintained constant externally (e.g., constant supersaturation of the gas phase), then (13.12) can be written as

$$I_D(D_p, t) = \frac{A}{D_p} \tag{13.13}$$

where $A = 4 D_i M_i (p_i - p_{\mathrm{eq},i})/RT\,\rho_p$ is a constant. This case therefore corresponds to a growth law $I_D(D_p) \approx D_p^{-1}$.

13.2.2 Solution of the Condensation Equation

The condensation equation assuming continuum regime growth, unity accommodation coefficient, and constant gas-phase supersaturation can be written using (13.10) and (13.13) as

$$\frac{\partial n_D(D_p, t)}{\partial t} + \frac{\partial}{\partial D_p}\left(\frac{An_D(D_p, t)}{D_p}\right) = 0 \tag{13.14}$$

or, after expanding the derivative, as

$$\frac{\partial n_D(D_p, t)}{\partial t} + \frac{A}{D_p}\frac{\partial n_D(D_p, t)}{\partial D_p} = \frac{A}{D_p^2} n_D(D_p, t) \tag{13.15}$$

This above partial differential equation (PDE) is to be solved subject to the following initial and boundary conditions:

$$n_D(D_p, 0) = n_0(D_p) \tag{13.16}$$
$$n_D(0, t) = 0 \tag{13.17}$$

Equation (13.16) provides simply the initial size distribution for the aerosol population, while (13.17) implies that there are no particles of zero diameter.

If we define

$$F(D_p, t) = \frac{A}{D_p} n_D(D_p, t) \tag{13.18}$$

then (13.14) can be rewritten as

$$\frac{\partial F(D_p, t)}{\partial t} + \frac{A}{D_p}\frac{\partial F(D_p, t)}{\partial D_p} = 0 \tag{13.19}$$

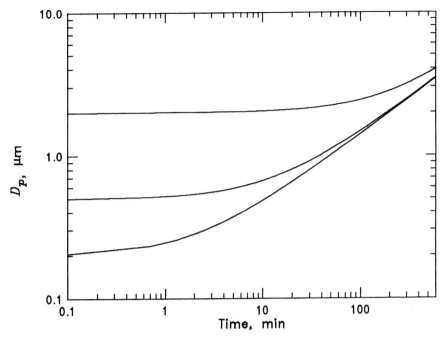

FIGURE 13.2 Growth of particles of initial diameters 0.2, 0.5, and $2\,\mu m$ assuming $D_i = 0.1\,cm^2\,s^{-1}$, $M_i = 100\,g\,mol^{-1}$, $(p_i - p_{eq,i}) = 10^{-9}$ atm (1 ppb), $T = 298\,K$, and $\rho_p = 1\,g\,cm^{-3}$.

Using the method of characteristics for partial differential equations, $F(D_p, t)$ will be constant along the characteristic lines given by

$$\frac{dD_p}{dt} = \frac{A}{D_p} \qquad (13.20)$$

or equivalently along the curves given by the solution of (13.20) with $D_p(t = 0) = D_{p0}$,

$$D_p^2 = D_{p0}^2 + 2At \qquad (13.21)$$

where D_{p0} is a constant corresponding to the initial diameter for a given characteristic. The characteristic curves for a typical set of parameters are given in Figure 13.2. Note that if $2At \ll D_{p0}^2$, then $D_p \simeq D_{p0}$; therefore for a timescale of 1 h, particles larger than $0.6\,\mu m$ will not grow appreciably under these conditions. The smaller particles grow much faster as the characteristics are moving rapidly to larger diameters. It is also interesting to note that as the diameters of the small particles change much faster than the diameters of the large ones, condensation tends to "shrink" the aerosol distribution. Note that as $t \to \infty$ the distribution tends to become more and more monodisperse.

Physically, the characteristics of the partial differential equation give us the trajectories in the time–diameter coordinate system of the various particle sizes. If, for a moment, we think of the particle population not as a continuous mathematical function but as groups of particles of specific diameters, the characteristic equations describe quantitatively the evolution of the sizes of these particles with time. The characteristic curves in Figure 13.2 are not the complete solution of the PDE, but they still convey information about the evolution of the size distribution for this case.

Returning to the solution of the condensation equation, the constant value of $F(D_p, t)$ along each characteristic curve can be determined by its value of $t = 0$, that is

$$F(D_p, t) = F(D_{p0}, 0) = \frac{A}{D_{p0}} n_0(D_{p0}) \tag{13.22}$$

and recalling the definition of $F(D_p, t)$ in (13.18), we obtain

$$n(D_p, t) = \frac{D_p}{D_{p0}} n_0(D_{p0}) \tag{13.23}$$

But D_p, D_{p0}, and t are related through the characteristic equation (13.21), so eliminating D_{p0} in (13.23) one gets

$$D_{p0} = (D_p^2 - 2At)^{1/2} \tag{13.24}$$

or substituting into (13.23)

$$n(D_p, t) = \frac{D_p}{(D_p^2 - 2At)^{1/2}} n_0[(D_p^2 - 2At)^{1/2}] \tag{13.25}$$

Note that the solution in (13.25) carries implicitly the constraint $D_p > \sqrt{2At}$.

Growth of a Lognormal Aerosol Size Distribution   Aerosol size distribution data are sometimes represented by the lognormal distribution function [see also (8.33)]

$$n(D_p) = \frac{N_0}{\sqrt{2\pi} D_p \ln \sigma_g} \exp\left(-\frac{\ln^2(D_p/\overline{D}_{pg})}{2 \ln^2 \sigma_g}\right) \tag{13.26}$$

For such an initial distribution using the solution obtained in (13.25), we get

$$n(D_p, t) = \frac{D_p}{(D_p^2 - 2At)} \frac{N_0}{\sqrt{2\pi} \ln \sigma_g} \exp\left(-\frac{\ln^2[(D_p^2 - 2At)^{1/2}/\overline{D}_{pg}]}{2 \ln^2 \sigma_g}\right) \tag{13.27}$$

For an aerosol population undergoing pure growth with no particle sources or sinks, the total particle number is preserved. If the solution in (13.27) is valid, it must satisfy the following integral relation at all times:

$$N_0 = \int_{\sqrt{2At}}^{\infty} n(D_p, t) dD_p \tag{13.28}$$

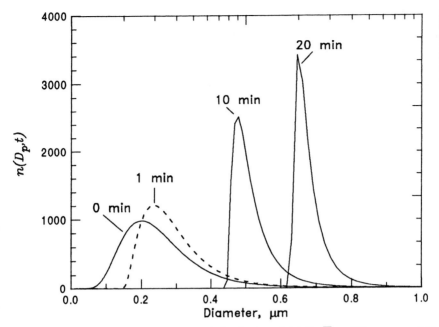

FIGURE 13.3 Evolution of a lognormal distribution (initially $\overline{D}_p = 0.2\,\mu\text{m}$, $\sigma_g = 1.5$) assuming $D_i = 0.1\,\text{cm}^2\,\text{s}^{-1}$, $M_i = 100\,\text{g mol}^{-1}$, $(p_i - p_{\text{eq},i}) = 10^{-9}\,\text{atm}$ (1 ppb), $T = 298\,\text{K}$, and $\rho_p = 1\,\text{g cm}^{-3}$.

Note that the lower limit of this integral is determined by the characteristic curve through the origin, since the solution is valid only for $D_p > \sqrt{2\,At}$. Equation (13.28) is readily simplified by using the substitutions

$$u = (D_p^2 - 2\,At)^{1/2} \tag{13.29}$$

and

$$du = \frac{D_p}{(D_p^2 - 2\,At)^{1/2}}\,dD_p \tag{13.30}$$

The lower and upper limits of the integral become 0 and ∞, respectively. The resulting integral is identical at all times in form with the integral of the original lognormal distribution, which we know is equal to the total number concentration.

  Figure 13.3 shows the growth of a lognormally distributed aerosol as a function of time. The difference in growth between the small and large particles, seen from the characteristic curves of Figure 13.2, is even more evident in Figure 13.3.

13.3 COAGULATION

Aerosol particles suspended in a fluid may come into contact because of their Brownian motion or as a result of their motion produced by hydrodynamic, electrical, gravitational, or other forces. Brownian coagulation is often referred to as *thermal coagulation*.

The theory of coagulation will be presented in two steps. At first, we will develop an expression describing the rate of collisions between two monodisperse particle populations consisting of N_1 particles with diameter D_{p1} and N_2 with diameter D_{p2}. In the next step a differential equation describing the rate of change of a full coagulating aerosol size distribution will be derived.

13.3.1 Brownian Coagulation

Let us start by considering the simplest coagulation problem, that of a monodisperse aerosol population of radius R_p at initial concentration N_0. Imagine one of the particles to be stationary with its center at the origin of the coordinate system. We wish to calculate the rate at which particles collide with this stationary particle as a result of their Brownian motion. Let us neglect for the time being the size and shape change of the particles as other particles collide with it. We will show later on that this assumption does not lead to appreciable error in the early stages of coagulation. Moreover, since spherical particles come into contact when the distance between their centers is equal to the sum of their radii, from a mathematical point of view, we can replace our stationary particle of radius R_p with an absorbing sphere of radius $2R_p$ and the other particles by point masses (Figure 13.4).

At this point, we need to describe mathematically the distribution of aerosols around the fixed particle. We need to consider, as with mass transfer to a particle, the continuum, free molecular, and transition regimes.

Continuum Regime   Assuming that the distribution of particles around our fixed particle can be described by the continuum diffusion equation, then that distribution $N(r, t)$ satisfies

$$\frac{\partial N(r,t)}{\partial t} = D\left(\frac{\partial^2 N(r,t)}{\partial r^2} + \frac{2}{r}\frac{\partial N(r,t)}{\partial r}\right) \tag{13.31}$$

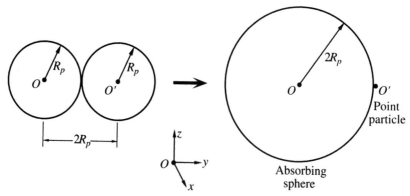

FIGURE 13.4   The collision of two particles of radii R_p is equivalent geometrically to collisions of point particles with an absorbing sphere of radius $2R_p$. Also shown in the corresponding coordinate system.

where r is the distance of the particles from the center of the fixed particle, N is the number concentration, and D is the Brownian diffusion coefficient for the particles. The initial and boundary conditions for (13.31) are

$$N(r,0) = N_0$$
$$N(\infty,t) = N_0$$
$$N(2R_p,t) = 0 \tag{13.32}$$

The initial condition assumes that the particles are initially homogeneously distributed in space with a number concentration N_0. The first boundary condition requires that the number concentration of particles infinitely far from the particle "absorbing sphere" not be influenced by it. Finally, the boundary condition at $r = 2R_p$ expresses the assumption that the fixed particle is a perfect absorber, that is, that particles adhere at every collision. Although little is known quantitatively about the sticking probability of two colliding aerosol particles, their low kinetic energy makes bounce-off unlikely. We shall therefore assume here a unity sticking probability.

The solution of (13.31) and (13.32) is

$$N(r,t) = N_0 \left[1 - \frac{2R_p}{r} + \frac{4R_p}{r\sqrt{\pi}} \int_0^{(r-2R_p)/2\sqrt{Dt}} e^{-\eta^2} d\eta \right]$$
$$= N_0 \left[1 - \frac{2R_p}{r} \operatorname{erfc}\left(\frac{r - 2R_p}{2\sqrt{Dt}} \right) \right] \tag{13.33}$$

and the rate (particles s^{-1}) at which particles collide with our stationary particle, which has an effective surface area of $16\pi R_p^2$, is

$$J_{\text{col}} = (16\pi R_p^2)D\left(\frac{\partial N}{\partial r}\right)_{r=2R_p} = 8\pi R_p D N_0 \left(1 + \frac{2R_p}{\sqrt{\pi Dt}}\right) \tag{13.34}$$

The collision rate is initially extremely fast (actually it starts at infinity) but for $t \gg 4R_p^2/\pi D$, it approaches a steady-state value of $J_{\text{col}} = 8\pi R_p D N_0$. Physically, at $t = 0$, other particles in the vicinity of the absorbing one collide with it, immediately resulting in a mathematically infinite collision rate. However, these particles are soon absorbed by the stationary particle and the concentration profile around our particle relaxes to its steady-state profile with a steady-state collision rate. One can easily calculate, given the Brownian diffusivities in Table 9.5, that such a system reaches steady state in 10^{-4} s for particles of diameter 0.1 μm and in roughly 0.1 s for 1 μm particles. Therefore neglecting the transition to this steady state is a good assumption for atmospheric applications.

Our analysis so far includes two major assumptions, that our particle is stationary and that all particles have the same radius. Let us relax these two assumptions by allowing our particle to undergo Brownian diffusion and also let it have a radius R_{p1} and the others in the fluid have radii R_{p2}. Our first challenge as we want to maintain the diffusion framework of (13.31) is to calculate the diffusion coefficient that characterizes the diffusion of particles of radius R_{p2} relative to those of radius R_{p1}.

As both particles are undergoing Brownian motion, suppose that in a time interval dt they experience displacements $d\mathbf{r}_1$ and $d\mathbf{r}_2$. Then their mean square relative displacement is

$$\langle |d\mathbf{r}_1 - d\mathbf{r}_2|^2 \rangle = \langle dr_1^2 \rangle + \langle dr_2^2 \rangle - 2\langle d\mathbf{r}_1 \cdot d\mathbf{r}_2 \rangle \tag{13.35}$$

Since the motions of the two particles are assumed to be independent, $\langle d\mathbf{r}_1 \cdot d\mathbf{r}_2 \rangle = 0$. Thus

$$\langle |d\mathbf{r}_1 - d\mathbf{r}_2|^2 \rangle = \langle dr_1^2 \rangle + \langle dr_2^2 \rangle \tag{13.36}$$

Referring back to our discussion of Brownian motion, we can identify coefficient D_{12} by (9.72):

$$\langle |d\mathbf{r}_1 - d\mathbf{r}_2|^2 \rangle = 6D_{12}dt \tag{13.37}$$

On the other hand, the two individual Brownian diffusion coefficients are defined by $\langle dr_1^2 \rangle = 6D_1\, dt$ and $\langle dr_2^2 \rangle = 6D_2 dt$. Consequently, we find that

$$D_{12} = D_1 + D_2 \tag{13.38}$$

Taking advantage of this result, we can now once more regard the particle with radius R_{p1} as stationary and those with radius R_{p2} as diffusing toward it. The diffusion equation (13.31) then governs $N_2(r,t)$, the concentration of particles with radius R_{p2}, with the diffusion coefficient D_{12}. The system is then described mathematically by

$$\frac{\partial N_2(r,t)}{\partial t} = D_{12}\left(\frac{\partial^2 N_2(r,t)}{\partial r^2} + \frac{2}{r}\frac{\partial N_2(r,t)}{\partial r}\right) \tag{13.39}$$

with conditions

$$N_2(r,0) = N_{20}$$
$$N_2(\infty,t) = N_{20}$$
$$N_2(R_{p1} + R_{p2},t) = 0 \tag{13.40}$$

Note that the boundary condition at $r = 2R_p$ is now applied at $r = R_{p1} + R_{p2}$. The solution of (13.39) and (13.40) is

$$N_2(r,t) = N_{20}\left[1 - \frac{R_{p1} + R_{p2}}{r}\,\text{erfc}\left(\frac{r - (R_{p1} + R_{p2})}{2\sqrt{D_{12}t}}\right)\right] \tag{13.41}$$

and the rate (s^{-1}) at which #2 (secondary) particles arrive at the surface of the absorbing sphere at $r = R_{p1} + R_{p2}$ is

$$J_{\text{col}} = 4\pi(R_{p1} + R_{p2})D_{12}N_{20}\left(1 + \frac{R_{p1} + R_{p2}}{\sqrt{\pi D_{12}t}}\right) \tag{13.42}$$

The steady-state rate of collision is

$$J_{col} = 4\pi(R_{p1} + R_{p2})D_{12}N_{20} \tag{13.43}$$

The collision rate we have derived is the rate, expressed as the number of #2 particles per second, at which the #2 particles collide with a single #1 (primary) particle. When there are N_{10} #1 particles, the total collision rate between #1 and #2 particles per volume of fluid is equal to the collision rate derived above multiplied by N_{10}. Thus the steady-state coagulation rate $(cm^{-3}\,s^{-1})$ between #1 and #2 particles is

$$J_{12} = 4\pi(R_{p1} + R_{p2})D_{12}N_{10}N_{20} \tag{13.44}$$

At this point there is no need to retain the subscript 0 on the number concentrations so the coagulation rate can be expressed as

$$J_{12} = K_{12}N_1N_2 \tag{13.45}$$

where

$$K_{12} = 2\pi(D_{p1} + D_{p2})(D_1 + D_2) \tag{13.46}$$

In the continuum regime the Brownian diffusivities are given by (9.73)

$$D_i = \frac{kT}{3\pi\mu D_{pi}} \tag{13.47}$$

where the slip correction C_c is set equal to 1, μ is the viscosity of air, and D_{pi} is the particle diameter. Using the expressions for D_1 and D_2, (13.46) becomes

$$K_{12} = \frac{2kT}{3\mu}\frac{(D_{p1} + D_{p2})^2}{D_{p1}D_{p2}} \tag{13.48}$$

The coagulation coefficient K_{12} can be expressed in terms of particle volume as

$$K_{12} = \frac{2kT}{3\mu}\frac{(v_1^{1/3} + v_2^{1/3})^2}{(v_1 v_2)^{1/3}} \tag{13.49}$$

The development described above is based on the assumption that continuum diffusion theory is a valid description of the concentration distribution of particles surrounding a central absorbing sphere.

Transition and Free Molecular Regime   When the mean free path λ_p of the diffusing aerosol particle is comparable to the radius of the absorbing particle, the boundary

TABLE 13.1 Fuchs Form of the Brownian Coagulation Coefficient K_{12}

$$K_{12} = 2\pi(D_1 + D_2)(D_{p1} + D_{p2})\left(\frac{D_{p1} + D_{p2}}{D_{p1} + D_{p2} + 2(g_1^2 + g_2^2)^{1/2}} + \frac{8(D_1 + D_2)}{(\bar{c}_1^2 + \bar{c}_2^2)^{1/2}(D_{p1} + D_{p2})}\right)^{-1}$$

where

$$\bar{c}_i = \left(\frac{8\,kT}{\pi\,m_i}\right)^{1/2}$$

$$\ell_i = \frac{8\,D_i^{\,a}}{\pi\,\bar{c}_i}$$

$$g_i = \frac{1}{3\,D_{pi}\ell_i}[(D_{pi} + \ell_i)^3 - (D_{pi}^2 + \ell_i^2)^{3/2}] - D_{pi}$$

$$D_i = \frac{kTC_c}{3\,\pi\mu D_{pi}}$$

^{a}The mean free path ℓ_i is defined slightly differently than in (9.88).

condition at the absorbing particle surface must be corrected to account for the nature of the diffusion process in the vicinity of the surface. The physical meaning of this can be explained as follows. As we have seen the diffusion equations can be applied to Brownian motion only for time intervals that are large compared to the relaxation time τ of the particles or for distances that are large compared to the aerosol mean free path λ_p. Diffusion equations cannot describe the motion of particles inside a layer of thickness λ_p adjacent to an absorbing wall. If the size of the absorbing sphere is comparable to λ_p, this layer has a substantial effect on the kinetics of coagulation.

Fuchs (1964) suggested that this effect can be corrected by the introduction of a generalized coagulation coefficient

$$K_{12} = 2\pi(D_{p1} + D_{p2})(D_1 + D_2)\beta \tag{13.50}$$

where D_i is given by (9.73). The Fuchs form of K_{12} is defined in Table 13.1. It is instructive to compare the values of the corrected coagulation coefficient K versus that neglecting the kinetic effect, K_0, corresponding to $\beta = 1$ (Table 13.2) for monodisperse aerosols in air. The correction factor β varies from 0.014 for particles of 0.001 μm radius, to practically 1 for 1 μm particles.

Let us examine the two asymptotic limits of this coagulation rate. In the continuum limit, Kn $\rightarrow$ 0, $\beta = 1$ and the collision rate reduces to that given by (13.48). In the free-molecule limit Kn $\rightarrow \infty$ and one can show that

$$K_{12} = \pi(R_{p1} + R_{p2})^2(\bar{c}_1^2 + \bar{c}_2^2)^{1/2} \tag{13.51}$$

This collision rate is, of course, equal to that obtained directly from kinetic theory.

Values of K_{12} are given in Table 13.3 and Figure 13.5. In using Figure 13.5, we find the smaller of the two particles as the abscissa and then locate the line corresponding to the larger particle.

TABLE 13.2 Coagulation Coefficients of Monodisperse Aerosols in Air[c]

D_p, μm	K_0, $cm^3 s^{-1a}$	K, $cm^3 s^{-1b}$
0.002	690×10^{-10}	8.9×10^{-10}
0.004	340×10^{-10}	13×10^{-10}
0.01	140×10^{-10}	19×10^{-10}
0.02	72×10^{-10}	24×10^{-10}
0.04	38×10^{-10}	23×10^{-10}
0.1	18×10^{-10}	15×10^{-10}
0.2	11×10^{-10}	11×10^{-10}
0.4	8.6×10^{-10}	8.2×10^{-10}
1.0	7.0×10^{-10}	6.9×10^{-10}
2.0	6.5×10^{-10}	6.4×10^{-10}
4.0	6.3×10^{-10}	6.2×10^{-10}

[a]Coagulation coefficient neglecting kinetic effects.
[b]Coagulation coefficient including the kinetic correction.
[c]Parameter values given in Figure 13.5.

Coagulation Rates Table 13.3 indicates that the smallest value of the coagulation coefficient occurs when both particles are of the same size. The coagulation coefficient rises rapidly when the ratio of the particle diameters increases. This increase is due to the synergism between the two particles. A large particle may be sluggish from the Brownian motion perspective but, because of its large surface area, provides an ample target for the small, fast particles. Collisions between large particles are slow because both particles move slowly. On the other hand, whereas very small particles have relatively high velocities, they tend to miss each other because their cross-sectional area for collision is small. Note that target area is proportional to D_p^2, whereas the Brownian diffusion coefficient decreases only as D_p.

We see from Figure 13.5 that for collisions among equal-sized particles a maximum coagulation coefficient is reached at about 0.02 μm. In the continuum regime, Kn ≪ 1 or $D_p > 1$ μm, for $D_{p1} = D_{p2}$ using (13.48), the coagulation coefficient is

$$K_{11} = \frac{8kT}{3\mu} \tag{13.52}$$

and is independent of the particle diameter.

TABLE 13.3 Coagulation Coefficients ($cm^3 s^{-1}$) of Atmospheric Particles[a]

	D_{p1} (μm)					
D_{p2} μm	0.002	0.01	0.1	1.0	10	20
0.002	8.9×10^{-10}	5.7×10^{-9}	3.4×10^{-7}	7.8×10^{-6}	8.5×10^{-5}	17×10^{-5}
0.01	5.7×10^{-9}	19×10^{-10}	2.5×10^{-8}	3.4×10^{-7}	3.5×10^{-6}	7.0×10^{-6}
0.1	3.4×10^{-7}	2.5×10^{-8}	15×10^{-10}	5.0×10^{-9}	4.5×10^{-8}	9.0×10^{-8}
1	7.8×10^{-6}	3.4×10^{-7}	5.0×10^{-9}	6.9×10^{-10}	2.1×10^{-9}	3.9×10^{-9}
10	8.5×10^{-5}	3.5×10^{-6}	4.5×10^{-8}	2.1×10^{-9}	6.1×10^{-10}	6.8×10^{-10}
20	17×10^{-5}	7.0×10^{-6}	9.0×10^{-8}	3.9×10^{-9}	6.8×10^{-10}	6.0×10^{-10}

[a]Parameter values given in Figure 13.5.

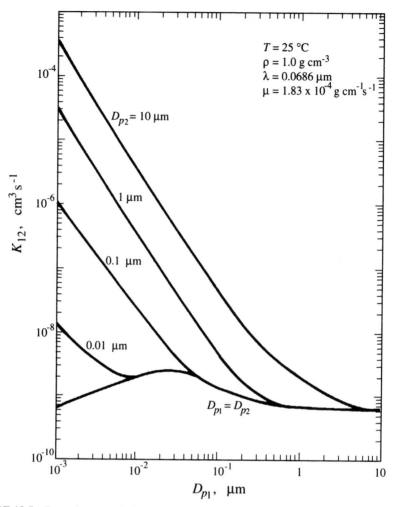

FIGURE 13.5 Brownian coagulation coefficient K_{12} for coagulation in air at 25°C of particles of diameters D_{p1} and D_{p2}. The curves were calculated using the correlation of Fuchs in Table 13.1. To use this figure, find the *smaller* of the two particles as the abscissa and then locate the line corresponding to the larger particle.

On the other hand, in free molecular regime coagulation with $D_{p1} = D_{p2}$, using (13.51), one can show that

$$K_{11} = 4 \left(\frac{6kT}{\rho_p} \right)^{1/2} D_{p1}^{1/2} \tag{13.53}$$

and the coagulation coefficient increases with increasing diameter. The maximum in the coagulation coefficient for equal-sized particles reflects a balance between particle mobility and cross-sectional area for collision.

In the continuum regime if $D_{p2} \gg D_{p1}$, the coagulation coefficient approaches the limiting value

$$\lim_{D_{p2} \gg D_{p1}} K_{12} = \frac{2kT}{3\mu} \frac{D_{p2}}{D_{p1}} \tag{13.54}$$

In the free molecular regime if $D_{p2} \gg D_{p1}$, the coagulation coefficient has the following asymptotic behavior:

$$\lim_{D_{p2} \gg D_{p1}} K_{12} = \left(\frac{3kT}{\rho_p}\right)^{1/2} \frac{D_{p2}^2}{D_{p1}^{3/2}} \tag{13.55}$$

By comparing (13.54) and (13.55) we see that, with D_{p1} fixed, K_{12} increases more rapidly with D_{p2} for free molecular regime than for continuum regime coagulation.

Collision Efficiency   In our discussion so far, we have assumed that every collision results in coagulation of the two colliding particles. Let us relax this assumption by denoting the coagulation efficiency (fraction of collisions that result in coagulation) by α. Then Fuchs (1964) showed that the coagulation coefficient becomes

$$K_{12} = 2\pi(D_1 + D_2)(D_{p1} + D_{p2})$$

$$\times \left[\frac{D_{p1} + D_{p2}}{D_{p1} + D_{p2} + 2(g_1^2 + g_2^2)^{1/2}} + \frac{8\alpha(D_1 + D_2)}{(\bar{c}_1^2 + \bar{c}_2^2)^{1/2}(D_{p1} + D_{p2})}\right]^{-1} \tag{13.56}$$

in place of the Table 13.1 expression.

One can show that in the free molecular regime the coagulation rate is proportional to the collision efficiency. In the continuum regime, coagulation is comparatively insensitive to changes in α. For example, for $R_p = 1\ \mu m$, the coagulation rate decreases by only 5% as α goes from 1.0 to 0.25, while at $R_p = 0.1\ \mu m$ a 5% decrease of the coagulation rate needs an $\alpha = 0.60$ (Fuchs 1964).

Our development has assumed that all particles behave as spheres. It has been found that, when coagulating, certain particles form chain-like aggregates whose behavior can no longer be predicted on the basis of ideal, spherical particles. Lee and Shaw (1984), Gryn and Kerimov (1990), and Rogak and Flagan (1992), among others, have studied the coagulation of such particles.

13.3.2   The Coagulation Equation

In the previous sections, we focused our attention on calculation of the instantaneous coagulation rate between two monodisperse aerosol populations. In this section, we will develop the overall expression describing the evolution of a polydisperse coagulating aerosol population. A good place to start is with a discrete aerosol distribution.

The Discrete Coagulation Equation   A spatially homogeneous aerosol of uniform chemical composition can be fully characterized by the number densities of particles of various monomer contents as a function of time, $N_k(t)$. The dynamic equation governing

$N_k(t)$ can be developed as follows. We have seen that the rate of collision between particles of two types (sizes) is $J_{12} = K_{12}N_1N_2$, expressed in units of collisions per cm^3 of fluid per second. A k-mer can be formed by collision of a $(k-j)$-mer with a j-mer. The overall rate of formation, J_{fk}, of the k-mer will be the sum of the rates that produce the k-mer or

$$J_{fk}(t) = \sum_{j=1}^{k-1} K_{k-j,j}N_{k-j}(t)N_j(t)$$

To investigate the validity of this formula let us apply it to a trimer, which can be formed only by collision of a monomer with a dimer

$$J_{f3} = K_{21}N_2N_1 + K_{12}N_1N_2$$

Note that because $K_{12} = K_{21}$, the overall rate is

$$J_{f3} = 2K_{12}N_1N_2$$

This expression appears to be incorrect, as we have counted the collisions twice. To correct for that we need to multiply the rate by a factor of $\frac{1}{2}$ so that collisions are not counted twice. Therefore

$$J_{fk} = \frac{1}{2} \sum_{j=1}^{k-1} K_{k-j,j}N_{k-j}(t)N_j(t) \tag{13.57}$$

What about equal-sized particles? For a dimer our corrected formation rate is

$$J_{f2} = \frac{1}{2}K_{11}N_1^2$$

Why do we need to divide by a factor of 2 here? Let us try to persuade ourselves that this factor of $\frac{1}{2}$ is always needed. Assume that we have particles consisting only of $(k/2)$ monomers. If one-half of these particles are painted red and one-half painted blue, then the rate of coagulation of red and blue particles is

$$J_{\text{red, blue}} = K_{k/2,k/2}\left(\frac{1}{2}N_{k/2}\right)\left(\frac{1}{2}N_{k/2}\right) = \frac{1}{4}K_{k/2,k/2}N_{k/2}^2$$

To calculate the overall coagulation rate we need to add to this rate the collision rate of red particles among themselves and blue particles among themselves. Let us now paint one-half the red particles as green and one-half the blue particles as yellow. The rate of coagulation within each of the initial particle categories is

$$J_{\text{red, green}} + J_{\text{blue, yellow}} = K_{k/2,k/2}\left[\left(\frac{1}{4}N_{k/2}\right)\left(\frac{1}{4}N_{k/2}\right) + \left(\frac{1}{4}N_{k/2}\right)\left(\frac{1}{4}N_{k/2}\right)\right]$$
$$\text{(red)}\quad\text{(green)} + \text{(blue)(yellow)}$$

We can continue this painting process indefinitely (provided that we do not run out of colors), and the result for the total rate of coagulation occurring in the population of equal-sized particles is

$$K_{k/2, k/2} N_{k/2}^2 \left[\frac{1}{2^2} + \frac{1}{2^3} + \frac{1}{2^4} + \cdots \right]$$

Since

$$\sum_{n=2}^{\infty} 2^{-n} = \frac{1}{2}$$

the total coagulation rate is

$$\frac{1}{2} K_{k/2, k/2} N_{k/2}^2$$

The factor of $\frac{1}{2}$ in the equal-sized particles is thus a result of the indistinguishability of the coagulating particles.

The rate of depletion of a k-mer by agglomeration with all other particles is

$$J_{dk}(t) = N_k(t) \sum_{j=1}^{\infty} K_{kj} N_j(t) \tag{13.58}$$

Question: Is this rate correct? Why do we *not* divide here by $\frac{1}{2}$ if $j = k$?

Answer: We do need to divide by 2, due to the indistinguishability argument. However, because each collision removes 2 k-mers, we also need to multiply by 2. These factors cancel each other out and therefore the expression above is correct.

Combining the coagulation production and depletion terms, we get the *discrete coagulation equation*:

$$\frac{dN_k(t)}{dt} = \frac{1}{2} \sum_{j=1}^{k-1} K_{j, k-j} N_j N_{k-j} - N_k \sum_{j=1}^{\infty} K_{k,j} N_j, \quad k \geq 2 \tag{13.59}$$

In this formulation, it is assumed that the smallest particle is one of size $k = 2$. In reality, there is a minimum number of monomers in a stable nucleus g^* and generally $g^* \gg 2$ (see Chapter 11). This simply means that

$$N_2 = N_3 = \cdots = N_{g^*-1} = 0$$

in (13.59), and being rigorous, we can write the following equation, replacing the summation limits,

$$\frac{dN_k(t)}{dt} = \frac{1}{2} \sum_{j=g^*}^{k-g^*} K_{j,k-j} N_j N_{k-j} - N_k \sum_{j=g^*}^{\infty} K_{k,j} N_j, \quad k \geq 2 \tag{13.60}$$

The Continuous Coagulation Equation   Although (13.59) and (13.60) are rigorous representations of the coagulating aerosol population, they are impractical because of the enormous range of k associated with the equation set above. It is customary to replace $N_k(t)\,(\text{cm}^{-3})$ with the continuous number distribution function $n(v, t)$ $(\mu\text{m}^{-3}\,\text{cm}^{-3})$, where $v = kv_1$ is the particle volume with v_1 the volume of the monomer. If we let $v_0 = g^*v_1$, then (13.60) becomes, in the limit of a continuous distribution of sizes

$$\frac{\partial n(v, t)}{\partial t} = \frac{1}{2} \int_{v_0}^{v - v_0} K(v - q, q)n(v - q, t)n(q, t)dq$$

$$- n(v, t) \int_{v_0}^{\infty} K(q, v)n(q, t)dq \tag{13.61}$$

The first term in (13.61) corresponds to the production of particles by coagulation of the appropriate combinations of smaller particles, while the second term corresponds to their loss with collisions with all available particles.

13.3.3 Solution of the Coagulation Equation

While the coagulation equation cannot be solved analytically in its most general form, solutions can be obtained assuming constant coagulation coefficients. While these solutions have limited applicability to ambient atmospheric conditions, they still provide useful insights and can be used as benchmarks for the development of numerical methods for the solution of the full coagulation equation. Such situations are also applicable to the early stages of coagulation of a monodisperse aerosol in the continuous regime when $K(v, q) = 8\,kT/3\mu$.

Discrete Coagulation Equation   Assuming $K_{k,j} = K$ in (13.59), we obtain

$$\frac{dN_k(t)}{dt} = \frac{1}{2}K \sum_{j=1}^{k-1} N_j(t)N_{k-j}(t) - K N_k(t) \sum_{j=1}^{\infty} N_j(t) \tag{13.62}$$

Noting that the last term in this equation is the total number of particles (the subscript t on $N(t)$ can be omitted)

$$N(t) = \sum_{j=1}^{\infty} N_j(t)$$

(13.62) is simplified to

$$\frac{dN_k(t)}{dt} = \frac{1}{2}K \sum_{j=1}^{k-1} N_j(t)N_{k-j}(t) - K N_k(t)N(t) \tag{13.63}$$

To solve (13.63), we need to know $N(t)$. By summing equations (13.63) from $k = 1$ to ∞, we obtain

$$\frac{dN(t)}{dt} = \frac{1}{2}K \sum_{k=1}^{\infty} \sum_{j=1}^{k-1} N_{k-j}(t)N_j(t) - K N^2(t) \tag{13.64}$$

The double summation on the right-hand side is equal to $N^2(t)$. So (13.64) becomes

$$\frac{dN}{dt} = -\frac{1}{2} K N^2(t) \tag{13.65}$$

If $N(0) = N_0$, the solution of (13.65) is

$$N(t) = \frac{N_0}{1 + (t/\tau_c)} \tag{13.66}$$

where τ_c is the characteristic time for coagulation given by

$$\tau_c = \frac{2}{K N_0} \tag{13.67}$$

Characteristic Timescale for Coagulation

$$\tau_c = \frac{2}{K N_0}$$

At $t = \tau_c$, $N(\tau_c) = \frac{1}{2} N_0$. Thus, τ_c is the time necessary for reduction of the initial number concentration to half its original value. The timescale shortens as the initial number concentration increases. Consider an initial population of particles of about 0.2 μm diameter, for which $K = 10 \times 10^{-10} \, \text{cm}^3 \, \text{s}^{-1}$. The coagulation timescales for $N_0 = 10^4 \, \text{cm}^{-3}$ and $10^6 \, \text{cm}^{-3}$ are

$$
\begin{array}{ll}
N_0 = 10^4 \, \text{cm}^{-3} & \tau_c \cong 55 \, \text{h} \\
N_0 = 10^6 \, \text{cm}^{-3} & \tau_c \cong 33 \, \text{min}
\end{array}
$$

We can now solve (13.63) inductively assuming that $N_1(0) = N_0$; that is, at $t = 0$ all particles are of size $k = 1$. Let us start from $k = 1$

$$\frac{dN_1}{dt} = -K N_1 N \tag{13.68}$$

to be solved subject to $N_1(0) = N_0$. Using (13.66) and solving (13.68) gives

$$N_1(t) = \frac{N_0}{[1 + (t/\tau_c)]^2}$$

Similarly, solving

$$\frac{dN_2}{dt} = \frac{1}{2} K N_1^2 - K N_2 N \tag{13.69}$$

subject to $N_2(0) = 0$ gives

$$N_2(t) = \frac{N_0(t/\tau_c)}{[1 + (t/\tau_c)]^3}$$

(13.70)

Continuing, we find

$$N_k(t) = \frac{N_0(t/\tau_c)^{k-1}}{[1 + (t/\tau_c)]^{k+1}}, \qquad k = 1, 2, \ldots$$

(13.71)

Note that when $t/\tau_c \gg 1$, it follows that

$$N_k(t) \simeq N_0 \left(\frac{\tau_c}{t}\right)^2$$

so the number concentration of each k-mer decreases as t^{-2}. In the short time limit $t/\tau_c \ll 1$, so

$$N_k(t) \simeq N_0 \left(\frac{t}{\tau_c}\right)^{k-1}$$

and N_k increases as t^{k-1}.

Continuous Coagulation Equation   We now consider the solution to the continuous coagulation equation (13.61) assuming a constant coagulation coefficient $K(q, v) = K$, assuming that $v_0 \simeq 0$:

$$\frac{\partial n(v, t)}{\partial t} = \frac{1}{2} K \int_0^v n(v - q, t)n(q, t)dq - Kn(v, t)N(t)$$

(13.72)

In order to solve (13.72) we also need to know $N(t)$. We have seen that the total number concentration $N(t)$ for any aerosol distribution assuming constant coagulation coefficient is given by (13.66). Substituting this expression for $N(t)$ into (13.72), we find that the continuous coagulation equation becomes

$$\frac{\partial n(v, t)}{\partial t} + \frac{K N_0}{1 + t/\tau_c} n(v, t) = \frac{1}{2} K \int_0^v n(v - q, t)n(q, t)\, dq$$

(13.73)

Let us solve (13.73) assuming the initial distribution

$$n(v, 0) = A \exp(-Bv)$$

(13.74)

where A and B are parameters. For this particular initial distribution, an analytical solution of (13.73) can be determined. In this case the total number and volume concentrations are initially

$$N_0 = \int_0^\infty n(v, 0)\, dv = \frac{A}{B}$$

(13.75)

and

$$V_0 = \int_0^\infty v n(v,0)\, dv = \frac{A}{B^2} \tag{13.76}$$

Combining (13.75) and (13.76) we find that the parameters A and B are related to the initial aerosol number and volume concentrations by $A = N_0^2/V_0$ and $B = N_0/V_0$; therefore the distribution given by (13.74) is equivalent to

$$n(v,0) = \frac{N_0^2}{V_0} \exp\left(\frac{-v N_0}{V_0}\right) \tag{13.77}$$

To solve (13.73) with the initial condition given by (13.77), we can either use an integrating factor and an assumed form of the solution or attack it directly with the Laplace transformation. Both approaches are illustrated in Appendix 13.2. The desired solution is then

$$n(v,t) = \frac{N_0^2}{V_0(1 + t/\tau_c)^2} \exp\left(-\frac{v N_0}{V_0(1 + t/\tau_c)}\right) \tag{13.78}$$

This size distribution is shown in Figure 13.6. The exponential shape of the distribution persists, but while the number of small particles decreases with time, the availability of larger particles slowly increases.

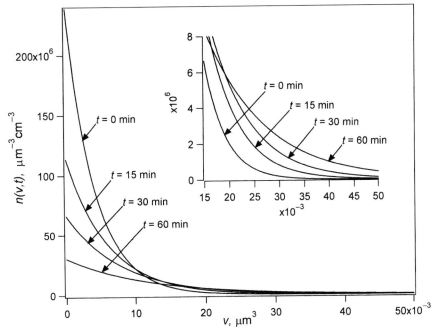

FIGURE 13.6 Solution of the continuous coagulation equation for an exponential initial distribution given by (13.78). The following parameters are used: $N_0 = 10^6\,\text{cm}^{-3}, V_0 = 4189\,\mu\text{m}^3\,\text{cm}^{-3}$, and $\tau_c = 2000\,\text{s}$.

For additional solutions of the discrete and continuous coagulation equations, the interested reader may wish to consult Drake (1972), Mulholland and Baum (1980), Tambour and Seinfeld (1980), and Pilinis and Seinfeld (1987).

13.4 THE DISCRETE GENERAL DYNAMIC EQUATION

A spatially homogeneous aerosol of uniform chemical composition can be fully characterized by the number concentrations of particles of various sizes $N_k(t)$ as a function of time. These particle concentrations may undergo changes due to coagulation, condensation, evaporation, nucleation, emission of fresh particles, and removal. The dynamic equation governing $N_k(t)$ for $k \geq 2$ can be developed as follows.

Consider first the contribution of coagulation to this concentration change. We showed in Section 13.3.2 that the production rate of k-mers due to coagulation of j-mers with $(k - j)$-mers is given by

$$\frac{1}{2} \sum_{j=1}^{k-1} K_{j,k-j} N_j N_{k-j}$$

where $K_{j,k-j}$ is the corresponding coagulation coefficient for these collisions. The loss rate of k-mers due to their coagulation with all other particles is given by (13.58) as

$$N_k \sum_{j=1}^{\infty} K_{k,j} N_j$$

To calculate the rate of change of $N_k(t)$ due to evaporation, let γ'_k be the flux of monomers per unit area leaving a k-mer. Then, the overall rate of escape of monomers from a k-mer, γ_k, with surface area a_k will be given by

$$\gamma_k = \gamma'_k a_k \tag{13.79}$$

The rate of loss of k-mers from evaporation can then be written as

$$\gamma_k N_k, \qquad k \geq 2$$

The rate of formation of k-mers by evaporation of $(k + 1)$-mers is then just

$$\gamma_{k+1} N_{k+1}, \qquad k \geq 2$$

Note that if a dimer dissociates, then two monomers are formed and the formation rate of monomers will be equal to $2\gamma_2 N_2$. We can account for this introducing the Krönecker delta, $\delta_{j,k}$, defined by

$$\delta_{j,k} = 1, \quad j = k$$
$$\delta_{j,k} = 0, \quad j \neq k$$

and generalizing the formation rate of k-mers by evaporation as

$$(1 + \delta_{1,k}) \gamma_{k+1} N_{k+1}$$

To account for the process of accretion of monomers by other particles, we define $p_k N_k$ as the rate of gain of $(k + 1)$-mers due to collisions of k-mers with monomers, where p_k (s^{-1}) is the frequency with which a monomer collides with a k-mer. This is the process we commonly refer to as condensation. Note that then

$$p_k = K_{1,k} N_1 \tag{13.80}$$

and the rate of gain of k-mers will be

$$p_{k-1} N_{k-1} = K_{1,k-1} N_1 N_{k-1}$$

The loss rate of k-mers due to the addition of a monomer and subsequent growth will then be

$$p_k N_k = K_{1,k} N_1 N_k$$

Summarizing, the rate of change of the k-mer concentration accounting for coagulation, condensation, and evaporation will be given by

$$\frac{dN_k}{dt} = \frac{1}{2} \sum_{j=1}^{k-1} K_{j,k-j} N_j N_{k-j} - N_k \sum_{j=1}^{\infty} K_{k,j} N_j$$
$$+ p_{k-1} N_{k-1} - (p_k + \gamma_k) N_k + \gamma_{k+1} N_{k+1} (1 + \delta_{1,k}) \tag{13.81}$$

In the preceding formulation of the discrete general dynamic equation it is implicitly assumed that the smallest particle is of size $k = 2$. No distinction has yet been made among the processes of coagulation, homogeneous nucleation, and heterogeneous condensation; they are all included in (13.81). In reality, there is a minimum number of monomers in a stable nucleus g^*, and generally $g^* \geq 2$. In the presence of a supersaturated vapor, stable clusters of size g^* will form continuously at a rate given by the theory of homogeneous nucleation. Let us denote the rate of formation of stable clusters containing g^* monomers from homogeneous nucleation as $J_0(t)$. Then coagulation, heterogeneous condensation of vapor on particles of size $k \geq g^*$, and nucleation of g^*-mers become distinct processes in (13.81). By changing the smallest particle from 2 to g^* and adding emission and removal terms, we find that (13.81) becomes

$$\frac{dN_k}{dt} = \frac{1}{2} \sum_{j=g^*}^{k-g^*} K_{j,k-j} N_j N_{k-j} - N_k \sum_{j=g^*}^{\infty} K_{k,j} N_j + p_{k-1} N_{k-1}$$
$$- (p_k + \gamma_k) N_k + \gamma_{k+1} N_{k+1} + J_0(t) \delta_{g^*,k} + S_k - R_k \tag{13.82}$$
$$k = g^*, g^* + 1, \dots$$

where S_k is the emission rate of k-mers by sources and R_k is their removal rate. Note that in (13.82) the nucleation term is used only in the equation for the g^*-mers, and it is zero for all other particles.

13.5   THE CONTINUOUS GENERAL DYNAMIC EQUATION

Although (13.82) is a rigorous representation of the system, it is impractical to deal with discrete equations because of the enormous range of k. Thus it is customary to replace the discrete number concentration $N_k(t)$ (cm^{-3}) by the continuous size distribution function $n(v,t)$ $(\mu\text{m}^{-3}\,\text{cm}^{-3})$, where $v = kv_1$, with v_1 being the volume associated with a monomer. Thus $n(v,t)dv$ is defined as the number of particles per cubic centimeter having volumes in the range from v to $v + dv$. If we let $v_0 = g^* v_1$ be the volume of the smallest stable particle, then (13.82) becomes in the limit of a continuous distribution of sizes

$$\frac{\partial n(v,t)}{\partial t} = \frac{1}{2}\int_{v_0}^{v-v_0} K(v-q,q)n(v-q,t)n(q,t)dq - n(v,t)\int_{v_0}^{\infty} K(q,v)n(q,t)dq$$

$$- \frac{\partial}{\partial v}[I_0(v)n(v,t)] + \frac{\partial^2}{\partial v^2}[I_1(v)n(v,t)]$$

$$+ J_0(v)\delta(v-v_0) + S(v) - R(v) \tag{13.83}$$

where

$$I_0(v) = v_1(p_k - \gamma_k) \tag{13.84}$$

$$I_1(v) = \frac{v_1^2}{2}(p_k + \gamma_k) \tag{13.85}$$

are the condensation/evaporation-related terms. Since $p_k - \gamma_k$ is the frequency with which a k-mer experiences a net gain of one monomer, $I_0(v)$ is the rate of change of the volume of a particle of size $v = kv_1$. The sum $p_k + \gamma_k$ is the total frequency by which monomers enter and leave a k-mer. $I_1(v)$ assumes the role of a diffusion coefficient from kinetic theory. Brock (1972) and Ramabhadran et al. (1976) have shown that the term in (13.83) involving I_1 can ordinarily be neglected. We shall do so and simply write I_0 as I, which can be calculated using (13.4).

In (13.83), a stable particle has been assumed to have a lower limit of volume of v_0. From the standpoint of the solution of (13.83), it is advantageous to replace the lower limits v_0 of the coagulation integrals by zero. Ordinarily this does not cause any difficulty, since the initial distribution $n(v,0)$ can be specified as zero for $v < v_0$, and no particles of volume $v < v_0$ can be produced for $t > 0$. Homogeneous nucleation provides a steady source of particles of size v_0 according to the rate defined by $J_0(t)$. Then the full equation governing $n(v,t)$ is as follows:

$$\frac{\partial n(v,t)}{\partial t} = \frac{1}{2}\int_{0}^{v} K(v-q,q)n(v-q,t)n(q,t)dq$$

$$- n(v,t)\int_{0}^{\infty} K(q,v)n(q,t)dq - \frac{\partial}{\partial v}[I(v)n(v,t)]$$

$$+ J_0(v)\delta(v-v_0) + S(v) - R(v) \tag{13.86}$$

TABLE 13.4   Properties of Coagulation, Condensation, and Nucleation

Process	Number Concentration	Volume Concentration
Coagulation	Decreases	No change
Condensation	No change	Increases
Nucleation	Increases	Increases
Coagulation and condensation	Decreases	Increases

This equation is called the *continuous general dynamic equation* for aerosols (Gelbard and Seinfeld 1979). Its initial and boundary conditions are

$$n(v,0) = n_0(v) \tag{13.87}$$
$$n(0,t) = 0 \tag{13.88}$$

In the absence of nucleation $(J_0(v) = 0)$, sources $(S(v) = 0)$, sinks $(R(v) = 0)$, and growth $[I(v) = 0]$, we have the continuous coagulation equation (13.61). If particle concentrations are sufficiently small, coagulation can be neglected. If there are no sources or sinks of particles then the general dynamic equation is simplified to the condensation equation (13.7).

Table 13.4 compares the properties of coagulation and condensation with respect to the total number of particles and the total particle volume. Coagulation reduces particle number but the total volume remains constant. Condensation, since it involves gas-to-particle conversion, increases total particle volume, but the total number of particles remains constant. If both processes are occurring simultaneously, total particle number decreases and total particle volume increases. A detailed study of the interplay between coagulation and condensation has been presented by Ramabhadran et al. (1976) and Peterson et al. (1978) to which we refer the interested reader.

APPENDIX 13.1   ADDITIONAL MECHANISMS OF COAGULATION

13.A.1   Coagulation in Laminar Shear Flow

Particles in a fluid in which a velocity gradient du/dy exists have a relative motion that may bring them into contact and cause coagulation (Figure 13.A.1). Smoluchowski in 1916 first studied this coagulation type assuming a uniform shear flow, no fluid dynamic interactions between the particles, and no Brownian motion. This is a simplification of the actual physics since the particles affect the shear flow and the streamlines have a curvature around the particles.

Smoluchowski showed that for a velocity gradient $\Gamma = du/dy$ perpendicular to the x axis the coagulation rate is

$$J_{\text{coag}} = K_{12}^{LS} N_1 N_2 = \frac{\Gamma}{6} (D_{p1} + D_{p2})^3 N_1 N_2 \tag{13.A.1}$$

The relative magnitudes of the laminar shear and Brownian coagulation coefficients for equal-sized particles in the continuum regime are given by the ratio $K_{12}^{LS}/K_{12} = (\Gamma \mu / 2kT) D_p^3$. Good agreement between this shear coagulation theory and experiments was reported by Swift and Friedlander (1964). From this ratio, we find that high shear rates are required to make the shear coagulation rate comparable to the Brownian even for particles

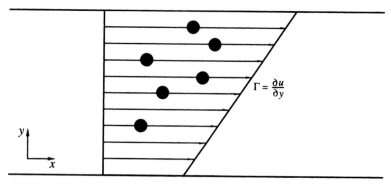

FIGURE 13.A.1   Schematic illustrating collisions of particles in shear flow.

of size around 1 μm. For example, for $D_p = 2\,\mu m$, a $\Gamma \simeq 60\,s^{-1}$ is necessary to make the two rates equal.

13.A.2   Coagulation in Turbulent Flow

Velocity gradients in turbulent flows cause relative particle motion and induce coagulation just as in laminar shear flow. The difficulty in analyzing the process rests with identifying an appropriate velocity gradient in the turbulence. In fact, a characteristic turbulent shear rate at the small lengthscales applicable to aerosols is ε_k/ν, where ε_k is the rate of dissipation of kinetic energy per unit mass and ν is the kinematic viscosity of the fluid (Tennekes and Lumley 1972). The analysis by Saffman and Turner (1956) gives the coagulation coefficient for turbulent shear as

$$K_{12}^{TS} = \left(\frac{\pi\varepsilon_k}{120\nu}\right)^{1/2}(D_{p1} + D_{p2})^3 \qquad (13.A.2)$$

The relative magnitudes of the turbulent shear and Brownian coagulation coefficients for equal-sized particles in the continuum regime is given by the ratio

$$\frac{K_{12}^{TS}}{K_{12}} = \frac{3\mu(\pi\varepsilon_k/120\nu)^{1/2}D_p^3}{kT} \qquad (13.A.3)$$

Typical measured values of $(\varepsilon_k/\nu)^{1/2}$ are on the order of $10\,s^{-1}$, so turbulent shear coagulation is significantly slower than Brownian for submicrometer particles, and the two rates become approximately equal for particles of about 5 μm in diameter (Figure 13.A.2). The calculations indicate that coagulation by Brownian motion dominates the collisions of submicrometer particles in the atmosphere. Turbulent shear contributes to the coagulation of large particles under conditions characterized by intense turbulence.

13.A.3   Coagulation from Gravitational Settling

Coagulation results in a settling particle population when heavier particles move faster, catching up with lighter particles, and collide with them. This process will be discussed again in Chapter 20, where we will be interested in the scavenging of particles by falling raindrops. Suffice it to say at this point that the coagulation coefficient is proportional to

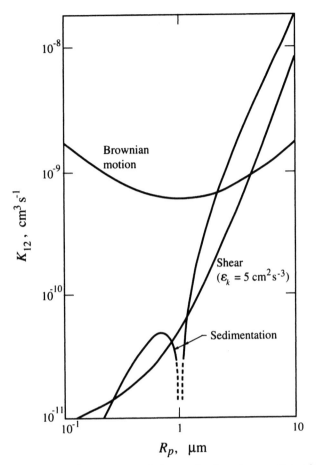

FIGURE 13.A.2 Comparison between coagulation mechanisms for a particle of 1 μm radius as a function of the particle radius of the second interacting particle.

the product of the target area, the relative distance swept out by the larger particle per unit time, and the collision efficiency between particles of sizes D_{p1} and D_{p2}

$$K_{12}^{GS} = \frac{\pi}{4}(D_{p1} + D_{p2})^2(v_{t1} - v_{t2})E(D_{p1}, D_{p2}), \tag{13.A.4}$$

where v_{t1} and v_{t2} are the terminal settling velocities for the large and small particles, respectively. The collision efficiency $E(D_{p1}, D_{p2})$ is discussed in Sections 17.8.2 and 20.3. Particles with equal diameters have the same terminal settling velocities and K_{12}^{GS} approaches zero as D_{p1} approaches D_{p2} (Figure 13.A.2). Gravitational coagulation can be neglected for submicrometer particles but becomes significant for particle diameters exceeding a few micrometers.

13.A.4 Brownian Coagulation and External Force Fields

In our discussion of Brownian motion and the resulting coagulation, we did not include the effect of external force fields on the particle motion. Here we extend our treatment of coagulation to include interparticle forces.

The force $\mathbf{F}_{12}$ between particles #1 and #2 can be written as the negative of the gradient of a potential Φ

$$\mathbf{F}_{12} = -\nabla\Phi \qquad (13.A.5)$$

Now let us assume that $\Phi = \Phi(r)$, where r is the distance between the particle centers. Then one can show that the steady-state coagulation flux is given by

$$J_{12} = \frac{1}{W} K_{12}^{\text{Brownian}} N_1 N_2 \qquad (13.A.6)$$

where the correction factor W is a result of the interparticle force as represented by the potential $\Phi(r)$ and can be calculated by

$$W = (R_{p1} + R_{p2}) \int_{R_{p1}+R_{p2}}^{\infty} \frac{1}{x^2} \exp\left(\frac{\Phi(x)}{kT}\right) dx \qquad (13.A.7)$$

Let us now apply the above theory to two specific types of interactions: van der Waals forces and Coulomb forces.

Van der Waals Forces   Van der Waals forces are the result of the formation of momentary dipoles in uncharged, nonpolar molecules. They are caused by fluctuations in the electron cloud, which can attract similar dipoles in other molecules. The potential of the attractive force can be expressed as

$$\Phi_v = -4\,\phi_1\left(\frac{\phi_2}{r}\right)^6 \qquad (13.A.8)$$

where ϕ_1 and ϕ_2 are constants with units of energy and length, respectively, that depend on the particular species involved. Values of these constants have been tabulated by Hirschfelder et al. (1954).

The van der Waals potential between two spherical particles of radii R_{p1} and R_{p2} whose centers are separated by a distance r is (Hamaker 1937)

$$\Phi_v(r) = -\frac{4\pi^2\phi_1\phi_2^6}{6v_m^2}\left[\frac{2R_{p1}R_{p2}}{r^2-(R_{p1}+R_{p2})^2} + \frac{2R_{p1}R_{p2}}{r^2-(R_{p1}-R_{p2})^2} + \ln\left(\frac{r^2-(R_{p1}+R_{p2})^2}{r^2-(R_{p1}-R_{p2})^2}\right)\right] \qquad (13.A.9)$$

where v_m is the molecular volume.

Let us investigate the case of equal-sized particles, $R_{p1} = R_{p2}$. Then, if $s = 2R_p/r$, we have

$$\Phi_v(s) = -\frac{4\pi^2\phi_1\phi_2^6}{6v_m^2}\left(\frac{s^2}{2(1-s^2)} + \frac{s^2}{2} + \ln(1-s^2)\right) \qquad (13.A.10)$$

Using (13.A.7) the correction factor, W_v, for the van der Waals force between equal-sized particles is

$$W_v = \int_0^1 \exp\left[-\frac{2\pi^2 \phi_1 \phi_2^6}{3 v_m^2 kT} \left(\frac{x^2}{2(1-x^2)} + \frac{x^2}{2} + \ln(1-x^2) \right) \right] dx \qquad (13.A.11)$$

The correction factor W_v is independent of particle size for identically sized particles and depends only on the van der Waals parameters ϕ_1 and ϕ_2 and the temperature T. W_v can be evaluated as a function of

$$\frac{Q}{kT} = \frac{4\phi_1 \phi_2^6}{v_m^2 kT} \qquad (13.A.12)$$

by numerically evaluating the integral of (13.A.11). The results of this integration are shown in Figure 13.A.3. Note that a 50% correction is necessary for a Q/kT value of roughly 10. The Hamaker constant is defined as

$$A = \frac{4\pi^2 \phi_1 \phi_2^6}{v_m^2} \qquad (13.A.13)$$

For particles of radii R_{p1} and R_{p2}, the continuum correction factor is given by (13.A.7) (Friedlander 1977; Schmidtt-Ott and Burtscher 1982)

$$W_v = (R_{p1} + R_{p2}) \int_{R_{p1}+R_{p2}}^{\infty} \frac{1}{r^2} \exp\left(\frac{\Phi_v(r)}{kT} \right) dr \qquad (13.A.14)$$

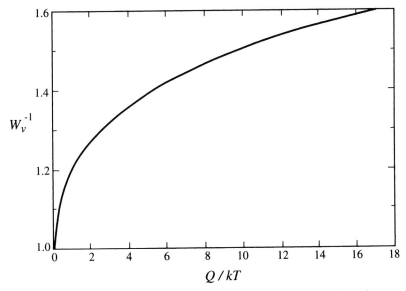

FIGURE 13.A.3 Correction factor to the Brownian coagulation coefficient for equal-sized particles in the continuum regime from van der Waals forces.

Marlow (1981) has extended this development to the transition and free-molecule regimes and determined that the effect of van der Waals forces on aerosol coagulation rates can be considerably more pronounced in these size ranges than in the continuum regime.

Coulomb Forces   Charged particles may experience either enhanced or retarded coagulation rates depending on their charges. The potential energy of interaction between two particles containing z_1 and z_2 charges (including sign) whose centers are separated by a distance r is

$$\Phi_c = \frac{z_1 z_2 e^2}{4\pi\varepsilon_0\varepsilon r} \tag{13.A.15}$$

where e is the electronic charge, ε_0 is the permittivity of vacuum ($8.854 \times 10^{-12}\ \mathrm{F\,m^{-1}}$), and ε the dielectric constant of the medium. For air at 1 atm, $\varepsilon = 1.0005$. Substituting (13.A.15) into (13.A.7) and evaluating the integral

$$W_c = \frac{e^\kappa - 1}{\kappa} \tag{13.A.16}$$

where

$$\kappa = \frac{z_1 z_2 e^2}{4\pi\varepsilon_0\varepsilon (R_{p1} + R_{p2})kT} \tag{13.A.17}$$

The constant κ can be interpreted as the ratio of the electrostatic potential energy at contact to the thermal energy kT. For like charge, κ is greater than zero, so $W_c > 1$ and the coagulation is retarded from that of pure Brownian motion. Conversely, for $\kappa < 0$ (unlike charges), $W_c < 1$ and the coagulation rate is enhanced.

If we consider a situation where there are equal numbers of positively and negatively charged particles, then we can estimate the coagulation correction factor as the average of the like—and unlike—correction factors. This estimate is shown in Table 13.A.1. We see that an aerosol consisting of an equal number of positively and negatively charged particles will exhibit an overall enhanced rate of coagulation. The enhanced rate of unlike charged particles more than compensates for the retarded rate of like charged particles.

TABLE 13.A.1   Coagulation Correction Factors for Coulomb Forces between Equal-Sized Particles

κ	W_c^{-1} (like)	W_c^{-1} (unlike)	$\frac{1}{2}[W_c^{-1}\ (\text{like}) + W_c^{-1}\ (\text{unlike})]$
0.1	0.9508	1.0508	1.0008
0.25	0.8802	1.1302	1.0052
0.5	0.7708	1.2708	1.0208
0.75	0.6714	1.4214	1.0464
1.0	0.5820	1.5820	1.0820
5.0	0.0339	5.0339	2.5339
10.0	0.00045	10.00045	5.00045

Pruppacher and Klett (1997) show that the average number of charges on an atmospheric particle at equilibrium $\bar{z}$, regardless of sign, is given by

$$\bar{z} = \frac{1}{e}\left(\frac{D_p kT}{\pi}\right) \tag{13.A.18}$$

resulting in fewer than 10 charges even for a 10-μm particle. For these particles κ is on the order of 10^{-11} or smaller so the correction factor W_c is equal to unity. Consequently, it is not expected that Coulomb forces will be important in affecting the coagulation rates of atmospheric particles. Only if an aerosol is charged far in excess of its equilibrium charge described by (13.A.18) will there be any effect on the coagulation coefficient.

Hydrodynamic Forces   Fluid mechanical interactions between particles arise because a particle in motion in a fluid induces velocity gradients in the fluid that influence the motion of other particles when they approach its vicinity. Because the fluid resists being "squeezed" out from between the approaching particles, the effect of so-called viscous forces is to retard the coagulation rate from that in their absence.

In our calculations of the common diffusivity D_{12} as $D_{12} = D_1 + D_2$, we have implicitly assumed that each particle approaches the other oblivious of the other's existence. In reality the velocity gradients around each particle influence the motion of an approaching particle. Spielman (1970) proposed that these effects can be considered by introducing a corrected diffusion coefficient D'_{12}. Alam (1987) has obtained an analytical solution for the ratio D_{12}/D'_{12} that closely approximates Spielman's solution:

$$\frac{D_{12}}{D'_{12}(r)} = 1 + \frac{2.6R_{p1}R_{p2}}{(R_{p1}+R_{p2})^2}\left(\frac{R_{p1}R_{p2}}{(R_{p1}+R_{p2})(r-R_{p1}-R_{p2})}\right)^{1/2}$$
$$+ \frac{R_{p1}R_{p2}}{(R_{p1}+R_{p2})(r-R_{p1}-R_{p2})} \tag{13.A.19}$$

Note that $D'_{12} \simeq D_{12}$ for sufficiently large separation r and D'_{12} decreases as the particle separation decreases. Figure 13.A.4 shows that the enhancement due to van der Waals forces is decreased when viscous forces are included in the calculations. For some values of the Hamaker constant there is an overall retardation of the coagulation rate due to the viscous forces.

Alam (1987) has proposed the following interpolation formula for hydrodynamic and van der Waals forces that is a function of a particle Knudsen number Kn_p and attains the proper limiting forms in the continuum and kinetic regimes. The result is

$$K_{12} = 4\pi(R_{p1}+R_{p2})D_{12}\frac{W_c^{-1}}{1+W_k Kn_p\delta/W_c}$$

$$D_{12} = D_1 + D_2, \qquad \delta = \frac{4D_{12}}{\bar{c}(\lambda_1^2+\lambda_2^2)^{1/2}}$$

$$Kn_p = \frac{(\lambda_1^2+\lambda_2^2)^{1/2}}{R_{p1}+R_{p2}}, \qquad \bar{c} = \left(\frac{8kT}{\pi m_1}+\frac{8kT}{\pi m_2}\right)^{1/2} \tag{13.A.20}$$

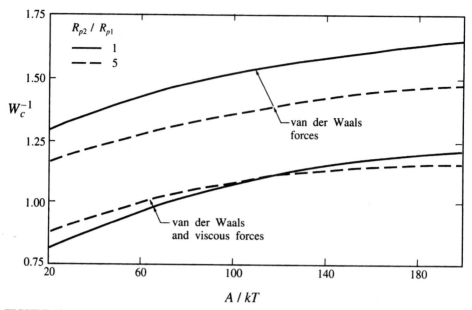

FIGURE 13.A.4 Enhancement to Brownian coagulation when van der Waals and viscous forces are included in the calculation.

W_c^{-1} and W_k^{-1} are the enhancement factors for van der Waals and viscous forces in the continuum and kinetic regimes, respectively. Figure 13.A.5 shows the coagulation coefficient K at 300 K, 1 atm, $\rho_p = 1\,\mathrm{g\,cm}^{-3}$, and $A/kT = 20$ as a function of Kn_p for particle radii ratios of 1, 2, and 5 both in the presence and in the absence of interparticle forces. Figure 13.A.6 shows the enhancement over pure Brownian coagulation as a function of particle Knudsen number for $A/kT = 20$. In the continuum regime $(Kn \ll 1)$ there is a retardation of coagulation because of the viscous forces. As shown in Figure 13.A.6 for sufficiently large Hamaker constant A, the effect of van der Waals forces overtakes that of viscous forces to lead to an enhancement factor greater than 1. In the kinetic regime, coagulation is always enhanced due to the absence of viscous forces.

APPENDIX 13.2   SOLUTION OF (13.73)

We solve (13.73) subject to (13.77) using the integrating factor method. The integrating factor will be given by the exponential of the integral of the term multiplying $n(v, t)$ on the left-hand side of (13.73):

$$\exp\left(\int_0^t \frac{KN_0}{1 + t'/\tau_c}\,dt'\right) = \left(1 + \frac{t}{\tau_c}\right)^2 \tag{13.A.21}$$

If we multiply both sides of (13.73) with the factor calculated in (13.A.21) and combine the resulting terms of the left-hand side, we find that

$$\frac{\partial}{\partial t}\left[\left(1 + \frac{1}{\tau_c}\right)^2 n(v, t)\right] = \frac{1}{2} K \left(1 + \frac{t}{\tau_c}\right)^2 \int_0^v n(v - q, t) n(q, t)\,dq \tag{13.A.22}$$

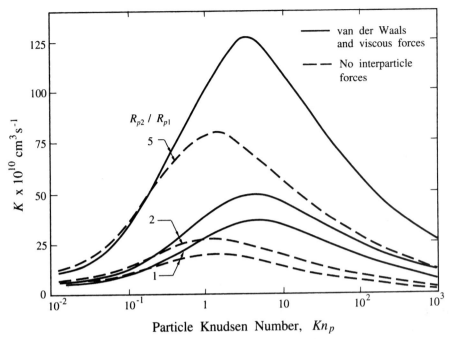

FIGURE 13.A.5 Coagulation coefficient at 300 K, $\rho_p = 1 \, \text{g cm}^{-3}$, $A/kT = 20$ as a function of Knudsen number (Kn_p) for a particle radii ratios of 1, 2, and 5 in both presence and absence of interparticle forces.

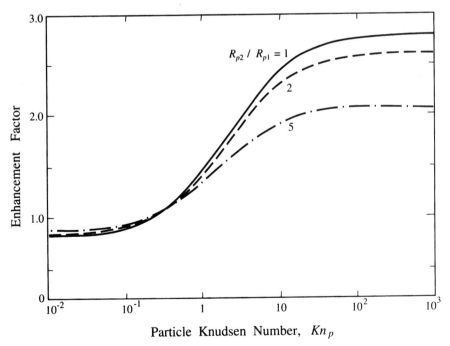

FIGURE 13.A.6 Enhancement over Brownian coagulation as a function of particle Knudsen number for $A/kT = 20$. In the continuum regime there is a retardation of coagulation because of the viscous forces.

If we let $y = N_0^{-1}(1 + t/\tau_c)^{-1}$ and $w = (1 + t/\tau_c)^2 n(v, t)$ then (13.A.22) can be transformed to

$$\frac{\partial w(v, y)}{\partial y} = -\int_0^v w(v - q, y) w(q, y) dq \qquad (13.A.23)$$

As a solution let us try $w(v, y) = a \exp(-bv)$, where a and b are to be determined, and we find that

$$w(v, y) = a \exp(-ayv) \qquad (13.A.24)$$

so returning to the original variables

$$(1 + t/\tau_c)^2 n(v, t) = a \exp\left(-\frac{av}{N_0(1 + t/\tau_c)}\right) \qquad (13.A.25)$$

For $t = 0$ (13.A.25) is simplified to

$$n(v, 0) = a \exp\left(-\frac{av}{N_0}\right) \qquad (13.A.26)$$

and comparing it with the initial condition given by (13.77) we find that $a = N_0^2/V_0$. Therefore the solution of our problem is (13.78)

$$n(v, t) = \frac{N_0^2}{V_0(1 + t/\tau_c)^2} \exp\left(-\frac{vN_0}{V_0(1 + t/\tau_c)}\right) \qquad (13.A.27)$$

PROBLEMS

13.1_A Derive the general dynamic equation describing the evolution of the aerosol size distribution based on diameter $n(D_p, t)$.

13.2_A By what factor does the average particle size of tobacco smoke increase as a result of coagulation during the 2 s that it takes for the smoke to travel from the cigarette to the smoker's lungs? Assume that the inhaled concentration is 10^{10} cm^3 and that the initial aerosol diameter is 20 nm.

13.3_B Estimate the expected coagulation lifetime of a particle in an urban area (see Chapter 8 for typical size distributions) as a function of its diameter.

13.4_B Show that for an aerosol population undergoing growth by condensation only with no sources or sinks:

a. The total number of particles is constant.

b. The average particle size increases with time.

c. The total aerosol volume increases with time.

13.5$_B$ Show that the solution of the coagulation equation with a constant coagulation coefficient, including first-order particle removal

$$\frac{\partial n(v,t)}{\partial t} = \frac{1}{2}K \int_0^v n(v - \bar{v},t)n(\bar{v},t)d\bar{v} - Kn(v,t)N(t) - R(t)n(v,t)$$

with $n_0(v) = (N_0/v_0)\exp(-v/v_0)$ is (Williams 1984)

$$n(v,t) = \frac{N_0}{v_0}e^{-v/v_0}\exp\left\{ -\int_0^t \left[P(t') - \frac{vN_0K}{2v_0}\exp\left(-\int_0^{t'} P(t'')dt'' \right) \right]dt' \right\}$$

where $P(t) = R(t) + KN(t)$.

13.6$_B$ Given particle growth laws expressed in terms of particle diameter for the three modes of gas-to-particle conversion

$$\frac{dD_p}{dt} = \begin{cases} h_d D_p^{-1} & \text{diffusion} \\ h_s & \text{surface reaction} \\ h_v D_p & \text{volume reaction} \end{cases}$$

show that an aerosol having an initial size distribution $n_0(D_p)$ evolves under the three growth mechanisms according to

$$n(D_p,t) = \begin{cases} n_0[(D_p^2 - 2h_dt)^{1/2}]\dfrac{D_p}{(D_p^2 - 2h_dt)^{1/2}} \\ n_0(D_p - h_st) \\ n_0(D_pe^{-h_vt})e^{-h_vt} \end{cases}$$

13.7$_B$ Consider the steady-state size distribution of an aerosol subject to growth by condensation, sources, and first-order removal. The steady-state size distribution in this situation is governed by

$$\frac{d}{dv}(I(v)n(v)) = S(v) - R(v)n(v)$$

a. Show that

$$n(v) = \frac{1}{I(v)} \int_0^v S(v')\exp\left(-\int_{v'}^v \frac{R(v'')}{I(v'')}dv'' \right)dv'$$

b. Taking $I(v) = h_a v^a$, $R(v) = R_m v^m$, and $S(v) = S_0 \delta(v - v_0)$, show that

$$n(v) = \frac{S_0}{v^a h_a} \exp\left[-\frac{R_m}{h_a(m+1-a)} \left(v^{m+1-a} - v_0^{m+1-a} \right) \right]$$

valid for $v \geq v_0$.

13.8$_B$ It is of interest to compare the relative magnitudes of processes that remove or add particles from one size regime to another in an aerosol size spectrum. Let us assume that a "typical" atmospheric aerosol size distribution can be divided into three size ranges, $0.01\,\mu m < D_p < 0.1\,\mu m$, $0.1\,\mu m < D_p < 1.0\,\mu m$, and $1\,\mu m < D_p < 10\,\mu m$, in which the particle number concentrations are assumed to be as follows:

$0.01\,\mu m < D_p < 0.1\,\mu m$	$0.1\,\mu m < D_p < 1.0\,\mu m$	$1.0\,\mu m < D_p < 10\,\mu m$
$10^5\,cm^{-3}$	$10^2\,cm^{-3}$	$10^{-1}\,cm^{-3}$

It is desired to estimate the relative contributions of the processes listed below to the rates of change of the aerosol number concentrations in each of the three size ranges:

1. Coagulation
   a. Brownian
   b. Turbulent shear
   c. Differential sedimentation
2. Heterogeneous condensation

For the condition of the calculation assume air at $T = 298\,K$, $p = 1\,atm$. For the purposes of the coagulation calculation, assume that the particles in the three size ranges have diameters $0.05\,\mu m$, $0.5\,\mu m$, and $5\,\mu m$. For turbulent shear coagulation, assume $\varepsilon_k = 1000\,cm^2\,s^{-3}$. In the differential sedimentation calculation, assume that the particles with which those in the three size ranges collide have $D_p = 10\,\mu m$ with a number concentration of $10^{-1}\,cm^{-3}$. For heterogeneous condensation, the flux of particles into and out of the size ranges can be estimated by calculating the rate of change of the diameter, dD_p/dt, assuming zero vapor pressure of the condensing vapor species and assuming that the number concentration in each size range is uniform across the range. In addition, assume that the condensing species has the properties of sulfuric acid:

$$D = 0.07\,cm^2\,s^{-1}$$
$$M_A = 98\,g\,mol^{-1}$$
$$p_A^\circ = 1.3 \times 10^{-4}\,Pa$$
$$\rho_p = 1.87\,g\,cm^{-3}$$

Examine saturation ratio values, $S = p_A/p_A^\circ(T)$, of 1.0, 2.0, and 10.0. Show all calculations and carefully note any assumptions you make. Present your results in Table 13.P.1.

TABLE 13.P.1 Estimated Rates of change of Aerosol Number Densities, $cm^{-3}s^{-1}$

	Size Range		
Process	$0.01\,\mu m < D_{p_i} < 0.1\,\mu m^a$	$0.1\,\mu m \leq D_{p_i} \leq 1.0\,\mu m^b$	$1.0\,\mu m < D_{p_i} < 10\,\mu m^c$
Coagulation			
Brownian			
Turbulent			
Sedimentation			
Heterogeneous			
condensation			
$S = 1.0$			
$S = 2.0$			
$S = 10.0$			

[a] Assuming that $D_{p_i} = 0.05\,\mu m$ for all particles, and $N = 10^5\,cm^{-3}$.
[b] Asssuming that $D_{p_i} = 0.5\,\mu m$ for all particles, and $N = 10^2\,cm^{-3}$.
[c] Assuming that $D_{p_i} = 5\,\mu m$ for all particles, and $N = 10^{-1}\,cm^{-3}$.

13.9$_C$ We want to explore the dynamics of aerosol size distributions undergoing simultaneous growth by condensation and removal at a rate dependent on the aerosol concentration, with a continuous source of new particles. The size distribution function in such a case is governed by

$$\frac{\partial n(v,t)}{\partial t} + \frac{\partial}{\partial v}\left(I(v,t)n(v,t)\right) = S(v,t) - R(v,t)n(v,t)$$

where $R(v,t)$ is the first-order removal constant. Let us assume that $I(v,t) = h_a v^a$, $R(v,t) = R_m v^m$, $S(v,t) = S_0 \delta(v - v_0)$, and $n_0(v) = N_0 \delta(v - v_*)$.

a. Show that under these conditions

$$n(v,t) = \frac{S_0}{v^a h_a}\exp\left[-\frac{R_m}{h_a(m+1-a)}\left(v^{m+1-a} - v_0^{m+1-a}\right)\right]$$

$$+ N_0 \exp\left\{-\frac{R_m}{h_a(m+1-a)}\right.$$

$$\times \left[[v_*^{1-a} + (1-a)h_a t]^{m+1-a/1-a} - v_*^{m+1-a}\right]\Big\}$$

$$\times \delta[v - (v_*^{1-a} + (1-a)h_a t)^{1/1-a}]$$

[*Hint:* The equation for $n(v,t)$ can be solved by assuming a solution of the form $n(v,t) = n_\infty(v) + n^0(v,t)$, where $n_\infty(v)$ is the steady-state solution of the equation (see Problem 12.7) and $n^0(v,t)$ is the transient solution corresponding to the intial condition $n_0(v)$ in the absence of the source $S(v,t)$.]

b. Express the solution determined in part (a) in terms of the following dimensionless groups:

$$\theta = \frac{(1-a)h_a t}{v_*^{1-a}} \qquad y = \frac{v}{v_0} \qquad \tilde{n} = \frac{n}{N}$$

$$\chi = \frac{v_*}{v_0} \qquad \lambda = \frac{R_m v_0^{m+1-a}}{h_a(m+1-a)} \qquad \rho = \frac{N_0 h_a v_*^a}{v_0 S_0}$$

where N is the total number of particles.

c. Examine the limiting cases of $\rho \to \infty$ and $\rho \to 0$. Show that the proper forms of the solution are obtained in these two cases.

d. Plot the steady-state solution $n_\infty(y)$ for $a = \frac{1}{3}$ and $\frac{2}{3}$, $m = \frac{2}{3}$, and $\lambda = 0.001$, 0.01, 0.1, 1.0, 10.0. Discuss the behavior of the solutions.

e. Plot $\tilde{n}(y, \theta)$ as a function of y for $a = \frac{1}{3}$, $m = \frac{2}{3}$, $\lambda = 0.01$, $\rho = 100$, and $\chi = 1.1$ for $\theta = 0.1$, 1, 10, and 100. Discuss.

REFERENCES

Alam, M. K. (1987) The effect of van der Waals and viscous forces on aerosol coagulation. *Aerosol Sci. Technol.* **6**, 41–52.

Brock, J. R. (1972) Condensational growth of atmospheric aerosols, *J. Colloid Interface Sci.* **39**, 32–36.

Drake, R. L. (1972) A general mathematical survey of the coagulation equation, in *Topics in Current Aerosol Research* (Part 2), G. M. Hidy and J. R. Brock, Pergamon, New York, pp. 201–376.

Friedlander, S. K. (1977) *Smoke, Dust, and Haze: Fundamentals of Aerosol Behavior*, Wiley, New York.

Fuchs, N. A. (1964) *Mechanics of Aerosols*, Pergamon, New York.

Fuchs, N. A. and Sutugin, A. G. (1971) High dispersed aerosols, in *Topics in Current Aerosol Research* (Part 2), G. M. Hidy and J. R. Brock, eds., Pergamon, New York, pp. 1–200.

Gelbard, F., and Seinfeld J. H. (1979) The General Dynamic Equation for aerosols—theory and application to aerosol formation and growth, *J. Colloid Interface Sci.* **68**, 363–382.

Gryn, V. I., and Kerimov, M. K. (1990) Integrodifferential equations for nonspherical aerosol coagulation, *USSR Comput. Math. Math. Phys.* **30**, 221–224.

Hamaker, H. C. (1937) The London–van der Waals attraction between spherical particles, *Physica* **4**, 1058–1072.

Hirschfelder, J. O., Curtiss, C. F., and Bird, R. B. (1954) *Molecular Theory of Gases and Liquids*, Wiley, New York.

Lee, P. S., and Shaw, D. T. (1984) Dynamics of fibrous type particles: Brownian coagulation and the charge effect, *Aerosol Sci. Technol.* **3**, 9–16.

Marlow, W. H. (1981) Size effects in aerosol particle interactions: The van der Waals potential and collision rates, *Surf. Sci.* **106**, 529–537.

Mulholland, G. W., and Baum, H. R. (1980) Effect of initial size distribution on aerosol coagulation, *Phys. Rev. Lett.* **45**, 761–763.

Peterson, T. W., Gelbard, F., and Seinfeld, J. H. (1978) Dynamics of source-reinforced, coagulating, and condensing aerosols, *J. Colloid Interface Sci.* **63**, 426–445.

Pilinis, C., and Seinfeld, J. H. (1987) Asymptotic solution of the aerosol general dynamic equation for small coagulation, *J. Colloid Interface Sci.* **115**, 472–479.

Pruppacher, H. R., and Klett, J. O. (1997) *Microphysics of Clouds and Precipitation*, 2nd ed., Kluwer, Boston.

Ramabhadran, T. E., Peterson T. W., and Seinfeld J. H. (1976) Dynamics of aerosol coagulation and condensation, *AIChE J.* **22**, 840–851.

Rogak, S. N., and Flagan, R. C. (1992) Coagulation of aerosol agglomerates in the transition regime, *J. Colloid Interface Sci.* **151**, 203–224.

Saffman, P. G., and Turner, J. S. (1956). On the collision of drops in turbulent clouds, *J. Fluid Mech.* **1**, 16–30.

Schmidt-Ott, A., and Burtscher H. (1982) The effect of van der Waals forces on aerosol coagulation, *J. Colloid. Interface. Sci.* **89**, 353–357.

Seinfeld, J. H., and Ramabhadran, T. E. (1975) Atmospheric aerosol growth by heterogeneous condensation, *Atmos. Environ.* **9**, 1091–1097.

Spielman, L. (1970) Viscous interactions in Brownian coagulation, *J. Colloid Interface Sci.* **33**, 562–571.

Swift, D. L., and Friedlander, S. K. (1964) The coagulation of hydrosols by Brownian motion and laminar shear flow, *J. Colloid Sci.* **19**, 621–647.

Tambour, Y., and Seinfeld, J. H. (1980) Solution of the discrete coagulation equation, *J. Colloid Interface Sci.* **74**, 260–272.

Tennekes, H., and Lumley, J. O. (1972) *A First Course in Turbulence*, MIT Press, Cambridge, MA.

Williams, M. M. R. (1984) On some exact solutions of the space- and time-dependent coagulation equation for aerosols, *J. Colloid. Interface. Sci.* **101**, 19–26.

14 Organic Atmospheric Aerosols

14.1 ORGANIC AEROSOL COMPONENTS

The carbonaceous fraction of ambient particulate matter consists of elemental carbon and a variety of organic compounds (organic carbon). Elemental carbon (EC), also called *black carbon* or *graphitic carbon*, has a chemical structure similar to impure graphite and is emitted directly into the atmosphere predominantly during combustion. Organic carbon (OC) is either emitted directly by sources (primary OC) or can be formed in situ by condensation of low-volatility products of the photooxidation of hydrocarbons (secondary OC). Note that organic carbon refers to only a fraction of the mass of the organic material (the rest is hydrogen, oxygen, nitrogen, etc.), but because the carbon fraction is measured directly the designation is used routinely. Additional quantities of aerosol carbon, usually small, may exist either as carbonates (e.g., $CaCO_3$) or CO_2 adsorbed onto particulate matter such as soot (Appel et al. 1989; Clarke and Karani 1992). Measured EC and OC values are based on operational definitions that reflect the method and purpose of the measurement. Definitions of EC and OC have been the subject of debate and use of different definitions may lead to some confusion. Measurement methods for OC and EC are summarized in Appendix 14.

14.2 ELEMENTAL CARBON

14.2.1 Formation of Soot and Elemental Carbon

Carbonaceous particles are a byproduct of the combustion of liquid or gaseous fuels. Particles formed this way consist of both EC and OC and are known as *soot*. Soot particles are agglomerates of small roughly spherical elementary carbonaceous particles. While the size and morphology of the clusters vary widely, the small spherical elementary particles are remarkably consistent from one to another. These elementary particles vary in size from 20 to 30 nm and cluster with each other, forming straight or branched chains. These chains agglomerate and form visible soot particles that have sizes up to a few micrometers.

Soot formed in combustion processes is not a unique substance. It consists mainly of carbon atoms but also contains up to 10% moles of hydrogen as well as traces of other elements. The composition of soot that has been aged in the high-temperature region of a flame is typically C_8H (Palmer and Cullis 1965; Flagan and Seinfeld 1988), but soot usually contains more hydrogen earlier in the flame. Furthermore, soot particles absorb organic vapors when the combustion products cool down, frequently accumulating significant quantities of organic compounds. Therefore soot can be viewed as a mixture of

---

Atmospheric Chemistry and Physics: From Air Pollution to Climate Change, Second Edition, by John H. Seinfeld and Spyros N. Pandis.   Copyright © 2006 John Wiley & Sons, Inc.

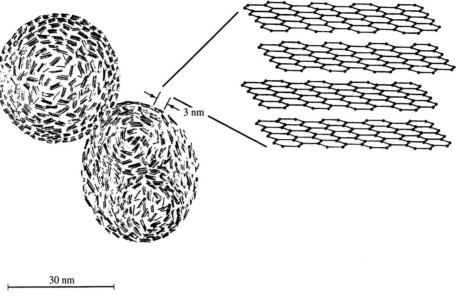

FIGURE 14.1 Schematic of soot microstructure.

EC, OC, and small amounts of other elements such as oxygen, nitrogen, and hydrogen incorporated in its graphitic structure (Chang et al. 1982). The EC found in atmospheric particles is not chunks of highly structured pure graphite, but rather is a related, more complex, three-dimensional array of carbon with small amounts of other elements. EC contains a number of crystallites 2 to 3 nm in diameter, with each crystallite consisting of several carbon layers having the hexagonal structure of graphite (Figure 14.1). The density of these soot particles is around $2 \, \text{g cm}^{-3}$.

The formation of soot depends critically on the carbon/oxygen ratio in the hydrocarbon–air mixture. Let us assume for the moment that the mixture has insufficient oxygen to form CO_2 and that CO is the combustion product of the fuel C_mH_n. Then the combustion stoichiometry is

$$C_mH_n + a\,O_2 \rightarrow 2a\,\text{CO} + 0.5n\,H_2 + (m - 2a)\,C_s$$

where C_s is the soot formed and the ratio of carbon to oxygen (C/O) is $m/2a$. When the C/O ratio is unity, that is, $m = 2a$, there is sufficient oxygen to tie up all the available carbon as CO and no soot is formed. If there is even more oxygen, the C/O ratio is lower than unity as $m < 2a$, and then the extra oxygen will be used for the conversion of CO to CO_2. On the other hand, if there is less oxygen, the C/O ratio is >1, $m > 2a$, and soot will start forming. This argument indicates that stoichiometrically one expects soot formation when the C/O ratio in the fuel–air mixture exceeds the critical value of unity. Actually both CO and CO_2 are formed in the combustion even at low C/O ratios. As oxygen is tied up in the stable CO_2 molecules, less of it is available for CO formation and soot starts forming at C/O ratios lower than 1. Wagner (1981) reported, for example, that soot formation is experimentally observed at C/O ratios close to 0.5. This critical C/O ratio is only weakly dependent on pressure or dilution with inert gases. Increasing temperature tends to generally inhibit the onset of soot formation.

Soot forms in a flame as the result of a chain of events starting with the oxidation and/or pyrolysis of the fuel into small molecules. Acetylene, C_2H_2, and polycyclic aromatic hydrocarbons (PAHs) are considered the main molecular intermediates for soot formation and growth (McKinnon and Howard 1990). The growth of soot particles involves first the formation of soot nuclei and then their rapid growth due to surface reactions (Harris and Weiner 1983a,b).

The soot nuclei themselves represent only a small fraction of the overall soot mass produced. However, the final soot mass depends critically on the number of nuclei formed, since the growth rate of soot particles is not a strong function of the fuel composition (Harris and Weiner 1983a,b). Unfortunately, despite numerous studies, this crucial nuclei formation step remains poorly understood. This step does not appear to involve homogeneous nucleation, but rather a series of gas-phase polymerization reactions producing large polycyclic aromatic hydrocarbons and eventually soot particles (Wang et al. 1983; Flagan and Seinfeld 1988). There is strong experimental evidence for this picture. The concentration of PAHs in a sooting flame rises rapidly prior to soot formation and reaches a peak nearly coincident with the zone of soot nucleation (McKinnon and Howard 1990). Acetylene is the most abundant hydrocarbon in rich flames (Harris and Weiner 1983a) and has been regarded for more than a century as the species responsible for mass addition to the soot particles. A series of mechanisms have been proposed in which acetylene molecules fuse to form an aromatic ring and then continue to react with the aromatic compound leading to PAH formation (Bittner and Howard 1981; Bockhorn et al. 1983; Frenklach et al. 1985). Reactive collisions between the large PAH molecules can accelerate their growth. The soot nuclei grow through these chemical reactions, reaching diameters larger than 10 nm, at which point they begin coagulating to form chain aggregates. The carbon/hydrogen ratio in the nuclei starts close to unity, but as soot ages in the flame it loses hydrogen, eventually exiting the flame with a carbon/hydrogen ratio of about 8.

The formation of soot depends strongly on the fuel composition. The rank ordering of sooting tendency of fuel components is: naphthalenes > benzenes > aliphatics. However, the order of sooting tendencies of the aliphatics (alkanes, alkenes, alkynes) varies dramatically depending on the flame type. The difference between the sooting tendencies of aliphatics and aromatics is thought to result mainly from the different routes of formation. Aliphatics appear to first form acetylene and polyacetylenes; aromatics can form soot both by this route and also by a more direct pathway involving ring condensation or polymerization reactions building on the existing aromatic structure (Graham et al. 1975; Flagan and Seinfeld 1988).

14.2.2 Emission Sources of Elemental Carbon

Fuel-specific emission rates of elemental carbon are shown in Table 14.1. These rates can vary significantly as they depend strongly on the conditions under which combustion takes place. Woodburning fireplaces and diesel automobiles are effective sources of EC per unit of fuel burned (Mulhbaier and Williams 1982; Brown et al. 1989; Dod et al. 1989; Mulhbaier and Cadle 1989; Hansen and Rosen 1990; Burtscher 1992).

Emissions of EC by diesel-burning motor vehicles are easily visible and have been the focus of many studies (Pierson and Brackaczek 1983; Raunemaa et al. 1984; Hildemann et al. 1991b; U.S. EPA 1996). Photomicrographs of particles collected from the exhaust of a diesel automobile indicate that the emitted aerosols are similar to the soot particles described above. Volume mean diameters of the aggregates tend to range from 0.05 to

TABLE 14.1   Estimates of Fuel-Specific Particulate Carbon Emission Rates, g(C) (kg fuel)$^{-1}$

Source	Organic Carbon	Elemental Carbon
Fireplace		
Hardwood	4.7	0.4
Softwood	2.8	1.3
Motor vehicles		
Noncatalytic	0.04–0.24	0.01–0.13
Catalytic	0.01–0.03	0.01–0.03
Diesel	0.7–1.0	2.1–3.4
Furnace (natural gas)		
Normal	0.0004	0.0002
Rich	0.007	0.12

Sources: Muhlbaier and Williams (1982) and Muhlbaier and Cadle (1989).

0.25 μm. More than 90% of the emitted EC mass is in submicrometer particles. Measured size distributions of passenger automobile exhaust aerosol burning leaded and unleaded gasoline are shown in Figure 14.2 for the Federal Test Procedure (FTP) driving cycle, and under steady-state conditions at 40 and 95 km h^{-1} (Hildemann et al. 1991b). The

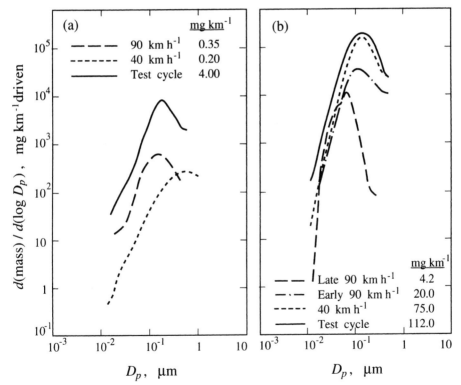

FIGURE 14.2 Mass distributions of primary automobile exhaust produced under cruise conditions as a function of speed during combustion of (a) unleaded gasoline and (b) leaded gasoline as reported by Hildemann et al. (1991b). (Reprinted from *Aerosol Sci. Technol.* **14**, Hildemann et al., 138–152, Copyright 1991, with kind permission from Elsevier Science Ltd., The Boulevard, Langford Lane, Kidlington OX5 1GB, UK.)

emitted aerosol volume distributions are characterized by a mode centered at a diameter of approximately 0.1 μm. Emissions during the driving cycle simulating traffic conditions in a city were almost a factor of 10 higher than those measured under steady-state conditions. The data in Figure 14.2 represent a very limited sample and should not be viewed as representative of all motor vehicles. Nevertheless, they do exhibit the general features of the aerosol distributions emitted by automobiles.

A series of tracer techniques have been developed for calculation of the source contribution to EC concentrations, including use of potassium (K) as a woodsmoke tracer (Currie et al. 1994) and use of the carbon isotopic tracers ^{14}C and ^{12}C (Klouda et al. 1988; Lewis et al. 1988; Currie et al. 1989). By this procedure 47% of the EC in Detroit, 93% in Los Angeles, and 30–60% in a rural area in Pennsylvania have been attributed to motor vehicle sources (Wolff and Korsog 1985; Pratsinis et al. 1988; Keeler et al. 1990). Global emissions of EC were estimated by Penner et al. (1993) to be 12.6–24 Tg (C) yr^{-1}, while the EC emission rate for the United States was 0.4–1.1 Tg yr^{-1} and for the rest of North America 0.2 Tg yr^{-1}. Bond et al. (2004) estimated that 8 Tg (C) yr^{-1} were emitted globally in 1996, much of it (42%) from open burning.

The EC percentages of the aerosol mass emitted by automobiles and other sources based on a series of different studies are shown in Table 14.2. All of these values are based on small samples of sources and show significant differences from sample to sample and from study to study.

14.2.3   Ambient Elemental Carbon Concentrations

EC concentrations in rural and remote areas usually vary from 0.2 to 2.0 μg m^{-3} and from 1.5 to 20 μg m^{-3} in urban areas. The concentration of EC over the remote oceans is approximately 5-20 ng m^{-3} (Clarke 1989). Average PM$_{10}$ EC values exceeding 10 μg m^{-3}

TABLE 14.2   Elemental and Organic Carbon Percentages in Emissions by Different Sources

Fuel	Elemental Carbon	Organic Carbon	References
Diesel	74 ± 21[a]	23 ± 8[a]	Watson et al. (1990)
	52 ± 5[a]	36 ± 3[a]	Cooper et al. (1987)
	43 ± 8[b]	49 ± 13[b]	Houck et al. (1989)
Unleaded gasoline	18 ± 11[a]	76 ± 29[a]	Watson et al. (1990)
	5 ± 7[c]	93 ± 52[c]	Cooper et al. (1987)
	39 ± 10[a]	49 ± 10[a]	Cooper et al. (1987)
Leaded gasoline	16 ± 7[a]	67 ± 23[a]	Watson et al. (1990)
	13 ± 1[c]	52 ± 4[c]	Cooper et al. (1987)
	15 ± 2[a]	51 ± 20[a]	Cooper et al. (1987)
Mixed vehicles	28 ± 19	50 ± 24	Watson et al. (1990)
(tunnel and roadside)	38 ± 5	38 ± 6	Cooper et al. (1987)
	36 ± 11	39 ± 19	Watson et al. (1990)
Dust (< 2.5 μm)	2	12	Watson and Chow (1994)
Woodburning	15	70	Watson and Chow (1994)
Coal-fired power plant	2	—	Olmez et al. (1988)

[a]Measured during dynamometer tests following the modified Federal Test Procedure (FTP).
[b]Roof monitoring at inspection station.
[c]Steady-speed (55 km h^{-1}) tests.

TABLE 14.3 Measured Daily Average Elemental and Organic Carbon Concentration in a Series of U.S. Studies

Location	Date	Samples	EC $\mu g\,m^{-3}$	EC %	OC $\mu g(C)\,m^{-3}$	OC %
Eastern USA						
Philadelphia	August 1994	21	0.76	2.4	4.51	14
Roanoke	Winter 1988/1989	—[a]	1.5	7.5	7.3	36.7
Western USA						
Los Angeles	Summer 1987	44	2.37	5.8	8.27	20.1
Los Angeles	Fall 1987	24	7.28	8.1	18.46	20.5
San Joaquin	1988–1989	35	3.24	10.8	4.87	16.3
Phoenix	Winter 1989/1990	200	7.47	25.4	10.10	34.4
Central USA						
Albuquerque	Winter 1984/1985	—[a]	2.1	10.2	13.2	64.6
Denver	Winter 1987/1988	136	4.41	22.4	7.25	36.9
Chicago	July 1994	16	1.31	9.6	5.39	43.5

[a]Not reported.

Source: U.S. EPA (1996).

are common for some urban locations (Chow et al. 1994). EC measurements from a series of major U. S. studies are shown in Table 14.3.

Annual average nonurban EC concentrations in the United States are shown in Figure 14.3 as absolute concentrations and as fractions of the PM_{10} aerosol mass. These data are based on regional background monitoring sites, away from urban industrial activities. These maps are useful for showing overall concentration trends and for illustrating the average effect of anthropogenic emissions on relatively remote continental areas (Sisler et al. 1993). The remote data indicate that the EC concentrations are higher over the Northwest, southern California, and locally in the eastern United States. In the northwestern United States EC exceeds 10% of the aerosol mass, but over most of the United States it is less than 5%.

14.2.4 Ambient Elemental Carbon Size Distribution

The mass distribution of EC emitted by automobiles is unimodal with a peak around 0.1 μm. Measurements inside tunnels have confirmed that over 85% of the emitted EC mass by vehicles is in particles smaller than 0.2 μm aerodynamic diameter (Venkataraman and Friedlander 1994a) (Figure 14.4). However, the ambient distribution of EC in polluted areas is bimodal with peaks in the 0.05–0.12 μm (mode I) and 0.5–1.0 μm (mode II) size ranges (Nunes and Pio 1993; Venkataraman and Friedlander 1994b) (Figure 14.5). In polluted urban areas mode I usually dominates, containing almost 75% of the total EC. The EC distribution shown in Figure 14.5b is an extreme case corresponding to well-aged aerosol measured during winter in an inland site in southern California. The first mode in the ambient EC size distribution is a result of the contribution of primary EC sources from combustion. The creation of mode II is mainly the result of accumulation of secondary aerosol products on primary aerosol particles and subsequent growth. This conclusion is supported by the existence of liquid aerosol components in the mode II particles.

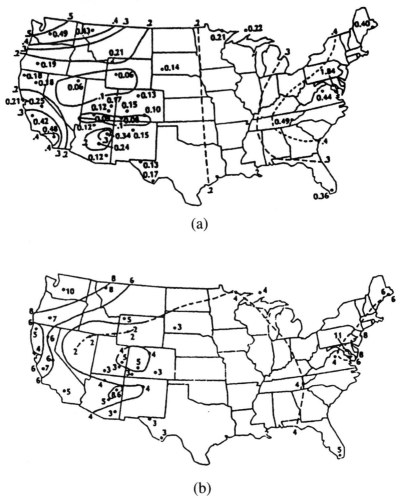

(a)

(b)

FIGURE 14.3 Annual mean EC concentration distribution in the United States based on samples from remote stations: (a) $\mu g\, m^{-3}$; (b) percent of total aerosol mass. The influence of the urban centers has been eliminated to a large extent and the results depict the regional distributions (U.S. EPA 1996).

14.3 ORGANIC CARBON

The organic component of ambient particles in both polluted and remote areas is a complex mixture of hundreds of organic compounds (Hahn 1980; Simoneit and Mazurek 1982; Graedel 1986; Rogge et al. 1993a–c). Compounds identified in the ambient aerosol include n-alkanes, n-alkanoic acids, n-alkanals, aliphatic dicarboxylic acids, diterpenoid acids and retene, aromatic polycarboxylic acids, polycyclic aromatic hydrocarbons, polycyclic aromatic ketones and quinones, steroids, N-containing compounds, regular steranes, pentacyclic triterpanes, and iso- and anteiso-alkanes. Sampling and analysis of ambient carbon pose a series of challenges that are discussed in Appendix 14.

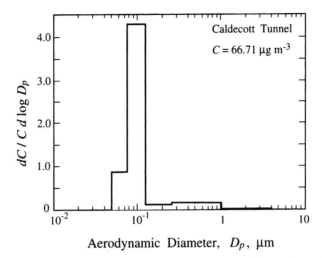

FIGURE 14.4 EC size distribution measured inside a tunnel in southern California. The distribution corresponds to the primary particles emitted by automobiles. (Reprinted with permission from Venkataraman, C. and Friedlander, S. K. Size distributions of polycyclic hydrocarbons and elemental carbon. 1. Sampling, measurement methods and source characterization. *Environ. Sci. Technol.* **28**. Copyright 1994 American Chemical Society.)

14.3.1 Ambient Aerosol Organic Carbon Concentrations

Most investigators have reported OC concentration as simply the concentration of carbon in $\mu g(C) m^{-3}$ and as such do not include the contribution to the aerosol mass of the other elements (namely, oxygen, hydrogen, and nitrogen) of the organic aerosol compounds. Wolff et al. (1991) suggested that measured OC values should be multiplied by a factor of 1.5 for calculation of the total organic mass associated with the OC, but values of 1.2–2.0 have also been used in the past by various investigators. Values that include the estimated total organic mass are usually given in $\mu g m^{-3}$.

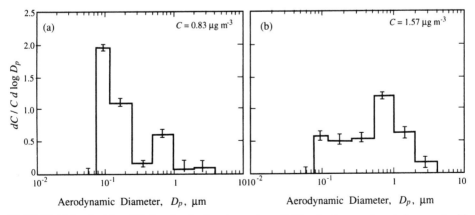

FIGURE 14.5 Ambient EC size distributions in southern California. A second mode is present with a mean of roughly 1 μm. (Reprinted with permission from Venkataraman, C., and Friedlander, S. K. Size distributions of polycyclic hydrocarbons and elemental carbon. 2. Ambient measurements and effects of atmospheric processes. *Environ. Sci. Technol.* **28**. Copyright 1994 American Chemical Society.)

TABLE 14.4 Concentration of Aerosol Organic Carbon from Remote Marine Areas

Location	Organic Carbon, $\mu g(C) \, m^{-3}$
Northern Hemisphere	
Enewetak Atoll	0.82 ± 0.17
Northern Atlantic	0.40 ± 0.39
West Ireland	0.57 ± 0.29
Sargasso Sea	0.44 ± 0.04
Hawaii	0.39 ± 0.03
Eastern Atlantic	0.35 ± 0.15
Equatorial Pacific	0.21 ± 0.11
Puerto Rico	0.66
Southern Hemisphere	
New Zealand	0.13 ± 0.02
Tropical Atlantic	0.23 ± 0.08
Samoa	0.11 ± 0.03
Peru	0.16 ± 0.07
Equatorial Pacific	0.15 ± 0.05
Amsterdam Island	0.15

Source: Penner (1995).

The concentration of OC is around $3.5 \, \mu g(C) \, m^{-3}$ in rural locations (Stevens et al. 1984) and $5-20 \, \mu g(C) \, m^{-3}$ in polluted atmospheres (Wolff et al. 1991). Table 14.3 presents a set of observed concentrations of OC measured at various sites in the United States. One should note that the data account only for the OC mass and that the total organic matter concentration ranges from 1.2 to 2.0 times the reported concentration (Duce et al. 1983; Countess et al. 1980). Concentrations of OC in aerosols sampled in remote marine areas of the globe are given in Table 14.4. Concentrations in the Northern Hemisphere are around $0.5 \, \mu g(C) \, m^{-3}$, while in the southern latitudes they are roughly a factor of 2 lower. In rural areas of the western United States particulate OC concentrations are comparable to sulfate, and organic carbon represents 30–50% of the PM_{10} mass (White and Macias 1989) (Figure 14.6). In more polluted areas OC contributes 10–40% of the $PM_{2.5}$ and PM_{10} mass (Table 14.3).

Organic compounds accumulate mainly in the submicrometer aerosol size range (McMurry and Zhang 1989), and their mass distribution is typically bimodal with the first peak around diameter $0.2 \, \mu m$ and the second around $1 \, \mu m$ (Pickle et al. 1990; Mylonas et al. 1991) (see Figure 14.7 as an example).

14.3.2 Primary versus Secondary Organic Carbon

Particulate organic carbon comprises a large number of compounds having significant variations in volatility; as a result, a number of these compounds can be present in both the gas and particulate phases. The ability of such "semivolatile" compounds to coexist in both phases complicates the distinction between primary and secondary OC. Strictly speaking, secondary OC starts its atmospheric life in the gas phase as a VOC, undergoes one or more chemical transformations in the gas phase to a less volatile compound, and finally transfers to the particulate phase by condensation or nucleation. Therefore the term

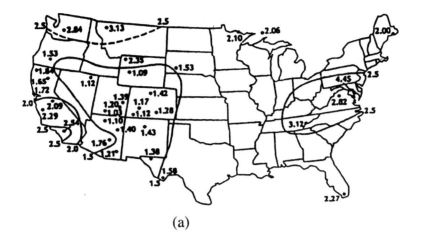

(a)

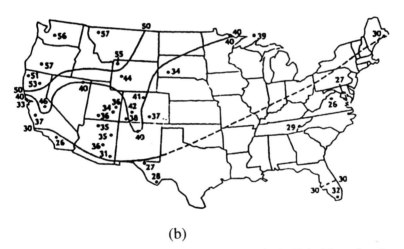

(b)

FIGURE 14.6 Annual mean OC concentration distribution in the United States based on samples from remote stations: (a) $\mu g\, m^{-3}$; (b) percent of total aerosol mass. The influence of the urban centers has been eliminated to a large extent and the results depict the regional distributions (U.S. EPA 1996).

secondary OC implies both a gas-phase chemical transformation and a change of phase. Organic compounds that are vapors at the high-temperature conditions of the exhaust of a combustion source and subsequently condense to the particulate phase as the emissions cool are considered as primary because they have not undergone chemical reaction in the gas phase. By the same token, organic compounds in an emitted particle are considered primary even if the compounds eventually undergo chemical change in the particulate phase because these compounds have not undergone a change in phase.

Evaluation of the primary and secondary components of aerosol OC have been difficult. Lack of a direct chemical analysis method for the identification of either of these OC components has led to several indirect methods. The simplest approach to estimating the contributions of primary and secondary sources to measured particulate OC is the EC

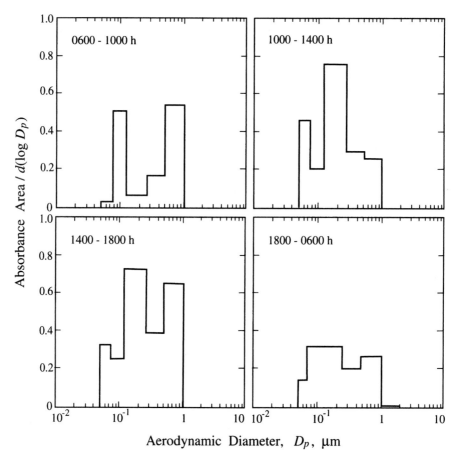

FIGURE 14.7 Diurnal variation of organic nitrate size distributions in southern California. Organic nitrates are secondary organic aerosol compounds. (Reprinted from *Atmos. Environ.* **25A**, Mylonas et al., 2855–2861. Copyright 1991, with kind permission from Elsevier Science, Ltd., The Boulevard, Langford Lane, Kidlington OX5 GB1, UK.)

tracer method (Chu and Macias 1981; Wolff et al. 1982; Novakov 1984; Gray et al. 1986; Huntzicker et al. 1986; Turpin et al. 1991; Cabada et al. 2004). If $[OC]_P$ and $[OP]_S$ are the primary and secondary OC concentrations, respectively, then the total OC concentration in a measured sample is just their sum:

$$[OC] = [OC]_P + [OC]_S \qquad (14.1)$$

Primary OC is emitted mainly by combustion or combustion-related sources (e.g., resuspension of combustion particles) and is accompanied by EC, but there is generally also a noncombustion component mainly from biogenic sources:

$$[OC]_P = [OC]_C + [OC]_{NC} \qquad (14.2)$$

where $[OC]_C$ and $[OC]_{NC}$ are the corresponding contributions of combustion and noncombustion sources to primary OC. If $[OC/EC]_C$ is the ratio of the OC to EC concentration for the combustion sources affecting the site at which the sample is measured, then using (14.1) and (14.2) yields

$$[OC]_S = [OC] - [OC/EC]_C[EC] - [OC]_{NC} \qquad (14.3)$$

The secondary OC concentration can be calculated from (14.3) and measurements of [OC] and [EC] if $[OC/EC]_C$ and $[OC]_{NC}$ are known. In general, these parameters are time-dependent and differ for different sites but can be determined from high temporal resolution measurements of OC and EC (Turpin et al. 1991; Cabada et al. 2004). The determination requires excluding from the dataset measurements corresponding to periods when there is a significant secondary OC component (e.g., periods of high photochemical activity) (Cabada et al. 2004). For the remaining data, the OC is dominated by the primary contributions, $[OC]_S \approx 0$ and from (14.3),

$$[OC] = [OC/EC]_C[EC] + [OC]_{NC} \qquad (14.4)$$

Therefore the $[OC/EC]_C$ and $[OC]_{NC}$ can be calculated from the slope and the intercept of the remaining [OC] versus [EC] data (Figure 14.8). For the data set of Figure 14.8, $[OC/EC]_C = 1.7$ and $[OC]_{NC} = 0.9\,\mu g\,m^{-3}$. Substituting these values into (14.3) and using all the measurements of OC and EC, one can estimate the secondary organic concentration as a function of time (Figure 14.9). If the measured ambient OC exceeds the value predicted by (14.4), then the additional OC can be considered to be secondary in origin. This analysis, when applied to atmospheric data, indicates that the secondary contribution to OC is quite variable, ranging from more than 50% during stagnation periods characterized by high aerosol loadings to almost zero during relatively clean periods. The secondary organic aerosol estimates from the EC tracer method are quite sensitive to the value of the $[OC/EC]_C$ used in the analysis. This ratio varies by source and therefore will

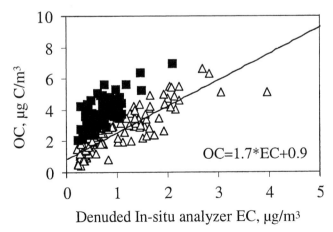

FIGURE 14.8 Determination of $[OC/EC]_C$ and $[OC]_{NC}$ for Pittsburgh, Pennsylvania during July 2001 (Cabada et al. 2004). The filled squares correspond to periods when there was significant photochemical activity and were excluded from the analysis. A linear fit of the remaining measurements provides estimates of $[OC/EC]_C$ (slope) and $[OC]_{NC}$ (intercept).

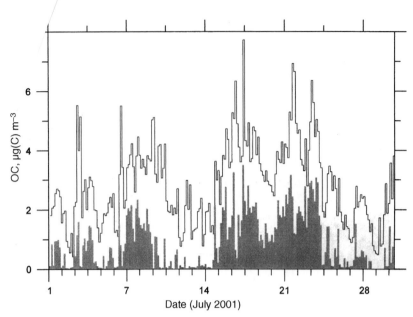

FIGURE 14.9 Measured OC concentrations (solid line) and estimated secondary organic aerosol concentrations (shaded area) during July 2001 in Pittsburgh, Pennsylvania (Cabada et al. 2004).

be influenced by meteorology, diurnal and seasonal variations in emissions, and local sources, This method should be used during periods for which the $[OC/EC]_C$ is approximately constant and with high-temporal-resolution data sets (1–4 h sampling time). For example, it can be applied with month-long data sets in areas with relatively uniform spatial distribution of sources and relatively small local emissions.

While the EC tracer method can supply useful information, its prediction can be quite uncertain, so additional approaches have been suggested. These include the use of models describing the emission and dispersion of primary OC (Gray et al. 1986; Larson et al. 1989; Hildemann 1990), use of models describing the formation of secondary OC (Pilinis and Seinfeld 1988; Pandis et al. 1992; Bowman et al. 1997; Kanakidou et al. 2005), and use of primary OC tracers (Rogge et al. 1996).

On average, primary OC tends to dominate in most polluted areas, but the secondary OC contribution is maximum and can exceed the primary contribution during peak photochemical air pollution episodes. For example, during periods of high photochemical smog in the South Coast Air Basin of California (Los Angeles air basin), as much as 80% of the OC can be secondary, with secondary concentrations as high as $15 \, \mu g(C) \, m^{-3}$ (Turpin and Huntzicker 1995). The yearly averaged contribution of secondary OC to total OC in that air basin is 10–40%.

14.4 PRIMARY ORGANIC CARBON

14.4.1 Sources

Primary carbonaceous particles are produced by combustion (pyrogenic), chemical (commercial products), geologic (fossil fuels), and natural (biogenic) sources. Rogge et al.

TABLE 14.5   Automobile Emissions of Fine Particles ($D_p < 2\,\mu m$)

Vehicle	Sample Size	Fuel Consumed FTP, mi gal^{-1}	OC Emission, mg (C) km^{-1}	EC Emission, mg (C) km^{-1}	Total Particulate Emission, mg km^{-1}
Noncatalyst	6	15.7	38.9	4.8	59.4
Catalyst	7	23.3	9.0	5.3	18.0
Diesel truck	2	7.6	132.9	163.2	408.0

Source: Hildemann et al. (1991a).

(1996) calculated that on average 29.8 tons of OC per day are emitted in the Los Angeles basin from 32 groups of sources. The most important of these include meat cooking operations (21.2%), paved road dust (15.9%), fireplaces (14%), noncatalyst vehicles (11.6%), diesel vehicles (6.2%), surface coatings (4.8%), forest fires (2.9%), catalyst-equipped automobiles (2.9%), and cigarettes (2.7%). Each of these sources emits a wide variety of chemical compounds. For example, more than 80 chemical compounds were identified in the OC emitted from meat cooking operations (Rogge et al. 1991).

Global primary OC emissions for 1996 were estimated as 17–77 Tg yr^{-1} (Bond et al. 2004), with open burning contributing around 75% of this total, while fossil fuel combustion was estimated at 7%.

Hildemann et al. (1991a) measured fine-particle emissions from 15 different automobiles and trucks using the cold-start Federal Test Procedure (FTP) to simulate urban driving conditions (Table 14.5). The reader should note that this is a very limited sample and such results should be extrapolated with caution to other situations. The noncatalyst automobiles tested by Hildemann et al. (1991a) and Rogge et al. (1993a) had consistently higher emissions than previously reported values.[1]

Source tests conducted by Hildemann et al. (1991a) suggested that meat cooking can be a significant source of primary organic aerosols, with broiling of regular hamburger meat resulting in fine- ($D_p \leq 2\,\mu m$) particle emissions of approximately 40 g per kg of meat cooked. The emissions from meat cooking operations depend strongly on the cooking method used, fat content of the meat, and the type of grease eliminator system that may be in operation above the cooking surface (Table 14.6).

TABLE 14.6   Fine ($D_p < 2\,\mu m$) Particle Mass Emissions during Meat Cooking

Source	Aerosol Emission, g kg (meat cooked)$^{-1}$
Frying extra lean (10% fat) meat	1
Frying regular (21% fat) meat	1
Charbroiling extra lean (10% fat) meat	7
Charbroiling regular (21% fat) meat	40

Source: Hildemann et al. (1991a).

---

[1]Rogge et al. (1993a) explained this discrepancy by the fact that they tested the automobiles in as-received condition, while other studies tuned the automobiles and changed the lubricant oil before testing to improve the consistency of the measurements. Discrepancies like this are indicative of the difficulties of obtaining accurate emission estimates, as these vary depending on the automobile and driving conditions.

TABLE 14.7   Fine ($D_p < 2 \mu$m) Particle Mass Emissions during Woodburning

Wood Type	Emission, g kg (wood)$^{-1}$
Oak	6.2 ± 0.3
Pine	13 ± 4.1
Synthetic log	12

Source: Hildemann et al. (1991a).

Wood combustion has been identified as a source that can contribute significantly to atmospheric particulate levels (Core et al. 1984; Ramdahl et al. 1984b; Sexton et al. 1984; Standley and Simoneit 1987; Hawthorne et al. 1992). Fine-particle emission rates reported by Hildemann et al. (1991a) are shown in Table 14.7.

There is strong evidence that plant leaves contribute a significant amount of leaf wax as primary organic aerosol particles (Simoneit 1977, 1986, 1989; Gagosian et al. 1982; Simoneit and Mazurek 1982; Wils et al. 1982; Doskey and Andren 1986; Standley and Simoneit 1987; Sicre et al. 1990; Simoneit et al. 1990).

14.4.2   Chemical Composition

Several hundred organic compounds have been identified in primary organic aerosol emissions. In spite of this these studies have been able to identify compounds representing only 10–40% of the emitted organic mass depending on the source. Even if our knowledge of the molecular composition of OC has increased significantly, the complexity of the mixture is such that tracer compounds are still necessary to unravel the contributions of the various sources.

Table 14.8 presents chemical compounds that have been identified as primary organic aerosol components (Rogge et al. 1991, 1993a–e) together with their annual average

TABLE 14.8   Compounds Identified in Primary Organic Aerosol Emissions from a Series of Sources and Ambient Aerosol Concentrations of Fine Aerosol Organic Compounds Measured in Pasadena, California, during 1982

Compound	Sources[a]	Concentration, ng m^{-3}
Alkanes		
n-Tricosane (C23)	M, A, RD, V, G, C, AS, B, W	5.4
n-Tetracosane (C24)	M, A, RD, V, G, C, AS, B, W	4.7
n-Pentacosane (C25)	M, A, RD, V, G, C, AS, B, W	9.5
n-Hexacosane (C26)	M, A, RD, V, G, C, AS, B, W	4.3
n-Heptacosane (C27)	M, A, RD, V, G, C, AS, B, W	5.6
n-Octacosane (C28)	M, A, RD, V, G, C, AS, B, W	2.5
n-Nonacosane (C29)	M, A, RD, V, G, C, AS, B, W	4.7
n-Triacontane (C30)	A, RD, V, G, C, AS, B, W	2.5
n-Hentriacontane (C31)	A, RD, V, G, C, AS, B, W	9.6
n-Dotriacontane (C32)	A, RD, V, G, C, AS, B, W	1.5
n-Tritriacontane (C33)	RD, V, G, C, B, W	2.3
n-Tetratriacontane (C34)	RD, V, C, W	0.7
Total alkanes		53.3

TABLE 14.8 (*Continued*)

Compound	Sources[a]	Concentration, ng m^{-3}
Alkanoic acids		
n-Nonanoic acid (C9)	M, A, RD, V, G, C, AS, B, W	5.3
n-Decanoic acid (C10)	M, A, RD, V, G, C, AS, B, W	2.4
n-Undecanoic acid (C11)	M, A, RD, V, G, C, AS, B, W	6.0
n-Dodecanoic acid (C12)	M, A, RD, V, G, C, AS, B, W	7.0
n-Tridecanoic acid (C13)	M, A, RD, V, G, C, AS, B, W	4.9
n-Tetradecanoic (myristic) acid (C14)	M, A, RD, V, G, C, AS, B, W	22.2
n-Pentadecanoic acid (C15)	M, A, RD, V, G, C, AS, B, W	6.1
n-Hexadecanoic acid (palmitic) (C16)	M, A, RD, V, G, C, AS, B, W	127.4
n-Heptadecanoic acid (C17)	M, A, RD, V, G, C, AS, B, W	5.2
n-Octadecanoic acid (stearic) (C18)	M, A, RD, V, G, C, AS, B, W	50.0
n-Nonadecanoic acid (C19)	A, RD, V, G, C, AS, B, W	1.1
n-Eikosanoic acid (C20)	A, RD, V, G, C, AS, B, W	6.1
n-Heneicosanoic acid (C21)	A, RD, V, G, C, AS, B, W	2.3
n-Docosanoic acid (C22)	RD, V, G, C, AS, B, W	9.9
n-Tricosanoic acid (C23)	RD, V, G, C, AS, B, W	2.5
n-Tetracosanoic acid (C24)	RD, V, G, C, AS, B, W	16.5
n-Pentacosanoic acid (C25)	RD, V, G, C, B, W	1.6
n-Hexacosanoic acid (C26)	RD, V, G, C, B, W	9.3
n-Heptacosanoic acid (C27)	RD, V, G, C, W	0.8
n-Octacosanoic acid (C28)	RD, V, G, C, W	4.9
n-Nonacosanoic acid (C29)	RD, V, G, C, W	0.6
n-Triacontanoic acid (C30)	RD, V, G, C, W	2.2
Total alkanoic acids		294.3
Alkenoic acids		
Octadecenoic (oleic) acid	M, A, RD, V, C, AS, W	26.0
Alkanals and alkenals		
Nonanal (C9)	M	26.0
Dicarboxylic acids		
Propanedioic (malonic) acid	W	44.4
2-Butenedioic acid		1.3
Butanedioic (succinic) acid	M, C, W	51.2
Methylbutanedioic (methylsuccinic) acid	C, W	15.0
Pentanedioic (glutaric) acid	M, C, W	28.3
Methylpentanedioic (methylglutaric) acid		16.6
Hydroxybutanedioic (hydroxysuccinic) acid		16.0
Hexanedioic (adipic) acid	M, W	14.1
Octanedioic (suberic) acid	M	4.1
Nonanedioic (azelaic) acid		22.8
Total dicarboxylic acids		213.8
Diterpenoic acids and retene		
Dehydroabetic acid	RD, W	22.6
13-Isopropyl-5 α-podocarpa-6,8, 11,13-tetraen-16-oic acid	W	1.2
8,15-Pimaradien-18-oic acid	W	0.6
Pimaric acid	W	4.8
Isopimaric acid	W	2.3

(*continued*)

TABLE 14.8   (*Continued*)

Compound	Sources[a]	Concentration, ng m^{-3}
7-Oxodehydroabetic acid	W	4.1
Sandaracopimaric acid	W	2.2
Retene	W	0.1
Total woodsmoke markers		37.6
Aromatic polycarboxylic acids		
1,2-Benzenedicarboxylic (phtalic) acid		55.7
1,3-Benzenedicarboxylic (isophthalic) acid	B	2.9
1,4-Benzenedicarboxylic (terephthalic) acid	B	1.5
4-Methyl-1,2-benzenedicarboxylic acid		28.8
1,2,4-Benzenetricarboxylic (trimellitic) acid		0.8
1,3,5-Benzenetricarboxylic (trimesic) acid		17.2
1,2,4,5-Benzenetricarboxylic (pyromellitic) acid		0.8
Total		107.7
Polycyclic aromatic hydrocarbons		
Fluoranthene	M, A, RD, G, C, AS, B, W	0.13
Pyrene	M, A, RD, G, C, AS, B, W	0.17
Benz[*a*]anthracene	M, A, RD, G, C, AS, B, W	0.25
Cyclopenta[*cd*]pyrene	A, W	0.41
Benzo[*ghi*]fluoranthene	A, RD, G, AS, B, W	0.30
Chrysene	M, A, RD, G, AS, B, W	0.43
Benzo[*k*]fluoranthene	M, A, RD, G, AS, B, W	1.20
Benzo[*b*]fluoranthene	M, A, RD, G, AS, B, W	0.85
Benzo[*e*]pyrene	M, A, RD, G, AS, B, W	0.93
Benzo[*a*]pyrene	M, A, RD, AS, B, W	0.44
Indeno[1,2,3-*cd*]pyrene	A, W	0.42
Perylene	M, A, RD, W	—
Indeno[1,2,3-*cd*]fluoranthene	A, RD, W	1.09
Benzo[*ghi*]perylene	M, A, RD, W	4.43
Coronene	A, W	2.41
Total		11.04
Polycyclic aromatic ketones (PAKs) and quinones (PAQs)		
7H-Benz[*de*]anthracen-7-one	A, RD, G, B, W	0.8
Benz[*a*]anthracene-7,12-dione	G	0.3
6H-benzo[*cd*]pyren-6-one	A, RD, W	1.2
Sterols		
Cholesterol	M, C	1.9
N-containing and other compounds		
3-Methoxypyridine		1.4
Isoquinoline	A	1.1
1-Methylisoquinoline		0.2
1,2-Dimethoxy-4-nitro-benzene		3.9
Total		6.6

TABLE 14.8 (*Continued*)

Compound	Sources[a]	Concentration, ng m^{-3}
Regular steranes		
20S&R-5 α(H), 14 β(H), 17 β(H)-Cholestanes	A, RD, AS, B	0.6
20R-5 α(H), 14 α(H), 17 α(H)-Cholestane	A, RD, AS	0.8
20S&R-5 α(H), 14 β(H), 17 β(H)-Ergostanes	A, RD, AS, B	0.8
20S&R-5 α(H), 14 β(H), 17 β (H)-Sitostanes	A, RD, AS, B	1.0
Total		3.2
Pentacyclic triterpanes		
22, 29, 30-Trisnorheohopane	A, RD, AS, B	0.4
17 α(H), 21 β(H)-29-Norhopane	A, RD, AS, B	1.0
17 α(H), 21 β(H)-Hopane	A, RD, AS, B	1.5
22S-17 α(H), 21 β(H)-30-Homohopane	A, RD, AS, B	0.6
22R-17α(H), 21 β(H)-30-Homohopane	A, RD, AS, B	0.4
22S-17 α(H), 21 β(H)-30-Bishomohopane	A, RD, B	0.4
22R-17 α(H), 21 β(H)-30-Bishomohopane	A, RD, B	0.3
Total		4.6
Iso-and anteisoalkanes		
Isotriacontane	V, C	0.3
Isohentriacontane	V, C	1.3
Isodotriacontane	C	0.1
Anteisotriacontane	V, C	0.2
Anteisohentriacontane	V, C	0.1
Anteisodotriacontane	V, C	0.9
Total		2.9

[a]*Key*: M, meat cooking; A, automobiles (leaded, unleaded and trucks); RD, road dust (includes tire wear and brake lining particles); V, vegetation; G, natural gas home appliances; C, cigarette smoke; AS, asphalt; B, boilers; W, wood burning (oak, pine).

ambient mass concentrations in the Los Angeles air basin. These ambient concentrations are the cumulative results of a variety of primary sources and secondary aerosol production. Normal alkanoic acids, aliphatic dicarboxylic acids, and aromatic polycarboxylic acids are the major constituents of the resolved urban aerosol mass with annual average concentrations of 0.25–0.3, 0.2–0.3, and around 0.1 µg m^{-3}, respectively.

A set of tracer compounds have been proposed for the identification and quantification of the contributions of a number of sources to ambient aerosol concentration levels. These sets of compounds are summarized in Table 14.9. For example, cholesterol concentrations can be used to estimate the contribution of meat cooking operations to the ambient OC aerosol in a given area (see Problem 14.1).

14.4.3 Primary OC Size Distribution

The mass distribution of primary particles emitted from boilers, fireplaces, automobiles, diesel trucks, and meat cooking operations is dominated by a mode around 0.1–0.2 µm in diameter (Hildemann et al. 1991b). Representative distributions measured by these authors using a dilution stack system are shown in Figures 14.2 and 14.10. Most of the observed

TABLE 14.9   Tracer Compounds for Atmospheric Primary Organic Aerosol Sources

Source	Compounds	Reference
Meat cooking	Cholesterol	Rogge et al. (1991)
	Myristic acid (*n*-tetradecanoic acid) ⎫	
	Palmitic acid (*n*-hexanoic acid) ⎪	
	Stearic acid (*n*-octadecanoic acid) ⎬ Rogge et al. (1991)	
	Oleic acid (*cis*-9-octadecenoic acid) ⎪	
	Nonanal ⎪	
	2-Decanone ⎭	
Tire wear	Benzothiazole	Kim et al. (1990)
		Rogge et al. (1993b)
Automobiles	Steranes ⎫	
	Pentacyclic triterpanes (hopanes) ⎬	Rogge et al. (1993a)
Biogenics	C27 *n*-alkane ⎫	
	C29 *n*-alkane ⎪	Mazurek and Simoneit (1984)
	C31 *n*-alkane ⎬	Simoneit (1984)
	C33 *n*-alkane ⎭	Rogge et al. (1993c)
Cigarette smoke	Iso-alkanes ⎫	
	Anteisoalkanes ⎬	Rogge et al. (1994)

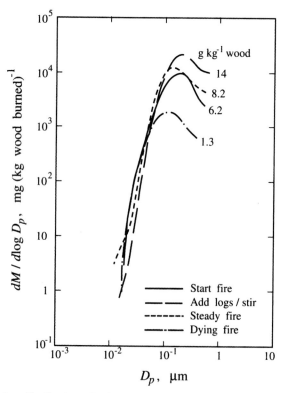

FIGURE 14.10   Mass distribution of primary organic aerosol emitted by fireplaces during the various stages of fire development (Hildemann et al. 1991b). (Reprinted from *Aerosol Sci. Technol.* **14**, Hildemann et al., 138–152. Copyright 1991, with kind permission from Elsevier Science Ltd., The Boulevard, Langford Lane, Kidlington OX5 1GB, UK.)

mass distributions are unimodal and 90% or more of the emitted mass is in the form of submicrometer particles (U.S. EPA 1996).

14.5 SECONDARY ORGANIC CARBON

14.5.1 Overview of Secondary Organic Aerosol Formation Pathways

Secondary organic aerosol material is formed in the atmosphere by the mass transfer to the aerosol phase of low vapor pressure products of the oxidation of organic gases. As organic gases are oxidized in the gas phase by species such as the hydroxyl radical (OH), ozone (O_3), and the nitrate radical (NO_3), their oxidation products accumulate. Some of these products have low volatilities and condense on the available particles in an effort to establish equilibrium between the gas and aerosol phases.

Let us consider as an example the reaction of cyclohexene with ozone in the atmosphere. This reaction has been studied in laboratory chamber experiments by Kalberer et al. (2000). A potential reaction mechanism is depicted in Figure 14.11. The first steps of the reaction are the formation of an initial molozonide M, its transformation to a peroxy radical intermediate, and then to a dioxyrane-type intermediate, and finally to an excited Criegee biradical $[CHO(CH_2)_4CHO\dot{O}]^*$. A series of reactions then lead from the Criegee biradical to stable products, some with five and some with six carbon atoms (Figure 14.11). This rather complicated series of reactions leads to the stable gas-phase products listed in Table 14.10.

The mass transfer flux of these products to the aerosol phase, J will be proportional to the difference between their gas-phase concentration c_g and their concentration in the gas phase at the particle surface c_{eq}:

$$J \sim c_g - c_{eq} \tag{14.5}$$

TABLE 14.10 Products of the Cyclohexene–Ozone Reaction

Compound	Total Average Molar Yield, %
Oxalic acid	6.16
Malonic acid	6.88
Succinic acid	0.63
Glutaric acid	5.89
Adipic acid	2.20
4-Hydroxybutanal	2.60
Hydroxypentanoic acid	1.02
Hydroxyglutaric acid	2.33
Hydroxyadipic acid	1.19
4-Oxobutanoic acid	6.90
5-Oxopentanoic acid	4.52
6-Oxohexanoic acid	4.16
1,4-butanedial	0.53
1,5-pentanedial	0.44
1,6-hexanedial	1.64
Pentanal	17.05
Total	64.14

Source: Kalberer et al. (2000).

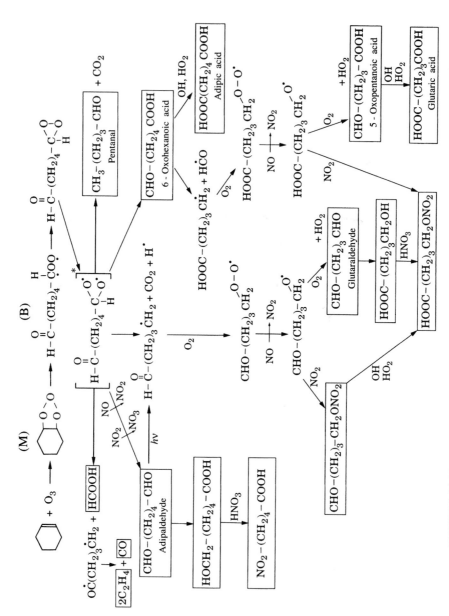

FIGURE 14.11 Proposed gas-phase reaction mechanism for the cyclohexene-ozone reaction.

If the two concentrations are equal, the organic compound will be in equilibrium, and there will be no net transfer between the two phases. If $c_g > c_{eq}$, then the gas phase is supersaturated with the compound and some of it will condense on the available aerosol. The equilibrium concentration c_{eq} depends not only on the properties of the species but also on its ability to form solutions with compounds already present in the aerosol phase (other organics, water, etc.). Let us assume for the moment that the preexisting aerosol particles are solid inorganic salts like ammonium sulfate and the organic products cannot form a solution. In that case c_{eq} will be equal to the pure component saturation concentration and if the concentration of such a product is smaller than c_{eq}, the species will remain in the gas phase. For cyclohexene, for example, some of its oxidation products like pentanal and formic acid have very high vapor pressures (Table 14.10) and will remain in the gas phase. Their gas-phase concentrations increase as more and more cyclohexene reacts, but the capacity of the gas phase to "carry" these species does not become saturated; products like pentanal will tend to reside virtually exclusively in the gas phase. On the other end of the volatility spectrum, products like adipic acid have low vapor pressures. Let us assume a molar yield for adipic acid equal to 0.02: that is, for every 100 molecules of cyclohexene that react, two molecules of adipic acid are formed. As cyclohexene keeps reacting with ozone, the concentration of adipic acid will start increasing. After the reaction of 1 ppb of cyclohexene in a closed system, there will be 0.02 ppb of adipic acid present. This concentration is lower than the saturation mixing ratio of adipic acid (0.08 ppb), so all adipic acid will still remain in the gas phase. If more cyclohexene reacts, the adipic acid gas-phase concentration will increase further. When the adipic acid mixing ratio reaches 0.08 ppb (i.e., after 4 ppb of cyclohexene has reacted), the gas phase will become saturated with the acid. Further cyclohexene reaction will lead to supersaturation of the gas phase in adipic acid and the excess will condense onto any available aerosol particles or homogeneously nucleate, leading to the production of secondary organic aerosol material. This simplified mechanism of secondary organic aerosol formation is depicted schematically in Figure 14.12.

The preceding description of aerosol formation from the cyclohexene–ozone reaction reflected the presumed mechanism of secondary organic aerosol formation for a number of years. In actuality, when organic species are already present in particles, as they frequently

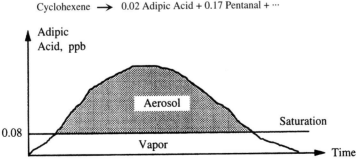

FIGURE 14.12 Schematic of the formation of secondary organic aerosol during the cyclohexene oxidation by mass transfer of adipic acid to the aerosol phase. This case corresponds to a single-component secondary organic aerosol phase without any interactions with the rest of the aerosol species.

are, organic vapors tend to dissolve in the particle-phase organics. As a result, the gas–particle partitioning of organic vapor species is determined by the molecular properties of the species and the organic phase and the amount of the particulate organic phase available for the vapor-phase compound to dissolve into (Pankow 1994a,b; Odum et al. 1996, 1997a,b; Bowman et al. 1997). This dissolution exhibits no threshold concentration. In the absence of an organic particulate phase, adsorption can serve as the mechanism to initiate gas–particle partitioning (Yamasaki et al. 1982; Pankow 1987; Ligocki and Pankow 1989; Pankow and Bidleman 1991).

Summarizing, there are two separate steps involved in the production of secondary organic aerosol. First, the organic aerosol compound is produced in the gas phase during the reaction of parent organic gases. Then, the organic compound partitions between the gas and particulate phases, forming secondary organic aerosol. The organic compound gas-phase production depends on the gas-phase chemistry of the organic aerosol precursor, while the partitioning is a physicochemical process that may involve interactions among the various compounds present in both phases. We will focus on the partitioning processes in the subsequent sections as the gas-phase oxidation of organic compounds has been discussed in Chapter 6.

14.5.2   Dissolution and Gas–Particle Partitioning of Organic Compounds

The thermodynamic principles introduced in Chapter 10 can be used to investigate the partitioning of organic compounds between the gas and aerosol phases. We will first focus on the analysis of a series of idealized scenarios and then discuss their applicability to the atmosphere.

Noninteracting Secondary Organic Aerosol Compounds   The simplest case is that of a compound that does not interact with already existing aerosol components. Such a compound does not adsorb on the aerosol surface or form solutions with the existing aerosol and gas-phase species. Let us assume that a reactive organic gas (ROG) undergoes atmospheric oxidation to produce products, P_1, P_2, and so on

$$\text{ROG} \rightarrow a_1\,P_1 + a_2\,P_2 + \cdots$$

where a_i is the molar yield (moles produced per mole of ROG reacted) of product P_i.

If $c_{t,i}$, $c_{g,i}$, and $c_{\text{aer},i}$, are the total, gas-phase, and aerosol-phase concentrations in $\mu\mathrm{g\,m^{-3}}$ of product i, then

$$c_{t,i} = c_{g,i} + c_{\text{aer},i} \tag{14.6}$$

With noninteracting aerosol components, P_i exists in the aerosol phase as pure P_i. Thus gas–aerosol equilibrium is reached for P_i when the gas-phase concentration, $c_{g,i}$, is that corresponding to the vapor pressure of pure i at that temperature, $c_{\text{eq},i} = p_i^\circ M_i / RT$, where M_i is its molecular weight. Thus at equilibrium

$$c_{g,i} = c_{\text{eq},i} = \frac{p_i^\circ M_i}{RT} \tag{14.7}$$

and

$$c_{t,i} = c_{eq,i} + c_{aer,i} \qquad (14.8)$$

If $c_{t,i} < c_{eq,i}$ the equilibrium condition of (14.8) cannot be satisfied because it would require negative aerosol concentrations. Therefore the aerosol concentration of species i will be

$$c_{aer,i} = 0 \qquad \text{if } c_{t,i} \le c_{eq,i} \qquad (14.9)$$
$$c_{aer,i} = c_{t,i} - c_{eq,i} \qquad \text{if } c_{t,i} > c_{eq,i} \qquad (14.10)$$

The total concentration of a product can be calculated using the reacted precursor concentration ΔROG (in $\mu g \, m^{-3}$) and the reaction stoichiometry

$$c_{t,i} = a_i \frac{M_i}{M_{ROG}} \Delta ROG \qquad (14.11)$$

where M_{ROG} is the molecular weight of the parent compound. Combining (14.10) with (14.11), we find that

$$
\begin{aligned}
c_{aer,i} &= 0 \qquad \text{if } \Delta ROG \le \Delta ROG_i^* \\
c_{aer,i} &= a_i \frac{M_i}{M_{ROG}} \Delta ROG - c_{eq,i} \qquad \text{if } \Delta ROG > \Delta ROG_i^*
\end{aligned}
\qquad (14.12)
$$

where ΔROG_i^* is the threshold reacted precursor concentration for formation of the aerosol compound i calculated from (14.7), (14.11) and (14.12) as

$$\Delta ROG_i^* = \frac{p_i^\circ M_{ROG}}{a_i RT} \qquad (14.13)$$

The lower the vapor pressure of the compound and the higher its molar yield, the smaller the amount of reacted ROG needed to saturate the gas phase and initiate the formation of secondary organic aerosol. For a compound with a vapor pressure of 10^{-10} atm (0.1 ppb), produced by a ROG with $M_{ROG} = 100 \, g \, mol^{-1}$, and a yield equal to 0.1; $\Delta ROG^* = 4.1 \, \mu g \, m^{-3}$.

Partitioning of Noninteracting SOA Compounds Find the mass fraction of the secondary organic aerosol species i that will exist in the particulate phase, $X_{p,i}$, as a function of reacted ROG. Plot the $X_{p,i}$ versus ΔROG curves for $a_i = 0.05$, $T = 298 \, K$, and $M_{ROG} = 150 \, g \, mol^{-1}$ for saturation mixing ratios equal to 0.1, 1, and 5 ppb.

Defining $X_{p,i}$ as the mass fraction of the species in the aerosol phase

$$X_{p,i} = \frac{c_{aer,i}}{c_{t,i}} \qquad (14.14)$$

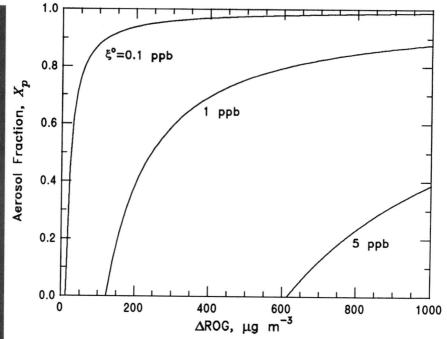

FIGURE 14.13 Mass fraction of ROG oxidation product in the aerosol phase as a function of the reacted ROG concentration and the product saturation mixing ratio. The product is assumed to be insoluble in the aerosol phase. Other conditions are $a_i = 0.05$, $T = 298$ K, and $M_{ROG} = 150$ g mol^{-1}.

we find that for this case

$$X_{p,i} = 0 \qquad \text{if } \Delta ROG \leq \Delta ROG_i^*$$

$$X_{p,i} = 1 - \frac{p_i^\circ M_{ROG}}{a_i RT \Delta ROG} = 1 - \frac{\Delta ROG_i^*}{\Delta ROG} \qquad \text{if } \Delta ROG > \Delta ROG_i^* \qquad (14.15)$$

This fraction is shown as a function of ΔROG for various saturation mixing ratios and molar yields in Figure 14.13. The organic compound will exist virtually exclusively in the aerosol phase if $\Delta ROG \gg \Delta ROG_i^*$.

Another useful quantity for atmospheric aerosol formation studies is the aerosol mass yield defined as

$$Y_i = \frac{c_{aer,i}}{\Delta ROG} \qquad (14.16)$$

This yield should be contrasted with the molar yield a_i that refers to the total concentration of i. For this case the yield Y_i can be calculated combining (14.16) and (14.12),

$$Y_i = 0 \qquad \text{if } \Delta ROG \leq \Delta ROG_i^*$$

$$Y_i = a_i \frac{M_i}{M_{ROG}} - \frac{p_i^\circ M_i}{RT \Delta ROG} \qquad (14.17)$$

The aerosol yield is therefore a function of the compound concentration that has reacted, starting from zero and approaching asymptotically the mass yield of the compound $a_i M_i / M_{ROG}$.

Aerosol Yield Calculation Find the aerosol mass yield Y_i as a function of the concentration of i in the particulate phase $c_{aer,i}$. What happens when $c_{aer,i} \to \infty$? Plot the aerosol mass yield as a function of ΔROG for the conditions of the previous example.

The aerosol yield can be expressed as a function of the organic aerosol concentration by solving (14.12) for ΔROG and substituting into (14.17) to obtain

$$Y_i = \frac{c_{aer,i} a_i M_i RT}{(c_{aer,i} RT + p_i^\circ M_i) M_{ROG}} \qquad (14.18)$$

If $c_{aer,i} \to \infty$, then from (14.18), $Y_i \to a_i M_i / M_{ROG}$, that is, it approaches the stoichiometric mass yield of the compound. In this case, the mass yield is $a_i M_i / M_{ROG} = 0.06$, so all yield curves approach asymptotically this value (Figure 14.14). For low-vapor-pressure products this asymptotic limit is reached as soon as a small amount of the parent ROG reacts. For more volatile products the actual yield will be less for all conditions relevant to the atmosphere.

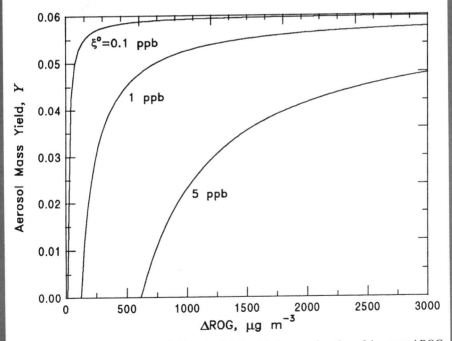

FIGURE 14.14 Aerosol mass yield during ROG oxidation as a function of the reacted ROG concentration and the product saturation mixing ratio. The product is assumed to be insoluble in the aerosol phase. Other conditions are $a_i = 0.05$, $T = 298$ K, $M_{ROG} = 150$ g mol^{-1}, and $M_i = 180$ g mol^{-1}.

Formation of Binary Ideal Solution with Preexisting Aerosol   Let us repeat the analysis assuming that now the organic vapor i produced during the ROG oxidation can form an ideal solution with at least some of the preexisting organic aerosol material. The preexisting organic aerosol mass concentration is m_0 (in $\mu g\,m^{-3}$) and has an average molecular weight M_0. Using the same notation as above, the mole fraction of the species in the aerosol phase x_i will be

$$x_i = \frac{c_{aer,i}/M_i}{(c_{aer,i}/M_i) + (m_0/M_0)} \tag{14.19}$$

If the solution is ideal, then the vapor pressure of i over the solution p_i will satisfy

$$p_i = x_i p_i^\circ \tag{14.20}$$

and converting to concentration units ($\mu g\,m^{-3}$)

$$c_{g,i} = \frac{x_i p_i^\circ M_i}{RT} \tag{14.21}$$

Combining (14.19), (14.21), and (14.11) we can eliminate x_i and $c_{g,i}$ to obtain an expression for the aerosol concentration of i,

$$c_{aer,i} + \frac{M_0 M_i p_i^\circ c_{aer,i}}{RT(M_0 c_{aer,i} + M_i m_0)} = \frac{a_i M_i \,\Delta ROG}{M_{ROG}} \tag{14.22}$$

This equation can be simplified assuming

$$c_{aer,i} \ll m_0 \tag{14.23}$$

that is, compound i is only a minor component of the overall organic phase. This assumption is valid for most atmospheric conditions where no single organic compound dominates the composition of organic aerosol. Using (14.23) and (14.22), we obtain

$$c_{aer,i} = \frac{a_i\,RT}{M_{ROG}} \left(\frac{M_i m_0}{m_0 RT + M_0 p_i^\circ} \right) \Delta ROG \tag{14.24}$$

Note that in this case if $\Delta ROG > 0$, then $c_{aer,i} > 0$, that is, *secondary aerosol is formed as soon as the ROG starts reacting.* In this case the threshold ΔROG_i^* is zero. The aerosol concentration of i is inversely related to its vapor pressure and also depends on the preexisting organic material concentration.

The mass fraction of i in the aerosol phase can be calculated combining (14.24), (14.14), and (14.11) as follows:

$$X_{p,i} = \frac{m_0 RT}{m_0 RT + p_i^\circ M_0} \tag{14.25}$$

For high-organic-aerosol loadings, or very low vapor pressures, $X_{p,i} \simeq 1$, and virtually all i will be in the aerosol phase.

The aerosol mass yield of the ROG reaction defined by (14.16) is then

$$Y_i = \frac{a_i\,RT}{M_{\mathrm{ROG}}} \left(\frac{M_i m_0}{m_0 RT + M_0 p_i^\circ} \right) \qquad (14.26)$$

As expected, for high-organic-aerosol loadings or products of very low volatility, the yield Y_i becomes equal to the stoichiometric mass yield of i in the reaction $a_i M_i / M_{\mathrm{ROG}}$.

Formation of Binary Ideal Solution with Other Organic Vapor If the products of a given ROG cannot form a solution with the existing aerosol material, but they can dissolve in each other, then the formation of secondary aerosol can be described by a modification of the above theories. Assuming that two products are formed

$$\mathrm{ROG} \rightarrow a_1 \mathrm{P}_1 + a_2 \mathrm{P}_2$$

with molecular weights M_1 and M_2, then each compound satisfies (14.6) and (14.11), resulting in

$$c_{g,1} + c_{\mathrm{aer},1} = \frac{a_1 M_1}{M_{\mathrm{ROG}}} \Delta\mathrm{ROG} \qquad (14.27)$$

$$c_{g,2} + c_{\mathrm{aer},2} = \frac{a_2 M_2}{M_{\mathrm{ROG}}} \Delta\mathrm{ROG} \qquad (14.28)$$

Assuming that the two species form an ideal organic solution, their gas-phase concentrations will satisfy

$$c_{g,1} = x_1 \frac{p_1^\circ M_1}{RT} = x_1 c_1^\circ \qquad (14.29)$$

$$c_{g,2} = (1 - x_1) \frac{p_2^\circ M_2}{RT} = x_2 c_2^\circ \qquad (14.30)$$

where x_1 is the mole fraction of the first compound in the binary solution, and c_i° is the saturation concentration in $\mu\mathrm{g\,m^{-3}}$ of pure i. The mole fraction of the second compound is simply $x_2 = 1 - x_1$. The mole fraction x_1 will be given by

$$x_1 = \frac{c_{\mathrm{aer},1}/M_1}{(c_{\mathrm{aer},1}/M_1) + (c_{\mathrm{aer},2}/M_2)} \qquad (14.31)$$

The five equations (14.27) to (14.31) form an algebraic system for the five unknowns $c_{\mathrm{aer},1}$, $c_{\mathrm{aer},2}$, $c_{g,1}$, $c_{g,2}$, and x_1. Combining them we are left with one equation for x_1, describing the equilibrium organic solution composition. Assuming that $M_1 \simeq M_2 \simeq M_{\mathrm{ROG}}$, we have

$$x_1^2 + \frac{(a_1 + a_2)\Delta\mathrm{ROG} - c_2^\circ + c_1^\circ}{c_2^\circ - c_1^\circ} x_1 - \frac{a_1\,\Delta\mathrm{ROG}}{c_2^\circ - c_1^\circ} = 0 \qquad (14.32)$$

One can show that (14.32) always has a real positive solution. However, this solution should not only exist but should satisfy the constraints

$$c_{aer,1} \geq 0 \quad \text{and} \quad c_{aer,2} \geq 0 \tag{14.33}$$

or equivalently

$$a_1 \Delta ROG - x_1 c_1^\circ \geq 0 \tag{14.34}$$

$$a_2 \Delta ROG - c_2^\circ + x_1 c_2^\circ \geq 0 \tag{14.35}$$

Instead of finding the most restrictive constraint let us add (14.34) and (14.35) to simplify the algebraic calculations. The result of this simplification may be a slight underestimation of the threshold ΔROG^*. After summation, we obtain

$$(a_1 + a_2) \Delta ROG \geq x_1 c_1^\circ + c_2^\circ - x_1 c_2^\circ \tag{14.36}$$

For small ΔROG the inequality will not be satisfied and the organic aerosol phase will not exist. However, when $\Delta ROG = \Delta ROG^*$ the equality will hold, or equivalently

$$(a_1 + a_2) \Delta ROG^* = x_1 c_1^\circ + c_2^\circ - x_1 c_2^\circ \tag{14.37}$$

Substituting (14.37) into (14.32) and simplifying, we find the following relationship

$$x_1^* c_1^\circ = a_1 \Delta ROG^* \tag{14.38}$$

which is satisfied at the point of formation of the organic aerosol phase and which relates the threshold concentration of the hydrocarbon with the composition of the organic phase at this point. Combining (14.38) with (14.37) and eliminating the composition variable x_1, we obtain

$$\frac{1}{\Delta ROG^*} = \frac{a_1}{c_1^\circ} + \frac{a_2}{c_2^\circ} \tag{14.39}$$

or in terms of vapor pressures

$$\frac{1}{\Delta ROG^*} = \left(\frac{a_1}{p_1^\circ M_1} + \frac{a_2}{p_2^\circ M_2} \right) RT \tag{14.40}$$

Odum et al. (1996) showed that secondary organic aerosol production by m-xylene at $40°C$ can be represented by the formation of two products with molar yields $a_1 = 0.03$ and $a_2 = 0.17$ and saturation concentrations $c_1^\circ = 31.2\,\mu g\,m^{-3}$ and $c_2^\circ = 525\,\mu g\,m^{-3}$. If the two species do not interact with each other, then (14.13) would be applicable, and assuming that $M_{ROG} \simeq M_i$, the corresponding thresholds would be $\Delta ROG_1^* = 1040\,\mu g\,m^{-3}$ and $\Delta ROG_2^* = 3090\,\mu g\,m^{-3}$, respectively. If the two species form a solution, then from (14.40), $\Delta ROG^* = 780\,\mu g\,m^{-3}$. Therefore, after the reaction of ΔROG^*, a binary solution will be formed containing both compounds with a composition given by (14.38) as $x_1^* = 0.75$. The solution formation lowers the corresponding thresholds significantly and facilitates the creation of the secondary organic aerosol phase.

Note that because we have not used the most restrictive constraint from (14.35) and (14.36) our solution could underestimate the threshold ΔROG^*. An exact calculation would involve repeating the above algorithm first using only (14.35) and then only (14.36), calculating two different ΔROG^* and then selecting the higher value (see Problem 14.4). The value calculated from (14.40) is an accurate approximation for most atmospheric applications.

The aerosol mass yield Y for such a reaction will be given by

$$Y = \frac{c_{aer,1} + c_{aer,2}}{\Delta ROG} \tag{14.41}$$

or defining the total organic aerosol concentration produced during the ROG oxidation, c_{aer}, as

$$c_{aer} = c_{aer,1} + c_{aer,2} \tag{14.42}$$

and assuming that $M_{ROG} \simeq M_1 \simeq M_2$ and combining (14.27) to (14.31) with (14.41) and (14.42), one can show that

$$Y = c_{aer} \left(\frac{a_1}{c_{aer} + c_1^\circ} + \frac{a_2}{c_{aer} + c_2^\circ} \right) \tag{14.43}$$

This equation, first proposed by Odum et al. (1996) in a slightly different form, indicates that for low-organic-mass concentrations the aerosol yield will be directly proportional to the total aerosol organic mass concentration c_{aer}. For very nonvolatile products and/or for large organic mass concentrations, the overall aerosol yield will be independent of the organic mass concentrations and equal to $a_1 + a_2$. Finally, (14.43) suggests that the aerosol yields will be sensitive to temperature since c_1° and c_2° depend exponentially on temperature.

Odum et al. (1996, 1997a,b) showed that laboratory chamber data for the oxidation of aromatic and biogenic hydrocarbons agree remarkably well with the theoretical model of (14.43) (Figure 14.15). This agreement indicates that the formation of secondary organic aerosol can be described by assuming the formation of solutions among the products of the various secondary organic aerosol components. Application of this model to the atmosphere requires knowledge of two parameters for each secondary organic aerosol product, its gas-particle partitioning coefficient $K_{p,i}$, and its stoichiometric yield a_i.

Deviations from the theory outlined above may occur if the organic solution is not ideal. In this case the mole fraction x_i should be replaced by the product $\gamma_i x_i$, where γ_i is the activity coefficient of the component i in the given solution. Saxena et al. (1995) have shown that a series of organic aerosol compounds can interact also with the aqueous phase in atmospheric particles. Therefore, for high relative humidities, the secondary organic aerosol compounds may exist in three phases—namely, the gas, the organic aerosol material, and the aqueous phase.

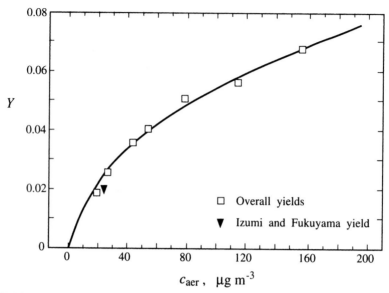

FIGURE 14.15 Measured and predicted [equation (14.43)] aerosol mass yields as a function of the produced aerosol mass for 1,2,4-trimethylbenzene oxidation (Odum et al. 1996). The theoretical line corresponds to $a_1 = 0.03$, $a_2 = 0.17$, $c_1^\circ = 18.9\,\mu g\,m^{-3}$, and $c_2^\circ = 500\,\mu g\,m^{-3}$.

14.5.3 Adsorption and Gas–Particle Partitioning of Organic Compounds

Most particles can adsorb vapor molecules on their surfaces. The interactions between adsorbed molecules and the particle surface are complex, involving both physical and chemical forces. The physics and chemistry of adsorption are quite complicated, and we refer the interested reader to Masel (1996) for detailed coverage of the subject.

The adsorption process involves first the partial covering of the particle surface with vapor molecules, leading to the formation of a monolayer, occasionally followed by the formation of additional layers. Mono- and multilayer adsorption processes often have different characteristics, as in the former the adsorbed molecules interact directly with the particle surface, whereas in the latter they interact mainly with already adsorbed molecules.

The adsorption behavior of a surface is generally characterized by an adsorption isotherm, that is, the functional dependence of the amount of gas adsorbed on the gas partial pressure at constant temperature. The frequently used Langmuir isotherm describes the equilibrium adsorption for the formation of a monolayer; it is based on the following assumptions:

1. All adsorption sites on the surface are equivalent.
2. There are no horizontal interactions among adsorbed molecules.
3. The heat of adsorption for all molecules to any site is the same.

A useful form of this equation is

$$V = V_m \frac{bp}{1 + bp} \qquad (14.44)$$

where V is the volume of gas adsorbed at equilibrium with a gas partial pressure p, V_m is the gas volume necessary to form a monolayer, and b is a constant depending on the surface properties of the adsorbing material. For low partial pressure p, (14.44) predicts that the amount adsorbed will be proportional to the partial pressure. As the partial pressure p increases it eventually exceeds $1/b$ and then as $bp \gg 1$, $V \simeq V_m$ and the monolayer is complete.

One of the most widely used isotherms is based on the BET theory, named from the initials of Brunauer, Emmett, and Teller. The BET theory is an extension of the Langmuir isotherm to include the adsorption of two or more molecular layers assuming, in addition to assumptions 1–3 listed above, that

4. Each molecule adsorbed in a particular layer can be a site for adsorption during the formation of the next layer.
5. The heat of adsorption is equal to the latent heat of evaporation for the bulk condensed gas for all adsorbed layers except the first.

Under these assumptions the BET isotherm can be written as

$$\frac{V}{V_m} = \frac{cS}{(1-S)[1+(c-1)S]} \tag{14.45}$$

where $S = p/p°$ is the gas-phase saturation ratio, p is the partial pressure of the adsorbing vapor, $p°$ is the saturation vapor pressure, and c is a constant for the surface depending on its properties. Note that according to the BET theory, as the gas-phase saturation approaches unity the volume of the adsorbed vapor increases rapidly as multiple layers are formed. The BET theory actually predicts that for $p = p°$ the adsorbed volume approaches infinity. For such conditions one should avoid the BET isotherm and use a more appropriate isotherm, for example, the FHH isotherm based on the work of Frenkel, Halsey, and Hill (Frenkel 1946). This isotherm is given by the relation

$$\ln\left(\frac{p°}{p}\right) = \frac{A}{(V/V_m)^B} \tag{14.46}$$

where A and B are constants for the adsorbing surface. Typical isotherms for the various theories are shown in Figure 14.16. Note that adsorption according to all the theories listed above can lead to transfer of a fraction of the vapor to the aerosol phase, even if the gas phase is not yet saturated with the vapor. These amounts, though, are often very small, consisting of only one or two monolayers.

The above theories are generally not useful to describe a priori adsorption of a vapor on atmospheric aerosols, as they rely on experimental information for determination of which theory is most suitable for describing the substrate–vapor adsorption pair and determination of the appropriate parameters of that theory.

In vapor adsorption on atmospheric particles, one often parametrizes the partitioning of a species between the gas and aerosol phase by introducing the partitioning coefficient defined as (Yamasaki et al. 1982; Pankow 1987)

$$K_p = \frac{c_{\text{aer}}}{M_t c_g} \tag{14.47}$$

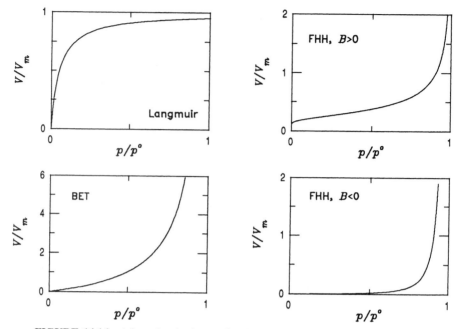

FIGURE 14.16   Adsorption isotherms for the Langmuir, BET, and FHH theories.

where K_p is a temperature-dependent partitioning coefficient ($m^3 \, \mu g^{-1}$), M_t is the total ambient aerosol mass concentration ($\mu g \, m^{-3}$), and c_{aer} and c_g are the aerosol- and gas-phase concentrations of the species partitioned between the two phases ($\mu g \, m^{-3}$). Note that this equation does not assume anything about the mechanism leading to the gas-to-particle partitioning, which could be adsorption, absorption, or a combination of both. The ratio c_{aer}/c_g is that of the particulate-associated organic compound to the gas-phase organic compound. From (14.47) $c_{aer}/c_g = K_p M_t$, so the particle-to-gas partitioning is the product of a partitioning constant and the total particulate mass concentration. The physical reasoning behind (14.47) is that organic compounds are adsorbed onto and/or absorbed into particulate matter and the amount adsorbed or absorbed depends on the total amount of particulate matter present. A larger value of K_p implies more of the organic compound resides in the particulate phase relative to the gas phase, at a given ambient aerosol mass concentration. As the aerosol mass concentration increases, more of the organic compound will be in the particulate phase because of the greater surface and volume of particulate matter available.

Pankow (1987) has shown that for physical adsorption the partitioning coefficient K_p is given by

$$K_p = \frac{N_s a_s T}{1600 \, p^\circ} \exp\left(\frac{Q_l - Q_v}{RT}\right) \qquad (14.48)$$

where N_s is the surface concentration of sorption sites (sites cm^{-2}), a_s is the specific area of the aerosol ($m^2 \, g^{-1}$), Q_l is the enthalpy of desorption from the surface (kJ mol^{-1}), Q_v is the enthalpy of vaporization of the compound as a liquid (kJ mol^{-1}), R is the gas constant, T is the temperature (in K), and p° (in Torr) is the vapor pressure of the compound as a liquid (subcooled if the organic compound is solid at temperature T). On the basis of this theoretical framework, one can fit experimentally determined partitioning data by expressions of the form

$$\log_{10}(K_p) = m \log_{10}(p^\circ) + b_r \tag{14.49}$$

Note that according to (14.48) $m = -1$, behavior that has been observed in several field and laboratory studies of specific compound classes (Pankow and Bidleman 1991). These authors suggest that $b_r = -7.8$ for PAHs, and $b_r = -8.5$ for organochlorines.

Gas–particle partitioning is described by the gas–particle partitioning coefficient K_p, given by (14.47). One question that arises with respect to the ambient atmosphere is what is the mechanism that controls gas–particle partitioning of typical semivolatile organic compounds in the atmosphere–adsorption or absorption. Liang et al. (1997) measured K_p values in an outdoor smog chamber for a group of polycyclic aromatic hydrocarbons (PAHs) and n-alkanes sorbing to three types of model aerosol materials: solid $(NH_4)_2SO_4$, liquid dioctyl phthalate, and secondary organic aerosol generated from the photooxidation of whole gasoline vapor. K_p values were also measured for n-alkanes sorbing to ambient aerosol during summertime smog episodes in Pasadena, CA. They showed that gas/particle partitioning of semivolatile organic compounds to urban aerosol is dominated by absorption into the organic fraction of the aerosol. Thus, even for a purely inorganic aerosol, once an organic layer is initiated, gas–particle partitioning is governed by absorption. Moreover, the absorption properties of the ambient smog aerosol and that generated in the outdoor smog chamber were found to be nearly identical. Similarities observed between ambient smog aerosol and chamber-generated secondary organic aerosol support the use of secondary organic aerosol yield data from smog chamber studies to predict ambient secondary organic aerosol formation. Dioctyl phthalate was found to be a good surrogate for ambient aerosol consisting mainly of organic compounds from primary emissions. The measured K_p values for nine n-alkanes and eight PAHs sorbing to the three classes of aerosol conformed to (14.49) with $m = -1$.

14.5.4 Precursor Volatile Organic Compounds

The ability of a given volatile organic compound (VOC) to produce during its atmospheric oxidation secondary organic aerosol depends on three factors:

1. The volatility of its oxidation products
2. Its atmospheric abundance
3. Its chemical reactivity

Oxidation of VOCs leads to the formation of more highly substituted and therefore lower volatility reaction products. The reduction in volatility is due mainly to the fact that adding oxygen and/or nitrogen to organic molecules reduces volatility (Seinfeld and Pankow 2003). Addition of carboxylic acid, alcohol, aldehyde, ketone, alkyl nitrate, and nitro groups to the precursor VOC can reduce its volatility by several orders of magnitude (see Section 14.5.1). The reactions of VOCs with O_3, OH, and NO_3 can all lead to SOA formation in the atmosphere.

The ability of almost all such oxidation products to dissolve in the organic particulate phase suggests that, in theory at least, every VOC contributes to SOA formation [see, e.g., (14.24), where, for any $\Delta ROG > 0$, some nonzero SOA is formed]. However, the contributions of small VOCs to atmospheric SOA are negligible owing to the relatively high vapor pressures of their products. 1-Butene is a good example. Its least volatile oxidation product is propionic acid with a saturation mixing ratio at 298 K of several hundred ppm. For a compound that is so volatile, all the mechanisms discussed above lead to only negligible amounts in the particulate phase, and therefore propionic acid will remain almost exclusively in the gas phase (see Problem 14.2). As a result, oxidation of 1-butene does not form organic aerosol even at its highest atmospheric concentrations. Organic aerosol precursors should also react sufficiently rapidly (at least as fast as their atmospheric dilution rate) so that their products can accumulate. Usually this is not a critical constraint as species that produce low-vapor-pressure products are generally reactive. Finally, enough precursor should be available in the atmosphere so that its products can reach sufficiently high concentrations in both the gas and particulate phases.

As a result of the above constraints, VOCs that for all practical purposes do not produce organic aerosol in the atmosphere include all alkanes with up to six carbon atoms (from methane to hexane isomers), all alkenes with up to six carbon atoms (from ethene to hexene isomers), and most other low-molecular-weight compounds. An important exception is isoprene, a five-carbon atom molecule with two double bonds (see Section 6.11); isoprene photooxidation produces SOA in laboratory chambers (Kroll et al. 2005, 2006) and its oxidation products have been detected in ambient aerosols (Claeys et al., 2004a; Edney et al. 2005). In general, large VOCs containing one or more double bonds are expected to be good SOA precursors. A set of structure–SOA formation potential relationships have been proposed by Keywood et al. (2004) for cycloalkenes:

- Yields increase as the number of carbon atoms in the cycloalkene ring increase (Figure 14.17).
- Yield is enhanced by the presence of a methyl group at a double-bonded carbon site.
- Yield is reduced by the presence of a methyl group at a non-double-bonded carbon site.
- The presence of an exocyclic double bond results in a reduction in yield compared with the equivalent methylated (at the double bond) cycloalkene.

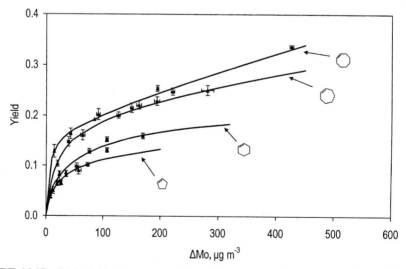

FIGURE 14.17 SOA yield data as a function of the produced organic aerosol mass for four cycloalkenes (Keywood et al. 2004).

Aromatic hydrocarbons are expected to be the most significant anthropogenic SOA precursors (Pandis et al. 1992) and the dominant component of SOA in large urban areas. Odum et al. (1997) showed that SOA formation during the photooxidation of gasoline vapors was determined by the aromatic content of the fuel. Monoterpene, and potentially sesquiterpene, and isoprene oxidation products are estimated to be the major contributors to SOA concentrations in rural forested areas and on a global scale. A list of the major SOA precursors that have been investigated in laboratory chamber studies is given in Table 14.11. Additional detailed information about the emissions, gas-phase chemistry, and kinetics of the oxidation of the major SOA precursors can be found in review papers by Kanakidou et al. (2005), Seinfeld and Pankow (2003), and Jacobson et al. (2000) and references cited therein.

TABLE 14.11 SOA Precursors and Laboratory Studies

	Type of Reaction	References
Aromatics		
Toluene	Photooxidation	Kleindienst et al. (1999, 2004), Jang and Kamens (2001), Forstner et al. (1997b), Edney et al. (2000, 2001), Hurley et al. (2001)
m-Xylene	Photooxidation	Cocker et al. (2001a), Odum et al. (1997a,b), Forstner et al. (1997b)
o-Xylene	Photooxidation	Odum et al. (1997a,b)
p-Xylene	Photooxidation	Kleindienst et al. (1999), Forstner et al. (1997b)
Ethylbenzene	Photooxidation	Forstner et al. (1997b)
1,2,4-Trimethylbenzene	Photooxidation	Odum et al. (1997a,b), Forstner et al. (1997b)
1,3,5-Trimethylbenzene	Photooxidation	Kleindienst et al. (1999), Cocker et al. (2001a)
n-Propylbenzene	Photooxidation	Odum et al. (1997a,b)
1-Methyl-3-n-propylbenzene	Photooxidation	Odum et al. (1997a,b)
1,2-Dimethyl-4-Ethylbenzene	Photooxidation	Odum et al. (1997a,b)
1,4-Dimethyl-2-Ethylbenzene	Photooxidation	Odum et al. (1997a,b)
1,2,4,5-Tetramethylbenzene	Photooxidation	Odum et al. (1997a,b)
p-Diethylbenzene	Photooxidation	Odum et al. (1997a,b)
m-Ethyltoluene	Photooxidation	Odum et al. (1997a,b), Forstner et al. (1997b)
o-Ethyltoluene	Photooxidation	Odum et al. (1997a,b)
p-Ethyltoluene	Photooxidation	Odum et al. (1997a,b), Forstner et al. (1997b)
Monoterpenes		
α-Pinene	O_3	Hoffmann et al. (1998) Jang and Kamens (1999), Hoffmann et al. (1997), Cocker et al. (2001b), Yu et al. (1998, 1999) Larsen et al. (2001), Bonn et al. (2002), Presto et al. (2005a,b)
	OH	Griffin et al. (1999)
	Photooxidation	Odum et al. (1996) Kamens and Jaoui (2001)

(continued)

TABLE 14.11   (*Continued*)

	Type of Reaction	References
β-Pinene	OH	Larsen et al. (2001)
	O_3	Yu et al. (1998, 1999)
		Bonn et al. (2002)
	Photooxidation	Griffin et al. (1999)
Δ^3-Carene	OH	Larsen et al. (2001)
	O_3	Yu et al. (1998, 1999)
		Bonn et al. (2002)
	Photooxidation	Griffin et al. (1999)
Sabinene	OH	Larsen et al. (2001)
	O_3	Yu et al. (1998, 1999)
		Bonn et al. (2002)
	Photooxidation	Griffin et al. (1999)
Myrcene	Photooxidation	Griffin et al. (1999)
Ocimene	Photooxidation	Griffin et al. (1999)
Limonene	OH	Larsen et al. (2001)
	Photooxidation	Griffin at al. (1999)
α-Terpinene	Photooxidation	Griffin at al. (1999)
Terpinolene	Photooxidation	Griffin et al. (1999)
10 Terpenes	O_3	Lee et al. (2006)
Other biogenic VOCs		
Isoprene	Photooxidation	Pandis et al. (1991),
		Claeys et al. (2004b),
		Kroll et al. (2005a, 2006)
Linalool	Photooxidation	Griffin et al. (1999)
Terpinene-4-ol	Photooxidation	Griffin et al. (1999)
Sesquiterpenes		
β-Caryophyllene	Photooxidation	Griffin et al. (1999)
		Jaoui et al. (2003)
α-Humulene	Photooxidation	Griffin et al. (1999)
Alkenes		
1-Tetradecene	O_3	Tobias and Ziemann (2000)
Cyclohexene	O_3	Kalberer et al. (2000)
1-Octene	Photooxidation	Forstner et al. (1997a)
1-Decene	Photooxidation	Forstner et al. (1997b)
Cyclic alkenes	O_3	Ziemann (2003)
		Keywood et al. (2004)

14.5.5   SOA Yields

SOA formation in the atmosphere can be viewed as a three step process:

1. Gas-phase chemical reactions producing the SOA compounds from the parent VOC
2. Partitioning of the SOA compounds between the gas and particulate phases
3. Particulate-phase reactions that can convert the SOA species to other chemical compounds.

The gas-phase chemistry is accelerated and therefore the total (gas and particulate phase) concentrations of the SOA compounds increase as the temperature and the sunlight intensity increase. NO_x significantly affects the pathways of reaction of the parent VOC by changing the fractions of the VOC reacting with O_3, OH, and NO_3, but also by affecting various intermediate steps in these pathways (Kroll et al. 2005a; 2006; Presto et al. 2005a,b). For example, under high-NO_x conditions, NO and NO_2 react with organo-peroxy radicals (RO_2) that would otherwise react with other peroxy radicals (RO_2 and HO_2). Relative humidity also affects the rates of important chemical pathways influencing the product distribution. Finally, other VOCs may also affect the product distributions of the parent VOC.

After the SOA compounds are formed, their gas–particle partitioning determines their concentrations in the particulate phase. This partitioning is affected by temperature (lower temperatures shift the partitioning toward the particulate phase), existence of other organics (in general higher organic aerosol concentrations help dissolve a larger fraction of the SOA compounds), and relative humidity (increases in RH lead to higher aerosol liquid water content and increased dissolution of water soluble organic vapors).

The SOA compounds may undergo further reactions in the particulate phase. These may lead to larger molecules (oligomerization) or more oxidized compounds (hetero-geneous oxidation) and a subsequent decrease in volatility and increase in SOA concentrations. However, heterogeneous reactions may also result in breaking of the original SOA compounds to smaller molecules and transfer of this material back to the gas phase. These reactions will be discussed in Section 14.5.8.

The complexity of the three steps outlined above suggests that the same parameter can have multiple effects on the ultimate aerosol yield. For example, increases in temperature may either increase or decrease the aerosol yield depending on the interplay between kinetics and thermodynamics. Strader et al. (1999) suggested that for the same scenario (same emissions and atmospheric dilution) there may be an optimum temperature for SOA formation for each area. In general, aerosol yields are expected to increase as additional amounts of the parent VOC react in the same air parcel and in cases where there is significant preexisting organic aerosol. This link between SOA precursors or between primary and secondary organic aerosol components has some interesting consequences. For example, increases of the anthropogenic organic aerosol concentrations (both primary and secondary) can facilitate the transfer of semivolatile biogenic vapors to the organic aerosol phase, increasing indirectly the biogenic SOA concentration (Kanakidou et al. 2000).

The two-product model of Odum et al. (1997) [equation (14.43)] has been used to fit laboratory smog chamber yield data for over 50 parent VOCs. Some typical sets of parameters are given in Table 14.12. While the model combines simplicity with the ability to reproduce accurately the laboratory measurements, it also has some limitations. Oxidation of even the simplest SOA precursor leads to tens of different products, each with different properties. It is therefore expected that changes of the oxidant levels, precursor concentrations, temperature, relative humidity, and other parameters will lead to a different product distribution than that for which the model parameters were derived. While the two-product model is expected to be relatively accurate for the experimental conditions used for its derivation, its predictions are expected to be uncertain by a factor of ≥ 2 for significantly different scenarios.

14.5.6 Chemical Composition

SOA compounds formed by gas-phase photochemical reactions of hydrocarbons, ozone, and nitrogen oxides have been identified in both urban and rural atmospheres. Most of these species are di- or polyfunctionally substituted compounds. These compounds

TABLE 14.12   Parameters for SOA Production by Selected VOCs

Precursor	Oxidant	Yield[a]		Saturation Mixing Ratio ($\mu g\,m^{-3}$)[b]	
		1	2	1	2
Toluene	OH	0.071	0.138	18.9	526
Xylenes	OH	0.038	0.167	23.8	714
α-Pinene	OH	0.038	0.326	5.8	250
	O_3	0.125	0.102	11.4	12.7
β-Pinene	OH	0.13	0.041	22.7	204
	O_3	0.21	0.49	5.1	333
Δ^3-Carene	OH	0.054	0.517	23.2	238
	O_3	0.128	0.068	7.8	14.7
	NO_3	0.743	0.257	113	120
Caryophyllene	OH	1.0		23	
Humulene	OH	1.0		20	

[a]Molar stoichiometric yields derived assuming that the precursor and the SOA species have the same molecular weight.
[b]Based on experiments at around 310 K.
Source: Griffin et al. (1999).

include dicarboxylic acids, aliphatic organic nitrates, carboxylic acids derived from aromatic hydrocarbons (benzoic and phenylacetic acids), polysubstituted phenols, and nitroaromatics from aromatic hydrocarbons (Kawamura et al. 1985; Satsumakayashi et al. 1989, 1990). Some species that have been identified as SOA components in laboratory and ambient aerosol are listed in Table 14.13. Despite the abovementioned studies, the available information about the molecular composition of atmospheric SOA remains incomplete. Reaction mechanisms leading to the observed products are to a great extent speculative at present for most precursors. Studies that have focused on specific hydrocarbons can be found in Table 14.11.

14.5.7   Physical Properties of SOA Components

Calculation of the partitioning of SOA components requires knowledge of their saturation concentrations and their dependence on temperature. However, there are only a few direct measurements of these saturation concentrations (Table 14.14). Asher et al. (2002) have proposed a UNIFAC-based group contribution method for estimating the supercooled liquid vapor pressure of oxygen-containing compounds of intermediate to low volatility. Such approaches are expected to have uncertainties of a factor 2-3. Saturation concentrations of condensable products from the oxidation of some aromatic hydrocarbons (toluene, *m*-xylene, and 1,3,5-trimethylbenzene) were estimated to lie in the range of 3–30 ppt (Seinfeld et al. 1987).

14.5.8   Particle-Phase Chemistry

The traditional view of SOA formation was that once the VOC oxidation products have condensed into the particulate phase, no further chemical reactions occur. Recently, it has

TABLE 14.13   Secondary Organic Aerosol Precursors and Aerosol Products[a]

Precursor	Major Products[b]		

α-Pinene   Pinonic acid   Pinic acid   Pinonaldehyde

β-Pinene   Norpinonic acid   Pinic acid   3-Oxo-pina ketone

Limonene   Limonic acid   Limonaldehyde   7-Hydroxylimonic acid

β-Carophyllene   β-Carophyllonic acid   β-Carophyllon aldehyde   Keto-β-carophyllon aldehyde

Toluene   Methyl glyoxal   2-Hydroxy-3-oxo butanal   2-Oxo propane-1,3-dial

Cyclohexene   Adipic acid   Glutaric acid   2-Hydroxyglutaric acid

[a]Vertices represent carbon atoms and bonds between carbon atoms are shown. Hydrogen atoms bonded to carbons are not explicitly indicated. Stereochemistry is omitted for simplicity.

[b]A sampling of products are listed here; actual product distribution depends on oxidant type and concentration, precursor concentration, and atmospheric conditions.

TABLE 14.14   Vapor Pressures of Selected SOA Compounds

Compound	Saturation Mixing Ratio, ppb	Temperature, K	Reference
Tetradecanoic acid	1.9	298	Tao and McMurry (1989)
Pentadecanoic acid	1.0	298	Tao and McMurry (1989)
Hexadecanoic acid	0.1	298	Tao and McMurry (1989)
Heptadecanoic acid	0.1	298	Tao and McMurry (1989)
Malonic acid (C3)	5.2	298	Bilde et al. (2003)
Succinic acid (C4)	0.4	298	Bilde et al. (2003)
Glutaric acid (C5)	7.6	296	Tao and McMurry (1989)
	6.1	296	Bilde and Pandis (2001)
	6.6	298	Bilde et al. (2003)
Adipic acid (C6)	0.14	298	Tao and McMurry (1989)
	0.14	298	Bilde et al. (2003)
Pimelic acid (C7)	1.0	298	Bilde et al. (2003)
Suberic acid (C8)	0.02	298	Bilde et al. (2003)
Azelaic acid (C9)	0.09	298	Bilde et al. (2003)
Octadecanoic acid	0.005	298	Tao and McMurry (1989)
Pinonaldehyde	42×10^3	296	Hallquist et al. (1997)
Pinic acid	0.3	296	Bilde and Pandis (2001)
Cis-pinonic acid	0.7	296	Bilde and Pandis (2001)
Caronaldehyde	23×10^3	296	Hallquist et al. (1997)
Trans-norpinic acid	1.3	296	Bilde and Pandis (2001)
Methyl-malonic acid	8.1	298	Monster et al. (2004)
Dimethyl-malonic acid	2.6	298	Monster et al. (2004)
2-Methyl succinic acid	15.5	298	Monster et al. (2004)
2,2-Dimethyl succinic acid	15.9	298	Monster et al. (2004)
3-Methyl glutaric acid	7	298	Monster et al. (2004)
3,3-Dimethyl glutaric acid	11.6	298	Monster et al. (2004)
3-Methyl adipic acid	1.3	298	Monster et al. (2004)

been shown that the particulate SOA phase often contains high-molecular-weight species that have the nature of oligomers (dimers, trimers, tetramers, etc. that form from the VOC oxidation products) (Kalberer et al. 2004; Gao et al. 2004a,b; Tolocka et al. 2004). These compounds have been shown by mass spectrometry to have molecular weights as high as 1000. Oligomerization takes place during the oxidation of both anthropogenic and biogenic VOCs and appears to be enhanced by the presence of strong acids like sulfuric acid (Jang et al., 2002, 2003; Gao et al. 2004a; Iinuma et al., 2004). However, oligomers are formed even in the absence of inorganic acids. The overall mechanism of SOA formation now must include particle-phase reactions involving the condensed SOA products; one effect of this is that the SOA products that are initially relatively nonvolatile are converted in the particle phase into high-molecular-weight species of virtually zero volatility (Kroll and Seinfeld, 2005b). The exact mechanisms of particulate-phase SOA reactions are under study (Barsanti and Pankow 2004, 2005; Tong et al., 2006).

Figure 14.18 shows two important features of SOA formation. Both panels present mass spectra of SOA. The upper panel is that for SOA formed from photooxidation of isoprene. Note the presence of compounds with molecular weight of ≤ 800. As indicated, about 19% of the quantified mass is oligomeric species. The lower panel is the mass

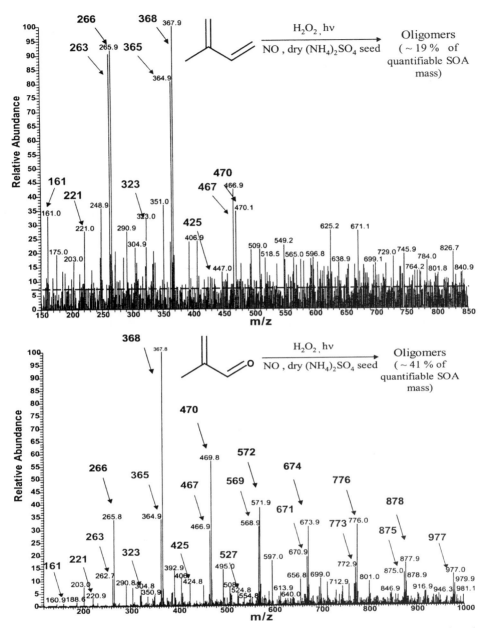

FIGURE 14.18   Mass spectra of SOA formed from photooxidation of isoprene (upper panel) and methacrolein (lower panel). Hydroxyl radicals were produced by photodissociation of H_2O_2. Experiment carried out in laboratory chamber at the California Institute of Technology.

spectrum of SOA formed from the photooxidation of methacrolein, one of the two main products of isoprene oxidation (see Section 6.11). The close correspondence between the mass spectra of isoprene and methacrolein SOA shows that the formation of isoprene SOA proceeds first to methacrolein, as the first-generation product, and then to more condensable species through the subsequent oxidation of methacrolein. (Oxidation of methyl vinyl ketone, the other main product of isoprene photooxidation, does not lead to SOA.) Therefore, besides the presence of oligomeric species, Figure 14.18 shows that the route to SOA formation often proceeds through a volatile first-generation oxidation product that itself must undergo further oxidation to produce the low volatility species that condense to form SOA (Ng et al. 2006).

14.6   POLYCYCLIC AROMATIC HYDROCARBONS (PAHs)

Polycyclic aromatic hydrocarbons (PAHs) were one of the first atmospheric species to be identified as being carcinogenic. Formed during the incomplete combustion of organic matter, for example, coal, oil, wood, and gasoline fuel, PAHs consist of two or more fused benzene rings in linear, angular, or cluster arrangements and, by definition, they contain only carbon and hydrogen (Table 14.15).

This class of compounds is ubiquitous in the atmosphere with more than 100 PAH compounds having been identified in urban air. PAHs observed in the atmosphere range from bicyclic species such as naphthalene, present mainly in the gas phase, to PAHs containing seven or more fused rings, such as coronene, which are present exclusively in the aerosol phase. Intermediate PAHs such as pyrene and anthracene are distributed in both the gas and aerosol phases. The major compounds that have been identified in the atmosphere with their abbreviations and some physical properties are shown in Table 14.15 (Baek et al. 1991). The ambient concentrations of PAHs vary from a few $ng\,m^{-3}$, with values as high as $100\,ng\,m^{-3}$ reported close to their sources (traffic, combustion sources). Concentrations are usually higher during the winter compared to the summer months (Table 14.16).

TABLE 14.15   Properties of Some PAHs

PAH	Molecular Weight	Solubility in Water, $\mu g\,L^{-1}$
Naphthalene (NAP)	128.06	31,700
Acenaphthylene (ACE)	152.06	3,930
9H-Fluorene (FLN)	166.08	1,980
Phenanthrene (PHEN)	178.08	1,290
Anthracene (ANTHR)	178.08	73
Pyrene (PYR)	202.08	135
Fluoranthene[a] (FLUR)	202.08	260
Cyclopenta[cd]pyrene[a] (CPY)	226.08	
Benz[a]anthracene[a] (BaA)	228.09	14
Triphenylene (TRI)	228.09	43
Chrysene[a] (CHRY)	228.09	2
Benzo[a]pyrene[a] (BaP)	252.09	0.05
Benzo[e]pyrene[a] (BeP)	252.09	3.8
Perylene (PER)	252.09	0.4
Benzo[ghi]perylene[a] (BghiP)	276.09	0.3
Coronene[a] (COR)	300.09	0.1

[a]Carcinogenic activity.

TABLE 14.16   Concentrations (ng m^{-3}) of Selected PAHs in the Ambient Air in Los Angeles (United States) and Munich (Germany)

	Los Angeles (1981–1982) (Grosjean 1983)		Munich (1978–1979) (Steinmetzer et al. 1984)
Compound	Summer	Winter	Annual
ANTHR	0.0	0.8	—
FLUR	0.8	1.0	1.8–2.0
PYR	1.5	1.7	1.8–2.1
BaA	0.2	0.6	1.5–1.8
CHRY	0.6	1.2	4.3–5.0
BeP	0.1	0.6	2.3–2.7
BbF[a]	0.4	1.2	—
BkF[b]	0.2	0.4	5.8–7.0
BaP	0.2	0.6	2.0–2.2
BghiP	0.3	4.5	1.6–2.0
COR	1.4	4.7	3.5–4.5

[a]Benzo[b]fluoranthene
[b]Benzo[k]fluoranthene

14.6.1   Emission Sources

Whereas some atmospheric PAHs result from natural forest fires and volcanic eruptions (Nikolaou et al. 1984), anthropogenic emissions are the predominant source. Stationary sources (residential heating, industrial processes, open burning, power generation) are estimated to account for roughly 80% of the annual total PAH emissions in the United States with the remainder produced by mobile sources (Peters et al. 1981; Ramdahl et al. 1983). Mobile sources, however, are the major contributors in urban areas (National Academy of Sciences 1983; Freeman and Cattel 1990; Baek et al. 1991). Estimates of the PAH emissions in the United States and Sweden based on the work of Ramdahl et al. (1983) and Peters et al. (1981) are shown in Table 14.17. These are rough estimates and their uncertainty can be as much as an order of magnitude. This uncertainty is evident by the comparison of the two different estimates for the United States.

14.6.2   Size Distributions

Measurements of the size distribution of PAHs indicate that while they are found in the 0.01–0.5-μm-diameter mode of fresh combustion emissions they exhibit a bimodal distribution in ambient urban aerosol, with an additional mode in the 0.5–1.0-μm-diameter range (Venkataraman and Friedlander 1994b). Distributions of PAHs measured inside a tunnel and in the ambient air in Southern California are shown in Figure 14.19.

14.6.3   Atmospheric Chemistry

PAHs adsorbed on the surfaces of combustion-generated particles may undergo chemical transformations leading to formation of products more polar than the parent PAHs (National Academy of Sciences 1983). Several studies have focused on the rates and

TABLE 14.17   Estimated Atmospheric Emissions of Total PAHs by Source Type (tons yr^{-1})

	United States		Sweden
Source Type	Peters et al. (1981)	Ramdahl et al. (1983)	Ramdahl et al. (1983)
Residential heating			
Coal and wood	3939	450	96
Oil and gas	17	930	36
Industrial			
Coke manufacturing	632	2490	277
Asphalt production	5	4	—
Carbon black	3	3	—
Aluminium	3	1000	35
Others	2	—	—
Incineration			
Municipal	—	—	—
Commercial	56	50	2
Open burning			
Coal refuse fires	29	100	—
Agricultural fires	1190	400	1
Forest fires	1478	600	1
Others	1328	—	—
Power generation			
Power plants	13	1	—
Industrial boilers	75	400	7
Mobile sources			
Gasoline engines	2161[a]	2100[a]	33
Diesel engines	105	70	14
Total	11031	8598	502

[a]This figure was not corrected for the cars with emission control devices (estimated reduction of 50%).

products of reactions of PAHs adsorbed on specific substrates and exposed in the dark or in the light to other species (Pitts et al. 1985b; Valerio and Lazzarotto 1985; Atkinson and Aschmann 1986; Behymer and Hites 1988; Coutant et al. 1988; Alebic-Juretic et al. 1990; De Raat et al. 1990; Kamens et al. 1990; Baek et al. 1991).

Benzo[*a*]pyrene (BaP) and other PAHs on a variety of aerosol substrates react with gaseous NO_2 and HNO_3 to form mono- and dinitro-PAHs (Finlayson-Pitts and Pitts 1986). The presence of HNO_3 along with NO_2 is necessary for PAH nitrification. The reaction rate depends strongly on the nature of the aerosol substrate (Ramdahl et al. 1984a), but the qualitative composition of the products does not. Aerosol water is also a favorable medium for heterogeneous PAH nitration reactions (Nielsen et al. 1983). Nielsen (1984) proposed a reactivity classification of PAHs based on chemical and spectroscopic parameters (Table 14.18).

14.6.4   Partitioning between Gas and Aerosol Phases

PAH vapor pressures vary over seven orders of magnitude (Table 14.19). As a result, at 25°C at equilibrium, naphthalene exists exclusively in the gas phase, while BaP, chrysene, and other PAHs with five and six rings exist predominantly in the aerosol phase. The

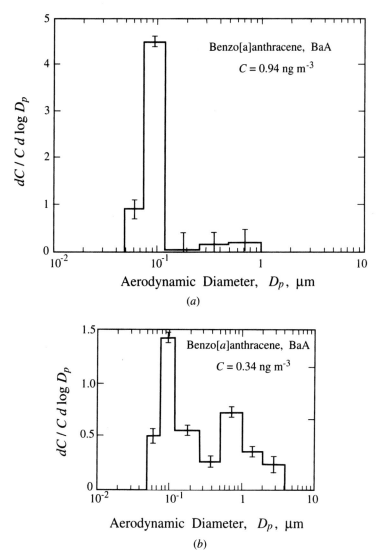

FIGURE 14.19 Size distributions of benzo[*a*]anthracene (a) measured inside a tunnel and (b) in the ambient atmosphere of southern California. (Reprinted with permission from Venkataraman, C., and Friedlander, S. K. Size distributions of polycyclic hydrocarbons and elemental carbon. 1. Sampling, measurement methods and source characterization. *Environ. Sci. Technol.* **28**. Copyright 1994 American Chemical Society.)

distribution of PAHs between the gas and aerosol phases has been parametrized using the partitioning constant K_p given by (14.47). Because of the semiempirical nature of the constant and the fact that the values are based on actual measurements, K_p can be used for estimation of PAH partitioning without actually knowing the controlling partitioning process (adsorption or absorption). For Osaka, Japan, for example, partitioning of all measured PAHs was described by

$$\log K_p = -1.028 \log_{10}(p^\circ) - 8.11 \qquad (14.50)$$

TABLE 14.18   Reactivity Scale for Electrophilic Reactions of PAHs

(Reactivity Decreases in the Order I to VI)

I	Benzo[*a*]tetracene, dibenzo[*a,h*]pyrene, pentacene, tetracene
II	Anthanthrene, anthracene, benzo[*a*]pyrene, cyclopenta[*cd*]pyrene, dibenzo[*a,l*]pyrene, dibenzo[*a,i*]pyrene, dibenzo[*a,c*]tetracene, perylene
III	Benz[*a*]anthracene, benzo[*g*]chrysene, benzo[*ghi*]perylene, dibenzo[*a,e*]pyrene, picene, pyrene
IV	Benzo[*c*]chrysene, benzo[*c*]phenanthrene, benzo[*e*]pyrene, chrysene, coronene, dibezanthracene, dibenzo[*e,l*]pyrene
V	Acenaphthylene, benzofluoranthenes, fluranthene, indeno[1,2,3-*cd*]fluoranthene, indeno[1,2,3-*cd*]pyrene, naphthalene, phenanthrene, triphenylene
VI	Biphenyl

Source: Nielsen (1984).

where $p°$ is the PAH vapor pressure in torr and K_p is in $m^3 \, \mu g^{-1}$ (Pankow and Bidleman 1992).

The partitioning coefficient K_p for PAHs is found to depend on temperature according to

$$\log K_p = \frac{A_p}{T} - 18.48 \qquad (14.51)$$

where the constant A_p depends on the chemical compound (Pankow 1991) with values given in Table 14.19.

Baek et al. (1991) proposed as a rule of thumb that two- and three-ring PAHs (naphthalene, fluorene, phenanthrene, anthracene, etc.) are predominantly in the gas phase, four-ring PAHs (pyrene) exist in both phases, while five- and six-ring PAHs

TABLE 14.19   Partitioning Estimation for Selected PAHs Based on Data Collected in Osaka, Japan, Assuming a Total Aerosol Mass of 100 $\mu g \, m^{-3}$

Compound	log $p°$ (torr) at 20°C	A_p(K)	K_p(20°C)	c_{aer}/c_g
Fluorene	−2.72	—	4.8×10^{-6}	0.00048
Phenanthrene	−3.50	4124	3.1×10^{-5}	0.0031
Anthracene	−3.53	4124	3.3×10^{-5}	0.0033
Fluoranthene	−4.54	4412	3.6×10^{-4}	0.0036
Pyrene	−4.73	4451	5.6×10^{-4}	0.056
Benzo[*a*]fluorene	−5.24	4549	1.9×10^{-3}	0.19
Benzo[*b*]fluorene	−5.22	4549	1.8×10^{-3}	0.18
Benz[*a*]anthracene	−6.02	4836	0.012	1.2
Chrysene	−6.06	4836	0.013	1.3
Triphenylene	−6.06	4836	0.013	1.3
Benzo[*b*]fluoranthene	−7.12	5180	0.16	16
Benzo[*k*]fluoranthene	−7.13	5180	0.17	17
Benzo[*a*]pyrene	−7.33	5301	0.27	27
Benzo[*e*]pyrene	−7.37	5301	0.29	29

Source: Pankow and Bidleman (1992).

exist primarily in the aerosol phase. Calculations of the ratios of the aerosol- : gas-phase concentrations for a series of PAHs based on data collected in Osaka and assuming an aerosol mass concentration of $100 \, \mu g \, m^{-3}$ are presented in Table 14.19. Such estimates should be used with caution as there is considerable variability in the coefficients of (14.51) even when obtained by the same researchers sampling in the same location.

APPENDIX 14   MEASUREMENT OF ELEMENTAL AND ORGANIC CARBON

A variety of methods have been applied to the measurement of EC and OC in aerosol samples with the thermal, thermal optical reflectance (TOR), and thermal manganese oxidation (TMO) methods being the most popular. Understanding the operational principles of these methods is often necessary for the interpretation of reported EC and OC data.

The TOR method was developed by Huntzicker et al. (1982) and is used for the quantification of OC and EC in aerosol samples collected on quartz filters. In this method the filter is first heated gradually from ambient temperature to 550°C in a pure helium atmosphere, resulting in volatilization of organic compounds in the sample. Then, the filter is exposed to a 2% oxygen, 98% helium oxidizing atmosphere and the temperature is ramped from 550 to 800°C in several steps (Figure 14.A.1). The carbon that evolves at each temperature in both steps is converted to methane and measured by a flame ionization detector. The filter sample reflectance is monitored throughout the process. This reflectance usually decreases during volatilization in the helium atmosphere due to the pyrolysis of the organic material. When oxygen is added the reflectance increases as the light-absorbing EC is oxidized and removed from the sample. OC is defined as the material that evolves from the beginning of the process, until the sample reflectance, after passing through its minimum, returns to its original value. The organic material that evolves after this point is defined as the EC (Figure 14.A.1). Note that in this example the separation of the total aerosol carbon into OC and EC is not straightforward, and that there is some uncertainty in the EC/OC split. The OC measured by this method is actually the OC that does not absorb light at the $0.63 \, \mu m$ wavelength used for the reflectance measurements, and EC is the light-absorbing carbon (Chow et al., 1993).

Simple thermal methods do not use the optical correction of the TOR method, but rather define as OC the carbon that evolves during heating in the helium atmosphere, and EC as the carbon that is produced during further heating in the oxidizing atmosphere.

In the TMO method (Fung 1990) manganese dioxide (MnO_2) is contacted with the sample throughout the process. Manganese dioxide serves as the oxidizing agent and the OC and EC are distinguished based on the temperature at which they volatilize. Carbon evolving at 525°C is classified as OC, and carbon evolving at 850°C as EC.

Comparisons among the results of these methods show that they agree within 5–15% on the total measured carbon from ambient aerosol and source samples (Kusko et al. 1989; Countess 1990). However, the individual EC and OC values are often quite different (Countess 1990; Hering et al. 1990). Finally, we should note that because these methods oxidize the organic compounds they measure organic carbon (OC) concentration (in $\mu g(C) \, m^{-3}$) and not organic aerosol mass directly. To convert the measured OC value to aerosol mass one needs to account for the other atoms in the OC molecules and multiplying factors ranging from 1.2 to 1.8 are used to do so.

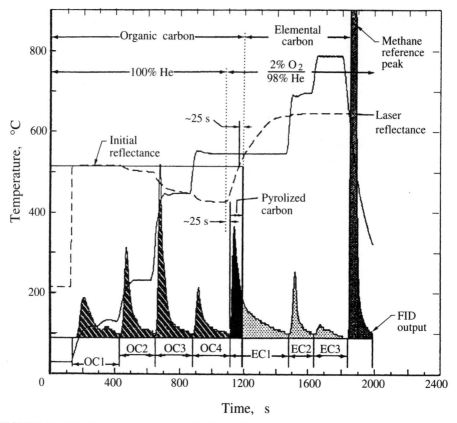

FIGURE 14.A.1 Example of a thermal/optical reflectance carbon analyzer thermogram for an ambient sample collected in Yellowstone National Park (Chow et al. 1993). Reflectance and FID output are in relative units. Reflectance is normalized to initial reflectance and FID output is normalized to the area of the reference peak. (Reprinted from *Atmos. Environ.* **27**, Chow et al., 1185–1201. Copyright 1993, with kind permission from Elsevier Science Ltd., The Boulevard, Langford Lane, Kidlington OX5 1GB, UK.)

Ambient organic material is usually collected on glass or quartz-fiber filters that have been specially treated to achieve low baseline carbon concentrations. OC can also be collected on a variety of particle-sizing devices, including low-pressure impactors, microorifice uniform deposit impactors (MOUDIs), and diffusion-based samplers (Eatough et al. 1993; Tang et al. 1994). The task of sampling organic compounds in airborne particles is complicated by the fact that many of these compounds have relatively high equilibrium vapor pressures, which are comparable to their ambient gas-phase concentrations. As a result, the distribution of these species between the gas and aerosol phases is sensitive to temperature and concentration changes. Also, artifacts may occur during the sampling process, including volatilization of sampled material and/or adsorption of vapors on the sampling surface and the sampled particles. Volatilization of sampled OC will lead to underestimation of the particle-phase concentration of organics (Arey et al. 1987; Coutant et al. 1988). Conversely, the adsorption of organic vapors on deposited particles or the sampling surface itself leads to overestimation of the aerosol concentration (Bidleman et al. 1986; Ligocki and Pankow 1989; McDow and Huntzicker

1990). In addition, several studies have suggested that chemical degradation of some organics may occur during sampling (Lindskog et al. 1985; Arey et al. 1988; Parmar and Grosjean 1990). For example, Wolff et al. (1991) suggested that the OC measured values in Los Angeles should be reduced by roughly 20% to correct for the sampling bias and then multiplied by 1.5 to account for the noncarbon mass of the organic aerosol compounds. After these corrections the organic aerosol mass is roughly 1.3 times the measured OC mass.

Partitioning of semivolatile organic compounds has been measured using a filter to collect particles followed by a solid adsorbent trap to collect the vapor fraction (Ligocki and Pankow 1989; Foreman and Bidleman 1990; Cotham and Bidleman 1992; Kaupp and Umlauf, 1992; Pankow 1992; Turpin et al. 1993).

PROBLEMS

14.1$_A$ The ambient fine-particle concentration of cholesterol in West Los Angeles was found to be $14.6 \, \text{ng} \, \text{m}^{-3}$ (Rogge et al. 1991). The same authors reported that the mean OC concentration for the same area for the same period was $7.5 \, \mu\text{g(C)} \, \text{m}^{-3}$ and that the emission rates from meat cooking (charbroiling and frying) in the area were:

Compound	Emission Rate
Cholesterol	$15.3 \, \text{kg} \, \text{day}^{-1}$
Fine organic carbon	$1400 \, \text{kg(C)} \, \text{day}^{-1}$

What fraction of the OC in the area, based on the measured concentrations, is due to meat cooking? Assume that the organic aerosol concentration can be calculated by multiplying the OC concentration by 1.2.

14.2$_B$ Calculate the maximum propionic acid concentrations in the aerosol phase after the reaction of 100 ppb of 1-butene considering the following scenarios:

a. Propionic acid does not adsorb or dissolve in the aerosol phase.

b. Propionic acid can form an ideal solution with the organic aerosol phase.

c. Propionic acid can form an ideal solution with the aqueous aerosol phase.

d. Propionic acid adsorbs on the aerosol particles following (14.49).

Assume extreme conditions (aerosol surface, aerosol liquid water content, aerosol organic carbon concentration) for these estimates and that the propionic acid vapor pressure is 0.005 atm.

14.3$_A$ Plot the gas- and aerosol-phase concentrations of the products P_1 and P_2 of α-pinene as a function of the amount of α-pinene that has reacted (0–1000 ppb) assuming that

$$\alpha\text{-pinene} \rightarrow 0.05 \, P_1 + 0.15 \, P_2$$

and that the saturation mixing ratios at 298 K are 0.15 and 6 ppb, respectively. Neglect the formation of solutions with other organic species or water.

14.4$_B$ Equation (14.40) was derived after summation of the constraints (14.34) and (14.35). Find ΔROG^* for m-xylene exactly solving (14.32) under the constraints (14.34) and (14.35) and compare with the approximate solution. What do you observe?

14.5$_B$ Turpin and Huntzicker (1995) proposed that during the summer in southern California primary OC is related to the EC concentration by

$$OC(primary) = 1.9 + 2.1 \times EC$$

Estimate the secondary OC fraction based on the following measurements collected during an air pollution episode in Los Angeles on July 11, 1987:

Time (PST)	OC ($\mu g(C) \, m^{-3}$)	EC ($\mu g(C) \, m^{-3}$)
2	4.0	1.0
4	5.2	1.1
6	5.4	1.1
8	8.1	1.7
10	8.2	1.7
12	13.6	2.7
14	14.1	2.4
16	14.9	2.1
18	14.0	1.8
20	12.5	1.9
22	8.1	1.4
24	6.5	1.3

Plot the secondary OC as a function of time and try to explain the observed behavior.

14.6$_A$ Repeat the calculation of Problem 14.3 assuming that the α-pinene products dissolve into $10 \, \mu g \, m^{-3}$ of preexisting organic aerosol, forming an ideal solution.

REFERENCES

Alebic-Juretic, A., Cvitas, T., and Klasmic, L. (1990) Heterogeneous polycyclic aromatic hydrocarbon degradation with ozone on silica-gel carrier, *Environ. Sci. Technol.* **24**, 62–66.

Appel, B. R., Cheng, W., and Salaymeh, F. (1989) Sampling of carbonaceous particles in the atmosphere. II, *Atmos. Environ.* **23**, 2167–2175.

Arey, J., Zielinska, B., Atkinson, R., and Winer, A. M. (1987) Polycyclic aromatic hydrocarbon and nitroarene concentrations in ambient air during a wintertime high NO$_x$ episode in the Los Angeles basin, *Atmos. Environ.* **21**, 1437–1444.

Arey, J., Zielinska, B., Atkinson, R., and Winer, A. M. (1988) Formation of nitroarenes during ambient high-volume sampling, *Environ. Sci. Technol.* **22**, 457–462.

Asher, W. E., Pankow, J. F., Erdakos, G. B., and Seinfeld, J. H. (2002) Estimating the vapor pressures of multi-functional oxygen-containing organic compounds using group contribution methods, *Atmos. Environ.* **36**, 1483–1498.

Atkinson, R., and Aschmann, R. M. (1986) Kinetics of the reactions of naphthalene, 2-methyl-naphthalene, and 2,3-dimethylnaphthalene with OH radicals and O_3, *Int. J. Chem. Kinet.* **18**, 569–573.

Baek, S. O., Field, R. A., Goldstone, M. E., Kirk, P. W., Lester, J. N., and Perry, R. (1991) A review of atmospheric polycyclic hydrocarbons: Sources, fate and behavior, *Water Air Soil Pollut.* **60**, 279–300.

Barsanti, K. C., and Pankow, J. F. (2004) Thermodynamics of the formation of atmospheric organic particulate matter by accretion reactions—Part 1: Aldehydes and ketones, *Atmos. Environ.* **38**, 4371–4382.

Barsanti, K. C., and Pankow, J. F. (2005) Thermodynamics of the formation of atmospheric organic particulate matter by accretion reactions—2. Dialdehydes, methylglyoxal, and diketones, *Atmos. Environ.* **39**, 6597–6607.

Behymer, T. D., and Hites, R. A. (1988) Photolysis of polycyclic aromatic hydrocarbons, *Environ. Sci. Technol.* **22**, 1311–1319.

Bidleman, T. F., Billings, W. N., and Foreman, W. T. (1986) Vapor-particle partitioning of semi-volatile organic compounds: Estimates from field collections, *Environ. Sci. Technol.* **20**, 1038–1043.

Bilde M., and Pandis, S. N. (2001) Evaporation rates and vapor pressures of individual aerosol species formed in the atmospheric oxidation of α- and β-pinene, *Environ. Sci. Technol.* **35**, 3344–3349.

Bilde M., Svenningsson, B., Monster, J., and Rosenorn, T. (2003) Even–odd alternation of evapo-ration rates and vapor pressures of C3–C9 dicarboxylic acid aerosols, *Environ. Sci. Technol.* **37**, 1371–1378.

Bittner, J. D., and Howard, J. B. (1981) Pre-particulate chemistry in soot formation, in *Particulate Carbon: Formation During Combustion*, D. C. Siegla and G. W. Smith, eds., Plenum Press, New York, pp. 109–142.

Bockhorn, H., Fetting, F., and Wenz, H. W. (1983) Chemistry of intermediate species in rich combustion of benzene, in *Soot Formation in Combustion Systems and Its Toxic Properties*, J. Lahaye, ed., Plenum Press, New York, pp. 57–94.

Bond, T. C., Streets, D. G., Yarber, K. F., Nelson, S. M., Woo, J. H., and Klimont, Z. (2004) A technology-based global inventory of black and organic carbon emissions from combustion, *J. Geophys. Res.* **109**, D14203 (doi: 10.1029/2003JD003697).

Bonn, B., Schuster, G., and Moortgat, G.K. (2002) Influence of water vapor on the process of new particle formation during monoterpene ozonolysis, *J. Phys. Chem. A.* **106**, 2869–2881.

Bowman, F. M., Odum, J. R., Pandis, S. N., and Seinfeld, J. H. (1997) Mathematical model for gas/particle partitioning of secondary organic aerosols, *Atmos. Environ.* **31**, 3921–3931.

Brown, N. J., Dod R. L., Mowrer, F. W., Novakov, T., and Williamson, R. B. (1989) Smoke emission factors from medium scale fires. Part 1, *Aerosol Sci. Technol.* **10**, 2–19.

Burtscher, H. (1992) Measurement and characteristics of combustion aerosols with special con-sideration of photoelectric charging by flame ions, *J. Aerosol Sci.* **23**, 549–595.

Cabada, J. C., Pandis, S. N., Subramanian, R., Robinson, A. L., Polidori, A., and Turpin, B. (2004) Estimating the secondary organic aerosol contribution to $PM_{2.5}$ using the EC tracer method, *Aerosol Sci. Technol.* **38**, 140–155.

Chang, S. G., Brodzinsky, R., Gundel, L. A., and Novakov, T. (1982) Chemical and catalytic properties of elemental carbon, in *Particulate Carbon: Atmospheric Life Cycle*, G. T. Wolff and R. L. Klimsch, eds., Plenum, New York, pp. 158–181.

Chow, J. C., Watson, J. G., Fujita, E. M., Lu, Z. Q., Lawson, D. R., and Ashbaugh, L. L. (1994) Temporal and spatial variations of $PM_{2.5}$ and PM_{10} aerosol in the Southern California Air Quality Study, *Atmos. Environ.* **28**, 2061–2080.

Chow, J. C., Watson, J. G., Pritchett, L. C., Pierson, W. R., Frazier, C. A., and Purcell, R. G. (1993) The DRI thermal/optical reflectance carbon analysis system: Description, evaluation and applications in U.S. air quality studies, *Atmos. Environ.* **27**, 1185–1201.

Chu, L. C., and Macias, E. S. (1981) Carbonaceous urban aerosol: Primary or secondary? in *Atmospheric Aerosol: Source/Air Quality Relationships*, E. S. Macias and P. K. Hopke, eds., American Chemical Society, Washington, DC, pp. 251–268.

Claeys, M. et al. (2004a) Formation of secondary organic aerosol through photooxidation of isoprene, *Science* **303**, 1173–1176.

Claeys, M., Wang, W., Ion, A. C., Kourtchev, I., Gelencser, A., and Maenhaut, W. (2004b) Formation of secondary organic aerosols from isoprene and its gas-phase oxidation products through reaction with hydrogen peroxide, *Atmos. Environ.* **38**, 4093–4098.

Clarke, A. D. (1989) Aerosol light absorption by soot in remote environments, *Aerosol Sci. Technol.* **10**, 161–171.

Clarke, A. G., and Karani, G. N. (1992) Characterization of the carbonate content of atmospheric aerosols, *J. Atmos. Chem.* **14**, 119–128.

Cocker, D. R. III, Mader, B. T., Kalberer, M., Flagan, R. C., and Seinfeld, J. C. (2001a) The effect of water on gas-particle partitioning of secondary organic aerosol, II. *m*-Xylene and 1,3,5-trimethylbenzene photooxidation systems, *Atmos. Environ.* **35**, 6073–6085.

Cocker, D. R. III, Clegg, S. L., Flagan, R. C., and Seinfeld, J. H. (2001b) The effect of water on gas-particle partitioning of secondary organic aerosol, I: α-pinene/ozone system, *Atmos. Environ.* **35**, 6049–6072.

Cooper, J. A., Redline, D. C., Sherman, J. R., Valdovinos, L. M., Pollard, W. L., Scavone, L. C., and Badgett-West, C. (1987) *PM$_{10}$ Source Composition Library for the South Coast Air Basin*: Vol. I. *Source Profile Development Documentation Final Report*, South Coast Air Quality Management District, Diamond Bar, CA.

Core, J. E., Cooper, J. A., and Newlicht, R. M. (1984) Current and projected impacts of residential wood combustion on Pacific Northwest air quality, *J. Air Pollut. Control Assoc.* **34**, 138–143.

Cotham, W. E., and Bidleman, T. F. (1992) Laboratory investigations of the partitioning of organochlorine compounds between the gas phase and atmospheric aerosols on glass fiber filters, *Environ. Sci. Technol.* **26**, 469–478.

Countess, R. J. (1990) Interlaboratory analyses of carbonaceous aerosol samples, *Aerosol Sci. Technol.* **12**, 114–121.

Countess, R. J., Wolff, G. T., and Cadle, S. H. (1980) The Denver winter aerosol: A comprehensive chemical characterization, *J. Air Pollut. Control Assoc.* **30**, 1194–1200.

Coutant, R. W., Brown, L., Chaung, J. C., Riggin, R. M., and Lewis, R. G. (1988) Phase distribution and artifact formation in ambient air sampling for polynuclear aromatic hydrocarbons, *Atmos. Environ.* **22**, 403–409.

Currie, L. A., Sheffield, A. E., and Riederer, C. (1994) Improved atmospheric understanding through exploratory data analysis and complementary modeling of the urban K-Pb-C system, *Atmos. Environ.* **28**, 1359–1369.

Currie, L. A., Stafford, T. W., Sheffield, A. E., Klouda, G. A., Wise, S. A., and Fletcher, R. A. (1989) Microchemical and molecular dating, *Radiocarbon* **31**, 448–463.

De Raat, W. K., Bakher, G. L., and De Meijere, F. A. (1990) Comparison of filter materials used for sampling of mutagens and polycyclic aromatic hydrocarbons in ambient airborne particles, *Atmos. Environ.* **24A**, 2875–2887.

Dod, R. L., Brown, N. J., Mowrer, F. W., Novakov, T., and Williamson, R. B. (1989) Smoke emission factors from medium scale fires. Part 2, *Aerosol Sci. Technol.* **10**, 20–27.

Doskey, P. V., and Andren, A. W. (1986) Particulate and vapor-phase *n*-alkanes in the northern Wisconsin atmosphere, *Atmos. Environ.* **20**, 1735–1744.

Duce, R. A., Mohnen, V. A., Zimmerman, P. R., Grosjean, D., Cautreels, W., Chatfield, R., Jaenicke, R., Ogren, J. A., Pellizzari, E. D., and Wallace, G. T. (1983) Organic material in the global troposphere, *Rev. Geophys. Space Phys.* **21**, 921–952.

Eatough, D. J., Sedar, B., Lewis, L., Hansen, L. D., Lewis, E. A., and Farber, R. J. (1993) Determination of the semivolatile organic compounds in particles in the Grand Canyon area, *Aerosol Sci. Technol.* **10**, 438–449.

Edney, E. O, Driscoll, D. J., Weathers, W. S., Kleindienst, T. E., Conver, T. S., McIver, C. D., and Li, W. (2001) Formation of polyketones in irradiated toluene/propylene/NO_x/air mixtures, *Aerosol Sci Technol.* **35**, 998–1008.

Edney, E. O., Driscoll, D. J., Speer, R. E., Weathers, W. S., Kleindienst, T. E., Li, W., Smith, D. F. (2000) Impact of aerosol liquid water on secondary organic aerosol yields of irradiated toluene/propylene/NO_x/$(NH_4)_2SO_4$/air mixtures, *Atmos. Environ.* **34**, 3907–3919.

Edney, E. O., Kleindienst, T. E., Jaoui, M., Lewandowski, M., Offenberg. J. H., Wang, W., and Claeys, M. (2005) Formation of 2-methyl tetrols and 2-methylglyceric acid in secondary organic aerosol from laboratory irradiated isoprene/NO_x/SO_2/air mixtures and their detection in ambient $PM_{2.5}$ samples collected in the eastern United States, *Atmos. Environ.* **39**, 5281–5289.

Finlayson-Pitts, B. J., and Pitts, J. N. Jr. (1986) *Atmospheric Chemistry*, Wiley, New York.

Flagan, R. C., and Seinfeld, J. H. (1988) *Fundamentals of Air Pollution Engineering*, Prentice-Hall, Englewood Cliffs, NJ.

Foreman, W. T., and Bidleman, T. F. (1990) Semivolatile organic compounds in the ambient air of Denver, Colorado, *Atmos. Environ.* **24A**, 2405–2416.

Forstner, H. J. L., Seinfeld, J. H., and Flagan, R. C. (1997a) Secondary organic aerosol formation from the photooxidation of aromatic hydrocarbons. Molecular composition, *Environ. Sci. Technol.* **31**, 1345–1358.

Forstner, H. J. L., Seinfeld, J. H., and Flagan, R. C. (1997b) Molecular speciation of secondary organic aerosol from the higher alkenes: 1-Octene and 1-decene, *Atmos. Environ.* **31**, 1953–1964.

Freeman, D. J., and Cattell, F. C. R. (1990) Wood burning as a source of atmospheric polycyclic aromatic hydrocarbons, *Environ. Sci. Technol.* **24**, 1581–1585.

Frenkel, J. (1946) *Kinetic Theory of Liquids*, Oxford Univ. Press, Oxford, UK.

Frenklach, M., Ramachandra, M. K., and Matula, R. A. (1985) Soot formation in shock tube oxidation of hydrocarbons, *Proc. 20th Int. Symp. Combustion*, The Combustion Institute, Pittsburgh, PA, pp. 97–104.

Fung, K. (1990) Particulate carbon speciation by MnO_2 oxidation, *Aerosol Sci. Technol.* **12**, 122–127.

Gao, S. et al. (2004a) Particle phase acidity and oligomer formation in secondary organic aerosol, *Environ. Sci. Technol.* **38**, 6582–6589.

Gao, S., Keywood, M., Ng, N.L., Surratt, J., Varutbangkul, V., Bahreini, R., Flagan, R.,C., and Seinfeld, J. H. (2004b) Low-molecular-weight and oligomeric components in secondary organic aerosol from the ozonolysis of cycloalkenes and α-pinene, *J. Phys. Chem. A.* **108**, 10147–10164.

Gagosian, R. B., Zafiriou, O. C., Peltzer, E. T., and Alford, J. B. (1982) Lipids in aerosols from the tropical North Pacific: Temporal variability, *J. Geophys. Res.* **87**, 11133–11144.

Graedel, T. E. (1986) *Atmospheric Chemical Compounds: Sources, Occurrence and Bioassay.* Academic Press, Orlando, FL.

Graham, S. C., Homer, J. B., and Rosenfeld, J. L. J. (1975) The formation and coagulation of soot aerosols generated in pyrolysis of aromatic hydrocarbons, *Proc. Roy. Soc. Lond.* **A344**, 259–285.

Gray, H. A., Cass, G. R., Huntzicker, J. J., Heyerdahl, E. K., and Rau, J. A. (1986) Characteristics of atmospheric organic and elemental carbon particle concentrations in Los Angeles, *Environ. Sci. Technol.* **20**, 580–589.

Griffin R. J., Cocker III, D. R., Flagan, R. C., Seinfeld, J. H. (1999) Organic aerosol formation from the oxidation of biogenic hydrocarbons, *J. Geophys. Res.* **104**, 3555–3567.

Grosjean, D. (1983) Polycyclic aromatic hydrocarbons in Los Angeles from samples collected on teflon, glass and quartz filters, *Atmos. Environ.* **17**, 2565–2573.

Grosjean, D. (1984a) Particulate carbon in Los Angeles air, *Sci. Total Environ.* **32**, 133–145.

Grosjean, D. (1984b) Photooxidation of 1-heptene, *Sci. Total Environ.* **37**, 195–211.

Grosjean, D. (1985) Reactions of *o*-cresol and nitrocresol with NO$_x$ in sunlight and with ozone–nitrogen dioxide mixtures in the dark, *Environ. Sci. Technol.* **19**, 968–974.

Hahn, J. (1980) Organic constituents of natural aerosols, *Ann. NY Acad. Sci.* **338**, 359–376.

Hallquist, M., Wangberg, I., and Ljungstrom, E. (1997) Atmospheric fate of carbonyl oxidation products originating from α-pinene and δ-carene: Determination of rate of reaction with OH and NO$_3$ radicals, UV absorption cross sections, and vapor pressures, *Environ. Sci. Technol.* **31**, 3166–3172.

Hansen, A. D. A., and Rosen, H. (1990) Individual measurements of the emission factor of aerosol black carbon in automobile plumes, *J. Air Waste Manag. Assoc.* **40**, 1654–1657.

Harris, S. J., and Weiner, A. M. (1983a) Surface growth of soot particles in premixed ethylene/air flames, *Combust. Sci. Technol.* **31**, 155–167.

Harris, S. J., and Weiner, A. M. (1983b) Determination of the rate constant for soot surface growth, *Combust. Sci. Technol.* **32**, 267–275.

Hawthorne, S. B., Miller, D. J., Langenfeld, J. J., and Krieger, M. S. (1992) PM$_{10}$ high volume collection and quantification of semi- and nonvolatile phenols, methoxylated phenols, alkanes, and polycyclic aromatic hydrocarbons from winter urban air and their relationship to wood smoke, *Environ. Sci. Technol.* **26**, 2251–2283.

Hering, S. V. et al. (1990) Comparison of sampling methods for carbonaceous aerosols in ambient air, *Aerosol Sci. Technol.* **12**, 200–213.

Hildemann, L. M. (1990) *A Study of the Origin of Atmospheric Organic Aerosols*, Ph.D. thesis, California Institute of Technology, Pasadena.

Hildemann, L. M., Markowski, G. R., and Cass, G. R. (1991a) Chemical composition of emissions from urban sources of fine organic aerosol, *Environ. Sci. Technol.* **25**, 744–759.

Hildemann, L. M., Markowski, G. R., Jones, M. C., and Cass, G. R. (1991b) Submicrometer aerosol mass distributions of emissions from boilers, fireplaces, automobiles, diesel trucks, and meat cooking operations, *Aerosol Sci. Technol.* **14**, 138–152.

Hoffmann, T., Bandur, R., Marggraf, U., and Linscheid, M. (1998) Molecular composition of organic aerosols formed in the α-pinene/O$_3$ reaction: Implications for new particle formation processes, *Atmos. Environ.* **32**, 1657–1661.

Hoffmann, T., Odum, J. R., Bowman, F., Collins, D., Klockow, D., Flagan, R. C., and Seinfeld, J. H. (1997) Formation of organic aerosols from the oxidation of biogenic hydrocarbons, *J. Atmos. Chem.* **26**, 189–222.

Houck, J. E., Chow, J. C., Watson, J. G., Simons, C. A., Pritchett, L. C., Goulet, J. M., and Frazier, C. A. (1989) *Determination of Particle Size Distribution and Chemical Composition of Particulate Matter from Selected Sources in California*, California Air Resources Board, Sacramento, CA.

Huntzicker, J. J., Heyerdahl, E. K., Rau, J. A., Griest, W. H., and MacDougall, C. S. (1986) Carbonaceous aerosol in the Ohio Valley, *J. Air Pollut. Control Assoc.* **36**, 705–709.

Huntzicker, J. J., Johnson, R. L., Shah, J. J., and Gary, R. A. (1982) Analysis of organic and elemental carbon in ambient aerosols by the thermal-optical method, in *Particulate Carbon—Atmospheric Life Cycle*, G. T. Wolff and R. L. Klimisch, eds., Plenum, New York, pp. 79–88.

Hurley, M. D., Sokolov, O., Wallington, T. J., Takekawa, H., Karasawa, M., Klotz, B., Barnes, I., and Becker, K. H. (2001) Organic aerosol formation during the atmospheric degradation of toluene, *Environ. Sci. Technol.* **35**, 1358–1366.

Iinuma, Y., Boge, O., Gnauk, T., and Hermann, H. (2004) Aerosol chamber study of the α-pinene/O₃ reaction: Influence of particle acidity on aerosol yields and products, *Atmos. Environ.* **38**, 761–773.

Izumi, K., and Fukuyama, T. (1990) Photochemical aerosol formation from aromatic hydrocarbons in the presence of NO$_x$, *Atmos. Environ.* **24A**, 1433–1441.

Izumi, K., Murano, K., Mizuochi, M., and Fukuyama, T. (1988) Aerosol formation by the photooxidation of cyclohexene in the presence of nitrogen oxides, *Environ. Sci. Technol.* **22**, 1207–1215.

Jacobson, M. C., Hansson, H. C., Noone, K. J., and Charlson, R. J. (2000) Organic atmospheric aerosols: Review and state of science, *Rev. Geophys.* **38**, 267–294.

Jang, M., and Kamens, R. M. (1999) Newly characterized products and composition of secondary aerosols from reaction of α-pinene with ozone, *Atmos. Environ.* **33**, 459–474.

Jang, M., and Kamens, R. M. (2001) Atmospheric secondary organic aerosol formation by heterogeneous reactions of aldehydes in the presence of a sulfuric acid catalyst, *Environ. Sci. Technol.* **35**, 4758–4766.

Jang, M., Czoschke, N. M., Lee, S., and Kamens, R. M. (2002) Heterogeneous atmospheric aerosol production by acid-catalyzed particle-phase reactions, *Science* **298**, 814–817.

Jang M., Carroll, B., Chandramouli, B., and Kamens, R. M. (2003) Particle growth by acid-catalyzed heterogeneous reactions of organic carbonyls on preexisting aerosols, *Environ. Sci. Technol.* **37**, 3828–3837.

Jaoui, M., Leungsakul, S., and Kamens, R. M. (2003) Gas and particle products distribution from the reaction of β-caryophyllene with ozone, *J. Atmos. Chem.* **45**, 261–287.

Kalberer, M., Yu, J. Cocker, D. R. III, Flagan, R. C., and Seinfeld, J. H. (2000) Aerosol formation in the cyclohexene-ozone system, *Environ. Sci. Technol.* **34**, 4894–4901.

Kalberer, M., Paulsen, D., Sax, M., Steinbacher, M., Dommen, J., Prevot, A. S. H., Fisseha, R., Weingartner, E., Frankevich, V., Zenobi, R., and Baltensperger, U. (2004) Identification of polymers as major components of atmospheric organic aerosols, *Science* **303**, 1659–1662.

Kamens, R. M., Guo, J., Guo, Z., and McDow, S. R. (1990) Polynuclear aromatic hydrocarbon degradation by heterogeneous reactions with N₂O₅ on atmospheric particles, *Atmos. Environ.* **24A**, 1161–1173.

Kamens, R. M. and Jaoui M. (2001) Modelling aerosol formation from α-pinene + NO$_x$ in the presence of natural sunlight using gas-phase kinetics and gas-particle partitioning theory, *Environ. Sci. Technol.* **35**, 1394–1405.

Kanakidou et al. (2005) Organic aerosol and climate modeling: A review (2005) *Atmos. Chem. Phys.* **5**, 1–70.

Kanakidou, M., Tsigaridis, K., Dentener, F. J., and Crutzen, P. J. (2000) Human-activity-enhanced formation of organic aerosols by biogenic hydrocarbon oxidation, *J. Geophys. Res.* **105**, 9243–9254.

Kaupp, H., and Umlauf, G. (1992) Atmospheric gas–particle partitioning of organic compounds: Comparison of sampling methods, *Atmos. Environ.* **26A**, 2259–2267.

Kawamura, K., Ng, L., and Kaplan, I. R. (1985) Determination of organic acids ($C_1 - C_{10}$) in the atmosphere, motor exhausts, and engine oils, *Environ. Sci. Technol.* **19**, 1082–1086.

Keeler, G. J., Japar, S. M., Brackaczek, W. W., Gorse, R. A., Norbeck, J. M., and Pierson, W. R. (1990) The sources of aerosol elemental carbon at Allegheny Mountain, *Atmos. Environ.* **24**, 2795–2805.

Keywood, M. D., Varutbangkul, V., Bahreini, R., Flagan, R. C., and Seinfeld, J. H. (2004) Secondary organic aerosol formation from the ozonolysis of cycloalkenes and related compounds, *Environ. Sci. Technol.* **38**, 4157–4164.

Kim, M. G., Yakawa, K., Inoue, H., Lee, Y. K., and Shirai, T. (1990) Measurement of tire tread in urban air by pyrolysis–gas chromatography with flame photometric detection, *Atmos. Environ.* **24A**, 1417–1422.

Kleindienst, T. E., Smith, D. F., Li, W., Edney, E. O., Driscoll, D. J., Speer, R. E., and Weathers, W. S. (1999) Secondary organic aerosol formation from the oxidation of aromatic hydrocarbons in the presence of dry submicron ammonium sulfate, *Atmos. Environ.* **33**, 3669–3681.

Kleindienst, T. E., Conver, T. S., McIver, C. D., and Edney, E. O. (2004) Determination of secondary organic aerosol products from the photooxidation of toluene and their implications in ambient $PM_{2.5}$, *J. Atmos. Chem.* **47**, 79–100.

Klouda, G. A., Currie, L. A., Verkouteren, R. M., Eifeld, W., and Zak, B. D. (1988) Advances in microradiocarbon dating and the direct tracing of environmental carbon, *J. Radioanal. Nucl. Chem.* **123**, 191–197.

Kroll, J. H., Ng, N. L., Murphy, S. M., Flagan, R. C., and Seinfeld, J. H. (2005a) Secondary organic aerosol formation from isoprene photooxidation under high-NO_x conditions, *Geophys. Res. Lett.* **32**, L18808 (doi: 10.1029/2005GL023637).

Kroll, J. H. Ng, N. L., Murphy, S. M., Flagan, R. C., and Seinfeld, J. H., (2006) Secondary organic aerosol formation from isoprene photooxidation, *Environ. Sci. Technol.* **40**, 1869–1877.

Kroll, J. H., and Seinfeld, J. H. (2005b) Representation of secondary organic aerosol (SOA) laboratory chamber data for the interpretation of mechanisms of particle growth, *Environ. Sci. Technol.* **39**, 4149–4165.

Kusko, B., Cahill, T. A., Eldred, R. A., Matsuda, Y., and Miyabe, H. (1989) Nondestructive analysis of total nonvolatile carbon by forward alpha scattering technique (FAST), *Aerosol Sci. Technol.* **10**, 390–396.

Larsen, B. R., Di Bella, D., Glasius, M., Winterhalter, R., Jensen, N. R., Hjorth, J. (2001) Gas-phase OH oxidation of monoterpenes: Gaseous and particulate products, *J. Atmos. Chem.* **38**, 231–276.

Larson, S. M., Cass, G. R., and Gray, H. A. (1989) Atmospheric carbon particles and the Los Angeles visibility problem, *Aerosol Sci. Technol.* **10**, 118–130.

Lee, A., Goldstein, A. H., Keywood, M. D., Gao, S., Varutbangkul, V., Bahreini, R., Flagan, R. C., and Seinfeld, J. H. (2006) Gas-phase products and secondary aerosol yields from the ozonlysis of ten different terpenes, *J. Geophys. Res*, **111**, D07302 (doi: 10.1029/2005JD006437).

Lewis, C. W., Baumgardner, R. E., Stevens, R. K., Claxton, L. D., and Lewtas, J. (1988) Contribution of woodsmoke and motor vehicle emissions to ambient aerosol, *Environ. Sci. Technol.* **22**, 968–971.

Liang, C., Pankow, J. F., Odum, J. R., and Seinfeld, J. H. (1997) Gas/particle partitioning of semivolatile organic compounds to model inorganic, model organic, and ambient smog aerosols, *Environ. Sci. Technol.* **31**, 3086–3092.

Ligocki, M. P., and Pankow, J. F. (1989) Measurements of the gas/particle distributions of atmospheric organic compounds, *Environ. Sci. Technol.* **23**, 75–83.

Lindskog, A., Brorstrom-Lunden, E., and Sjodin, A. (1985) Transformation of reactive PAH on particles by exposure to oxidized nitrogen compounds and ozone, *Environ. Int.* **11**, 125–130.

Masel, R. I. (1996) *Principles of Adsorption and Reaction on Solid Surfaces*, Wiley, New York.

Mazurek, M. A., and Simoneit, B. R. T. (1984) Characterization of biogenic and petroleum-derived organic matter in aerosols over remote, rural and urban areas, in *Identification and Analysis of Organic Pollutants in Air*, Ann Arbor Science/Butterworth, Stoneham, MA, pp. 353–370.

McDow, S. R., and Huntzicker, J. J. (1990) Vapor adsorption artifact in the sampling of organic aerosol: Face velocity effects, *Atmos. Environ.* **24A**, 2563–2571.

McKinnon, J. T., and Howard, J. B. (1990) Application of soot formation model—effects of chlorine, *Combust. Sci. Technol.* **74**, 175–197.

McMurry, P. H., and Zhang, X. Q. (1989) Size distributions of ambient organic and elemental carbon, *Aerosol Sci. Technol.* **10**, 430–437.

Monster, J., Rosenorn, T., Svenningsson, B., and Bilde, A. (2004) Evaporation of methyl- and dimethyl-substituted malonic, succinic, glutaric, and adipic acid particles at ambient temperatures, *J. Aerosol Sci.* **35**, 1453–1465.

Mulhbaier, D. J., and Cadle, S. H. (1989) Atmospheric carbon particles in Detroit urban area: Wintertime sources and sinks, *Aerosol Sci. Technol.* **10**, 237–248.

Mulhbaier, J. L., and Williams, R. L. (1982) Fireplaces, furnaces, and vehicles as emission sources of particulate carbons, in *Particulate Carbon: Atmospheric Life Cycle*, G. T. Wolff and R. L. Klimsch, eds., Plenum, New York, pp. 185–205.

Mylonas, D. T., Allen, D. T., Ehrman, S. H., and Pratsinis, S. E. (1991) The sources and size distributions of organonitrates in the Los Angeles aerosol, *Atmos. Environ.* **25A**, 2855–2861.

National Academy of Sciences (1983) *Polycyclic Aromatic Hydrocarbons: Evaluation of Sources and Effects*, National Academy Press, Washington, DC.

Ng, N. L., Kroll, J. H., Keywood, M.D., Bahreini, R., Varutbangkul, V., Flagan, R. C., Seinfeld, J. H., Lee, A., and Goldstein, A. H. (2006) Contribution of first-versus second-generation products to secondary organic aerosols formed in the oxidation of biogenic hydrocarbons, *Environ. Sci. Technol.* **40**, 2283–2297.

Nielsen, T. (1984) Reactivity of polycyclic aromatic hydrocarbons towards nitrating species, *Environ. Sci. Technol.* **18**, 157–163.

Nielsen, T., Ramdahl, T., and Bjorseth, A. (1983) The fate of airborne polycyclic organic matter, *Environ. Health Perspect.* **47**, 103–114.

Nikolaou, K., Masclet, P., and Mouvier, G. (1984) Sources and chemical reactivity of polynuclear aromatic hydrocarbons in the atmospherie. A critical review, *Sci. Total Environ.* **32**, 103–132.

Novakov, T. (1984) The role of soot and primary oxidants in atmospheric chemistry, *Sci. Total Environ.* **36**, 1–10.

Nunes, T. V., and Pio, C. A. (1993) Carbonaceous aerosols in industrial and coastal atmospheres, *Atmos. Environ.* **27**, 1339–1346.

Odum, J. R., Hoffmann, T., Bowman, F., Collins, T., Flagan, R. C., and Seinfeld, J. H. (1996) Gas-particle partitioning and secondary organic aerosol yields, *Environ. Sci. Technol.* **30**, 2580–2585.

Odum, J. R., Jungkamp, T. P. W., Griffin, R. J., Flagan, R. C., and Seinfeld, J. H. (1997a) The atmospheric aerosol-forming potential of whole gasoline vapor, *Science* **276**, 96–99.

Odum, J. R., Jungkamp, T. P. W., Griffin, R. J., Forstner, H. J. L., Flagan, R. C., and Seinfeld, J. H. (1997b) Aromatics, reformulated gasoline, and atmospheric organic aerosol formation, *Environ. Sci. Technol.* **31**, 1890–1897.

Olmez, I., Sheffield, A. E., Gordon, G. E., Houck, J. E., Pritchett, L. C., Cooper, J. A., Dzubay, T. G., and Bennett, R. L. (1988) Compositions of particles from selected sources in Philadelphia for receptor modeling applications, *J. Air Pollut. Control Assoc.* **38**, 1392–1402.

Palmer, H. B., and Cullis, C. F. (1965) The formation of carbon from gases, in *Chemistry and Physics of Carbon*, Vol. 1, P. L. Walker, ed., Marcel Dekker, New York, pp. 265–325.

Pandis, S. N., Harley, R. H., Cass, G. R., and Seinfeld, J. H. (1992) Secondary organic aerosol formation and transport, *Atmos. Environ.* **26A**, 2269–2282.

Pandis, S. N., Paulson, S. E., Seinfeld, J. H., and Flagan, R. C. (1991) Aerosol formation in the photooxidation of isoprene and β-pinene, *Atmos. Environ.* **25A**, 997–1008.

Pankow, J. F. (1987) Review and comparative analysis of the theories of partitioning between the gas and aerosol particulate phases in the atmosphere, *Atmos. Environ.* **21**, 2275–2283.

Pankow, J. F. (1991) Common y-intercept and single compound regressions of gas–particle partitioning data vs. $1/T$, *Atmos. Environ.* **25A**, 2229–2239.

Pankow, J. F. (1992) Application of common y-intercept regression parameters for log K_p vs $1/T$ for predicting gas–particle partitioning in the urban environment, *Atmos. Environ.* **26A**, 2489–2497.

Pankow, J. F. (1994a) An absorption model of gas/particle partitioning of organic compounds in the atmospheric, *Atmos. Environ.* **28**, 185–188.

Pankow, J. F. (1994b) An absorption model of the gas/aerosol partitioning involved in the formation of secondary organic aerosol, *Atmos. Environ.* **28**, 189–193.

Pankow, J. F., and Bidleman, T. F. (1992) Interdependence of the slopes and intercepts from log-log correlations of measured gas–particle partitioning and vapor pressure. I. Theory and analysis of available data, *Atmos. Environ.* **26A**, 1071–1080.

Pankow, J. F., and Bidleman, T. F. (1991) Effects of temperature, TSP and percent non-exchangeable material in determining the gas–particle partitioning of organic compounds, *Atmos. Environ.* **25A**, 2241–2249.

Parmar, S. S., and Grosjean, D. (1990) Laboratory tests of KI and alkaline annular denuders, *Atmos. Environ.* **24A**, 2695–2698.

Penner, J. E. (1995) Carbonaceous aerosols influencing atmospheric radiation: Black and organic carbon, in *Aerosol Forcing of Climate: Report of the Dahlem Workshop on Aerosol Forcing of Climate, Berlin 1994*, eds., R. J. Charlson and J. Heintzenberg, Wiley, New York, pp. 91–108.

Penner, J. E., Eddleman, H., and Novakov, T. (1993) Towards the development of a global inventory for black carbon emissions, *Atmos. Environ.* **27**, 1277–1295.

Peters, J. A., Deangelis, D. G., and Hughes, T. W. (1981) in *Chemical Analysis and Biological Fate: Polynuclear Aromatic Hydrocarbons*, M. Cooke and A. J. Dennis, eds., Battelle Press, Columbus, OH, pp. 571–582.

Pickle, T., Allen, D. T., and Pratsinis, S. E. (1990) The sources and size distributions of aliphatic and carbonyl carbon in Los Angeles aerosol, *Atmos. Environ.* **24**, 2221–2228.

Pierson, W. R., and Brachaczek, W. W. (1983) Particulate matter associated with vehicles on the road: 2, *Aerosol Sci. Technol.* **2**, 1–40.

Pilinis, C., and Seinfeld, J. H. (1988) Development and evaluation of an Eulerian photochemical gas-aerosol model, *Atmos. Environ.* **22**, 1985–2001.

Pitts, J. N. Jr., Paur, H. R., Zielinska, B., Sweetman, J. A., Winer, A. M., Ramdahl, T., and Mejia, V. (1986) Factors influencing the reactivity of polycyclic aromatic hydrocarbons adsorbed on model substrates and in ambient POM with ambient levels of ozone, *Chemosphere* **15**, 675–685.

Pitts, J. N. Jr., Sweetman, J. A., Zielinska, B., Winer, A. M., and Atkinson, R. (1985a) Determination of 2-nitrofluoranthene and 2-nitropyrene in ambient particulate organic matter: Evidence for atmospheric reactions, *Atmos. Environ.* **19**, 1601–1608.

Pitts, J. N. Jr., Sweetman, J. A., Zielinska, B., Winer, A. M., Atkinson, R., and Harger, W. P. (1985b) Formation of nitroarenes from the reaction of polycyclic aromatic hydrocarbons with dinitrogen pentoxide, *Environ. Sci. Technol.* **19**, 1115–1121.

Pratsinis, S. E., Zeldin, M. D., and Ellis, E. C. (1988) Source resolution of fine carbonaceous aerosol by principal component–stepwise regression analysis, *Environ. Sci. Technol.* **22**, 212–216.

Presto, A. A., Huff Hartz, K. E., and Donahue, N. M. (2005a) Secondary organic aerosol production from terpene ozonolysis: 1. Effect of UV radiation, *Environ. Sci. Technol.* **39**, 7036–7045.

Presto, A. A., Huff Hartz, K. E., and Donahue, N. M. (2005b) Secondary organic aerosol production from terpene ozonolysis: 2. Effect of NO_x concentration, *Environ. Sci. Technol.* **39**, 7046–7054.

Ramdahl, T., Alfheim, I., and Bjorseth, A. (1983) in *Mobil Source Emissions Including Polycyclic Organic Species*, D. Rondia, M. Cooke, and R. K. Haroz, eds., Reidel, Dordrecht, The Netherlands, pp. 278–298.

Ramdahl, T., Bjorseth, A., Lokensgard, D., and Pitts, J. N. Jr. (1984a) Nitration of polycyclic aromatic hydrocarbons adsorbed to different carriers in a fluidized bed reactor, *Chemosphere* **13**, 527–534.

Ramdahl, T., Schjoldager, J., Currie, L. A., Hanssen, J. E., Moller, M., Klouda, G. A., and Alfheim, I. (1984b) Ambient impact of residential wood combustion in Elverum, Norway, *Sci. Total Environ.* **36**, 81–90.

Raunemaa, T., Hyvonen, V., and Kauppinen, E. (1984) Submicron size particle growth and chemical transformation in gasoline exhaust, 1. Size distribution and transformation, *J. Aerosol Sci.* **15**, 335–341.

Rogge, W. F., Hildemann, L. M., Mazurek, M. A., and Cass, G. R. (1994) Sources of fine organic aerosol, 6. Cigarette smoke in the urban atmosphere, *Environ. Sci. Technol.* **28**, 1375–1388.

Rogge, W. F., Hildemann, L. M., Mazurek, M. A., Cass, G. R., and Simoneit, B. R. T. (1991) Sources of fine organic aerosol 1. Charbroiler and meat cooking operations, *Environ. Sci. Technol.* **25**, 1112–1125.

Rogge, W. F., Hildemann, L. M., Mazurek, M. A., Cass, G. R., and Simoneit, B. R. T. (1993a) Sources of fine organic aerosol, 2. Non catalyst and catalyst equipped automobiles and heavy duty diesel trucks, *Environ. Sci. Technol.* **27**, 636–651.

Rogge, W. F., Hildemann, L. M., Mazurek, M. A., Cass, G. R., and Simoneit, B. R. T. (1993b) Sources of fine organic aerosol, 3. Road dust, tire debris and organometallic brake lining dust, *Environ. Sci. Technol.* **27**, 1892–1904.

Rogge, W. F., Hildemann, L. M., Mazurek, M. A., Cass, G. R., and Simoneit, B. R. T. (1993c) Sources of fine organic aerosol, 4. Particulate abrasion products from leaf surfaces or urban plants, *Environ. Sci. Technol.* **27**, 2700–2711.

Rogge, W. F., Hildemann, L. M., Mazurek, M. A., Cass, G. R., and Simoneit, B. R. T. (1993d) Sources of fine organic aerosol, 5. Natural gas home appliances, *Environ. Sci. Technol.* **27**, 2736–2744.

Rogge, W. F., Mazurek, M. A., Hildemann, L. M., Cass, G. R., and Simoneit, B. R. T. (1993e) Quantification of urban organic aerosols at a molecular level. Identification, abundance and seasonal variation, *Atmos. Environ.* **27**, 1309–1330.

Rogge, W. F., Hildemann, L. M., Mazurek, M. A., Cass, G. R., and Simoneit, B. R. T. (1996) Mathematical modeling of atmospheric fine particle associated primary organic compound concentrations, *J. Geophys. Res.* **101**, 19379–19394.

Satsumakayashi, H., Kurita, H., Yokouchi, Y., and Ueda, H. (1989) Mono and dicarboxylic acids under long-range transport of air pollution in central Japan, *Tellus* **41B**, 219–229.

Satsumakayashi, H., Kurita, H., Yokouchi, Y., and Ueda, H. (1990) Photochemical formation of particulate dicarboxylic acids under long range transport in central Japan, *Atmos. Environ.* **24A**, 1443–1450.

Saxena, P., Hildemann, L. M., McMurry, P. H., and Seinfeld, J. H. (1995) Organics alter hygroscopic behavior of atmospheric particles, *J. Geophys. Res.* **100**, 18755–18767.

Seinfeld, J. H., Flagan, R. C., Petti, T. B., Stern, J. E., and Grosjean, D. (1987) *Aerosol Formation in Aromatic Hydrocarbon–NO_x Systems*, California Institute of Technology, Pasadena, CA, Final Report to the Coordinating Research Council under project AP-6.

Seinfeld, J. H., and Pankow, J. E. (2003) Organic atmospheric particulate material, *Annu. Rev. Phys. Chem.* **54**, 121–140.

Sexton, K., Spengler, J. D., Treitman, R. D., and Turner, W. A. (1984) Effects of residential wood combustion on indoor air quality—a case study in Waterbury, Vermont, *Atmos. Environ.* **18**, 1357–1370.

Shah, J. J., Johnson, R. L., Heyerdahl, E. K., and Huntzicker, J. (1986) Carbonaceous aerosol at urban and rural sites in the United States, *J. Air Pollut. Control Assoc.* **36**, 254–257.

Sicre, M. A., Marty, J. C., and Saliot, A. (1990) n-Alkanes, fatty acid esters, and fatty acid salts in size fractionated aerosols collected over the Mediterranean Sea, *J. Geophys. Res.* **95**, 3649–3657.

Simoneit, B. R. T. (1977) Organic matter in eolian dusts over the Atlantic Ocean, *Mar. Chem.* **5**, 443–464.

Simoneit, B. R. T. (1984) Organic matter in the troposphere: III. Characterization and sources of petroleum and pyrogenic residues in aerosols over the Western United States, *Atmos. Environ.* **18**, 51–67.

Simoneit, B. R. T. (1986) Characterization of organic constituents in aerosols in relation to their origin and transport: A review, *Int. J. Environ. Anal. Chem.* **23**, 207–237.

Simoneit, B. R. T. (1989) Organic matter in the troposphere—V. Application of molecular marker analysis to biogenic emissions into the troposphere for source reconciliations, *J. Atmos. Chem.* **8**, 251–278.

Simoneit, B. R. T., Cardoso, J. N., and Robinson, N. (1990) An assessment of the origin and composition of higher molecular weight organic matter in aerosols over Amazonia, *Chemosphere* **21**, 1285–1301.

Simoneit, B. R. T., and Mazurek, M. A. (1982) Organic matter in the troposphere II. Natural background of biogenic lipid matter in aerosols over the rural Western United States, *Atmos. Environ.* **16**, 2139–2159.

Sisler, J. F., Huffman, D., Lattimer, D. A., Malm, W. C., and Pitchford, M. L. (1993) Spatial and temporal patterns and the chemical composition of the haze in the United States: An analysis of data from the IMPROVE network, 1988–1991. Colorado State University, Cooperative Institute for Research in the Atmosphere (CIRA), Fort Collins, CO.

Standley, L. J., and Simoneit, B. R. T. (1987) Characterization of extractable plant wax, resin, and thermally matured components in smoke particles from prescribed burns, *Environ. Sci. Technol.* **21**, 163–169.

Steinmetzer H. C., Baumeister W., and Vierle O. (1984) Analytical investigation on the contents of polycyclic aromatic hydrocarbons in airborne particulate matter from 2 Bavarian cities, *Sci. Total Environ.* **36**, 91–96.

Strader R., Lurmann, F., and Pandis, S. N. (1999) Evaluation of secondary organic aerosol formation in winter, *Atmos. Environ.* **33**, 4849–4863.

Stevens, R. K., Dzubay, T. G., Lewis, C. W., and Shaw, R. W. Jr. (1984) Source apportionment methods applied to the determination of the origin of ambient aerosols that affect visibility in forested areas, *Atmos. Environ.* **18**, 261–272.

Tang, H., Lewis, E. A., Eautough, D. J., Burton, R. M., and Farber, R. J. (1994) Determination of the particle size distribution and chemical composition of semi-volatile organic compounds in atmospheric fine particles with a diffusion denuder sampling system, *Atmos. Environ.* **29**, 939–947.

Tao, Y., and McMurry, P. H. (1989) Vapor pressures and surface free energies of C_{14}-C_{18} monocarboxylic acids and C_5-dicarboxylic and C_6-dicarboxylic acids, *Environ. Sci. Technol.* **23**, 1519–1523.

Tobias H. J., Docherty, K. S., Beving, D. E., and Ziemann, P. J. (2000) Effect of relative humidity on the chemical composition of secondary organic aerosol formed from reactions of 1-tetradecene and O_3, *Environ. Sci. Technol.* **34**, 2116–2125.

Tolocka, M. P., Jang, M., Ginter, J., Cox, F., Kamens, R., and Johnston, M. (2004) Formation of oligomers in secondary organic aerosol, *Environ. Sci. Technol.* **38**, 1428–1434.

Tong, C., Blanco, M., Goddard III, W. A., and Seinfeld J. H. (2006) Secondary organic aerosol formation by heterogeneous reactions of aldehydes and ketones: A quantum mechanical study, *Environ. Sci. Technol.* **40**, 2333–2338.

Turpin, B. J., and Huntzicker, J. J. (1991) Secondary formation of organic aerosol in the Los Angeles Basin: A descriptive analysis of organic and elemental carbon concentrations, *Atmos. Environ.* **25A**, 207–215.

Turpin, B. J., Huntzicker, J. J. (1995) Identification of secondary organic aerosol episodes and quantitation of primary and secondary organic aerosol concentration during SCAQS, *Atmos. Environ.* **29**, 3527–3544.

Turpin, B. J., Huntzicker, J. J., Larson, S. M., and Cass, G. R. (1991) Los Angeles summer midday particulate carbon—primary and secondary aerosol, *Environ. Sci. Technol.* **25**, 1788–1793.

Turpin, B. J., Liu, S. P., Podolske, K. S., Gomes, M. S. P., Eisenreich, S. J., and McMurry, P. H. (1993) Design and evaluation of a novel diffusion separator for measuring gas/particle distributions of semivolatile organic compounds, *Environ. Sci. Technol.* **27**, 2441–2449.

United States Environmental Protection Agency (1996) *Air Quality Criteria for Particulate Matter*, EPA/600/P-95/001.

Valerio, F., and Lazzarotto, A. (1985) Photochemical degradation of polycylic aromatic hydrocarbons (PAH) in real and laboratory conditions, *Int. J. Environ. Anal. Chem.* **23**, 135–151.

Venkataraman, C., and Friedlander, S. K. (1994a) Size distributions of polycyclic aromatic hydrocarbons and elemental carbon 1. Sampling, measurement methods and source characterization, *Environ. Sci. Technol.* **28**, 555–562.

Venkataraman, C., and Friedlander, S. K. (1994b) Size distributions of polycyclic aromatic hydrocarbons and elemental carbon 2. Ambient measurements and effects of atmospheric processes, *Environ. Sci. Technol.* **28**, 563–572.

Wagner, H. G. (1981) Soot formation—an overview, in *Particulate Carbon Formation During Combustion*, D. C. Siegla and G. W. Smith, eds., Plenum, New York, pp. 1–29.

Wang, T. S., Matula, R. A., and Farmer, R. C. (1983) Combustion kinetics of soot formation from toulene, *Proc. 19th Int. Symp. Combustion*, The Combustion Institute, Pittsburgh, PA, pp. 1149–1158.

Watson, J. G., and Chow, J. C. (1994) Clear sky visibility as a challenge for society, *Annu. Rev. Energy Environ.* **19**, 241–266.

Watson, J. G., Chow, J. C., Lowenthal, L. C., Pritchett, C. A., Frazier, C. A., Neuroth, G. R., and Robbins, R. (1994) Differences in the carbon composition of source profiles for diesel- and gasoline-powered vehicles, *Atmos. Environ.* **28**, 2493–2505.

Watson, J. G., Chow, J. C., Pritchett, L. C., Houck, J. A., Ragassi, R. A., and Burns, S. (1990) Chemical source profiles for particulate motor vehicle exhaust under cold and high altitude operating conditions, *Sci. Total Environ.* **93**, 183–190.

White, W. H., and Macias, E. S. (1989) Carbonaceous particles and regional haze in the Western United States, *Aerosol Sci. Technol.* **10**, 106–110.

Wils, E. R. J., Hulst, A. G., and Hartog, J. C. (1982) The occurrence of plant wax constituents in airborne particulate matter in an urban area, *Chemosphere* **11**, 1087–1096.

Wolff, G. T., Groblicki, P. J., Cadle, S. H., and Countess, R. J. (1982) Particulate carbon at various locations in the United States, in *Particulate Carbon: Atmospheric Life Cycle*, G. T. Wolff and R. L. Klimsch, eds., Plenum, New York, pp. 297–315.

Wolff, G. T., and Korsog, P. E. (1985) Estimates of the contribution of sources to inhalable particulate concentrations in Detroit, *Atmos. Environ.* **19**, 1399–1409.

Wolff, G. T., Ruthkosky, M. S., Stroup, D. P., and Korsog, P. E. (1991) A characterization of the principal PM-10 species in Claremont (summer) and Long Beach (fall) during SCAQS, *Atmos. Environ.* **25A**, 2173–2186.

Yamasaki, H., Kuwata, K., and Miyamoto, H. (1982) Effects of temperature on aspects of airborne polycyclic aromatic hydrocarbons, *Environ. Sci. Technol.* **16**, 189–194.

Yu, J., Flagan, R. C., and Seinfeld, J. H. (1998) Identification of products containing —COOH, —OH, and — C=O in atmospheric oxidation of hydrocarbons, *Environ. Sci. Technol.* **32**, 2357–2370.

Yu, J., Cocker D. R. III, Griffin, R. J., Flagan, R. C., and Seinfeld, J. H. (1999) Gas-phase oxidation of monoterpenes: Gaseous and particulate products, *J. Atmos. Chem.* **34**, 207–258.

Ziemann, P. J. (2003) Formation of alkoxyhydroperoxy aldehydes and cyclic peroxyhemiacetals from reactions of cyclic alkenes with O_3 in the presence of alcohols, *J. Phys. Chem. A.* **107**, 2048–2060.

15 Interaction of Aerosols with Radiation

15.1 SCATTERING AND ABSORPTION OF LIGHT BY SMALL PARTICLES

When a beam of light impinges on a particle, electric charges in the particle are excited into oscillatory motion. The excited electric charges reradiate energy in all directions (scattering) and may convert a part of the incident radiation into thermal energy (absorption). Electromagnetic radiation transports energy. The amount crossing an area of a detector perpendicular to its direction of propagation is its *intensity*, measured in units of $W\,m^{-2}$. We give the incident intensity of radiation the symbol F_0.

The energy scattered by a particle is proportional to the incident intensity

$$\widetilde{F}_{scat} = C_{scat}F_0 \tag{15.1}$$

where C_{scat}, in units of m^2, is the single-particle scattering cross section. For absorption, the energy absorbed is described analogously

$$\widetilde{F}_{abs} = C_{abs}F_0 \tag{15.2}$$

where $C_{abs}(m^2)$ is the single-particle absorption cross section.

Conservation of energy requires that the light removed from the incident beam by the particle is accounted for by scattering in all directions and absorption in the particle. The combined effect of scattering and absorption is referred to as *extinction*, and an extinction cross section (C_{ext}) can be defined by

$$C_{ext} = C_{scat} + C_{abs} \tag{15.3}$$

C_{ext} has the units of area; in the language of geometric optics, one would say that the particle casts a "shadow" of area C_{ext} on the radiative energy passing the particle. This shadow, however, can be considerably greater, or much less, than the particle's geometric shadow. The dimensionless *scattering efficiency* of a particle Q_{scat}, is C_{scat}/A, where A is the cross-sectional area of the particle. Defining Q_{abs} and Q_{ext} in the same way, we obtain

$$Q_{ext} = Q_{scat} + Q_{abs} \tag{15.4}$$

Atmospheric Chemistry and Physics: From Air Pollution to Climate Change, Second Edition, by John H. Seinfeld and Spyros N. Pandis. Copyright © 2006 John Wiley & Sons, Inc.

The ratio of Q_{scat} to Q_{ext} is called the *single-scattering albedo*,

$$\omega = \frac{Q_{scat}}{Q_{ext}} = \frac{C_{scat}}{C_{ext}} \tag{15.5}$$

Thus the fraction of light extinction that is scattered by a particle is ω, and the fraction absorbed is $1 - \omega$.

Light scattering mechanisms of particles can be divided into three categories:

Elastic scattering—the wavelength (frequency) of the scattered light is the same as that of the incident beam λ_0.

Quasi-elastic scattering—the wavelength (frequency) shifts owing to Doppler effects and diffusion broadening.

Inelastic scattering—the emitted radiation has a wavelength different from that of the incident radiation.

Figure 15.1 depicts the various processes that can occur when radiation of wavelength λ_0 interacts with a particle. Inelastic scattering processes include Raman scattering and fluorescence. For the interaction of solar radiation with atmospheric aerosols, elastic light scattering is the process of interest.

The absorption and elastic scattering of light by a spherical particle is a classical problem in physics, the mathematical formalism of which is called Mie theory (sometimes

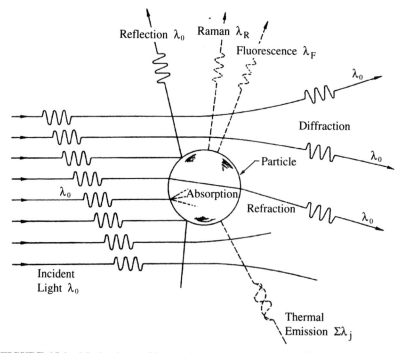

FIGURE 15.1 Mechanisms of interaction between incident radiation and a particle.

also Mie–Debye–Lorenz theory). It is a beautiful and elegant theory, which is the subject of entire treatises. In particular, we refer the reader to Kerker (1969) and Bohren and Huffman (1983). The key parameters that govern the scattering and absorption of light by a particle are (1) the wavelength λ of the incident radiation; (2) the size of the particle, usually expressed as a dimensionless size parameter

$$\alpha = \frac{\pi D_p}{\lambda} \tag{15.6}$$

the ratio of the circumference of the particle to the wavelength of light; and (3) the particle optical property relative to the surrounding medium, the *complex refractive index*:

$$N = n + ik \tag{15.7}$$

Both the real part n and the imaginary part k of the refractive index are functions of λ. The real and imaginary parts of the refractive index represent the nonabsorbing and absorbing components, respectively. Refractive index N is usually normalized by the refractive index of the medium, N_0, and denoted by m

$$m = \frac{N}{N_0} \tag{15.8}$$

where the medium of interest to us is air. Since the refractive index of air is effectively unity, for example, $N_0 = 1.00029 + 0i$ at $\lambda = 589\,\text{nm}$, for all practical purposes N and m are identical. Henceforth the refractive index will be denoted as $m = n + ik$, with the understanding that it is referenced to that of air.

Table 15.1 gives the real and imaginary parts of the refractive index of water. In the visible range of the spectrum, 400–700 nm, because of the negligibly small value of k, water is totally nonabsorbing. Table 15.2 summarizes the refractive indices of a number of other substances of atmospheric interest. The most significant absorbing component in atmospheric particles is elemental carbon (soot); this is reflected in the relatively large size of the imaginary component k.

The angular distribution of light intensity scattered by a particle at a given wavelength is called the *phase function*, or the *scattering phase function*; it is the scattered intensity at a particular angle θ relative to the incident beam and normalized by the integral of the scattered intensity at all angles

$$P(\theta, \alpha, m) = \frac{F(\theta, \alpha, m)}{\int_0^\pi F(\theta, \alpha, m) \sin\theta \, d\theta} \tag{15.9}$$

where $F(\theta, \alpha, m)$ is the intensity scattered into angle θ. (Note that the assumption of a spherical particle allows removal of the azimuthal ϕ dependence of P.) The integral of the phase function over the unit sphere centered on the particle is 4π:

$$\int_0^{2\pi} \int_0^\pi P(\theta, \alpha, m) \sin\theta \, d\theta \, d\phi = 4\pi \tag{15.10}$$

TABLE 15.1   Refractive Index, Real n, and Imaginary k Part for Liquid Water

λ, μm	n	k
0.25	1.362	3.35×10^{-8}
0.30	1.349	1.6×10^{-8}
0.35	1.343	6.5×10^{-9}
0.40	1.339	1.86×10^{-9}
0.45	1.337	1.02×10^{-9}
0.50	1.335	1.00×10^{-9}
0.55	1.333	1.96×10^{-9}
0.60	1.332	1.09×10^{-8}
0.65	1.331	1.64×10^{-8}
0.70	1.331	3.35×10^{-8}
0.75	1.330	1.56×10^{-7}
0.80	1.329	1.25×10^{-7}
0.85	1.329	2.93×10^{-7}
0.90	1.328	4.86×10^{-7}
0.95	1.327	2.93×10^{-6}
1.0	1.327	2.89×10^{-6}
1.2	1.324	9.89×10^{-6}
1.4	1.321	1.38×10^{-4}
1.6	1.317	8.55×10^{-5}
1.8	1.312	1.15×10^{-4}
2.0	1.306	1.10×10^{-3}
2.2	1.296	2.89×10^{-4}
2.4	1.279	9.56×10^{-4}
2.6	1.242	3.17×10^{-3}
2.8	1.142	0.115
3.0	1.371	0.272
3.2	1.478	0.0924
3.4	1.420	0.0195
3.6	1.385	0.00515
3.8	1.364	0.00340
4.0	1.351	0.00460
4.2	1.342	0.00688
4.4	1.334	0.00103
4.6	1.330	0.0147
4.8	1.330	0.0150
5.0	1.325	0.0124
5.5	1.298	0.0116
6.0	1.265	0.107
6.5	1.339	0.0392
7.0	1.317	0.0320
7.5	1.304	0.0326
8.0	1.291	0.0343
9.0	1.262	0.0399
10.0	1.218	0.0508
11.0	1.153	0.0968
12.0	1.111	0.199
14.0	1.210	0.370
16.0	1.325	0.422
18.0	1.423	0.426
20.0	1.480	0.393

Source: Hale and Querry (1973).

TABLE 15.2 Refractive Indices of Atmospheric Substances at
$\lambda = 589\,\text{nm}$ **(Unless Otherwise Indicated)**

		$m = n + ik$	
Substance	n	k	
Water	1.333	0 (see Table 15.1)	
Water (ice)	1.309		
NaCl	1.544	0	
H_2SO_4	1.426^a	0	
NH_4HSO_4	1.473^b	0	
$(NH_4)_2SO_4$	1.521^b	0	
SiO_2	1.55	0	($\lambda = 550\,\text{nm}$)
Carbon^c	1.95	−0.79	($\lambda = 550\,\text{nm}$)
Mineral dust^d	1.56	−0.006	($\lambda = 550\,\text{nm}$)

^{a}Stelson (1990), assuming a 97% pure (by mass) mixture of H_2SO_4 with H_2O.
^{b}Weast (1987).
^{c}Bond and Bergstrom (2006) report a narrow range of refractive indices of light-absorbing carbon. The value in the table represents the upper limit.
^{d}Tegen et al. (1996).

Several useful parameters can be derived from the phase function to express the distribution of scattered intensity. Henceforth we will not indicate explicitly the dependence on α and m, but it is understood. The *asymmetry parameter g* is defined as the intensity-weighted average of the cosine of the scattering angle;

$$
\begin{aligned}
g &= \frac{1}{2} \frac{\int_0^\pi \cos\theta \, F(\theta) \sin\theta \, d\theta}{\int_0^\pi F(\theta) \sin\theta \, d\theta} \\
&= \frac{1}{2} \int_0^\pi \cos\theta \, P(\theta) \sin\theta \, d\theta
\end{aligned}
\tag{15.11}
$$

The factor of $\frac{1}{2}$ ensures that $g = 1$ for light scattered totally at $\theta = 0°$ (the forward direction) and $g = -1$ for light scattered completely at $\theta = 180°$ (the backward direction). For a particle that scatters light isotropically (the same in all directions), $g = 0$. A positive value of g indicates that the particle scatters more light in the forward as opposed to the backward direction; a negative g signifies the reverse.

Another measure of the distribution of scattered intensity is the *hemispheric backscatter ratio b*, often referred to simply as the backscatter ratio. It is the fraction of the scattered intensity that is redirected into the backward hemisphere of the scattering particle:

$$
b = \frac{\int_{\pi/2}^\pi P(\theta) \sin\theta \, d\theta}{\int_0^\pi P(\theta) \sin\theta \, d\theta}
\tag{15.12}
$$

An advantage of the backscatter ratio is that it can be measured directly with an instrument known as a nephelometer; determination of g requires knowledge of the entire phase function. There is no general one-to-one relationship between g and b, but in certain instances one-to-one relationships can be derived (Wiscombe and Grams 1976; Marshall et al., 1995).

It is possible to derive expressions for the cross sections of a spherical particle exactly (Bohren and Huffman 1983). The formulas for Q_{scat} and Q_{ext} are

$$Q_{scat}(m, \alpha) = \frac{2}{\alpha^2} \sum_{k=1}^{\infty} (2k + 1) \left[|a_k|^2 + |b_k|^2 \right] \tag{15.13}$$

$$Q_{ext}(m, \alpha) = \frac{2}{\alpha^2} \sum_{k=1}^{\infty} (2k + 1) \mathrm{Re}[a_k + b_k] \tag{15.14}$$

where

$$a_k = \frac{\alpha \psi_k'(y) \psi_k(\alpha) - y \psi_k'(\alpha) \psi_k(y)}{\alpha \psi_k'(y) \zeta_k(\alpha) - y \zeta_k'(\alpha) \psi_k(y)}$$

$$b_k = \frac{y \psi_k'(y) \psi_k(\alpha) - \alpha \psi_k'(\alpha) \psi_k(y)}{y \psi_k'(y) \zeta_k(\alpha) - \alpha \zeta_k'(\alpha) \psi_k(y)}$$

with $y = \alpha m$. The functions $\psi_k(z)$ and $\zeta_k(z)$ are the Riccati–Bessel functions:

$$\psi_k(z) = \left(\frac{\pi z}{2} \right)^{1/2} J_{k+1/2}(z) \tag{15.15}$$

$$\zeta_k(z) = \left(\frac{\pi z}{2} \right)^{1/2} [J_{k+1/2}(z) + i(-1)^k J_{-k-1/2}(z)] \tag{15.16}$$

where $J_{k+1/2}$ and $J_{-k-1/2}$ are Bessel functions of the first kind. A procedure for evaluating the Mie formulas is given in Appendix 15.

Mie theory can serve as the basis of a computational procedure to calculate the scattering and absorption of light by any sphere as a function of wavelength (Bohren and Huffman 1983). There are, in addition, approximate expressions, valid in certain limiting cases, that provide insight into the physics of the problem. On the basis of the α value, light scattering can be divided into three domains:

$\alpha \ll 1$ *Rayleigh* scattering (particle small compared with the wavelength)
$\alpha \simeq 1$ *Mie* scattering (particle of about the same size as the wavelength)
$\alpha \gg 1$ *Geometric* scattering (particle large compared with the wavelength)

15.1.1 Rayleigh Scattering Regime

When the particle is much smaller than the wavelength of incident light, this is called the *Rayleigh scattering regime*, in which a closed form solution of the scattering problem is possible. With respect to light in the visible range of the spectrum, particles of diameter $\leq 0.1\ \mu m$ lie in the Rayleigh scattering regime. In this regime the pattern of scattered light intensity is symmetrical in the forward and backward directions and more or less independent of particle shape (Figure 15.2). With reference to Figure 15.2, if θ is the angle between the incident beam and the scattered beam, the scattering phase function is

$$P(\theta) = \frac{\lambda^2}{8\pi^2} \left(\frac{\pi D_p}{\lambda} \right)^6 \left| \frac{m^2 - 1}{m^2 + 2} \right| (1 + \cos^2 \theta) F_0 \tag{15.17}$$

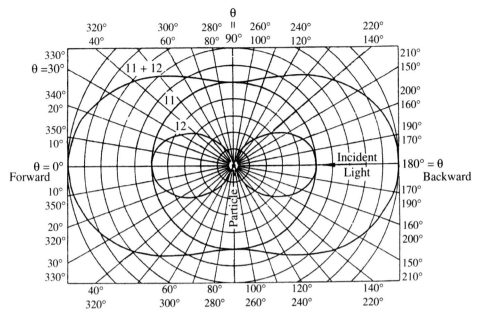

FIGURE 15.2 Pattern of light scattering (scattering phase function) by a particle in the Rayleigh regime. The scattered light intensity pattern is symmetric in the forward and backward directions, totally polarized at 90°, and independent of particle shape. Incident beam enters from the right. 11 indicates the circular component independent of θ; 12 is the θ-dependent term.

$P(\theta)$ consists of the product of a circular component independent of θ and the $(1 + \cos^2 \theta)$ term. These two terms are shown in Figure 15.2. Line 11 depicts the circularly symmetric part, and line 12 is the θ-dependent part. The total scattering intensity is the product of the two. If the term $|(m^2 - 1)/(m^2 + 2)|$ is weakly dependent on wavelength (this is not always the case), the irradiance scattered by a sphere small compared with the wavelength, or, indeed, any sufficiently small particle regardless of its shape, is proportional to $1/\lambda^4$. Such behavior is often referred to as *Rayleigh scattering*. Thus, at increasingly longer wavelengths, the scattered intensity falls off as the fourth power of λ. As a result, small particles scatter light of short wavelengths more effectively than they do light of long wavelengths. A consequence of this phenomenon is the reddening of white light on passing through a population of very small particles. The shorter-wavelength blue component is preferentially scattered out of the line of sight, leaving the redder components to reach the observer. This can be demonstrated by putting a few drops of milk into a glass of water. A collimated beam of white light assumes a reddish tint after transmission through this suspension because the shorter-wavelength blue light is extinguished more effectively than the longer-wavelength red light. Increasing extinction toward shorter wavelengths is a general characteristic of nonabsorbing particles that are small compared with the wavelength of light. This effect is responsible for the reddish hue of sunsets. Lord Rayleigh observed that the sky is bluest when air is purest. (The ideal atmosphere, one totally devoid of particles, is sometimes referred to as a *Rayleigh atmosphere*.)

The extinction and scattering efficiencies for particles in the Rayleigh scattering regime are (to terms of order α^4) (Bohren and Huffman 1983)

$$Q_{ext}(m, \alpha) = 4\alpha \, \text{Im}\left\{\frac{m^2 - 1}{m^2 + 2}\left[1 + \frac{\alpha^2}{15}\left(\frac{m^2 - 1}{m^2 + 2}\right)\frac{m^4 + 27m^2 + 38}{2m^2 + 3}\right]\right\}$$
$$+ \frac{8}{3}\alpha^4 \text{Re}\left\{\left(\frac{m^2 - 1}{m^2 + 2}\right)^2\right\} \tag{15.18}$$

$$Q_{scat}(m, \alpha) = \frac{8}{3}\alpha^4\left|\frac{m^2 - 1}{m^2 + 2}\right|^2 \tag{15.19}$$

If $|m|\alpha \ll 1$, the coefficient (in brackets) of $(m^2 - 1)/(m^2 + 2)$ in the first term of (15.18) is approximately unity. As a result, the absorption efficiency is

$$Q_{abs}(m, \alpha) = 4\alpha \, \text{Im}\left\{\frac{m^2 - 1}{m^2 + 2}\right\}\left[1 + \frac{4}{3}\alpha^3 \, \text{Im}\left\{\frac{m^2 - 1}{m^2 + 2}\right\}\right] \tag{15.20}$$

Thus, if $(4\alpha^3/3) \, \text{Im}\{(m^2 - 1)/(m^2 + 2)\} \ll 1$, a condition satisfied for sufficiently small α, the absorption efficiency is approximately

$$Q_{abs}(m, \alpha) = 4\alpha \, \text{Im}\left\{\frac{m^2 - 1}{m^2 + 2}\right\} \tag{15.21}$$

In this case, the absorption cross section $C_{abs} = (\pi D_p^2/4)Q_{abs}$, is proportional to D_p^3 and therefore to the volume of the particle, whereas the absorption efficiency Q_{abs} is proportional to D_p. Since $\alpha \ll 1$, $Q_{abs} > Q_{scat}$ in this regime.

In summary, for sufficiently small particles, if $(m^2 - 1)/(m^2 + 2)$ is a weak function of wavelength, the wavelength dependences of the scattering and absorption efficiencies are

$$Q_{scat} \approx \lambda^{-4} \qquad Q_{abs} \approx \lambda^{-1}$$

In both scattering and absorption, shorter wavelengths are extinguished more effectively than longer wavelengths. This is the "reddening" of the spectrum referred to above.

15.1.2 Geometric Scattering Regime

Particles for which $\alpha \gg 1$ fall into the so-called geometric scattering regime. In this case the scattering can be determined on the basis of the geometrical optics of reflection, refraction, and diffraction. Scattering is strongly dependent on particle shape and orientation relative to the incoming beam. Let us consider a large, weakly absorbing sphere. We do so because water droplets are the most important class of "large" particles in the atmosphere and they are, for all practical purposes, nonabsorbing. Using geometric optics it can be shown for a weakly absorbing sphere ($4\alpha k \ll 1$) that (Bohren and Huffman 1983)

$$Q_{abs}(m, \alpha) = \frac{8}{3}\alpha\frac{k}{n}\left[n^3 - (n^2 - 1)^{3/2}\right] \tag{15.22}$$

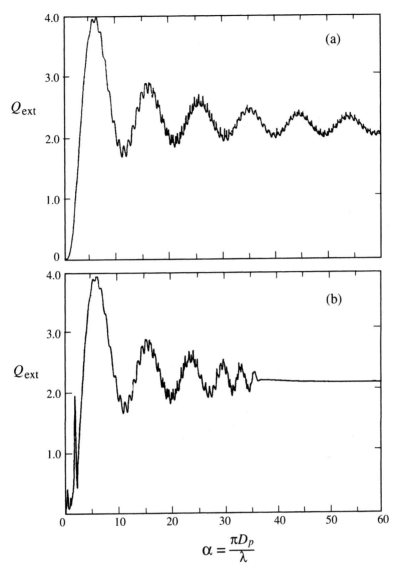

FIGURE 15.3 Extinction efficiency Q_{ext} for a water droplet: (a) wavelength of the radiation is constant at $\lambda = 0.5 \, \mu m$ and diameter is varied; (b) diameter $= 2 \, \mu m$ and wavelength is varied (Bohren and Huffman 1983).

The extinction efficiency of a water droplet as a function of the size parameter α is shown in Figure 15.3. We note that Q_{ext} approaches the limiting value 2 as the size parameter increases:

$$\lim_{\alpha \to \infty} Q_{ext}(m, \alpha) = 2 \qquad (15.23)$$

This is twice as large as predicted by geometric optics, which is the so-called extinction paradox (Bohren and Huffman 1983). In qualitative terms, the incident wave is influenced beyond the physical boundaries of the sphere; the edge of the sphere deflects rays in its neighborhood, rays that, from the viewpoint of geometric optics, would have passed unimpeded. Also, all the geometrically incident light that is not externally reflected enters the sphere and is absorbed, as long as the absorptive part of the refractive index is not identically zero.

15.1.3 Scattering Phase Function

The scattering phase function describes the angle-dependent scattering of light incident on a particle. Figure 15.4 shows the scattering phase functions for $(NH_4)_2SO_4$ aerosol at 80% RH for several particle sizes at $\lambda = 550\,nm$. At small diameter, the phase function is symmetric in the forward and backward directions (Figure 15.2). Note that, for all but the very smallest particles, the scattering is highly peaked in the forward direction. The directional asymmetry becomes more and more pronounced as the particle size increases. The small scattering lobes for $\theta > 90°$ become almost imperceptible compared with the strong forward-scattering lobes for the larger particles. The consequence of strong forward scattering is the reason why driving into a bright setting sun can produce a blinding glare. A similar phenomenon occurs at night in a fog when light from oncoming automobile headlights is scattered in the forward direction.

15.1.4 Extinction by an Ensemble of Particles

The foregoing discussion relates to the scattering of light by a single particle. A rigorous treatment of the scattering by an ensemble of particles is very complicated. If, however, the average distance between particles is large compared to the particle size, it can be assumed that the total scattered intensity is the sum of the intensities scattered by individual particles, and single-particle scattering theory can be used. Such an assumption is easily obeyed for even the most concentrated atmospheric conditions. Consider, for example, a total particle concentration of $10^6\,cm^{-3}$, consisting of monodisperse 1-μm-diameter spheres, a number density well above that of most ambient conditions. Even at this extreme aerosol loading, the volume fraction occupied by particles per cm^3 of air is only $(\pi/6) \times 10^{-6}$.

Consider the solar beam traversing an atmospheric layer containing aerosol particles. Light extinction occurs by the attenuation of the incident light by scattering and absorption as it traverses the layer. Recalling Section 4.3, the fractional reduction in intensity over an incremental depth of the layer dz can be expressed as $dF = -b_{ext}F\,dz$, where b_{ext} is the extinction coefficient, having units of inverse length (m^{-1})

$$b_{ext} = C_{ext}N \tag{15.24}$$

where N is the total particle number concentration (particles m^{-3}). Equation (15.24) is written, for the moment, for a collection of monodisperse particles. Thus b_{ext} is the fractional loss of intensity per unit pathlength, and

$$\frac{dF}{dz} = -b_{ext}F \tag{15.25}$$

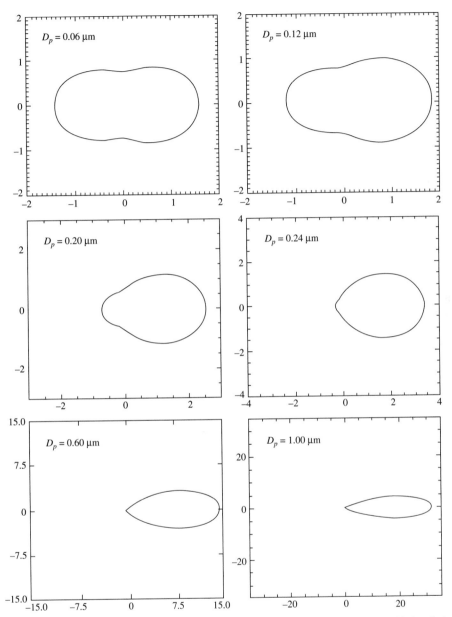

FIGURE 15.4 Aerosol scattering phase function for $(NH_4)_2SO_4$ particles at 80% relative humidity at $\lambda = 550\,nm$ (Nemesure et al. 1995). Incident beam enters from the left.

If F_0 is the incident flux at $z = 0$, taken to represent, say, the top of the layer, then intensity at any distance z into the layer is

$$\frac{F}{F_0} = \exp(-b_{ext}z) \qquad (15.26)$$

As noted in Chapter 4, the dimensionless product $\tau = b_{ext}\, z$ is called the *optical depth* of the layer, and (15.26) is called the Beer–Lambert law. In the visible portion of the spectrum, optical depth of tropospheric aerosols can range from less than 0.05 in remote, pristine environments to close to 1.0 near the source of intense particulate emissions such as in the plume of a forest fire.

For a population of monodisperse, spherical particles at a number concentration of N, the extinction coefficient is related to the dimensionless extinction efficiency by

$$b_{ext} = \frac{\pi D_p^2}{4} N Q_{ext} \qquad (15.27)$$

The extinction coefficient can be expressed as the sum of a scattering coefficient b_{scat} and an absorption coefficient b_{abs}

$$b_{ext} = b_{scat} + b_{abs} \qquad (15.28)$$

with similar relations to Q_{scat} and Q_{abs}.

That extinction is the sum of scattering and absorption, and that either of the two phenomena can dominate extinction, is illustrated nicely by the classroom demonstration described by Bohren and Huffman (1983):

> Two transparent containers (Petri dishes serve quite well) are filled with water, placed on an overhead projector, and their images focused on a screen. To one container, a few drops of milk are added; to the other, a few drops of India ink. The images can be changed from clear to a reddish hue to black by increasing the amount of milk or ink. Indeed, both images can be adjusted so that they appear equally dark; in this instance it is not possible, judging solely by the light transmitted to the screen, to distinguish one from the other: the amount of extinction is about the same. But the difference between the two suspensions immediately becomes obvious if one looks directly at the containers: the milk is white whereas the ink is black. Milk is a suspension of very weakly absorbing particles which therefore attenuate light primarily by scattering; India ink is a suspension of very small carbon particles which attenuate light primarily by absorption. Although this demonstration is not meant to be quantitative, and its complete interpretation is complicated somewhat by multiple scattering, it clearly shows the difference between extinction by scattering and extinction by absorption. Moreover, it shows that merely by observing transmitted light it is not possible to determine the relative contributions of absorption and scattering to extinction; to do so requires an additional independent observation.

We can decompose b_{abs} and b_{scat} into contributions from the gas and particulate components of the atmosphere

$$b_{abs} = b_{ag} + b_{ap} \qquad (15.29)$$
$$b_{scat} = b_{sg} + b_{sp} \qquad (15.30)$$

where

b_{sg} = scattering coefficient due to gases (the so-called Rayleigh scattering coefficient)
b_{sp} = scattering coefficient due to particles
b_{ag} = absorption coefficient due to gases
b_{ap} = absorption coefficient due to particles

A ratio b_{scat}/b_{sg} of unity indicates the cleanest possible air $(b_{sp} = 0)$. Thus the higher this ratio, the greater is the contribution of particulate scattering to total light scattering.

15.2   VISIBILITY

Visibility degradation is probably the most readily perceived impact of air pollution. The term "visibility" is generally used synonymously with "visual range," meaning the farthest distance at which one can see a large, black object against the sky at the horizon. Even if no distant objects are within view, subjective judgments about visual range can be made based on the coloration and light intensity of the sky and nearby objects. For example, one perceives reduced visual range if a distant mountain that is usually visible cannot be seen, if nearby objects look "hazy" or have diminished contrast, or if the sky is white, gray, yellow, or brown, instead of blue.

Several factors determine how far one can see through the atmosphere, including optical properties of the atmosphere, amount and distribution of light, characteristics of the objects observed, and properties of the human eye. Visibility is reduced by the absorption and scattering of light by both gases and particles. Absorption of certain wavelengths of light by gas molecules and particles is sometimes responsible for atmospheric colorations. However, light scattering by particles is the most important phenomenon responsible for impairment of visibility. Visibility is reduced when there is significant scattering because particles in the atmosphere between the observer and the object scatter light from the sun and other parts of the sky through the line of sight of the observer. This light decreases the contrast between the object and the background sky, thereby reducing visibility. These effects are depicted in Figure 15.5.

To examine the effect of atmospheric constituents on visibility reduction, we consider the case in which a black object is being viewed against a white background. We first

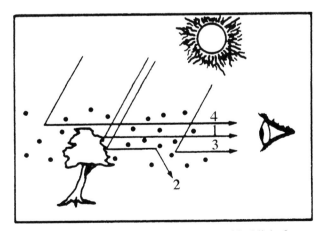

FIGURE 15.5   Contributions to atmospheric visibility: (1) residual light from target reaching the observer; (2) light from the target scattered out of the observer's line of sight; (3) airlight from intervening atmosphere scattered into the observer's line of sight; and (4) airlight constituting the horizon sky.

define the visual contrast at a distance x from the object, $C_V(x)$, as the relative difference between the light intensity of the background and the target

$$C_V(x) = \frac{F_B(x) - F(x)}{F_B(x)} \qquad (15.31)$$

where $F_B(x)$ and $F(x)$ are the intensities of the background and the object, respectively. At the object $(x = 0)$, $F(0) = 0$, since the object is assumed to be black and therefore absorbs all light incident on it. Thus $C_V(0) = 1$. Over the distance x between the object and the observer, $F(x)$ will be affected by two phenomena: (1) absorption of light by gases and particles and (2) addition of light that is scattered into the line of sight. Light can reenter the beam by multiple scattering; that is, light scattered by particles outside the beam may ultimately contribute to the irradiance at the target. This is because scattered light, in contrast to absorbed light, is not irretrievably lost from the system—it merely changes direction and is lost from a beam propagating in a particular direction—but contributes to other directions. The greater the individual particle scattering coefficient, number concentration of particles, and depth of the beam, the greater will be the multiple scattering contribution to the irradiance at x. Over a distance dx the intensity change dF is the result of these two effects. The fraction of F diminished is assumed to be proportional to dx, since dx is a measure of the amount of suspended gases and particles present. The fractional reduction in F is written $dF = -b_{ext}F\,dx$. In addition, the intensity F can be increased over the distance dx by scattering of light from the background into the line of sight. The increase can be expressed as $b'F_B(x)dx$, where b' is a constant. The net change in intensity is given by

$$dF(x) = [b'F_B(x) - b_{ext}F(x)]dx \qquad (15.32)$$

By its definition as background intensity, F_B must be independent of x. Thus, along any other line of sight,

$$dF_B(x) = 0 = [b'F_B(x) - b_{ext}F_B(x)]dx \qquad (15.33)$$

We see that $b' = b_{ext}$. Thus we find that the contrast $C_V(x)$ itself obeys the Beer–Lambert law

$$\frac{dC_V(x)}{dx} = -b_{ext}C_V(x) \qquad (15.34)$$

and therefore that the contrast decreases exponentially with distance from the object,

$$C_V(x) = \exp(-b_{ext}x) \qquad (15.35)$$

The lowest visually perceptible brightness contrast is called the *liminal contrast* or *threshold contrast*. The threshold contrast has been the object of considerable interest since it determines the maximum distances at which various components of a scene can be discerned. Laboratory experiments indicate that for most daylight viewing conditions, contrast ratios as low as 0.018–0.03 are perceptible. Typical observers can detect a 0.02 or greater contrast between large, dark objects and the horizon sky. A threshold contrast value of 2% $(C_V = 0.02)$ is usually employed for visual range calculations.

Equation (15.35) can be evaluated at the distance at which a black object has a standard 0.02 contrast ratio against a white background. When the contrast in (15.35) becomes the threshold contrast, the distance becomes the visual range. If $C_V = 0.02$, then

$$x_v = \frac{3.912}{b_{ext}} \tag{15.36}$$

with x_v and b_{ext} in similar units (i.e., x in m and b_{ext} in m^{-1}). This is called the *Koschmeider equation*. Thus the visual range can be expressed equivalently in terms of an extinction coefficient (b_{ext}) or a distance (x_v). If the extinction coefficient is measured along a sight path, then x_v is the visual range. If the extinction coefficient is measured at a point, then x_v is taken to be the local visual range. The two values of x_v are equal in a homogenous atmosphere.

At sea level the Rayleigh atmosphere has an extinction coefficient b_{ext} of approximately 13.2×10^{-6} m^{-1} at $\lambda = 520$ nm wavelength, limiting visibility in the cleanest possible atmosphere to about 296 km. Rayleigh scattering decreases with altitude and is proportional to air density. At $\lambda = 520$ nm wavelength:

Altitude above Sea Level, km	$b_{sg} \times 10^6$, m^{-1}
0	13.2
1.0	11.4
2.0	10.6
3.0	9.7
4.0	8.8

For convenience, a megameter, denoted Mm and equal to 1000 km, is often used as the unit of distance; in this unit the sea-level Rayleigh extinction coefficient is 13 Mm^{-1}.

Rayleigh scattering thus represents an irreducible level of extinction against which other extinction components (e.g., anthropogenic pollutants) can be compared. Figure 15.6 shows median midday visual range in the United States in the mid-1990s. The greatest visual range in the United States is present in the western states. Indeed, maintenance of the visual range in this region of the country, where there are many national parks, has been a stated objective in the U.S. Clean Air Acts.

Scattering by particles of sizes comparable to the wavelength of visible light (the Mie scattering range) is mostly responsible for visibility reduction in the atmosphere. Scattering by particles accounts for 50–95% of extinction, depending on location, with urban sites in the 50–80% range and nonurban sites in the 80–95% range. Particle absorption is on the order of 5–10% of particle extinction in remote areas; its contribution can rise to 50% in urban areas. Values of 10–25% are typical for suburban and rural locations. As we will see shortly, particles in the range 0.1–1 μm in diameter are the most effective, per unit aerosol mass, in reducing visibility. The scattering coefficient b_{scat} is more or less directly dependent on the atmospheric aerosol concentration in this size range. Scattering by air molecules usually has a minor influence on urban visibility. For visual ranges exceeding 30 km the effect of air molecules must be taken into account. The addition of small amounts of submicrometer particles throughout the viewing distance tends to whiten the horizon sky, making distant dark objects and intervening airlight

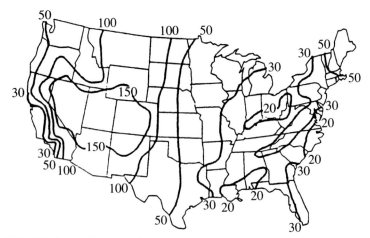

FIGURE 15.6 Median midday visual range (km) in the United States [based on data given in Malm et al. (1994)]. The visibility isopleths are the result of a national visibility and aerosol monitoring network of 36 stations established in 1988 to track spatial and temporal trends of visibility and visibility-reducing particles in the United States. The major visibility-reducing aerosol species—sulfates, nitrates, organics, carbon, and windblown dust—are monitored. Sulfates and organics are responsible for most of the extinction at most locations throughout the United States. In the eastern United States, sulfates contribute about two-thirds of the extinction. In almost all cases, extinction is highest in the summer and lowest in the winter months.

appear more gray. On the whole, as we have seen, particles generally scatter more light in the forward direction than in other directions (see Figure 15.4); thus haze appears bright in the forwardscatter mode and dark in the backscatter mode.

Nitrogen dioxide (NO_2) is the only light-absorbing atmospheric gas present in optically significant quantities in the troposphere. NO_2 absorbs light selectively and is strongly blue-absorbing; as a result, its presence will color plumes red, brown, or yellow. In spite of the coloration properties of NO_2, nevertheless, the brown haze characteristic of smoggy atmospheres is largely a result of aerosol scattering rather than NO_2 absorption (Charlson and Ahlquist 1969).

Table 15.3 presents estimated contributions of chemical species to the wintertime extinction coefficient in Denver, Colorado about 1980. We note, for example, that in

TABLE 15.3 Contribution of Chemical Species to the Extinction Coefficient b_{ext} in Denver Wintertime

Fine-Particle Species	Mean Percent Contribution
$(NH_4)_2SO_4$	20.2
NH_4NO_3	17.2
Organic C	12.5
Elemental C (scattering)	6.5
Elemental C (absorption)	31.2
Other	6.6
NO_2	5.7
Total	100

Source: Groblicki et al. (1981).

Denver during winter, elemental carbon was responsible for about 40% of visibility reduction. The disproportionately large influence of black carbon on light extinction, relative to its concentration, arises in part, as we will see, because black carbon is more effective than nonabsorbing aerosol particles (such as sulfates and nitrates) of the same size in attenuating light (Faxvog and Roessler 1978).

15.3 SCATTERING, ABSORPTION, AND EXTINCTION COEFFICIENTS FROM MIE THEORY

Aerosol scattering, absorption, and extinction coefficients are functions of the particle size, the complex refractive index of the particles m, and the wavelength λ of the incident light. The extinction coefficient for a monodisperse ensemble of particles was given in terms of the dimensionless extinction efficiency by (15.27). With a population of different-sized particles of identical refractive index m with a number size distribution function of $n(D_p)$, the extinction coefficient is given by[1]

$$b_{\text{ext}}(\lambda) = \int_0^{D_p^{\max}} \frac{\pi D_p^2}{4} Q_{\text{ext}}(m, \alpha) n(D_p) \, dD_p \qquad (15.37)$$

where $D_p^{\max}$ is an upper limit diameter for the particle population. Similar expressions can be written for $b_{\text{scat}}(\lambda)$ and $b_{\text{abs}}(\lambda)$ in terms of Q_{scat} and Q_{abs}.

It is frequently useful to express b_{ext} in terms of the aerosol mass distribution function

$$n_M(D_p) = \rho_p \frac{\pi D_p^3}{6} n(D_p) \qquad (15.38)$$

where ρ_p is the density of the particle. The result is

$$b_{\text{ext}}(\lambda) = \int_0^{D_p^{\max}} \frac{3}{2 \rho_p D_p} Q_{\text{ext}}(m, \alpha) n_M(D_p) \, dD_p \qquad (15.39)$$

which can be written as

$$b_{\text{ext}}(\lambda) = \int_0^{D_p^{\max}} E_{\text{ext}}(D_p, \lambda, m) n_M(D_p) \, dD_p \qquad (15.40)$$

where $E_{\text{ext}}(D_p, \lambda, m)$ is the *mass extinction efficiency*:

$$E_{\text{ext}}(D_p, \lambda, m) = \frac{3}{2 \rho_p D_p} Q_{\text{ext}}(m, \alpha) \qquad (15.41)$$

---

[1] Although $b_{\text{abs}}, b_{\text{scat}}$, and b_{ext} can be decomposed into contributions from gases and particles as in (15.29) and (15.30), it is common practice to use these terms to refer to the particulate component only.

Similarly, the *mass scattering* and *mass absorption efficiencies* are

$$E_{\text{scat}}(D_p, \lambda, m) = \frac{3}{2\rho_p D_p} Q_{\text{scat}}(m, \alpha) \tag{15.42}$$

$$E_{\text{abs}}(D_p, \lambda, m) = \frac{3}{2\rho_p D_p} Q_{\text{abs}}(m, \alpha) \tag{15.43}$$

The mass scattering efficiencies of spherical particles of water, ammonium nitrate and sulfate, silica (SiO_2), and carbon (expressed in units of $m^2 g^{-1}$) are shown as a function of particle diameter in Figure 15.7 at a wavelength of $\lambda = 550 \, nm$. We see that particles between 0.1 and 1.0 μm diameter scatter light most efficiently. The mass scattering efficiencies of the four substances in Figure 15.7 exhibit peaks as follows: H_2O, $\sim 0.85 \, \mu m$; NH_4NO_3, $\sim 0.5 \, \mu m$; $(NH_4)_2SO_4$, $\sim 0.5 \, \mu m$; carbon, $\sim 0.2 \, \mu m$. Figure 15.8 shows how the mass scattering and absorption efficiencies vary with systematic variation of the real and imaginary parts of the refractive index. Figures 15.8a–d show how the imaginary component k influences the mass scattering efficiency at a constant value of the real part n.

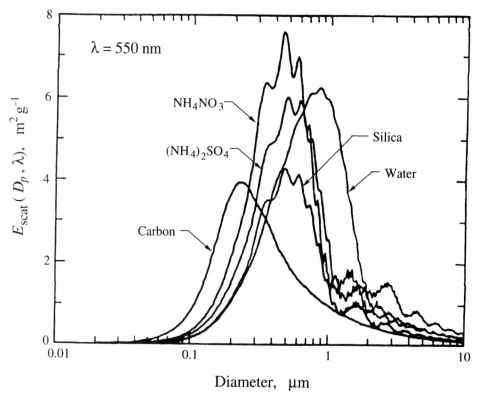

FIGURE 15.7 Mass scattering efficiencies of homogeneous spheres of $(NH_4)_2SO_4$, NH_4NO_3, carbon, H_2O, and silica at $\lambda = 550 \, nm$.

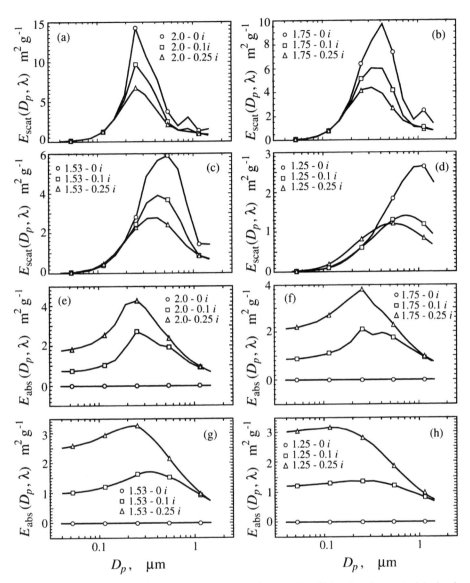

FIGURE 15.8 Mass scattering (a–d) and absorption (e–h) efficiencies for materials having refractive indices $m = n + ik$: $n = 2.00$, 1.75, 1.53, 1.25, and $k = -0$, -0.1, -0.25 [Note that $m = 1.53 - 0k$ is the refractive index of $(NH_4)_2SO_4$.]

We note that, at any given value of n, increasing k lowers the peak value of the mass scattering efficiency and decreases the diameter at which the scattering peak occurs. Both reflect the fact that, as k increases, more of the total photon energy is being transformed into absorption rather than scattering. We also note that the larger the value of n, the higher the peak value of E_{scat} and the smaller the diameter at which peak scattering occurs. Figures 15.8e–h show the corresponding behavior of the mass absorption efficiency, E_{abs}. At any given value of n, increasing k raises the overall E_{abs} curve. The peak value of E_{abs} is weakly affected by n (larger n, larger E_{abs} at the peak). More importantly, for the lower values of n, E_{abs} remains close to its peak value well into the sub-0.1 μm range. The mass absorption efficiency remains high for absorbing particles less than 0.1 μm in diameter, and, as a result, overall light extinction by particles with diameters smaller than 0.1 μm is due primarily to absorption. The black plume of soot from an oil burner exemplifies this behavior, where most of the particles are smaller than 0.1 μm in diameter.

It is of interest to examine the dependence of the mass scattering and absorption coefficients on particle diameter in the two extremes of $\alpha \ll 1 (D_p \ll \lambda)$ and $\alpha \gg 1 (D_p \gg \lambda)$. Consider first $E_{scat}(D_p, \lambda, m)$. In the Rayleigh regime $(D_p \ll \lambda)$, from (15.19) $Q_{scat} \approx D_p^4$, whereas in the regime $D_p \gg \lambda$, Q_{scat} becomes independent of particle size. Thus from (15.42) the mass scattering efficiency varies with particle size in the Rayleigh and large-particle regimes according to

$$
\begin{aligned}
E_{scat}(D_p, \lambda, m) &\approx D_p^3 \quad D_p \ll \lambda \\
&\approx D_p^{-1} \quad D_p \gg \lambda
\end{aligned}
$$

Thus the mass scattering efficiency increases as D_p^3 for the smallest particles and falls off as D_p^{-1} for large particles. This behavior is most clearly evident from the E_{scat} curve for carbon in Figure 15.7. The mass absorption coefficient behavior for $D_p \ll \lambda$ is seen from (15.21), from which $Q_{abs} \approx D_p$. There is no straightforward way to see the dependence of Q_{abs} on D_p for $D_p \gg \lambda$, but $Q_{abs} \approx D_p^0$ in this region. Then, from (15.43), we obtain the dependence of mass absorption efficiency on particle size in the two extremes as

$$
\begin{aligned}
E_{abs}(D_p, \lambda, m) &\approx D_p^0 \quad D_p \ll \lambda \\
&\approx D_p^{-1} \quad D_p \gg \lambda
\end{aligned}
$$

Therefore E_{abs} is independent of particle size for small particles and descreases as D_p^{-1} for large particles. Any of the curves for $k = -0.25$ in Figures 15.8e–h illustrate this behavior.

In the case of a multicomponent particle, the index of refraction m reflects the mixture of species in the particle and can be approximated by the volume average of the indices of refraction of the individual components. Doing so assumes implicitly that the particle is a homogeneous mixture. This is not quite accurate if the particle, in fact, contains a core of one type of material surrounded by a shell of another. Previous studies have shown, however, that the error in refractive index incurred by the homogenous particle assumption, as compared with a calculation of refractive index that explicitly accounts for an insoluble core, is less that 20% (Sloane 1983; Larson et al. 1988). The volume-average index of refraction $\bar{m}$ for an aerosol containing n components is calculated from

$$
\bar{m} = \sum_{i=1}^{n} m_i f_i \tag{15.44}
$$

where m_i is the index of refraction of component i and f_i is the volume fraction of component i.

Ångstrom Exponent   It has proved to be useful in some cases to represent the wavelength dependence of the aerosol extinction coefficient b_{ext}, by $b_{ext} \approx \lambda^{-\mathring{a}}$. The exponent, $\mathring{a}$, is called the *Ångstrom exponent*. The Ångstrom exponent is calculated from measured values of the extinction coefficient as a function of

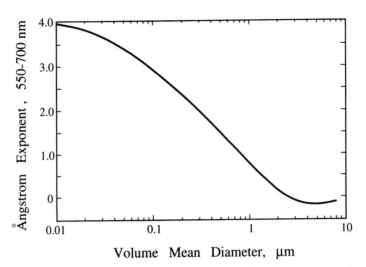

FIGURE 15.9 Ångstrom exponent for a lognormally distributed water aerosol $(\sigma_g = 2.0)$ with refractive index $m = 1.33 - 0i$ in the wavelength range $\lambda = 550$ to 700 nm. (Courtesy of J. A. Ogren.)

wavelength by

$$\overset{\circ}{a} = -\frac{d \log b_{\text{ext}}}{d \log \lambda}$$
$$\cong -\frac{\log \left(b_{\text{ext}_1}/b_{\text{ext}_2}\right)}{\log \left(\lambda_1/\lambda_2\right)} \tag{15.45}$$

If the aerosol number distribution is represented in a certain size range by a relation of the form

$$\frac{dN}{dD_p} \approx D_p^{-v} \tag{15.46}$$

then $\overset{\circ}{a} = v - 3$. Figure 15.9 shows the Ångstrom exponent for a log-normally distributed water aerosol $(\sigma_g = 2.0)$ with refractive index $m = 1.33 - 0i$, as a function of volume mean diameter in the range $\lambda = 550$ to 700 nm. In the Rayleigh regime $(D_p \leq 0.1\ \mu\text{m})$, the extinction coefficient varies with wavelength to a power between -3 and -4, reflecting the contributions of both scattering and absorption. In the large-particle regime, the Ångstrom exponent ranges between 1 and 0, again reflecting the wavelength dependence of Q_{abs} and Q_{scat} in this region.

15.4 CALCULATED VISIBILITY REDUCTION BASED ON ATMOSPHERIC DATA

Table 15.4 presents typical aerosol size distributions and concentrations based on measurements in many locations, including mass median diameters D_{pg}, geometric standard deviations σ_g, and volume concentrations $V(\mu\text{m}^3\ \text{cm}^{-3})$ of the nuclei,

TABLE 15.4 Size Distributions (Lognormal) of Different Classes of Atmospheric Aerosol

	Nuclei Mode			Accumulation Mode			Coarse-Particle Mode		
	D_{pg}, μm	σ_g	V, μm³ cm⁻³	D_{pg}, μm	σ_g	V, μm³ cm⁻³	D_{pg}, μm	σ_g	V, μm³ cm⁻³
Marine background	0.019	1.6[a]	0.0005	0.3	2.0	0.10	12.0	2.7	12.0
Clean continental background	0.03	1.6	0.006	0.35	2.1	1.5	6.2	2.2	5.0
Average background	0.034	1.7	0.037	0.32	2.0	4.45	6.04	2.16	25.9
Background and aged urban plume	0.028	1.6	0.029	0.36	1.84	44.0	4.51	2.12	27.4
Background and local sources	0.021	1.7	0.62	0.25	2.11	3.02	5.6	2.09	39.1
Urban average	0.038	1.8	0.63	0.32	2.16	38.4	5.7	2.21	30.8
Urban and freeway	0.032	1.74	9.2	0.25	1.98	37.5	6.0	2.13	42.7
Labadie plume (1976)[b]	0.015	1.5	0.1	0.18	1.96	12.0	5.5	2.5	24.0

[a]Assumed.

[b]Typical distribution observed in the plume from the Labadie coal-fired power plant near St. Louis, Missouri.

accumulation, and coarse-particle modes of eight classifications of aerosol. We note only a small variability in the accumulation and coarse-mode size distributions for the variety of aerosol classes (excluding the marine aerosol). The average specifications for the accumulation and coarse modes are as follows:

Mode	D_{pg}, μm	σ_g
Accumulation	0.29 ± 0.06	2.0 ± 0.1
Coarse	6.3 ± 2.3	2.3 ± 0.2

The contribution to the total scattering coefficient b_{scat} can be calculated for each of the modes of the different aerosol classes in Table 15.4 from the theory in Section 15.3. The accumulation mode is found to be the dominant scattering mode, with the coarse mode contributing a small amount and the nuclei mode a negligible amount. For the two background aerosol cases:

Type of Aerosol	b_{scat}, m^{-1}	Nuclei	Accumulation	Coarse	Rayleigh
			Fractional Modal Contribution to b_{scat} from		
Clean continental background	0.23×10^{-4}	0.01	0.39	0.17	0.43
Average background	0.57×10^{-4}	0.01	0.46	0.36	0.17

These calculations indicate that for background conditions the accumulation mode is a larger contributor to light scattering than the coarse mode but that the coarse mode is a non-negligible component of the scattering coefficient. This situation contrasts with that of polluted urban atmospheres in which the accumulation mode is responsible for more than 90% of the total scattering coefficient.

Larson et al. (1984) analyzed atmospheric composition and visibility in Los Angeles on two days in 1983: one day, April 7, representing relatively clean conditions, and the second, August 25, characterized by heavy smog. Table 15.5 gives the ambient measured concentrations of the species sampled on the two days. In preparing Table 15.5 Larson et al. (1984) used the procedure of Stelson and Seinfeld (1981) for developing a material balance on the chemical composition of the aerosol samples. Stelson and Seinfeld (1981) showed that urban aerosol mass can generally be accounted for from measurements of SO_4^{2-}, Cl^-, Br^-, NO_3^-, NH_4^+, Na^+, K^+, Ca^{2+}, Fe, Mg, Al, Si, Pb, carbonaceous material, and aerosol water. Their method assumes that trace metals are present in the form of common oxides:

Element	Oxide Form
Al	Al_2O_3
Ca	CaO
Fe	Fe_2O_3
Si	SiO_2
Mg	MgO
Pb	PbO
Na	Na_2O
K	K_2O

TABLE 15.5   Atmospheric Concentrations Measured on 2 Days in Los Angeles

Component (Concentration)	Clear Day April 7, 1983	Heavy Smog August 25, 1983
$(NH_4)_2SO_4$, $\mu g\,m^{-3}$	3.54	14.90
NH_4NO_3, $\mu g\,m^{-3}$	1.23	1.79
$NaNO_3$, $\mu g\,m^{-3}$	—	12.80
Na_2SO_4, $\mu g\,m^{-3}$	—	—
Elemental carbon, $\mu g\,m^{-3}$	0.99	6.37
Organic carbon, $\mu g\,m^{-3}$	6.78	32.74
Al_2O_3, $\mu g\,m^{-3}$	2.82	9.34
SiO_2, $\mu g\,m^{-3}$	4.53	15.54
K_2O, $\mu g\,m^{-3}$	0.36	1.45
CaO, $\mu g\,m^{-3}$	0.45	1.83
Fe_2O_3, $\mu g\,m^{-3}$	0.91	4.58
PbO, $\mu g\,m^{-3}$	0.06	0.72
NO_2, ppm	0.05	0.10
O_3, ppm	0.05	0.21
CO, ppm	1.6	3.39
T, °C	23.0	29.8
RH, %	23.6	50.5
b_{sp}, m^{-1}	0.259×10^{-4}	4.08×10^{-4}
b_{sg}, m^{-1} (Rayleigh)	0.111×10^{-4}	0.107×10^{-4}
b_{ag}, m^{-1} (NO_2)	0.012×10^{-4}	0.030×10^{-4}
b_{ap}, m^{-1} (carbon)	0.093×10^{-4}	0.787×10^{-4}
Calculated b_{ext}, m^{-1}	0.475×10^{-4}	5.00×10^{-4}

Source: Larson et al. (1984).

In Table 15.5 this procedure was followed with the exception of Na, which was assumed to be in the form of an ionic solid. Mg was present at negligible levels, and thus its chemical form is unimportant to the aerosol mass balance. To account for hydrogen and oxygen present in the hydrocarbons, the mass of organic carbonaceous material was taken to be 1.2 times the organic carbon mass measured (Countess et al. 1980). The ionic material was assumed to be distributed as follows:

Na^+ was associated with Cl^-.

NH_4^+ was associated with SO_4^{2-}.

NH_4^+ remaining, if any, was associated with NO_3^-.

Na^+ remaining, if any, was associated with remaining NO_3^-, if any.

Na^+ remaining, if any, was associated with remaining SO_4^{2-}, if any.

Figure 15.10 shows the aerosol volume distributions and gives the total extinction coefficients on the 2 days at $\lambda = 550$ nm. The estimated contributions to b_{ext} on the 2 days are given at the bottom of Table 15.5. Note that the total calculated b_{ext} on the smoggy day was somewhat smaller than that measured since not all atmospheric constituents were included in the calculation.

A comprehensive review of visibility is given by Watson (2002).

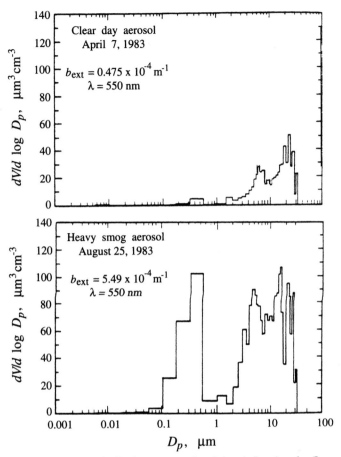

FIGURE 15.10   Aerosol volume distributions measured on 2 days in Los Angeles (Larson et al. 1984).

APPENDIX 15   CALCULATION OF SCATTERING AND EXTINCTION COEFFICIENTS BY MIE THEORY

This appendix outlines the evaluation of the coefficients (Pilinis 1989) a_k and b_k in (15.13) and (15.14). If we define

$$A_k(y) = \psi'_k(y)/\psi_k(y) \tag{15.A.1}$$

then (Wickramasinghe 1973)

$$a_k = \frac{\left(\dfrac{A_k(y)}{m} + \dfrac{k}{\alpha}\right) \mathrm{Re}[\zeta_k(\alpha)] - \mathrm{Re}[\zeta_{k-1}(\alpha)]}{\left(\dfrac{A_k(y)}{m} + \dfrac{k}{\alpha}\right)\zeta_k(\alpha) - \zeta_{k-1}(\alpha)} \tag{15.A.2}$$

$$b_k = \frac{\left(A_k(y)m + \dfrac{k}{\alpha}\right) \mathrm{Re}[\zeta_k(\alpha)] - \mathrm{Re}[\zeta_{k-1}(\alpha)]}{\left(A_k(y)m + \dfrac{k}{\alpha}\right)\zeta_k(\alpha) - \zeta_{k-1}(\alpha)} \tag{15.A.3}$$

To generate $A_k(y)$, we use the recurrence relation

$$A_k(y) = -\frac{k}{y} + \left(\frac{k}{y} - A_{k-1}(y)\right)^{-1} \tag{15.A.4}$$

with $A_0(y) = \cos y / \sin y$. For $\zeta_k(\alpha)$, we use the relation

$$\zeta_k(\alpha) = \frac{2k-1}{p}\zeta_{k-1}(\alpha) - \zeta_{k-2}(\alpha) \tag{15.A.5}$$

with

$$\zeta_{-1}(\alpha) = \cos\alpha - i\sin\alpha \tag{15.A.6}$$

$$\zeta_0(\alpha) = \sin\alpha + i\cos\alpha \tag{15.A.7}$$

Using (15.A.1)–(15.A.7) one may calculate the sums in (15.13) and (15.14) by adding terms until a desired accuracy is achieved. Convergence is rapid and, for most cases, the number of terms required is less than 20.

PROBLEMS

15.1$_A$ Rayleigh scattering is the irreducible minimum of light scattering owing to air molecules along an atmospheric sight path. The amount of scattered light varies as λ^{-4}. The Rayleigh scattering coefficient at three wavelengths is:

$$\begin{aligned} b_{sg} &= 25.9\left(\frac{293}{T}\right)p & \lambda = 450\,\text{nm} \\[2mm] &= 11.4\left(\frac{293}{T}\right)p & \lambda = 550\,\text{nm} \\[2mm] &= 5.8\left(\frac{293}{T}\right)p & \lambda = 650\,\text{nm} \end{aligned}$$

where T is in K and p is in atmospheres. Calculate b_{sg} at 550 nm at sea level, 1 km, 2 km, and 3 km altitude.

15.2$_A$ Light extinction can be divided into the sum of its scattering and absorption components as follows

$$b_{ext} = b_{sg} + b_{ag} + b_{sp} + b_{ap}$$

where b_{ext} = light extinction coefficient, b_{sg} = Rayleigh scattering (light scattering by molecules of air), b_{ag} = light absorption because of gases (mainly NO_2), b_{sp} = light scattering by particles, and b_{ap} = light absorption by particles. The scattering and absorption by gases (b_{sg} and b_{ag}) can be calculated knowing the air pressure (altitude) and temperature, and the concentration of NO_2, respectively. To deal with the particle-related extinction (b_{sp} and b_{ap}), a standard approach is to allocate portions of the extinction to each species of the mixture and then summarize the contributions to arrive at the total particle-caused extinction. With this approach, one sums up the individual contributions

$$b_{sp} = e_{\text{sulfate}}[\text{sulfate}] + e_{\text{nitrate}}[\text{nitrate}] + e_{\text{OC}}[\text{organic carbon}]$$
$$+ e_{\text{soil}}[\text{soil}] + e_{\text{coarse}}[\text{coarse}]$$

and, for absorption of light by particles

$$b_{ap} = e_{\text{BC}}[\text{black carbon}]$$

where the e values are the extinction efficiencies. The units of e are M m^{-1} per μg m^{-3}; hence m^2 g^{-1}. Extinction efficiencies depend on the size distribution and the molecular composition of the sulfates, nitrates, and organic carbon. The extinction efficiencies for hygroscopic substances (sulfate, nitrate, and organic carbon) are dependent on the relative humidity. Values are usually reported for dry particles; the uptake of water can multiply the given sulfate and nitrate efficiencies manyfold at high relative humidities. The effect of water uptake on the organic carbon efficiency is not as well established as that for the inorganic salts. Ranges of dry extinction efficiencies are

$$e_{\text{sulfate}} = 1.5 - 4\,\text{m}^2\,\text{g}^{-1}$$
$$e_{\text{nitrate}} = 2.5 - 3\,\text{m}^2\,\text{g}^{-1}$$
$$e_{\text{OC}} = 1.8 - 4.7\,\text{m}^2\,\text{g}^{-1}$$
$$e_{\text{soil}} = 1 - 1.25\,\text{m}^2\,\text{g}^{-1}$$
$$e_{\text{coarse}} = 0.3 - 0.6\,\text{m}^2\,\text{g}^{-1}$$
$$e_{\text{BC}} = 8 - 12\,\text{m}^2\,\text{g}^{-1}$$

Calculate the visual range of an atmosphere at a 0.02 contrast ratio for which $e_{\text{sulfate}} = 3\,\text{m}^2\,\text{g}^{-1}$, $e_{\text{nitrate}} = 3\,\text{m}^2\,\text{g}^{-1}$, $e_{\text{OC}} = 4\,\text{m}^2\,\text{g}^{-1}$, and $e_{\text{BC}} = 10\,\text{m}^2\,\text{g}^{-1}$, and for which

[Sulfate]	$= 20\,\mu\text{g m}^{-3}$
[Nitrate]	$= 5\,\mu\text{g m}^{-3}$
[Organic carbon]	$= 25\,\mu\text{g m}^{-3}$
[Black carbon]	$= 7.5\,\mu\text{g m}^{-3}$

15.3$_B$ Consider a droplet size distribution given by the modified gamma distribution

$$n(r) = \frac{N_0}{\Gamma(\beta)r_n}\left(\frac{r}{r_n}\right)^{\beta-1}\exp\left(-\frac{r}{r_n}\right)$$

where N_0 is the total number concentration of particles, $\Gamma(\beta)$ is the gamma function, with β as integer, and r_n is related to the mean radius r_m of the population by $r_m = (\beta + 1)r_n$. The transmission of a light beam of wavelength λ through the droplet population is

$$T(\lambda) = \exp[-\tau(\lambda)]$$

where the optical thickness for pathlength ℓ is given by

$$\tau(\lambda) = \ell \int \pi r^2 Q_{ext}(r, \lambda) n(r) dr$$

N_0 typically ranges between 50 and 500 cm^{-3}. r_m ranges between 5 and 25 μm, and values of $\beta = 2$ and 5 have been used for tropospheric clouds. Consider $r_m = 20$ μm and an optical pathlength of 1 m. Calculate and plot (a) the size distribution and (b) the transmission of light through a droplet population with this modified gamma distribution for both $\beta = 2$ and 5. For the plot of $T(\lambda)$, consider N_0 ranging from 0 to 800 cm^{-3}. Identify the sensitivity of the number concentration at which $T = 0.5$ to the two values of β. [The size distribution can be plotted as $n(r)/N_0$.]

15.4$_D$ Calculate the scattering coefficient b_{scat} for a lognormally distributed carbon aerosol of mass concentration 20 μg m^{-3}, with $\overline{D}_{pg} = 0.02$ μm and $\sigma_g = 3.0$ at $\lambda = 550$ nm. Note that it will be necessary to numerically evaluate the appropriate integral based on the data in Figure 15.7. What is the visibility under these conditions?

REFERENCES

Bohren, C. F., and Huffman, D. R. (1983) *Absorption and Scattering of Light by Small Particles*, Wiley, New York.

Bond, T. C., and Bergstrom, R. W. (2006) Light absorption by carbonaceous particles: An investigative review, *Aerosol Sci. Technol.*, **40**, 27–67.

Charlson, R. J., and Ahlquist, N. C. (1969) Brown haze: NO_2 or aerosol? *Atmos. Environ.* **3**, 653–656.

Countess, R. J., Wolff, G. T., and Cadle, S. H. (1980) The Denver winter aerosol: A comprehensive chemical characterization, *J. Air Pollut. Control Assoc.* **30**, 1194–1200.

Faxvog, F. R., and Roessler, D. M. (1978) Carbon aerosol visibility versus particle size distribution, *Appl. Opt.* **17**, 2612–2616.

Groblicki, P. J., Wolff, G. T., and Countess, R. J. (1981) Visibility-reducing species in the Denver "brown cloud"—I. Relationships between extinction and chemical composition, *Atmos. Environ.* **15**, 2473–2484.

Hale, G. M., and Querry, M. R. (1973) Optical constants of water in the 200 nm to 200 μm wavelength region, *Appl. Opt.* **12**, 555–563.

Kerker, M. (1969) *The Scattering of Light and Other Electromagnetic Radiation*, Academic Press, New York.

Larson, S., Cass, G., Hussey, K., and Luce, F. (1984) *Visibility Model Verification by Image Processing Techniques*, Final Report to State of California Air Resources Board under Agreement A2-077-32, Sacramento, CA.

Larson, S. M., Cass, G. R., Hussey, K. J., and Luce, F. (1988) Verification of image processing–based visibility models, *Environ. Sci. Technol.* **22**, 629–637.

Malm, W. C., Sisler, J. F., Huffman, D., Eldred, R. A., and Cahill, T. A. (1994) Spatial and seasonal trends in particle concentration and optical extinction in the United States, *J. Geophys. Res.* **99**, 1347–1370.

Marshall, S. F., Covert, D. S., and Charlson, R. J. (1995) Relationship between asymmetry parameter and hemispheric backscatter ratio: Implications for climate forcing by aerosols, *Appl. Opt.* **34**, 6306–6311.

Nemesure, S., Wagener, R., and Schwartz, S. E. (1995) Direct shortwave forcing of climate by the anthropogenic sulfate aerosol: Sensitivity to particle size, composition, and relative humidity, *J. Geophys. Res.* **100**, 26105–26116.

Pilinis, C. (1989) Numerical simulation of visibility degradation due to particulate matter: Model development and evaluation, *J. Geophys. Res.* **94**, 9937–9946.

Sloane, C. S. (1983) Optical properties of aerosols: Comparison of measurements with Model calculation, *Atmos. Environ.* **17**, 409–419.

Stelson, A. W. (1990) Urban aerosol refractive index prediction by partial molar refraction approach, *Environ. Sci. Technol.* **24**, 1676–1679.

Stelson, A. W., and Seinfeld, J. H. (1981) Chemical mass accounting of urban aerosol, *Environ. Sci. Technol.* **15**, 671–679.

Tegen, I., Lacis, A. A., and Fung, I. (1996) The influence of climate forcing of mineral aerosols from disturbed soils, *Nature* **380**, 419–423.

Watson, J. G. (2002) Visibility: Science and regulation, *J. Air Waste Manage. Assoc.* **52**, 628–713.

Weast, R. C. (1987) Physical constants of organic compounds, in *CRC Handbook of Chemistry and Physics*, 68th ed., R. C. Weast, ed., CRC Press, Boca Raton, FL, pp. B67–B146.

Wickramasinghe, N. C. (1973) *Light Scattering Functions for Small Particles with Applications in Astronomy*, Wiley, New York.

Wiscombe, W., and Grams, G. (1976) The backscattered fraction in two-stream approximations, *J. Atmos. Sci.* **33**, 2440–2451.

16 Meteorology of the Local Scale

Meteorology is the study of the atmosphere, its motion, and its phenomena. The word *meteorology* comes from the Greek word *meteoros* meaning "suspended in air." The term itself was introduced by Aristotle, who wrote *Meteorologica* in 340 B.C., summarizing what was known at the time about weather and climate including discussions of clouds, rain, snow, wind, hail, and thunder. This first treatment of atmospheric phenomena by Aristotle was speculative and based on qualitative observations and philosophy. There was little progress in the next 2000 years until the invention of the thermometer (end of sixteenth century), the barometer (in 1643) and the hygrometer (eighteenth century) that allowed progress in the science of meteorology especially in the nineteenth and twentieth centuries.

The air in our atmosphere is continuously in motion, creating a complicated three-dimensional flow field that varies in scales ranging from a few meters (around a building or a small hill) to thousands of kilometers (a major storm). At the same time, the temperature and relative humidity of the atmosphere are highly variable; there are clouds forming, evaporating, raining, and snowing at different altitudes. Meteorology is intimately connected with air quality, determining the concentration levels of locally emitted primary pollutants, the formation of secondary pollutants, their transport to other areas, and their ultimate removal from the atmosphere. Even if anthropogenic emissions are the source of air quality degradation, meteorology determines the magnitude of the problems that will be caused by these emissions as well as where and when these problems will occur. The importance of meteorology for air quality in a given area is clear from Figure 16.1, showing the $PM_{2.5}$ concentrations in a large urban area. Both local and regional anthropogenic emissions of the major pollutants (SO_2, NO_x, total VOCs) in the eastern United States are relatively constant from day to day, and they vary by a factor of <2 from month to month. However, the measured $PM_{2.5}$ concentrations shown in Figure 16.1 vary by a factor of ≤ 10 from day to day. This variability of pollutant concentrations and the resulting clean and "polluted" days in an area with more or less constant emissions are determined by meteorology. Our goal in this chapter is to study the basic concepts in meteorology and to understand their links to air pollution. We will focus on the *microscale* (phenomena occurring on scales of the order of 1 km). Phenomena occurring on larger scales and the global circulation of the atmosphere will be discussed in Chapter 21.

Each of these scales of atmospheric motion plays a role in air pollution, although over different periods of time. For example, the microscale meteorological effects determine the dispersion of a plume from an industrial stack or a highway over timescales on the order of minutes to a few hours. On the other hand, mesoscale phenomena take place over hours or days and influence the transport and dispersal of pollutants to areas that are hundreds of kilometers from their sources.

Atmospheric Chemistry and Physics: From Air Pollution to Climate Change, Second Edition, by John H. Seinfeld and Spyros N. Pandis.   Copyright © 2006 John Wiley & Sons, Inc.

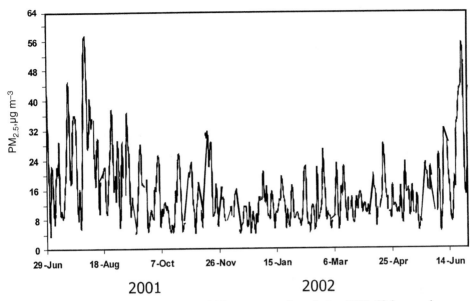

FIGURE 16.1 Measured daily average $PM_{2.5}$ concentrations during 2001–02 in an urban area (Pittsburgh, PA).

Most of the emissions of gases and vapors by either anthropogenic activities or nature itself enter the atmosphere in the lower level of the troposphere, the planetary boundary layer. After their emission they may be either dispersed and diluted rapidly, resulting in low concentrations; in other periods they may be concentrated in a relatively small volume leading to an air pollution episode. The degree of mixing of pollutants in the first kilometer or so of the atmosphere is determined to a large extent by its temperature profile (variation of temperature with altitude) and the windspeed. The three-dimensional temperature and wind fields are related to each other in a rather complicated way so we will first investigate their effects separately and then study their interplay.

In the troposphere the temperature normally decreases with increasing altitude because of the decrease in pressure with height (see Chapter 1). If the atmosphere is at its equilibrium state, an air parcel has no tendency to either rise or fall. The temperature for this *neutral* state decreases slowly with height and is considered the reference temperature profile for the atmosphere. The atmosphere is, however, very seldom in such delicate equilibrium; the influence of surface heating and phenomena at larger scales usually result in a temperature profile different from the reference profile. If the atmospheric temperature decreases faster with height than the reference profile, then the atmosphere is *unstable*. If an air parcel is displaced either upward or downward, it will continue moving in this direction. An unstable atmosphere is therefore characterized by strong vertical currents in both the upward and downward directions, leading to rapid mixing of emitted pollutants. On the other hand, if the temperature decreases more slowly with height than the reference profile (or even increases), the atmosphere is *stable*. In stable atmospheres air parcels are inhibited from either upward or downward motion, resulting in weak mixing of emitted pollutants. If the temperature of an atmospheric layer actually increases with altitude, it is called an *inversion layer*. An inversion occurs when warm air lies above cool air, and this condition is extremely stable. One type of stable atmosphere occurs during nights with clear skies and low winds. Then the ground and the air in the lower atmosphere can become cooler than the air above them. This is known as a *surface* (or *radiation inversion*). The temperature profiles in the atmosphere and atmospheric stability are discussed in Sections 16.1 and 16.2.

The air around us is continuously in motion, and circulations of all dimensions exist in the atmosphere. Small whirls form inside larger whirls, which in turn take place inside even larger whirls. One way to think about these motions is by applying the concept of *turbulent eddies*: relatively small volumes of air that are spinning, have their own life history, and behave differently from the larger flow in which they exist. Small turbulent eddies with dimensions of centimeters or a few meters can be seen in the smoke from a fire or a cigarette. These eddies coexist with those of larger sizes on up to the major dimensions of the motion (hundreds of meters or more in the atmosphere), the so-called large eddies. This atmospheric turbulence is responsible for the dispersion and transport of material from the sources to the rest of the atmosphere. The basic phenomena that influence atmospheric turbulence will be discussed in Sections 16.3 and 16.4.

16.1   TEMPERATURE IN THE LOWER ATMOSPHERE

The variation of temperature with altitude in the atmosphere is a key variable in determining the degree to which material will mix vertically. The study of the atmospheric temperature profile is facilitated by the concept of an *air parcel*: a hypothetical mass of air that may deform as it moves in the atmosphere but remains as a single unit without exchanging mass with its surroundings. Think of it as an invisible, extremely elastic balloon enclosing a mass of air and not allowing it to escape. If we assume that the air parcel does not exchange any heat with its surroundings, then we call it an *adiabatic air parcel*.

If an adiabatic air parcel rises in the atmosphere, it will enter a region where the surrounding air pressure is lower. Taking advantage of this lower-pressure environment, the air molecules will move the imaginary walls of the balloon outward, causing the expansion of the air parcel and the lowering of its pressure. The expansion of the air parcel, because there are no other sources of energy for this system, will consume some of the energy of the air molecules inside the parcel. The molecules will slow down, and the temperature of the air parcel will decrease. As a result, air that rises expands and cools. The opposite will happen if the air parcel sinks, its temperature increases and its volume decreases. An adiabatic air parcel and its surroundings can be at different temperatures (but not different pressures). This difference in temperature will determine whether a moving air parcel will continue moving in the same direction or if it will reach equilibrium and stop.

The concept of an air parcel is a tenable one as long as the parcel is of such size that the exchange of air molecules across its boundary is small when compared to the total number of molecules in the parcel. The process of vertical mixing in the atmosphere can, to a first degree, be approximated as one involving a large number of air parcels rising and falling.

16.1.1   Temperature Variation in a Neutral Atmosphere

Let us assume that the atmosphere is in a state in which adiabatic air parcels that move vertically are always in equilibrium with their surroundings. This atmospheric equilibrium state is defined as neutral, and the corresponding temperature profile will be used as the reference profile for our discussion of atmospheric stability. This neutral state corresponds to that of a ball lying on a flat surface. If the ball is moved from its original position where it was in equilibrium to a new position, it will still be in equilibrium.

We would like to calculate the temperature profile $T(z)$ that corresponds to a neutral atmosphere. If an air parcel of mass m moves infinitesimally in the atmosphere, its change of internal energy dU satisfies the first law of thermodynamics

$$dU = dQ + dW \tag{16.1}$$

where dQ is the heat transferred to the air parcel from its surroundings and dW is the work done on the system. The infinitesimal work done on the air parcel is $dW = -p\,dV$, where p is the system pressure and dV its volume change. The air parcel is assumed to be adiabatic therefore $dQ = 0$. The change in internal energy is finally $dU = m\hat{c}_{v,air}dT$, where $\hat{c}_{v,air}$ is the specific heat capacity of air at constant volume and dT its temperature change. Therefore, the first law of thermodynamics for this adiabatic air parcel results in

$$m\hat{c}_{v,air}dT = -p\,dV \tag{16.2}$$

Note that the adiabatic air parcel is assumed to be always in equilibrium (mechanical and thermal) with its surroundings so T and p are the same for the air parcel and the neutral atmosphere around it. Also for a rising air parcel, $dT < 0$, and the left-hand side (LHS) is negative while $dV > 0$, and the right-hand side (RHS) is also negative. Physically, this equation says that the change in energy because of the expansion work is equal to the energy reduction because of the temperature decrease. The opposite arguments can be made for a sinking air parcel.

To replace the term dV in (16.2) with terms that depend on temperature and pressure, we can use the ideal-gas law as $pV = mRT/M_{air}$ for the air parcel. Taking the derivatives of both sides and recognizing that the mass of the air parcel by definition remains constant,

$$d(pV) = \frac{mR}{M_{air}}dT \tag{16.3}$$

The LHS is equal to $d(pV) = p\,dV + V\,dp$ and therefore combining this with (16.3) and using the ideal-gas law for V, we obtain

$$p\,dV = \frac{mR}{M_{air}}dT - \frac{mRT}{M_{air}}\frac{dp}{p} \tag{16.4}$$

Substituting (16.4) into (16.2), the first law of thermodynamics reduces to

$$m\hat{c}_{v,air}\,dT = \frac{mRT}{M_{air}}\frac{dp}{p} - \frac{mR}{M_{air}}dT \tag{16.5}$$

Rearranging and dividing both sides by dz, we have

$$\left(m\hat{c}_{v,air} + \frac{mR}{M_{air}}\right)\frac{dT}{dz} = \frac{mRT}{pM_{air}}\frac{dp}{dz} \tag{16.6}$$

We have already derived the change of pressure with height in the atmosphere using the hydrostatic equation to derive (1.3). Substituting the dp/dz term and simplifying, we get

$$\frac{dT}{dz} = -\frac{g}{\hat{c}_{v,\text{air}} + \frac{R}{M_{\text{air}}}} \tag{16.7}$$

We note that $\hat{c}_{v,\text{air}} + (R/M_{\text{air}}) = \hat{c}_{p,\text{air}}$, the heat capacity at constant pressure per unit mass of air. Thus

$$\frac{dT}{dz} = -\frac{g}{\hat{c}_{p,\text{air}}} \tag{16.8}$$

and the vertical temperature gradient for a neutral atmosphere is equal to a constant. The heat capacity at constant pressure of air is $\hat{c}_{p,\text{air}} = 1005\,\text{J}\,\text{kg}^{-1}\,\text{K}^{-1}$ and therefore $g/\hat{c}_{p,\text{air}} = 0.976 \times 10^{-2}\,\text{K}\,\text{m}^{-1}$ or $9.76\,\text{K}\,\text{km}^{-1}$. This value refers to water-free air and is called the *dry adiabatic lapse rate*, denoted by Γ. Therefore, for a neutral, water-free atmosphere the temperature decreases by approximately 1 K per 100 m. If the temperature at the surface is T_0, then, integrating (16.8), we find that the temperature profile to be $T = T_0 - \Gamma z$.

Adiabatic Lapse Rate of an Air Parcel Containing Water Vapor   The atmosphere contains also water vapor with heat capacity at constant pressure equal to $1952\,\text{J}\,\text{kg}^{-1}\,\text{K}^{-1}$. Calculate the adiabatic lapse rate Γ' for an air parcel as a function of its water vapor content. What is the maximum expected difference between Γ' and Γ?

Equation (16.8) is still applicable if we replace the heat capacity of air $\hat{c}_{p,\text{air}}$ with the heat capacity of the mixture of air and water vapor $\hat{c}'_p$ and

$$\Gamma' = g/\hat{c}'_p$$

If w_v is the water vapor mass mixing ratio (ratio of the mass of water vapor to the mass of dry air) in the air parcel, then

$$c'_p = (1 - w_v)\hat{c}_{p,\text{air}} + w_v\,\hat{c}_{p,\text{water vapor}} = 1005 + 947\,w_v \quad (\text{J}\,\text{kg}^{-1}\,\text{K}^{-1})$$

Combination of these equations allows the calculation of the adiabatic lapse rate Γ' as a function of w_v. The higher the water vapor content, the larger the heat capacity of the air parcel, and the smaller the adiabatic lapse rate. The atmospheric concentration of water vapor is a function of temperature and relative humidity (see, e.g., Figure 7.1) ranging from a few $\text{g}\,\text{m}^{-3}$ at low temperatures to as much as $40\,\text{g}\,\text{m}^{-3}$ at 40°C. The mass concentration of air for these conditions ($p = 1$ atm, $T = 313$ K) is, using the ideal-gas law, $1.1 \times 10^3\,\text{g}\,\text{m}^{-3}$. The water vapor mass mixing ratio is then $w_v = 40/(40 + 1100) = 0.035$ and $\hat{c}'_p = 1038\,\text{J}\,\text{kg}^{-1}\,\text{K}^{-1}$.

In this extreme case, $\Gamma' = 0.945 \times 10^{-2} \, \text{K m}^{-1}$ or $9.45 \, \text{K km}^{-1}$. Therefore the difference between Γ' and Γ is always less than 3% and the water vapor content of a cloud free air parcel can be neglected for all practical purposes.

The existence of clouds in the atmosphere can significantly change the temperature profile. As an air parcel containing water vapor rises, its temperature decreases and its relative humidity increases (recall Figure 7.1). If the relative humidity exceeds 100%, water will begin to condense on the available particles, and the corresponding latent heat of water condensation will be released. The heat released during water condensation offsets some of the cooling due to expansion of the rising parcel. The air no longer cools at the dry adiabatic rate but at a lower rate called the *moist adiabatic lapse rate*, Γ_s.

The latent heat of evaporation of water per gram ΔH_v is a weak function of temperature and is 2.5 kJ g^{-1} at 0°C and 2.25 kJ g^{-1} at 100°C. If m_w is the mass of water vapor in the air parcel, then the latent heat released for the infinitesimal motion of the air parcel will be equal to $-\Delta H_v \, dm_w$. This term needs to be added to the RHS of the parcel energy balance given by (16.2):

$$m\hat{c}_{v,\text{air}} \, dT = -p \, dV - \Delta H_v \, dm_w \tag{16.9}$$

Even if the air parcel is saturated with water vapor, the thermodynamic properties of the gas phase are, for all practical purposes, the same as that of air. We can therefore substitute (16.4) into (16.9) using the molecular weight of air:

$$m\hat{c}_{v,\text{air}} \, dT = \frac{mRT}{M_{\text{air}}} \frac{dp}{p} - \frac{mR}{M_{\text{air}}} dT - \Delta H_v \, dm_w \tag{16.10}$$

Dividing all terms by the air parcel mass and recognizing the heat capacity under constant pressure term, we obtain

$$\hat{c}_{p,\text{air}} \, dT = \frac{RT}{M_{\text{air}}} \frac{dp}{p} - \Delta H_v \frac{dm_w}{m} \tag{16.11}$$

The term m_w/m is equal to the mass mixing ratio of water in the air parcel, w_v. Dividing this expression by dz and once more using (1.3), we find the expression for the moist adiabatic lapse rate:

$$\Gamma_s = -\frac{dT}{dz} = \frac{g}{\hat{c}_{p,\text{air}}} + \frac{\Delta H_v}{\hat{c}_{p,\text{air}}} \frac{dw_v}{dz} \tag{16.12}$$

If the water vapor concentration in the air parcel remains constant (no condensation or evaporation), then $(dw_v/dz) = 0$ and we recover (16.8). For an air parcel saturated with water vapor that is rising, the water vapor concentration will be decreasing due to condensation and $(dw_v/dz) < 0$. Thus the temperature gradient inside a cloud is less than that for cloud-free air or $\Gamma_s < \Gamma$.

One can relate the gradient dw_v/dz to dw_v/dT to show (see Problem 16.1) that

$$\Gamma_s = \frac{g}{\hat{c}_{p,\text{air}} + \Delta H_v \dfrac{dw_v}{dT}} \tag{16.13}$$

Since the saturation vapor pressure of water is a strong function of temperature, (dw_v/dT) also depends on temperature. Note that (dw_v/dT) is proportional to the slope of the saturation line (100% RH in Figure 7.1). Therefore the moist lapse rate is not a constant but rather a strong function of temperature. For $p = 1$ atm, Γ_s is equal to 3 K km^{-1} at 40°C and 9.5 K km^{-1} at -40°C. At high temperatures the atmosphere contains more water vapor (see Figure 7.1), resulting in larger derivatives (dw_v/dT), and a lower moist lapse rate. As the temperature decreases, so does (dw_v/dT), and Γ_s approaches Γ. In warm tropical air the wet adiabatic lapse rate is roughly one-third of the dry adiabatic lapse rate, whereas in cold polar regions there is little difference between the two.

Moist Adiabatic Lapse Rate Calculation   Calculate the moist adiabatic lapse rate at 0°C and 1 atm. An expression for the water vapor saturation vapor pressure as a function of temperature is given in Table 17.2.

The calculation requires the estimation of dw_v/dT at 0°C. Differentiating the expression of Table 17.2 with respect to temperature,

$$dp°/dT = a_1 + 2\,a_2 T + 3\,a_3 T^2 + 4\,a_4 T^3 + 5\,a_5 T^4 + 6\,a_6 T^5$$

where T is in °C and $p°$ in mbar. For $T = 0$°C, $dp°/dT = a_1 = 0.44$ mbar K$^{-1} = 4.4 \times 10^{-4}$ atm K^{-1}. Using the ideal-gas law we can convert the partial pressure to mass concentration $(c = pM_w/RT)$, and the derivative of the water saturation concentration is 0.35 g m^{-3} K^{-1}. Finally, we need to calculate the airmass concentration under these conditions. Note that this concentration depends not only on temperature but also on pressure. This means that the wet adiabatic lapse rate also depends on pressure. Using the ideal-gas law once more but for air this time, we find that the air concentration in the air parcel is 1290 g m^{-3}. Therefore, $dw_v/dT = (0.35/1290)$ K$^{-1} = 2.7 \times 10^{-4}$ K^{-1}. Substituting into (16.13), we find that $\Gamma_s = 9.81/(1005 + 2.7 \times 10^{-4} \times 2.5 \times 10^6)$ K m$^{-1} = 5.8$ K km^{-1}. This value is only 60% of the dry adiabatic lapse rate.

16.1.2 Potential Temperature

Let us assume that an air parcel somewhere in the atmosphere has a temperature T and pressure p and is initially in equilibrium (same T and p) with the surrounding atmosphere. If this air parcel is moved dry adiabatically to the surface with a pressure of $p_0 = 1000$ mbar, it will attain a temperature θ called the *potential temperature*. The potential temperature can be calculated from (16.5) integrating from initial conditions (T, p) to the final state (θ, p_0) to find that

$$\theta = T\left(\frac{p_0}{p}\right)^{R/(c_p M_{air})} = T\left(\frac{p_0}{p}\right)^{0.286} \tag{16.14}$$

For $p = p_0$ the potential temperature is equal to the surface temperature T_0, and the potential temperature profile of the atmosphere starts at the surface temperature. The potential temperature is used in meteorology to compare the temperature of air parcels under identical conditions. As we will see subsequently, it is also useful in the stability analysis of the atmosphere.

If the adiabatic dry air parcel is always in equilibrium with the atmosphere during its motion from the original position to the surface, the atmosphere by definition is neutral and its temperature profile satisfies (16.8). No matter where the air parcel starts in this atmosphere, it will always attain the same temperature when brought to the surface at pressure p_0. In other words, the potential temperature of the air parcel will not change during its motion and will always be equal to θ. The equilibrium of the air parcel with the surrounding environment means that the neutral atmosphere (or a neutral atmospheric layer) has the same potential temperature at all heights z and therefore $d\theta/dz = 0$. Plots of altitude versus potential temperature for a neutral (adiabatic) atmosphere are vertical lines at $\theta = T_0$.

The results described above can be demonstrated mathematically by differentiating (16.14) with respect to z to get

$$\frac{1}{\theta}\frac{d\theta}{dz} = \frac{1}{T}\frac{dT}{dz} - \frac{R}{\hat{c}_{p,\text{air}}M_{\text{air}}}\frac{1}{p}\frac{dp}{dz} \tag{16.15}$$

Replacing (1.3) for dp/dz and simplifying, recognizing that $\Gamma = g/\hat{c}_p$, we find that

$$\frac{1}{\theta}\frac{d\theta}{dz} = \frac{1}{T}\left(\frac{dT}{dz} + \Gamma\right) \tag{16.16}$$

But in a neutral atmosphere $dT/dz = -\Gamma$, and therefore $d\theta/dz = 0$.

For a neutral atmosphere $\theta = \text{const} = T_0$ and also $T = T_0 - \Gamma z$. Combining these two results and solving for the potential temperature, we find that

$$\theta = T + \Gamma z \tag{16.17}$$

16.1.3 Buoyancy of a Rising (or Falling) Air Parcel in the Atmosphere

In the previous sections we have assumed that an adiabatic air parcel is always in equilibrium with the surrounding atmosphere. As a result, if the parcel moves to another altitude it would still be in equilibrium and would have no tendency to keep moving. It turns out that the atmosphere rarely has an adiabatic temperature profile. A number of processes, for example the nighttime radiative cooling of the Earth's surface and the influx of cold air brought in by the wind (cold advection), can cool the air next to the surface and decrease the lapse rate. The atmospheric lapse rate can also decrease if the air aloft is replaced by warmer air (warm advection). On the other hand, daytime heating of the surface and influx of warm air by the wind next to the surface can increase the air temperature near the ground and the atmospheric lapse rate. Finally, the atmospheric lapse rate can also increase owing to cold advection aloft and radiative cooling of the clouds. This dynamic behavior of the atmosphere resulting from winds, solar heating, or radiative cooling usually dominates the behavior of the vertical temperature profile in the atmosphere.

An adiabatic air parcel moving in such an atmosphere may become more or less buoyant than the surrounding air. To study this motion, let us assume that an adiabatic air parcel is rising in such a nonadiabatic atmosphere with a lapse rate Λ. Let us assume that the atmospheric temperature T_a profile is

$$T_a(z) = T_0 - \Lambda z \tag{16.18}$$

The change of pressure with altitude in this environment is then given by (1.3), or using this notation, $dp/dz = -M_{air} g \, p/R T_a(z)$.

Now consider an adiabatic air parcel at $z = 0$ with initial temperature T_0. Its adiabatic motion as it is rising will still satisfy (16.6). Combining this with the pressure versus altitude of the atmosphere that we have just seen, we find that

$$\frac{dT}{dz} = \frac{-\Gamma T(z)}{T_0 - \Lambda z} \tag{16.19}$$

Integrating this equation from the initial position of the air parcel ($z = 0$ and T_0) to the final position, we obtain

$$T(z) = T_0 \left(\frac{T_0 - \Lambda z}{T_0} \right)^{\Gamma/\Lambda} = T_0 \left(\frac{T_a(z)}{T_0} \right)^{\Gamma/\Lambda} \tag{16.20}$$

If $\Lambda = \Gamma$, then (16.20) simplifies to $T(z) = T_a(z) = T_0 - \Gamma z$, the adiabatic temperature profile.

If the density of the air parcel is ρ and that of the atmosphere around it ρ_a, the acceleration experienced by the air parcel a will be proportional to the density difference between the atmosphere and the air parcel:

$$a = \frac{g(\rho_a - \rho)}{\rho} \tag{16.21}$$

If the density of the air parcel is lower than that of the air surrounding it, the parcel will be more buoyant and will rise. If the air has a lower density, the parcel will decelerate and buoyancy will oppose its motion. The air parcel has the same pressure as the surrounding atmosphere, so using the ideal-gas law (16.21) becomes

$$a = \frac{g(T_a - T)}{T_a} \tag{16.22}$$

Substituting (16.20) into (16.22), we obtain

$$a = g \left[\left(\frac{T_a(z)}{T_0} \right)^{(\Gamma/\Lambda) - 1} - 1 \right] \tag{16.23}$$

If $\Lambda > \Gamma$, that is if the atmospheric temperature decreases faster than that of the adiabatic rising parcel (Figure 16.2a), then $T_a(z) < T_0$ and $(\Gamma/\Lambda) - 1 < 0$. Therefore the term in brackets in (16.23) is positive, and $a > 0$. As the parcel rises it will accelerate, rise more rapidly, accelerate more, and so on. Its vertical motion is enhanced by the atmosphere surrounding it and the resulting buoyancy force.

If $\Lambda < \Gamma$ (Figure 16.2b), then one can show (considering the cases $\Lambda > 0$ and $\Lambda < 0$) that the acceleration given by (16.23) is negative, or $a < 0$. The rising air parcel will

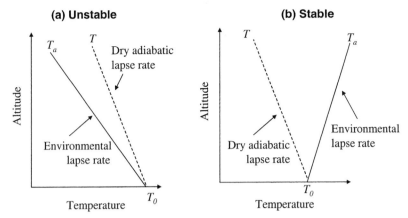

FIGURE 16.2 Temperature profiles for (a) an unstable atmosphere and (b) a stable atmosphere. The dry adiabatic lapse rate is also shown.

decelerate as it rises in an atmosphere with a lapse rate smaller than the adiabatic. The deceleration will lead to a stop of the air parcel motion.

Our previous discussion focused on a rising air parcel. The same conclusions are applicable to a sinking air parcel. If $\Lambda > \Gamma$, then the atmosphere enhances its motion, whereas if $\Lambda < \Gamma$, the atmosphere suppresses it. Finally, if the air parcel is saturated with water vapor, one would need to use the moist adiabatic lapse rate Γ_s instead of Γ in the discussion above.

16.2 ATMOSPHERIC STABILITY

The lapse rate in the lower portion of the atmosphere has a great influence on the vertical motion of air. Buoyancy can resist or enhance vertical air motion of airmasses, thus affecting the mixing of pollutants. Comparing the environmental lapse rate Λ with the adiabatic lapse rate Γ, we can define three regimes of atmospheric stability (Figure 16.2):

$$\Lambda > \Gamma : \quad \text{unstable}$$
$$\Lambda = \Gamma : \quad \text{neutral}$$
$$\Gamma > \Lambda : \quad \text{stable}$$

If $\Lambda > \Gamma$, the atmospheric lapse rate is *superadiabatic* and the atmosphere is unstable. A rising adiabatic air parcel will be warmer than its surroundings (Figure 16.2a) and will continue to rise away from its original position. Vertical motions are enhanced by buoyancy and pollutants are mixed rapidly in an unstable atmosphere. The mechanical analog in this case is that of a rock resting on top of a hill; any perturbation of the rock left or right will lead to continuous motion of the rock in that direction. The atmosphere becomes more unstable as the temperature decreases more rapidly with altitude. This can result from either warming of the air near the ground or cooling of the air aloft. The surface air warms up as a result of the daytime solar heating of the ground surface, air moving over a warm surface, or advection of a warm airmass next to the ground. Radiative cooling of clouds and advection of colder airmasses aloft can also make the atmosphere unstable.

If the atmospheric lapse rate is adiabatic, an air parcel displaced vertically is always at equilibrium with its surroundings. For this atmospheric state, vertical displacements are not affected by buoyancy forces. The mechanical analog of a neutral atmosphere is that of a rock lying on a flat surface. If the rock is displaced, it will still be in equilibrium. The atmosphere may be close to neutral when the sky is heavy with clouds and there is a moderate to high wind. Under these conditions, the clouds prevent the radiative heating or cooling of the ground and the wind mixes the air, smoothing out temperature deviations from the adiabatic profile. This atmospheric state is rather rare.

Finally, if $\Lambda < \Gamma$, a rising air parcel cools more rapidly with height than does its environment (Figure 16.2b). The rising air will therefore be cooler and heavier than the air surrounding it. Stable air strongly resists vertical motion, and if it is forced to rise, it will tend to spread horizontally instead. The mechanical analog of this case is that of a rock lying in a valley; if it is moved, it will return to its original position. The atmosphere is stable when the lapse rate is small, that is, when the temperature difference between the surface and aloft is relatively small. As a result, the atmosphere tends to become more stable owing to processes causing either cooling of the air near the ground or warming of the air aloft. The air near the ground can cool as a result of nighttime radiative cooling of the surface, movement of air next to a cool surface, or advection of cool air by the wind close to the ground. Replacement of the air aloft with warmer air masses can also contribute to the formation of a stable atmosphere. As noted earlier, if the temperature of the atmosphere increases with altitude in a layer in the atmosphere, then this layer is very stable and is called an *inversion*.

The arguments presented above are strictly applicable to a cloud-free atmosphere. For a cloudy atmosphere $\Gamma_s < \Gamma$ and one can then distinguish the following regimes of behavior (Figure 16.3):

$$\Lambda > \Gamma : \quad \text{absolutely unstable atmosphere}$$
$$\Gamma > \Lambda > \Gamma_s : \quad \text{conditionally stable atmosphere}$$
$$\Gamma_s > \Lambda : \quad \text{absolutely stable atmosphere}$$

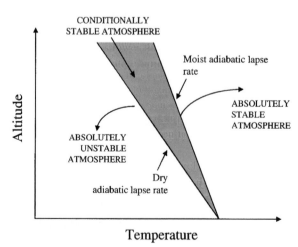

FIGURE 16.3   Regimes of absolute stability, absolute instability, and conditional stability in the atmosphere.

If the environmental lapse rate lies between the dry and moist values, then the stability of the atmosphere depends on whether the rising air is saturated. When an air parcel is not saturated, then the dry adiabatic lapse rate is the relevant reference state and the atmosphere is stable. For a saturated air parcel inside a cloud, the moist adiabatic rate is the applicable criterion for comparison and the atmosphere is unstable. Therefore a cloudy atmosphere is inherently less stable than the corresponding dry atmosphere with the same lapse rate.

The radiative cooling of the ground during the night and its heating by solar radiation during the day cause diurnal changes in the stability of the lower atmosphere. During the night (especially if there are no clouds aloft and the winds are light), the radiative cooling of the ground surface often leads to surface air that is colder than the air above it. This leads to an inversion near the ground known as a *radiation* (or *surface*) *inversion*. A stable layer thus exists in the lower hundred or so meters in the atmosphere that prevents the ventilation of emissions during the night. Pollutants emitted during the night inside this shallow layer get trapped and can reach relatively high concentrations (Figure 16.4a).

As the sun rises, the ground and the air next to it start warming up and a temperature profile corresponding to an unstable atmosphere is established. This change occurs over a period of a few hours in the morning and results in breaking of the inversion usually before noon. The changing atmospheric stability, from stable in the early morning to unstable in the afternoon, has a significant effect on pollutant concentrations. In large urban areas

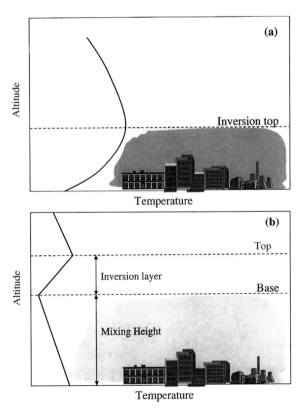

FIGURE 16.4   Temperature profile and pollutant mixing for (a) nighttime radiation inversion and (b) subsidence inversion.

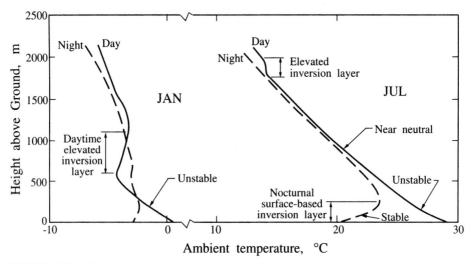

FIGURE 16.5 Monthly average diurnal and seasonal variations of the vertical thermal structure of the planetary boundary layer at a rural site near St. Louis, MO, based on 1976 data for January and July.

pollutants emitted mainly by transportation (carbon monoxide, oxides of nitrogen, black carbon) have much higher concentrations during the morning rush-hour than in the afternoon, even if the traffic is the same. This unstable layer in direct contact with the surface is called the *mixed layer* and its height the *mixing height* (or *mixing depth*). The mixing height is usually of the order of 1 km varying from a few hundred meters to 3 km (Figure 16.4b). The daytime mixed layer is characterized by vigorous turbulent mixing.

Subsidence inversions form as the air above slowly sinks and warms. A typical temperature profile for such an inversion is shown in Figure 16.4b. Usually the air below the inversion is unstable, and pollutants mix relatively rapidly up to the inversion. The stable layer of the inversion acts as a lid and does not allow penetration. Some of the most dangerous air pollution episodes are related to subsidence inversions that persist for several days. The mixing height can be as low as a few hundred meters in such air pollution episodes.

Monthly average diurnal and seasonal variations of the vertical temperature profiles near St. Louis in 1976 are shown in Figure 16.5. The winter nights are characterized by a weak radiation inversion layer extending up to 200 m, while during the day a strong inversion layer is formed at 500 m above the ground extending up to 1000 m. During the summer nights a much stronger radiation inversion layer is present in the lower 100 m, while the atmosphere is unstable during the day. An elevated inversion layer is present with a base close to 1700 m.

16.3 MICROMETEOROLOGY

In this section we will concentrate on air motion in the lowest layers of the atmosphere. Such air motion, taking place adjacent to a solid boundary of variable temperature and roughness, is virtually always turbulent. This atmospheric turbulence is responsible for the transfer of heat, water vapor, and trace gases and aerosols between the surface and the atmosphere as a whole. Our objective will be to understand the basic phenomena that influence atmospheric turbulence. Our treatment here will, of necessity, be rather limited. The interested reader may pursue this area in greater depth in Monin and Yaglom (1971, 1975), Plate (1971, 1982), Nieuwstadt and van Dop (1982), and Panofsky and Dutton (1984).

16.3.1 Basic Equations of Atmospheric Fluid Mechanics

We will derive first the equations that govern the fluid density, temperature, and velocities in the lowest layer of the atmosphere. These equations will form the basis from which we can subsequently explore the processes that influence atmospheric turbulence. In our discussion we shall consider only a shallow layer adjacent to the surface, in which case we can make some rather important simplifications in the equations of continuity, motion, and energy.

The equation of continuity for a compressible fluid is

$$\frac{\partial \rho}{\partial t} + \frac{\partial}{\partial x}(\rho u_x) + \frac{\partial}{\partial y}(\rho u_y) + \frac{\partial}{\partial z}(\rho u_z) = 0 \qquad (16.24)$$

where ρ is the fluid density and u_x, u_y, u_z are the fluid velocity components in the directions x, y, and z, respectively. To make the subsequent equations a little more compact, we will use first the *index notation*, that is, denote the coordinate axes as x_1, x_2, and x_3 and the corresponding velocities as u_1, u_2, and u_3. With this notation the continuity equation becomes

$$\frac{\partial \rho}{\partial t} + \frac{\partial}{\partial x_1}(\rho u_1) + \frac{\partial}{\partial x_2}(\rho u_2) + \frac{\partial}{\partial x_3}(\rho u_3) = \frac{\partial \rho}{\partial t} + \sum_{i=1}^{3} \frac{\partial}{\partial x_i}(\rho u_i) = 0 \qquad (16.25)$$

Equation (16.25) can be simplified further if we use the *summation convention*, that is, if we omit the summation symbol from the equation, assuming that repeated symbols are summed from 1 to 3. As a result, the continuity equation becomes

$$\frac{\partial \rho}{\partial t} + \frac{\partial}{\partial x_i}(\rho u_i) = 0 \qquad (16.26)$$

Because we are interested only in processes taking place on limited spatial and temporal scales over which the air motion is not influenced by the rotation of the Earth, we will neglect the Coriolis acceleration (see Chapter 21) and write the equation of motion of a compressible, Newtonian fluid in a gravitational field as

$$\rho \left(\frac{\partial u_i}{\partial t} + u_j \frac{\partial u_i}{\partial x_j} \right) = \frac{\partial}{\partial x_k} \left[\mu \left(\frac{\partial u_i}{\partial x_k} + \frac{\partial u_k}{\partial x_i} \right) - \left(p + \frac{2}{3}\mu \frac{\partial u_j}{\partial x_j} \right) \delta_{ik} \right] - \rho g \delta_{3i} \qquad (16.27)$$

where μ is the fluid viscosity and δ_{ij} is the Krönecker delta, defined by $\delta_{ij} = 1$ if $i = j$, and $\delta_{ij} = 0$ if $i \neq j$. For example, in the x direction (16.27) becomes

$$\rho \left(\frac{\partial u_x}{\partial t} + u_x \frac{\partial u_x}{\partial x} + u_y \frac{\partial u_x}{\partial y} + u_z \frac{\partial u_x}{\partial z} \right) = -\frac{\partial p}{\partial x} + \frac{\partial}{\partial x} \left[2\mu \frac{\partial u_x}{\partial x} - \frac{2}{3}\mu \left(\frac{\partial u_x}{\partial x} + \frac{\partial u_y}{\partial y} + \frac{\partial u_z}{\partial z} \right) \right]$$
$$+ \frac{\partial}{\partial y} \left[\mu \left(\frac{\partial u_x}{\partial y} + \frac{\partial u_y}{\partial x} \right) \right] + \frac{\partial}{\partial z} \left[\mu \left(\frac{\partial u_x}{\partial z} + \frac{\partial u_z}{\partial x} \right) \right]$$

$$(16.28)$$

Finally the energy equation, assuming that the contribution of viscous dissipation to the energy balance of the atmosphere is negligible, is

$$\rho \hat{c}_v \left(\frac{\partial T}{\partial t} + u_j \frac{\partial T}{\partial x_j} \right) = k \frac{\partial^2 T}{\partial x_j \partial x_j} - p \frac{\partial u_j}{\partial x_j} + Q \tag{16.29}$$

where $\hat{c}_v$ is the heat capacity of air at constant volume per unit mass, k is the thermal conductivity (assumed constant) and Q represents the heat generated by any sources in the fluid.

Equations (16.26), (16.27), and (16.29) represent five equations for the six unknowns u_1, u_2, u_3, p, ρ, and T. The sixth equation necessary for closure is the ideal-gas law:

$$p = \frac{\rho R T}{M_{\text{air}}} \tag{16.30}$$

These six equations can therefore be solved, in principle, subject to appropriate boundary and initial conditions to yield velocity, pressure, density, and temperature profiles in the atmosphere.

Atmosphere at Equilibrium   Solve the system of the equations given above to calculate the pressure, density, and temperature profiles in the atmosphere for $u_i = 0$, assuming that there no heat sources and that the temperature is zero at $x_3 = H$.

When the atmosphere is at rest, its density, pressure, and temperature are constant with time, and (16.27) and (16.29) become

$$\frac{\partial p_e}{\partial x_1} = 0 \qquad \frac{\partial p_e}{\partial x_2} = 0 \qquad \frac{\partial p_e}{\partial x_3} = -\rho_e g \tag{16.31}$$

$$\frac{\partial^2 T}{\partial x_1^2} + \frac{\partial^2 T}{\partial x_2^2} + \frac{\partial^2 T}{\partial x_3^2} = 0 \tag{16.32}$$

where the subscript e denotes equilibrium. From (16.31) it is obvious that the pressure and density depend only on altitude: $p_e = p(x_3)$, $\rho_e = \rho(x_3)$. This means, given the ideal-gas law, that the temperature is only be a function of altitude and (16.32) becomes

$$\frac{\partial^2 T}{\partial x_3^2} = 0 \tag{16.33}$$

To solve (16.33) we need two boundary conditions. If T_0 is the ground temperature and zero at $x_3 = H$, then, integrating, we obtain

$$T_e = T_0 \left(1 - \frac{x_3}{H} \right) \tag{16.34}$$

Substituting (16.30) and (16.34) into (16.31) and integrating, we find that

$$p_e = p_0 \left(1 - \frac{x_3}{H} \right)^{(gHM_{\text{air}}/RT_0)} \tag{16.35}$$

where p_0 is the surface pressure. Finally, from (16.34), (16.35), and the ideal-gas law, we obtain

$$\rho_e = \rho_0 \left(1 - \frac{x_3}{H}\right)^{(gHM_{\text{air}}/RT_0)-1} \tag{16.36}$$

where $\rho_0 = p_0 M_{\text{air}}/RT$ is the air density at the surface.

The lapse rate for the atmosphere is in this case equal to $\Lambda = T_0/H$. The adiabatic atmosphere (see Section 16.1.1) is a special case of this more general example.

Equations (16.26), (16.27), and (16.29) can be simplified for a shallow atmospheric layer next to the ground using the *Boussinesq* approximations. The conditions of their validity have been examined by Spiegel and Veronis (1960), Calder (1968), and Dutton (1976). The fundamental idea is of these approximations is to first express the equilibrium profiles of pressure, density, and temperature as functions of x_3 only as follows:

$$\begin{aligned} p_e &= p_0 + p_m(x_3) \\ \rho_e &= \rho_0 + \rho_m(x_3) \\ T_e &= T_0 + T_m(x_3) \end{aligned} \tag{16.37}$$

We consider only a shallow layer, so that p_m/p_0, ρ_m/ρ_0, and T_m/T_0 are all smaller than unity. When there is motion, we can express the actual pressure, density, and temperature as the sum of the equilibrium values and a small correction due to the motion (denoted by a tilde). Thus we write

$$\begin{aligned} p(x_1, x_2, x_3, t) &= p_0 + p_m(x_3) + \tilde{p}(x_1, x_2, x_3, t) \\ \rho(x_1, x_2, x_3, t) &= \rho_0 + \rho_m(x_3) + \tilde{\rho}(x_1, x_2, x_3, t) \\ T(x_1, x_2, x_3, t) &= T_0 + T_m(x_3) + \tilde{T}(x_1, x_2, x_3, t) \end{aligned} \tag{16.38}$$

where we assume that the deviations induced by the motion are sufficiently small that the quantities $\tilde{p}/p_0$, $\tilde{\rho}/\rho_0$, and $\tilde{T}/T_0$ are also small compared with unity. Substituting (16.38) into (16.26), (16.27), and (16.29) and simplifying using the assumptions presented above, (see also Appendix 16), one can derive the basic equations of atmospheric fluid mechanics that are applicable for a shallow atmospheric layer

$$\frac{\partial u_i}{\partial x_i} = 0$$

$$\frac{\partial u_i}{\partial t} + u_j \frac{\partial u_i}{\partial x_j} = -\frac{1}{\rho_0}\frac{\partial \tilde{p}}{\partial x_i} + \frac{\mu}{\rho_0}\frac{\partial^2 u_i}{\partial x_j \partial x_j} + \frac{g\tilde{T}}{T_0}\delta_{i3} \tag{16.39}$$

$$\rho_0 \hat{c}_p \left(\frac{\partial \theta}{\partial t} + u_j \frac{\partial \theta}{\partial x_j}\right) = k\left(\frac{\partial^2 \theta}{\partial x_j \partial x_j}\right) + Q$$

where θ is the potential temperature. The Boussinesq approximations lead to a considerable simplification of the original equations. The first is the continuity equation for an incompressible fluid. The equation of motion is identical to the incompressible form of the equation with the exception of the last term, which accounts for the acceleration due to buoyancy forces. Finally, the energy equation is just the usual heat conduction equation with T replaced by θ. The complete set of equations consists now of the five equations in (16.39) for the five unknowns, u_1, u_2, u_3, ρ, and θ. The ideal-gas equation of state is no longer required as it has been incorporated into the equations.

Although ρ_0 and T_0 in (16.39) refer to the constant surface values, equations of precisely the same form can be derived in which ρ_0 and T_0 are replaced by ρ_e and T_e, the reference profiles. The equations written in this form will be useful later when we consider the dynamics of potential temperature in the atmosphere.

16.3.2 Turbulence

Equations (16.39) govern the fluid velocity and temperature in the lower atmosphere. Although these equations are at all times valid, their solution is impeded by the fact that the atmospheric flow is turbulent (as opposed to laminar). Turbulence is a characteristic of flows and not of fluids themselves. It is difficult to define turbulence; instead we can cite a number of characteristics of turbulent flows. Turbulent flows are irregular and random, so that the velocity components at any location vary randomly with time. Since the velocities are random variables, their exact values can never be predicted precisely. Thus (16.39) become partial differential equations, the dependent variables of which are random functions. We cannot expect to solve any of these equations exactly; rather, we must be content to determine some statistical properties of the velocities and temperature. The random fluctuations in the velocities result in rates of momentum, heat, and mass transfer in turbulence that are many orders of magnitude greater than the corresponding rates due to pure molecular transport. Turbulent flows are dissipative in the sense that there is a continuous conversion of kinetic to internal energy. Thus, unless energy is continuously supplied, turbulence will decay. The usual source of energy for turbulence is shear in the flow field, although in the atmosphere buoyancy can also be a source of energy.

Let us consider a situation of turbulent pipe flow. If the same pipe and pressure drop are used each time the experiment is repeated, the velocity field would always be different no matter how carefully the conditions of the experiment are reproduced. A particular turbulent flow in the pipe can be envisioned as one of an infinite ensemble of flows from identical macroscopic boundary conditions. The mean or average velocity, for instance, as a function of radial position in the pipe could be determined, in principle, by averaging the readings made over an infinite ensemble of identical experiments. Figure 16.6 shows a hypothetical record of the ith velocity component at a certain point in a turbulent flow. The specific features of a second velocity record taken under the same conditions would be different but there might well be decided resemblance in some of the characteristics of the record. In practice, it is rarely possible to repeat measurements, particularly in the atmosphere. To compute the mean value of u_i at location $\mathbf{x}$ and time t, we would need to average the values of u_i at $\mathbf{x}$ and time t for all the similar records. This ensemble mean is denoted by $< u_i(t,\mathbf{x}) >$. If the ensemble mean does not change with time t, we can substitute a time average for the ensemble average. The time-average velocity is defined by

$$\bar{u}_i = \lim_{\tau \to \infty} \frac{1}{\tau} \int_{t_0}^{t_0+\tau} u_i(t)\,dt$$

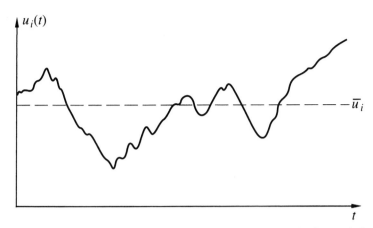

FIGURE 16.6   Typical record of the velocity in direction i at a point in a turbulent flow.

In practice, the statistical properties of u_i depend on time. For example, the flow may change with time. However, we still wish to define a mean velocity; this is done by defining:

$$\bar{u}_i(t) = \frac{1}{\tau} \int_{t-\tau/2}^{t+\tau/2} u_i(t')dt'$$

Clearly, $\bar{u}_i(t)$ will depend on the averaging interval τ. We need to choose τ large enough so that an adequate number of fluctuations are included, but yet not so large that important macroscopic features of the flow would be masked. For example, if τ_1 and τ_2 are timescales associated with fluctuations and macroscopic changes in the flow, respectively, we would want $\tau_2 \gg \tau \gg \tau_1$.

It is customary to represent the instantaneous value of the wind velocity u_i as the sum of a mean and a fluctuating component, $\bar{u}_i + u_i'$. The mean values of the velocities tend to be smooth and slowly varying. The fluctuations $u_i' = u_i - \bar{u}_i$ are characterized by extreme spatial and temporal variations. In spite of the severity of fluctuations, it is observed experimentally that turbulent spatial and temporal inhomogeneities still have considerably greater sizes than molecular scales. The viscosity of the fluid prevents the turbulent fluctuations from becoming too small. Because the smaller scales (or eddies) are still many orders of magnitude larger than molecular dimensions, the turbulent flow of a fluid is described by the basic equations of continuum mechanics in Section 16.3.1. In general, the largest scales of motion in turbulence (the so-called large eddies) are comparable to the major dimensions of the flow and are responsible for most of the transport of momentum, heat, and mass. Large scales of motion have comparatively long timescales, where the small scales have short timescales and are often statistically independent of the large-scale flow. A physical picture that is often used to describe turbulence involves the transfer of energy from the larger to the smaller eddies, which ultimately dissipate the energy as heat.

16.3.3   Equations for the Mean Quantities

In the equations of motion and energy (16.39), the dependent variables u_i, p, and θ are random variables, making the equations virtually impossible to solve. We modify our goal from trying to calculate the actual variables to trying to estimate their mean values. We decompose the velocities, temperature, and pressure into a mean and a fluctuating component:

$$u_i = \bar{u}_i + u'_i$$
$$\theta = \bar{\theta} + \theta' \tag{16.40}$$
$$p = \bar{p} + p'$$

By definition, the mean of a fluctuating quantity is zero:

$$\overline{u'_i} = \overline{\theta'} = \overline{p'} = 0 \tag{16.41}$$

Our objective is to determine equations for $\bar{u}_i, \bar{\theta}$, and $\bar{p}$. To obtain these equations, we first substitute (16.40) into (16.39). We then average each term in the resulting equations with respect to time. The result, employing (16.41), is

$$\frac{\partial \bar{u}_i}{\partial x_i} = 0 \tag{16.42}$$

$$\frac{\partial \bar{u}_i}{\partial t} + \bar{u}_j \frac{\partial \bar{u}_i}{\partial x_j} + \overline{u'_j \frac{\partial u'_i}{\partial x_j}} = -\frac{1}{\rho_0}\frac{\partial \bar{p}}{\partial x_i} + \frac{\mu}{\rho_0}\frac{\partial^2 \bar{u}_i}{\partial x_j \partial x_j} + \frac{g\bar{\theta}}{T_0}\delta_{i3} \tag{16.43}$$

$$\rho_0 \hat{c}_p \left(\frac{\partial \bar{\theta}}{\partial t} + \bar{u}_j \frac{\partial \bar{\theta}}{\partial x_j} + \overline{u'_j \frac{\partial \theta'}{\partial x_j}} \right) = k \left(\frac{\partial^2 \bar{\theta}}{\partial x_j \partial x_j} \right) \tag{16.44}$$

It is customary to employ the relation

$$\frac{\partial u'_i}{\partial x_i} = 0 \tag{16.45}$$

obtained by subtracting (16.42) from the first equation (16.39) to transform the third terms on the LHS of (16.43) and (16.44) to $\partial \overline{u'_i u'_j}/\partial x_j$ and $\partial \overline{u'_j \theta'}/\partial x_j$. Then (16.43) and (16.44) are written in the form

$$\frac{\partial}{\partial t}(\rho_0 \bar{u}_i) + \frac{\partial}{\partial x_j}(\rho_0 \bar{u}_i \bar{u}_j) = -\frac{\partial \bar{p}}{\partial x_i} + \frac{\partial}{\partial x_j}\left(\mu \frac{\partial \bar{u}_i}{\partial x_j} - \rho_0 \overline{u'_i u'_j} \right) + \frac{g\bar{\theta}}{T_0}\delta_{i3}$$

$$\rho_0 \hat{c}_p \left(\frac{\partial \bar{\theta}}{\partial t} + \bar{u}_j \frac{\partial \bar{\theta}}{\partial x_j} \right) = \frac{\partial}{\partial x_j}\left(k \frac{\partial \bar{\theta}}{\partial x_j} - \rho_0 c_p \overline{u'_j \theta'} \right) \tag{16.46}$$

The good news is that these equations, now time-averaged, contain only smoothly time-varying average quantities, so that the difficulties associated with the stochastic nature of the original equations have been alleviated. However, there is also some bad news. We note the emergence of new dependent variables, $\overline{u'_i u'_j}$ and $\overline{u'_j \theta'}$ for $i, j = 1, 2, 3$. When the equations are written in the form of (16.45) and (16.46), we can see that $\rho_0 \overline{u'_i u'_j}$ represents a new contribution to the total stress tensor and that $\rho_0 \hat{c}_p \overline{u'_j \theta'}$ is a new contribution to the heat flux vector.

The terms $\overline{\rho_0 u'_i u'_j}$, called the *Reynolds stresses*, indicate that the velocity fluctuations lead to a transport of momentum from one volume of fluid to another. Let us consider the physical interpretation of the Reynolds stresses by using the Cartesian notation for simplicity. We envision the situation of a steady mean wind in the x direction near the ground. Let the y direction be the horizontal direction perpendicular to the mean wind and z the vertical direction. A sudden increase or gust in the mean wind would result in a positive u'_x, whereas a lull would lead to a negative u'_x. Left- and right-hand swings of the wind direction from its mean direction can be described by positive and negative u'_y, respectively, and upward and downward vertical gusts by positive and negative u'_z. The air needed to sustain a gust in the x direction must come from somewhere and usually it comes from faster moving air from above. Therefore, we expect positive values of u'_x to be correlated with negative values of u'_z. Similarly, a lull will result when air is transported upward rather than forward, so that we would expect negative u'_x to be associated with positive u'_z. As a result of both effects, $\overline{u'_x u'_z}$ is not zero and the corresponding Reynolds stress will play an important role in the transport of momentum.

Equations (16.42) and (16.46) have, as dependent variables, $\overline{u}_i, \overline{p}, \overline{\theta}, \overline{u'_i u'_j}$, and $\overline{u'_j \theta'}$. We have thus 14 dependent variables (note that $\overline{u'_i u'_j} = \overline{u'_j u'_i}$, so there are six Reynolds stresses plus three $\overline{u'_j \theta'}$ variables, three velocities, pressure, and temperature), but we have only five equations from (16.46). We need eight additional equations to close the system. We could attempt to write conservation equations for the new dependent variables. For example, we can derive such an equation for the variables $\overline{u'_i u'_j}$ by first subtracting (16.43) from the second equation in (16.39), leaving an equation for u'_i. We then multiply by u'_j and average all terms. Although we can arrive at an equation for $\overline{u'_i u'_j}$ this way, we unfortunately have at the same time generated still more dependent variables $\overline{u'_i u'_j u'_l}$. This problem, arising in the description of turbulence, is called the *closure problem*, for which no general solution has yet been found. At present we must rely on models and estimates based on intuition and experience to obtain a closed system of equations. Since mathematics by itself will not provide a solution, we will resort to quasi-physical models to obtain additional equations for the Reynolds stresses and the turbulent heat fluxes. The next section is devoted to the most popular semiempirical models for the turbulent momentum and energy fluxes.

16.3.4 Mixing-Length Models for Turbulent Transport

The simplest approaching to closing the equations (16.46) is based on an appeal to the physical picture of the actual nature of turbulent momentum transport. We can envision the turbulent fluid as comprising lumps of fluid, which, for a short time, retain their integrity before being destroyed. These lumps or eddies transfer momentum, heat, and mass from one location to another. Thus it is possible to imagine an eddy, originally at one level in the fluid, breaking away and conserving some or all of its momentum until it mixes with the mean flow at another level.

Let us assume a steady turbulent shear flow in which $\overline{u}_1 = \overline{u}_1(x_2)$ and $\overline{u}_2 = \overline{u}_3 = 0$. We first consider turbulent momentum transport, that is, the Reynolds stresses. The mean flux of x_1 momentum in the x_2 direction due to turbulence is $\overline{\rho u'_1 u'_2}$. Let us see if we can derive an estimate for this flux.

We can assume that the fluctuation in u_1 at any level x_2 is due to the arrival at that level of a fluid lump or eddy that originated at some other location where the mean velocity was different from that at x_2. We illustrate this idea in Figure 16.7, in which a fluid lump that is at $x_2 = x_2 + l_a$ at $t - \tau_a$ with an original velocity equal to the mean velocity at that level,

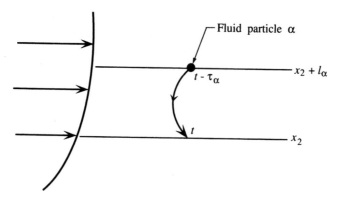

FIGURE 16.7   Eddy transfer in a turbulent shear flow.

namely, $\bar{u}_1(x_2 + l_a)$, arrives at x_2 at t. Let u'_{1a} be the fluctuation in u_1 at x_2 at time t due to the αth eddy. If the eddy maintains its x_1 momentum during its sojourn, this fluctuation can be written

$$u'_{1\alpha} = u_{1\alpha}(x_2, t) - \bar{u}_1(x_2) \tag{16.47}$$

As long as the x_1 momentum of the eddy is conserved, then

$$u_{1\alpha}(x_2, t) = \bar{u}_1(x_2 + l_\alpha, t - \tau_\alpha) \tag{16.48}$$

Substituting (16.48) into (16.47) and expanding $\bar{u}_1(x_2 + l_\alpha, t - \tau_\alpha)$ in a Taylor series about the point (x_2, t), we get

$$u'_{1\alpha} = l_\alpha \frac{\partial \bar{u}_1}{\partial x_2} - \tau_\alpha \frac{\partial \bar{u}_1}{\partial t} + \frac{1}{2} l_\alpha^2 \frac{\partial^2 \bar{u}_1}{\partial x_2^2} + \frac{1}{2} \tau_\alpha^2 \frac{\partial^2 \bar{u}_1}{\partial t^2} + \cdots \tag{16.49}$$

First, we note that, since the flow has been assumed to be steady, $\bar{u}_1$ does not vary with time and the corresponding derivatives are zero. Thus (16.49) becomes

$$u'_{1\alpha} = l_\alpha \frac{\partial \bar{u}_1}{\partial x_2} + \frac{1}{2} l_\alpha^2 \frac{\partial^2 \bar{u}_1}{\partial x_2^2} + \cdots \tag{16.50}$$

The second- and higher-order terms in (16.50) can be truncated if the distance l_α over which the eddy maintains its integrity is small compared to the characteristic lengthscale of the $\bar{u}_1$ field, leaving

$$u'_{1\alpha} = l_\alpha \frac{\partial \bar{u}_1}{\partial x_2} \tag{16.51}$$

Multiplying (16.51) by $u'_{2\alpha}$, the turbulent fluctuation in the x_2 direction associated with the same αth eddy is

$$u'_{1\alpha} u'_{2\alpha} = u'_{2\alpha} l_\alpha \frac{\partial \bar{u}_1}{\partial x_2} \tag{16.52}$$

To obtain the turbulent flux $\overline{u'_1 u'_2}$, we need to consider an ensemble of eddies that pass the point x_2. If we average (16.52) over all these eddies, we find that

$$\overline{u'_1 u'_2} = \overline{u'_2 l} \frac{\partial \overline{u}_1}{\partial x_2} \tag{16.53}$$

The term $\overline{u'_2 l}$ represents the correlation between the fluctuating x_2 velocity at x_2 and the distance of travel of the eddy. Let us try to estimate the order of the term $\overline{u'_2 l}$. First note that if $\partial \overline{u}_1 / \partial x_2 > 0$, then from (16.53) $\overline{u'_2 l}$ will have the same sign as $\overline{u'_1 u'_2}$, that is, it will be negative (see discussion in Section 16.3.3). If L_e is the maximum distance over which an eddy maintains its integrity and $\hat{u}_2$ is the turbulent intensity $\left(\overline{u'^2_2} \right)^{1/2}$, then the term $\overline{u'_2 l}$ will be proportional to $L_e \hat{u}_2$ or if c is a positive constant of proportionality, then

$$\overline{u'_2 l} = -c L_e \hat{u}_2 \tag{16.54}$$

Substituting (16.54) into (16.53), we find that

$$\overline{u'_1 u'_2} = -c L_e \hat{u}_2 \frac{\partial \overline{u}_1}{\partial x_2} \tag{16.55}$$

The *mixing length L_e* is a measure of the maximum distance in the fluid over which the velocity fluctuations are correlated, or in some sense, of the eddy size. The experimental determination of L_e simply involves measuring the velocities at two points separated by larger and larger distances until their correlation approaches zero. From (16.55), we define the *eddy viscosity* or *turbulent momentum diffusivity* K_M as $K_M = c L_e \hat{u}$, so that (16.55) becomes:

$$\overline{u'_1 u'_2} = -K_M \frac{\partial \overline{u}_1}{\partial x_2} \tag{16.56}$$

We can extend the mixing-length concept to the turbulent heat flux. We consider the same shear flow as above, in which buoyancy effects are for the moment neglected. By analogy to the definition of the eddy viscosity, we can define an eddy viscosity for heat transfer by

$$\overline{u'_2 \theta'} = -K_T \frac{\partial \overline{\theta}}{\partial x_2} \tag{16.57}$$

Equations (16.56) and (16.57) provide a solution to the closure problem inasmuch as the turbulent fluxes have been related directly to the mean velocity and potential temperature. However, we have essentially exchanged our lack of knowledge of $\overline{u'_1 u'_2}$ and $\overline{u'_2 \theta'}$ for K_M and K_T, respectively. In general, both K_M and K_T are functions of location in the flow field and are different for transport in different coordinate directions. The variation of these coefficients, as we will see later, is determined by a combination of scaling arguments and experimental data.

The result of the mixing-length idea used to derive the expressions (16.56) and (16.57) is that the turbulent momentum and energy fluxes are related to the gradients of the mean quantities. Substitution of these relations into (16.46) leads to closed equations for the mean quantities. Thus, except for the fact that K_M and K_T vary with position and direction, these models for turbulent transport are analogous to those for molecular transport of momentum and energy.

The use of a diffusion equation model implies that the lengthscale of the transport processes is much smaller than the characteristic length over which the mean velocity and temperature profiles are changing. In molecular diffusion in a gas at normal densities, the lengthscale of the diffusion process, the mean free path of a molecule, is many orders of magnitude smaller than the distances over which the mean properties of the gas (e.g., velocity or temperature) vary. In turbulence, on the other hand, the motions responsible for the transport of momentum, heat, and material are usually roughly the same size as the characteristic lengthscale for changes in the mean fields of velocity, temperature, and concentration. In the atmosphere, for example, the characteristic vertical dimension of eddies is of the same order as is the characteristic scale for changes of the velocity profile. Thus, in general, a diffusion model for transport like the one given by (16.55) and (16.56) has some serious weaknesses for the description of atmospheric turbulence. The success in using these equations depends on two factors: (1) they should ideally be employed in situations in which the lengthscale for changes in the mean properties is considerably greater than that of the eddies responsible for transport; and (2) the values and functional forms of K_M and K_T should be determined empirically for situations similar to those in which (16.56) and (16.57) are applied.

16.4   VARIATION OF WIND WITH HEIGHT IN THE ATMOSPHERE

The atmosphere near the surface of the Earth can be divided into four layers: the free atmosphere; the Ekman layer; the surface layer; and immediately adjacent to the ground surface, the laminar sublayer. The thickness of the laminar sublayer is typically less than a centimeter, and this layer can be neglected for all practical purposes in the present discussion. (The fluid viscosity becomes important in this thin layer; we will examine the processes in this layer in our discussion of dry deposition in Chapter 19.) The surface layer extends practically from the surface to a height of ≤ 30–50 m. Within this layer, the vertical turbulent fluxes of momentum and heat are assumed constant with respect to height, and indeed they define the extent of this region. The Ekman layer extends to a height of 300–500 m depending on the terrain type, with the greatest thickness corresponding to the more uneven terrain. In this layer, the wind direction is affected by the rotation of the Earth (see Chapter 21). The windspeed in the Ekman layer generally increases rapidly with height; however, the rate lessens near the free atmosphere.

In this section we consider the variation of wind with height in the surface and Ekman layers, which constitute the so-called *planetary boundary layer*. Most of our attention will be devoted to the surface layer, the region in which pollutants are usually first released. The exact vertical distribution of wind velocity depends on a number of parameters, including the surface roughness and the atmospheric stability.

In meteorological applications, the surface roughness is usually characterized by the height of the roughness elements (buildings, trees, bushes, grass, etc) ε. These elements are usually so closely distributed that only their height and spacing are important. In general, *smooth surfaces* allow the establishment of a laminar sublayer in which they are submerged. On the other hand, a rough surface is one in which the roughness elements are high enough to prevent the formation of a laminar sublayer, so that the flow is turbulent down to the roughness elements. The depth of the laminar sublayer, and hence the classification of the surface as smooth or rough, depends on the Reynolds number of the flow.

Knowledge of the governing physics of flows in the surface layer is not sufficiently complete to derive the vertical mean velocity profiles based on first principles. Useful empirical relationships have been developed using approximate theories and experimental measurements. Similarity theories, based on dimensional analysis, provide a convenient method for grouping of the system variables into dimensionless parameters and then the derivation of universal similarity relationships.

16.4.1   Mean Velocity in the Adiabatic Surface Layer over a Smooth Surface

Consider first the steady, ground-parallel flow of air over a flat homogeneous surface. We assume that the wind flow is in the x direction ($\bar{u}_y = 0$) and that $\bar{u}_x = \bar{u}_x(z)$. Our goal is to determine $\bar{u}_x(z)$ assuming that the vertical temperature profile is adiabatic.

Let us see what can be determined about the functional dependence of $\bar{u}_x(z)$ employing dimensional analysis. For this analysis, we need to define a characteristic velocity for the flow u_*. This characteristic velocity depends on the turbulence flux $\overline{u'_x u'_z}$ and is called the *friction velocity u_**. It is actually equal to $\sqrt{\tau_0/\rho}$, where τ_0 is the shear stress at the surface. The friction velocity can be calculated from an actual measurement of the velocity at a given height. With the addition of the friction velocity, we have five variables in our problem, $\bar{u}_x$, ν, ρ, z, and u_*, involving three dimensions (mass, length, and time). We can facilitate the following steps by replacing in this set of variables the velocity $\bar{u}_x$ with its gradient, so that the variable set becomes $d\bar{u}_x/dz$, ν, ρ, z, and u_*.

The Buckingham π theorem states that in this system there are only $5 - 3 = 2$ independent dimensionless groups, π_1 and π_2, relating the five variables, so that $F(\pi_1, \pi_2) = 0$. The π theorem does not tell us what the groups are, only how many exist. As the first group we select

$$\pi_1 = \frac{zu_*}{\nu} \tag{16.58}$$

which is essentially a Reynolds number for the flow. For the second dimensionless parameter a convenient choice is a dimensionless velocity gradient

$$\pi_2 = \frac{d\bar{u}_x}{dz}\frac{z}{u_*} \tag{16.59}$$

and according to the π theorem, $F(\pi_1, \pi_2) = 0$ or equivalently $\pi_2 = G(\pi_1)$. Using (16.58) and (16.59), we obtain

$$\frac{d\bar{u}_x}{dz} = \frac{u_*}{z}G\left(\frac{zu_*}{\nu}\right) \tag{16.60}$$

This is as far as dimensional analysis will bring us. We now need some physical insight to determine the unknown function G. The kinematic viscosity ν is important only in the laminar sublayer, and $d\bar{u}_x/dz$ should not depend on ν in the region of interest above the sublayer. Thus we can set the function G equal to a constant $(1/\kappa)$ obtaining

$$\frac{d\bar{u}_x}{dz} = \frac{1}{\kappa}\frac{u_*}{z} \tag{16.61}$$

On integration, we obtain

$$\bar{u}_x(z) = \frac{u_*}{\kappa}\ln z + \text{const}$$

This equation is usually written in the dimensionless form:

$$\frac{\bar{u}_x(z)}{u_*} = \frac{1}{\kappa}\ln\frac{u_*z}{\nu} + \text{const} \tag{16.62}$$

The integration constant in (16.62) has been evaluated experimentally for smooth surfaces and has been found to be equal to 5.5. The constant κ is known as the *von Karman constant* with a value of ~ 0.4. In spite of an uncertainty of the order of 5% in empirical estimates of κ for different surfaces, it is considered a universal constant. The mean velocity profile can therefore be calculated by

$$\bar{u}_x(z) = u_*\left(\frac{1}{\kappa}\ln\frac{u_*z}{\nu} + 5.5\right) \tag{16.63}$$

and the velocity $\bar{u}_x$ is predicted to increase according to the logarithm of z for a perfectly smooth surface. Inspecting (16.63), we see that $\bar{u}_x$ becomes zero when $\ln(u_*z/\nu) = -5.5\kappa = -2.2$, that is, when $z = z^* = 0.11\,\nu/u_*$. Typical values of u_* and ν for the atmosphere are 100 cm s^{-1} and $0.1\text{ cm}^2\text{s}^{-1}$, so $\bar{u}_x$ vanishes at around 10^{-4} cm and becomes negative for even smaller z. Obviously, (16.63) holds only for values of z greater than z^*. Equation (16.63) has limited utility in the real atmosphere because all actual surfaces have some roughness.

16.4.2 Mean Velocity in the Adiabatic Surface Layer over a Rough Surface

For the case of a surface with roughness elements of height ε, one can repeat the analysis used in the previous section for a smooth surface. In this case there is no laminar sublayer so we do not need to use the kinematic viscosity in our analysis. On the other hand, we do need to add the height of the roughness elements. Thus the five parameters for the dimensional analysis are now $d\bar{u}_x/dz$, ε, ρ, z, and u_*. One can then show that

$$\frac{\bar{u}_x(z)}{u_*} = \frac{1}{\kappa}\ln\frac{z}{\varepsilon} + Q \tag{16.64}$$

where Q is an integration constant. One can incorporate the integration constant in the logarithm by defining the *roughness length* z_0 so that

$$\frac{1}{\kappa}\ln\frac{z}{z_0} = \frac{1}{\kappa}\ln\frac{z}{\varepsilon} + Q \tag{16.65}$$

TABLE 16.1 Roughness Lengths for Various Surfaces

Surface	z_0, m
Very smooth (ice, mud flats)	10^{-5}
Snow	10^{-3}
Smooth sea	10^{-3}
Level desert	10^{-3}
Lawn	10^{-2}
Uncut grass	0.05
Full-grown root crops	0.1
Tree covered	1
Low-density residential	2
Central business district	5–10

Source: McRae et al. (1982).

Using (16.65), the velocity profile becomes

$$\bar{u}_x(z) = \frac{u_*}{\kappa} \ln \frac{z}{z_0} \quad z > z_0 \tag{16.66}$$

The roughness length z_0 is related to the height of the roughness elements according to (16.65) or, after some algebra, $z_0 = \varepsilon \exp(-\kappa Q)$. It has been found experimentally that $z_0 \approx \varepsilon/30$. Values of the roughness length for typical surfaces are given in Table 16.1. Note that according to (16.66), $\bar{u}_x = 0$ for $z = z_0$, and that the equation is valid only for heights significantly greater than the roughness length.

Commonly, the friction velocity u_* is obtained from a measurement of the velocity at some reference height h_r, often equal to 10 m. Then, if $\bar{u}_x(h_r)$ is known,

$$u_* = \frac{\kappa \bar{u}_x(h_r)}{\ln(h_r/z_0)} \tag{16.67}$$

A better way to obtain u_* would be a direct measurement of the surface shear stress, but this requires elaborate experimental measurements and is not routinely available. Equation (16.67) works satisfactorily in adiabatic boundary layers (Plate 1971).

Calculation of the Mean Velocity Profile The mean wind velocity at 4 m over a surface with $z_0 = 0.0015$ m was measured at $7.8 \, \text{m s}^{-1}$ during a period of the Wangara experiment (Deardoff 1978). The atmosphere was adiabatic. Calculate the velocity at 0.5 and 16 m and compare with the observed values of 5.7 and $9.1 \, \text{m s}^{-1}$.

Using (16.67), $u_* = 0.4 \, \text{m s}^{-1}$. Using this value and (16.66), we estimate that $\bar{u}_x(0.5 \, \text{m}) = 5.8 \, \text{m s}^{-1}$ and $\bar{u}_x(16 \, \text{m}) = 9.3 \, \text{m s}^{-1}$. Both of these agree within 2% of the measured values in the field study.

16.4.3   Mean Velocity Profiles in the Nonadiabatic Surface Layer

The basic logarithmic velocity profile (16.66) is applicable only to adiabatic conditions. However, the atmosphere is seldom adiabatic, and the velocity profiles for stable and unstable conditions deviate from this logarithmic law. For the more frequently encountered nonadiabatic atmosphere (also called *stratified*), the *Monin–Obukhov similarity theory* is usually employed (Monin and Obukhov 1954).

According to this theory, the turbulence characteristics in the surface layer are governed in general by the following seven variables: $d\bar{u}_x/dz$, z, z_0, u_*, ρ, (g/T_0), and $\bar{q}_z = \rho \hat{c}_p \overline{u_z'\theta'}$, where the first five were also used in the previous section, (g/T_0) is a parameter related to buoyancy, and $\bar{q}_z$ is the vertical mean turbulent flux. Assuming that variations in the roughness length z_0 do not affect the form of the velocity profiles but only shift them, we can neglect this parameter and reduce the list to six variables. There are four dimensions in the problem (mass, length, time, and temperature), so, according to the Buckingham π theorem, there are two dimensionless groups governing the behavior of the system. The first group is called the *flux Richardson number* (Rf) and is defined as

$$\text{Rf} = -\frac{\kappa g z \bar{q}_z}{\rho \hat{c}_p T_0 u_*^3} \tag{16.68}$$

where $\kappa = 0.4$ is the von Karman constant. One can show that the flux Richardson number is equal to the ratio of the production of turbulent kinetic energy by buoyancy to its production by shear stresses. Rf can be positive or negative depending on the sign of the vertical mean turbulent flux $\bar{q}_z = \rho \hat{c}_p \overline{u_z'\theta'}$. There are three cases:

Case 1. If $\overline{u_z'\theta'} > 0$, then $\bar{q}_z > 0$ and Rf < 0. For this case, positive values of u_z' tend to be associated with positive values of θ' and vice versa. This case corresponds to an unstable atmosphere (decreasing potential temperature with height). If an air parcel moves upward because of a positive fluctuation in its velocity u_z', it rises to a region of lower potential temperature. However, the air parcel temperature changes adiabatically during this small rapid fluctuation, and its potential temperature remains constant. As a result, the air parcel potential temperature will exceed the potential temperature of its surroundings and will cause a positive potential temperature fluctuation θ' in its new position.

Case 2. If $\overline{u_z'\theta'} < 0$, then $\bar{q}_z < 0$ and Rf > 0. In this case, positive values of u_z' occur with negative values of θ', and the atmosphere (using the same arguments as above) is stable.

Case 3. If $\overline{u_z'\theta'} = 0$ then $\bar{q}_z = 0$ and Rf $= 0$. This case corresponds to an adiabatic (neutral) atmosphere.

The flux Richardson number according to (16.68) is a function of the distance from the ground. Because it is dimensionless, it can actually be viewed as a dimensionless length

$$\text{Rf} = \frac{z}{L} \tag{16.69}$$

TABLE 16.2   Monin–Obukhov Length L with Respect to Atmospheric Stability

L		Stability Condition
Very large negative	$L < -10^5$ m	Neutral
Large negative	-10^5 m $\leq L \leq -100$ m	Unstable
Small negative	-100 m $< L < 0$	Very unstable
Small positive	$0 < L < 100$ m	Very stable
Large positive	100 m $\leq L \leq 10^5$ m	Stable
Very large positive	$L > 10^5$ m	Neutral

where L is the Monin–Obukhov length and according to (16.68) and (16.69) is given by

$$L = -\frac{\rho \hat{c}_p T_0 u_*^3}{\kappa g \bar{q}_z} \tag{16.70}$$

By definition the *Monin–Obukhov length* is the height at which the production of turbulence by both mechanical and buoyancy forces is equal. The parameter L, like the flux Richardson number, provides a measure of the stability of the surface layer. As we discussed, when $Rf > 0$ and therefore according to (16.69) $L > 0$ the atmosphere is stable. On the other hand, when the atmosphere is unstable, $Rf < 0$ and then $L < 0$. Because of the inverse relationship between Rf and L, an adiabatic atmosphere corresponds to very small (positive or negative) values of Rf and to very large (positive or negative) values of L. Typical values of L for different atmospheric stability conditions are given in Table 16.2.

The second dimensionless group is the dimensionless velocity gradient

$$\frac{\kappa L}{u_*} \frac{\partial \bar{u}_x}{\partial z}$$

Let us now redefine the first dimensionless group as

$$\zeta = \frac{z}{L} \tag{16.71}$$

The dimensionless length ζ is equal to the flux Richardson number, $\zeta = Rf$, only in the surface layer. Using once more the π theorem, we can write the second dimensionless group as a function of the first or

$$\frac{\kappa L}{u_*} \frac{\partial \bar{u}_x}{\partial z} = g(\zeta) \tag{16.72}$$

Our next step is to find the function $g(\zeta)$. Replacing L from (16.71) into (16.72) and rearranging, we obtain

$$\frac{\partial \bar{u}_x}{\partial z} = \frac{u_*}{\kappa z} \phi(\zeta) \tag{16.73}$$

where we have defined $\phi(\zeta) = \zeta g(\zeta)$. For $\zeta \to 0$ the atmosphere becomes neutral, therefore (16.73) should become (16.60) at this limit. Comparing these two equations, we find that

$$\lim_{\zeta \to 0} \phi(\zeta) = 1 \qquad (16.74)$$

Generally accepted forms of the universal function $\phi(\zeta)$ are those of Businger et al. (1971):

$$
\begin{array}{ll}
\text{Stable } \zeta > 0 & \phi(\zeta) = 1 + 4.7\,\zeta \\
\text{Neutral } \zeta = 0 & \phi(\zeta) = 1 \\
\text{Unstable } \zeta < 0 & \phi(\zeta) = (1 - 15\,\zeta)^{-1/4}
\end{array}
\qquad (16.75)
$$

The velocity $\bar{u}_x(z)$ can be determined by integrating (16.73) from $z = z_0$ and $\bar{u}_x = 0$ using (16.75):

$$\bar{u}_x(z) = \frac{u_*}{\kappa} \int_{z_0/L}^{z/L} \frac{\phi(\zeta)}{\zeta} \, d\zeta \qquad (16.76)$$

For neutral conditions, (16.76) results in the logarithmic law consistent with (16.66):

$$\bar{u}_x(z) = \frac{u_*}{\kappa} \ln \frac{z}{z_0} \qquad (16.77)$$

For stable conditions, combining (16.76) and the first equation in (16.75) we find that

$$\bar{u}_x(z) = \frac{u_*}{\kappa} \ln \frac{z}{z_0} + 4.7 \frac{u_*}{\kappa} \frac{z - z_0}{L} \qquad (16.78)$$

This profile relation may be viewed as a correction to the logarithmic law. For an almost neutral atmosphere, L is a large positive number, and the relationship between velocity and height is logarithmic. As stability increases, the positive L decreases and the deviation from the logarithmic behavior increases. Finally, for very stable conditions, L approaches zero, the second term in (16.78) dominates, and the velocity profile becomes linear.

For unstable conditions, the velocity profile given by (16.76) and the third equation in (16.75) is

$$\bar{u}_x(z) = \frac{u_*}{\kappa} \int_{z_0/L}^{z/L} \frac{d\zeta}{\zeta(1 - 15\,\zeta)^{1/4}} \qquad (16.79)$$

Use of (16.76) or (16.77)–(16.79) requires the calculation of the friction velocity u_*. This requires once more the measurement of the velocity at a height h_r (say, 10 m). So if $\bar{u}_x(h_r)$ is known, we can integrate (16.73) between z_0 and h_r and solve for u_* assuming that $\bar{u}_x(z_0) = 0$

$$u_* = \frac{\kappa \bar{u}_x(h_r)}{\int_{z_0}^{h_r} (\phi(z/L)/z)\,dz} \tag{16.80}$$

For an adiabatic atmosphere, (16.80) reduces to (16.67). For a stable atmosphere, we find that

$$u_* = \frac{\kappa \bar{u}_x(h_r)}{\ln\left(\dfrac{h_r}{z_0}\right) + \dfrac{4.7(h_r - z_0)}{L}} \tag{16.81}$$

Finally, for an unstable atmosphere, Benoit (1977) derived the following approximate relationship after integration of (16.80):

$$u_* = \frac{\kappa \bar{u}_x(h_r)}{\ln\left(\dfrac{h_r}{z_0}\right) + \ln\left[\dfrac{(n_0^2 + 1)(n_0 + 1)^2}{(n_1^2 + 1)(n_1 + 1)^2}\right] + 2(\tan^{-1} n_1 - \tan^{-1} n_0)}$$

$$n_0 = \left[1 - 15\frac{z_0}{L}\right]^{1/4} \qquad n_1 = \left[1 - 15\frac{h_r}{L}\right]^{1/4} \tag{16.82}$$

Use of the equations derived in this section requires estimation of the Monin–Obukhov length L. A number of approaches are available, including the profile and gradient methods using available measurements (Arya 1999). The simplest approach based on the Pasquill stability classes will be discussed in the next section.

The applicability of the velocity profiles discussed in this section is limited to only the lowest 10–15% of the planetary boundary layer, in which wind direction does not change significantly with height. Observations suggest that for unstable conditions the wind speed is rather uniform above this surface layer.

16.4.4 The Pasquill Stability Classes—Estimation of L

The Monin–Obukhov length L is not a parameter that is routinely measured. Recognizing the need for a readily usable way to define atmospheric stability based on routine observations, Pasquill (1961) proposed a discrete atmospheric stability classification scheme that was later modified by Turner (1969). The scheme relies on observations of near-surface (10 m) wind, solar radiation, and cloudiness. If these

TABLE 16.3   Estimation of Pasquill Stability Classes[a]

Surface (10 m) Windspeed, $m s^{-1}$	Daytime[c]			Nighttime[c]	
	Incoming Solar Radiation[b]			Cloud Cover Fraction	
	Strong	Moderate	Slight	$\geq \frac{4}{8}$	$\leq \frac{3}{8}$
<2	A	A–B	B	—	—
2–3	A–B	B	C	E	F
3–5	B	B–C	C	D	E
5–6	C	C–D	D	D	D
>6	C	D	D	D	D

[a]*Key*: A—extremely unstable; B—moderately unstable; C—slightly unstable; D—neutral; E—slightly stable; F—moderately stable.
[b]Solar radiation: strong ($>700 \, W \, m^{-2}$), moderate ($350 - 700 \, W \, m^{-2}$), slight ($<350 \, W \, m^{-2}$)
[c]The neutral category D should be used, regardless of windspeed, for overcast conditions during day or night.
Source: Turner (1969).

variables are known, then the corresponding atmospheric stability class can be found using Table 16.3.

Golder (1972) established a relation among the Pasquill stability classes, the roughness length z_0, and L shown in Figure 16.8. Using this figure, one can estimate the Monin–Obukhov length for an area with a given surface roughness after estimating the stability

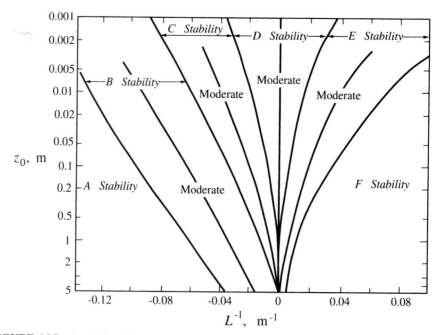

FIGURE 16.8   A relationship between Monin–Obukhov length L and roughness height z_0 for various Pasquill stability classes (Myrup and Ranzieri 1976).

TABLE 16.4 Correlation Parameters for the Estimation of L using (16.83)

Pasquill Stability Class	Coefficients	
	a	b
A (extremely unstable)	-0.096	0.029
B (moderately unstable)	-0.037	0.029
C (slightly unstable)	-0.002	0.018
D (neutral)	0	0
E (slightly stable)	0.004	-0.018
F (moderately stable)	0.035	-0.036

class using Table 16.3. To simplify calculation of L, Golder (1972) proposed the following correlation

$$\frac{1}{L} = a + b \log z_0 \qquad (16.83)$$

with the corresponding coefficients a and b given in Table 16.4 for the different stability classes.

Estimation of the Monin–Obukhov Length   Estimate L for an agricultural area with windspeed at 10 m equal to $2.5\,\mathrm{m\,s^{-1}}$ and solar flux $500\,\mathrm{W\,m^{-2}}$.

Assuming that the area has full-grown crops, we select, using Table 16.1, a roughness length z_0 equal to 0.1 m. The corresponding stability class for these atmospheric conditions is B, which is moderately unstable. Using Figure 16.8, L ranges from -30 to -10 m for this roughness length and stability class. Using (16.83), we obtain

$$(1/L) = -0.037 + 0.029 \log(0.1) = -0.066\,\mathrm{m}$$

and therefore $L = -15$ m. The correlations (16.83) correspond to the "moderate" lines in the middle of the stability classes of Figure 16.8 and therefore provide intermediate values of L.

The most important advantages of Pasquill's stability classification scheme are its simplicity and reliance on measurements that are readily available. As a result, it is the most frequently used scheme for characterization of atmospheric turbulence in pollutant dispersion calculations. Its most important limitation is its discrete nature, covering a wide range of atmospheric conditions with only six classes. As a result, a wide range of conditions is covered by each class, resulting in a range of potential values of L and other parameters that can be calculated for given atmospheric conditions. For example, the turbulence intensity approximately doubles as one moves from the lower to the upper end of stability class E (Arya 1999). It should be recognized that the estimation of L using this scheme introduces considerable uncertainty. Reviews of more accurate methods to calculate the surface fluxes of momentum and heat using micrometeorological measurements can be found in Stull (1988) and Arya (1988). The simplest approaches require temperature measurements at two heights h_1 and h_2 in the surface layer, or temperature and windspeed measurements at h_1 and h_2 (Arya 1999).

16.4.5   Empirical Equation for the Mean Wind Speed

The mean wind speed generally increases with height at least in the lower half of the planetary boundary layer. The wind vertical profiles based on the Monin–Obukhov similarity approach discussed in Section 16.4.3 are useful approximations of these velocities. An alternative empirical approach used often in air pollutant dispersion calculations is the power law wind velocity profile

$$\bar{u}_x(z) = \bar{u}_x(h_r)\left(\frac{z}{h_r}\right)^p \tag{16.84}$$

where h_r is the reference height at which a measurement of the windspeed is available. The exponent p is less than or equal to unity and should be determined from the atmospheric conditions. The exponent generally increases with increasing surface roughness and increasing stability. Equation (16.84) is empirical and lacks a sound theoretical basis; nonetheless, it often provides a reasonable fit to the observed wind speed profiles in the lower part of the planetary boundary layer.

Huang (1979) provided estimates of p as a function of z_0 and L shown in Figure 16.9. The value of p that should be used in (16.84) depends on the reference height h_r. The values of p for $h_r = 10$ m are shown in Figure 16.9a. The upper curve of the diagram (small positive L) corresponds to very stable conditions, the lower curve to unstable conditions, while the neutral conditions are somewhere in the middle. For very smooth surfaces, values of p are in the 0.05–0.15 range unless the atmosphere is very stable. Over a moderately rough surface p may range from 0.1 for very unstable conditions to near unity for very stable conditions. Finally for very rough surfaces p begins to approach unity and the velocity profile has an almost linear dependence on altitude. Clearly, p increases as the surface roughness increases for all atmospheric conditions.

APPENDIX 16   DERIVATION OF THE BASIC EQUATIONS OF SURFACE LAYER ATMOSPHERIC FLUID MECHANICS

Equation (16.26) can be written as

$$\frac{\partial \rho}{\partial t} + u_i\frac{\partial \rho}{\partial x_i} + \rho\frac{\partial u_i}{\partial x_i} = 0 \tag{16.A.1}$$

or equivalently

$$\frac{\partial u_i}{\partial x_i} = -\frac{1}{\rho}\left(\frac{\partial \rho}{\partial t} + u_j\frac{\partial \rho}{\partial x_j}\right) \tag{16.A.2}$$

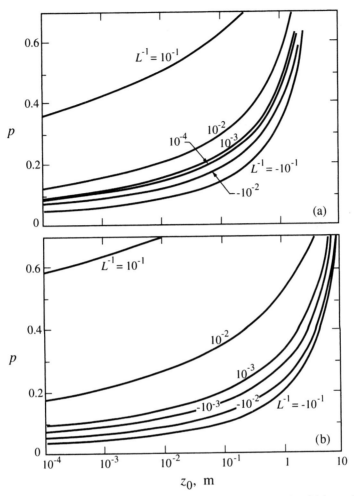

FIGURE 16.9 Exponent in the power-law expression for windspeed $\bar{u}_x(z)/\bar{u}_r = (z/h_r)^p$ as a function of the roughness length z_0 and the Monin–Obukhov length L: (a) $h_r = 10$ m; (b) $h_r = 30$ m (Huang 1979).

The term in parenthesis is by definition the substantial derivative $D\rho/Dt$ and therefore

$$\frac{\partial u_i}{\partial x_i} = -\frac{1}{\rho}\frac{D\rho}{Dt} = -\frac{D\ln\rho}{Dt} \tag{16.A.3}$$

Using the density expression from (16.38), we obtain

$$\frac{\partial u_i}{\partial x_i} = -\frac{D}{Dt}\ln[\rho_0 + \rho_m + \tilde{\rho}] = -\frac{D}{Dt}\ln\left[\rho_0\left(1 + \frac{\rho_m}{\rho_0} + \frac{\tilde{\rho}}{\rho_0}\right)\right] \tag{16.A.4}$$

For a shallow atmospheric layer $\rho_m/\rho_0 \ll 1$ and $\tilde{\rho}/\rho_0 \ll 1$, therefore, the expression in brackets on the RHS of (16.A.4) is just equal to $\ln\rho_0$ and its derivative is zero. The continuity equation then becomes

$$\frac{\partial u_i}{\partial x_i} = 0 \tag{16.A.5}$$

which, of course, is the continuity equation for an incompressible fluid.

To derive the equation of motion, we subtract the reference equilibrium state (16.31) from the equation of motion (16.27) and also use (16.A.5) to obtain

$$\frac{\partial u_i}{\partial t} + u_j \frac{\partial u_i}{\partial x_j} = -\frac{1}{\rho}\frac{\partial \tilde{p}}{\partial x_i} + \frac{\mu}{\rho}\frac{\partial^2 u_i}{\partial x_j \partial x_j} - \frac{g\tilde{\rho}}{\rho}\delta_{i3} \tag{16.A.6}$$

Let us examine the first and third terms on the RHS of (16.A.6). For a shallow layer the air density ρ will be approximately equal to the density at the surface. Therefore

$$\frac{1}{\rho}\frac{\partial \tilde{p}}{\partial x_i} \approx \frac{1}{\rho_0}\frac{\partial \tilde{p}}{\partial x_i} \tag{16.A.7}$$

The last term on the RHS of (16.A.6) expresses the vertical acceleration on a fluid element as a result of the density fluctuations. Assuming that the fluctuations in density from the surface value are small, we can expand ρ in a Taylor series about ρ_0 as follows:

$$\rho = \rho_0\left(1 + \frac{p_m}{p_0} - \frac{T_m}{T_0}\right) + \rho_0\left(\frac{\tilde{p}}{p_0} - \frac{\tilde{T}}{T_0}\right) \tag{16.A.8}$$

In the atmosphere the relative magnitude of pressure deviations from the reference pressure as a result of motion, that is, $\tilde{p}$, is small compared with temperature fluctuations:

$$\frac{\tilde{p}}{p_0} \ll \frac{\tilde{T}}{T_0} \tag{16.A.9}$$

Thus (16.A.8) after some algebraic manipulation, becomes

$$\tilde{\rho} = -\rho_0\frac{\tilde{T}}{T_0} \tag{16.A.10}$$

that is, the deviations in density from the reference state can be attributed solely to temperature deviations. The final form of the approximate equation of motion can be obtained by replacing (16.A.7) and (16.A.10) in (16.A.6):

$$\frac{\partial u_i}{\partial t} + u_j \frac{\partial u_i}{\partial x_j} = -\frac{1}{\rho_0}\frac{\partial \tilde{p}}{\partial x_i} + \frac{\mu}{\rho_0}\frac{\partial^2 u_i}{\partial x_j \partial x_j} + \frac{g\tilde{T}}{T_0}\delta_{i3} \tag{16.A.11}$$

We now consider the energy equation (16.29), and we subtract once more the corresponding equation for the equilibrium state, (16.32), to obtain the equation governing the temperature fluctuations

$$\rho\hat{c}_v\left(\frac{\partial(T_e + \tilde{T})}{\partial t} + u_j\frac{\partial(T_e + \tilde{T})}{\partial x_j}\right) = k\frac{\partial^2 \tilde{T}}{\partial x_j \partial x_j} - p\frac{\partial u_j}{\partial x_j} + Q \tag{16.A.12}$$

where we have retained the term $p(\partial u_j / \partial x_j)$ since, although $\partial u_j / \partial x_j \approx 0$, the pressure p is so large that $p(\partial u_j / \partial x_j)$ is of the same order of magnitude as the other terms in the equation. We can actually estimate this term as follows. From the continuity equation and the ideal-gas law, we have

$$
\begin{aligned}
p \frac{\partial u_j}{\partial x_j} &= -\frac{p}{\rho} \frac{D\rho}{Dt} \\
&= -\frac{RT}{M_{air}} \frac{D\rho}{Dt} \\
&= \frac{\rho_0 R}{M_{air}} \frac{D}{Dt}(T_e + \tilde{T}) - \frac{Dp_e}{Dt} \\
&= \frac{\rho_0 R}{M_{air}} \frac{D}{Dt}(T_e + \tilde{T}) + u_3 g \rho_0
\end{aligned}
\tag{16.A.13}
$$

where in the last step we have employed (16.31). Using the relation $\hat{c}_p - \hat{c}_v = R/M_{air}$, and on substituting (16.A.13) into (16.A.12), we obtain the approximate form of the energy equation:

$$
\rho_0 \hat{c}_p \left(\frac{\partial \tilde{T}}{\partial t} + u_j \frac{\partial \tilde{T}}{\partial x_j} \right) + \rho_0 \hat{c}_p u_3 \left(\frac{\partial T_e}{\partial x_3} + \frac{g}{c_p} \right) = k \frac{\partial^2 \tilde{T}}{\partial x_j \, \partial x_j} + Q
\tag{16.A.14}
$$

This equation holds regardless of the choice of reference equilibrium atmosphere. Usually, however, the equilibrium reference condition is chosen to be the adiabatic case, in which

$$
\frac{\partial T_e}{\partial x_3} = -\frac{g}{\hat{c}_p}
$$

Let us assume that at some initial time $\tilde{T} = 0$ relative to the adiabatic atmosphere. Then, from (16.A.14) we see that if $Q = 0$, the condition of $\tilde{T} = 0$ is preserved for $t > 0$ even though there may be motion of air. Also, the equation of motion (16.A.11) reduces to the usual form of the Navier–Stokes equation for the dynamics of an incompressible fluid under the influence of a motion-induced pressure fluctuation $\tilde{p}$ with no contribution from buoyancy forces since $\tilde{T} = 0$. Therefore, for an atmosphere with no sources of heat and initially having an adiabatic lapse rate, the temperature profile is unaltered if the atmosphere is set in motion. As a result, the adiabatic condition can be envisioned as one in which a large number of parcels are rising and falling, a sort of "convective" equilibrium. Thus, we have been able to derive the relation for the adiabatic lapse rate here from the full equation of continuity, motion, and energy, in contrast with the derivation presented in Section 16.1.1, which is based on thermodynamic arguments.

With the choice of the adiabatic atmosphere as the equilibrium state, (16.A.14) becomes

$$
\rho_0 \hat{c}_p \left(\frac{\partial \tilde{T}}{\partial t} + u_j \frac{\partial \tilde{T}}{\partial x_j} \right) = k \frac{\partial^2 \tilde{T}}{\partial x_j \, \partial x_j} + Q
\tag{16.A.15}
$$

which is the classic form of the heat conduction equation for an incompressible fluid with constant physical properties.

Now, using (16.16), we obtain

$$\frac{1}{\theta}\frac{\partial\theta}{\partial x_i} = \frac{1}{T}\left(\frac{\partial T}{\partial x_i} + \frac{g}{\hat{c}_p}\delta_{3i}\right) \tag{16.A.16}$$

where the quantity in parentheses is the difference between the actual and the adiabatic lapse rate:

$$\frac{1}{\theta}\frac{\partial\theta}{\partial x_i} = \frac{1}{T}\frac{\partial\tilde{T}}{\partial x_i} \tag{16.A.17}$$

Since θ is quite close to T, we can replace (16.A.17) by

$$\frac{\partial\theta}{\partial x_i} \approx \frac{\partial\tilde{T}}{\partial x_i} \tag{16.A.18}$$

Now, using (16.A.18), we can rewrite (16.A.15) as

$$\rho_0\hat{c}_p\left(\frac{\partial\theta}{\partial t} + u_j\frac{\partial\theta}{\partial x_j}\right) = k\left(\frac{\partial^2\theta}{\partial x_j\,\partial x_j}\right) + Q \tag{16.A.19}$$

PROBLEMS

16.1$_A$ Use the chain rule and (16.2) to derive the moist adiabatic lapse rate equation (16.13).

16.2$_B$ Calculate the moist adiabatic lapse rate at $0°C$ and 0.6 atm. Compare with the value at 1 atm.

16.3$_A$ Show that if the atmosphere is isothermal, the temperature change of a parcel of air rising adiabatically is

$$T(z) = T_0 e^{-\Gamma z/T_0'}$$

where T_0 and T_0' are the temperatures of the parcel at the surface and of the air at the surface, respectively.

16.4$_A$ A rising parcel of air will come to rest when its temperature T equals that of the surrounding air, T'. Show that the height z where this occurs is given by

$$z = \frac{1}{\Lambda}\left[T_0' - \left(\frac{T_0'^{\,\Gamma}}{T_0^{\Lambda}}\right)^{1/(\Gamma-\Lambda)}\right]$$

What condition must hold for this result to be valid?

16.5$_A$ Show that the condition that the density of the atmosphere does not change with height is

$$\frac{dT}{dz} = -3.42 \times 10^{-2}\,°C\,m^{-1}$$

16.6$_B$ It has been proposed that air pollution in Los Angeles can be abated by drilling large tunnels in the mountains surrounding the basin and pumping the air out into the surrounding deserts. You are to examine the power requirements in displacing the volume of air over the Los Angeles basin. Assume that the basin has an area of 4000 km^2 and that the polluted air is confined below an elevated inversion with a mean height of 400 m. The coefficient of friction for air moving over the basin is assumed to be 0.5, and the minimum energy needed to sustain airflow is equal to the energy dissipated by ground friction. Determine the power required to move the airmass with a velocity of 7 km h^{-1}. Compare your result with the capacity of Hoover Dam: 1.25×10^6 kW.

16.7$_B$ Elevated inversion layers are a prime factor responsible for the incidence of community air pollution problems. It is interesting to consider the feasibility of eliminating an elevated inversion layer. In principle, this could be done either by cooling all the air from the inversion base upward to a temperature below that at the inversion base or by heating all the air below the top of the inversion to a temperature higher than that at the inversion top. Show that the energy E required to destroy an elevated inversion by heating from below is given by

$$E = \rho \hat{c}_p [\Gamma(H_T - H_B) + (T_T - T_B)]\frac{H_T + H_B}{2}$$

where

$$\rho = \text{average density of air}$$
$$\hat{c}_p = \text{heat capacity of air}$$
$$H_B, H_T = \text{heights of base and top of inversion}$$
$$T_B, T_T = \text{temperatures of base and top of inversion}$$

Assume that the lapse rate below the base of the inversion is adiabatic and that the rate of temperature increase with height in the inversion is linear. Estimate the value of E for typical September conditions at Long Beach, California, at 7 a.m. Use

$$H_B = 475\,m \qquad T_B = 14.1°C$$
$$H_T = 1055\,m \qquad T_T = 22.4°C$$

If the area of the Los Angeles basin is 4000 km^2 and the energy produced by oil burning with 100% efficiency is 1.04×10^7 cal/kg(oil)$^{-1}$, what is the amount of oil required in order to destroy the inversion over the entire basin?

16.8$_C$ Consider the prediction of the diurnal atmospheric temperature profile under stagnant conditions. In this circumstance it is necessary to consider spatial variations in temperature in only the vertical direction. If it is asumed that absorption of radiation by the atmosphere can be neglected, the potential temperature θ satisfies

$$\frac{\partial \theta}{\partial t} = \frac{\partial}{\partial z}\left(K \frac{\partial \theta}{\partial z}\right) \tag{A}$$

assuming that K may be taken as constant. It may also be assumed that, at sufficiently high altitudes, the temperature profile should approach the adiabatic lapse rate:

$$\theta \to 0 \qquad \text{as } z \to \infty \tag{B}$$

The ground $(z = 0)$ temperature is governed by solar heating during the day and radiational cooling during the night. Therefore $\theta(0, t)$ may be expressed as

$$\theta(0, t) = A \cos \omega t \tag{C}$$

where A is the amplitude of the diurnal surface temperature variation and $\omega = 7.29 \times 10^{-5} \text{ s}^{-1}$.

a. Show that a solution satisfying (A) to (C) is

$$\theta(z, t) = A e^{-\beta z} \cos(\omega t - \beta z)$$

where $\beta = \sqrt{\omega/2K}$. This is the so-called long-time solution or steady-state solution, which represents the temperature dynamics corresponding to the influence of the surface forcing function (C).

b. Show that the elevation H that marks either the base or top of an inversion is found from

$$\sin(\omega t - \beta H) - \cos(\omega t - \beta H) = \frac{g}{A\beta \hat{c}_p} e^{\beta H}$$

Can there be more than one inversion layer?

c. Consider the evolution of the temperature profile from an initial profile

$$\theta(z, 0) = f(z) \tag{D}$$

Show that the solution of (A) to (D) is

$$\theta(z, t) = A e^{-\beta z} \cos(\omega t - \beta z) - \frac{A}{\pi} \int_0^\infty e^{-\eta t} \sin\left(z\sqrt{\frac{\eta}{K}}\right) \frac{\eta}{\eta^2 + \omega^2} d\eta$$
$$+ \frac{1}{2\sqrt{\pi K t}} \int_0^\infty f(\eta)[e^{-(z-\eta)^2/4Kt} - e^{-(z+\eta)^2/4Kt}] d\eta$$

d. Plot the steady-state solution for $K = 10^4$ cm^2 s^{-1} and $A = 4°$C. Note that a surface inversion is predicted at night (take $t = 0$ to be midnight), followed by weak, elevated, probably multiple inversions during the day.

REFERENCES

Arya, S. P. S., and Plate, E. J. (1969) Modeling of the stably stratified atmospheric boundary layer, *J. Atmos. Sci.* **26**, 656–665.

Arya, S. P. (1988) *Introduction to Micrometeorology*, Academic Press, San Diego.

Arya, S. P. (1999) *Air Pollution Meteorology and Dispersion*, Oxford Univ. Press, New York.

Benoit, R. (1977) On the integral of the surface layer profile-gradient functions, *J. Appl. Meteorol.* **16**, 859–860.

Blackadar, A. K., and Tennekes, H. (1968) Asymptotic similarity in neutral barotropic planetary boundary layers, *J. Atmos. Sci.* **25**, 1015–1020.

Businger, J. A., Wyngaard, J. C, Izumi, Y., and Bradley, E. F. (1971) Flux profile relationships in the atmospheric surface layer, *J. Atmos. Sci.* **28**, 181–189.

Calder, K. L. (1968) In clarification of the equations of shallow-layer thermal convection for a compressible fluid based on the Boussinesq approximation, *Quart. J. Roy. Meteorol. Soc.* **94**, 88–92.

Deardoff, J. W. (1978) Observed characteristics of the outer layer, in *Short Course on the Planetary Boundary Layer*, A. K. Blackadar, ed., American Meteorological Society, Boston.

Dutton, J. A. (1976) *The Ceaseless Wind*, McGraw-Hill, New York.

Golder, D. (1972) Relations among stability parameters in the surface layer, *Boundary Layer Meteorol.* **3**, 47–58.

Huang, C. H. (1979) Theory of dispersion in turbulent shear flow, *Atmos. Environ.* **13**, 453–463.

McRae, G. J., Goodin, W. R., and Seinfeld, J. H. (1982) Development of a second-generation mathematical model for urban air pollution 1. Model formulation, *Atmos. Environ.* **16**, 679–696.

Monin, A. S., and Obukhov, A. M. (1954) Basic turbulent mixing laws in the atmospheric surface layer, *Tr. Geofiz. Inst. Akad. Nauk SSSR* **24**, 163–187.

Monin, A. S., and Yaglom, A. M. (1971) *Statistical Fluid Mechanics*, Vol. 1, MIT Press, Cambridge, MA.

Monin, A. S., and Yaglom, A. M. (1975) *Statistical Fluid Mechanics*, Vol. 2, MIT Press, Cambridge, MA.

Myrup, L. O., and Ranzieri, A. J. (1976) *A Consistent Scheme for Estimating Diffusivities to Be Used in Air Quality Models*, Report CA-DOT-TL-7169-3-76-32, California Department of Transportation, Sacramento.

Nieuwstadt, F. T. M., and van Dop, H., eds.(1982) *Atmospheric Turbulence and Air Pollution Modeling*, Reidel Dordrecht, The Netherlands.

Panofsky, H. A., and Dutton, J. A. (1984) *Atmospheric Turbulence*, Wiley, New York.

Pasquill, F. (1961) The estimation of the dispersion of windborne material, *Meteorol. Mag.* **90**, 33–49.

Plate, E. J. (1971) *Aerodynamic Characteristics of Atmospheric Boundary Layers*, U.S. Atomic Energy Commission, Oak Ridge, TN.

Plate, E. J. (Ed.) (1982) *Engineering Meteorology*, Elsevier, New York.

Record, F. A., and Cramer, H. E. (1966) Turbulent energy dissipation rates and exchange processes above a nonhomogeneous surface, *Quart. J. Roy. Meteorol. Soc.* **92**, 519–532.

Spiegel, E. A., and Veronis, G. (1960) On the Boussinesq approximation for a compressible fluid, *Astrophys. J.* **131**, 442–447.

Stull, R. B. (1988) *An Introduction to Boundary Layer Meteorology*, Kluwer Academic, Boston.

Turner, D. B. (1969) *Workbook of Atmospheric Diffusion Estimates*, U.S. Environmental Protection Agency Report 999-AP-26, Washington, DC.

Zilitinkevich, S. S. (1972) On the determination of the height of the Ekman boundary layer, *Boundary Layer Meteorol.* **3**, 141–145.

17 Cloud Physics

Clouds are one of the most significant elements of the atmospheric system, playing several key roles:

1. Clouds are a major factor in the Earth's radiation budget, reflecting sunlight back to space or blanketing the lower atmosphere and trapping infrared radiation emitted by the Earth's surface.
2. Clouds deliver water from the atmosphere to the Earth's surface as rain or snow and are thus a key step in the hydrologic cycle.
3. Clouds scavenge gaseous and particulate materials and return them to the surface (wet deposition).
4. Clouds provide a medium for aqueous-phase chemical reactions and production of secondary species.
5. Clouds significantly affect vertical transport in the atmosphere. Updrafts and downdrafts associated with clouds determine in a major way the vertical redistribution of trace species in the atmosphere.

Despite their great importance, clouds still remain one of the least understood components of the weather and climate system. We begin our discussion of clouds by summarizing the properties of their basic constituent, water. We then investigate the formation of droplets in a cooling air parcel. The microphysics of a droplet population and the dynamics of cloud formation are then examined. Finally, we revisit the chemical processes taking place in clouds and fogs using the material already developed in Chapter 7. A comprehensive discussion of cloud physics, beyond the scope of this book, can be found in Pruppacher and Klett (1997).

17.1 PROPERTIES OF WATER AND WATER SOLUTIONS

Liquid water, H_2O, is characterized by the strong hydrogen bonds between its molecules, which give rise to a number of unique properties. Because of the strength of these bonds, a relatively large amount of energy is required to evaporate a unit mass of water. Similarly, the latent heat of freezing is also relatively large, as a result of further strong bonding in ice crystals. The surface tension (surface free energy) is also large. Table 17.1 summarizes these physical properties of water. In the following sections we discuss the atmospherically relevant properties of water and and its solutions.

TABLE 17.1 Properties of Water

Property	Phase/Temperature	Symbol	Value
Specific heat at constant pressure	Vapor	$\hat{c}_{pv}$	$1.952\,\mathrm{J\,g^{-1}\,K^{-1}}$
	Liquid (0°C)	$\hat{c}_{pw}$	$4.218\,\mathrm{J\,g^{-1}\,K^{-1}}$
Specific heat at constant volume	Vapor	$\hat{c}_{vw}$	$1.463\,\mathrm{J\,g^{-1}\,K^{-1}}$
Latent heat of evaporation	0°C	ΔH_v	$2.5\,\mathrm{kJ\,g^{-1}}$
	100°C		$2.25\,\mathrm{kJ\,g^{-1}}$
Latent heat of fusion	0°C	ΔH_m	$0.33\,\mathrm{kJ\,g^{-1}}$
Surface tension	Water/air (20°C)	σ_{w0}	$0.073\,\mathrm{J\,m^{-2}}$

17.1.1 Specific Heat of Water and Ice

The specific heat of liquid water $\hat{c}_{pw}$ varies with temperature and can be described by the semiempirical relationships

$$\hat{c}_{pw} = 4.218 + 3.47 \times 10^{-4}(T - 273)^2, \qquad 233 \le T \le 273\,\mathrm{K} \qquad (17.1)$$
$$= 4.175 + 1.3 \times 10^{-5}(T - 308)^2$$
$$+ 1.6 \times 10^{-8}(T - 308)^4, \qquad 273 \le T \le 308\,\mathrm{K} \qquad (17.2)$$

where T is in K and $\hat{c}_{pw}$ is in $\mathrm{J\,g^{-1}\,K^{-1}}$. This heat capacity refers to pure water. Most ions lower the heat capacity of water, but this change is negligible for solute concentrations smaller than 0.1 M. Therefore, for cloud applications, we will assume that the heat capacity of water is independent of the droplet concentration and depends only on temperature.

17.1.2 Latent Heats of Evaporation and of Melting for Water

Empirical fits to the specific latent heats of evaporation ΔH_v and of melting ΔH_m are as follows for temperature T in K:

$$\Delta H_v(\mathrm{kJ\,g^{-1}}) = 2.5 \left(\frac{273.15}{T} \right)^{0.167 + 3.67 \times 10^{-4}\,T} \qquad (17.3)$$

$$\Delta H_m(\mathrm{J\,g^{-1}}) = 333.5 + 2.03\,T - 0.0105\,T^2 \qquad (17.4)$$

These enthalpies refer to pure water-phase changes and are expected to differ for solutions. However, even at NaCl concentrations of 5 M, the enthalpy of evaporation changes by less than 0.2% (Pruppacher and Klett 1997). Thus, for our purposes, these enthalpies will also be assumed to depend on temperature only.

17.1.3 Water Surface Tension

The surface tension of pure water decreases with increasing temperature. Pruppacher and Klett (1997) recommend use of the following function

$$\sigma_{w0} = 0.0761 - 1.55 \times 10^{-4}(T - 273) \qquad (17.5)$$

for the temperature range from -40 to $40°C$, where σ_{w0} is in $J\,m^{-2}$. The surface tension of water is $76.1 \times 10^{-3}\,J\,m^{-2}$ at $0°C$, and decreases by $1.55 \times 10^{-3}\,J\,m^{-2}$ for every $10°C$. Note that these values can be used for supercooled water also, that is, liquid water existing at temperatures below $0°C$.

The dissolution of other compounds in water alters its surface tension. Experimental values of the variation of water solution surface tension with the solution concentration are tabulated in the *Handbook of Physics and Chemistry*. For salts like NaCl and $(NH_4)_2SO_4$, the dependence of the solution surface tension, σ_w, on the solution molarity is practically linear over the range of atmospheric interest

$$\sigma_w(m_{NaCl}, T) = \sigma_{w0}(T) + 1.62 \times 10^{-3}\,m_{NaCl}$$

$$\sigma_w(m_{(NH_4)_2SO_4}, T) = \sigma_{w0}(T) + 2.17 \times 10^{-3}\,m_{(NH_4)_2SO_4}$$

$$(17.6)$$

where $\sigma_{w0}(T)$ is, as above, the surface tension of pure water and m_{NaCl} and $m_{(NH_4)_2SO_4}$ are the molarities of NaCl and $(NH_4)_2SO_4$ in M, respectively.[1]

A last issue is the dependence of the water surface tension on the size of the droplet. One would expect that as the droplet surface tension is the result of attractive forces between water molecules near the surface, a change in droplet diameter would change the number of molecules interacting with the molecules at the surface, thus changing the surface tension. However, because of the small range of molecular interaction, this dependence of σ_{w0} on size is significant only for extremely small drops, consisting merely of a few thousands of molecules, and the exact dependence is still a subject of debate. The change is probably smaller than 1% for water drops as small as $0.1\,\mu m$ and becomes significant only at drop sizes less than $0.01\,\mu m$. Therefore the dependence of surface tension on droplet size can be neglected for atmospheric cloud applications.

17.2 WATER EQUILIBRIUM IN THE ATMOSPHERE

Water in the atmosphere exists in the gas phase as water vapor and in the aqueous phase as water droplets and wet aerosol particles. In this section we will investigate the conditions for water equilibrium between the gas and aqueous phases. This equilibrium is complicated by two effects: the curvature of the particles and the formation of aqueous solutions. We will start from the simplest case—the equilibrium between a flat pure water surface and the atmosphere. Then the equilibrium of a pure water droplet will be investigated, followed by a flat water solution surface. Finally, these effects will be

---

[1] Because solutes alter the surface tension of water, one would expect variations of the concentrations of these species near the droplet–air interface. For example, as nature tries to reach states of lower energies, if a solute increases the surface tension of water, this species at equilibrium should have lower concentrations at the interface than in the bulk solution. The opposite should happen for a surface-active compound that lowers the surface tension. One would expect higher concentrations of this species near the interface than in the bulk. For a 1 M NaCl solution this surface tension effect results in a NaCl deficiency at the interface of less than 1% (Pruppacher and Klett, 1997). The same authors suggested that for drops with diameters larger than $0.2\,\mu m$ and NaCl concentrations lower than 1 M, this concentration gradient due to surface tension is less than 1% and can safely be ignored. The effect of solution inhomogeneity due to surface tension will therefore be neglected for our discussion of atmospheric droplet formation.

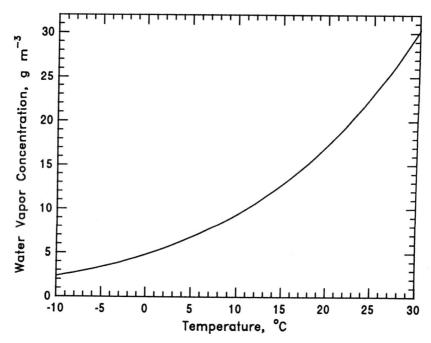

FIGURE 17.1 Saturation concentration of water over a flat water surface as a function of temperature. Values below 0°C correspond to supercooled (metastable) water.

integrated, and the desired equilibrium conditions for an aqueous solution droplet will be derived.

17.2.1   Equilibrium of a Flat Pure Water Surface with the Atmosphere

When a pure substance is at equilibrium with its vapor, its gas-phase partial pressure is equal by definition to its saturation vapor pressure, $p°$. The gas-phase molar saturation concentration, $c° = p°/RT$, of a compound is determined by its chemical structure and for water at atmospheric conditions is, on a mass basis, the order of a few $\mathrm{g\,m^{-3}}$ (Figure 17.1).

The change of the saturation pressure with temperature can be calculated, as we showed in Chapter 10, by the Clausius–Clapeyron equation:

$$\frac{dp°}{dT} = \frac{\Delta H_v(T)M_w}{T(v_v - v_w)} \tag{17.7}$$

where ΔH_v is the latent heat for water evaporation, M_w is the molecular weight of water, and v_v and v_w are the molar volumes of water vapor and liquid water correspondingly. Assuming that $v_v \gg v_w$ and that water vapor satisfies the ideal-gas law ($p°v_v = RT$), then (17.7) becomes (10.68)

$$\frac{dp°}{dT} \simeq \frac{\Delta H_v(T)p°M_w}{RT^2} \tag{17.8}$$

Replacing in this equation a function describing the temperature dependence of the latent heat of evaporation [e.g., (17.3)], one can integrate and obtain an explicit expression for

TABLE 17.2   Saturation Vapor Pressure of Water Vapor over a Flat Pure Water or Ice Surface

$p°(\text{mbar}) = a_0 + a_1T + a_2T^2 + a_3T^3 + a_4T^4 + a_5T^5 + a_6T^6 (T \text{ is in } °C)$	
Water (-50 to $+50°C$)	Ice (-50 to $0°C$)
$a_0 = 6.107799961$	$a_0 = 6.109177956$
$a_1 = 4.436518521 \times 10^{-1}$	$a_1 = 5.034698970 \times 10^{-1}$
$a_2 = 1.428945805 \times 10^{-2}$	$a_2 = 1.886013408 \times 10^{-2}$
$a_3 = 2.650648471 \times 10^{-4}$	$a_3 = 4.176223716 \times 10^{-4}$
$a_4 = 3.031240396 \times 10^{-6}$	$a_4 = 5.824720280 \times 10^{-6}$
$a_5 = 2.034080948 \times 10^{-8}$	$a_5 = 4.838803174 \times 10^{-8}$
$a_6 = 6.136820929 \times 10^{-11}$	$a_6 = 1.838826904 \times 10^{-10}$

Source: Lowe and Ficke (1974).

$p°(T)$. A series of such expressions exist in the literature (see Problem 1.1), and that proposed by Lowe and Ficke (1974) is given in Table 17.2.

17.2.2   Equilibrium of a Pure Water Droplet

In Chapter 10 we showed that the vapor pressure over a curved interface always exceeds that of the same substance over a flat surface. The dependence of the water vapor pressure on the droplet diameter is given by the Kelvin equation (10.86) as

$$\frac{p_w(D_p)}{p°} = \exp\left(\frac{4M_w\sigma_{w0}}{RT\rho_w D_p}\right) \tag{17.9}$$

where $p_w(D_p)$ is the water vapor pressure over the droplet of diameter D_p, $p°$ is the water vapor pressure over a flat surface at the same temperature, M_w is the molecular weight of water, σ_{w0} is the air–water surface tension, and ρ_w is the water density. The equilibrium water vapor concentrations at $0°C$ and $20°C$ are shown in Figure 17.2 as a function of the droplet diameter. Note that the effect of curvature for water droplets becomes important only for $D_p < 0.1\ \mu m$.

Since $p_w(D_p) > p°$, for equilibrium of a pure water droplet with the environment the air needs to be supersaturated with water vapor. For the equilibration of a large pure water droplet, a modest supersaturation is necessary, but a large supersaturation is necessary for a small droplet. Let us investigate the stability of such an equilibrium state. We assume that a droplet of pure water with diameter D_p is in equilibrium with the atmosphere. Maintaining the temperature constant at T, the atmosphere will have a water vapor partial pressure $p_w = p_w(D_p)$. Let us assume that a few molecules of water vapor collide with the droplet causing its diameter to increase infinitesimally to D_p'. This will cause a small decrease in the water vapor pressure required for the droplet equilibration to $p_w(D_p')$. The water vapor concentration in the environment has not changed and therefore $p_w > p_w(D_p')$, causing more water molecules to condense on the droplet and the droplet to grow even more. The opposite will happen if a few water molecules leave the droplet. The droplet diameter will decrease, the water vapor pressure at the droplet surface will exceed that of the environment, and the droplet will continue evaporating. These simple arguments indicate that *the equilibrium of a pure water droplet is unstable.* A minor

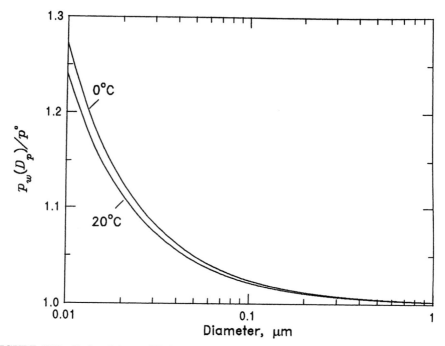

FIGURE 17.2   Ratio of the equilibrium vapor pressure of water over a droplet of diameter D_p, $p_w(D_p)$, to that over a flat surface, p°, as a function of droplet diameter for 0°C and 20°C.

perturbation is sufficient either for the complete evaporation of the droplet or for its uncontrollable growth.

Droplets in the atmosphere never consist exclusive of water; they always contain dissolved compounds. However, understanding the behavior of a pure water droplet is necessary for understanding the behavior of an aqueous droplet solution.

17.2.3   Equilibrium of a Flat Water Solution

Let us consider a water solution (flat surface) at constant temperature T and pressure p in equilibrium with the atmosphere. Water equilibrium between the gas and aqueous phases requires equality of the corresponding water chemical potentials in the two phases (see Chapter 10):

$$\mu_w(g) = \mu_w(aq) \tag{17.10}$$

Water vapor behaves in the atmosphere as an ideal gas, so its gas-phase chemical potential is

$$\mu_w(g) = \mu_w^\circ(T) + RT \ln p_s^\circ \tag{17.11}$$

where p_s° is the water vapor partial pressure over the solution. The chemical potential of liquid water is given by

$$\mu_w(aq) = \mu_w^* + RT \ln \gamma_w x_w \tag{17.12}$$

where γ_w is the water activity coefficient and x_w the mole fraction of water in solution. Combining (17.10) and (17.12), we obtain

$$\frac{p_s^{\circ}}{\gamma_w x_w} = \exp\left(\frac{\mu_w^{\circ} - \mu_w^{*}}{RT}\right) = K(T) \tag{17.13}$$

This equation describes the behavior of the system for any solution composition. Note that the right-hand side (RHS) is a function of temperature only and therefore will be equal to a constant, K, for constant temperature. Considering the case of pure water (no solute) we note that when $x_w = 1$, $\gamma_w \to 1$ and $p_s^{\circ} = p^{\circ} = K(T)$, where p° is the vapor pressure of water over pure water. Therefore (17.13) can be rewritten as

$$p_s^{\circ} = \gamma_w x_w p^{\circ} \tag{17.14}$$

Equation (17.14) is applicable for any solution and does not assume ideal behavior. Nonideality is accounted for by the activity coefficient γ_w.

The mole fraction of water in a solution consisting of n_w water moles and n_s solute moles is given by

$$x_w = \frac{n_w}{n_w + n_s} \tag{17.15}$$

and therefore the vapor pressure of water over its solution is given by

$$p_s^{\circ} = \frac{n_w}{n_w + n_s} \gamma_w p^{\circ} \tag{17.16}$$

If the solution is dilute, then γ_w approaches its infinite dilution limit $\gamma_w \to 1$. Therefore, at high dilution, the vapor pressure of water is given by Raoult's law

$$p_s^{\circ} = x_w p^{\circ} \tag{17.17}$$

and the solute causes a reduction of the water equilibrium vapor pressure over the solution.

The vapor pressure of water over NaCl and $(NH_4)_2SO_4$ solutions is shown in Figure 17.3. Also shown is the ideal solution behavior. Note that because NaCl dissociates into two ions, the number of equivalents in solution is twice the number of moles of NaCl. For $(NH_4)_2SO_4$, the number of ions in solution is three times the number of dissolved salt moles. In calculating the number of moles in solution, n_s, a dissociated molecule that has dissociated into i ions is treated as i molecules, whereas an undissociated molecule is counted only once. A similar diagram is given in Figure 17.4, using now the concentration of salt as the independent variable. Solutes that dissociate (e.g., salts) reduce the vapor pressure of water more than do solutes that do not dissociate, and this reduction depends strongly on the type of salt.

17.2.4 Atmospheric Equilibrium of an Aqueous Solution Drop

In the previous sections we developed expressions for the water vapor pressure over a pure water droplet and a flat solution. Atmospheric droplets virtually always contain

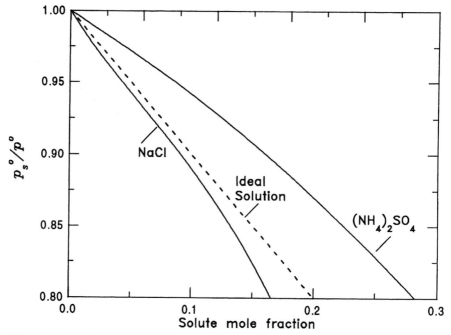

FIGURE 17.3 Variation of water vapor pressure ratio (p_s°/p°) as a function of the solute mole fraction at 25°C for solution of NaCl and $(NH_4)_2SO_4$ and an ideal solution. The mole fraction of the salts has been calculated taking into account their complete dissociation.

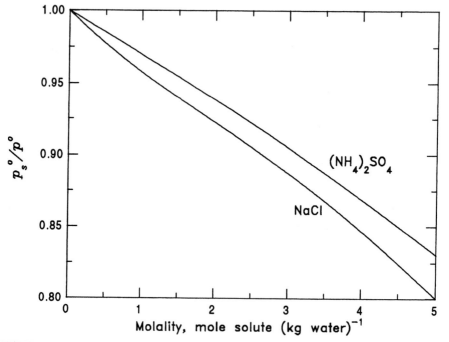

FIGURE 17.4 Variation of water vapor pressure ratio (p_s°/p°) as a function of the salt molality (mol of salt per kg of water) of NaCl and $(NH_4)_2SO_4$ at 25°C.

dissolved solutes, so we need to combine these two previous results to treat this general case.

Let us consider a droplet of diameter D_p containing n_w moles of water and n_s moles of solute (e.g., a nonvolatile salt). If the solution were flat, the water vapor pressure over it would satisfy (17.14). Substituting this expression into (17.9), one obtains

$$\frac{p_w(D_p)}{p^\circ \gamma_w x_w} = \exp\left(\frac{4\bar{v}_w \sigma_w}{RTD_p}\right) \tag{17.18}$$

$\bar{v}_w$ and $\bar{v}_s$ are the partial molar volumes of the two components in the solution, and the total drop volume will satisfy

$$\frac{1}{6}\pi D_p^3 = n_w \bar{v}_w + n_s \bar{v}_s \tag{17.19}$$

Note that since we have assumed that the solute is nonvolatile, n_s is the same regardless of the value of D_p. Using (17.19) and (17.15), we find that

$$\frac{1}{x_w} = 1 + \frac{n_s}{n_w} = 1 + \frac{n_s \bar{v}_w}{(\pi/6)D_p^3 - n_s \bar{v}_s} \tag{17.20}$$

Replacing the water mole fraction appearing in (17.18) by (17.20), we obtain

$$\ln\left(\frac{p_w(D_p)}{p^\circ}\right) = \frac{4\bar{v}_w \sigma_w}{RTD_p} + \ln \gamma_w - \ln\left(1 + \frac{n_s \bar{v}_w}{(\pi/6)D_p^3 - n_s \bar{v}_s}\right) \tag{17.21}$$

In the derivation of (17.21) we have implicitly assumed that temperature, pressure, and number of solute moles n_s remain constant as the droplet diameter changes. This equation relates the water vapor pressure over the droplet solution of diameter D_p to that of water over a flat surface p° at the same temperature.

If the solution is dilute, the volume occupied by the solute can be neglected relative to the droplet volume; that is,

$$n_s \bar{v}_s \ll \frac{\pi}{6}D_p^3 \tag{17.22}$$

and the molar volume of water approximately equal to M_w/ρ_w, and (17.21) can be simplified to

$$\ln\left(\frac{p_w(D_p)}{p^\circ}\right) = \frac{4M_w \sigma_w}{RT\rho_w D_p} + \ln \gamma_w - \ln\left(1 + \frac{6n_s \bar{v}_w}{\pi D_p^3}\right) \tag{17.23}$$

For dilute solutions one can also assume that $\gamma_w \to 1$ and $6n_s \bar{v}_w / \pi D_p^3 \to 0$; so recalling that $\ln(1+x) \simeq x$ as $x \to 0$, we find

$$\ln\left(\frac{p_w(D_p)}{p^\circ}\right) = \frac{4M_w \sigma_w}{RT\rho_w D_p} - \frac{6n_s \bar{v}_w}{\pi D_p^3} \tag{17.24}$$

Recall that $\bar{v}_w$ is the molar volume of water in the solution, which for a dilute solution is equal to the molar volume of pure water, that is

$$\bar{v}_w \simeq \frac{M_w}{\rho_w} \tag{17.25}$$

where M_w is the molecular weight of water and ρ_w its density. Replacing (17.25) in (17.24), we obtain

$$\ln\left(\frac{p_w(D_p)}{p^\circ}\right) = \frac{4M_w\sigma_w}{RT\rho_w D_p} - \frac{6n_s M_w}{\pi\rho_w D_p^3} \tag{17.26}$$

It is customary to write

$$A = \frac{4M_w\sigma_w}{RT\rho_w} \qquad B = \frac{6n_s M_w}{\pi\rho_w}$$

and (17.26) as

$$\ln\left(\frac{p_w(D_p)}{p^\circ}\right) = \frac{A}{D_p} - \frac{B}{D_p^3} \tag{17.27}$$

Equations (17.21), (17.24), (17.26), and (17.27) are different forms of the *Köhler equations* (Köhler 1921, 1926). These equations express the two effects that determine the vapor pressure over an aqueous solution droplet—the Kelvin effect that tends to increase vapor pressure and the solute effect that tends to decrease vapor pressure. For a pure water drop there is no solute effect and the Kelvin effect results in higher vapor pressures compared to a flat interface. By contrast, the vapor pressure of an aqueous solution drop can be larger or smaller than the vapor pressure over a pure water surface depending on the magnitude of the *solute effect* term B/D_p^3 relative to the *curvature term* A/D_p. Note that both effects increase with decreasing droplet size but the solute effect increases much faster. One should also note that a droplet may be in equilibrium in a subsaturated environment if $D_p^2 A < B$.

Figure 17.5 shows the water vapor pressure over NaCl and $(NH_4)_2SO_4$ drops. The A term in the Köhler equations can be approximated by

$$A = \frac{4M_w\sigma_w}{RT\rho_w} \simeq \frac{0.66}{T} \qquad \text{(in μm)} \tag{17.28}$$

where T is in K, and the solute term

$$B = \frac{6n_s M_w}{\pi\rho_w} \simeq \frac{3.44 \times 10^{13}\, v m_s}{M_s} \qquad \text{(in μm}^3) \tag{17.29}$$

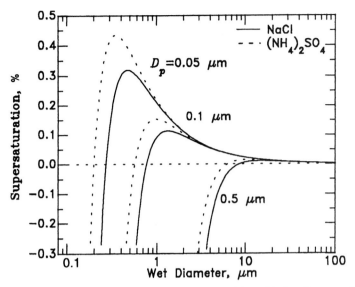

FIGURE 17.5 Köhler curves for NaCl and $(NH_4)_2SO_4$ particles with dry diameters 0.05, 0.1, and 0.5 μm at 293 K (assuming spherical dry particles). The supersaturation is defined as the saturation minus one. For example, a supersaturation of 1% corresponds to a relative humidity of 101%.

where m_s is the solute mass (in g) per particle, M_s the solute molecular weight (in g mol^{-1}), and v is the number of ions resulting from the dissociation of one solute molecule. For example, $v = 2$ for NaCl and NaNO$_3$, while $v = 3$ for $(NH_4)_2SO_4$.

All the curves in Figure 17.5 pass through a maximum. These maxima occur at the *critical droplet diameter* D_{pc}

$$D_{pc} = \left(\frac{3B}{A}\right)^{1/2} \tag{17.30}$$

and at this diameter (denoted by the subscript c)

$$\left(\ln\frac{p_w}{p^\circ}\right)_c = \left(\frac{4A^3}{27B}\right)^{1/2} \tag{17.31}$$

The ratio p_w/p° is the saturation relative to a flat pure water surface required for droplet equilibrium and therefore at the critical diameter, the critical saturation, S_c, is

$$\ln S_c = \left(\frac{4A^3}{27B}\right)^{1/2} \tag{17.32}$$

The steeply rising portion of the Köhler curves represents a region where solute effects dominate. As the droplet diameter increases, the relative importance of the Kelvin effect over the solute effect increases, and finally beyond the critical diameter, domination of the Kelvin effect is evident. In this range all Köhler curves approach the Kelvin equation, represented by the equilibrium of a pure water droplet. Physically, the solute concentration is so small in this range (recall that each Köhler curve refers to fixed solute amount) that the droplet becomes similar to pure water.

The Köhler curves also represent the equilibrium size of a droplet for different ambient water vapor concentrations (or relative humidity values). If the water vapor partial pressure in the atmosphere is p_w, a droplet containing n_s moles of solute and having a diameter D_p satisfying the Köhler equations should be at equilibrium with its surroundings. Realizing that the Köhler curves can be viewed as size–RH equilibrium curves poses a number of interesting questions:

> What happens if the atmosphere is supersaturated with water vapor and the supersaturation exceeds the critical supersaturation for a given particle?
>
> What happens if for a given atmospheric saturation there are two diameters for which the droplet can satisfy the Köhler equations? Are there two equilibrium states?

To answer these questions and understand the cloud and fog creation processes in the atmosphere, we need to investigate the stability of the equilibrium states given by the Köhler equation.

Stability of Atmospheric Droplets   We have already seen in the previous section that a pure water droplet cannot be at stable equilibrium with its surroundings. A small perturbation of either the droplet itself or its surroundings causes spontaneous droplet growth or shrinkage.

Let us consider first a drop lying on the portion of the Köhler curve for which $D_p < D_{pc}$. We assume that the atmospheric saturation is fixed at S. A drop will constantly experience small perturbations caused by the gain or loss of a few molecules of water. Say that the drop grows slightly with the addition of a few molecules of water. At its momentary larger size, its equilibrium vapor pressure is larger than the fixed ambient value and the drop will evaporate water, eventually returning its original equilibrium state. The same phenomenon will be observed if the droplet loses a few molecules of water. Its equilibrium vapor pressure will decrease, becoming less than the ambient, and water will condense on the droplet, returning it to its original size. Therefore drops in the rising part of the Köhler curve are in stable equilibrium with their environment.

Now consider a drop on the portion of the curve for which $D_p > D_{pc}$ that experiences a slight perturbation, causing it to grow by a few molecules of water. At its slightly larger size its equilibrium vapor pressure is lower than the ambient. Thus water molecules will continue to condense on the drop and it will grow even larger. Conversely, a slight shrinkage leads to a drop that has a higher equilibrium vapor pressure than the ambient so the drop continues to evaporate. If it is a drop of pure water, it will evaporate completely. If it contains a solute, it will diminish in size until it intersects the ascending branch of the Köhler curve that corresponds to stable equilibrium. In conclusion, the descending branches of the curves describe unstable equilibrium states.

If the ambient saturation ratio S is lower than the critical saturation S_c for a given particle, then the particle will be in equilibrium described by the ascending part of the curve. If $1 < S < S_c$, then there are two equilibrium states (two diameters corresponding to S). One of them is a stable state, and the other is unstable. The particle can reach stable equilibrium only at the state corresponding to the smaller diameter.

If the ambient saturation ratio S happens to exceed the particle critical saturation S_c, there is no feasible equilibrium size for the particle. For any particle diameter the ambient saturation will exceed the saturation at the particle surface (equilibrium saturation), and the particle will grow indefinitely. In such a way a droplet can grow to a size much larger than the original size of the dry particle. It is, in fact, through this process that particles as small as $0.01\,\mu m$ in diameter can grow one billion times in mass to become $10\,\mu m$ cloud or fog droplets. Moreover, in cloud physics a particle is not considered to be a cloud droplet unless its diameter exceeds its critical diameter D_{pc}.

The critical saturation S_c of a particle is an important property. If the environment has reached a saturation larger than S_c, the particle is said to be *activated* and starts growing rapidly, becoming a cloud droplet. For a spherical aerosol particle of diameter d_s (dry diameter), density ρ_s, and molecular weight M_s, the number of moles (after complete dissociation) in the particle is given by

$$n_s = \frac{\nu\pi\,d_s^3\rho_s}{6M_s} \tag{17.33}$$

and combining this result with (17.32), we find that

$$\ln S_c = \left(\frac{4A^3\rho_w M_s}{27\nu\rho_s M_w d_s^3}\right)^{1/2} \tag{17.34}$$

This equation gives the critical saturation for a dry particle of diameter d_s. Note that the smaller the particle, the higher its critical saturation. When a fixed saturation S exists, all particles whose critical saturation S_c exceeds S come to a stable equilibrium size at the appropriate point on their Köhler curve. All particles whose S_c is below S become activated and grow indefinitely as long as $S > S_c$.

Figure 17.6 shows the critical supersaturation for spherical salt particles as a function of their diameters. One should note that critical saturations are always higher than unity, and one often defines $s_c = S_c - 1$ as the critical supersaturation.

Figure 17.7 presents the Köhler curves for $(NH_4)_2SO_4$ for complete dissociation and no dissociation. Note that dissociation lowers the critical saturation ratio of the particle (the particle is activated more easily) and increases the drop critical diameter (the particle absorbs more water).

17.2.5 Atmospheric Equilibrium of an Aqueous Solution Drop Containing an Insoluble Substance

Our analysis so far has assumed that the aerosol particle consists of a soluble salt that dissociates completely as the RH exceeds 100%. Most atmospheric particles contain both water-soluble and water-insoluble substances (dust, elemental carbon, etc.). Our goal here is to extend the results of the previous section to account for the existence of insoluble material. Our assumption is that the original particle contains soluble material with mass fraction ε_m, and the rest is insoluble. We also assume that the insoluble portion does not interact at all with water or the salt ions.

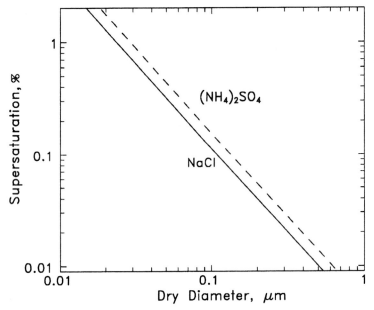

FIGURE 17.6   Critical supersaturation for activating aerosol particles composed of NaCl and $(NH_4)_2SO_4$ as function of the dry particle diameters (assuming spherical particles) at 293 K.

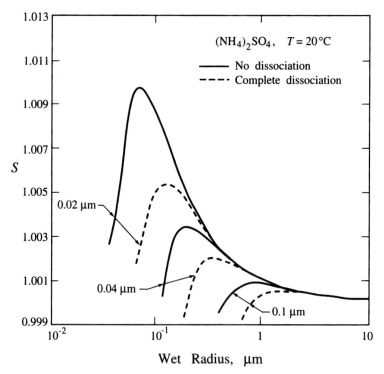

FIGURE 17.7   Köhler curves for $(NH_4)_2SO_4$ assuming complete dissociation and no dissociation of the salt in solution for dry radii of 0.02, 0.04, and 0.1 μm.

Our analysis follows exactly Section 17.2.4, until the derivation of the mole fraction of water x_w in (17.19) and (17.20). The existence of the insoluble material needs to be included in these equations. If the insoluble particle fraction is equivalent to a sphere of diameter d_u, then the droplet volume will be

$$\frac{1}{6}\pi D_p^3 = n_w \bar{v}_w + n_s \bar{v}_s + \frac{1}{6}\pi d_u^3 \tag{17.35}$$

and the mole fraction of water x_w in the solution will be given by

$$\frac{1}{x_w} = 1 + \frac{n_s}{n_w} = 1 + \frac{n_s \bar{v}_w}{(\pi/6)(D_p^3 - d_u^3) - n_s \bar{v}_s} \tag{17.36}$$

Substituting this expression into (17.18), we find

$$\ln\left(\frac{p_w(D_p)}{p^\circ}\right) = \frac{4M_w \sigma_w}{RT\rho_w D_p} + \ln\gamma_w - \ln\left(1 + \frac{n_s \bar{v}_w}{(\pi/6)(D_p^3 - d_u^3) - n_s \bar{v}_s}\right) \tag{17.37}$$

For a dilute solution, we may simplify (17.37) as before and obtain expressions analogous to (17.26) and (17.27) with the same definitions of A and B. For example, in this case (17.27) becomes

$$\ln\left(\frac{p_w(D_p)}{p^\circ}\right) = \frac{A}{D_p} - \frac{B}{(D_p^3 - d_u^3)} \tag{17.38}$$

The effect of the insoluble material is to increase in absolute terms the solute effect. Physically, the insoluble material is responsible for part of the droplet volume, displacing the equivalent water. Therefore, for the same overall droplet diameter, the solution concentration will be higher and the solute effect more significant.

An alternative expression can be developed replacing the equivalent diameter of the insoluble material d_u with the mass fraction of soluble material ε_m, assuming that the insoluble material has a density ρ_u. Then for a dry particle of diameter d_s the following relationship exists between the insoluble core diameter d_u and the mass fraction ε_m

$$d_u^3 = \frac{d_s^3}{\left(\dfrac{\rho_u}{\rho_s}\dfrac{\varepsilon_m}{1 - \varepsilon_m} + 1\right)} \tag{17.39}$$

and (17.38) can be rewritten as a function of the particle soluble fraction and the initial particle diameter provided that the densities of soluble and insoluble material are known. The number of moles of solute are in this case given not by (17.33) but by the following expression:

$$n_s = \frac{\varepsilon_m}{M_s} \frac{v\pi d_s^3}{6\left(\dfrac{\varepsilon_m}{\rho_s} + \dfrac{1 - \varepsilon_m}{\rho_u}\right)} \tag{17.40}$$

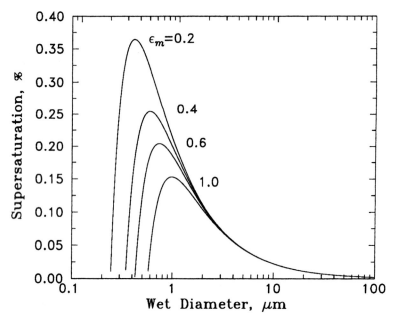

FIGURE 17.8   Variation of the equilibrium vapor pressure of an aqueous solution drop containing $(NH_4)_2SO_4$ and insoluble material for an initial dry particle diameter of 0.1 μm for soluble mass fractions 0.2, 0.4, 0.6, and 1.0 at 293 K.

Köhler curves for a particle consisting of various combinations of $(NH_4)_2SO_4$ and insoluble material are given in Figure 17.8. We see that the smaller the water-soluble fraction the higher the supersaturation needed for activation of the same particle, and the lower the critical diameter. Critical supersaturation as a function of the dry particle diameter is given in Figure 17.9.

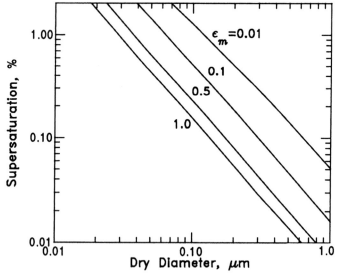

FIGURE 17.9   Critical supersaturation as a function of the particle dry diameter for different contents of insoluble material. The soluble material is $(NH_4)_2SO_4$.

17.3   CLOUD AND FOG FORMATION

The ability of a given particle to become activated depends on its size and chemical composition and on the maximum supersaturation experienced by the particle. If, for example, the ambient RH does not exceed 100%, no particle will be activated and a cloud cannot be formed.[2] In this section we will examine the mechanisms by which clouds are created in the atmosphere. A necessary condition for this cloud formation is the increase in the RH of an air parcel to a value exceeding 100%. This RH increase is usually the result of cooling of a moist air parcel. Even if the water mass inside the air parcel does not change, its saturation water vapor concentration decreases as its temperature decreases, and therefore its RH increases.

There are several mechanisms by which air parcels can cool in the atmosphere. It is useful to separate these mechanisms into two groups: isobaric cooling and adiabatic cooling. Isobaric cooling is the cooling of an air parcel under constant pressure. It usually is the result of radiative losses of energy (fog and low stratus formation) or horizontal movement of an airmass over a colder land or water surface or colder airmass. If an air parcel ascends in the atmosphere, its pressure decreases, the parcel expands, and its temperature drops. The simplest model of such a process assumes that during this expansion there is no heat exchanged between the rising parcel of air and the environment. This idealized process is usually termed "adiabatic cooling." In reality there is always some heat and mass exchange between the parcel and its environment. Let us briefly examine the basic thermodynamic relations for these two mechanisms of cloud formation.

17.3.1   Isobaric Cooling

Let us consider a volume of moist air that is cooled isobarically. Assuming that we can neglect mass exchange between the air parcel and its surroundings, the water vapor partial pressure (p_w) will remain constant. A temperature decrease from an initial value T_0 to T_d will lead to a decrease of the saturation vapor pressure from $p^°(T_0)$ to $p^°(T_d)$.

We would like to calculate when the parcel will become saturated. In other words, if a parcel initially has a temperature T_0 and relative humidity RH (from 0 to 1), what is the temperature T_d at which it will become saturated (RH = 1)? The temperature T_d is called the *dew temperature* or *dewpoint*.

T_d can be calculated recognizing that by definition RH $= p_w/p^°(T_0)$ and that at the dew point $p_w = p^°(T_d)$. The dependence of the saturation vapor pressure on temperature is given by the Clausius–Clapeyron equation (17.8). Integration between T_0, $p^°(T_0)$, and T_d, p_w yields

$$\int_{p^°(T_0)}^{p_w} d\ln p = \int_{T_0}^{T_d} \frac{\Delta H_v M_w}{RT^2}\, dT \qquad (17.41)$$

Assuming that the latent heat ΔH_v is approximately constant from T_0 to T_d, we get

$$\ln\left(\frac{p_w}{p^°(T_0)}\right) = \frac{\Delta H_v M_w}{R}\left(\frac{T_d - T_0}{T_0 T_d}\right) \qquad (17.42)$$

---

[2] The classic Köhler formulation does not consider the cases of solutes that are not completely soluble or soluble gases, both of which influence the solute effect term in (17.27). In such cases cloud droplets can exist at $S < 1$ since $B/D_p^3 > A/D_p$.

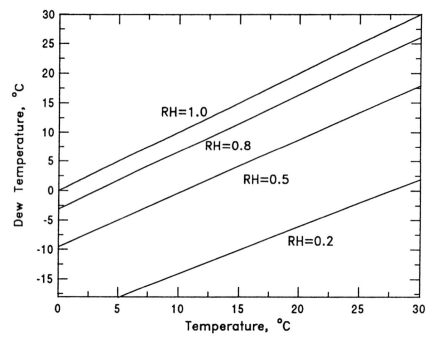

FIGURE 17.10 Dew temperature of an air parcel as a function of its temperature and relative humidity.

Noting that the ratio on the LHS is equal to the initial relative humidity, and assuming that the temperature change is small enough so that $T_0 T_d \simeq T_0^2$, we get

$$T_d \simeq T_0 + \frac{RT_0^2}{\Delta H_v M_w} \ln (\text{RH}) \qquad (17.43)$$

For an air parcel initially at 283 K with a relative humidity of 80% (RH = 0.8), a temperature reduction of 3.3 K is required to bring the parcel to saturation. The dew point of a subsaturated air parcel is always lower than its actual temperature (Figure 17.10). The two become equal only when the relative humidity reaches 100%.

17.3.2 Adiabatic Cooling

The thermodynamic behavior of a rising moist air parcel can be examined in two steps: the cooling of the air parcel from its initial condition to saturation followed by the cooling of the saturated air.

Let us consider first a moist unsaturated air parcel. Assuming that the rise is adiabatic (no heat exchange with its surroundings) and reversible, then it will also be isentropic. Recall that for a reversible process $dQ = T \, dS$ and therefore when $dQ = 0$, $dS = 0$ also. Under these conditions we have shown in Chapter 16 that if the air parcel is dry (no water vapor), its temperature will vary linearly with height according to (16.8)

$$\frac{dT}{dz} = -\Gamma \qquad (17.44)$$

where $\Gamma = g/\hat{c}_p = 9.76°C\,km^{-1}$ is the dry adiabatic lapse rate. If the air is moist, then, as we saw in Chapter 16, the heat capacity of the air parcel $\hat{c}_p$ must be corrected. The changes are small as the water vapor mass fraction is usually less than 3%. Even at this rather extreme condition, the lapse rate of the moist parcel is $9.71°C\,km^{-1}$, an essentially negligible change. Setting the moist adiabatic lapse rate equal to the dry adiabatic lapse rate Γ results in negligible error for all practical purposes.

We can use the preceding information about temperature changes of a rising unsaturated air parcel to calculate the height at which the parcel will become saturated. This height is called the *lifting condensation level* (LCL) and is usually very close to the cloud base. To calculate the temperature T_L at the lifting condensation level, we need to recall (see (16.14)) that the temperature of an air parcel undergoing adiabatic cooling varies as function of pressure p_a according to

$$T = \alpha p_a^{\kappa_a} \tag{17.45}$$

where α is a constant and $\kappa_a = (\gamma - 1)/\gamma \simeq 0.286$ ($\gamma = \hat{c}_p/\hat{c}_v$ is the ratio of the air heat capacity under constant pressure to its heat capacity under constant volume). We also need to recall that because both the air and water vapor masses are conserved during the air parcel rise, the water vapor mass mixing ratio w_v will remain constant. Because both gases are ideal, it follows that

$$w_v = \frac{M_w p_w}{M_a p_a} \tag{17.46}$$

The water vapor and liquid water mixing ratios w_v and w_L are very useful quantities in cloud physics. They do not have units and can be defined on a molar, a volume, or a mass basis. The mass basis will be used in this chapter unless otherwise indicated. Several rather complicated expressions can be simplified considerably using these mixing ratios. If necessary, (17.46) can be used to convert the water vapor mixing ratio to water vapor partial pressure.

At the LCL the air is saturated and therefore $p_w = p°(T_L)$. The air pressure at this height will be, according to (17.46),

$$p_a = \frac{M_w p°(T_L)}{M_a w_v}$$

and substituting into (17.45), we obtain

$$T_L = \alpha \left(\frac{M_w p°(T_L)}{M_a w_v}\right)^{\kappa_a} \tag{17.47}$$

The constant α can be eliminated noting that initially $T_0 = \alpha p_{a0}^{\kappa_a}$, where p_{a0} is the initial air pressure of the parcel, to get

$$T_L = T_0 \left(\frac{M_w p°(T_L)}{M_a w_v p_{a0}}\right)^{\kappa_a} \tag{17.48}$$

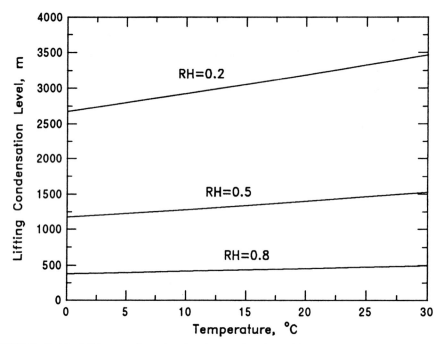

FIGURE 17.11 Lifting condensation level as a function of initial temperature and relative humidity of the air parcel assuming that the air parcel starts initially at the ground at $p = 1$ atm.

The temperature at the lifting condensation level can be calculated by solving (17.48) numerically for T_L. Once T_L is determined, the height at the LCL, h_{LCL} can be found simply by

$$h_{LCL} = \frac{T_0 - T_L}{\Gamma} \tag{17.49}$$

The lifting condensation level h_{LCL} is shown in Figure 17.11 as a function of the initial temperature and relative humidity of the air parcel.

If the air parcel is lifted beyond the LCL, water will start condensing on the available particles, and latent heat of condensation $(-\Delta H_v)$ will be released. The lapse rate Γ_s in this case can be calculated from (16.13).

The latent heat release during cloud formation makes clouds warmer than the surrounding cloud-free air. This higher temperature enhances the buoyancy of clouds, as will have been noticed by anyone who has flown through clouds in an airplane.

17.3.3 Cooling with Entrainment

So far we have treated a rising air parcel as a closed system. This is rarely the case in real clouds, where air from the rising parcel is mixed with the surrounding air. If m is the mass of the air parcel one defines the *entrainment rate e* as

$$e = \frac{1}{m}\frac{dm}{dz} \tag{17.50}$$

The entrainment rate is often written as $e = 1/l$, where l is the lengthscale characteristic of the entrainment process. If the water vapor mixing ratio of the environment around the rising parcel is w_v' and its temperature T', one can show that the lapse rate in the cloud Γ_c is given by (Pruppacher and Klett 1997)

$$\Gamma_c = -\frac{dT}{dz} = \frac{g + e[(\Delta H_v(w_v - w_v') + \hat{c}_p(T - T')]}{\hat{c}_p + \Delta H_v(dw_{vs}/dT)}$$

$$= \frac{g}{\hat{c}_p} + \frac{\Delta H_v}{\hat{c}_p}\frac{dw_{vs}}{dz} + \frac{e}{\hat{c}_p}[\Delta H_v(w_v - w_v') + \hat{c}_p(T - T')]$$

(17.51)

Γ_c exceeds Γ_s because $e > 0$, $w_v > w_v'$, and $T > T'$. Observations show that use of (16.13) usually overestimates the temperature differences between updrafts in cumulus clouds and the environment. The correction for entrainment represented by (17.51) increases the lapse rate by 1–2°C and describes cloud formation more realistically.

17.3.4 A Simplified Mathematical Description of Cloud Formation

Let us revisit the rising moist air parcel assuming that we are in a Lagrangian reference frame, moving with the air parcel. The air parcel is characterized by its temperature T, water vapor mixing ratio w_v, liquid water mixing ratio w_L, and velocity W. At the same time we need to know the temperature T', pressure p, and water vapor mixing ratio w_v' of the air around it (Figure 17.12). The pressure of the air parcel is assumed to be equal to its environment.

Let us assume that the air parcel has mass m and air density ρ (excluding the liquid water). The velocity of the air parcel will be the result of buoyancy forces and the gravitational force due to liquid water. The buoyancy force is proportional to the volume of the air parcel m/ρ and the density difference between the air parcel and its surroundings,

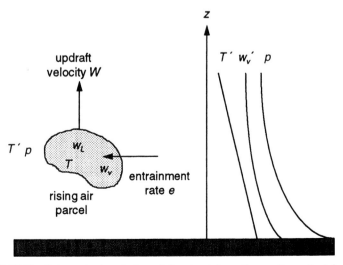

FIGURE 17.12 Schematic description of the cloud formation mathematical framework.

$\rho' - \rho$. The liquid water mass is mw_L and the corresponding gravitational force gmw_L. The equation for conservation of momentum is

$$\frac{d}{dt}(mW) = gm\left(\frac{\rho' - \rho}{\rho} - w_L\right) \tag{17.52}$$

where ρ' is the density of the surrounding air.

As the air parcel is moving, it causes the acceleration of surrounding airmasses, resulting in a decelerating force on the air parcel. The deceleration force is proportional to the mass of the displaced air, m', and the corresponding deceleration, $-dW/dt$. Pruppacher and Klett (1997) show that this effect is actually equivalent to an acceleration of an "induced" mass $m/2$ and therefore a term $-\frac{1}{2}m\,dW/dt$ should be added on the right-hand side of (17.52). Using the ideal-gas law $(\rho' - \rho)/\rho = (T - T')/T'$, and the modified (17.52) can be rewritten as

$$\frac{3}{2}\frac{dW}{dt} + \frac{W}{m}\frac{dm}{dt} = g\left(\frac{T - T'}{T'} - w_L\right) \tag{17.53}$$

and by employing the definition of the entrainment rate e, we obtain

$$e|W| = \frac{1}{m}\frac{dm}{dt} \tag{17.54}$$

Therefore the velocity of the air parcel is described by

$$\frac{dW}{dt} = \frac{2}{3}g\left(\frac{T - T'}{T'} - w_L\right) - \frac{2}{3}eW^2 \tag{17.55}$$

The rate of change of temperature can be calculated using (17.51), noting that $dT/dt = W\,dT/dz$, and also that w_{vs} should be replaced by w_v to allow the creation of supersaturations. The final result is

$$-\frac{dT}{dt} = \frac{gW}{\hat{c}_p} + \frac{\Delta H_v}{\hat{c}_p}\frac{dw_v}{dt} + e\left[\frac{\Delta H_v}{\hat{c}_p}(w_v - w_v') + (T - T')\right]W \tag{17.56}$$

The condensed water is related to w_v through the water mass balance for the entraining parcel. If airmass dm enters the parcel from the outside, then the water vapor and liquid water mixing ratios will change according to

$$m(w_v + w_L) + w_v'dm = (w_v + dw_v + w_L + dw_L)(m + dm)$$

Neglecting products of differentials and dividing by dt, we find that

$$\frac{dw_v}{dt} = -\frac{dw_L}{dt} - eW(w_v + w_L - w_v') \tag{17.57}$$

Let us assume that the environmental profiles of temperature and water vapor are constant with time at $T'(z)$ and $w'_v(z)$ and are known. Then because the air parcel is moving with speed W, it will appear to an observer moving with the air parcel that the surrounding conditions are changing according to

$$\frac{dT'}{dt} = W\frac{dT'}{dz} \qquad (17.58)$$

and

$$\frac{dw'_v}{dt} = W\frac{dw'_v}{dz} \qquad (17.59)$$

For a given environment and given entrainment rate e, we need one more equation to close the system, namely, an equation describing the liquid water mixing ratio w_L of the drop population. This liquid water content can be calculated if the droplet size distribution is known, such as a simple integral over the distribution. We thus need to derive differential equations for the droplet diameter rate of change dD_p/dt. These equations will link the cloud dynamics discussed here with the cloud microphysics discussed in the following sections.

17.4   GROWTH RATE OF INDIVIDUAL CLOUD DROPLETS

When cloud and fog droplets have diameters significantly larger than 1 μm, mass transfer of water to a droplet can be expressed by the mass transfer equation for the continuum regime (see Chapter 12)

$$\frac{dm}{dt} = 2\pi D_p D_v(c_{w,\infty} - c_w^{eq}) \qquad (17.60)$$

where m is the droplet mass, D_p its diameter, D_v the water vapor diffusivity, $c_{w,\infty}$ (in mass per volume of air) the concentration of water vapor far from the droplet, and c_w^{eq} (in mass per volume of air) the equilibrium water vapor concentration of the droplet.

The diffusivity of water vapor in air is given as a function of temperature and pressure by

$$D_v = \frac{0.211}{p}\left(\frac{T}{273}\right)^{1.94} \qquad (17.61)$$

where D_v is in $cm^2\,s^{-1}$, T is in K, and p is in atm.

Equation (17.60) neglects noncontinuum effects that may influence very small cloud droplets. These effects can be included in this equation by introducing a modified diffusivity D'_v, where (Fukuta and Walter 1970)

$$D'_v = \frac{D_v}{1 + \frac{2D_v}{\alpha_c D_p}\left(\frac{2\pi M_w}{RT}\right)^{1/2}} \qquad (17.62)$$

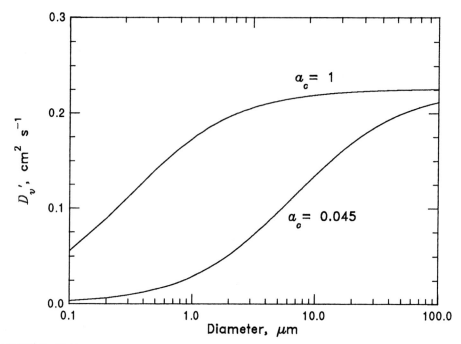

FIGURE 17.13 Water vapor diffusivity corrected for noncontinuum effects and imperfect accommodation as a function of the droplet diameter at $T = 283\,\text{K}$ and $p = 1$ atm.

where α_c is the water accommodation coefficient (often called condensation coefficient). The corrected diffusivity is plotted as a function of droplet diameter in Figure 17.13. The magnitude of the correction depends strongly on the value of the water accommodation coefficient used. For a value equal to unity, the correction should be less than 25% for particles larger than $1\,\mu\text{m}$ and less than 5% for droplet diameters larger than $5\,\mu\text{m}$. However, the correction increases significantly if the α_c value is lower than unity. The value of α_c has been the subject of debate. A value of 0.045 was used by Pruppacher and Klett (1997), while ambient measurements seem to suggest a value closer to unity (Leaitch et al. 1986). Laaksonen et al. (2005) argued that only values near unity yield theoretical predictions consistent with the most accurate growth rate measurements.

The equilibrium vapor concentration in (17.60) corresponds to the concentration at the droplet surface and has been derived in Section 17.2 as a function of temperature. However, during water condensation, heat is released at the droplet surface, and the droplet temperature is expected to be higher than the ambient temperature.

Let us calculate the droplet temperature by deriving an appropriate energy balance. If T_a is the temperature at the drop surface and T_∞ the temperature of the environment, an energy balance gives [see also (12.21)]

$$2\pi D_p k_a'(T_\infty - T_a) = -\Delta H_v \left(\frac{dm}{dt}\right) \tag{17.63}$$

where k_a' is the effective thermal conductivity of air corrected for noncontinuum effects. Equation (17.63) simply states that at steady state the heat released during water condensation is equal to the heat conducted to the droplet surroundings. The temperature at the droplet surface is then

$$T_a = T_\infty + \frac{\Delta H_v \rho_w}{4k_a'} D_p \frac{dD_p}{dt} = T_\infty(1+\delta) \tag{17.64}$$

where we have used the mass balance

$$\frac{dm}{dt} = \frac{1}{2}\pi\rho_w D_p^2 \frac{dD_p}{dt} \tag{17.65}$$

and we have defined

$$\delta = \frac{\Delta H_v \rho_w}{4k_a' T_\infty} D_p \frac{dD_p}{dt} \tag{17.66}$$

For atmospheric cloud droplet growth $\delta \ll 1$; however, let us continue the dervation without assuming that $T_a = T_\infty$. Combining (17.60) and (17.65) and using the ideal-gas law, we find that

$$D_p \frac{dD_p}{dt} = \frac{4D_v' M_w p^\circ(T_\infty)}{\rho_w RT} \left(S_{v,\infty} - \frac{p_w(D_p, T_a)}{p^\circ(T_\infty)} \right) \tag{17.67}$$

where $S_{v,\infty} = p_{w\infty}/p^\circ(T_\infty)$ is the environmental saturation ratio. Recall that for a relative humidity equal to 100% the partial pressure of water in the atmosphere, $p_{w\infty}$ is equal to the saturation vapor pressure $p^\circ(T_\infty)$ and $S_{v,\infty}$ is equal to unity. The ratio of the water saturation pressures at T_a and T_∞ is given by the Clausius–Clapeyron equation as

$$\frac{p^\circ(T_a)}{p^\circ(T_\infty)} = \exp\left[\frac{\Delta H_v M_w}{R}\left(\frac{T_a - T_\infty}{T_a T_\infty}\right)\right] \tag{17.68}$$

Combining (17.64), (17.67), (17.38), and (17.68) we finally get

$$D_p \frac{dD_p}{dt} = \frac{4D_v' M_w p^\circ(T_\infty)}{\rho_w RT_\infty}\left(S_{v,\infty} - \frac{1}{1+\delta} \right.$$
$$\left. \times \exp\left[\frac{\Delta H_v M_w}{RT_\infty}\frac{\delta}{1+\delta} + \frac{4M_w \sigma_w}{RT\rho_w D_p(1+\delta)} - \frac{6n_s M_w}{\pi\rho_w(D_p^3 - d_u^3)}\right]\right) \tag{17.69}$$

This result can be simplified since $\delta \ll 1$ and

$$\exp[\Delta H_v M_w \delta/RT_\infty(1+\delta)] \simeq 1 + \Delta H_v M_w \delta/RT_\infty$$

After some algebra the implicit dependence on δ can be resolved to obtain

$$D_p \frac{dD_p}{dt} = \frac{S_{v,\infty} - \exp\left(\dfrac{4M_w \sigma_w}{RT_\infty \rho_w D_p} - \dfrac{6n_s M_w}{\pi\rho_w(D_p^3 - d_u^3)}\right)}{\dfrac{\rho_w RT_\infty}{4p^\circ(T_\infty)D_v' M_w} + \dfrac{\Delta H_v \rho_w}{4k_a' T_\infty}\left(\dfrac{\Delta H_v M_w}{T_\infty R} - 1\right)} \tag{17.70}$$

This equation describes the growth/evaporation rate of an atmospheric droplet. The numerator is the driving force for the mass transfer of water, namely, the difference between the ambient saturation $S_{v,\infty}$ and the equilibrium saturation for the droplet (or equivalently the water vapor saturation at the droplet surface). The equilibrium saturation includes, as we saw in Section 17.2.4, the contributions of the Kelvin effect (first term in the exponential) and the solute effect (second term in the exponential). When the ambient saturation exceeds the equilibrium saturation, the cloud droplets grow and vice versa. The numerator is qualitatively equivalent to the term $c_{w,\infty} - c_w^{eq}$ in (17.60). The first term in the denominator corresponds to the diffusivity of water vapor (compare with (17.60)), while the second accounts for the temperature difference between the droplet and its surroundings. Note that if no heat were released during condensation, $\Delta H_v = 0$, and this term would be zero.

The thermal conductivity of air k_a is given by

$$k_a = 10^{-3}(4.39 + 0.071T) \tag{17.71}$$

where k_a is in $J\,m^{-1}\,s^{-1}\,K^{-1}$ and T is in K. The modified form for the thermal conductivity k_a' accounting for non-continuum effects is given by

$$k_a' = k_a \bigg/ \left[1 + \frac{2k_a}{\alpha_T D_p \rho \hat{c}_p}\left(\frac{2\pi M_a}{RT_a}\right)^{1/2}\right] \tag{17.72}$$

where α_T is the thermal accommodation coefficient. The value of α_T is also uncertain and it is often set equal to the value of the mass accommodation coefficient α_c.

For a cloud droplet larger than $10\,\mu m$, using $\alpha_c = \alpha_T = 1$, at 283 K, we obtain

$$\frac{\rho_w RT_\infty}{p^\circ(T_\infty)D_v' M_w} = 4.85 \times 10^5\,s\,cm^{-2}$$

$$\frac{\Delta H_v \rho_w}{k_a' T_\infty}\left(\frac{\Delta H_v M_w}{T_\infty R} - 1\right) = 6.4 \times 10^5\,s\,cm^{-2}$$

and defining $S_{v,eq}$ as the equilibrium saturation of the droplet, (17.70) can be rewritten as

$$D_p\frac{dD_p}{dt} = 3.5 \times 10^{-6}(S_{v,\infty} - S_{v,eq})$$

where D_p is in cm and t in s. The growth of an aerosol size distribution under constant supersaturation of 1% is shown in Figure 17.14. The rate of growth of droplets is inversely proportional to their diameters so smaller droplets grow faster than larger ones. As a result, small droplets catch up in size with larger ones during the growth stage of the cloud.

17.5   GROWTH OF A DROPLET POPULATION

The growth of an aerosol population to cloud droplets can be investigated using the growth equation derived in the previous section. In general, one would need to integrate simultaneously the differential equations derived in Section 17.3.4 for the air parcel

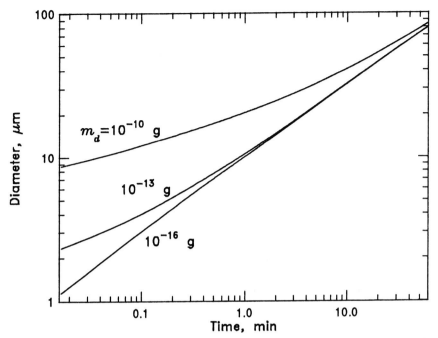

FIGURE 17.14 Diffusional growth of individual drops with different dry masses as a function of time. The drops are initially at equilibrium at 80% RH.

updraft velocity, temperature, water vapor mixing ratio, and environmental temperature and water vapor mixing ratio, coupled with a set of droplet growth equations, one for each droplet size class. The liquid water mixing ratio of the population consisting of N_i droplets per volume of air of diameters D_{pi} will then be

$$w_L = \frac{\rho_w}{\rho_a} \frac{\pi}{6} \sum_{i=1}^{n} N_i D_{pi}^3 \qquad (17.73)$$

where we have assumed that there are n groups of droplets.

It is instructive, before examining the interactions between cloud dynamics and microphysics, to focus our attention on the microphysics. Figure 17.15 presents the results of the integration of these equations for a polluted urban aerosol population (Pandis et al. 1990a) for an aerosol distribution consisting of seven size sections. At time zero the relative humidity is assumed to be 100%. At this time the particles have grown several times from their dry size as a result of water absorption and are assumed to be in equilibrium with the surrounding environment. The temperature of the air parcel is assumed to decrease with a constant rate of $2\,\mathrm{K\,h^{-1}}$. As the temperature decreases, the saturation of the air parcel increases. The particles absorb water vapor, but the cooling rate is too rapid compared to mass transfer and the air parcel becomes supersaturated. After a few minutes the particles start becoming activated. The larger particles become activated first (Section 17.2) and the smaller soon follow. As particles become activated, they are able to grow much faster. Note that based on the Köhler curves as a particle grows the driving force for growth $(c_{w,\infty} - c_w^{eq})$ becomes larger for almost constant

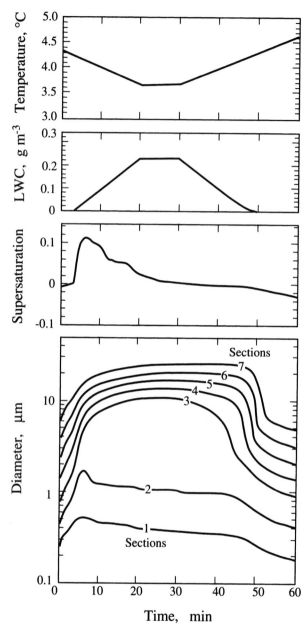

FIGURE 17.15 Simulated evolution of temperature, liquid water content, supersaturation, and particle diameters during the lifetime of a cloud (Pandis et al. 1990a,b). The sections denote seven sizes of initial particles.

$c_{w,\infty}$, because c_w^{eq} decreases rapidly with size. Therefore, as more and more particles get activated, the rate of transport of water from the vapor to the particulate phase increases, while the rate of supersaturation increase due to the cooling remains approximately constant. The result is that the supersaturation increase slows down and after 6 min reaches a maximum value of 0.1%.

Let us describe this situation quantitatively, by deriving the equation for the rate of change of the supersaturation s_v. The water vapor mixing ratio w_v is related to the water vapor partial pressure p_w by (17.46), while by definition

$$1 + s_v = \frac{p_w}{p^\circ}$$

Therefore combining these two relationships, one gets

$$s_v = \frac{M_a p_a}{M_w p^\circ} w_v - 1 \tag{17.74}$$

Differentiating this expression with respect to time and rearranging the terms, we obtain

$$\frac{ds_v}{dt} = \frac{M_a p_a}{M_w p^\circ} \frac{dw_v}{dt} - (1 + s_v)\left(\frac{1}{p^\circ}\frac{dp^\circ}{dt} - \frac{1}{p_a}\frac{dp_a}{dt}\right) \tag{17.75}$$

The change of the air pressure with time can be calculated assuming that the environment is in hydrostatic equilibrium so that

$$\frac{dp_a}{dt} = -\frac{g p_a M_a}{RT} W \tag{17.76}$$

where we have assumed that $T' \simeq T$. The change of the parcel saturation pressure with time can be calculated using the chain rule and the Clausius–Clapeyron equation:

$$\frac{dp^\circ}{dt} = \frac{dp^\circ}{dT}\frac{dT}{dt} = \frac{\Delta H_v M_w p^\circ}{RT^2}\frac{dT}{dt} \tag{17.77}$$

Substituting (17.76) and (17.77) into (17.75), one gets

$$\frac{ds_v}{dt} = \frac{M_a p_a}{M_w p^\circ}\frac{dw_v}{dt} - (1 + s_v)\left(\frac{\Delta H_v M_w}{RT^2}\frac{dT}{dt} + \frac{g M_a}{RT} W\right) \tag{17.78}$$

If we assume that there is no entrainment ($e = 0$) and substitute (17.54) and (17.55) into (17.78), we obtain

$$\frac{ds_v}{dt} = \left(\frac{\Delta H_v M_w g}{\hat{c}_p R T^2} - \frac{g M_a}{RT}\right) W - \left(\frac{p_a M_a}{p^\circ M_w} + \frac{\Delta H_v^2 M_w}{\hat{c}_p R T^2}\right)\frac{dw_L}{dt} \tag{17.79}$$

where we have assumed that $1 + s_v \simeq 1$, as $s_v \simeq 0.01$ in clouds. Equation (17.79) reveals that in the absence of condensation, the saturation varies linearly with the updraft velocity and is decreased by water condensation. One could replace in (17.79) the updraft velocity with the cooling rate

$$W \simeq \frac{\hat{c}_p}{g}\left(-\frac{dT}{dt}\right) \tag{17.80}$$

where we have assumed that the updraft velocity is almost constant, and therefore the air parcel cooling rate is also constant. Equation (17.79) is the mathematical representation of our previous qualitative theoretical arguments. It suggests that the supersaturation inside the cloud is the result of a balance between the cooling rate and the liquid water increase. The latter is limited by the mass transport to particles, which in turn depends on the particle size distribution and on their state of activation.

The maximum supersaturation reached inside a cloud/fog is an important parameter. Particles with critical supersaturations lower than this value will become activated and become cloud droplets. The rest remain close to equilibrium but never grow enough to be considered droplets and are called *interstitial aerosol*. The aerosol population inside a cloud is therefore separated into two groups: interstitial aerosols that contain significant amounts of water but are not activated (their sizes are usually smaller than 2 μm) and cloud droplets, with size increases corresponding to mass changes of three orders of magnitude.

In our example, the maximum supersaturation of 0.1% was sufficient to activate aerosols with dry diameters larger than approximately 0.3 μm, corresponding to size sections 3–7 in Figure 17.15. Note in Figure 17.15 that particles in section 1 (dry size 0.1 μm) grow up to 0.5 μm for the maximum saturation and then evaporate slowly, following the relative humidity. These particles remain in equilibrium throughout the cloud lifetime. Particles in the second size section (dry diameter 0.2 μm) exhibit a more interesting behavior. Their critical supersaturation is slightly lower than the maximum supersaturation, so they activate growing to a size of 1.5 μm. However, as the supersaturation decreases, they deactivate after a few minutes and do not have the time to grow more in size. The rest of the particles (dry sizes larger than 0.3 μm) all get activated and grow to droplets larger than 10 μm in size.

Direct measurements of ambient supersaturations in clouds have been extremely challenging. Not only does one try to measure a small deviation from saturation, but clouds are frequently patchy with supersaturated regions next to subsaturated ones corresponding to "dry" entrained masses. Previous measurements have indicated that ambient supersaturations are usually less than 1% and almost never exceed 2% (Warner 1968). A median value of 0.1% was reported in these measurements. Most of our knowledge of these supersaturations is based on theoretical calculations using measurements of atmospheric conditions and are rather similar to that presented here as an example. Ranges of supersaturations expected in various cloud types are given in Table 17.3.

Note that supersaturations depend both on the macroscale cloud dynamics represented by the cloud updraft velocities (larger updrafts result in higher supersaturations) and on the

TABLE 17.3   Updraft Velocities and Maximum Supersaturations for Clouds and Fogs

Cloud Type	Updraft Velocity, m s^{-1}	Maximum Supersaturation, %	Reference
Continental cumulus	~1–17	0.25–0.7	Pruppacher and Klett (1997)
Maritime cumulus	~1–2.5	0.3–0.8	Pruppacher and Klett (1997) Mason (1971)
Stratiform	~0–1	~0.05	Pruppacher and Klett (1997)
Fog	—	~0.1	Pandis and Seinfeld (1989)

microphysics (the details of the aerosol size distribution). Cleaner environments, with lower aerosol concentrations, usually result in higher supersaturations.

17.6   CLOUD CONDENSATION NUCLEI

Supersaturations of several hundred percent are necessary for the formation of water droplets in particle free air (see Chapter 11). The need for such high supersaturations indicates the necessity of particles for cloud formation in the ambient atmosphere. The ability of a given particle to serve as a nucleus for water droplet formation, as we have seen in the previous sections, will depend on its size, chemical composition, and the local supersaturation.

Particles that can activate at a given supersaturation are defined as *cloud condensation nuclei* (CCN) for this supersaturation. In the cloud physics literature one often defines as *condensation nuclei* (CN) those particles that form droplets at supersaturations of $\geq 400\%$ and therefore CN include all the available particles. One can therefore assume that the CN concentration is equal to the total aerosol number concentration. This CN definition should be contrasted with the CCN definition where supersaturations often well less than 2% are used. Therefore CCN represent the particles that can form cloud droplets under reasonable atmospheric supersaturations. We caution the reader that CCN concentrations always refer to a specific supersaturation, for example, CCN(1%) or CCN(0.5%) and one should be careful when comparing CCN concentrations measured or estimated at different supersaturations.

The CCN concentration of a given supersaturation corresponds under ideal cloud formation conditions (e.g., spatial uniformity) to the number concentrations of droplets if the cloud had the same supersaturation. We will use the symbol CCN(s) for CCN at $s\%$ supersaturation.

For a given aerosol population CCN(s) depends on both the size and composition of the particles. In the simple case of an aerosol population that has uniform size-independent composition, by definition,

$$\mathrm{CCN}(s) = \int_{D_s}^{\infty} n(D_p)dD_p \qquad (17.81)$$

where $n(D_p)$ is the number distribution of the aerosol population, and D_s the activation diameter for $s\%$ supersaturation of these particles. Note that CN = CCN(∞) according to the above notation. Therefore, if all particles had the same composition, one needs to know only the activation diameter and the size distribution of these particles to estimate the corresponding CCN concentration (Figure 17.16a). However, for aerosol populations where chemical composition is size-dependent, CCN concentration is a more complicated function,

$$\mathrm{CCN}(s) = \int_{0}^{\infty} f_s(D_p)n(D_p)dD_p \qquad (17.82)$$

where $f_s(D_p)$ is the fraction of the aerosol particles of diameter D_p that are activated at supersaturation $s\%$ (Figure 17.16b). Note that if the aerosol particles are internally mixed

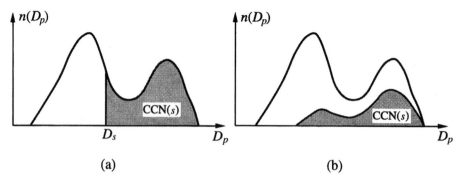

FIGURE 17.16 Schematic of an aerosol size distribution with the shaded area indicating the CCN for (a) uniform chemical composition and (b) a typical multicomponent aerosol population that is neither externally nor internally mixed and has a size-dependent composition.

(all particles of the same diameter have the same chemical composition), $f_s(D_p)$ will be either zero or one.

Because of the difficulties associated with measurement of the aerosol size composition distribution, and then calculating from that their CCN properties, a series of empirical parametrizations have been developed based on atmospheric measurements. One of the most popular relates CCN(s) concentration to atmospheric supersaturation s (percent) with a power law

$$CCN(s) = cs^k \qquad (17.83)$$

where CCN(s) is in particles cm^{-3} and s is the supersaturation expressed as a percentage (Twomey 1959). The constant c corresponds to the CCN(1%), the particles active at 1% supersaturation. It should be noted that knowledge of the parameters c and k is often sufficient for many cloud microphysics applications. Information about the size/composition of the aerosol population is embedded in the empirical parameters c and k. Values for c and k based on ambient measurements are given in Table 17.4.

Since ambient supersaturations rarely exceed 1%, the CCN measurements in Table 17.4 indicate that cloud droplet concentrations should range from 200 to 1000 cm^{-3} over continents and from 10 to 200 cm^{-3} over oceans.

The link between aerosol chemical composition and CCN behavior is still not completely understood. Whereas behavior of soluble inorganic aerosols is relatively well established, less is known about the ability of organic aerosols (alone or mixed with inorganic components) to serve as CCN. A number of studies have measured the CCN behavior of carbonaceous species, either by themselves or mixed together with inorganic salts (Lammel and Novakov, 1995; Cruz and Pandis, 1997; Carrigan and Novakov, 1999; Hegg et al., 2001; Raymond and Pandis, 2002; Bilde and Svenningson, 2004; Abbatt et al., 2005; VanReken et al., 2005).

Köhler theory was originally developed for inorganic salt solutes and, as in Section 17.2.5, can be readily adapted to an insoluble substance as well. Recently Köhler theory has been extended to soluble trace gases, slightly soluble substances, and solutes that

TABLE 17.4   Empirical Parameters for the CCN Concentration Dependence on the Supersaturation s

c, cm^{-3}	k	Location
125	0.3	Maritime (Australia)
53–105	0.5–0.6	Maui (Hawaii)
100	0.5	Atlantic, Pacific Oceans
190	0.8	Pacific
250	1.3–1.4	North Atlantic
145–370	0.4–0.9	North Atlantic
100–1000	—	Arctic
140	0.4	Cape Grim (Australia)
250	0.5	North Atlantic
25–128	0.4–0.6	North Pacific
27–111	1.0	North Pacific
400	0.3	Polluted North Pacific
100	0.4	Equatorial Pacific
600	0.5	Continental
2000	0.4	Continental (Australia)
3500	0.9	Continental (Buffalo, NY)

Source: Hegg and Hobbs (1992).

change the surface tension of water. The latter two cases begin to address theoretically the effect of organic species on particle activation. Appendix 17 presents these extensions of Köhler theory.

The CCN behavior of ambient particles can be measured by drawing an air sample into an instrument in which the particles are subjected to a known supersaturation, a so-called CCN counter (Nenes et al. 2001). If the size distribution and chemical composition of the ambient particles are simultaneously measured, then the measured CCN behavior can be compared to that predicted by Köhler theory on the basis of their size and composition. Such a comparison can be termed a *CCN closure*, that is, an assessment of the extent to which measured CCN activation can be predicted theoretically [see, for example, VanReken et al. (2003), Ghan et al. (2006), and Rissman et al. (2006)]. The next level of evaluation is an *aerosol-cloud drop closure*, in which a cloud parcel model, which predicts cloud drop concentration using observed ambient aerosol concentration, size distribution, cloud updraft velocity, and thermodynamic state, is evaluated against direct airborne measurements of cloud droplet number concentration as a function of altitude above cloud base. The predicted activation behavior can also be evaluated by independent measurements by a CCN instrument on board the aircraft. Such an aerosol-cloud drop closure was carried out by Conant et al. (2004) for warm cumulus clouds in Florida.

Figure 17.17 shows the results of a CCN closure study carried out during the 2003 Coastal Stratocumulus Imposed Perturbation Experiment (CSTRIPE), in which an aircraft-based CCN counter measured activation by marine boundary layer aerosols at 3 supersaturations. Observed CCN concentrations are compared to CCN concentrations predicted on the basis of the measured size distribution of the aerosol and assuming the aerosol was pure $(NH_4)_2SO_4$. Most of the data indicate that actual activation was less than that predicted. This result is consistent with the presence of insoluble or partially soluble material in the particles or externally mixed particles. Note the difference in concentration scales between the marine and continental airmasses; this shows that the continental

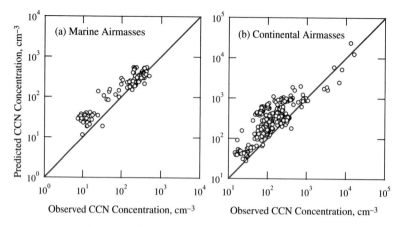

FIGURE 17.17   CCN closure study carried out during July 2003 in the Pacific Ocean marine boundary layer off the coast of Monterey, California. A CCN counter on board an aircraft measured activation of the boundary layer aerosol at 3 supersaturations; data for 0.1% supersaturation are shown here. Panels (a) and (b) show data when the airmasses sampled were of marine and continental origins, respectively. The predicted CCN concentrations are based on the assumption that the aerosol was pure ammonium sulfate. Source: California Institute of Technology

aerosol likely contained a large fraction of insoluble or partially soluble material and/or was largely externally mixed. Other examples of atmospheric CCN closure studies are given by VanReken et al. (2003), Ghan et al. (2006), and Rissman et al. (2006).

17.7   CLOUD PROCESSING OF AEROSOLS

During the processing of an air parcel by a nonraining cloud the aerosol size–composition distribution is transformed by a variety of processes. First, a fraction of the aerosol distribution is activated and becomes cloud droplets while the rest remains as interstitial particles. This process, often described as *nucleation scavenging* of aerosols, determines the initial composition of the cloud droplets. If this were the only process taking place, after cloud evaporation the aerosol distribution would return to its original form. However, a series of additional processes can modify the distribution, including chemical reaction in the aqueous phase, collisions between interstitial aerosols and clouddrops, and coalescence among clouddrops.

If the cloud is raining, there are additional interactions between the raindrops and the aerosols both in and around clouds, leading to removal of material from the atmosphere. Finally, there are other processes that can occur around clouds that may lead to the formation of new particles.

17.7.1   Nucleation Scavenging of Aerosols by Clouds

Nucleation scavenging of aerosols in clouds refers to activation and subsequent growth of a fraction of the aerosol population to cloud droplets. This process is described by (17.70) and has been discussed in Section 17.5.

If $C_{i,0}$ is the concentration (in mass per volume of air) of an aerosol species in clear air before cloud formation (e.g., at the cloud base), and $C_{i,\text{cloud}}$ and $C_{i,\text{int}}$ are its concentrations again in mass per volume of air in the aqueous phase and in the interstitial aerosol, respectively, one can define the cloud *mass scavenging ratio* for species $i(F_i)$ as

$$F_i = \frac{C_{i,0} - C_{i,\text{int}}}{C_{i,0}} \qquad (17.84)$$

Note that if there is no production or removal of i in the cloud, then $C_{i,0} = C_{i,\text{int}} + C_{i,\text{cloud}}$. The mass scavenging ratio defined above may vary from zero to unity. The number scavenging ratio F_N can be defined as

$$F_N = \frac{N_0 - N_{\text{int}}}{N_0} \qquad (17.85)$$

where N_0 is the aerosol number concentration before cloud formation and N_{int} is the number concentration of interstitial aerosol.

Theoretically, as particles larger than $0.5\,\mu m$ or so become cloud droplets in a typical cloud, and these particles represent most of the aerosol mass, one would expect mass activation efficiencies close to unity. Junge (1963) predicted sulfate scavenging ratios from nucleation scavenging alone to range from 0.5 to 1.0. Since then all theoretical studies have predicted high mass nucleation scavenging efficiencies for all aerosol species. For example, Flossmann et al. (1985, 1987) reported calculated aerosol scavenging efficiencies exceeding 0.9 in typical cloud environments. Pandis et al. (1990a) estimated scavenging efficiencies of 0.7 for sulfate and 0.8 for nitrate and ammonia in polluted clouds. In other numerical studies, Flossmann (1991) reported mass scavenging efficiencies of 0.9 or higher for warm clouds over the Atlantic.

These theoretical estimates are in good agreement with the high mass scavenging efficiencies measured in the atmosphere. Ten Brink et al. (1987) observed nearly complete scavenging of aerosol sulfate in clouds. The data of Daum et al. (1984) also showed that the bulk of the sulfate mass is incorporated into cloud droplets. Hegg and Hobbs (1988) reported scavenging ratios for sulfate of 0.5 ± 0.2.

On the contrary, low number scavenging efficiencies are expected in clouds influenced by anthropogenic sources because of the prevalence of fine aerosol particles; number scavenging efficiencies of a few percent or less are expected in most such situations. Only in clouds in the remote marine atmosphere does the total number scavenging efficiency exceed 0.1.

17.7.2 Chemical Composition of Cloud Droplets

During the droplet growth stage of a cloud, droplets of different sizes dilute at different rates; in particular, smaller droplets grow faster and therefore dilute faster than the larger ones. This can be shown by the following argument (Noone et al. 1988). Assume that the aerosol population consists of i particle groups of dry diameters $D_{s,i}$. These particles grow and become aqueous droplets of diameters D_i. Assuming for simplicity that the density of the dry particles is $1\,\text{g\,cm}^{-3}$, the solute mass fraction of droplets in section i is $X_i = (D_{s,i}/D_i)^3$. Defining the dilution rate DR_i of section i as the normalized rate of change of the solute mass fraction in the droplet, we obtain

$$\text{DR}_i = -\frac{1}{X_i}\frac{dX_i}{dt} \qquad (17.86)$$

and assuming that the mass of scavenged aerosol in droplets of group i remains constant with time (i.e., neglecting processes like scavenging of gases, coagulation), then, for two aerosol groups with $D_1 < D_2$

$$\frac{DR_1}{DR_2} = \frac{D_2}{D_1} \frac{dD_1/dt}{dD_2/dt} \qquad (17.87)$$

For sufficiently large droplets, the growth rate is approximately proportional to the inverse of the droplet diameter [see (17.70)] and

$$\frac{dD_i}{dt} \simeq \frac{K}{D_i} \qquad (17.88)$$

where K is a constant that is only a very weak function of D_i. Then, combining (17.87) and (17.88), one finds that

$$\frac{DR_1}{DR_2} \simeq \left(\frac{D_2}{D_1}\right)^2 > 1 \qquad (17.89)$$

and the smaller droplets, D_1, are diluted at a faster rate than the larger D_2 ones if growth by water diffusion is the dominant process occurring.

The above argument suggests that, with time, the total solute concentrations of droplets for an aerosol population would tend to increase with particle size. These arguments are supported by calculations like those by Pandis et al. (1990a) shown in Figure 17.18. Note that initially, before cloud creation, solute concentration is high across the size spectrum and larger particles have slightly lower concentrations as they contain hydrophilic components such as NaCl. As aerosols become activated, their solute concentrations decrease. For a mature cloud solute concentration shows a minimum at around 10 μm. This minimum is a result of existence of nonactivated particles for which solute concentration decreases with increasing size, and droplets for which the solute concentration increases with increasing size. Note that for a mature cloud droplets of diameter around 10 μm have a solute concentration of roughly $100\,\text{mg}\,\text{L}^{-1}$, whereas drops of diameter 24 μm have a solute concentration that is 3.4 times higher. During the evaporation stage of a cloud, smaller droplets get deactivated and evaporate first (Figure 17.18b). Therefore the minimum gradually disappears and the system returns close to its original state.

The predictions presented above agree with measured concentration/size dependencies measured in clouds that are not heavily influenced by anthropogenic sources. Noone et al. (1988) sampled droplets from a marine stratus cloud and calculated that the volumetric mean solute concentration of the 9–18-μm droplets was a factor of 2.7 smaller than in the 18–23-μm droplets. Ogren et al. (1989) reported similar results for a cloud in Sweden. On the other hand, similar measurements for cloud and fog droplets in heavily polluted environments suggest that solute concentrations decrease with increasing droplet size (Munger et al. 1989; Ogren et al. 1992). No satisfactory explanation exists for such behavior.

The discussion above concerns total solute concentration. For individual species, because their aerosol concentrations are also generally size-dependent, there is an additional reason for size-dependent droplet concentrations. Concentrations of some major aerosol species measured in small and large droplets in a cloud are shown in Figure 17.19. The droplet population in these measurements was separated into only two samples with significant overlap and therefore the concentration deviations shown probably underestimate the actual differences.

The above differences in solute concentrations are accompanied by differences in acidity among droplets of different sizes. Figure 17.20 summarizes measurements of

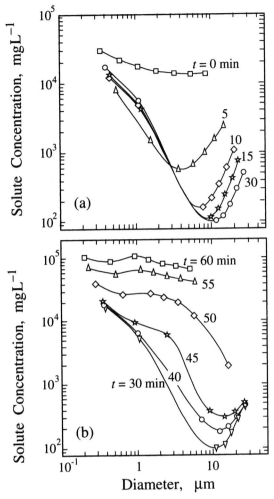

FIGURE 17.18 Predicted dependence of the total solute concentration on droplet diameter during the lifetime of a cloud (Pandis et al. 1990a).

Collett et al. (1994) in a variety of environments. Significant pH differences are observed. Once more, because of significant mixing among droplets of different sizes during sampling, the actual differences are probably even greater than those depicted in Figure 17.20.

Measurements of bulk cloudwater concentrations have been presented by a number of investigators. These concentrations vary significantly because of both the aerosol loading (degree of anthropogenic influence) and the liquid water content of the cloud.

17.7.3 Nonraining Cloud Effects on Aerosol Concentrations

Significant production of sulfate has been detected and/or predicted in clouds and fogs in different environments (Hegg and Hobbs 1987, 1988; Pandis and Seinfeld 1989; Husain et al. 1991; Pandis et al. 1992; Swozdiak and Swozdiak 1992; Develk 1994; Liu et al. 1994). Detection of sulfate-producing reactions is often hindered by variability of cloud

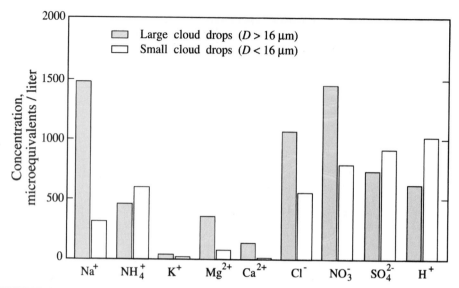

FIGURE 17.19   Measured composition of the small and large cloud droplets collected in coastal stratus clouds at La Jolla Peak, California, in July 1993 (Collett et al. 1994).

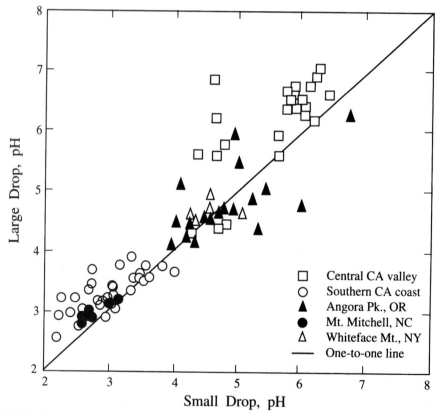

FIGURE 17.20   Measured pH of small and large droplets in a series of clouds and fogs in typical environments (Collett et al. 1994).

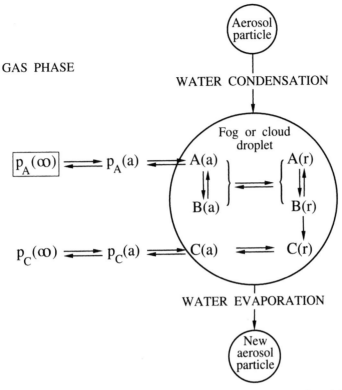

FIGURE 17.21   Schematic of the cloud processing of an aerosol particle.

liquid water content and temporal instability and spatial variability in concentrations of reagents and product species (Kelly et al. 1989).

During cloud formation, aerosols that serve as cloud condensation nuclei (CCN) become activated and grow freely by vapor diffusion. Soluble gases such as nitric acid, ammonia, and sulfur dioxide dissolve into the droplets. The cloudwater serves as the reacting medium for a series of aqueous-phase reactions, most importantly the transformation of dissolved SO_2, S(IV), to sulfate, S(VI). The sulfate formed is not volatile and remains in the particulate phase. Other reactions—for example, the oxidation of formaldehyde to formic acid—result in volatile products that return to the gas phase (Figure 17.21). During the cloud evaporation stage, several species that were dissolved in the cloudwater evaporate. Others, like sulfate, remain in the aerosol phase. Ammonia often accompanies the sulfate formed as the neutralizing cation. Species like nitrate or chloride that may have existed in the original particle can be displaced by the sulfate produced and forced to return to the gas phase. The result of these aqueous-phase processes is usually an overall increase in particle mass and size. Chemical composition of the particles may also change, with sulfate and ammonium concentrations generally increasing and nitrate and chloride decreasing (Pandis et al. 1990b).

Available evidence suggests that the single most important reaction during aerosol processing by clouds is the oxidation of HSO_3^- by H_2O_2. This reaction, as we saw in Chapter 7, is particularly fast with rates often exceeding 100% $SO_2\,h^{-1}$. Daum et al. (1984) showed that SO_2 and H_2O_2 did not coexist in interstitial cloud air (Figure 17.22). If

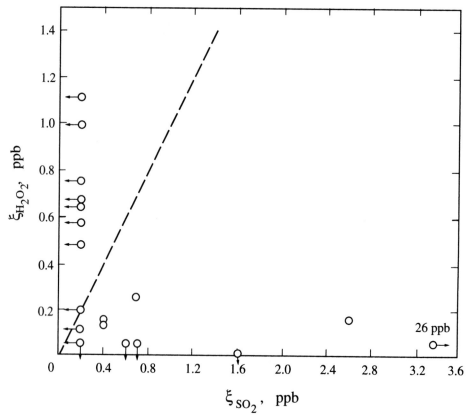

FIGURE 17.22   Measurements of the gas-phase partial pressures of H_2O_2 versus the SO_2 partial pressure for interstitial cloud air (Daum et al. 1984). Arrows signify that the mixing ratio was below the detection limit.

H_2O_2 was high then interstitial SO_2 was low and vice versa, a pattern consistent with the rapid and quantitative reaction of SO_2 with peroxide in which either SO_2 or peroxide acts as limiting reagent. Hydrogen peroxide has been reported to dominate aqueous sulfate formation in the northeastern United States. Measured H_2O_2 gas-phase mixing ratios over the northeastern and central United States vary from 0.2 to 6.7 ppb (Sakugawa et al. 1990) with the highest values during the summer and the lowest during the winter months. Availability of hydrogen peroxide is often limiting to sulfate formation in clouds, a limitation more pronounced near SO_2 sources and during winter months. The seasonal contribution of clouds to sulfate levels depends on both availability of oxidants and on cloud cover. In cases where sulfate production is oxidant-limited, changes in aerosol sulfate levels will be less than proportional to SO_2 emission changes, with the relationship being more nonlinear in winter than in spring or summer (U.S. NAPAP 1991).

Cloud processing is a major source of sulfate and aerosol mass in general on regional and global scales. Walcek et al. (1990) calculated that, during passage of a midlatitude storm system, over 65% of tropospheric sulfate over the northeastern United States was formed in cloud droplets via aqueous-phase reactions. The same authors estimated that, during a 3-day springtime period, chemical reactions in clouds occupying 1–2% of the tropospheric volume were responsible for sulfate production comparable to the gas-phase

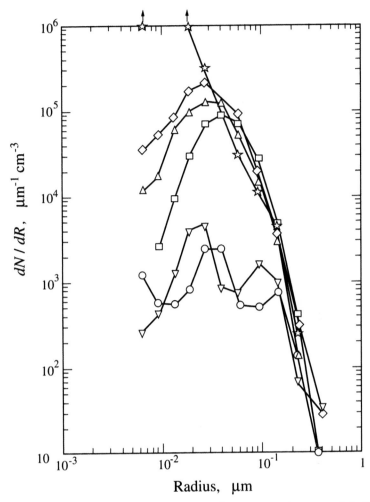

FIGURE 17.23 Size distribution measurements of remote marine aerosol indicating the formation of an additional peak as the airmass is advected off North America. (Reprinted from *Atmos. Environ.* **24A**, Hoppel, W. A., and Frick, G. M., 645–649. Copyright 1990, with kind permission from Elsevier Science Ltd., The Boulevard, Langford Lane, Kidlington OX5 1GB, UK.)

reactions throughout the entire tropospheric volume under consideration. McHenry and Dennis (1994) proposed that annually more than 60% of the ambient sulfate in central and eastern United States is produced in mostly nonprecipitating clouds. Similar conclusions were reached by Dennis et al. (1993) and Karamachandani and Venkatram (1992). Aqueous-phase SO_2 oxidation in clouds is predicted to be the most important pathway for the conversion of SO_2 to sulfate on a global scale (Hegg 1985; Liao et al., 2003).

Effect of cloud processing of aerosols in the remote marine atmosphere has been demonstrated in a series of field studies (Hoppel et al. 1986; Frick and Hoppel 1993). Figure 17.23 shows the formation of a second peak in the accumulation mode as an airmass is advected off North America to the Atlantic and the Pacific Oceans. Note that the two modes observed in the number distribution should not be confused with modes of the

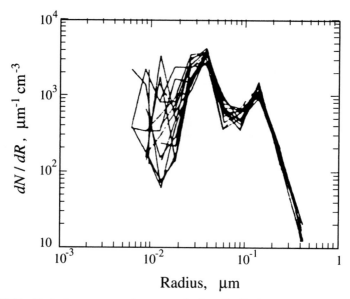

FIGURE 17.24   Typical remote marine aerosol size distributions. (Reprinted from *Atmos. Environ.* **24A**, Hoppel, W. A., and Frick, G. M., 645–649. Copyright 1990, with kind permission from Elsevier Science Ltd., The Boulevard, Langford Lane, Kidlington OX5 1GB, UK.)

mass distribution. Hoppel et al. (1986) proposed that cloud processing of aerosol is an efficient mechanism for accumulating mass in the 0.08–0.5 μm size range in the marine atmosphere. The gap observed in most remote marine aerosol number distributions (Figure 17.24) is, as a result, often referred to as the "Hoppel gap." These effects on the aerosol number distribution are expected to be important only in the remote atmosphere, where the number of CCN is a significant fraction of the total aerosol number. However, significant effects on the aerosol mass distribution are expected under all circumstances.

Measurements of the urban aerosol mass distribution have shown that two distinct modes often exist in the 0.1 to 1.0 μm diameter range (Hering and Friedlander 1982; McMurry and Wilson 1983; Wall et al. 1988; John et al. 1990). These are referred to as the *condensation mode* (approximate aerodynamic diameter 0.2 μm) and the *droplet mode* (aerodynamic diameter around 0.7 μm). These two submicrometer mass distribution modes have also been observed in nonurban continental locations (McMurry and Wilson 1983; Hobbs et al. 1985; Radke et al. 1989). Hering and Friedlander (1982) and John et al. (1990) proposed that the larger mode could be the result of aqueous-phase chemical reactions. Meng and Seinfeld (1994) showed that growth of condensation mode particles by accretion of water vapor or by gas-phase or aerosol-phase sulfate production cannot explain existence of the droplet mode. Activation of condensation mode particles, formation of cloud/fog drops, followed by aqueous-phase chemistry, and droplet evaporation were shown to be a plausible mechanism for formation of the aerosol droplet mode.

Simulations of the aerosol–cloud–aerosol cycle have shown that sulfate formed during cloud/fog processing of an airmass favors aerosol particles that have access to most of the cloud liquid water content, which are those with diameters in the 0.5–1.0 μm range (Pandis et al. 1990a). Figure 17.25 depicts simulation of fog processing of an urban aerosol population. Note that the shape of the aerosol distribution changes with the creation of an

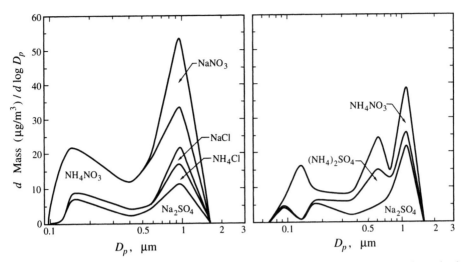

FIGURE 17.25 Predicted aerosol size–composition distributions before and after a fog episode (Pandis et al. 1990b).

extra mode, resulting mainly from the formation of $(NH_4)_2SO_4$. Note also that there are significant changes in the aerosol chemical composition before and after the fog with sulfates replacing nitrate and chloride salts.

17.7.4 Interstitial Aerosol Scavenging by Cloud Droplets

Interstitial aerosol particles collide with cloud droplets and are removed from cloud interstitial air. The coagulation theory of Chapter 13 can be used to quantify the rate and effects of such removal. If $n(D_p, t)$ is the aerosol number distribution and $n_d(D_p, t)$ the droplet number distribution at time t, the loss rate of aerosol particles per unit volume of air due to scavenging by cloud drops is governed by

$$-\frac{\partial n(D_p, t)}{\partial t} = n(D_p, t) \int_0^\infty K(D_p, x) n_d(x, t)\, dx \qquad (17.90)$$

where $K(D_p, x)$ is the collection coefficient for collisions between an interstitial aerosol particle of diameter D_p and a droplet of diameter x. One can define the scavenging coefficient $\Lambda(D_p, t)$ for the full droplet population:

$$\Lambda(D_p, t) = -\frac{1}{n(D_p, t)} \frac{\partial n(D_p, t)}{\partial t} = \int_0^\infty K(D_p, x) n_d(x, t)\, dx \qquad (17.91)$$

If the scavenging coefficient does not vary with time and is equal to $\Lambda(D_p)$, then the evolution of the number distribution is given by

$$n(D_p, t) = n(D_p, 0) \exp[-\Lambda(D_p)t]$$

Assuming for the time being that cloud droplets are stationary, then particles are captured by Brownian diffusion. The collection of particles by a falling drop will be

**TABLE 17.5   Estimated Lifetimes of Aerosols in a Nonraining Cloud
($N = 955\,\mathrm{cm}^{-3}$, $D_p = 10\,\mu\mathrm{m}$, $w_L = 0.5\,\mathrm{g\,cm}^{-3}$) at Standard Conditions**

D_p, μm	$K(D_p, 10\,\mu\mathrm{m})$, $\mathrm{cm}^{-3}\,\mathrm{s}^{-1}$	$\Lambda(D_p)$, s^{-1}	Lifetime $(= 1/\Lambda)$
0.002	4×10^{-5}	0.038	0.4 min
0.01	1.6×10^{-6}	1.5×10^{-3}	11 min
0.05	1.1×10^{-7}	1×10^{-4}	2.8 h
0.1	2.2×10^{-8}	2.1×10^{-5}	13 h
1.0	1.03×10^{-9}	9.8×10^{-7}	11.8 days
10.0	3×10^{-10}	2.9×10^{-7}	40 days

discussed when we consider wet deposition in Chapter 20. The collection coefficient $K(D_p, x)$ can then be estimated by (13.54). Let us estimate this collection rate assuming that the cloud has a liquid water content of $0.5\,\mathrm{g\,m}^{-3}$ and that all drops have diameters of $10\,\mu\mathrm{m}$, resulting in a number concentration of $N_d = 955\,\mathrm{cm}^{-3}$. For such a monodisperse droplet population (17.91) simplifies to

$$\Lambda(D_p) = N_d K(D_p, 10\,\mu\mathrm{m}) \tag{17.92}$$

and $K(D_p, 10\,\mu\mathrm{m})$ can be calculated from (13.54) (see also Figure 13.5 and Table 13.3). Corresponding particle collision efficiencies and lifetimes are shown in Table 17.5. Nuclei smaller than $10\,\mathrm{nm}$ will be scavenged in a few minutes in such a cloud whereas particles larger than $0.1\,\mu\mathrm{m}$ will not be collected by droplets during the cloud lifetime.

The above results indicate that for average residence times of air parcels in clouds (on the order of an hour) only the very fine aerosol particles will be collected by cloud droplets. These particles often represent a significant fraction of the aerosol number, so the total number may be reduced significantly, and the shape of the aerosol number distribution may change dramatically. However, these particles contain little of the mass and therefore the mass distribution effectively will not change. These qualitative conclusions are in agreement with detailed simulations of aerosol processing by clouds (Flossmann and Pruppacher 1988; Flossmann 1991).

17.7.5   Aerosol Nucleation Near Clouds

Enhanced aerosol number concentrations in the vicinity of clouds have been observed (Saxena and Hendler, 1983; Hegg et al., 1990, 1991; Radke and Hobbs, 1991; Perry and Hobbs, 1994; Weber et al., 2001). Assuming that nucleation in the vicinity of clouds in the free troposphere is a result of H_2SO_4–H_2O binary nucleation, two effects could be responsible for enhanced particle nucleation near cloud boundaries: (1) regions in the immediate vicinity of clouds have high relative humidity (Kerminen and Wexler, 1994; Lu et al., 2002, 2003); (2) high actinic fluxes near cloudtops resulting from upward scattering of solar radiation could lead to enhanced OH concentrations and rapid oxidation of SO_2 to H_2SO_4 (Hegg et al., 1991). Note that nucleation in the vicinity of clouds produces little aerosol mass but a large number of particles that could be a contributor to maintenance of aerosol number concentrations in the free troposphere.

17.8 OTHER FORMS OF WATER IN THE ATMOSPHERE

Our discussion in the preceding sections has focused on "warm" nonraining tropospheric clouds. Water in the atmosphere can also exist as ice, rain, snow, and so on. We summarize here aspects of the formation and removal of these water forms that are most associated with atmospheric chemistry. The interested reader is referred for more information to Pruppacher and Klett (1997) and references therein.

17.8.1 Ice Clouds

Atmospheric observations indicate that water readily supercools, and water clouds are frequently found in the atmosphere at temperatures below 0°C. Figure 17.26 shows that supercooled clouds are quite common in the atmosphere, especially if cloudtop temperature is warmer than −10°C. However, the likelihood of ice increases with decreasing temperature, and at −20°C only about 10% of clouds consist entirely of water drops. At these low temperatures, ice particles coexist with water drops in the same cloud.

The temperature dependence of the equilibrium between water vapor and ice can be described by the Clausius–Clapeyron equation. Following (17.7), one finds that

$$\frac{dp_{\text{sat},i}}{dT} = \frac{\Delta H_s}{T(v_v - v_i)} \tag{17.93}$$

where $p_{\text{sat},i}$ is the saturation vapor pressure of water over ice, ΔH_s is the molar enthalpy for ice sublimation, and v_i and v_v are the molar volumes of ice and water vapor. If we assume that $v_v \gg v_i$ and that water vapor behaves as an ideal gas, then

$$\frac{d \ln p_{\text{sat},i}}{dT} \simeq \frac{\Delta H_s}{RT^2} \tag{17.94}$$

Note that (17.94) is similar to the Clausius–Clapeyron equation for water vapor–water equilibrium (17.8), with the enthalpy of sublimation replacing the enthalpy of evaporation.

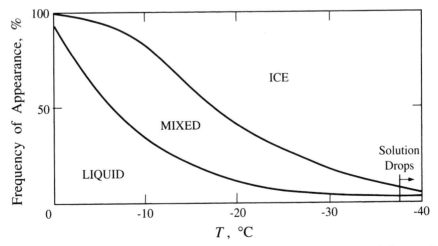

FIGURE 17.26 Average frequencies of appearance of supercooled water, mixed phase, and ice clouds as a function of temperature in layer clouds over Russia (Boronikov et al. 1963).

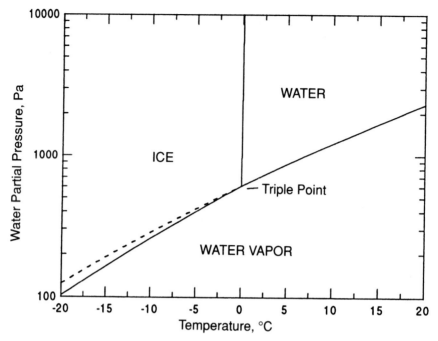

FIGURE 17.27   Pressure–temperature phase diagram for water. The dashed line corresponds to supercooled water and its metastable equilibrium with water vapor.

Finally, if the Clausius–Clapeyron equation is applied to the equilibrium between ice and water, we get

$$\frac{dp_m}{dT} = \frac{\Delta H_m}{T(v_w - v_i)} \tag{17.95}$$

where p_m is the melting pressure of ice, ΔH_m the molar enthalpy of melting, and v_w the molar volume of water. Equations (10.66), (17.94), and (17.95) can be integrated and plotted to produce the $p - T$ phase diagram for pure water shown in Figure 17.27. Note that there is only one point ($T = 0°C$, $p = 6.1$ mbar), the *water triple point* at which all phases coexist. Another interesting observation is that for temperatures below $0°C$ the water vapor pressure over liquid water is higher than water vapor pressure over ice:

$$p_{sat,w} > p_{sat,i} \qquad (T < 0°C)$$

So, if air is saturated with respect to ice, it is subsaturated with respect to water. As a result, supercooled water droplets cannot coexist in equilibrium with ice crystals. Figure 17.26 refers to the equilibrium of bulk water (curvature is neglected), without any impurities (zero solute concentration). Curvature and solute effects cause the behavior of water in the atmosphere to deviate significantly from Figure 17.27. We discuss these effects briefly below.

Freezing-Point Depression   Dissolution of a salt in water lowers the vapor pressure over the solution. A direct result of this is the depression of the freezing point of water. In order to quantify this change, assume that our system has constant pressure and contains air and an aqueous salt solution in equilibrium with ice. According to thermodynamic equilibrium the chemical potential of water in the aqueous and ice phases will be the same, $\mu_i = \mu_w$. If a_w is the activity of water in solution, then

$$\mu_i = \mu_w^\circ(T) + RT \ln a_w$$

Dividing by T and differentiating with respect to T under constant pressure p, we have

$$\left(\frac{\partial(\mu_i/T)}{\partial T}\right)_p = \left(\frac{\partial(\mu_w^\circ/T)}{\partial T}\right)_p + R\left(\frac{\partial \ln a_w}{\partial T}\right)_p \qquad (17.96)$$

But the chemical potential is related to the enthalpy by

$$\left(\frac{\partial(\mu_k/T)}{\partial T}\right)_p = -\frac{H_k}{T^2} \qquad (17.97)$$

and therefore

$$\left(\frac{\partial \ln a_w}{\partial T}\right)_p = \frac{H_w - H_i}{RT^2} = \frac{\Delta H_m}{RT^2} \qquad (17.98)$$

This equation describes the equilibrium between a salt solution and pure ice. This can be integrated as follows, noting that for pure water $a_w = 1$, and denoting the pure water freezing temperature by T_0, yields

$$\int_{a_w=1}^{a_w} d \ln a_w = \int_{T_0}^{T_e} \frac{\Delta H_m}{RT^2} dT \qquad (17.99)$$

assuming that ΔH_m is approximately constant in the temperature interval from T_0 to T_e:

$$T_0 - T_e = \frac{RT_0 T_e}{\Delta H_m}(-\ln a_w) \qquad (17.100)$$

Because $a_w \leq 1$, $T_0 - T_e \geq 0$ and the new freezing temperature T_e is lower than the pure water freezing temperature. The difference $\Delta T_f = T_0 - T_e$ is the equilibrium freezing point depression. To get an estimate of this depression, we can assume that the solution is ideal so that $a_w = x_w$ and that $T_0 T_e \simeq T_0^2$. Noting that $\ln a_w = \ln x_w \simeq -n_s/n_w$, we obtain the estimate

$$\boxed{\Delta T_f = \frac{RT_0^2 M_w}{1000 \, \Delta H_m} m \qquad (17.101)}$$

where m is the solution molarity. For ideal solutions this results in a depression of $1.86°C \, M^{-1}$ of solute, and for a salt that dissociates into two ions this results in a depression of $3.72°C\,M^{-1}$ of the salt. As a result of the nonideality of real salts, actual depression at 1 M concentration is $3.35°C$ for NaCl.

Curvature Effects Ice crystals in the atmosphere have a variety of shapes, with hexagonal prismatic being the basic one (Pruppacher and Klett 1997). For illustration purposes, let us ignore this complexity and concentrate on the behavior of a spherical ice particle of diameter D_p. Our analysis for water droplets and the Kelvin effect is directly applicable here and vapor pressure of water over the ice particle surface p_i is

$$p_i = p_{\text{sat},i} \exp\left(\frac{4M_w\sigma_{ia}}{RT\rho_i D_p}\right) \tag{17.102}$$

where $p_{\text{sat},i}$ is the vapor pressure over a flat ice surface, σ_{ia} the surface energy of ice in air, and ρ_i the ice density. Pruppacher and Klett (1997) reviewed theoretical and experimental values for σ_{ia} and suggested that it varies from $0.10–0.11\,N\,m^{-1}$. This surface tension that is higher than the water/air value results in relative vapor increases (p/p_{sat}) higher than those for a water droplet of the same size.

Pruppacher and Klett (1997) show that the freezing temperature decreases with decreasing size of the ice particle. This decrease becomes particularly pronounced for crystal diameters smaller than 20 nm and is further enhanced by the solute effect for solute concentrations larger than 0.1 M.

Ice Nuclei Ice particles can be formed through a variety of mechanisms. All of these require the presence of a particle, which is called an *ice nucleus* (IN). These mechanisms are (1) water vapor adsorption onto the IN surface and transformation to ice (deposition mode), (2) transformation of a supercooled droplet to an ice particle (freezing mode), and (3) collision of a supercooled droplet with an IN and initiation of ice formation (contact mode).

Formation of ice particles in the absence of IN is possible only at very low temperatures, below $-40°C$ (Hobbs 1995). The presence of IN allows ice formation at higher temperatures. Aerosols that can serve as IN are rather different from those that serve as CCN. Ice-forming nuclei are usually insoluble in water and have chemical bonding and crystallographic structures similar to ice. Larger particles are more efficient than smaller ones. While our understanding of the ice nucleating abilities of aerosols remains incomplete, particles that are known to serve as IN include dust particles (especially clay particles such as haolinite) and combustion particles (containing metal oxides). Experiments have shown that the ice-nucleating active fraction of an aerosol population increases with decreasing temperature.

IN concentrations in the atmosphere are quite variable. A proposed empirical relation for their concentration as a function of temperature is (Pruppacher and Klett 1997)

$$IN(L^{-1}) = \exp[0.6(253 - T)] \tag{17.103}$$

This relation suggests that ice nucleation in the atmosphere is a very selective process. For example, at temperatures as low as $-20°C$ the atmosphere typically contains $1\,IN\,L^{-1}$ and 10^6 particles L^{-1}. So, at most, one out of a million particles can serve as an IN.

Measurements of ice particle concentrations for cloud top temperatures below $-10°C$ have given concentrations varying from 0.1 to $200\,L^{-1}$. For temperatures below $-20°C$, the concentrations vary from 10 to $300 \, L^{-1}$. This number of ice crystals observed in clouds often exceeds by several orders of magnitude the IN concentration. The enhancement of ice particle concentration over the IN concentrations has been addressed (Rangno and Hobbs 1991). Proposed explanations include breakup of primary ice particles, ice splinter production during droplet freezing, and unusually high supersaturations. The reader is referred to Mason (1971), Pruppacher and Klett (1997), and Rogers and De Mott (1991) for further discussion of the nature, origins, and concentrations of atmospheric IN.

17.8.2 Rain

For a cloud to generate precipitation, some drops need to grow to precipitable size around 1 mm. Growth of droplets can proceed via a series of mechanisms: (1) water vapor condensation, (2) droplet coalescence, and (3) ice processes.

We have already examined the growth of droplets by water vapor condensation. While this is a very effective mechanism for the initial growth of aerosol particles from a fraction of a micrometer to 10 μm in a few minutes, further growth of a drop is a slow process. Figure 17.14 demonstrates that an hour is necessary under a constant supersaturation of 1% for drop growth to 100 μm. After an entire hour the drop has collected only 0.1% of the water mass of an average raindrop.

For warm clouds (i.e., containing no ice), large drops can grow to precipitation size by collecting smaller droplets that lie in their fallpath. Therefore the largest drops in the cloud drop spectrum are of particular importance because they are the ones that initiate precipitation. The typical concentration of large drops required to initiate precipitation is one per liter of air, or only one out of a million cloud drops. These large drops form on giant particles, which are again the one in a million nuclei. The dramatic dependence of the precipitation on one out of a million droplets is indicative of the difficulty of quantitatively describing the overall process. Clearly a broad cloud drop size distribution is more conducive to precipitation development than is a narrow distribution. Remote marine clouds tend to have broad size distribution and tend to produce precipitation more effectively than similar continental clouds.

Larger drops fall faster than smaller drops, resulting in the larger drops overtaking the smaller drops, colliding and coalescing with them. This process is growth by accretion, sometimes also called gravitational coagulation, coalescence, or collisional growth.

The collision process is illustrated in Figure 17.28 for viscous flow around a sphere of diameter D_p. As the large droplet approaches small drops of diameter d_p, the viscous forces exerted by the flow field around the large droplet push the droplets away from the center of flow, modifying their trajectories. In Figure 17.28 droplet b is collected by the raindrop while droplet a is not. Therefore the falling raindrop will in general collect fewer drops than those existing in the cylinder of diameter D_p below it. Droplets in the cylinder of diameter y will be collected. The distance y is defined by the grazing trajectory a and is a function of the raindrop size D_p and drop size d_p. One defines the collision efficiency E as the ratio of the actual collision cross section to the geometric cross section, or

$$E = \frac{y^2}{\left(D_p + d_p\right)^2} \tag{17.104}$$

Note that all drops of diameter d_p in the cylinder with diameter y, below the falling drop with diameter D_p, will be collected by it. Because small drops tend to move away from the falling raindrop, E is expected to be smaller than unity for most cases.

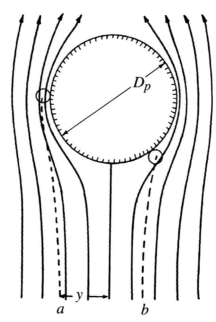

FIGURE 17.28   Schematic of the flow around a falling drop. The dashed lines are the trajectories of small drops considered as mass points. Trajectory a is a grazing trajectory, while b is a collision trajectory.

Figure 17.29 shows theoretically estimated collision efficiencies among drops as functions of the radii of the small and large drops. There is a rapid increase in collision efficiency as the two drops approach equal size. This is a result of fluid mechanical interactions that accelerate the upper drop more than the lower one but have little importance for the atmosphere where the probability of collision of equal-sized drops is extremely small. Figure 17.29 shows that drops with diameters below 50 μm have a small collision efficiency and growth of drops to larger size is required for the acceleration of accretional growth.

Calculations for larger drops are complicated by phenomena such as shape deformation, wake oscillations, and eddy shedding, making theoretical estimates of E difficult. The overall process of rain formation is further complicated by the fact that drops on collision trajectories may not coalesce but bounce off each other. The principal barrier to coalescence is the cushion of air between the two drops that must be drained before they can come into contact. An empirical coalescence efficiency E_c suggested by Whelpdale and List (1971) to address droplet bounce-off is

$$E_c = \left(\frac{D_p}{D_p + d_p} \right)^2 \tag{17.105}$$

which is applicable for $D_p > 400\,\mu m$. Equation (17.105) can probably be viewed as an upper limit for E_c. Note that while (17.104) describes the area swept by a falling drop, (17.105) quantifies the probability of coalescence of two colliding droplets.

The growth of a falling drop of mass m as a result of accretion of drops can be described by

$$\frac{dm}{dt} = \frac{\pi}{4} E_t (D_p + d_p)^2 w_L (v_D - v_d) \tag{17.106}$$

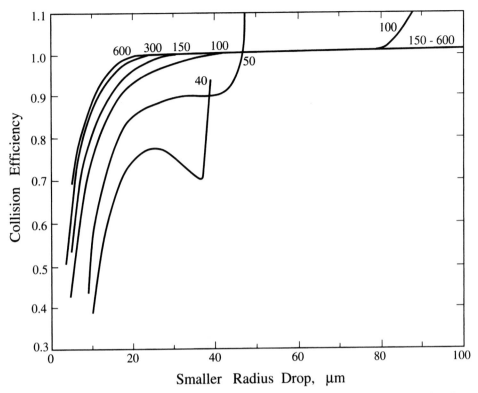

FIGURE 17.29 Theoretically estimated collision efficiencies between cloud droplets as functions of their radii. The radius of the smaller drop is on the x axis and the radius of the larger drop is on the curve. (From H. G. Houghton, *Physical Meteorology*, The MIT Press, Cambridge, MA, 1985, p. 265.)

where E_t is the overall accretion efficiency (collision efficiency times coalescence efficiency, $E_t = EE_c$), w_L is the liquid water content of the small drops, and v_D and v_d are, respectively, the fall speeds of the larger and smaller particles. Equation (17.106) is called the *continuous-accretion equation* and assumes that the smaller drops are uniformly distributed in space. The description of the size change of droplets falling through a cloud by (17.106) implicitly assumes that all droplets of the same size will grow in the same way. According to the continuous equation, if several drops with the same initial diameter fall through a cloud, they would maintain the same size at all times. In reality, each droplet–droplet collision is a discrete event. Droplets that collide first with others grow faster than do droplets that had initially the same size. As a result, the simple continuous equation can seriously underestimate the collision frequency and the growth rate of falling drops, especially when the collector and collected drops have the same size. The coagulation equation [see (13.59)] is a better mathematical description of these discrete collisions. Equation (13.59) (often called the *stochastic accretion equation* in cloud microphysics) describes implicitly individual collisions and predicts that even if all falling droplets initially had the same size, some will grow more than others.

The role of ice in rain formation was first addressed by Bergeron in 1933, based on the calculations of Wegener. Using thermodynamics, Wegener showed in 1911 that at temperatures below 0°C supercooled water drops and ice crystals cannot exist in equilibrium. Using this result, Bergeron proposed that in cold clouds the ice crystals grow by vapor diffusion at the expense of the water droplets until either all drops have been consumed or all ice crystals have fallen out of the cloud as precipitation. Findeisen later produced additional observations supporting the mechanism described above, which is often called the Wegener–Bergeron–Findeisen mechanism. Mathematically the description of the mechanism requires solution of the growth equations by vapor diffusion for both ice crystals and water drops in a supersaturated environment.

Raindrop Distributions   A number of empirical formulas have been proposed for the raindrop spectrum. The distribution proposed by Best is often used to describe the fraction of rainwater comprised of raindrops smaller than D_p, $F(D_p)$

$$F(D_p) = 1 - \exp\left[-\left(\frac{D_p}{1.3\,p_0^{0.232}}\right)^{2.25}\right] \tag{17.107}$$

where p_0 is the rainfall intensity in $(\mathrm{mm\,h}^{-1})$ and D_p is in mm.

Probably the most widely used is the Marshall–Palmer (MP) distribution, where

$$n(D_p) = n_0 \exp(-\psi D_p) \tag{17.108}$$

where $n(D_p) = dn/dD_p$ is the number distribution in drops $\mathrm{m}^{-3}\,\mathrm{mm}^{-1}$, $n_0 = 8000\,\mathrm{m}^{-3}\,\mathrm{mm}^{-1}$, and $\psi = 4.1\,p_0^{-0.21}\,\mathrm{mm}^{-1}$. The MP distribution is often not sufficiently general to describe observed rain spectra. Sekhon and Srivastava (1971) among others noted that n_0 is not a constant but rather depends on the rainfall intensity. They suggested use of $n_0 = 7000\,p_0^{0.37}\,\mathrm{m}^{-3}\,\mathrm{mm}^{-1}$ and $\psi = 3.8\,p_0^{-0.14}\,\mathrm{mm}^{-1}$. Raindrop spectra and a fitted MP distribution with variable n_0 are shown in Figure 17.30. The MP distribution may overestimate by as much as 50% the number of small droplets in the 0.02–0.12 cm range and it is strictly applicable for $D_p \geq 0.12\,\mathrm{cm}$.

APPENDIX 17   EXTENDED KÖHLER THEORY

Derivation of the Köhler equation is based on a combination of two expressions: the Kelvin equation, which governs the increase in water vapor pressure over a curved surface; and modified Raoult's law, which describes the water equilibrium over a flat solution:

$$\frac{p_w(D_p)}{x_w \gamma_w p_w^{\mathrm{sat}}} = \exp\left(\frac{4M_w \sigma_w}{RT\rho_w D_p}\right) \tag{17.A.1}$$

The total volume of the droplet is given by (17.19)

$$\frac{\pi}{6}D_p^3 = n_w \bar{v}_w + v_s n_s \bar{v}_s \tag{17.A.2}$$

where v_s is the number of ions generated by the dissociation of a molecule of salt that forms the cloud condensation nucleus. Using

$$\frac{1}{x_w} = 1 + \frac{v_s n_s}{n_w} = 1 + \frac{v_s n_s \bar{v}_w}{\dfrac{\pi}{6}D_p^3 - v_s n_s \bar{v}_s} \tag{17.A.3}$$

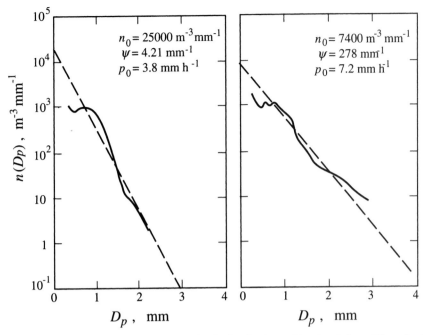

FIGURE 17.30 Raindrop spectra and fitted MP distributions. (Reprinted from *Microphysics of Clouds and Precipitation*, Pruppacher, H. R., and Klett, J. D., 1997, with kind permission from Kluwer Academic Publishers.)

we get (17.26):

$$\ln\left(\frac{p_w(D_p)}{p^\circ}\right) = \frac{4M_w\sigma_w}{RT\rho_w D_p} - \frac{6v_s n_s M_w}{\pi\rho_w D_p^3} \tag{17.A.4}$$

We now define the saturation ratio S and the two constants A_w and B_s as

$$S = \frac{p_w(D_p)}{p^\circ} \qquad A_w = \frac{4M_w\sigma_w}{RT\rho_w} \qquad B_s = \frac{6v_s n_s M_w}{\pi\rho_w}$$

and then arrive at the Köhler equation (17.27):

$$\ln S = \frac{A_w}{D_p} - \frac{B_s}{D_p^3} \tag{17.A.5}$$

The subscripts w and s will come in handy as we add more phenomena.

17.A.1 Modified Form of Köhler Theory for a Soluble Trace Gas

Let us consider how droplet activation changes when a soluble trace gas is present (Laaksonen et al. 1998). We can first rewrite (17.A.1) as

$$S_w = x_w \gamma_w \exp\left(\frac{A_w}{D_p}\right) \tag{17.A.6}$$

When a soluble gas is present, an equation analogous to (17.A.6) governs the gas equilibrium

$$S_a = x_a \gamma_a \exp\left(\frac{A_a}{D_p}\right) \tag{17.A.7}$$

where A_a is defined analogously as A_w, and the subscript a refers to the soluble gas:

$$A_a = \frac{4M_a \sigma}{RT \rho_a} \tag{17.A.8}$$

To approximate the right-hand side (RHS) of (17.A.7), the activity coefficient of a completely dissociated gas is given by

$$x_a \gamma_a = \frac{(x_a \gamma_\pm)^{(\nu_+ + \nu_-)} \nu_+^{\nu_+} \nu_-^{\nu_-}}{p_a^{\text{sat}} H_a} \tag{17.A.9}$$

where $\gamma_\pm$ is the mean activity coefficient; ν_+ and ν_- are the number of cations and anions produced by a molecule of dissolved gas, respectively; and H_a is the Henry's law constant of the gas. We again make the assumption of dilute solution, for which $\gamma_\pm$ is unity. We also specify that the soluble gas dissociates into exactly one cation and one anion, such as in the case of HNO_3 and HCl. Now, to write the left-hand side (LHS) of (17.A.7), we express S_a explicitly as the ratio of the partial pressure of the soluble gas to its saturation pressure:

$$S_a = \frac{p_a}{p_a^{\text{sat}}}$$

The partial pressure of the soluble gas can be expressed in terms of its molar concentration, N_a (moles per volume of air) by

$$p_a = N_a RT$$

The amount of soluble gas that is dissolved in droplets per unit volume of air can be expressed as Cn_a, where C is the number of droplets per unit volume of air and n_a is the moles of soluble gas per droplet. If the total amount of available soluble gas per unit volume of air is expressed as $n_{\text{tot}}C$, where n_{tot} is the total moles of soluble gas available per droplet, then the soluble gas remaining in the gas phase at any time is $(n_{\text{tot}} - n_a)C$, and

$$S_a = \frac{(n_{\text{tot}} - n_a)CRT}{p_a^{\text{sat}}} \tag{17.A.10}$$

Combining (17.A.7), (17.A.9), and (17.A.10), we get

$$\frac{(n_{\text{tot}} - n_a)CRT}{p_a^{\text{sat}}} = \frac{x_a^2}{p_a^{\text{sat}} H_a} \exp\left(\frac{A_a}{D_p}\right)$$

or

$$x_a^2 = H_a CRT(n_{\text{tot}} - n_a) \exp\left(-\frac{A_a}{D_p}\right) \tag{17.A.11}$$

The exponential term is approximately unity for $D_p \geq 0.2\,\mu\text{m}$; since cloud droplets exceed this size easily, we will take this term as unity. For a dilute drop, we can approximate the aqueous-phase mole fraction of soluble gas as $x_a \approx n_a/n_w$, giving

$$n_a^2 = n_w^2 H_a CRT(n_{\text{tot}} - n_a) \tag{17.A.12}$$

We can then write the number of moles of water in the drop as

$$n_w = \frac{m_w}{M_w} = \frac{\pi D_p^3 \rho_w}{6M_w}$$

$$= \frac{\nu_s n_s D_p^3}{B_s} = \frac{\nu_a n_a D_p^3}{B_a} \tag{17.A.13}$$

where B_a is defined analogously to B_s as

$$B_a = \frac{6\nu_a n_a M_w}{\pi \rho_w} \tag{17.A.14}$$

In order to simplify (17.A.12), we will define β as

$$\beta = H_a RT \left(\frac{\pi \rho_w}{6M_w}\right)^2 \tag{17.A.15}$$

and (17.A.12) can then be rewritten in terms of this β:

$$n_a^2 = C\beta D_p^6 (n_{\text{tot}} - n_a)$$

$$0 = n_a^2 + C\beta D_p^6 n_a - C\beta D_p^6 n_{\text{tot}} \tag{17.A.16}$$

The positive root of this quadratic equation is

$$n_a = \frac{C\beta D_p^6}{2}\left(-1 + \sqrt{1 + \frac{4n_{\text{tot}}}{C\beta D_p^6}}\right) \tag{17.A.17}$$

or

$$n_a = \frac{2n_{\text{tot}}}{1 + \sqrt{1 + \dfrac{4n_{\text{tot}}}{C\beta D_p^6}}} \tag{17.A.18}$$

We now turn to (17.A.6). We write the mole fraction of water as

$$x_w = \frac{n_w}{n_w + \nu_s n_s + \nu_a n_a}$$

$$\approx 1 - \frac{\nu_s n_s + \nu_a n_a}{n_w} \tag{17.A.19}$$

Recalling the expression for n_w from (17.A.13), we rewrite x_w as

$$x_w = 1 - \frac{B_s}{D_p^3} - \frac{B_a}{D_p^3} \qquad (17.A.20)$$

Substituting this back into (17.A.6), and again assuming that $\gamma_w \approx 1$, we obtain

$$S_w = \left(1 - \frac{B_s}{D_p^3} - \frac{B_a}{D_p^3}\right) \exp\left(\frac{A_w}{D_p}\right) \qquad (17.A.21)$$

The exponential term can then be expanded in a Taylor series. If we collect the lowest-order terms of D_p, we then have

$$S_w = 1 + \frac{A_w}{D_p} - \frac{B_s}{D_p^3} - \frac{B_a}{D_p^3} \qquad (17.A.22)$$

In this expression for saturation ratio, the second term describes the Kelvin effect, while the third term describes the solute effect as in the classical form of the Köhler equation. The fourth term is the added one that accounts for the dissolved gas. In evaluating S_w, we keep in mind that B_a is defined in terms of n_a, which is given by (17.A.18) such that B_a is also a function of D_p.

If the soluble gas concentration in the vapor phase is maintained at a constant value, that is, is not depleted as a result of absorption into the droplets, then the foregoing result simplifies. Equation (17.A.11) becomes

$$x_a^2 = p_a H_a \exp\left(-\frac{A_a}{D_p}\right) \qquad (17.A.23)$$

where, as above, we can neglect the exponential term for D_p greater than $\sim 0.2\,\mu m$. And approximating x_a as n_a/n_w, we get

$$\frac{n_a}{n_w} = \sqrt{p_a H_a} \qquad (17.A.24)$$

Using (17.A.24), the form of the Köhler equation for a constant soluble gas concentration is

$$S_w = 1 + \frac{A_w}{D_p} - \frac{B_s}{D_p^3} - v_a\sqrt{p_a H_a} \qquad (17.A.25)$$

In summary, the gaseous solute effect term, B_a/D_p^3, in (17.A.22) is replaced by $v_a\sqrt{p_a H_a}$ when the soluble gas concentration in the vapor is constant.

17.A.2 Modified Form of the Köhler Theory for a Slightly Soluble Substance

We take an approach similar to that above; here

$$x_w = \frac{n_w}{n_w + v_s n_s + v_a n_a + v_{ss} n_{ss}}$$

$$\approx 1 - \frac{v_s n_s + v_a n_a + v_{ss} n_{ss}}{n_w} \qquad (17.A.26)$$

where the subscript ss refers to the slightly soluble substance. Since now the number of moles of water in the drop is written to exclude the volume of the insoluble substance, we obtain

$$n_w = \frac{\pi(D_p^3 - D_{ss}^3)\rho_w}{6M_w}$$

$$= \frac{v_s n_s(D_p^3 - D_{ss}^3)}{B_s} = \frac{v_a n_a(D_p^3 - D_{ss}^3)}{B_a}$$

(17.A.27)

The corresponding water mole fraction becomes

$$x_w = 1 - \frac{B_s}{(D_p^3 - D_{ss}^3)} - \frac{B_a}{(D_p^3 - D_{ss}^3)} - \frac{v_{ss} n_{ss}}{n_w}$$

(17.A.28)

To calculate the number of moles of the slightly soluble species that actually dissolves into the aqueous phase, we use the solubility Γ, expressed in kg of the substance per kg of water:

$$n_{ss} = \frac{M_w n_w \Gamma}{M_{ss}}$$

(17.A.29)

Thus, we write the concentration of the slightly soluble substance in the aqueous phase C_{ss} as

$$C_{ss} = \frac{v_{ss} n_{ss}}{n_w} = \frac{v_{ss} M_w \Gamma}{M_{ss}}$$

(17.A.30)

This expression for C_{ss} is valid only when a portion of the slightly soluble core has been dissolved. When the entire core has dissolved into the droplet, then

$$C_{ss} = \frac{v_{ss} M_w \Gamma n_w^0}{M_{ss} n_w}$$

(17.A.31)

where n_w^0 refers to the number of moles of water required to dissolve the core. Finally, we have

$$x_w = 1 - \frac{B_s}{(D_p^3 - D_{ss}^3)} - \frac{B_a}{(D_p^3 - D_{ss}^3)} - C_{ss}$$

(17.A.32)

and thus we get an equation analogous to (17.A.22):

$$S_w = 1 + \frac{A_w}{D_p} - \frac{B_s}{(D_p^3 - D_{ss}^3)} - \frac{B_a}{(D_p^3 - D_{ss}^3)} - C_{ss}$$

(17.A.33)

This expression is similar to (17.A.22) except for the new final term that accounts for the extra solute effect from the slightly soluble substance, and all terms involving of D_p^3 in (17.A.22) are replaced by $(D_p^3 - D_{ss}^3)$, reflecting the subtracted volume owing to the undissolved core. This replacement also holds in the expression for n_a as in (17.A.18): D_p^6 is replaced by $(D_p^3 - D_{ss}^3)^2$. Note that the case of completely insoluble core is included as the limiting case of this equation where $C_{ss} = 0$.

17.A.3 Modified Form of the Köhler Theory for a Surface-Active Solute

Because surface-active compounds are generally organic solutes, we can modify (17.A.33) to include the effects of surface-active, slightly soluble material (Shulman et al. 1996). Note that this correction affects only the Kelvin term. First, we assume that the mixture surface tension can be approximated as the mole fraction weighted average of the surface tensions of pure water and pure organic solute:

$$\sigma_{mix} = x_w \sigma_w + (1 - x_w)\sigma_{ss}$$
$$= x_w(\sigma_w - \sigma_{ss}) + \sigma_{ss} \tag{17.A.34}$$

We can then use this mixture surface tension σ_{mix} in place of σ_w in the A_w term. The resulting expression can be written in shorthand as

$$S_w = \frac{4M_w}{RT\rho_w D_p}(x_w(\sigma_w - \sigma_{ss}) + \sigma_{ss}) + x_w \tag{17.A.35}$$

where x_w is given by (17.A.32).

17.A.4 Examples

We consider first the effect of a soluble trace gas on droplet activation, in this case HNO_3. The Köhler curves are generated with (17.A.22); these are shown in Figure 17.A.1 for a system in which the HNO_3 is depleted and one in which HNO_3 is held at constant concentration, both with an initial HNO_3 mixing ratio of 10 ppb. The salt assumed to be present is $(NH_4)_2SO_4$ with a dry diameter of 0.05 μm.

We see that in the case of constant HNO_3 concentration, the resulting Köhler curve has a shape similar to that of the original curve in the absence of soluble gas. The critical saturation ratio is reduced by about 0.006, translating to easier droplet activation. The case in which the HNO_3 is depleted exhibits a similar behavior as the droplet begins to take on water, but as the HNO_3 in the gas phase is depleted, the curve approaches the classical curve. The result is a Köhler curve with two maxima and a local minimum. The barrier for activation in this case is set by the second, higher maximum, because after S reaches the first peak, the equilibrium droplet size is given by the point at the same supersaturation on the other ascending branch. As S increases further, the droplet finally activates at the second maximum. Note that if S were to be decreased from some point on the second ascending branch, the change in droplet diameter would be the reverse—it would reach the local minimum, and jump over to the left branch, thereby bypassing the local maximum. In general, the soluble trace gas, regardless of whether its concentration changes or remains constant, helps to accelerate activation by lowering the critical saturation ratio. The higher the soluble gas concentration, the greater the decrease in S_c.

Figure 17.A.2 shows another set of curves similar to those in Figure 17.A.1, but with a larger $(NH_4)_2SO_4$ dry diameter of 0.1 μm. Here we note that with a sufficiently large dry

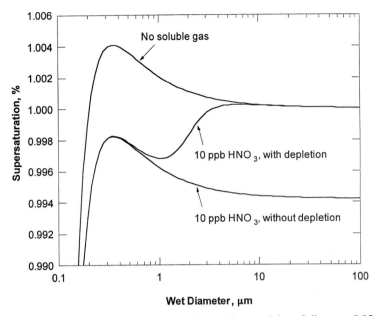

FIGURE 17.A.1 Köhler curves for dry ammonium sulfate particles of diameter 0.05 μm in the absence and presence of HNO₃ at 20°C and 1 atm. Ammonium sulfate properties are $v_s = 3$ ions molecule^{-1} and $\rho_s = 1770 \, \text{kg m}^{-3}$. Droplet number concentration $C = 1000 \, \text{cm}^{-3}$. Surface tension of water $\sigma_w = 0.073 \, \text{N m}^{-1}$.

diameter, a local minimum does not appear in the case in which the HNO₃ is depleted, and the particle growth occurs normally. In both cases described above, when the concentration of HNO₃ is held constant, the critical diameter is not changed from that in the absence of soluble gas. When depletion occurs, however, a larger critical diameter results. In this case, larger particles can exist in equilibrium without being activated.

Now we investigate the effect on the Köhler curves of slightly soluble organic matter in the particle core in both presence and absence of HNO₃. We choose adipic acid (C₆H₁₀O₄) as the slightly soluble species because of its moderate solubility and ability to lower surface tension. First, we ignore the effect of adipic acid on the surface tension of the solution and approximate the surface tension as that of pure water. Figure 17.A.3 shows the modified curves for the slightly soluble adipic acid at various HNO₃ concentrations, as predicted by (17.A.33). Again, the general effect is a downward shift of the Köhler curve, which translates to accelerated activation at a lower saturation ratio. A sharp minimum occurs, which is produced by the final dissolution of the slightly soluble core. If a larger core of adipic acid exists initially, say, 0.5 μm, as in Figure 17.A.4, the sharp minimum occurs at a later stage, as expected. Since the minimum now occurs at a larger diameter, the effect of the HNO₃ depletion that precedes it can be seen in the form of a smooth minimum. For a sufficiently high HNO₃ concentration, the effect of HNO₃ depletion dominates and masks the smaller sharp minimum. To compare the new critical parameters owing to the slightly soluble organic for the case with no HNO₃, we note that from Figure 17.A.3 (smaller initial core), $S_c = 1.0025$ and $D_{pc} = 0.2 \, \mu\text{m}$, while that from Figure 17.A.4 (larger initial core), $S_c = 1.001$ and $D_{pc} = 0.6 \, \mu\text{m}$. Thus, a larger amount of the slightly soluble matter decreases the critical saturation ratio, while the critical diameter is increased mainly because the initial particle size is larger.

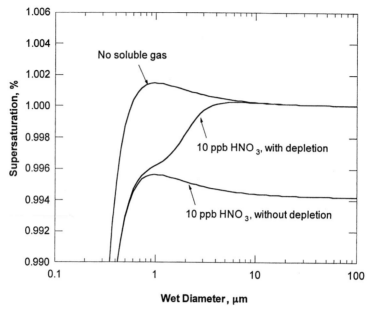

FIGURE 17.A.2   Same conditions as Figure 17.A.1 except dry diameter of ammonium sulfate particles in 0.1 μm.

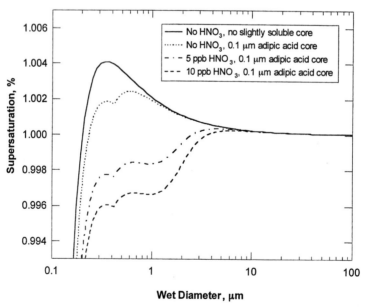

FIGURE 17.A.3   Köhler curves for particles consisting of a 0.1 μm adipic acid core coated with an equivalent of 0.05 μm ammonium sulfate with 0, 5, and 10 ppb of HNO_3 at 20°C and 1 atm. For adipic acid, a slightly soluble C_6 dicarboxylic acid, the following parameters are used: $\nu_{ss} = 1$ ions molecule^{-1}, $M_{ss} = 0.146$ kg mol^{-1}, $\rho_{ss} = 1360$ kg m^{-3} (Cruz and Pandis 2000), and $\Gamma = 0.018$ kg kg water^{-1} (Raymond and Pandis 2002). Surface tension is taken as that of water.

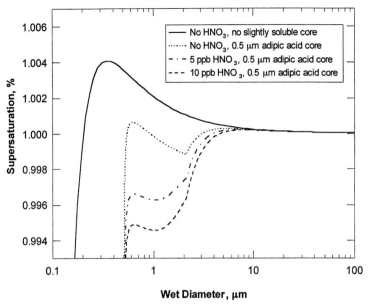

FIGURE 17.A.4   Same conditions as Figure 17.A.3 except 0.5 μm adipic acid core.

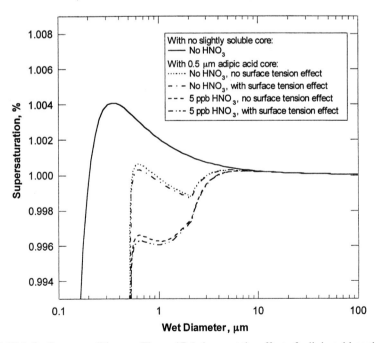

FIGURE 17.A.5   Same conditions as Figure 17.A.4 except the effect of adipic acid on the surface tension of the solution is accounted for. The surface tension of the aqueous solution of adipic acid is calculated by interpolation using experimental data (Shulman et al. 1996) as a function of the molar concentration of adipic acid in the droplet.

Finally, we take into account the effect of decreased surface tension as a result of the dissolved adipic acid. The effective surface tension for adipic acid is calculated from literature data (Shulman et al. 1996), and we use (17.A.35) to generate the curves in Figure 17.A.5. Once more, the curve is shifted down (activation is accelerated) as a result of the decreased surface tension. We also note that the surface-active solute effect lowers the critical S appreciably only when there is very little or no soluble gas. This is because in the presence of sufficient soluble gas, the S_c is determined by the second maximum rather than the first, and the downward shift of this point owing to surface tension effects is quite small.

PROBLEMS

17.1$_A$ Show equations (17.39) and (17.40).

17.2$_A$ Lawrence (2005) proposed the following simple formula for the dewpoint of an air parcel:

$$T_d \cong T - 0.2(100 - \text{RH})$$

where T_d and T are in °C and RH is in percent. Compare the results of this equation with those of (17.43) for $T = 15\,°C$ and RH varying from 50 to 100%.

17.3$_B$ Compare the values predicted by (Lawrence 2005)

$$h_{\text{LCL}} = (20 + 0.2\,T)(100 - \text{RH})$$

where T is in °C with (17.49) for $T = 20\,°C$ and RH $= 70\%$.

17.4$_A$ You dissolve 5 g of NaCl in a glass containing $200\,\text{cm}^3$ of water. The glass is in a room with constant temperature equal to 20°C and relative humidity 80%. Calculate the volume of water that will be left in the glass after several days of residence in this environment. Repeat the calculation for a relative humidity of 95%.

17.5$_A$ Your grandmother and grandfather are upset because it has just started raining. After listening to the weather prediction for a cloudy day but without rain, they planned to spend the day working in the garden. They have started criticizing the local weather forecaster who, despite the impressive gadgets (radar maps, 3D animated maps), still cannot reliably predict if it will rain tomorrow. They turn to you and ask why, if we can send people to the Moon, we still cannot tell whether it is going to rain. Explain, avoiding scientific terminology.

17.6$_B$ What is the activation diameter at 0.3% supersaturation for particles consisting of 50% $(NH_4)_2SO_4$, 30% NH_4NO_3, and 20% insoluble material?

17.7$_C$ A 100-nm-diameter dry $(NH_4)_2SO_4$ particle is suddenly exposed to an atmosphere with a 0.3% water supersaturation. How long will it take for the particle to reach a diameter of 5 µm? Assume a unity accommodation coefficient.

17.8$_C$ Repeat Problem 17.7 neglecting the correction to the diffusion coefficients and thermal conductivities for noncontinuum effects (assume that $D'_v = D_v, k'_a = k_a$). Compare with Problem 17.7 and discuss your observations.

17.9$_C$ Repeat Problem 17.7 assuming that $\alpha_c = 0.045$. Compare with Problem 17.7 and discuss your observations.

17.10$_A$ Dry activation diameters of roughly 0.5 μm have been observed in fogs in urban areas.

    a. If the particles consisted entirely of $(NH_4)_2SO_4$, what would be the maximum supersaturation inside the fog layer?

    b. Assuming that the maximum supersaturation was 0.05% and the particles contained $(NH_4)_2SO_4$ and insoluble material, calculate the mass fraction of the insoluble material.

REFERENCES

Abbatt, J. P. D., Broekhuizen, K., and Kumal, P. P. (2005) Cloud condensation nucleus activity of internally mixed ammonium sulfate/organic acid aerosol particles, *Atmos. Environ.* **39**, 4767–4778.

Bilde, M., and Svenningsson, B. (2004) CCN activation of slightly soluble organics: The importance of small amounts of inorganic salt and particle phase, *Tellus B* **56**, 128–134.

Collett, J. L. Jr., Bator, A., Rao, X., and Demoz, B. (1994) Acidity variations across the cloud drop size spectrum and their influence on rates of atmospheric sulfate production, *Geophys. Res. Lett.* **21**, 2393–2396.

Conant, W. C., VanReken, T. M., Rissman, T. A., Varutbangkul, V., Jonsson, H. H., Nenes, A., Jimenez, J. L., Delia, A. E., Bahreini, R., Roberts, G. C., Flagan, R. C., and Seinfeld, J. H. (2004) Aerosol-cloud drop concentration closure in warm cumulus, *J. Geophys. Res.* **109**, D13204 (doi:10.1029/2003JD004324).

Corrigan, C. E., and Novakov, T. (1999) Cloud condensation nucleus activity of organic compounds: A laboratory study, *Atmos. Res.* **33**, 2661–2668.

Cruz, C., and Pandis, S. N. (1997) The CCN activity of secondary organic aerosol compounds, *Atmos. Environ.* **31**, 2205–2214.

Cruz, C. N., and Pandis, S. N. (2000) Deliquescence and hygroscopic growth of mixed inorganic-organic atmospheric aerosol, *Environ. Sci. Technol.* **34**, 4313–4319.

Daum, P. H., Schwartz, S. E., and Newman, L. (1984) Acidic and related constituents in liquid water stratiform clouds, *J. Geophys. Res.* **89**, 1447–1458.

Dennis, R. L., McHenry, J. N., Barchet, W. R., Binkowski, F. S., and Byun, D. W. (1993) Correcting RADM sulfate underprediction—discovery and correction of errors and testing the correction through comparisons with field data, *Atmos. Environ.* **27**, 975–997.

Develk, J. P. (1994) A model for cloud chemistry—a comparison between model simulations and observations in stratus and cumulus, *Atmos. Environ.* **28**, 1665–1678.

Flossmann, A. I. (1991) The scavenging of two different types of marine aerosol particles calculated using a two dimensional detailed cloud model, *Tellus* **43B**, 301–321.

Flossmann, A. I., Hall, W. D., and Pruppacher, H. R. (1985) A theoretical study of the wet removal of atmospheric pollutants. Part I: The redistribution of aerosol particles captured through nucleation and impaction scavenging by growing cloud drops, *J. Atmos. Sci.* **42**, 583–606.

Flossmann, A. I., Hall, W. D., and Pruppacher, H. R. (1987) A theoretical study of the wet removal of atmospheric pollutants. Part II: The uptake and redistribution of $(NH_4)_2SO_4$ particles and SO_2 gas simultaneously scavenged by growing cloud drops, *J. Atmos. Sci.* **44**, 2912–2923.

Flossmann, A. I. and Pruppacher, H. R. (1988) A theoretical study of the wet removal of atmospheric pollutants 3. The uptake, redistribution, and deposition of $(NH_4)_2SO_4$ particles by a convective cloud using a two-dimensional cloud dynamics model, *J. Atmos. Sci.* **45**, 1857–1871.

Frick, G. M., and Hoppel, W. A. (1993) Airship measurements of aerosol size distributions, cloud droplet spectra, and trace gas concentrations in the marine boundary layer, *Bull. Am. Meteorol. Soc.* **74**, 2195–2202.

Ghan, S. J. et al. (2006) Use of in situ cloud condensation nuclei, extinction, and aerosol size distribution measurements to test a method for retrieving cloud condensation nuclei profiles from surface measurements, *J. Geophys. Res.* **111**, D05S10 (doi: 10.1029/2004JD005752).

Hegg, D. A. (1985) The importance of liquid-phase oxidation of SO_2 in the troposphere, *J. Geophys. Res.* **90**, 3773–3779.

Hegg, D. A., and Hobbs, P. V. (1987) Comparisons of sulfate production due to ozone oxidation in clouds with a kinetic rate equation, *Geophys. Res. Lett.* **14**, 719–721.

Hegg, D. A., and Hobbs, P. V. (1988) Comparisons of sulfate and nitrate production in clouds on the mid-Atlantic and Pacific Northwest coast of the United States, *J. Atmos. Chem.* **7**, 325–333.

Hegg, D. A., Radke, L. F., and Hobbs, P. V. (1990) Particle production associated with marine clouds, *J. Geophys. Res.* **95**, 13917–13926.

Hegg, D. A., Radke, L. F., and Hobbs, P. V. (1991) Measurements of Aitken nuclei and cloud condensation nuclei in the marine atmosphere and their relationship to the DMS–cloud–climate hypothesis, *J. Geophys. Res.* **96**, 18727–18733.

Hegg, D. A., and Hobbs, P. V. (1992) Cloud condensation nuclei in the marine atmosphere: A review, in *Nucleation and Atmospheric Aerosols*, N. Fukuta and P. E. Wagner, eds., Deepak Publishing, Hampton, VA, pp. 181–192.

Hegg, D. A., Gao, S., Hoppel, W., Frick, G., Caffrey, P., Leaitch, W. R., Shantz, N., Ambrusko, J., and Albrechcinski, T. (2001) Laboratory studies of the efficiency of selected organic aerosols as CCN, *Atmos. Res.* **58**, 155–166.

Hering, S. V., and Friedlander, S. K. (1982) Origins of aerosol sulfur distributions in the Los Angeles basin, *Atmos. Environ.* **16**, 2647–2656.

Hobbs, P. V., Bowdle, D. A., and Radke, L. F. (1985) Particles in the lower troposphere over the high plains of the United States, 1. Size distributions, elemental compositions and morphologies, *J. Climat. Appl. Meteorol.* **24**, 1344–1356.

Hobbs, P. V. (1995) Aerosol–cloud interactions, in *Aerosol–Cloud–Climate Interactions*, P. V. Hobbs, ed., Academic Press, San Diego, pp. 33–73.

Hoppel, W. A., Frick, G. M., and Larson, R. E. (1986) Effects of non–precipitating clouds on the aerosol size distribution in the marine boundary layer, *Geophys. Res. Lett.* **13**, 125–128.

Hoppel, W. A., and Frick, G. M. (1990) Submicron aerosol size distributions measured over the tropical and south Pacific, *Atmos. Environ.* **24A**, 645–659.

Houghton H. G. (1985) *Physical Meterology*, MIT Press, Cambridge, MA.

Hudson, J. G., and Frisbie, P. R. (1991) Cloud condensation nuclei near marine stratus, *J. Geophys. Res.* **96**, 20795–20808.

Husain, L., Dutkiewicz, V. A., Hussain, M. M., Khwaja, H. A., Burkhard, E. G., Mehmood, G., Parekh, P. P., and Canelli, E. (1991) A study of heterogeneous oxidation of SO_2 in summer clouds, *J. Geophys. Res.* **96**, 18789–18805.

John, W., Wall, S. M., Ondo, J. L., and Winklmayr, W. (1990) Modes in the size distributions of atmospheric inorganic aerosol, *Atmos. Environ.* **24A**, 2349–2359.

Junge, C. E. (1963) *Air Chemistry and Radioactivity*, Academic Press, New York.

Karamachandani, P., and Venkatram, A. (1992) The role of non–precipitating clouds in producing ambient sulfate during summer—results from simulation with the Acid Deposition and Oxidant Model (ADOM), *Atmos. Environ.* **26**, 1041–1052.

Kelly, T. J., Schwartz, S. E., and Daum, P. H. (1989) Detection of acid producing reactions in natural clouds, *Atmos. Environ.* **23**, 569–583.

Kerminen, V. M., and Wexler, A. S. (1994) Post–fog nucleation of H_2SO_4–H_2O particles in smog, *Atmos. Environ.* **28**, 2399–2406.

Köhler, H. (1921) Zur Kondensation des Wasserdampfe in der Atmosphäre, *Geofys. Publ.* **2**, 3–15.

Köhler, H. (1926) Zur Thermodynamic der Kondensation an hygroskopischen Kernen und Bemerkungen über das Zusammenfliessen der Tropfen, *Medd. Met. Hydr. Anst. Stockholm* **3** (8).

Laaksonen, A., Korhonen, P., Kulmala, M., and Charlson, R. J. (1998) Modification of the Köhler equation to include soluble trace gases and slightly soluble substances, *J. Atmos. Sci.* **55**, 853–862.

Laaksonen, A., Vesala, T., Kulmala, M., Winkler, P. M., and Wagner, P. E. (2005) Commentary on cloud modelling and the mass accommodation coefficient of water, *Atmos. Chem. Phys.* **5**, 461–464.

Lammel, G., and Novakov, T. (1995) Water nucleation properties of carbon black and diesel soot particles, *Atmos. Environ.* **29**, 817–823.

Lawrence, M. G. (2005) The relationship between relative humidity and the dewpoint temperature in moist air, *Bull. Am. Meteorol. Soc.* **86**, 225–233.

Leaitch, W. R., Strapp, J. W., Isaac, G. A., and Hudson, J. G. (1986) Cloud droplet nucleation and scavenging of aerosol sulphate in polluted atmospheres, *Tellus* **38B**, 328–344.

Liao, H., Adams, P. J., Chung, S. H., Seinfeld, J. H., Mickley, L. J., and Jacob, D. J. (2003) Interactions between tropospheric chemistry and aerosols in a unified general circulation model, *J. Geophys. Res.* **108**, 4001 (doi: 10.1029/2001JD001260).

Liu, P. S. K., Leaitch, W. R., McDonald, A. M., Isaac, G. A., Strapp, J. W., and Wiebe, H. A. (1994) Sulfate production in a summer cloud over Ontario, Canada, *Tellus* **45B**, 368–389.

Lowe, P. R., and Ficke, J. M. (1974) Technical Paper 4–74, Environmental Prediction Research Facility, Naval Postgraduate School, Monterey, CA.

Lu, M. L., McClatchey, R. A., and Seinfeld, J. H. (2002) Cloud halos: Numerical simulation of dynamical structure and radiative impact, *J. Appl. Meteorol.* **41**, 832–848.

Lu, M. L., Wang, J., Freedman, A., Jonsson, H. H., Flagan, R. C., McClatchey, R. A., and Seinfeld, J. H. (2003) Analysis of humidity halos around trade wind cumulus clouds, *J. Atmos. Sci* **60**, 1041–1059.

Mason, B. J. (1971) *The Physics of Clouds*, Oxford Univ. Press, Oxford, UK.

McHenry, J. N., and Dennis, R. L. (1994) The relative importance of oxidation pathways and clouds to atmospheric ambient sulfate production as predicted by the regional acid deposition model, *J. Appl. Meteorol.* **33**, 890–905.

McMurry, P. H., and Wilson, J. C. (1983) Droplet phase (heterogeneous) and gas-phase (homogeneous) contributions to secondary ambient aerosol formation as functions of relative humidity, *J. Geophys. Res.* **88**, 5101–5108.

Meng, Z., and Seinfeld, J. H. (1994) On the source of the submicrometer droplet mode of urban and regional aerosols, *Aerosol Sci. Technol.* **20**, 253–265.

Munger, J. W., Collett, J. Jr., Daube, B., and Hoffmann, M. R. (1989) Chemical composition of coastal clouds: Dependence on droplet size and distance from the coast, *Atmos. Environ.* **23**, 2305–2320.

Nenes, A., Chuang, P. Y., Flagan, R. C., and Seinfeld, J. H. (2001) A theoretical analysis of cloud condensation nucleus (CCN) instruments, *J. Geophys. Res.* **106**, 3449–3474.

Noone, K. J., Charlson, R. J., Covert, D. S., and Heintzenberg, J. (1988) Cloud droplets: Solute concentration is size dependent, *J. Geophys. Res.* **93**, 9477–9482.

Ogren, J. A., Heitzenberg, J., Zuber, A., Noone, K. J., and Charlson, R. J. (1989) Measurements of the size-dependence of solute concentrations in cloud droplets, *Tellus* **41B**, 24–31.

Ogren, J. A. et al. (1992) Measurements of the size dependence of the concentration of non-volatile material in fog droplets, *Tellus* **44B**, 570–580.

Pandis, S. N., and Seinfeld, J. H. (1989) Mathematical modeling of acid deposition due to radiation fog, *J. Geophys. Res.* **94**, 12156–12176.

Pandis, S. N., Seinfeld, J. H., and Pilinis, C. (1990a) Chemical composition differences among droplets of different sizes, *Atmos. Environ.* **24A**, 1957–1969.

Pandis, S. N., Seinfeld, J. H., and Pilinis, C. (1990b) The smog–fog–smog cycle and acid deposition, *J. Geophys. Res.* **95**, 18489–18500.

Pandis, S. N., Seinfeld, J. H., and Pilinis, C. (1992) Heterogeneous sulfate production in an urban fog, *Atmos. Environ.* **26A**, 2509–2522.

Perry, K. D., and Hobbs, P. V. (1994) Further evidence for particle nucleation in clean air adjacent to marine cumulus clouds, *J. Geophys. Res.* **99**, 22803–22818.

Pruppacher, H. R., and Klett, J. D. (1997) *Microphysics of Clouds and Precipitation*, Kluwer, Dordrecht, The Netherlands.

Radke, L. F., and Hobbs, P. V. (1991) Humidity and particle fields around some small cumulus clouds, *J. Atmos. Sci.* **48**, 1190–1193.

Rangno, A. L., and Hobbs, P. V. (1991) Ice particle concentrations and precipitation development in small polar maritime cumuliform clouds, *Quart. J. Roy. Meteorol. Soc.* **117**, 207–241.

Raymond, T. M., and Pandis, S. N. (2002) Cloud activation of single-component organic aerosol particles, *J. Geophys. Res.*, **107**(D24), 4787 (doi: 10.1029/2002JD002159).

Rissman, T. A., VanReken, T. M., Wang, J., Gasparini, R., Collins, D. R., Jonsson, H. H., Brechtel, F. J., Flagan, R. C., and Seinfeld, J. H. (2006) Characterization of ambient aerosol from measurements of cloud condensation nuclei during the 2003 Atmospheric Radiation Measurement Aerosol Intensive Observational Period at the Southern Great Plains site in Oklahoma, *J. Geophys. Res.* **111**, D05S11 (doi: 10.1029/2004JD005695).

Rogers, D. C., and DeMott, P. J. (1991) Advances in laboratory cloud physics 1987–1990, *Rev. Geophys. Suppl.* 80–87.

Sakugawa, H., Kaplan, I. R., Tsai, W., and Cohen, Y. (1990) Atmospheric hydrogen peroxide, *Environ. Sci. Technol.* **24**, 1452–1462.

Saxena, V. K., and Hendler, A. H. (1983) In–cloud scavenging and resuspension of cloud active aerosols during winter storms over Lake Michigan, in *Precipitation Scavenging, Dry Deposition and Resuspension*, H. R. Pruppacher, R. G. Semonin, and W. G. N. Slinn, eds., Elsevier, New York, pp. 91–102.

Sekhon, R. S., and Srivastava, R. C. (1971) Doppler observations of drop size distributions in a thunderstorm, *J. Atmos. Sci.* **28**, 983–994.

Shulman, M. L., Jacobson, M. C., Charlson, R. J., Synovec, R. E., and Young, T. E. (1996) Dissolution behavior and surface tension effects of organic compounds in nucleating cloud droplets, *Geophys. Res. Lett.* **23**, 277–280.

Swozdziak, J. W., and Swozdziak, A. B. (1992) Sulfate aerosol production in the Sudely Range, Poland, *J. Aerosol Sci.* S369–S372.

Ten Brink, H. M., Schwartz, S. E., and Daum, P. H. (1987) Efficient scavenging of aerosol sulfate by liquid water clouds, *Atmos. Environ.* **21**, 2035–2052.

Twomey, S. (1959) The nuclei of natural clouds formation. Part II: The supersaturation in natural clouds and the variation of cloud droplet concentration, *Geofis. Pura Appl.* **43**, 243–249.

United States National Acid Precipitation Assessment Program (U.S. NAPAP) (1991) *Acidic Deposition: State of Science and Technology*, Vol.1, *Emissions, Atmospheric Processes and Deposition*, P. M. Irving, ed., U.S. Government Printing Office, Washington, DC.

VanReken, T. M., Rissman, T. A., Roberts, G. C., Varutbangkul, V., Jonsson, H. H., Flagan, R. C., and Seinfeld, J. H. (2003) Toward aerosol/cloud condensation nuclei (CCN) closure during CRYSTAL-FACE, *J. Geophys. Res.* **108**, 4633 (doi: 10.1029/2003JD003582).

VanReken, T. M., Ng, N. L., Flagan, R. C., and Seinfeld, J. H. (2005) Cloud condensation nucleus activation properties of biogenic secondary organic aerosol, *J. Geophys. Res.* **110**, D07206 (doi: 10.1029/2004JD005465).

Walcek, C. J., Stockwell, W. R., and Chang, J. S. (1990) Theoretical estimates of the dynamic radiative and chemical effects of clouds on tropospheric gases, *Atmos. Res.* **25**, 53–69.

Wall, S. M., John, W., and Ondo, J. L. (1988) Measurements of aerosol size distributions for nitrate and major ionic species, *Atmos. Environ.* **22**, 1649–1659.

Warner, J. (1968) The supersaturation in natural clouds, *J. Appl. Meteorol.* **7**, 233–237.

Weber, R. J., Chen, G., Davis, D. D., Mauldin III, R. L., Tanner, D. J., Eisele, F. L., Clarke, A. D., Thornton, D. C., and Bandy, A. R. (2001) Measurements on enhanced H_2SO_4 and 3-4 nm particles near a frontal cloud during the first Aerosol Characterization Experiment (ACE 1), *J. Geophys. Res.* **106**, 24107–24117.

Whelpdale, D. M., and List, R. (1971) The coalescence process in droplet growth, *J. Geophys. Res.* **76**, 2836–2856.

18 Atmospheric Diffusion

A major goal of our study of the atmospheric behavior of trace species is to be able to describe mathematically the spatial and temporal distribution of contaminants released into the atmosphere. This chapter is devoted to the subject of atmospheric diffusion and its description. It is common to refer to the behavior of gases and particles in turbulent flow as turbulent "diffusion" or, in this case, as atmospheric "diffusion," although the processes responsible for the observed spreading or dispersion in turbulence are not the same as those acting in ordinary molecular diffusion. A more precise term would perhaps be "atmospheric dispersion," but to conform to common terminology we will use "atmospheric diffusion." This chapter is devoted primarily to developing the two basic ways of describing turbulent diffusion. The first is the *Eulerian* approach, in which the behavior of species is described relative to a fixed coordinate system. The Eulerian description is the common way of treating heat and mass transfer phenomena. The second approach is the *Lagrangian*, in which concentration changes are described relative to the moving fluid. As we will see, the two approaches yield different types of mathematical relationships for the species concentrations that can, ultimately, be related. Each of the two modes of expression is a valid description of turbulent diffusion; the choice of which approach to adopt in a given situation will be seen to depend on the specific features of the situation.

18.1 EULERIAN APPROACH

Let us consider N species in a fluid. The concentration of each must, at each instant, satisfy a material balance taken over a volume element. Thus any accumulation of material over time, when added to the net amount of material convected out of the volume element, must be balanced by an equivalent amount of material that is produced by chemical reaction in the element, that is emitted into it by sources, and that enters by molecular diffusion. Expressed mathematically, the concentration of each species c_i must satisfy the continuity equation

$$\frac{\partial c_i}{\partial t} + \frac{\partial}{\partial x_j}(u_j c_i) = D_i \frac{\partial^2 c_i}{\partial x_j \, \partial x_j} + R_i(c_1, \ldots, c_N, T) + S_i(\mathbf{x}, t) \tag{18.1}$$
$$i = 1, 2, \ldots, N$$

Atmospheric Chemistry and Physics: From Air Pollution to Climate Change, Second Edition, by John H. Seinfeld and Spyros N. Pandis. Copyright © 2006 John Wiley & Sons, Inc.

where u_j is the jth component of the fluid velocity, D_i is the molecular diffusivity of species i in the carrier fluid, R_i is the rate of generation of species i by chemical reaction (which depends in general on the fluid temperature T), and S_i is the rate of addition of species i at location $\mathbf{x} = (x_1, x_2, x_3)$ and time t.[1]

In addition to the requirement that the c_i satisfy (18.1), the fluid velocities u_j and the temperature T, in turn, must satisfy the Navier–Stokes and energy equations, and which themselves are coupled through the u_j, c_i, and T with the total continuity equation and the ideal-gas law (we restrict our attention to gaseous systems). In general, it is necessary to carry out a simultaneous solution of the coupled equations of mass, momentum, and energy conservation to account properly for the changes in u_j, T, and c_i and the effects of the changes of each of these on each other. In dealing with atmospheric trace gases, occurring at parts per million and smaller mixing ratios, it is quite justifiable to assume that the presence of these species does not affect the meteorology to any detectable extent; thus the equation of continuity (18.1) can be solved independently of the coupled momentum and energy equations.[2] Consequently, the fluid velocities u_j and the temperature T can be considered independent of the c_i. From this point on we will not explicitly indicate the dependence of R_i on T.

The complete description of atmospheric gas behavior rests with the solution of (18.1). Unfortunately, because the flows of interest are turbulent, the fluid velocities u_j are random functions of space and time. As was done in Chapter 16, it is customary to represent the wind velocities u_j as the sum of a deterministic and stochastic component, $\bar{u}_j + u_j'$.

To illustrate the importance of the definition of the deterministic and stochastic velocity components $\bar{u}_j$ and u_j', let us suppose a puff of species of known concentration distribution $c(\mathbf{x}, t_0)$ at time t_0. In the absence of chemical reaction and other sources, and assuming molecular diffusion to be negligible, the concentration distribution at some later time is described by the following *advection equation*:

$$\frac{\partial c}{\partial t} + \frac{\partial}{\partial x_j}(u_j c) = 0 \qquad (18.2)$$

If we solve this equation with $u_j = \bar{u}_j$ and compare the solution with observations, we would find in reality that the material spreads out more than predicted. This extra spreading is, in fact, what is referred to as turbulent diffusion and results from the influence of the random component u_j', which we have ignored. Now let us solve this equation with the precise velocity field u_j. We should then find that the solution agrees exactly with the observations (assuming, of course, that molecular diffusion is negligible), implying that if we knew the velocity field precisely at all locations and times, there would be no such phenomenon as turbulent diffusion. Thus turbulent diffusion is an artifact of our lack of complete knowledge of the true velocity field. Consequently, one of the fundamental tasks in turbulent diffusion theory is to define the deterministic and stochastic components of the velocity field.

---

[1] We will use the convention of (x_1, x_2, x_3) as coordinate axes and (u_1, u_2, u_3) as fluid velocity components when dealing with the transport equations in their most general form. Eventually it will be easier to use an (x, y, z), (u, v, w) system when dealing with specific problems and we will do so at that time.

[2] Two effects could, in principle, serve to invalidate this assumption. Highly unlikely in the troposphere is that sufficient heat would be generated by chemical reactions to influence the temperature. Absorption, reflection, and scattering of radiation by trace gases and particles could result in alterations of the fluid behavior.

Replacing u_j by $\bar{u}_j + u_j'$ in (18.1) gives

$$\frac{\partial c_i}{\partial t} + \frac{\partial}{\partial x_j}[(\bar{u}_j + u_j')c_i] = D_i \frac{\partial^2 c_i}{\partial x_j \partial x_j} + R_i(c_1, \ldots, c_N) + S_i(\mathbf{x}, t) \qquad (18.3)$$

Since the u_j' are random variables, the c_i resulting from the solution of (18.3) must also be random variables; that is, because the wind velocities are random functions of space and time, the airborne species concentrations are themselves random variables in space and time. Thus the determination of the c_i, in the sense of being a specified function of space and time, is not possible, just as it is not possible to determine precisely the value of any random variable in an experiment. We can at best derive the probability that at some location and time the concentration of species i will lie between two closely spaced values. Unfortunately, the specification of the probability density function for a random process as complex as atmospheric diffusion is almost never possible. Instead, we must adopt a less desirable but more feasible approach, the determination of certain statistical properties of the c_i, most notably the mean $\langle c_i \rangle$.

The mean concentration can be interpreted in the following way. Let us imagine an experiment in which a puff of material is released at a certain time and concentrations are measured downwind at subsequent times. We would measure $c_i(\mathbf{x}, t)$, which would exhibit random characteristics because of the wind. If it were possible to repeat this experiment under identical conditions, we would again measure $c_i(\mathbf{x}, t)$, but because of the randomness in the wind field we could not reproduce the first $c_i(\mathbf{x}, t)$. Theoretically we could repeat this experiment an infinite number of times. We would then have a so-called ensemble of experiments. If at every location $\mathbf{x}$ and time t we averaged all the concentration values over the infinite number of experiments, we would have computed the theoretical mean concentration $\langle c_i(\mathbf{x}, t) \rangle$.[3] Experiments like this cannot, of course, be repeated under identical conditions, and so it is virtually impossible to measure $\langle c_i \rangle$. Thus a measurement of the concentration of species i at a particular location and time is more suitably envisioned as one sample from a hypothetically infinite ensemble of possible concentrations. Clearly, an individual measurement may differ considerably from the mean $\langle c_i \rangle$.

It is convenient to express c_i as $\langle c_i \rangle + c_i'$, where, by definition, $\langle c_i' \rangle = 0$. Averaging (18.3) over an infinite ensemble of realizations of the turbulence yields the equation governing $\langle c_i \rangle$:

$$\frac{\partial \langle c_i \rangle}{\partial t} + \frac{\partial}{\partial x_j}(\bar{u}_j \langle c_i \rangle) + \frac{\partial}{\partial x_j}\langle u_j' c_i' \rangle = D_i \frac{\partial^2 \langle c_i \rangle}{\partial x_j \partial x_j} + \langle R_i(\langle c_1 \rangle + c_1', \ldots, \langle c_N \rangle + c_N') \rangle + S_i(\mathbf{x}, t)$$

$$(18.4)$$

Let us consider the case of a single inert species, that is, $R = 0$. We note that (18.4) contains dependent variables $\langle c \rangle$ and $\langle u_j' c' \rangle$, $j = 1, 2, 3$. We thus have more dependent variables than equations. Again, this is the closure problem of turbulence. For example, if we were to derive an equation for the $\langle u_j' c' \rangle$ by subtracting (18.4) from (18.3), multiplying

---

[3]We have used different notations for the mean values of the velocities and the concentrations, that is, $\bar{u}_j$ versus $\langle c_i \rangle$, in order to emphasize the fact that the mean fluid velocities are normally determined by a process involving temporal and spatial averaging, whereas the $\langle c_i \rangle$ always represent the theoretical ensemble average.

the resulting equation by u_j' and then averaging, we would obtain

$$\frac{\partial}{\partial t}\langle u_j'c'\rangle + \frac{\partial}{\partial x_k}(\bar{u}_k\langle u_j'c'\rangle) + \langle u_j'u_k'\rangle\frac{\partial\langle c\rangle}{\partial x_k} = \frac{\partial}{\partial x_k}\left(D\frac{\partial}{\partial x_k}\langle u_j'c'\rangle - \langle u_j'u_k'c'\rangle\right) \tag{18.5}$$
$$j = 1, 2, 3$$

Although we have derived the desired equations, we have at the same time generated new dependent variables $\langle u_j'u_k'c'\rangle$, $j,k = 1,2,3$. If we generate additional equations for these variables, we find that still more dependent variables appear. The closure problem becomes even worse if a nonlinear chemical reaction is occurring. If the single species decays by a second-order reaction, then the term $\langle R \rangle$ in (18.4) becomes $-k(\langle c\rangle^2 + \langle c'^2\rangle)$, where $\langle c'^2\rangle$ is a new dependent variable. If we were to derive an equation for $\langle c'^2\rangle$, we would find the emergence of new dependent variables $\langle u_j'c'^2\rangle$, $\langle c'^3\rangle$, and $\langle \partial c'/\partial x_j \partial c'/\partial x_j\rangle$. It is because of the closure problem that an Eulerian description of turbulent diffusion will not permit exact solution even for the mean concentration $\langle c\rangle$.

18.2 LAGRANGIAN APPROACH

The Lagrangian approach to turbulent diffusion is concerned with the behavior of representative fluid particles.[4] We therefore begin by considering a single particle that is at location $\mathbf{x}'$ at time t' in a turbulent fluid. The subsequent motion of the particle can be described by its trajectory, $\mathbf{X}[\mathbf{x}', t'; t]$, that is, its position at any later time t. Let $\psi(x_1, x_2, x_3, t)dx_1\,dx_2\,dx_3 = \psi(\mathbf{x}, t)d\mathbf{x}$ = probability that the particle at time t will be in volume element x_1 to $x_1 + dx_1$, x_2 to $x_2 + dx_2$, and x_3 to $x_3 + dx_3$, that is, that $x_1 \le X_1 < x_1 + dx_1$, and so on. Thus $\psi(\mathbf{x}, t)$ is the probability density function (pdf) for the particle's location at time t. By the definition of a probability density function

$$\int_{-\infty}^{\infty}\int_{-\infty}^{\infty}\int_{-\infty}^{\infty} \psi(\mathbf{x}, t)d\mathbf{x} = 1$$

The probability density of finding the particle at $\mathbf{x}$ at t can be expressed as the product of two other probability densities:

1. The probability density that if the particle is at $\mathbf{x}'$ at t', it will undergo a displacement to $\mathbf{x}$ at t. Denote this probability density $Q(\mathbf{x}, t\,|\,\mathbf{x}', t')$ and call it the *transition probability density* for the particle.
2. The probability density that the particle was at $\mathbf{x}'$ at t', $\psi(\mathbf{x}', t')$, integrated over all possible starting points $\mathbf{x}'$. Thus

$$\psi(\mathbf{x}, t) = \int_{-\infty}^{\infty}\int_{-\infty}^{\infty}\int_{-\infty}^{\infty} Q(\mathbf{x}, t\,|\,\mathbf{x}', t')\psi(\mathbf{x}', t')d\mathbf{x}' \tag{18.6}$$

---

[4]By a "fluid particle" we mean a volume of fluid large compared with molecular dimensions but small enough to act as a point that exactly follows the fluid. The "particle" may contain fluid of a different composition than the carrier fluid, in which case the particle is referred to as a "marked particle."

The density function $\psi(\mathbf{x}, t)$ has been defined with respect to a single particle. If, however, an arbitrary number m of particles are initially present and the position of the ith particle is given by the density function $\psi_i(\mathbf{x}, t)$, it can be shown that the ensemble mean concentration at the point $\mathbf{x}$ is given by

$$\langle c(\mathbf{x}, t) \rangle = \sum_{i=1}^{m} \psi_i(\mathbf{x}, t) \tag{18.7}$$

By expressing the probability density function (pdf) $\psi_i(\mathbf{x}, t)$ in (18.7) in terms of the initial particle distribution and the spatiotemporal distribution of particle sources $S(\mathbf{x}, t)$, say, in units of particles per volume per time, and then substituting the resulting expression into (18.6), we obtain the following general formula for the mean concentration:

$$\langle c(\mathbf{x}, t) \rangle = \int_{-\infty}^{\infty} \int_{-\infty}^{\infty} \int_{-\infty}^{\infty} Q(\mathbf{x}, t | \mathbf{x}_0, t_0) \langle c(\mathbf{x}_0, t_0) \rangle d\mathbf{x}_0$$
$$+ \int_{-\infty}^{\infty} \int_{-\infty}^{\infty} \int_{-\infty}^{\infty} \int_{t_0}^{t} Q(\mathbf{x}, t | \mathbf{x}', t') S(\mathbf{x}', t') dt' \, d\mathbf{x}' \tag{18.8}$$

The first term on the right-hand side (RHS) represents those particles present at t_0, and the second term on the RHS accounts for particles added from sources between t_0 and t.

Equation (18.8) is the fundamental Lagrangian relation for the mean concentration of a species in turbulent fluid. The determination of $\langle c(\mathbf{x}, t) \rangle$, given $\langle c(\mathbf{x}_0, t_0) \rangle$ and $S(\mathbf{x}', t)$, rests with the evaluation of the transition probability $Q(\mathbf{x}, t | \mathbf{x}', t')$. If Q were known for $\mathbf{x}, \mathbf{x}', t$, and t', the mean concentration $\langle c(\mathbf{x}, t) \rangle$ could be computed by simply evaluating (18.8). However, there are two substantial problems with using (18.8): (1) it holds only when the particles are not undergoing chemical reactions, and (2) such complete knowledge of the turbulence properties as would be needed to know Q is generally unavailable except in the simplest of circumstances.

18.3 COMPARISON OF EULERIAN AND LAGRANGIAN APPROACHES

The techniques for describing the statistical properties of the concentrations of marked particles, such as trace gases, in a turbulent fluid can be divided into two categories: Eulerian and Lagrangian. The Eulerian methods attempt to formulate the concentration statistics in terms of the statistical properties of the Eulerian fluid velocities, that is, the velocities measured at fixed points in the fluid. A formulation of this type is very useful not only because the Eulerian statistics are readily measurable (as determined from continuous-time recordings of the wind velocities by a fixed network of instruments) but also because the mathematical expressions are directly applicable to situations in which chemical reactions are taking place. Unfortunately, the Eulerian approaches lead to a serious mathematical obstacle known as the "closure problem," for which no generally valid solution has yet been found.

By contrast, the Lagrangian techniques attempt to describe the concentration statistics in terms of the statistical properties of the displacements of groups of particles released in the fluid. The mathematics of this approach is more tractable than that of the Eulerian methods, in that no closure problem is encountered, but the applicability of the resulting equations is limited because of the difficulty of accurately determining the required particle statistics. Moreover, the equations are not directly applicable to problems involving nonlinear chemical reactions.

Having demonstrated that exact solution for the mean concentrations $\langle c_i(\mathbf{x}, t) \rangle$ even of inert species in a turbulent fluid is not possible in general by either the Eulerian or Lagrangian approaches, we now consider what assumptions and approximations can be invoked to obtain practical descriptions of atmospheric diffusion. In Section 18.4 we shall proceed from the two basic equations for $\langle c_i \rangle$, (18.4) and (18.8), to obtain the equations commonly used for atmospheric diffusion. A particularly important aspect is the delineation of the assumptions and limitations inherent in each description.

18.4 EQUATIONS GOVERNING THE MEAN CONCENTRATION OF SPECIES IN TURBULENCE

18.4.1 Eulerian Approaches

As we have seen, the Eulerian description of turbulent diffusion leads to the so-called closure problem, as illustrated in (18.4) by the new dependent variables $\langle u_j' c_i' \rangle, j = 1, 2, 3$, as well as any that might arise in $\langle R_i \rangle$ if nonlinear chemical reactions are occurring. Let us first consider only the case of chemically inert species, that is, $R_i = 0$. The problem is to deal with the variables $\langle u_j' c' \rangle$ if we wish not to introduce additional differential equations.

The most common means of relating the turbulent fluxes $\langle u_j' c' \rangle$ to the $\langle c \rangle$ is based on the mixing-length model of Chapter 16. In particular, it is assumed that (summation implied over k)

$$\langle u_j' c' \rangle = -K_{jk} \frac{\partial \langle c \rangle}{\partial x_k} \qquad j = 1, 2, 3 \tag{18.9}$$

where K_{jk} is called the *eddy diffusivity*. Equation (18.9) is called both *mixing-length theory* and *K theory*. Since (18.9) is essentially only a definition of the K_{jk} which are, in general, functions of location and time, we have, by means of (18.9), replaced the three unknowns $\langle u_j' c' \rangle, j = 1, 2, 3$, with the six unknowns $K_{jk}, j, k = 1, 2, 3$ ($K_{jk} = K_{kj}$). If the coordinate axes coincide with the principal axes of the eddy diffusivity tensor $\{K_{jk}\}$, then only the three diagonal elements K_{11}, K_{22}, and K_{33} are nonzero, and (17.9) becomes[5]

$$\langle u_j' c' \rangle = -K_{jj} \frac{\partial \langle c \rangle}{\partial x_j} \tag{18.10}$$

---

[5]No summation is implied in this term, for example

$$\langle u_1' c' \rangle = -K_{11} \frac{\partial \langle c \rangle}{\partial x_1}$$

It is beyond our scope to consider the conditions under which $\{K_{jk}\}$ may be taken as diagonal. For our purposes we note that ordinarily there is insufficient information to assume that $\{K_{jk}\}$ is not diagonal.

In using (18.4), two other assumptions are ordinarily invoked:

1. Molecular diffusion is negligible compared with turbulent diffusion:

$$D \frac{\partial^2 \langle c \rangle}{\partial x_j \partial x_j} \ll \frac{\partial}{\partial x_j} \langle u_j' c' \rangle$$

2. The atmosphere is incompressible:

$$\frac{\partial \bar{u}_j}{\partial x_j} = 0$$

With these assumptions and (18.10), (18.4) becomes

$$\frac{\partial \langle c \rangle}{\partial t} + \bar{u}_j \frac{\partial \langle c \rangle}{\partial x_j} = \frac{\partial}{\partial x_j} \left(K_{jj} \frac{\partial \langle c \rangle}{\partial x_j} \right) + S(\mathbf{x}, t) \qquad (18.11)$$

This equation is termed the "semiempirical equation of atmospheric diffusion," or just the *atmospheric diffusion equation*, and will play an important role in what is to follow.

Let us return to the case in which chemical reactions are occurring, for which we refer to (18.4). Since R_i is almost always a nonlinear function of the c_i, we have already seen that additional terms of the type $\langle c_i' c_j' \rangle$ will arise from $\langle R_i \rangle$. The most obvious approximation we can make regarding $\langle R_i \rangle$ is to replace $\langle R_i(c_1, \ldots, c_N) \rangle$ by $R_i(\langle c_1 \rangle, \ldots, \langle c_N \rangle)$, thereby neglecting the effect of concentration fluctuations on the rate of reaction. Invoking this approximation, as well as those inherent in (18.11), we obtain for each species i,

$$\frac{\partial \langle c_i \rangle}{\partial t} + \bar{u}_j \frac{\partial \langle c_i \rangle}{\partial x_j} = \frac{\partial}{\partial x_j} \left(K_{jj} \frac{\partial \langle c_i \rangle}{\partial x_j} \right) + R_i(\langle c_1 \rangle, \ldots, \langle c_N \rangle) + S_i(\mathbf{x}, t) \qquad (18.12)$$

A key issue is whether we can develop the conditions under which (18.12) is a valid description of atmospheric diffusion and chemical reaction.

It can be shown that (18.14) is a valid description of turbulent diffusion and chemical reaction as long as the reaction processes are slow compared with turbulent transport and the characteristic lengthscale and timescale for changes in the mean concentration field are large compared with the corresponding scales for turbulent transport.

18.4.2 Lagrangian Approaches

We now wish to consider the derivation of usable expressions for $\langle c_i(\mathbf{x}, t) \rangle$ based on the fundamental Lagrangian expression (18.8). As we have seen, the utility of (18.8) rests on the ability to evaluate the transition probability $Q(\mathbf{x}, t \mid \mathbf{x}', t')$. The first question, then, is: "Are there any circumstances under which the form of Q is known?"

To attempt to answer this question let us go back to the advection equation, (18.2), for a one-dimensional flow with a general source term $S(x,t)$:

$$\frac{\partial c}{\partial t} + \frac{\partial}{\partial x}(uc) = S(x,t) \tag{18.13}$$

Since the velocity u is a random quantity, the concentration c that results from the solution of (18.13) is also random. What we want to do is to solve (18.13) for a particular choice of the velocity u and find the concentration c corresponding to that choice of u. For simplicity, let us assume that u is independent of x and depends only on time t. Thus the velocity $u(t)$ is a random variable depending on time. Since $u(t)$ is a random variable, we need to specify the probability density for $u(t)$. A reasonable assumption for the probability distribution of $u(t)$ is that it is Gaussian

$$p_u(u) = \frac{1}{(2\pi)^{1/2}\sigma_u}\exp\left(-\frac{(u-\bar{u})^2}{2\sigma_u^2}\right) \tag{18.14}$$

where the mean value is $\langle u(t)\rangle = \bar{u}$, and the variance is σ_u^2. In the process of solving (18.13) we will need an expression for the term $\langle(u(t)-\bar{u})(u(\tau)-\bar{u})\rangle$. We will assume that $u(t)$ is a stationary random process with the correlation

$$\langle(u(t)-\bar{u})(u(\tau)-\bar{u})\rangle = \sigma_u^2\exp(-b|t-\tau|) \tag{18.15}$$

This expression states that the maximum correlation between the velocities at two times occurs when those times are equal, and is equal to σ_u^2. As the time separation between $u(t)$ and $u(\tau)$ increases, the correlation decays exponentially with a characteristic decay time of $1/b$. Stationarity implies that the statistical properties of u at two different times t and τ depend only on $t-\tau$ and not on t and τ individually.

Let us now solve (18.13) for a time-varying source of strength $S(t)$ at $x=0$. The solution, obtained by the method of characteristics, is

$$c(x,t) = \int_0^t \delta(x-X(t,\tau))S(\tau)d\tau \tag{18.16}$$

where

$$X(t,\tau) = \int_\tau^t u(t')dt' \tag{18.17}$$

is just the distance a fluid particle travels between times τ and t. Since $X(t,\tau)$ is a random variable, so is $c(x,t)$. We are really interested in the mean, $\langle c(x,t)\rangle$. Thus, taking the expected value of (18.16), we obtain

$$\langle c(x,t)\rangle = \int_0^t \langle\delta(x-X(t,\tau))\rangle S(\tau)d\tau \tag{18.18}$$

Since the pdf of $u(t)$ is Gaussian, that for $X(t, \tau)$ is also Gaussian

$$p_X(X; t, \tau) = \frac{1}{(2\pi)^{1/2}\sigma_x} \exp\left(-\frac{(X - \bar{X})^2}{2\sigma_x^2}\right) \tag{18.19}$$

where $\bar{X}(t, \tau)$ and $\sigma_x^2(t, \tau)$ are the mean and variance, respectively, of $X(t, \tau)$. The expression for $\langle c(x, t)\rangle$ can be written in terms of p_X as

$$\langle c(x, t)\rangle = \int_0^t S(t') \int_{-\infty}^{\infty} \delta(x - X(t, t')) p_X(X; t, t') dX \, dt'$$
$$= \int_0^t S(t') p_X(x; t, t') dt' \tag{18.20}$$

By comparing (18.20) with (18.8), we see that $p_X(x; t, t')$ is precisely $Q(x, t \,|\, x', t')$, except that there is no dependence on x' in this case. Let us compute $\bar{X}(t, \tau)$ and $\sigma_x^2(t, \tau)$ based on the definition of $X(t, \tau)$ and the properties of u. We note that

$$\bar{X}(t, \tau) = (t - \tau)\bar{u} \tag{18.21}$$

and

$$\sigma_x^2(t, \tau) = \langle(X(t, \tau) - \bar{X})^2\rangle$$
$$= \langle X(t, \tau)^2\rangle - \bar{X}^2 \tag{18.22}$$

Since

$$\langle X(t, \tau)^2\rangle = \left\langle \int_\tau^t u(t') dt' \int_\tau^t u(t'') dt'' \right\rangle$$
$$= \int_\tau^t \int_\tau^t \langle u(t')u(t'')\rangle dt' \, dt'' \tag{18.23}$$

using (18.15), (18.21), (18.22), and (18.23), we obtain

$$\sigma_x^2(t, \tau) = \frac{2\sigma_u^2}{b^2}[b(t - \tau) + e^{-b(t-\tau)} - 1] \tag{18.24}$$

so $\sigma_x^2(t, \tau) = \sigma_x^2(t - \tau)$.

If the source is just a pulse of unit strength at $t = 0$, that is, $S(t) = \delta(t)$, then (18.20) becomes

$$\langle c(x, t)\rangle = \frac{1}{(2\pi)^{1/2}\sigma_x(t)} \exp\left(-\frac{(x - \bar{u}t)^2}{2\sigma_x^2(t)}\right) \tag{18.25}$$

Thus we have found that the mean concentration of a tracer released in a flow where the velocity is a stationary, Gaussian random process has a distribution that is, itself, Gaussian. This is an important result.

The mean position of the distribution at time t is $\bar{u}t$, just the distance a tracer molecule has traveled over a time t at the mean fluid velocity $\bar{u}$. The variance of the mean concentration distribution, $\sigma_x^2(t)$, is just the variance of $X(t,\tau)$. This result makes sense since $X(t,\tau)$ is just the random distance that a fluid particle travels between times τ and t, and this distance is precisely that which a tracer molecule travels.

Let us examine the limits of (18.24) for large and small values of t. For large t, that is, $t \gg b^{-1}$, the characteristic time for velocity correlations, (18.24) reduces to

$$\sigma_x^2 = \frac{2\sigma_u^2 t}{b} \tag{18.26}$$

For $t \ll b^{-1}$, $\exp(-bt) \simeq 1 - bt + (bt)^2/2$, and

$$\sigma_x^2 = \sigma_u^2 t^2 \tag{18.27}$$

Thus we find that the variance of the mean concentration distribution varies with time according to

$$\sigma_x^2 \approx \begin{cases} t^2 & \text{small } t \\ t & \text{large } t \end{cases}$$

This example can readily be generalized to three dimensions. If we continue to assume that there is a mean flow only in the x direction, then the expression for the mean concentration resulting from an instantaneous point source of unit strength at the origin is

$$\langle c(x,y,z,t) \rangle = \frac{1}{(2\pi)^{3/2}\sigma_x(t)\sigma_y(t)\sigma_z(t)}$$
$$\times \exp\left(-\frac{(x - \bar{u}t)^2}{2\sigma_x^2(t)} - \frac{y^2}{2\sigma_y^2(t)} - \frac{z^2}{2\sigma_z^2(t)} \right) \tag{18.28}$$

where σ_y^2 and σ_z^2 are given by equations analogous to that for σ_x^2.

To summarize again, we have shown through a highly idealized example that the mean concentration in a stationary, homogeneous Gaussian flow field is itself Gaussian. If the turbulence is stationary and homogeneous, the transition probability density Q of a particle depends only on the displacements in time and space and not on where or when the particle was introduced into the flow. Thus, in that case, $Q(\mathbf{x},t \mid \mathbf{x}',t') = Q(\mathbf{x} - \mathbf{x}'; t - t')$. The Gaussian form of Q turns out to play an important role in atmospheric diffusion theory, as we will see.

18.5   SOLUTION OF THE ATMOSPHERIC DIFFUSION EQUATION FOR AN INSTANTANEOUS SOURCE

We begin in this section to obtain solutions for atmospheric diffusion problems. Let us consider, as we did in the previous section, an instantaneous point source of strength S at the origin in an infinite fluid with a velocity $\bar{u}$ in the x direction. We desire to solve the atmospheric diffusion equation, (18.11), in this situation. Let us assume, for lack of anything better at the moment, that K_{xx}, K_{yy}, and K_{zz} are constant. Then (18.11) becomes

$$\frac{\partial \langle c \rangle}{\partial t} + \bar{u} \frac{\partial \langle c \rangle}{\partial x} = K_{xx} \frac{\partial^2 \langle c \rangle}{\partial x^2} + K_{yy} \frac{\partial^2 \langle c \rangle}{\partial y^2} + K_{zz} \frac{\partial^2 \langle c \rangle}{\partial z^2} \tag{18.29}$$

to be solved subject to

$$\langle c(x, y, z, 0) \rangle = S \delta(x) \delta(y) \delta(z) \tag{18.30}$$

$$\langle c(x, y, z, t) \rangle = 0 \qquad x, y, z \rightarrow \pm\infty \tag{18.31}$$

The solution of (18.29) to (18.31) is given in Appendix 18.A.1. It is

$$\langle c(x, y, z, t) \rangle = \frac{S}{8(\pi t)^{3/2}(K_{xx}K_{yy}K_{zz})^{1/2}}$$

$$\times \exp\left(-\frac{(x - \bar{u}t)^2}{4K_{xx}t} - \frac{y^2}{4K_{yy}t} - \frac{z^2}{4K_{zz}t}\right) \tag{18.32}$$

Note the similarity of (18.32) and (18.28). In fact, if we define $\sigma_x^2 = 2K_{xx}t$, $\sigma_y^2 = 2K_{yy}t$, and $\sigma_z^2 = 2K_{zz}t$, we note that the two expressions are identical. There is, we conclude, evidently a connection between the Eulerian and Lagrangian approaches embodied in a relation between the variances of spread that arise in a Gaussian distribution and the eddy diffusivities in the atmospheric diffusion equation. We will explore this relationship further as we proceed.

18.6 MEAN CONCENTRATION FROM CONTINUOUS SOURCES

We just obtained expression (18.32) for the mean concentration resulting from an instantaneous release of a quantity S of material at the origin in an infinite fluid with stationary, homogeneous turbulence and a mean velocity $\bar{u}$ in the x direction. We now wish to consider a continuously emitting source under the same conditions. The source strength is $q(\text{g s}^{-1})$.

18.6.1 Lagrangian Approach

A continuous source is viewed conceptually as one that began emitting at $t = 0$ and continues as $t \rightarrow \infty$. The mean concentration achieves a steady state, independent of time, and $S(x, y, z, t) = q\delta(x)\delta(y)\delta(z)$. The basic Lagrangian expression (18.8) becomes

$$\langle c(x, y, z) \rangle = \int_0^t Q(x, y, z, t|0, 0, 0, t')q \, dt' \tag{18.33}$$

The steady-state concentration is given by

$$\langle c(x,y,z)\rangle = \lim_{t\to\infty}\langle c(x,y,z,t)\rangle = \lim_{t\to\infty}\int_0^t Q(x,y,z,t|0,0,0,t')q\,dt' \qquad (18.34)$$

The transition probability density Q has the general Gaussian form of (18.28)

$$Q(x,y,z,t|0,0,0,t') = (2\pi)^{-3/2}[\sigma_x(t-t')\sigma_y(t-t')\sigma_z(t-t')]^{-1}$$

$$\times \exp\left(-\frac{(x-\bar u(t-t'))^2}{2\sigma_x^2(t-t')}-\frac{y^2}{2\sigma_y^2(t-t')}-\frac{z^2}{2\sigma_z^2(t-t')}\right) \qquad (18.35)$$

which can be expressed as $Q(x,y,z,t-t'|0,0,0,0)$. Thus

$$\langle c(x,y,z)\rangle = \lim_{t\to\infty}\int_0^t Q(x,y,z,\tau|0,0,0,0)q\,d\tau \qquad (18.36)$$

Therefore the steady-state concentration resulting from a continuous source is obtained by integrating the unsteady-state concentration over all time from 0 to ∞:

$$\langle c(x,y,z)\rangle = \int_0^\infty \frac{q}{(2\pi)^{3/2}\sigma_x\sigma_y\sigma_z}\exp\left(-\frac{(x-\bar ut)^2}{2\sigma_x^2}-\frac{y^2}{2\sigma_y^2}-\frac{z^2}{2\sigma_z^2}\right)dt \qquad (18.37)$$

We need to evaluate the integral in (18.37). To do so, we need to specify $\sigma_x(t)$, $\sigma_y(t)$, and $\sigma_z(t)$. For the moment let us assume simply that $\sigma_x(t) = \sigma_y(t) = \sigma_z(t) = \sigma(t)$, where we do not specify how σ depends on t. We note that the term $\exp[(x-\bar ut)^2/2\sigma_x^2]$ is peaked at $t = x/\bar u$ and falls off exponentially for longer or shorter values of t. Thus the major contribution of this term to the integral comes from values of t close to $x/\bar u$. Therefore let us perform a Taylor series expansion of

$$G(t) = \exp\left(-\frac{(x-\bar ut)^2 + y^2 + z^2}{2\sigma^2(t)}\right)$$

about $t = x/\bar u$. Using

$$G(t) = G(x/\bar u) + \left(\frac{dG}{dt}\right)_{t=x/\bar u}(t-x/\bar u)$$

we find

$$\left(\frac{dG}{dt}\right)_{t=x/\bar u} = -\exp\left(-\frac{y^2+z^2}{2\sigma^2(t)}\right)\left[\frac{1}{\sigma^3(t)}\frac{d\sigma}{dt}(y^2+z^2)\right]_{t=x/\bar u}$$

If we let $E = (y^2+z^2)/2\sigma^2(t)$, then

$$\left(\frac{dG}{dt}\right)_{t=x/\bar u} = -\exp(-E)\left[\frac{2}{\sigma}\frac{d\sigma}{dt}E\right]_{t=x/\bar u}$$

Thus the Taylor series approximation of the exponential is

$$\exp\left(-\frac{(x-\bar{u}t)^2 + y^2 + z^2}{2\sigma^2(t)}\right) = e^{-E}\left[1 - \left(t - \frac{x}{\bar{u}}\right)\left(E\frac{2}{\sigma}\frac{d\sigma}{dt}\right)_{t=x/\bar{u}}\right] \qquad (18.38)$$

Now assume that the major contribution to the integral of (18.37) comes from values of t in the range

$$\frac{x}{\bar{u}} - a\frac{\sigma}{\bar{u}} \le t \le \frac{x}{\bar{u}} + a\frac{\sigma}{\bar{u}}$$

Substituting (18.38) into (18.37) yields

$$\langle c(x,y,z)\rangle = \int_{(x-a\sigma)/\bar{u}}^{(x+a\sigma)/\bar{u}} \frac{q}{(2\pi)^{3/2}\sigma^3(t)} e^{-E}\left[1 - \left(t - \frac{x}{\bar{u}}\right)\left(\frac{2}{\sigma}\frac{d\sigma}{dt}E\right)_{t=x/\bar{u}}\right] dt \qquad (18.39)$$

If we now neglect the $(t - x/\bar{u})$ term as being of order $a\sigma/\bar{u}$ and assume that within the limits of integration $\sigma(t) \simeq \sigma(x/\bar{u})$, we get

$$\langle c(x,y,z)\rangle = \frac{2qe^{-E}(a\sigma/\bar{u})}{(2\pi)^{3/2}\sigma^3} \qquad (18.40)$$

To obtain a, if we impose the condition of conservation of mass, that is, that the total flow of material through the plane at any x is q, we obtain $a = (\pi/2)^{1/2}$. Thus (18.40) becomes

$$\langle c(x,y,z)\rangle = \frac{q}{2\pi\bar{u}\sigma^2(x/\bar{u})}\exp\left(-\frac{y^2+z^2}{2\sigma^2(x/\bar{u})}\right) \qquad (18.41)$$

The assumptions made in deriving (18.41) require that $a\sigma/\bar{u} \ll x/\bar{u}$. Thus the condition for validity of the result is

$$\frac{\sigma(x/\bar{u})}{x} \ll 1$$

Physically, the mean concentration emanating from a point source is a plume that can be visualized to be composed of many puffs, each of whose concentration distributions is sharply peaked about its centroid at all travel distances. Thus the spread of each puff is small compared to the downwind distance it has traveled. This assumption is called the *slender plume approximation*.

The procedure we have followed to obtain (18.41) can readily be generalized to $\sigma_x \ne \sigma_y \ne \sigma_z$. The result is

$$\langle c(x,y,z)\rangle = \frac{q}{2\pi\bar{u}\sigma_y\sigma_z}\exp\left(-\frac{y^2}{2\sigma_y^2} - \frac{z^2}{2\sigma_z^2}\right) \qquad (18.42)$$

This equation for the mean concentration from a continuous point source occupies a key position in atmospheric diffusion theory and we will have occasion to refer to it again and again.

An Alternate Derivation of (18.42) Our derivation of (18.42) was based on physical reasoning concerning the amount of spreading of a puff compared to its downwind distance. We can obtain (18.42) in a slightly different manner by assuming specific functional forms for σ_x, σ_y, and σ_z. Specifically, let us select the "long-time" form [recall (18.26)]:

$$\sigma_x^2 = a_x t, \qquad \sigma_y^2 = a_y t, \qquad \sigma_z^2 = a_z t \qquad (18.43)$$

Thus it is desired to evaluate

$$\langle c(x, y, z)\rangle = \int_0^\infty \frac{q}{(2\pi)^{3/2}(a_x a_y a_z)^{1/2}t^{3/2}} \exp\left(-\frac{(x - \bar{u}t)^2}{2a_x t} - \frac{y^2}{2a_y t} - \frac{z^2}{2a_z t}\right) dt$$

This integral can be expressed as

$$\langle c(x, y, z)\rangle = \frac{q}{(2\pi)^{3/2}(a_x a_y a_z)^{1/2}} \int_0^\infty t^{-3/2} \exp\left(-\frac{(r^2 - 2\bar{u}xt + \bar{u}^2 t^2)}{2a_x t}\right) dt$$

where $r^2 = x^2 + (a_x/a_y)y^2 + (a_x/a_z)z^2$. Thus

$$\langle c(x, y, z)\rangle = \frac{q}{(2\pi)^{3/2}(a_x a_y a_z)^{1/2}} e^{\bar{u}x/a_x} \int_0^\infty t^{-3/2} \exp\left[-\left(\frac{r^2}{2a_x t} + \frac{\bar{u}^2 t}{2a_x}\right)\right] dt$$

Let $\eta = t^{-1/2}$ and

$$\langle c(x, y, z)\rangle = \frac{2q}{(2\pi)^{3/2}(a_x a_y a_z)^{1/2}} e^{\bar{u}x/a_x} \int_0^\infty \exp\left[-\left(\frac{r^2\eta^2}{2a_x} + \frac{\bar{u}^2}{2a_x \eta^2}\right)\right] d\eta$$

The integral is of the general form

$$\int_0^\infty e^{-(a\eta^2 + b/\eta^2)} d\eta = \frac{1}{2}\left(\frac{\pi}{a}\right)^{1/2} e^{-2(ab)^{1/2}}$$

so finally

$$\langle c(x, y, z)\rangle = \frac{q}{2\pi(a_y a_z)^{1/2}r} \exp\left(-\frac{\bar{u}}{a_x}(r - x)\right) \qquad (18.44)$$

is the expression for the mean concentration from a continuous point source of strength q at the origin in an infinite fluid with the variances given by (18.43). If advection dominates plume dispersion so that only the concentrations close to the plume centerline are of importance, we will be interested in the solution only for values of x, y, and z that satisfy

$$\frac{\left(\frac{a_x}{a_y}\right)y^2 + \left(\frac{a_x}{a_z}\right)z^2}{x^2} \ll 1$$

which can be viewed as the result of two assumptions, that

$$\frac{y^2 + z^2}{x^2} \ll 1$$

and that

$$\frac{a_x}{a_y} = O(1); \qquad \frac{a_x}{a_z} = O(1)$$

The latter assumption implies that the variances of the windspeeds are of the same order of magnitude. Since

$$r = x\{1 + [(a_x/a_y)y^2 + (a_x/a_z)z^2]/x^2\}^{1/2} \tag{18.45}$$

using $(1 + \zeta)^p = 1 + \zeta p + \cdots$, (18.45) can be approximated by[6]

$$r = x\left(1 + \frac{(a_x/a_y)y^2 + (a_x/a_z)z^2}{2x^2}\right) \tag{18.46}$$

In (18.44) we approximate r by x and $r - x$ by the expression above. Thus (18.44) becomes

$$\langle c(x, y, z)\rangle = \frac{q}{2\pi(a_y a_z)^{1/2}x}\exp\left\{-\frac{\bar{u}}{2a_x x}\left[\left(\frac{a_x}{a_y}\right)y^2 + \left(\frac{a_x}{a_z}\right)z^2\right]\right\} \tag{18.47}$$

---

[6]The expression

$$(1 + \zeta)^p = 1 + \binom{p}{1}\zeta + \binom{p}{2}\zeta^2 + \cdots$$

is valid for all p if $-1 < \zeta < 1$. Thus

$$(1 + \zeta)^{1/2} = 1 + \frac{1}{2}\zeta + \frac{1}{8}\zeta^2 + \cdots$$

for $-1 < \zeta < 1$. Note that

$$\binom{p}{v} = \frac{p!}{v!(p - v)!}$$

where for noninteger p, $p! = \Gamma(p + 1)$.

If we relate time and distance from the source x by $t = x/\bar{u}$, then we can use (18.43) to write

$$\sigma_y^2 = \frac{a_y x}{\bar{u}}, \qquad \sigma_z^2 = \frac{a_z x}{\bar{u}} \tag{18.48}$$

Then (18.47) becomes identical to (18.42).

Still Another Derivation of (18.42)   We can use the relation

$$\lim_{\sigma \to 0} \frac{1}{(2\pi)^{1/2}\sigma} \exp\left(-\frac{(x - x')^2}{2\sigma^2}\right) = \delta(x - x') \tag{18.49}$$

to take the limit of the x term in (18.38) as $\sigma_x \to 0$. Using (18.49) and letting $\xi = \bar{u}t$, we get

$$\langle c(x, y, z) \rangle = \lim_{\xi \to \infty} \int_0^{\xi/\bar{u}} \frac{q}{2\pi\bar{u}\sigma_y\sigma_z} \exp\left(-\frac{y^2}{2\sigma_y^2} - \frac{z^2}{2\sigma_z^2}\right) \delta(x - \xi)\,d\xi$$

where σ_y and σ_z are now functions of $\xi/\bar{u}$. This last formula can be simplified to give (18.42). This derivation is the shortest, but we delayed it after those that more clearly illustrate the physical assumptions.

18.6.2 Eulerian Approach

The problem of determining the concentration distribution resulting from a continuous source of strength q at the origin in an infinite isotropic fluid with a velocity $\bar{u}$ in the x direction can be formulated by the Eulerian approach as follows:

$$\bar{u}\frac{\partial \langle c \rangle}{\partial x} = K\left(\frac{\partial^2 \langle c \rangle}{\partial x^2} + \frac{\partial^2 \langle c \rangle}{\partial y^2} + \frac{\partial^2 \langle c \rangle}{\partial z^2}\right) + q\delta(x)\delta(y)\delta(z) \tag{18.50}$$

$$\langle c(x, y, z) \rangle = 0 \qquad x, y, z \to \pm\infty \tag{18.51}$$

The solution of (18.50) and (18.51) is given in Appendix 18.A.2. It is

$$\langle c(x, y, z) \rangle = \frac{q}{4\pi Kr} \exp\left(-\frac{\bar{u}(r - x)}{2K}\right) \tag{18.52}$$

where $r^2 = x^2 + y^2 + z^2$, or

$$r = x\left(1 + \frac{y^2 + z^2}{x^2}\right)^{1/2} \tag{18.53}$$

If we invoke the slender plume approximation, we are interested only in the solution close to the plume centerline. Thus, as in (18.45), (18.53) can be approximated by

$$r \simeq x\left(1 + \frac{y^2 + z^2}{2x^2}\right) \tag{18.54}$$

If in (18.52) r is approximated by x and $r - x$ by $(y^2 + z^2)/2x$, (18.52) becomes

$$\langle c(x, y, z) \rangle = \frac{q}{4\pi Kx} \exp\left[-\left(\frac{\bar{u}}{4Kx}\right)(y^2 + z^2)\right] \tag{18.55}$$

An Alternate Derivation of (18.55)   Equation (18.55) is based on the slender plume approximation as expressed by (18.54). We will now show that the slender plume approximation is equivalent to neglecting diffusion in the direction of the mean flow in the atmospheric diffusion equation. Thus $\langle c \rangle$ is governed by

$$\bar{u}\frac{\partial \langle c \rangle}{\partial x} = K\left(\frac{\partial^2 \langle c \rangle}{\partial y^2} + \frac{\partial^2 \langle c \rangle}{\partial z^2}\right) + q\delta(x)\delta(y)\delta(z) \tag{18.56}$$

$$\langle c(0, y, z) \rangle = 0 \tag{18.57}$$

$$\langle c(x, y, z) \rangle = 0 \qquad y, z \rightarrow \pm\infty \tag{18.58}$$

Before we proceed to solve (18.56), a few comments about the boundary conditions are useful. When the x diffusion term is dropped in the atmospheric diffusion equation, the equation becomes first-order in x, and the natural point for the single boundary condition on x is at $x = 0$. Since the source is also at $x = 0$ we have an option of whether to place the source on the RHS of the equation, as in (18.56), or in the $x = 0$ boundary condition. If we follow the latter course, then the $x = 0$ boundary condition is obtained by equating material fluxes across the plane at $x = 0$. The result is

$$\bar{u}\frac{\partial \langle c \rangle}{\partial x} = K\left(\frac{\partial^2 \langle c \rangle}{\partial y^2} + \frac{\partial^2 \langle c \rangle}{\partial z^2}\right) \tag{18.59}$$

$$\langle c(0, y, z) \rangle = \frac{q}{\bar{u}}\delta(y)\delta(z) \tag{18.60}$$

$$\langle c(x, y, z) \rangle = 0 \qquad y, z \rightarrow \pm\infty \tag{18.61}$$

The sets of (18.56) to (18.58) and (18.59) to (18.61) are entirely equivalent. We present the solution of (18.59) in Appendix 18.A.3. The solution is

$$\langle c(x, y, z) \rangle = \frac{q}{4\pi Kx} \exp\left(-\frac{\bar{u}}{4Kx}(y^2 + z^2)\right) \tag{18.62}$$

If we allow K_{yy} to be different from K_{zz}, the analogous result is

$$\langle c(x, y, z) \rangle = \frac{q}{4\pi(K_{yy} K_{zz})^{1/2}x} \exp\left[-\frac{\bar{u}}{4x}\left(\frac{y^2}{K_{yy}} + \frac{z^2}{K_{zz}}\right)\right] \tag{18.63}$$

18.6.3   Summary of Continuous Point Source Solutions

Table 18.1 presents a summary of the solutions obtained in this section. Of primary interest at this point is a comparison of the forms of the Lagrangian and Eulerian expressions, in particular, the relationships between eddy diffusivities and the plume dispersion variances. For the slender plume cases, for example, the Lagrangian and Eulerian

TABLE 18.1   Expressions for the Mean Concentration from a Continuous Point Source in an Infinite Fluid in Stationary, Homogeneous Turbulence

Approach	Full Solution	Solution Employing Slender Plume Approximation
Lagrangian $\sigma_x^2 = a_x t$	$\dfrac{q}{2\pi(a_y a_z)^{1/2} r} \exp\left(-\dfrac{\bar{u}}{a_x}(r-x)\right)$	$\dfrac{q}{2\pi\bar{u}\sigma_y\sigma_z} \exp\left[-\left(\dfrac{y^2}{2\sigma_y^2} + \dfrac{z^2}{2\sigma_z^2}\right)\right]$
$\sigma_y^2 = a_y t$ $\sigma_z^2 = a_z t$	$r^2 = x^2 + (a_x/a_y)y^2 + (a_x/a_z)z^2$	
Eulerian	$\dfrac{q}{4\pi(K_{yy}K_{zz}x^2 + K_{xx}K_{zz}y^2 + K_{xx}K_{yy}z^2)^{1/2}}$	$\dfrac{q}{4\pi(K_{yy}K_{zz})^{1/2}x}$
	$\times \exp\left\{-\dfrac{\bar{u}}{2K_{xx}}\left[\left(\dfrac{x^2}{K_{xx}} + \dfrac{y^2}{K_{yy}} + \dfrac{z^2}{K_{zz}}\right)^{1/2} - x\right]\right\}$	$\times \exp\left[-\dfrac{\bar{u}}{4x}\left(\dfrac{y^2}{K_{yy}} + \dfrac{z^2}{K_{zz}}\right)\right]$

expressions are identical if

$$\sigma_y^2 = \frac{2K_{yy}x}{\bar{u}} \qquad \sigma_z^2 = \frac{2K_{zz}x}{\bar{u}} \qquad (18.64)$$

In most applications of the Lagrangian formulas, the dependences of σ_y^2 and σ_z^2 on x are determined empirically rather than as indicated in (18.64). Thus the main purpose of the formulas in Table 18.1 is to provide a comparison between the two approaches to atmospheric diffusion theory.

18.7   STATISTICAL THEORY OF TURBULENT DIFFUSION

Up to this point in this chapter we have developed the common theories of turbulent diffusion in a purely formal manner. We have done this so that the relationship of the approximate models for turbulent diffusion, such as the K theory and the Gaussian formulas, to the basic underlying theory is clearly evident. When such relationships are clear, the limitations inherent in each model can be appreciated. We have in a few cases applied the models obtained to the prediction of the mean concentration resulting from an instantaneous or continuous source in idealized stationary, homogeneous turbulence. In Section 18.7.1 we explore further the physical processes responsible for the dispersion of a puff or plume of material. Section 18.7.2 can be omitted on a first reading of this chapter; that section goes more deeply into the statistical properties of atmospheric dispersion, such as the variances $\sigma_i^2(t)$, which are needed in the actual use of the Gaussian dispersion formulas.

18.7.1   Qualitative Features of Atmospheric Diffusion

The two idealized source types commonly used in atmospheric turbulent diffusion are the instantaneous point source and the continuous point source. An instantaneous point source

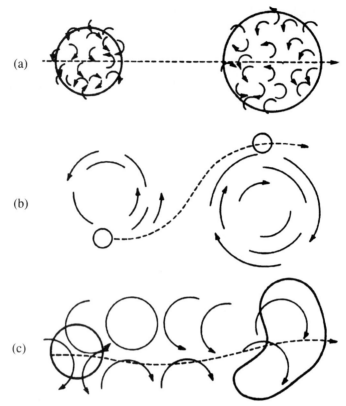

FIGURE 18.1   Dispersion of a puff of material under three turbulence conditions: (a) puff embedded in a field in which the turbulent eddies are smaller than the puff, (b) puff embedded in a field in which the turbulent eddies are larger than the puff, and (c) puff embedded in a field in which the turbulent eddies are comparable in size to the puff.

is the conventional approximation to a rapid release of a quantity of material. Obviously, an "instantaneous point" is a mathematical idealization since any rapid release has finite spatial dimensions. As the puff is carried away from its source by the wind, it will disperse under the action of turbulent velocity fluctuations. Figure 18.1 shows the dispersion of a puff under three different turbulence conditions. Figure 18.1a shows a puff embedded in a turbulent field in which all the turbulent eddies are smaller than the puff. The puff will disperse uniformly as the turbulent eddies at its boundary entrain fresh air. In Figure 18.1b, a puff is embedded in a turbulent field all of whose eddies are considerably larger than the puff. In this case the puff will appear to the turbulent field as a small patch of fluid that will be convected through the field with little dilution. Ultimately, molecular diffusion will dissipate the puff. Figure 18.1c shows a puff in a turbulent field of eddies of size comparable to the puff. In this case the puff will be both dispersed and distorted. In the atmosphere, a cloud of material is always dispersed since there are almost always eddies of size smaller than the cloud. From Figure 18.1 we can see that the dispersion of a puff relative to its center of mass depends on the initial size of the puff relative to the lengthscales of the turbulence. In order to describe such relative dispersion, we must consider the statistics of the separation of two representative fluid particles in the puff. The

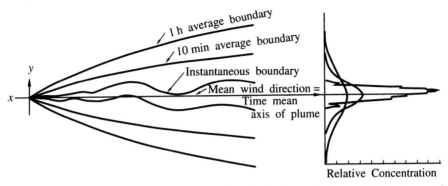

FIGURE 18.2 Plume boundaries and concentration distributions of a plume at different averaging times.

analysis of the wandering of a single particle is insufficient to tell us about the dispersion of a cloud (Csanady 1973).

A continuous source emits a plume that might be envisioned, as we have noted, as an infinite number of puffs released sequentially with an infinitesimal time interval between them. The quantity of material released is expressed in terms of a rate, say, grams per minute. The dimensions of a plume perpendicular to the plume axis are generally given in terms of the standard deviation of the mean concentration distribution since the mean cross-sectional distributions are often nearly Gaussian. Figure 18.2 shows the plume "boundaries" and concentration distributions as might be seen in an instantaneous snapshot and exposures of a few minutes and several hours. An instantaneous picture of a plume reveals a meandering behavior with the width of the plume gradually growing downwind of the source. Longer-time averages give a more regular appearance to the plume and a smoother concentration distribution.

If we were to take a time exposure of the plume at large distances from the source, we would find that the boundaries of the time-averaged plume would begin to meander, because the plume would come under the influences of larger and larger eddies, and the averaging time, say, several hours, would still be too brief to time average adequately the effect of these larger eddies. Eddies larger in size than the plume dimension tend to transport the plume intact, whereas those that are smaller tend to disperse it. As the plume becomes wider, larger and larger eddies become effective in dispersing the plume and the smaller eddies become increasingly ineffective.

The theoretical analysis of the spread of a plume from a continuous point source can be achieved by considering the statistics of the diffusion of a single fluid particle relative to a fixed axis. The actual plume would then consist of a very large number of such identical particles, the average over the behavior of which yields the ensemble statistics of the plume.

18.7.2 Motion of a Single Particle Relative to a Fixed Axis

Let us consider, as shown in Figure 18.3, a single particle that is at position $\mathbf{x}_0$ at time t_0 and is at position $\mathbf{x}$ at some later time t in a turbulent field. The complete statistical properties of the particle's motion are embodied in the transition probability density $Q(\mathbf{x}, t \mid \mathbf{x}', t')$. An analysis of this problem for stationary, homogeneous turbulence was presented by Taylor (1921) in one of the classic papers in the field of turbulence. If the

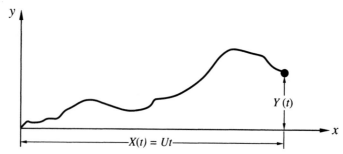

FIGURE 18.3   Motion of a single marked particle in a turbulent flow.

turbulence is stationary and homogeneous, $Q(\mathbf{x}, t \mid \mathbf{x}', t') = Q(\mathbf{x} - \mathbf{x}'; t - t')$; that is, Q depends only on the displacements in space and time and not on the initial position or time. The single particle may be envisioned as one of a very large number of particles that are emitted sequentially from a source located at $\mathbf{x}_0$. The distribution of the concentration of marked particles in the fluid is known once the statistical behavior of one representative marked particle is known. For convenience we assume that the particle is released at the origin at $t = 0$, so that its displacement corresponds to its coordinate location at time t.

The most important statistical quantity is the mean-square displacement of the particle from the source after a time t, since the mean displacement from the axis parallel to the flow direction will be zero. If we envision this particle as one being emitted from a continuous source, we can see that the mean-square displacement of the particle from the axis of the plume will tell us the width of the plume and hence the variances $\sigma_i^2(t)$.

The mean displacement of the particle along the ith coordinate is defined by

$$\langle X_i(t) \rangle = \int_0^t x_i Q(\mathbf{x}; t) d\mathbf{x} \tag{18.65}$$

where the braces ($\langle \rangle$) indicate an ensemble average over an infinite number of identical marked particles. If the velocity of the particle in the ith direction at any time is $v_i(t)$, the position of the particle at time t is given by [recall (18.17)]

$$X_i(t) = \int_0^t v_i(t') dt' \tag{18.66}$$

where the velocity of the particle at any instant is equal, by definition, to the fluid velocity at the spot where the particle happens to be at that instant:

$$v_i(t) = u_i[\mathbf{X}(t), t] \tag{18.67}$$

Let us consider a situation in which there is no mean velocity, so that $v_i(t) = u_i'[\mathbf{X}(t), t]$. If there is a mean velocity, say, in the x_i direction, we will be interested in the dispersion about a point moving with the mean velocity. Therefore the influence of the fluctuating Eulerian velocities on the wanderings of the marked particle from its axis is the key issue here, not its translation in the mean flow. We might also note that, in discussing the statistics of particle motion, all averages are conceptually ensemble averages, carried out over a very large number of similar particle releases. For this reason, we denote mean quantities by braces, as in (18.65), as opposed to overbars, which have been reserved for time averages. It is understood, however, that when discussing mean properties of the velocity field itself, due to the condition of stationarity, the ensemble average and the time average are identical. The mean displacement can also be computed by ensemble averaging of (18.66),

$$\langle X_i(t) \rangle = \int_0^t \langle v_i(t') \rangle \, dt' \tag{18.68}$$

where the averaging can be taken inside the integral.

The variance of the displacements is the expected value of the product of X_i and X_j, which is expressed in terms of Q by

$$P_{ij}(t) = \langle X_i(t) X_j(t) \rangle = \int_{-\infty}^{\infty} x_i x_j Q(\mathbf{x}; t) d\mathbf{x} \tag{18.69}$$

The diagonal elements $\langle X_i^2(t) \rangle$ are of principal interest since they describe the rate of spreading along each axis. $P_{ii}(t)$ is just $\sigma_i^2(t)$. Using (18.66) in (18.69), we obtain [recall (18.23)]

$$P_{ij}(t) = \left\langle \int_0^t v_i(t') dt' \int_0^t v_j(t'') dt'' \right\rangle$$

$$= \int_0^t \int_0^t \langle v_i(t') v_j(t'') \rangle \, dt' \, dt'' \tag{18.70}$$

The integrand of (18.70) is defined as the Lagrangian correlation function $R_{ij}(t' - t'')$, so that (18.70) may be rewritten

$$P_{ij}(t) = \int_0^t \int_0^t R_{ij}(t' - t'') \, dt' \, dt''$$

$$= \int_0^t \int_0^{t-t'} R_{ij}(\zeta) \, d\zeta \, dt' \tag{18.71}$$

By definition $R_{ij}(\zeta) = R_{ji}(-\zeta)$, so that

$$P_{ij}(t) = \int_0^t (t - \zeta)[R_{ij}(\zeta) + R_{ji}(\zeta)] d\zeta \tag{18.72}$$

We now consider the form of $P_{ij}(t)$ in the two limiting situations, $t \to 0$ and $t \to \infty$. First, for $t \to 0$, we have $R_{ij}(\zeta) \simeq R_{ij}(0) = \langle u_i' u_j' \rangle = \overline{u_i' u_j'}$. Thus

$$P_{ij}(t) = \overline{u_i' u_j'} t^2 \qquad t \to 0 \tag{18.73}$$

and the dispersion increases as t^2. Next, as $t \to \infty$ we expect the Lagrangian correlation function $R_{ij}(t)$ to approach zero as the motion of the particle becomes uncorrelated with its original velocity. We expect convergence of the following integral

$$\int_0^{\infty} [R_{ij}(\zeta) + R_{ji}(\zeta)] d\zeta = I_{ij}$$

where I_{ij} is proportional to the Lagrangian timescale of the turbulence. Defining

$$\int_0^\infty \zeta[R_{ij}(\zeta) + R_{ji}(\zeta)]\,d\zeta = J_{ij}$$

as $t \to \infty$, $P_{ij}(t)$ becomes proportional to t:

$$P_{ij}(t) = I_{ij}t - J_{ij} \tag{18.74}$$

We can obtain these results by a slightly different route. Let us, for example, compute the rate of change of the dispersion $\langle X_i^2(t) \rangle$:

$$\frac{d}{dt}\langle X_i^2(t)\rangle = 2\langle X_i(t)v_i(t)\rangle$$

$$= 2\left\langle v_i(t) \int_0^t v_i(t')\,dt' \right\rangle$$

$$= 2 \int_0^t \langle v_i(t)v_i(t')\rangle\,dt'$$

$$= 2 \int_0^t R_{ii}(t - t')\,dt' \tag{18.75}$$

Integration of (18.75) with respect to t gives (18.71).

In summary, we have found that mean-square dispersion of a particle in stationary, homogeneous turbulence has the following dependence on time

$$P_{ii}(t) = \begin{cases} \overline{u_i^2}t^2 & t \to 0 \\ 2K_{ii}t & t \to \infty \end{cases} \tag{18.76}$$

as we had anticipated by (18.26) and (18.27), where

$$K_{ii} = \lim_{t\to\infty} \int_0^t R_{ii}(t - t')\,dt' \tag{18.77}$$

Since $\overline{u_i'^2}$ is proportional to the total turbulent kinetic energy, the total energy of the turbulence is important in the early dispersion. After long times the largest eddies will contribute to R_{ii} and R_{ii} will not go to zero until the particle can escape the influence of the largest eddies. From its definition, K_{ii} has the dimensions of a diffusivity, since as $t \to \infty$

$$\frac{1}{2}\frac{d\langle X_i^2(t)\rangle}{dt} = K_{ii} \tag{18.78}$$

Now, if we return to Section 18.5 we see that this K_{ii} is precisely the constant eddy diffusivity used in the atmospheric diffusion equation for stationary, homogeneous turbulence. Since the Lagrangian timescale is defined by

$$T_L = \max_i \left(\frac{1}{\overline{u_i'^2}} \int_0^\infty R_{ii}(t)\,dt \right) \tag{18.79}$$

it becomes clear that K theory, with constant K values, should apply only when the diffusion time t is much greater than T_L. Because of large atmospheric eddies, T_L might be quite large. In general, large eddies dominate atmospheric diffusion when diffusion is measured relative to a fixed coordinate system.

It has been recognized that the simple exponential function, $\exp(-t/T_L)$, where t is travel time from the source, appears to approximate $R_{ii}(t)$ rather well (Neumann 1978; Tennekes 1979). If

$$R_{ii}(t) = \overline{u_i^2} \exp(-t/T_L) \tag{18.80}$$

then the mean-square particle displacement is given by

$$\langle X_i^2(t) \rangle = 2\overline{u_i^2} \int_0^t (t - t') e^{-t'/T_L} \, dt'$$

$$= 2\overline{u_i^2} T_L^2 \left(\frac{t}{T_L} - \left(1 - e^{-t/T_L} \right) \right) \tag{18.81}$$

This result is identical to (18.25) if $b = T_L^{-1}$, which provides a nice connection between the statistical theory of turbulent diffusion and the basic example considered earlier.

18.8 SUMMARY OF ATMOSPHERIC DIFFUSION THEORIES

Turbulent diffusion is concerned with the behavior of individual particles that are supposed to faithfully follow the airflow or, in principle, are simply marked minute elements of the air itself. Because of the inherently random character of atmospheric motions, one can never predict with certainty the distribution of concentration of marked particles emitted from a source. Although the basic equations describing turbulent diffusion are available, there does not exist a single mathematical model that can be used as a practical means of computing atmospheric concentrations over all ranges of conditions.

There are two basic ways of considering the problem of turbulent diffusion: the so-called Eulerian and Lagrangian approaches. The Eulerian method is based on carrying out a material balance over an infinitesimal region fixed in space, whereas the Lagrangian approach is based on considering the meandering of marked fluid particles in the flow. Each approach can be shown to have certain inherent difficulties that render impossible an exact solution for the mean concentration of particles in turbulent flow. For the purposes of practical computation, several approximate theories have been used for calculating mean concentrations of species in turbulence. Two are the K theory, based on the atmospheric diffusion equation, and the statistical theory, based on the behavior of individual particles in stationary, homogeneous turbulence.

The basic issues of interest with respect to the K theory are (1) under what conditions on the source configuration and the turbulent field can this theory be applied and (2) to what extent can the eddy diffusivities be specified in an a priori manner from measured properties of the turbulence. In summary, the spatial and temporal scales of the turbulence should be small in comparison with the corresponding scales of the concentration field.

The statistical theory is concerned with the actual velocities of individual particles in stationary, homogeneous turbulence. Under this assumption the statistics of the motion of one typical particle provides a statistical estimate of the behavior of all particles and that of two particles, an estimate of the behavior of a cluster of particles. In the atmosphere one may expect the crosswind component (v) of turbulence to be nearly homogeneous since the variations in the scale and intensity of v with height are often small. On the other hand, the vertical velocity component (w) is decidedly inhomogeneous, since characteristically w increases with height above the ground. Thus the statistical theory should be suitable for describing the spread of a plume in the crosswind direction regardless of the height but for vertical spread only in the early stages of travel from a source considerably elevated above the ground.

The deciding factor in judging the validity of a theory for atmospheric diffusion is the comparison of its predictions with experimental data. It must be kept in mind, however, that the theories we have discussed are based on predicting the ensemble mean concentration $\langle c \rangle$, whereas a single experimental observation constitutes only one sample from the hypothetically infinite ensemble of observations from that identical experiment. Thus it is not to be expected that any one realization should agree precisely with the predicted mean concentration even if the theory used is applicable to the set of conditions under which the experiment has been carried out. Nevertheless, because it is practically impossible to repeat an experiment more than a few times under identical conditions in the atmosphere, one must be content with at most a few experimental realizations when testing any available theory.

18.9   ANALYTICAL SOLUTIONS FOR ATMOSPHERIC DIFFUSION: THE GAUSSIAN PLUME EQUATION AND OTHERS

18.9.1   Gaussian Concentration Distributions

We have seen that under certain idealized conditions the mean concentration of a species emitted from a point source has a Gaussian distribution. This fact, although strictly true only in the case of stationary, homogeneous turbulence, serves as the basis for a large class of atmospheric diffusion formulas in common use. The collection of Gaussian-based formulas is sufficiently important in practical application that we devote a portion of this chapter to them. The focus of these formulas is the expression for the mean concentration of a species emitted from a continuous, elevated point source, the so-called Gaussian plume equation.

The basic Lagrangian expression for the mean concentration is (18.8)

$$
\begin{aligned}
\langle c(x,y,z,t) \rangle = {} & \int_{-\infty}^{\infty} \int_{-\infty}^{\infty} \int_{-\infty}^{\infty} Q(x,y,z,t \mid x_0, y_0, z_0, t_0) \langle c(x_0, y_0, z_0, t_0) \rangle \\
& \times dx_0 \, dy_0 \, dz_0 + \int_{-\infty}^{\infty} \int_{-\infty}^{\infty} \int_{-\infty}^{\infty} \int_{t_0}^{t} Q(x,y,z,t \mid x', y', z', t') \\
& \times S(x', y', z', t') dt' \, dx' \, dy' \, dz'
\end{aligned}
\tag{18.82}
$$

The transition probability density Q expresses physically the probability that a tracer particle that is at x', y', z' at t' will be at x, y, z at t. We showed that under conditions of stationary, homogenous turbulence Q has a Gaussian form. For example, in the case of a mean wind directed along the x axis, that is, $\bar{v} = \bar{w} = 0$, and an infinite domain, Q is

$$Q(x, y, z, t \mid x', y', z', t')$$
$$= \frac{1}{(2\pi)^{3/2} \sigma_x \sigma_y \sigma_z} \exp\left(-\frac{(x - x' - \bar{u}(t - t'))^2}{2\sigma_x^2} - \frac{(y - y')^2}{2\sigma_y^2} - \frac{(z - z')^2}{2\sigma_z^2}\right) \quad (18.83)$$

where the variances σ_x^2, σ_y^2, and σ_z^2 are functions of the travel time, $t - t'$.

Up to this point we have considered an infinite domain. For atmospheric applications a boundary at $z = 0$, the Earth, is present. Because of the barrier to diffusion at $z = 0$ it is necessary to modify the z dependence of Q to account for this fact. We can separate out the z dependence in (18.83) by writing

$$Q(x, y, z, t \mid x', y', z', t')$$
$$= \frac{1}{2\pi \sigma_x \sigma_y} \exp\left(-\frac{(x - x' - \bar{u}(t - t'))^2}{2\sigma_x^2} - \frac{(y - y')^2}{2\sigma_y^2}\right) Q_z(z, t \mid z', t') \quad (18.84)$$

To determine the form of $Q_z(z, t \mid z', t')$, we enumerate the following possibilities:

1. Form of upper boundary
   (a) $0 \leq z \leq \infty$
   (b) $0 \leq z \leq H$ with no diffusion across $z = H$ (i.e., inversion layer)
2. Type of interaction between the diffusing material and the surface
   (a) Total reflection
   (b) Total absorption
   (c) Partial absorption

For the moment let us continue to consider the vertical domain to be $0 \leq z \leq \infty$.

Total Reflection at z = 0   We assume that the presence of the surface at $z = 0$ can be accounted for by adding the concentration resulting from a hypothetical source at $z = -z'$ to that from the source at $z = z'$ in the region $z \geq 0$. Then Q_z assumes the form

$$Q_z(z, t \mid z', t') = \frac{1}{(2\pi)^{1/2} \sigma_z} \left[\exp\left(-\frac{(z - z')^2}{2\sigma_z^2}\right) + \exp\left(-\frac{(z + z')^2}{2\sigma_z^2}\right)\right] \quad (18.85)$$

Total Absorption at z = 0   If the Earth is a perfect absorber, the concentration of material at $z = 0$ is zero. The form of Q_z can be obtained by the same method of an image source at $-z'$, with the change that we now *subtract* the distribution from the source at $-z'$ from that for the source at $+z'$. The result is

$$Q_z(z, t \mid z', t') = \frac{1}{(2\pi)^{1/2} \sigma_z} \left[\exp\left(-\frac{(z - z')^2}{2\sigma_z^2}\right) - \exp\left(-\frac{(z + z')^2}{2\sigma_z^2}\right)\right] \quad (18.86)$$

The case of partial absorption at $z = 0$ cannot be treated by the same image source approach since some particles are reflected and some are absorbed. We will consider this case shortly.

We now turn to the case of a continuous source. The mean concentration from a continuous point source of strength q at height h above the (totally reflecting) Earth is given by (it is conventional to let h denote the source height, and we do so henceforth)

$$\langle c(x, y, z) \rangle = \lim_{t \to \infty} \int_0^t \frac{q}{(2\pi)^{3/2} \sigma_x \sigma_y \sigma_z} \exp\left(-\frac{(x - \bar{u}t')^2}{2\sigma_x^2} - \frac{y^2}{2\sigma_y^2}\right)$$
$$\times \left[\exp\left(-\frac{(z - h)^2}{2\sigma_z^2}\right) + \exp\left(-\frac{(z + h)^2}{2\sigma_z^2}\right)\right] dt' \qquad (18.87)$$

As usual, we will be interested in the slender plume case, so we evaluate the integral in the limit of $\sigma_x \to 0$. [Recall (18.49).] The result is

$$\langle c(x, y, z) \rangle = \frac{q}{2\pi \bar{u} \sigma_y \sigma_z} \exp\left(-\frac{y^2}{2\sigma_y^2}\right)$$
$$\times \left[\exp\left(-\frac{(z - h)^2}{2\sigma_z^2}\right) + \exp\left(-\frac{(z + h)^2}{2\sigma_z^2}\right)\right] \qquad (18.88)$$

the *Gaussian plume equation*.

For a totally absorbing surface at $z = 0$, we obtain

$$\langle c(x, y, z) \rangle = \frac{q}{2\pi \bar{u} \sigma_y \sigma_z} \exp\left(-\frac{y^2}{2\sigma_y^2}\right)$$
$$\times \left[\exp\left(-\frac{(z - h)^2}{2\sigma_z^2}\right) - \exp\left(-\frac{(z + h)^2}{2\sigma_z^2}\right)\right] \qquad (18.89)$$

18.9.2   Derivation of the Gaussian Plume Equation as a Solution of the Atmospheric Diffusion Equation

We saw that by assuming constant eddy diffusivities K_{xx}, K_{yy}, and K_{zz}, the solution of the atmospheric diffusion equation has a Gaussian form. Thus it should be possible to obtain (18.88) or (18.89) as a solution of an appropriate form of the atmospheric diffusion equation. More importantly, because of the ease in specifying different physical situations in the boundary conditions for the atmospheric diffusion equation, we want to include those situations that we were unable to handle easily in Section 18.9.1, namely, the existence of an inversion layer at height H and partial absorption at the surface. Readers not concerned with the details of this solution may skip directly to Section 18.9.3.

Let us begin with the atmospheric diffusion equation with eddy diffusivities that are functions of time:

$$\frac{\partial \langle c \rangle}{\partial t} + \bar{u} \frac{\partial \langle c \rangle}{\partial x} = K_{xx} \frac{\partial^2 \langle c \rangle}{\partial x^2} + K_{yy} \frac{\partial^2 \langle c \rangle}{\partial y^2} + K_{zz} \frac{\partial^2 \langle c \rangle}{\partial z^2} + S(x,y,z,t)$$

$$\langle c(x,y,z,0) \rangle = 0$$ (18.90)

$$\langle c(x,y,z,t) \rangle = 0 \qquad x, y \to \pm\infty$$

For the boundary conditions on z we assume that an impermeable barrier exists at $z = H$:

$$\frac{\partial \langle c \rangle}{\partial z} = 0 \qquad z = H$$ (18.91)

The case of an unbounded region $z \geq 0$ is simply obtained by letting $H \to \infty$. To include the case of partial absorption at the surface, we write the $z = 0$ boundary condition as

$$K_{zz} \frac{\partial \langle c \rangle}{\partial z} = v_d \langle c \rangle \qquad z = 0$$ (18.92)

where v_d is a parameter that is proportional to the degree of absorptivity of the surface, the so-called deposition velocity. We will study the properties and specification of v_d in Chapter 19; for now let us treat it merely as a parameter. For total reflection, $v_d = 0$, and for total absorption, $v_d = \infty$.

Solution of (18.90) to (18.92)   The solution of (18.90) can be expressed in terms of the Green's function $G(x,y,z,t \mid x',y',z',t')$ as

$$\langle c(x,y,z,t) \rangle = \int_0^H \int_{-\infty}^{\infty} \int_{-\infty}^{\infty} \int_0^t G(x,y,z,t \mid x',y',z',t')$$

$$\times S(x',y',z',t) dt' \, dx' \, dy' \, dz'$$ (18.93)

where (18.93) is identical to (18.82) and where G satisfies

$$\frac{\partial G}{\partial t} + \bar{u} \frac{\partial G}{\partial x} = K_{xx} \frac{\partial^2 G}{\partial x^2} + K_{yy} \frac{\partial^2 G}{\partial y^2} + K_{zz} \frac{\partial^2 G}{\partial z^2}$$ (18.94)

$$G(x,y,z,t' \mid x',y',z',t') = \delta(x - x')\delta(y - y')\delta(z - z')$$

$$G = 0 \qquad x, y \to \pm\infty$$

$$\frac{\partial G}{\partial z} = \beta G \qquad z = 0$$

$$\frac{\partial G}{\partial z} = 0 \qquad z = H$$

where $\beta = v_d / K_{zz}$. Physically, G represents the mean concentration at (x,y,z) at time t resulting from a unit source at (x',y',z') at time t'. First we remove the convection term by the coordinate transformation, $\xi = x - \bar{u}(t - t')$, which converts (18.94) to

$$\frac{\partial G}{\partial t} = K_{xx}\frac{\partial^2 G}{\partial \xi^2} + K_{yy}\frac{\partial^2 G}{\partial y^2} + K_{zz}\frac{\partial^2 G}{\partial z^2}$$

To obtain G, we let

$$G(\xi,y,z,t \,|\, \xi',y',z',t') = A(\xi,y,t \,|\, \xi',y',t')B(z,t \,|\, z',t')$$

where

$$\frac{\partial A}{\partial t} = K_{xx}\frac{\partial^2 A}{\partial \xi^2} + K_{yy}\frac{\partial^2 A}{\partial y^2} \tag{18.95}$$

$$A(\xi,y,t' \,|\, \xi',y',t') = \delta(\xi - \xi')\delta(y - y')$$
$$A(\xi,y,t \,|\, \xi',y',t') = 0 \qquad \xi,y \to \pm\infty$$

$$\frac{\partial B}{\partial t} = K_{zz}\frac{\partial^2 B}{\partial z^2} \tag{18.96}$$

$$B(z,t' \,|\, z',t') = \delta(z - z')$$

$$\frac{\partial B}{\partial z} = \beta B \qquad z = 0$$

$$\frac{\partial B}{\partial z} = 0 \qquad z = H$$

We begin with the solution of (18.95). Using separation of variables, $A(\xi,y,t \,|\, \xi',y',t') = A_x(\xi,t \,|\, \xi',t')A_y(y,t \,|\, y',t')$. The solutions are symmetric

$$A_x(\xi,t \,|\, \xi',t') = \frac{a}{(\bar{K}_{xx})^{1/2}}\exp\left(-\frac{(\xi - \xi')^2}{4\bar{K}_{xx}}\right)$$

$$A_y(y,t \,|\, y',t') = \frac{b}{(\bar{K}_{yy})^{1/2}}\exp\left(-\frac{(y - y')^2}{4\bar{K}_{yy}}\right)$$

where

$$\bar{K}_{ii} = \int_{t'}^{t} K_{ii}(\tau - t')d\tau$$

Thus

$$A(\xi,y,t \,|\, \xi',y',t') = \frac{a'}{(\bar{K}_{xx}\bar{K}_{yy})^{1/2}}\exp\left(-\frac{(\xi - \xi')^2}{4\bar{K}_{xx}} - \frac{(y - y')^2}{4\bar{K}_{yy}}\right)$$

We determine a' from the initial condition

$$A(\xi,y,t' \,|\, \xi',y',t') = \delta(\xi - \xi')\delta(y - y')$$

or

$$\frac{a'}{(\bar{K}_{xx}\bar{K}_{yy})^{1/2}} \int_{-\infty}^{\infty} \int_{-\infty}^{\infty} \exp\left(-\frac{(\xi-\xi')^2}{4\bar{K}_{xx}}\right) \exp\left(-\frac{(y-y')^2}{4\bar{K}_{yy}}\right) d\xi\, dy$$

$$= \int_{-\infty}^{\infty} \int_{-\infty}^{\infty} \delta(\xi-\xi')\delta(y-y')d\xi\, dy$$

which reduces to $a' = 1/4\pi$. Thus

$$A(\xi,y,t\,|\,\xi',y',t') = \frac{1}{4\pi(\bar{K}_{xx}\bar{K}_{yy})^{1/2}} \exp\left(-\frac{(\xi-\xi')^2}{4\bar{K}_{xx}} - \frac{(y-y')^2}{4\bar{K}_{yy}}\right) \tag{18.97}$$

Now we proceed to solve (18.96) to obtain

$$B(z,t\,|\,z',t') = 2\sum_{n=1}^{\infty} \frac{(\lambda_n^2+\beta^2)\cos[\lambda_n(H-z')]\cos[\lambda_n(H-z)]}{H(\lambda_n^2+\beta^2)+\beta} \exp(-\lambda_n^2\bar{K}_{zz}) \tag{18.98}$$

where the λ_n are the roots of

$$\lambda_n \tan \lambda_n H = \beta$$

In the case of a perfectly reflecting surface, $\beta = 0$, and

$$B(z,t\,|\,z',t') = \frac{2}{H}\left(1 + \sum_{n=2}^{\infty} \cos[\lambda_n(H-z')]\cos[\lambda_n(H-z)]\exp[-\lambda_n^2\bar{K}_{zz}]\right)$$

where $\sin \lambda_n H = 0$. This result can be simplified somewhat to

$$B(z,t\,|\,z',t') = \frac{2}{H}\left\{1 + \sum_{n=1}^{\infty} \cos\left(\frac{n\pi z}{H}\right)\cos\left(\frac{n\pi z'}{H}\right)\exp\left[-\left(\frac{n\pi}{H}\right)^2\bar{K}_{zz}\right]\right\} \tag{18.99}$$

The desired solution for $G(x,y,z,t\,|\,x',y',z',t')$ is obtained by combining the expressions for $A(\xi,y,t\,|\,\xi',y',t')$ and $B(z,t\,|\,z',t')$. In the case of a totally reflecting Earth, we have

$$G(\xi,y,z,t\,|\,\xi',y',z',t') = \frac{1}{2\pi H(\bar{K}_{xx}\bar{K}_{yy})^{1/2}}$$

$$\times \left\{1 + \sum_{n=1}^{\infty} \cos\left(\frac{n\pi z}{H}\right)\cos\left(\frac{n\pi z'}{H}\right)\exp\left[-\left(\frac{n\pi}{H}\right)^2\bar{K}_{zz}\right]\right\}$$

$$\times \exp\left(-\frac{(\xi-\xi')^2}{4\bar{K}_{xx}} - \frac{(y-y')^2}{4\bar{K}_{yy}}\right) \tag{18.100}$$

As usual, we are interested in neglecting diffusion in the x direction as compared with convection, that is, the slender plume approximation. We could return to the original problem, neglecting the term $\partial^2\langle c\rangle/\partial x^2$ in (18.90) and repeat the solution. We can also work with (18.100) and let $\bar{K}_{xx} \to 0$. To do so return to (18.97), and let $\sigma_x^2 = 2\bar{K}_{xx}$. Using (18.49) to take the limit of the x term as $\sigma_x \to 0$:

$$G(x,y,z,t \mid x',y',z',t') = \left\{1 + \sum_{n=1}^{\infty} \cos\left(\frac{n\pi z}{H}\right)\cos\left(\frac{n\pi z'}{H}\right)\exp\left[-\left(\frac{n\pi}{H}\right)^2 \bar{K}_{zz}\right]\right\}$$

$$\times \frac{1}{2(\pi\bar{K}_{yy})^{1/2}}\exp\left(-\frac{(y-y')^2}{4\bar{K}_{yy}}\right)\delta(x - x' - \bar{u}(t - t')) \quad (18.101)$$

Now we consider a continuous point source of strength q at $(0,0,h)$. The continuous source solution is obtained from the unsteady solution from

$$\langle c(x,y,z)\rangle = \lim_{t\to\infty} \int_0^H \int_{-\infty}^{\infty} \int_{\infty}^{\infty} \int_0^t G(x,y,z,t \mid x',y',z',t')$$

$$\times q\delta(x')\delta(y')\delta(z' - h)dt' \, dx' \, dy' \, dz'$$

The solution is illustrated for the case of a totally reflecting earth. Using (18.101) and carrying out the integration, we obtain

$$\langle c(x,y,z)\rangle = q\int_0^t \left\{1 + \sum_{n=1}^{\infty} \cos\left(\frac{n\pi z}{H}\right)\cos\left(\frac{n\pi h}{H}\right)\exp\left[-\left(\frac{n\pi}{H}\right)^2 \bar{K}_{zz}\right]\right\}$$

$$\times \frac{1}{2(\pi\bar{K}_{yy})^{1/2}}\exp\left(-\frac{y^2}{4\bar{K}_{yy}}\right)\delta(x - \bar{u}(t - t'))dt'$$

Evaluating the integral yields

$$\langle c(x,y,z)\rangle = \frac{q}{(\pi\bar{K}_{yy})^{1/2}H\bar{u}}\left\{1 + \sum_{n=1}^{\infty} \cos\left(\frac{n\pi z}{H}\right)\cos\left(\frac{n\pi h}{H}\right)\exp\left[-\left(\frac{n\pi}{H}\right)^2 \bar{K}_{zz}\right]\right\}$$

$$\times \exp\left(-\frac{y^2}{4\bar{K}_{yy}}\right) \quad (18.102)$$

This is the expression for the steady-state concentration resulting from a continuous point source located at $(0,0,h)$ between impermeable, nonabsorbing boundaries separated by a distance H when diffusion in the direction of the mean flow is neglected.

Finally, we wish to obtain the result when $H \to \infty$, that is, only one bounding surface at a distance h from the source. Let

$$K_{yy} = \frac{1}{2}\frac{d\sigma_y^2}{dt} \qquad K_{zz} = \frac{1}{2}\frac{d\sigma_z^2}{dt}$$

and (18.102) can be expressed as

$$\langle c(x,y,z) \rangle = \frac{2q}{(2\pi)^{1/2}\bar{u}\sigma_y H} \left\{ 1 + \sum_{n=1}^{\infty} \cos\left(\frac{n\pi z}{H}\right) \right.$$

$$\left. \times \cos\left(\frac{n\pi h}{H}\right) \exp\left[-\left(\frac{n\pi}{H}\right)^2 \frac{\sigma_z^2}{2}\right] \right\} \exp\left(-\frac{y^2}{2\sigma_y^2}\right)$$

Now let $H \to \infty$:

$$\langle c(x,y,z) \rangle = \frac{2q}{(2\pi)^{1/2}\bar{u}\sigma_y} \exp\left(-\frac{y^2}{2\sigma_y^2}\right)$$

$$\times \int_0^{\infty} \cos\left(\frac{\pi z}{H}\right) \cos\left(\frac{\pi h}{H}\right) \exp\left(-\frac{(\pi\sigma_x)^2}{2H^2}\right) d(1/H)$$

Now

$$\cos\left(\frac{\pi z}{H}\right)\cos\left(\frac{\pi h}{H}\right) = \frac{1}{2}\left[\cos\left(\frac{\pi(z+h)}{H}\right) + \cos\left(\frac{\pi(z-h)}{H}\right)\right]$$

and

$$\langle c(x,y,z) \rangle = \frac{q}{(2\pi)^{1/2}\bar{u}\sigma_y} \exp\left(-\frac{y^2}{2\sigma_y^2}\right)$$

$$\times \left[\int_0^{\infty} \cos\left(\frac{\pi(z+h)}{H}\right) \exp\left(-\frac{(\pi\sigma_z)^2}{2H^2}\right) d(1/H) \right.$$

$$\left. + \int_0^{\infty} \cos\left(\frac{\pi(z-h)}{H}\right) \exp\left(-\frac{(\pi\sigma_z)^2}{2H^2}\right) d(1/H) \right]$$

The integrals can be evaluated[7] to produce (18.88).

18.9.3 Summary of Gaussian Point Source Diffusion Formulas

The various point source diffusion formulas we have derived are summarized in Table 18.2.

18.10 DISPERSION PARAMETERS IN GAUSSIAN MODELS

We have derived several Gaussian-based models for estimating the mean concentration resulting from point source releases of material. We have noted that the conditions under

[7]Note that

$$\int_0^{\infty} e^{-\alpha x^2} \cos\beta x\, dx = \frac{1}{2}\left(\frac{\pi}{\alpha}\right)^{1/2} \exp\left(-\frac{\beta^2}{4\alpha}\right)$$

TABLE 18.2 Point Source Gaussian Diffusion Formulas

Mean Concentration	Assumptions
Gaussian puff formula $$\langle c(x,y,z,t)\rangle = \frac{q}{(2\pi)^{3/2}\sigma_x\sigma_y\sigma_z}\exp\left(-\frac{(x-x'-\bar{u}(t-t'))^2}{2\sigma_x^2}-\frac{(y-y')^2}{2\sigma_y^2}\right)$$ $$\times\left[\exp\left(-\frac{(z-z')^2}{2\sigma_z^2}\right)+\exp\left(-\frac{(z+z')^2}{2\sigma_z^2}\right)\right]$$	Total reflection at $z = 0$ $\bar{\mathbf{u}} = (\bar{u},0,0)$ $S = q\delta(x-x')\delta(y-y')\delta(z-z')\delta(t-t')$ $0 \le z \le \infty$
Gaussian puff formula $$\langle c(x,y,z,t)\rangle = \frac{q}{(2\pi)^{3/2}\sigma_x\sigma_y\sigma_z}\exp\left(-\frac{(x-x'-\bar{u}(t-t'))^2}{2\sigma_x^2}-\frac{(y-y')^2}{2\sigma_y^2}\right)$$ $$\times\left[\exp\left(-\frac{(z-z')^2}{2\sigma_z^2}\right)-\exp\left(-\frac{(z+z')^2}{2\sigma_z^2}\right)\right]$$	Total absorption at $z = 0$ $\bar{\mathbf{u}} = (\bar{u},0,0)$ $S = q\delta(x-x')\delta(y-y')\delta(z)\delta(z-z')\delta(t-t')$ $0 \le z \le \infty$
Gaussian puff formula $$\langle c(x,y,z,t)\rangle = \frac{q}{2\pi H\sqrt{K_{xx}K_{yy}}}\exp\left(-\frac{(x-x'-\bar{u}(t-t'))^2}{4\bar{K}_{xx}}-\frac{(y-y')^2}{4\bar{K}_{yy}}\right)$$ $$\times\left\{1+\sum_{n=1}^{\infty}\cos\lambda_n z\cos\lambda_n z'\exp[-\lambda_n^2\bar{K}_{zz}]\right\}$$ $$\lambda_m=\frac{n\pi}{H}\quad \bar{K}_{xx}=\frac{1}{2}\sigma_x^2\quad \bar{K}_{yy}=\frac{1}{2}\sigma_y^2\quad \bar{K}_{zz}=\frac{1}{2}\sigma_z^2$$	Total reflection at $z = 0$ $\bar{\mathbf{u}} = (\bar{u},0,0)$ $S = q\delta(x-x')\delta(y-y')\delta(z-z')\delta(t-t')$ $0 \le z \le H$

$$\langle c(x,y,z,t)\rangle = \frac{q}{2\pi\sqrt{\bar{K}_{xx}\bar{K}_{yy}}}\exp\left(-\frac{(x-x'-\bar{u}(t-t'))^2}{4\bar{K}_{xx}}-\frac{(y-y')^2}{4\bar{K}_{yy}}\right)$$

$$\times\sum_{n=1}^{\infty}\frac{(\lambda_n^2+\beta^2)\cos[\lambda_n(H-z')]\cos[\lambda_n(H-z)]}{H(\lambda_n^2+\beta^2)+\beta}\times\exp(-\lambda_n^2\bar{K}_{zz})$$

$$\lambda_n\tan\lambda_n H=\beta\quad\beta=v_d/K_{zz}$$

Partial absorption at z = 0
$\bar{\mathbf{u}}=(\bar{u},0,0)$
$S=q\delta(x-x')\delta(y-y')\delta(z-z')\delta(t-t')$
$0\leq z\leq H$

Gaussian plume formula

$$\langle c(x,y,z)\rangle=\frac{q}{2\pi\bar{u}\sigma_y\sigma_z}\exp\left(-\frac{y^2}{2\sigma_y^2}\right)\left[\exp\left(-\frac{(z-h)^2}{2\sigma_z^2}\right)+\exp\left(-\frac{(z+h)^2}{2\sigma_z^2}\right)\right]$$

Total reflection at z = 0
$\bar{\mathbf{u}}=(\bar{u},0,0)$
$S=q\delta(x)\delta(y)\delta(z-h)$
Slender plume approximation
$0\leq z\leq\infty$

Gaussian plume formula

$$\langle c(x,y,z)\rangle=\frac{q}{2\pi\bar{u}\sigma_y\sigma_z}\exp\left(-\frac{y^2}{2\sigma_y^2}\right)\left[\exp\left(-\frac{(z-h)^2}{2\sigma_z^2}\right)-\exp\left(-\frac{(z+h)^2}{2\sigma_z^2}\right)\right]$$

Total absorption at z = 0
$\bar{\mathbf{u}}=(\bar{u},0,0)$
$S=q\delta(x)\delta(y)\delta(z-h)$
Slender plume approximation
$0\leq z\leq\infty$

Gaussian plume formula

$$\langle c(x,y,z)\rangle=\frac{2q}{\sqrt{2\pi}\bar{u}\sigma_y H}\left\{1+\sum_{n=1}^{\infty}\cos\left(\frac{n\pi z}{H}\right)\cos\left(\frac{n\pi h}{H}\right)\right.$$

$$\left.\times\exp\left[-\left(\frac{n\pi}{H}\right)^2\frac{\sigma_z^2}{z}\right]\right\}\exp\left(-\frac{y^2}{2\sigma_y^2}\right)$$

Total reflection at z = 0
$\bar{\mathbf{u}}=(\bar{u},0,0)$
$S=q\delta(x)\delta(y)\delta(z-h)$
$0\leq z\leq H$

which the equation is valid are highly idealized and therefore that it should not be expected to be applicable to very many actual ambient situations. Because of its simplicity, however, the Gaussian plume equation has been applied widely (U.S. Environmental Protection Agency 1980). The justification for these applications is that the dispersion parameters σ_y and σ_z used have been derived from concentrations measured in actual atmospheric diffusion experiments under conditions approximating those of the application. This section is devoted to a summary of several results available for estimating Gaussian dispersion coefficients.

18.10.1   Correlations for σ_y and σ_z Based on Similarity Theory

As we noted in Section 18.7, the variances of the mean plume dimensions can be expressed in terms of the motion of single particles released from the source. (At a *particular instant* the plume outline is defined by the statistics of the trajectories of two particles released simultaneously at the source. We have not considered the two-particle problem here.) In an effort to overcome the practical difficulties associated with using (18.72) to obtain results for σ_y and σ_z, Pasquill (1971) suggested an alternate definition that retained the essential features of Taylor's statistical theory but that is more amenable to parametrization in terms of readily measured Eulerian quantities. As adopted by Draxler (1976), the American Meteorological Society (1977), and Irwin (1979), the Pasquill representation leads to

$$\sigma_y = \sigma_v t F_y \tag{18.103}$$

$$\sigma_z = \sigma_w t F_z \tag{18.104}$$

where σ_v and σ_w are the standard deviations of the wind velocity fluctuations in the y and z directions, respectively, and F_y and F_z are universal functions of a set of parameters that specify the characteristics of the atmospheric boundary layer. The exact forms of F_y and F_z are to be determined from data.

The variables on which F_y and F_z are assumed to depend are the friction velocity u_*, the Monin–Obukhov length L, the Coriolis parameter f, the mixed-layer depth z_i, the convective velocity scale w_*, the surface roughness z_0, and the height of pollutant release above the ground h.[8]

The variances σ_y^2 and σ_z^2 are therefore treated as empirical dispersion coefficients, the functional forms of which are determined by matching the Gaussian solution to data. In that way, σ_y and σ_z actually compensate for deviations from stationary, homogeneous conditions that are inherent in the assumed Gaussian distribution.

Of the two standard deviations, σ_y and σ_z, more is known about σ_y. First, most of the experiments from which σ_y and σ_z values are inferred involve ground-level measurements. Such measurements provide an adequate indication of σ_y, whereas vertical concentration distributions are needed to determine σ_z. Also, the Gaussian expression for vertical concentration distribution is known not to be obeyed for ground-level releases, so the fitting of a measured vertical distribution to a Gaussian form is considerably more difficult than that for the horizontal distribution where lateral symmetry and an approximate Gaussian form are good assumptions.

---

[8]It is conventional when referring to unstable conditions to represent the depth of the unstable, or mixed, layer by z_i. From a mathematical point of view in terms of the equations for mean concentration, z_i is identical to H, the height of an elevated layer impermeable to diffusion. The Coriolis parameter is discussed in Chapter 21.

For lateral dispersion, Irwin (1979) developed expressions for σ_v based on the work of Deardorff and Willis (1975), Draxler (1976), and Nieuwstadt and van Duuren (1979):

$$\sigma_v = \begin{cases} 1.78\,u_*[1 + 0.059(-z_i/L)]^{1/3} & z_i/L < 0 \\ 1.78u_* & z_i/L \geq 0 \end{cases} \qquad (18.105)$$

Irwin (1979), based on the work of Panofsky et al. (1977) and Nieuwstadt (1980), proposed the following form for F_y:

$$F_y = \begin{cases} \left[1 + \left(\dfrac{t}{T_i}\right)^{1/2}\right]^{-1}; T_i^{-1} = \dfrac{2.5u_*}{z_i}\left[1 + 0.0013\left(-\dfrac{z_i}{L}\right)\right]^{1/3} & \dfrac{z_i}{L} \leq 0 \\[4mm] \left[1 + 0.9\left(\dfrac{t}{T_0}\right)\right]^{-1}; T_0^{-1} = 1.001 & \dfrac{z_i}{L} > 0 \end{cases} \qquad (18.106)$$

We note that (18.103) when combined with (18.105)–(18.106) possesses the same limiting behavior as the statistical theory; that is, $\sigma_y \approx t$ as $t \to 0$ and $\sigma_y \approx t^{1/2}$ as $t \to \infty$.

The case of unstable or convective conditions is a special one in determining atmospheric dispersion. Since under convective conditions the most energetic eddies in the mixed layer scale with z_i, the timescale relevant to dispersion is z_i/w_*. This timescale is therefore roughly the time needed after release for material to become well mixed through the depth of the mixed layer.

A wide range of field and laboratory measurements on vertical and velocity fluctuations under unstable conditions can be represented by $\sigma_w = w_* G(z/z_i)$ (Irwin 1979), where

$$G(z/z_i) = \begin{cases} 1.342(z/z_i)^{0.333} & z/z_i < 0.03 \\ 0.763(z/z_i)^{0.175} & 0.03 < z/z_i < 0.40 \\ 0.722(1 - z/z_i)^{0.207} & 0.40 < z/z_i < 0.96 \\ 0.37 & z/z_i > 0.96 \end{cases} \qquad (18.107)$$

Under neutral and stable conditions the formulation for σ_w developed by Binkowski (1979) can be used

$$\sigma_w = u_*\left(\frac{\phi_m(z/L) - z/L}{3\kappa f_m}\right)^{1/3} \qquad z/L \geq 0 \qquad (18.108)$$

where from (16.75)

$$\phi_m(z/L) = 1 + 4.7z/L \qquad (18.109)$$

and where

$$f_m = \begin{cases} 0.4[1 + 3.9z/L - 0.25(z/L)^2] & z/L \leq 2 \\ 0.4[6.78 + 2.39(z/L - 2)] & z/L > 2 \end{cases} \qquad (18.110)$$

The next step to complete parametrization of the vertical dispersion coefficients is to specify F_z. Between neutral and very stable conditions, Irwin (1979) gives an interpolation formula that we do not reproduce here. Draxler (1976) developed the following results for F_z under neutral and stable conditions:

$$F_z = \begin{cases} [1 + 0.9(t/T_0)^{1/2}]^{-1} & z < 50\,\text{m} \\ [1 + 0.945(t/T_0)^{0.8}]^{-1} & z \geq 50\,\text{m} \end{cases} \tag{18.111}$$

Both expressions require specification of the characteristic time T_0. While an initial estimate of 50 s was given by Draxler, Irwin (1979) proposed additional values of T_0.

18.10.2 Correlations for σ_y and σ_z Based on Pasquill Stability Classes

The correlations for σ_y and σ_z in the previous subsection require knowledge of atmospheric variables that may not be available. In that case, one needs correlations for σ_y and σ_z based on readily available ambient data. The Pasquill stability categories A through F introduced in Chapter 16 provide a basis for such correlations.

The most widely used σ_y and σ_z correlations based on the Pasquill stability classes have been those developed by Gifford (1961). The correlations, commonly referred to as the "Pasquill–Gifford curves," appear in Figures 18.4 and 18.5.

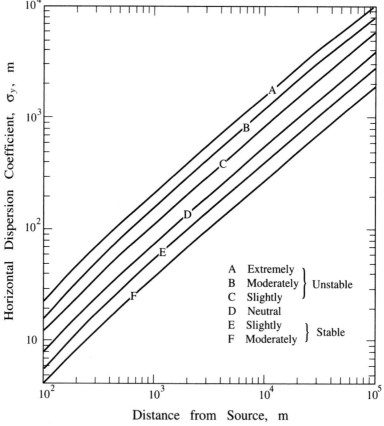

FIGURE 18.4 Correlations for σ_y based on the Pasquill stability classes A to F (Gifford 1961). These are the so-called Pasquill–Gifford curves.

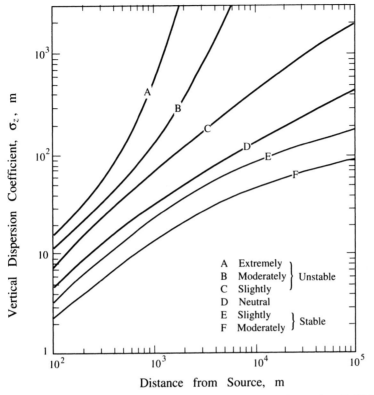

FIGURE 18.5 Correlations for σ_z based on the Pasquill stability classes A to F (Gifford 1961). These are the so-called Pasquill–Gifford curves.

For use in dispersion formulas it is convenient to have analytical expressions for σ_y and σ_z as functions of x. Many of the empirically determined forms can be represented by the power-law expressions

$$\sigma_y = R_y x^{r_y} \tag{18.112}$$

$$\sigma_z = R_z x^{r_z} \tag{18.113}$$

where R_y, R_z, r_y, and r_z depend on the stability class and the averaging time. Some commonly used dispersion coefficients, including those of Pasquill–Gifford (PG), are summarized in Table 18.3. Both the ASME and Klug dispersion coefficients are expressible by (18.112) and (18.113). Although the σ_y correlation for the PG coefficients is expressible in the form (18.112), that for σ_z requires a three-parameter form:[9]

$$\sigma_z = \exp[I_z + J_z \ln x + K_z (\ln x)^2] \tag{18.114}$$

---

[9]There are points of uncertainty surrounding the PG coefficients, including the stack heights at which they apply and the downwind distances to which they extend. Regardless of their technical merits, the PG coefficients have been incorporated (with and without modifications) into computer programs developed and endorsed by the U.S. Environmental Protection Agency. Note that some of these programs conservatively assume that the PG coefficients are for an averaging time of 1 h despite the fact that the true PG averaging time is about 10 min.

TABLE 18.3 Coefficients in Gaussian Plume Dispersion Parameter Correlations[a]

$$\sigma_y(x) = R_y x^{r_y} \qquad \sigma_z(x) = R_z x^{r_z}$$
$$\sigma_y(x) = \exp[I_y + J_y \ln x + K_y (\ln x)^2] \qquad \sigma_z(x) = \exp[I_z + J_z \ln x + K_z (\ln x)^2]$$

Source	Averaging Time, min	Coefficient	Stability Class					
			A	B	C	D	E	F
Pasquill–Gifford (Turner 1969;	10	R_y	0.443	0.324	0.216	0.141	0.105	0.071
Martin 1976)		r_y	0.894	0.894	0.894	0.894	0.894	0.894
ASME (1973)	60	R_y	0.40	0.36		0.32		0.31
		r_y	0.91	0.86		0.78		0.71
		R_z	0.40	0.33		0.22		0.06
		r_z	0.91	0.86		0.78		0.71
Klug (1969)	10	R_y	0.469	0.306	0.230	0.219	0.237	0.273
		r_y	0.903	0.885	0.855	0.764	0.691	0.594
		R_z	0.017	0.072	0.076	0.140	0.217	0.262
		r_z	1.380	1.021	0.879	0.727	0.610	0.500
Pasquill–Gifford (Turner 1969)	10	I_y	−1.104	−1.634	−2.054	−2.555	−2.754	−3.143
		J_y	0.9878	1.0350	1.0231	1.0423	1.0106	1.0148
		K_y	−0.0076	−0.0096	−0.0076	−0.0087	−0.0064	−0.0070
		I_z	4.679	−1.999	−2.341	−3.186	−3.783	−4.490
		J_z	−1.7172	0.8752	0.9477	1.1737	1.3010	1.4024
		K_z	0.2770	0.0136	−0.0020	−0.0316	−0.0450	−0.0540

[a]Application restricted to downwind distances not exceeding 10 km (Hanna et al. 1982).

For completeness we also give σ_y in this form in Table 18.3 for the PG correlations. The three sets of coefficients in Table 18.3 are based on different data. In choosing a set for a particular application, one should attempt to use that set most representative of the conditions of interest [see Gifford (1976), Weber (1976), AMS Workshop (1977), Doran et al. (1978), Sedefian and Bennett (1980), and Hanna et al. (1982)].

18.11   PLUME RISE

In order to ensure that waste gases emitted from stacks will rise above the top of the chimney, the gases are usually released at temperatures above that of the ambient air and are emitted with considerable initial momentum. Since the maximum mean ground-level concentration of effluents from an elevated point source depends roughly on the inverse square of the effective stack height, the amount of plume rise obtained is an important factor in reducing ground-level concentrations. The effective stack height is taken to be the sum of the actual stack height h_s and the plume rise Δh, defined as the height at which the plume becomes passive and subsequently follows the ambient air motion:

$$h = h_s + \Delta h \qquad (18.115)$$

The behavior of a plume is affected by a number of parameters, including the initial source conditions (exit velocity and difference between the plume temperature and that of the air), the stratification of the atmosphere, and the wind speed. Based on the initial source conditions, plumes can be categorized in the following manner:

Buoyant plume	Initial buoyancy $\gg$ initial momentum
Forced plume	Initial buoyancy $\simeq$ initial momentum
Jet	Initial buoyancy $\ll$ initial momentum

We shall deal here with buoyant and forced plumes only. Our interest is in predicting the rise of both buoyant and forced plumes in calm and windy, thermally stratified atmospheres.

Chracterization of plume rise in terms of the exhaust gas properties and the ambient atmospheric state is a complex problem. The most detailed approach involves solving the coupled mass, momentum, and energy conservation equations. This approach is generally not used in routine calculations because of its complexity. An alternate approach, introduced by Morton et al. (1956), is to consider the integrated form of the conservation equations across a section normal to the plume trajectory [e.g., see Fisher et al. (1979) and Schatzmann (1979)].

Table 18.4 (see also Table 18.5) presents a summary of several available plume rise formulas expressed in the form

$$\Delta h = \frac{Ex^b}{\bar{u}^a} \qquad (18.116)$$

TABLE 18.4   Summary of Several Plume Rise Formulas Expressed in the Form^a $\Delta h = Ex^b/\bar{u}^a$

Atmospheric Stability	a	b	E	Conditions	Reference
Plumes Dominated by Buoyancy Forces					
Neutral and unstable	1	0	$7.4(Fh_s^2)^{1/3}$		ASME (1973)
Stable^b	$\frac{1}{3}$	0	$29(F/S_1)^{1/3}$		
Neutral and unstable	1	$\frac{2}{3}$	$1.6F^{1/3}$	$F < 55, x < 49F^{5/8}$	Briggs (1969,
	1	0	$21.4\,F^{3/4}$	$F < 55, x \geq F^{5/8}$	1971, 1974)
	1	$\frac{2}{3}$	$1.6\,F^{1/3}$	$F \geq 55, x < 119F^{2/5}$	
	1	0	$38.7\,F^{3/5}$	$F \geq 55, x \geq 119\,F^{2/5}$	
Stable^{b,c}	$\frac{1}{3}$	0	$2.4(F/S_2)^{1/3}$		
	0	0	$5F^{1/4}S_2^{-3/8}$		
	1	$\frac{2}{3}$	$1.6F^{1/3}$		
Plumes Dominated by Momentum Forces					
All	1.4	0	$dV_s^{1.4}$	$V_s > 10\,\mathrm{m\,s}^{-1}$	ASME (1973)
				$V_s > \bar{u}$	
				$\Delta T < 50\,\mathrm{K}$	
Neutral^d	$\frac{2}{3}$	$\frac{1}{3}$	$1.44(dV_s)^{2/3}$	$V_s/\bar{u} \geq 4$	Briggs (1969)
	1	0	$3dV_s$	$V_s/\bar{u} \geq 4$	

Nomenclature for Table 18.4:
d = stack diameter, m
F = buoyancy flux parameter, $gd^2V_s(T_s - T_a)/4T_s$, $\mathrm{m^4\,s^{-3}}$
g = acceleration due to gravity, $9.807\,\mathrm{m\,s}^{-2}$
p = atmospheric pressure, kPa
p_0 = 101.3 kPa
$S_1 = (g\partial\theta/\partial z/T_a)(p/p_0)^{0.29}$, s^{-2}
$S_2 = (g\partial\theta/\partial z)/T_a$, s^{-2}
T_a = ambient temperature at stack height, K
T_s = stack exit temperature at stack height, K
$\Delta T = T_s - T_a$
V_s = stack exit velocity, $\mathrm{m\,s}^{-1}$
^{a}For further information we refer the reader to Hanna et al. (1982).
^{b}If the appropriate field data are not available to estimate S_1 and S_2. Table 18.5 can be used.
^{c}Of these formulas for stable conditions, use the one that predicts the least plume rise.
^{d}Of the two formulas for neutral conditions, use the one that predicts the least plume rise.

TABLE 18.5   Relationship between Pasquill–Gifford Stability Classes and Temperature Stratification

Stability Class	Ambient Temperature Gradient $\partial T/\partial z$, °C $100\,\mathrm{m}^{-1}$	Potential Temperature Gradient^a $\partial\theta/\partial z$, °C $100\,\mathrm{m}^{-1}$
A (extremely unstable)	< -1.9	< -0.9
B (moderately unstable)	-1.9 to -1.7	-0.9 to -0.7
C (slightly unstable)	-1.7 to -1.5	-0.7 to -0.5
D (neutral)	-1.5 to -0.5	-0.5 to 0.5
E (slightly stable)	-0.5 to 1.5	0.5 to 2.5
F (moderately stable)	> 1.5	> 2.5

^{a}Calculated by assuming $\partial\theta/\partial z = \partial T/\partial z + \Gamma$, where Γ is the adiabatic lapse rate, 0.986°C $100\,\mathrm{m}^{-1}$.

18.12 FUNCTIONAL FORMS OF MEAN WINDSPEED AND EDDY DIFFUSIVITIES

While the Gaussian equations have been widely used for atmospheric diffusion calculations, the lack of ability to include changes in windspeed with height and nonlinear chemical reactions limits the situations in which they may be used. The atmospheric diffusion equation provides a more general approach to atmospheric diffusion calculations than do the Gaussian models, since the Gaussian models have been shown to be special cases of that equation when the windspeed is uniform and the eddy diffusivities are constant. The atmospheric diffusion equation in the absence of chemical reaction is

$$
\frac{\partial \langle c \rangle}{\partial t} + \bar{u}\frac{\partial \langle c \rangle}{\partial x} + \bar{v}\frac{\partial \langle c \rangle}{\partial y} + \bar{w}\frac{\partial \langle c \rangle}{\partial z}
$$
$$
= \frac{\partial}{\partial x}\left(K_{xx}\frac{\partial \langle c \rangle}{\partial x}\right) + \frac{\partial}{\partial y}\left(K_{yy}\frac{\partial \langle c \rangle}{\partial y}\right) + \frac{\partial}{\partial z}\left(K_{zz}\frac{\partial \langle c \rangle}{\partial z}\right) \tag{18.117}
$$

The key problem in the use of (18.117) is to choose the functional forms of the windspeeds, $\bar{u}$, $\bar{v}$, and $\bar{w}$, and the eddy diffusivities, K_{xx}, K_{yy}, and K_{zz}, for the particular situation of interest.

18.12.1 Mean Windspeed

The mean windspeed, usually taken as that coinciding with the x direction, is often represented as a power-law function of height by (16.84)

$$
\frac{\bar{u}}{\bar{u}_r} = \left(\frac{z}{z_r}\right)^p \tag{18.118}
$$

where p depends on atmospheric stability and surface roughness (Section 16.4.5).

18.12.2 Vertical Eddy Diffusion Coefficient K_{zz}

The expressions available for K_{zz} are based on Monin–Obukhov similarity theory coupled with observational or computationally generated data. It is best to organize the expressions according to the type of stability.

In the surface layer K_{zz} can be expressed as

$$
K_{zz} = \frac{\kappa u_* z}{\phi(z/L)} \tag{18.119}
$$

where $\phi(z/L)$ is given by

$$
\phi(z/L) = \begin{cases} 1 + 4.7\,z/L & z/L > 0 \quad \text{stable} \\ 1 & z/L = 0 \quad \text{neutral} \\ (1 - 15\,z/L)^{-1/2} & z/L < 0 \quad \text{unstable} \end{cases} \tag{18.120}
$$

We note that for stable and neutral conditions $\phi(z/L)$ is identical to that for momentum transfer, $\phi_m(z/L)$, given by (16.75). For unstable conditions, $\phi(z/L) = (\phi_m(z/L))^2$ (Galbally 1971; Crane et al. 1977).

Since we generally need expressions for K_{zz} that extend vertically beyond the surface layer, we now consider some available correlations for the entire Ekman layer.

Unstable Conditions   In unstable conditions there is usually an inversion base height at $z = z_i$ that defines the extent of the mixed layer. The two parameters that are key in determining K_{zz} are the convective velocity scale w_* and z_i. We expect that a dimensionless profile $\tilde{K}_{zz} = K_{zz}/w_* z_i$, which is a function only of z/z_i, should be applicable. This form should be valid as long as $\tilde{K}_{zz}$ is independent of the nature of the source distribution. Lamb and Duran (1977) determined that $\tilde{K}_{zz}$ does depend on the source height. With the proviso that the result be applied when emissions are at or near ground level, Lamb et al. (1975) and Lamb and Duran (1977) derived an empirical expression for $\tilde{K}_{zz}$ under unstable conditions, using the numerical turbulence model of Deardorff (1970):

$$\frac{K_{zz}}{w_* z_i} = \begin{cases} 2.5\left(\kappa\dfrac{z}{z_i}\right)^{4/3}(1 - 15\,z/L)^{1/4} & 0 \leq \dfrac{z}{z_i} < 0.05 \\[2ex] \begin{aligned}&0.021 + 0.408\left(\dfrac{z}{z_i}\right) + 1.351\left(\dfrac{z}{z_i}\right)^2 \\ &-4.096\left(\dfrac{z}{z_i}\right)^3 + 2.560\left(\dfrac{z}{z_i}\right)^4\end{aligned} & 0.05 \leq \dfrac{z}{z_i} \leq 0.6 \\[2ex] 0.2\,\exp\left[6 - 10\left(\dfrac{z}{z_i}\right)\right] & 0.6 < \dfrac{z}{z_i} \leq 1.1 \\[2ex] 0.0013 & \dfrac{z}{z_i} > 1.1 \end{cases} \tag{18.121}$$

Equation (18.121) is shown in Figure 18.6. The maximum value of K_{zz} occurs at $z/z_i \simeq 0.5$ and has a magnitude $\simeq 0.21 w_* z_i$. For typical meteorological conditions this corresponds to a magnitude of $O(100\,\mathrm{m}^2\,\mathrm{s}^{-1})$ and a characteristic vertical diffusion time, z_i^2/K_{zz}, of $O(5 z_i/w_*)$.

Other expressions for K_{zz} in unstable conditions exist, notably those of O'Brien (1970) and Myrup and Ranzieri (1976), that are similar in nature to (18.121).

Neutral Conditions   Under neutral conditions $K_{zz} = \kappa u_* z$ in the surface layer. Since K_{zz} will not continue to increase without limit above the surface layer, it is necessary to specify its behavior at the higher elevations. Myrup and Ranzieri (1976) proposed the following empirical form for K_{zz} under neutral conditions:

$$K_{zz} = \begin{cases} \kappa u_* z & z/z_i < 0.1 \\ \kappa u_* z(1.1 - z/z_i) & 0.1 \leq z/z_i \leq 1.1 \\ 0 & z/z_i > 1.1 \end{cases} \tag{18.122}$$

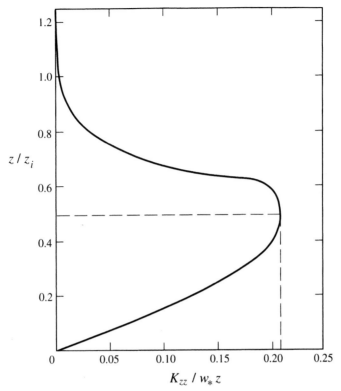

FIGURE 18.6   Vertical turbulent eddy diffusivity K_{zz} under unstable conditions derived by Lamb and Duran (1977).

Shir (1973) developed the following relationship for K_{zz} under neutral conditions from a one-dimensional turbulent transport model:

$$K_{zz} = \kappa u_* z \exp\left(-\frac{8fz}{u_*}\right)$$

(18.123)

We note that Myrup and Ranzieri chose the mixed-layer depth z_i as the characteristic vertical lengthscale, whereas Shir uses the Ekman layer height, u_*/f. (Since $L = \infty$ under neutral conditions, the Monin–Obukhov length cannot be used as a characteristic length scale.)

Lamb et al. (1975) calculated K_{zz} under neutral conditions from a numerical turbulence model and obtained

$$K_{zz} = \begin{cases} \dfrac{u_*^2}{f}\left[7.396 \times 10^{-4} + 6.082 \times 10^{-2}\left(\dfrac{zf}{u_*}\right) + 2.532\left(\dfrac{zf}{u_*}\right)^2 \right. \\ \left. -12.72\left(\dfrac{zf}{u_*}\right)^3 + 15.17\left(\dfrac{zf}{u_*}\right)^4\right], \qquad 0 \le \left(\dfrac{zf}{u_*}\right) \le 0.45 \\ \simeq 0 \qquad\qquad\qquad\qquad\qquad\qquad\qquad \left(\dfrac{zf}{u_*}\right) < 0.45 \end{cases}$$

(18.124)

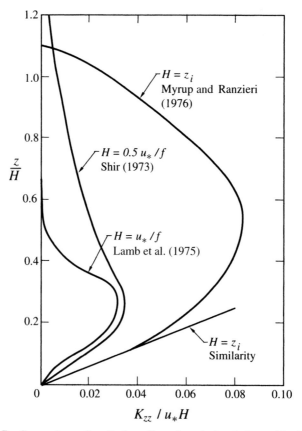

FIGURE 18.7 Comparison of vertical profiles of vertical turbulent eddy diffusivity K_{zz}.

The predictions of the three expressions (18.122) to (18.124) are compared in Figure 18.7.

Note that the scale height H in the figure differs for each equation, in particular

$$H = \begin{cases} z_i & \text{for (18.122)} \\ 0.5\,u_*/f & \text{for (18.123)} \\ u_*/f & \text{for (18.124)} \end{cases}$$

It is clear that substantial differences exist in the magnitude of K_{zz} predicted by the three expressions. This is due to the lack of knowledge of the form of K_{zz} above the surface layer.

Stable Conditions   Under stable conditions the appropriate characteristic vertical length scale is L. Businger and Arya (1974) proposed a modification of surface layer similarity theory to extend its vertical range of applicability:

$$K_{zz} = \frac{\kappa u_* z}{0.74 + 4.7(z/L)} \exp\left(-\frac{8fz}{u_*}\right) \tag{18.125}$$

For typical meteorological conditions the maximum value of K_{zz} under stable conditions is in the range $0.5-0.5 \, \text{m}^2 \, \text{s}^{-1}$.

18.12.3 Horizontal Eddy Diffusion Coefficients K_{xx} and K_{yy}

We have seen that when σ^2 varies as t the crosswind eddy diffusion coefficient K_{yy} is related to the variance of plume spread by

$$K_{yy} = \frac{1}{2} \frac{d\sigma_y^2}{dt} = \frac{1}{2} \bar{u} \frac{d\sigma_y^2}{dx} \qquad (18.126)$$

for $t \gg T_L$, where T_L is the Lagrangian timescale.

Measurements of T_L in the atmosphere are extremely difficult to perform, and it is difficult to establish whether the condition $t \gg T_L$ holds for urban scale flows. Csanady (1973) indicates that a typical eddy that is generated by shear flow near the ground has a Lagrangian timescale on the order of 100 s. Lamb and Neiburger (1971), in a series of measurements in the Los Angeles basin, estimated the Eulerian timescale T_E to be about 50 s. In a discussion of some field experiments, Lumley and Panofsky (1964) suggested that $T_L < 4T_E$. If the averaging interval is selected to be equal to the travel time, then an approximate value for K_{yy} can be deduced from the measurements of Willis and Deardorff (1976). Their data indicate that for unstable conditions ($L > 0$) and a travel time $t = 3z_i/w_*$

$$\frac{\sigma_y^2}{z_i^2} \simeq 0.64 \qquad (18.127)$$

Employing the previous travel time estimate and combining this result with (18.126) gives

$$K_{yy} = \frac{1}{6} \frac{\sigma_y^2}{z_i^2} w_* z_i \simeq 0.1 \, w_* z_i \qquad (18.128)$$

This latter result can be expressed in terms of the friction velocity u_* and the Monin–Obukhov length L as

$$K_{yy} \simeq 0.1 \, z_i^{3/4} (-\kappa L)^{-1/3} u_* \qquad (18.129)$$

For a range of typical meteorological conditions this formulation results in diffusivities under unstable conditions of $O(50-100 \, \text{m}^2 \, \text{s}^{-1})$. In practical applications it is usually assumed that $K_{xx} = K_{yy}$.

18.13 SOLUTIONS OF THE STEADY-STATE ATMOSPHERIC DIFFUSION EQUATION

The Gaussian expressions are not expected to be valid descriptions of turbulent diffusion close to the surface because of spatial inhomogeneities in the mean wind and the turbulence. To deal with diffusion in layers near the surface, recourse is generally made to the atmospheric diffusion equation, in which, as we have noted, the key problem is proper specification of the spatial dependence of the mean velocity and eddy diffusivities. Under steady-state conditions, turbulent diffusion in the direction of the mean wind is usually neglected (the slender plume approximation), and if the wind direction coincides with the x axis, then $K_{xx} = 0$. Thus it is necessary to specify only the lateral, K_{yy}, and vertical, K_{zz}, coefficients. It is generally assumed that horizontal homogeneity exists so that $\bar{u}$ and K_{yy} are independent of y. Hence (18.117) becomes

$$\bar{u}\frac{\partial \langle c \rangle}{\partial x} = \frac{\partial}{\partial y}\left(K_{yy}\frac{\partial \langle c \rangle}{\partial y}\right) + \frac{\partial}{\partial z}\left(K_{zz}\frac{\partial \langle c \rangle}{\partial z}\right) \tag{18.130}$$

Section 18.12 has been devoted to expressions for $\bar{u}$, K_{zz}, and K_{yy} based on atmospheric boundary-layer theory. Because of the rather complicated dependence of $\bar{u}$ and K_{zz} on z,(18.130) must generally be solved numerically. However, if they can be found, analytical solutions are advantageous for studying the behavior of the predicted mean concentration. For solutions beyond those presented here we refer the reader to Lin and Hildemann (1996, 1997).

18.13.1 Diffusion from a Point Source

A solution of (18.130) has been obtained by Huang (1979) in the case when the mean windspeed and vertical eddy diffusivity can be represented by the power-law expressions

$$\bar{u}(z) = az^p \tag{18.131}$$
$$K_{zz}(z) = bz^n \tag{18.132}$$

and when the horizontal eddy diffusivity is related to σ_y^2 by (18.126).

For a point source of strength q at height h above the ground, the solution of (18.130) subject to (18.126), (18.131), and (18.132) is

$$\langle c(x,y,z) \rangle = \frac{q}{(2\pi)^{1/2}\sigma_y}\exp\left(-\frac{y^2}{2\sigma_y^2}\right)\frac{(zh)^{(1-n)/2}}{b\alpha x}$$

$$\times \exp\left(-\frac{a(z^\alpha + h^\alpha)}{b\alpha^2 x}\right)I_{-v}\left(\frac{2a(zh)^{\alpha/2}}{b\alpha^2 x}\right) \tag{18.133}$$

where $\alpha = 2 + p - n$, $v = (1 - n)/\alpha$, and I_{-v} is the modified Bessel function of the first kind of order $-v$.

Equation (18.133) can be used to obtain some special cases of interest. If it is assumed that $p = n = 0$, then (18.133) reduces to

$$\langle c(x,y,z) \rangle = \frac{q}{(2\pi)^{1/2}\sigma_y}\exp\left(-\frac{y^2}{2\sigma_y^2}\right)\frac{(zh)^{1/2}}{\sigma_z^2\bar{u}}\exp\left(-\frac{z^2 + h^2}{2\sigma_z^2}\right)I_{-1/2}\left(\frac{zh}{\sigma_z^2}\right) \tag{18.134}$$

where

$$\sigma_y^2 = 2K_{yy}x/\bar{u} \qquad \sigma_z^2 = 2K_{zz}x/\bar{u} \qquad (18.135)$$

Using the asymptotic form

$$I_{-1/2}(x) = \left(\frac{2}{\pi x}\right)^{1/2} \cosh x \qquad x \to \infty \qquad (18.136)$$

(18.134) reduces to the Gaussian plume equation. Note that the asymptotic condition in (18.136) corresponds to $zh \gg \sigma_z^2$.

The case of a point source at or near the ground can also be examined. We can take the limit of (18.133) as $h \to 0$ using the asymptotic form of $I_\nu(x)$ as $x \to 0$

$$I_\nu(x) = \frac{x^\nu}{2^\nu \Gamma(1+\nu)} \qquad x \to 0 \qquad (18.137)$$

to obtain

$$\langle c(x,y,z) \rangle = \frac{q}{(2\pi)^{1/2}\sigma_y x^{(1+p)/\alpha}} \frac{\alpha}{a^\nu (b\alpha^2)^{(1+p)/\alpha}\Gamma((1+p)/\alpha)}$$

$$\times \exp\left(-\frac{y^2}{2\sigma_y^2}\right) \exp\left(-\frac{a(z^\alpha + h^\alpha)}{b\alpha^2 x}\right) \qquad (18.138)$$

18.13.2   Diffusion from a Line Source

The mean concentration downwind of a continuous, crosswind line source at a height h emitting at a rate $q_l (\mathrm{g\,m^{-1}\,s^{-1}})$ is governed by

$$\bar{u}\frac{\partial \langle c \rangle}{\partial x} = \frac{\partial}{\partial z}\left(K_{zz}\frac{\partial \langle c \rangle}{\partial z}\right) \qquad (18.139)$$

$$\langle c(0,z) \rangle = (q_l/\bar{u}(h))\delta(z-h)$$

$$-K_{zz}(0)\left(\frac{\partial \langle c \rangle}{\partial z}\right)_{z=0} = 0 \qquad z = 0$$

$$\langle c(x,z) \rangle = 0 \qquad z \longrightarrow \infty$$

The solution of (18.139) for the power-law profiles (18.131) and (18.132) is

$$\langle c(x,z) \rangle = \frac{q_l (zh)^{(1-n)/2}}{b\alpha x}\exp\left(-\frac{a(z^\alpha + h^\alpha)}{b\alpha^2 x}\right)I_{-\nu}\left(\frac{2a(zh)^{\alpha/2}}{b\alpha^2 x}\right) \qquad (18.140)$$

For a ground-level line source, $h = 0$, and

$$\langle c(x, z) \rangle = \frac{\alpha q_l}{a\Gamma((p+1)/\alpha)} \left(\frac{a}{\alpha^2 bx}\right)^{(p+1)/\alpha} \exp\left(-\frac{az^\alpha}{\alpha^2 bx}\right) \qquad (18.141)$$

The General Structure of Multiple-Source Plume Models   Multiple-source plume (and particularly Gaussian plume) models (Calder 1977) are commonly used for predicting concentrations of inert pollutants over urban areas. Although there are many special-purpose computational algorithms currently in use, the basic element that is common to most is the single-point-source release. The spatial concentration distribution from such a source is the underlying component and the multiple-source model is then developed by simple superposition of the individual plumes from each of the sources.

The starting point for predicting the mean concentration resulting from a single point source is the assumption of a quasi-steady state. In spite of the obvious long-term variability of emissions, and of the meteorological conditions affecting transport and dispersion, it is assumed as a working approximation that the pollutant concentrations can be treated as though they resulted from a time sequence of different steady states. In urban modeling the time interval is normally relatively short and on the order of 1 h. The time sequence of steady-state concentrations is regarded as leading to a random series of (1 h) concentration values, from which the frequency distribution and long-term average can be calculated as a function of receptor location. An assumption normally made is that the dispersion of material is "horizontally homogeneous" and independent of the horizontal location of the source—that is, the source–receptor relation is invariant under arbitrary horizontal translation of the source–receptor pair. Furthermore, the dispersion in a simple pointsource plume is assumed to be "isotropic" with respect to direction, and independent of the actual direction of the wind transport over the urban area. In principle, the short-term multiple-source plume model then only involves simple summation or integration over all point, line, and area sources and for the different meteorological conditions of plume dispersion.

In most short-term plume models in use at the present time, it is assumed that for each time interval of the quasi-steady-state sequence, transport by the wind over the urban area can be characterized in terms of a single "wind direction." Also, for each time interval the three meteorological variables that influence the transport and dispersion are taken to be the mean wind speed, the atmospheric stability category, and the mixing depth. Although the wind direction θ is evidently a continuous variable that may assume any value ($0° < \theta < 360°$), for reasons of practical simplicity the three dispersion variables are usually considered as discrete, with a relatively small number of possible values for each. The short-term, steady-state, three-dimensional spatial distribution of mean concentration for an individual point source plume must then be expressed in terms of a plume dispersion equation of fixed functional form, such as the Gaussian plume equation, although the dispersion parameters that appear in its arguments will, of course, depend on actual conditions. Many models assume the well-known Gaussian form, and the horizontal and vertical standard deviation functions, σ_y and σ_z, are expressed as functions of the downwind distance in the plume and of the stability category only.

To illustrate these ideas we consider a fixed rectangular coordinate system, with the plane $z = 0$ at ground level, and also assume that the pollutant transport by the wind over the urban area can be characterized in terms of a single horizontal direction that makes an angle, say, θ, with the positive x axis of the coordinate system. We thus explicitly separate the wind direction variable from all other meteorological variables that influence transport and dispersion, for example, windspeed, atmospheric stability category, and mixing depth. These latter variables may be regarded as defining a meteorological dispersion index $j(j = 1, 2, \ldots)$ that will be a function of t_n, the interval of time over which meteorological conditions are constant; that is, $j = j(t_n)$. The index number j simply designates a given meteorological state out of the totality of possible combinations, that is, a specified combination of windspeed, stability, and mixing depth. Then from the additive property of concentration it follows that the mean concentration $\langle c(x, y, z; t_n) \rangle$ for the time interval t_n that results from the superposition can be written as a summation (actually representing a three-dimensional integral)

$$\langle c(x, y, z; t_n) \rangle = \sum_{(x', y', z')} Q\{x, y, z; x', y', z', \theta_n, j(t_n)\} S(x', y', z'; t_n) \Delta V$$

where $S(x', y', z'; t_n)$ is the steady emission rate per unit volume and unit time at position (x', y', z') for time interval $t = t_n$; $Q\{x, y, z; x', y', z'; \theta_n, j(t_n)\}$ is the mean concentration at (x, y, z) produced by a steady point source of unit strength located at (x', y', z'), and for time interval t_n when $\theta = \theta_n$; and the summation is taken over all the elemental source locations of the entire region. The function Q is a meteorological dispersion function that is explicitly dependent on the wind direction θ_n and implicitly on the other meteorological variables that will be determined by the time sequence t_n. In all simple urban models now in use the function Q is taken to be a deterministic function, although its variables θ_n and $j(t_n)$ may be regarded as random. It is usually assumed that the dispersion function Q is independent of the horizontal location of the source, so that Q is invariant under arbitrary horizontal translations of the source–receptor pair, reducing Q to a function of only two independent horizontal spatial variables, so that the preceding equation may be rewritten as

$$\langle c(x, y, z; t_n) \rangle = \sum_{(X, Y, z')} Q\{X, Y, z, z'; \theta_n, j(t_n)\} S(x - X, y - Y, z'; t_n) \Delta V$$

where $X = x - x'$ and $Y = y - y'$, and the summation extends over the entire source distribution. Finally, it is normally assumed that the dispersion function Q depends on θ only through a simple rotation of the horizontal axes, and that it is independent of the actual value of the single horizontal wind direction that is assumed to characterize the pollutant transport over the urban area, provided the x axis of the coordinate system is taken along this direction. If this is done then the dispersion function is directionally isotropic.

The long-term average concentration, which we denote by an uppercase C, is given by

$$C(x,y,z) = \frac{1}{N} \sum_{n=1}^{N} \langle c(x,y,z;t_n) \rangle$$

When the emission intensities S and the dispersion functions Q, that is, the meteorological conditions, can be assumed to be independent or uncorrelated, the average value of the product of S and Q is equal to the product of their average values, and we thus have

$$C(x,y,z) = \sum_{(X,Y,z')} \bar{Q}(X,Y,z,z') \bar{S}(x-X, y-Y, z') \Delta V$$

where $\bar{S}$ denotes the time-average source strength and $\bar{Q}$ the average value of the meteorological dispersion function. Thus if $P(\theta, j)$ denotes the joint frequency function for wind direction θ and dispersion index j, so that

$$\sum_{\theta} \sum_{j} P(\theta, j) = 1$$

then

$$\bar{Q}(X,Y,z,z') = \sum_{\theta} \sum_{j} Q\{X,Y,z,z';\theta,j\} P(\theta,j)$$

APPENDIX 18.1 FURTHER SOLUTIONS OF ATMOSPHERIC DIFFUSION PROBLEMS

18.A.1 Solution of (18.29)–(18.31)

To solve this problem we let $\langle c(x,y,z,t) \rangle = c_x(x,t) c_y(y,t) c_z(z,t)$ with $c_x(x,0) = S^{1/3} \delta(x)$, $c_y(y,0) = S^{1/3} \delta(y)$, and $c_z(z,0) = S^{1/3} \delta(z)$. Then (18.29) becomes

$$\frac{\partial c_x}{\partial t} + \bar{u} \frac{\partial c_x}{\partial x} = K_{xx} \frac{\partial^2 c_x}{\partial x^2} \tag{18.A.1}$$

$$\frac{\partial c_y}{\partial t} = K_{yy} \frac{\partial^2 c_y}{\partial y^2} \tag{18.A.2}$$

$$\frac{\partial c_z}{\partial t} = K_{zz} \frac{\partial^2 c_z}{\partial z^2} \tag{18.A.3}$$

Each of these equations may be solved by the Fourier transform. We illustrate with (18.A.1). The Fourier transform of $c_x(x, t)$ is

$$C(\alpha, t) = F\{c_x(x, t)\} = \frac{1}{(2\pi)^{1/2}} \int_{-\infty}^{\infty} c_x(x, t) e^{-i\alpha x} dx$$

and thus transforming, we obtain

$$\frac{\partial C}{\partial t} + i\alpha \bar{u} C = -\alpha^2 K_{xx} C$$

$$C(\alpha, 0) = \frac{S^{1/3}}{(2\pi)^{1/2}} \tag{18.A.4}$$

The solution of (18.A.4) is

$$C(\alpha, t) = \frac{S^{1/3}}{(2\pi)^{1/2}} \exp[-(\alpha^2 K_{xx} + i\alpha \bar{u})t]$$

The inverse transform is

$$c_x(x, t) = \frac{1}{(2\pi)^{1/2}} \int_{-\infty}^{\infty} C(\alpha, t) e^{i\alpha x} d\alpha$$

Thus

$$c_x(x, t) = \frac{S^{1/3}}{2\pi} \int_{-\infty}^{\infty} \exp\{-[\alpha^2 K_{xx} t - i\alpha(x - \bar{u}t)]\} d\alpha$$

Completing the square in the exponent, we obtain

$$\alpha^2 K_{xx} t - i\alpha(x - \bar{u}t) - \frac{(x - \bar{u}t)^2}{4K_{xx}t} + \frac{(x - \bar{u}t)^2}{4K_{xx}t}$$

$$= \left(\alpha(K_{xx}t)^{1/2} - \frac{i(x - \bar{u}t)}{2(K_{xx}t)^{1/2}} \right)^2 - \frac{(x - \bar{u}t)^2}{4K_{xx}t}$$

Let $\eta = \alpha(K_{xx}t)^{1/2} - i(x - \bar{u}t)/2(K_{xx}t)^{1/2}$ and $d\eta = (K_{xx}t)^{1/2} d\alpha$. Then

$$c_x(x, t) = \frac{S^{1/3}}{2\pi(K_{xx}t)^{1/2}} \exp\left(-\frac{(x - \bar{u}t)}{4K_{xx}t} \right) \int_{-\infty}^{\infty} e^{-\eta^2} d\eta$$

the integral equals $\pi^{1/2}$, so

$$c_x(x,t) = \frac{S^{1/3}}{2(\pi K_{xx}t)^{1/2}} \exp\left(-\frac{(x-\bar{u}t)^2}{4K_{xx}t}\right)$$

By the same method

$$c_y(y,t) = \frac{S^{1/3}}{2(\pi K_{yy}t)^{1/2}} \exp\left(-\frac{y^2}{4K_{yy}t}\right)$$

$$c_z(z,t) = \frac{S^{1/3}}{2(\pi K_{zz}t)^{1/2}} \exp\left(-\frac{z^2}{4K_{zz}t}\right)$$

and thus the mean concentration is given by

$$\langle c(x,y,z,t)\rangle = \frac{S}{8(\pi t)^{3/2}(K_{xx}K_{yy}K_{zz})^{1/2}}$$

$$\times \exp\left(-\frac{(x-\bar{u}t)^2}{4K_{xx}t} - \frac{y^2}{4K_{yy}t} - \frac{z^2}{4K_{zz}t}\right) \tag{18.A.5}$$

18.A.2 Solution of (18.50) and (18.51)

To solve (18.50) we begin with the transformation

$$f(x,y,z) = \langle c(x,y,z)\rangle e^{-kx}$$

where $k = \bar{u}/2K$, and (18.50) becomes

$$\nabla^2 f - k^2 f = \frac{-q}{K}e^{-kx}\delta(x)\,\delta(y)\,\delta(z)$$

$$f(x,y,z) = 0 \qquad x,y,z \to \pm\infty$$

First we will solve

$$\nabla_r^2 f - k^2 f = 0$$

where $r^2 = x^2 + y^2 + z^2$. To do so, let $f(r) = g(r)/r$ and

$$\nabla_r^2 f = \frac{1}{r^2}\frac{d}{dr}\left(r^2\frac{df}{dr}\right) = \frac{1}{r}\frac{d^2g}{dr^2}$$

so that the equation to be solved is

$$\frac{d^2g}{dr^2} - k^2 g = 0$$

The solution is

$$g(r) = A_1 e^{kr} + A_2 e^{-kr}$$

Then $A_1 = 0$ satisfies the condition that g is finite as $r \to \infty$, and

$$f(r) = \frac{A}{r} e^{-kr}$$

To determine A, we will evaluate

$$\int_V (\nabla_r^2 f - k^2 f) dV = - \int_V \frac{q}{K} e^{-kx} \delta(x) \, \delta(y) \, \delta(z) \, dV$$

on the sphere with unit radius. The right-hand side is simply $-q/K$. Using Green's theorem, the left-hand side is

$$\int_V (\nabla_r^2 f - k^2 f) dV = \int_S \frac{\partial f}{\partial n} dS - k^2 \int_V f \, dV$$

On a sphere of unit radius

$$\int_S \frac{\partial f}{\partial n} dS = \int_0^{2\pi} \int_{-\pi}^{\pi} \cos\theta \left(\frac{\partial f}{\partial r}\right)_{r=1} d\theta \, d\phi$$
$$= -4\pi A e^{-k}(k+1)$$

and

$$-\int_V f \, dV = \int_0^1 4\pi r^2 f \, dr = 4\pi A[e^{-k}(k+1) - 1]$$

Thus we obtain $4\pi A = q/K$. Finally

$$\langle c(x, y, z) \rangle = \frac{q}{4\pi K r} \exp\left[-\frac{\bar{u}(r-x)}{2K}\right]$$

18.A.3   Solution of (18.59)–(18.61)

The solution can be carried out by Fourier transform, first with respect to the y direction and then with respect to the z direction. If $C(x, \alpha, z) = F_y\{\langle c(x, y, z)\rangle\}$ and $C'(x, \alpha, \beta) = F_z\{C(x, \alpha, z)\}$, then (18.59) becomes

$$\bar{u}\frac{\partial C'}{\partial x} = -K(\alpha^2 + \beta^2)C'(x, \alpha, \beta) \tag{18.A.6}$$

The $x = 0$ boundary condition, when transformed doubly, is

$$C'(0, \alpha, \beta) = \frac{q}{2\pi\bar{u}} \qquad (18.A.7)$$

The solution of (18.A.6) subject to (18.A.7) is

$$C'(x, \alpha, \beta) = \frac{q}{2\pi\bar{u}} \exp\left(-\frac{Kx}{\bar{u}}(\alpha^2 + \beta^2)\right) \qquad (18.A.8)$$

We must now invert (18.A.8) twice to return to $\langle c(x, y, z)\rangle$. First

$$C(x, \alpha, z) = \frac{1}{(2\pi)^{1/2}} \int_{-\infty}^{\infty} C'(x, \alpha, \beta) e^{i\beta z} d\beta$$

Thus

$$C(x, \alpha, z) = \frac{q}{(2\pi)^{3/2}\bar{u}} e^{-\alpha^2 Kx/\bar{u}} \int_{-\infty}^{\infty} \exp\left(i\beta z - \frac{Kx\beta^2}{\bar{u}}\right) d\beta \qquad (18.A.9)$$

It is now necessary to express the exponential in the integrand as

$$-(Kx\beta^2/\bar{u} - i\beta z) = -[(Kx/\bar{u})^{1/2}\beta - (iz/2)(\bar{u}/Kx)^{1/2}]^2 - z^2\bar{u}/4Kx$$

and let $\eta = (Kx/\bar{u})^{1/2}\beta - (iz/2)(\bar{u}/Kx)^{1/2}$. Then (18.A.9) becomes

$$C(x, \alpha, z) = \frac{q}{(2\pi)^{3/2}\bar{u}}\left(\frac{\bar{u}}{Kx}\right)^{1/2} \exp\left(-\frac{\alpha^2 Kx}{\bar{u}} - \frac{z^2\bar{u}}{4Kx}\right) \int_{-\infty}^{\infty} e^{-\eta^2} d\eta$$

Proceeding through identical steps to invert $C(x, \alpha, z)$ to $\langle c(x, y, z)\rangle$, we obtain

$$\langle c(x, y, z)\rangle = \frac{q}{4\pi Kx} \exp\left(-\frac{\bar{u}}{4Kx}(y^2 + z^2)\right) \qquad (18.A.10)$$

APPENDIX 18.2   ANALYTICAL PROPERTIES OF THE GAUSSIAN PLUME EQUATION

When material is emitted from a single elevated stack, the resulting ground-level concentration exhibits maxima with respect to both downwind distance and windspeed. Both directly below the stack, where the plume has not yet touched the ground, and far downwind, where the plume has become very dilute, the concentrations approach zero; therefore a maximum ground-level concentration occurs at some intermediate distance. Both at very high wind speeds, when the plume is rapidly diluted, and at very low windspeeds, when plume rise proceeds relatively unimpeded, the contribution to the ground-level concentration is essentially zero; therefore a maximum occurs at some intermediate wind speed.

To investigate the properties of the maximum ground-level concentration with respect to distance from the source and windspeed, we begin with the Gaussian plume equation evaluated along the plume centerline ($y = 0$) at the ground ($z = 0$):

$$\langle c(x,0,0) \rangle = \frac{q}{\pi \bar{u} \sigma_y \sigma_z} \exp\left(-\frac{h^2}{2\sigma_z^2}\right) \tag{18.A.11}$$

The maximum in $\langle c(x,0,0) \rangle$ arises because of the x dependence of σ_y, σ_z, and h, as parametrized, for example, by (18.112), (18.113), (18.115), and (18.116). As a buoyant plume rises from its source, the effect of the wind is to bend it over until it becomes level at some distance x_f from the source. Windspeed $\bar{u}$ enters the expression for the ground-level concentration through two terms having opposite effects: (1) $\langle c(x,0,0) \rangle$ is proportional to $\bar{u}^{-1}$, such that the lighter the wind, the greater is the ground-level concentration; and (2) the plume rise Δh is inversely proportional to $\bar{u}^a$, such that the lighter the wind, the higher the plume rise and the smaller is the ground-level concentration. Thus a "worst" windspeed exists at which the ground-level concentration is a maximum at any downwind location. This maximum is different from the largest ground-level concentration that is achieved as a function of downwind distance for any given wind speed. The fact that $\langle c(x,0,0) \rangle$ depends on both x and $\bar{u}$ suggests that one may find simultaneously the windspeed and distance at which the highest possible ground-level concentration may occur, the so-called critical concentration. In computing these various maxima, there are two regimes of interest. The first is the region of $x < x_f$, where the plume has not reached its final height, and the second is $x \geq x_f$, when the plume has reached its final height. The latter case corresponds to $b = 0$ in the formula for the plume rise.

We want to calculate the location x_m of the maximum ground-level concentration for any given wind speed. By differentiating (18.117) with respect to x and setting the resulting equation equal to zero, we find

$$-\frac{1}{\sigma_y}\frac{d\sigma_y}{dx} - \frac{1}{\sigma_z}\frac{d\sigma_z}{dx} + \frac{h^2}{\sigma_z^3}\frac{d\sigma_z}{dx} - \frac{h}{\sigma_z^2}\left(\frac{dh}{dx}\right) = 0 \tag{18.A.12}$$

Now using the expressions (18.112), (18.113), (18.115), and (18.116) for h and σ_y and σ_z as a function of x, we obtain the following implicit equation for x_m

$$\sigma_z^2(x_m) = \frac{-hb\Delta h + h^2 r_z}{r_y + r_z} \tag{18.A.13}$$

where if (18.114) and its analogous relation for σ_y are used, r_y and r_z in (18.A.13) are replaced by $J_y + 2K_y \ln x_m$ and $J_z + 2K_z \ln x_m$, respectively.

In the special case in which $\Delta h = 0$ and σ_y and σ_z are given by (18.112) and (18.113), the expression for x_m reduces to (Ragland 1976)

$$x_m = \left(\frac{h_s^2 r_z}{R_z^2(r_y + r_z)}\right)^{1/2r_z} \tag{18.A.14}$$

The value of the maximum ground-level concentration at x_m in the case of σ_y and σ_z given by (18.112) and (18.113) and when $b = 0$ is

$$\langle c \rangle_m = \frac{q \gamma^{\gamma/2} R_z^{\gamma-1}}{\pi e^{\gamma/2} R_y h^\gamma \bar{u}} \tag{18.A.15}$$

where $\gamma = 1 + r_y/r_z$.

Now we consider the effect of wind speed $\bar{u}$ on the maximum ground-level concentration. The highest concentration at any downwind distance can be determined as a function of $\bar{u}$. Differentiating (18.A.11) with respect to $\bar{u}$, we obtain

$$\bar{u}^{2a} - \frac{h_s a E x^b}{\sigma_z^2} \bar{u}^a - \frac{a E^2 x^{2b}}{\sigma_z^2} = 0 \tag{18.A.16}$$

from which we find that the "worst" windspeed, $\bar{u}_w$, is

$$\bar{u}_w = \left(\frac{2 E x^b}{h_s (\zeta - 1)} \right)^{1/a} \tag{18.A.17}$$

where $\zeta = (1 + 4\sigma_z^2/ah_s^2)^{1/2}$.

The value of the ground-level concentration at this windspeed is

$$\langle c(x) \rangle_w = \frac{q 2^{-1/a} h_s^{1/a} (\zeta - 1)^{1/a}}{\pi \sigma_y \sigma_z E^{1/a} x^{b/a}} \exp\left(-\frac{h_s^2 (\zeta + 1)^2}{8 \sigma_z^2} \right) \tag{18.A.18}$$

When the plume has reached its final height, h becomes independent of x, and the effective plume height can be written as

$$h = h_s + \frac{E}{\bar{u}^a} \tag{18.A.19}$$

In that case the "worst" windspeed is (Roberts 1980; Bowman 1983)

$$\bar{u}_w = \left(\frac{2E}{h_s (\zeta - 1)} \right)^{1/a} \tag{18.A.20}$$

Finally, we can find the combination of location and windspeed that produces the highest possible ground-level concentration, the so-called critical concentration. Considering σ_y and σ_z given by (18.112) and (18.113), the critical downwind distance x_c is given by

$$x_c = \left[\left(\frac{h_s}{R_z} \right) \left(\frac{r}{r - 1/a} \right) \left(\frac{1}{r} \right)^{1/2} \right]^{1/r_z} \tag{18.A.21}$$

where $r = (r_y + r_z + b/a)/r_z$. The critical windspeed $\bar{u}_c$ is

$$\bar{u}_c = (aE)^{1/a} \left(\frac{r^{1/2}}{R_z}\right)^{b/ar_z} \left(\frac{r - 1/a}{h_s}\right)^{(1/a - b/ar_z)} \tag{18.A.22}$$

and the critical concentration $\langle c \rangle_c$ is

$$\langle c \rangle_c = \frac{qR_z^{r-1}(r - 1/a)^{r-1/a}}{\pi(aE)^{1/a}R_y h_s^{r-1/a} e^{r/2} r^{r/2}} \tag{18.A.23}$$

All terms in (18.A.23), other than exponents, are positive except for $r - 1/a$. If this term is zero or negative, a critical concentration does not exist. Thus for a critical concentration to exist, it is necessary that $r > 1/a$, or

$$a > \frac{r_z - b}{r_y + r_z} \tag{18.A.24}$$

Once the plume has reached its final height, $b = 0$, and (18.A.24) becomes simply $a > r_z/(r_y + r_z)$.

Let us now apply these general results to some specific cases. Figures 18.A.1 and 18.A.2 show the critical distance x_c as a function of the source height h_s and stability class for the cases in which the plume has reached its final height and before it has reached its final height, respectively. For the level plume, that is, one that has reached its final height, the critical downwind distance is

$$x_c = \left(\frac{a^2 r_z (r_y + r_z)}{R_z^2 [a(r_y + r_z) - r_z]^2}\right)^{1/2r_z} h_s^{1/r_z} \tag{18.A.25}$$

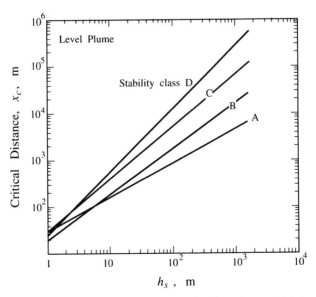

FIGURE 18.A.1   Critical downwind distance x_c as a function of source height h_s, and stability class for a plume that has reached its final height.

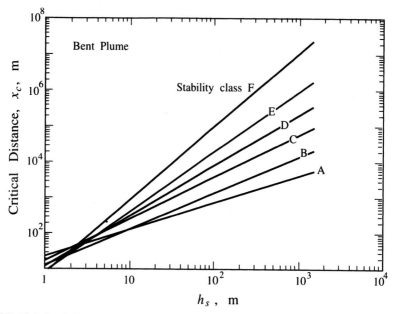

FIGURE 18.A.2 Critical downwind distance x_c as a function of source height h_s, and stability class for a plume that has not yet reached its final height.

For a plume that has not yet reached its final height, $a = 1$, and

$$x_c = \left(\frac{3r_z(3r_z + 3r_y + 2)}{R_z^2(3r_y + 2)^2} \right)^{1/2r_z} h_s^{1/r_z} \qquad (18.A.26)$$

For the level plume, $a = 1$ in neutral and unstable conditions and $a = \frac{1}{3}$ under stable conditions. Coefficients for different stability conditions are given in Table 18.3.

We see that as h_s increases, the downwind distance at which the critical concentration occurs increases. The distance also increases as the atmosphere becomes more stable. When $a = 1$, (18.A.24) holds regardless of r_y and r_z (neutral and unstable conditions); for stability classes E and F, however, $a = \frac{1}{3}$, and from the r_y and r_z values in Table 18.3, we see that (18.A.24) is not satisfied. Thus for a level plume, a critical concentration does not exist for stability classes E and F.

PROBLEMS

18.1$_A$ A power plant burns $10^4 \, \text{kg h}^{-1}$ of coal containing 2.5% sulfur. The effluent is released from a single stack of height 70 m. The plume rise is normally about 30 m, so that the effective height of emission is 100 m. The wind on the day of interest, which is a sunny summer day, is blowing at $4 \, \text{m s}^{-1}$. There is no inversion layer. Use the Pasquill–Gifford dispersion parameters from Table 18.3.

a. Plot the ground-level SO_2 concentration at the plume centerline over distances from 100 m to 10 km. (Use log–log coordinates.)

b. Plot the ground-level SO_2 concentration versus crosswind distance at downwind distances of 200 m and 1 km.

c. Plot the vertical centerline SO_2 concentration profile from ground level to 500 m at distances of 200 m, 1 km, and 5 km.

18.2$_A$ Repeat the calculation of Problem 18.1 for an overcast day with the same windspeed.

18.3$_A$ Repeat the calculation of Problem 18.1 assuming that an inversion layer is present at a height of 300 m.

18.4$_A$ A power plant continuously releases from a 70-m stack a plume into the ambient atmosphere of temperature 298 K. The stack has diameter 5 m, the exit velocity is $25 \, \mathrm{m \, s^{-1}}$, and the exit temperature is 398 K.

a. For neutral conditions and a $20 \, \mathrm{km \, h^{-1}}$ wind, how high above the stack is the plume 200 m downwind?

b. If a ground-based inversion layer 150 m thick is present through which the temperature increases 0.3°C and if the wind is blowing at $10 \, \mathrm{km \, h^{-1}}$, will the plume penetrate the inversion layer?

18.5$_C$ Mathematical models for atmospheric diffusion are generally derived for either instantaneous or continuous releases. Short-term releases of material sometimes occur and it is of interest of develop the appropriate diffusion formulas to deal with them (Palazzi et al. 1982). The mean concentration resulting from a release of strength q at height h that commenced at $t = 0$ is given by

$$
\langle c(x,y,z,t) \rangle = \int_0^t \frac{q}{(2\pi)^{3/2} \sigma_x \sigma_y \sigma_z} \exp\left(-\frac{[x - \bar{u}(t - t')]^2}{2\sigma_x^2} - \frac{y^2}{2\sigma_y^2} \right)
$$
$$
\times \left[\exp\left(-\frac{(z - h)^2}{2\sigma_z^2} \right) + \exp\left(-\frac{(z + h)^2}{2\sigma_z^2} \right) \right] dt' \tag{A}
$$

The duration of the release is t_r.

a. Show that if the slender plume approximation is invoked, the mean concentration resulting from the release is

$$
\langle c(x,y,z,t) \rangle = \begin{cases} \langle c_G(x,y,z) \rangle \displaystyle\int_0^t \frac{\bar{u}}{(2\pi)^{1/2} \sigma_x} \exp\left(-\frac{[x - \bar{u}(t - t')]^2}{2\sigma_x^2} \right) dt' & t \le t_r \\[4mm] \langle c_G(x,y,z) \rangle \displaystyle\int_0^{t_r} \frac{\bar{u}}{(2\pi)^{1/2} \sigma_x} \exp\left(-\frac{[x - \bar{u}(t - t')]^2}{2\sigma_x^2} \right) dt' & t > t_r \end{cases}
$$

where $\langle c_G(x,y,z) \rangle$ is the steady-state Gaussian plume result. Show that this reduces to

$$
\langle c(x,y,z,t) \rangle = \begin{cases} \dfrac{1}{2} \langle c_G(x,y,z) \rangle \left[\mathrm{erf}\left(\dfrac{x}{\sqrt{2}\sigma_x} \right) - \mathrm{erf}\left(\dfrac{x - \bar{u}t}{\sqrt{2}\sigma_x} \right) \right] & t \le t_r \\[4mm] \dfrac{1}{2} \langle c_G(x,y,z) \rangle \left[\mathrm{erf}\left(\dfrac{x - \bar{u}(t - t_r)}{\sqrt{2}\sigma_x} \right) - \mathrm{erf}\left(\dfrac{x - \bar{u}t}{\sqrt{2}\sigma_x} \right) \right] & t > t_r \end{cases}
$$

b. For $t < t_r$ show that $\langle c \rangle$ reaches its maximum at $t = t_r$. After $t = t_r$, the puff continues to flow downwind. Show that at locations where $x \le \bar{u}t_r/2$, the concentration decreases with increasing t and the maximum concentration is given by that at $t = t_r$. For $x > \bar{u}t_r/2$, show that the maximum concentration occurs when $\partial\langle c \rangle/\partial t = 0$, which gives $t = t_r/2 + x/\bar{u}$ and is

$$\langle c(x, y, z, t) \rangle_{\text{max}} = \langle c_G(x, y, z) \rangle \text{erf}\left(\frac{\bar{u}t_r}{2\sqrt{2}\sigma_x}\right) \quad \begin{matrix} t > t_r \\ x > \bar{u}t_r/2 \end{matrix}$$

c. Compare $\langle c \rangle_{\text{max}}$ and $\langle c_G \rangle$ as a function of $\bar{u}t_r/2\sqrt{2}\sigma_x$. What do you conclude?

d. We can define the *dosage* over a time interval from t to $t_0 + t_e$ as

$$\bar{c}(x, y, z) = \frac{1}{t_e}\int_{t_0}^{t_0+t_e} \langle c(x, y, z, t) \rangle dt \quad \text{(B)}$$

Three different situations can be visualized, according to when the release stops:

$$\begin{matrix} t_r \le t_0 & \text{release stops before exposure period begins} \\ t_0 \le t_r \le t_0 + t_e & \text{release stops during exposure period} \\ t_r \ge t_0 + t_e & \text{release stops after the exposure period} \end{matrix}$$

Show that substituting the appropriate form of $\langle c(x, y, z, t) \rangle$ into (B) leads to the following formulas for the concentration dosage:

$$\bar{c}(x, y, z) = \begin{cases} \begin{aligned} & \frac{\langle c_G(x, y, z) \rangle \sqrt{2}\sigma_x}{2t_e\bar{u}}\left[F\left(\frac{x - \bar{u}(t_0 - t_r)}{\sqrt{2}\sigma_x}\right) \right. \\ & - F\left(\frac{x - \bar{u}(t_0 + t_e - t_r)}{\sqrt{2}\sigma_x}\right) \\ & \left. + F\left(\frac{x - \bar{u}(t_0 + t_e)}{\sqrt{2}\sigma_x}\right) - F\left(\frac{x - \bar{u}t_0}{\sqrt{2}\sigma_x}\right)\right] \quad t_r \le t_0 \end{aligned} \\[2em] \begin{aligned} & \frac{\langle c_G(x, y, z) \rangle \sqrt{2}\sigma_x}{2t_e\bar{u}}\left[F\left(\frac{x}{\sqrt{2}\sigma_x}\right) - F\left(\frac{x - \bar{u}(t_0 + t_e - t_r)}{\sqrt{2}\sigma_x}\right) \right. \\ & + F\left(\frac{x - \bar{u}(t_0 + t_e)}{\sqrt{2}\sigma_x}\right) - F\left(\frac{x - \bar{u}t_0}{\sqrt{2}\sigma_x}\right) \\ & \left. - \frac{\bar{u}(t_r - t_0)}{\sqrt{2}\sigma_x}\text{erfc}\left(\frac{x}{\sqrt{2}\sigma_x}\right)\right] \quad t_0 \le t_r \le t_0 + t_e \end{aligned} \\[2em] \begin{aligned} & \frac{\langle c_G(x, y, z) \rangle \sqrt{2}\sigma_x}{2t_e\bar{u}}\left[F\left(\frac{x - \bar{u}(t_0 + t_e)}{\sqrt{2}\sigma_x}\right) \right. \\ & \left. - F\left(\frac{x - \bar{u}t_0}{\sqrt{2}\sigma_x}\right) - \frac{\bar{u}t_e}{\sqrt{2}\sigma_x}\text{erfc}\left(\frac{x}{\sqrt{2}\sigma_x}\right)\right] \quad t_r \ge t_0 + t_e \end{aligned} \end{cases}$$

where $F(x) = \int_x^{\infty} \text{erfc}\,(\xi)\, d\xi$.

e. The dosage reaches its maximum value in the interval defined by $t_0 > t_r - t_e$. Show that when $x \geq \bar{u}(t_r + t_e)/2$, the maximum dosage corresponds to

$$t_0 = \frac{x}{\bar{u}} + \frac{t_r - t_e}{2}$$

and that it can be found from

$$c_{\max} = \frac{\langle c_G(x,y,z) \rangle \sqrt{2}\sigma_x}{2t_e\bar{u}} \left[\bar{u}\frac{t_r + t_e}{\sqrt{2}\sigma_x} - \bar{u}\frac{|t_e - t_r|}{\sqrt{2}\sigma_x} + 2F\left(\frac{\bar{u}(t_r + t_e)}{2\sqrt{2}\sigma_x}\right) - 2F\left(\frac{\bar{u}|t_e - t_r|}{2\sqrt{2}\sigma_x}\right) \right]$$

18.6$_B$ Chlorine is released at a rate of $30\,\mathrm{kg\,s^{-1}}$ from an emergency valve at a height of $20\,\mathrm{m}$. We need to evaluate the maximum ground-level dosage of Cl_2 for different durations of release and exposure. Assume a $5\,\mathrm{m\,s^{-1}}$ windspeed and neutral stability. We may use the dispersion parameters for a continuous emission (they are a good approximation for exposure times exceeding about 2 min). Plot the Cl_2 dosage in ppm versus exposure time in minutes on a log–log scale over a range of 1 to 100 min exposure time for release durations of 2, 10, and 30 min. (The results needed to solve this problem can be taken from the statement of Problem 18.5.)

18.7$_B$ SO_2 is emitted from a stack under the following conditions:

$$\text{Stack diameter} = 3\,\mathrm{m}$$
$$\text{Exit velocity} = 10\,\mathrm{m\,s^{-1}}$$
$$\text{Exit temperature} = 430\,\mathrm{K}$$
$$\text{Emission rate} = 10^3\,\mathrm{g\,s^{-1}}$$
$$\text{Stack height} = 100\,\mathrm{m}$$

For Pasquill stability categories A, B, C, and D, calculate (Appendix 18.2 is needed):

a. The distance from the source at which the maximum ground-level concentration is achieved as a function of windspeed.

b. The windspeed at which the ground-level concentration is a maximum as a function of downwind distance.

c. The plume rise at the wind speed in part (b).

d. The critical downwind distance, windspeed, and concentration.

18.8$_C$ Gifford (1968, 1980) has discussed the determination of lateral and vertical dispersion parameters σ_y and σ_z from smoke plume photographs. Let us consider a single plume emanating from a continuous point source that will be assumed to be at ground level. Thus the mean concentration divided by the source strength is given by the Gaussian plume equation with $h = 0$:

$$\chi(x,y,z) = \frac{\langle c(x,y,z) \rangle}{q} = (\pi\sigma_y\sigma_z\bar{u})^{-1} \exp\left(-\frac{y^2}{2\sigma_y^2} - \frac{z^2}{2\sigma_z^2}\right) \qquad \text{(A)}$$

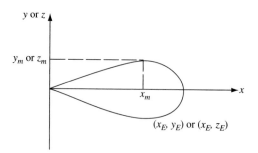

FIGURE 18.P.1   Plume boundaries.

A view of the plume from either above or the side is sketched in Figure 18.P.1, where y_E and z_E represent the coordinates of the visible edge of the plume, and y_m and z_m denote the maximum values of y_E and z_E with respect to downwind distance x. Let us consider the visible edge in the y direction, that is, the plume as observed from above. Integrating (A) over z from 0 to ∞ and evaluating y at y_E gives

$$
\begin{aligned}
\chi_E(x) &= \int_0^\infty \chi(x, y_E, z)\,dz \\
&= (\pi \sigma_y \bar{u})^{-1} \exp\left(-\frac{y_E^2}{2\sigma_y^2}\right) \int_0^\infty \frac{1}{\sigma_z} e^{-z^2/2\sigma_z^2}\,dz \\
&= (2\pi)^{-1/2}(\sigma_y \bar{u})^{-1} \exp\left(-\frac{y_E^2}{2\sigma_y^2}\right)
\end{aligned}
\tag{B}
$$

$\chi_E(x)$ represents the mean concentration normalized by the source strength, evaluated at the value of y corresponding to the visible edge of the plume and integrated through the depth of the plume.

a. Because the visible edge of the plume is presumably characterized by a constant integrated concentration, $\chi_E(x)$ is a constant χ_E. Show that, at $y = y_m$

$$
y_m^2 = \sigma_z^2(x_m) = \sigma_{y_m}^2
\tag{C}
$$

b. Since χ_E is a constant, (B) is invariant whether evaluated at any x or at x_m. Thus show that at any x

$$
\chi_E = (2\pi)^{-1/2}(\sigma_y \bar{u})^{-1} \exp\left(-\frac{y_E^2}{2\sigma_y^2}\right)
$$

and at x_m

$$
\chi_E = (2\pi e)^{-1/2}(y_m \bar{u})^{-1}
$$

Thus show that

$$\sigma_y^2 = y_E^2 \left[\ln \left(\frac{ey_m^2}{\sigma_y^2} \right) \right]^{-1}$$

18.9$_C$ Derive (18.89) as a solution of the atmospheric diffusion equation.

18.10$_C$ Murphy and Nelson (1983) have shown that the Gaussian plume equation can be modified to include dry deposition with a deposition velocity v_d by replacing the source strength q, usually taken as constant, with the depleted source strength $q(t)$ as a function of travel time t, where

$$q(t) = q_0 \exp \left\{ - \left(\frac{2}{\pi} \right)^{1/2} v_d \int_0^t \frac{1}{\sigma_z(t')} \exp \left(- \frac{h^2}{2\sigma_z^2(t')} \right) dt' \right\}$$

Verify this result. Show that, if $\sigma_z = (2K_{zz}t)^{1/2}$, then

$$q(t) = q_0 \exp \left(- \frac{2v_d t^{1/2}}{(\pi K_{zz})^{1/2}} \right)$$

18.11$_A$ An eight-lane freeway is oriented so that the prevailing wind direction is usually normal to the freeway. During a typical day the average traffic flow rate per lane is 30 cars per minute, and the average speed of vehicles in both directions is $80 \, \text{km} \, \text{h}^{-1}$. The emission of CO from an average vehicle is $90 \, \text{g} \, \text{km}^{-1}$ traveled.

a. Assuming a $5 \, \text{km} \, \text{h}^{-1}$ wind and conditions of neutral stability, with no elevated inversion layer present, determine the average ground-level CO concentration as a function of downwind distance, using the appropriate Gaussian plume formula.

b. A more accurate estimate of downwind concentrations can be obtained by taking into account the variation of wind velocity and turbulent mixing with height. For neutral conditions we have seen that

$$\bar{u}(z) = u_0 z^{1/7}$$

and

$$K_{zz} = \kappa u_* z = 0.4 \, u_* z$$

should be used. Assume that the $5 \, \text{km} \, \text{h}^{-1}$ wind reading was taken at a $10 \, \text{m}$ height and that for the surface downwind of the freeway $u_* = 0.6 \, \text{km} \, \text{h}^{-1}$ (grassy field). Repeat the calculation of (a). Discuss your result.

18.12_B It is desired to estimate the ground-level, centerline concentration of a contaminant downwind of a continous, elevated point source. The source height including plume rise is 100 m. The wind speed at a reference height of 10 m is $4\,m\,s^{-1}$. The atmospheric diffusion equation with the power-law correlations (18.131) and (18.132) is to be used. The roughness length is 0.1 m.

    a. Calculate the ground-level, centerline mean concentration under the following conditions:

       (1) Stable (Pasquill stability class F)
       (2) Neutral (Pasquill stability class D)
       (3) Unstable (Pasquill stability class B)

    b. Compare your predicitions to those of the Gaussian plume equation assuming that $\bar{u} = 4\,m\,s^{-1}$ at all heights. Discuss the reasons for any differences in the two results.

18.13_B To account properly for terrain variations in a region over which the transport and diffusion of pollutants are to be predicted, the following dimensionless coordinate transformation is used

$$\rho = \frac{z - h(x, y)}{Z} \qquad Z = H - h(x, y)$$

where $h(x, y)$ is the ground elevation at point (x, y) and H is the assumed extent of vertical mixing. Likewise, a similar change of variables for the horizontal coordinates may be performed

$$\xi = \frac{x - x_W}{X} \qquad X = x_E - x_W$$

$$\eta = \frac{y - y_S}{Y} \qquad Y = y_N - y_S$$

where x_E, x_W, y_N, and y_S are the coordinates of the horizontal boundaries of the region. Show that the form of (18.117) in ξ, η, ρ, t coordinates is

$$\frac{\partial \langle c \rangle}{\partial t} + \frac{\bar{u}}{X}\frac{\partial \langle c \rangle}{\partial \xi} + \frac{\bar{v}}{Y}\frac{\partial \langle c \rangle}{\partial \eta} + \frac{W}{Z}\frac{\partial \langle c \rangle}{\partial \rho} = \frac{1}{X^2}\frac{\partial}{\partial \xi}\left[K_{xx}\left(\frac{\partial \langle c \rangle}{\partial \xi} - \Lambda_\xi \frac{\partial \langle c \rangle}{\partial \rho}\right)\right]$$

$$- \frac{\Lambda_\xi}{X}\frac{\partial}{\partial \rho}\left[\frac{K_{xx}}{X}\left(\frac{\partial \langle c \rangle}{\partial \xi} - \Lambda_\xi \frac{\partial \langle c \rangle}{\partial \rho}\right)\right]$$

$$+ \frac{1}{Y^2}\frac{\partial}{\partial \eta}\left[K_{yy}\left(\frac{\partial \langle c \rangle}{\partial \eta} - \Lambda_\eta \frac{\partial \langle c \rangle}{\partial \rho}\right)\right]$$

$$- \frac{\Lambda_\eta}{Y}\frac{\partial}{\partial \rho}\left[\frac{K_{yy}}{Y}\left(\frac{\partial \langle c \rangle}{\partial \eta} - \Lambda_\eta \frac{\partial \langle c \rangle}{\partial \rho}\right)\right]$$

$$+ \frac{1}{Z^2}\frac{\partial}{\partial \rho}\left(K_{zz}\frac{\partial \langle c \rangle}{\partial \rho}\right)$$

where $W = \bar{w} - (\bar{u}/X)\Lambda_\xi Z - (\bar{v}/Y)\Lambda_\eta Z$ and where

$$\Lambda_\xi = \frac{1}{Z}\left(\frac{\partial h}{\partial \xi} + \rho \frac{\partial Z}{\partial \xi}\right) \cdot \Lambda_\eta = \frac{1}{Z}\left(\frac{\partial h}{\partial \eta} + \rho \frac{\partial Z}{\partial \xi}\right)$$

18.14$_C$ You have been asked to formulate the problem of determining the best location for a new power plant from the standpoint of minimizing the average yearly exposure of the population to its emissions of SO_2. Assume that the long-term average concentration of SO_2 downwind of the plant can be fairly accurately represented by the Gaussian plume equation. Let (X', Y', Z') be the location of the source in the (x', y', z') coordinate system. Assume that

$$\sigma_y^2 = a\bar{u}^\alpha(x' - X')^\beta$$
$$\sigma_z^2 = b\bar{u}^\alpha(x' - X')^\beta \qquad (A)$$

where a, b, α, and β depend on the atmospheric stability. The wind $\bar{u}$ is assumed to be in the x' direction. A conventional fixed (x, y, z) coordinate system with x and y pointing in the east and north directions, respectively, will not necessarily coincide with the (x', y', z') system, which is always chosen so that the wind direction is parallel to x'. The vertical coordinate is the same in each system, that is, $z = z'$. Given that the (x', y') coordinate system is at an angle θ to the fixed (x, y) system, show that

$$x' = x\cos\theta + y\sin\theta$$
$$y' = -x\sin\theta + y\cos\theta \qquad (B)$$

It is necessary to convert the ground-level concentration $\langle c(x', y', 0)\rangle$ to $\langle c(x, y, 0)\rangle$, the average concentration at $(x, y, 0)$ from a point source at (X, Y, Z) with the wind in a direction at an angle of θ to the x axis. Show that the result is

$$\langle c(x, y, 0)\rangle = \frac{q}{\pi(ab)^{1/2}\bar{u}^{1+\alpha}[(x - X)\cos\theta + (y - Y)\sin\theta]^\beta}$$
$$\times \exp\left(-\frac{b[(X - x)\sin\theta + (y - Y)\cos\theta]^2 + aZ^2}{2\bar{u}^\alpha ab[(x - X)\cos\theta + (y - Y)\sin\theta]^\beta}\right) \qquad (C)$$

and that this equation is valid as long as

$$(x - X)\cos\theta + (y - Y)\sin\theta > 0 \qquad (D)$$

Now let $P(\bar{u}_j, \theta_k)$ be the fraction of time over a year that the wind blows with speed $\bar{u}_j$ in direction θ_k. Thus, if you consider J discrete windspeed classes and K directions:

$$\sum_{j=1}^{J}\sum_{k=1}^{K} P(\bar{u}_j, \theta_k) = 1 \qquad (E)$$

The yearly average concentration at location $(x, y, 0)$ from the source at (X, Y, Z) is given by

$$c(x, y, 0) = \sum_{j=1}^{J} \sum_{k=1}^{K} \langle c(x, y, 0) \rangle P(\bar{u}_j, \theta_k) \tag{F}$$

where $\langle c(x, y, 0) \rangle$ is given by (C). The total exposure of the region is defined by

$$E = \iint C(x, y, 0) dx \, dy \tag{G}$$

The optimal source location problem is then to choose (X, Y) (assuming that the stack height Z is fixed) to minimize E subject to the constraint that (X, Y) lies in the region. Carry through the solution for the optimal location X of a ground-level cross-wind line source (parallel to the y axis) on a region $0 \leq x \leq L$. Let $P_0(\bar{u}_j)$ and $P_1(\bar{u}_j)$ be the fractions of time that the wind blows in the $+x$ and $-x$ directions, respectively. Show that the value of X to minimize E subject to $0 \leq X \leq L$ is

$$X_{\text{opt}} = \begin{cases} 0 & \alpha_0 < \alpha_1 \\ L & \alpha_0 > \alpha_1 \end{cases}$$

$$\alpha_i = \frac{q}{\pi^{1/2} a} \frac{L^{(1-\beta/2)}}{1 - \beta/2} \sum_{j=1}^{J} \bar{u}_j^{-1-\alpha/2} P_i(\bar{u}_j) \qquad i = 0, 1$$

18.15$_B$ When a cloud of pollutant is deep enough to occupy a substantial fraction of the Ekman layer, its lateral spread will be influenced or possibly dominated by variations in the direction of the mean velocity with height. The mean vertical position of a cloud released at ground level increases at a velocity proportional to the friction velocity u_*. If the thickness of the Ekman layer can be estimated as $0.2u_*/f$, where f is the Coriolis parameter, estimate the distance that a cloud must travel from its source in midlatitudes for crosswind shear effects to become important.

18.16$_B$ Consider a continuous, ground-level crosswind line source of finite length b and strength $S(\text{g km}^{-1} \text{s}^{-1})$. Assume that conditions are such that the slender plume approximation is applicable.

a. Taking the origin of the coordinate system as the center of the line, show that the mean ground-level concentration of material at any point downwind of the source is given by

$$\langle c(x, y, 0) \rangle = \frac{S}{(2\pi)^{1/2} \sigma_z \bar{u}} \left[\text{erf}\left(\frac{b/2 - y}{\sqrt{2} \sigma_y} \right) + \text{erf}\left(\frac{b/2 + y}{\sqrt{2} \sigma_y} \right) \right]$$

b. For large distances from the source, show that the ground-level concentration along the axis of the plume $(y = 0)$ may be approximated by

$$\langle c(x,0,0)\rangle = \frac{Sb}{\pi \sigma_y \sigma_z \bar{u}}$$

c. The width w of a diffusing plume is often defined as the distance between the two points where the concentration drops to 10% of the axial value. For the finite line source, at large enough distances from the source, the line source can be considered a point source. Show that under these conditions σ_y may be determined from a measurement of w from

$$\sigma_y = \frac{w}{4.3}$$

18.17$_C$ In this problem we wish to examine two aspects of atmospheric diffusion theory: (1) the slender plume approximation and (2) surface deposition. To do so, consider an infinitely long, continuously emitting, ground-level crosswind line source of strength q_l. We will assume that the mean concentration is described by the atmospheric diffusion equation,

$$\bar{u}\frac{\partial \langle c\rangle}{\partial x} = K\left(\frac{\partial^2 \langle c\rangle}{\partial x^2} + \frac{\partial^2 \langle c\rangle}{\partial z^2}\right) + q_l\delta(x)\delta(z)$$

$$K\frac{\partial \langle c\rangle}{\partial z} = v_d\langle c\rangle \qquad z = 0$$

$$\langle c(x,z)\rangle = 0 \qquad z \to +\infty \qquad x = \pm L$$

where the mean velocity $\bar{u}$ and eddy diffusivity K are independent of the height, v_d is a parameter that is proportional to the degree of absorptivity of the surface (the so-called deposition velocity), and L is an arbitrary distance from the source at which the concentration may be assumed to be at background levels. (L is a convenience in the solution and can be made as large as desired to approximate the condition $\langle c\rangle = 0$ as $x \to \pm\infty$.)

a. It is convenient to place this problem in dimensionless form by defining $X = x/L$, $Z = z/L$, $C(X,Z) = \langle c(x,z)\rangle K/q_l$, $Pe = \bar{u}L/K$, and $Sh = v_d L/K$.

Show that the dimensionless solution is

$$C(X,Z) = e^{Pe(x/2)}\sum_{n=1}^{\infty}\left[(\theta_n + Sh)\left(1 + \frac{\sin 2\lambda_n}{2\lambda_n}\right)\right]^{-1} e^{-\theta_n Z}\cos\lambda_n X$$

where $\theta_n^2 = \lambda_n^2 + (Pe)^2/4$ and $\lambda_n = [(2n-1)/2]\pi$.

b. To explore the slender plume approximation, let us now consider

$$\bar{u}\frac{\partial \langle c \rangle}{\partial x} = K\frac{\partial^2 \langle c \rangle}{\partial z^2} + q_l \delta(x)\delta(z)$$

$$K\frac{\partial \langle c \rangle}{\partial z} = v_d \langle c \rangle \qquad z = 0$$

$$\langle c(x,z) \rangle = 0 \qquad z \to +\infty$$

$$\langle c(0,z) \rangle = 0$$

First, express this problem with the source q_l in the $x = 0$ boundary condition and then in the $z = 0$ boundary condition. Solve any one of the three formulations to obtain

$$C(X,Z) = \frac{1}{Pe}\left\{ \left(\frac{Pe}{\pi X}\right)^{1/2} \exp\left(-\frac{(Pe)Z^2}{4X}\right) - Sh \, \exp\left((Sh)Z + \frac{(Sh)^2 X}{Pe} \right) \right.$$
$$\left. \times \, \mathrm{erfc}\left[\left(\frac{X}{Pe}\right)^{1/2} \left(Sh + \frac{(Pe)Z}{2X} \right) \right] \right\}$$

(Note that even though the length L does not enter into this problem, we can use it merely to obtain the same dimensionless variables as in the previous problem.)

c. We want to compare the two solutions numerically for parameters of interest in air pollution to explore (1) the effect of diffusion in the X direction and (2) the effect of surface deposition. Thus evaluate the two solutions for the following parameter values:

$$\bar{u} = 1 \text{ and } 5 \, \mathrm{m \, s^{-1}}$$
$$K = 1 \text{ and } 10 \, \mathrm{m^2 \, s^{-1}}$$
$$v_d = 0 \text{ and } 1 \, \mathrm{cm \, s^{-1}}$$
$$L = 1000 \, \mathrm{m}$$

Plot $C(X,Z)$ against Z at a couple of values of X to examine the two effects. Discuss your results.

REFERENCES

American Meteorological Society (1977) Workshop on Stability Classification Schemes and Sigma Curves—summary of recommendations, *Bull. Am. Meteorol. Soc.* **58**, 1305–1309.

American Society of Mechanical Engineers (ASME) (1973) *Recommended Guide for the Prediction of the Dispersion of Airborne Effluents*, 2nd ed., ASME, New York.

Binkowski, F. S. (1979) A simple semi-empirical theory for turbulence in the atmospheric surface layer, *Atmos. Environ.* **13**, 247–253.

Bowman, W. A. (1983) Characteristics of maximum concentrations, *J. Air Pollut. Control Assoc.* **33**, 29–31.

Briggs, G. A. (1969) *Plume Rise*, U.S. Atomic Energy Commission Critical Review Series T/D 25075.

Briggs, G. A. (1971) Some recent analyses of plume rise observations, *Proc. 2nd Int. Clean Air Congress*, H. M. Englund and W. T. Beery, eds., Academic Press, New York, pp. 1029–1032.

Briggs, G. A. (1974) Diffusion estimation for small emissions, in *Environmental Research Laboratories Air Resources Atmospheric Turbulence and Diffusion Laboratory 1973 Annual Report*, USAEC Report ATDL-106, National Oceanic and Atmospheric Administration, Washington, DC.

Businger, J. A., and Ayra, S. P. S. (1974) Height of the mixed layer in the stably stratified planetary boundary layer, *Adv. Geophys.* **18A**, 73–92.

Calder, K. L. (1977) Multiple-source plume models of urban air pollution—their general structure, *Atmos. Environ.* **11**, 403–414.

Crane, G., Panofsky, H., and Zeman, O. (1977) A model for dispersion from area sources in convective turbulence, *Atmos. Environ.* **11**, 893–900.

Csanady, G. I. (1973) *Turbulent Diffusion in the Environment*, Reidel, Dordrecht, The Netherlands.

Deardorff, J. W. (1970) A three-dimensional numerical investigation of the idealized planetary boundary layer, *Geophys. Fluid Dyn.* **1**, 377–410.

Deardorff, J. W., and Willis, G. E. (1975) A parameterization of diffusion into the mixed layer, *J. Appl. Meteorol.* **14**, 1451–1458.

Doran, J. C., Horst, T. W., and Nickola, P. W. (1978) Experimental observations of the dependence of lateral and vertical dispersion characteristics on source height, *Atmos. Environ.* **12**, 2259–2263.

Draxler, R. R. (1976) Determination of atmospheric diffusion parameters, *Atmos. Environ.* **10**, 99–105.

Fisher, H. B., List, E. J., Koh, R. C. Y., Imberger, J., and Brooks, N. H. (1979) *Mixing in Inland and Coastal Waters*, Academic Press, New York.

Galbally, I. E. (1971) Ozone profiles and ozone fluxes in the atmospheric surface layer, *Quart. J. Roy. Meteorol. Soc.* **97**, 18–29.

Gifford, F. A. (1961) Use of routine meteorological observations for estimating atmospheric dispersion, *Nucl. Safety* **2**, 47–51.

Gifford, F. A. (1968) An outline of theories of diffusion in the lower layers of the atmosphere, in *Meteorology and Atomic Energy*, D. Slade, ed., USAEC TID-24190, Chapter 3, U.S. Atomic Energy Commission, Oak Ridge, TN.

Gifford, F. A. (1976) Turbulent diffusion-typing schemes: A review, *Nucl. Safety* **17**, 68–86.

Gifford, F. A. (1980) Smoke as a quantitative atmospheric diffusion tracer, *Atmos. Environ.* **14**, 1119–1121.

Hanna, S. R., Briggs, G. A., and Hosker, R. P. Jr. (1982) *Handbook on Atmospheric Diffusion*, U.S. Department of Energy Report DOE/TIC-11223, Washington, DC.

Huang, C. H. (1979) Theory of dispersion in turbulent shear flow, *Atmos. Environ.* **13**, 453–463.

Irwin, J. S. (1979) *Scheme for Estimating Dispersion Parameters as a Function of Release Height*, EPA-600/4-79-062, U.S. Environmental Protection Agency, Washington, DC.

Klug, W. (1969) A method for determining diffusion conditions from synoptic observations, *Staub-Reinhalt. Luft* **29**, 14–20.

Lamb, R. G., Chen, W. H., and Seinfeld, J. H. (1975) Numerico-empirical analyses of atmospheric diffusion theories, *J. Atmos. Sci.* **32**, 1794–1807.

Lamb, R. G., and Duran, D. R. (1977) Eddy diffusivities derived from a numerical model of the convective boundary layer, *Nuovo Cimento* **1C**, 1–17.

Lamb, R. G., and Neiburger, M. (1971) An interim version of a generalized air pollution model, *Atmos. Environ.* **5**, 239–264.

Lin, J. S., and Hildemann, L. M. (1996) Analytical solutions of the atmospheric diffusion equation with multiple sources and height-dependent wind speed and eddy diffusivities, *Atmos. Environ.* **30**, 239–254.

Lin, J. S., and Hildemann, L. M. (1997) A generalized mathematical scheme to analytically solve the atmospheric diffusion equation with dry deposition, *Atmos. Environ.* **31**, 59–72.

Lumley, J. L., and Panofsky, H. A. (1964) *The Structure of Atmospheric Turbulence*, Wiley, New York.

Martin, D. O. (1976) Comment on the change of concentration standard deviations with distance, *J. Air Pollut. Control Assoc.* **26**, 145–146.

Morton, B. R., Taylor, G. I., and Turner, J. S. (1956) Turbulent gravitational convection from maintained and instantaneous sources, *Proc. Roy. Soc. Lond. Ser. A*, **234**, 1–23.

Murphy, B. D., and Nelson, C. B. (1983) The treatment of ground deposition, species decay and growth and source height effects in a Lagrangian trajectory model, *Atmos. Environ.* **17**, 2545–2547.

Myrup, L. O., and Ranzieri, A. J. (1976) *A Consistent Scheme for Estimating Diffusivities to Be Used in Air Quality Models*, Report CA-DOT-TL-7169-3-76-32, California Department of Transportation, Sacramento.

Neumann, J. (1978) Some observations on the simple exponential function as a Lagrangian velocity correlation function in turbulent diffusion, *Atmos. Environ.* **12**, 1965–1968.

Nieuwstadt, F. T. M. (1980) Application of mixed layer similarity to the observed dispersion from a ground level source, *J. Appl. Meteorol.* **19**, 157–162.

Nieuwstadt, F. T. M., and van Duuren, H. (1979) Dispersion experiments with SF_6 from the 213 m high meteorological mast at Cabau in The Netherlands, *Proc. 4th Symp. Turbulence, Diffusion and Air Pollution*, Reno, NV, American Meteorological Society, Boston, pp. 34–40.

O'Brien, J. (1970) On the vertical structure of the eddy exchange coefficient in the planetary boundary layer, *J. Atmos. Sci.* **27**, 1213–1215.

Palazzi, E., DeFaveri, M., Fumarola, G., and Ferraiolo, G. (1982) Diffusion from a steady source of short duration, *Atmos. Environ.* **16**, 2785–2790.

Panofsky, H. A., Tennekes, H., Lenschow, D. H., and Wyngaard, J. C. (1977) The characteristics of turbulent velocity components in the surface layer under convective conditions, *Boundary Layer Meteorol.* **11**, 355–361.

Pasquill, F. (1971) Atmospheric diffusion of pollution, *Quart. J. Roy. Meteorol. Soc.* **97**, 369–395.

Pasquill, F. (1974) *Atmospheric Diffusion*, 2nd ed., Halsted Press, Wiley, New York.

Ragland, K. W. (1976) Worst case ambient air concentrations from point sources using the Gaussian plume model, *Atmos. Environ.* **10**, 371–374.

Roberts, E. M. (1980) Conditions for maximum concentration, *J. Air Pollut. Control Assoc.* **30**, 274–275.

Schatzmann, M. (1979) An integral model of plume rise, *Atmos. Environ.* **13**, 721–731.

Sedefian, L., and Bennett, E. (1980) A comparison of turbulence classification schemes, *Atmos. Environ.* **14**, 741–750.

Shir, C. C. (1973) A preliminary numerical study of atmospheric turbulent flows in the idealized planetary boundary layer, *J. Atmos. Sci.* **30**, 1327–1339.

Taylor, G. I. (1921) Diffusion by continuous movements, *Proc. Lond. Math. Soc. Ser. 2* **20**, 196.

Tennekes, H. (1979) The exponential Lagrangian correlation function and turbulent diffusion in the inertial subrange, *Atmos. Environ.* **13**, 1565–1567.

Turner, D. B. (1969) *Workbook of Atmospheric Diffusion Estimates*, USEPA 999-AP-26, U.S. Environmental Protection Agency, Washington, DC.

U.S. Environmental Protection Agency (1980) *OAQPS Guideline Series, Guidelines on Air Quality Models*, Research Triangle Park, NC.

Weber, A. H. (1976) *Atmospheric Dispersion Parameters in Gaussian Plume Modeling*, EPA-600/4-76-030A, U.S. Environmental Protection Agency, Washington, DC.

Willis, G. E., and Deardorff, J. W. (1976) A laboratory model of diffusion into the convective boundary layer, *Quart. J. Roy. Meteorol. Soc.* **102**, 427–447.

19 Dry Deposition

Wet deposition and dry deposition are the ultimate paths by which trace gases and particles are removed from the atmosphere. The relative importance of dry deposition, as compared with wet deposition, for removal of a particular species depends on the following factors:

- Whether the substance is present in the gaseous or particulate form
- The solubility of the species in water
- The amount of precipitation in the region
- The terrain and type of surface cover

Dry deposition is, broadly speaking, the transport of gaseous and particulate species from the atmosphere onto surfaces in the absence of precipitation. The factors that govern the dry deposition of a gaseous species or a particle are the level of atmospheric turbulence, the chemical properties of the depositing species, and the nature of the surface itself. The level of turbulence in the atmosphere, especially in the layer nearest the ground, governs the rate at which species are delivered down to the surface. For gases, solubility and chemical reactivity may affect uptake at the surface. For particles, size, density, and shape may determine whether capture by the surface occurs. The surface itself is a factor in dry deposition. A nonreactive surface may not permit absorption or adsorption of certain gases; a smooth surface may lead to particle bounce-off. Natural surfaces, such as vegetation, whereas highly variable and often difficult to describe theoretically, generally promote dry deposition.

19.1 DEPOSITION VELOCITY

It is generally impractical, in terms of the atmospheric models within which such a description is to be embedded, to simulate, in explicit detail, the microphysical pathways by which gases and particles travel from the bulk atmosphere to individual surface elements where they adhere. In the universally used formulation for dry deposition, it is assumed that the dry deposition flux is directly proportional to the local concentration C of the depositing species, at some reference height above the surface (e.g., 10 m or less)

$$F = -v_d C \tag{19.1}$$

where F represents the vertical dry deposition flux, the amount of material depositing to a unit surface area per unit time. The proportionality constant between flux and concentration v_d has units of length per unit time and is known as the *deposition velocity*. Because C is a function of height z above the ground, v_d is also a function of z and must be related to a reference height at which C is specified. By convention, a downward flux is negative, so that v_d is positive for a depositing substance.

The advantage of the deposition velocity representation is that all the complexities of the dry deposition process are bundled in a single parameter, v_d. The disadvantage is that, because v_d contains a variety of physical and chemical processes, it may be difficult to specify properly. The flux F is assumed to be constant up to the reference height at which C is specified. Equation (19.1) can be readily adapted in atmospheric models to account for dry deposition and is usually incorporated as a surface boundary condition to the atmospheric diffusion equation.

The process of dry deposition of gases and particles is generally represented as consisting of three steps: (1) aerodynamic transport down through the atmospheric surface layer to a very thin layer of stagnant air just adjacent to the surface; (2) molecular (for gases) or Brownian (for particles) transport across this thin stagnant layer of air, called the *quasi-laminar sublayer*, to the surface itself; and (3) uptake at the surface. Each of these steps contributes to the value of the deposition velocity v_d.

Transport through the atmospheric surface layer down to the quasi-laminar sublayer occurs by turbulent diffusion. Both gases and particles are subject to the same eddy transport in the surface layer. Sedimentation of particles may also contribute to the downward flux for larger particles.

As a consequence of the no-slip boundary condition for airflow at a surface, the air at an infinitesimal distance above a stationary surface will also be stationary. The surface of the Earth consists generally of a large number of irregularly shaped obstacles, such as leaves and buildings. Adjacent to each surface is a boundary layer, of thickness on the order of millimeters, in which the air is more or less stationary. As noted above, this layer is called the quasi-laminar sublayer. Transport across the quasi-laminar sublayer occurs by diffusion (gases and particles) and sedimentation (particles). Particle removal can actually occur within the quasi-laminar sublayer by interception, when particles moving with the mean air motion pass sufficiently close to an obstacle to collide with it. Particle deposition by impaction may also occur when particles cannot follow rapid changes of direction of the mean flow, and their inertia carries them across the sublayer to the surface. Like interception, impaction occurs when there are changes in the direction of airflow; unlike interception, a particle subject to impaction leaves its air streamline and crosses the sublayer with inertia derived from the mean flow (recall Figure 9.12).

The final step in the dry deposition process is actual uptake of the vapor molecules or particles by the surface. Gaseous species may absorb irreversibly or reversibly into the surface; particles simply adhere. The amount of moisture on the surface and its stickiness are important factors at this step. For moderately soluble gases, such as SO_2 and O_3, the presence of surface moisture can have a marked effect on whether the molecule is actually removed. For highly soluble and chemically reactive gases, such as HNO_3, deposition is rapid and irreversible on almost any surface. Solid particles may bounce off a smooth surface; liquid particles are more likely to adhere upon contact.

19.2   RESISTANCE MODEL FOR DRY DEPOSITION

It has proved useful to interpret the deposition process in terms of an electrical resistance analogy, in which the transport of material to the surface is assumed to be governed by three resistances in series: the aerodynamic resistance r_a, the quasi-laminar layer resistance r_b, and the surface or canopy (the term "canopy" refers to the vegetation canopy) resistance r_c. The total resistance, r_t, to deposition of a gaseous species is the sum of the three individual resistances and is, by definition, the inverse of the deposition velocity:

$$v_d^{-1} = r_t = r_a + r_b + r_c \tag{19.2}$$

(The overall resistance for particles includes the effect of sedimentation and will be discussed shortly.) This basic conceptual model is depicted schematically in Figure 19.1.

Equation (19.2) is derived as follows. By reference to Figure 19.1, let C_3 be the concentration at the top of the surface layer [the reference height referred to in (19.1)]; C_2, that at the top of the quasi-laminar sublayer; C_1, at the bottom of the quasi-laminar sublayer, and $C_0 = 0$, in the surface itself. The difference between C_1 and C_0 accounts for

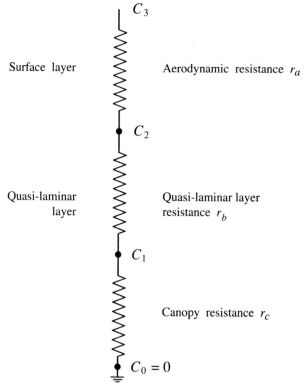

FIGURE 19.1   Resistance model for dry deposition.

the resistance offered by the surface itself. At steady state the overall flux of a vapor species is related to the concentration differences and resistances across the layers by

$$F = \frac{C_3 - C_2}{r_a} = \frac{C_2 - C_1}{r_b} = \frac{C_1 - C_0}{r_c} = \frac{C_3 - C_0}{r_t} \tag{19.3}$$

where $C_0 = 0$. Thus

$$\frac{C_3}{r_t} = \frac{C_3 - C_2}{r_a} = \frac{C_2 - C_1}{r_b} = \frac{C_1}{r_c} \tag{19.4}$$

By solving the three independent equations for the three unknowns, C_1, C_2, C_3, one obtains (19.2).

For particles, the model is identical to that for gases except that particle settling operates in parallel with the three resistances in series. The sedimentation flux is equal to the particle settling velocity, v_s, multiplied by the particle concentration. It is usually assumed that particles adhere to the surface on contact so that the surface or canopy resistance $r_c = 0$. In this case the vertical flux is

$$F = \frac{C_3 - C_2}{r_a} + v_s C_3 = \frac{C_2 - C_1}{r_b} + v_s C_2 = \frac{C_3 - C_1}{r_t} \tag{19.5}$$

where, because of assumed perfect removal, $C_1 = 0$. Thus

$$\frac{C_3}{r_t} = \frac{C_3 - C_2}{r_a} + v_s C_3 = \frac{C_2}{r_b} + v_s C_2 \tag{19.6}$$

Solving the two independent equations to eliminate C_2 and C_3 yields the resistance relation analogous to (19.2) for particle dry deposition:

$$v_d = \frac{1}{r_t} = \frac{1}{r_a + r_b + r_a r_b v_s} + v_s \tag{19.7}$$

Therefore the deposition velocity of particles may be viewed as the reciprocal of three resistances in series (r_a, r_b, and $r_a r_b v_s$) and one in parallel ($1/v_s$). The third resistance in series is a virtual resistance since it is a mathematical artifact of the equation manipulation and not an actual physical resistance.

The aerodynamic and quasi-laminar resistances are affected by windspeed, vegetation height, leaf size, and atmospheric stability. In general, $r_a + r_b$ decreases with increasing windspeed and vegetation height. Thus smaller resistances and hence higher deposition rates are expected over tall forests than over short grass. Also, smaller resistances are expected under unstable than under stable and neutral conditions. Typical aerodynamic layer resistances for a $4 \, \text{m s}^{-1}$ windspeed are as follows:

$$
\begin{array}{llll}
z_0 = 0.1 \, \text{m} & \text{Grass} & & r_a \simeq 60 \, \text{s m}^{-1} \\
 = 0.1 \, \text{m} & \text{Crop} & & \simeq 20 \, \text{s m}^{-1} \\
 = 10 \, \text{m} & \text{Conifer forest} & & \simeq 10 \, \text{s m}^{-1}
\end{array}
$$

In daytime, values of r_a tend to be relatively low,[1] and hence the precise value is often not important (an exception is very reactive species such as HNO_3, HCl, and NH_3 that deposit efficiently onto any surface with which they come into contact). At night, r_a is considerably higher over land and is often the controlling factor in the overall rate of deposition. r_a values up to $150 \, s \, m^{-1}$ can occur at night over land when turbulent mixing is reduced. When r_a is high, it is usually difficult to evaluate. Fortunately, in this case, the resulting deposition fluxes are low, and hence the error in predicting the dry deposition rate tends to be relatively less important. Over water, r_a lacks the strong diurnal cycle typical of that over land. The larger the body of water and the farther from shore, the smaller the diurnal cycle in r_a.

The absolute magnitude of r_b does not change significantly as atmospheric conditions vary. Using perfect-sink surfaces to provide zero surface resistance, Wu et al. (1992a) showed that r_b for SO_2 in the ambient atmosphere reached a maximum value of about $100 \, s \, cm^{-1}$ for a range of windspeeds for smooth anthropogenic (human-made) surfaces. Much smaller values would be expected for rough natural vegetation. This suggest that r_b can be of comparable value to r_a during the day and is probably much smaller than r_a at night. Thus r_b is not often rate-limiting for gases.

Under stable conditions, especially with low winds, r_a dominates dry deposition, whereas under neutral and unstable conditions, r_a is seldom the controlling resistance for dry deposition. Over some surfaces (especially more barren landscapes), r_c can be quite large, thereby dominating the dry deposition process. The greatest deposition velocities are achieved under highly unstable conditions over transpiring vegetation, when the three resistances are all small and of approximately the same order of magnitude. Table 19.1 summarizes typical dry deposition velocities for some atmospheric trace gases.

For gaseous species, solubility and reactivity are the major factors affecting surface resistance and overall deposition velocity. For particles, the factor most strongly influencing the deposition velocity is the particle size. Particles are transported by turbulent diffusion toward the surface identically to gaseous species, a process enhanced for larger particles by gravitational settling. Across the quasi-laminar layer ultrafine particles ($< 0.05 \, \mu m$ diameter) are transported primarily by Brownian diffusion, analogous to the molecular diffusion of gases. Larger particles possess inertia, which may enhance the flux through the quasi-laminar sublayer (Davidson and Wu 1990).

TABLE 19.1   Typical Dry Deposition Velocities for Some Atmospheric Gases

Species	$v_d(cm \, s^{-1})$ over		
	Continent	Ocean	Ice/Snow
CO	0.03	0	0
N_2O	0	0	0
NO	0.016	0.003	0.002
NO_2	0.1	0.02	0.01
HNO_3	4	1	0.5
O_3	0.4	0.07	0.07
H_2O_2	0.5	1	0.32

Source: Hauglustaine et al. (1994).

[1]Remember that a small resistance equates to a large deposition velocity.

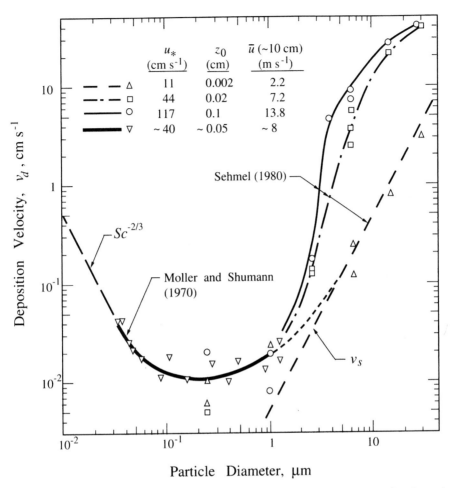

FIGURE 19.2 Particle dry deposition velocity data for deposition on a water surface in a wind tunnel (Slinn et al. 1978).

Figure 19.2 shows deposition velocity data for particles as a function of particle size for a water surface in a wind tunnel. Zufall et al. (1999) have shown that deposition to bodies of water exposed to the atmosphere can have as much as double the deposition velocity compared to a flat water surface owing to the presence of waves. Furthermore, growth of hygroscopic particles in the humid boundary layer just above the air–water interface can increase deposition velocities slightly compared with deposition to dry surfaces (Zufall et al. 1998).

The data in Figure 19.2 indicate a characteristic minimum in deposition velocity in the particle diameter range between 0.1 and 1.0 μm. The reason for this behavior is that small particles behave much like gases and are efficiently transported across the quasi-laminar layer by Brownian diffusion. Brownian diffusion ceases to be an effective transport mechanism for particles with diameters above $\sim$0.05 μm. Moderately large particles in the 2–20 μm diameter range are efficiently transported across the laminar sublayer by inertial impaction (see Figure 9.12); the deposition of even larger particles ($D_p > 20\,\mu$m) is governed principally by gravitational settling since the settling velocity increases with the square of the particle diameter. The transport mechanisms for particles in the 0.05–2 μm diameter range are less effective than those of smaller or larger particles. Uncertainties in ambient particle size distributions thus have a marked effect on the overall predicted particle dry deposition velocity (Ruijgrok et al. 1995).

19.3 AERODYNAMIC RESISTANCE

Turbulent transport is the mechanism that brings material from the bulk atmosphere down to the surface and therefore determines the aerodynamic resistance. The turbulence intensity is dependent principally on the lower atmospheric stability and the surface roughness and can be determined from micrometeorological measurements and surface characteristics such as windspeed, temperature, and radiation and the surface roughness length (see Chapter 16). During daytime conditions, the turbulence intensity is typically large over a reasonably thick layer (i.e., the well-mixed layer), thus exposing a correspondingly ample reservoir of material to potential surface deposition. During the night, stable stratification of the atmosphere near the surface often reduces the intensity and vertical extent of the turbulence, effectively diminishing the overall dry deposition flux. The aerodynamic resistance is independent of species or whether a gas or particle is involved except that gravitational settling must be taken into account for large particles.

The aerodynamic component of the overall dry deposition resistance is typically based on gradient transport theory and mass transfer/momentum transfer similarity (or mass transfer/heat transfer similarity). It is presumed that turbulent transport of species through the surface layer (i.e., constant-flux layer) is expressible in terms of an eddy diffusivity multiplied by a concentration gradient, that turbulent transport of material occurs by mechanisms that are similar to those for turbulent heat and/or momentum transport, and that measurements obtained for one of these entities thus can be applied, using scaling parameters, to calculate the corresponding behavior of another (see Chapter 16). Expressions for the aerodynamic resistance are most easily obtained by integrating the micrometeorological flux–gradient relationships. Applications of similarity theory to turbulent transfer through the surface layer suggest that the eddy diffusivity should be proportional to the friction velocity and the height above the ground. Under diabatic conditions the eddy diffusivity is modified from its neutral form by a function dependent on the dimensionless height scale, $\zeta = z/L$, where L is the Monin–Obukhov length.

The vertical turbulent flux of a quantity, say, C, through the (constant-flux) surface layer is expressed as

$$F_a = K \frac{\partial C}{\partial z} \tag{19.8}$$

where K is the appropriate eddy diffusivity and F_a is, by definition, constant across the layer. From dimensional analysis and micrometeorological measurements, the eddy momentum (K_M) and heat diffusivities (K_T) can be expressed by

$$K_M = \frac{\kappa u_* z}{\phi_M(\zeta)} \tag{19.9}$$

$$K_T = \frac{\kappa u_* z}{\phi_T(\zeta)} \tag{19.10}$$

where κ is the von Karman constant, u_* is the friction velocity, and ϕ_M and ϕ_T, respectively, are empirically determined dimensionless momentum and temperature profile functions.

If (19.8) is integrated across the depth of the constant-flux (i.e., surface) layer from z_3 down to z_2, the flux F_a may be written as

$$F_a = (C_3 - C_2) \left(\int_{z_2}^{z_3} \frac{\phi(\zeta)}{\kappa u_* z} dz \right)^{-1} \tag{19.11}$$

where, as above, C_3 and C_2 refer to concentrations at the top and bottom of the constant-flux layer and $\phi(\zeta)$ denotes either $\phi_M(\zeta)$ or $\phi_T(\zeta)$, whichever is deemed analogous to the species profile function. The aerodynamic resistance is thus given by

$$r_a = \int_{z_2}^{z_3} \frac{\phi(\zeta)}{\kappa u_* z} dz \tag{19.12}$$

The integral in (19.12) is evaluated from the bottom of the constant-flux layer (at z_0, the roughness length) to the top (z_r, the reference height implicit in the definition of v_d).

Explicit Expressions for r_a If the stability-dependent temperature profile function is given by (16.75), then

$$\phi_T(\zeta) = \begin{cases} 1 + 4.7\zeta & \text{for } 0 < \zeta < 1 \quad \text{(stable)} \\ 1 & \text{for } \zeta = 0 \quad \text{(neutral)} \\ (1 - 15\zeta)^{-1/4} & \text{for } -1 < \zeta < 0 \quad \text{(unstable)} \end{cases} \tag{19.13}$$

the corresponding aerodynamic resistance is

$$r_a = \begin{cases} \dfrac{1}{\kappa u_*} \left[\ln\left(\dfrac{z}{z_0}\right) + 4.7(\zeta - \zeta_0) \right] & \text{(stable)} \\[2ex] \dfrac{1}{\kappa u_*} \ln\left(\dfrac{z}{z_0}\right) & \text{(neutral)} \\[2ex] \dfrac{1}{\kappa u_*} \left[\ln\left(\dfrac{z}{z_0}\right) + \ln\left(\dfrac{(\eta_0^2 + 1)(\eta_0 + 1)^2}{(\eta_r^2 + 1)(\eta_r + 1)^2}\right) + 2(\tan^{-1}\eta_r - \tan^{-1}\eta_0) \right] & \text{(unstable)} \end{cases} \tag{19.14}$$

where $\eta_0 = (1 - 15\zeta_0)^{1/4}$, $\eta_r = (1 - 15\zeta_r)^{1/4}$, $\zeta_0 = z_0/L$.

The theory is applicable only in the surface layer where the flux is nondivergent, that is, $-3 \lesssim Rf \lesssim 2$. An approximate maximum vertical extent is $\sim$100 m.

19.4 QUASI-LAMINAR RESISTANCE

The resistance model for dry deposition postulates that adjacent to the surface exists a quasi-laminar layer, across which the resistance to transfer depends on molecular properties of the substance and surface characteristics. This layer does not usually correspond to a laminar boundary layer in the classical sense; rather, it is the consequence of many viscous layers adjacent to the obstacles constituting the overall, effective surface seen by the atmosphere. The depth of this layer constantly changes in response to turbulent shear stresses adjacent to the surface or surface elements. In fact, the layer may only exist intermittently on such surfaces as plant leaves, which are often in continuous motion. Whether a quasi-laminar layer actually exists physically depends on the smoothness and the shape of the surface or surface elements, and to some extent the variability of the near-surface turbulence, but, in terms of the theory, it is considered to exist.

19.4.1 Gases

A viscous boundary layer adjacent to the surface of some obstacle on which deposition is occurring is an impediment to all depositing species, regardless of the orientation of the target surface. Molecular (and Brownian) diffusion occurs independently of direction; molecular diffusion can occur to the underside of a leaf just as easily as it can to the top surface. The flux across the quasi-laminar sublayer adjacent to the surface is expressed in terms of a dimensionless transfer coefficient, B, multiplying the concentration difference across the layer, $C_2 - C_1$. Since, under steady-state conditions, this flux is equal to that across each layer, we write

$$F_b = Bu_*(C_2 - C_1) \tag{19.15}$$

where C_1 is the concentration at the surface, and, by convention, the transfer coefficient is dimensionalized by u_*. The quasi-laminar layer resistance is then given by

$$r_b = \frac{1}{Bu_*} \tag{19.16}$$

The quasi-laminar resistance r_b depends on the molecular diffusivity of the gas being considered. This dependence can be accounted for through the dimensionless Schmidt number, $Sc = v/D$, where v is the kinematic viscosity of air and D is the molecular diffusivity of the species. Measurements over canopies have shown r_b to be relatively insensitive to the canopy roughness length z_0. A useful expression for r_b for gases in terms of the Schmidt number is,

$$r_b = \frac{5\,Sc^{2/3}}{u_*} \tag{19.17}$$

19.4.2 Particles

For gases, molecules are transported across the quasi-laminar sublayer by molecular diffusion, and then whether the gas molecule is actually taken up by the surface depends on the nature of the surface (e.g. whether it is wet) and the solubility and reactivity of the gas. Across the quasi-laminar sublayer, particles are transported analogously by Brownian diffusion. Particles also possess inertia, which leads to deposition on surface elements by impaction. In addition, because of their finite size, particles are also removed by interception on elements of the surface. All of these particle processes can be considered as occurring in the laminar sublayer. In so doing, the surface resistance r_c can be assumed to be zero. (Some references consider these processes to constitute the surface resistance for particles; whether this resistance is called the *quasi-laminar* or *surface* resistance is just a question of terminology.)

The overall particle dry deposition velocity is given by (19.7). The particle settling velocity v_s is given by (9.42)

$$v_s = \frac{\rho_p D_p^2 g C_c}{18\mu} \tag{19.18}$$

where ρ_p is the density of the particle, D_p is particle diameter, μ is the viscosity of air, and C_c is the slip correction coefficient, given by (9.34). The viscosity of air as a function of temperature is $\mu = 1.8 \times 10^{-5} \, (T/298)^{0.85} \, (\text{kg m}^{-1}\text{s}^{-1})$, where T is in K.

An expression for the overall quasi-laminar resistance for particles has been developed by Zhang et al. (2001)[2]

$$r_b = \frac{1}{\varepsilon_0 u_*(E_B + E_{IM} + E_{IN})R_1} \tag{19.19}$$

where

E_B = collection efficiency from Browian diffusion

E_{IM} = collection efficiency from impaction

E_{IN} = collection efficiency from interception

and where R_1 is a correction factor representing the fraction of particles that stick to the surface and ε_0 is an empirical constant = 3.0.

Transport of particles by Brownian diffusion depends on the Schmidt number, $Sc = v/D$, where D is now the Brownian diffusivity of particles, given by (9.73):

$$D = \frac{kTC_c}{3\pi\mu D_p} \tag{19.20}$$

E_B is represented as

$$E_B = Sc^{-\gamma} \tag{19.21}$$

where γ lies between $\frac{1}{2}$ and $\frac{2}{3}$, with larger values for rougher surfaces. Zhang et al. (2001) suggest different values of γ corresponding to different land-use categories (to be presented shortly).

---

[2]Zhang et al. (2001) refer to this as the surface resistance. Their model includes only the aerodynamic and surface resistances.

The dimensionless group governing impaction is the Stokes number (St), given by (9.101). A number of authors have proposed expressions for E_{IM} involving St. The particular form adapted by Zhang et al. (2001) is that of Peters and Eiden (1992)

$$E_{IM} = \left(\frac{St}{\alpha + St}\right)^{\beta} \tag{19.22}$$

where α depends on the land use category and $\beta = 2$. The Stokes number in (19.22) is defined as follows:

$$St = \frac{v_s u_*^2}{g\nu} \quad \text{smooth surfaces or surfaces with bluff} \tag{19.23}$$
$$\text{roughness elements}$$

$$St = \frac{v_s u_*}{gA} \quad \text{vegetated surfaces} \tag{19.24}$$

Here A is the characteristic radius of collectors, the value of which depends on land-use categories.

The efficiency of particle collection by interception depends on the particle diameter and the characteristic dimension of the collectors. Again, a variety of expressions have been proposed for E_{IN}. The one adapted here is

$$E_{IN} = \frac{1}{2}\left(\frac{D_p}{A}\right)^2 \tag{19.25}$$

Finally, the parameter R_1 represents the fraction of particles, once in contact, that stick to the surface. Generally, small particles do not possess sufficient inertia to bounce off a surface, and R_1 is simply unity. We noted this earlier in arguing that the surface resistance r_c for particles is usually taken to be zero. Whether a particle possesses sufficient inertia to rebound from a surface depends on its Stokes number. Zhang et al. (2001) use the following form for R_1 suggested by Slinn (1982):

$$R_1 = \exp\left(-St^{1/2}\right) \tag{19.26}$$

As $St \to 0$, $R_1 = 1$. If the surface is wet, $R_1 = 1$, that is, particles are assumed to stick to a wet surface regardless of their size.

Summarizing, the expression for the quasi-laminar resistance for particle dry deposition is

$$r_b = \frac{1}{3u_*\left[Sc^{-\gamma} + \left(\dfrac{St}{\alpha + St}\right)^2 + \dfrac{1}{2}\left(\dfrac{D_p}{A}\right)^2\right]R_1} \tag{19.27}$$

where specification of γ, α, and A is based on land-use categories (Table 19.2). The three terms in (19.27) account for transfer by Brownian diffusion and collection by impaction and interception. The data of Table 19.2 must be considered rough averages over many

TABLE 19.2   Parameters for Selected Land Use Categories in the Dry Deposition model of Zhang et al. (2001)

Land use categories (LUC)
   1 Evergreen–needleleaf trees
   4 Deciduous broadleaf trees
   6 Grass
   8 Desert
  10 Shrubs and interrupted woodlands

Seasonal categories (SC)
   1 Midsummer with lush vegetation
   2 Autumn with cropland not harvested
   3 Late autumn after frost, no snow
   4 Winter, snow on ground
   5 Transitional

	LUC	1	4	6	8	10
z_0, m	SC1	0.8	1.05	0.1	0.04	0.1
	SC2	0.9	1.05	0.1	0.04	0.1
	SC3	0.9	0.95	0.05	0.04	0.1
	SC4	0.9	0.55	0.02	0.04	0.1
	SC5	0.8	0.75	0.05	0.04	0.1
A, mm	SC1	2.0	5.0	2.0	N/A	10.0
	SC2	2.0	5.0	2.0	N/A	10.0
	SC3	2.0	10.0	5.0	N/A	10.0
	SC4	2.0	10.0	5.0	N/A	10.0
	SC5	2.0	5.0	2.0	N/A	10.0
α		1.0	0.8	1.2	50.0	1.3
γ		0.56	0.56	0.54	0.54	0.54

types of vegetation that make up the landscape; data for individual plants are available for some species [e.g., Davidson et al. (1982)].

Figure 19.3 shows $1/r_b$ as a function of particle diameter for particles of density $1\,\mathrm{g\,cm^{-3}}$ at $u_* = 1$ and $10\,\mathrm{m\,s^{-1}}$. If the aerodynamic resistance r_a can be neglected, then the overall particle dry deposition velocity will be just $v_d = (1/r_b) + v_s$. Thus, $1/r_b$ is the component of the overall dry deposition velocity that accounts for processes occurring in the surface layer other than settling. The value of $1/r_b$ is large for very small particles because of the efficiency of Brownian diffusion as a mechanism for transporting such particles across the surface layer. For very large particles, impaction and interception lead to effective removal. The result is that $1/r_b$ achieves a minimum in the range between 0.1 and 1.0 μm diameter where none of the removal process is especially effective. Since all the removal terms in the denominator of (19.19) are multiplied by u_*, the value of $1/r_b$ scales directly with u_*; the larger the value of u_*, the larger the deposition velocity.

19.5   SURFACE RESISTANCE

The surface resistance r_c for gases depends critically on the nature of the surface. It is useful to divide potential surfaces into three categories: water, ground, and vegetative

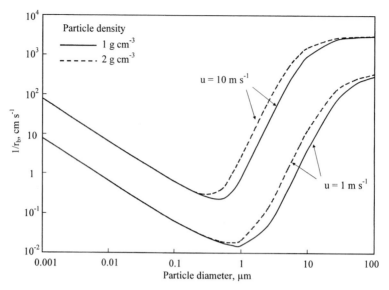

FIGURE 19.3 $1/r_b$ for particle dry deposition.

canopy. Figure 19.4 depicts the various resistance paths, including those for particles to these three types of surfaces. For gases, depending on the nature of the surface, one of the following three resistances applies:

r_{cw}	Water resistance
r_{cg}	Ground resistance
r_{cf}	Foliar resistance

When a vegetative canopy is present, the surface resistance r_c is often referred to as the *canopy resistance*. The canopy resistance is the most difficult of the three to evaluate because of the complexity of vegetative surfaces. The capture of gases by vegetation depends ultimately on the accessibility of the gas to sites within the plant capable of absorbing the gas. Numerous field studies show, for example, that r_c is the dominant resistance for SO_2 deposition to a plant canopy (Matt et al. 1987). Studies of pathways of deposition of gases to vegetation have constituted a major area of dry deposition research [e.g., Hicks et al. (1987), Baldocchi (1988), Hicks and Matt (1988), Meyers and Baldocchi (1988), Meyers and Hicks (1988), Wesely (1989), Erisman and Baldocchi (1994)]. There is a sizable body of work devoted to estimating r_c as a function of chemical species, canopy type, and meteorological conditions.

A number of pathways are available from the quasi-laminar layer to the vegetation canopy or ground, including uptake by plant tissue inside leaf pores (stomata), the waxy skin of some leaves (cuticle), and mesophyll of leaves, deposition to the soil, and reactions with wetted surfaces. Each of these pathways can be represented by an appropriate resistance to transport; the total canopy resistance r_c is then determined by summing these resistances in parallel

$$\frac{1}{r_c} = \frac{1}{r_{cf}} + \frac{1}{r_{cg}}$$

(19.28a)

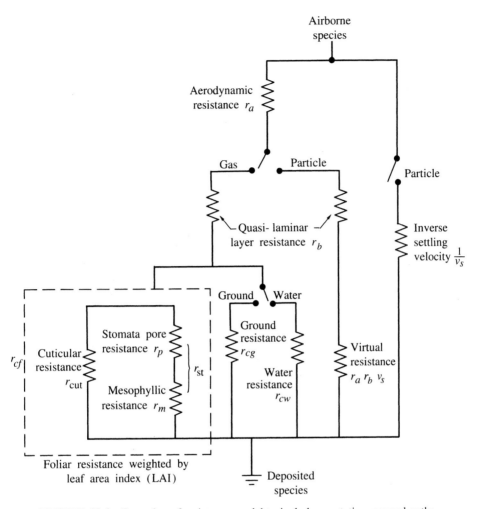

FIGURE 19.4 Extension of resistance model to include vegetative removal paths.

or, if a water surface is present

$$\frac{1}{r_c} = \frac{1}{r_{cf}} + \frac{1}{r_{cw}} \qquad (19.28b)$$

where r_{cf} is the foliar resistance, r_{cg} is the resistance to uptake by the ground (e.g., soil or snow), and r_{cw} is the resistance to uptake by water surfaces. These resistances act in parallel since the material can take only one deposition pathway, either to vegetation or to the Earth's surface. The foliar resistance r_{cf} can itself be divided into resistance to uptake at the cuticle (epidermal) surfaces (r_{cut}) and at the plant's stomata (r_{st}). As depicted in Figure 19.4, these resistances also act in parallel, so

$$r_{cf} = \left(\frac{1}{r_{cut}} + \frac{1}{r_{st}}\right)^{-1} (\text{LAI})^{-1} \qquad (19.29)$$

where these two resistances are weighted by LAI, the leaf area index, that is, the ratio of the total leaf area to the area of the ground, so that r_{cf} receives the proper magnitude relative to r_{cg} or r_{cw}.

The resistance to absorption of a species at the cuticle surfaces is frequently referenced to that of SO_2, $r_{cut} = r_{cut(0)}A_0/A$, where $r_{cut(0)}$ and A_0 are the values of cuticle resistance and the gas reactivity of SO_2, and A is the reactivity of other specific gases. Transfer of gases through the cuticle is generally less important than that through the stomata and can often be neglected. Typical values for r_{cut} for water vapor diffusion through leaf surfaces are 30 to 200 s cm^{-1}, as compared with values of r_{st} in the range 1–20 s cm^{-1}. For SO_2, cuticle resistance far exceeds the stomatal resistance (van Hove 1989). This resistance is observed to decrease as relative humidity increases.

As shown in Figure 19.4, the stomatal resistance (r_{st}) is then assumed to be composed of two components acting in series—the resistance to diffusion through the stomatal pores, r_p, and the resistance to dissolution at the mesophyllic tissue (i. e., the leaf interior), r_m. The specific locations of gaseous removal often depend on the plant's biological activity level. Sulfur dioxide, for example, enters plants through the stomata. As gas molecules enter the leaf, deposition occurs as molecules react with the moist cells in the substomatal chamber and the mesophyll. For example, the diurnally variable stomatal resistance often influences the net SO_2 and O_3 deposition to plants (Wesely and Hicks 1977). Daytime opening of the stomatal pores exposes more reactive plant tissues (the mesophyllic tissue) to the species, thus decreasing the surface resistance. During periods when stomatal openings are closed, the normally higher cuticular resistance becomes important in overall surface resistance. Higher stomatal resistance occurs at night and under conditions of severe water stress, when the stomata remain closed to minimize transpiration. Stomatal resistance decreases with increasing light. Low and high temperatures cause stomatal closure, and moderate temperature promotes stomatal opening. Values of r_p for SO_2 during daytime range between 30 and 300 s m^{-1} for a range of herbaceous annuals and woody perennials (Baldocchi et al. 1987; Matt et al. 1987). The mesophyll resistance r_m depends on the solubility of the gas (Meyers 1987). Readily soluble gases, such as SO_2, NH_3, and HNO_3, are assumed to experience no resistance at the mesophyll. Moderately soluble gases such as NO, NO_2, and O_3 have values of r_m in the range of 90 s cm^{-1} (NO) and 5 s cm^{-1} (NO_2 and O_3).

19.5.1   Surface Resistance for Dry Deposition of Gases to Water

Recapitulating, the dry deposition velocity for vapor species i is related to the three resistances by

$$v_{d,i} = \frac{1}{r_a + r_{b,i} + r_{c,i}} \tag{19.30}$$

where no subscript is needed on r_a since the aerodynamic resistance is the same for all gases. The expression for $r_{b,i}$ is given by (19.17). We consider first the surface resistance for dry deposition of gases to water, $r_{c,i} = r_{cw,i}$.

Transfer of a species A from the gas phase to a liquid phase can be depicted as shown in Figure 19.5. The gas phase is assumed to be well mixed by turbulence down to a thin stagnant film just above the air–water interface. All the resistance to mass transfer in the

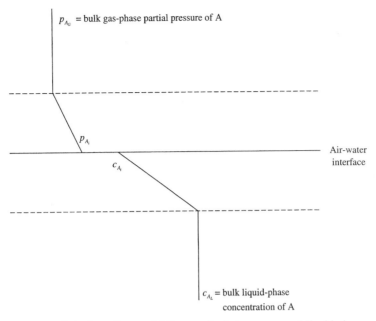

p_{A_G} = bulk gas-phase partial pressure of A

p_{A_i}

c_{A_i}

Air-water
interface

c_{A_L} = bulk liquid-phase
concentration of A

FIGURE 19.5 Two-film model for transfer between gas and liquid phases.

gas phase is assumed to occur in this thin film adjacent to the interface. Likewise, it is assumed that all the resistance to mass transfer of the dissolved gas away from the interface into the bulk liquid is confined to a thin stagnant layer of liquid just below the air–water interface. Right at the interface, the partial pressure of A in the gas phase is in equilibrium with the concentration of A in the liquid phase. This traditional representation of mass transfer between two phases is called the *two-film model*. The gas-phase flux of A across the thin film to the interface is represented in terms of a gas-phase mass transfer coefficient k_G as

$$F = k_G(p_{A_G} - p_{A_i}) \tag{19.31}$$

where p_{A_G} is the partial pressure of A in the bulk gas and p_{A_i} is the partial pressure of A just above the gas–liquid interface. At steady state, this flux must be equal to that of dissolved A away from the interface into the bulk liquid phase; this flux is written in terms of a liquid-phase mass transfer coefficient k_L as

$$F = k_L(c_{A_i} - c_{A_L}) \tag{19.32}$$

where c_{A_L} is the concentration of A in the bulk liquid. The interfacial partial pressure of A in the gas phase p_{A_i} must be in equilibrium with the interfacial concentration of A in the liquid phase c_{A_i}. Usually the interfacial partial pressure and concentration are not known, so it is customary to express the flux in terms of *overall* mass transfer coefficients K_G and K_L as follows

$$F = K_G(p_{A_G} - p_A^*) = K_L(c_A^* - c_{A_L}) \tag{19.33}$$

where p_A^* is the gas-phase partial pressure that would be in equilibrium with the bulk liquid-phase concentration c_{A_L}. p_A^* and c_{A_L} are related by Henry's law (see Chapter 7)

$$p_A^* = c_{A_L}/H_A \tag{19.34}$$

where H_A is the Henry's law coefficient for species A (expressed in units of $M\,atm^{-1}$). Similarly, c_A^* is the liquid-phase concentration that would be in equilibrium with the bulk gas-phase partial pressure p_{A_G}:

$$c_A^* = p_{A_G}H_A \tag{19.35}$$

To ensure a net flux of A from the gas phase to the liquid phase, the bulk liquid-phase concentration of A must not be in equilibrium with the bulk gas-phase partial pressure p_{A_G} and $c_{A_L} < H_A p_{A_G}$. (If the reverse is true, then the net flux of A is from the liquid to the gas.) The absence of equilibrium can be a result of aqueous-phase chemical reaction of A or simply the fact that water is undersaturated relative to the atmospheric concentration of the gas. (We do not consider here the enhancement of deposition that arises from chemical reaction in the liquid phase.)

Thus, using (19.34) and (19.35), equation (19.33) becomes

$$F = K_G\left(p_{A_G} - \frac{c_{A_L}}{H_A}\right) = K_L(H_A p_{A_G} - c_{A_L}) \tag{19.36}$$

From (19.36), we see that the two overall mass transfer coefficients are related through the Henry's law coefficient:

$$K_L = \frac{K_G}{H_A} \tag{19.37}$$

We now want to relate K_G and K_L to the individual phase mass transfer coefficients, k_G and k_L. Since the interface must be at equilibrium, so $c_{A_i} = H_A p_{A_i}$. Using this relation together with (19.31), (19.32), and (19.36) produces the following relations:

$$\frac{1}{K_G} = \frac{1}{k_G} + \frac{1}{k_L H_A} \tag{19.38}$$

$$\frac{1}{K_L} = \frac{H_A}{k_G} + \frac{1}{k_L} \tag{19.39}$$

We see that there is a critical value of the Henry's law constant, $H_{A,crit} = k_G/k_L$, for which the gas- and liquid-phase resistances are equal. For $H_A \ll H_{A,crit}$ (i.e., a slightly soluble gas), liquid-phase mass transport is controlling. If $H_A \gg H_{A,crit}$ (i.e., a very soluble gas), gas-phase mass transport controls the deposition process, and the transport is independent of the value of H_A. The two limiting cases are

$$\frac{1}{K_G} \cong \frac{1}{k_L H_A} \qquad H_A \ll H_{A,crit} \tag{19.40}$$

$$\frac{1}{K_G} \cong \frac{1}{k_G} \qquad H_A \gg H_{A,crit} \tag{19.41}$$

Up to this point we have expressed the gas-phase concentration of A in terms of partial pressure, in order to employ the usual Henry's law coefficient in units of $M\,atm^{-1}$. If we express the gas-phase concentration in units of $mol\,cm^{-3}$, then k_G, k_L, K_G, and K_L all have units of $cm\,s^{-1}$. The only change is that the Henry's law constant must now be dimensionless, that is, moles cm^{-3} (gas)/moles cm^{-3} (liq). Thus, if we denote the dimensionless Henry's law constant as $\widetilde{H}_A$, then

$$\widetilde{H}_A = H_A RT \, \frac{mol\,A\,cm^{-3}\ (gas)}{mol\,A\,cm^{-3}\ (liquid)} \tag{19.42}$$

and H_A is replaced by $\widetilde{H}_A$ in (19.38) and (19.39). At 298 K, for example, $\widetilde{H}_A = 24.44 H_A$.

To determine the surface resistance for dry deposition of a gas to a water surface, it is necessary to determine values of k_G and/or k_L, depending on the solubility of the gas in water. The key variable on which these coefficients depend is the windspeed. For a highly soluble gas, such as HNO_3, H_2SO_4, HCl, and NH_3, only k_G is important. Hicks and Liss (1976) show that k_G is about 0.13% of the windspeed at a reference height of 10 m. For windspeeds of $3-15\,m\,s^{-1}$, values of k_G range from 0.4 to $2\,cm\,s^{-1}$.

For a slightly soluble gas, such as CO_2 or O_3, (19.40) applies, and the resistance to transfer depends on the Henry's law coefficient and k_L. Considerable effort has gone into determining the relationship between k_L and windspeed. The two most widely used relationships are those of Liss and Merlivat (1986) and Wanninkhof (1992):

Liss and Merlivat (1986)

$$k_{L,A} = \frac{1}{3600} \begin{cases} 0.17\,u_{10}S_A^{2/3} & u_{10} \le 3.6\,m\,s^{-1} \\ 0.612\,S_A^{2/3} + (2.85\,u_{10} - 10.26)S_A^{1/2} & 3.6 < u_{10} \le 13\,m\,s^{-1} \\ 0.612\,S_A^{2/3} + (5.9\,u_{10} - 49.9)S_A^{1/2} & u_{10} > 13\,m\,s^{-1} \end{cases} \tag{19.43}$$

where $k_{L,A}$ is in units of $cm\,s^{-1}$ and u_{10} is the windspeed $(m\,s^{-1})$ at 10 m above water level. (The factor 3600 converts $cm\,h^{-1}$, the original units given by Liss and Merlivat, to $cm\,s^{-1}$.) S_A is the ratio of the Schmidt number of CO_2 at 293 K to that of the species in question at the temperature of interest

$$S_A = \frac{Sc_{CO_2}(293\,K)}{Sc_A(T)} \tag{19.44}$$

where $Sc_A = \nu/D_A$, ν the kinematic viscosity of water, and D_A the molecular diffusivity of A in water. The Schmidt number of CO_2 in seawater over the range 0–30°C has been correlated as (Wanninkhof 1992)

$$Sc_{CO_2} = 2073.1 - 125.62\,T + 3.6276\,T^2 - 0.043219\,T^3 \tag{19.45}$$

where T is in °C. Its value at 293 K is 660.

Wanninkhof (1992)

$$k_{L,A} = \frac{0.31\,u_{10}^2 S_A^{1/2}}{3600} \qquad (cm\,s^{-1}) \tag{19.46}$$

Figure 19.6 shows the two expressions for $k_{L,A}$ as a function of windspeed for CO_2 at 293 K (i.e., $S_A = 1$).

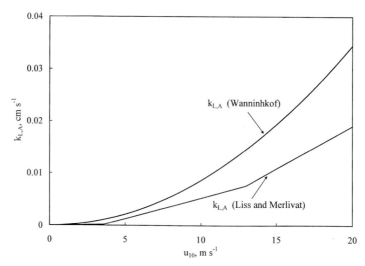

FIGURE 19.6 $k_{L,A}$ as a function of windspeed.

In summary, we have the following general result for the surface resistance to deposition of a gas to a water surface

$$r_{cw,A} = \frac{1}{k_G} + \frac{1}{k_{L,A}\widetilde{H}_A} \qquad (19.47)$$

and the two limiting cases

$$r_{cw,A} \cong \frac{1}{k_{L,A}\widetilde{H}_A} \quad \text{slightly soluble gas} \qquad (19.48)$$

$$r_{cw,A} \cong \frac{1}{k_G} \quad \text{highly soluble gas} \qquad (19.49)$$

For highly soluble gases, the surface resistance tends to be small, and dry deposition is generally controlled by the aerodynamic and laminar sublayer resistances. For slightly soluble gases, $\widetilde{H}_A$ is relatively small, $r_{cw,A}$ is large, and dry deposition tends to be controlled by the surface resistance, $r_{cw,A}$. If the dissolving gas dissociates in water, the effect of dissociation can be accounted for by using the effective Henry's law constant $\widetilde{H}_A^*$ in the relations presented above.

19.5.2   Surface Resistance for Dry Deposition of Gases to Vegetation

The approach adopted here is based primarily on the methodology developed by Wesely (1989) for regional-scale modeling over a range of species, land-use types, and seasons. For all land-use categories, the surface resistance is divided into component resistances

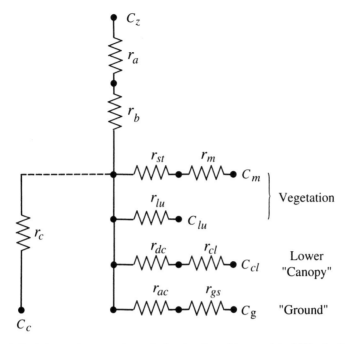

FIGURE 19.7   Resistance schematic for dry deposition model of Wesely (1989).

that are somewhat more detailed than we have discussed above. The surface resistance, shown in Figure 19.7, is calculated from the individual resistances by

$$r_c = \left(\frac{1}{r_{st} + r_m} + \frac{1}{r_{lu}} + \frac{1}{r_{dc} + r_{cl}} + \frac{1}{r_{ac} + r_{gs}} \right)^{-1} \qquad (19.50)$$

where the first term includes the leaf stomatal (r_{st})[3] and mesophyll (r_m) resistances; the second term is outer surface resistance in the upper canopy (r_{lu}), which includes the leaf cuticular resistance in healthy vegetation and the other outer surface resistances; the third term is resistance in the lower canopy, which includes the resistance to transfer by buoyant convection (r_{dc}) and the resistance to uptake by leaves, twigs, and other exposed surfaces (r_{cl}); and the fourth term is resistance at the ground, which includes a transfer resistance (r_{ac}) for processes that depend only on canopy height and a resistance for uptake by the soil, leaf litter, and so on at the ground surface (r_{gs}). Table 19.3 provides tabulated values of seven components of baseline resistances $(r_j, r_{lu}, r_{ac}, r_{gsS}, r_{gsO}, r_{clS}, r_{clO})$ for five categories of seasons and for eleven land-use types. r_j is the minimum bulk canopy stomatal resistance for water vapor, and the subscripts O and S refer to O_3 and SO_2, respectively. All the resistances needed to apply (19.50) to an individual gas can be calculated from the baseline resistance values in Table 19.3 using the following formulas.

---

[3]In Wesely's (1989) formulation what we have called the "stomatal pore resistance" r_p is called just the "stomatal resistance" and given the symbol r_{st}. As in Figure 19.4, it is assumed to act in series with the mesophyll resistance.

TABLE 19.3   Input Resistances[a] (s m^{-1}) for Computations of Surface Resistances (r_c)

Resistance Component	Land Use Type[b]										
	1	2	3	4	5	6	7	8	9	10	11
	Seasonal Category 1: Midsummer with Lush Vegetation										
r_j	9999	60	120	70	130	100	9999	9999	80	100	150
r_{lu}	9999	2000	2000	2000	2000	2000	9999	9999	2500	2000	4000
r_{ac}	100	200	100	2000	2000	2000	0	0	300	150	200
r_{gsS}	400	150	350	500	500	100	0	1000	0	220	400
r_{gsO}	300	150	200	200	200	300	2000	400	1000	180	200
r_{clS}	9999	2000	2000	2000	2000	2000	9999	9999	2500	2000	4000
r_{clO}	9999	1000	1000	1000	1000	1000	9999	9999	1000	1000	1000
	Seasonal Category 2: Autumn with Unharvested Cropland										
r_j	9999	9999	9999	9999	250	500	9999	9999	9999	9999	9999
r_{lu}	9999	9000	9000	9000	4000	8000	9999	9999	9000	9000	9000
r_{ac}	100	150	100	1500	2000	1700	0	0	200	120	140
r_{gsS}	400	200	350	500	500	100	0	1000	0	300	400
r_{gsO}	300	150	200	200	200	300	2000	400	800	180	200
r_{clS}	9999	9000	9000	9000	2000	4000	9999	9999	9000	9000	9000
r_{clO}	9999	400	400	400	1000	600	9999	9999	400	400	400
	Seasonal Category 3: Late Autumn after Frost, No Snow										
r_j	9999	9999	9999	9999	250	500	9999	9999	9999	9999	9999
r_{lu}	9999	9999	9000	9000	4000	8000	9999	9999	9000	9000	9000
r_{ac}	100	10	100	1000	2000	1500	0	0	100	50	120
r_{gsS}	400	150	350	500	500	200	0	1000	0	200	400
r_{gsO}	300	150	200	200	200	300	2000	400	1000	180	200
r_{clS}	9999	9999	9000	9000	3000	6000	9999	9999	9000	9000	9000
r_{clO}	9999	1000	400	400	1000	600	9999	9999	800	600	600
	Seasonal Category 4: Winter, Snow on Ground and Subfreezing										
r_j	9999	9999	9999	9999	400	800	9999	9999	9999	9999	9999
r_{lu}	9999	9999	9999	9999	6000	9000	9999	9999	9000	9000	9000
r_{ac}	100	10	10	1000	2000	1500	0	0	50	10	50
r_{gsS}	100	100	100	100	100	100	0	1000	100	100	50
r_{gsO}	600	3500	3500	3500	3500	3500	2000	400	3500	3500	3500
r_{clS}	9999	9999	9999	9000	200	400	9999	9999	9000	9999	9000
r_{clO}	9999	1000	1000	400	1500	600	9999	9999	800	1000	800
	Seasonal Category 5: Transitional Spring with Partially Green Short Annuals										
r_j	9999	120	240	140	250	190	9999	9999	160	200	300
r_{lu}	9999	4000	4000	4000	2000	3000	9999	9999	4000	4000	800
r_{ac}	100	50	80	1200	2000	1500	0	0	200	60	120
r_{gsS}	500	150	350	500	500	200	0	1000	0	250	400
r_{gsO}	300	150	200	200	200	300	2000	400	1000	180	200
r_{clS}	9999	4000	4000	4000	2000	3000	9999	9999	4000	4000	8000
r_{clO}	9999	1000	500	500	1500	700	9999	9999	600	800	800

[a]Entries of 9999 indicate that there is no air–surface exchange via that resistance pathway.
[b](1) Urban land, (2) agricultural land, (3) range land, (4) deciduous forest, (5) coniferous forest, (6) mixed forest including wetland, (7) water, both salt and fresh, (8) barren land, mostly desert, (9) nonforested wetland, (10) mixed agricultural and range land, and (11) rocky open areas with low-growing shrubs.
Source: Wesely (1989).

The bulk canopy stomatal resistance is calculated from tabulated values of r_j, the solar radiation (G in $W\,m^{-2}$), and surface air temperature (T_s in $°C$ between 0 and 40°C) using

$$r_{st} = r_j \left[1 + \left(\frac{200}{G + 0.1} \right)^2 \left(\frac{400}{T_s(40 - T_s)} \right) \right] \qquad (19.51)$$

Outside this range, the stomata are assumed to be closed and r_{st} is set to a large value. The combined minimum stomatal and mesophyll resistance is calculated from

$$r_{sm}^i = r_{st}^i + r_m^i = r_{st} \left(\frac{D_{H_2O}}{D_i} \right) + \frac{1}{3.3 \times 10^{-4} H_i^* + 100 f_0^i} \qquad (19.52)$$

where D_{H_2O}/D_i is the ratio of the molecular diffusivity of water to that of the specific gas, H_i^* is the effective Henry's law constant ($M\,atm^{-1}$) for the gas, and f_0^i is a normalized (0 to 1) reactivity factor for the dissolved gas (Table 19.4). The second term on the RHS of (19.52) is the mesophyll resistance for the gas of interest.

The resistance of the outer surfaces in the upper canopy for a specific gas is computed from

$$r_{lu}^i = r_{lu} \left(\frac{1}{10^{-5} H_i^* + f_0^i} \right) \qquad (19.53)$$

where r_{lu} is tabulated for each season and land-use category in Table 19.3.

TABLE 19.4   Relevant Properties of Gases for Dry Deposition Calculations

Species	Ratio of Molecular Diffusivities $(D_{H_2O}/D_{species})$	Henry's Law Constant[b] (H^*), $M\,atm^{-1}$	Henry's Law Exponent[a] (A)	Normalized Reactivity (f_0)
Sulfur dioxide	1.89	1×10^5	-3020	0
Ozone	1.63	1×10^{-2}	$+2300$	1
Nitrogen dioxide	1.6	1×10^{-2}	-2500	0.1
Nitric oxide	1.29	2×10^{-3}	-1480	0
Nitric acid	1.87	1×10^{14}	-8650	0
Hydrogen peroxide	1.37	1×10^5	-6800	1
Acetaldehyde	1.56	15	-6500	0
Propionaldehyde	1.8	15	-6500	0
Formaldehyde	1.29	6×10^3	-6500	0
Methyl hydroperoxide	1.6	220	-5600	0.3
Formic acid	1.6	4×10^6	-5740	0
Acetic acid	1.83	4×10^6	-5740	0
Ammonia	0.97	2×10^4	-3400	0
Petroxyacetyl nitrate	2.59	3.6	-5910	0.1
Nitrous acid	1.62	1×10^5	-4800	0.1
Pernitric acid	2.09	2×10^4	-1500	0
Hydrochloric acid	1.42	2.05×10^6	-2020	0

[a]The exponent A is used in the expression $H(T) = H \exp \{A[1/298 - 1/T]\}$ to calculate H at the surface temperature.
[b]Effective Henry's law constant assuming a pH of about 6.5.

The resistance r_{dc} is determined by the effects of mixing forced by buoyant convection (as a result of surface heating of the ground and/or lower canopy) and by penetration of winds into canopies on the sides of hills. The resistance (in $s\,m^{-1}$) is estimated from

$$r_{dc} = 100\left(1 + \frac{1000}{G + 10}\right)\left(\frac{1}{1 + 1000\,\theta}\right) \tag{19.54}$$

where θ is the slope of the local terrain in radians. The resistance of the exposed surfaces in the lower portions of structures (canopies or buildings) is computed from

$$r_{cl}^i = \left(\frac{10^{-5} H_i^*}{r_{clS}} + \frac{f_0^i}{r_{clO}}\right)^{-1} \tag{19.55}$$

where r_{clS} and r_{clO} are given for each season and land-use category in Table 19.3. Similarly, at the ground, the resistances are computed from

$$r_{gs}^i = \left(\frac{10^{-5} H_i^*}{r_{gsS}} + \frac{f_0^i}{r_{gsO}}\right)^{-1} \tag{19.56}$$

where r_{gsS} and r_{gsO} are likewise given for each season and land-use category in Table 19.3.

Table 19.4 lists relevant properties needed to calculate gaseous deposition layer and surface resistances for this model. It is important to recognize that the reactivity factors assigned to the depositing species are approximate and may vary significantly with the vegetation type or state. The Henry's law constants can be adjusted for temperature using the expression given in the footnote to the table and adjusted for pH on the leaf surface (see Chapter 7). The Henry's law constants for species that dissociate in the aqueous phase (SO_2, HNO_3, NH_3) are calculated using the expressions shown in Table 19.5.

Wesely (1989) recommends an alternate surface resistance equation when the ground surface is wet from rain or dew. For surfaces covered with dew, the upper-canopy resistances for SO_2 and O_3 are calculated from

$$r_{lu}^{SO_2} = 100\,s\,m^{-1} \tag{19.57}$$

$$r_{lu}^{O_3} = \left(\frac{1}{3000} + \frac{1}{3r_{lu}}\right)^{-1} \tag{19.58}$$

TABLE 19.5   Correction Factors for Henry's Law Coefficients for Species that Dissociate on Absorption

Species	Henry's Law Expression[a]
SO_2	$H(T)\{1 + 1.23 \times 10^{-2} \exp(-2010\gamma)\}/[H^+]$ $+\{1.23 \times 10^{-2} \exp(-2010\gamma)6.6 \times 10^{-8} \exp(5.1 \times 10^{-4}\gamma)\}/[H^+]^2$
HNO_3	$H(T)\{(1 + 2.2 \times 10^{-12} \exp(3730\gamma)\}/[H^+]$
NH_3	$H(T)\{1 + 1.75 \times 10^{-5} \exp(450\gamma)[H^+]/[1 \times 10^{-14} \exp(6710\gamma)]\}$

[a] $\gamma = \frac{1}{298} - \frac{1}{T}.$

When it is raining, the SO_2 and ozone upper-canopy resistance is calculated from

$$r_{lu}^{SO_2} = \left(\frac{1}{5000} + \frac{1}{3r_{lu}}\right)^{-1} \qquad \text{for nonurban land use} \qquad (19.59)$$

$$r_{lu}^{SO_2} = 50\,\text{s m}^{-1} \qquad \text{for urban land use} \qquad (19.60)$$

$$r_{lu}^{O_3} = \left(\frac{1}{1000} + \frac{1}{3r_{lu}}\right)^{-1} \qquad (19.61)$$

The upper-canopy resistance for other species is calculated from

$$r_{lu}^i = \left(\frac{1}{3r_{lu}} + 10^{-7} H_i^* + \frac{f_0^i}{r_{luO}}\right)^{-1} \qquad (19.62)$$

when the surface is covered by rainwater or dew.

All the formulas described above are for unstressed vegetation, which is the default vegetation status. Optionally, for vegetation stress due to lack of water, the stomatal resistance is increased by a factor of 10 and for inactive vegetation (winter deciduous), a stomatal resistance of 10,000 s m^{-1} should be used, indicating a complete shutdown of this pathway.

19.6   MEASUREMENT OF DRY DEPOSITION

A broad range of techniques have been used to measure dry deposition (Businger 1986). Applicability to different spatial and temporal scales, chemical species, and complexities of terrain varies among the different techniques.

Dry deposition measurements can be divided into two general categories: direct and indirect. In the direct methods, an explicit determination is made of the flux of material to the surface, either by collecting material deposited on the surface itself or by measuring the vertical flux in the air near the surface. Indirect methods derive flux values by measurements of secondary quantities, such as the mean concentration or vertical gradients of the mean concentration of the depositing material, and relating these quantities to the flux. Direct methods require that fewer assumptions be made in analyzing the data but generally demand considerably more effort and relatively sophisticated instrumentation. A detailed treatment of micrometeorological measurement techniques is provided by Businger (1986). Erisman (1993a,b) and Baldocchi et al. (1988) discuss dry deposition measurement methods.

19.6.1   Direct Methods

Surrogate Surfaces   Whereas one can use surrogate surfaces, such as filter substrates, to collect depositing material, the approach is most suitable when the aerodynamic resistance r_a is the dominant resistance. If r_b or r_c is significant, a surrogate surface is unlikely to mimic the correct behavior of the natural surface of interest. The surrogate surface technique is used primarily for particles where the particular nature of the surface is less important than for gases. Aerodynamically designed surrogate surfaces in the shape of symmetric airfoils have proven useful for measuring the deposition of large particles (Wu et al. 1992b).

Natural Surfaces   Analysis of material deposited on natural surfaces represents a method to infer dry deposition rates especially when micrometeorological methods are difficult to apply (complex terrain, forest edges). Foliar extraction (e.g., leaf washing and analysis of snow) can provide a specific measure of the amount of material removed from the air by an individual element of a plant canopy. Such an approach is generally ineffective for gases because of chemical binding to the surface. The throughfall technique measures the total material flux below the canopy. Hicks (1986) points out that significant differences have been found between predictions made for forest canopies, based on extrapolations from individual elements, and field measurements.

Chamber Method   In the chamber method gas uptake by the depositing surface is measured. In open-top or in closed chambers the factors thought to influence deposition are controlled. Deposition to the surfaces in the chamber (soil or vegetation) can be calculated by measuring the fluxes in and out of the chamber over a given time interval (Granat and Johansson 1983; Taylor et al. 1983).

Eddy Correlation   Eddy correlation is the most common of the "direct micrometeorological" techniques to measure dry deposition rates (Wesely et al. 1982). Statistical correlations of the fluctuations in wind and concentration fields are measured to directly obtain values of the associated vertical fluxes. In the case of eddy correlation, high-speed measurements of vertical velocity and concentration are used to obtain time series of the corresponding fluctuating components, $w'(t)$ and $C'(t)$. These time series are used to derive the time-averaged vertical turbulent flux, $\overline{w'C'}$. Under the assumptions that vertical turbulent transport dominates and that chemical reactions are absent, this is a direct measure of the local vertical flux at the measurement point, $F = -\overline{w'C'}$. For measurement techniques that rely on obtaining statistically representative samples of the local turbulence and species concentration, measured over finite averaging times, representative samples can be obtained as long as their average properties do not change over periods of time that are long compared to the measurement time. The deposition velocity can be obtained from the measured value of F simply by dividing by the mean concentration at a suitably chosen reference height.

To infer a dry deposition rate from an eddy correlation measurement, a nondivergent vertical species flux should exist. Nondivergence essentially stipulates that quasi-one-dimensional transport exists. The nondivergence assumption is, in fact, equivalent to the constant-flux-layer assumption of the surface layer; in practical terms, nondivergence is best satisfied in relatively flat topography for which a substantial fetch over the terrain exists.

Eddy correlation measurements require fast-response instrumentation to resolve the turbulent fluctuations that contribute primarily to the vertical flux. These requirements are particularly severe under stable conditions where response times on the order of 0.2 s or less may be required. In practice, it is often possible to use somewhat slower instruments and apply various corrections to the computed fluxes as compensation. The eddy correlation technique has been used in aircraft (Pearson and Steadman 1980; Lenschow et al. 1982) as well as with tower-mounted instruments.

Eddy Accumulation   Eddy accumulation depends on essentially the same conditions and assumptions as those for eddy correlation. In this method, air is collected on two separate filters (or in containers), with the vertical velocity determining which filter (container) receives the sampled air (Hicks and McMillen, 1984). One filter (container) is used for positive vertical velocities, and the second is used for negative vertical velocities; the instantaneous sampling rate for each filter (container) is proportional to the magnitude of the velocity. The filter (collected air) is then analyzed for the species of interest, and the results are used to calculate the net flux (Businger 1986).

19.6.2   Indirect Methods

Gradient Method   In the gradient method the deposition velocity is determined by measuring the vertical gradient of the depositing substance and using gradient transport theory to infer the associated deposition flux. This is usually done by combining (19.1) and (19.8) and eliminating the flux from the two expressions, resulting in $v_d = (\kappa u_* z/\phi)(1/\overline{C})(\partial\overline{C}/\partial z)$, where $\overline{C}$ is the average concentration over the height interval used to determine the gradient. Whereas the gradient method is theoretically straightforward, it requires relatively accurate concentration values at two or more heights, since the difference between such values can be very small if the deposition rate is small. For example, for a deposition velocity of $0.2\,\mathrm{cm\,s^{-1}}$ and u_* of $0.4\,\mathrm{m\,s^{-1}}$, the concentrations at 2 and 4 m above the surface will differ by less than 1% under neutral conditions. The difficulty of achieving such relative accuracy can be addressed by using a single detector for the species of interest, thereby eliminating interinstrument differences, to sample the air at different heights, for example, with a movable sample probe or with a mechanism that switches between sampling lines. The gradient method tends to be impractical over extremely rough surfaces, because the measuring heights should satisfy the criterion $z/z_0 \gg 1$; however, that condition can actually place the measurement above the constant-flux layer. In such a case the turbulent diffusivity based on the gradient transport assumption may then be poorly known.

Inferential Method   The inferential technique for determining dry deposition rates is based on the direct application of (19.1). Measured ambient concentrations at a particular reference height are multiplied by a deposition velocity assumed to be representative of the local surface to compute the dry deposition rate. This approach is most suited when routine monitoring data are available, but the values of the derived flux values are clearly dependent on the validity of the estimates for v_d. Detailed canopy models using information about the surface and meteorology surrounding the concentration monitor can be used to calculate the deposition velocity.

19.6.3   Comparison of Methods

Direct and indirect methods are distinguished, in part, by the spatial scales over which they are representative. Surface methods yield dry deposition fluxes representative of the spatial scales of the foliage or surrogate surface element sampled, typically on the order of fractions of a square meter. If these surface elements dominate the overall surface, they may account for more area than just the local area where the measurement was made. The micrometeorological methods most commonly used provide data that are representative of the flux over a larger spatial scale than that associated with a surface method. This is because the turbulent eddies responsible for the flux inherently include a degree of spatial averaging of the conditions in the vicinity of the measurement site. Satisfaction of the nondivergence assumption depends on the degree of variability within this vicinity.

In short, micrometeorological methods generally meet the nondivergence criterion more frequently than do surface sampling methods satisfy representativeness for the surrounding area.

19.7   SOME COMMENTS ON MODELING AND MEASUREMENT OF DRY DEPOSITION

Implicit in the use of the deposition velocity v_d is the assumption of a constant-flux layer (nondivergence). This assumption is met in certain circumstances, typically if the reference height for measurement is on the order of a few meters or less and the fetch upwind of a measuring site is flat and uniform for a sufficient distance. Slinn (1983) has provided scaling arguments to estimate the fetches that are required but suggests somewhat more stringent requirements. In any event, it is generally acknowledged that a fetch: measurement height ratio on the order of 100 or more is required for micrometeorological measurements of heat and momentum fluxes; even larger ratios may be desirable for slowly depositing materials.

To ensure a constant-flux layer, one can simply move the measurement height closer to the surface. For the eddy correlation method, however, the response time of the instrument must be faster as the measurement height approaches the surface, because high-frequency turbulent eddies then contribute proportionally more to the concentration fluxes than at higher levels. On the other hand, fluxes measured very close to the surface may be less representative of those over the entire area for which the measurement is intended. For the gradient method, the requirement that $z/z_0 \gg 1$ (based on the requirements of similarity theory) constrains the minimum measurement height. Under very stable conditions, when turbulence may be intermittent, turbulent fluxes may become very small, and the constant-flux layer may be very shallow. Under conditions such as these, it can be quite difficult to determine the aerodynamic resistance term r_a.

A complicating factor in dry deposition measurements is the presence of sources of the depositing substance in the "footprint" of the measurement. Whereas the flux of SO_2 is nearly always unidirectional (downward) and the surface is a sink for SO_2, gases such as H_2S, NH_3, and NO_x may have surface sources. It may be possible in some cases to specify a surface emission rate. For NO_2, the situation appears to be even more complex than just adding a surface emission rate to the resistance model. As much as 50% of the NO_2 initially removed at the surface can reappear as NO as a result of surface emissions (Meyers and Baldocchi 1988).

Bidirectional fluxes of particles can also occur. One important example is the deposition and subsequent resuspension of automobile-emitted lead. During the decades of leaded gasoline use, deposition onto soil in urban areas exceeded resuspension and there was a net buildup of lead in surface soil. Now that leaded gasoline has been largely eliminated, resuspension exceeds deposition, and the soil surface is becoming gradually depleted in lead. Harris and Davidson (2005) have shown that the lifetime for depletion of lead in the soil is more than a century.

The assumptions of similarity weaken as one moves from flat and uniform surfaces into hilly terrain and associated natural surface covers (Doran et al. 1989). Fluxes are likely to change substantially over rather short distances (1 km or less), and it may be extremely difficult to establish a representative flux value for more extended regions from measurements at a single site. Some measurements have been carried out over sloping terrain, and methods have been developed to take into account the effects of such slopes in the determination of flux values (McMillen 1988). However, there is presently no generally useful method for establishing flux values over rough terrain.

The dimensionless concentration gradient that is a function of atmospheric stability, ϕ_C (or, more accurately, ϕ_T or ϕ_M), has been determined from numerous measurement programs over the years at carefully selected sites and conditions. Measurement sites have satisfied the fetch conditions noted earlier, and the meteorological conditions have been chosen to be relatively constant over the measurement period, typically 30–60 min. During transition periods, which occur for several hours each day around sunrise and sunset, the assumption of stationarity is not met, and functions such as ϕ_C are not well defined. Moreover, the surface characteristics may change rapidly during transition periods as plant stomata open or close, dew is formed or evaporated, and so on. The resulting changes in surface resistance values r_c may not be well represented by assumed steady-state values.

The constant-flux approximation also presumes that the species being transported is conservative. Since most atmospheric species, especially those subject to dry deposition, undergo some sort of chemical reactions in the atmosphere, the characteristic time of reaction should be either very long or very short when compared to the characteristic time for vertical turbulent diffusion in the surface layer for the constant-flux assumption to hold. If the characteristic time for reaction is long compared to the transport time, then during the course of transport down through the surface layer any changes in the species flux from chemical reaction will be negligible. At the other extreme, if the characteristic reaction time is very short compared to the transport time, the species will be in a local chemical equilibrium at any point in the surface layer. Then the constant-flux assumption will hold for the chemical family that is involved in the fast reactions. This situation is considered in detail by Pandis and Seinfeld (1990).

PROBLEMS

19.1$_A$ **a.** Show the steps followed to derive (19.2).
   b. Show the steps followed to derive (19.7).

19.2$_A$ Determine the overall dry deposition velocity of particles of diameter 1 μm and density of 1 g cm^{-3}, over a desert surface at 298 K. Assume $u_* = 10\,\text{m s}^{-1}$ with neutral stability conditions. What are the relative contributions of the aerodynamic and quasi-laminar resistances to the overall dry deposition velocity?

19.3$_A$ Prepare a plot of $1/r_b$ for dry deposition of particles over grass as a function of particle diameter from 0.01 to 10.0 μm. Assume $T = 298$ K, $u_* = 1\,\text{m s}^{-1}$ and seasonal category 2 in Table 19.2. Prepare lines for particle densities of $\rho_p = 1$ and 2 g cm^{-3}.

19.4$_A$ Determine the surface resistance to dry deposition of O_3 to surface water at 298 K as a function of windspeed at 10 m above the water surface for wind speeds ranging from 1 to 10 m s^{-1}. The Henry's law coefficient for O_3 in water at 298 K is 0.0113 M atm^{-1}. The Schmidt number for ozone in water is 571.

19.5$_A$ Calculate the surface resistance to dry deposition of SO_2 to a flat deciduous forest at 298 K under midsummer conditions with lush vegetation. Assume solar radiation $G = 1000\,\text{W m}^{-2}$. How does the resistance change in the case when the ground surface is wet?

19.6$_A$ Calculate the dry deposition velocity of SO_2 under the following conditions: neutral stability, $u_* = 1 \text{ m s}^{-1}$, $\bar{u}(10\,\text{m}) = 1.5 \text{ m s}^{-1}$, and $z_0 = 1\,\text{m}$. Consider surface resistance values of $r_c = 0, 0.3$, and $2.0\,\text{s cm}^{-1}$.

19.7$_C$ Bolin et al. (1974) have presented a one-dimensional steady-state model based on the atmospheric diffusion equation to be able to estimate the effect of dry deposition on vertical species concentration distribution. In the model the mean concentration is governed by

$$\frac{d}{dz}\left(K_{zz}\frac{dc}{dz}\right) + q\delta(z - h) = 0$$

$$\left(K_{zz}\frac{dc}{dz}\right)_{z=z_r} = v_d c \qquad z = z_v$$

$$c = 0 \qquad z \longrightarrow \infty$$

with

$$K_{zz} = \begin{cases} \kappa u_* z & z_v \leq z \leq H \\ \kappa u_* H & z > H \end{cases}$$

Thus the source is taken at a height h and of strength q. Horizontal inhomogeneity is neglected, and neutral stability is assumed up to a layer at height H, thereafter remaining constant. The object of the model is to be able to study the effect of v_d on the vertical concentration profiles and thereby to assess the degree of importance of dry deposition.

a. When the source height is in the constant diffusivity layer, that is, $h > H$, show that

$$c(z) = \begin{cases} \dfrac{1 + \dfrac{v_d}{\kappa u_*}\ln\left(\dfrac{z}{z_v}\right)}{1 + \dfrac{v_d}{\kappa u_*}\left[\dfrac{h - H}{H} + \ln\left(\dfrac{H}{z_v}\right)\right]} & z_v \leq z \leq H \\[4ex] 1 - \dfrac{v_d(h - z)}{\kappa u_* H\left\{1 + \dfrac{v_d}{\kappa u_*}\left[\dfrac{h - H}{H} + \ln\left(\dfrac{H}{z_v}\right)\right]\right\}} & H \leq z < h \end{cases}$$

b. When the source height $h < H$, show that

$$c(z) = \frac{1 + \dfrac{v_d}{\kappa u_*}\ln\left(\dfrac{z}{z_v}\right)}{1 + \dfrac{v_d}{\kappa u_*}\ln\left(\dfrac{h}{z_v}\right)}$$

c. Calculate and plot the vertical concentration distribution for the following conditions:

$$h = 50, 200 \, \text{m}$$
$$v_d = 0.1, 10, 10^7 \, \text{cm s}^{-1}$$
$$z_v = 0.5 \, \text{cm}$$
$$u_* = 1, 4, 8 \, \text{m s}^{-1}$$

Discuss your results.

REFERENCES

Baldocchi, D. D. (1988) A multi-layer model for estimating sulfur dioxide deposition to a deciduous oak forest canopy, *Atmos. Environ.* **22**, 869–884.

Baldocchi, D. D., Hicks, B. B., and Camara, P. (1987) A canopy stomatal resistance model for gaseous deposition to vegetated surfaces, *Atmos. Environ.* **21**, 91–101.

Baldocchi, D. D., Hicks, B. B., and Meyers, T. P. (1988) Measuring biosphere–atmosphere exchanges of biologically related gases with micrometeorological methods, *Ecology* **69**, 1331–1340.

Bolin, B., Aspling, G., and Persson, C. (1974) Residence time of atmospheric pollutants as dependent on source characteristics, atmospheric diffusion processes, and sink mechanisms, *Tellus* **26**, 185–194.

Businger, J. A. (1986) Evaluation of the accuracy with which dry deposition can be measured with current micrometeorological techniques, *J. Appl. Meteorol.* **25**, 1100–1124.

Davidson, C. I., Miller, J. M., and Pleskow, M. A. (1982) The influence of surface structure on predicted particle dry deposition to natural grass canopies, *Water, Air, Soil Pollut.* **18**, 25–44.

Davidson, C. I., and Wu, Y. L. (1990) Dry deposition of particles and vapors, in *Acid Precipitation*, Vol. 3, S. E. Lindberg, A. L. Page, and S. A. Norton, eds., Springer-Verlag, Berlin, pp. 103–209.

Doran, J. C., Wesely, M. L., McMillen, R. T., and Neff, W. D. (1989) Measurements of turbulent heat and momentum fluxes in a mountain valley, *J. Appl. Meteorol.* **28**, 438–444.

Erisman, J. W. (1993a) Acid deposition onto nature areas in the Netherlands: Part I. Methods and results, *Water Air Soil Pollut.* **71**, 51–80.

Erisman, J. W. (1993b) Acid deposition onto nature areas in the Netherlands: Part II. Throughfall measurements compared to deposition estimates, *Water Air Soil Pollut.* **71**, 81–99.

Erisman, J. W., and Baldocchi, D. (1994) Modelling dry deposition of SO_2 *Tellus* **46B**, 159–171.

Granat, L., and Johansson, C. (1983) Dry deposition of SO_2 and NO_x in winter, *Atmos. Environ.* **17**, 191–192.

Harris, A. R., and Davidson, C. I. (2005) The role of resuspended soil in lead flows in the California South Coast Air Basin, *Environ. Sci. Technol.* **39**, 7410–7415.

Hauglustaine, D. A., Granier, C., Brasseur, G. P., and Megie, G. (1994) The importance of atmospheric chemistry in the calculation of radiative forcing on the climate system, *J. Geophys. Res.* **99**, 1173–1186.

Hicks, B. B. (1986) Measuring dry deposition: A re-assessment of the state of the art, *Water Air Soil Pollut.* **30**, 75–90.

Hicks, B. B, Baldocchi, D. D., Meyers, T. P., Hosker, R. P., and Matt, D. R. (1987) A preliminary multiple resistance routine for deriving dry deposition velocities from measured quantities, *Water Air Soil Pollut.* **36**, 311–330.

Hicks, B. B., and Liss, P. S. (1976) Transfer of SO_2 and other reactive gases across the air-sea interface, *Tellus* **28**, 348–354.

Hicks, B. B., and Matt, D. R. (1988) Combining biology, chemistry, and meteorology in modeling and measuring dry deposition, *J. Atmos. Chem.* **6**, 117–131.

Hicks, B. B., and McMillen, R. T. (1984) A simulation of the eddy accumulation method for measuring pollutant fluxes, *J. Climate Appl. Meteorol.* **23**, 637–643.

Lenschow, D. H., Pearson, R. Jr., and Stankov, B. B. (1982) Measurement of ozone vertical flux to ocean and forest, *J. Geophys. Res.* **87**, 8833–8837.

Liss, P. S., and Merlivat, L. (1986) Air-sea gas exchange rates: Introduction and synthesis, in *The Role of Air-Sea Exchange in Geochemical Cycling*, P. Buat-Menard, ed., Reidel, Hingham, MA, pp. 113–127.

Matt, D. R., McMillen, R. T., Womack, J. D., and Hicks, B. B. (1987) A comparison of estimated and measured SO_2 deposition velocities, *Water Air Soil Pollut.* **36**, 331–347.

McMillen, R. T. (1988) An eddy correlation technique with extended applicability to non-simple terrain, *Boundary-Layer Meteorol.* **43**, 231–245.

Meyers, T. P. (1987) The sensitivity of modelled SO_2 fluxes and profiles to stomatal and boundary layer resistance, *Water Air Soil Pollut.* **35**, 261–278.

Meyers, T. P., and Baldocchi, D. D. (1988) A comparison of models for deriving dry deposition fluxes of O_3 and SO_2 to a forest canopy, *Tellus* **40B**, 270–284.

Meyers, T. P., and Hicks, B. B. (1988) Dry deposition of O_3, SO_2, and HNO_3 to different vegetation in the same exposure environment, *Environ. Pollut.* **53**, 13–25.

Moller, U., and Schumann, G. (1970) Mechanisms of transport from the atmosphere to the Earth's surface, *J. Geophys. Res.* **75**, 3013–3019.

Pandis, S. N., and Seinfeld, J. H. (1990) On the interaction between equilibration processes and wet or dry deposition, *Atmos. Environ.* **24A**, 2313–2327.

Pearson, R. Jr., and Stedman, D. H. (1980) Instrumentation for fast-response ozone measurements from aircraft, *Atmos. Technol.* **12**, 51–55.

Peters, K., and Eiden, R. (1992) Modeling the dry deposition velocity of aerosol particles to a spruce forest, *Atmos. Environ.* **26**, 2555–2564.

Ruijgrok, W., Davidson, C. I., and Nicholson, K. (1995) Dry deposition of particles: Implications and recommendations for mapping of deposition over Europe, *Tellus* **47B**, 587–601.

Sehmel, G. A. (1980) Particle and gas deposition, a review, *Atmos. Environ.* **14**, 983–1011.

Slinn, W. G. N. (1982) Predictions for particle deposition to vegetative surfaces, *Atmos. Environ.* **16**, 1785–1794.

Slinn, W. G. N. (1983) A potpourri of deposition and resuspension questions, in *Precipitation Scavenging, Dry Deposition, and Resuspension*, Vol. 2, *Dry Deposition and Resuspension*, Proc. 4th Int. Conf., Santa Monica, CA, Nov. 29–Dec. 3, 1982 (coordinators: H. R. Pruppacher, R. G. Semonin, and W. G. N. Slinn), Elsevier, New York.

Slinn, W. G. N., Hasse, L., Hicks, B. B., Hogan, A. W., Lai, D., Liss, P. S., Munnich, K. O., Sehmel, G. A., and Vittori, O. (1978) Some aspects of the transfer of atmospheric trace constituents past the air–sea interface, *Atmos. Environ.* **12**, 2055–2087.

Taylor, G. E., McLaughlin, S. B., Shriner, D. S., and Selvidge, W. J. (1983) The flux of sulphur containing gases to vegetation, *Atmos. Environ.* **17**, 789–796.

van Hove, L. W. A. (1989) *The Mechanism of NH_3 and SO_2 Uptake by Leaves and Its Physiological Effects*, thesis, Univ. Wageningen, The Netherlands.

Wanninkhof, R. (1992) Relationship between wind speed and gas exchange over the ocean, *J. Geophys. Res.* **97**, 7373–7382.

Wesely, M. L. (1989) Parameterizations of surface resistance to gaseous dry deposition in regional-scale, numerical models, *Atmos. Environ.* **23**, 1293–1304.

Wesely, M. L., Eastman, J. A., Stedman, D. H., and Yalvac, E. D. (1982) An eddy-correlation measurement of NO_2 flux to vegetation and comparison to O_3 flux, *Atmos. Environ.* **16**, 815–820.

Wesely, M. L., and Hicks, B. B. (1977) Some factors that affect the deposition rates of sulfur dioxide and similar gases on vegetation, *J. Air Pollut. Control Assoc.* **27**, 1110–1116.

Wu, Y. L., Davidson, C. I., Dolske, D. A., and Sherwood, S. I. (1992a) Dry deposition of atmospheric contaminants to surrogate surfaces and vegetation, *Aerosol Sci. Technol.* **16**, 65–81.

Wu, Y. L., Davidson, C. I., Lindberg, S. E., and Russell, A. G. (1992b) Resuspension of particulate chemical species at forested sites, *Environ. Sci. Technol.* **26**, 2428–2435.

Zhang, L., Gao, S., Padro, J., and Barrie, L. (2001) A size-segregated dry deposition scheme for an atmospheric aerosol module, *Atmos. Environ.* **35**, 549–560.

Zufall, M. J., Bergin, M. H., and Davidson, C. I. (1998) Effects of non-equilibrium hygroscopic growth of $(NH_4)_2SO_4$ on dry deposition to water surfaces, *Environ. Sci. Technol.* **32**, 584–590.

Zufall, M. J., Dai, W., Davidson, C. I., and Etyemezian, V. (1999) Dry deposition of particles to wave surfaces: I. Mathematical modeling, *Atmos. Environ.* **33**, 4273–4281.

20 Wet Deposition

Wet deposition refers to the natural processes by which material is scavenged by atmospheric hydrometeors (cloud and fog drops, rain, snow) and is consequently delivered to the Earth's surface. A number of different terms are used more or less synonymously with wet deposition including precipitation scavenging, wet removal, washout, and rainout. *Rainout* usually refers to in-cloud scavenging and *washout*, to below-cloud scavenging by falling rain, snow, and so on.

1. Precipitation scavenging, that is, the removal of species by a raining cloud
2. Cloud interception, the impaction of cloud droplets on the terrain usually at the top of tall mountains
3. Fog deposition, that is, the removal of material by settling fog droplets
4. Snow deposition, removal of material during a snowstorm

In all of these processes three steps are necessary for wet removal of a material. Specifically, the species (gas or aerosol) must first be brought into the presence of condensed water. Then, the species must be scavenged by the hydrometeors, and finally it needs to be delivered to the Earth's surface. Furthermore, the compound may undergo chemical transformations during each one of the above steps. These wet deposition steps are depicted in Figure 20.1. Note that almost all processes are reversible. For example, rain may scavenge particles below cloud, but raindrops that evaporate produce new aerosols. Several of the microphysical steps in the wet deposition process have already been discussed in previous chapters (nucleation scavenging during cloud formation, dissolution into aqueous droplets, etc.). In this chapter, we begin by developing a general mathematical framework for wet deposition processes and then discuss in detail the scavenging of material below a cloud. Then, we will integrate these processes into an overall framework. Our discussion will focus mainly on precipitation scavenging. The chapter will end with an overview of the acid deposition problem.

20.1 GENERAL REPRESENTATION OF ATMOSPHERIC WET REMOVAL PROCESSES

Wet removal pathways depend on multiple and composite processes, involve numerous physical phases, and are influenced by phenomena on a variety of physical scales. Figure 20.2 indicates the variety of lengthscales that influence wet removal. The challenge of understanding processes that operate on the microscale (10^{-6} m) and the macroscale (10^6 m) makes wet deposition one of the most complex atmospheric processes.

Atmospheric Chemistry and Physics: From Air Pollution to Climate Change, Second Edition, by John H. Seinfeld and Spyros N. Pandis. Copyright © 2006 John Wiley & Sons, Inc.

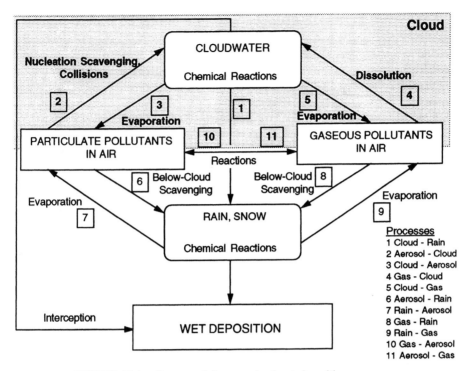

FIGURE 20.1 Conceptual framework of wet deposition processes.

The first challenge concerns the involvement of multiple phases in wet deposition. Not only does one deal with the three usual phases (gas, aerosol, and aqueous), but the aqueous phase can be present in several forms (cloudwater, rain, snow, ice crystals, sleet, hail, etc.), all of which have a size resolution. To complicate matters even further, different processes operate inside a cloud, and others below it. Our goal will initially be to create a mathematical framework for this rather complicated picture. To simplify things as much as possible we consider a "warm" raining cloud without the complications of ice and snow. There are four "media" or "phases" present, namely, air, cloud droplets, aerosol particles, and rain droplets. A given species may exist in each of these phases; for example, nitrate may exist in air as nitric acid vapor, dissolved in rain and cloud droplets as nitrate, and in various salts in the aerosol phase. Nonvolatile species like metals exist only in droplets and aerosols, while gases like HCHO exist only in the gas phase and the droplets. The size distribution of cloud droplets, rain droplets, and aerosols provides an additional complication. Let us initially neglect this feature. For a species i, one needs to describe mathematically its concentration in air $C_{i,\text{air}}$, cloudwater $C_{i,\text{cloud}}$, rainwater $C_{i,\text{rain}}$, and the aerosol phase $C_{i,\text{part}}$. We assume that all concentrations are expressed as moles of i per volume of air (e.g., mol m^{-3} of air). These concentrations will be a function of the location (x,y,z) and time and can be described by the atmospheric diffusion equation

$$\frac{\partial C_{im}}{\partial t} = -\mathbf{v}_m \cdot \nabla C_{im} + \nabla \cdot (\mathbf{K}\,\nabla C_{im}) + \sum_n W^i_{n/m} + R_{im} + E_{im} \qquad (20.1)$$

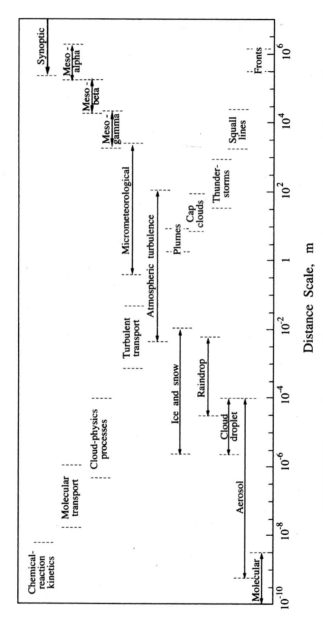

FIGURE 20.2 Lengthscales significant to wet scavenging processes (U.S. NAPAP, 1991)

where C_{im} is the concentration of species i in medium (phase) m (air, rain, cloud, aerosol), $\mathbf{v}_m$ is the velocity of medium m, K is the turbulent diffusivity, $W^i_{n/m}$ is the transport flux from medium n to medium m, R_{im} is the production rate of species i in medium m, and E_{im} is the corresponding emission rate. The terms on the right-hand side (RHS) of (20.1) correspond to advection, dispersion, transport from other media, reactions, and emissions. The velocities $\mathbf{v}_m$ can be different from medium to medium, for example, $\mathbf{v}_{\text{rain}}$ contains the precipitation fall speed of raindrops and the wind-induced horizontal velocity. It should be noted that the distinction of velocity components for individual media, the existence of multiple phase and species combinations, and the intercorporation of interphase transport terms are key ingredients distinguishing wet deposition from dry removal.

The terms $W^i_{n/m}$ correspond to the rates of transport of species i from medium n to medium m. For this simplified case with only four media involved, we need to consider a total of 11 transformation pathways (Figure 20.1). Of these we have already discussed the transformations inside the cloud (rain formation, aerosol–cloud interactions, gas–cloud interactions, and aerosol–gas interactions). Interactions between rain and gaseous compounds will be discussed next to complete the picture.

The complexity of the wet removal process led early investigators to attempt to quantify the relationship between airborne species concentrations, meteorological conditions, and wet deposition rates by lumping the effects of these processes into a few parameters. We will summarize below the definition of these parameters to create a historical perspective, but more importantly to stress the assumptions involved in use of such semiempirical parameters.

The rate of transfer of a soluble gas or a particle into rain droplets below a cloud can be approximated by first-order relationships

$$W^i_{\text{gas/rain}} = \Lambda_{ig} C_{i,\text{gas}}$$
$$W^i_{\text{part,rain}} = \Lambda_{ip} C_{i,\text{part}}$$
(20.2)

where Λ_{ig} and Λ_{ip} are the *scavenging coefficients* for species i in the gas and particulate phases, respectively (Chamberlain 1953). Scavenging coefficients in general are a function of location, time, rainstorm characteristics, and the aerosol size distribution of species i. Use of (20.2) is allowable only if scavenging is irreversible and only if it is independent of the quantity of material scavenged previously.

If $C_g(z,t)$ is the concentration of a species in a horizontally homogeneous atmosphere, "washed out" by rain, the below-cloud scavenging rate F_{bc} will be equal to

$$F_{\text{bc}}(t) = \int_0^h \Lambda_g(z,t) C_g(z,t)\,dz$$
(20.3)

where h is the cloud base height and Λ_g the height-dependent scavenging coefficient for the species. Note that Λ_g has units of inverse time, so the overall below-cloud scavenging rate F_{bc} has units of (mass area^{-1} time^{-1}).

The overall wet flux of a species is the sum of transfer of the species from the cloud to rain plus the below-cloud scavenging. The rate of removal of a species from the cloud is often referred to as the "rainout" rate and the rate of below-cloud scavenging, as the "washout" rate.

If the atmosphere below the cloud is homogeneous, then one can also define an average scavenging coefficient $\overline{\Lambda}_g$ so that

$$F_{bc}(t) = C_g(t) \int_0^h \Lambda_g(z,t)dz = \overline{\Lambda}_g h C_g(t) \tag{20.4}$$

If the species exists only in the gas phase (not in the aerosol), and if the contribution of "rainout" is negligible compared to "washout," the wet deposition flux of the species will be equal to the below-cloud scavenging rate as given by (20.4).

> **Below-Cloud Scavenging Coefficient**   The concentration of gas A is $10\,\mu g\,m^{-3}$ below a raining cloud. Assuming a constant scavenging coefficient of $3.3\,h^{-1}$, cal-culate the concentration of A in the atmosphere after 30 min of rain, and the overall wet deposition flux. Assume a cloud base at 2 km.
>
> Assuming that the atmosphere below the cloud is homogeneous before and during the precipitation event, the concentration of A is given by
>
> $$\frac{\partial C}{\partial t} = -W_{air/rain} + R + E$$
>
> If A is not reacting ($R = 0$) in any of the phases (gas or aqueous) and there are no emissions ($E = 0$), then
>
> $$\frac{\partial C}{\partial t} = -\Lambda C$$
>
> Assuming that the scavenging coefficient remains constant with time, we can solve the above equation to get
>
> $$C = C_0 e^{-\Lambda t}$$
>
> and after 0.5 h, $C = 0.19 C_0 = 1.9\,\mu g\,m^{-3}$. The amount of A removed per volume of air is equal to $C_0 - C = 8.1\,\mu g\,m^{-3}$, and for a column of 2 km height, this is equivalent to $8.1 \times 2000 = 16.2\,mg\,m^{-2}$ deposited. This will be equal to the overall wet deposition flux of A, only if the initial rain-water concentration of A is zero.

This example illustrates that the scavenging coefficient can be a useful parameter provided that it can be estimated reliably and one is aware of its physical meaning and limitations. Use of a scavenging coefficient in (20.2) implies linear, irreversible transport of a species into droplets.

Another empirical parameter that has been used in wet removal studies is the *scavenging ratio* ζ_i, defined as the ratio of species concentration in collected preci-pitation divided by that in air:

$$\zeta_i = \frac{C_{i,rain}}{C_{i,air}(\text{entering storm})} \tag{20.5}$$

$C_{i,\text{air}}$ is the gas-phase concentration of i aloft at the point where it enters the storm, while $C_{i,\text{rain}}$ is its aqueous-phase concentration (per unit volume of water) in precipitation arriving at the surface. Owing to difficulties of obtaining measurements aloft at the region where species enter the storm, it has been usual practice to presume that collocated ground-level measurements of $C_{i,\text{rain}}$ and $C_{i,\text{air}}$ suffice. This practice has led to problems in the application of measured scavenging ratios to calculate wet deposition. Measured scavenging ratios for the same species often vary by orders of magnitude. Barrie (1992), in his article "Scavenging ratios a useful tool or black magic," cautioned that use of incorrectly measured scavenging ratios or extrapolation of measured values to other situations may result in erroneous results.

The *washout ratio* is defined as

$$w_r = \frac{\text{concentration of material in surface-level precipitation}}{\text{concentration of material in surface-level air}}$$
$$= \frac{C_{i,\text{rain}}(x, y, 0, t)}{C_{i,\text{air}}(x, y, 0, t)} \tag{20.6}$$

The net wet deposition flux F_w can be expressed in terms of $C_{i,\text{rain}}$ $(x, y, 0, t)$ by

$$F_w = C_{i,\text{rain}}(x, y, 0, t)p_0 \tag{20.7}$$

where p_0 is the precipitation intensity, usually reported in mm h^{-1}. Typical values of p_0 are $0.5 \, \text{mm h}^{-1}$ for drizzle and $25 \, \text{mm h}^{-1}$ for heavy rain.

On the basis of F_w we can define the *wet deposition velocity* as

$$u_w = \frac{F_w}{C_{i,\text{air}}(x, y, 0, t)} \tag{20.8}$$

where $C_{i,\text{air}}$ $(x, y, 0, t)$ is the concentration of the species measured at the ground level.

Note that according to the definitions of (20.7) and (20.8) the relation between the wet deposition velocity u_w and the washout ratio w_r is

$$u_w = \frac{F_w}{C_{i,\text{air}}(x, y, 0, t)} = \frac{C_{i,\text{rain}} p_0}{C_{i,\text{air}}(x, y, 0, t)} = w_r p_0 \tag{20.9}$$

20.2 BELOW-CLOUD SCAVENGING OF GASES

During rain, soluble species that exist below clouds dissolve into falling raindrops and are removed from the atmosphere. We would like to estimate the rate of removal of these species based on rain event characteristics (rain intensity, raindrop size) and species physical and chemical properties.

The rate of transfer of a gas to the surface of a stationary or falling drop can be calculated by

$$W_t(z, t) = K_c(C_g(z, t) - C_{eq}(z, t)) \tag{20.10}$$

where K_c is the species mass transfer coefficient (m s^{-1}), C_g is the bulk concentration of the species in the gas phase, and C_{eq} is the concentration of the species at the droplet surface, that is, in equilibrium with the aqueous-phase concentration of the dissolved gas.

Both C_g and C_{eq} are gas-phase concentrations and therefore have units of moles of A per unit volume of air, whereas W_t has units of moles of A transferred per unit surface area per time. Using Henry's law, $C_{eq} = C_{aq}/(H^*RT)$, where H^* is the effective Henry's law coefficient of the species, and C_{aq} is its aqueous-phase concentration (mol A per unit volume of droplet). Therefore (20.10) can be written as

$$W_t(z, t) = K_c \left(C_g(z, t) - \frac{C_{aq}(z, t)}{H^*RT} \right) \qquad (20.11)$$

The mass transfer coefficient of a gaseous molecule to a sphere can be calculated by the empirical correlation (Bird et al. 1960)

$$K_c = \frac{D_g}{D_p} \left[2 + 0.6 \left(\frac{\rho_{air} U_t D_p}{\mu_{air}} \right)^{1/2} \left(\frac{\mu_{air}}{\rho_{air} D_g} \right)^{1/3} \right] \qquad (20.12)$$

where D_p is the droplet diameter; D_g is the gas-phase diffusivity of the species, ρ_{air} and μ_{air} are the air density and viscosity, respectively; and U_t is the droplet velocity. Note that the group Sh $= K_c D_p/D_g$ is the Sherwood number, Re $= \rho_{air} U_t D_p/\mu_{air}$ is the Reynolds number, and Sc $= \mu_{air}/\rho_{air} D_g$ is the Schmidt number.

The major challenge in using (20.11) is that the gas- and aqueous-phase concentrations $C_g(z, t)$ and $C_{aq}(z, t)$ for a horizontally uniform atmosphere are a function of time and altitude. Therefore one needs to estimate the evolution of both variables in the general case. We shall consider two cases, the simplified one of an irreversibly soluble gas and then the more general case of a reversibly soluble one.

20.2.1   Below-Cloud Scavenging of an Irreversibly Soluble Gas

Equations (20.10) and (20.11) apply to the general case where a gas can be transferred both from the gas to the aqueous phase (when $C_g > C_{eq}$), and vice versa (when $C_g < C_{eq}$). However, in the limiting case when $C_g \gg C_{eq}$, one can neglect the flux from the aqueous to the gas phase and assume that

$$W_t(z, t) \simeq K_c C_g(z, t) \qquad (20.13)$$

This simplification frees us from the need to estimate the aqueous-phase concentration $C_{aq}(z, t)$. This will be a rather good approximation for a very soluble species, that is, a species with sufficiently high effective Henry's law constant H^*. Nitric acid, with a Henry's law constant of 2.1×10^5 M atm^{-1}, is a good example of such a species. Recalling that dissolved nitric acid dissociates to produce nitrate, we obtain

$$HNO_3(aq) \rightleftharpoons NO_3^- + H^+$$

and the effective Henry's law coefficient $H^*_{HNO_3}$, for the equilibrium between nitric acid vapor and total dissolved nitrate $N(V) = [HNO_3] + [NO_3^-]$, is given by (7.59)

$$H^*_{HNO_3} = H_{HNO_3} \left(1 + \frac{K_n}{[H^+]} \right) \tag{20.14}$$

where $H_{HNO_3} = 2.1 \times 10^5 \, M \, atm^{-1}$ at 298 K, and K_n is the dissociation constant of HNO_3 (aq) equal to 15.4 M. Therefore the effective Henry's law coefficient is $H^*_{HNO_3} \simeq 3.23 \times 10^6/[H^+]$ and for a reasonable ambient pH range from 2 to 6 it varies from 3.2×10^8 to $3.2 \times 10^{12} \, M \, atm^{-1}$. If the gas-phase HNO_3 mixing ratio is 10 ppb (10^{-8} atm), then the corresponding equilibrium aqueous-phase concentration is 3–30,000 M. This is an extremely high concentration (rainwater concentrations of nitrate are several orders of magnitude less) and the assumption $C_g \gg C_{aq}/H^*_{HNO_3}$ appears reasonable. To view it from a different perspective, nitrate rainwater concentrations C_{aq} are in the 10–300 μM range (U.S. NAPAP 1991), so the corresponding equilibrium concentrations $C_{eq} = C_{aq}/H^*_{HNO_3} < 0.001$ ppb and as $C_g \simeq 10$ ppb the assumption $C_g \gg C_{eq}$ should hold under most ambient conditions.

Let us calculate the scavenging rate of $HNO_3(g)$ during a rain event. There are two ways to approach this problem:

1. To follow a falling raindrop calculating the amount of $HNO_3(g)$ removed by it.
2. To simulate the evolution of the gas-phase concentration of HNO_3 as a function of height.

Let us use the falling drop approach. The rate of increase of the concentration C_{aq} of an irreversibly scavenged gas in a droplet can be estimated by a mass balance between the rate of increase of the mass of species in the droplet and the rate of transport of species to the drop

$$\frac{1}{6} \pi D_p^3 \frac{dC_{aq}}{dt} = \pi D_p^2 W_t \tag{20.15}$$

or, using the irreversible flux given by (20.13):

$$\frac{dC_{aq}}{dt} = \frac{6 K_c}{D_p} C_g \tag{20.16}$$

For a falling raindrop with a terminal velocity U_t we can change the independent variable in (20.16) from time to height noting that, from the chain rule

$$\frac{dC_{aq}}{dt} = \frac{dC_{aq}}{dz} \frac{dz}{dt} = U_t \frac{dC_{aq}}{dz} \tag{20.17}$$

where z is the fall distance. Combining (20.16) and (20.17), we obtain

$$\frac{dC_{aq}}{dz} = \frac{6 K_c}{D_p U_t} C_g \tag{20.18}$$

Assuming that the droplet diameter remains constant as the drop falls and that its initial concentration is C_{aq}^0, we get the following, by integrating (20.18):

$$C_{aq} = C_{aq}^0 + \frac{6 K_c}{U_t D_p} \int_0^z C_g(y) \, dy \tag{20.19}$$

If the atmosphere is homogeneous, and C_g is constant with height, then

$$C_{aq} = C_{aq}^0 + \frac{6 K_c C_g}{U_t D_p} z \tag{20.20}$$

and the concentration of the irreversibly scavenged species in the drop varies linearly with the fall distance. Note that $z = 0$ corresponds to cloud base level. If the drop falls a distance h corresponding to the cloud base, the amount scavenged below cloud per droplet will be

$$m_{scav} = \left(\frac{1}{6} \pi D_p^3 \right)(C_{aq} - C_{aq}^0) = \frac{\pi D_p^2 K_c C_g h}{U_t} \tag{20.21}$$

Let us assume that the rain intensity is p_0 (mm h^{-1}) and all rain droplets have diameter D_p (m). The volume of water deposited per unit surface area will then be $10^{-3} p_0$ (m^3 m^{-2} h^{-1}). The number of droplets falling per surface area per hour will be equal to $6 \times 10^{-3} p_0 / \pi D_p^3$ and the rate of wet removal of the material below cloud F_{bc} will be equal to [using (20.21)]

$$F_{bc} = \frac{6 \times 10^{-3} p_0 K_c h}{U_t D_p} C_g \tag{20.22}$$

A mass balance on the species in a 1 m^2 column of air below the cloud suggests that

$$F_{bc} = -h \frac{dC_g}{dt} \tag{20.23}$$

and after integration

$$C_g = C_g^0 e^{-\Lambda t} \tag{20.24}$$

where the scavenging coefficient is

$$\Lambda = \frac{6 \times 10^{-3} p_0 K_c}{U_t D_p} \tag{20.25}$$

The parameters K_c and U_t are given in Table 20.1 as a function of the droplet diameter. The scavenging coefficients in Table 20.1 have been estimated for $p_0 = 1$ mm h^{-1} assuming that all the raindrops have the same size D_p. As the scavenging coefficient according to (20.25) depends linearly on p_0, one just needs to multiply the Λ values in Table 20.1 with

TABLE 20.1 Estimation of the Scavenging Coefficient Λ for Irreversible Scavenging in a Homogeneous Atmosphere ($p_0 = 1\,\mathrm{mm\,h^{-1}}$)

D_p, cm	U_t, cm s^{-1}	K_c, cm s^{-1}	Λ, h^{-1}	$1/\Lambda$, h
0.001	0.3	220	4.4×10^5	2.3×10^{-6}
0.01	26	32	73.8	0.01
0.1	300	13	0.26	3.8
1.0	1000	6	0.0036	278

the appropriate rainfall intensity. Table 20.1 demonstrates that the scavenging coefficient depends dramatically on raindrop diameter. Very small drops are very efficient in scavenging soluble gases for two reasons: (1) they fall more slowly, so they have more time in their transit to "clean" the atmosphere; and (2) mass transfer is more efficient for these drops (high K_c). Note that the scavenging rate varies over eight orders of magnitude when the drop diameter increases by three orders. Also note that the scavenging time scale ($1/\Lambda$) can vary from less than a second to several hours depending on the raindrop size distribution. Droplets smaller than 2 mm are responsible for most of the scavenging in general.

Typical scavenging rates are in the range of 1–3% min^{-1} for irreversibly soluble gases such as HNO$_3$. These rates indicate that HNO$_3$ and other very soluble gases are significantly depleted during a typical 30-min rainfall.

An assumption inherent in our preceding analysis is that the diameter of a raindrop remains approximately constant during its fall. Raindrops usually evaporate during their fall, as they pass through a subsaturated environment. Finally, we can now revisit our initial assumption that HNO$_3$ scavenging is practically irreversible. Figure 20.3 shows the

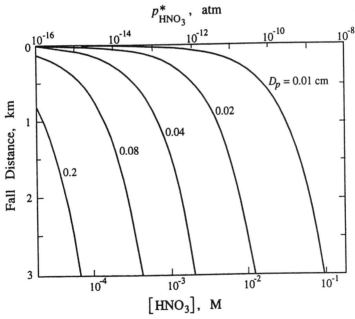

FIGURE 20.3 Concentration of dissolved HNO$_3$ and corresponding equilibrium vapor pressure over drops of different diameters as a function of fall distance for $p_{\mathrm{HNO_3}} = 10^{-8}$ atm (10 ppb) (Levine and Schwartz 1982).

HNO$_3$ concentration in the aqueous phase for droplets of different sizes, falling through an atmosphere containing 10 ppb of HNO$_3$. The equilibrium gas-phase mixing ratio (plotted on the upper x axis) remains below 10 ppb (10^{-8} atm) even for the smaller droplets, providing more support for the validity of our assumption.

20.2.2   Below-Cloud Scavenging of a Reversibly Soluble Gas

For a reversibly soluble gas, one needs to retain both the flux from the gas to the aqueous phase and the inverse flux. The mass balance of (20.15) is still valid, so combining it with (20.11), we get

$$\frac{dC_{aq}(z,t)}{dt} = \frac{6 K_c}{D_p}\left(C_g(z,t) - \frac{C_{aq}(z,t)}{H^* RT}\right) \tag{20.26}$$

where H^* is the effective Henry's law coefficient for the species (it can be a function of the pH) and C_{aq} is the net aqueous-phase species concentration (including both the associated and dissociated forms). Equation (20.26) can be rewritten using height as an independent variable as

$$\frac{dC_{aq}(z,t)}{dz} = \frac{6 K_c}{D_p U_t}\left(C_g(z,t) - \frac{C_{aq}(z,t)}{H^* RT}\right) \tag{20.27}$$

This equation indicates that the driving force for scavenging is the difference between the gas-phase concentration of the species and the concentration at the drop surface (droplet equilibrium concentration). As the aqueous-phase concentration increases during the droplet fall, the driving force will tend to decrease.

Let us assume that the rain droplet size and droplet pH remain constant during its fall and that the atmosphere is homogeneous so that $C_g(z,t) = C_g(t)$. Then (20.27) can be integrated using $C_{aq}(0,t) = C_{aq}^0$ to give

$$C_{aq}(z,t) = C_g(t)H^* RT - [C_g(t)H^* RT - C_{aq}^0]\exp\left(\frac{-6 K_c z}{D_p U_t H^* RT}\right) \tag{20.28}$$

This solution exhibits several interesting limiting cases. If the droplet starts at equilibrium with the below-cloud atmosphere ($C_{aq}^0 = H^* RT C_g(t)$), then its aqueous-phase concentration will remain constant and the saturated droplet will not scavenge any of the species from the below-cloud region. We note also that for $z \gg D_p U_t H^* RT/(6 K_c)$, $C_{aq} \simeq H^* RT C_g(t)$ and the drop after a certain fall distance reaches equilibrium with the environment and stops scavenging additional material from the gas phase.

The scavenging rate $s(t)$ for one droplet of diameter D_p can be calculated by

$$s(t) = \frac{1}{6}\pi D_p^3 \frac{dC_{aq}}{dt} = \frac{1}{6}\pi D_p^3 U_t \frac{dC_{aq}}{dz} \tag{20.29}$$

Differentiating (20.28) and substituting into (20.29) we get

$$s(t) = \frac{\pi D_p^2 K_c}{H^* RT}[C_g(t)H^* RT - C_{aq}^0]\exp\left(-\frac{6 K_c z}{D_p U_t H^* RT}\right) \tag{20.30}$$

The overall below-cloud scavenging rate, W_{bc}, can be calculated once more by multiplying the per droplet scavenging rate by the number of droplets per volume of air $(6 \times 10^{-3} p_0/\pi D_p^3 U_t)$:

$$W_{bc} = \frac{6 \times 10^{-3} p_0 K_c}{H^* R T D_p U_t} [C_g(t) H^* R T - C_{aq}^0] \exp\left(-\frac{6 K_c z}{D_p U_t H^* R T}\right) \tag{20.31}$$

Because of the reversibility of the scavenging process, one can define a scavenging coefficient only if $C_g H^* R T \gg C_{aq}^0$. This is valid when the initial raindrop species concentration is much lower than the equilibrium concentration corresponding to the below-cloud atmospheric conditions. If this is valid, then

$$\Lambda(z) = \frac{6 \times 10^{-3} p_0 K_c}{D_p U_t} \exp\left(-\frac{6 K_c z}{D_p U_t H^* R T}\right) \tag{20.32}$$

and, as expected, the local scavenging coefficient is a function of fall distance. As the droplet keeps falling, its aqueous-phase concentration approaches equilibrium and the local scavenging rate approaches zero.

When the drop reaches the ground after a fall through an atmospheric layer of thickness h, it will have a concentration according to (20.28) equal to

$$C_{aq}(h, t) = C_g(t) H^* R T - [C_g(t) H^* R T - C_{aq}^0] \exp\left(-\frac{6 K_c h}{D_p U_t H^* R T}\right) \tag{20.33}$$

and will have scavenged a mass of species equal to

$$\begin{aligned} m_{scav} &= \frac{1}{6} \pi D_p^3 [C_{aq}(h, t) - C_{aq}^0] \\ &= \frac{1}{6} \pi D_p^3 [C_g(t) H^* R T - C_{aq}^0] \left[1 - \exp\left(-\frac{6 K_c h}{D_p U_t H^* R T}\right)\right] \end{aligned} \tag{20.34}$$

The total scavenging rate by the rain from the below-cloud atmosphere (equal to the wet deposition rate due to below-cloud scavenging) is

$$F_{bc} = 10^{-3} p_0 [C_g(t) H^* R T - C_{aq}^0] \left[1 - \exp\left(-\frac{6 K_c h}{D_p U_t H^* R T}\right)\right] \tag{20.35}$$

It is interesting to note that this overall scavenging rate can be negative if $C_g(t) H^* R T < C_{aq}^0$. This case corresponds to a droplet falling through an atmosphere that is below saturation compared to the species concentration in the droplet. In this case, the species will evaporate from the droplet during its fall and rain will result in a net gain of species mass for the below-cloud atmosphere. Note also that the maximum scavenging rate is reached for $6 K_c h/D_p U_t H^* R T \gg 1$ and is equal to

$$(F_{bc})_{max} = 10^{-3} p_0 [C_g(t) H^* R T - C_{aq}^0] \tag{20.36}$$

Our analysis above applies to a highly idealized scavenging scenario. In general, during a drop's fall, the dissolution of species changes the pH and therefore the effective Henry's law constant of dissociating species. At the same time, aqueous-phase reactions may take place inside the raindrops, usually resulting in an acceleration of the scavenging process. Study of the interaction of these processes requires solution of a system of differential equations numerically. A representative example is given next.

Fall of a Droplet through a Layer Containing CO_2, SO_2, HNO_3, O_3, and H_2O_2
We want to calculate the evolution of the aqueous-phase concentrations in a drop of diameter D_p ($D_p = 2\,mm$ and $5\,mm$), falling through an $h = 1000\,m$ deep atmospheric layer, as a function of its fall distance. Assume that the atmospheric gas-phase concentrations of all species remain constant with values given in Table 20.2.

Case A represents the base-case scenario. In case B, we remove the HNO_3, so by comparing cases A and B, we can determine the HNO_3 effect on droplet pH. In case C we remove the oxidants (H_2O_2 and O_3), to study the effect of S(IV) on the drop composition. Finally, case D has no NH_3, so comparing cases A and D we can see the effect of the neutralization by NH_3. We assume that initally the drop is in equilibrium with the ambient CO_2 and NH_3.

The droplet composition will be characterized by the concentrations of the following species: $CO_2 \cdot H_2O$, HCO_3^-, CO_3^{2-}, $SO_2 \cdot H_2O$, HSO_3^-, SO_3^{2-}, NH_4OH, NH_4^+, $HNO_3\,(aq)$, NO_3^-, O_3, H_2O_2, HO_2^-, H^+, OH^-, H_2SO_4, HSO_4^{2-}, and SO_4^{2-}. As we saw in our discussion of aqueous-phase chemistry, instead of describing all the above concentrations, we can define the total species concentrations as

$$[CO_2^T] = [CO_2 \cdot H_2O] + [HCO_3^-] + [CO_3^{2-}]$$

$$[S(IV)] = [SO_2 \cdot H_2O] + [HSO_3^-] + [SO_3^{2-}]$$

$$[HNO_3^T] = [HNO_3(aq)] + [NO_3^-]$$

$$[NH_3^T] = [NH_3 \cdot H_2O] + [NH_4^+]$$

$$[H_2O_2^T] = [H_2O_2] + [HO_2^-]$$

$$[S(VI)] = [H_2SO_4] + [HSO_4^-] + [SO_4^{2-}]$$

(20.37)

and adding O_3 we have reduced the problem size from 18 variables to 7, plus the pH that can be calculated from the electroneutrality equation. Note that sulfate will be the product of the S(IV) oxidation.

TABLE 20.2 Species Mixing Ratios (ppb) for Example

Case	$\xi_{H_2O_2}$	ξ_{O_3}	ξ_{HNO_3}	ξ_{NH_3}	ξ_{SO_2}	$\xi_{CO_2}{}^a$
A	10	50	10	5	20	300
B	10	50	0	5	20	300
C	0	0	10	5	20	300
D	10	50	10	0	20	300

^{a}In ppm.

The mass balance for total aqueous-phase nitric acid (assuming reversible transfer between the droplets and the surrounding atmosphere) will then be

$$\left(\frac{1}{6}\pi D_p^3\right) U_t \frac{d[\text{HNO}_3^T]}{dz} = (\pi D_p^2) K_{c,\text{HNO}_3}\left(\frac{p_{\text{HNO}_3}}{RT} - \frac{[\text{HNO}_3^T]}{H_{\text{HNO}_3}^* RT}\right) \tag{20.38}$$

where $H_{\text{HNO}_3}^*$ is the effective Henry's law coefficient for nitric acid given by (20.14)

$$H_{\text{HNO}_3}^* = H_{\text{HNO}_3}\left(1 + \frac{K_n}{[\text{H}^+]}\right) \tag{20.39}$$

with H_{HNO_3} and K_n the nitric acid Henry's law coefficient and the first dissociation constant, respectively. Equation (20.38) indicates that the total dissolved nitric acid changes due to mass transfer, as described by the product of the mass transfer coefficient for HNO_3 transport to a falling sphere of diameter D_p in air K_{c,HNO_3} and the "driving force," which is the difference between the ambient HNO_3 concentration p_{HNO_3}/RT and vapor pressure of HNO_3 just above the drop surface.

Using the definition of the effective Henry's law coefficient in (20.39), we can rewrite (20.38) as

$$\frac{d[\text{HNO}_3^T]}{dz} = \frac{6\,K_{c,\text{HNO}_3}}{U_t D_p RT}\left(p_{\text{HNO}_3} - \frac{[\text{HNO}_3^T][\text{H}^+]}{H_{\text{HNO}_3}([\text{H}^+] + K_n)}\right) \tag{20.40}$$

Note that only $[\text{HNO}_3^T]$ and $[\text{H}^+]$ appear on the RHS of the equation. Equation (20.40) is an example of a material balance obtained on a drop for a gas that dissociates on absorption but does not participate in further aqueous-phase chemistry.

Consider now O_3, a gas that does not dissociate on absorption but that participates in aqueous-phase chemistry reacting with S(IV). The material balance in this case is

$$\frac{d[\text{O}_3]}{dz} = \frac{6\,K_{c,\text{O}_3}}{U_t D_p RT}\left(p_{\text{O}_3} - \frac{[\text{O}_3]}{H_{\text{O}_3}}\right) + \frac{1}{U_t}\left(\frac{d[\text{O}_3]}{dt}\right)_{\text{react}} \tag{20.41}$$

The aqueous-phase reaction rate between O_3 and S(IV) can be calculated using the kinetic expressions given in Chapter 7.

Dissolved SO_2 both dissociates and reacts with O_3 and H_2O_2. The mass balance in this case is

$$\frac{d[\text{S(IV)}]}{dz} = \frac{6\,K_{c,\text{SO}_2}}{U_t D_p RT}\left(p_{\text{SO}_2} - \frac{[\text{S(IV)}][\text{H}^+]^2}{H_{\text{SO}_2}([\text{H}^+]^2 + K_{s1}[\text{H}^+] + K_{s1}K_{s2})}\right)$$
$$+ \frac{1}{U_t}\left(\frac{d[\text{S(IV)}]}{dt}\right)_{\text{react}} \tag{20.42}$$

Sulfate, [S(VI)], is an example of a species that is produced in the droplet but is not transferred between the gas and aqueous phases. Its material balance is simply

$$\frac{d[\text{S(VI)}]}{dz} = -\frac{1}{U_t}\left(\frac{d[\text{S(IV)}]}{dt}\right)_{\text{react}} \tag{20.43}$$

The mass balances for the remaining species can be written similarly, recalling that $[CO_2^T]$ and $[NH_3^T]$ are species that are transferred between the two phases, dissociate, but do not participate in aqueous-phase reactions, while H_2O_2 has a behavior similar to that of S(IV).

All these differential equations are coupled through the electroneutrality equation

$$[H^+] + [NH_4^+] = [OH^-] + [HCO_3^-] + 2[CO_3^{2-}] + [HSO_3^-] + 2[SO_3^{2-}]$$
$$+ [NO_3^-] + 2[SO_4^{2-}] + [HSO_4^-] + [HO_2^-] \qquad (20.44)$$

Each term in (20.44) can be related to $[NH_3^T]$, $[CO_2^T]$, $[S(IV)]$, $[HNO_3^T]$, $[H_2O_2^T]$, and $[S(VI)]$. Mathematically the problem is equivalent to the solution of seven differential equations (for $[CO_2^T]$, $[S(IV)]$, $[HNO_3^T]$, $[NH_3^T]$, $[H_2O_2^T]$, $[S(VI)]$, and $[O_3]$) coupled with one algebraic equation, (20.44). The problem could be simplified depending on the conditions by using appropriate equilibrium assumptions. For example, if one assumes that CO_2 is in equilibrium between the gas and aqueous phases and its gas-phase concentration remains constant, one of the differential equations is eliminated. The reader is referred back to Chapter 12 for a discussion of the validity of similar assumptions for the other species.

In the solutions presented below we have assumed that CO_2, NH_3, and O_3 are in equilibrium between the two phases. Figure 20.4 shows $[H^+]$ as a function of drop fall distance in cases A, B, C, and D. As expected, the pH is lowest in case D in the absence of NH_3. The removal of HNO_3 in case B relative to A leads to only a slight increase in pH, primarily because the large drop does not absorb an appreciable amount of HNO_3 during its fall. The corresponding sulfate profiles are given in

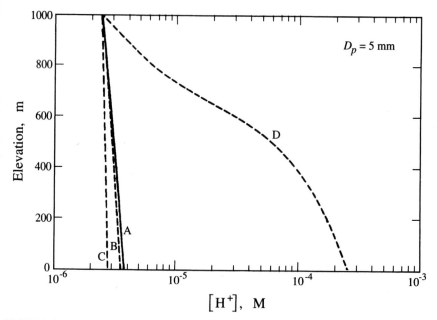

FIGURE 20.4 Fall of a drop through a layer containing CO_2, SO_2, HNO_3, NH_3, O_3, and H_2O_2. $[H^+]$ as a function of fall distance in cases A–D in text.

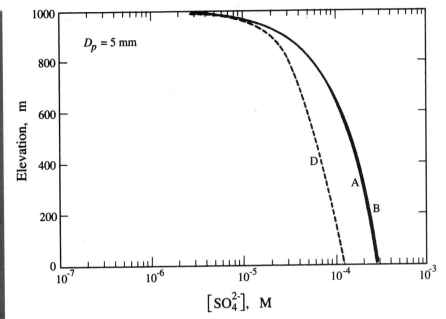

FIGURE 20.5 Fall of a drop through a layer containing CO_2, SO_2, HNO_3, NH_3, O_3, and H_2O_2. $[SO_4^{2-}]$ as a function of fall distance in cases A to D in text.

Figure 20.5. Sulfate concentrations increase to $300\,\mu M$ in cases A and B, since the absence of HNO_3 for the 5 mm drop was seen in the previous figure to have only a negligible effect on pH. Absence of NH_3 in case D leads to a much lower sulfate concentration of a little more than $100\,\mu M$, due to the significant pH reduction.

20.3 PRECIPITATION SCAVENGING OF PARTICLES

As a raindrop falls through air, it collides with airborne particles and collects them. As the droplet falls, it sweeps per unit time the volume of a cylinder equal to $\pi D_p^2 U_t/4$, where U_t is its falling velocity. As a first approximation, one would be tempted to conclude that the droplet would collect all the particles that are in this volume. Actually, if the aerosol particles have a diameter d_p, a collision will occur if the center of the particle is inside the cylinder with diameter $D_p + d_p$. Also, the particles are themselves settling with a velocity $u_t(d_p)$. So the "collision volume" per time is actually $\pi(D_p + d_p)^2[U_t(D_p) - u_t(d_p)]/4$. The major complication associated with this simple picture, as we saw in Chapter 17, arises because the falling drop perturbs the air around it, creating the flow field depicted in Figure 17.27. The flow streamlines diverge around the drop. As the raindrop approaches the particle, it exerts a force on it via the air as a medium and modifies its trajectory. Whether a collision will in fact occur depends on the sizes of the drop and the particle and their relative locations. Prediction of the actual trajectory of the particle is a complicated fluid mechanics problem. Solutions are expressed as the collision efficiency $E(D_p, d_p)$, which is defined in exactly the same manner as is the drop-to-drop collision efficiency. Therefore $E(D_p, d_p)$ is the fraction of particles of diameter d_p contained within the collision volume of a drop of diameter D_p that are collected. Thus $E(D_p, d_p)$ can be viewed as a correction factor accounting for the interaction between the falling raindrop and the aerosol particle. If the aerosol size distribution is given by $n(d_p)$, the number of collisions between particles of diameters in the range $[d_p, d_p + dd_p]$ and a drop of diameter D_p is

$$\frac{\pi}{4}(D_p + d_p)^2[U_t(D_p) - u_t(d_p)]E(D_p, d_p)n(d_p)dd_p \tag{20.45}$$

The rate of mass accumulation of these particles by a single falling drop can be calculated by simply replacing the number distribution $n(d_p)$ by the mass distribution $n_M(d_p)$ to obtain

$$\frac{\pi}{4}(D_p + d_p)^2[U_t(D_p) - u_t(d_p)]E(D_p, d_p)n_M(d_p)dd_p \tag{20.46}$$

The total rate of collection of mass of all particles of diameter d_p is obtained by integrating (20.46) over the size distribution of collector drops

$$W_{bc}(d_p) = n_M(d_p)dd_p \int_0^\infty \frac{\pi}{4}(D_p + d_p)^2[U_t(D_p) - u_t(d_p)] \\ \times E(D_p, d_p)N(D_p)dD_p \tag{20.47}$$

where $N(D_p)$ is the raindrop number distribution.

Two approximations can generally be made in (20.47):

1. $U_t(D_p) \gg u_t(d_p)$.
2. $(D_p + d_p)^2 \simeq D_p^2$.

Using these approximations, (20.47) becomes

$$W_{bc}(d_p) = n_M(d_p)dd_p \int_0^\infty \frac{\pi}{4}D_p^2 U_t(D_p)E(D_p, d_p)N(D_p)dD_p \tag{20.48}$$

Therefore the below-cloud scavenging (rainout) rate of aerosol particles of diameter d_p can be written as

$$\frac{dn_M(d_p)}{dt} = -\Lambda(d_p)n_M(d_p) \tag{20.49}$$

where the scavenging coefficient $\Lambda(d_p)$ is given by

$$\Lambda(d_p) = \int_0^\infty \frac{\pi}{4}D_p^2 U_t(D_p)E(D_p, d_p)N(D_p)dD_p \tag{20.50}$$

Calculation therefore of the aerosol scavenging rate, for a given aerosol diameter d_p, requires knowledge of the droplet size distribution $N(D_p)$ and the scavenging efficiency $E(D_p, d_p)$.

The total aerosol mass scavenging rate can be calculated by integrating (20.50) over the aerosol size distribution to get

$$\frac{dM_{\text{aer}}}{dt} = \frac{d}{dt} \int_0^\infty n_M(d_p) dd_p = -\int_0^\infty \Lambda(d_p) n_M(d_p) dd_p \qquad (20.51)$$

Finally, the rainfall rate p_0 (mm h^{-1}) is related to the raindrop size distribution by

$$p_0 = \int_0^\infty \frac{\pi}{6} D_p^3 U_t(D_p) N(D_p) dD_p \qquad (20.52)$$

While routine measurements of p_0 are available, knowledge of reliable raindrop size distributions remains a problem because of their variability from event to event as well as during the same rainstorm.

20.3.1 Raindrop–Aerosol Collision Efficiency

The collision efficiency $E(D_p, d_p)$ is by definition equal to the ratio of the total number of collisions occurring between droplets and particles to the total number of particles in an area equal to the droplet's effective cross-sectional area. A value of $E = 1$ implies that all particles in the geometric volume swept out by a falling drop will be collected. Usually $E \ll 1$, although E can exceed unity under certain conditions (charged particles). Experimental data suggest that all particles that hit a hydrometeor stick, and therefore, a sticking efficiency of unity is assumed.

Theoretical solution of the Navier–Stokes equation for prediction of the collision efficiency, $E(D_p, d_p)$, for the general raindrop–aerosol interaction case is a difficult undertaking. Complications arise because the aerosol size varies over orders of magnitude, and also because the large raindrop size results in complicated flow patterns (drop oscillations, wake creation, eddy shedding, etc.) Pruppacher and Klett (1997) present a critical overview of the theoretical attempts for the solution of the problem. A detailed discussion of these efforts is outside our scope. However, it is important to understand at least qualitatively the various processes involved.

Particles can be collected by a falling drop as a result of their Brownian diffusion. This random motion of the particles will bring some of them in contact with the drop, and will tend to increase the collection efficiency E. Because the Brownian diffusion of particles decreases rapidly as particle size increases, we expect that this removal mechanism will be most important for the smaller particles ($d_p < 0.2 \mu m$). Inertial impaction occurs when a particle is unable to follow the rapidly curving streamlines around the falling spherical drop and, because of its inertia, continues to move toward the drop and is eventually captured by it. Inertial impaction increases in importance as the aerosol size increases and accelerates scavenging of particles with diameters larger than 1 μm. The preceding arguments indicate that whereas the scavenging of small and large particles is expected to be efficient, scavenging of particles in the 0.1 to 1 μm size range is expected to be relatively slow. In the literature, this minimum is sometimes referred to as the "Greenfield gap" after S. Greenfield who first identified it. Finally, interception takes place when a particle, following the streamlines of flow around an obstacle, comes into contact with the raindrop, because of its size. If the streamline on which the particle center lies is within a distance $d_p/2$ or less from the drop surface, interception will occur.

Interception and inertial impaction are closely related, but interception occurs as a result of particle size neglecting its mass, while inertial impaction is a result of its mass neglecting its size.

An alternative to exact solution of the Navier–Stokes equation is the use of dimensional analysis coupled with experimental data. To formulate a correlation for E based on dimensional analysis, we must identify first the parameters that influence E. There are eight such variables: aerosol diameter (d_p), raindrop diameter (D_p), velocity of the aerosol and raindrop (u_t, U_t), viscosity of water and air (μ_w, μ_{air}), aerosol diffusivity (D), and air density (ρ_a). We assume here that the aerosol has density ρ_p equal to $1\,g\,cm^{-3}$, so this density is not included in the list. Water viscosity appears in the list because internal circulations may be established in the drop, affecting the flow field around it and thus the capture efficiency. These eight variables have three fundamental dimensions (mass, length, and time). Thus, by the Buckingham π theorem, there are five (eight minus three) independent dimensionless groups. These groups can be obtained by nondimensionalizing the equations of motion for air and for the particles. They are (Slinn 1983):

$$Re = D_p U_t \rho_a / 2\mu_a \qquad \text{(Reynolds number of raindrop based on its radius)}$$
$$Sc = \mu_a / \rho_a D \qquad \text{(Schmidt number of collected particle)}$$
$$St = 2\tau(U_t - u_t)/D_p \qquad \text{(Stokes number of collected particle, where τ is its characteristic relaxation time)}$$
$$\phi = d_p / D_p \qquad \text{(ratio of diameters)}$$
$$\omega = \mu_w / \mu_a \qquad \text{(viscosity ratio)}$$

On the basis of these equations, Slinn (1983) proposed the following correlation for E that fits experimental data:

$$
\begin{aligned}
E = \frac{4}{Re\,Sc} &[1 + 0.4\,Re^{1/2}Sc^{1/3} + 0.16\,Re^{1/2}Sc^{1/2}] \\
&+ 4\phi[\omega^{-1} + (1 + 2\,Re^{1/2})\phi] + \left(\frac{St - S^*}{St - S^* + \frac{2}{3}}\right)^{3/2}
\end{aligned}
\tag{20.53}
$$

where

$$
S^* = \frac{1.2 + \frac{1}{12}\ln(1 + Re)}{1 + \ln(1 + Re)}
\tag{20.54}
$$

For particles of density different from $1\,g\,cm^{-3}$, the last term in (20.53) should be scaled by $(\rho_w/\rho_p)^{1/2}$. The first term in (20.53) is the contribution from Brownian diffusion, the second is due to interception, and the third represents impaction.

The third term on the RHS of (20.53) requires some comment; this term represents the effect of impaction. If $S^* - St$ is smaller than $\frac{2}{3}$, the term inside the parentheses is < 0. This corresponds to the case in which the diameter of the particle being collected is less than $1\,\mu m$. Physically, impaction for a particle of this size is likely to be less important than diffusion. From a practical point of view, if $S^* - St < \frac{2}{3}$, the third term on the RHS of (20.53) should be set equal to zero. For very large particles, this term approaches 1.0,

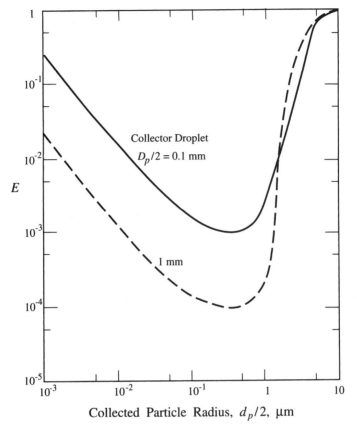

FIGURE 20.6 Semiempirical correlation for the collection efficiency E of two drops (Slinn 1983) as a function of the collected particle size. The collected particle is assumed to have unit density.

potentially leading to a predicted overall efficiency exceeding unity; in this case, the overall efficiency is taken as 1.0.

Figure 20.6 shows E calculated from (20.53) as a function of the collected particle radius ($d_p/2$), for raindrop radius of 0.1 mm and 1 mm. As expected, Brownian diffusion dominates for $d_p < 0.1\ \mu m$, whereas impaction and interception control removal for large d_p. The characteristic minimum in E occurs in the regime where the particles are too large to have an appreciable Brownian diffusivity yet too small to be collected effectively by either impaction or interception.

For very large aerosol particles ($d_p > 20\ \mu m$) and extremely small ones, the collection efficiency approaches unity. However, for particles close to the minimum ($d_p \simeq 1\ \mu m$) the drops collect only particles that are extremely close to the center of the volume swept by the raindrop.

20.3.2 Scavenging Rates

Using expressions obtained for the collision efficiency for $E(D_p, d_p)$ in (20.50) and (20.51), one can estimate the scavenging coefficient, and the scavenging rate for a rain event. The calculation requires knowledge of the size distributions of the raindrops and the below-cloud aerosols.

The scavenging coefficient $\Lambda(d_p)$ given by (20.50) describes the rate of removal of particles of diameter d_p by rain with a raindrop size distribution $N(D_p)$. If one assumes that all raindrops have the same diameter D_p, and a number concentration N_D, then (20.50) simplifies to

$$\Lambda(d_p) = \frac{\pi}{4} D_p^2 U_t(D_p) E(D_p, d_p) N_D \qquad (20.55)$$

The only term depending on the aerosol size d_p is the collection efficiency E. The number concentration of drops can be estimated using (20.52) as a function of the rainfall intensity by

$$p_0 = \frac{\pi}{6} D_p^3 U_t(D_p) N_D \qquad (20.56)$$

and for monodisperse aerosols and raindrops, (20.55) becomes

$$\Lambda(d_p) = \frac{3}{2} \frac{E(D_p, d_p) p_0}{D_p} \qquad (20.57)$$

The scavenging coefficient for this simplified scenario is shown in Figure 20.7 for $p_0 = 1 \text{ mm h}^{-1}$ for drops of diameters $D_p = 0.2$ and 2 mm. This figure indicates the sensitivity of the scavenging coefficient to sizes of both aerosols and raindrops, suggesting the need for realistic size distributions for both aerosol and drops in order to obtain useful estimates for ambient scavenging rates.

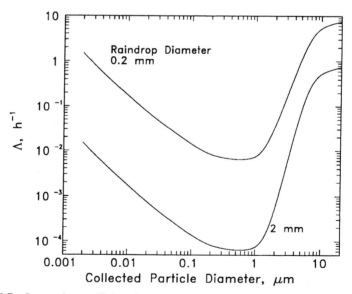

FIGURE 20.7   Scavenging coefficient for monodisperse particles as a function of their diameter collected by monodisperse raindrops with diameters 0.2 and 2 mm assuming a rainfall intensity of 1 mm h^{-1}.

Total aerosol mass scavenging rate is given by (20.51). Replacing the aerosol mass distribution with the number distribution, the overall mass scavenging rate is given by

$$\frac{dM_{\text{aer}}}{dt} = -\int_0^\infty \frac{\pi}{6} d_p^3 n(d_p) \rho_p \Lambda(d_p) dd_p \qquad (20.58)$$

Our calculations can be simplified by defining a mean mass scavenging coefficient Λ_m so that

$$\frac{\pi}{6} \rho_p \int_0^\infty d_p^3 n(d_p) \Lambda(d_p) dd_p = \Lambda_m \frac{\pi}{6} \rho_p \int_0^\infty d_p^3 n(d_p) dd_p \qquad (20.59)$$

or equivalently by

$$\Lambda_m = \frac{\int_0^\infty \Lambda(d_p) d_p^3 n(d_p) dd_p}{\int_0^\infty n(d_p) d_p^3 dd_p} \qquad (20.60)$$

Note that the denominator is equal to $6V_{\text{aer}}/\pi$, where V_{aer} is the particle volume concentration and assuming a particle density ρ_p the denominator is equal to $6M_{\text{aer}}/\pi\rho_p$. Therefore, by definition, we have

$$\frac{dM_{\text{aer}}}{dt} = -\Lambda_m M_{\text{aer}} \qquad (20.61)$$

and all the scavenging information has effectively been incorporated into one parameter.

20.4   IN-CLOUD SCAVENGING

Species can be incorporated into cloud and raindrops inside the raining cloud. These processes determine the initial concentration C_{aq}^0 of the raindrops, before they start falling below the cloud base, and have been discussed previously. Let us summarize the rates of in-cloud scavenging for gases and aerosols.

Gases like HNO_3, NH_3, and SO_2 can be removed from interstitial cloud air by dissolution into clouddrops. For a droplet size distribution $N(D_p)$, the local rate of removal of an irreversibly soluble gas like HNO_3 with a concentration C_g is given by

$$W_{\text{ic}} = C_g \int_0^\infty K_c \pi D_p^2 N(D_p) dD_p = \Lambda C_g \qquad (20.62)$$

where Λ is the scavenging coefficient. Levine and Schwartz (1982) used the cloud droplet distribution of Battan and Reitan (1957)

$$N(D_p) = a e^{-bD_p} \qquad (20.63)$$

with the parameters $a = 2.87\,\text{cm}^{-4}$, $b = 2.65\,\text{cm}^{-1}$, and $D_p = 5 - 40\,\mu\text{m}$, to calculate $\Lambda = 0.2\,\text{s}^{-1}$ for a cloud with 288 drops cm^{-3} and $L = 0.17\,\text{g}\,\text{m}^{-3}$. This value of Λ corresponds to a characteristic time of approximately 5 s for the scavenging of a highly soluble gas like HNO_3 in a typical cloud. In this case the uptake process in a cloud is rapid compared with the characteristic times of air movement through the cloud and the change of cloud liquid water content amounts by condensation and precipitation.

For less soluble gases like SO_2, uptake is a complex function not only of species properties and cloud droplet distribution, but also of other species that may participate in aqueous-phase reactions (e.g., H_2O_2) or control the cloud droplet pH (e.g., NH_3).

Finally, aerosol in-cloud scavenging is the result of two processes. First, nucleation scavenging (growth of CCN to cloud drops) and then collection of a fraction of the remaining aerosols by cloud or rain droplets. Nucleation scavenging as we have seen is an efficient process, often incorporating in cloud drops most of the aerosol mass. On the other hand, as we saw in Chapter 17, interstitial aerosol collection by cloud droplets is a rather slow process and often scavenges a negligible fraction of the interstitial aerosol mass.

20.5 ACID DEPOSITION

The atmosphere is a potent oxidizing medium. In Chapters 6 and 7 we have seen that, once emitted to the atmosphere, SO_2 and NO_x become oxidized to sulfate and nitrate through both gas- and aqueous-phase processes. Organic acids can be produced during the oxidation of emitted organic compounds. The result of these reactions is the existence in the atmosphere of acids in the gas phase (HNO_3, HCl, $HCOOH$, CH_3COOH, etc.), in the aerosol phase (sulfate, nitrate, chloride, organic acids, etc.), and in the aqueous phase. These acidic species are removed from the atmosphere by both wet and dry deposition, and the overall process is termed *acid deposition*. The removal of acidic material by rain has traditionally been termed *acid rain*. Therefore acid deposition includes acid rain, dry deposition of both acid vapor and acid particles, and other forms of wet removal (acid fogs, cloud interception, etc.). Because of the historical focus on the composition of rainwater, this entire process is often called simply acid rain. Even though the term acid rain implies removal only by wet deposition, it is important to keep in mind that the effects attributable to acid rain are in fact a result of the combination of wet and dry deposition.

20.5.1 Acid Rain Overview

A raindrop in a pollutant-free atmosphere has, as we have seen, a pH of 5.6 as a result of the dissolution of CO_2. However, emissions of SO_2 and NO_x lead to conversion of these species during transport from their sources to acidic sulfate and nitrate, and their incorporation into cloud and rainwater. Addition of these acids lowers the rainwater pH, and the rain reaching the ground is acidic.

Historical Perspective   The phenomenon of acid rain appears to have been present at least three centuries ago. Sulfur dioxide emitted from industry in the United Kingdom was traveling far downwind and even causing bleaching of dyed cloth in France (U.S. NAPAP 1991). In 1692 Robert Boyle published his book, *A General History of the Air*, recognizing the presence of sulfur compounds and acids in air and rain. Boyle referred to them as "nitrous or salino-sulfurous spirits." In 1853 Robert Angus Smith, an English chemist, published a report on the chemistry of rain in and around the city of Manchester, "discovering" the phenomenon. Smith published *Air and Rain: The Beginnings of Chemical Climatology* in 1872, introducing the term "acid rain." Little attention was paid to acid rain for almost a century. In 1961, Svante Odin, a Swedish chemist, established a Scandinavian network to monitor surface water chemistry. On the basis of his measurements, Odin showed that acid rain was a large-scale regional phenomenon in much of Europe with well-defined source and sink regions, that precipitation and surface waters were becoming more acidic, and that there were marked seasonal trends in the deposition of major ions and acidity. Odin also hypothesized long-term ecological effects of acid rain, including decline of fish population, leaching of toxic metals from soils into surface waters, and decreased forest growth.

The major foundations for our present understanding of acid rain and its effects were laid by Eville Gorham. On the basis of his research in England and Canada, Gorham showed as early as 1955 that much of the acidity of precipitation near industrial regions can be attributed to combustion emissions, that progressive acidification of surface waters can be traced to precipitation, and that the free acidity in soils receiving acid precipitation results primarily from sulfuric acid.

Scandinavian and European studies, including the Norwegian Interdisciplinary Research Programme "Acid Precipitation—Effects on Forest and Fish," and the study by the European Organization for Economic Cooperation and Development, elucidated the effects of acid rain on fish and forests and long-distance transport of pollutants in Europe.

Studies of acid rain in North America by individual researchers started as early as 1923. Concern about acid rain developed first in Canada and later in the United States with the true scope of the problem not being appreciated until the 1970s. In 1978 the United States and Canada established a Bilateral Research Consultation Group on the Long Range Transport of Pollutants and in 1980 signed a memorandum of intent to develop bilateral agreements on transboundary air pollution, including acid deposition. Both countries have instituted long-term programs for the chemical analysis of precipitation. The National Acid Precipitation Assessment Program (NAPAP) served from 1980 to 1990 as the umbrella of one of the most comprehensive acid rain studies (U.S. NAPAP 1991).

Brimblecombe (1977) and Cowling (1982) have provided excellent historical perspectives associated with the discovery of and attempts to deal with acid rain.

Definition of the Problem   The global pattern of precipitation acidity as determined by the Background Air Pollution Monitoring Program of the World Meteorological Organization (Whelpdale and Miller 1989) is shown in Figure 20.8. Average pH measured in background sites varied from 3.8 to 6.3. All of these values are below the absolutely neutral precipitation pH, which has a value of 7. The "natural" acidity of rainwater is often taken to be pH 5.6, which is that of pure water in equilibrium with the global atmospheric concentration of CO_2 (about 350 ppm). Thus a pH value of 5.6 has been considered as the demarcation line of acidic precipitation.

Figure 20.8 demonstrates that the answers to "What is acid rain?" and "Who is responsible for it—ourselves or nature?" are complicated. Note first that rainwater in several areas has average pH values higher than 6 (China, SE Australia), indicating an alkaline tendency. High pH values observed in several regions of the world are the result of alkaline dust emissions, usually from deserts. These alkaline particles have significant buffering capacity, and even after the addition of anthropogenic acidity, pH of the resulting rain may be over 6. A more puzzling observation is pH values measured in precipitation over remote oceans that are often acidic and sometimes very acidic (Figure 20.8). These observations indicate that considering rainwater pH below 5.6 as an indication of anthropogenic acid rain may be misleading.

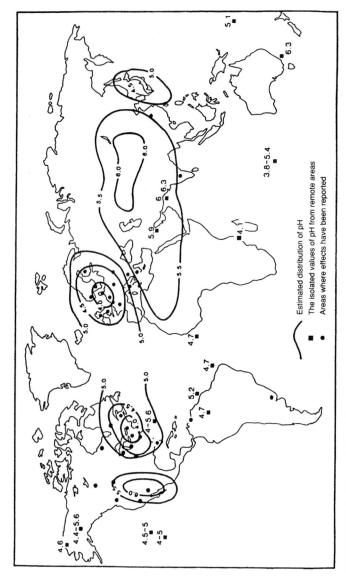

FIGURE 20.8 Map of the global pattern of precipitation acidity as determined by the Background Air Pollution Monitoring Program of the World Meteorological Organization [after Whelpdale and Miller (1989)].

Charlson and Rodhe (1982) have shown that, in the absence of common basic compounds such as NH_3 and $CaCO_3$, rainwater pH values as a result of natural anthropogenic compounds alone could be expected to be around 5.0. These authors suggested that in some extreme cases pH values of 4 or even less can be produced without any contribution from anthropogenic sources. Galloway et al. (1982) reported precipitation chemistry data from remote areas of the globe, with average values around 5.0. Analysis of eastern U.S. precipitation data by Stensland and Semonin (1982) also indicated background values of around 5.0. Keene et al. (1983) showed that organic acids (mainly formic and acetic) can play a significant role in lowering rainwater pH in some remote areas. Their measurements in Katherine, Australia showed that these acids accounted for more than half the rainwater acidity. While it is undeniable that human activities contribute, sometimes overwhelmingly, to rain acidity, it is often difficult to quantify this contribution from rain pH alone.

We can conclude from the discussion above that rain of pH higher than 5.6 has not been influenced by humans, or if it has, it has sufficient buffering capacity so that acidification did not occur. Rain with pH between 5.0 and 5.6 may have been influenced by anthropogenic (human-induced) emissions, but not to an extent exceeding that of natural background sulfur compounds. Again, anthropogenic influences, if present in these cases, may have been mitigated by natural buffering. Although the issues of what acid rain is and who is responsible for it are complex, it is reasonable to consider precipitation with pH < 5.0 as acid rain.

20.5.2 Acid Rain Data and Trends

In this section we summarize recent understanding of acid rain occurrences in various areas of the world. Driven mainly by data availability, we focus on wet deposition in remote areas of the world, in Europe, and in North America.

Precipitation measurements are usually reported as volume-weighted means. The volume-weighted mean is computed by first multiplying the concentration of a species by either the depth of rainfall or the liquid equivalent depth of snowfall. Summing over all precipitation events and dividing by the total liquid depth for all events gives the volume-weighted mean. The volume-weighted mean, assuming that the species is conserved during mixing, is the concentration that would result if the samples from all events were physically mixed and the resulting concentration measured. The volume-weighted pH is computed by first converting the individual pH measurement to H^+ concentration, averaging using the sample volume as a weighting factor and reconverting to pH.

Remote Areas   Table 20.3 shows rain composition in five remote areas in 1980–81 based on the study of Galloway et al. (1982).

These results indicate that indeed no single pH value is universally representative of precipitation in the absence of anthropogenic influences. The same is true for the individual compounds in precipitation. This is a direct result of the variability in natural sources of acid-forming compounds and alkaline materials and meteorological variability (wet and dry years, changes in wind flow patterns, low volume rainfall samples, etc.).

Europe   Spatial and temporal wet deposition data are available for Europe from the European Monitoring and Evaluation Program (EMEP), the European Air Chemistry Network (EACN) in NW Europe, several national programs, and individual studies. Spatial deposition patterns based on EMEP network data for 1985 are shown in Figure 20.9. In general, sulfate concentrations in precipitation have a maximum [$\sim 2.8\,mg\,(S)\,L^{-1}$]

TABLE 20.3   Mean Rain Composition in Remote Areas of the World Based on 1980–1981 Data

Species	Amsterdam Island (Indian Ocean)	Poker Flat (Alaska)	Katherine (Australia)	San Carlos (Venezuela)	St. Georges (Bermuda)
	(all concentration values are in μ equiv L^{-1})				
pH	4.92	4.96	4.78	4.81	4.79
SO_4^{2-}	52.6	10.5	8.3	3.3	48.8
$ex - SO_4^{2-a}$	11.5	10.2	6.9	3.0	21.6
NO_3^-	2.7	2.4	5.9	3.5	7.9
Cl^-	406	4.8	20.6	4.3	264
Mg^{2+}	72.8	0.5	3.6	0.8	49.3
Na^+	334	2.1	11.3	2.7	221
K^+	7.2	1.2	1.2	1.1	6.5
Ca^{2+}	15.5	0.5	5.7	0.5	14.4
NH_4^+	5.1	2.0	2.8	2.3	4.8
H^+	18.2	11.8	19.2	17.0	21.1

[a]Excess sulfate calculated by substracting the seasalt sulfate from the total.

Source: Galloway et al. (1982).

in eastern Europe, while the highest concentrations of nitrate [~ 1.0 mg (N) L^{-1}] in precipitation are found along a line from France to the NE, to the Baltic Sea. Significant ammonia deposition is shown in England, the Netherlands, and Belgium. The pH of precipitation is lowest in southern Scandinavia and in central Europe. For the most part, European precipitation and sulfate averages showed sharp declines by the mid-1980s.

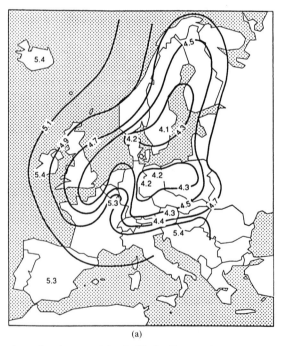

(a)

FIGURE 20.9   Annual volume-weighted pH, for Europe for 1985 (Schaug et al. 1987).

North America In North America, most of the northeastern United States, portions of southeastern Canada, and the western coast of the United States have received acidic precipitation (U.S. NAPAP 1991). The spatial distribution of volume-weighted pH for eastern and western North America is shown in Figure 20.10.

Although precipitation has been most acidic in the NE United States and SE Canada, the geographic extent of the problem encompasses all states east of the Mississippi (Likens et al. 1979; U.S. NAPAP 1991). In California mean precipitation pH values varied from 4.7 to 5.6 even if individual events may have lower pH values (Amar et al. 1988).

In both the western and eastern United States, the summer and winter pH spatial patterns are distinctly different. In the East, summer pH values are lower (median in summer is 4.45 while in the winter it is 4.64) based on 1987 data. The center of the lowest pH values moves east from NE Ohio to eastern Pennsylvania during the summer. In the East, summer sulfate rainwater concentration is nearly twice that in the winter, but overall sulfate deposition is just slightly greater in summer. Summer nitrate and ammonium wet deposition, for the same area, is approximately twice winter deposition. This rather counterintuitive behavior is a result of much higher concentrations observed in rainwater during the summer and the fact that water precipitation during summer is almost equal in magnitude to water precipitation during winter in the eastern and central United States. In the western United States, summer sulfate deposition (median $1.4 \, kg \, ha^{-1}$)[1] is twice that of winter (median $0.6 \, kg \, ha^{-1}$).

Historical Trends Examination of historical trends of anthropogenic emissions and accompanying changes in rain acidity can provide insights into the complex nature of the acid rain problem. Figure 20.11 shows the temporal trends in precipitation acidity in North America and Europe for the period from the mid-1950s to 1970. Total SO_2 emissions in the eastern United States doubled from 1950 to 1970 and then decreased by about 8% from 1970 to 1980 (Gschwandtner et al. 1983). Estimates of Canadian SO_2 emissions indicate a 20% increase from 1955 to 1976. Total NO_x emissions in the eastern United States increased by a factor of 2.4 from 1950 to 1980, and Canadian NO_x emissions tripled between 1955 and 1976. Effects of these emission increases are evident in rainwater pH during these periods. Similar behavior was observed in Europe.

The situation changed during the 1980s. Mean annual precipitation pH for North America based on an average over 39 sites remained practically constant during the 1978 to 1987 period with a median value of around 4.4. Similar conclusions were reached by the analysis of precipitation pH based on data from 148 sites during the 1982–1987 period. Most of these sites were in the eastern United States, so the average is not representative of the whole country. Most of the sites had lower sulfate concentration values in 1987 than in 1979. The total SO_2 emission rate was reduced from 24.8 to $20.9 \, Tg \, yr^{-1}$ during the 1979–1988 period. Emissions of NO_x were decreased from approximately 21 to $18 \, Tg \, yr^{-1}$.

20.5.3 Effects of Acid Deposition

Effects of acidic deposition include the following:

1. Acidification of surface waters (lakes, rivers, etc.) and subsequent damage to aquatic ecosystems
2. Damage to forests and vegetation
3. Damage of materials and structures

---

[1]One hectare (ha) is an area equal to $10^4 \, m^2$.

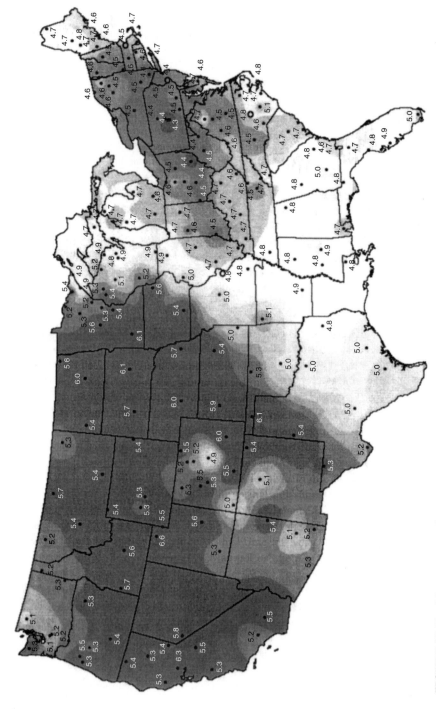

FIGURE 20.10 Measurements of average precipitation pH in the United States during 2004 (data from National Deposition Program 2005).

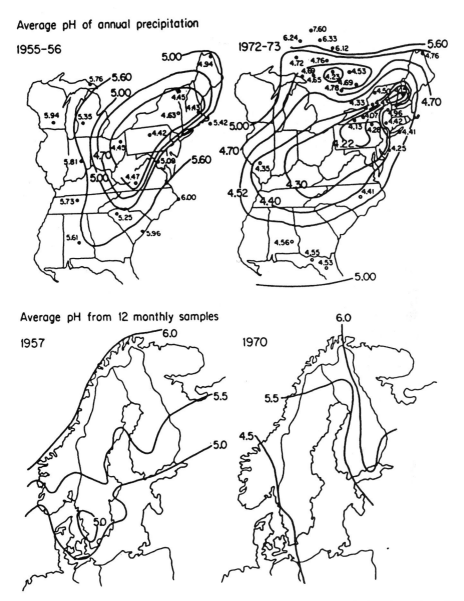

FIGURE 20.11 Historical trends in precipitation acidity in eastern North America and northern Europe (*Chemical and Engineering News*, Nov. 22, 1976).

Lake Acidification   Lakes acidify when they lose alkalinity (Roth et al. 1985). The total alkalinity or acid-neutralizing capacity is the reservoir of bases in solution. The acid-neutralizing capacity of a lake buffers it against large changes in pH. In natural clean waters, most of the acid-neutralizing capacity consists of bicarbonate ion, HCO_3^-. Carbonate alkalinity is defined by

$$Alkalinity = [HCO_3^-] + 2[CO_3^{2-}] + [OH^-] - [H^+]$$

When strong acids such as sulfuric acid enter bicarbonate waters, the additional acidity can be neutralized by reacting with the bicarbonate. In other terms, addition of acidity can be balanced by a shifting of the equilibrium,

$$H^+ + HCO_3^- \rightleftharpoons CO_2 \cdot H_2O$$

toward $CO_2 \cdot H_2O$. If input of hydrogen ion continues, the supply of buffering ions is eventually exhausted and the lake pH decreases. Watersheds vary in their ability to neutralize acidity, so some are more susceptible than others. Soils around the surface water help buffer the pH of the water. Many soils take up hydrogen ions from acidic deposition by cation exchange with Ca^{2+}, Mg^{2+}, Na^+, or K^+. Surface water acidification can thus be avoided if the supply of base cations is sufficient and if precipitation contacts soil long enough before flowing into surface waters.

The hydrology and chemistry of lakes and streams are highly individualistic. Lakes surrounded by poorly buffered soil and underlain by granitic bedrock appear to be more susceptible to acidification when exposed to acid rain (Havas et al. 1984). Other lake characteristics such as dominance of atmospheric input or surface/subsurface runoff as the major source of water, type and depth of soil, bedrock characteristics, lake size and depth, area of the drainage basin, and residence time of water in the lake are all features that influence the response of a lake to acid rain.

As the pH of a lake decreases, its chemistry changes and this may have profound effects on the biota inhabiting the lake. As pH decreases, concentrations of several potentially toxic metals, such as aluminum, iron, manganese, copper, nickel, zinc, lead, cadmium, and mercury increase (Dickson 1980; Havas et al. 1984). Abundant data show a significant correlation between increasing acidity and decreasing fish populations. Morphological deformities, spawning failure, and changes in blood chemistry have all been linked to acidification both in field observations and laboratory experiments (Havas et al. 1984). At first it was thought that acidity alone was killing fish, but it appears that aluminum dissolved from the soil by acid rain is primarily responsible. Aluminum ions irritate fish gills and cause the gills to produce a protective mucus. This initiates a process that erodes the gill filament until the fish suffocates. Understanding of the mechanisms of damage to other forms of aquatic life, from algae to amphibians, is incomplete compared to fish.

Quantification of the relationship between atmospheric acidity input and surface water acidification is difficult. First, certain lakes, known as *dystrophic lakes*, are naturally acidic. They usually have brown to yellow water caused by humic-derived organic acids. However, organic acids in these naturally acidic lakes reduce metal toxicity over that in clear water lakes (Baker and Schofield 1980). Second, acid deposition consists of both dry and wet deposition, and "wet" includes not only rain but also fog and cloud deposition. Because measurements of only the precipitation component of the atmospheric acid deposition input are widely available, measurements of acid deposition have been correlated with the acid rain component only. Moreover, often the only measurements available include the pH alone. Evidence derived from studies in the eastern United States and Canada indicate that damage has probably not occurred yet in areas receiving precipitation of pH greater than 4.7, or wet sulfate deposition of less than $14\,kg\,ha^{-1}\,yr^{-1}$. Damage is most probably occurring with wet sulfate loadings exceeding $20\,kg\,ha^{-1}\,yr^{-1}$, and almost certainly occurs for loadings over $30\,kg\,ha^{-1}\,yr^{-1}$. Lakes with alkalinities less than $200\,\mu equiv\,L^{-1}$ are sensitive to current levels of acid precipitation.

20.5.4   Cloudwater Deposition

Materials scavenged by cloud droplets (in-cloud scavenging) can be removed by wet deposition without rain formation if the cloud comes in contact with the Earth's surface. This wet deposition of intercepted cloudwater can be locally important, especially for slopes of mountains that regularly intercept clouds. In general, the concentrations of SO_4^{2-}, NO_3^-, H^+, and NH_4^+ in cloudwater are 5–10 times greater than in precipitation at the same sites in both the eastern and western United States. This process is relevant to high-altitude ecosystems and will be distinguished here from in situ fog formation, which can as easily occur at low altitudes. Areas most susceptible to wet deposition by cloud droplet interception and impaction are mountain summits above the average cloud base. The deposition of cloudwater will obviously be dependent on the height of the site. For example, the site at 1500 m in Whiteface Mountain spends roughly 4 times more time immersed in clouds than a location at the same mountain at 1000 m (Mohnen 1988). Land areas in the United States that are above the mean cloud base include the Appalachian Mountains and Rocky Mountains.

Most cloud droplets have diameters over 5 μm and their main removal mechanism for flat terrain at low wind speeds is gravitational settling. However, surfaces of interest are usually forest canopies on complex terrain with relatively high turbulence. Wet removal rate is therefore a function not only of the cloud droplet size distribution, but also the size, shape, and distribution of elements protruding into the airflow. While there are several empirical descriptions of particle deposition to vegetative canopies (Thorne et al. 1982; Grant 1983), the actual deposition process is complex. Turbulent transport over and among inhomogeneous surfaces, the driving force for cloudwater interception, is also poorly described.

Droplet deposition velocities of 2–5 cm s^{-1} have been measured on grass, while forests under orographic cloud may have droplet deposition velocities of 10–20 cm s^{-1} (U.S. NAPAP 1991). These values can be contrasted with the settling velocity of 10-μm-diameter droplet of 0.3 cm s^{-1}, of a 20 μm droplet of 1.2 cm s^{-1}, and of a 50 μm droplet of 7.5 cm s^{-1}. Volume mean diameters are on the order of 10 μm, indicating that turbulence has the potential to significantly increase the deposition velocity of cloud drops even over a flat surface in mountaintops characterized by high windspeeds and high roughness lengths.

Modeling studies (Lovett 1984; Lovett and Reiners 1986; Mueller 1990) have suggested that cloud droplet removal rate depends critically on characteristics of the forest canopy. Mueller (1990) calculated that cloudwater deposition at the edge of a forest was four to five times greater than in a closed forest. The cloud LWC and the size distribution of cloud droplets (especially the concentration of large droplets) also influence significantly the overall wet deposition rate.

A combination of measurements and simulations have indicated that cloudwater sulfate, acidity, nitrate, and ammonium deposition exceeds wet deposition via precipitation for some NE U.S. sites located above 400 m. While the current large uncertainties that are associated with cloudwater deposition models make firm estimates impossible, the available data indicate that cloudwater deposition is a substantial contributor to wet deposition inputs to high elevation sites in the eastern United States. Similar conclusions have been reached by European studies.

Cloud droplets are typically far more acidic than precipitation droplets collected at the ground. In essence, cloud drops are small and have not been subjected to the dilution associated with growth to the size of raindrops, snowflakes, and so on, nor the neutralization associated with the capture of surface-derived NH_3 and alkaline particles held in layers at lower altitudes. Interception of these droplets therefore provides a route by which concentrated solutions of sulfate and nitrate can be transferred to foliage in high-elevation areas that are exposed to clouds. Only limited areas of the eastern part of the United States are frequently exposed to such deposition, but for these sensitive areas cloud interception is an important acid deposition pathway.

20.5.5 Fogs and Wet Deposition

Fogs can be viewed as clouds that are in contact with the Earth's surface. Fogs are created during cooling of air next to the Earth's surface either by radiation to space (radiation fogs) or by contact with a surface (advection fogs) and in general can be distinguished from clouds that are just intercepted by a mountain slope. The distinction between certain types of clouds created as a result of flow of air up a mountain (orographic clouds and cap clouds) is a little more difficult.

Fogs can be created in heavily polluted, densely populated regions. Fogs were associated with one of the more severe air pollution episodes, the 1952 London "killer fog," during which approximately 4000 excess deaths were recorded. The fog was accompanied by very low inversion heights (as low as 50 m), very low visibility, and extremely high SO_2 and particulate concentration levels. The actual species that were responsible for the excess deaths have not been identified.

Urban fogs scavenge acidic particles and gases by the same mechanisms as clouds and are often characterized by low pH values. The most extreme event observed in southern California was a relatively light fog at Corona del Mar during which the fog pH reached a low of 1.7. The nitrate level was roughly 6 times the sulfate level in that particular fog (Jacob et al. 1985). Areas most influenced by such fogs in the United States include the western coast, as far south as Los Angeles, and northeastern coastal zones (Court and Gerston 1966).

Composition and pH of fogs are a function of time. At the beginning of a fog, the concentrations of its major constituents are high; as the fog develops, liquid water content rises, droplets are diluted, and acidity drops, and then as the air is heated, evaporation takes place, relative humidity decreases, and the pH is lowered again. While fogs have liquid water contents $(0.1–0.2 \text{ g m}^{-3})$ and droplet sizes ($\sim 10 \, \mu m$) similar to clouds, they are usually much more concentrated. This should be expected because they are in general created in areas of higher particulate and gaseous concentrations than those in clouds. Table 20.4 shows for comparison composition of fogs, clouds, and rain in southern California.

TABLE 20.4 Aqueous-Phase Concentrations (μequiv L^{-1}) of Clouds, Fog, and Rainwater in Pasadena, California

Species	Rain (Mean for 1984–1987)	Cloud	Fog (1981 Data)
pH	4.72	4.4	2.92–4.85
Na^+	19.8	120	320–500
K^+	1.1	6	33–53
Ca^{2+}	7.6	45	140–530
Mg^{2+}	5.1	30	90–360
NH_4^+	18.8	500	1290–2380
Cl^-	21.7	150	480–730
NO_3^-	23.8	500	1220–2350
SO_4^{2-}	18.7	150	480–950

Fogs appear to accelerate removal of the major aerosol species by 5–20 times compared to nonfoggy periods (Waldman and Hoffmann 1987). The contribution of fogs to overall wet deposition flux has been investigated in only a few areas in the United States; California is the most studied area. In this area, because of low annual precipitation, fogs are a major pathway for overall acid deposition.

20.6 ACID DEPOSITION PROCESS SYNTHESIS

Perhaps the single word that best characterizes the acid deposition phenomenon is "competition." The nature of acid deposition depends on competition between gas-phase and liquid-phase chemistry, competition between airborne transport and removal, and competition between dry and wet deposition. The key questions in acid deposition are related to identifying the essential processes involved and then understanding their interactions and quantifying their contributions. In this section we summarize our current understanding of the answers to these questions.

20.6.1 Chemical Species Involved in Acid Deposition

The gas/particulate/aqueous forms of sulfuric and nitric acid are the major contributors in polluted areas. Smaller contributions in these areas are made by hydrochloric acid and organic acids (mainly formic and acetic). These latter contributions have usually been neglected in most acid deposition studies, with the exception of remote areas. However, atmospheric bases, mainly NH_3 and Ca^{2+}, and other crustal elements play as important a role as do the acids, neutralizing to various extents the atmospheric acidity. The net acidity is the overall result of the acid and basic contributions. NAPAP suggested that changes in the emissions of these bases may often be more important than changes in SO_2 and NO_x emissions. For example, precipitation pH in North America remained constant during the 1980s despite reduction in sulfate and nitrate deposition. These reductions were apparently accompanied by reductions in Ca^{2+} emissions, resulting in a constant net acidity input to surface waters, vegetation, and so on. Emissions of bases are natural to a large extent and our understanding of their rates remains incomplete.

In characterizing acid deposition, much attention has focused on free acidity, expressed either as H^+ concentration or as pH. However, pH is often an inherently misleading measure of acid concentration because at low pH values substantial changes in H^+ concentrations are masked by only slight changes in pH, whereas at high pH values, slight changes in H^+ concentrations are exaggerated when expressed as pH. Expressing acid deposition in terms of pH would also diminish the apparent accomplishment of any emissions reduction program. Quoting Lodge, "If whatever control strategy is hit upon is successful in cutting the acidity in half, an evil conspiracy of the chemists will only allow the pH of precipitation to increase by 0.3" (Schwartz 1989).

20.6.2 Dry versus Wet Deposition

One can often be misled by the relative slow removal of material by dry deposition and rapid removal by wet deposition into underestimating the significance of the dry removal pathway that operates continuously.

Let us follow the simple calculation of Schwartz (1989) to compare the magnitude of the various processes. The average emission fluxes of SO_2 and NO_x in the northeastern United States are 130 mmol m^{-2} per year and 120 mmol m^{-2} per year, respectively. For comparison, total annual fluxes in Ohio are 360 mmol m^{-2} yr^{-1} and 210 mmol m^{-2} yr^{-1}. Dry deposition monitoring is extremely difficult as the fluxes determined with passive collectors (e.g., plates or coated surfaces) are often significantly different from fluxes on trees, lakes, or mountains. Let us for the purposes of this calculation assume a dry deposition velocity of SO_2 of 1 cm s^{-1} as a representative value and use available SO_2 concentration measurements to get a range of annual average SO_2 concentrations over the United States. Schwartz (1989) estimated an annual average value of SO_2 equal to 0.16–1.25 µmol m^{-3}, leading to dry deposition fluxes of 16–125 mmol m^{-2} yr^{-1}. Measured annual average sulfate wet deposition fluxes range from 10 to 40 mmol m^{-2} yr^{-1} in this region. Such estimates indicate that, on an annual basis, dry deposition of sulfur exceeds wet deposition in near-source regions (corresponding to high SO_2 concentrations) and is comparable to wet deposition farther from source regions. These conclusions are in agreement with the results of three-dimensional acid deposition models and the available deposition measurements (U.S. NAPAP 1991).

Note that impaction of cloud or fog droplets can lead locally to significant fluxes. Lovett et al. (1982) reported impaction fluxes at a high elevation forest of 280, 160, and 240 mmol m^{-2} yr^{-1} for sulfate, nitrate, and H$^+$, respectively.

20.6.3 Chemical Pathways for Sulfate and Nitrate Production

Atmospheric reactions modify the physical and chemical properties of emitted materials, changing removal rates and exerting a major influence on acid deposition rates. Sulfur dioxide can be converted to sulfate by reactions in gas, aerosol, and aqueous phases. As we noted in Chapter 17, the aqueous-phase pathway is estimated to be responsible for more than half of the ambient atmospheric sulfate concentrations, with the remainder produced by the gas-phase oxidation of SO_2 by OH (Walcek et al. 1990; Karamachandani and Venkatram 1992; Dennis et al. 1993; McHenry and Dennis 1994). These results are in agreement with box model calculations suggesting that gas-phase daytime SO_2 oxidation rates are $\sim$1–5% per hour, while a representative in-cloud oxidation rate is $\sim$10% per minute for 1 ppb of H_2O_2.

Fogs in polluted environments have the potential to increase aerosol concentrations by droplet-phase reactions but, at the same time, to cause reductions because of the rapid deposition of larger fog droplets compared to smaller particles (Pandis et al. 1990). Pandis et al. (1992) estimated that more than half of the sulfate in a typical aerosol air pollution episode was produced inside a fog layer the previous night.

The low amount of liquid water associated with particles (volume fraction 10^{-10}, compared to clouds, for which the volume fraction is on the order of 10^{-7}) precludes significant aqueous-phase conversion of SO_2 in such droplets. These particles can contribute to sulfate formation only for very high relative humidities (90% or higher) and in areas close to emissions of NH_3 or alkaline dust. Seasalt particles can also serve as the sites of limited sulfate production (Sievering et al. 1992), as they are buffered by the alkalinity of seawater. The rate of such a reaction as a result of the high pH of fresh seasalt particles is quite rapid, 60 µM min^{-1}, corresponding to 8% h^{-1} for the remote oceans (SO_2 = 0.05 ppb). Despite this initial high rate of the reaction, the extent of such production may be quite limited. For a seasalt concentration of 100 nmol m^{-3}, the alkalinity of seasalt is around 0.5 nmol m^{-3}. Consequently, after 0.25 nmol m^{-3} of SO$_2$ is taken up in solution and oxidized, the initial alkalinity would be exhausted and the reaction rapidly quenched as the pH would immediately decrease.

Snider and Vali (1994) reported studies of SO$_2$ oxidation in winter orographic clouds in which SO$_2$ was released and the increased concentrations of sulfate in cloudwater relative to the unperturbed cloud were compared to decreased concentrations of H$_2$O$_2$. Despite considerable scatter, the data fell fairly close to the one-to-one line, indicative of the expected stoichiometry of the reaction.

By analogy to the sulfur system, atmospheric nitrate sources can be distinguished into primary, gas phase, aqueous phase, and aerosol phase. Primary nitric acid emissions are considered to be small (U.S. EPA 1996) and can be neglected.

Gas-phase production of HNO$_3$ by reaction of OH with NO$_2$ is well established, and the only uncertainty in predicting the reaction rate is the concentration of OH. Note that the reaction of OH with NO$_2$ is approximately 10 times faster than the OH reaction with SO$_2$. The peak daytime conversion rate of NO$_2$ to HNO$_3$ in the gas phase is $\sim$10–50% per hour.

A second pathway for formation of nitrate is the reaction sequence (see Chapter 6)

$$NO_2 + O_3 \rightarrow NO_3 + O_2$$
$$NO_3 + NO_2 \rightleftharpoons N_2O_5$$
$$N_2O_5 + H_2O(aq) \rightarrow 2\,HNO_3(aq)$$

Note that this reaction pathway will operate only in the nighttime, as during the daytime NO$_3$ photolyzes rapidly. The NO$_3$ radical formed during the nighttime also reacts with a series of organic compounds, producing organic nitrates. Aqueous-phase production of nitrate by NO and NO$_2$ reactions is negligible under most conditions, including reactions of NO$_2$ with O$_2$, O$_3$, H$_2$O$_2$, and H$_2$O (see Chapter 7). This leaves the gas-phase NO$_2$ + OH reaction as the dominant daytime nitrate production pathway and the NO$_3$ heterogeneous reaction as the main nighttime production pathway.

Following the overview by Hales (1991), the chemical and removal processes for SO$_2$ and NO$_x$ are recapped below. SO$_2$ is emitted primarily from point sources. It is moderately soluble in water, and its solubility decreases with increasing acidity of the solution. SO$_2$ is not scavenged efficiently from fresh plumes, but this efficiency improves as the plumes age and dilute. It is essentially insoluble in ice and cold snow but can be scavenged by wet slushy snow and snow composed of graupel formed by rimming of supercooled cloud-water. Only a small fraction of the emitted SO$_2$ is removed as unreacted S(IV), mainly during the winter in the United States (significantly in the form of HMSA ions), and virtually none in the summer because of the high droplet acidity. Roughly one-third of the sulfur emitted annually in North America is believed to be removed by precipitation.

Point sources are a relatively small contributor of NO$_x$ emissions compared to SO$_2$, but still substantial. Both NO and NO$_2$ have low solubility in water. Virtually no NO$_x$ is removed from fresh plumes. HNO$_3$ formed by gas-phase oxidation of NO$_2$ is very soluble in water and the principal source of nitrate in precipitation. Since the secondary products are much more easily scavenged than NO$_x$, its scavenging increases with plume dilution and oxidation. Mesoscale studies show much variation in the efficiency of wet scavenging of SO$_x$ and NO$_x$, depending on the storm type and plume history. About one-third of the anthropogenic NO$_x$ emissions in the United States are estimated to be removed by wet deposition. The distinct seasonal character of sulfur wet deposition is absent in the case of nitrogen wet deposition for a series of reasons: HNO_3 has a strong affinity for ice as well as liquid water; its formation has no direct dependence on hydrogen peroxide, which peaks in summer; and nitrate can be formed in low winter sunlight.

20.6.4   Source–Receptor Relationships

In developing control strategies to deal with acid deposition, one seeks to establish relationships between emissions and depositional fluxes, so-called source-receptor relationships. For sulfur, for example, it is desired to relate deposition at site j, as $mmol\,m^{-2}$ per year, to the strength of source i, as measured, say, also in $mmol\,m^{-2}$ per year. Clearly, the amount of deposited sulfur will vary from day to day depending on the prevailing meteorological conditions, such as the windspeed and direction, the presence of clouds and precipitation, mixing state and chemical reactivity of the atmosphere, and so forth. Such source–receptor relationships on a day-to-day basis or averaging over a year can be derived only from mathematical models that include transport, chemical reactions, and removal processes. Sulfur and nitrogen oxides emitted into the atmosphere are necessarily returned to the surface of the Earth. For example, the entire northeastern United States (the region bounded by and including North Carolina and Tennessee on the south and the Mississippi River on the west) yields an average flux for sulfur of about $130\,mmol\,m^{-2}$ per year and for nitrogen of about $120\,mmol\,m^{-2}$ per year. A fraction of these emissions are transported out of the region mainly toward the Atlantic. For example, at Bermuda the 1980–1981 annual wet deposition of nonseasalt sulfate was $11\,mmol\,m^{-2}$ per year, and for nitrate it was $22\,mmol\,m^{-2}$ per year (Church et al. 1982). Estimates of the fraction of North American emissions exported off the continent by the prevailing winds are 25–35% for sulfur oxides and 15–25% for nitrogen oxides (Galloway and Whelpdale 1987). The remaining 65–75% of the emitted sulfur and 75–85% of nitrogen oxides are deposited in various chemical forms (primary and oxidized gases) and by a variety of deposition pathways on the NE United States.

A key measure of the source–receptor relation for acid deposition is the mean transport distance $\bar{x}$, the distance of travel between source and receptor averaged over all emitted molecules (Schwartz 1989). This quantity defines the region of influence of a specific source and is a measure of the decay of the deposition flux as one moves away from the source. One approach to getting a rough estimate for $\bar{x}$ is by multiplying the transport velocity with the mean residence time of the emitted material in the atmosphere. Climatologically averaged transport distances in the atmospheric mixed layer indicate a mean transport velocity of around 400 km per day. Atmospheric lifetimes of SO_2, NO_x, and their oxidation products are 1–3 days (Schwartz 1979; Levine and Schwartz 1982), suggesting mean transport distances of 400–1200 km.

If the receptor region is, for example, the Adirondack Mountains region of New York State, a possible specific question of the overall source–receptor problem would be which states contribute to acid deposition in the area. One could make the question even more specific, by asking what fraction of the sulfate deposition in the Adirondacks is emitted as SO_2 in the state of Ohio. The development of source–receptor relationships is a key policy question associated with acid deposition.

Development of reliable source–receptor relationships remains a challenging task. The magnitude of the task may be appreciated by envisioning North America, an area of about 2500×2500 km gridded in cells of 250×250 km. This results in some 100 cells, each of which is in principle both a source and a receptor of acidity. The source–receptor relation is the contribution of each source to acid deposition at each receptor. Therefore one needs to quantify 10,000 elements of the source–receptor matrix.

One approach for the determination of these relationships is to perform carefully planned field experiments. Possible experiments include the release of passive tracers; use of tracers of opportunity, such as transition metal ions; mass balance experiments; large-scale change of emissions; and releases of stable isotopes of sulfur and nitrogen. Unfortunately, each of these approaches has major problems starting with the shear magnitude of the undertaking (Hidy 1984). Passive tracers cannot mimic the acid deposition processes, nor can reactive tracers or tracers of opportunity. Mass balance experiments are subject to large uncertainties. Large-scale changes of emissions have technical constraints and are also quite costly. Use of stable isotopes requires large quantities of the isotopes, because similar isotopes are already emitted by the various sources. Use of radioactive isotopes would probably not be accepted by the public. All these difficulties leave us with one choice, atmospheric modeling.

Derivation of source–receptor relationships include statistical, Lagrangian, and Eulerian approaches (see Chapters 25 and 26). Despite significant progress during the last two decades, limitations of these models leave uncertainties of a factor of ≥ 2 in the most substantial property of source to receptor relationships—the mean transport distance (Schwartz 1989).

20.6.5 Linearity

The source–receptor relationships we just discussed, if available, tell us the fraction of acid deposition at a receptor that results from emissions of a particular source over a given averaging time. While this information is valuable, we would like to know something more in order to design emission control strategies. What we need to calculate is how much deposition of, say, sulfate at a receptor site will be reduced if SO_2 emissions by a certain source are reduced by a certain amount. Let us use as an example the estimate presented in the previous section. Assume that the utility SO_2 emissions in the Lower Ohio Valley are reduced by 50% (cut in half). This area as of about two decades ago (according to the RADM results) appeared to contribute on average $1.8 \, kg(S) \, ha^{-1} \, yr^{-1}$ to the sulfur deposition on the Adirondacks. What would be the contribution after the emission reductions?

One could argue that if the emissions are reduced by 50%, and because "Whatever goes up must eventually come down," that the sum of sulfate acid deposition (wet and dry) averaged over a year would also decrease by 50%. Such an answer implies that sulfate deposition varies linearly with changes in SO_2 emissions. We know enough about the processes and pathways connecting the SO_2 emissions with the sulfate deposition (Figure 20.12) to critically evaluate the validity of such an assumption.

Let us take a step back and synthesize what we know about the behavior of the system and its response to an SO_2 emission change. A reasonable assumption for our discussion is that this local change of emissions will not affect the meteorological component of the acid deposition process (windspeed and direction, mixing, cloud occurrence and pollutant processing, rainfall, etc.). This leaves us free to concentrate on the changes of the chemical component of acid deposition. Simplifying the problem this way, we can now focus on the two components of the acid deposition—the clean-air and the cloud-related pathways. The clean-air processes include emissions of SO_2, atmospheric transport, conversion to sulfate by reaction with the OH radical, and dry deposition of SO_2 and sulfate. If we follow

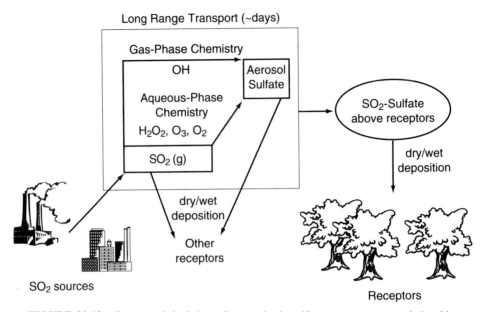

FIGURE 20.12   Conceptual depiction of atmospheric sulfur source–receptor relationship.

an air parcel moving from the source to the receptor, then all the clean-air process rates depend for all practical purposes linearly on the SO_2 concentration. For example, the gas-phase conversion rate of SO_2 to sulfate is

$$R_r = k[OH][SO_2]$$

The OH radical concentration is not sensitive to the levels of SO_2 present as it is determined by the available organic and NO_x concentrations present. Therefore a doubling of the SO_2 gas-phase concentration would lead to a doubling of the conversion rate. The same is true for the dry removal rate, which for a well-mixed box boundary layer is equal to

$$R_d = \frac{v_d}{H_m}[SO_2]$$

where v_d is the dry deposition velocity of SO_2 and H_m is the characteristic mixing height. Again, because the SO_2 concentration does not affect the meteorological conditions and because the deposition velocity and mixing height will remain constant even if SO_2 changes, a doubling of SO_2 concentrations will lead to a doubling of the dry removal rate. These arguments show that the clean air component of the sulfur deposition process is practically linear. This leaves us with the aqueous-phase chemistry component.

Let us first consider the conversion of SO_2 to sulfate in clouds by H_2O_2. As we saw, this reaction often dominates the overall process as its rate can exceed 10% SO_2 per minute. Let us assume a scenario where an air parcel contains 0.8 ppb of SO_2 and 0.5 ppb of H_2O_2. One would expect that, in this case, all the available H_2O_2 will react with 0.5 ppb SO_2, producing roughly $2\,\mu g\,m^{-3}$ of sulfate. Let us assume that the cloud pH is low enough that no other reaction is taking place with an appreciable rate. Then assume that SO_2 emissions increase by a factor of 2 and the SO_2 concentration entering the cloud doubles to 1.6 ppb.

Let us assume that this emission increase does not significantly affect the H_2O_2 concentration, which will remain close to 0.5 ppb. Again, the H_2O_2 will control the reaction and again roughly $2\,\mu g\,m^{-3}$ of sulfate will be produced. This example shows that if the sulfate production in the aqueous phase is controlled by the availability of oxidants, then the sulfate concentrations may respond nonlinearly to SO_2 emission and concentration changes. Our example here has oversimplified the situation by assuming that the conversion of SO_2 to sulfate by H_2O_2 is instantaneous, but it still illustrates the problems of the linearity assumption. Therefore we would expect nonlinear effects if the SO_2 conversion to sulfate is controlled by the availability of H_2O_2. This can be the case in areas very close to sources, where SO_2 concentrations are high, and during the winter, when H_2O_2 concentrations are lower. In these scenarios, there may not be adequate oxidants to permit all the sulfur dioxide to be transformed to sulfate. Note that in this case a 50% reduction in sulfur emissions will not result in a reduction of acid deposition by 50% but will have a lesser effect. In our simple example, a 50% reduction of emissions would lead to a reduction of the SO_2 concentration to 0.4 ppb, and the sulfate produced would be $1.6\,\mu g\,m^{-3}$, corresponding to a reduction by only 20%.

In addition to the chemical nonlinearities described above, there are other nonlinearities associated with other chemical processes. For example, one can argue that cloud droplets may become saturated with SO_2, and this can occur quite rapidly if the pH of the droplets is low. As a consequence, wet deposition of dissolved SO_2 may be nonlinearly related to emissions of areas with high SO_2.

The important policy question is, however, not whether there are reasons for nonlinear behavior of the system, but whether these processes dominate the overall behavior or whether their effects are small in comparison with the linear processes. We have already seen that the clean-air component is close to linear for all practical purposes. If gas-phase and aerosol-phase processes dominate compared to aqueous-phase chemistry, the system will be practically linear.

PROBLEMS

20.1$_A$ Calculate the aerosol scavenging coefficient Λ as a function of the diameter d_p of the scavenged particle over a size range of 0.01–0.1 μm diameter for drops of diameter $D_p = 5\,mm$ falling at their terminal velocity in air at 25°C.

20.2$_B$ Calculate the mean aerosol number scavenging coefficient for the same conditions as Problem 20.1, where the aerosol being scavenged has a lognormal size distribution with $\bar{d}_{pg} = 0.01\,\mu m$ and $1.0\,\mu m$, each case having $\sigma_g = 2.0$.

20.3$_B$ For a more accurate determination of the scavenging rate of a gas, we need to include in the calculation the full droplet size spectrum. Show that if the droplet size distribution is given by the function $N(D_p)$, the rate of removal of an irreversibly soluble gas is given by

$$W_{bc} = \int_0^\infty K_c \pi D_p^2 C_g N(D_p)\,dD_p$$

Using this, calculate the scavenging coefficient for the Marshall–Palmer distribution (17.108) (Marshall and Palmer 1948)

$$N(D_p) = 0.08 \exp(-41 D_p p_0^{-0.21})$$

for $p_0 = 1\,\text{mm}\,\text{h}^{-1}$ and $25\,\text{mm}\,\text{h}^{-1}$.

20.4$_B$ Prepare a plot analogous to Figure 20.3 for the scavenging of NH_3 for an ammonia mixing ratio of 10 ppb. Will any droplet get saturated in ammonia after falling 3 km?

20.5$_B$ Calculate the scavenging coefficient for below-cloud scavenging of NH_3 based on the Marshall–Palmer raindrop size distribution for rainfall intensities of $p_0 = 1, 5, 15$, and $25\,\text{mm}\,\text{h}^{-1}$. Assume a minimum raindrop diameter of 0.02 cm.

20.6$_B$ Consider a raindrop falling through a layer containing a uniform concentration of SO_2. Show how to calculate the rate of approach of the dissolved S(IV) to its equilibrium value as a function of time of fall. Apply your result to a drop with 5 mm diameter, $T = 25°C$, $\xi_{SO_2} = 10$ ppb to compute how long it takes for the drop to reach equilibrium with the surrounding gas phase. Repeat the calculation for H_2O_2 at a background mixing ratio of 1 ppb.

20.7$_C$ Assuming a rainfall with intensity $p_0 = 10\,\text{mm}\,\text{h}^{-1}$ and a monodisperse raindrop distribution with $D_p = 4\,\text{mm}$, calculate the mass scavenging coefficients for the model continental, marine, and urban distribution of Chapter 8.

20.8$_C$ Calculate the number scavenging efficiencies of the aerosol populations for the conditions of Problem 20.1. Compare with the mass scavenging efficiencies and discuss your results.

20.9$_A$ Calculate the effect on the pH of atmospheric liquid water of dissolved aerosol SO_4^{2-} and the liquid water content L, assuming that the mixing ratios of CO_2 and SO_2 are constant and equal to 350 ppm and 100 ppt, respectively. Ammonia is assumed to be absent. In particular, fill in the pH values in the table below:

	$L(\text{g m}^{-3})$		
SO_4^{2-} $(\mu\text{g m}^{-3})$	0.1	0.5	2.5
0.04			
0.2			
1.0			

20.10$_B$ The probability of an SO_2 molecule being transformed to sulfate and wet-deposited on land can be expressed as the product of the probabilities of the molecule being processed through precipitating air parcels over land and of being absorbed and deposited (Oppenheimer 1983a). Show that if t_{SO_2} is the regional mean lifetime of SO_2 and $f(t)$ is the probability of a dry period lasting a time of length t, that is, of a wet event occurring at time t, then the probability of an SO_2 molecule being processed through precipitation is equal to

$$\int_0^\infty f(t) e^{-t/t_{SO_2}}\, dt$$

Assuming that the frequency distribution of rain events is given by

$$f(t) = \frac{1}{T_D} e^{-t/T_D}$$

show that this probability is equal to $t_{SO_2}/(t_{SO_2} + T_D)$. If we identify t_{SO_2} with the characteristic time for SO_2 removal by dry deposition, we obtain an upper limit on t_{SO_2} because some SO_2 will be lost by gas-phase conversion into material that is not wet-deposited. Let us use $t_{SO_2} = 60\,h$. A value of T_D typical of the eastern United States is $T_D = 138\,h$. Thus we see that the probability of an emitted SO_2 molecule being processed through precipitation before dry deposition is about 0.3. Considering the region of interest to be the eastern United States and Canada, Oppenheimer (1983a) estimated the following sulfur emissions and wet deposition for the 1977–1979 period:

$$\text{Emissions} = 10.26 \times 10^9\,kg(S)yr^{-1}$$
$$\text{Wet deposition} = 2.25 \times 10^9\,kg(S)yr^{-1}$$

Calculate the probability of an SO_2 molecule being absorbed and deposited in this region. Discuss your result.

20.11$_B$ In order to evaluate potential emission control strategies for acid rain, one would like to understand the relationship of decreases in SO_2 emissions to changes in precipitation sulfate levels. A lower limit on the change in precipitation sulfate can be obtained by assuming that all sulfate formation occurs in the liquid phase under oxidant-limited circumstances. Assume that airborne SO_2 occurs in two streams, a part that is processed through precipitating air parcels during its airborne lifetime and a part that is not, labeled A and B, respectively. The part that is passed through precipitating air parcels, A, is then subdivided into two streams, a portion that is absorbed, oxidized, and deposited as sulfate and a part that is not oxidized, labeled A_1 and A_2, respectively. Assume that an emissions reduction is not reflected in A_1, but only in A_2 and B, until A_2 is exhausted. Thereafter, the emissions reduction reduces A_1 on a proportional basis. Show that the response of wet deposition to a fractional emissions reduction $X/(A + B)$ is

$$\delta = \frac{1}{\beta} \left(\frac{X}{A + B} - \frac{A_2}{A + B} \right)$$

where β is the probability of an SO_2 molecule being absorbed and deposited. [Note that if $\beta = 1$, $A_2 = 0$, and $\delta = X/(A + B)$.] Using the value of β determined in Problem 20.10, evaluate δ for a 50% reduction in emissions.

20.12$_C$ In this problem, following Oppenheimer (1983b), we wish to explore the scenario that the atmospheric path leading to precipitation sulfate occurs only by gas-phase oxidation of SO_2 to sulfate aerosol, followed by either incorporation of the aerosol in droplets or dry deposition. Let

$g_6(t) =$ probability density that an air parcel is subjected to a dry period of length t, followed by precipitation beginning in the interval $(t, t + dt)$

p_6 = probability that aerosol processed through a precipitating air parcel is scavenged and deposited

p_1, p_5 = probabilities that SO_2 or sulfate aerosol, respectively, are not lost from the regional boundary layer before wet or dry deposition

$f_2(t), f_7(t)$ = probabilities that SO_2 or sulfate aerosol, respectively, escape dry deposition for time t

$g_4(t)$ = probability density that an SO_2 molecule remains unoxidized for a period of length t, followed by oxidation in the interval $(t, t + dt)$

Show that the probability of an SO_2 molecule being oxidized to sulfate aerosol in the regional boundary layer before passing through a precipitating air parcel is

$$p_{ox} = p_1 \int_0^\infty g_4(t) f_2(t) dt$$

and the probability of sulfate aerosol being wet deposited is

$$p_{wd} = p_5 p_6 \int_0^\infty g_6(t) f_7(t) dt$$

and thus that the probability that an SO_2 molecule is both oxidized to sulfate and deposited is

$$p_{dep} = p_{ox} p_{wd}$$

Let us assume that $f_7(t) = e^{-t/\tau_7}$, with $\tau_7 = 280$ h. Based on $v_d = 0.1 \, \mathrm{cm \, s^{-1}}$ and a 1 km boundary layer, $p_1 = p_5 = 0.75$, $g_6(t) = \tau_6^{-1} e^{-t/\tau_6}$, with $\tau_6 = 138$ h. Also, we take $p_{dep} = 0.22$. Finally, values of p_6 ranging from $\frac{1}{3}$ to 1.0 have been assumed. Essentially, p_6 is very uncertain because the cloud microphysics is unknown. Using the above values show that

$$p_{wd} = 0.5 p_6$$

and consequently that $p_{ox} = 0.44 p_6^{-1}$. Discuss this result. Assuming that all SO_2 emission reductions reduce the unoxidized SO_2 fraction before any of the oxidized part is reduced, and using $p_6 = 0.5$, determine the percentage reduction in wet sulfate deposition resulting from a 50% reduction in SO_2 emissions.

20.13$_C$ To quantify source–receptor relationships, long-range transport models have been developed [e.g., see Gislason and Prahm (1983)]. Many of these models are of the so-called trajectory type in which the concentration changes in a parcel of air are computed as the parcel is advected by the wind field. In some of the trajectory models used in the SO_2/sulfate system, all transformations and removal processes are represented as first-order. (The use of trajectory models to describe long-range transport presumes that we can estimate the location of an air parcel as a fuction of time. It is clearly an approximation to assume that an air parcel maintains its integrity over a period of many hours to days.

Nevertheless, the concept of an integral air parcel has been useful in allowing the formulation of the trajectory model.) The first-order dry deposition rate constants for SO_2 and SO_4^{2-} for use in a trajectory model are $k_{SO_2}^d = v_{dSO_2}/H$ and $k_{SO_4}^d = v_{dSO_4}/H$, where v_d are the deposition velocities and H is the height of the vertical extent of the model. The height of the mixed layer H varies both diurnally and seasonally, although it is common for regional-scale, single-layer trajectory models to consider only the seasonal variation. Using the following parameter values, characteristic of noontime, summertime conditions in the northeastern United States (Samson and Small 1982):

$$H = 1500\,\text{m}$$
$$k_c = 0.03\,\text{h}^{-1}$$
$$v_{dSO_2} = 1\,\text{cm}\,\text{s}^{-1}$$
$$v_{dSO_4} = 0.4\,\text{cm}\,\text{s}^{-1}$$
$$\left.\begin{array}{l} k_{wSO_2} = 5 \times 10^4\, p_0/H \\[4pt] k_{wSO_4} = 2.32 \times 10^5\, p_0^{0.625}/H \end{array}\right\} \begin{array}{l} p_0 \text{ in mm h}^{-1} \\ H \text{ in mm} \end{array}$$

carry out a trajectory model simulation of long-range transport of the sulfur species in that area. Note that the model does not include cloud removal of SO_2. Start a trajectory at the Indiana/Ohio/Kentucky three-state intersection and run it on a line to Albany, New York, at a speed of $10\,\text{km}\,\text{h}^{-1}$. Continue the calculation for 96 h and record the complete sulfur material balance for the parcel, including airborne SO_2 and SO_4^{2-}, dry-deposited SO_2 and SO_4^{2-}, and wet-deposited sulfur. The SO_2 emission rates can be taken as those for 1980:

Ohio	$2602 \times 10^3\,\text{Mt}\,\text{yr}^{-1}$
Pennsylvania	$1971 \times 10^3\,\text{Mt}\,\text{yr}^{-1}$
New York	$901 \times 10^3\,\text{Mt}\,\text{yr}^{-1}$

Assume that three emissions are distributed uniformly over each state. The areas of the states are

Ohio	$115,750\,\text{km}^2$
Pennsylvania	$119,316\,\text{km}^2$
New York	$137,976\,\text{km}^2$

and the distance as the crow flies from the Indiana/Ohio/Kentucky intersection to Albany, New York, is 966 km. Assume that it rains during the last hour of each 24 h period at a rate of $p_0 = 10\,\text{mm}\,\text{h}^{-1}$. Calculate the total quantity SO_2 emitted along the trajectory, together with the quantity remaining at the end, the quantities lost by wet and dry deposition, and the quantity converted to sulfate. Perform the same material balance on sulfate. For the purpose of calculation assume the moving air parcel to have a base of $10^4\,\text{m}^2$.

20.14$_D$ Clouds act like giant pumps through which large quantities of air are processed. Aside from transporting boundary-layer air into the free troposphere, clouds act as filters and remove soluble gases and aerosols from that air. Although the physics of the interaction among cloud droplets, gases, and aerosols is very complex, insights can be gained by representing in-cloud processes as a flowthrough reactor in which boundary-layer air containing SO_2, HNO_3, NO_x, NH_3, O_3, H_2O_2, CO_2, and aerosol flows through the well-mixed region (reactor) representing the cloud (Hong and Carmichael 1983). We say that the cloud consists of cloud droplets of radius R_c and number concentration N_c. Gases are absorbed and aerosol particles are captured, and inside the droplets chemical reactions occur. Show that, subject to these assumptions, within the gas phase the concentration of a species i obeys

$$V\frac{dc_i}{dt} = q(c_i^0 - c_i) - Vk_{ni}c_i - k_{gi}V4\pi R_c^2 N_c(c_i - c_i^l/H_iRT)$$

where V is cloud volume, q is volumetric flow rate of air through the cloud, c_i^0 is the concentration of species i in the entrained air, k_{ni} is the first-order rate constant for gas-phase conversion of species i, k_{gi} is the mass transfer coefficient, and c_i^l is the liquid-phase concentration of species i. Then show that the liquid-phase concentration of species i is governed by

$$V_l\frac{dc_i^l}{dt} = k_{gi}V4\pi R_c^2 N_c(c_i - c_i^l/H_iRT) - q_lc_i^l + V_lR_i$$

where V_l is the liquid water volume, q_l is the volumetric flow rate of water out of the cloud by rainfall, and R_i is the rate of generation of species i by liquid-phase reactions. The microphysics of the cloud will be represented in terms of the liquid water content w_L, rainfall intensity p_0, updraft velocity w, cloud depth h_c, and cloud cross-sectional area A. We wish to use this model to investigate sulfate production in clouds. The conditions to be simulated are as follows (take $T = 273$ K):

Case	[SO$_2$] (ppb)	[O$_3$] (ppb)	[NH$_3$] (ppb)	[HNO$_3$] (ppb)	[H$_2$O$_2$] (ppb)	R_c (cm)	p_0 (mm h^{-1})	w (m s^{-1})	h_c (km)	w_L (cm^3 m^{-3})
1	50	100	1	10	10	0.003	1	2.5	5	0.3
2	50	100	1	10	50	0.003	1	2.5	5	0.3
3	50	100	1	10	50	0.003	10	2.5	5	0.3
4	50	100	1	10	50	0.03	1	2.5	5	0.3

The in-cloud S(IV) oxidation reactions to be considered are those involving O_3 and H_2O_2. The necessary equilibrium and kinetic data can be obtained from Chapter 7. The dynamic calculations should be done by numerically integrating the differential equations for c_i and c_i^l together with the relevant electroneutrality relation. The results of the computation should be presented in terms of the ratio of c_i at $t = 25$ min to c_i at $t = 0$. Discuss your results.

REFERENCES

Amar, P. et al. (1988) *The Fifth Annual Report to the Governor and the Legislature on the Air Resources Board's Acid Deposition Research and Monitoring Program*, State of California Air Resources Board, Research Division, Sacramento.

Baker, J. P., and Schofield, C. L. (1980) *Proc. Int. Conf. Ecological Impact of Acid Precipitation*, Sandefjord, Norway, eds., D. Drablos and A. Tollau, SNSF Project, Oslo. pp. 292–293.

Barrie, L. (1992) Scavenging ratios a useful tool or black magic, in *Precipitation Scavenging and Atmosphere–Surface Exchange Processes*, Hemisphere Publishing, Washington, DC.

Battan, L. J., and Reitan, C. H. (1957) Droplet size measurements in convective clouds, in *Artificial Stimulation of Rain*, Pergamon Press, New York, pp. 184–191.

Bird, R. B., Steward, W. E., and Lightfoot, E. N. (1960) *Transport Phenomena*, Wiley, New York.

Brimblecombe, P. (1977) London air pollution, 1500–1900, *Atmos. Environ.* **11**, 1157–1162.

Chamberlain, A. C. (1953) *Aspects of Travel and Deposition of Aerosols and Vapor Clouds*, AERE Harwell, Report R1261 HMSO, London.

Charlson, R. J., and Rodhe, H. (1982) Factors controlling the acidity of natural rainwater, *Nature* **295**, 683–685.

Church, T. M., Galloway, J. N., Jickels, T. D., and Knap, A. H. (1982) The chemistry of western Atlantic precipitation at the mid-Atlantic coast and on Bermuda, *J. Geophys. Res.* **87**, 1013–1018.

Court, A., and Gerston, R. D. (1966) Fog frequency in the United States, *Geogr. Rev.* **56**, 543–550.

Cowling, E. B. (1982) Acid precipitation in historical perspective, *Environ. Sci. Technol.* **16**, 110A–123A.

Dennis, R. L., McHenry, J. N., Barchet, W. R., Binkowski, F. S., and Byun, D. W. (1993) Correcting RADM's sulfate underprediction: discovery and correction of model errors and testing the corrections through comparisons against field data, *Atmos. Environ.* **27A**, 975–997.

Dickson, W. (1980) *Proc. Int. Conf. Ecological Impact of Acid Precipitation*, Sandefjord, Norway, D. Drablos and A. Tollau, eds., SNSF Project, Oslo. pp. 75–83.

Galloway, J. N., Likens, G. E., Keene, W. C., and Miller, J. M. (1982) The composition of precipitation in remote areas of the world, *J. Geophys. Res.* **89**, 1447–1458.

Galloway, J. N., and Whelpdale, D. M. (1987) WATOX-86 overview and western North Atlantic Ocean S and N atmospheric budgets, *Global Biogeochem. Cycles* **1**, 261–281.

Gislason, K. B., and Prahm, L. P. (1983) Sensitivity study of air trajectory long-range transport modeling, *Atmos. Environ.* **17**, 2463–2472.

Grant, R. H. (1983) The scaling of flows in vegetative structures, *Boundary-Layer Meteorol.* **27**, 171–184.

Gschwandtner, G., Gschwandtner, K. C., and Elridge, K. (1983) *Historic Emissions of Sulfur and Nitrogen Oxides in the United States from 1890 to 1980*, report on EPA contract 68-02-3311, prepared by Pacific Environmental Services, Inc., Durham, NC.

Hales, J. M. (1991) Atmospheric process research and process model development, in *Acidic Deposition: State of Science and Technology*, Vol. I: *Emissions, Atmospheric Processes, and Deposition*, U. S. National Acid Precipitation Program, Washington, DC.

Havas, M., Hutchinson, T. C., and Likens, G. E. (1984) Red herrings in acid rain research, *Environ. Sci. Technol.* **18**, 176A–186A.

Hidy, G. M. (1984) Source–receptor relationships for acid deposition: Pure and simple? *J. Air Pollut. Control Assoc.* **34**, 518–531.

Hong, M. S., and Carmichael, G. R. (1983) An investigation of sulfate production in clouds using a flow through chemical reactor model approach, *J. Geophys. Res.* **88**, 733–743.

Jacob, D. J., Waldman, J. M., Munger, J. W., and Hoffmann, M. R. (1985) Chemical composition of fogwater collected along the California coast, *Environ. Sci. Technol.* **19**, 730–736.

Karamachandani, P., and Venkatram, A. (1992) The role of non-precipitating clouds in producing ambient sulfate during summer: results from simulations with the Acid Deposition and Oxidant Model (ADOM), *Atmos. Environ.* **26A**, 1041–1052.

Keene, W. C., Galloway, J. N., and Holden, J. D. Jr. (1983) Measurements of weak organic acidity in precipitation from remote areas of the world, *J. Geophys. Res.* **88**, 5122–5130.

Levine, S. Z., and Schwartz, S. E. (1982) In-cloud and below-cloud scavenging of nitric acid vapor, *Atmos. Environ.* **16**, 1725–1734.

Likens, G. E., Wright, R. G., Galloway, J. N., and Butler, T. J. (1979) Acid rain, *Sci. Am.* **241**, 43–51.

Lovett, G. M., Reiners, W. A., and Olson, R. K. (1982) Cloud droplet deposition in subalpine balsam fir forests: Hydrological and chemical inputs, *Science* **218**, 1303–1304.

Lovett, G. M. (1984) Rates and mechanisms of cloud water deposition to a subalpine balsam fir forest, *Atmos. Environ.* **18**, 361–371.

Lovett, G. M., and Reiners, W. A. (1986) Canopy structure and cloud water deposition in subalpine coniferous forests, *Tellus* **38B**, 319–327.

Marshall, J. S., and Palmer, M. W. M. (1948) The distribution of raindrops with size, *J. Meteorol.* **5**, 165–166.

McHenry, J. N., and Dennis, R. L. (1994) The relative importance of oxidation pathways and clouds to atmospheric ambient sulfate production as predicted by the Regional Acid Deposition Model, *J. Appl. Meteorol.* **33**, 890–905.

Mohnen, V. A. (1988) *Mountain Cloud Chemistry Project: Wet, Dry, and Cloud Water Deposition*, Report to EPA Contract CR-813934-01-2, US EPA, Research Triangle Park, NC.

Mueller, S. F. (1990) Estimating cloud water deposition to subalpine spruce-fir forests, 1. Modification of an existing model, *Atmos. Environ.* **25**, 1093–1104.

Oppenheimer, M. (1983a) The relationship of sulfur emissions to sulfate in precipitation, *Atmos. Environ.* **17**, 451–460.

Oppenheimer, M. (1983b) The relationship of sulfur emissions to sulfate in precipitation—II. Gas phase processes, *Atmos. Environ.* **17**, 1489–1495.

Pandis, S. N., Seinfeld, J. H., and Pilinis, C. (1990) The smog–fog–smog cycle and acid deposition, *J. Geophys. Res.* **95**, 18489–18500.

Pandis, S. N., Seinfeld, J. H., and Pilinis, C. (1992) Heterogeneous sulfate production in an urban fog, *Atmos. Environ.* **26A**, 2509–2522.

Pruppacher, H. R., and Klett, J. D. (1997) *Microphysics of Clouds and Precipitation*, Kluwer, Dordrecht, The Netherlands.

Roth, R., Blanchard, C., Harte, J., Michaels, H., and El-Ashry, M. T. (1985) *The American West's Acid Rain Test*, World Resources Institute, Washington, DC.

Samson, P. J., and Small, M. J. (1982) The use of atmospheric trajectory models for diagnosing the sources of acid precipitation, American Chemical Society, 183rd National Meeting, Las Vegas.

Schaug, J., Hanssen J. E., Nodop K., Ottar B., and Pacyna J. M. (1987) *Co-operative Programme for Monitoring and Evaluation of the Long Range Transmission of Air Pollutants in Europe (EMEP)*, EMEP-CCC-Report 3/87, Norwegian Institute for Air Research, Lillestrom, Norway.

Schwartz, S. E. (1979) Residence times in reservoirs under non-steady-state conditions: Application to atmospheric SO_2 and aerosol sulfate, *Tellus* **31**, 520–547.

Schwartz, S. E. (1989) Acid deposition: Unraveling a regional phenomenon, *Science* **243**, 753–763.

Sievering, H., Boatman, J., Gorman, E., Kim, Y., Anderson, L., Ennis, G., Luria, M., and Pandis, S. N. (1992) Removal of sulfur from the marine boundary layer by ozone oxidation in sea-salt, *Nature* **360**, 571–573.

Slinn, W. G. N. (1983) Precipitation scavenging, in *Atmospheric Sciences and Power Production—1979*, Chapter 11, Division of Biomedical Environmental Research, U.S. Department of Energy, Washington, DC.

Snider, J. R., and Vali, G. (1994) Sulfur oxidation in winter orographic clouds, *J. Geophys. Res.* **99**, 18713–18733.

Stensland, G. J., and Semonin, R. G. (1982) Another interpretation of the pH trend in the United States, *Bull. Am. Meteorol. Soc.* **63**, 1277–1284.

Thorne, P. G., Lovett, G. M., and Reiners, G. M. (1982) Experimental determination of droplet impaction on canopy components of balsam fir, *J. Appl. Meteorol.* **21**, 1413–1416.

U.S. Environmental Protection Agency (1983) National air pollutant emission estimates, 1940–1967, EPA-450/4-88-022, Research Triangle Park, NC.

U.S. Environmental Protection Agency (1996) *Air Quality Criteria for Particulate Matter*, Vol. I, Office of Research and Development, EPA/600/P-95/001aF.

U.S. National Acid Precipitation Assessment Program (NAPAP) (1991) *The U.S. National Acid Precipitation Assessment Program*, Vol. I, U.S. Government Printing Office, Washington, DC.

Walcek, C. J., Stockwell, W. R., and Chang, J. S. (1990) Theoretical estimates of the dynamic, radiative and chemical effects of clouds on tropospheric gases, *Atmos. Res.* **25**, 53–69.

Waldman, J. M., and Hoffmann, M. R. (1987) Depositional aspects of pollutant behavior in fog and intercepted clouds, in *Sources and Fates of Aquatic Pollutants*, Advances in Chemistry Series, no. 216, R. A. Hites and S. J. Eisenreich, eds., Wiley, New York, pp. 79–129.

Whelpdale, D. M., and Miller, J. W. (1989) *GAW and Precipitation Chemistry Measurement Activities*, Fact Sheet 5, Background Air Pollution Monitoring Program, World Meteorological Organization, Geneva, Switzerland.

21 General Circulation of the Atmosphere

The fundamental process driving the global scale circulation of the Earth's atmosphere is the uneven heating of the Earth's surface by solar radiation. Although the total energy received by the Earth from the Sun is balanced by the total energy radiated back to space, this balance does not hold for every location on Earth. The tropics, for example, receive more energy from the Sun than is radiated back to space; the polar regions receive less energy than they emit.

Figure 21.1 shows the zonally annual averaged absorbed solar and emitted infrared fluxes, as observed from satellites. We note a net gain of radiative energy between about 40°N and 40°S, and a net loss of energy in the polar regions. This pattern results largely from the decrease in insolation to the polar regions in winter and from the high surface albedo in the polar regions. The outgoing infrared flux displays only a modest latitudinal dependence. As a result of the net gain of radiative energy in the tropics and the net loss in the polar regions, an equator-to-pole temperature gradient is generated.

The Earth's atmospheric circulation is one mechanism (about 60%) for redistributing energy from areas of the globe with an energy excess to those with an energy deficit; the Earth's ocean circulation is the other mechanism (about 40%). The circulation of the global atmosphere is one of the most complex, and theoretically deep, areas in fluid dynamics. It is the subject of several texts [e.g., Holton (1992), Salby (1996)]. To understand atmospheric flows at a truly fundamental level requires an advanced fluid mechanics background. The approach taken here is to describe the general nature of large-scale flow in the atmosphere at a level corresponding to a basic course in fluid mechanics.

Imagine, for a moment, that the Earth is not rotating and that the atmosphere can be represented as being confined to a huge rectangular enclosure that is heated at one end (tropics) and cooled at the other end (poles). Air at the warm end rises because of its decreased density, and air at the cool end falls because of its increased density. The buoyant rise of air along the heated end creates a decreased pressure at the surface, while the sinking of cold air at the other end leads to a higher pressure. This pressure difference creates a flow along the bottom of the enclosure from the cold end to the warm end. To conserve mass, rising air at the warm end must travel along the top of the enclosure to the cold end. The result is a single large circulation. Applied to the atmosphere, this single-cell model leads to a single circulation between the equator and the poles, in which air rises in the equatorial regions, spreads poleward at the tropopause, converges, and sinks in the polar regions, finally moving back toward the equator at the Earth's surface. Such a

Atmospheric Chemistry and Physics: From Air Pollution to Climate Change, Second Edition, by John H. Seinfeld and Spyros N. Pandis. Copyright © 2006 John Wiley & Sons, Inc.

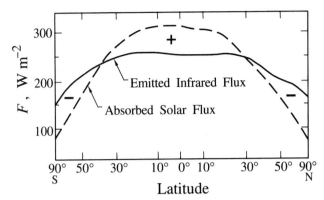

FIGURE 21.1   Zonally averaged components of the absorbed solar flux and emitted thermal infrared flux at the top of the atmosphere. The + and − signs denote energy gain and loss, respectively. (From *Radiation and Cloud Processes in the Atmosphere: Theory Observation and Modeling* by Kuo-Nan Liou. Copyright © 1992 by Oxford University Press, Inc. Used by permission of Oxford University Press, Inc.)

circulation, proposed in the early eighteenth century by the British meteorologist George Hadley (1685–1768), is called a *Hadley cell* (also the "tropical cell") (Hadley 1735). Although the Hadley cell captures the concept of rising and falling masses of air resulting from differential heating and cooling, it fails to describe the large-scale circulation of the atmosphere because it does not take into account the rotation of the Earth.

The fact that the Earth is rotating gives rise to another force, besides pressure gradients, that profoundly influences circulation of the atmosphere, the *Coriolis force*. Because of this force, air cannot flow in one unhindered cell from pole to equator and back. It turns out that a good approximation is that of a three-cell model (Figure 21.2). (There must be an odd number of cells so that air rising and falling in adjacent cells is oriented in the proper direction.) The cell in the equatorial region retains the name of the Hadley cell. The reverse cell in midlatitudes is called the *Ferrel cell* (also called the "midlatitude cell"). The third cell at high latitudes, called the *polar cell*, contains the sinking motion over the poles. The three-cell model, while highly oversimplified, is a reasonable qualitative description of the large-scale pattern of atmospheric circulation, especially that associated with surface flow.

21.1 HADLEY CELL

The Hadley cell in the Northern Hemisphere is characterized by ascent at the equator, northward movement along the tropopause, convergence, and descent at about 30°N and southward movement toward the equator along the surface. In the Southern Hemisphere, flow is southward along the tropopause, with convergence and descent at about 30°S, and northward movement toward the equator. Air over the equator is warm, and horizontal pressure gradients are weak. Little wind exists in this region, which has become known as the *doldrums*. The excess radiative heating in the tropics is compensated partially by evaporation, but even with evaporation, the air is still warmer than that to the north and south. The warm, moist air over the tropics is unstable; as soon as a small uplift occurs, the

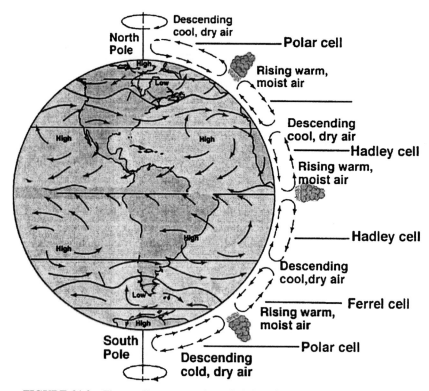

FIGURE 21.2   Three-cell representation of global circulation of the atmosphere.

air saturates with water vapor and continues to move upward. Large amounts of latent heat are released as towering cumulus clouds form, providing even more energy for the upward motion. At the tropical tropopause ($\sim$15 km), the atmosphere is stable and the upward motion is largely brought to a halt. Air that does not enter the stratosphere is deflected to the north and south.

The Hadley cell accomplishes a large fraction of global heat transport toward the higher latitudes. While the Hadley cell also transports latent heat, in the form of water vapor, from higher latitudes toward the equator, on net balance the transport of sensible heat (the warm air itself) poleward is larger. In the Northern Hemisphere, air converges as it moves northward and cools. At latitudes of about 30°N, cooling and convergence produce a large, heavy mass of air above the Earth's surface. This large, heavy mass of air sinks and produces a widespread, semipermanent high-pressure system known as a *subtropical high*. While sinking, the air warms by compression, producing clear skies, warm, dry climates, and weak surface winds.[1] These zones are where the great deserts of the world are found. At the surface, some of the descending air moves back toward the equator and is deflected to the right in the Northern Hemisphere as a result of the Coriolis force (which we will discuss shortly) as it moves. On both sides of the equator, there exists a wide region where

---

[1]In the days of sailing, winding up in these regions meant serious delays in the voyage. They were referred to as *horse latitudes*, because it is said that the horses that were carried on board had to be thrown off to lighten the load or to conserve drinking water.

winds blow from east to west (easterlies) with a slight equatorward tilt (see Figure 21.2).[2] This region is called the *trade wind belt*. Such winds were well known to early mariners, taking sailing ships from east to west.

Near the equator, northeasterly winds from the Northern Hemisphere converge with southeasterly winds from the Southern Hemisphere. This region is known as the *intertropical convergence zone* (ITCZ). The convergence aids the ascent of air resulting in a widespread area of low pressure known as the *equatorial low*. The predominant rising motion in the ITCZ leads to a band of extensive convective clouds and high rainfall. Major tropical rainforests are found in this zone.

21.2   FERRELL CELL AND POLAR CELL

At 30°N, the air descending to the surface that does not move south toward the equator in the Hadley cell begins to move north. As it does so, it is deflected toward the right by the Coriolis force. In these regions north and south of the trade wind belt (in the Northern and Southern Hemispheres, respectively), winds tend to blow from west to east (westerlies); these are referred to as *westerly wind belts*.

When the temperature gradient and upper level winds are sufficiently large, the westerly flow breaks down into large-scale eddies. As a result, day-to-day weather patterns do not exhibit the zonal symmetry of the large-scale climatological Ferrell cell. The breakdown into eddies is termed *baroclinic instability* (*baros* means "weight" and *clini* means "sloping"). The stronger the flow, the more likely it is to be unstable. Baroclinic eddies, the eddies resulting from baroclinic instability, are the source of midlatitude weather. They are characterized by trains of alternating low- and high-pressure systems moving eastward, which are the source of frontal systems that produce the clouds and storminess of the midlatitudes. Baroclinic eddies push warm air poleward and cold air southward, cooling the subtropics and warming the polar latitudes. Baroclinic eddy heat fluxes are so important that their description lies at the heart of any theory of the general circulation.

Figure 21.3 sketches four conditions of midlatitude flow in the Northern Hemisphere. Panel (a) is the typical westerly flow under baroclinically stable conditions, when the energy imbalance between tropics and poles is not excessively large. Panel (a) shows a distinct band of relatively strong winds (usually exceeding $30\,\mathrm{m\,s^{-1}}$) called *jetstreams*. They are located above areas of especially strong temperature gradients. Since westerly flow parallel to latitude circles does not transport much energy poleward, the equator to pole temperature gradient will continue to increase as long as the atmosphere is in this state. When a critical north–south temperature gradient is reached, meanders begin to form in the jetstream and the atmosphere becomes baroclinically unstable [panel (b)]. Strong waves form in the upper airflow [panel (c)] and relative highs and lows are formed along each latitude band. Finally, the flow returns to a pattern of flatter flow aloft [panel (d)]. In short, baroclinic instability arises when inevitable small perturbations in the velocities grow rather than decay.

The overall direction of winds between 30° and 60° is from the west. Above ~500 mbar, these winds actually flow in wavelike patterns, with troughs and ridges. Troughs occur where the flow dips equatorward, and ridges occur where the flow moves poleward [panels (b) and (c) in Figure 21.3]. The flow in these upper-level waves leads to storms that move warm air poleward and cold air equatorward. The Northern Hemisphere is typically

---

[2] Recall that winds are identified by the direction they are coming from, not heading to.

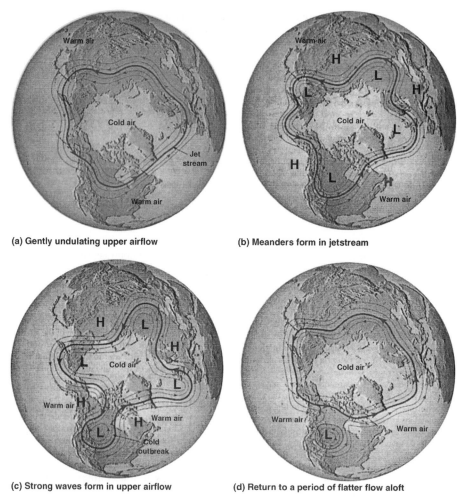

(a) Gently undulating upper airflow

(b) Meanders form in jetstream

(c) Strong waves form in upper airflow

(d) Return to a period of flatter flow aloft

FIGURE 21.3 Baroclinic stability and instability: (a) gently undulating upper airflow; (b) meanders form in jetstream; (c) strong waves form in upper airflow; (d) return to a period of flatter flow aloft.

encircled by several of these waves at any given time. These long-wavelength waves, which drift slowly eastward, are referred to as *Rossby waves*, after Carl-Gustav Rossby, the Swedish meteorologist who first identified them. When the waves have large amplitude with deep troughs and peaked ridges, cold air flows equatorward and warm air flows poleward.

The high-latitude end of the Ferrel cell is characterized by rising warm, moist air. Some of the rising air, on reaching the tropopause, moves back southward toward the region of convergence above 30°N. This circulation, northward along the surface, upward at the polar front, back to the south along the tropopause, and sinking toward the surface at the horse latitudes, constitutes the Ferrel cell. The portion of the air aloft that does not move equatorward moves toward the poles where it converges and sinks. This circulation, which results in high pressure at the pole (the *polar high*), is known as the *polar cell*; it is the weakest of the three cells (Figure 21.2). As the cold air sinks over the poles, it must flow equatorward at the surface. The Coriolis force turns these winds toward the west, forming the *polar easterlies* at 60° latitude and above.

Mild air moving north and east along the Earth's surface in the midlatitude Northern Hemisphere encounters much cooler air moving south and west in the polar easterlies. These airmasses of differing temperature and density result in the formation of the *polar front* along the boundary where they meet. The polar front is a zone of strong convergence and rapidly rising air that results in a widespread surface low-pressure system known as the *subpolar low*. This region of rapid, widespread ascent produces conditions conducive to storm formation. Polar air, which is cooler and more dense than the air moving up from the south, can push the location of the polar front further south, especially in the wintertime, producing a cold polar outbreak.

21.3 CORIOLIS FORCE

Because the Earth is a rotating sphere, points on the Earth's surface move at different speeds depending on their latitude. Consider a point completing a circle of radius R in time t. Its tangential speed is $v = 2\pi R/t$. The angular velocity (Ω) in radians per second, is $2\pi/t$, since 2π radians is the angle covered in t seconds. Thus, $v = \Omega R$. The angular velocity of the Earth is $2\pi/(24 \times 60 \times 60) = 7.27 \times 10^{-5}$ rad s^{-1}. Since the Earth's radius at the equator is about 6370 km, the tangential speed of a point on the surface at the equator is $v = 463$ m s^{-1}. At latitude ϕ, the radius of the circle described by a point on the surface is $R \cos \phi$, so that the tangential speed is

$$v = \Omega R \cos \phi \qquad (21.1)$$

which is just $\cos \phi$ times the speed at the equator. At latitude 30°, for example, the tangential speed is 401 m s^{-1}. Thus, at 30°N, a parcel of upper-level air transported from the equator has a "surplus" momentum corresponding to a velocity of 62 m s^{-1}. In a simplistic view, this surplus momentum forms the westerly subtropical jetstream, the average velocity of which is about 60 m s^{-1}. The difference in tangential speeds between points on the Earth's surface at different latitudes gives rise to the *Coriolis force*, named for the French engineer who first described it.

Consider in Figure 21.4 a parcel of air moving northward from point X to point Y. As the parcel heads north toward Y, it carries with it the eastward momentum from the Earth's rotation at X. Point X has a greater tangential speed than Y. As a result, as the parcel moves slowly northward, it appears to an observer on the Earth to gain speed toward the east. The combination of the northward and eastward velocities results in a bending path that deflects to the right of its destination. Now imagine the air parcel moving southward from point Y to X. Since the parcel has the lower tangential velocity of point Y as it heads southward, it appears to an observer on the surface to lag behind the spinning Earth, and its path also deflects to the right of its direction of travel. We conclude that the Coriolis force deflects moving air to the right (in its direction of travel) in the Northern Hemisphere. If point Y is in the Southern Hemisphere, the air parcel, in going from X to Y, deflects toward the left. Air, traveling from Y to X, also deflects to the left. Thus, in the Southern Hemisphere, the Coriolis force deflects moving air to the left (in its direction of travel).

The magnitude of the Coriolis force per unit mass of air is proportional to the distance from the equator and the windspeed. This force is expressed as fv, where f is the Coriolis

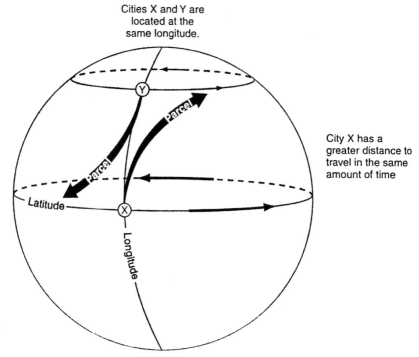

FIGURE 21.4 Illustration of the effect of Coriolis force on air parcels in the Northern Hemisphere (Ackerman and Knox 2003).

parameter:

$$f = 2\Omega \sin \phi \tag{21.2}$$

The Coriolis parameter $f > 0$ in the Northern Hemisphere ($\phi > 0$) and $f < 0$ in the Southern Hemisphere ($\phi < 0$). Note that the Coriolis parameter f equals zero at the equator and increases in magnitude as $\sin \phi$ toward the poles; thus, winds at higher latitudes are more strongly affected by the Coriolis force.

The momentum of an object moving in a straight line is equal to the product of its mass and its velocity. A rotating body has angular momentum, defined as the product of its mass, its rotation velocity, and the perpendicular distance from the axis of rotation. If undisturbed, a rotating object conserves its angular momentum. Thus, if the distance from the axis of rotation decreases, then its velocity must accordingly increase. (This is why a figure skater spins faster as his or her arms are pulled closer to the body.) As an air parcel moves north or south from the equator, its distance from the Earth's axis of rotation decreases, and therefore its angular velocity must increase to conserve angular momentum.

As air flows northward in the upper level of the Hadley cell, the Coriolis force turns it to the right (eastward). The magnitude of the Coriolis force increases as the air flows toward the poles, and by the time the air reaches about 30°N latitude, the Coriolis force has turned the flow entirely from west to east (see Figure 21.2). To conserve angular momentum, the speed of the wind also increases as the air moves toward the pole.

Order of Magnitude of Atmospheric Quantities   It is useful to estimate the order of magnitude of quantities in the equations of motion on the synoptic scale (systems having a spatial scale of ~ 1000 km). Typical orders of magnitude of quantities are as follows (McIlveen 1992):

Horizontal length scale	L	1000 km	10^6 m
Vertical length scale	H	10 km	10^4 m
Time scale	t	1 day	10^5 s
Horizontal pressure	Δp	10 mbar	10^3 Pa
Vertical pressure change	P	1000 mbar	10^5 Pa
Air density	ρ		$1\,\mathrm{kg\,m}^{-3}$
Earth's angular velocity	Ω		$10^{-4}\,\mathrm{rad\,s}^{-2}$
Gravitational constant	g		$10\,\mathrm{m\,s}^{-2}$

L is the horizontal distance over which there is a substantial change in the pressure or wind field. Δp is a typical horizontal pressure gradient for synoptic weather systems. The vertical pressure change is just the hydrostatic head of air from the surface to the tropopause. (In the spirit of order of magnitude, we estimate this as 1000 mbar rather than the accurate 900 mbar.) The time scale of 1 day is set by the Earth's rate of rotation. Moreover, the observed timescale of synoptic scale weather systems is about one day; such systems take days rather than hours to form and dissipate.

From the orders of magnitude above, we can estimate the following characteristic values:

Horizontal windspeed	$u \sim L/t$	$10\ \mathrm{m\,s}^{-1}$
Vertical windspeed	$w \sim H/t$	$10^{-1}\ \mathrm{m\,s}^{-1}$
Horizontal acceleration	$u/t(\text{or } u^2/L)$	$10^{-4}\ \mathrm{m\,s}^{-2}$
Vertical acceleration	$w/t(\text{or } w^2/H)$	$10^{-6}\ \mathrm{m\,s}^{-2}$
Coriolis acceleration	Ωu	$10^{-3}\ \mathrm{m\,s}^{-2}$
Horizontal pressure gradient	$\Delta p/L$	$10^{-3}\ \mathrm{Pa\,m}^{-1}$

21.4   GEOSTROPHIC WINDSPEED

At an altitude sufficiently far above the Earth's surface (about 500 m) where the effect of the surface can be neglected, the airflow is influenced only by horizontal pressure gradients and the Coriolis force. This layer is called the *geostrophic layer*, and the windspeed attained is the *geostrophic windspeed*. In the geostrophic layer the flow may be assumed to be inviscid (molecular viscosity effects are negligible). We can compute the geostrophic windspeed and direction at any latitude as a function of the prevailing pressure gradient from the equations of continuity and motion.

It can be shown that the acceleration experienced by an object on the surface of the Earth (or in the atmosphere) moving with a velocity vector **u** consists of two components, $-\Omega \times (\Omega \times \mathbf{r})$ and $-2(\Omega \times \mathbf{u})$, where Ω is the angular rotation vector for the Earth and **r** is the radius vector from the center of the Earth to the point in question. The first term is simply the centrifugal force, in a direction that acts normal to the Earth's surface and is counterbalanced by gravity. The second term, $\Omega \times \mathbf{u}$, is the Coriolis force. This force arises only when an object, such as an air parcel, is moving; that is, $\mathbf{u} \neq 0$. Even though the Coriolis force is of much smaller magnitude than the centrifugal force, only the Coriolis force has a horizontal component. Since the winds are horizontal in the geostrophic layer, the Coriolis acceleration is given by the horizontal component of the Coriolis term, namely, $2u_g \Omega \sin \phi$. The direction of the Coriolis force is perpendicular to the wind velocity. Windspeed u_g at latitude ϕ lies in the horizontal plane.

In the geostrophic layer it may be assumed that the atmosphere is inviscid (frictionless). The equations of continuity and motion for such a fluid are

$$\frac{\partial u}{\partial x} + \frac{\partial v}{\partial y} + \frac{\partial w}{\partial z} = 0 \tag{21.3}$$

and

$$\frac{\partial u}{\partial t} + u\frac{\partial u}{\partial x} + v\frac{\partial u}{\partial y} + w\frac{\partial u}{\partial z} = -\frac{1}{\rho}\frac{\partial p}{\partial x} + F_{cx}$$

$$\frac{\partial v}{\partial t} + u\frac{\partial v}{\partial x} + v\frac{\partial v}{\partial y} + w\frac{\partial v}{\partial z} = -\frac{1}{\rho}\frac{\partial p}{\partial y} + F_{cy} \tag{21.4}$$

$$\frac{\partial w}{\partial t} + u\frac{\partial w}{\partial x} + v\frac{\partial w}{\partial y} + w\frac{\partial w}{\partial z} = -\frac{1}{\rho}\frac{\partial p}{\partial z} + F_{cz}$$

where u, v, and w are the three components of the velocity and F_{cx}, F_{cy}, and F_{cz} are the three components of the Coriolis force.

Let the axes be fixed in the Earth, with the x axis horizontal and extending to the east, the y axis horizontal and extending to the north, and the z axis normal to the Earth's surface. As before, Ω is the angular velocity of rotation of the Earth and ϕ is the latitude. The components of the Coriolis force in the x, y, and z directions on an air parcel are the components of $\mathbf{F}_c = -2(\Omega \times \mathbf{u})$

$$F_{cx} = -2\Omega(w \cos \phi - v \sin \phi)$$

$$F_{cy} = -2\Omega u \sin \phi \tag{21.5}$$

$$F_{cz} = 2\,\Omega u \cos \phi$$

In the geostrophic layer, the vertical velocity component w can usually be neglected relative to the horizontal components u and v. Therefore, substituting (21.5) into (21.4), we obtain, for steady motion

$$u\frac{\partial u}{\partial x} + v\frac{\partial u}{\partial y} = 2\Omega v \sin \phi - \frac{1}{\rho}\frac{\partial p}{\partial x}$$

$$u\frac{\partial v}{\partial x} + v\frac{\partial v}{\partial y} = -2\Omega u \sin \phi - \frac{1}{\rho}\frac{\partial p}{\partial y} \tag{21.6}$$

We see that the air moves so that a balance is achieved between the pressure gradient and the Coriolis force. Let us consider the situation in which the velocity vector is oriented in the x direction, and so $v = 0$; then

$$u\frac{\partial u}{\partial x} = -\frac{1}{\rho}\frac{\partial p}{\partial x} \tag{21.7}$$

$$0 = -2\Omega u \sin \phi - \frac{1}{\rho}\frac{\partial p}{\partial y} \tag{21.8}$$

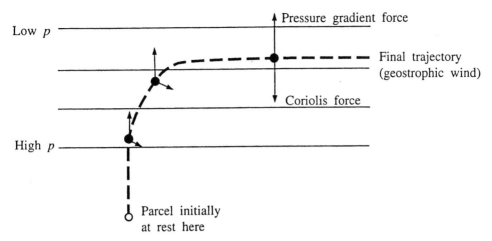

FIGURE 21.5 Approach to geostrophic equilibrium.

From the continuity equation (21.3), we see that $\partial u/\partial x = 0$, since $v = w = 0$. Thus from (21.7) $\partial p/\partial x = 0$, and we are left with only (21.8). From (21.8), we see that the component of the Coriolis force, $-fu$, is exactly balanced by the pressure gradient, $(1/\rho)\,\partial p/\partial y$, and the direction of flow is perpendicular to the pressure gradient $\partial p/\partial y$. Therefore the *geostrophic windspeed* u_g is given by

$$u_g = -\frac{\partial p/\partial y}{2\rho\Omega\sin\phi} = -\frac{1}{\rho f}\frac{\partial p}{\partial y} \qquad (21.9)$$

The approach to the geostrophic equilibrium for an air parcel starting from rest, accelerated by the pressure gradient and then affected by the Coriolis force, is shown in Figure 21.5.

21.4.1 Buys Ballot's Law

In the geostrophic equilibrium, the pressure gradient must be parallel to the y axis, with pressure decreasing in the positive y direction. The fact that, for a flow in the easterly direction, the isobars (lines of constant pressure) must themselves lie east–west was noted empirically by the nineteenth-century Dutch meteorologist Buys Ballot, and is called *Buys Ballot's law*. The law states that "Low pressure in the Northern Hemisphere is on the left when facing downwind" (on the right in the Southern Hemisphere). Because the pressure gradient is directly proportional to the geostrophic windspeed, the spacing of the isobars must decrease as the windspeed increases and vice versa. Instead of flowing in the direction of decreasing pressure gradient, the geostrophic wind flows perpendicular to the pressure gradient.

For an area of low pressure in the Northern Hemisphere that is completely surrounded by areas of high pressure, the isobars are concentric circles with a low pressure at the center. The pressure gradient force causes the air to flow toward the center of this low-pressure system, but because of the Coriolis force the air is forced to deflect to the right. As

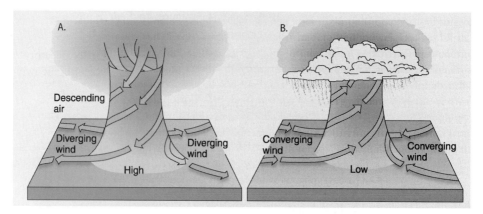

FIGURE 21.6   Northern Hemisphere cyclones and anticyclones.

a result, the Northern Hemisphere *low* or *cyclone* is characterized by a counterclockwise flow that spirals inward with rising air at the center [see Figure 21.6 as well as panel (c) of Figure 21.3]. In the case of an area of high pressure completely surrounded by areas of low pressure, the flow outward from the center will also be deflected to the right. This *high* or *anticyclone* exhibits a clockwise flow that spirals outward with descending air at its center (Figure 21.6). In the Southern Hemisphere, while the pressure gradient force acts in the direction from high pressure to low pressure, the Coriolis force now causes the deflection to be directed to the left of the path of motion. Thus a cyclone in the Southern Hemisphere is characterized by a clockwise flow of air that spirals inward with rising air at the center, while an anticyclone exhibits a counterclockwise flow that spirals outward with descending air in the center. Note that the only difference between the two hemispheres is the direction of rotation.

21.4.2   Ekman Spiral

The geostrophic balance determines the wind direction at altitudes above about 500 m. In order to describe the air motions at lower levels, we must take into account the friction of the Earth's surface. The presence of the surface induces a shear in the wind profile, as in a turbulent boundary layer over a flat plate generated in a laboratory wind tunnel. In analyzing the geostrophic windspeed, we found that for steady flow a balance exists between the pressure force and the Coriolis force. Consequently, steady flow of air at levels near the ground leads to a balance of three forces: pressure force, Coriolis force, and friction force due to the Earth's surface. Thus, as shown in Figure 21.7, the net result of these three forces must be zero for a nonaccelerating air parcel. Since the pressure gradient force F_p must be directed from high to low pressure, and the frictional force F_f must be directed opposite to the velocity u, a balance can be achieved only if the wind is directed at some angle toward the region of low pressure. This angle between the wind direction and the isobars increases as the ground is approached since the frictional force increases. At the ground, over open terrain, the angle of the wind to the isobars is usually between $10°$ and $20°$. Because of the relatively smooth boundary existing over this type of terrain, the windspeed at a 10 m height (the height at which the so-called surface wind is usually measured) is already almost 90% of the geostrophic windspeed. Over built-up areas, on the other hand, the speed at a 10 m height may be only 50% of the geostrophic windspeed,

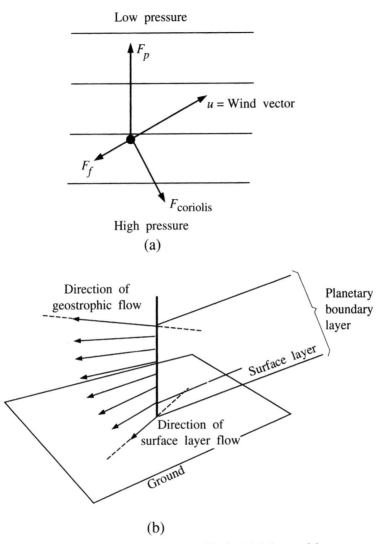

FIGURE 21.7 Variation of wind direction with altitude: (a) balance of forces among pressure gradient, Coriolis force, and friction; (b) the Ekman spiral.

owing to the mixing induced by the surface roughness. In this case the surface wind may be at an angle of 45° to the isobars.

As a result of these frictional effects, the wind direction commonly turns with height, as shown in Figure 21.7. The variation of wind direction with altitude is known as the *Ekman spiral*. Derivation of the expression for the Ekman spiral is the subject of Problem 21.2.

Horizontal and Vertical Flows   The relationship between horizontal and vertical velocities is embodied in the mass continuity equation

$$\frac{\partial \rho}{\partial t} + \frac{\partial}{\partial x}(\rho u) + \frac{\partial}{\partial y}(\rho v) + \frac{\partial}{\partial z}(\rho w) = 0 \qquad (21.10)$$

Atmospheric flows are such that the horizontal mass flux divergence, the second and third terms on the LHS of (21.10) are essentially balanced by the fourth term, the vertical mass flux divergence. Thus

$$\frac{\partial}{\partial x}(\rho u) + \frac{\partial}{\partial y}(\rho v) + \frac{\partial}{\partial z}(\rho w) \cong 0 \tag{21.11}$$

When flow converges at lower levels, it flows vertically upward in such a way so as to satisfy (21.11). If we consider a column of air and integrate the last term on the LHS of (21.11) from the bottom (b) of the column to its top (t), we get

$$\int_b^t \frac{\partial}{\partial z}(\rho w)dz = \rho w \big|_b^t = (\rho w)_t \tag{21.12}$$

since there is no mass flux of air into the Earth's surface. Thus, from (21.11), we have

$$w_t = -\frac{1}{\rho_t} \int_b^t \left[\frac{\partial}{\partial x}(\rho u) + \frac{\partial}{\partial y}(\rho v) \right] dz \tag{21.13}$$

Atmospheric measurements in a variety of circumstances show that the RHS of (21.13) generally has a magnitude of order $10\,\mathrm{cm\,s^{-1}}$ in the midtroposphere. (Updraft rates in cumulonimbus clouds can, however, reach values of $\sim 5\,\mathrm{m\,s^{-1}}$.)

A simple depiction of a tropospheric air column with surface-level convergent flow is given in Figure 21.8, wherein horizontal convergence at the surface is matched by horizontal divergence at upper levels. The situation sketched in Figure 21.8 especially represents that in cloudy weather systems.

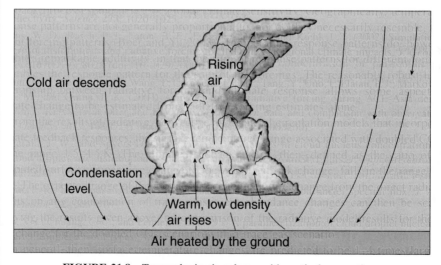

FIGURE 21.8   Tropospheric air column with vertical convection.

21.5   THE THERMAL WIND RELATION[3]

We have seen that the horizontal wind components in the atmosphere obey the geostrophic equations:

$$u_g = -\frac{1}{\rho f}\frac{\partial p}{\partial y} \qquad v_g = \frac{1}{\rho f}\frac{\partial p}{\partial x} \tag{21.14}$$

A basic question is how the geostrophic windspeed varies with height, the so-called vertical wind shear. If one can obtain an expression for the gradients of u_g and v_g with altitude, this relation is the key to understanding how the flow in the atmosphere behaves at all levels. The two fundamental relations that govern the large-scale flow in the atmosphere are the geostrophic and hydrostatic balances, and these two relations can be combined to produce the desired relation for the vertical gradient of the windspeed.

We desire the expressions for the variation of u_g and v_g with height; it turns out to be easier to use pressure as the vertical coordinate rather than height. So we seek expressions for $-\partial u_g/\partial p$ and $-\partial v_g/\partial p$, where the minus sign allows the derivative with respect to the vertical coordinate to be taken upward (since pressure decreases with altitude). The quantities

$$\frac{\partial u_g}{\partial p} \qquad \text{and} \qquad \frac{\partial v_g}{\partial p}$$

are called the *thermal wind*, and the relation we will obtain is called the *thermal wind relation*. We note, despite the name, that these quantities are vertical gradients of the windspeed, not windspeeds.

The hydrostatic relation is usually expressed as the gradient of pressure with respect to height,

$$\frac{\partial p}{\partial z} = -g\rho \tag{21.15}$$

but it can be written with pressure as the coordinate and z as the dependent variable by simply inverting:

$$\frac{\partial z}{\partial p} = -\frac{1}{g\rho} \tag{21.16}$$

We can substitute for ρ using the ideal-gas law ($\rho = p/RT$), and we get

$$-\frac{\partial z}{\partial p} = \frac{RT}{pg} \tag{21.17}$$

This equation can be nicely interpreted in finite difference form. Consider two pressure surfaces, $p_0 + \Delta p/2$ and $p_0 - \Delta p/2$. The two pressure surfaces, at any location, must

---

[3]This section presents somewhat more advanced material.

occur at different heights, with a separation Δz. This distance between the two pressure surfaces is called its *thickness*. Thus

$$\Delta z \cong -\frac{\Delta \rho}{\rho} \frac{RT}{g} \tag{21.18}$$

Then, combining (21.14) and (21.16) gives

$$u_g = -\frac{g}{f} \frac{\partial z}{\partial y} \qquad v_g = \frac{g}{f} \frac{\partial z}{\partial x} \tag{21.19}$$

Taking the p derivative of the x component of the geostrophic wind in pressure coordinates, we have

$$\frac{\partial u_g}{\partial p} = -\frac{g}{f} \frac{\partial^2 z}{\partial p \, \partial y}$$

$$= -\frac{g}{f} \left(\frac{\partial}{\partial y} \left[\frac{\partial z}{\partial p} \right] \right)_p$$

$$= \frac{1}{f} \frac{\partial}{\partial y} \left(\frac{1}{\rho} \right)_p \tag{21.20}$$

Since $1/\rho = RT/p$, it follows that

$$\frac{\partial}{\partial y} \left(\frac{1}{\rho} \right)_p = \frac{R}{p} \left(\frac{\partial T}{\partial y} \right)_p \tag{21.21}$$

Thus, we obtain the *thermal wind relations*:

$$\frac{\partial u_g}{\partial p} = \frac{R}{fp} \left(\frac{\partial T}{\partial y} \right)_p \qquad \frac{\partial v_g}{\partial p} = -\frac{R}{fp} \left(\frac{\partial T}{\partial x} \right)_p \tag{21.22}$$

This relation shows that horizontal temperature gradients must be accompanied by vertical gradients of geostrophic wind speed.

Imagine that in the Northern Hemisphere temperature decreases in the northward direction, that is, $\partial T/\partial y < 0$. At a certain height z at lower latitude, say, at point A, the air is warm and therefore has relatively low density, while beneath, say, point B, at higher latitude, the air is colder and more dense. From the hydrostatic balance, $p_A > p_B$, since the air density is less at point A than at point B and it must be p_0 at the surface at both places. The x component of the geostrophic wind (in the Northern Hemisphere) is westerly since $\partial p/\partial y < 0$ (lower pressure on its left). Because $\partial T/\partial y < 0$, $\partial u_g/\partial p < 0$, and thus u_g increases with height. In summary, in the Northern Hemisphere, a northward decrease of temperature leads to westerly winds that increase in speed with altitude. In the Northern Hemisphere, $(\partial T/\partial y)_p$ is negative, and in the Southern Hemisphere, $(\partial T/\partial y)_p$ is positive. The Coriolis parameter f is positive in the NH and negative in the SH. The negative signs cancel, and u_g flows in the same direction (west to east) in each hemisphere.

As we noted earlier, the jetstreams are bands of relatively strong winds (usually exceeding $30\,\text{m}\,\text{s}^{-1}$) located above areas of especially strong temperature gradients. The occurrence of jetstreams is a consequence of the thermal wind relation. In such areas, the pressure gradients and resulting windspeeds increase with height all the way up to the tropopause. Jetstreams occur in the upper troposphere, at altitudes between 9 and 18 km. Earlier, we mentioned the polar front jetstream. The polar front is the boundary between polar and midlatitude air. In winter this boundary can extend all the way down to $30°$, while in summer it tends to reside at $50°$–$60°$. The temperature gradient $(\partial T/\partial y)_p$ is stronger in winter than in summer. Thus, the polar front jetstream is located closer to the equator in winter and is stronger during that time. The polar front jets move west to east but meander with the general upper-air waves, serving to affect the movement of major low-level airmasses.

Another important jetstream is located at the boundary between the Hadley and Ferrel cells, the poleward limit of the equatorial tropical air above the transition zone between tropical and midlatitude air (about $30°$–$40°$ latitude). This jet is not necessarily characterized by especially large surface temperature gradients but rather by strong temperature gradients in the midtroposphere. The subtropical jet is generally less intense than the polar jet and is found at higher altitudes. Finally, a high-level jet called the "tropical easterly jetstream" is found in the tropical latitudes of the NH. Such jets are located about $15°$N over continental regions; they result from latitudinal heating contrasts over tropical land masses that do not exist over the tropical oceans. Jetstreams in the SH generally resemble those in the NH, although they exhibit less day-to-day variability because of the presence of less land mass.

The thermal wind relation also describes the prevailing flows in the stratosphere. In contrast to the troposphere, the midlatitude winds in the stratosphere blow from west to east in the winter and from east to west in the summer. The explanation of the differing large-scale flows between the troposphere and stratosphere lies in the temperature structure of the stratosphere. Above $\sim$30 mbar, stratospheric temperatures have their maximum at the summer pole, not the equator, and decrease uniformly from that point to the winter pole. Thus, $(\partial T/\partial y)_p > 0$ in the summer NH. When the NH is in summer, the SH is in winter, and $(\partial T/\partial y)_p > 0$ in the SH. Since f changes sign in the two hemispheres, but $(\partial T/\partial y)_p$ does not, $\partial u_g/\partial p$ has different signs in the two hemispheres. In the summer hemisphere, $\partial u_g/\partial p > 0$ [in the NH, both $(\partial T/\partial y)_p$ and f are positive; in the SH both are negative]. In the winter hemisphere, $\partial u_g/\partial p < 0$. Thus, in the summer hemisphere of the stratosphere, winds become more easterly as altitude increases; in the winter hemisphere, they become more westerly. As the stratospheric temperature structure changes over the course of a year, the direction of the jets changes. In contrast to the troposphere, stratospheric jets, therefore, change direction seasonally and flow in opposite directions in the two hemispheres.

Application of the Thermal Wind Relation   The mean temperature in the layer between 65 and 60 kPa decreases in the eastward direction by $4°$C per 100 km. If the geostrophic wind at 65 kPa is from the southeast at $20\,\text{m}\,\text{s}^{-1}$, what are the geostrophic windspeed and direction at 60 kPa. Assume $f = 10^{-4}\,\text{s}^{-1}$.

Winds from the southeast are midway between easterly and southerly, so the geostrophic windspeed components at 65 kPa altitude are

$$u_g = (-20\,\text{m}\,\text{s}^{-1})(\cos 45°) = -14.14\,\text{m}\,\text{s}^{-1}$$
$$v_g = (20\,\text{m}\,\text{s}^{-1})(\sin 45°) = 14.14\,\text{m}\,\text{s}^{-1}$$

Now we need to determine the thickness Δz of the layer between 65 and 60 kPa. To do so we need the mean density of air in the layer. Taking $T = 248$ K at the average pressure of 62.5 kPa, we obtain

$$\rho = \frac{p}{RT} = \frac{6.25 \times 10^4 \, \text{Pa}}{(287 \, \text{J kg}^{-1} \, \text{K}^{-1})(248 \, \text{K})} = 0.88 \, \text{kg m}^{-3}$$

Then the thickness of the layer between 65 and 60 kPa is found from the hydrostatic relation

$$\Delta z = \frac{-\Delta p}{\rho g} = \frac{5000 \, \text{Pa}}{(0.88 \, \text{kg m}^{-3})(9.8 \, \text{m s}^{-2})} = 581 \, \text{m}$$

Now the thermal wind relation will allow us to calculate the change in geostrophic wind over this layer. The only temperature gradient is in the eastward (x) direction. From (21.22), there is no change u_g with altitude. From (21.22), we can express $\partial v_g / \partial p$ as

$$\frac{\partial v_g}{\partial z} = \frac{g}{fT} \left(\frac{\partial T}{\partial x} \right)_p$$

In finite difference form, this is

$$\Delta v_g = \Delta z \frac{g}{fT} \frac{\Delta T}{\Delta x}$$

$$= (581 \, \text{m}) \left[\frac{9.8 \, \text{m s}^{-2}}{(10^{-4} \, \text{s}^{-1})(248 \, \text{K})} \right] \left(\frac{-4 \, \text{K}}{10^5 \, \text{m}} \right) = -9.2 \, \text{m s}^{-1}$$

With $v_g = 14.14 \, \text{m s}^{-1}$ at 65 kPa, the meridional wind component at 60 kPa is $14.14 - 9.2 = 4.9 \, \text{m s}^{-1}$. The zonal component is $-14.14 \, \text{m s}^{-1}$, so the net windspeed at 60 kPa $= [(4.9)^2 + (-14.14)^2]^{1/2} = 15 \, \text{m s}^{-1}$. The wind direction at 60 kPa is out of the southeast at a direction of $\tan \theta = 4.9/14.14$. Thus, $\theta = 19°$, and the wind is from the southeast at $19°$ north of west.

21.6 STRATOSPHERIC DYNAMICS

The classic depiction of stratospheric transport is that material enters the stratosphere in the tropics, is transported poleward and downward, and finally exits the stratosphere at middle and high latitudes. The mean meridional stratospheric circulation, known as the *Brewer–Dobson circulation*, is generated by stratospheric wave forcing, with the circulation at any level being controlled by the wave forcing above that level (recall Figure 5.25). This process is also called the "extratropical pump." The composition of the lowermost stratosphere varies with season, suggesting a seasonal dependence in the balance between the downward transport of stratospheric air and the horizontal transport of air of upper tropospheric character. The stratosphere and troposphere are actually coupled by more dynamically complex mechanisms than the traditional model of large-scale circulation-driven exchange. Waves generated in the troposphere propagate into the stratosphere, where they can exert forces on circulation, which then can extend back down into the troposphere. A coupling exists between the variability of the stratosphere and troposphere through what are termed the *northern* and *southern annular modes* (NAM and SAM), also called the *Arctic* and *Antarctic oscillations* (AO and AAO). The extreme states of this mode of variability correspond to strong and weak polar vortices.

21.7 THE HYDROLOGIC CYCLE

The hydrologic cycle refers to the fluxes of water, in all its states and forms, over the Earth. Table 21.1 gives the estimated volume and percentage of total water in all the major global reservoirs, and Table 21.2 shows the estimated annual fluxes between major reservoirs. We note that 97% of the Earth's water resides in the oceans. Of the remainder, about $\frac{3}{4}$ is sequestered in the ice caps of Greenland and Antarctica, and $\frac{1}{4}$ is in underground aquifers and lakes and rivers. Only a fraction of 10^{-5} is in the atmosphere, almost all of which is in the form of water vapor. If the total water vapor content of the atmosphere were suddenly precipitated, it would cover the Earth with a rainfall of only 3 cm (see Problem 21.1).

The hydrosphere is thought to have been formed by outgassing of steam from volcanoes during the early life of Earth. Although the mass of the current hydrosphere is effectively unchanging, water rapidly cycles between reservoirs (Table 21.2). More water is evaporated from the oceans than falls on them as precipitation, and more water falls as precipitation on the land masses than is evaporated. The system is balanced by river runoff. Most of the evaporation occurs from the tropical oceans, where the water vapor is carried to the ITCZ. Here it is uplifted in rising convective systems, where it precipitates in intense

TABLE 21.1 Water in Major Global Reservoirs

Reservoir	Volume of water, km^3	Percentage of total
Oceans	1,370,000,000	97.25
Glaciers and ice sheets	29,000,000	2.05
Underground aquifers	9,565,000	0.69
Lakes	125,000	0.01
Rivers	1,700	0.0001
Atmosphere	13,000	0.001
Biosphere	600	0.00001
Total	1,408,705,300	100

TABLE 21.2 Water Fluxes between Reservoirs

Reservoirs	Process	Flux, km^3 yr^{-1}
Ocean–atmosphere	Evaporation	400,000
	Precipitation	370,000
Land masses–atmosphere	Evaporation	60,000
	Precipitation	90,000
Land masses–ocean	Runoff	30,000

downpours. This intense convection over tropical land masses accounts for much of the world's rainfall and much of the flux of water from the oceans to land masses.

The annual flux of water through the atmosphere is about $460,000\,\mathrm{km^3\,yr^{-1}}$, about 35 times greater than the amount of water stored in the atmosphere at any one time. Thus, the average residence time of a molecule of water in the atmosphere can be estimated as $13,000\,\mathrm{km^3}/460,000\,\mathrm{km^{-3}\,yr^{-1}}$, or about 10 days. By contrast, the ocean reservoir is over 3000 times larger than the annual flux to the atmosphere, so the average residence time of a molecule of water in the oceans is very long (about 3400 years).

APPENDIX 21   OCEAN CIRCULATION

The atmospheric circulation is driven by temperature differences over the globe. The atmosphere is a fluid heated from below, and fluids that expand when heated, like air, rise when heated from below. The world's oceans differ from the atmosphere in two important respects: (1) the oceans are heated from above; and (2) water is densest at $4°C$. It expands both when cooled below $4°C$ and heated above $4°C$.

These two facts have important implications for ocean circulation. Because water becomes less dense when heated above $4°C$, surface heating forms a warm, buoyant, and stable layer at the surface. If surface water is cooled below $4°C$, it also becomes less dense, forming a cold, buoyant, and stable layer on the surface. Water also undergoes a dramatic expansion on freezing, increasing in volume by 9%, so pack ice also forms a stable upper layer. Thus, in each of these circumstances, vertical motions are suppressed. Only where ocean surface waters with temperatures over $4°C$ undergo cooling can surface temperature changes cause sinking motions, and these tend not to extend to very great depths.

Temperature, however, is not the only consideration; ocean-water density is also influenced by salinity. Seawater contains dissolved ions, mainly sodium and chlorine, and, as a result, is denser than freshwater (typically $1035\,\mathrm{kg\,m^{-3}}$ compared with $1000\,\mathrm{kg\,m^{-3}}$). The salinity of seawater is not uniform everywhere in the ocean. Evaporation at the sea surface increases the salinity (as water is evaporated off but dissolved ions are left behind), whereas inputs from rain or river runoff lead to decreased salinity. Salinity is also increased by the formation of sea ice; as ice forms, salt is rejected from the freezing water, and the remaining liquid water is enriched in salt.

The density of seawater thus depends both on its temperature and salinity. Highly saline waters at about $4°C$ are the most dense and will tend to sink, whereas warm, less saline water is less dense, and will tend to rise. This behavior has important implications for vertical motions in the sea. Descent of surface waters to the deep ocean is possible only where saline waters are cooled. This actually occurs in only a few locations, most notably in the North Atlantic off Greenland, and in the Weddell Sea, part of the Antarctic Ocean south of the Atlantic. North Atlantic deep water has a temperature of $2.5°C$ and a salinity of 35 g salt per kg of seawater. Deep water forming in the Weddell Sea has a temperature of $-1.0°C$ and a salinity of 34.6 g salt per kg seawater. In both of these locations, saline surface waters are cooled and sink, a process known as *downwelling*. The surface waters are made saline by evaporation nearer the equator in the case of the North Atlantic, and by the formation of sea ice in the case of the Weddell Sea. Shallower downwelling, to intermediate depths in the oceans, also occurs in areas of high evaporation.

A term that has been widely used to describe the ocean's circulation is the *thermohaline circulation* (Wunsch 2002). Ocean circulation is defined most clearly in terms of the mass of water transported; other properties such as heat or salt content are transported with the mass. The upper layers of the ocean are wind-driven; mass fluxes in the upper several hundred meters of the ocean are controlled by the stress of wind on the surface. The ocean is heated and cooled within about 100 m of its surface; below this, it has a stable stratification. The ocean's mass flux is sustained primarily by the wind. Surface buoyancy boundary conditions strongly influence the transport of heat and salinity, because the fluid must become sufficiently dense to sink, but these boundary conditions do not actually drive the circulation. The wind field not only controls the near-surface wind-driven components of the mass flux but also influences the turbulence at depth, which controls the deep stratification. The term "thermohaline circulation" can be reserved for the resulting circulations of heat and salt.

Deep downwelling in the polar oceans draws in surface waters to replace the sinking waters; at the same time the downwelling water reaches the seabed, then travels equatorward. Deep waters are formed by downwelling in the North Atlantic and Weddell Sea; the latter is the area of ocean between South America and Antarctica. Saline surface waters are conveyed to the North Atlantic, whereupon cooling causes downwelling and the formation of deep waters; sinking in the North Atlantic occurs at a rate of about $17,000,000 \, \text{m}^3 \, \text{s}^{-1}$ in winter, 20 times the flow volume of all the rivers on Earth. At a depth of 2–3 km, the cold water begins its trip around the globe: southward through the western Atlantic basin to the Antarctic circumpolar current and from there into the Indian and Pacific Oceans. In upwelling regions, seawater rises once again to the surface. Since deep-water currents move very slowly, at most $0.36 \, \text{km} \, \text{h}^{-1}$, it takes centuries for water to circulate around the entire globe on these global conveyor belts.

Water evaporated from the oceans returns to its ocean of origin mostly in river flows. In some areas, however, atmospheric motions and drainage patterns result in water evaporated from one ocean being returned to another. In particular, some of the water evaporated from the Atlantic is returned to the Pacific and Indian Oceans following transport in clouds across Central America and Asia. In this process, the Atlantic loses water and is left more saline, whereas the Indian and Pacific Oceans gain water and become less saline. These imbalances are redressed by the large scale circulation. A similar cycle occurs between the Mediterranean and Atlantic. Enhanced evaporation from the Mediterranean produces more saline water than in the neighboring Atlantic. Mediterranean water flows out through the Straits of Gibraltar, where it sinks to intermediate depths in the Atlantic. Atlantic surface waters flow back into the Mediterranean to replace the saline water and that lost to evaporation. No deep water forms today in the North Pacific, mainly because the North Pacific surface waters are too fresh, even at the freezing point, to sink lower than a few hundred meters. In the North Atlantic, surface salinity remains high even in regions of net precipitation.

Changes in the ocean's overturning are one of the most important feedbacks in greenhouse warming. Manabe and Stouffer (1993), employing a coupled ocean–atmosphere model, predicted that a fourfold increase of CO_2 from its preindustrial level would lead to a cessation of the ocean's overturning within a period of 200 years; the cause was the lowering of the salinity of the North Atlantic as a result of increased precipitation in a warmer, greenhouse world.

The main force driving the oceanic circulation is the wind. As wind blows across the sea surface, part of the wind's momentum is transferred to the ocean, setting up currents.[4]

---

[4]In contrast to winds, ocean currents are defined in the direction to which they flow or, more descriptively, by their thermal characteristics. For example, an ocean current that flows from south to north in the Northern Hemisphere is called a *northerly current* or a *warm current*.

Just as winds are deflected by the Coriolis force, so are ocean currents. Wind-driven currents are also influenced by the topography of the oceans, both the form of the seabed and the position of the coasts. Currents driven onshore are forced to flow along the coast and, as a result, ocean currents in many parts of the world flow along the boundaries of the oceans. Currents driven offshore result in vertical motions. For example, a wind blowing parallel to a coast in the Northern Hemisphere, with the ocean on the right, will establish a surface current flowing to the right (i.e., offshore). This transports water away from the coast, which must eventually be replaced by upwelling deeper waters. Upwelling brings nutrients from depth to the surface; since most marine organisms live near the surface where light can penetrate and promote photosynthesis (the photic zone), this transport promotes biological productivity.

The midlatitude westerlies and the northeast trade winds flow in opposite directions around 30°N. Thus, a wind stress is introduced that produces a clockwise circulation or *gyre* around 30°N. This same situation exists in the Southern Hemisphere, except that the gyre circulates counter-clockwise. At 60°N, the polar easterlies and mid-latitude westerlies meet to produce a counter-clockwise gyre. A similar gyre, albeit rotating clockwise, would be produced in the Southern Hemisphere except that there is no land at 60°S to produce a bounded basin. Thus, the Southern Ocean (often used to refer to the Atlantic and Pacific Oceans south of about 50°S) circles Antarctica in a west-to-east flow. Since the ocean currents are deflected to the right in the Northern Hemisphere and to the left in the Southern Hemisphere, the ocean currents flow from east to west in both hemispheres in the Tropical Pacific. Close to the Equator is an eastward-flowing current, the Equatorial Countercurrent. This exists because the trade winds drive water to the western Pacific, where it is blocked by land masses. Sea surfaces rises westward (due to the wind stress), so in the area of light winds near the Equator, water is able to flow back eastward. This flow is dramatically accentuated during El Niño events.

Warm currents flowing poleward transport heat to colder regions, whereas cold currents flowing equatorward have the opposite effect. One of the most important examples of poleward heat transport by an ocean current is the Gulf Stream in the North Atlantic. The Gulf Stream, which moves 500 times as much water as the Amazon River, and flows at speeds of $\leq 9\,\mathrm{km\,h^{-1}}$, is a tongue of warm surface waters that enters the Atlantic from the Gulf of Mexico. These warm surface waters give up heat to the atmosphere, which is then transported in the midlatitude westerlies. Heat from the Gulf Stream makes an important contribution to climate in northwest Europe. In the western Pacific, the Kuroshio current has a similar warming effect on Japan, whereas in the eastern Pacific, the California and Humboldt currents exert a cooling influence.

PROBLEMS

21.1_A Compute the magnitude of the geostrophic windspeed (westerly) at 40°N latitude and 5000 m altitude assuming that the pressure at this latitude and altitude decreases from south to north by 400 Pa over a distance of 200 km.

21.2_A The total mass of water in the atmosphere is $1.3 \times 10^{16}\,\mathrm{kg}$, and the global mean rate of precipitation is estimated to be $0.2\,\mathrm{cm\,day^{-1}}$. Estimate the global mean residence time of a molecule of water in the atmosphere.

21.3_A Accounting for the order of magnitude of atmospheric quantities, estimate the time needed to mix a species throughout a hemisphere.

21.4$_C$ The Ekman spiral describes the variation of wind direction with altitude in the planetary boundary layer. The analytical form of the Ekman spiral can be derived by considering a two-dimensional wind field (no vertical component), the two components of which satisfy

$$\frac{\partial}{\partial z}\left(K_M \frac{\partial \bar{u}}{\partial z}\right) - \frac{1}{\rho}\frac{\partial p}{\partial x} + f\bar{v} = 0$$

$$\frac{\partial}{\partial z}\left(K_M \frac{\partial \bar{v}}{\partial z}\right) - \frac{1}{\rho}\frac{\partial p}{\partial y} - f\bar{u} = 0$$

where K_M is a constant eddy viscosity and f is the Coriolis parameter. At some height the first terms in each of these equations are expected to become negligible, leading to the geostrophic wind field. Take the x axis to be oriented in the direction of the geostrophic wind, in which case

$$f\bar{u}_g = -\frac{1}{\rho}\frac{\partial p}{\partial y}$$

Show that the solutions for $\bar{u}(z)$ and $\bar{v}(z)$ are

$$\bar{u}(z) = \bar{u}_g(1 - e^{\alpha z}\cos \alpha z)$$
$$\bar{v}(z) = \bar{u}_g e^{-\alpha z}\sin \alpha z$$

where $\alpha = \sqrt{f/2K_M}$. In what way is this solution oversimplified?

21.5$_B$ The outer boundary of the planetary boundary layer can be defined as the point at which the component in the Ekman spiral disappears. From the solution of Problem 21.4, it is seen that this occurs when $\alpha z_g = \pi$.

a. Using representative values of f and K_M, estimate the depth of the planetary boundary layer.

b. The magnitude of the total turbulent stress at the surface is given by

$$\tau_0 = \frac{\rho K_M}{\sqrt{2}}\left(\left.\frac{\partial \bar{u}}{\partial z}\right|_{z=0} + \left.\frac{\partial \bar{v}}{\partial z}\right|_{z=0}\right)$$

Show that for a planetary boundary layer

$$\tau_0 = \frac{z_g}{\sqrt{2\pi}}f\rho\bar{u}_g$$

Estimate the magnitude of τ_0.

c. The surface layer can be estimated as that layer in which τ_0 changes by only 10% of its value at the surface. In the planetary boundary layer the turbulent stress terms in the equations of motion are of the same order as the Coriolis acceleration terms. Estimate $\partial\tau/\partial z$, and from this estimate the thickness of the surface layer.

REFERENCES

Ackerman, S. A., and Knox, J. A. (2003) *Meteorology: Understanding the Atmosphere*, Brooks/Cole, Pacific Grove, CA.

Hadley, G. (1735) Concerning the cause of the general trade winds, *Philos. Trans.* **39**.

Holton, J. R. (1992) *An Introduction to Dynamic Meterology*, 3rd ed., Academic Press, San Diego.

Manabe, S., and Stouffer, R. J. (1993) Century-scale effects of increased atmospheric CO_2 on the ocean-atmosphere system, *Nature* **364**, 215–218.

McIlveen, R. (1992) *Fundamentals of Weather and Climate*, Chapman & Hall, London.

Salby, M. L. (1996) *Fundamentals of Atmospheric Physics*, Academic Press, San Diego.

Wunsch, C. (2002) What is the thermohaline circulation? *Science* **298**, 1179–1180.

22 Global Cycles: Sulfur and Carbon

Biogeochemical cycles describe the exchange of molecules containing a given atom between various reservoirs within the Earth system. Important biogeochemical cycles in the Earth system include oxygen, nitrogen, sulfur, carbon, and phosphorus. Here we consider the global biogeochemical cycles of sulfur and carbon, two of the most important elements from the standpoint of atmospheric chemistry and climate.

22.1 THE ATMOSPHERIC SULFUR CYCLE

Figure 22.1 depicts the major reservoirs in the biogeochemical cycle of sulfur, with estimated quantities [in Tg(S)] in each reservoir. Directions of fluxes between the reservoirs are indicated by arrows. The major pathways of sulfur compounds in the atmosphere are depicted in Figure 22.2. The numbers on each arrow refer to the description of the process given in the caption to the figure (not to fluxes). Note the small amount of sulfur in the atmosphere relative to that in the other reservoirs. Also note the significant amount of sulfur in the marine atmosphere; this is the result of dimethyl sulfide (DMS) emissions from the sea.

We wish now to analyze that portion of the global sulfur cycle involving atmospheric SO_2 and sulfate shown in Figure 22.2. We will denote the natural and anthropogenic emissions of SO_2 as $P^n_{SO_2}$ and $P^a_{SO_2}$, respectively. $P^n_{SO_2}$ includes a contribution from the oxidation of reduced sulfur species to SO_2. SO_2 is removed by dry and wet deposition and oxidized to sulfate by chemical reaction. Sulfate is also removed from the atmosphere by dry and wet deposition. Our goal is to obtain estimates for the lifetimes of SO_2 and SO_4^{2-}.

Writing (2.15) for both SO_2 and SO_4^{2-}, we obtain

$$P^n_{SO_2} + P^a_{SO_2} - (k^d_{SO_2} + k^w_{SO_2} + k^c_{SO_2})Q_{SO_2} = 0 \qquad (22.1)$$

$$k^c_{SO_2}Q_{SO_2} - (k^d_{SO_4} + k^w_{SO_4})Q_{SO_4} = 0 \qquad (22.2)$$

The mean residence time of SO_2 is

$$\begin{aligned}
\tau_{SO_2} &= \frac{Q_{SO_2}}{P^n_{SO_2} + P^a_{SO_2}} \\
&= \frac{1}{k^d_{SO_2} + k^w_{SO_2} + k^c_{SO_2}}
\end{aligned} \qquad (22.3)$$

Atmospheric Chemistry and Physics: From Air Pollution to Climate Change, Second Edition, by John H. Seinfeld and Spyros N. Pandis.   Copyright © 2006 John Wiley & Sons, Inc.

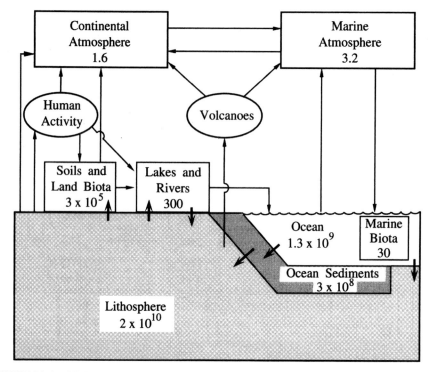

FIGURE 22.1   Major reservoirs and burdens of sulfur, in Tg(S) (Charlson et al. 1992). (Reprinted by permission of Academic Press.)

whereas that for sulfate is

$$\tau_{SO_4} = \frac{Q_{SO_4}}{k^c_{SO_2} Q_{SO_2}}$$
$$= \frac{1}{k^d_{SO_4} + k^w_{SO_4}} \tag{22.4}$$

The mean residence time of a sulfur atom is

$$\tau_S = \frac{Q_{SO_2} + Q_{SO_4}}{P^n_{SO_2} + P^a_{SO_2}} \tag{22.5}$$

Which can be expressed in terms of the two previous mean residence times as

$$\tau_S = \tau_{SO_2} + b\tau_{SO_4} \tag{22.6}$$

where

$$b = \frac{k^c_{SO_2} Q_{SO_2}}{P^n_{SO_2} + P^a_{SO_2}} \tag{22.7}$$

the fraction of S converted to SO_4^{2-} before being removed.

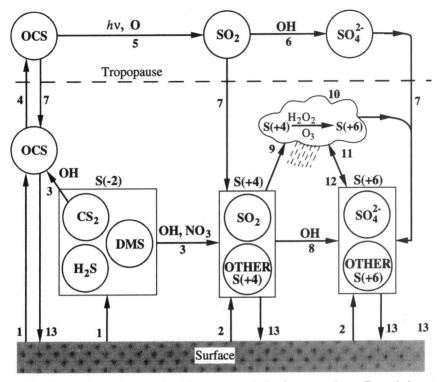

FIGURE 22.2 Major pathways of sulfur compounds in the atmosphere (Berresheim et al. 1995). The paths are labeled according to the processes: (1) emission of DMS, H_2S, CS_2, and OCS; (2) emission of S(+4) and S(+6); (3) oxidation of DMS, H_2S, and CS_2 by OH, and DMS, by NO_3 in the troposphere; (4) transport of OCS into the stratosphere; (5) photolysis of OCS or reaction with O atoms to form SO_2 in the stratosphere; (6) oxidation of SO_2 in the stratosphere; (7) transport of stratospheric OCS, SO_2, and sulfate back into the troposphere; (8) oxidation of SO_2 and other S(+4) products by OH in the troposphere; (9) absorption of S(+4), mainly SO_2, into hydrosols (cloud/fog/rain droplets, moist aerosol particles); (10) liquid phase oxidation of S(+4) by H_2O_2(aq) in hydrosols (and by O_2 in the presence of elevated levels of catalytic metal ions); (11) absorption/growth of S(+6) aerosol—mainly sulfate—into hydrosols; (12) evaporation of cloud-water leaving residual S(+6) aerosol; (13) deposition of OCS, S(+4), and S(+6).

We can also define individual characteristic times such as

$$\tau^d_{SO_2} = (k^d_{SO_2})^{-1}, \qquad \tau^w_{SO_2} = (k^w_{SO_2})^{-1}, \qquad \tau^d_{SO_4} = (k^d_{SO_4})^{-1}$$

so that

$$\frac{1}{\tau_{SO_2}} = \frac{1}{\tau^d_{SO_2}} + \frac{1}{\tau^w_{SO_2}} + \frac{1}{\tau^c_{SO_2}} \tag{22.8}$$

$$\frac{1}{\tau_{SO_4}} = \frac{1}{\tau^d_{SO_4}} + \frac{1}{\tau^w_{SO_4}} \tag{22.9}$$

We can also define the mean residence time of a sulfur atom before surface removal or precipitation scavenging by

$$\tau_S^{d,w} = \frac{Q_{SO_2} + Q_{SO_4}}{k_{SO_2}^{d,w} Q_{SO_2} + k_{SO_4}^{d,w} Q_{SO_4}} \tag{22.10}$$

so that using

$$\frac{P_{SO_2}^n + P_{SO_2}^a}{Q_{SO_2} + Q_{SO_4}} = (k_{SO_2}^d + k_{SO_2}^w)\frac{Q_{SO_2}}{Q_{SO_2} + Q_{SO_4}} + (k_{SO_4}^d + k_{SO_4}^w)\frac{Q_{SO_4}}{Q_{SO_2} + Q_{SO_4}} \tag{22.11}$$

we get

$$\frac{1}{\tau_S} = \frac{1}{\tau_S^d} + \frac{1}{\tau_S^w} \tag{22.12}$$

The mean residence times for a sulfur atom before surface removal (or precipitation scavenging) can be related to the mean surface removal residence times for SO_2 and SO_4^{2-} as follows. Noting that

$$\begin{aligned}
\tau_S^d &= \frac{Q_{SO_2} + Q_{SO_4}}{k_{SO_2}^d Q_{SO_2} + k_{SO_4}^d Q_{SO_4}} \\
&= \frac{Q_{SO_2} + Q_{SO_4}}{\dfrac{Q_{SO_2}}{\tau_{SO_2}^d} + \dfrac{Q_{SO_4}}{\tau_{SO_4}^d}} \\
&= \frac{1}{\dfrac{Q_{SO_2}}{Q_{SO_2} + Q_{SO_4}}\dfrac{1}{\tau_{SO_2}^d} + \dfrac{Q_{SO_4}}{Q_{SO_2} + Q_{SO_4}}\dfrac{1}{\tau_{SO_4}^d}}
\end{aligned} \tag{22.13}$$

we have

$$\frac{1}{\tau_S^d} = \frac{c}{\tau_{SO_2}^d} + \frac{(1-c)}{\tau_{SO_4}^d} \tag{22.14}$$

where

$$c = \left[1 + \frac{b(k_{SO_2}^d + k_{SO_2}^w + k_{SO_2}^c)}{k_{SO_4}^d + k_{SO_4}^w}\right]^{-1} \tag{22.15}$$

By virtue of (22.14), c is the fraction of the total sulfur that is SO_2.

Rodhe (1978) has estimated values of the sulfur residence times (in hours):

$\tau_{SO_2}^d$	$\tau_{SO_2}^w$	$\tau_{SO_2}^c$	τ_{SO_2}	$\tau_{SO_4}^d$	$\tau_{SO_4}^w$	τ_{SO_4}
60	100	80	25	>400	80	80

Assuming $c = 0.5$, the sulfur atom residence times are (in hours):

τ_S^d	τ_S^w	τ_S
120	90	50

Because of the uneven spatial distribution of anthropogenic sources and the relatively short residence time of sulfur in the atmosphere, global averages do not provide an accurate description of human influence on the sulfur cycle in populated parts of the world.

22.2 THE GLOBAL CARBON CYCLE

22.2.1 Carbon Dioxide

In terms of total mass of emissions, carbon dioxide is the single most important waste product of our industrialized society. The primary effect of CO_2 in the atmosphere is climatic (see Chapter 23). Charles D. Keeling of the Scripps Institution of Oceanography began measurement of atmospheric CO_2 levels at Mauna Loa, Hawaii in 1958 (Keeling 1983; Keeling and Whorf 2003). Figure 22.3 shows CO_2 mixing ratios measured at Mauna Loa from 1958 to 2004.

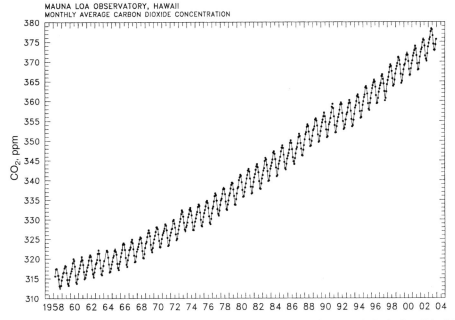

FIGURE 22.3 CO_2 mixing ratio measured at Mauna Loa, Hawaii, since 1958 (Carbon Dioxide Research Group, Scripps Institution of Oceanography).

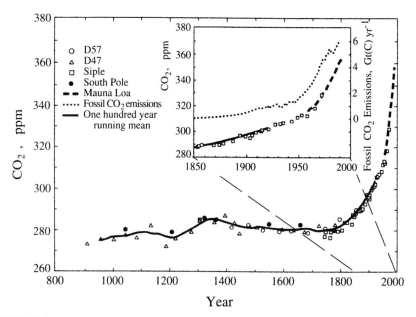

FIGURE 22.4 CO_2 mixing ratio over the past 1000 years from the recent ice core record and (since 1958) from the Mauna Loa measurement site. The inset shows the period from 1850 in more detail, including CO_2 emissions from fossil fuel. Data sources are given in IPCC (1995). The smooth curve is based on a 100-year running mean. All ice core measurements were taken in Antarctica.

The preindustrial (pre–Industrial Revolution) level of CO_2 over the last 1000 years, as determined from ice core records, was about 280 ppm (Figure 22.4).[1] By 2003, the global mean atmospheric CO_2 level had reached 375 ppm, based on an average of values measured at Barrow, Alaska, Mauna Loa, Hawaii, American Samoa, and the South Pole. The CO_2 record exhibits a seasonal cycle evident in Figure 22.3. Each year during the Northern Hemisphere growing season in spring and summer, atmospheric CO_2 mixing ratios decrease as carbon is incorporated into leafy plants. From October through January, photosynthesis is largely confined to the tropics and the relatively small land mass of the Southern Hemisphere continents. At this time of year, plant respiration and decay dominate and CO_2 levels increase. At Mauna Loa, the peak-to-trough CO_2 seasonal cycle has remained between 5 and 6 ppm since 1958. The amplitude of the seasonal cycle increases to about 15 ppm in the boreal forest zone ($55°–65°N$). In the Southern Hemisphere the mean amplitude of the CO_2 seasonal cycle is only about 1 ppm.

Over the period 1850–2000 it is estimated that 282 Pg C (1 Pg = 1 Gt = 10^{15} g) were released to the atmosphere by fossil fuel combustion, and an additional 5.5 Pg C from cement manufacture (Figure 22.5). (The 2000 global, fossil-fuel emission estimate of 6.6 Pg C represents a 1.8% increase from 1999. The average annual fossil-fuel release of CO_2 over the decade 1990 to 1999 was 6.35 Pg C.) In addition, land-use changes are estimated to have resulted in a net transfer of 154 Pg C to the atmosphere since 1850. This totals

---

[1] Estimates of the pre-industrial CO_2 mixing ratio vary from 280 to 290 ppm. Records generated both by proxy techniques and recovery of ancient air trapped in glacial ice show the CO_2 abundance cycling between 200 and 300 ppm as the climate cycled from glacial to interglacial conditions (Barnola et al., 1987; Genthon et al., 1987).

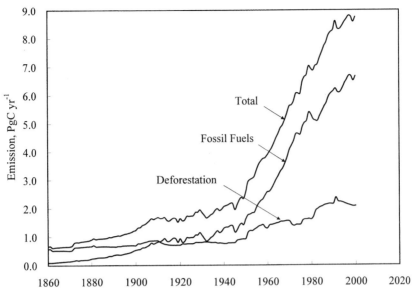

FIGURE 22.5 Global CO_2 emissions from fossil fuel and deforestation. Data from Carbon Dioxide Information Analysis Center, Oak Ridge National Laboratory; http://cdiac.esd.ornl.gov/trends/emis/tre-glob.htm.

441.5 Pg C, 64% of which is due to fossil fuel combustion. The atmospheric CO_2 mixing ratio rose from a preindustrial value of about 280 ppm to 370 ppm in 2000, an increase of 90 ppm. Each ppm of CO_2 in the atmosphere corresponds to 2.1 Pg C (see Problem 1.6), so the increase in the atmospheric burden of carbon from 1850 to 2000 was 189 Pg C. Thus, about 43% (189/441.5) of the carbon added to the atmosphere since 1850 has remained in the atmosphere; the other 57% has been transferred to the oceans and the terrestrial biosphere. The 370 ppm of CO_2 translates into 777 Pg C, of which 189 Pg C has been added since 1850. We noted above that 64% of that addition can be attributed to fossil fuel combustion.

Of the values for all the quantities of interest in the global carbon cycle, such as the exact preindustrial CO_2 mixing ratio, the total fossil fuel emissions of CO_2, the transfer of CO_2 to the atmosphere as a result of deforestation, and the amounts of carbon in the preindustrial reservoirs, only the fossil fuel emissions are known with significant accuracy. As a result, different studies employ somewhat different procedures or databases to estimate these quantities, leading to a range of estimates in the literature. For example, the parameters in the global carbon cycle model that we will employ here have been determined on the basis of a value for preindustrial atmospheric carbon loading somewhat different from the 588 Pg C corresponding to a CO_2 mixing ratio of 280 ppm. Nonetheless, the important features of the carbon cycle, such as the percentage of carbon emissions remaining in the atmosphere, are reflected in most models.

22.2.2 Compartmental Model of the Global Carbon Cycle

Carbon is naturally contained in all of Earth's compartments; the global carbon cycle is a description of how carbon moves among those compartments in response to perturbations, such as emissions from fossil fuel burning and deforestation. Compartmental models of

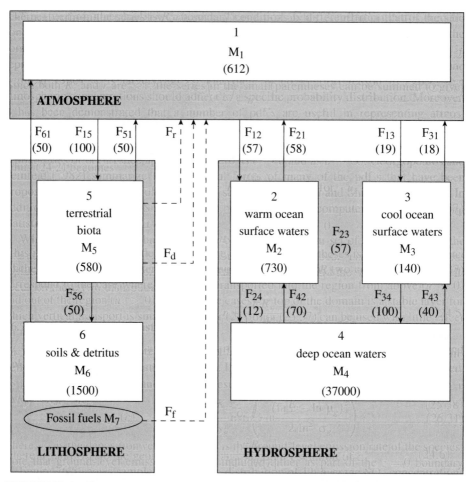

FIGURE 22.6 Six-compartment model of the global carbon cycle (Schmitz 2002). Values shown for reservoir masses (M_i, in Pg C) and fluxes (F_{ij}, in Pg C yr^{-1}) represent those for the preindustrial steady state.

the global carbon cycle provide a means to simulate the flux of carbon between the various reservoirs and to estimate future levels corresponding to different CO_2 emission scenarios.

Figure 22.6 shows a six-compartment model of the carbon cycle due to Schmitz (2002). The quantities shown in parentheses in each compartment are estimates of the pre-industrial ($\sim$1850) amount of carbon, indicated by the M_i symbols, measured in petagrams (Pg C) in each reservoir. Since the amount of carbon in the aquatic biosphere and in rivers, streams, and lakes is negligible compared with those in the other reservoirs in Figure 22.6, these are omitted. The fossil fuel reservoir affects the global carbon cycle only as a source of carbon. Sediments are actually the largest carbon reservoir of all, but the fluxes of carbon into and out of sediments are so small that sediments can be neglected as a compartment over any realistic timescale.[2] Estimates of the pre-industrial flux of carbon

---

[2] Some of the carbon sequestration approaches under consideration are aimed at transferring some of the CO_2 that would otherwise be released to the atmosphere to the sediments or underground deep-water reservoirs by direct injection of CO_2 emitted by power plants.

from reservoir i to reservoir $j(F_{ij})$ are given in parentheses next to the arrow indicating the flux; at the preindustrial steady state, the net flow of carbon into or out of any reservoir was zero.

Some of the fluxes shown in Figure 22.6 can be elaborated on:

1. F_{12}, F_{21}, F_{13}, and F_{31} are rates of transfer of CO_2 between the atmosphere and surface waters of the ocean.
2. F_{23} is the flux of carbon from warm to cool surface ocean reservoirs. This flow, which accounts for much of mixing in the oceans, results from sinking of cool surface water at high latitudes and upwelling at low latitudes.
3. F_{15} is the uptake of atmospheric CO_2 by terrestrial vegetation (photosynthesis).
4. F_{56} is the flux of carbon in litter fall, such as dead leaves.
5. F_{51} and F_{61} are the fluxes of carbon to the atmosphere by respiration of biota.

The carbon content of the oceans is more than 50 times greater than that of the atmosphere. Over 95% of the oceanic carbon is in the form of inorganic dissolved species, bicarbonate (HCO_3^-) and carbonate (CO_3^{2-}) ions; the remainder exists in various forms of organic carbon (Druffel et al. 1992). Oceanic uptake of CO_2 involves three steps: (1) transfer of CO_2 across the air–sea interface, (2) chemical interaction of dissolved CO_2 with seawater constituents, and (3) transport to the deeper ocean by vertical mixing processes. Steps 2 and 3 are the rate-determining processes in the overall transfer of CO_2 from the atmosphere to the ocean, and oceanic transport and mixing processes are the primary uncertainties in predicting the rate of oceanic uptake of CO_2.

The ocean is both a source and sink for atmospheric CO_2. Two fundamental effects control the ocean distribution of CO_2. First, the cold waters, which fill the deep ocean from the high latitudes, are rich in CO_2; this results because the solubility of CO_2 increases as temperature decreases. In the sunlit uppermost 100 m of the ocean, photosynthesis is a sink for CO_2 and a source of O_2. Dying marine organisms decay as they settle into the deeper layers of the ocean, consuming dissolved O_2 and giving off CO_2. Thus, these layers have higher CO_2 concentrations and lower O_2 concentrations than the surface waters. This process of transporting CO_2 to deeper water is referred to as the *biological pump*.

Since preindustrial conditions, the rise in atmospheric CO_2 has led to an increased air-to-sea CO_2 flux everywhere on Earth. From tens of thousands of individual measurements made in every corner of the world's ocean in every season of the year, a global map of the present-day flux of CO_2 across the sea surface has emerged (Takahashi et al. 1997). While the net global flux of CO_2 is into the ocean, there are regions in which CO_2 is outgassed. The equatorial Pacific is the strongest continuous global natural source of CO_2 to the atmosphere, the cause of which is vigorous upwelling along the equator, driven by divergence of surface currents. The upwelling water comes from only a few hundred meters' depth but is warmed by several degrees as it ascends, decreasing the solubility of CO_2. Even though the upwelling water also brings abundant nutrients to the surface, the plankton are not able to make full use of these because of the lack of sufficient amounts of the nutrient iron. The absence of biological production in these surface waters means that the CO_2 brought to the surface is released to the atmosphere rather than taken up by photosynthesis.

The most intense oceanic sink region for CO_2 is the North Atlantic. The Gulf Stream and North Atlantic drift transport warm water rapidly northward. As the water cools, CO_2 solubility increases. In addition, the North Atlantic is biologically the most productive ocean region on Earth, because of an abundant supply of nutrients, including iron from Saharan dust. Other important ocean sink areas are all associated with zones where the ocean is transferring heat to the atmosphere and cooling.

Carbon in the form of CO_2 is removed from the atmosphere and incorporated into the living tissue of green plants by photosynthesis. Respiration returns carbon to the atmosphere (Schimel 1995). Photosynthesis responds essentially instantaneously to an increase in atmospheric CO_2. Respiration does not respond directly to atmospheric CO_2 concentrations but exponentially increases with increasing temperature.

Carbon is stored in terrestrial ecosystems in two main forms: carbon in biomass $\left(\sim\frac{1}{3}\right)$, principally as wood in forests; and carbon in soil $\left(\sim\frac{2}{3}\right)$. The lifetime of carbon in these reservoirs can be quite different. If elevated CO_2 leads to increased growth of a tree, this results in a CO_2 sink. The tree might typically live for a century before it dies and decomposes, releasing CO_2 back into the atmosphere via heterotrophic respiration. Large terrestrial carbon reservoirs in the form of trees or forest floor litter are susceptible to rapid return to the atmosphere through fire.

Figure 22.6 shows three fluxes denoted by dashed lines:

F_f Emissions of carbon from fossil-fuel burning

F_d Flux of carbon from renewable terrestrial biota to the atmosphere (deforestation)

F_r Flux of carbon from the atmosphere to terrestrial biota (reforestation)

These three fluxes are the direct anthropogenic perturbations to the carbon cycle and are assumed to be negligibly small in preindustrial times. This is an excellent assumption for F_f and F_r, but it also presumes that natural forest fires (F_d) were responsible for small emissions of CO_2.

The preindustrial compartmental masses of carbon in Figure 22.6 are those assumed by Schmitz (2002). The preindustrial atmospheric mass of CO_2 is taken to be 612 Pg C, which corresponds to a CO_2 mixing ratio of 291 ppm, toward the high end of the range of 280–290 ppm usually assumed as the preindustrial value.

The carbon mass balances for each of the six compartments follow the development in Chapter 2. For example, for the carbon in the atmosphere, we have

$$\frac{dM_1}{dt} = F_{21} - F_{12} + F_{31} - F_{13} + F_{51} - F_{15} + F_{61} + F_f + F_d - F_r \qquad (22.16)$$

At the preindustrial steady state, the perturbation fluxes, $F_f = F_d = F_r = 0$, and $F_{21} - F_{12} + F_{31} - F_{13} + F_{51} - F_{15} + F_{61} = 0$, so that $dM_1/dt = 0$. The total amount of carbon stored in fossil fuels affects the carbon cycle only through the amount actually released to the atmosphere. Falkowski et al. (2001) have estimated that the fossil fuel reservoir contains 4130 Pg C. (The ratio of the estimated fossil fuel reservoir to the current atmospheric reservoir, 4130/788 = 5.2. Thus, if the entire estimated fossil fuel reservoir were added to the atmosphere with no carbon taken up by other reservoirs, the atmospheric concentration of CO_2 would increase by a factor of 6.2. This can be considered as an upper-limit estimate for increase of atmospheric CO_2.)

As a first approximation, the intercompartmental fluxes can be assumed to be proportional to the mass in the compartment from which the flux is taking place

$$F_{ij} = k_{ij}M_i \qquad (22.17)$$

where the k_{ij} values are first-order exchange coefficients. These k_{ij} values can be determined from the fluxes and masses in Figure 22.6 using (22.17), assuming that the exchange coefficients between compartments have remained the same from preindustrial times to the present. There are three exceptions: F_{21} (warm ocean waters to atmosphere), F_{31} (cold ocean waters to atmosphere), and F_{15} (atmosphere to terrestrial biota).

Once CO_2 dissolves in water, an equilibrium is established among dissolved CO_2 ($CO_2 \cdot H_2O$), bicarbonate ions (HCO_3^-), and carbonate ions (CO_3^{2-}) (see Chapter 7). Thus, while the sea-to-air fluxes F_{21} and F_{31} can be linearly related to dissolved CO_2 ($CO_2 \cdot H_2O$), they are not linearly related to the total dissolved carbon (M_2 and M_3), which is the sum of $CO_2 \cdot H_2O$, HCO_3^-, and CO_3^{2-}. This relationship depends on seawater temperature and pH, the latter of which depends in a complex way on salinity and concentrations of dissolved salts. Seawater pH varies between ~ 7.5 and 8.4, and therefore most of the CO_2 that dissolves in the ocean is not in the $CO_2 \cdot H_2O$ form (Chapter 7). For the purpose of a global carbon cycle model, one wishes to avoid the complication of an explicit ocean chemistry model to relate F_{21} and F_{31} to M_2 and M_3, respectively. The following empirical relationship has been developed (Ver et al., 1999)

$$F_{21} = k_{21} M_2^{\beta_2} \qquad F_{31} = k_{31} M_3^{\beta_3} \qquad (22.18)$$

where β_1 and β_2 are positive empirical constants. Given numerical values of β_2 and β_3, the values of k_{21} and k_{31} can be determined from the preindustrial conditions shown in Figure 22.6. Note that the units of k_{21} and k_{31} are no longer simply yr^{-1}, but are Pg $C^{(1-\beta_2)} yr^{-1}$ and Pg $C^{(1-\beta_3)} yr^{-1}$, respectively. As M_2 and M_3 increase with more dissolved carbon in the oceans, the fluxes of carbon back to the atmosphere F_{21} and F_{31} increase so that the net flux from atmosphere to ocean decreases. In this way, the fraction of CO_2 added to the atmosphere that is taken up by the oceans decreases with increasing CO_2 burden because of reduced buffer capacity of the $CO_2 \cdot H_2O/HCO_3^-/CO_3^{2-}$ system.

F_{15} expresses the rate of photosynthetic uptake of atmospheric CO_2 by vegetation. Physically, F_{15} increases with increasing CO_2 but eventually saturates. Lenton (2000) has suggested the following empirical form for F_{15}:

$$F_{15} = \begin{cases} 0 & M_1 \leq \gamma \\ k_{15}\, G\, \dfrac{M_1 - \gamma}{M_1 + \Gamma} & M_1 > \gamma \end{cases} \qquad (22.19)$$

γ is a threshold value for M_1, below which no CO_2 uptake occurs ($\gamma = 62$ Pg C). Γ is a saturation parameter ($\Gamma = 198$ Pg C). At sufficiently high CO_2 (in the range of 800–1000 ppm), increasing carbon assimilation via photosynthesis would cease to increase. $G(t)$ is a scaling factor that corrects the flux from the atmosphere to the terrestrial biota for the permanent effects of deforestation and reforestation that have taken place from preindustrial times (1850) until time t. For example, if forested areas are cleared and converted to urban use, then those forested areas are forever removed from the Earth's supply of terrestrial biota, and, even in the absence of any fossil fuel burning, the new carbon cycle steady state will be different from the preindustrial state because the CO_2 uptake capacity of terrestrial biota will have been permanently altered. The preindustrial value of G is taken to be 1.0.

$G(t)$ can be expressed by

$$G(t) = 1 + \int_0^t \frac{a_r F_r - a_d F_d}{M_5(0)} dt \qquad (22.20)$$

where time zero is taken as preindustrial (1850), a_d is the fraction of forested area that is not available for regrowth after deforestation, and a_r is the fraction of reforested area that increases the Earth's terrestrial biota. If no deforestation or reforestation occurs, then the integral is zero, and $G = 1$. The integral accounts for the permanent effects of changes to terrestrial vegetation. For example, if only deforestation occurs, then $G < 1$, and F_{15}, the flux of CO_2 from the atmosphere to vegetation, as given by (22.19), is less than that if deforestation had not taken place.

Equation (22.20) can be replaced by its corresponding differential equation:

$$\frac{dG}{dt} = \frac{a_r F_r - a_d F_d}{M_5(0)}; \qquad G(0) = 1 \qquad (22.21)$$

Returning to (22.19), the numerical value of k_{15} can be calculated from the preindustrial values in Figure 22.6, using $G = 1$ and the values of γ and Γ.

The full compartmental model and numerical values of the coefficients are given in Table 22.1. It is assumed that all reforested land increases the terrestrial biota, so $a_r = 1.0$. The value of $a_d = 0.23$ is that suggested by Schmitz (2002) [Lenton (2000) proposed 0.27]. Initial values for all M_i are the preindustrial values given in Figure 22.6. The preindustrial fossil fuel reservoir is assumed to have contained 5300 Pg carbon; the actual value is not important, only the emission rate $F_f(t)$. The model requires as input $F_f(t), F_d(t)$, and $F_r(t)$ from preindustrial times to the present. $F_f(t)$ is obtained from the historical record of carbon emissions from fossil fuels; $F_d(t)$ is that for deforestation, expressed also in units of Pg C yr^{-1}. Until very recently, reforestation, F_r, can be assumed to have been negligibly small.

The differential equations for $M_1, M_2, M_3, M_4, M_5, M_6, M_7$, and G can be integrated numerically to predict burdens of atmospheric CO_2 from 1850 to the present using the emissions from Figure 22.5; the result is shown in Figure 22.7, and agreement with observed mixing ratios over this period is close. Table 22.2 compares the predicted quantities of carbon in each of the seven compartments in 1990 with their preindustrial values. We also note that G is predicted to have decreased from 1.0 to 0.952 over this time period. Of the six reservoirs, only the atmosphere has shown a substantial increase in carbon loading since preindustrial times. Carbon in terrestrial biota (M_5) is predicted to have decreased by only 3 Pg, from 580 to 577 Tg; decreases resulting from deforestation and increased uptake owing to higher ambient CO_2 levels virtually offset each other.

22.2.3 Atmospheric Lifetime of CO_2

As we know, the mean lifetime of a species at steady state can be computed as the ratio of its total abundance (e.g., Tg) to its total rate of removal (e.g., Tg yr^{-1}). CO_2 can be presumed to have been in steady state under preindustrial conditions, in which, according to Figure 22.6, the total atmospheric abundance was 612 Pg and the total rate of removal was 176 Pg yr^{-1}. Thus, the overall mean lifetime of a molecule of CO_2 in the preindustrial atmosphere was 3.5 years. For the deep-ocean reservoir, the mean carbon residence time was 37,000 Pg/112 Pg $yr^{-1} = 330$ years. At these steady-state conditions the overall

TABLE 22.1   Compartmental Model for the Global Carbon Cycle

Governing Equations	Symbol	Value	Units
$\dfrac{dM_1}{dt} = -(k_{12} + k_{13})M_1 - k_{15}G\dfrac{M_1 - \gamma}{M_1 + \Gamma} + k_{21}M_2^{\beta_2}$	k_{12}	0.0931	yr^{-1}
$\quad + k_{31}M_3^{\beta_3} + k_{51}M_5 + k_{61}M_6 + F_f(t) + F_d(t) - F_r(t)$	k_{13}	0.0311	yr^{-1}
$\dfrac{dM_2}{dt} = k_{12}M_1 - (k_{23} + k_{24})M_2 - k_{21}M_2^{\beta_2} + k_{42}M_4$	k_{15}	147	yr^{-1}
$\dfrac{dM_3}{dt} = k_{13}M_1 + k_{23}M_2 - k_{34}M_3 - k_{31}M_3^{\beta_3} + k_{43}M_4$	k_{21}	$58(730^{-\beta_2})$	$PgC^{(1-\beta_2)}yr^{-1}$
$\dfrac{dM_4}{dt} = k_{24}M_2 + k_{34}M_3 - (k_{42} + k_{43})M_4$	k_{23}	0.0781	yr^{-1}
$\dfrac{dM_5}{dt} = k_{15}G\dfrac{M_1 - \gamma}{M_1 + \Gamma} - (k_{51} + k_{56})M_5 - F_d(t) + F_r(t)$	k_{24}	0.0164	yr^{-1}
$\dfrac{dM_6}{dt} = k_{56}M_5 - k_{61}M_6$	k_{31}	$18(140^{-\beta_3})$	$PgC^{(1-\beta_3)}yr^{-1}$
$\dfrac{dM_7}{dt} = -F_f(t)$	k_{34}	0.714	yr^{-1}
$\dfrac{dG}{dt} = -\left[\dfrac{a_d F_d(t) - a_r F_r(t)}{M_5(0)}\right]$	k_{42}	0.00189	yr^{-1}
	k_{43}	0.00114	yr^{-1}
	k_{51}	0.0862	yr^{-1}
	k_{56}	0.0862	yr^{-1}
	k_{61}	0.0333	yr^{-1}
	β_2	9.4	
	β_3	10.2	
	γ	62.0	PgC
	Γ	198	PgC
	a_d	0.230	
	a_r	1.0	

Source: Schmitz (2002).

TABLE 22.2   Pre–Industrial Revolution and 1990 Predicted Compartmental Burdens of Carbon (Pg C)

	Preindustrial	1990
M_1	612	753
M_2	730	744
M_3	140	143
M_4	37000	37071
M_5	580	577
M_6	1500	1489
M_7	5300	5086
G	1.0	0.952

Source: Schmitz (2002).

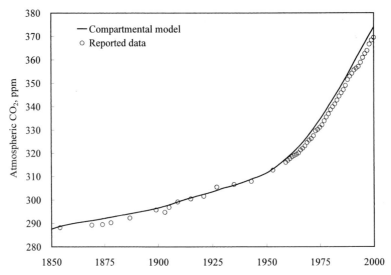

FIGURE 22.7   Prediction of atmospheric carbon loading from preindustrial time to the year 2000 from numerical integration of the six-compartment model given in Table 22.1 with initial conditions based on the carbon amounts shown in Figure 22.6. Data points correspond to global average CO_2 data.

sources and sinks of carbon into and out of any reservoir are equal; for example, the overall rate of removal from the atmosphere of 176 Pg yr^{-1} was balanced by an equal overall emission rate into the atmosphere. The removal of CO_2 from the atmosphere occurs for three major reservoirs: terrestrial biota and warm and cool ocean surface waters; each of these reservoirs exchanges carbon not just with the atmosphere but with additional reservoirs, soils and detritus, and the deep ocean, where the soil and detritus reservoir is also a direct source of carbon to the atmosphere. Under preindustrial steady-state conditions, flows into and out of each of these reservoirs were in balance. Note, however, that the mean carbon lifetime at steady state in each of these reservoirs is different. For terrestrial biota, that lifetime is 11.6 years; for warm ocean surface waters, 5.7 years; for cool ocean surface waters, 1.2 years; and for soils and detritus, 30 years. And, as noted above, the carbon residence time in the deep-ocean reservoir was 330 years.

Starting under preindustrial conditions, increasing amounts of CO_2 have been emitted year by year to the atmosphere (Figure 22.5). The dynamic response of the atmospheric compartment is represented in simplified form by (2.14)

$$\frac{dQ}{dt} = P - \frac{Q}{\tau} \tag{22.22}$$

where Q is the atmospheric CO_2 abundance (in Pg C) above the initial preindustrial steady-state value and P is the anthropogenic emission rate (Pg C yr^{-1}). With these definitions of Q and P, the initial condition for (22.22) is $Q = 0$ at $t = 0$; because we are starting from a steady state, it suffices to consider the dynamics of the perturbed system to reach its new steady state.

A critical issue is how to interpret the value of τ in (22.22). Equation (22.22) only represents a balance on the amount of atmospheric carbon over the preindustrial steady state. It does not account for how carbon levels in the other reservoirs are changing and how these changes, in turn, affect the dynamics of the atmospheric carbon. In (22.22), then, τ represents essentially a relaxation time for the atmospheric carbon system to reach a new steady state corresponding to P. The mean atmospheric carbon lifetime derived from the preindustrial steady state of Figure 22.6 is not the same as τ in (22.22) because the atmospheric carbon is no longer in steady state once the anthropogenic perturbation occurs. Because $Q, dQ/dt$, and P are relatively well known, one can actually use these quantities in (22.22) to solve for the relaxation time τ from

$$\tau = \frac{Q}{P - dQ/dt} \qquad (22.23)$$

Because $Q, dQ/dt$, and P vary yearly, the relaxation time τ inferred from this equation also varies from year to year.

Let us estimate τ, for example, in year 2000 based on (22.23). To determine dQ/dt, consider the period 1995–2000, during which CO_2 increased at a rate of 1.8 ppm yr^{-1} (Keeling and Whorf 2003). This equates to $(1.8 \text{ ppm yr}^{-1})(2.1 \text{ Pg C ppm}^{-1}) = 3.78 \text{ Pg yr}^{-1}$. The fossil fuel CO_2 emission rate in 2000 has been estimated as 6.6 Pg C yr^{-1} (Marland et al. 2003) and that from biomass burning as ranging between 1.5 and 2.7 Pg C yr^{-1} (Jacobson 2004). This produces a CO_2 emission range of 8.1–9.3 Pg C yr^{-1}. The global average CO_2 mixing ratio in 2000 was 370 ppm. With an assumed preindustrial mixing ratio of 280 ppm, the anthropogenic burden of CO_2 in 2000 was $(90 \text{ ppm})(2.1 \text{ Pg C ppm}^{-1}) = 189 \text{ Pg C}$. On the basis of the upper and lower estimates of the CO_2 emission rate in year 2000, we obtain the following inferred range for the mean atmospheric relaxation time of CO_2 in year 2000:

$$\tau \sim 34 - 44 \text{ yr}$$

Similar calculations are presented by Gaffin et al. (1995) and Jacobson (2005).

Figure 22.8 shows the value of τ as inferred from $Q, dQ/dt$, and P from the emissions in Figure 22.5 and the atmospheric loading in Figure 22.7 from 1860 to 2000. The fact that the value of τ is not the same as the 3.5-year steady-state preindustrial CO_2 lifetime is clearly evident from the fact that the relaxation time corresponding to the initial anthropogenic perturbation in 1860 is about 17 years. In other words, in 1860 it would have taken 17 years for the atmosphere to reach a new steady state corresponding to the initial anthropogenic emissions. This roughly 35-year relaxation time reflects the time needed for a quasi-equilibrium to be reached between atmospheric CO_2 and the surface ocean waters and terrestrial biosphere.

Only about half of the annual CO_2 emissions remain in the atmosphere; the remainder partition primarily into the near-surface ocean layer. The equilibrium ratio is about 0.7 : 1.0 for CO_2 partitioning between the atmosphere and the near-surface ocean water. At present, the total amount of CO_2 in the atmosphere is about 750 Pg C, with approximately 900 Pg C in the near-surface ocean layer. If humans add 700 Pg C by the year 2050, this CO_2 will partition by equilibration (with the roughly 35-year relaxation time) between the atmosphere and the surface ocean layer. Neglecting any significant uptake by biomass, the total C in the atmosphere/surface ocean will be $750 + 900 + 700 = 2350$ Pg C. At an equilibrium partitioning ratio of 0.7 : 1.0, this will lead to ~ 950 Pg C in the atmosphere and ~ 1400 Pg C in the near-surface ocean. The atmospheric CO_2 mixing ratio will be ~ 475 ppm.

Mixing of the surface ocean waters with the deep ocean will eventually reduce these carbon burdens in the atmosphere and surface ocean, but the characteristic time for this

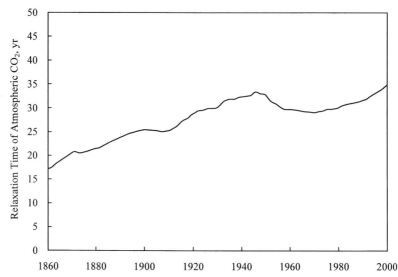

FIGURE 22.8   Relaxation time of atmospheric CO_2 to adjust to the perturbations induced by anthropogenic emissions over the 1860–2000 period. Note that this relaxation time is based on achieving a quasi-equilibrium with the surface ocean waters and the terrestrial biosphere and does not reflect mixing with the deep ocean, the characteristic time of which is several hundred years.

mixing is several hundred years. *For this reason, if global CO₂ emissions were abruptly stopped today, it would take almost a millennium for atmospheric CO₂ to return to a preindustrial level.*

22.3   ANALYTICAL SOLUTION FOR A STEADY-STATE FOUR-COMPARTMENT MODEL OF THE ATMOSPHERE

Consider Figure 22.9, in which four natural atmospheric reservoirs are depicted, the NH and SH troposphere and the NH and SH stratosphere.[3] We use this model of four interconnected atmospheric compartments as a vehicle to derive balance equations from which global biogeochemical cycles can be analyzed.

We will assume that the substance of interest is removed by different first-order processes in the troposphere and stratosphere, characterized by first-order rate constants k_T and k_S. Thus the rates of removal of the substance in the two tropospheres are $k_T Q_{NH}^T$ and $k_T Q_{SH}^T$; the rates of removal in the two stratospheres are $k_S Q_{NH}^S$ and $k_S Q_{SH}^S$. k_T could represent, for example, OH reaction in the troposphere and k_S could represent photolysis in the stratosphere. k_T can also include removal at the Earth's surface.

---

[3] While the division of the troposphere into two compartments, NH and SH, is reasonable in terms of estimating lifetimes of relatively long-lived tropospheric constituents, such a division is less applicable in the stratosphere. In the case of stratospheric transport and mixing, a better compartmental division would be between the tropical and midlatitude stratosphere. With such a division, the stratosphere would actually be represented by three compartments, the tropical stratosphere and the NH and SH midlatitude to polar stratospheres. The development in this section can be extended to such a five-compartment model, if desired.

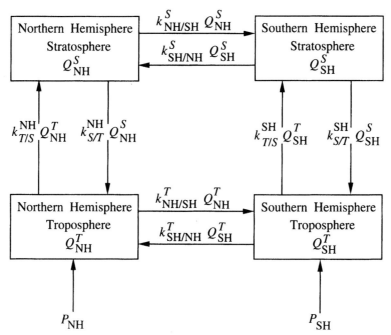

FIGURE 22.9 Four-compartment model of the atmosphere (NH = Northern Hemisphere, SH = Southern Hemisphere, S = stratosphere, T = troposphere). P_{NH} and P_{SH} are source emission rates of the compound in the NH and SH troposphere, respectively. The quantities of the species of interest in the four reservoirs are denoted $Q_{NH}^T, Q_{NH}^S, Q_{SH}^T,$ and Q_{SH}^S. The fluxes between the reservoirs are proportional to the content of the compound in the reservoir where the flux originates. The flux of material from the NH troposphere to the SH troposphere is $k_{NH/SH}^T Q_{NH}^T$ and the reverse flux is $k_{SH/NH}^T Q_{SH}^T$. Other intercompartmental fluxes are defined similarly.

A dynamic balance on the mass of substance in the NH troposphere component is

$$\frac{dQ_{NH}^T}{dt} = k_{SH/NH}^T Q_{SH}^T - k_{NH/SH}^T Q_{NH}^T$$

(exchange between NH and SH tropospheres)

$$+ k_{S/T}^{NH} Q_{NH}^S - k_{T/S}^{NH} Q_{NH}^T$$

(exchange between NH tropospheres and stratosphere)   (22.24)

$$- k_T^{NH} Q_{NH}^T$$

(removal in troposphere)

$$+ P_{NH}$$

(source emission into NH troposphere)

At steady state, the sources and sinks balance:

$$0 = k_{SH/NH}^T Q_{SH}^T - k_{NH/SH}^T Q_{NH}^T + k_{S/T}^{NH} Q_{NH}^S - k_{T/S}^{NH} Q_{NH}^T$$
$$- k_T^{NH} Q_{NH}^T + P_{NH}$$

(22.25)

This equation can be rearranged as follows:

$$0 = -(k^T_{NH/SH} + k^{NH}_{T/S} + k^{NH}_T)Q^T_{NH} + k^T_{SH/NH}Q^T_{SH} + k^{NH}_{S/T}Q^S_{NH} + P_{NH} \tag{22.26}$$

A similar steady-state balance on the SH troposphere yields

$$0 = -(k^T_{SH/NH} + k^{SH}_{T/S} + k^{SH}_T)Q^T_{SH} + k^T_{NH/SH}Q^T_{NH} + k^{SH}_{S/T}Q^S_{SH} + P_{SH} \tag{22.27}$$

Comparable balances on the two stratosphere compartments yield

$$0 = -(k^S_{NH/SH} + k^{NH}_{S/T} + k^{NH}_S)Q^S_{NH} + k^S_{SH/NH}Q^S_{SH} + k^{NH}_{T/S}Q^T_{NH} \tag{22.28}$$

$$0 = -(k^S_{SH/NH} + k^{SH}_{S/T} + k^{SH}_S)Q^S_{SH} + k^S_{NH/SH}Q^S_{NH} + k^{SH}_{T/S}Q^T_{SH} \tag{22.29}$$

Equations (22.26) to (22.29) constitute four equations in the four unknowns $Q^T_{NH}, Q^T_{SH}, Q^S_{NH}$, and Q^S_{SH}. It is useful to rewrite these equations as

$$0 = -\alpha_1 Q_1 + \alpha_2 Q_2 + \alpha_3 Q_3 + P_1 \tag{22.30}$$

$$0 = -\alpha_4 Q_2 + \alpha_5 Q_1 + \alpha_6 Q_4 + P_2 \tag{22.31}$$

$$0 = -\alpha_7 Q_3 + \alpha_8 Q_4 + \alpha_9 Q_1 \tag{22.32}$$

$$0 = -\alpha_{10} Q_4 + \alpha_{11} Q_3 + \alpha_{12} Q_2 \tag{22.33}$$

where $Q_1 = Q^T_{NH}, Q_2 = Q^T_{SH}, Q_3 = Q^S_{NH}, Q_4 = Q^S_{SH}, P_1 = P_{NH}$, and $P_2 = P_{SH}$. Also

$$\alpha_1 = k^T_{NH/SH} + k^{NH}_{T/S} + k^{NH}_T \qquad \alpha_2 = k^T_{SH/NH} \qquad \alpha_3 = k^{NH}_{S/T}$$

$$\alpha_4 = k^T_{SH/NH} + k^{SH}_{T/S} + k^{SH}_T \qquad \alpha_5 = k^T_{NH/SH} \qquad \alpha_6 = k^{SH}_{S/T}$$

$$\alpha_7 = k^S_{NH/SH} + k^{NH}_{S/T} + k^{NH}_S \qquad \alpha_8 = k^S_{SH/NH} \qquad \alpha_9 = k^{NH}_{T/S}$$

$$\alpha_{10} = k^S_{SH/NH} + k^{SH}_{S/T} + k^{SH}_S \qquad \alpha_{11} = k^S_{NH/SH} \qquad \alpha_{12} = k^{SH}_{T/S}$$

Equations (22.32) and (22.33) can be solved simultaneously to obtain Q_3 and Q_4 in terms of Q_1 and Q_2

$$Q_3 = \beta_1 Q_2 + \beta_2 Q_1 \tag{22.34}$$

$$Q_4 = \beta_3 Q_2 + \beta_4 Q_1 \tag{22.35}$$

where

$$\beta_1 = \frac{\dfrac{\alpha_8\,\alpha_{12}}{\alpha_7\,\alpha_{10}}}{1 - \dfrac{\alpha_8\,\alpha_{11}}{\alpha_7\,\alpha_{10}}} \qquad\qquad \beta_2 = \frac{\dfrac{\alpha_9}{\alpha_7}}{1 - \dfrac{\alpha_8\,\alpha_{11}}{\alpha_7\,\alpha_{10}}}$$

$$\beta_3 = \frac{\dfrac{\alpha_{11}}{\alpha_{10}}\dfrac{\alpha_8\,\alpha_{12}}{\alpha_7\,\alpha_{10}}}{1 - \dfrac{\alpha_8\,\alpha_{11}}{\alpha_7\,\alpha_{10}}} + \frac{\alpha_{12}}{\alpha_{10}} \qquad \beta_4 = \frac{\dfrac{\alpha_9\,\alpha_{11}}{\alpha_7\,\alpha_{10}}}{1 - \dfrac{\alpha_8\,\alpha_{11}}{\alpha_7\,\alpha_{10}}}$$

The resulting equations for Q_1 and Q_2 are

$$0 = (\alpha_3\beta_2 - \alpha_1)Q_1 + (\alpha_2 + \alpha_3\beta_1)Q_2 + P_1 \tag{22.36}$$

$$0 = (\alpha_5 + \alpha_6\beta_4)Q_1 + (\alpha_6\beta_3 - \alpha_4)Q_2 + P_2 \tag{22.37}$$

the solutions of which are

$$Q_1 = \frac{(\alpha_2 + \alpha_3\beta_1)P_2 - (\alpha_6\beta_3 - \alpha_4)P_1}{(\alpha_3\beta_2 - \alpha_1)(\alpha_6\beta_3 - \alpha_4) - (\alpha_5 + \alpha_6\beta_4)(\alpha_2 + \alpha_3\beta_1)} \tag{22.38}$$

$$Q_2 = \frac{(\alpha_5 + \alpha_6\beta_4)P_1 - (\alpha_3\beta_2 - \alpha_1)P_2}{(\alpha_3\beta_2 - \alpha_1)(\alpha_6\beta_3 - \alpha_4) - (\alpha_5 + \alpha_6\beta_4)(\alpha_2 + \alpha_3\beta_1)} \tag{22.39}$$

Equations (22.38) and (22.39) give the steady-state concentrations of the substance in the NH and SH troposphere, respectively ($Q_1 = Q_{NH}^T$ and $Q_2 = Q_{SH}^T$), as a function of the source rates into the two hemispheres and all the transport and removal parameters of the four compartments. Steady-state concentrations in the two stratospheric reservoirs, $Q_3 = Q_{NH}^S$ and $Q_4 = Q_{SH}^S$, are then obtained from (22.34) and (22.35). These equations provide a general, steady-state analysis of a four-compartment model of a substance that is emitted into the troposphere and removed by separate first-order processes in each of the four compartments.

The mass of the substance in the entire atmosphere is the sum of the masses in the four compartments,

$$Q_{total} = Q_{NH}^T + Q_{SH}^T + Q_{NH}^S + Q_{SH}^S \tag{22.40}$$

The overall average residence time of a molecule of the substance in the atmosphere can be obtained by dividing its total quantity in the atmosphere by its total rate of introduction from sources,

$$\tau = \frac{Q_{total}}{P_{NH} + P_{SH}} \tag{22.41}$$

Average residence times in any of the four compartments can be calculated from the quantity in the compartment divided by the net source rate in that compartment.

The foregoing analysis can be simplified to fewer than four atmospheric compartments. For a substance that is completely removed in the troposphere, only the two tropospheric hemispheric components need be considered. If the lifetime of such a substance is shorter than the time needed to mix throughout the entire global troposphere, then a two-compartment model, NH troposphere and SH troposphere, is called for. If the substance's lifetime is long compared to the interhemispheric mixing time, then the entire troposphere can be considered as a single compartment. Because horizontal mixing in the stratosphere is so much faster than vertical mixing, for many substances that reach the stratosphere, the stratosphere can be considered as a single compartment.

Application of the Four-Compartment Model to Methyl Chloroform (CH$_3$CCl$_3$)
Methyl chloroform is an anthropogenic substance, the total emissions of which to the atmosphere are reasonably well known. Its atmospheric degradation occurs almost entirely by hydroxyl radical reaction. The CH$_3$CCl$_3$ mixing ratio in the atmosphere is well established; thus the global steady-state budget of CH$_3$CCl$_3$ can be used as a means of estimating a global average OH radical concentration.

Such estimates are usually accomplished with a three-dimensional atmospheric model (Prinn et al. 1992), but here we apply the simple four-compartment model to analyze the global budget of CH_3CCl_3.

The following emissions data for CH_3CCl_3 are available (Prinn et al. 1992):

$$P_{NH} = 5.647 \times 10^{11}\,g\,yr^{-1} \quad (1978 - 1990\ average)$$
$$P_{SH} = 2.23 \times 10^{10}\,g\,yr^{-1}$$

CH_3CCl_3 is removed by OH reaction

$$CH_3CCl_3 + OH \rightarrow CH_2CCl_3 + H_2O$$

with a rate constant $k = 1.6 \times 10^{-12} \exp(-1520/T)$ (see Table B.1). To evaluate average values of the rate constant in the troposphere and stratosphere, we use the tropospheric average temperature of 277 K. An average temperature for the stratosphere is taken as that at 12 km in the U.S. Standard Atmosphere, $T = 216.7$ K. A global average tropospheric OH concentration is taken as 8.7×10^5 molecules cm^{-3} (Prinn et al. 1992). An average stratospheric OH mixing ratio in the midlatitude is 1 ppt (see Chapter 5), which translates, at 12 km, into a concentration of 6.48×10^6 molecules cm^{-3}. CH_3CCl_3 is also degraded by photolysis in the stratosphere, but at a rate almost 4000 times slower than OH reaction, so we will neglect it here. Finally, CH_3CCl_3 is lost by deposition to the Earth's surface with a first-order loss coefficient of 0.012 yr^{-1} (Prinn et al. 1992).

For the exchange rates among the four atmospheric compartments, we use the following values:

$$k^T_{SH/NH} = k^T_{NH/SH} = 1.0\,yr^{-1}$$

$$k^S_{SH/NH} = k^S_{NH/SH} = 0.25\,yr^{-1}$$

$$k^{NH}_{S/T} = k^{SH}_{S/T} = 0.4\,yr^{-1} \quad \text{(van Velthoven and Kelder 1996)}$$

$$k^{NH}_{T/S} = k^{SH}_{T/S} = 0.063\,yr^{-1} \quad \text{(van Velthoven and Kelder 1996)}$$

The four-compartment model can be used, with the parameter values and yearly emission rates given above, to estimate the steady-state mixing ratios of CH_3CCl_3 in the troposphere and stratosphere and the overall atmospheric residence time of CH_3CCl_3. The results are given below.

	Calculated by Four-Compartment Model	Observed Mixing Ratio
Tropospheric mixing ratio, ppt	129	160[b]
Stratospheric mixing ratio, ppt	75	
Atmospheric lifetime, yr	5.0[a]	

[a]The CH_3CCl_3 lifetime estimated by IPCC (2001) is 4.8 years based on three-dimensional model calculations.
[b]See Table 2.15.

PROBLEMS

22.1$_A$ In the simplified calculation of the atmospheric sulfur cycle, if the value of c, the SO_2 fraction of the total sulfur, is taken as 0.5, a sulfur atom residence time of 50 h is estimated. What is the value of b, the fraction of sulfur converted to SO_4^{2-} before being removed, that is consistent with this choice of c?

22.2$_A$ It is desired to compute the global emissions of carbon that resulted from oil and gas production in 1986. Worldwide oil production in 1986 was 5.518×10^7 barrels day^{-1}, and worldwide gas production in the same year was 6.22×10^{13}ft^{3}yr^{-1} (Katz and Lee 1990). For oil a density of 0.85 g cm^{-3} can be assumed. Assume that oil has an average composition of $C_{10}H_{22}$. A barrel has volume of 0.16 m^3. Assume that the gas produced is entirely methane. Calculate global emissions from each in g(C) yr^{-1}.

22.3$_A$ The current population of the Earth is about 6.5 billion people. Each person, on average, exhales about 1 kg of CO_2 per day. How does this compare with current fossil fuel emissions? This amount is not included in inventories of CO_2 emissions. Why?

22.4$_A$ **a.** Plot the relative amount of CO_2 remaining in the atmosphere from an instantaneous emission in year zero. Use (22.22) with $\tau = 40$ years.

   b. Repeat part (a) for a sustained emission of $P = 0.01$ Pg C yr^{-1} that begins in year zero.

22.5$_B$ Derive the balance equations for a substance that is completely removed in the troposphere, but for which two tropospheric reservoirs should be considered. Apply the balance to CO using the source and sink data from Table 2.14. As a first approximation, assume that anthropogenic sources are totally concentrated in the Northern Hemisphere and that biomass burning sources are totally in the Southern Hemisphere. The CH_4 oxidation source can be apportioned according to the NH/SH ratio of CH_4 concentration. NMHC oxidation can be assumed to be entirely in the NH. Biogenic CO sources can be equally apportioned between the NH and SH, and the ocean source according to a 2 : 1 ratio SH to NH. Soil uptake can be apportioned in the reverse ratio. The NH/SH exchange rates in Section 22.3 can be used. A global mean OH concentration of 8.7×10^5 molecules cm^{-3} can be assumed, and the CO–OH reaction rate constant is given in Table B.1. Calculate the CO mixing ratios in the NH and SH, using the mean values of the ranges in Table 2.14 and compare with those observed.

22.6$_D$ **a.** Use the compartmental model for the global carbon cycle (Table 22.1) to estimate the concentration of CO_2 in the atmosphere assuming that

$$F_r(t) = 0$$
$$F_d(t) = 0.3 + 0.01\,t \qquad\qquad (t = 0 \text{ for } 1850)$$
$$F_f(t) = \begin{cases} 0.014\,t & \text{from 1850 to 1950 } (t = 0 \text{ for } 1850) \\ 1.4 + (4.6/40)\,t & \text{from 1950 to 1990 } (t = 0 \text{ for } 1950) \end{cases}$$

   where the fluxes are in Pg(C) yr^{-1} and t is in yr. Compare this value to the current measurements.

b. Assume that both deforestation and fossil fuel combustion emissions continue to increase at the same rate [0.01 Pg(C) yr^{-1} for deforestation and 46/40 Pg(C) yr^{-1}] for fossil fuel combustion. When will the concentration of CO_2 in the atmosphere double?

c. Assume that global emissions of CO_2 from fossil fuel combustion and deforestation are stabilized at their 1990 levels. What will be the CO_2 concentration in 2050 and 2100?

22.7$_D$ One approach to slowing down the increase of the CO_2 concentration in the atmosphere and stabilizing it at a lower level is carbon sequestration, that is, the transfer of some of the atmospheric CO_2 (or preferably of the CO_2 produced by a source like power plants) to one of the reservoirs in the hydrosphere or lithosphere. Let us investigate the effectiveness of sequestration in the deep-ocean waters as an example using the compartmental model for the global cycle (Table 22.1). Assume that $t = 0$ corresponds to year 2010, and use the 1990 values of Table 22.2 as the initial conditions for the various reservoirs.

a. Calculate the CO_2 concentration from 2010 to 2150 assuming that

$$F_r(t) = 0$$
$$F_d(t) = 1.8 + 0.01\,t \ \ (t = 0 \text{ for } 2010)$$
$$F_f(t) = 6 + (4.6/40)\,t \ \ (t = 0 \text{ for } 2010)$$

This is the "business as usual" scenario that assumes that we will continue to increase the CO_2 emissions at the present rate.

b. Let us assume that starting in 2010, F_0 Pg(C)yr^{-1} are transferred from the atmosphere to the deep ocean. Extend the model to account for this additional flux and repeat the calculations of part (a) for $F_0 = 1.3$ Pg(C)yr^{-1} (this corresponds to 20% of the current emissions from fossil fuel combustion).

c. What should be the value of F_0 to avoid reaching a CO_2 concentration of 500 ppm by 2150?

REFERENCES

Barnola, J. M., Raynaud, D., Korotkevich, Y. S., and Lorius, C. (1987) Vostok ice core provides 160,000-year record of atmospheric CO_2, *Nature* **329**, 408–414.

Berresheim, H., Wine, P. H., and Davis, D. D. (1995) Sulfur in the atmosphere, in *Composition, Chemistry, and Climate of the Atmosphere*, H. B. Singh, ed., Van Nostrand Reinhold, New York, pp. 251–307.

Charlson, R. J., Anderson, T. L., and McDuff, R. E. (1992) The sulfur cycle, in *Global Biogeochemical Cycles*, S. S. Butcher, R. J. Charlson, G. H. Oriana, and G. V. Wolfe, eds., Academic Press, New York, pp. 285–300.

Druffel, E. R. M., Williams, P. M., Bauer, J. E., and Ertel, J. R. (1992) Cycling of dissolved particulate organic matter in the open ocean, *J. Geophys. Res.* **97**, 15639–15659.

Falkowski, P. et al. (2001) The global carbon cycle: A test of our knowledge of earth as a system, *Science* **290**, 291–296.

Gaffin, S. R., O'Neill, B. C, and Oppenheimer, M. (1995) Comment on "The lifetime of excess atmospheric carbon dioxide," by Berrien Moore III and B. H. Braswell, *Global Biochem. Cycles* **9**, 167–169.

Genthon, C., Barnola, J. M., Raynaud, D., Lorius, C., Jouzel, J., Barkov, N. I., Korotkevich, Y. S., and Kotlyakov, V. M. (1987) Vostok ice core: Climate response to CO_2 and orbital forcing changes over the last climatic cycle, *Nature* **329**, 414–418.

Intergovernmental Panel on Climate Change (IPCC) (1995) *Climate Change 1994: Radiative Forcing of Climate Change and an Evaluation of the IPCC IS92 Emission Scenarios*. Cambridge Univ. Press, Cambridge, UK.

Intergovernmental Panel on Climate Change (IPCC) (2001) *Climate Change 2001: The Scientific Basis*, Cambridge Univ. Press, Cambridge, UK.

Jacobson, M. Z. (2004) The short-term cooling but long-term global warming due to biomass burning, *J. Climate* **17**(l5), 2909–2926.

Jacobson, M. Z. (2005) Correction to "Control of fossil-fuel particulate black carbon and organic matter, possibly the most effective method of slowing global warming," *J. Geophys. Res.* **110**, D14105 (doi: 10.1029/2005JD005888).

Katz, D. L., and Lee, R. L. (1990) *Natural Gas Engineering*, McGraw-Hill, New York.

Keeling, C. D. (1983) The global carbon cycle: What we know and could know from atmospheric, biospheric, and oceanic observations, *Proc. CO_2 Research Conf.: Carbon Dioxide, Science and Consensus*, DOE Conf-820970, II.3-II.62. U.S. Department of Energy, Washington, DC.

Keeling, C. D., and Whorf, T. P. (2003) Atmospheric CO_2 concentrations (ppmv) derived from in situ air samples collected at Mauna Loa Observatory, Hawaii, http://cdiac.esd.ornl.gov/ftp/maunaloa-co2/maunaloa.co2.

Lenton, T. M. (2000) Land and ocean carbon cycle feedback effects on global warming in a simple earth system model, *Tellus B* **52**, 1159–1188.

Marland, G., Boden, T. A., and Andres, R. J. (2003) Global CO_2 emissions from fossil-fuel burning, cement manufacture, and gas flaring: 1751–2000, in *Trends Online: A Compendium of Data on Global Change*, Carbon Dioxide Information Analysis Center, Oak Ridge National Laboratory, U.S. Department of Energy, Oak Ridge, TN, USA.

Prinn, R. et al. (1992) Global average concentration and trend for hydroxyl radicals deduced from ALE/GAGE trichloroethane (methyl chloroform) data from 1978–1990, *J. Geophys. Res.* **97**, 2445–2461.

Rodhe, H. (1978) Budgets and turn-over times of atmospheric sulfur compounds, *Atmos. Environ.* **l2**, 671–680.

Schimel, D. S. (1995) Terrestrial ecosystems and the carbon cycle, *Global Change Biol.* **1**, 77–91.

Schmitz, R. A. (2002) The Earth's carbon cycle, *Chem. Eng. Educ.* 296.

Takahashi, T., Feely, R. A., Weiss, R. F., Wanninkhof, R. H., Chipman, D. W., Sutherland, S. C, and Takahashi, T. T. (1997) Global air-sea flux of CO_2: An estimate based on measurements of sea-air pCO_2 difference, *Proc. Natl. Acad. Sci.* **94**, 8292–8299.

van Velthoven, P. F. J., and Kelder, H. (1996) Estimates of stratosphere–troposphere exchange: Sensitivity to model formulation and horizontal resolution, *J. Geophys. Res.* **101**, 1429–1434.

Ver, L. M. B., Mackenzie, F. T., and Lerman, A. (1999) Biogeochemical responses of the carbon cycle to natural and human perturbations: Past, present, and future, *Am. J. Sci.* **299**, 762–801.

23 Climate and Chemical Composition of the Atmosphere

Climate refers to the mean behavior of the weather over some appropriate averaging time. The actual averaging time to use when defining climate is not immediately obvious; times too short are insufficient to average out normal year-to-year fluctuations; times too long may incur on timescales of climate change itself. Timescales of relevance for climate change are the order of decades to centuries. Balancing these considerations, a traditional averaging time for defining climate is 30 years. Perhaps the most fundamental climate variable is the global annual mean surface temperature, although other variables such as the frequency and amount of rainfall can also be considered. Climate change involves changes not only in the mean values of these variables but also in their variances.

The planetary annual average surface temperature is controlled by the solar energy input and the surface albedo of the planet. The surface temperature adjusts so that the planetary emission of heat to space balances the solar energy input. As we saw in Chapter 4, annually and globally averaged, about $342 \, \mathrm{W \, m^{-2}}$ of solar energy strikes the Earth[1] and 30% is reflected back to space, leaving about $235 \, \mathrm{W \, m^{-2}}$ to be absorbed. Of the $107 \, \mathrm{W \, m^{-2}}$ reflected back to space, clouds account for about two-thirds of the loss and the surface for about one-eighth, with atmospheric Rayleigh scattering reflecting the rest. The average surface temperature of the Earth is about 288 K or 15°C. Were the atmosphere transparent to infrared radiation, the surface temperature would simply adjust to balance the absorbed insolation and would be about 255 K.

Although the total absorbed and emitted energy of the surface–atmosphere system balances at $235 \, \mathrm{W \, m^{-2}}$, the fluxes of processes transferring energy within the system, between the surface and atmosphere and within the atmosphere itself, can exceed the net $235 \, \mathrm{W \, m^{-2}}$ solar input. Actually, about 70 of the $235 \, \mathrm{W \, m^{-2}}$ in absorbed insolation is absorbed by trace constituents, aerosols, and clouds in the atmosphere. About $390 \, \mathrm{W \, m^{-2}}$ is radiated upward from the surface; thus trace atmospheric constituents are responsible for trapping the excess surface-emitted infrared radiation. The Earth–atmosphere system must be in balance with the solar energy input, so the effective radiating temperature of the Earth and its atmosphere must be lower than the surface temperature. As noted in Chapter 4, this is accomplished by the climate system in two steps: (1) the atmosphere is not transparent to outgoing infrared radiation by virtue of H_2O, CO_2, clouds, and other trace constituents; and (2) the temperature of the atmosphere decreases with altitude in the troposphere.

---

[1] Think of this as 3.4 100-W lightbulbs for each square meter of the Earth's surface.

Atmospheric Chemistry and Physics: From Air Pollution to Climate Change, Second Edition, by John H. Seinfeld and Spyros N. Pandis.   Copyright © 2006 John Wiley & Sons, Inc.

Any factor that alters the Earth's radiation balance, the radiation received from the Sun or lost to space, or that alters the redistribution of energy in the atmosphere–land–ocean system can affect climate. Changes in the solar energy reaching the Earth are the main external forcing mechanism on climate. The Sun's output of energy varies by small amounts over the 11-year cycle associated with sunspots, and the solar output may vary by larger amounts over longer time periods. As mentioned in Chapter 4, subtle variations in the Earth's orbit, over periods of multiple decades to thousands of years, have led to changes in the solar radiation reaching the Earth that are thought to have played an important role in past climate changes, including the formation of ice ages.

It is important to distinguish between *climate forcing* and *climate response*. Climate forcing is a change imposed on the planetary energy balance that has the potential to alter global temperature, for example, a change in solar radiation incident on Earth or change in atmospheric CO_2 abundance. Climate forcing is measured in watts per square meter $(W\,m^{-2})$. Climate response is the meteorological result of these forcings, such as global temperature change, rainfall changes, or sea-level changes.

The concept that as human activity results in ever more CO_2 being released the temperature of the globe can increase was first enunciated by Svante Arrhenius in 1896 (Arrhenius 1896). After Arrhenius' pioneering paper almost half a century passed before concern was initially expressed about possible global warming. In 1938 British meteorologist Guy Callendar asserted that atmospheric CO_2 levels had increased by 10% since the 1890s, but his paper received little attention (Weart 1997). In fact, it was believed that the oceans would absorb the vast majority of any CO_2 that was put into the atmosphere. Moreover, it was thought that CO_2 infrared absorption lines lay right on top of those for water vapor and thus addition of CO_2 would not affect radiation that was already being absorbed by H_2O. In 1955, G. Plass accurately calculated the infrared absorption spectrum of CO_2 and showed that the CO_2 lines do not lie right on top of water vapor absorption lines as had previously been thought; this result meant that adding CO_2 to the atmosphere would lead to interception of more infrared radiation. Then in 1956 R. Revelle showed that, because of the chemistry of seawater, the ocean surface layer did not have the virtually unlimited capacity for CO_2 absorption as previously believed. Cracks were opening up in the "theory" that CO_2 could not affect climate. Finally, Charles Keeling began his measurements of CO_2 in the atmosphere in late 1957; by 1960 he was able to detect a rise after only 2 years of measurements. His data, the standard of atmospheric CO_2, have chronicled the increase in CO_2 ever since (see Figure 22.3). That the concentrations of greenhouse gases (GHGs) have been increasing over the past century is indisputable. As the concentrations of GHGs increase, the net trapping of long-wave, infrared radiation is enhanced. The climate change that will occur as a result depends on a number of factors, including the amount of increase of concentration of each GHG, the radiative properties of each, interactions with other radiatively important atmospheric constituents, and, importantly, climate feedbacks.

Climate varies naturally over all temporal and spatial scales. To distinguish anthropogenic effects on climate from natural variations, it is necessary to identify the human-induced signal in climate against the background noise of natural climate variability. Although the physics of the effects of GHGs and aerosols (see Chapter 24) on atmospheric radiation is beyond question, controversy surrounding the "greenhouse effect" has resulted because it has been difficult to extract an absolutely unambiguous greenhouse warming signal from the climate record. One of the prime reasons for this is because, in addition to GHGs, aerosols can influence climate by absorbing and reflecting solar radiation and by altering cloudiness and cloud reflectivity. There is evidence that the cooling effect of aerosols over the industrialized regions of the Northern Hemisphere may be masking the warming effect of GHGs over that part of the globe. When the effect of aerosols is taken into account in analyzing temperature records, the observed temperature changes are more consistent with what is observed than if only GHGs are considered. Because we need to consider the effect of aerosols in addition to GHGs in comparing observed temperature records to those predicted by climate models, we postpone such an analysis until the next chapter.

23.1   THE GLOBAL TEMPERATURE RECORD

Global temperature data are available from about 1860 to the present. The record of surface temperature over this period, shown in Figure 23.1, reveals a warming since the late nineteenth century of 0.4–0.8°C. Temperatures before the establishment of a global measurement network are determined from a combination of ice core, ocean sediment, tree-ring, fossil, and other geologic data. Annually, deposited layers on ice sheets contain bubbles of air trapped at the time when the ice sheet was formed, providing a means to determine atmospheric concentrations of CO_2 and other trace gases. The layers in the ice cores also contain a record of ocean temperature in the form of the ratio of isotopically "heavy" water (enriched in ^{18}O) to "light" water in the ice. When temperatures are higher, heavy water is evaporated preferentially. From the ocean temperature estimate, the air temperature for large regions can be estimated.

Global mean surface air temperature warmed by about 0.6°C during the twentieth century. The warming is consistent with the global retreat of mountain glaciers,[2] reduction in snow-cover extent, thinning of Arctic sea ice, the accelerated rise of sea level relative to the past few thousand years, and the increase in upper tropospheric water vapor and rainfall rates over most regions. The ocean, the largest heat reservoir in the climate system, has warmed by about 0.05°C averaged over the layer extending from the surface down to about 3 km.

The stratosphere has cooled markedly since the 1960s. This trend is partially a result of stratospheric ozone depletion and partially a result of the accumulation of greenhouse gases, which warm the troposphere but cool the stratosphere. Stratospheric circulation has responded to the radiatively induced temperature changes by concentrating the effects at high altitudes of the winter hemisphere, where cooling of up to 5°C has been observed.

Figure 23.1 also shows the global temperature record of the last 1000 years, as estimated from a variety of sources. The so-called "Medieval Warm Period" (MWP), extending from 1000 to about 1400, was followed by a span of considerably colder climate, the "Little Ice Age," (LIA) from 1450 to 1890, when the global mean temperature may have been 0.5–1.0°C lower than today. During the LIA glaciers moved into lower elevations, and rivers that rarely freeze today were often ice-covered in winter.

The variations in regional surface temperatures for the last 18,000 years are shown in Figure 23.2. In Figure 23.2, we see that the MWP and the LIA are little more than ripples on the longer and more significant warming trend of 4–5°C of the last 15,000 years, which marks the recovery of the Earth from the Ice Age. The present warm epoch, beginning

---

[2] Glaciers are present on every continent except Australia; as such, they constitute an excellent global indicator of climate change.

Variations of the Earth's surface temperature for:

(a) the past 140 years

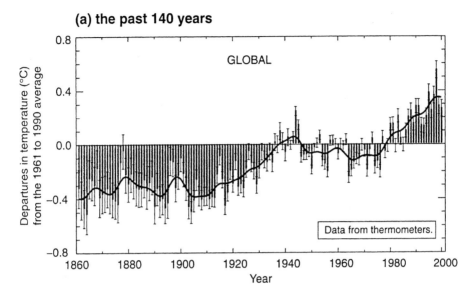

(b) the past 1,000 years

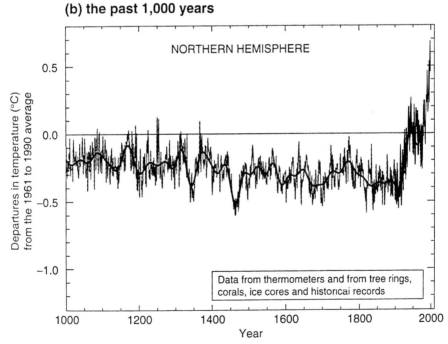

FIGURE 23.1   Variations of the Earth's surface temperature over the last 140 years (a) and the past 1000 years (b) (IPCC 2001).

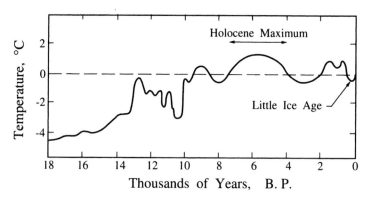

FIGURE 23.2   Variations in surface temperatures for the last 18,000 years, estimated from a variety of sources. Shown are changes (in °C), relative to the value for 1900 [Crowley (1996); based on Bradley and Eddy (1991)].

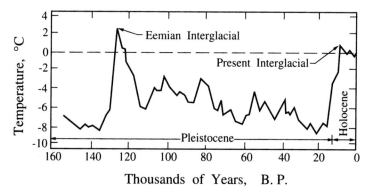

FIGURE 23.3   Air temperature near Antarctica for the last 150,000 years. Temperatures given are inferred from hydrogen/deuterium ratios measured in an ice core from the Antarctic Vostok station, with reference to the value for 1900 [Crowley (1996); based on Bradley and Eddy (1991) and Jouzel et al. (1987)].

roughly 10,000 years ago, is known as the "Holocene interglacial".[3] The "inter" is used because such warm intervals are relatively infrequent on timescales of a million years or so and have lasted on average about 10,000 years before a return to colder climates. The onset and recovery from the Ice Age is now attributed to the Milankovitch cycle (see Chapter 4), slow changes in the Earth's orbit that modify the seasonal cycle of solar radiation at the Earth's surface.

Figure 23.3 shows the air temperature near Antarctica for the last 150,000 years. We see the Holocene Interglacial following the Ice Age that lasted about 100,000 years. The interglacial period that began about 130,000 years ago is called the "Eemian Interglacial." Temperatures during the Eemian Interglacial were 1–2°C higher than those during the Holocene Interglacial. At the time of the coldest portion of the last Ice Age, about

---

[3] The *Holocene* refers to the last 11,000 years of Earth's history—the time since the last major glacial epoch. In general, the Holocene has been a relatively warm period. Another name for the Holocene that is sometimes used is the *Anthropocene*, the "Age of Man."

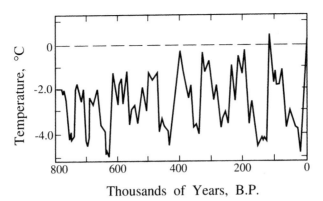

FIGURE 23.4 Estimate of surface temperature for the last 800,000 years, inferred from measurements of the ratio of ^{16}O to ^{18}O in fossil plankton that settled to the sea floor. The use of oxygen isotope ratios is based on the assumption that changes in global temperature approximately track changes in the global ice volume. Detailed studies for the last glacial maximum provide the temperature scale. Shown are changes in temperature (in °C) from the modern value. [Based on data from Imbrie et al. (1984) and presented by Crowley (1996).]

15,000–23,000 years ago, ice sheets more than 2 km thick extended as far south as New York City and London. The massive amount of water contained in the ice sheets required evaporation of about 50×10^6 km^3 of water from the oceans; as a result, sea level was about 105 m below present levels. The sea level retreat exposed most of the continental shelves, and allowed early humans to migrate across the Bering Strait, to populate North and South America. The last Ice Age was also marked by an equatorward expansion of sea ice in both hemispheres, a significant reduction in the area of tropical rainforests, and an expansion of savanna vegetation typical of drier climates. There was also a worldwide increase of the amount of dust in the atmosphere.

Figure 23.4 shows an estimate of the Earth's temperature for the last 800,000 years, as inferred from measurements of the ratio of ^{16}O to ^{18}O in fossil plankton that settled to the ocean floor. The valleys are Ice Ages; the warm peaks are interglacial periods, including, at the right-hand end, the Eemian and Holocene. What is immediately obvious is the rarity, over this roughly million-year period, of periods as warm as the present. Ancestral *Homo sapiens* did not appear until the middle part of the period shown, and modern *Homo sapiens* (Cro-Magnon Man) not until the last 100,000 years. As noted above, periodic changes in the Earth's orbit trigger the waxing and waning of Ice Ages. However, these slow variations alone may not be sufficient to account for glacial–interglacial fluctuations of the last million years. Changes in greenhouse gas levels are thought to amplify the effects of changes in solar radiation.

Finally, Figure 23.5 is a schematic reconstruction of global temperature over the last 100 million years, based on analysis of various marine and terrestrial deposits. The Earth's climate, prior to the last million years, was considerably warmer than present. The warmest temperatures occurred in the Cretaceous Period, about 100 million years ago, when the global mean surface temperature may have been as much as 6–8°C above that of the present. During most of the long period of time up to about 1 million years ago, there is little evidence for ice sheets of continental scale. Subtropical plants and animals existed up to 55°–60° latitude, as compared with their present extent of about 30°N, the latitude of

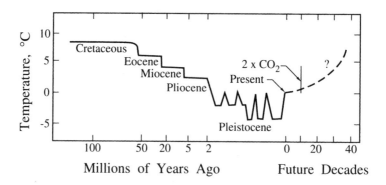

FIGURE 23.5   Schematic reconstruction of mean global surface temperature through the last 100 million years, based on analyses of various marine and terrestrial deposits. Predictions of future trends represent an assumption of substantial utilization of the fossil fuel reservoir. [Modified from Crowley (1990) and presented by Crowley (1996).]

northern Florida. The time of dinosaurs, which ended about 65 million years ago, overlays most of this warm period, and their fossils have been found on the North Slope of Alaska. Climate model simulations suggest that large increases in atmospheric CO_2 are needed to explain the high temperatures of the Cretaceous Period. There is growing evidence from the geologic record to support these conclusions.

23.2   SOLAR VARIABILITY

Burning steadily in stable, middle age, the Sun, now about 5 billion years old, provides the Earth's light and energy. Like other stars of similar age, size, and composition, the Sun exhibits variability in its output. Most pronounced, and most familiar, is a cycle of about 11 years in the number of dark spots (sunspots) on its surface (Figure 23.6). As seen in Figure 23.6, the number of sunspots at the peak of the sunspot cycles has been increasing over the last 100 years. Although the total radiation over all wavelengths received from the

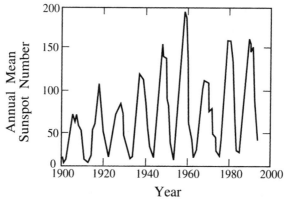

FIGURE 23.6   Annual averages of *sunspot number*—a measure of how many spots appear on the Sun—during the twentieth century (Lean and Rind 1996).

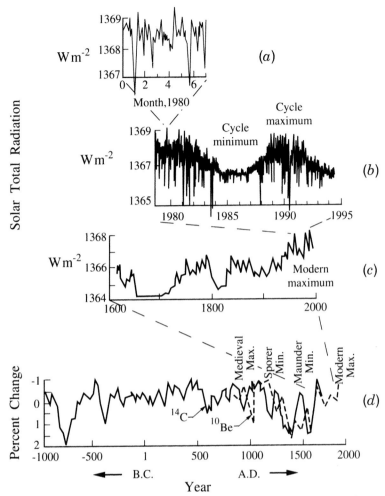

FIGURE 23.7 Variations in solar total radiation incident on the Earth (in W m^{-2}), on different timescales (Lean and Rind 1996). (a) Recorded day-to-day changes for a period of 7 months at a time of high solar activity. The largest dips of up to 0.3% persist for about a month and are the result of large sunspot groups that are carried across the face of the Sun with solar rotation. (b) Observed changes for the 15-year period over which direct measurements have been made, showing the 11-year cycle of amplitude about 0.1%. (c) A reconstruction of variations in solar radiation since about 1600, based on historical records of sunspot numbers and postulated solar surface brightness during the 70-year Maunder Minimum. Estimated variations are of larger amplitude than have yet been observed. (d) A longer record of solar activity based on postulated changes in solar radiation that are derived from measured variations in ^{14}C and ^{10}Be.

Sun is called the solar "constant," in fact the solar constant varies. Spacecraftborne measurements made since 1979 provide a precise record of solar output, on timescales from minutes to decades (Figure 23.7).

There are no direct measurements of solar radiation extending back beyond the last century or so. Much longer records of solar activity are derived from the abundance of atomic isotopes that are produced in the atmosphere by the impact of cosmic rays, the rate of incidence of which on the Earth is affected by conditions on the Sun. When the Sun is more active, its own magnetic fields deflect some of the cosmic rays that would otherwise impact on the Earth; conversely, when the Sun is less active, the Earth receives more cosmic rays. Isotopes ^{14}C, found in tree rings, and ^{10}Be, sequestered in ice deposits, are sensitive to the cosmic ray influx. Records of both of these isotopes exist for thousands of years. They exhibit cyclic variations of 2300, 210, and 88 years, as well as that of the 11-year sunspot cycle, all of which are ascribed to the Sun. Even longer-term changes in orbital features of 19,000, 24,000, 41,000, and 100,000 years exist induced by the changing gravitational pull of the other planets and the Moon. The prominent 100,000-year periodicity in the climate record (see Figure 23.4) is associated with this 100,000-year cycle of the eccentricity of the Earth's orbit, which oscillates between circular and slightly elliptical, altering the Sun–Earth distance.

The total radiation from the Sun is continually changing, with variations of $\leq 0.2\%$ from one month to the next. The timing and nature of these shorter-term fluctuations are consistent with the Sun's 27-day period of rotation. They occur because brighter or darker areas on the solar surface alter the amount of sunlight received at the Earth. This "rotational" variation of total solar radiation (see top panel in Figure 23.7) is superimposed on the 11-year sunspot cycle, which had an amplitude of about $1 \, \text{W m}^{-2}$ (0.1%) in the two most recent cycles (Haigh, 1996). The 11-year sunspot cycle is clearly evident in the second panel from the top of Figure 23.7.

Changes in the Sun's total radiation occur primarily as a result of two phenomena that alter the surface brightness and modulate the outward flow of energy. The first of these are the dark spots (sunspots), which, because they are cooler than surrounding regions, block some of the radiation the Sun would otherwise emit. The second are *faculae*, areas brighter than the surrounding surface, that add to the overall radiative output. The radiation that is emitted from the Sun varies continually in response to the push and pull of these two competing and constantly changing features. During the Maunder Minimum, which occurred for the latter half of the seventeenth century (see third panel of Figure 23.7), it has been postulated that the Sun's surface was not only largely devoid of spots and faculae but also less bright overall. The Sun's output seems to be building up over the last several hundred years, approaching levels last experienced in the twelfth-century Medieval Maximum (third panel of Figure 23.7). In fact, the present production of ^{14}C and ^{10}Be appears to be near historically low levels, as a result of persistently high solar activity that inhibits the rate at which these isotopes are produced. There has been an increase in total solar radiation of about 0.25% over the past 300 years.

The Earth has warmed by $\sim 0.8°$C since the seventeenth century. Estimates of Northern Hemisphere surface temperatures from 1600 to 1800 correlate well with a reconstruction of changes in total solar radiation, suggesting a predominant solar influence on climate during this 200-year, preindustrial period. Figure 23.8 shows trends in total solar irradiance, atmospheric dust loading, CO_2 level, and temperature over the last 400 years. The top and bottom panels of Figure 23.8, matching solar radiation and temperature trends over this 200 year period, show an increase in solar radiation of 0.14% and a concomitant warming of 0.28°C. The implied climate sensitivity from this record is 2°C per 1% change of solar output. Applying this sensitivity to the period since 1850, the 0.13% increase in total solar radiation in the last 140 years should have produced a warming of 0.26°C. This is about half of what has been observed. If we apply the same sensitivity to the last 25 years, solar changes can account for less than one-third of the observed warming (Figure 23.9).

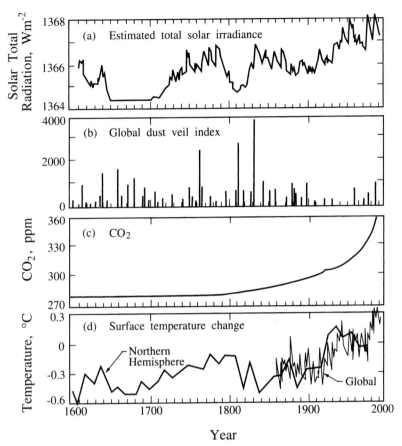

FIGURE 23.8 Trends in natural and human influences relevant to Earth's climate during the past four centuries (Lean and Rind 1996). Compared are annual averages of (a) estimates of the solar total radiation, from Figure 23.7c; (b) variations in the amount of volcanic aerosols derived from an index (global dust veil index) of known eruptions; and (c) the concentration of CO_2 in the atmosphere. The decade-averaged estimate of the Earth's surface temperature, shown in (d), combines continuous instrumental records from the most recent 150 years with a less certain reconstruction based on various climatic indicators.

23.3 RADIATIVE FORCING

A *radiative forcing* is a change imposed on the Earth's radiation balance. Examples of the causes of such perturbations are increases in the concentrations of radiatively active species (e.g., CO_2, aerosols), changes in the solar irradiance incident on Earth, volcanic eruptions, and changes in the surface reflection properties of the planet. Radiative forcing is measured by the change in net radiative flux (down minus up), at some level in the atmosphere, predicted to occur in response to the perturbation. Several definitions of radiative forcing are possible, depending on the atmospheric level at which the net flux change is defined and whether the stratospheric temperature profile is allowed to adjust to the perturbation. A climate *feedback* is an internal climate process that amplifies or dampens the climate response to an initial forcing. An example of a climate feedback is the increase in atmospheric water vapor that results from the initial warming due to rising greenhouse gas levels, which then acts to amplify the warming since water vapor itself is a greenhouse gas.

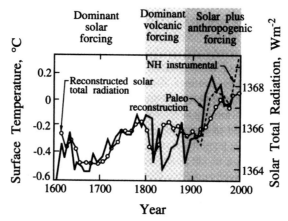

FIGURE 23.9   A comparison of reconstructed solar total radiation with two curves of Northern Hemisphere (NH) summer temperature for the period since 1600 (Lean and Rind 1996). The shorter run of temperature shown as a dashed line is taken from direct meteorological measurements that commence in about 1850. The solid line is a longer and less certain reconstruction of NH temperature, derived primarily from tree-ring data, that has been scaled to match the direct instrumental data during the period of overlap. The temperature curves are in °C and show departures from an arbitrary mean value. All data have been averaged over 10 years. Periods in which different climate forcing mechanisms appear to have predominated are distinguished by shading.

Direct radiative forcings affect directly the Earth's radiative balance; for instance, added CO_2 absorbs infrared radiation. *Indirect* forcings lead to a radiative imbalance by first altering some component of the climate system that then leads to a change in radiative fluxes. An example of an indirect effect is increasing aerosol levels that produce clouds with smaller drops; clouds with smaller drops are not as prone to produce precipitation, so the clouds persist longer and reflect and absorb more radiation.

Radiative forcing defined as a change in the energy flux at the tropopause is referred to as *instantaneous forcing* (Figure 23.10a). The *adjusted radiative forcing*, which is the measure now commonly used, is the radiative forcing at the top of the atmosphere (TOA) in which the stratospheric temperature profile is allowed to adjust to the perturbation and

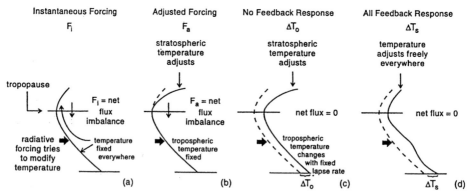

FIGURE 23.10   Assumptions used in calculating (a) ΔF_i, the instantaneous radiative forcing; (b) ΔF_a, the adjusted radiative forcing; (c) ΔT_0, the surface temperature response with no feedbacks; and (d) ΔT_s, the surface temperature response with feedbacks (Hansen et al. 1997). (Reprinted by permission of American Geophysical Union.)

with the tropospheric temperature held fixed at the unperturbed value (Figure 23.10b). The rationale for using the adjusted forcing is that the stratospheric temperature profile will, in response to a perturbation, relax to a new equilibrium in a matter of months, as compared to several decades for the tropospheric temperature. Thus, the adjusted forcing is a somewhat better measure of the expected climate response to forcings that persist for at least several months. The adjusted forcing can be calculated at TOA because the net radiative flux is constant throughout the stratosphere at radiative equilibrium.

Figure 23.10 illustrates the differences between the no-feedback surface temperature response ΔT_0 and the ultimate equilibrium response ΔT_s. ΔT_0 tends to be directly proportional to the adjusted forcing; this proportionality provides the rationale for using the adjusted forcing as a measure of the expected climate response to the perturbation. This proportionality assumes that the lapse rate in the troposphere is fixed (and that in the stratosphere is determined by radiative equilibrium). Ultimately, the question is whether ΔT_s is proportional to the adjusted forcing, when the tropospheric lapse rate is allowed to change in response to climate feedbacks; such feedbacks include changes in clouds and precipitation.

A standard scenario used to judge climate change is a doubling of the preindustrial atmospheric concentration of CO_2, so called $2 \times CO_2$. If the amount of CO_2 in the atmosphere is doubled, the atmosphere becomes more opaque, temporarily reducing longwave emission to space. More recent estimates of adjusted forcing at the tropopause resulting from $2 \times CO_2$ are between 3.5 and $+4.1 \, W \, m^{-2}$ (IPCC 2001). (This estimate includes the negative contribution for the surface–troposphere system owing to the extra shortwave absorption due to stratospheric CO_2, about 5% of the total radiative forcing.) The IPCC (2001) "best estimate" for $2 \times CO_2$ is $+3.7 \, W \, m^{-2}$. The increment of forcing attributable to CO_2 since preindustrial times to the present is $1.46 \, W \, m^{-2}$.

Figure 23.11 shows a way to view the greenhouse effect. The mean atmospheric temperature profile is shown starting from the current mean surface temperature of 288 K. If we assume a global average rate of decrease of temperature with height of $5.5 \, K \, km^{-1}$, the temperature of 255 K, the blackbody temperature corresponding to the mean emitted

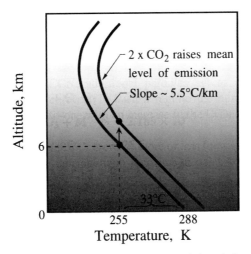

FIGURE 23.11 Greenhouse effect. Doubling CO_2 increases infrared absorption, raising by about 200 m the mean level from which thermal energy escapes to space. Because it is colder, the higher up one goes in the troposphere, the energy emitted to space is temporarily reduced and the Earth radiates less energy than it absorbs. The surface temperature must rise by 1.2 K to restore the energy balance, if the temperature gradient and other factors are held constant.

power of the Earth–atmosphere system, is reached at about 6 km altitude. A doubling of CO_2 from the preindustrial level diminishes the mean emitted power of the Earth–atmosphere system. The result is that the mean altitude from which thermal energy escapes to space rises by about 200 m. Because it is colder 200 m higher, the energy emitted to space is temporarily reduced and the Earth radiates less energy than it absorbs. The surface temperature must rise by 1.2 K to restore the energy balance, if the vertical temperature gradient and other factors are held fixed.

Figure 23.12 is the estimated global, annual mean radiative forcings for the period from preindustrial times to about 2000, as presented by the Intergovernmental Panel on Climate

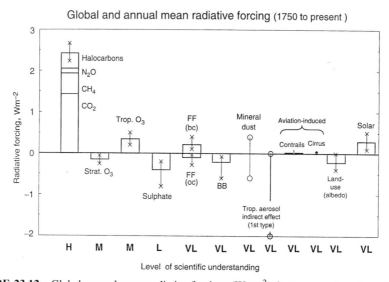

FIGURE 23.12   Global, annual mean radiative forcings ($W\,m^{-2}$) due to a number of agents for the period from pre–Industrial Revolution (1750) to present (late 1990s; about 2000) (IPCC, 2001) The height of the rectangular bar denotes a central or best estimate value, while its absence denotes that no best estimate is possible. The vertical line about the rectangular bar with "x" delimiters indicates an estimate of the uncertainty range, guided by the spread in the published values of the forcing and physical understanding. A vertical line without a rectangular bar and with "o" delimiters denotes a forcing for which no central estimate can be given owing to large uncertainties. The uncertainty range specified here has no statistical basis. A "level of scientific understanding" (LOSU) index is accorded to each forcing, with H, M, L and VL denoting high, medium, low, and very low levels, respectively. This represents the IPCC judgment about the reliability of the forcing estimate, involving factors such as the assumptions necessary to evaluate the forcing, the degree of knowledge of the physical/chemical mechanisms determining the forcing, and the uncertainties surrounding the quantitative estimate of the forcing. The well-mixed greenhouse gases are grouped together into a single rectangular bar with the individual mean contributions due to CO_2, CH_4, N_2O, and halocarbons shown; "halocarbons" refers to all halogen-containing compounds. "FF" denotes fossil fuel burning while "BB" denotes biomass burning aerosol. Fossil fuel burning is separated into the "black carbon" (bc) and "organic carbon" (oc) components with its separate best estimate and range. The sign of the effects due to mineral dust is itself an uncertainty. Only the *first* type of indirect effect due to aerosols (see Chapter 24) as applicable in the context of liquid clouds is considered here. All the forcings shown have distinct spatial and seasonal features such that the global, annual means appearing on this plot do not yield a complete picture of the radiative perturbation. They are only intended to give, in a relative sense, a first-order perspective on a global, annual mean scale, and cannot be readily employed to obtain the climate response to the total natural and/or anthropogenic forcings. The positive and negative global mean forcings cannot be added up and viewed a priori as providing offsets in terms of the complete global climate impact.

Change (IPCC) 2001 assessment document. The radiative forcing attributable to all well-mixed greenhouse gases was estimated by IPCC as $+2.43\,\text{W m}^{-2}$. The overall uncertainty estimated for this value is 10%, with the uncertainty for CO_2 less than those for the other GHGs. The individual forcings for CO_2, CH_4, and N_2O are

CO_2	$+1.46\,\text{W m}^{-2}$
CH_4	$+0.48$
N_2O	$+0.15$

Methane oxidation leads to a net loss of OH in the atmosphere, thereby lengthening the lifetime of CH_4 itself (we will discuss this later in this chapter). It is estimated that this longer lifetime increases the radiative forcing of CH_4 by 25–35% over that in the absence of this feedback effect. Methane oxidation also leads to tropospheric O_3; this indirectly increases the greenhouse effect by another 30–40% through the effect of the added O_3 itself. Finally, increases in CH_4 also indirectly lead to further climate forcing by increasing stratospheric H_2O (about 7% of CH_4 is oxidized in the upper troposphere).

Ozone absorbs UV, visible, and infrared radiation and acts as both a direct radiative forcing agent and as a climate feedback. Changes in O_3 driven by changes in anthropogenic emissions of VOCs and NO_x represent a direct forcing. Ozone levels also respond to changes in temperature and UV radiation and natural emissions from vegetation and lightning; the forcing induced by these responses represents a climate feedback. Anthropogenic tropospheric O_3 was estimated by IPCC (2001) to cause a radiative forcing of about $+0.4\,\text{W m}^{-2}$ at the tropopause, a portion of which is linked to CH_4 emissions.

Figure 23.12 does not provide information about the relative timescales associated with the different forcing agents. The GHGs have atmospheric lifetimes of decades or longer; the various aerosols are removed in days to weeks. Nor does the globally averaged forcing reveal information about the spatial distribution of the forcing over the Earth.

23.4 CLIMATE SENSITIVITY

The usefulness of radiative forcing is based on the assumption that the change in global annual mean surface temperature is proportional to the imposed global annual mean forcing. The surface and troposphere are strongly coupled by convective heat-transfer processes, that is, heat transfer by vertical motions of air. The vertical temperature profile within the troposphere (the lapse rate) is determined largely by this vertical convective heat transport, while the vertically averaged surface–troposphere temperature is regulated by radiative flux equilibrium at the tropopause. This situation is referred to as *radiative–convective equilibrium* (Manabe and Strickler 1964; Manabe and Wetherald 1967).

Climate models predict an approximately linear relationship between global mean radiative forcing ΔF and the equilibrium global mean surface temperature change.

$$\Delta T_s = \lambda \, \Delta F \qquad (23.1)$$

where λ is a *climate sensitivity parameter* [in $\text{K}(\text{W m}^{-2})^{-1}$]. This relationship is relatively independent of the nature of the forcing, for example, whether the change in forcing is a result of a change in GHG concentration or a change in solar output. All climate feedback processes are implicitly included in λ, such as, for example, water vapor feedback, cloud feedback, and ice–albedo feedback. An example of a positive feedback to a climate forcing that causes warming is melting of some of the sea ice. Because the darker ocean absorbs more sunlight than the ice it replaced, even more warming occurs; this is a positive feedback. Water vapor and cloud droplets are the dominant atmospheric absorbing species, and how these substances respond to climate forcings is a principal determinant of climate sensitivity. The major source of atmospheric water vapor is evaporation from the oceans, the response of which to climate changes can be slow because the ocean requires time to warm (or cool). The response time depends on how rapidly the ocean circulation transmits changes in surface temperature into the deep ocean. For $2 \times CO_2$, for example, for which a midrange global average temperature increase of about 3°C is predicted, several centuries are required for the full climate response (see Chapter 22).

If feedbacks are not considered, the forcing–temperature relation can be written as [recall 4.9]

$$\Delta T_0 = \lambda_0 \, \Delta F \qquad (23.2)$$

where ΔT_0 is the surface temperature response to an imposed radiative forcing ΔF and λ_0 is the climate sensitivity in the absence of feedbacks (see Figure 23.10). Thus water vapor, clouds, and surface albedo are not allowed to change in response to ΔF. The radiative forcing for a doubling of CO_2 is around 4 W m^{-2}, and the estimated surface temperature increase in the absence of any climate feedback processes is calculated to be in the range 1.2–1.3 K at equilibrium.

The parameter λ can be calculated in two ways. One is to employ a general circulation model (GCM) to simulate the response to a change in forcing. GCMs predict values of λ ranging from 0.3 to $1.4 \text{ K} \left(\text{W m}^{-2} \right)^{-1}$, differing primarily because of different model treatments of cloud feedback. The second approach is to identify in the climate record historical periods in which both changes in forcing and surface temperature can be deduced. The constant of proportionality λ is a measure of the overall strength of climate feedback processes and hence of global climate sensitivity. Geographically, temperature response patterns are not generally proportional to, nor do they necessarily resemble, their parent forcing patterns (Boer and Yu 2003). Temperature response patterns do, however, exhibit a remarkable additivity, in that the sum of response patterns for different forcings resembles the response pattern for the sum of the forcings. The reasonably robust linear relationship between radiative forcing and climate response allows some aspects of climate change to be estimated from radiative forcing estimates alone.

From the results of radiative–convective and general circulation models that incorporate climate feedback responses, the surface temperature change associated with doubled CO_2 is in the range 1.5–4.5 K. The climate feedback factor, then, defined as the ratio of the computed surface temperature change to the zero-feedback change, falls in the range 1.2–3.75. The expected range of the global surface temperature change from the direct radiative effects on any combination of trace constituent abundance changes can then be scaled, applying the results given above, by comparison of the radiative model results for the net flux change for the doubled CO_2 scenario with a trace gas scenario.

In general, then, surface temperature changes are predicted to be 1–4 times larger if climate feedbacks are included. The increase in atmospheric water vapor provides the strongest positive feedback. Snow and/or ice in the polar regions will decrease in a warmer globe, and the albedo of the polar regions will consequently decrease. Clouds produce positive feedback in some GCMs at low and middle latitudes; they are a negative feedback in the polar regions. Feedback resulting from advective energy transports (sensible and latent heat, geopotential energy) tend to be opposite in sign to the cloud feedbacks.

The relation between ΔT_s and ΔF given by (23.1) does not hold in all cases of forcing (National Research Council 2005). The nature of the climatic response to forcing can depend critically on the vertical distribution of the forcing. In the next chapter, we will consider the radiative forcing associated with tropospheric aerosols. When particles that absorb solar radiation are in the troposphere, the atmosphere gains shortwave radiative energy while the surface experiences a deficit. Complex feedbacks involving changes in the atmospheric lapse rate and cloudiness from particles near the Earth's surface could alter the climate sensitivity substantially from that for a similar magnitude of perturbation at other altitudes. In the case of volcanic eruptions that inject particles directly into the lower stratosphere, the lower stratosphere is radiatively warmed while the surface-troposphere cools. This contrasts with the effects from CO_2 increases, wherein the surface–troposphere heats and the stratosphere cools. Therefore, the vertical partitioning of forcing can, in general, affect changes of climate variables in such a way that, for two identical forcings, the same value of λ may not apply even though the overall values of ΔF are the same.

Because the assumption of a constant, linear relationship between changes in global mean surface temperature and global mean TOA radiative forcing breaks down for absorbing aerosols (which may have only small TOA forcing, but larger surface forcing due to their absorption of solar radiation), the concept of *efficacies* of different forcing agents has been introduced (Joshi et al. 2003; Hansen and Nazarenko 2004) by the following equation

$$E_i = \frac{\lambda_i}{\lambda_{CO_2}} \tag{23.3}$$

which is the ratio of the climate sensitivity parameter λ_i for a given forcing agent to that for a doubling of CO_2. The efficacy can then be used to define an effective forcing:

$$\Delta F_e = \Delta F_i E_i \tag{23.4}$$

A value of $E_i > 1$ indicates that the agent leads to a larger effective forcing than does that of $2 \times CO_2$; $E_i < 1$ means that the agent is less efficient than CO_2 in changing the surface temperature for a given amount of forcing.

Climate simulations have revealed that radiative forcing occurring predominantly in one portion of the globe can induce climate changes in other parts of the world substantially separated from the site where the forcing occurs; these have been referred to as *teleconnections*. Teleconnections occur, for example, through the transport of energy by atmospheric waves. Regional weather patterns have been associated with sea surface temperature changes. Land-use changes can produce changes in atmospheric circulation patterns at long distances. Deforestation in the tropics has been found to impact rainfall in other regions (Werth and Avissar 2005); for example, deforestation in the Amazon could lead to rainfall decreases in the U.S. Midwest during spring and summer. The explanation for this effect is that tropical deforestation alters the latent and sensible heat fluxes to the atmosphere, and the associated change in pressure distribution modifies zones of atmospheric convergence and divergence, which shifts the polar jetstream and its associated precipitation.

23.5   RELATIVE RADIATIVE FORCING INDICES

The concept of a relative radiative forcing index was devised to place the various GHGs on a common scale. Factors on which such an index depends are (1) the strength with which a given species absorbs infrared radiation and the spectral location of its absorbing wavelengths and (2) the atmospheric lifetime (or response time) of the species in the atmosphere. In addition, it is necessary to specify the time period over which the radiative effects of the compound are to be considered since a number of the climate responses to radiative forcing are long. The radiative forcing index measures the cumulative radiative forcing of a species.

Figure 23.13 shows radiative forcings as a function of time as a result of the emissions of 1 kg of several GHGs. Note that after 1 year, instantaneous radiative forcings can differ by four orders of magnitude. Figure 23.14 shows the time integrals of the instantaneous forcings in Figure 23.13 up to 500 years. The time integral of the forcing index $(W\,m^{-2}\,kg^{-1})$ can be called the *absolute global warming potential* $(W\,m^{-2}\,kg^{-1}\,yr)$.

The potential of 1 kg of a compound A to contribute to radiative forcing relative to that of 1 kg of a reference compound R can be called the *global warming potential* (GWP)

$$GWP = \frac{\int_0^{t_f} a_A [A(t)]\,dt}{\int_0^{t_f} a_R [R(t)]\,dt} \qquad (23.5)$$

where t_f is the time horizon, a_A is the radiative forcing resulting from 1 kg increase of compound A, $[A(t)]$ is the time decay of a pulse of compound A, and a_R and $[R(t)]$ are the comparable quantities for the reference compound. The reference compound generally used is CO_2. The values of a_A and a_R are derived from radiative transfer models. The time decays of A and R are based on their atmospheric lifetimes. We see that the previously mentioned absolute global warming potential (AGWP) is just

$$AGWP_A = \int_0^{t_f} a_A [A(t)]\,dt \qquad (23.6)$$

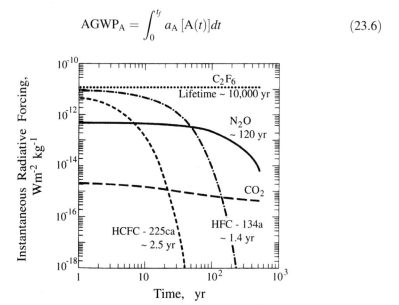

FIGURE 23.13   Radiative forcing $(W\,m^{-2}\,kg^{-1})$ versus time after a pulse release for several different greenhouse gases (IPCC 1995).

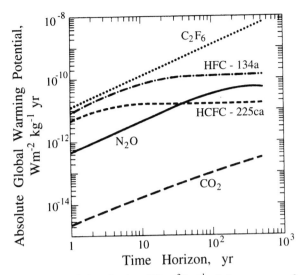

FIGURE 23.14 Integrated radiative forcing ($W\,m^{-2}\,kg^{-1}\,yr$) for a range of greenhouse gases, calculated from Figure 23.13 (IPCC 1995).

An advantage of the GWP over the AGWP is that uncertainties that enter into the calculation of a_A, like the role of clouds, also enter into the calculation of a_R and thus are normalized out of the calculation.

Figure 23.15 shows GWPs for C_2F_6, HFC-134a, HCFC-225ca, and N_2O relative to CO_2. For the purpose of generating these GWPs, the global lifetime of CO_2 has been assumed to be 150 years, although, as noted earlier, a single lifetime is generally not adequate to characterize CO_2. This is viewed as a major weakness of the GWP concept that uses CO_2 as the reference compound. Note that the GWP for N_2O is relatively

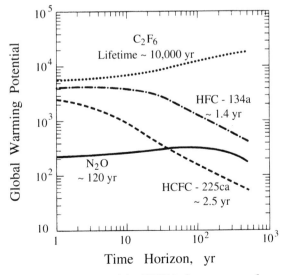

FIGURE 23.15 Global warming potentials (GWPs) for a range of greenhouse gases with differing lifetimes, as in Figures 23.13 and 23.14, using CO_2 as the reference gas (IPCC 1995).

TABLE 23.1   Greenhouse gases: Abundances, lifetimes, and GWPs.

Chemical Species	Formula	Abundance[a], ppt		Lifetime, yr	100-yr GWP
		1998	1750		
Methane	CH_4 (ppb)	1745	700	$8.4/12^a$	23
Nitrous oxide	N_2O (ppb)	314	270	$120/114^a$	296
Perfluoromethane	CF_4	80	40	>50,000	5,700
Perfluoroethane	C_2F_6	3.0	0	10,000	11,900
Sulfur hexafluoride	SF_6	4.2	0	3,200	22,200
HFC-23	CHF_3	14	0	260	12,000
HFC-134a	CF_3CH_2F	7.5	0	13.8	1,300
HFC-152a	CH_3CHF_2	0.5	0	1.40	120

Important Greenhouse Halocarbons under Montreal Protocol and Its Amendments

CFC-11	$CFCl_3$	268	0	45	4,600
CFC-12	CF_2Cl_2	533	0	100	10,600
CFC-13	CF_3Cl	4	0	640	14,000
CFC-113	$CF_2ClCFCl_2$	84	0	85	6,000
CFC-114	CF_2ClCF_2Cl	15	0	300	9,800
CFC-115	CF_3CF_2Cl	7	0	1700	7,200
Carbon tetrachloride	CCl_4	102	0	35	1,800
Methyl chloroform	CH_3CCl_3	69	0	4.8	140
HCFC-22	CHF_2Cl	132	0	11.9	1,700
HCFC-141b	CH_3CFCl_2	10	0	9.3	700
HCFC-142b	CH_3CF_2Cl	11	0	19	2,400
Halon-1211	CF_2ClBr	3.8	0	11	1,300
Halon-1301	CF_3Br	2.5	0	65	6,900
Halon-2402	CF_2BrCF_2Br	0.45	0	<20	

[a] Species with chemical feedbacks that change the duration of the atmospheric response. The first number is the global mean atmospheric lifetime; for example, the global mean atmospheric lifetime of CH_4 against reaction with OH is 8.4 years. The second number is the perturbation lifetime, which is the time required for a perturbation to decay back to its initial state. For example, an increase in CH_4 leads to a decrease in the OH level in the atmosphere and this reduced OH level, in turn, leads to a slower rate of removal of CH_4; thus the perturbation lifetime of CH_4 is 12 years.

Source: IPCC (2001).

insensitive to the choice of time horizon, whereas those for the fluorocarbons depend on that choice. C_2F_6, with an estimated lifetime of 10,000 years, has a GWP that continues to increase as the time horizon is increased. GWPs for the other two compounds also depend on the time horizon, with that for the shorter-lived of the two, HCFC-225ca, falling off the most rapidly with time horizon. Table 23.1 summarizes GWPs for the important greenhouse gases.

It is worth reiterating that the GWP concept is relevant only for compounds that have sufficiently long lifetimes to become globally well-mixed. Gases with vertical profiles or latitudinal variations pose difficulties in the simple GWP framework. Also, the CFCs, in addition to their role as GHGs, have, of course, a stratospheric ozone-depleting effect, and the change in stratospheric O_3 will itself exert a complicated radiative forcing that will vary with latitude. These effects are difficult to include in the GWP calculation without a full climate–chemistry model.

23.6 UNREALIZED WARMING

We saw in Chapter 22 that the time needed for a perturbation in atmospheric CO_2 concentration to lead to a new equilibrium level is of the order of about 100 years (assuming about two relaxation times) and much longer if the deep ocean is accounted for. The large heat capacity of the ocean also causes a time lag to occur between a change in external radiative forcing and the ultimate temperature change that results when the ocean achieves its new equilibrium. This time lag is also of the order of a century or more.

We can define the equilibrium warming responding to a given increase in GHG emissions as ΔT_e, which is that when the atmosphere–ocean system comes to its new equilibrium. If the warming actually realized to date is ΔT_r, then the additional warming expected to occur is $\Delta T_e - \Delta T_r$. This difference has been called *unrealized warming* or the *warming commitment*. If ΔF is the current forcing due to GHGs and ΔF_r is the forcing corresponding to an equilibrium warming of ΔT_r, then $\Delta F - \Delta F_r$ is the current radiation imbalance; this quantity is essentially the flux of heat that must be absorbed by the oceans.

Given estimates for the forcing and temperature change associated with a doubling of CO_2, one can estimate $\Delta T_e - \Delta T_r$ by

$$\Delta T_e - \Delta T_r = (\Delta F - \Delta F_r)\left(\frac{\Delta T_{2\times CO_2}}{\Delta F_{2\times CO_2}}\right) \tag{23.7}$$

where $\Delta F_{2\times CO_2}$ is the radiative forcing for a CO_2 doubling ($\sim 3.7\,\mathrm{W\,m^{-2}}$) and $\Delta T_{2\times CO_2}$ is the corresponding equilibrium global mean warming. Taking $\Delta F \sim 1.7\,\mathrm{W\,m^{-2}}$ and $\Delta T_r \sim 0.7°C$, with $\Delta T_{2\times CO_2} \sim 2.6°C$, this gives an unrealized warming of $0.5°C$, with $\Delta F - \Delta F_r \sim 0.7\,\mathrm{W\,m^{-2}}$ (Wigley 2005).

Temperature increases are predicted to be greatest at high northern latitudes, primarily because the melting ice will allow more solar energy to be absorbed. Warming will also be greatest over large land masses and least over oceans, because of the comparatively low thermal mass of the land and its relative inability to absorb heat. A detailed study of 244 glacier fronts on the Antarctic peninsula over the past 60 years or so shows that 87% of them have retreated, behavior broadly consistent with that expected by atmospheric warming (Cook et al. 2005).

An increase in the rate of cycling of water in the hydrologic cycle is expected in response to higher global average temperatures. Higher evaporation rates will lead to more rapid drying of soils after rain. The drier soils will warm more intensely during daytime, resulting in higher temperatures, even faster evaporation, and an increase in the diurnal temperature range (highest daytime temperature minus lowest nighttime temperature). The faster recycling of water will lead to higher rainfall rates and an increase in the frequency of heavy precipitation. Data indicate that during 1970–2000 the hydrological cycle accelerated at high northern latitudes (Stocker and Raible 2005). The resulting increase in freshwater flow into the Arctic Ocean may have long-range effects. Finally, there have been a number of studies on the effect of global warming on sea-level rise [e.g., Meehl et al. (2005) and Wigley (2005)].

23.7 ATMOSPHERIC CHEMISTRY AND CLIMATE CHANGE

A changing climate influences atmospheric chemistry through not only temperature and precipitation changes but also changes in atmospheric transport processes, changes in the budgets of species with biological sources (which respond to temperature and moisture changes), changes in vegetative cover (which would alter dry deposition rates), changes in the rate of export of pollutants from the urban/regional environment to the global, and so on. Changes in atmospheric chemical composition itself will lead to climate change. Responses of these various subsystems may be highly coupled to both each other and climate change itself. For example, climate change has the potential to alter water vapor concentrations; changes in the concentration of water vapor would alter many species concentrations by impacting atmospheric OH concentrations. Changes in H_2O will also impact tropospheric O_3 concentrations. Because O_3 itself is a GHG, these changes will feed back to alter the climate. Climate change also has the potential to alter biological sources and sinks of radiatively active species, changes that will feed back to alter the climate response.

From the standpoint of regional tropospheric chemistry—which involves near-surface abundances of ozone, wet and dry deposition of acidic species, and transport and lifetimes of trace atmospheric constituents—the climate variables of interest include the variability of distributions of temperature, precipitation, clouds, and boundary-layer meteorology. In the global sense, these variables are controlled by surface and atmospheric temperature and water content. The distributions of temperature and water vapor are in turn controlled by solar and longwave radiation transfer involving the surface and the atmosphere.

Many potentially important climate feedback processes have been identified (IPCC 2001). Because water vapor and clouds are the principal constituents responsible for reflecting, absorbing, and emitting radiation in the atmosphere, while snow and ice cover on the surface contribute significantly to surface albedo at higher latitudes, most feedbacks involve processes affecting the hydrologic cycle. These include the dependence of absolute humidity on temperature, the extent of cloud cover and cloud properties (optical depth and radiating temperature), the dependence of the tropospheric temperature profile on humidity, and the temperature dependence of the extent of surface coverage by snow and ice. The abundance of water vapor in the stratosphere appears to be controlled by the temperature of the tropical tropopause, through which air entering the stratosphere passes, and the stratospheric oxidation of CH_4. Increases in low-latitude tropopause temperature may result from an altered surface–atmosphere radiative balance and may increase stratospheric water vapor, with subsequent feedback on surface temperatures and stratospheric photochemistry. Some of these feedback processes are poorly quantified; in the case of some cloud feedbacks, even the signs are in doubt.

23.7.1   Indirect Chemical Impacts

In addition to changes in direct radiative forcing by changes in trace species abundances, changes in the atmospheric radiation balance can be forced through chemical modification of atmospheric composition. Many of the radiatively important atmospheric trace constituents also participate in chemical processes that control ozone abundances and lifetimes and budgets of trace species. Whereas small direct changes in net radiative forcings are additive with respect to their impact on the surface–troposphere system, chemical processes affecting atmospheric composition are in general coupled and nonlinear, and their effects on atmospheric composition are not additive. Estimation of the net radiative impact of changes mediated by chemistry is usually specific to the details of a scenario of coupled trace gas abundance and emission trends.

Given the importance of H_2O, CH_4, CO, NO, O_3, and tropospheric solar fluxes to tropospheric OH abundance, it is clear that, to the extent that climate would impact the concentrations of these species and photolysis rate constants, a change in climate would alter OH levels. The hydroxyl radical is the major chemical scavenger for most tropospheric species, such as CO and CH_4. Reactions between OH and CO and CH_4 cycle OH to other forms in the HO_x family, modifying the ratio of OH to HO_x. The importance of these interactions in HO_x chemistry implies that increases in CO and CH_4 abundances can lead to decreases in global OH. Similarly, increases in tropospheric NO and O_3 can lead to increases in OH by enhancing cycling reactions that convert HO_2 to OH. The extent to which these effects on OH will offset each other is a problem requiring analysis by tropospheric chemical models.

Because water vapor is a parent compound for OH and other HO_x species, changes to its concentration alter the concentration of tropospheric OH. Since tropospheric water vapor is in balance with an evaporation and transpiration source from the oceans, soils, and plants and the precipitation sink, global increases in temperature are expected to lead to changes in the tropospheric water vapor concentration. If the sources of water vapor are not perturbed by changes in vegetative cover and if circulation patterns do not lead to more frequent precipitation events, then the concentration of H_2O might be expected to increase.

Tropospheric O_3 concentrations appear to be increasing as a result of increased direct emissions of NO_x, hydrocarbons, and CH_4. Because of the role of NO in partitioning HO_x and because of the large variation in the concentration of NO between remote oceanic areas and continental areas, an increase in O_3 by a factor of 2 has been estimated to increase OH by perhaps 10% over ocean areas and by probably more than 10% over the continents (Thompson et al. 1989). These changes might, of course, feed back on the concentration perturbation of O_3.

Increases in the concentrations of halocarbons and N_2O are calculated to lead to a net decrease in stratospheric O_3. Stratospheric cooling, as driven by increases in CO_2 and other infrared emitters, acts to increase stratospheric O_3. The direct effects of emissions and indirect effects of climate change are calculated to have caused only small changes in stratospheric O_3 at present. Because most of the ozone column resides in the stratosphere and because ozone is responsible for much of the atmospheric opacity below 290 nm, these changes would alter tropospheric OH by changing photolysis rates for any species with significant absorption in this range. Of most importance, absorption by ozone leading to the formation of $O(^1D)$ would change, creating a change in the direct source of HO_x and OH (Madronich and Granier 1992, 1994). Of lesser importance, an increase in the photolysis rate of H_2O_2 could affect HO_x loss rates. A decrease in column O_3 of 20% is estimated to lead to an OH increase of roughly 15% over continental areas (Thompson et al. 1989).

Another mechanism for climate impact on OH concerns the magnitude of the source of OH created by the oxidation of nonmethane hydrocarbons (NMHCs). Trainer et al. (1987) have estimated that in areas of low NO_x concentration, the concentration of OH might be decreased by a factor of ≤ 2 in the first 100 m above ground level as a result of biogenic emissions of isoprene and terpenes from that in their absence. An increase of 5 K could lead to an increase in biogenic hydrocarbon emissions by a factor of > 3. An estimate of the impact on global OH concentrations of such a change is difficult to make because the importance of the role of NMHCs in the global budget for OH is not well defined. However, it is clear that locally, at least, these changes would lead to decreases in OH.

Temperature increases can impact bacterial sources of CH_4. Reaction of CH_4 with OH is a net sink for OH in a low NO_x environment (Chapter 6), although the net effect of CH_4 oxidation on HO_x depends on the exact oxidation pathway. Because the bacterial source is about 50% of the total source of CH_4 and because CH_4 emissions are exponentially dependent on temperature, an increase of 5 K could lead to almost a 50% increase in the total (bacterial plus anthropogenic) source strength of CH_4. The estimated increase in atmospheric abundance of CH_4 would actually be larger than the increase in its source strength because the concentration of OH would decrease as CH_4 is increased (see Section 23.7.2).

23.7.2 Atmospheric Lifetimes and Adjustment Times

In general, the timescale for a small perturbation of a species to disappear from the atmosphere—the *adjustment time*—is simply equal to the compound's mean lifetime τ. For two species, CH_4 and N_2O, their adjustment times are not equal to their mean lifetimes. The reason is that these trace gases interact strongly with the global atmospheric chemistry, such that their addition to the atmosphere actually perturbs the background chemical mixture that is responsible for their removal.

Increased CH_4, with its oxidation by OH, leads to additional CO formation; both the CH_4 itself and the CO formed suppress OH, the major sink for CH_4. As a result of the depletion of OH, the timescale for removal of the CH_4 lengthens over that prior to its addition. Global model calculations (Isaksen and Hov 1987; Prather 1994; see Table 23.1) indicate that the adjustment time for CH_4, based on atmospheric chemistry alone, is 12 years. This is to be compared to a mean global lifetime (τ), based on tropospheric OH reaction of 8.4 years.

The other major compound for which the lifetime and the adjustment time differ is N_2O. N_2O is the source of stratospheric NO_x. The stratospheric NO_x derived from N_2O controls, to some extent, stratospheric O_3 and hence the ultraviolet radiation field in the stratosphere, which, in turn, controls the rate of destruction of N_2O. Increases in N_2O lead to more NO_x, less O_3, and therefore less absorption of ultraviolet radiation. This enhanced radiation flux, in turn, increases the photolytic removal of N_2O. As a result, the adjustment time for N_2O is about 10% shorter than the N_2O lifetime.

Although we talk about lifetimes of short-lived trace gases such as NO_x (1–10 days) and CO (1–4 months), it is not possible to designate a global mean lifetime for such gases that have highly variable local loss rates. For example, the lifetime of CO varies from 1 month in the tropics to 4 months at midlatitudes in spring and autumn, to much longer lifetimes in high-latitude winter. The atmospheric circulation[4] does not mix CO from equator to pole in much less than a season. As a result, CO emitted in the tropics will tend to decay within a month in the tropics, whereas high-latitude emissions of CO in winter will survive to be mixed throughout the troposphere before substantial removal takes place, and the adjustment time does not depend on the location of the sources.

To illustrate the nonlinear chemical feedbacks in the atmospheric system, let us consider methane. One kilogram of CH_4 released from the surface becomes well mixed in the troposphere. A portion of this CH_4 is transported into the stratosphere. As we have just described, the added kilogram of CH_4 is removed with an adjustment time of about 12 years (and not with its global lifetime of 8.4 years due to OH reaction and stratospheric loss). That amount of the CH_4 perturbation that makes it into the stratosphere directly affects stratospheric chemistry that controls stratospheric O_3 abundance. More CH_4 will

---

[4] Recall that tropospheric mixing times are about 2 weeks from the ground to the tropopause, about 3 months from pole to tropics, and about 1 year between hemispheres.

sequester more active chlorine, slow down the ClO_x catalytic cycles, and increase O_3. At the same time, oxidation of this CH_4 increases stratospheric H_2O, leading to an enhanced efficiency of the HO_x catalytic cycle, depleting O_3. The additional H_2O might also aid heterogeneous chemistry in the stratosphere by providing more condensable water. Changes to stratospheric H_2O and O_3 have direct greenhouse impacts. Changes in stratospheric O_3 affect tropospheric chemistry through altered ultraviolet flux into the troposphere and through altered O_3 flux from the stratosphere to the troposphere. The additional CH_4 also impacts tropospheric chemistry by reduction of OH and increased production of CO and O_3. The resulting decreased OH increases the lifetimes, and hence concentrations, of those trace gases that are themselves removed by OH. Many of these trace gases are greenhouse gases. Some of these are, in fact, HCFCs, and their tropospheric concentration increase will lead to increased fluxes into the stratosphere with attendant effect of the increased stratospheric chlorine on stratospheric O_3. Again, it is clear that climate feedbacks involving atmospheric chemistry are quite complex and couple tropospheric and stratospheric chemistry.

23.8   RADIATIVE EFFECTS OF CLOUDS

One of the prominent uncertainties in climate modeling is how the cloud system reacts in response to increases in the levels of greenhouse gases. In general, high clouds act as a greenhouse and warm the Earth, whereas low clouds, by reflecting sunlight back to space, tend to cool the planet. Overall, clouds averaged together globally have a net cooling effect at present of about $-15\,W\,m^{-2}$. The main contribution to this overall cooling effect is from low clouds. Over regions of the Earth that are particularly prone to low boundary-layer stratus clouds, such as the northern Pacific Ocean, the net cooling effect from clouds can range from as much as -40 to $-60\,W\,m^{-2}$. It has been estimated that for each $1°C$ decrease in sea surface temperature, a 6% increase in low-level, boundary-layer clouds would result.

Water droplets or ice crystals possess absorption bands over virtually the entire terrestrial infrared spectrum. As a result, clouds considerably affect the infrared radiation escaping to space from the Earth's surface and atmosphere. Cloudtops generally have lower temperatures than the Earth's surface and the lower part of the troposphere, and thus the outgoing infrared radiation from cloudtops is less than that from the Earth's surface. By this means, clouds act to reduce outgoing infrared radiation from that in their absence. The higher the cloud, the lower its temperature, and the greater the reduction in outgoing infrared radiation. On the other hand, higher clouds, particularly ice clouds, tend to have a lower water content and optical thickness (see Chapter 24 for treatment of cloud optical thickness) and therefore are relatively more transparent than lower clouds. This reduces the infrared trapping effect.

PROBLEMS

23.1$_A$ The global mean precipitation rate over the Earth is one meter per year ($1\,m\,yr^{-1}$). Show that this corresponds to an atmospheric energy input of $80\,W\,m^{-2}$.

23.2$_A$ General circulation model (GCM) simulations suggest that relative humidity would stay constant in a warmer Earth. Thus, when atmospheric temperature increases, the ratio of the H_2O partial pressure to the saturation vapor pressure remains constant. Assuming a temperature increase of 4°C and ambient temperatures of 250, 275, and 300 K, calculate the increase in water vapor concentration in the atmosphere. The vapor pressure of water as a function of temperature is given in Table 17.2.

23.3$_A$ Show that if trace gas A has an atmospheric lifetime τ_A, its global warming potential (GWP) relative to that of CO_2 over a time period t_f is given by

$$\text{GWP} = \frac{a_A \tau_A (1 - e^{-t_f/\tau_A})}{a_{CO_2} \tau_{CO_2} (1 - e^{-t_f/\tau_{CO_2}})}$$

where τ_{CO_2} is an effective lifetime for CO_2. Use this formula together with greenhouse efficiencies and atmospheric residence times given in Table 23.1 to calculate GWP values for $t_f = 20$ yr for CH_4, $CFCl_3$, CF_2Cl_2, and CH_3CCl_3.

23.4$_C$ At steady state the atmospheric lifetime of a species is defined as the total amount of the species divided by its globally, annually averaged loss rate. The tropospheric CH_4–CO–OH system is highly coupled, and the apparent timescale for a perturbation in, say, CH_4 to decay is longer than the steady-state CH_4 lifetime as defined above (Prather 1994, 1996). In this problem we analyze this phenomenon. For our analysis we assume that the atmosphere can be represented as a single box. The chemical reactions of the CH_4–CO–OH system considered are:

$$CH_4 + OH \xrightarrow{1} CO + \text{products}$$

$$CO + OH \xrightarrow{2} CO_2 + H$$

$$OH + X \xrightarrow{3} \text{products}$$

The only product of CH_4 oxidation that we will be concerned with is CO. (This is an approximation because HCHO is also a product of CH_4 oxidation and HCHO is further oxidized itself to CO.) The general molecule X accounts for OH sinks that are independent of the CH_4–CO system. We include constant global source rates, S_{CH_4}, S_{CO}, and S_{OH}. Typical values of the rate constants and source rates are (Prather 1994):

$$k_1 = 5 \times 10^{-15} \text{ cm}^3 \text{ molecule}^{-1} \text{ s}^{-1}$$

$$k_2 = 2 \times 10^{-13} \text{ cm}^3 \text{ molecule}^{-1} \text{ s}^{-1}$$

$$k_3[X] = 1 \text{ s}^{-1}$$

$$S_{CH_4} = 1.6 \times 10^5 \text{ molecules cm}^{-3} \text{ s}^{-1}$$

$$S_{CO} = 2.4 \times 10^5 \text{ molecules cm}^{-3} \text{ s}^{-1}$$

$$S_{OH} = 11.2 \times 10^5 \text{ molecules cm}^{-3} \text{ s}^{-1}$$

a. Show that at steady state, a positive solution for [OH] results only if $S_{OH} > 2 S_{CH_4} + S_{CO}$, that is, the source of OH must be large enough to oxidize the combined sources of CH_4 and CO.

b. The rate equations for the three reactions

$$\frac{d[CH_4]}{dt} = S_{CH_4} - R_1$$

$$\frac{d[CO]}{dt} = S_{CO} + R_1 - R_2$$

$$\frac{d[OH]}{dt} = S_{OH} - R_1 - R_2 - R_3$$

where R_i is the rate of reaction i, can be expressed concisely as

$$\frac{d\mathbf{x}}{dt} = \mathbf{f}(\mathbf{x})$$

where

$$\mathbf{x} = \begin{bmatrix} x_1 \\ x_2 \\ x_3 \end{bmatrix} = \begin{bmatrix} CH_4 \\ CO \\ OH \end{bmatrix} \qquad \mathbf{f} = \begin{bmatrix} f_1 \\ f_2 \\ f_3 \end{bmatrix} = \begin{bmatrix} S_{CH_4} - R_1 \\ S_{CO} + R_1 - R_2 \\ S_{OH} - R_1 - R_2 - R_3 \end{bmatrix}$$

Consider a perturbation $\delta\mathbf{x}$ to all three species. Thus $\mathbf{x} = \bar{\mathbf{x}} + \delta\mathbf{x}$, where $\bar{\mathbf{x}}$ is the state before the perturbation. Show that the differential equations that govern the perturbation can be approximated as

$$\frac{d(\bar{\mathbf{x}} + \delta\mathbf{x})}{dt} \simeq \mathbf{f}(\bar{\mathbf{x}}) + \mathbf{J}\,\delta\mathbf{x}$$

where the 3×3 Jacobian matrix $\mathbf{J}$ is

$$\mathbf{J} = \begin{bmatrix} \dfrac{\partial f_1}{\partial x_1} & \dfrac{\partial f_1}{\partial x_2} & \dfrac{\partial f_1}{\partial x_3} \\ \dfrac{\partial f_2}{\partial x_1} & \dfrac{\partial f_2}{\partial x_2} & \dfrac{\partial f_2}{\partial x_3} \\ \dfrac{\partial f_3}{\partial x_1} & \dfrac{\partial f_3}{\partial x_2} & \dfrac{\partial f_3}{\partial x_3} \end{bmatrix}$$

Thus the decay of the perturbation in the concentrations changes according to

$$\frac{d\delta\mathbf{x}}{dt} = \mathbf{J}\,\delta\mathbf{x}$$

c. If the reactions were independent of each other, then $\mathbf{J}$ would have only diagonal elements and the three diagonal elements would just be the instantaneous loss rates of each species. Because the CH_4–CO–OH system is highly coupled, J contains important off-diagonal terms. Show that the Jacobian matrix corresponding to the steady-state solution for the parameters given above is (all elements have units s^{-1})

$$\begin{bmatrix} -2.80 \times 10^{-9} & 0 & -0.285714 \\ +2.80 \times 10^{-9} & -1.12 \times 10^{-7} & -0.428571 \\ -2.80 \times 10^{-9} & -1.12 \times 10^{-7} & -2.0 \end{bmatrix}$$

d. Show that the eigenvalues (s^{-1}) of J are

$$\lambda_1 = -1.769 \times 10^{-9} \qquad \lambda_2 = -8.863 \times 10^{-8} \qquad \lambda_3 = -2.0$$

and that any perturbation can be decomposed into a combination of the three eigenvectors of J, each of which decays with its own time constant. Show that the first eigenvector corresponds to perturbations that decay with a timescale of $1/\lambda_1 = 17.9$ years and represents largely the response to the CH_4 perturbation. The second eigenvector decays with a timescale $1/\lambda_2 = 0.36$ year and corresponds mainly to the CO–OH perturbation. The third eigenvector is solely the decay of the OH perturbation: $1/\lambda_3 = 0.5\,s$.

REFERENCES

Arrhenius, S. (1896) On the influence of carbonic acid in the air upon the temperature of the ground, *Philos. Mag.* **4**, 237–276.

Boer, G. J., and Yu, B. (2003) Climate sensitivity and response, *Climate Dyn.* **20**, 415–429.

Bradley, R. S., and Eddy, J. A. (1991) *Earth Quest*, **5**(1).

Cook, A. J., Fox, A. J., Vaughan, D. G., and Ferrigno, J. G. (2005) Retreating glacier fronts on the Antarctic Peninsula over the past half-century, *Science* **308**, 541–544.

Crowley, T. J. (1990) Are there any satisfactory geologic analogs for a future greenhouse warming, *Climate* **3**, 1282–1292.

Crowley, T. J. (1996) Remembrance of things past: Greenhouse lessons from the geologic record, *Consequences* **2**(1), 3–12.

Haigh, J. D. (1996) The impact of solar variability on climate, *Science* **272**, 981–984.

Hansen, J. Sato, M., and Ruedy, R. (1997) Radiative forcing and climate response, *J. Geophys. Res.* **102**, 6831–6864.

Hansen, J., and Nazarenko, L. (2004) Soot climate forcing via snow and ice albedos, *Proc. Natl. Acad. Sci.* **101** (doi: 10.1073/pnas.2237157100, 423–428).

Imbrie, J., Hays, J. D., Martinson, D. G., McIntyre, A., Mix, A. C., Morley, J. J., Pisias, N. G., Prell, W. L., and Shackleton, N. J. (1984), in *Milankovitch and Climate*, A. Berger, J. Imbrie, J. Hays, G. Kukla, and B. Saltzman, eds., Reidel, Dordrecht, The Netherlands, pp. 269–305.

Intergovernmental Panel on Climate Change (IPCC) (1995) *Climate Change 1994: Radiative Forcing of Climate Change and an Evaluation of the IPCC IS92 Emission Scenarios*, Cambridge Univ. Press, Cambridge, UK.

Intergovernmental Panel on Climate Change (IPCC) (2001) *Climate Change 2001: The Scientific Basis*, Cambridge Univ. Press, Cambridge, UK.

Isaksen, I. S. A., and Hov, O. (1987) Calculation of trends in the tropospheric concentration of O_3, OH, CH_4, and NO_x, *Tellus* **39B**, 271–285.

Joshi, M., Shine, K., Ponater, M., Stuber, M., Sausen, R., and Li, L. (2003) A comparison of climate response to different radiative forcings in three general circulation models: Towards an improved metric of climate change, *Climate Dyn.* **20**(7–8), 843–854.

Jouzel, J. et al. (1987) Vostok ice core—a continuous isotope temperature record over the last climatic cycle (160,000 years), *Nature* **329**, 403–408.

Lean, J., and Rind, D. (1996) The sun and climate, *Consequences* **2**(1), 27–36.

Madronich, S., and Granier, C. (1992) Impact of recent total ozone changes on tropospheric ozone photodissociation, hydroxyl radicals and methane trends, *Geophys. Res. Lett.* **19**, 465–467.

Madronich, S., and Granier, C. (1994) Tropospheric chemistry changes due to increased UV-B radiation, in *Stratospheric Ozone Depletion—UV-B Radiation in the Biosphere*, R. H. Biggs and M. E. B. Joyner, eds., Springer, Berlin, pp. 3–10.

Manabe, S., and Strickler, R. F. (1964) Thermal equilibrium of the atmosphere with a convective adjustment, *J. Atmos. Sci.* **21**(4), 361–385.

Manabe, S., and Weatherald, R. (1967) Thermal equilibrium of the atmosphere with a given distribution of relative humidity, *J. Atmos. Sci.* **24**, 241–259.

Meehl, G. A., Washington, W. M., Collins, W. D., Arblaster, J. M., Hu, A., Buja, L. E., Strand, W. G., and Teng, H. (2005) How much more global warming and sea level rise? *Science* **307**, 1769–1772.

National Research Council (2005) *Radiative Forcing of Climate Change*, National Academies Press, Washington, DC.

Prather, M. J. (1994) Lifetimes and eigenstates in atmospheric chemistry, *Geophys. Res. Lett.* **21**, 801–804.

Prather, M. J. (1996) Time scales in atmospheric chemistry: Theory, GWPs for CH_4 and CO, and runaway growth, *Geophys. Res. Lett.* **23**, 2597–2600.

Stocker, T. F. and Raible, C. C. (2005) Water cycle shifts gear, *Nature* **434**, 830–832.

Thompson, A. M., Stewart, R. W., Owens, M. A., and Herwehe, J. A. (1989) Sensitivity of tropospheric oxidants to global chemical and climate change, *Atmos. Environ* **23**, 519–532.

Trainer, M., Hsie, E. Y., KcKeen, S. A., Tallamraju, R., Parrish, D. D., Fehsenfeld, F. C., and Liu, S. C. (1987) Impact of natural hydrocarbons on hydroxyl and peroxy radicals at a remote site, *J. Geophys. Res.* **92**, 11879–11894.

Weart, S. R. (1997) The discovery of the risk of global warming, *Physics Today*, 34–40, Jan.

Werth, D. and Avissar, R. (2005) The local and global effects of African deforestation, *Geophys. Res. Lett.* **32**(12), L12704 (doi: 10.1029/2005GL022969).

Wigley, T. M. L. (2005) The climate change commitment, *Science* **307**, 1766–1769.

24 Aerosols and Climate

Aerosols influence climate directly by the scattering and absorption of solar radiation and indirectly through their role as cloud condensation nuclei. The magnitude of the direct forcing of aerosols (measured in $W\,m^{-2}$) at a particular time and location depends on the amount of radiation scattered back to space, which itself depends on the size, abundance, and optical properties of the particles and the solar zenith angle. The so-called indirect effect arises when increases in aerosol number concentrations from anthropogenic sources lead to increased concentrations of cloud condensation nuclei, which, in turn, lead to clouds with larger number concentrations of droplets with smaller radii, which, in turn, lead to higher cloud albedos.

Particles can both scatter and absorb radiation; as particles become increasingly absorbing versus scattering, a point is reached, depending on their size and the albedo of the underlying surface, where the overall effect of the particle layer changes from one of cooling to heating. In addition, if the particles consist of a mixture of purely scattering material, such as ammonium sulfate, and partially absorbing material, such as soot, the cooling–heating effect depends on the manner in which the two substances are mixed throughout the particle population. The two extremes in this regard are when every particle contains some absorbing material and when the absorbing material is distinct from the scattering particles. The effects are further complicated when a cloud is present. Particles exist both above and below clouds and, to some extent, even in the cloud itself. The amount of light scattered back to space depends on the properties of both the aerosols and the cloud.

The *direct* effect can be observed visually as the sunlight reflected upward from haze when viewed from above (e.g., from a mountain or an airplane). The result of the process of scattering of sunlight is an increase in the amount of light reflected by the planet and hence a decrease in the amount of solar radiation reaching the surface. The amount of light reflected upward by aerosol is roughly proportional to the total column mass burden of particles (typically reported in grams per square meter). The direct effect of aerosols on climate is a result of the same physics responsible for the reduction of visibility that occurs in airmasses laden with particles. The major difference is that, whereas visibility reduction is attributable to aerosol scattering in all directions, the direct climatic effect of aerosols results only from radiation that is scattered in the upward direction, that is, back to space.

Indirect climate effects of aerosols are more complex and more difficult to assess than direct effects because they depend on a chain of phenomena that connect aerosol levels to concentrations of cloud condensation nuclei, cloud condensation nuclei concentrations to cloud droplet number concentrations (and size), and these, in turn, to cloud albedo and cloud lifetime. Changes in the number concentration of aerosols are observed to cause variations in the population and sizes of cloud droplets, which are expected to cause changes in cloud albedo and areal extent. Other meteorological influences might occur as a result of perturbations in the number concentration of aerosols, such as changes in precipitation.

In contrast to GHGs, which act only on outgoing, infrared radiation, aerosol particles can influence both sides of the energy balance (Table 24.1). Particles of diameters less than 1 μm are highly effective at scattering incoming solar radiation, sending a portion of that scattered radiation back to space. In so doing, these particles reduce the amount of incoming solar energy as compared with that in their absence and consequently cool the Earth. Over industrialized parts of the world, sulfate particles produced by the oxidation of anthropogenically emitted SO_2 constitute much of this light-scattering aerosol. In the tropics, biomass burning of forests and savannas is a dominant source of airborne particles, which consist mainly of organic matter and soot. Mineral dust from wind acting on soils is always present in the atmosphere to some degree, although human activities, such as disruption of soils by changing use of land in arid and subarid regions, can increase the loading of dust over that present "naturally." Because of their size and composition, mineral dust particles can scatter and absorb both incoming and outgoing radiation. In the visible part of the spectrum, the light-scattering effect dominates, and mineral dust exerts an overall cooling effect; in the infrared region, mineral dust is an absorber and acts like a greenhouse gas.

Greenhouse gases such as CO_2, CH_4, N_2O, and the CFCs are virtually uniform globally; aerosol concentrations, on the other hand, are highly variable in space and time. With lifetimes of about a week, sulfate aerosols are most abundant close to their sources in the industrialized areas of the Northern Hemisphere. Biomass aerosols are emitted predominantly during the dry season in tropical areas. Mineral dust is at highest concentrations downwind of large arid regions. Moreover, greenhouse gas forcing operates day and night; aerosol forcing operates only during daytime. Aerosol radiative effects depend in a complicated way on the solar angle, relative humidity, particle size and composition, and the albedo of the underlying surface. When superimposed on each other, the spatial distribution of GHG warming and aerosol cooling do not occur at the same locations.

Aerosol residence times in the troposphere are roughly 1–2 weeks; if all SO_2 sources were shut off today, anthropogenic sulfate aerosols would disappear from the planet in 2 weeks. By contrast, not only are GHG residence times measured in decades to centuries, but because of the great inertia of the climate system, as noted in the previous chapter, the effect of GHG forcing takes decades to be fully transformed into equilibrium climate warming. As a result, if both CO_2 and aerosol emissions were to cease today, the Earth would continue to warm as the climate system continues to respond to the accumulated amount of CO_2 already in the atmosphere.

Analyses of global temperatures have, as we saw in the preceding chapter, concluded that the Earth's mean temperature has increased by about 0.6 K since preindustrial times. Until recently, the inability to reconcile the observed temperature trend since the preindustrial period with that predicted by general circulation models (GCMs) based on GHG increases alone led to nagging uncertainties about our understanding of the climate effects of GHG forcing. However, inclusion of aerosol effects in climate models has made a crucial difference in the ability to simulate observed temperature trends. Patterns of observed temperature changes show significant similarities with GCM predicted changes when aerosols are included but show less similarities if GHGs alone are considered.

TABLE 24.1 Comparison of Climate Forcing by Aerosols with Forcing by Greenhouse Gases (GHGs)

Factor	Long-Lived GHGs (CO_2, CH_4, CFCs)	Short-Lived GHGs (O_3, HCFCs, VOCs)	Aerosols
Optical properties	Infrared absorption is well quantified for all major and minor GHGs	Infrared absorption is reasonably well quantified	Refractive indices of pure substances are known, but size-dependent mixing of numerous species and the nature of mixing have optical effects difficult to quantify
Important electromagnetic spectrum	Almost entirely longwave ($\lambda > 1\ \mu m$)	For O_3, solar and longwave are important	For tropospheric aerosol, mainly solar; for stratospheric aerosol, solar and longwave contributions lead to stratospheric warming
Amounts of material	Well mixed; nearly uniform within the troposphere	Highly variable in space and time; concentrations may be estimated by chemical models, but with some uncertainty	Pronounced spatial and temporal variations
Determination of forcing	Well-posed problem in radiative transfer	Radiative aspects well posed; global networks provide some data to test model predictions of geographic and altitudinal distributions	*Direct:* Relatively well-posed problem, but dependent on empirical values for several key aerosol properties; dependent on models for geographical/temporal variations of forcing *Indirect:* Depends on aerosol number distribution; inadequacy of descriptions of aerosols and clouds restricts abilities to predict indirect forcing
Dependence of forcing on loading (at present)	Varies as weak function (square root or logarithm) of concentration	Generally nonlinear with concentration; for halocarbons, a linear dependence	*Direct:* Almost linear in the concentration of the particles *Indirect:* Nonlinear
Nature of forcing	Forcing is exerted at the surface and in the troposphere, operates night and day	Strongly dependent on geographic, altitudinal, and temporal variation (e.g., O_3); forcing is exerted at the surface and troposphere; operates night and day	Tropospheric forcing varies strongly with location and season and occurs only during daytime; maxima of forcing occur near sources and at the Earth's surface; stratospheric forcing includes some longwave effect but is dominated by shortwave radiation (daytime only); following major volcanic events, stratospheric mixing yields a forcing that is substantially global in nature For nonabsorbing tropospheric aerosols, forcing is almost entirely at surface; for stratospheric aerosols, there is a small heating resulting in a transient warming of stratosphere

Source: National Research Council (1996).

24.1   SCATTERING–ABSORBING MODEL OF AN AEROSOL LAYER

Consider a direct solar beam impinging on the layer shown in Figure 24.1. Assume, for the moment, that the beam is directly overhead, at a solar zenith angle of $\theta_0 = 0°$. The fraction of the incident beam transmitted through the layer is $e^{-\tau}$, where τ is the optical depth of the layer. The fraction reflected back in the direction on the beam is $r = (1 - e^{-\tau})\omega\beta$, where ω is the *single-scattering albedo* of the aerosol, and β is the *upscatter fraction*, the fraction of light that is scattered by a particle into the upward hemisphere.

The fraction of light absorbed within the layer is $(1 - \omega)(1 - e^{-\tau})$. The fraction scattered downward is $\omega(1 - \beta)(1 - e^{-\tau})$. The total fraction of radiation incident on the layer that is transmitted downward is

$$t = e^{-\tau} + \omega(1 - \beta)(1 - e^{-\tau}) \tag{24.1}$$

If the albedo of the underlying Earth's surface is R_s, then the fraction of the radiation incident on the surface that is reflected is $R_s t$.

Of the intensity $R_s t F_0$ reflected upward by the Earth back into the aerosol layer, some is backscattered back to the Earth, some is absorbed by the layer, and some is scattered upward. On the first downward pass through the layer the fraction of the intensity transmitted is t. Thus, on the first upward pass of the reflected beam, the fraction transmitted is also t. By turning the layer upside down in our minds, the fraction of the beam from the Earth reflected back downward to the Earth from the layer is just $rR_s t$. That beam reflected downward is itself reflected off the surface. So, for two complete passes, the total upward flux is

$$F_r = r + R_s t F_0 t + R_s r R_s t F_0 t$$
$$= (r + t^2 R_s + t^2 R_s^2 r) F_0$$

The process can be continued, giving

$$F_r = (r + t^2 R_s + t^2 R_s^2 r + \cdots) F_0$$
$$= [r + t^2 R_s (1 + R_s r + \cdots)] F_0$$

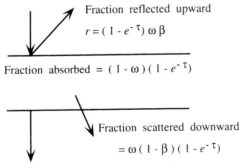

FIGURE 24.1   Scattering model of an aerosol layer above the Earth's surface. Total downward transmitted fraction is $t = e^{-\tau} + \omega(1 - \beta)(1 - e^{-\tau})$. Total reflected off surface $= R_s t$. ($R_s =$ surface albedo).

The next term in the series is $R_s^2 r^2$, so

$$F_r = [r + t^2 R_s(1 + R_s r + R_s^2 r^2 + R_s^3 r^3 + \cdots)]F_0 \tag{24.2}$$

Since both R_s and r are <1, the series in the small parentheses can be summed to give

$$1 + R_s r + R_s^2 r^2 + R_s^3 r^3 + \cdots = \frac{1}{1 - R_s r} \tag{24.3}$$

Thus (24.2) becomes

$$F_r = \left[r + \frac{t^2 R_s}{1 - R_s r} \right] F_0 \tag{24.4}$$

This is the total upward reflected flux. The quantity in parentheses is the total reflectance of the aerosol–surface system and can be given the symbol R_{as}:

$$R_{as} = r + \frac{t^2 R_s}{1 - R_s r} \tag{24.5}$$

In the absence of aerosol ($\tau = 0$), $t = 1$ and $r = 0$, and $F_r = R_s F_0$; that is, the only reflected beam is that from the surface.

The change in reflectance as the result of the presence of an aerosol layer is $\Delta R_p = R_{as} - R_s$:

$$\Delta R_p = \left[r + \frac{t^2 R_s}{1 - R_s r} \right] - R_s \tag{24.6}$$

If the fraction of the vertical column covered by clouds is A_c, as a first approximation we can assume that only that fraction free of clouds experiences the change in albedo and multiply (24.3) by $(1 - A_c)$:

$$\Delta R_p = (1 - A_c)\left[\left(r + \frac{t^2 R_s}{1 - R_s r} \right) - R_s \right] \tag{24.7}$$

Up to this point we have assumed that only the aerosol in the atmosphere produces scattering and absorption. Actually the atmosphere itself, even if totally devoid of particles, does not completely transmit the incident solar beam. Let T_a denote the fractional transmittance of the atmosphere. It is useful to think of the atmosphere as overlaying the aerosol layer. Then, instead of F_0, the incident flux on the aerosol layer is $T_a F_0$. The total upward reflected flux, given by (24.4), is now corrected as follows:

$$\left[r + \frac{t^2 R_s}{1 - R_s r} \right](T_a F_0)T_a = T_a^2 F_r \tag{24.8}$$

The term $(T_a F_0)$ accounts for the reduction in flux first impinging on the top of the aerosol layer and the second factor of T_a accounts for the fact that the upward reflected flux is itself reduced by the transmittance T_a before it exits the top of the atmosphere.

Thus the change in planetary albedo is

$$\Delta R_p = (1 - A_c) T_a^2 \left[\left(r + \frac{t^2 R_s}{1 - R_s r} \right) - R_s \right] \tag{24.9}$$

and the change in outgoing radiative flux as the result of an aerosol layer underlying an atmospheric layer is

$$\Delta F = F_0 (1 - A_c) T_a^2 \left[\left(r + \frac{t^2 R_s}{1 - R_s r} \right) - R_s \right] \tag{24.10}$$

Direct aerosol radiative forcing is defined as the difference in the net (down minus up) radiative flux between two cases: the first, with aerosol present; and the second, in which aerosol is not present. If $\Delta R_p > 0$, then the planetary albedo is increased in the presence of aerosols, and more radiation is reflected back to space than in their absence. In this case, the change in outgoing radiative flux $\Delta F > 0$. In terms of climate forcing, one is concerned with the net change in *incoming* radiative flux, so the net change in forcing is $-\Delta F$.

To summarize, ΔF depends on the following parameters:

F_0 = incident solar flux, $W\,m^{-2}$
A_c = fraction of the surface covered by clouds
T_a = fractional transmittance of the atmosphere
R_s = albedo of the underlying Earth surface
ω = single-scattering albedo of the aerosol
β = upscatter fraction of the aerosol
τ = aerosol optical depth

The single-scattering albedo ω depends on the aerosol size distribution and chemical composition and is wavelength-dependent. The upscatter fraction β depends on aerosol size and composition, as well as on the solar zenith angle θ_0. Aerosol optical depth depends largely on the mass concentration of aerosol.

As a first approximation, it was assumed that direct aerosol forcing occurs only in cloud-free regions. Actually, some amount of aerosol forcing does occur when clouds are present. Imagine a relatively uniform layer of particles extending from the surface up to a height of, say, 5 km. Consider first a cloud layer above the aerosol layer. Some solar radiation does penetrate the cloud layer to be intercepted by the aerosol layer beneath it. However, because the aerosols receive only a portion of the incoming radiation and because any radiation they do scatter upward is, in turn, scattered by the clouds, the aerosol forcing in this situation turns out not to be terribly significant. If the cloud layer exists within the aerosol layer, then those particles above the clouds scatter radiation back to space in the same manner as in the cloud-free case. A somewhat mitigating factor is that the radiation scattered upward by the cloud layer is intercepted by the aerosol layer and some of that radiation is scattered back down toward the cloud layer. Boucher and Anderson (1995) find, for example, in their global simulation that the ratio of aerosol forcing in the presence of clouds to that in the absence of clouds is $\Delta F_{cloud}/\Delta F_{clear} = 0.25$. When aerosol scattering within and below clouds was removed from the simulation, the resulting decrease in forcing was 7%. Thus aerosols above the cloud layer are responsible for most of the direct effect in cloudy regions.

24.2 COOLING VERSUS HEATING OF AN AEROSOL LAYER

The sign of the change in planetary albedo as the result of an aerosol layer ΔR_p determines whether the forcing is negative (cooling effect) or positive (heating effect). The key parameter governing the amount of cooling versus heating is the single-scattering albedo ω. The value of ω at which $\Delta R_p = 0$ defines the boundary between cooling and heating. The scattering versus absorption properties of an aerosol are measured by the value of the single-scattering albedo. Whether particle absorption actually results in an increase or decrease of the planetary albedo depends on the particle single-scattering albedo as well as on the albedo of the underlying surface. Over dark surfaces like the ocean, virtually all particles tend to increase the planetary albedo because upscattering of incident solar radiation by the particle layer exceeds that from the dark surface regardless of how much absorption is occurring. On the other hand, over high-albedo surfaces, such as snow or bright deserts, absorption by particles can reduce the solar flux reflected from the surface and result in a net reduction of radiation to space.

Let us derive an expression for the value of ω at which $\Delta R_p = 0$. To do so it is useful to employ the fact that for global tropospheric conditions optical depth values are usually about 0.1, so that the mathematical approximation $\tau \ll 1$ is a reasonable one. Thus

$$\begin{aligned} r &= (1 - e^{-\tau})\omega\beta \\ &\simeq \tau\omega\beta \end{aligned} \tag{24.11}$$

and

$$\begin{aligned} t &= e^{-\tau} + \omega(1 - \beta)(1 - e^{-\tau}) \\ &\simeq 1 - \tau + \omega(1 - \beta)\tau \end{aligned} \tag{24.12}$$

Thus, from (24.6), we obtain

$$\Delta R_p \simeq \tau\omega\beta + \frac{[1 - \tau + \omega(1 - \beta)\tau]^2 R_s}{1 - R_s \tau\omega\beta} - R_s \tag{24.13}$$

Rearranging the RHS of (24.13) and neglecting terms involving τ^2 yield

$$\Delta R_p \simeq \tau\omega\beta + \left[(1 - R_s)^2 - \frac{2R_s}{\beta}\left(\frac{1}{\omega} - 1\right)\right]$$

We can also neglect $\tau\omega\beta$ relative to the other term, giving

$$\Delta R_p = (1 - R_s)^2 - \frac{2R_s}{\beta}\left(\frac{1}{\omega} - 1\right) \tag{24.14}$$

The boundary between cooling and heating, that is, at $\Delta R_p = 0$, occurs for values of ω satisfying

$$\omega_{\text{crit}} = \frac{2R_s}{2R_s + \beta(1 - R_s)^2} \tag{24.15}$$

Values of $\omega > \omega_{\text{crit}}$ lead to cooling ($\Delta R_p > 0$).

Figure 24.2 shows the critical single-scattering albedo that defines the boundary between negative (cooling) and positive (heating) forcing regions as a function of surface albedo R_s and upscatter fraction β. The critical value of ω is practically independent of the actual value of τ. A typical global mean surface albedo R_s is about 0.15, and a representative value of the spectrally and solar zenith angle averaged β is about 0.29. For

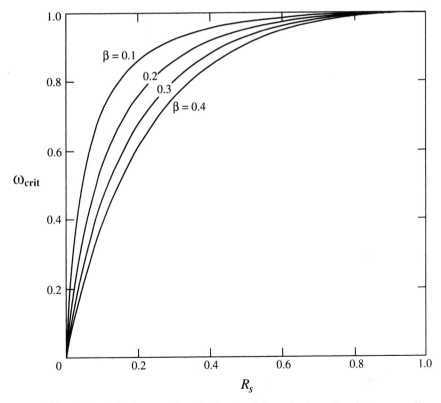

FIGURE 24.2 Critical single-scattering albedo that defines the boundary between cooling and heating regions.

these values of R_s and β, the critical value of ω that defines the crossover point between heating and cooling is about 0.6. At values of surface albedo approaching 1.0, the critical value of ω itself approaches 1.0, independently of the value of β. At high values of R_s the total reflectance of the aerosol–surface system is large to begin with and even a small amount of aerosol absorption leads to a heating effect. At values of R_s approaching zero, only a small amount of aerosol scattering is required to produce a cooling effect of the aerosol layer.

24.3 SCATTERING MODEL OF AN AEROSOL LAYER FOR A NONABSORBING AEROSOL

The aerosol layer scattering model can be simplified if a nonabsorbing aerosol, that is, $\omega = 1$, is considered and account is taken of the fact that the optical depth for most tropospheric conditions is usually about 0.1 or smaller, thereby allowing one to assume $\tau \ll 1$. We note that hydrated ammonium/sulfate particles ranging from pure $(NH_4)_2SO_4$ to pure H_2SO_4, a mixture frequently used to represent background global aerosol, are essentially nonabsorbing at wavelengths below 2 μm (Figure 24.3). This, coupled with the fact that the small amount ($\sim 5\%$) of solar intensity above $\lambda = 2$ μm is mostly absorbed by atmospheric gases, means that the approximation $\omega = 1$ is reasonable for global sulfate aerosols. The simplification occurs in the expression for the change in planetary reflectance (24.6):

$$\Delta R_p = R_{as} - R_s$$
$$= \left[r + \frac{t^2 R_s}{1 - R_s r} \right] - R_s$$

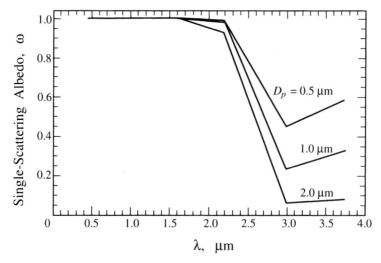

FIGURE 24.3 Single-scattering albedo for $(NH_4)_2SO_4$ particles as a function of wavelength (RH = 50%).

For a nonabsorbing aerosol, $\omega = 1$ and $r + t = 1$. Thus

$$\begin{aligned}
\Delta R_p &= r + \frac{(1-r)^2 R_s}{1 - R_s r} - R_s \\
&\simeq (r + R_s - 2\,r\,R_s)(1 + R_s r) - R_s
\end{aligned} \tag{24.16}$$

where we have used the condition that $1 \gg R_s^2 r^2$. Using this condition yet again, we find

$$\Delta R_p \simeq r(1 - R_s)^2 \tag{24.17}$$

Note that $\Delta R_p > 0$. This means that for a nonabsorbing aerosol layer, the change in planetary reflectance is always positive; that is, the presence of the aerosol layer leads to an increased planetary albedo over the that in its absence and, consequently, a cooling effect.

Since with $\omega = 1$, and $\tau \ll 1$, it follows that

$$\begin{aligned}
r &= (1 - e^{-\tau})\beta \\
&\simeq \tau\beta
\end{aligned} \tag{24.18}$$

so

$$\Delta R_p \simeq (1 - R_s)^2 \tau\beta \tag{24.19}$$

Adding the factors of $(1 - A_c)$ and T_a^2 as in (24.10), we obtain the change in outgoing radiative flux:

$$\Delta F \simeq F_0 T_a^2 (1 - A_c)(1 - R_s)^2 \beta\tau \tag{24.20}$$

For radiation at solar zenith angle θ_0 (see Chapter 4), the pathlength relative to a unit depth of the layer is (neglecting for the moment the correction given in Table 4.1) $\sec\theta_0$, so at any solar zenith angle θ_0, the flux along the path of the Sun is

$$\frac{F}{F_0} = \exp(-\tau \sec\theta_0) \tag{24.21}$$

where $\tau \sec\theta_0$ can be thought of as the slant path optical depth, which replaces τ in (24.20). But the incident solar flux F_0 at solar zenith angle θ_0 has an intensity, normal to the surface, of $F_0 \cos\theta_0$, which replaces F_0 in (24.20). Thus the two factors, $\sec\theta_0$ and $\cos\theta_0$, accounting for the solar zenith angle just cancel each other. The upscatter fraction β depends on aerosol particle size and solar zenith angle θ_0. For use of (24.20) in estimating radiative forcing, β can be replaced by its spectrally and solar zenith angle averaged value, $\bar{\beta}$.

Equation (24.20) can be used to estimate the globally and annually averaged forcing, $\Delta\bar{F}$, as a result of an aerosol layer. First F_0 is replaced by $F_0/2$ to reflect the fact that any point on the globe is illuminated by sunlight only one-half of the time over the course of a year. Finally, the optical depth, which is a function of wavelength λ, is replaced by its spectrally weighted value, $\bar{\tau}$. The result for the change in *incoming* solar flux (the negative of the outgoing flux) is

$$\Delta\bar{F} = -\tfrac{1}{2}F_0 T_a^2(1 - A_c)(1 - R_s)^2\,\bar{\beta}\bar{\tau} \qquad (24.22)$$

It is useful to reiterate the approximations inherent in (24.22): (1) the aerosol is assumed to be nonabsorbing, that is, $\omega = 1$; (2) the optical depth is considerably smaller than 1.0; (3) annual mean conditions are assumed; (4) a single, globally averaged value of the surface albedo R_s is used; (5) $\bar{\beta}$ is solar zenith angle and spectrally averaged; (6) a single, globally averaged value of atmospheric transmittance T_a is used; and (7) direct aerosol forcing occurs only in cloud-free regions. The expression for ΔF combines geophysical variables, aerosol microphysics, and optical depth:

$$\Delta\bar{F} = \underbrace{-\tfrac{1}{2}F_0 T_a^2(1 - A_c)(1 - R_s)^2\,\bar{\beta}}_{\text{geophysical variables}}\ \overset{\text{optical depth}}{\bar{\tau}}$$

$$\text{aerosol microphysics}$$

The general result for the net radiative forcing of an aerosol that both scatters and absorbs ($\omega < 1$) radiation is

$$\Delta\bar{F} = -\tfrac{1}{2}F_0 T_a^2(1 - A_c)\omega\bar{\beta}\bar{\tau}\left[(1 - R_s)^2 - \frac{2R_s}{\bar{\beta}}\left(\frac{1}{\omega} - 1\right)\right] \qquad (24.23)$$

This equation was first derived by Haywood and Shine (1995). We will now consider explicitly how upscatter fraction and optical depth may be calculated.

24.4 UPSCATTER FRACTION

The fraction of light that is scattered by a particle into the upward hemisphere relative to the local horizon is the fraction of the integral over the angular distribution of scattered radiation that is in the upward hemisphere. This *upscatter fraction* β therefore depends on particle size and solar zenith angle. Note that it is the *upward* hemisphere that is important in the effect of aerosols on the Earth's radiative balance, not the *back* hemisphere relative to the direction of direct of incident radiation (scattering angles of 90°–270° with respect to the incident ray). The two coincide exactly only when the solar zenith angle is directly overhead, $\theta_0 = 0°$. At any given solar zenith angle θ_0, the shift toward forward scattering associated with increasing particle size decreases the upscatter fraction (see Figure 24.4).

Radiative transfer theory to calculate the *upscatter fraction* β as a function of particle size and solar zenith angle θ_0 was developed by Wiscombe and Grams (1976). Figure 24.5 gives β as a function of $\mu_0 = \cos\theta_0$ and particle radius. For Sun at the horizon

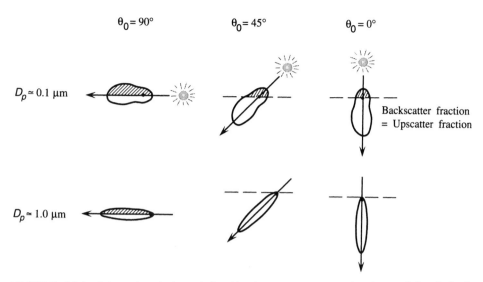

FIGURE 24.4 Schematic of the relationship between upscatter fraction and hemispheric backscatter ratio as a function of solar zenith angle for two particle sizes.

$(\theta_0 = 90°, \mu_0 = 0)\ \beta = 0.5$, independent of particle size because of the symmetry of the phase function (see Figure 24.4). For solar zenith angles decreasing from 90°, forward scatter leads to a decrease in the value of β, although the decrease is slight for the smallest particles because of the relative symmetry of the scattering function in the forward and backward directions for small particles.

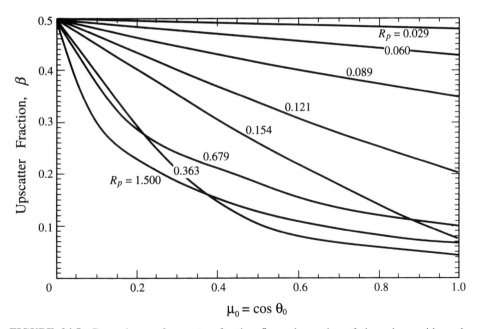

FIGURE 24.5 Dependence of upscatter fraction β on the cosine of the solar zenith angle, $\mu_0 = \cos\theta_0$, and particle radius (in μm) at $\lambda = 550$ nm and $m = 1.4 - 0i$ (Nemesure et al. 1995).

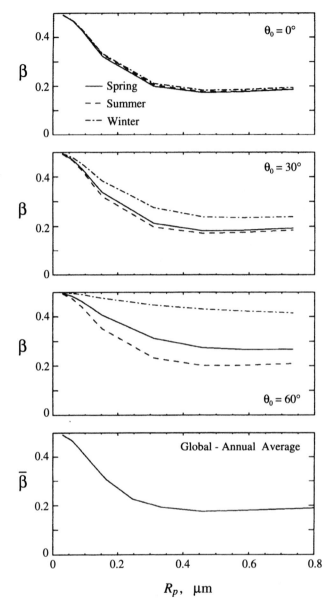

FIGURE 24.6 Diurnal (24-h) average upscatter fraction for $(NH_4)_2SO_4$ particles averaged over the solar spectrum, for three seasons and several latitudes (°N) as a function of particle radius (Nemesure et al. 1995). Here winter = December 21, spring = March 21, and summer = June 21. The bottom panel represents the global annual average β.

For models used to assess the effect of aerosols on radiative forcing, it is desirable to average the forcing over solar zenith angle. In doing so, it is necessary to average β over solar zenith angle. Figure 24.6 shows the global annual average β ($\bar{\beta}$) averaged over both diurnal solar zenith angle and the solar spectrum as calculated for a $(NH_4)_2SO_4$ aerosol. The global average $\bar{\beta}$ approaches 0.5 for the smallest particles and decreases to a value of about 0.2 for the largest particles.

24.5   OPTICAL DEPTH AND COLUMN FORCING

It is useful to define the aerosol optical depth as the product of coefficients per unit mass of aerosol and the aerosol mass concentration. For sulfate aerosol, for example, we can express the optical depth as (recall that the absorption component for sulfate aerosols is zero)

$$\tau = \alpha_{SO_4^{2-}} m_{SO_4^{2-}} H \tag{24.24}$$

where $\alpha_{SO_4^{2-}}$ is the light-scattering mass efficiency of the aerosol, expressed in units of $m^2 (g\, SO_4^{2-})^{-1}$, $m_{SO_4^{2-}}$ is the sulfate mass concentration ($g\, m^{-3}$), and H is the pathlength through the aerosol layer (m). Equation (24.24) defines $\alpha_{SO_4^{2-}}$, a quantity that, when multiplied by the mass concentration of sulfate, produces the sulfate aerosol scattering coefficient b_{sp}.

By reference to Chapter 15, we see that $\alpha_{SO_4^{2-}}$ is just the mass scattering efficiency $E_{scat}(m, D_p, \lambda)$, expressed in units of $m^2 (g\, SO_4^{2-})^{-1}$, for sulfate-containing particles. Figure 15.7 showed the mass scattering efficiency of a dry $(NH_4)_2SO_4$ aerosol at $\lambda = 550\, nm$. In that figure, the units of E_{scat} are $m^2\, g^{-1}$, referring to the total mass of $(NH_4)_2SO_4$. Expressed in units of $m^2 (g\, SO_4^{2-})^{-1}$, the mass scattering efficiency increases over that shown in Figure 15.7 by the ratio 132/96. Thus $\alpha_{SO_4^{2-}}$ for $(NH_4)_2SO_4$ aerosol reaches a maximum of about $9\, m^2 (g\, SO_4^{2-})^{-1}$ at a diameter just about equal to the wavelength of the radiation.

Because particle size and therefore $\alpha_{SO_4^{2-}}$ depend strongly on the ambient relative humidity (RH), it is customary to define $\alpha_{SO_4^{2-}}$ at a reference, low RH (RH_r) and then multiply it by a factor $f(RH)$ that is the ratio of the scattering cross section at any RH to that at RH_r

$$\alpha_{SO_4^{2-}} = \alpha_{SO_4^{2-}}^{RH_r} f(RH) \tag{24.25}$$

where $\alpha_{SO_4^{2-}}^{RH_r}$ represents the light-scattering efficiency of sulfate aerosol, including its associated cations, at RH_r, typically 30%. Its value depends on the aerosol size distribution and the chemical compounds associated with sulfate in the particles. An alternative to including all the associated cations and other substances along with sulfate in the definition of $\alpha_{SO_4^{2-}}^{RH_r}$ is to have the quantity refer only to the sulfate ion itself and treat the mass scattering efficiency of all substances separately and additively.

The dependence of scattering efficiency on RH is accounted for in (24.25) by the factor $f(RH)$. Most inorganic aerosols are hygroscopic. Sulfuric acid particles are always hydrated, but salts such as $(NH_4)_2SO_4$ and NH_4HSO_4 exist as dry particles at sufficiently low RH and experience an abrupt uptake of water at the relative humidity of deliquescence (DRH) (see Chapter 10). The DRH for these two ammonium sulfate salts are, for example, 80% for $(NH_4)_2SO_4$ (Shaw and Rood 1990; Tang and Munkelwitz 1994) and 39% for NH_4HSO_4 (Tang and Munkelwitz 1994). Figure 10.4 showed the particle size change that occurs with increasing RH for $(NH_4)_2SO_4$, NH_4HSO_4 and H_2SO_4. Sulfuric acid remains a liquid throughout the entire range of RH.

As we saw in Chapter 10, starting at an RH > DRH and decreasing RH, a particle does not crystallize when the DRH is reached but remains in a metastable equilibrium until a much lower humidity (RHC) at which crystallization finally occurs. This behavior gives rise to a hysteresis phenomenon in which a particle of a particular compound can be a different size at RHs between the RHC and DRH depending on whether it reached that RH through increasing or decreasing relative humidity (Figure 10.4). Crystallization relative humidities for the two ammonium sulfate salts are 37% for $(NH_4)_2SO_4$ and $<5\%$ for NH_4HSO_4 (Tang and Munkelwitz 1994). Metastable droplets at ambient RHs between the RHC and DRH of common salts have been found at urban and rural sites (Rood et al. 1989). Of course, it is not possible to know generally whether ambient aerosols have experienced a rising or falling RH history. The two hysteresis curves can be considered to represent the two extremes of particle size for deliquescent particles in computing optical properties.

Figure 24.7 shows the mass scattering efficiency of the three sulfate aerosols at four relative humidities averaged over the solar spectrum as a function of particle size. The primary label represents the moles of sulfate per particle; the secondary axes are the dry particle radius, given by

$$R_{p_0} = \left(\frac{3 N_{SO_4^{2-}} M}{4 \pi \rho_p} \right)^{1/3} \tag{24.26}$$

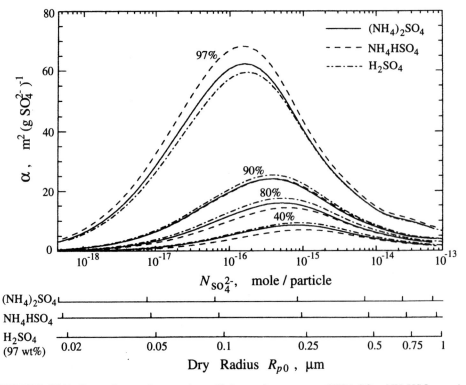

FIGURE 24.7 Dependence of scattering efficiency for aqueous $(NH_4)_2SO_4$, NH_4HSO_4, and H_2SO_4 aerosol on dry particle size expressed as amount of substance per particle $N_{SO_4^{2-}}$ for indicated values of RH, averaged over the solar spectrum (Nemesure et al. 1995). Particle dry radius, R_{p0} (evaluated as the radius of the sphere of equal volume) is also shown by means of the secondary axes.

where M is the molecular weight of the sulfur compound. In the case of H_2SO_4 the dry radius is taken at RH = 0. At each RH, $\alpha_{SO_4^{2-}}$ exhibits a maximum at about the coincidence of particle diameter with wavelength. Charlson et al. (1992) suggested, at $\lambda = 550$ nm, an average value of $\alpha_{SO_4^{2-}}$ of 8.5 m^2 (g SO_4^{2-})$^{-1}$, corresponding to $\alpha_{SO_4^{2-}}^{RH_r} = 5\,m^2(g\,SO_4^{2-})^{-1}$ and a growth factor $f(RH) = 1.7$.

Early studies employed (24.22) to estimate direct aerosol forcing by sulfate aerosols in a column of air extending from the surface to the top of the atmosphere (Charlson et al. 1991, 1992; Pilinis et al. 1995; Nemesure et al. 1995; Penner et al. 1994). The mean optical depth is computed as the product of the aerosol mass scattering coefficient $[m^2(g\,SO_4^{2-})^{-1}]$ and the column integrated mass of aerosol (g SO_4^{2-} m^{-2}):

$$\bar{\tau} = \alpha_{SO_4^{2-}}\bar{m}_{SO_4^{2-}} \tag{24.27}$$

The column burden of sulfate aerosol ($\bar{m}_{SO_4^{2-}}$) can be estimated from the global source strength of SO_2 (Q_{SO_2}), the average fractional conversion of SO_2 to sulfate in the atmosphere (y_{SO_2}), the mean residence time of sulfate aerosol in the atmosphere ($\tau_{SO_4^{2-}}$), and the area A of the geographic region over which the estimate is performed (e.g., the entire globe, the Northern Hemisphere) as follows:

$$\bar{m}_{SO_4^{2-}} = Q_{SO_2}\,y_{SO_2}\,\tau_{SO_4^{2-}}/A \tag{24.28}$$

Global mean direct radiative forcing from such an approach is thus given by

$$\Delta\bar{F}_{SO_4^{2-}} = -\frac{1}{2}F_0 T_a^2 (1 - A_c)(1 - R_s)^2 \underbrace{\bar{\beta}\alpha_{SO_4^{2-}}^{RH_r} f(RH)}_{\text{aerosol microphysics}} \overbrace{Q_{SO_2}\,y_{SO_2}\,\tau_{SO_4^{2-}}/A}^{\text{optical depth}} \tag{24.29}$$

geophysical variables   aerosol microphysics   sulfate column burden

Table 24.2 presents global mean values of the parameters in (24.29), along with estimates of the uncertainty associated with each parameter, as suggested by Penner et al. (1994).

TABLE 24.2   Parameters in the Estimation of Global Mean Direct Forcing of Climate by Anthropogenic Sulfate Aerosol

Parameter	Value	Units	Estimated Uncertainty Factor[a]
F_0	1370	W m^{-2}	—
$1 - A_c$	0.4	—	1.1
T_a	0.76	—	1.15
$1 - R_s$	0.85	—	1.1
$\bar{\beta}$	0.29	—	1.3
$\alpha_{SO_4^{2-}}^{RH_r}$	5[b]	$m^2(g\,SO_4^{2-})^{-1}$	1.5
$f(RH)$	1.7	—	1.2
Q_{SO_2}	80	Tg yr^{-1}	1.15
y_{SO_2}	0.4	—	1.5
$\tau_{SO_4^{2-}}$	0.02	yr	1.5
A	5×10^{14}	m^2	—

[a]The central value, divided/multiplied by the uncertainty factor, gives the estimated range of values of the parameter.
[b]Includes associated cations.

Source: Penner et al. (1994).

At any given RH and solar zenith angle θ_0 there is a maximum in forcing corresponding to the maximum in $\alpha_{SO_4^{2-}}$ as a function of particle size. Also, at any given particle dry radius and RH there is a maximum in forcing as a function of θ_0. Since the effective pathlength varies as sec θ_0, and the incident solar flux varies as cos θ_0. these two factors cancel each other out and the maximum forcing as a function of solar zenith angle cannot be a result of geometry alone. Now, from Figure 24.5, we know that β achieves its maximum at $\theta_0 = 90°(\mu_0 = 0)$, so at first glance it would appear that forcing should actually be a maximum when the Sun is at the horizon, but as the Sun approaches the horizon, Rayleigh scattering begins to become important because of the increasingly longer pathlength. As a result, the incident flux decreases. The competition between these two effects leads to a maximum in forcing at a particular zenith angle. In box model simulations Pilinis et al. (1995) determined that the maximum in forcing occurs at a solar zenith angle of about 78°. This result has implications for global aerosol forcing. All else being equal (aerosol optical properties, size, RH, etc.) for two regions at different latitudes, that at the higher latitude will experience a greater aerosol forcing because of the larger mean solar zenith angle.

Aerosol Radiative Forcing Efficiency   The radiative forcing of an aerosol species can be normalized by its column mass burden to produce a *radiative forcing efficiency per unit mass* $\Delta \bar{G}$. For sulfate aerosol, for example, one obtains

$$\Delta \bar{G}_{SO_4^{2-}} = \Delta \bar{F}_{SO_4^{2-}} / \bar{m}_{SO_4^{2-}} \qquad \left[W(g\,SO_4^{2-})^{-1} \right]$$

One may also define a related measure, the *radiative forcing efficiency per unit optical depth*, for example

$$\Delta \bar{G}^{\tau}_{SO_4^{2-}} = \Delta \bar{F}_{SO_4^{2-}} / \bar{\tau}_{SO_4^{2-}} \qquad (W\,m^{-2}\,\tau^{-1})$$

$\Delta \bar{G}^{\tau}$ depends on the single-scattering albedo of the aerosol, the albedo of the underlying surface, and the length of daylight.

Volcanic Eruptions   Massive increases of stratospheric aerosols from volcanic eruptions are natural climate "experiments." Because of the 1–2-year residence time of particles in the stratosphere, stratospheric aerosol injections become nearly global in extent. Observations following the eruption of Mt. Pinatubo in the Philippines in June 1991 have offered a wealth of details regarding radiative forcing and response characteristics associated with stratospheric aerosols (McCormick et al. 1995). The most optically thick portions of the aerosol were located between 20 and 25 km altitude and were confined to 10°S–30°N during the early period (see *Geophysical Research Letters* **19**, 149–218, 1992). Within 2–3 months, perturbed stratospheric optical depths were observed to at least 70°N, along with the enhancement in the SH. Figure 24.8 shows global mean aerosol optical depth at $\lambda = 550\,nm$ as computed in the Goddard Institute for Space Studies (GISS) General Circulation Model. Eruption of Mt. Pinatubo occurred in June 1991. Satellite observations indicate a global mean decrease of about 5 W m^{-2} in the absorbed solar radiation in the period immediately following the eruption. Thus the radiative forcing that resulted during the first 2 years after eruption of Mt. Pinatubo is estimated to have been of the same order as GHG forcing over the past century.

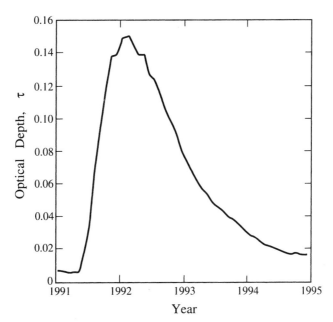

FIGURE 24.8   Global mean stratospheric aerosol optical depth at $\lambda = 550\,nm$ following the eruption of Mt. Pinatubo in June 1991 (Lacis 1996). Optical depth and effective particle size, which vary with time and latitude, were derived from multispectral SAGE observations assuming a monomodal aerosol size distribution.

24.6   INTERNAL AND EXTERNAL MIXTURES

The radiative effects of a population of particles depend on the composition of the particles, through their refractive indices. As we know, the atmospheric aerosol seldom consists exclusively of a single component; it is generally a mixture of species from a number of sources. How all the components are distributed among the particles constitutes the *mixing state* of the aerosol. Two extremes of mixing state are depicted in Figure 24.9. One extreme is termed an *external mixture*, where, in the aerosol population, each particle

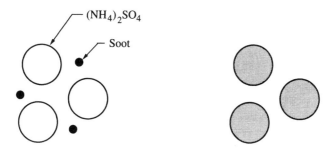

FIGURE 24.9   Mixing states of an aerosol—internal and external mixtures of $(NH_4)_2SO_4$ and soot.

arises from only one source. The example of an external mixture shown in Figure 24.9 is a collection of pure $(NH_4)_2SO_4$ particles mixed with a population of pure black carbon or soot particles. The other extreme is an *internal mixture*, in which all particles of a given size contain a uniform mixture of components from each of the sources. In Figure 24.9 the internally mixed particles are represented in gray to show that there is some soot in every particle. With respect to the absorption component of overall extinction, when a light-absorbing compound is present, all particles in an internal mixture exhibit some absorption, whereas in an external mixture, some particles exhibit absorption and others do not. Hygroscopic growth is also influenced by the mixing state of the aerosol. Assuming that both hygroscopic and non-hygroscopic components are present, in an internal mixture every particle exhibits some growth as RH is increased; in an external mixture, only the hygroscopic particles grow.

Let us examine the sensitivity of aerosol optical properties to the mixing state for an aerosol that is a mixture of soot and $(NH_4)_2SO_4$ only. For simplicity, we assume that all particles have diameter 0.5 μm. The relative humidity is 50%. We will examine the effect of mixing state for a fixed total aerosol mass of $30\,\mu g\,m^{-3}$, as the composition of the aerosol varies from pure $(NH_4)_2SO_4$ to pure soot. Refractive indices (at $\lambda = 530\,nm$) are taken as

$$\begin{array}{ll} \text{Soot} & 1.90\text{--}0.66i \\ (NH_4)_2SO_4 & 1.53\text{--}0i \end{array}$$

Figure 24.10 shows the total extinction b_{ext}, scattering b_{scat}, and absorption b_{abs}, coefficients as a function of the percentage of $(NH_4)_2SO_4$ for both external and internal mixtures. The scattering coefficient of the external mixture exceeds that of the internal mixture at all compositions, whereas the reverse is true for the absorption coefficient. As a result of this opposing behavior, the total extinction coefficients for the internal and external mixtures are virtually identical. The total extinction coefficient increases as the particles go from entirely soot to entirely $(NH_4)_2SO_4$, reflecting the stronger scattering per unit mass of sulfate particles as compared to soot particles.

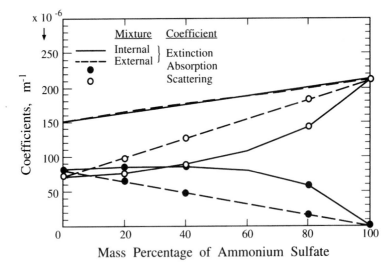

FIGURE 24.10   Total extinction, scattering, and absorption coefficients for internal versus external mixtures of soot and $(NH_4)_2SO_4$.

Let us consider the difference in behavior of the scattering and absorption coefficients for the internally and externally mixed aerosols. Consider first the scattering coefficient and the overall composition of 20% soot/80% $(NH_4)_2SO_4$. For 0.05 μm diameter particles, the scattering cross section of an internally mixed particle is $0.46 \, \mu m^2$. In the external mixture, the scattering cross section is an average of the sulfate value (0.71) and the soot value (0.26), weighted according to the number of particles of each present. That weighted average is $0.63 \, \mu m^2$/particle. Thus the scattering coefficient for the external mixture exceeds that for the internal mixture (0.63 > 0.46). This occurs because the decrease in E_{scat} that takes place when soot is mixed into every particle is proportionately more than the fraction of soot that has been added. (Figure 15.8 shows this behavior in terms of the refractive index.) Now consider the behavior of the absorption coefficients in the internal and external mixtures. The absorption coefficient increases with increasing soot content of the overall aerosol, regardless of whether the soot and sulfate are internally or externally mixed, but, contrary to the scattering coefficient, the absorption coefficient of the internal mixture always exceeds that of the external mixture. In the external mixture, some particles are pure soot and the others are pure $(NH_4)_2SO_4$, and all the absorption is concentrated in the pure soot particles. In the internal mixture, the absorbing component is spread equally over all particles. The absorption cross section of a given amount of soot in a particle that consists also of nonabsorbing material is greater than that of the pure soot particle in air. That is, in the internal mixture the soot exerts its influence in the absorption of every single particle, which exceeds that if all the soot is concentrated exclusively in pure soot particles (frequently the case) and that the soot particles are dispersed uniformly in the sulfate droplets. Simply from geometric optics, more light is incident on a small sphere when it is at the center of a much larger sphere than when it is in air by itself. On the whole, the increase in scattering coefficient with sulfate content outweighs the increase in absorption coefficient with soot content and the overall extinction coefficient increases with increasing sulfate content. The single-scattering albedo ω for the two mixtures is shown in Figure 24.11. The fact that ω for the external

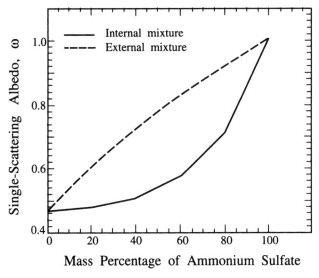

FIGURE 24.11   Single-scattering albedo for internal versus external mixture of soot and $(NH_4)_2SO_4$.

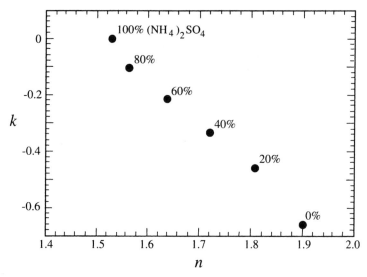

FIGURE 24.12   Real (n) and imaginary (k) parts of the refractive indices of internally mixed soot and $(NH_4)_2SO_4$ particles.

mixture exceeds that for the internal mixture at all compositions reflects the dominance of the scattering of pure $(NH_4)_2SO_4$ particles because of the absence of an absorbing component (recall Figure 15.8). The volume-average refractive indices of the internally mixed particles are shown in Figure 24.12.

24.7   TOP-OF-THE-ATMOSPHERE VERSUS SURFACE FORCING

Aerosol direct radiative forcing measured at the top of the atmosphere (TOA) is called *TOA aerosol forcing*; measured at the surface, it is called *surface aerosol forcing*. If the aerosol does not absorb radiation, the surface forcing is approximately equal to that at TOA. With black carbon and certain mineral dusts, which absorb solar radiation, the surface forcing can be a factor of 2–3 greater than the magnitude of the TOA forcing.

Of all particulate species, black carbon (BC) exerts the most complex effect on climate. Like all aerosols, BC scatters a portion of the direct solar beam back to space, which leads to a reduction in solar radiation reaching the surface of the Earth. This reduction is manifest as an increase in solar radiation reflected back to space at the top of the atmosphere, that is, a *negative* radiative forcing (cooling). A portion of the incoming solar radiation is absorbed by BC-containing particles in the air. This absorption leads to a further reduction in solar radiation reaching the surface. At the surface, the result of this absorption is cooling because solar radiation that would otherwise reach the surface is prevented from doing so. However, the absorption of radiation by BC-containing particles leads to a heating in the atmosphere itself. Thus, absorption by BC leads to a *negative* radiative forcing at the surface and a *positive* radiative forcing in the atmosphere. Finally, the BC-containing aerosol absorbs radiation from the diffuse upward beam of scattered radiation. This reduces the solar radiation that is reflected back to space, leading to a *positive* radiative forcing at the top of the atmosphere. This effect is particularly accentuated when BC aerosol lies above clouds.

TABLE 24.3   Scattering and Absorption Properties of East Asian Aerosol[a]

Type	D_m, nm	σ_g	α m^2/g (500 nm; 80% RH)	ω (500 nm; 80% RH)	$\dfrac{\alpha_e(500\,\text{nm}; 80\%\text{RH})}{\alpha_e(500\,\text{nm}; \text{dry})}$
Water-soluble (used for sulfate and organic carbon)	42	2.24	8.13	0.987	2.25
Black carbon	24	2.0	10.66	0.226	1
Seasalt (accumulated)	418	2.03	4.60	1.000	3.63
Seasalt (coarse)	3500	2.03	0.97	1.000	3.87
Mineral (fine)	140	1.95	2.79	0.986	1
Mineral (intermediate)	780	2.0	0.56	0.939	1
Mineral (coarse)	3800	2.15	0.23	0.864	1
Externally mixed model (sulfate + OC + BC)			8.28	0.928	2.15
Internally mixed model (sulfate + OC + BC)	70/150	1.4/1.6	9.52	0.905	2.36

[a]The external and internal mixtures correspond only to sulfate, organic carbon (OC), and black carbon (BC) for average concentrations predicted over land. D_m is geometric mean diameter, σ_g is geometric variance, α is mass scattering efficiency, α_e is mass extinction efficiency, ω is single-scattering albedo.

Source: Conant et al. (2003).

Top-of-atmosphere (TOA) forcing for BC is the sum of the negative forcing at the surface due to scattering of incoming radiation and the positive forcing from absorption of upward diffuse radiation. The absorption of incoming solar radiation by BC does not contribute to TOA forcing because it adds heat to the atmosphere and reduces solar heating at the surface by the same amount. The net effect of BC is to add heat to the atmosphere and decrease radiative heating of the surface.

Two regions of the world have been identified as having significant amounts of absorbing aerosol. As discovered in the Indian Ocean Experiment (INDOEX), a large plume of biomass and industrial emissions rich in organic and BC aerosol is carried from the Indian subcontinent by the northeast monsoon over the northern Indian Ocean each winter (Satheesh and Ramanathan 2000; Ramanathan et al. 2001; Collins et al. 2002). This plume is responsible for a large (from -15 to $-30\,\text{W m}^{-2}$) surface radiative forcing over a substantial fraction of the northern Indian Ocean. The Asian Pacific Regional Aerosol Characterization Experiment (ACE-Asia), carried out in spring 2001, characterized the aerosol that is carried from East Asia over the western Pacific (Huebert et al. 2003; Seinfeld et al. 2004). East Asia is one of the most concentrated aerosol source regions on the globe, with emissions from industrial sources, biomass burning, and dust storms. Conant et al. (2003) constructed an optical model of the East Asian aerosol (size, chemistry, single-scatter albedo, hygrosopocity, and mixing state) measured during ACE-Asia and predicted radiative forcing and forcing efficiency for all the species and their combinations (Table 24.3) during the period April 5–15, 2001, when a powerful dust storm carried significant levels of mineral dust mixed with anthropogenic sulfur, organic carbon, and BC over the North Pacific.

Figure 24.13 shows predicted column optical depth, surface and TOA forcing efficiency, and surface and TOA forcing for each of the aerosol types averaged over April 5–15,

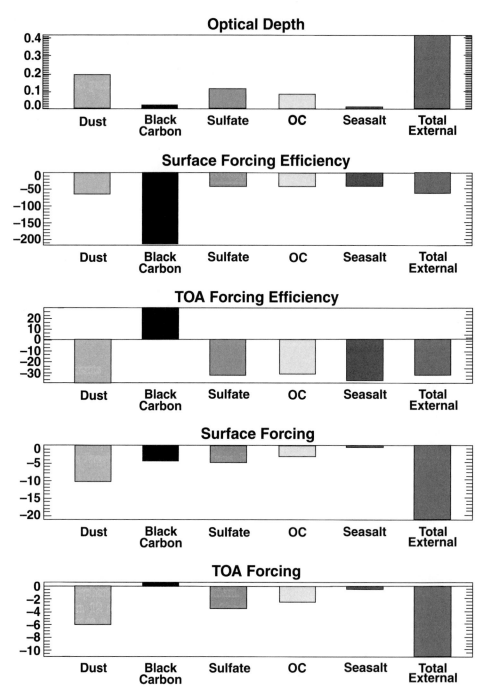

FIGURE 24.13   Radiative effects of East Asian aerosol during the period April 5–15, 2001 (Conant et al. 2003): Optical depth; surface forcing efficiency ($W\,m^{-2}\,\tau^{-1}$); TOA forcing efficiency ($W\,m^{-2}\,\tau^{-1}$); surface forcing ($W\,m^{-2}$); and TOA forcing ($W\,m^{-2}$). Values are for clear skies assuming that the aerosol components are externally mixed.

2001 over the area 20°N–50°N, 100°E–150°E. Mineral dust was the strongest single contributor to the aerosol optical depth and forcing over this period, contributing $-10 \, \text{W m}^{-2}$ to the surface forcing and $-6 \, \text{W m}^{-2}$ to the TOA forcing. Sulfate and organic carbon (treated together as a water-soluble component) exerted a combined forcing of $-8 \, \text{W m}^{-2}$ at the surface and $-7 \, \text{W m}^{-2}$ at TOA. Black carbon is estimated to have had a slight ($0.25 \, \text{W m}^{-2}$) warming effect at the TOA, yet exerted surface cooling of $-4 \, \text{W m}^{-2}$. Seasalt forcing contributed only $-0.5 \, \text{W m}^{-2}$ at the surface and TOA. Mean predicted forcing efficiency per unit optical depth by dust is $-65 \, \text{W m}^{-2}\tau^{-1}$. Sulfate and organic carbon are predicted to have comparable surface and TOA forcing efficiencies per unit optical depth of $-30 \, \text{W m}^{-2}\tau^{-1}$, whereas BC is far more efficient at reducing surface radiation with a surface forcing efficiency per unit optical depth of $-220 \, \text{W m}^{-2}\tau^{-1}$.

The predicted forcing efficiency is sensitive to whether the aerosol is externally or internally mixed and whether clouds are present. Figure 24.14 shows the overall predicted regional forcing for the externally mixed case (clear sky), the internally mixed case (clear sky), and the internally mixed all-sky case (clear and cloudy skies). As noted earlier, clouds generally decrease the magnitude of TOA forcing relative to clear-sky conditions because of multiple scattering effects. The effect of clouds on atmospheric absorption, however, depends on the vertical distribution of the cloud and aerosol layers. Aerosol absorption is reduced by high clouds, which shade the aerosol, but is enhanced by low clouds, which reflect radiation back up through the absorbing aerosol layers.

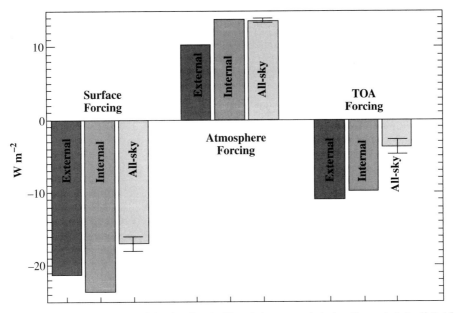

FIGURE 24.14 Direct aerosol forcing for the East Asian aerosol during the period April 5–15, 2001 at the TOA, the atmosphere, and at the surface for an externally mixed aerosol (clear sky), an internally mixed aerosol (clear sky), and the all-sky case (clear and cloudy skies) (Conant et al. 2003). The range bars for the all-sky cases reflect the sensitivity to cloud fraction.

24.8   INDIRECT EFFECTS OF AEROSOLS ON CLIMATE

If aerosol number concentrations are substantially increased as a result of anthropogenic emissions over that in the absence of such emissions, the number concentration of cloud droplets, which is governed by the number concentration of aerosol particles below cloud, may also be increased. An increased number concentration of cloud droplets leads, in turn, to an enhanced multiple scattering of light within clouds and to an increase in the optical depth and albedo of the cloud. The areal extent of the cloud and its lifetime may also increase. This is the essence of the indirect effect of aerosols on climate. A key measure of aerosol influences on cloud droplet number concentrations is the number concentration of cloud condensation nuclei (CCN).

It is well established that CCN concentrations are greater in continental airmasses than in the marine atmosphere; CCN concentrations in maritime air uninfluenced by anthropogenic emissions rarely exceed $100\,\mathrm{cm}^{-3}$, whereas concentrations in well-aged continental air generally exceed $1000\,\mathrm{cm}^{-3}$. Twomey et al. (1978) reported measurements of CCN concentrations exceeding $4500\,\mathrm{cm}^{-3}$ after relatively clean air (CCN levels as low as $50\,\mathrm{cm}^{-3}$) passed over an industrial area in southeastern Australia. Radke and Hobbs (1976) reported CCN concentrations of $1000\text{--}3500\,\mathrm{cm}^{-3}$ in air advecting off the eastern seaboard of the United States. Hudson (1991) reported CCN measurements along the western coast of the United States, indicating a background marine concentration of $20\text{--}40\,\mathrm{cm}^{-3}$, nonurban inland concentrations in Oregon of $100\text{--}200\,\mathrm{cm}^{-3}$, and urban concentrations in the vicinity of Santa Cruz, California, of $3000\text{--}5000\,\mathrm{cm}^{-3}$. Frisbie and Hudson (1993) report CCN concentrations upwind and downwind of Denver, Colorado, of 500 and $5000\,\mathrm{cm}^{-3}$, respectively.

It has long been recognized that continental clouds tend to exhibit greater cloud droplet number concentrations than do marine clouds. Numerous studies exist that link high cloud drop number concentrations to identified industrial sources (Mészáros 1992). Warner and Twomey (1967) found that the number concentration in clouds downwind of sugarcane fires averaged $510\,\mathrm{cm}^{-3}$, as compared with that in the upwind maritime air of $104\,\mathrm{cm}^{-3}$. Fitzgerald and Spyers-Duran (1973) observed the same effect downwind of St. Louis, Missouri. Studying cloud droplet size distributions upwind and downwind of Denver, Colorado, Alkezweeny et al. (1993) found that the droplet size distribution in the downwind air was shifted to a diameter smaller than that of the nonurban air (volume median diameter of $14\,\mu\mathrm{m}$ vs. $28\,\mu\mathrm{m}$) and the droplet concentration was increased by an order of magnitude ($22\,\mathrm{cm}^{-3}$ versus $226\,\mathrm{cm}^{-3}$).

"Ship tracks," linear features of high cloud reflectivity embedded in marine stratus clouds, result from aerosols emitted or formed from the exhaust of ships' engines (Coakley et al. 1987; Scorer 1987; Radke et al. 1989; King et al. 1993) (Figure 24.15). Aircraft observations have confirmed enhanced droplet concentrations and decreased drop sizes in the ship tracks themselves as compared with the adjacent, unperturbed regions of the clouds (Radke et al. 1989; King et al. 1993; Johnson et al. 1996).

The indirect effect of aerosols on climate is exemplified by the processes that link SO_2 emissions to cloud albedo. Sulfur dioxide is oxidized in gas and aqueous phases to aerosol sulfate. Although increased SO_2 emissions can be expected to lead to increased mass of sulfate aerosol, the relation between an increased *mass* of aerosol and the corresponding change of the *number* concentration of aerosol is not well established. Yet, it is the aerosol number concentration that is most closely related to the cloud drop number concentration. Aerosol mass is created by gas-to-particle conversion, which can occur by growth of existing particles or nucleation of fresh particles. Which route predominates will influence the aerosol number concentration; growth produces fewer larger particles, nucleation, many smaller ones. Particle lifetime depends on particle size; smaller particles are likely to have a shorter lifetime because of coagulation–scavenging processes. The formation of precipitation in clouds depends on the cloud drop size distribution; precipitation is accelerated, for a given liquid water content, with fewer, larger drops than with a greater number of smaller drops. One possible effect of reduced cloud droplet size is an increase in cloud lifetime resulting from a decreased tendency for precipitation (Albrecht 1989). It is the above-described chain of events that links anthropogenic SO_2 and particle emissions to cloud albedo changes.

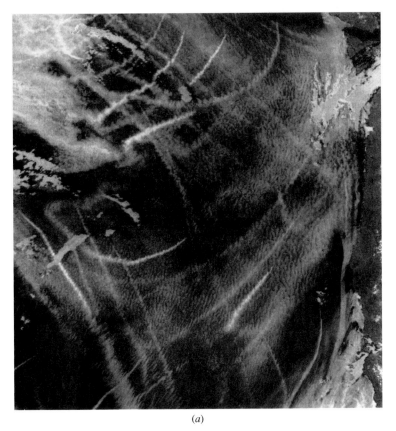

(a)

FIGURE 24.15   Ship tracks: (a) satellite image at 3.7 μm wavelength of ship tracks off the western coast of the United States (courtesy of P. A. Durkee); (b) schematic of processes leading to ship tracks in marine stratocumulus clouds; (c) cloud droplet number concentration (CDNC) and effective droplet radius (r_e) measured during two transects through a ship track in cloud 60 km and 70 km from the ship. The center of the ship track is at $\sim$ 16 km along the transect. (Reprinted from Johnson, D. W., Osborne, S. R., and Taylor, J. P., The effects of a localised aerosol perturbation on the microphysics of a stratocumulus cloud layer, in *Nucleation and Atmospheric Aerosols 1996*, M. Kulmala and P. E. Wagner, eds., p. 864, 1996, with kind permission from Elsevier Science Ltd. The Boulevard, Langford Lane, Kidlington OX5 1GB, UK.)

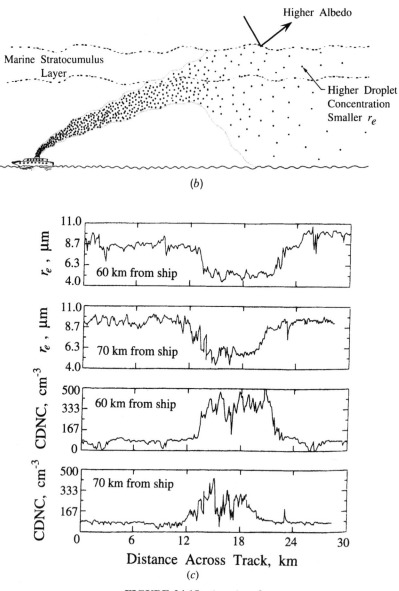

FIGURE 24.15   (*continued*)

24.8.1   Radiative Model for a Cloudy Atmosphere

Consider a spatially uniform cloud of depth h with a droplet number concentration distribution $n(r)$ based on droplet radius[1] r. Extinction of radiation by the cloud is given by (15.37),

$$b_{\text{ext}}(\lambda) = \int_0^{r^{\max}} \pi r^2 Q_{\text{ext}}(m, \alpha) n(r) \, dr \qquad (24.30)$$

---

[1]It is often customary when dealing with the microphysics of clouds to use radius as the size variable, rather than diameter.

The optical depth of the cloud, τ_c, is just the product,

$$\tau_c = b_{ext} h \tag{24.31}$$

Assume for the moment that the cloud droplet size distribution can be approximated as monodisperse, with a number concentration N and radius r_e. Then (24.30) and (24.31) become

$$\tau_c = h\, N\, \pi r_e^2 Q_{ext}\left(\frac{2\pi\, r_e}{\lambda}\right) \tag{24.32}$$

Recall from Figure 15.3 and (15.23) that at visible wavelengths for spheres the size of typical cloud drops ($r \approx 10\,\mu m$), $Q_{ext}(2\pi\, r_e/\lambda) \simeq 2$. Thus (24.32) becomes

$$\tau_c \simeq 2\, h\, N\pi r_e^2 \tag{24.33}$$

The liquid water content of the cloud, L (g H_2O/m^3 air) is

$$L = \tfrac{4}{3}\pi r_e^3 N\, \rho_w \tag{24.34}$$

where ρ_w is the density of water.

Combining (24.33) and (24.34), we can express the cloud optical thickness in terms of L as

$$\tau_c = \frac{3\, L h}{2\, r_e \rho_w} \tag{24.35}$$

Thus τ_c depends on the liquid water content of the cloud, its thickness, and the mean radius of the cloud droplets. τ_c can also be expressed in terms of the cloud droplet number concentration N. Since $r_e = (3L/4\pi N\rho_w)^{1/3}$, we obtain

$$\tau_c = h\left(\frac{9\,\pi\, L^2 N}{2\,\rho_w^2}\right)^{1/3} \tag{24.36}$$

Thus $\tau_c \approx hL^{2/3}N^{1/3}$.

The next step is to relate cloud albedo R_c to τ_c. The two-stream approximation for the reflectance (albedo) of a nonabsorbing, horizontally homogeneous cloud gives (Lacis and Hansen 1974)

$$R_c = \frac{\sqrt{3}(1-g)\tau_c}{2 + \sqrt{3}(1-g)\tau_c} \tag{24.37}$$

where g is the asymmetry factor (15.11). The value of g for cloud droplets of radius much greater than the wavelength of visible light is 0.85, for which (24.37) becomes

$$R_c = \frac{\tau_c}{\tau_c + 7.7} \tag{24.38}$$

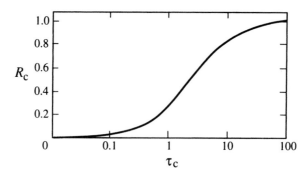

FIGURE 24.16   Cloud albedo as a function of cloud optical thickness.

Figure 24.16 shows R_c as a function of τ_c. As $\tau_c \to 0$, $R_c \to 0$, and for $\tau_c \gg 7.7$, $R_c \to 1$.

The essence of the potential effect of aerosols on cloud reflectively is embodied in Figure 24.17 which shows cloud albedo as a function of cloud droplet number concentration at various cloud thicknesses at a constant liquid water content of $L = 0.3\,\mathrm{g\,m^{-3}}$. At constant liquid water content and constant cloud thickness, cloud albedo increases with increasing CDNC. For a cloud 50 m thick with this liquid water content, an increase of CDNC from 100 to 1000 cm^{-3}, corresponding to going from remote marine to continental conditions, leads to almost a doubling of cloud albedo.

24.8.2   Sensitivity of Cloud Albedo to Cloud Droplet Number Concentration

We desire to calculate the sensitivity of cloud albedo R_c to changes in cloud droplet number concentration $N(dR_c/dN)$. R_c is a function of N, L, and h, so this derivative can be written as follows:

$$\begin{aligned}
\frac{dR_c}{dN} &= \frac{dR_c}{d\tau_c}\frac{d\tau_c}{dN} \\
&= \frac{dR_c}{d\tau_c}\left(\frac{\partial\tau_c}{\partial h}\frac{dh}{dN} + \frac{\partial\tau_c}{\partial L}\frac{dL}{dN} + \frac{\partial\tau_c}{\partial N}\right)
\end{aligned} \tag{24.39}$$

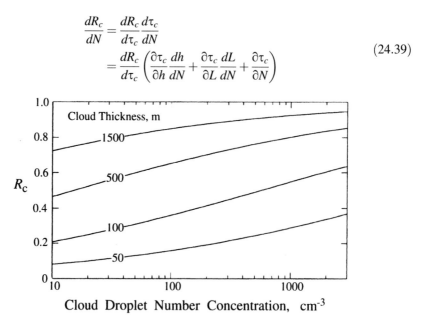

Cloud Droplet Number Concentration, cm^{-3}

FIGURE 24.17   Cloud albedo as a function of cloud droplet number concentration. Liquid water content $L = 0.3\,\mathrm{g\,m^{-3}}$ (Schwartz and Slingo 1996). (Reprinted by permission of Springer-Verlag.)

It is usually assumed that there is no dependence of liquid water content L on N. Thus $dL/dN = 0$. Likewise, it is assumed that there is no dependence of cloud thickness h on N, so $dh/dN = 0$. Thus we obtain

$$\frac{dR_c}{dN} = \frac{dR_c}{d\tau_c}\frac{\partial\tau_c}{\partial N}$$

$$= \left(\frac{R_c}{\tau_c}(1 - R_c)\right)\left(\frac{\tau_c}{3N}\right) \tag{24.40}$$

$$= \frac{R_c(1 - R_c)}{3N}$$

or, equivalently

$$\frac{dR_c}{d\ln N} = \frac{R_c(1 - R_c)}{3} \tag{24.41}$$

Twomey termed the quantity dR_c/dN, at constant liquid water content, the *susceptibility*; it is a measure of the sensitivity of cloud reflectance to changes in microphysics (Twomey 1991; Platnick and Twomey 1994). The susceptibility is inversely proportional to N such that when N is low, as in marine clouds, the susceptibility is high. For a fractional change in cloud droplet number concentration of $\Delta N/N$, the discrete version of (24.41) is

$$\Delta R_c = \frac{R_c(1 - R_c)}{3}\frac{\Delta N}{N}$$

$$= \tfrac{1}{3}R_c(1 - R_c)\Delta\ln N \tag{24.42}$$

The quantity $R_c(1 - R_c)$ achieves its maximum value of $\tfrac{1}{4}$ when $R_c = 0.5$. At this point ΔR_c attains its maximum value of $\tfrac{1}{12}\Delta N/N$. The approximation $\Delta R_c \simeq 0.075\Delta\ln N$ is accurate to within 10% for cloud albedos in the range $0.28 \le R_c \le 0.72$. A 10% relative increase in N ($\Delta N/N = 0.1$) leads to an increase in cloud albedo of 0.75%. This sensitivity of R_c to changes in N, together with the global importance of cloud albedo, is, in a nutshell, the source of the indirect climatic effects of aerosols.

Let us estimate the change in radiative flux resulting from a change in the cloud droplet number concentration. Assume a layer of marine stratus clouds underlying an atmospheric layer. For the ocean a surface albedo $R_s \simeq 0$ can be assumed. With a surface albedo of zero only a single-reflection radiative model is needed and the outgoing radiative flux at the top of the atmosphere is $F_0 A_c T_a^2 R_c$, where A_c is the fractional area by clouds and F_0 is the incoming radiative flux at the top of the atmosphere. The change in shortwave forcing resulting from a change in cloud albedo is

$$\Delta F_c = -F_0 A_c T_a^2 \Delta R_c$$

$$= -\tfrac{1}{3}F_0 A_c T_a^2 R_c(1 - R_c)\Delta\ln N \tag{24.43}$$

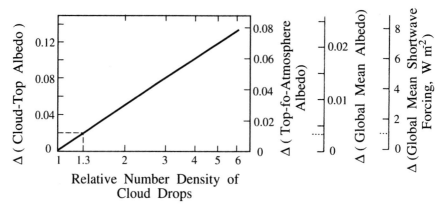

FIGURE 24.18 Calculated perturbation in cloudtop albedo (left ordinate), top-of-atmosphere albedo above marine stratus, global mean albedo, and global mean cloud radiative forcing (right ordinates) resulting from a uniform increase in cloud droplet number concentration by the factor indicated on the abscissa [Schwartz (1996), modified from Charlson et al. (1992)]. The global mean calculations were made with the assumption that the perturbation affects only nonoverlapping marine stratus and stratocumulus clouds having a fractional area of 30%; the fractional atmospheric transmittance of shortwave radiation above the cloud layer was taken as 76%. The dotted line indicates the perturbations resulting from a 30% increase in CDNC. (Reprinted from Schwartz, S. E., Cloud droplet nucleation and its connection to aerosol properties, in *Nucleation and Atmospheric Aerosols 1996*, M. Kulmala and P. E. Wagner, eds., p. 770, 1996, with kind permission from Elsevier Science Ltd, The Boulevard, Langford Lane, Kidlington OX5 1GB, UK.)

For numerical values typical of marine stratus clouds, $R_c = 0.5$, $A_c = 0.3$ (global average), and $T_a = 0.76$, a 30% increase in cloud droplet number concentration, $\Delta N / N = 0.3$, yields

$$\Delta F_c = -1.1 \, \text{W m}^{-2}$$

This estimate of forcing can be compared with the estimate derived from the Earth Radiation Budget Experiment (ERBE): compared with a total absence of clouds on the Earth, the presence of clouds increases the global average reflection of shortwave radiation by about 50 W m^{-2} (Hartmann 1993).

Figure 24.18 shows the change in cloudtop albedo, top-of-the-atmosphere albedo above marine stratus, global mean albedo, and global mean shortwave radiative forcing resulting from a uniform increase in cloud droplet number concentration N by the factor indicated on the abscissa. A global average cloud fractional area of $A_c = 0.3$ is assumed, and $T_a = 0.76$. The point corresponding to a 30% increase in N is indicated on the figure.

24.8.3 Relation of Cloud Droplet Number Concentration to Aerosol Concentrations

Clouds form when an air parcel is cooled sufficiently through vertical lifting that the water vapor in the parcel becomes supersaturated. Once water supersaturation is achieved, aerosol particles become activated to form cloud droplets (Chapter 17). The surface area of the growing drops provides an increasing sink for water vapor, causing the ambient super-saturation to reach a maximum and eventually decrease. Those particles that have been activated continue to grow; those smaller than their critical sizes for activation shrink and become interstitial particles.

Whether a particular particle becomes activated depends on its size and its content of soluble material, in addition to the degree of supersaturation achieved in the cloud and the rate at which the supersaturation changes. The magnitude of the critical supersaturation for activation decreases with increasing particle size and mass fraction of soluble material in the particle. As a result, larger particles activate sooner than do smaller particles, and because they contain a large fraction of soluble salts, marine aerosols tend to activate more readily than do continental aerosols of the same size for a given supersaturation. Concentrations of CCN as a function of supersaturation, so-called CCN spectra, can be measured by exposing samples of air to known supersaturations and then counting the concentration of droplets that form. Empirical expressions have been proposed for the activation spectrum of CCN (in cm^{-3}) as a function of super-saturation s (in %) (Twomey 1959; Ghan et al. 1993; Hobbs 1993) that are of the form [see (17.83)]

$$CCN(s) = cs^k \qquad (24.44)$$

Since s is measured in percent supersaturation, $s = 1$ corresponds to 1% supersaturation, and c is the CCN concentration activated at 1% supersaturation. k is an empirical parameter. Approximate values of c and k determined in different environments have been given in Table 17.4. The number of CCN that are activated to cloud droplets increases as the peak supersaturation in the air increases. Based on (24.44), the larger the value of c, the larger the ultimate droplet concentration. Since the liquid water contents of continental and marine clouds do not differ greatly, the average size of droplets in continental clouds should be less than that in marine clouds. Marine cumulus clouds have a median droplet concentration of about 45 cm^{-3} and a median droplet diameter of about 30 μm; continental cumulus clouds have a median droplet concentration of about 230 cm^{-3} and a median diameter in the vicinity of 10 μm (Hobbs 1993).

To a first approximation, the number concentration of cloud droplets might be expected to increase one-for-one with an increasing number of aerosol particles in the size range suitable for activation. However, the supersaturation time history of an air parcel itself depends on the number of CCN; additional particles act to depress the maximum supersaturation that can be achieved. In such a regime the number of cloud droplets formed does not increase one-to-one with an increase in the CCN concentration [e.g., see Gillani et al. (1992)].

At low CCN concentration, CDNCs increase more or less linearly as CCN concentration increases. Eventually, as CCN concentration increases, a point is reached where CDNC increases less than linearly; for example, in the work of Gillani et al. (1992) the transition occurred between 600 and 800 cm^{-3} for the continental stratiform clouds studied.

A direct relationship between CDNC and ambient aerosol mass is not necessarily to be expected. As we saw in Chapter 8, aerosol mass and number concentrations lie in different size regimes; aerosol mass peaks at much larger particle size than particle number. Thus aerosol mass may not accurately reflect aerosol number, and it is the latter that determines the CCN concentration. In addition, the lifetimes of particles at the peak in the mass and number distributions differ; the larger particles have a longer lifetime than do the smaller particles since the paths for removal of small particles are more efficient. Empirically, however, ambient data do show a general increase of CDNC with increasing sulfate aerosol mass. Boucher and Lohmann (1995) considered the data of Leaitch et al. (1992a,b), Berresheim et al. (1993), Quinn et al. (1993), and Hegg (1994). They fit all the data sets individually, separating continental and marine clouds, and also fit all the data sets to the single relationship

$$CDNC = 10^{2.21+0.41 \log{(m_{SO_4})}} \tag{24.45}$$

where m_{SO_4} is expressed in $\mu g(SO_4^{2-})m^{-3}$ and CDNC in cm^{-3}. This relationship is strongly sublinear; a doubling of m_{SO_4} leads to a 33% increase in CDNC.

Role of Dimethyl Sulfide (DMS) in Regulating Marine CCN Levels   Dimethyl sulfide (DMS) is emitted naturally by phytoplankton in the oceans (see Chapter 2). Once in the atmosphere, DMS reacts with OH and NO_3 radicals to produce SO_2 and methane sulfonic acid (MSA), among other products (see Chapter 6). Sulfur dioxide, itself, reacts with OH to produce H_2SO_4 (see Chapter 6), which can nucleate to produce H_2SO_4–H_2O particles. Whether H_2SO_4 formed from oxidation of DMS in the marine atmosphere nucleates homogeneously or condenses on existing particles depends on the surface area concentration of the existing particles (see Chapter 11). The SO_2 may also be absorbed into existing droplets, where it can be converted heterogeneously to sulfate. By these processes, sulfur emitted as DMS constitutes a major fraction of nonseasalt sulfate (NSS) and of marine cloud condensation nuclei. Charlson et al. (1987) suggested that DMS is a prime regulator of marine CCN concentrations and that a climate feedback mechanism could exist in which temperature changes influence phytoplankton productivity and DMS emissions, thereby affecting marine cloudiness. Figure 24.19 depicts the processes that connect DMS emissions to CCN development and cloud droplet formation. For quantitative modeling of the processes shown in Figure 24.19, we refer the reader to Pandis et al. (1994) and Russell et al. (1994).

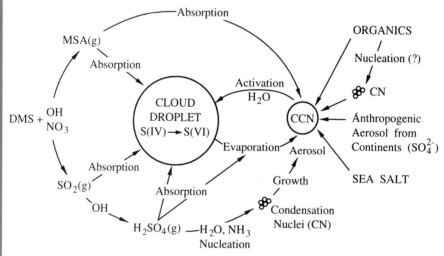

FIGURE 24.19   Relationship between dimethyl sulfide (DMS) emissions and CCN formation.

PROBLEMS

24.1$_A$ Determine the aerosol optical depth at $\lambda = 550\,nm$ of a uniform layer of dry ammonium sulfate aerosol of diameter $0.4\,\mu m$ and concentration $1\,\mu g\,m^{-3}$ extending from the surface to $10\,km$ altitude. The mass scattering efficiency of ammonium sulfate particles can be found in Figure 15.7.

24.2$_A$ What is the aerosol optical depth of the ammonium sulfate layer in Problem 24.1 at 80% RH averaged over the solar spectrum? (You will need to use Figure 24.7.)

24.3$_A$ Prepare a plot of annual mean direct aerosol forcing, $\Delta\bar{F}$ ($W\,m^{-2}$), versus aerosol optical depth $\bar{\tau}$ for different values of single-scattering albedo ω, using equation (24.23). Consider $0 \le \bar{\tau} \le 1.0$. Use values of F_0, T_a, $(1\text{-}A_c)$, $(1\text{-}R_s)$, and $\bar{\beta}$ from Table 24.2. Consider ω values of 1.0, 0.9, 0.8, 0.7, and 0.5.

24.4$_B$ Ghan et al. (1993) have developed a crude aerosol/cloud microphysical parameterization of the form

$$\text{CDNC} = \frac{N_a v}{c' N_a + v}$$

where CDNC is cloud drop number concentration (cm^{-3}), N_a is aerosol number concentration (cm^{-3}), v is the updraft velocity ($cm\,s^{-1}$), and c' is a correction factor ($cm^4\,s^{-1}$) derived using a detailed microphysical model that accounts for the aerosol size distribution and composition. This expression may be used with those developed in Section 24.7.1 to obtain a relation between aerosol concentration and cloud albedo and then between aerosol concentration and reflected radiation, measured in $W\,m^{-2}$. To complete the latter connection, the following approximate expression for the reflectance of a nonabsorbing, horizontally homogeneous cloud (Charlson et al. 1992) may be used

$$F = \tfrac{1}{2} R_c F_0 \mu_0 T_a^2$$

where F_0 is the solar flux at the top of the atmosphere, T_a is the transmittance of the atmosphere above the cloud, and μ_0 is the cosine of the solar zenith angle.

a. Using $\mu_0 = 0.5$, calculate and plot F (in $W\,m^{-2}$) as a function of N_a (in cm^{-3}) for N_a varying from 10^2 to $10^4\,cm^{-3}$ for constant values of

$$p = h\left(\frac{9\pi L^2}{2\,\rho_w^2}\right)^{1/3} = \frac{\tau_c}{\text{CDNC}^{1/3}}$$

equal to 1, 2, 5, and 10, for $v = 20\,cm\,s^{-1}$. For the calculation use $c' = 0.0034\,cm^4\,s^{-1}$, which is the value for a lognormal $(NH_4)_2SO_4$ aerosol with number mode radius $0.05\,\mu m$ and geometric standard deviation 2.0 at $800\,mbar$ altitude and $273\,K$.

b. Evaluate the sensitivity of the predicted reflected radiation to variation of the updraft velocity v over the range of v from 10 to $50\,cm\,s^{-1}$ at $p = 1$.

24.5$_B$ It is desired to estimate the radiative forcing resulting from natural dimethyl sulfide (DMS) emissions from the oceans. As we saw in Chapter 6, DMS is oxidized in the atmosphere to SO_2, which is further oxidized to sulfate. Use (24.29) with a global DMS flux of 19 Tg (S) yr^{-1}. It will be necessary to estimate a global average yield of SO_2 from DMS. To obtain bounds on the possible forcing, consider a 100% and a 70% yield of SO_2 from DMS. The parameters in Table 24.2 may be used.

REFERENCES

Albrecht, B. A. (1989) Aerosols, cloud microphysics, and fractional cloudiness, *Science* **245**, 1227–1230.

Alkezweeny, A. J., Burrows, A. D., and Grainger, A. C. (1993) Measurements of cloud-droplet size distributions in polluted and unpolluted stratiform clouds, *J. Appl. Meteorol.* **32**, 106–115.

Berresheim, H., Eisele, F. L., Tanner, D. J., McInnes, L. M., Ramsey-Bell, D. C., and Covert, D. S. (1993) Atmospheric sulfur chemistry and cloud condensation nuclei (CCN) concentrations over the Northeastern Pacific coast, *J. Geophys. Res.* **98**, 12701–12711.

Boucher, O., and Anderson, T. L. (1995) General circulation model assessment of the sensitivity of direct climate forcing by anthropogenic sulfate aerosols to aerosol size and chemistry, *J. Geophys. Res.* **100**, 26117–26134.

Boucher, O., and Lohmann, U. (1995) The sulfate-CCN-cloud albedo effect, *Tellus* **47B**, 281–300.

Charlson, R. J., Langner, J., Rodhe, H., Leovy, C. B., and Warren, S. G. (1991) Perturbation of the northern hemisphere radiative balance by backscattering from anthropogenic sulfate aerosols, *Tellus* **43AB**, 152–163.

Charlson, R. J., Lovelock, J. E., Andreae, M. O., and Warren, S. G. (1987) Oceanic phytoplankton, atmospheric sulfur, cloud albedo and climate, *Nature* **326**, 655–661.

Charlson, R. J., Schwartz, S. E., Hales, J. M., Cess, R. D., Coakley, J. A., Hansen, J. E., and Hofmann, D. J. (1992) Climate forcing by anthropogenic aerosols, *Science* **255**, 423–430.

Coakley, J. A. Jr., Bernstein, R. L., and Durkee, P. A. (1987) Effect of ship-track effluents on cloud reflectivity, *Science* **273**, 1020–1023.

Collins, W. D., Rasch, P. J., Eaton, B. E., Fillmore, D. W., and Kiehl, J. T. (2002) Simulation of aerosol distributions and radiative forcing for INDOEX: Regional climate impacts, *J. Geophys. Res.* **107**(D19), 8028 (doi: 10.1029/2000JD000032).

Conant, W. C., Seinfeld, J. H., Wang, J., Carmichael, G. R., Tang, Y., Uno, I., Flatau, P. J., Markowicz, K. M., and Quinn, P. K. (2003) A model for the radiative forcing during ACE-Asia derived from CIRPAS Twin Otter and R/V *Ronald H. Brown* data and comparison with observations, *J. Geophys. Res.* **108**(D23), 8661 (doi: 10.1029/2002JD003260).

Fitzgerald, J. W., and Spyers-Duran, P. A. (1973) Changes in cloud nucleus concentration and cloud droplet size distribution associated with pollution from St. Louis, *J. Appl. Meteorol.* **12**, 511–516.

Frisbie, P. R., and Hudson, J. G. (1993) Urban cloud condensation nuclei spectral flux, *J. Appl. Meteorol.* **32**, 666–676.

Ghan, S. J., Chuang, C. C., and Penner, J. E. (1993) A parameterization of cloud droplet nucleation. Part I: Single aerosol types, *Atmos. Res.* **30**, 198–221.

Gillani, N. V., Daum, P. H., Schwartz, S. E., Leaitch, W. R., Strapp, J. W., and Isaac, G. A. (1992) Fractional activation of accumulation-model particles in warm continental stratiform clouds, in *Precipitation Scavenging and Atmosphere–Surface Exchange*, Vol. 1, S. E. Schwartz and W. G. N. Slinn, eds., Hemisphere Publishing, Washington, DC, pp. 345–358.

Hansen, J., and Nazarenko, L. (2004) Soot climate forcing via snow and ice albedos, *Proc. Natl. Acad. Sci.* **101**(2), 423–428.

Hartmann, D. L. (1993) Radiative effects of clouds on Earth's climate, in *Aerosol–Cloud–Climate Interactions*, P. V. Hobbs, ed., Academic Press, San Diego, pp. 151–173.

Haywood, J. J., and Shine, K. P. (1995) The effect of anthropogenic sulfate and soot aerosol on the clear sky planetary radiation budget, *Geophys. Res. Lett.* **22**, 603–606.

Holben, B. N. et al. (2001) An emerging ground-based aerosol climatology: Aerosol optical depth from AERONET, *J. Geophys. Res.* **106**, 12067–12097.

Hegg, D. A. (1994) Cloud condensation nucleus–sulfate mass relationship and cloud albedo, *J. Geophys. Res.* **99**, 25903–25907.

Hobbs, P. V. (1993) Aerosol–cloud interactions, in *Aerosol–Cloud–Climate Interactions*, P. V. Hobbs, ed., Academic Press, San Diego, pp. 33–73.

Hudson, J. G. (1991) Observations of anthropogenic cloud condensation nuclei, *Atmos. Environ.* **25A**, 2449–2455.

Huebert, B. J., Bates, T., Russell, P. B., Shi, G., Kim, Y. J., Kawamura, K., Carmichael, G., and Nakajima, T. (2003) An overview of ACE-Asia: Strategies for quantifying the relationship between Asian aerosols and their climatic impacts, *J. Geophys. Res.* **108**(D23), 8633 (doi: 10.1029/2003JD003550).

Jacobson, M. Z. (2004) Climate response of fossil fuel and biofuel soot, accounting for soot's feedback to snow and sea ice albedo and emissivity, *J. Geophys. Res.* **109**, D21201 (doi: 10.1029/2004JD004945).

Johnson, D. W., Osborne, S. R., and Taylor, J. P. (1996) The effects of a localised aerosol perturbation on the microphysics of a stratocumulus cloud layer, in *Nucleation and Atmospheric Aerosols 1996*, M. Kulmala and P. E. Wagner, eds., Elsevier, Oxford, pp. 864–867.

Kaufman, Y. J., Tanré, D., and Boucher, O. (2002) A satellite view of aerosols in the climate system, *Nature* **419**, 215–223.

King, M. D., Kaufman, Y. J., Tanré, D., and Nakajima, T. (1999) Remote sensing of tropospheric aerosols from space: Past, present, and future, *Bull. Am. Meteorol. Soc.* **80**, 2229–2259.

King, M. D., Radke, L. F., and Hobbs, P. V. (1993) Optical properties of marine stratocumulus clouds modified by ships, *J. Geophys. Res.* **98**, 2729–2739.

Lacis, A. A. (1996) Radiative model of aerosols in the GISS GCM, presented at ACACIA Sulfate Aerosol Research Project Planning Meeting, National Center for Atmospheric Research, Boulder, CO, Feb. 27–28.

Lacis, A. A., and Hansen, J. E. (1974) A parameterization of the absorption of solar radiation in the Earth's atmosphere, *J. Atmos. Sci.* **31**, 118–133.

Leaitch, W. R., Isaac, G. A., Strapp, J. W., Banic, C. M., and Wiebe, H. A. (1992a) Concentrations of major ions in Eastern North American cloud water and their control of cloud droplet number concentrations, in *Precipitation Scavenging and Atmosphere–Surface Exchange*, Vol. 1, S. E. Schwartz and W. G. N. Slinn, eds., Hemisphere Publishing, Washington, DC, pp. 333–343.

Leaitch, W. R., Isaac, G. A., Strapp, J. W., Banic, C. M., and Wiebe, H. A. (1992b). The relationship between cloud droplet number concentrations and anthropogenic pollution: Observations and climatic implications, *J. Geophys. Res.* **97**, 2463–2474.

McCormick, M. P., Thomason, L. W., and Trepte, C. R. (1995) Atmospheric effects of the Mt. Pinatubo eruption, *Nature* **373**, 399–404.

Mészáros, E. (1992) Structure of continental clouds before the industrial era: A mystery to be solved, *Atmos. Environ.* **26A**, 2469–2470.

National Research Council (1996) *A Plan for a Research Program on Aerosol Radiative Forcing and Climate Change*, National Academy Press, Washington, DC.

Nemesure, S., Wagener, R., and Schwartz, S. E. (1995) Direct shortwave forcing of climate by the anthropogenic sulfate aerosol: Sensitivity to particle size, composition, and relative humidity, *J. Geophys. Res.* **100**, 26105–26116.

Pandis, S. N., Russell, L. M., and Seinfeld, J. H. (1994) The relationship between DMS flux and CCN concentration in remote marine regions, *J. Geophys. Res.* **99**, 16945–16957.

Penner, J. E., Charlson, R. J., Hales, J. M., Laulainen, N. S., Leifer, R., Novakov, T., Ogren, J., Radke, L. F., Schwartz, S. E., and Travis, L. (1994) Quantifying and minimizing uncertainty of climate forcing by anthropogenic aerosols, *Bull. Am. Meteorol. Soc.* **75**, 375–400.

Pilinis, C., Pandis, S. N., and Seinfeld, J. H. (1995) Sensitivity of direct climate forcing by atmospheric aerosols to aerosol size and composition, *J. Geophys. Res.* **100**, 18739–18754.

Platnick, S., and Twomey, S. (1994) Determining the susceptibility of cloud albedo to changes in droplet concentration with the advanced very high resolution radiometer, *J. Appl. Meteorol.* **33**, 334–347.

Quinn, P. K., Covert, D. S., Bates, T. S., Kapustin, V. N., Ramsey-Bell, D. C., and McInnes, L. M. (1993) Dimethylsulfide/cloud condensation nuclei/climate system: Relevant size-resolved measurements of the chemical and physical properties of the atmospheric aerosol particles, *J. Geophys. Res.* **98**, 10411–10427.

Radke, L. F., Coakley, J. A., and King, M. D. (1989) Direct and remote sensing observations of the effects of ships on clouds, *Science* **246**, 1146–1148.

Radke, L. F., and Hobbs, P. V. (1976) Cloud condensation nuclei on the Atlantic Seaboard of the United States, *Science* **193**, 999–1002.

Ramanathan, V., Crutzen, P. J., Kiehl, J. T., and Rosenfeld, D. (2001) Aerosols, climate, and the hydrological cycle, *Science* **294**, 2119–2124.

Satheesh, S. K., and Ramanathan, V. (2000) Large differences in tropical aerosol forcing at the top of the atmosphere and Earth's surface, *Nature* **405**, 60–63.

Rood, M. J., Shaw, M. A., Larson, T. V., and Covert, D. S. (1989) Ubiquitous nature of ambient metastable aerosol, *Nature* **357**, 537–539.

Russell, L. M., Pandis, S. N., and Seinfeld, J. H. (1994) Aerosol production and growth in the marine boundary layer, *J. Geophys. Res.* **99**, 20989–21003.

Schwartz, S. E. (1996) Cloud droplet nucleation and its connection to aerosol properties, in *Nucleation and Atmospheric Aerosols 1996*, M. Kulmala and P. E. Wagner, eds., Elsevier, Oxford, pp. 770–779.

Schwartz S. E., and Slingo, A. (1996) Enhanced shortwave cloud radiative forcing due to anthropogenic aerosols, in *Clouds, Chemistry and Climate–Proceedings of NATO Advanced Research Workshop*, P. Crutzen and V. Ramanathan, eds., Springer, Heidelberg, pp. 191–236.

Scorer, R. S. (1987) Ship trails, *Atmos. Environ.* **21**, 1417–1425.

Seinfeld, J. H. et al. (2004) ACE-Asia: Regional climatic and atmospheric chemical effects of Asian dust and pollution, *Bull. Am. Meteorol. Soc.* 367–380.

Shaw, M. A., and Rood, M. J. (1990) Measurement of the crystallization humidities of ambient aerosol particles, *Atmos. Environ.* **24A**, 1837–1841.

Tang, I. N., and Munkelwitz, R. H. (1994) Water activities, densities, and refractive indices of aqueous sulfate and nitrate droplets of atmospheric importance, *J. Geophys. Res.* **99**, 18801–18808.

Twomey, S. (1959) The nuclei of natural cloud formation, Part II: The supersaturation in natural clouds and the variation of cloud droplet concentration, *Geofis. Pura Appl.* **43**, 243–249.

Twomey, S. (1991) Aerosols, clouds, and radiation, *Atmos. Environ.* **25A**, 2435–2442.

Twomey, S., Davidson, A. K., and Seton, K. J. (1978) Results of five years' observations of cloud nucleus concentration at Robertson, New South Wales, *J. Atmos. Sci.* **35**, 650–656.

Warner, J., and Twomey, S. (1967) The production of cloud nuclei by cane fires and the effect on cloud droplet concentration, *J. Atmos. Sci.* **24**, 704–706.

Wiscombe, W., and Grams, G. (1976) The backscattered fraction in two-stream approximations, *J. Atmos. Sci.* **33**, 2440–2451.

25 Atmospheric Chemical Transport Models

25.1 INTRODUCTION

The atmosphere is an extremely complex reactive system in which numerous physical and chemical processes occur simultaneously. Ambient measurements give us only a snapshot of atmospheric conditions at a particular time and location. Such measurements are often difficult to interpret without a clear conceptual model of atmospheric processes. Moreover, measurements alone cannot be used directly by policymakers to establish an effective strategy for solving air quality problems. An understanding of individual atmospheric processes (chemistry, transport, removal, etc.) does not imply an understanding of the system as a whole. Mathematical models provide the necessary framework for integration of our understanding of individual atmospheric processes and study of their interactions. A combination of state-of-the-science measurements with state-of-the-science models is the best approach for making real progress toward understanding the atmosphere.

In urban and regional air quality studies one has often to address questions such as the following:

- What is the contribution of source A to the concentration of pollutants at site B?
- What is the most cost-effective strategy for reducing pollutant concentrations below an air quality standard?
- What will be the effect on air quality of the addition or the reduction of a specific air pollutant emission flux?
- Where should one place a future source (industrial complex, freeway, etc.) to minimize its environmental impacts?
- What will be the air quality tomorrow or the day after?

Addressing these questions requires understanding of the relationships between emission fluxes and ambient concentrations. A model involving descriptions of emission patterns, meteorology, chemical transformations, and removal processes is the essential tool for establishing such relationships. Such a model provides a link between emission changes from source control measures and resulting changes in airborne concentrations.

The components of an atmospheric model are depicted in Figure 25.1. The three basic components are species emissions, transport, and physicochemical transformations. We should note the close interaction among ambient monitoring, laboratory experiments, and modeling. Monitoring (routine or intensive) identifies the state of the atmosphere and provides

---

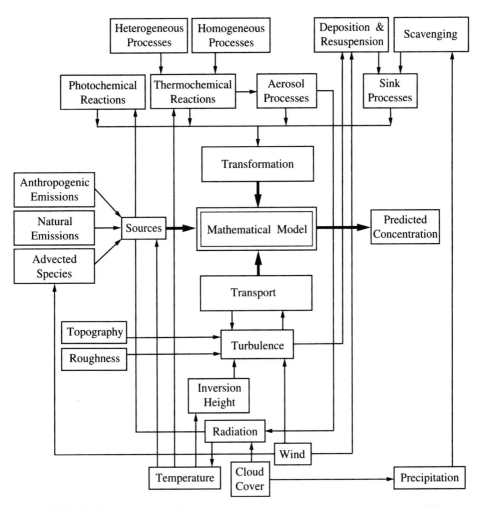

FIGURE 25.1 Elements of a mathematical atmospheric chemical transport model.

data needed for use and evaluation of atmospheric models. Laboratory studies generally focus on a single atmospheric process, providing parameters needed by atmospheric models. Models are the tools that integrate our understanding of atmospheric processes. Evaluation of models often points to gaps in our understanding, leading to more laboratory and field measurements and then further model development. This is, of course, just the application of the scientific method, where the atmospheric model represents the conceptual understanding that is in continuous interplay with the experimental evidence (field or laboratory studies).

25.1.1 Model Types

Atmospheric models can be divided, broadly speaking, into two types: physical and mathematical. Physical models are sometimes used to simulate atmospheric processes by means of a small-scale representation of the actual system, for example, a small-scale replica of an urban area or a portion thereof in a wind tunnel. Problems associated with properly duplicating the actual scales of atmospheric motion make physical models of this variety of limited usefulness. We henceforth focus our attention on mathematical models, and the term *model* will refer strictly to mathematical models.

Mathematical models of atmospheric behavior can broadly be classified into two types:

1. Models based on the fundamental description of atmospheric physical and chemical processes
2. Models based on statistical analysis of data

Most regions contain a number of monitoring stations operated by governmental authorities at which 1 h to daily average concentration levels are measured and reported. A great deal of information is potentially available in these enormous databases, and statistical analysis of such data can provide valuable insights. An example of how such data can be used is a simple forecast model, where, for a certain region, concentration levels in the next few hours are given as a statistical function of current concentrations and other variables from correlations among past measurements and concentration trends. Statistical models take advantage of the available databases and are relatively simple to apply. However, their reliance on past data is also their major weakness. Because these models do not explicitly describe causal relationships, they cannot be reliably extrapolated beyond the bounds of the data from which they were derived. As a result, statistically based models are not ideally suited to the task of predicting the impact of significant changes in emissions. Statistical analysis of air quality data and corresponding modeling approaches will be discussed in Chapter 26. Models based on fundamental description of atmospheric physics and chemistry are the subject of this chapter.

25.1.2   Types of Atmospheric Chemical Transport Models

A wide variety of atmospheric models have been proposed and used. Some of these simulate changes in the chemical composition of a given air parcel as it is advected in the atmosphere (Lagrangian models), while others describe the concentrations in an array of fixed computational cells (Eulerian models).

A Lagrangian modeling framework is one that moves with the local wind so that there is no mass exchange between the air parcel and its surroundings, with the exception of species emissions that are allowed to enter the parcel through its base (Figure 25.2). The

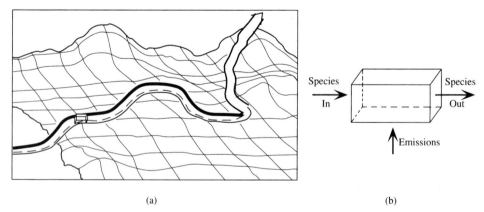

(a)   (b)

FIGURE 25.2   Schematic depiction of (a) a Lagrangian model and (b) a Eulerian model.

TABLE 25.1 Atmospheric Chemical Transport Models Defined According to Their Spatial Scale

Model	Typical Domain Scale	Typical Resolution
Microscale	$200 \times 200 \times 100\,m$	$5\,m$
Urban	$100 \times 100 \times 5\,km$	$4\,km$
Regional	$1000 \times 1000 \times 10\,km$	$20\,km$
Synoptic (continental)	$3000 \times 3000 \times 20\,km$	$80\,km$
Global	$65{,}000 \times 65{,}000 \times 20\,km$	$5° \times 5°$

air parcel moves continuously, so the model actually simulates concentrations at different locations at different times. On the contrary, an Eulerian modeling framework remains fixed in space (Figure 25.2). Species enter and leave each cell through its walls, and the model simulates the species concentrations at all locations as a function of time.

The domain of an atmospheric model—that is, the area that is simulated—varies from a few hundred meters to thousands of kilometers (Table 25.1). The computational domain usually consists of an array of computational cells, each having a uniform chemical composition. The size of these cells, that is, the volume over which the predicted concentrations are averaged, determines the spatial resolution of the model. Variation of concentrations at scales smaller than the model resolution cannot easily be resolved. For example, concentration variations over the Los Angeles basin cannot be described by a synoptic scale model that treats the entire area as one computational cell of uniform chemical composition.

Atmospheric models are also characterized by their dimensionality. The simplest is the box model (zero-dimensional), where the atmospheric domain is represented by only one box (Figure 25.3). In a box model concentrations are the same everywhere and

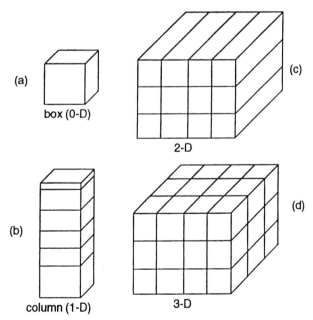

FIGURE 25.3 Schematic depiction of (a) a box model (zero-dimensional), (b) a column model (one-dimensional), (c) a two-dimensional model, and (d) a three-dimensional model.

therefore are functions of time only: $c_i(t)$. The next step in complexity are column models (one-dimensional) that assume that concentrations are functions of height and time: $c_i(z,t)$. The modeling domain consists of horizontally homogeneous layers. Two-dimensional models assume that species concentrations are uniform along one dimension and depend on the other two and time, for example: $c_i(x,z,t)$. Two-dimensional models have often been used in descriptions of global atmospheric chemistry, assuming that concentrations are functions of latitude and altitude but do not depend on longitude. Finally, three-dimensional models simulate the full concentration field: $c_i(x,y,z,t)$. Obviously, both model complexity and accuracy increase with dimensionality. Governing equations of these models will be derived subsequently.

25.2 BOX MODELS

We begin our discussion with the simplest modeling representation—the box model—and derive the governing equations in both Eulerian and Lagrangian frameworks.

25.2.1 The Eulerian Box Model

The Eulerian box enclosing a region of the atmosphere is assumed to have a height $H(t)$ equal, for example, to the mixing-layer height. This mixing height may be allowed to vary diurnally to simulate the evolution of the atmospheric mixing state. The model is based on the mass conservation of a species inside a fixed Eulerian box of volume $H\,\Delta x\,\Delta y$ (Figure 25.4). A mass balance for the concentration c_i of species i results in

$$\frac{d}{dt}(c_i\,\Delta x\,\Delta y\,H) = Q_i + R_i\,\Delta x\,\Delta y\,H - S_i + uH\,\Delta y(c_i^0 - c_i) \qquad (25.1)$$

where Q_i is the mass emission rate of i (kg h^{-1}), S_i the removal rate of i (kg h^{-1}), R_i its chemical production rate (kg m^{-3} h^{-1}), c_i^0 its background concentration, and u the windspeed with the wind assumed to have a constant direction. Equation (25.1) can be simplified by dividing by $\Delta x\,\Delta y$ to give

$$\frac{d}{dt}(c_i H) = q_i + R_i H - s_i + u\frac{H}{\Delta x}(c_i^0 - c_i) \qquad (25.2)$$

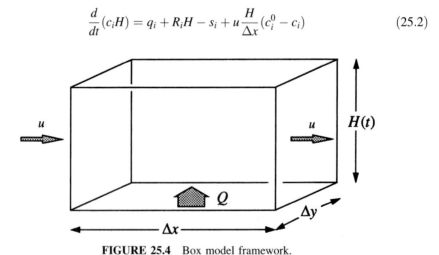

FIGURE 25.4 Box model framework.

with q_i and s_i the emission and removal rates of i per unit area ($\mathrm{kg\,m^{-2}\,h^{-1}}$). The removal rate due to dry deposition can be described using the dry deposition velocity of the species $v_{d,i}$ as

$$s_i = v_{d,i} c_i \tag{25.3}$$

and the box model governing equation assuming constant mixing height H becomes

$$\frac{dc_i}{dt} = \frac{q_i}{H} + R_i - \frac{v_{d,i}}{H} c_i + \frac{u}{\Delta x}(c_i^0 - c_i) \tag{25.4}$$

The terms on the right-hand side (RHS) of (25.4) correspond to the changes in the concentration of i as a result of emission, chemical reaction, dry deposition, and advection. The ratio of the length of the box to the prevailing windspeed u is equal to the residence time of air over the area τ_r, or

$$\tau_r = \frac{\Delta x}{u} \tag{25.5}$$

Note that (25.4) applies to a box with fixed height H and therefore neglects the effects of dilution that occur if the height is changing with time, entraining or detraining material into or out of the box, respectively.

Box Model   An inert species has as initial concentration $c_i(0)$ and is emitted at a rate $q_i = 200\,\mu\mathrm{g\,m^{-2}\,h^{-1}}$. Assuming that its background concentration is $c_i^0 = 1\,\mu\mathrm{g\,m^{-3}}$, calculate its steady-state concentration over a city characterized by an average windspeed of $3\,\mathrm{m\,s^{-1}}$. Assume that the city has dimensions $100 \times 100\,\mathrm{km}$ and a constant mixing height of $1000\,\mathrm{m}$.

The species of interest is inert and therefore $R_i = 0$. We do not have any information regarding its removal rate so we assume that its removal is dominated by advection, or

$$\frac{v_{d,i}}{H} c_i \ll \frac{c_i - c_i^0}{\tau_r} \tag{25.6}$$

and dry removal can be neglected. Equation (25.4) is then simplified to

$$\frac{dc_i}{dt} = \frac{q_i}{H} + \frac{c_i^0 - c_i}{\tau_r} \tag{25.7}$$

with initial condition at $t = 0$, $c_i = c_i(0)$. The solution of (25.7) is

$$c_i(t) = c_i(0)e^{-t/\tau_r} + \left(\frac{q_i \tau_r}{H} + c_i^0\right)(1 - e^{-t/\tau_r}) \tag{25.8}$$

The time τ_r is the *flushing time* for the box, that is, the time required by the airshed to clean itself after emissions are set to zero. Note that the concentration evolution of the inert species in (25.8) is the sum of two contributions: the initial condition contribution that decays exponentially with characteristic time $\tau_r \simeq 10\,h$, and the contribution due to emissions and advection. For $t \gg \tau_r$, $\exp(-t/\tau_r) \simeq 0$; and the inert species reaches a steady-state concentration

$$c_i^{ss} = \frac{q_i \tau_r}{H} + c_i^0 \simeq 3\,\mu g\,m^{-3} \tag{25.9}$$

which is $2\,\mu g\,m^{-3}$ above the background level. Equation (25.9) can be a reasonable first guess for slowly reacting material over an urban area.

Entrainment   Up to this point it has been assumed that the height of the mixing layer that defines the vertical extent of the box remains constant. Frequently, the mixing height varies diurnally with low values during the nighttime and higher values during the daytime. Let us assume that this variation is given by the function $H(t)$.

Physically, when the mixing height decreases, there is no direct change in the concentration c_i inside the mixed layer. As the mixing height decreases, air originally inside the box is left aloft above the box. Of course, later on, because the box will be smaller, surface sources and sinks will have a more significant effect. However, if the mixing height increases, the box entrains air from the layer above. This entrainment and subsequent dilution will change the concentration c_i, and this process should be explicitly included in the model. This dilution rate will depend on the concentration of the species aloft, c_i^a.

Assume that at a given moment the box has a height H, a concentration c_i, and the concentration above it is c_i^a. If after time Δt, the box height increases to $H + \Delta H$ and the concentration of i to $c_i + \Delta c_i$ then a mass balance for i gives

$$(c_i + \Delta c_i)(H + \Delta H) = c_i H + c_i^a \Delta H \tag{25.10}$$

which, after neglecting the second-order term $\Delta c_i \, \Delta H$, simplifies to

$$H\Delta c_i = (c_i^a - c_i)\Delta H \tag{25.11}$$

Dividing by Δt and taking the limit $\Delta t \to 0$, (25.11) becomes

$$\frac{dc_i}{dt} = \frac{c_i^a - c_i}{H}\frac{dH}{dt} \tag{25.12}$$

For increasing mixing height, if $c_i^a > c_i$, this entrainment term is positive and vice versa. The entrainment term given by (25.12) should be included in the box model only if the mixing height is increasing. Summarizing, the entraining Eulerian box model equations are

$$\frac{dc_i}{dt} = \frac{q_i}{H(t)} + R_i - \frac{v_{d,i}}{H(t)}c_i + \frac{c_i^0 - c_i}{\tau_r} \qquad \text{for } \frac{dH}{dt} \le 0 \tag{25.13}$$

$$\frac{dc_i}{dt} = \frac{q_i}{H(t)} + R_i - \frac{v_{d,i}}{H(t)}c_i + \frac{c_i^0 - c_i}{\tau_r} + \frac{c_i^a - c_i}{H(t)}\frac{dH}{dt} \qquad \text{for } \frac{dH}{dt} > 0 \tag{25.14}$$

Equations (25.13) and (25.14) describe mathematically the concentration of species above a given area assuming that the corresponding airshed is well mixed, accounting for emissions, chemical reactions, removal, advection of material in and out of the airshed, and entrainment of material during growth of the mixed layer. These equations cannot be solved analytically if one uses a realistic gas-phase chemical mechanism for the calculation of the R_i terms. Numerical solutions will be discussed subsequently.

25.2.2 A Lagrangian Box Model

The Eulerian box model developed in the previous section does not have any spatial resolution, in that the entire airshed is assumed to be well mixed. This weakness can be circumvented by selecting as the modeling framework a much smaller box that is allowed to move with the wind in the airshed (Figure 25.2). This box extends vertically up to the mixing height, while its horizontal dimensions can be selected arbitrarily. This model simulates advection of an air parcel over the airshed, its subsequent movement and collection of emissions, the buildup of primary and secondary species, and finally its exit from the airshed. The major assumption of the Lagrangian box model is that there is no horizontal dispersion; that is, material in the box is not removed by mixing and dilution with the surrounding air.

The Lagrangian box is assumed to behave like a point identically following the wind patterns. Knowing the position of the box at every moment $s(t)$, one can calculate the corresponding emission fluxes $E_i(t)$. For example, if such a box is located in the eastern suburbs of a city at 3 p.m. and in the downtown area at midnight, then the emissions into the box at 3 p.m. are those from the eastern suburbs at this time, and at midnight those from downtown.

The mass balance for the Lagrangian box is identical to (25.13) and (25.14) with the exception that the advection terms are absent:

$$\frac{dc_i}{dt} = \frac{q_i}{H(t)} + R_i - \frac{v_{d,i}}{H(t)}c_i \quad \text{for } \frac{dH}{dt} \leq 0 \tag{25.15}$$

$$\frac{dc_i}{dt} = \frac{q_i}{H(t)} + R_i - \frac{v_{d,i}}{H(t)}c_i + \frac{c_i^a - c_i}{H(t)}\frac{dH}{dt} \quad \text{for } \frac{dH}{dt} > 0 \tag{25.16}$$

The initial condition for the solution of (25.15) and (25.16) is

$$c_i(0) = c_i^0 \tag{25.17}$$

perhaps reflecting concentrations upwind of the area of interest.

While (25.13) and (25.14) and (25.15) and (25.16) appear to be similar, they are fundamentally different; these differences are demonstrated in the following example.

Urban SO$_2$ SO$_2$ is emitted in an urban area with a flux of 2000 μg m^{-2}h^{-1}. The mixing height over the area is 1000 m, the atmospheric residence time 20 h, and SO$_2$ reacts with an average rate of 3% h^{-1}. Rural areas around the city are characterized by a SO$_2$ concentration equal to 2 μg m^{-3}. What is the average SO$_2$ concentration in the urban airshed for these conditions? Assume an SO$_2$ dry deposition velocity of 1 cm s^{-1} and a cloud/fog-free atmosphere.

Let us first represent this situation with a Eulerian box model enclosing the entire airshed. The concentration of SO_2 over this area c_s will satisfy (25.13), or in the case of a first-order reaction

$$\frac{dc_s}{dt} = \frac{q_s}{H} - kc_s - \frac{v_s}{H}c_s + \frac{c_s^0 - c_s}{\tau_r} \tag{25.18}$$

with solution

$$c_s(t) = \frac{A}{B} + \left(c_s(0) - \frac{A}{B}\right)e^{-Bt} \tag{25.19}$$

where

$$A = \frac{q_s}{H} + \frac{c_s^0}{\tau_r} \tag{25.20}$$

$$B = k + \frac{v_s}{H} + \frac{1}{\tau_r} \tag{25.21}$$

The term B corresponds to the various sinks of SO_2, namely, chemical reaction $(k = 0.03\,h^{-1})$, dry removal $(v_s/H = 0.036\,h^{-1})$, and advection out of the urban airshed $(1/\tau_r = 0.05\,h^{-1})$. All sinks contribute significantly to SO_2 removal and $B = 0.116\ h^{-1}$. The term A corresponds to the sources of SO_2, that is, emissions $(q_s/H = 2\,\mu g\,m^{-3}\,h^{-1})$ and advection of background $SO_2(c_s^0/\tau_r = 0.1\,\mu g\,m^{-3}\,h^{-1})$. The urban emissions dominate in this case and $A = 2.1\,\mu g\,m^{-3}\,h^{-1}$. The concentration $c_s(0)$ is the initial SO_2 concentration for the airshed. Independently of our choice of this value, the predicted concentration $c_s(t)$ approaches a steady-state value for $t \gg 1/B$, or $t \gg 12.5\,h$, equal to A/B or $18.1\,\mu g\,m^{-3}$. The fundamental assumption of the Eulerian box model is that the airshed is homogeneous, so all areas even if they are close to the airshed boundary will have, according to this modeling approach, the same concentration of $18.1\,\mu g\,m^{-3}$.

Let us solve the same problem using a Lagrangian box advected from an area outside the urban center into the airshed and out of it. Equation (25.15) is applicable in this case and therefore

$$\frac{dc_s}{dt} = \frac{q_s}{H} - kc_s - \frac{v_s}{H}c_s \tag{25.22}$$

with initial condition

$$c_s(0) = c_s^0 \tag{25.23}$$

The solution of (25.22) is

$$c_s(t) = \frac{D}{E} + \left(c_s^0 - \frac{D}{E}\right)e^{-Et} \tag{25.24}$$

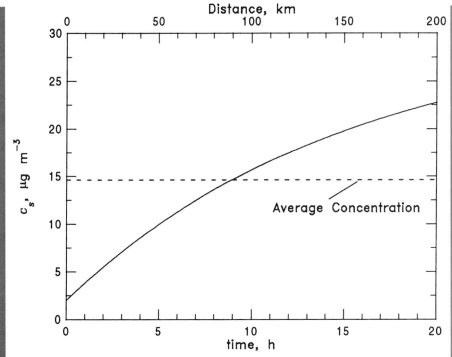

FIGURE 25.5 Concentration as a function of travel time across an urban area predicted by a Lagrangian box model assuming length 200 km and a constant windspeed of 2.8 m s⁻¹.

where

$$D = \frac{q_s}{H} \qquad E = k + \frac{v_d}{H} \tag{25.25}$$

Note that the SO_2 advection term does not appear explicitly in the D and E terms in this moving framework. Under the conditions of the example, $D = 2\,\mu g\,m^{-3}\,h^{-1}$ and $E \simeq 0.066\,h^{-1}$. Once more, the predicted concentration eventually reaches a steady-state value equal to $D/E = 30.3\,\mu g\,m^{-3}$. However, c_s approaches this steady-state value for $t \gg 33.3\,h$ and the air parcel stays in the urban airshed only $\tau_r = 20\,h$. After this period, (25.22) is no longer valid since $q_s = 0$. Therefore the steady state arising from (25.24) is irrelevant as it is not reached during the available residence time of the air parcel in the airshed. Equation (25.24) predicts that as the air parcel enters the urban area its SO_2 concentration increases rapidly from the background value of $2\,\mu g\,m^{-3}$ to $15.7\,\mu g\,m^{-3}$ in the center of the airshed, and to $22.7\,\mu g\,m^{-3}$ at the opposite end (Figure 25.5). The average concentration over the area is then $14.6\,\mu g\,m^{-3}$.

The above described results illustrate the strengths and weaknesses of the two box models. The Eulerian box model is easy to apply but oversimplifies everything by assuming a homogeneous airshed. The Lagrangian model can provide more information, such as, for example, a spatial distribution of concentrations, but by neglecting horizontal dispersion, it may predict higher concentrations downwind of emission sources.

25.3 THREE-DIMENSIONAL ATMOSPHERIC CHEMICAL TRANSPORT MODELS

The starting point of atmospheric chemical transport models is the mass balance equation (18.1) for a chemical species i[1]

$$\frac{\partial c_i}{\partial t} + \nabla \cdot (\mathbf{u} c_i) = R_i(c_1, c_2, \ldots, c_n) + E_i - S_i \qquad (25.26)$$

where $c_i(\mathbf{x}, t)$ is the concentration of i as a function of location $\mathbf{x}$ and time t, $\mathbf{u}(\mathbf{x}, t)$ is the velocity vector (u_x, u_y, u_z), R_i is the chemical generation term for i, and $E_i(\mathbf{x}, t)$ and $S_i(\mathbf{x}, t)$ are its emission and removal fluxes, respectively. This equation leads directly to the atmospheric diffusion equation (18.12) by splitting the atmospheric transport term into an advection and turbulent transport contribution[2]

$$\frac{\partial c_i}{\partial t} + u_x \frac{\partial c_i}{\partial x} + u_y \frac{\partial c_i}{\partial y} + u_z \frac{\partial c_i}{\partial z}$$
$$= \frac{\partial}{\partial x}\left(K_{xx} \frac{\partial c_i}{\partial x}\right) + \frac{\partial}{\partial y}\left(K_{yy} \frac{\partial c_i}{\partial y}\right) + \frac{\partial}{\partial z}\left(K_{zz} \frac{\partial c_i}{\partial z}\right) \qquad (25.27)$$
$$+ R_i(c_1, c_2, \ldots, c_n) + E_i(x, y, z, t) - S_i(x, y, z, t)$$

where $u_x(x, y, z, t), u_y(x, y, z, t)$, and $u_z(x, y, z, t)$ are the x, y, and z components of the wind velocity and $K_{xx}(x, y, z, t), K_{yy}(x, y, z, t)$, and $K_{zz}(x, y, z, t)$ are the corresponding eddy diffusivities. The turbulent fluctuations $\mathbf{u}'$ and c_i' of the velocity and concentration fields relative to their average values $\mathbf{u}$ and c_i have been approximated using the K theory (or mixing length or gradient transport theory) in (25.27) by

$$\langle u'c' \rangle = -\mathbf{K} \cdot \nabla c \qquad (25.28)$$

Equation (25.28) is the simplest solution to the closure problem and is currently used in the majority of chemical transport models. Higher-order closure approximations have been developed but are computationally expensive. Some more recent formulations have shown promise of becoming computationally competitive with the commonly employed K theory (Pai and Tsang 1993).

25.3.1 Coordinate System—Uneven Terrain

In our discussion so far we have implicitly assumed that the terrain is flat. Obviously, this is rarely the case. For example, severe air pollution problems often occur in regions that

---

[1]We showed in Chapter 18 that the molecular diffusion term on the RHS of (18.1) can be neglected in atmospheric flows. In Chapter 18 we used index notation, for example, $\partial(u_j c_i)/\partial x_j$. Here we use both vector and index notation. The three wind velocity components can be denoted $(u_1, u_2, u_3), (u_x, u_y, u_z)$, or (u, v, w) in Cartesian coordinates, and each of these notations has been used at various points in the book.

[2]For simplicity we drop the bracket notation on the concentration (c_i), which has been used in Chapter 18 to denote the ensemble mean concentration. All concentrations in this chapter are, however, understood to represent theoretical ensemble mean values. Likewise, we drop the overbars on the wind velocity components, but they are also understood to represent mean values.

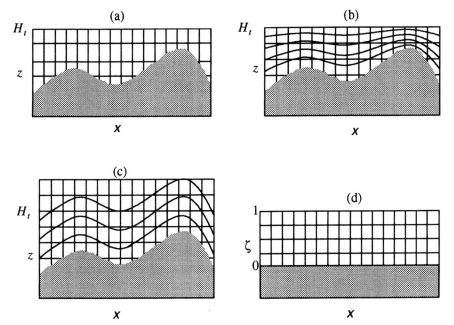

FIGURE 25.6 Coordinate transformation for uneven terrain: (a) two-dimensional terrain in $x - z$ space; (b) same as (a) but with contours of constant ζ superimposed; (c) same as (a) but with contours of constant z' superimposed; (d) two-dimensional terrain in $x - \zeta$ computational space (the terrain is indicated by the shaded region).

are at least partially bounded by mountains that restrict air movement. The surface of the Earth can be characterized by its topography:

$$Z = h(x, y) \tag{25.29}$$

The upper boundary of the modeling domain can be characterized by the mixing height $H_m(x, y, t)$ or an appropriately chosen constant height H_t. The presence of the topographic relief complicates the numerical solution of the atmospheric diffusion equation. Instead of using the actual height of a site (i.e., measured using sea level as a reference height) one usually transforms the domain into one of simpler geometry (Figure 25.6). This can be accomplished by a mapping that transforms points (x, y, z) from the physical domain to points (x, y, ζ) of the computational domain. The *terrain-following coordinate transformation* is commonly used, where

$$\zeta = \frac{z - h(x, y)}{H_t - h(x, y)} = \frac{z - h(x, y)}{\Delta H(x, y, t)} \tag{25.30}$$

that scales the vertical extent of the modeling domain into a new domain ζ that varies from 0 to 1. One may also simply subtract the surface height and define the new vertical coordinate by

$$z' = z - h(x, y) \tag{25.31}$$

and now z' has units of length and is not bounded between 0 and 1.

The above transformations result in changes of the components of the wind field from (u_x, u_y, u_z) to (u_x, u_y, w), where for the terrain-following coordinate system (McRae et al. 1982a)

$$w = \frac{1}{\Delta H}\left[u_z - u_x\left(\frac{\partial h}{\partial x} + z\frac{\partial \Delta H}{\partial x}\right) - u_y\left(\frac{\partial h}{\partial y} + z\frac{\partial \Delta H}{\partial y}\right) - z\frac{\partial \Delta H}{\partial t}\right] \qquad (25.32)$$

Note that even if the original vertical windspeed u_z is small, the new windspeed w may be significant if the terrain is rough (large derivatives $\partial h/\partial x$ and $\partial h/\partial y$). For the simple coordinate system transformation of (25.31), the vertical velocity is

$$w' = u_z - u_x\frac{\partial \Delta H}{\partial x} - u_y\frac{\partial \Delta H}{\partial y} \qquad (25.33)$$

The coordinate transformations discussed above result in changes of the eddy diffusivities. Initially, the eddy diffusivity tensor **K** was diagonal, but the transformed form is no longer diagonal. However, the off-diagonal terms contribution to turbulent transport in most urban flows is negligible (McRae et al. 1982a). If the terrain is extremely rugged the off-diagonal terms must be maintained, complicating significantly the overall problem. Note that even if these terms are neglected, if the terrain-following coordinate system is used, then

$$K_{\zeta\zeta} = \frac{K_{zz}}{\Delta H^2} \qquad (25.34)$$

No change is necessary for the simple transformation of (25.31). The changes required for the three coordinate systems are summarized in Table 25.2.

One should note that for the physical coordinate system boundary conditions must be applied at $z = h(x, y)$, while for the terrain-following systems application is at $\zeta = 0$ and $z' = 0$ (Figure 25.6). In the following discussion we will use the simple terrain-following coordinate transformation (25.31), which effectively "flattens out" the terrain, leading to the flat modeling domain shown in Figure 25.6d.

25.3.2   Initial Conditions

The solution of the full atmospheric diffusion equation requires the specification of the initial concentration field of all species:

$$c_i(x, y, z, 0) = c_i^*(x, y, z)$$

TABLE 25.2   Coordinate Systems for Solution of the Atmospheric Diffusion Equation

Coordinate System	Coordinates	Definitions	Vertical Windspeed	Vertical Eddy Diffusivity
Physical	x, y, z		u_z	K_{zz}
Terrain-following	x, y, ζ	$\zeta = \dfrac{z - h(x, y)}{\Delta H(x, y, t)}$	w^a	$\dfrac{K_{zz}}{\Delta H^2}$
Simple terrain-following	x, y, z'	$z' = z - h(x, y)$	w'^b	K_{zz}

[a]Given by (25.32).
[b]Given by (25.33).

A typical modeling domain at the urban or global scales contains on the order of 10^5 computational cells, and as a result one needs to specify the concentrations of all simulated species at all these points. There are practically never sufficient measurements (especially for the upper-lever computational cells), so one needs to extrapolate from the few available data to the rest of the modeling domain. Use of an inaccurate concentration field may introduce significant errors during the early part of the simulation. The sensitivity of the model to the specified initial conditions is qualitatively similar to that of (25.19). Initially, the model prediction is equal to c_i^*, the specified initial concentration, and will obviously be only as good as the specified value. As the simulation proceeds the initial condition term decays exponentially, while those due to emissions and advection increase in relative magnitude. Even if the specified initial conditions contain gross errors, their effect is eventually lost and the solution is dominated by the emissions q_i and the boundary conditions c_i^0. It is therefore common practice to start atmospheric simulations some period of time before that which is actually of interest. At the end of this "startup" period the model should have established concentration fields that do not seriously reflect the initial conditions and the actual simulation and comparison with observations may commence. The startup period for any atmospheric chemical transport model is determined by the residence time τ_r of an air parcel in the modeling domain.

25.3.3 Boundary Conditions

The atmospheric diffusion equation in three dimensions requires horizontal boundary conditions, two for each of the x, y, and z directions. The only exception are global-scale models simulating the whole Earth's atmosphere. One usually specifies the concentrations at the horizontal boundaries of the modeling domain as a function of time:

$$
\begin{aligned}
c(x,0,z,t) &= c_{x0}(x,z,t) \\
c(x,\Delta Y,z,t) &= c_{x1}(x,z,t) \\
c(0,y,z,t) &= c_{y0}(y,z,t) \\
c(\Delta X,y,z,t) &= c_{y1}(y,z;t)
\end{aligned}
\tag{25.35}
$$

Unfortunately, species concentration fields are practically never known at all points of the boundary of a modeling domain. Unlike initial conditions, boundary conditions, especially at the upwind boundaries, continue to affect predictions throughout the simulation. The simple box model solution given by (25.19) and (25.20) once more provides valuable insights as c_i^0 represents the corresponding boundary conditions. After the startup period the solution is the sum of the emission term and a term proportional to the concentration at the boundary. If $c_i^0 \ll q_i \tau_r / H$, errors in c_i^0 will have a small effect on the model predictions, which will be dominated by emissions. Therefore one should try to place the limits of the modeling domain in relatively clean areas (low c_i^0), where boundary conditions are relatively well known and have a relatively small effect on model predictions. A rule of thumb is to try to include all sources that have any effect on the air quality of a given region inside the modeling domain. If this cannot be done because of computational constraints, then the source effect has to be included implicitly in the boundary conditions. Uncertainty in urban air pollution model predictions as a result of uncertain side boundary conditions may be reduced by use of larger-scale models (e.g., regional models) to provide the boundary conditions to the urban-scale model. This technique is called *nesting* (the urban-scale model is "nested" inside the regional-scale model).

Treatment of upper and lower boundary conditions is different from that of the side conditions. One usually chooses a total reflection condition at the upper boundary of the computational domain (e.g., the top of the planetary boundary layer):

$$K_{zz}\left(\frac{\partial c}{\partial z}\right)_{z=H_t} = 0 \tag{25.36}$$

An alternative boundary condition can be formed using the vertical velocity u_z at the top of the modeling region (Reynolds et al. 1973)

$$|u_z|_{z=H_t}[c_i^a - c_i(z = H_t)] = K_{zz}\left(\frac{\partial c}{\partial z}\right)_{z=H_t} \qquad \text{for} \quad u_z \leq 0$$

$$K_{zz}\left(\frac{\partial c}{\partial z}\right)_{z=H_t} = 0 \qquad \text{for } u_z > 0 \tag{25.37}$$

where c_i^a is the concentration above the modeling region. The two conditions in (25.37) correspond to the case where material is transported into the region from above ($u_z \leq 0$) and out of the region ($u_z > 0$). In desirable cases the top of the domain is a stable layer for which vertical transport is small, and either (25.36) or (25.37) can be used with little effect on model predictions.

The boundary condition used at the Earth's surface accounts for surface sources and sinks of material

$$\left(v_{d,i}c_i - K_{zz}\frac{\partial c_i}{\partial z}\right)_{z=0} = E_i \tag{25.38}$$

where $v_{d,i}$ is the deposition velocity and E_i is the ground-level emission rate of the species. Note that ground-level emissions can be included either as part of the $z = 0$ boundary condition or directly in the differential equation as a source term in the ground-level cells.

25.4   ONE-DIMENSIONAL LAGRANGIAN MODELS

Whereas the atmospheric diffusion equation (25.27) is ideally suited for predicting the concentration distribution over extended areas, in many situations the atmosphere needs to be simulated only at a particular location. A Lagrangian model that simulates air parcels that eventually reach the receptor can often be used in this case. The air parcels are vertical columns of air extending from the ground up to a height H.

For example, assume that our goal is to simulate air quality in a given location, say, Claremont, California on a given day, for example, August 28, 1987. The first step using a Lagrangian modeling approach is to calculate, based on the wind field, the paths for the 24 air parcels that arrived at Claremont at 0:00, 1:00, ..., and 23:00 of the specific date. These "backward trajectories" can be calculated if we know the three-dimensional wind field $u_x(x, y, z, t), u_y(x, y, z, t)$, and $u_z(x, y, z, t)$ from

$$\frac{ds(t)}{dt} = \mathbf{u}(t) \tag{25.39}$$

where $\mathbf{s}(t)$ is the location of the air parcel at time t and $\mathbf{u}$ is the wind velocity vector. Integrating from t to t_0, we obtain

$$\mathbf{s}_0 - \mathbf{s}(t) = \int_t^{t_0} \mathbf{u}(\tau)d\tau \qquad (25.40)$$

or

$$\mathbf{s}(t) = \mathbf{s}_0 - \int_t^{t_0} \mathbf{u}(\tau)d\tau \qquad (25.41)$$

where we have assumed that at time t_0 the trajectory ends at the desired location $\mathbf{s}_0$. The location of the air parcel $\mathbf{s}(t)$ for a given moment t on this backward trajectory can be calculated by a straightforward integration according to (25.41). This integration is carried backward a couple of days until the air parcel is in a relatively clean area or in one with well-characterized atmospheric composition. In Figure 25.7 we show a calculated air parcel trajectory arriving at Claremont during August 28, 1987. The air parcel started over the Pacific Ocean, traversed the Los Angeles air basin picking up emissions from the various sources along their way, and eventually arrived at the receptor (Claremont).

After the trajectory path $\mathbf{s}(t)$ has been calculated from (25.41) the second step is the calculation of the emission fluxes corresponding to the trajectory path by interpolation of the emission field $E(x, y, z, t)$. For example, at 22:00 on August 27, 1987 the air parcel arriving in Claremont at 14:00 of the next day is over Hawthorne and therefore will pick up emissions from that area. Thus emission fluxes along the trajectory $E_t(t)$ are given by

$$E_t(t) = E(\mathbf{s}(t), t) \qquad (25.42)$$

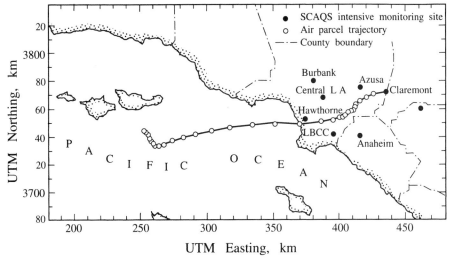

FIGURE 25.7 Air parcel trajectory arriving at Claremont, California, at 2 p.m. on August 28, 1987 (Pandis et al. 1992). SCAQS monitoring stations refer to those established in the 1987 Southern California Air Quality Study.

Let us now derive the simplified form of (25.27) corresponding to a moving coordinate system. Let x', y', and z' be the coordinates in the new system and u'_x, u'_y, and u'_z be the wind velocities with respect to this new moving coordinate system. The coordinate system moves horizontally with velocity equal to the wind speed and therefore $u'_x = u'_y = 0$, while $u'_z = u_z$. Therefore the second and third terms on the left-hand side [LHS] of (25.27) are zero in this case. Physically, the air parcel is moving with velocity equal to the windspeed, so there is no exchange of material with its surroundings by advection. The atmospheric diffusion equation then simplifies to

$$\frac{\partial c_i}{\partial t} + u_z \frac{\partial c_i}{\partial z} = \frac{\partial}{\partial x}\left(K_{xx}\frac{\partial c_i}{\partial x}\right) + \frac{\partial}{\partial y}\left(K_{yy}\frac{\partial c_i}{\partial y}\right) + \frac{\partial}{\partial z}\left(K_{zz}\frac{\partial c_i}{\partial z}\right)$$
$$+ R_i(c_1, c_2, \ldots, c_n) + E_{t,i}(t) - S_i(t) \tag{25.43}$$

A number of additional simplifying assumptions can be invoked at this stage. The first is that vertical advective transport is generally small compared to the vertical turbulent dispersion

$$\left|u_z \frac{\partial c_i}{\partial z}\right| \ll \frac{\partial}{\partial z}\left(K_{zz}\frac{\partial c_i}{\partial z}\right) \tag{25.44}$$

so that the former term can be neglected. This assumption can easily be relaxed, and the term may be retained if the vertical component of the wind field is large. According to the second assumption, the horizontal turbulent dispersion terms can be neglected:

$$\frac{\partial}{\partial x}\left(K_{xx}\frac{\partial c_i}{\partial x}\right) \simeq 0, \qquad \frac{\partial}{\partial y}\left(K_{yy}\frac{\partial c_i}{\partial y}\right) \simeq 0 \tag{25.45}$$

This assumption implies that horizontal concentration gradients are relatively small so that these terms make a negligible contribution to the overall mass balance. The error introduced by this assumption is small in areas with spatially homogeneous emissions but becomes significant in areas dominated by a few strong point sources. Finally, use of a Lagrangian trajectory model implies that the column of air retains its integrity during its transport. This assumption is equivalent to neglecting the wind shear:

$$u_x(x, y, z, t) \simeq u_x(x, y, t)$$
$$u_y(x, y, z, t) \simeq u_y(x, y, t) \tag{25.46}$$

This is a critical assumption and a major source of error in some trajectory model calculations, especially those that involve long transport times (Liu and Seinfeld 1975).

After employing the preceding three assumptions, the Lagrangian trajectory model equation is simplified to

$$\frac{\partial c_i}{\partial t} = \frac{\partial}{\partial z}\left(K_{zz}\frac{\partial c_i}{\partial z}\right) + R_i(c_1, c_2, \ldots, c_n) + E_{t,i}(t) - S_i(t) \tag{25.47}$$

Our previous discussion regarding the treatment of uneven terrain and boundary conditions is directly applicable to one-dimensional (1D) atmospheric models. For example, it is often difficult to specify appropriate initial conditions for a one-dimensional column model. In this modeling framework the characteristic species decay time is much longer than the residence time τ_r, because the only loss mechanism of inert chemical species is dry deposition [see (25.24) and (25.25)]. The characteristic time for a Lagrangian model can then be defined as

$$\tau_L = H/v_{d,i}$$

where H is the height of the column and $v_{d,i}$ the pollutant dry deposition velocity. For example, for $H \approx 1000\,\mathrm{m}$ and $v_{d,i} \approx 0.3\,\mathrm{cm\,s^{-1}}$, $\tau_L \approx 4$ days. This order-of-magnitude calculation suggests that column models will be relatively sensitive to the specified initial conditions in that an incorrectly specified value may persist for days after initiation of the simulation. The solution to this problem is to start the trajectory simulation far upwind of the source region in a relatively clean area with relatively well-known background concentrations.

One advantage of one-dimensional column models is that there are no side boundaries, and therefore no horizontal boundary conditions are necessary. One could argue, however, that the tradeoff is increased sensitivity to the initial conditions. Equations (25.36)–(25.38) can be used as the upper and lower boundary conditions in this case also.

25.5   OTHER FORMS OF CHEMICAL TRANSPORT MODELS

Our discussion in the previous sections has been based on a Cartesian coordinate system using x, y, and z as the coordinates and the molar concentration of species as the dependent variable. Other coordinate systems and concentration units used in chemical transport models are outlined below.

25.5.1   Atmospheric Diffusion Equation Expressed in Terms of Mixing Ratio

Atmospheric trace gas levels are frequently expressed in terms of mixing ratios. The volume mixing ratio of a species i (ξ_i) is identical to its mole fraction. We have developed forms of the atmospheric diffusion equation using the concentration c_i as the dependent variable. Let us take c_i as the molar concentration, expressed in units of mol i m^{-3}. Since the mass concentration m_i and the molar concentration c_i are related by $m_i = c_i M_i$, where M_i is the molecular weight of species i, the atmospheric diffusion equation applies equally well to the mass concentration.

Recall that the volume mixing ratio of species i at any point in the atmosphere is its mole fraction

$$\xi_i = \frac{c_i}{c_{\mathrm{air}}} \tag{25.48}$$

where c_{air} is the total molar concentration of air: $c_{\mathrm{air}} = p/RT$. The atmospheric mass density of air is

$$\rho = \frac{p\,M_{\mathrm{air}}}{RT} \tag{25.49}$$

To emphasize the altitude dependence of ρ, p, and T, we write (25.49) as

$$\rho(z) = \frac{p(z)M_{air}}{RT(z)} \tag{25.50}$$

Thus we can express the molar concentration of i in terms of its volume mixing ratio at any height z as

$$c_i(z) = \xi_i(z)\frac{\rho(z)}{M_{air}} \tag{25.51}$$

The basic question we address is: "What is the proper form of the atmospheric diffusion equation when written in terms of the mixing ratio?" Let us assume initially a horizontally homogeneous atmosphere. In this case, because only the vertical coordinate is affected by a changing density, we need consider only the form of equation for vertical transport. In what ensues we follow the development by Venkatram (1993). The mixing-length argument of Section 16.3 that led to the gradient transport equations applies to a variable that is conserved as an air parcel moves from one level to another. If, for some variable q, its substantial derivative satisfies

$$\frac{Dq}{Dt} = 0 \tag{25.52}$$

then the gradient transport relation in the z direction is

$$\langle u'_z q' \rangle = -K\frac{\partial\langle q\rangle}{\partial z} \tag{25.53}$$

To investigate whether c_i satisfies (25.52), we begin with the conservation equation using index notation

$$\frac{\partial c_i}{\partial t} + \frac{\partial}{\partial x_j}(u_j c_i) = 0 \tag{25.54}$$

where u_j is the instantaneous velocity in direction j. The substantial derivative

$$\frac{Dc_i}{Dt} \equiv \frac{\partial c_i}{\partial t} + u_j\frac{\partial c_i}{\partial x_j} \tag{25.55}$$

is then as follows, combining (25.54) and (25.55):

$$\frac{Dc_i}{Dt} = -c_i\frac{\partial u_j}{\partial x_j} \tag{25.56}$$

Previously we assumed an incompressible atmosphere, for which $\partial u_j/\partial x_j = 0$. This is a reasonable approximation for layers close to the Earth's surface and therefore for urban-scale atmospheric chemical transport models. The continuity equation for the compressible atmosphere is

$$\frac{\partial\rho}{\partial t} + \frac{\partial}{\partial x_j}(\rho u_j) = 0 \tag{25.57}$$

Then

$$\frac{\partial u_j}{\partial x_j} = -\frac{1}{\rho}\left(u_j\frac{\partial \rho}{\partial x_j} + \frac{\partial \rho}{\partial t}\right) \tag{25.58}$$

which is not zero in the vertical direction. Thus Dc_i/Dt is not zero in the vertical direction when vertical variations of density are taken into account.

To test if the mixing ratio $\xi_i = c_i M_{air}/\rho$ is conserved, we write (25.54) in terms of the mixing ratio as

$$\frac{\partial}{\partial t}\left(\frac{\xi_i \rho}{M_{air}}\right) + \frac{\partial}{\partial x_j}\left(u_j \frac{\xi_i \rho}{M_{air}}\right) = 0 \tag{25.59}$$

which gives

$$\rho\frac{\partial \xi_i}{\partial t} + \xi_i\frac{\partial \rho}{\partial t} + \rho u_j\frac{\partial \xi_i}{\partial x_j} + \xi_i\frac{\partial}{\partial x_j}(u_j\rho) = 0 \tag{25.60}$$

From (25.57), this reduces to

$$\frac{\partial \xi_i}{\partial t} + u_j\frac{\partial \xi_i}{\partial x_j} = 0 \tag{25.61}$$

or

$$\frac{D\xi_i}{Dt} = 0 \tag{25.62}$$

Thus the mixing ratio is a conserved quantity in the sense of the mixing-length arguments. This suggests that the proper mixing-length form is

$$\langle u_z'\xi_i'\rangle = -K_{zz}\frac{\partial\langle\xi_i\rangle}{\partial z} \tag{25.63}$$

The continuity equation for the mean concentration of species i, with a zero mean vertical velocity, is

$$\frac{\partial\langle c_i\rangle}{\partial t} = -\frac{\partial}{\partial z}\langle u_z'c_i'\rangle \tag{25.64}$$

Expressing c_i in terms of mixing ratio using (25.51), $c_i = \xi_i\rho/M_{air}$, the mean values and fluctuations of each quantity are

$$\langle c_i\rangle + c_i' = \langle\xi_i\rangle\bar{\rho} + \xi_i'\bar{\rho} + \langle\xi_i\rangle\rho' + \xi_i'\rho' \tag{25.65}$$

Taking the mean of both sides of the equation, we obtain

$$\langle c_i\rangle = \langle\xi_i\rangle\bar{\rho} \tag{25.66}$$

since $\langle \xi_i' \rho' \rangle$, which is proportional to $\bar{\rho}'^2/\bar{\rho}^2$, can be neglected. The vertical turbulent flux term in (25.64) becomes

$$u_z' c_i = u_z'(\xi_i'\bar{\rho} + \langle\xi_i\rangle\rho' + \xi_i'\rho') \tag{25.67}$$

which, on averaging and neglecting $\overline{u_z'\xi_i'\rho'}$, becomes

$$\langle u_z' c_i' \rangle = \bar{\rho}\langle u_z'\xi_i'\rangle + \langle\xi_i\rangle\overline{\rho'u_z'} \tag{25.68}$$

Substituting (25.66) and (25.68) into (25.64), we obtain

$$\frac{\partial\langle\xi_i\rangle}{\partial t} = \frac{1}{\bar{\rho}}\frac{\partial}{\partial z}(-\bar{\rho}\langle u_z'\xi_i'\rangle) - \frac{\partial\langle\xi_i\rangle}{\partial z}\frac{\overline{\rho'u_z'}}{\bar{\rho}} \tag{25.69}$$

where we have used the conservation equation for air density:

$$\frac{\partial\bar{\rho}}{\partial t} + \frac{\partial}{\partial z}\overline{(\rho'u_z')} = 0 \tag{25.70}$$

On the basis of the earlier analysis showing that the mixing ratio is conserved along a vertical fluctuation, we adopt (25.63), and (25.69) becomes

$$\frac{\partial\langle\xi_i\rangle}{\partial t} = \frac{1}{\bar{\rho}}\frac{\partial}{\partial z}\left(\bar{\rho}K_{zz}\frac{\partial\langle\xi_i\rangle}{\partial z}\right) - \frac{\partial\langle\xi_i\rangle}{\partial z}\frac{\overline{\rho'u_z'}}{\bar{\rho}} \tag{25.71}$$

which states that turbulent transport will not change the mean mixing ratio of a species $\langle\xi_i\rangle$ in an atmosphere in which $\langle\xi_i\rangle$ does not already vary with height. Venkatram (1993) has shown that the second term on the RHS of (25.71) involving the flux of air can generally be neglected.

Therefore the form of the atmospheric diffusion equation for vertical transport when vertical variations of density are taken into account is

$$\frac{\partial\langle\xi_i\rangle}{\partial t} = \frac{1}{\bar{\rho}}\frac{\partial}{\partial z}\left(\bar{\rho}K_{zz}\frac{\partial\langle\xi_i\rangle}{\partial z}\right) \tag{25.72}$$

If the atmosphere is not horizontally homogeneous, the same arguments can be used for the x and y directions to obtain

$$\frac{\partial\xi_i}{\partial t} + u_x\frac{\partial\xi_i}{\partial x} + u_y\frac{\partial\xi_i}{\partial y} + u_z\frac{\partial\xi_i}{\partial z} = \frac{1}{\rho}\frac{\partial}{\partial x}\left(\rho K_{xx}\frac{\partial\xi_i}{\partial x}\right) + \frac{1}{\rho}\frac{\partial}{\partial y}\left(\rho K_{yy}\frac{\partial\xi_i}{\partial y}\right)$$
$$+ \frac{1}{\rho}\frac{\partial}{\partial z}\left(\rho K_{zz}\frac{\partial\xi_i}{\partial z}\right) + R_i - S_i \tag{25.73}$$

where ξ_i and ρ are the mean mixing ratio and air density, respectively.

25.5.2 Pressure-Based Coordinate System

For modeling domains that cover the full troposphere and not only the lowest 1–2 km, it is useful to replace the height z, based coordinate system with the σ vertical coordinate system, defined as

$$\sigma = \frac{p - p_t}{p_s - p_t} \qquad (25.74)$$

where p_s is the surface pressure, p_t is the pressure that defines the top of the modeling domain (usually around 0.1 atm for the troposphere), and p is the pressure at the point where σ is evaluated. At the surface, even if the terrain is not flat, $\sigma = 1$, while at the top of the modeling domain $\sigma = 0$. An example of a σ-coordinate system is given in Table 25.3. The σ-coordinate system is similar to the height-based ζ system defined by (25.30). The corresponding form of the diffusion equation can then be developed, noting that from (25.74)

$$dp = (p_s - p_t)d\sigma \qquad (25.75)$$

and recalling that $dp = -\rho g\, dz$. Thus

$$dz = -\frac{p^*}{\rho g}d\sigma \qquad (25.76)$$

where

$$p^* = p_s - p_t \qquad (25.77)$$

Substituting (25.76) into (25.73), we obtain

$$\frac{\partial \xi_i}{\partial t} + u_x \frac{\partial \xi_i}{\partial x} + u_y \frac{\partial \xi_i}{\partial y} + u_\sigma \frac{\partial \xi_i}{\partial \sigma} = \frac{1}{\rho}\frac{\partial}{\partial x}\left(\rho K_{xx}\frac{\partial \xi_i}{\partial x}\right) + \frac{1}{\rho}\frac{\partial}{\partial y}\left(\rho K_{yy}\frac{\partial \xi_i}{\partial y}\right)$$
$$+ \frac{g^2}{(p^*)^2}\frac{\partial}{\partial \sigma}\left(\rho^2 K_{zz}\frac{\partial \xi_i}{\partial \sigma}\right) + R_i - S_i \qquad (25.78)$$

TABLE 25.3 Vertical Levels, σ Coordinates, Pressures, and Altitudes

Level	Coordinate, σ	Pressure, atm	z, km
1	0.995	0.995	0.04
2	0.97	0.972	0.22
3	0.90	0.905	0.75
4	0.75	0.763	2.0
5	0.60	0.620	3.6
6	0.45	0.478	5.5
7	0.30	0.355	8.2
8	0.15	0.193	12.4
9	0	0.050	22.5

where u_σ is the vertical velocity in the σ-coordinate system. The velocity u_σ can be computed using the mass continuity equation in σ coordinates

$$\frac{\partial p^*}{\partial t} = -\frac{\partial}{\partial x}(u_x p^*) - \frac{\partial}{\partial y}(u_y p^*) - \frac{\partial}{\partial \sigma}(u_\sigma p^*) - \frac{2u_z p^*}{R_e} \tag{25.79}$$

where R_e is the Earth's radius. The last term is much smaller than the other terms and can be ignored. The velocity u_σ can be calculated by integrating (25.79):

$$u_\sigma = -\frac{1}{p^*}\int_0^\sigma \left(\frac{\partial}{\partial x}(u_x p^*) + \frac{\partial}{\partial y}(u_y p^*)\right)d\sigma - \frac{\sigma}{p^*}\frac{\partial p^*}{\partial t} \tag{25.80}$$

25.5.3 Spherical Coordinates

For global atmospheric chemical transport models with a domain representing the entire atmosphere of the planet, spherical coordinate systems offer a series of advantages. The coordinates chosen are the latitude ϕ, the longitude l, and the σ-vertical coordinate described in the previous section. The coordinate-system-independent atmospheric diffusion equation is

$$\frac{\partial \xi_i}{\partial t} = -\mathbf{u} \cdot \nabla \xi_i + \frac{1}{\rho}\nabla \cdot (\rho \mathbf{K}\nabla \xi_i) - S_i + R_i \tag{25.81}$$

This equation can be written in spherical coordinates

$$
\begin{aligned}
&\frac{1}{p_s}\frac{\partial}{\partial t}(p_s \xi_i) + \frac{1}{p_s}\frac{\partial}{\partial l}(p_s u_l \xi_i) + \frac{1}{p_s \cos\phi}\frac{\partial}{\partial \phi}(p_s u_\phi \xi_i \cos\phi) + \frac{\partial}{\partial\sigma}(u_\sigma \xi_i) \\
&= \frac{1}{p_s}\frac{\partial}{\partial l}\left(p_s K_{ll}\frac{\partial \xi_i}{\partial l}\right) + \frac{1}{p_s \cos\phi}\frac{\partial}{\partial \phi}\left(p_s K_{\phi\phi}\cos\phi \frac{\partial \xi_i}{\partial \phi}\right) \\
&\quad + \left(\frac{g}{p_s}\right)^2 \frac{\partial}{\partial\sigma}\left(\rho^2 K_{zz}\frac{\partial \xi_i}{\partial\sigma}\right) - S_i + R_i
\end{aligned} \tag{25.82}
$$

where p_s is the atmospheric surface pressure at the corresponding point; $K_{ll}, K_{\phi\phi}$, and K_{zz} are the components of the diffusion tensor. The components of the wind velocity are given by

$$
\begin{aligned}
u_l &= \frac{1}{R_e \cos\phi} u_x \\
u_\phi &= \frac{1}{R_e} u_y \\
u_\sigma &= -\frac{\sigma}{p_s}\left(\frac{\partial p_s}{\partial t} + \frac{\partial p_s}{\partial x}u_x + \frac{\partial p_s}{\partial y}u_y\right) - \frac{\rho g}{p_s}u_z
\end{aligned} \tag{25.83}
$$

and

$$K_{ll} = \frac{1}{R_e^2 \cos^2 \phi} K_{xx}$$

$$K_{\phi\phi} = \frac{1}{R_e^2} K_{yy}$$

(25.84)

25.6 NUMERICAL SOLUTION OF CHEMICAL TRANSPORT MODELS

In this section we will discuss some aspects of numerical solution of chemical transport models. We do not attempt to specify which numerical method is best; rather, we point out some considerations in assessing the adequacy and appropriateness of numerical methods for chemical transport models. Our treatment is, by necessity, brief. We refer the reader to Peyret and Taylor (1983) and Oran and Boris (1987) for more study in this area.

Chemical transport models solve chemical species equations of the general form

$$\frac{\partial c_i}{\partial t} = \left(\frac{\partial c_i}{\partial t}\right)_{\text{adv}} + \left(\frac{\partial c_i}{\partial t}\right)_{\text{diff}}$$

$$+ \left(\frac{\partial c_i}{\partial t}\right)_{\text{cloud}} + \left(\frac{\partial c_i}{\partial t}\right)_{\text{dry}} + \left(\frac{\partial c_i}{\partial t}\right)_{\text{aeros}} + R_{gi} + E_i$$

(25.85)

where c_i is the concentration of species i; $(\partial c_i/\partial t)_{\text{adv}}$, $(\partial c_i/\partial t)_{\text{diff}}$, $(\partial c_i/\partial t)_{\text{cloud}}$, $(\partial c_i/\partial t)_{\text{dry}}$, and $(\partial c_i/\partial t)_{\text{aeros}}$ are the rates of change of c_i due to advection, diffusion, cloud processes (cloud scavenging, evaporation of cloud droplets, aqueous-phase reactions, wet deposition, etc.), dry deposition, and aerosol processes (transport between gas and aerosol phases, aerosol dynamics, etc), respectively; R_{gi} is the net production from gas-phase reactions; and E_i is the emission rate.

Species simulated can be in the gas, aqueous, or aerosol phases. Therefore chemical transport models are characterized by the following operators:

A	Advection operator
D	Diffusion operator
C	Cloud operator
G	Gas-phase chemistry operator
P	Aerosol operator
S	Source/sink operator

For example, the advection operator is $\mathbf{u} \cdot \nabla c_i$. If only the advection operator is applied to all species, then they will be advected following the wind patterns. If $\mathbf{c}(x, y, z; t)$ is the concentration vector of all species at time t, then its value at the next timestep $t + \Delta t$ will be the net result of the simultaneous application of all operators

$$\mathbf{c}(x, y, z; t + \Delta t) = \mathbf{c}(x, y, z; t) + [A(\Delta t) + D(\Delta t) + C(\Delta t) + G(\Delta t)$$
$$+ P(\Delta t) + S(\Delta t)]\mathbf{c}(x, y, z; t)$$

(25.86)

Each of these operators is distinctly different in basic character, each usually requiring quite different numerical techniques to obtain numerical solutions. We should note that no single numerical method is uniformly best for all chemical transport models. The relative contributions of each of the operators to the overall solution as well as other considerations, such as boundary conditions and wind fields, can easily change from application to application, leading to different numerical requirements.

25.6.1 Coupling Problem—Operator Splitting

Equation (25.85), the basis of every atmospheric model, is a set of time-dependent, nonlinear, coupled partial differential equations. Several methods have been proposed for their solution including global finite differences, operator splitting, finite element methods, spectral methods, and the method of lines (Oran and Boris 1987). Operator splitting, also called the *fractional step method* or *timestep splitting*, allows significant flexibility and is used in most atmospheric chemical transport models.

Finite Difference Methods   Finite difference methods approximate and solve the full equation (25.85). To illustrate the basic idea let us focus on the one-dimensional diffusion equation

$$\frac{\partial c}{\partial t} = \frac{\partial^2 c}{\partial x^2} \tag{25.87}$$

which is one of the simplest subcases of (25.85). First, a spatial grid $x_1, x_2, \ldots, x_k$ is defined over the spatial domain of interest $0 \leq x \leq L$. The concentration function $c(x, t)$ is approximated by its values $c_1, c_2, \ldots, c_k$ at the corresponding grid points (Figure 25.8). Partial spatial derivatives of concentration can be approximated by divided difference quotients; for example, if the grid spacing is uniform and equal to Δx [e.g., $x_i = x_1 + (i - 1)\Delta x$], then

$$\frac{\partial c}{\partial x} = \frac{c_i - c_{i-1}}{\Delta x} \tag{25.88}$$

$$\frac{\partial^2 c}{\partial x^2} = \frac{c_{i+1} - 2c_i + c_{i-1}}{(\Delta x)^2} \tag{25.89}$$

As the concentration $c(x, t)$ changes with time, at each moment there will be a different set of values c_i. Time can be discretized similarly to space by using a time interval Δt. So we are interested in the values of c_i at $t_0, t_0 + \Delta t, \ldots$. Defining

$$t_n = t_0 + n\,\Delta t \tag{25.90}$$

we would like to calculate the concentration values c_i^n corresponding to the grid point i at the time t_n. Time derivatives can be approximated similarly to space derivatives by

$$\frac{\partial c}{\partial t} = \frac{c_i^{n+1} - c_i^n}{\Delta t} \tag{25.91}$$

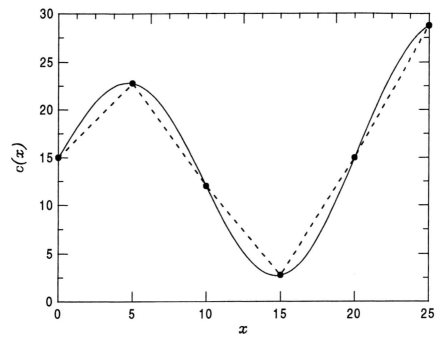

FIGURE 25.8   Example of a discrete approximation of the continuous concentration function $c(x)$.

By combining (25.89) and (25.91), a finite difference approximation of (25.87) is

$$\frac{c_i^{n+1} - c_i^n}{\Delta t} = \frac{c_{i+1}^n - 2\,c_i^n + c_{i-1}^n}{(\Delta x)^2} \tag{25.92}$$

or solving for c_i^{n+1}

$$c_i^{n+1} = c_i^n + \frac{c_{i+1}^n - 2c_i^n + c_{i-1}^n}{(\Delta x)^2}\,\Delta t \tag{25.93}$$

Initially (time zero) the values of c_i^0 are known and therefore the values of c_i^1 (concentrations at $t = \Delta t$) can be calculated by direct application of (25.93). This approach can be repeated and the values of c_i^{n+1} can be calculated from the previously calculated values of c_i^n. This is an example of an *explicit finite difference method*, where, if approximate solution values are known at time $t_n = n\Delta t$, then approximate values at time $t_{n+1} = (n+1)\Delta t$ may be explicitly and immediately calculated using (25.93). Typically, explicit techniques require that constraints be placed on the size of Δt that may be used to avoid significant numerical errors and for stable operation. In a stable method, unavoidable small errors in the solution are suppressed with time; in an unstable method a small initial error may increase significantly, leading to erroneous results or a complete failure of the method. Equation (25.93) is stable only if $\Delta t < (\Delta x)^2/2$, and therefore one is obliged to use small integration timesteps.

The derivative $\partial^2 c/\partial x^2$ can also be approximated by using concentration values at time t_{n+1} or

$$\frac{\partial^2 c}{\partial x^2} = \frac{c_{i+1}^{n+1} - 2c_i^{n+1} + c_{i-1}^{n+1}}{(\Delta x)^2} \tag{25.94}$$

leading to the following finite difference approximation of (25.87):

$$\frac{c_i^{n+1} - c_i^n}{\Delta t} = \frac{c_{i+1}^{n+1} - 2c_i^{n+1} + c_{i-1}^{n+1}}{(\Delta x)^2} \tag{25.95}$$

Now, even if values of $c_i^n (i = 1, 2, \ldots, k)$ are known, (25.95) cannot be solved explicitly, as c_i^{n+1} is also a function of the unknown c_{i+1}^{n+1} and c_{i-1}^{n+1}. However, all k equations of the form (25.95) for $i = 1, 2, \ldots, k$ form a system of linear algebraic equations with k unknowns, namely, $c_1^{n+1}, c_2^{n+1}, \ldots, c_k^{n+1}$. This system can be solved and the solution can be advanced from t_n to t_{n+1}. This is an example of an *implicit finite difference method*. In general, implicit techniques have better stability properties than explicit methods. They are often unconditionally stable and any choice of Δt and Δx may be used (the choice is ultimately based on accuracy considerations alone).

Similar global implicit formalisms can be developed to treat any form of partial differential equations in an atmospheric model. Accuracy of the solutions can be tested by increasing the temporal and spatial resolution (decreasing $\Delta x, \Delta y, \Delta z$, and Δt) and repeating the calculation. However, because the full problem is solved as a whole, implicit formalisms require solution of very large systems, are slow, and require huge computational resources. Even if finite difference methods are easy to apply, they are rarely used for solution of the full (25.85) in atmospheric chemical transport models.

Finite Element Methods   In using a finite element method, one typically divides the spatial domain in zones or "elements" and then requires that the approximate solutions have the form of a specified polynomial over each of the elements. Discrete equations are then derived by requiring that the error in the piecewise polynomial approximate solution (i.e., the residual when the polynomial is substituted into the basic partial differential equation) be minimum. The Galerkin approach is one of the most popular, requiring that the error be orthogonal to the piecewise polynomial space itself. Characteristically, finite element methods lead to implicit approximation equations to be solved, which are usually more complicated than analogous finite difference methods. Their computational expense is comparable to, or can exceed, global implicit methods.

Operator Splitting   Operator splitting is the most popular technique for the solution of (25.85). The basic idea is, instead of solving the full equation at once, to solve independently the pieces of the problem corresponding to the various processes and then couple the various changes resulting from the separate partial calculations (Yanenko 1971). Considering (25.86), each process contributes a part of the overall change in the concentration vector $\mathbf{c}$. Let us define the change over the timestep Δt as

$$\Delta \mathbf{c} = \mathbf{c}(t + \Delta t) - \mathbf{c}(t) \tag{25.96}$$

Then decoupling the operators

$$\begin{aligned}
\Delta \mathbf{c}^A &= A(\Delta t)\mathbf{c}(t) \\
\Delta \mathbf{c}^D &= D(\Delta t)\mathbf{c}(t) \\
\Delta \mathbf{c}^C &= C(\Delta t)\mathbf{c}(t) \\
\Delta \mathbf{c}^G &= G(\Delta t)\mathbf{c}(t) \\
\Delta \mathbf{c}^P &= P(\Delta t)\mathbf{c}(t) \\
\Delta \mathbf{c}^S &= S(\Delta t)\mathbf{c}(t)
\end{aligned} \tag{25.97}$$

where $\Delta \mathbf{c}^A$ is the change to the concentration vector because of advection, $\Delta \mathbf{c}^D$ is the change because of diffusion, and so on. The overall change for the timestep can then be found by summing all changes

$$\Delta \mathbf{c} = \Delta \mathbf{c}^A + \Delta \mathbf{c}^D + \Delta \mathbf{c}^C + \Delta \mathbf{c}^G + \Delta \mathbf{c}^P + \Delta \mathbf{c}^S \tag{25.98}$$

and the new concentration vector is

$$\mathbf{c}(t + \Delta t) = \mathbf{c}(t) + \Delta \mathbf{c} \tag{25.99}$$

Equations (25.97)–(25.99) suggest that each process can be simulated individually and then the results can be combined.

The approach of (25.97)–(25.99) can work quite well, but other alternatives exist and are often used. Instead of applying the operators in parallel to $\mathbf{c}$, one can apply them in series, or

$$\begin{aligned}
\mathbf{c}^1(t + \Delta t) &= A(\Delta t)\mathbf{c}(t) \\
\mathbf{c}^2(t + \Delta t) &= D(\Delta t)\mathbf{c}^1(t + \Delta t) \\
\mathbf{c}^3(t + \Delta t) &= C(\Delta t)\mathbf{c}^2(t + \Delta t) \\
\mathbf{c}^4(t + \Delta t) &= G(\Delta t)\mathbf{c}^3(t + \Delta t) \\
\mathbf{c}^5(t + \Delta t) &= P(\Delta t)\mathbf{c}^4(t + \Delta t) \\
\mathbf{c}(t + \Delta t) &= S(\Delta t)\mathbf{c}^5(t + \Delta t)
\end{aligned} \tag{25.100}$$

For example, $\mathbf{c}^3(t + \Delta t)$ includes changes in concentration because of advection, diffusion, and cloud processing. Operators can be combined; for example, diffusion and advection can be applied together as a transport operator T, or they may be split. One may apply first the transport operator in the x direction, T_x, then T_y, and finally the vertical transport operator T_z.

Order of operator application is another issue. McRae et al. (1982a) recommended using a symmetric operator splitting scheme for the solution of the atmospheric diffusion equation. They used the scheme

$$\begin{aligned}
\mathbf{c}(t + \Delta t) = {} & T_x\left(\frac{\Delta t}{2}\right) T_y\left(\frac{\Delta t}{2}\right) T_z\left(\frac{\Delta t}{2}\right) G(\Delta t) \\
& \times T_z\left(\frac{\Delta t}{2}\right) T_y\left(\frac{\Delta t}{2}\right) T_x\left(\frac{\Delta t}{2}\right) \mathbf{c}(t)
\end{aligned} \tag{25.101}$$

where T_x is the horizontal transport (advection and diffusion) operator in the east–west direction, T_y is the north–south transport operator, T_z is the vertical transport operator, and G is the gas-phase chemistry operator. Note that the transport operators are applied twice for $\Delta t/2$ and the gas-phase chemistry operator once for Δt during each cycle, to advance the solution for Δt.

The qualitative criterion for the validity of (25.97) or (25.100) is that the concentration values must not change too quickly over the applied timestep from any of the individual processes. The error because of the splitting becomes zero as $\Delta t \rightarrow 0$, but unfortunately the maximum allowable value of Δt cannot be easily determined a priori. The allowable Δt depends strongly on the simulated processes and scenario. For example, for the application of (25.101) in an urban airshed, McRae et al. (1982a) recommended a value of Δt smaller than 10 min.

To illustrate the advantages of operator splitting, consider the solution of the three-dimensional diffusion equation

$$\frac{\partial c}{\partial t} = \frac{\partial^2 c}{\partial x^2} + \frac{\partial^2 c}{\partial y^2} + \frac{\partial^2 c}{\partial z^2} \tag{25.102}$$

on a region having the shape of a cube. We divide the cube up with a uniform grid with N divisions in each direction; that is, we divide it up into N^3 smaller cubes. A global implicit finite difference method would require the solution of a linear system of equations with N^3 unknowns. Because the resulting matrix will have a bandwidth of about N^2 at each timestep, the cost of solving the linear system using conventional banded Gaussian elimination would be proportional to $(N^2)^2 N^3 = N^7$. Using a splitting technique for each step, one would solve discrete versions of the following three one-dimensional diffusion problems

$$\frac{\partial c}{\partial t} = \frac{\partial^2 c}{\partial x^2}$$
$$\frac{\partial c}{\partial t} = \frac{\partial^2 c}{\partial y^2} \tag{25.103}$$
$$\frac{\partial c}{\partial t} = \frac{\partial^2 c}{\partial z^2}$$

or using finite differences

$$\frac{c_i^{n+1/3} - c_i^n}{\Delta t} = \frac{c_{i+1}^{n+1/3} - 2c_i^{n+1/3} + c_{i-1}^{n+1/3}}{(\Delta x)^2}$$
$$\frac{c_j^{n+2/3} - c_j^{n+1/3}}{\Delta t} = \frac{c_{j+1}^{n+2/3} - 2c_j^{n+2/3} + c_{j-1}^{n+2/3}}{(\Delta y)^2} \tag{25.104}$$
$$\frac{c_k^{n+1} - c_k^{n+2/3}}{\Delta t} = \frac{c_{k+1}^{n+1} - 2c_k^{n+1} + c_{k-1}^{n+1}}{(\Delta z)^2}$$

where c^n are the values at $t_n = n\,\Delta t$ and c^{n+1} are the values at t_{n+1}. To accomplish the solution of (25.104) on the $N \times N \times N$ grid displayed above would require N^2 solutions of each of the three equations above, and the cost of each solution would be proportional to N.

So the total cost to advance over one timestep using splitting would be proportional to $2N^3$, which is considerably less than the N^7 cost for the fully implicit case. If each one-dimensional method is stable, then the overall splitting technique is usually stable. Splitting does not require exclusive use of finite difference methods; for example, finite element techniques could be used to solve the above one-dimensional problems.

Operator splitting methods require less computational resources but demand more thought. In general, stability is not guaranteed. However, operator splitting encourages modular models and allows the use of the best available numerical technique for each module. Thus advection, kinetics, cloud processes, and aerosol processes each reside in different modules. Techniques used for these subproblems will be outlined subsequently. For applications in other problems, the reader is referred to Oran and Boris (1987).

25.6.2 Chemical Kinetics

The gas-phase chemistry operator involves solution of a system of ordinary differential equations of the form

$$\frac{dc_i}{dt} = P_i - c_i L_i \tag{25.105}$$

where c_i are the species concentrations and P_i and $c_i L_i$ are production and loss terms, respectively. Equation (25.105) is a system of coupled nonlinear differential equations. The P_i and L_i terms are functions of the concentrations c_i and provide the coupling among the equations.

The simplest method for the solution of (25.105) is based on the finite difference approach discussed in the previous section. Writing

$$\frac{dc_i}{dt} = \frac{c_i^{n+1} - c_i^n}{\Delta t}$$

and solving for c_i^{n+1} yields

$$c_i^{n+1} = c_i^n + (P_i^n - c_i^n L_i^n)\Delta t \tag{25.106}$$

where P_i^n and $c_i^n L_i^n$ are the production and loss terms for $t = t_n$. This method is known as the *Euler method* and is explicit. Starting with the initial conditions c_i^0 one can apply (25.106) to easily find the concentrations to integrate (25.105), finding c_i^n for each timestep. However, integration of chemical kinetic equations poses a major difficulty as indicated by the following example.

Stiff ODEs   Consider the following two ordinary differential equations (ODEs)

$$\frac{dc_1}{dt} = -1001\, c_1 + 999\, c_2 + 2$$
$$\frac{dc_2}{dt} = 999\, c_1 - 1001\, c_2 + 2 \tag{25.107}$$

with initial conditions $c_1(0) = 3$ and $c_2(0) = 1$. Let us try to integrate this system numerically from $t = 0$ to $t = 1$ min using the Euler method. Then

$$
\begin{aligned}
c_1^{n+1} &= c_1^n + (2 - 1001\, c_1^n + 999\, c_2^n)\Delta t \\
c_2^{n+1} &= c_2^n + (2 + 999\, c_1^n - 1001\, c_2^n)\Delta t
\end{aligned}
\tag{25.108}
$$

If we select a step $\Delta t = 0.1$ min, then using $c_1^0 = 3$ and $c_2^0 = 1$ and (25.108), we calculate that $c_1^1 = 197$ and $c_2^1 = 201$. If we reapply the algorithm for $t = 0.2$ min, then $c_1^2 = 4 \times 10^4$ and $c_2^2 = -4 \times 10^4$, and both solutions approach infinity. The numerical solution of the problem has failed. Let us try once more using $\Delta t = 0.01$ min. Applying (25.108), we find that for $t = 0.01$ min, $c_1^1 = -17, c_2^1 = 21$. In the next step $c_1^2 = 362$ and $c_2^2 = -359$, and the algorithm once more fails. One may suspect that there is something peculiar going on with the solution of the problem. The system of differential equations has the solution

$$
\begin{aligned}
c_1(t) &= 1 + \exp(-2t) + \exp(-2000t) \\
c_2(t) &= 1 + \exp(-2t) - \exp(-2000t)
\end{aligned}
\tag{25.109}
$$

and each function is simply the sum of two exponential terms and a constant. The true $c_1(t)$ starts from its initial value $c_1(0) = 3$ and, as the two exponentials decay, approaches a steady-state value of $c_1 = 1$. The second function reaches a maximum and then decays back to 1.0. If the solution that we are seeking is so simple, why then is the Euler method having so much trouble reproducing it? Figure 25.9 and (25.109) suggest

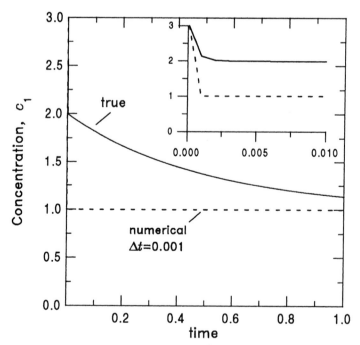

FIGURE 25.9 Numerical solution of the stiff ODE example using the explicit Euler method and $\Delta t = 0.001$. Also shown is the true solution.

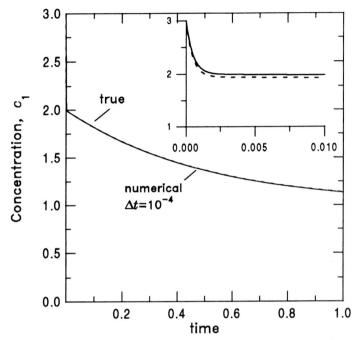

FIGURE 25.10   Numerical solution of the stiff ODE example using the explicit Euler method and $\Delta t = 0.0001$. Also shown is the true solution.

that the desired solution consists of two transients. The fast one, which could be due to a fast reaction, decays by $t = 0.002 \, \text{min}$. The slow transient affects the solution up to $t = 10 \, \text{min}$ or so. As a result of the fast transient, the concentration c_1 decays in less than $0.005 \, \text{min}$ by 33% from $c_1 = 3$ to $c_1 = 2$. Extremely small timesteps are necessary to resolve this rapid decay. If we use a timestep $\Delta t = 0.001 \, \text{min}$, our solution has significant errors throughout the integration interval but at least remains positive and does not explode (Figure 25.9). Once more the fast transient during the first $0.002 \, \text{min}$ is the cause of the numerical error. Finally, if $\Delta t = 0.0001 \, \text{min}$ is used, the algorithm is able to resolve both transients and reproduce the correct solution (Figure 25.10). Small errors exist only in the first few steps. Note that application of (25.109) 10,000 times is necessary to integrate the system of ordinary differential equations (ODEs).

In the example above, a term in the solution important for only $10^{-3} \, \text{min}$ dictated the choice of the timestep. If a step larger than $10^{-3} \, \text{min}$ is used, the solution is unstable. Similar problems almost always accompany the numerical solution of systems of ODEs describing a set of chemical reactions. These systems are characterized by timescales that vary over several orders of magnitude and are characterized as *numerically stiff*. For example, the differential equations (25.107) have timescales on the order of $10^{-3} \, \text{min}$ and $10 \, \text{min}$. Stiffness can be defined rigorously for a system of linear ODEs that can be expressed in the vector form

$$\frac{d\mathbf{c}}{dt} = \mathbf{Ac} \qquad (25.110)$$

with $\mathbf{c}$ the unknown concentration vector and $\mathbf{A}$ an $N \times N$ matrix based on the real part of the eigenvalues of $\mathbf{A}$, $\mathrm{Re}(\lambda_i)$. The system (25.110) is said to be stiff if the following two conditions are satisfied:

1. The real part of at least one of the eigenvalues is negative

$$\mathrm{Re}(\lambda_i) < 0$$

2. There are significant differences between the maximum and minimum eigenvalues

$$\frac{\max |\mathrm{Re}(\lambda_i)|}{\min |\mathrm{Re}(\lambda_i)|} \gg 1 \tag{25.111}$$

For a more general nonlinear system

$$\frac{d\mathbf{c}}{dt} = \mathbf{f}(\mathbf{c}) \tag{25.112}$$

the *Jacobian matrix* of the system is defined as

$$\mathbf{J} = \frac{\partial \mathbf{f}}{\partial \mathbf{c}} \tag{25.113}$$

and has elements

$$J_{ik} = \frac{\partial f_i}{\partial c_k} \tag{25.114}$$

Note that if the system is linear, $\mathbf{J} = \mathbf{A}$. For a nonlinear system the stiffness definition is applied to its Jacobian and its eigenvalues. For the previous example the corresponding Jacobian is

$$\mathbf{J} = \begin{pmatrix} -1001 & 999 \\ 999 & -1001 \end{pmatrix} \tag{25.115}$$

and the corresponding eigenvalues are $\lambda_1 = -2 \, \mathrm{min}^{-1}$ and $\lambda_2 = -2000 \, \mathrm{min}^{-1}$, with a ratio of $\lambda_2/\lambda_1 = 10^3$.

McRae et al. (1982a) calculated the eigenvalues and characteristic reaction times for a typical tropospheric chemistry mechanism. The 24 eigenvalues of the corresponding Jacobian span 12 orders of magnitude (from 10^{-5} to $10^7 \, \mathrm{min}^{-1}$). The inverse of these eigenvalues $(1/|\lambda_i|)$ corresponds rougly to the characteristic reaction times of the reactive species. For example, the rapidly reacting OH radical with a lifetime of $10^{-4} \, \mathrm{min}$ corresponds to an eigenvalue of approximately $10^4 \, \mathrm{min}^{-1}$. The system of the corresponding 24 differential equations is extremely stiff and its integration a formidable task.

One approach commonly used in the integration of such chemical kinetics problems is the *pseudo-steady-state* approximation (PSSA) (see Chapter 3). For example, instead of solving a differential equation for short-lived species like O, OH, and NO_3, one calculates and solves the corresponding PSSA algebraic equations. For example, McRae et al. (1982a) estimated that nine species (O, RO, OH, RO_2, NO_3, RCO, HO_2, HNO_4, and N_2O_5) with characteristic lifetimes less than $0.1 \, \mathrm{min}$ in the environment of interest could be assumed to be in pseudo–steady state. This assumption resulted in the replacement of the 24 ODE system by one consisting of 15 ODEs and 9 algebraic equations. The ratio of the larger to the smaller eigenvalues and the corresponding stiffness were reduced to 10^6. The choice of the species in pseudo–steady state is difficult, as it is an approximation and therefore introduces numerical errors. As a rule of thumb, the longer the lifetime of a species set in pseudo–steady state the larger the numerical error introduced.

Even after the use of the PSSA the remaining problem is stiff and its integration cannot be performed efficiently with an explicit Euler method. Fully implicit, stiffly stable integration techniques have been developed and are routinely used for such problems.

Backward Differentiation Methods   If the rate of change of c_i on the RHS of (25.105) is approximated not by its value at t_n but with the value at t_{n+1}, then

$$c_i^{n+1} = c_i^n + (P_i^{n+1} - c_i^{n+1} L_i^{n+1}) \, \Delta t \qquad (25.116)$$

This algorithm is the *backward Euler* or *fully implicit Euler method*. Because this formula is implicit, a system of algebraic equations must be solved to calculate c_i^{n+1} from c_i^n for $i = 1, 2, \ldots, N$. This step is expensive because the Jacobian of the ODE system needs to be inverted, a process that requires on the order of N^3 operations. This inversion should in principle be repeated in each step.

A number of other backward differentiation methods and numerical routines for their use can be found in Gear (1971). These algorithms have been rewritten and incorporated in the package ODEPACK (Hindmarsh 1983).

Asymptotic Methods   Equation (25.105) can be rewritten in the form

$$\frac{dc_i}{dt} = P_i - \frac{c_i}{\tau_i} \qquad (25.117)$$

where $\tau_i = 1/L_i$ is the characteristic time for the approach of c_i to equilibrium. Dropping the subscript i, (25.117) can be approximated numerically by

$$\frac{c^{n+1} - c^n}{\Delta t} = \frac{P^{n+1} + P^n}{2} - \frac{c^{n+1} + c^n}{\tau^{n+1} + \tau^n}$$

Solving for c^{n+1}, we obtain

$$c^{n+1} = \frac{c^n(\tau^{n+1} + \tau^n - \Delta t) + 0.5 \, \Delta t(P^{n+1} + P^n)(\tau^{n+1} + \tau^n)}{(\tau^{n+1} + \tau^n + \Delta t)} \qquad (25.118)$$

Equation (25.118) cannot be used directly as c^{n+1} is a function of the unknown τ^{n+1} and P^{n+1}. However, if we assume that the solution varies slowly so that $\tau^{n+1} \simeq \tau^n$ and $P^{n+1} \simeq P^n$, then

$$c_*^{n+1} \simeq \frac{c^n(2\tau^n - \Delta t) + 2 \, \Delta t \, \tau^n P^n}{2\tau^n + \Delta t} \qquad (25.119)$$

and the concentrations c_*^{n+1} can now be calculated explicitly. To improve the accuracy of the approximation (25.119) and (25.118) are combined in a *predictor–corrector* algorithm. First, the predictor (25.119) is used. Then, τ_*^{n+1} and P_*^{n+1} are calculated using the c_*^{n+1}. Finally, the corrector (25.118) is applied using τ_*^{n+1} and P_*^{n+1} instead of τ_*^{n+1} and P_*^{n+1}.

This *asymptotic method* is stable and does not require the solution of algebraic systems (no expensive Jacobian inversions) for the advancement of the solution to the next timestep. As a result it is very fast but moderately accurate. The described above asymptotic method does not necessarily conserve mass. This weakness provides a convenient check on accuracy. Deviation of the mass balance corresponds roughly to the numerical error that has been introduced by the method. Algorithms (CHEMEQ) have been developed for the asymptotic method (Young and Boris 1977) and are routinely used in atmospheric chemistry models. In these algorithms, equations are often split into stiff ones (integrated by the asymptotic method) and nonstiff ones (integrated with an inexpensive explicit Euler method). Oran and Boris (1987) and Dabdub and Seinfeld (1995) present additional methods for the integration of stiff ODEs in chemical kinetics.

25.6.3 Diffusion

From a numerical point of view the diffusion operator that solves

$$\frac{\partial c}{\partial t} = \frac{\partial}{\partial x}\left(K_{xx}\frac{\partial c}{\partial x}\right) + \frac{\partial}{\partial y}\left(K_{yy}\frac{\partial c}{\partial y}\right) + \frac{\partial}{\partial z}\left(K_{zz}\frac{\partial c}{\partial z}\right) \tag{25.120}$$

is probably the simplest to deal with of those involved in the full atmospheric diffusion equation. Because of its physical nature, diffusion tends to smooth out gradients and lend overall stability to the physical processes. Most typical finite difference or finite element procedures are quite adequate for solving diffusion-type operator problems.

In Section 25.6.1 we discussed finite difference schemes for the solution of the one-dimensional diffusion equation. This explicit scheme of (25.93) is stable only if $\Delta t < (\Delta x)^2/2$. If K is not equal to unity, the corresponding stability criterion is $K\Delta t < (\Delta x)^2/2$. Therefore explicit schemes cannot be used efficiently because stability considerations dictate relatively small timesteps. Thus implicit methods are used for the diffusion aspect of a problem. In higher dimensions this requirement implies that splitting will almost certainly have to be used.

Implicit algorithms that can be used include the globally implicit algorithm of (25.95) or the popular *Crank–Nicholson algorithm* that can be derived as follows. For the one-dimensional problem with constant diffusivity K, we obtain

$$\frac{\partial c}{\partial t} = K\frac{\partial^2 c}{\partial x^2} \tag{25.121}$$

If we approximate the spatial derivative by

$$\frac{\partial^2 c}{\partial x^2} = \frac{1}{2\Delta x^2}\left[(c_{i-1}^n + c_{i-1}^{n+1}) - 2(c_i^n + c_i^{n+1}) + (c_{i+1}^n + c_{i+1}^{n+1})\right] \tag{25.122}$$

and $\partial c/\partial t = (c_i^{n+1} - c_i^n)/\Delta t$, then (25.121) becomes

$$-\frac{K\Delta t}{2\Delta x^2} c_{i-1}^{n+1} + \left(1 + \frac{K\Delta t}{\Delta x^2}\right) c_i^{n+1} - \frac{K\Delta t}{2\Delta x^2} c_{i+1}^{n+1}$$

$$= \frac{K\Delta t}{2\Delta x^2} c_{i-1}^n + \left(1 - \frac{K\Delta t}{\Delta x^2}\right) c_i^n + \frac{K\Delta t}{2\Delta x^2} c_{i+1}^n \qquad (25.123)$$

The algebraic system of equations resulting from the Crank–Nicholson scheme in (25.123) is tridiagonal and can therefore be solved efficiently with specialized routines.

25.6.4 Advection

Advection problems tend to be more difficult to solve numerically than are diffusion problems. The advection operator included in

$$\frac{\partial c}{\partial t} + u_x \frac{\partial c}{\partial x} + u_y \frac{\partial c}{\partial y} + u_z \frac{\partial c}{\partial z} = 0 \qquad (25.124)$$

does not smooth or damp out gradients or solutions but simply transports them about intact. Consider, for example, the one-dimensional advection equation with constant velocity u

$$\frac{\partial c}{\partial t} + u \frac{\partial c}{\partial x} = 0 \qquad (25.125)$$

and let us assume that it is used to describe the movement of a homogeneous plume with concentration $c(x, t)$ initially given by

$$c(x, 0) = 1 \qquad 1 \leq x \leq 2$$
$$c(x, 0) = 0 \qquad x < 1 \text{ and } 2 < x \qquad (25.126)$$

This is the classic square-wave problem, in which the wave moves along the x axis with velocity u without changing shape. The center of the wave will be centered at $x = ut + 1.5$ as shown in Figure 25.11.

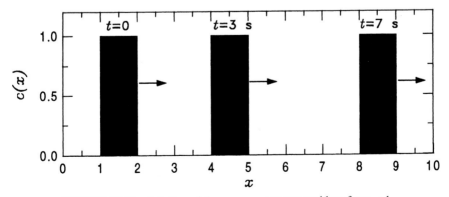

FIGURE 25.11   Solution of the square-wave test problem for $u = 1$.

Let us attempt to solve this problem numerically using finite differences. First, we need to discretize the time and space domains; let us select an equally spaced grid Δx and timesteps Δt. Then, using finite differences, both derivatives in (25.125) can be written as

$$\frac{\partial c}{\partial t} = \frac{c_i^{n+1} - c_i^n}{\Delta t} \qquad \frac{\partial c}{\partial x} = \frac{c_i^n - c_{i-1}^n}{\Delta x} \tag{25.127}$$

and (25.125) can be written, solving for c_i^{n+1}, as

$$c_i^{n+1} = c_i^n + \frac{u\,\Delta t}{\Delta x}\left(c_{i-1}^n - c_i^n\right) \tag{25.128}$$

The algorithm presented above uses the solution information for the grid point $i - 1$ to estimate the solution at i and is known as the *upwind*, or "one-sided", or "donor cell" algorithm. Consider the numerical solution of the square-wave problem assuming that the concentration initially satisfies (25.126) and using $\Delta x = 0.1\,\text{m}, \Delta t = 0.01\,\text{s}$, and $u = 1\,\text{m s}^{-1}$. The results of (25.128) after one step are shown in Figure 25.12. Note that for the 21st grid point corresponding to $x = 2.1\,\text{m}, c_{21}^1 = c_{21}^0 + 0.1(1 - 0) = 0.1$ and in the first step the concentration increases significantly. In reality, during this 0.01 s the wave

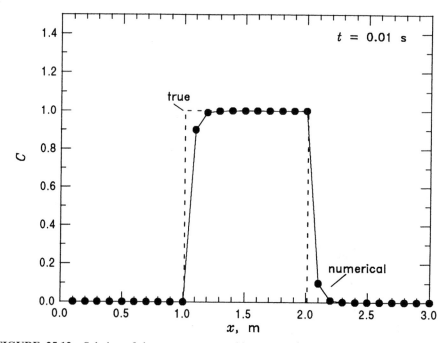

FIGURE 25.12   Solution of the square-wave problem using $\Delta x = 0.1\,\text{m}, \Delta t = 0.01\,\text{s}$, and $u = 1\,\text{m s}^{-1}$ with the upwind finite difference scheme. Numerical results and true solution after one timestep.

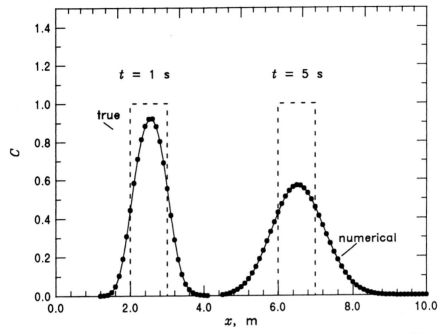

FIGURE 25.13 Solution of the square-wave problem using $\Delta x = 0.1\,\text{m}, \Delta t = 0.01\,\text{s}$, and $u = 1\,\text{m s}^{-1}$ with the upwind finite difference scheme. Numerical results and true solution for $t = 1\,\text{s}$ (100 steps) and $t = 5\,\text{s}$ (500 steps).

advances $1 \times 0.01 = 0.01\,\text{m}$ and occupies only 10% of the corresponding grid cell. Our grid structure is not sufficiently detailed to accurately describe this wave advancement, and the concentration is artificially spread over the whole grid cell. This numerical error is known as *numerical diffusion* because the solution "diffuses" artificially into the next grid cell. This diffusion takes place in every timestep of the upwind scheme and the error accumulates (Figure 25.13). After 100 steps ($t = 1\,\text{s}$) the predicted concentration peak is 10% less than the true one, and after 500 steps 50% less. Similar numerical diffusion is introduced by most advection algorithms. The upwind scheme is characterized by significant numerical diffusion and has therefore severe limitations.

If the spatial derivative is expressed using the values around the point of interest

$$\frac{\partial c}{\partial x} = \frac{c_{i+1}^n - c_{i-1}^n}{2\Delta x} \tag{25.129}$$

then after substitution into (25.125) the explicit algorithm

$$c_i^{n+1} = c_i^n + \frac{u\Delta t}{2\Delta x}\left(c_{i-1}^n - c_{i+1}^n\right) \tag{25.130}$$

is obtained. For this algorithm information from both upwind and downwind grid points is used to advance the solution for each grid point. Solution of the same square-wave problem using $\Delta x = 0.1\,\text{m}, \Delta t = 0.02\,\text{s}$, and $u = 1\,\text{m s}^{-1}$ results in the terrible solution shown in Figure 25.14. Oscillations that get larger with time are evident, and errors have

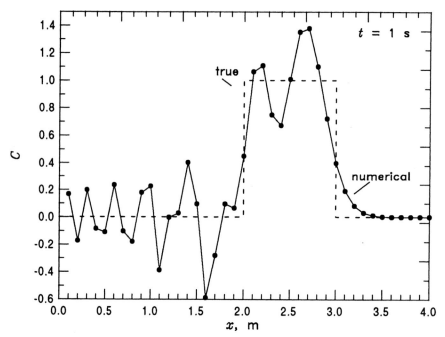

FIGURE 25.14 Solution of the square-wave problem using $\Delta x = 0.1\,\text{m}$, $\Delta t = 0.02\,\text{s}$, and $u = 1\,\text{m s}^{-1}$ with the explicit finite difference scheme. Numerical results and true solution for $t = 1\,\text{s}$ (50 steps).

even propagated backward to $x = 0$. This explicit algorithm is unstable for all timesteps Δt and the solution obtained has little resemblance to the true solution.

Stability is a problem associated with most explicit advection algorithms. The explicit algorithm of (25.130) has the worst behavior. The upwind scheme is stable if

$$\frac{u\Delta t}{\Delta x} \leq 1 \tag{25.131}$$

The preceding criterion is known as the *Courant limit*. For example, for our application of the upwind scheme discussed above, $\Delta t < 0.1\,\text{s}$ must be used to ensure stability of the solution.

A third problem plaguing numerical solution of advection problems can be seen if the upwind scheme is modified to include the updated information in the cell $i - 1$. Replacing c_{i-1}^n with c_{i-1}^{n+1} in (25.128), we get

$$c_i^{n+1} = c_i^n + \frac{u\Delta t}{\Delta x}\left(c_{i-1}^{n+1} - c_i^n\right) \tag{25.132}$$

This algorithm is still explicit, because, as applied sequentially for $i = 1, \ldots, N$, the value of c_{i-1}^{n+1} is known when we need to calculate c_i^{n+1}. Results from the application of the algorithm to the same square-wave problem are depicted in Figure 25.15. The numerical solution in this case not only diffuses but moves faster than the actual solution. Use of the

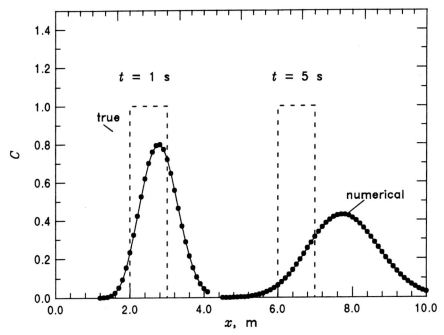

FIGURE 25.15 Solution of the square-wave problem using $\Delta x = 0.1\,\mathrm{m}, \Delta t = 0.02\,\mathrm{s}$, and $u = 1\ \mathrm{m\,s^{-1}}$ with the finite difference scheme of (25.132). Numerical results and true solution for $t = 1\,\mathrm{s}$ (100 steps) and $t = 5\,\mathrm{s}$ (500 steps).

updated information resulted in an artificial acceleration of the solution and the creation of a phase difference.

All the problems encountered during application of the algorithms presented above illustrate the difficulty of solution of the advection problem in atmospheric transport models. A number of techniques have been developed to treat advection accurately, including flux-corrected transport (FCT) algorithms (Boris and Book, 1973), spectral and finite element methods [for reviews, see Oran and Boris (1987), Rood (1987), and Dabdub and Seinfeld (1994). Bott (1989, 1992), Prather (1986), Yamartino (1992), Park and Liggett (1991), and others have developed schemes specifically for atmospheric transport models.

25.7 MODEL EVALUATION

Atmospheric chemical transport models are evaluated by comparison of their predictions against ambient measurements. A variety of statistical measures of the agreement or disagreement between predicted and observed values can be used. Although raw statistical analysis may not reveal the cause of the discrepancy, it can offer valuable insights about the nature of the mismatch. Three different classes of performance measures have been used for urban ozone models. These tests are heavily weighted toward the ability of the model to reproduce the peak ozone concentrations and include:

1. Analysis of predictions and observations paired in space and time. This is the most straightforward and demanding test.
2. Comparison of predicted and observed maximum concentrations.

3. Comparisons paired in space but not in time. These comparisons allow timing errors of model predictions.

Several numerical and graphical procedures are used routinely to quantify the performance of atmospheric models. Assume that there are n measurement stations each performing m measurements during the evaluation period. Let $PRED_{i,j}$ and $OBS_{i,j}$ be, respectively, the predicted and observed concentrations of a species for the station i during interval j. Tests that appear to be most helpful in determining the adequacy of a simulation include

1. The paired peak prediction accuracy
2. The unpaired peak prediction accuracy
3. The mean normalized bias (MNB) defined by

$$\text{MNB} = \frac{1}{nm} \sum_{i=1}^{n} \sum_{j=1}^{m} \frac{PRED_{i,j} - OBS_{i,j}}{OBS_{i,j}} \tag{25.133}$$

4. The mean bias (MB) defined by

$$\text{MB} = \frac{1}{nm} \sum_{i=1}^{n} \sum_{j=1}^{m} PRED_{i,j} - OBS_{i,j} \tag{25.134}$$

5. The mean absolute normalized gross error (MANGE) defined by

$$\text{MANGE} = \frac{1}{nm} \sum_{i=1}^{n} \sum_{j=1}^{m} \frac{|PRED_{i,j} - OBS_{i,j}|}{OBS_{i,j}} \tag{25.135}$$

6. The mean error (ME) defined by

$$\text{ME} = \frac{1}{nm} \sum_{i=1}^{n} \sum_{j=1}^{m} |PRED_{i,j} - OBS_{i,j}| \tag{25.136}$$

Diagnostic Analysis   A series of diagnostic tests are used to gain insights into the operation of complex atmospheric chemical transport models. Similar tests can often reveal model flaws or diagnose model problems. These tests for urban-scale models include use of zero emissions, zero initial conditions, zero boundary conditions, zero surface deposition, and mixing-height variations. Similar tests can be performed for regional- and global-scale models.

The zero emission test is intended to quantify the sensitivity of the model to emissions. The model should produce concentrations close to background or at least representative of upwind concentrations. Insensitivity to emissions raises questions about the utility of the simulated scenario for design of control strategies. The zero (or as close to zero as possible) initial and boundary condition tests reveal the effect of these often uncertain conditions on predicted concentrations. This initial condition effect should be small for the second or third day of an urban-scale simulation. The expected results of the zero deposition tests are increases of the primary pollutant concentrations downwind of their sources. However, concentrations of secondary pollutants may increase or decrease during this test because of nonlinear chemistry. The objective of the mixing height test is to quantify the sensitivity of predictions to the mixing height used in the simulations. Additional tests that are useful for atmospheric chemical transport models include overall mass balances and sensitivity analysis to model parameters and inputs.

PROBLEMS

25.1$_B$ SO_2 is emitted in an urban area at a rate of $3\,mg\,m^{-2}\,h^{-1}$. The hydroxyl radical concentration varies diurnally according to

$$[OH] = [OH]_{max} \sin[\pi(t-6)/12]$$

where t is the time in hours using midnight as a starting point. Assume that $[OH]$ is zero from $0 \le t \le 6$ and $18 \le t \le 24$. The mixing height changes at sunrise from $300\,m$ during the night to $1000\,m$ during the day. Assuming that, for this area, $[OH]_{max} = 10^7\,molecules\,cm^{-3}$, residence time $= 24\,h$, background $SO_2 = 0.2\,\mu g\,m^{-3}$, and background sulfate $= 1.0\,\mu g\,m^{-3}$, and using an Eulerian box model, calculate the following:

   a. The average and maximum SO_2 and sulfate concentrations.

   b. The time of occurrence of the maximum concentrations. Explain the results.

   c. Assume that during an air pollution episode the residence time increases to $36\,h$. Calculate the maximum SO_2 and sulfate concentration levels during the episode.

25.2$_B$ Repeat the calculations of Problem 25.1 using a Lagrangian box model. Discuss and explain the differences in predictions between the two modeling approaches.

25.3$_B$ Derive the atmospheric diffusion equations (25.82) to (25.84) in spherical coordinates. What happens when $\cos \phi = 0$? Assume constant surface pressure.

25.4$_D$ The following simplified reaction mechanism describes the main chemical processes taking place in an irradiated mixture of ethene/NO_x at a constant temperature of $298\,K$ and a pressure of $1\,atm$.

   (1) $NO_2 + h\nu \rightarrow NO + O$ $k_1 = 0.5$

   (2) $O + O_2 + M \rightarrow O_3 + M$ $k_2 = 2.183 \times 10^{-5}$

   (3) $O_3 + NO \rightarrow NO_2 + O_2$ $k_3 = 26.6$

   (4) $HO_2 + NO \rightarrow OH + NO_2$ $k_4 = 1.2 \times 10^4$

   (5) $OH + NO_2 \rightarrow HNO_3$ $k_5 = 1.6 \times 10^4$

   (6) $C_2H_4 + OH \rightarrow HOCH_2CH_2O_2$ $k_6 = 1.3 \times 10^4$

   (7) $HOCH_2CH_2O_2 + NO \rightarrow NO_2 + 0.72\,HCHO + 0.72\,CH_2OH$
$$+ 0.28\,HOCH_2CHO + 0.28\,HO_2 \quad k_7 = 1.1 \times 10^4$$

   (8) $CH_2OH + O_2 \rightarrow HCHO + OH$ $k_8 = 2.1 \times 10^3$

$$(9) \ C_2H_4 + O_3 \rightarrow HCHO + 0.4 \ H_2COO + 0.18 \ CO_2 + 0.42 \ CO$$
$$+ 0.12 \ H_2 + 0.42 \ H_2O + 0.12 \ HO_2 \quad k_9 = 2.6 \times 10^{-2}$$

$$(10) \ H_2COO + HO_2 \rightarrow HCOOH + H_2O \quad k_{10} = 0.034$$

The units of all the reaction constants are in the ppm-min unit system. The initial mixture contains C_2H_4, NO, NO_2, and air. Note that several additional reactions like photolysis of ozone, CO reactions, and carbonyl reactions have not been included in the preceding list to avoid unnecessary complications.

a. Simulate the system above without using the pseudo-steady-state assumption (PSSA) for a period of 12 h. Use the following values as initial conditions:

$$[NO_x]_0 = 0.5 \text{ ppm}, \quad [NO_2]_0/[NO_x]_0 = 0.25, \quad [C_2H_4]_0 = 3.0 \text{ ppm}$$

To account for the varying light intensity during the day assume that the reaction constant k_1 is

$$k_1 (\text{min}^{-1}) = 0.2053 \ t - 0.02053 \ t^2 \quad \text{for} \quad 0 \le t \le 10$$
$$k_1 = 0 \quad \text{for} \quad 10 \le t$$

where t is the time from the beginning of the simulation in hours. Therefore $t = 0$ corresponds to sunrise, $t = 10$ h corresponds to sunset, and the last 2 h cover the beginning of the night.

b. Prepare a list of the species participating in the chemical processes listed above and decide which will be treated as active and constant and which can be treated by using the PSSA in a typical polluted atmosphere. Explain your choices.

c. Use the PSSA to derive expressions for the concentrations of the steady-state species as function of the concentrations of the active and constant species.

d. Repeat part (a) using the PSSA. Compare the solution with (a). Discuss the computational requirements and timesteps for the two approaches.

e. Discuss the effect of the hydrocarbon on the maximum ozone value by repeating the preceding simulation for initial ethene values of 1, 2, and 4 ppm.

25.5$_D$ Integrate numerically the two differential equations of the stiff ODE (ordinary differential equation) using (a) the backward differentiation method given by (25.116) and (b) the asymptotic method given by (25.118) to (25.119). Compare your numerical solutions with the true solution of the system and discuss the necessary timesteps for accurate solution of the problem.

25.6$_D$ Solve the square-wave problem discussed in Section 25.6.4 using the Lax–Wendroff scheme given by

$$c_i^{n+1} = c_i^n + \left(\tfrac{1}{2} u \Delta t / \Delta x\right)\left(c_{i-1}^n - c_{i+1}^n\right) + \tfrac{1}{2}\left(u \Delta t / \Delta x\right)^2\left(c_{i-1}^n - 2c_i^n + c_{i+1}^n\right)$$

Compare your solution with the true solution and the results from other methods discussed in Section 25.6.4.

REFERENCES

Boris, J. P., and Book, D. L. (1973) Flux corrected transport I: SHASTA a fluid transport algorithm that works, *J. Comput. Phys.* **66**, 1–20.

Bott, A. (1989) A positive definite advection scheme obtained by non-linear renormalization of the advective fluxes, *Mon. Weather Rev.* **117**, 1006–1015.

Bott, A. (1992) Monotone flux limitation in the area-preserving flux-norm advection algorithm, *Mon. Weather Rev.* **120**, 2592–2602.

Dabdub, D., and Seinfeld, J. H. (1994) Numerical advective schemes used in air-quality models-sequential and parallel implementation, *Atmos. Environ.* **28**, 3369–3385.

Dabdub, D., and Seinfeld, J. H. (1995) Extrapolation techniques used in the solution of stiff ODEs associated with chemical kinetics of air quality models, *Atmos. Environ.* **29**, 403–410.

Gear, C. W. (1971) *Numerical Initial Value Problems in Ordinary Differential Equations*, Prentice-Hall, Englewood Cliffs, NJ.

Hindmarsh, A. C. (1983) ODEPACK, a systematic collection of ODE solvers, in *Numerical Methods for Scientific Computation*, R. S. Stepleman, ed., North Holland, New York, pp. 55–64.

Liu, M. K., and Seinfeld, J. H. (1975) On the validity of grid and trajectory models of urban air pollution, *Atmos. Environ.* **9**, 553–574.

McRae, G. J., Goodin, W. R., and Seinfeld, J. H. (1982a) Numerical solution of the atmospheric diffusion equation for chemically reacting flows, *J. Comput. Phys.* **45**, 1–42.

McRae, G. J., Goodin, W. R., and Seinfeld, J. H. (1982b) Development of a second generation mathematical model for urban air pollution, I. Model formulation, *Atmos. Environ.* **16**, 679–696.

Oran, E. S., and Boris, J. P. (1987) *Numerical Simulation of Reactive Flow*, Elsevier, New York.

Pai, P., and Tsang, T. H. (1993) On parallelization of time dependent three-dimensional transport equations in air pollution modeling, *Atmos. Environ.* **27A**, 2009–2015.

Pandis, S. N., Harley, R. H., Cass, G. R., and Seinfeld, J. H. (1992) Secondary organic aerosol formation and transport, *Atmos. Environ.* **26A**, 2269–2282.

Park, N. S., and Liggett, J. A. (1991) Application of a Taylor least–squares finite element to three dimensional advection–diffusion equation, *Int. J. Numer. Meth. Fluids* **13**, 769–773.

Peyret, R., and Taylor, T. D. (1983) *Computational Methods for Fluid Flow*, Springer-Verlag, New York.

Prather, M. J. (1986) Numerical advection by conservation of second-order moments, *J. Geophys. Res.* **91**, 6671–6681.

Reynolds, S. D., Roth, P. M., and Seinfeld, J. H. (1973) Mathematical modeling of photochemical air pollution. I. Formulation of the model, *Atmos. Environ.* **7**, 1033–1061.

Rood, R. B. (1987) Numerical advection algorithms and their role in atmospheric chemistry and transport models, *Rev. Geophys.* **25**, 71–100.

Venkatram, A. (1993) The parameterization of the vertical dispersion of a scalar in the atmospheric boundary layer, *Atmos. Environ.* **27A**, 1963–1966.

Yamartino, R. J. (1992) Non-negative, conserved scalar transport using grid-cell-centered, spectrally constrained Blackman cubics for applications on a variable thickness mesh, *Mon. Weather Rev.* **120**, 753–763.

Yanenko, N. N. (1971) *The Method of Fractional Steps*, Springer-Verlag, New York.

Young, T. R., and Boris, J. P. (1977) A numerical technique for solving stiff ordinary differential equations associated with chemical kinetics for reactive-flow problems, *J. Phys. Chem.* **81**, 2424–2427.

26 Statistical Models

The development of mathematical models based on fundamentals of atmospheric chemistry and physics has been discussed in Chapter 25. These models are essential tools in tracking emissions from many sources, their atmospheric transport and transformation, and finally their contribution to concentrations at a given location (receptor). A number of factors may often limit the application of these mathematical models including need for spatially resolved time-dependent emission inventories and meteorological fields. In some cases an alternative to the use of atmospheric chemical transport models is available. It is possible to attack the source contribution identification problem in reverse order, proceeding from concentrations at a receptor site backward to responsible emission sources. The corresponding tools, named *receptor models*, attempt to relate measured concentrations at a given site to their sources without reconstructing the dispersion patterns of the material. Sometimes, receptor and atmospheric chemical transport models are used synergistically. For example, receptor models can refine the input emission information used by atmospheric chemical transport models.

In the first part of this chapter a number of receptor modeling approaches will be discussed. These models are used for apportionment of the contributions of each source, identification of sources and their emission composition, and for determination of the spatial distribution of emission fluxes from a group of sources. In the second part, we will develop the tools needed to analyze the statistical character of air quality data.

26.1 RECEPTOR MODELING METHODS

Receptor models are based on measured mass concentrations and the use of appropriate mass balances. For example, assume that the total concentration of particulate iron measured at a site can be considered to be the sum of contributions from a number of independent sources

$$Fe_{total} = Fe_{soil} + Fe_{auto} + Fe_{coal} + \cdots \tag{26.1}$$

where Fe_{total} is the measured iron concentration, Fe_{soil} and Fe_{auto} are the concentrations contributed by soil emissions and automobiles, and so on. Let us start from a rather simple scenario illustrating the major concepts used in receptor modeling.

Atmospheric Chemistry and Physics: From Air Pollution to Climate Change, Second Edition, by John H. Seinfeld and Spyros N. Pandis.   Copyright © 2006 John Wiley & Sons, Inc.

Source Apportionment Assume that for a rural site the measured PM_{10} concentration is $32 \, \mu g \, m^{-3}$ containing $2.58 \, \mu g \, m^{-3}$ Si and $3.84 \, \mu g \, m^{-3}$ Fe. The two major sources contributing to the location's particulate concentration are a coal-fired power plant and soil-related dust. Analysis of the emissions of these sources indicates that the soil contains $200 \, mg(Si) \, g^{-1}$ (20% of the total emissions) and $32 \, mg(Fe) \, g^{-1}$ (3.2% of the total emissions), while the particles emitted by the power plant contain $10 \, mg(Si) \, g^{-1}$ (1%) and $150 \, mg(Fe) \, g^{-1}$ (15%). Neglecting Si and Fe contributions from other sources

$$Si_{total} = Si_{soil} + Si_{power} \tag{26.2}$$

$$Fe_{total} = Fe_{soil} + Fe_{power} \tag{26.3}$$

If S and P are the total aerosol contributions (in $\mu g \, m^3$) from dust and the power plant to the PM_{10} concentration in the receptor, then

$$PM_{10} = S + P + E \tag{26.4}$$

where E is the contribution from any additional sources. If the composition of the particles does not change during their transport from the sources to the receptor, then, using the initial composition of the emissions, we obtain

$$
\begin{aligned}
Si_{soil} &= 0.2 \, S \\
Fe_{soil} &= 0.032 \, S \\
Si_{power} &= 0.01 \, P \\
Fe_{power} &= 0.15 \, P
\end{aligned}
\tag{26.5}
$$

Substituting (26.5) into (26.2) and (26.3) yields

$$
\begin{aligned}
Si_{total} &= 0.2 \, S + 0.01 \, P \\
Fe_{total} &= 0.032 \, S + 0.15 \, P
\end{aligned}
\tag{26.6}
$$

The preceding is an algebraic system of two equations with two unknowns, the contributions of the two sources, S and P, to the receptor aerosol concentration. The solution of the system using the measured Si and Fe concentrations is $S = 12 \, \mu g \, m^{-3}$ and $P = 18 \, \mu g \, m^{-3}$. Using (26.4), we also find that $E = 2 \, \mu g \, m^{-3}$, and therefore the power plant is contributing 56.2%, the dust 37.5%, and the unknown sources 6.3% to the PM_{10} of the specific location. Recall that we have implicitly assumed that the unknown sources contribute negligible Si or Fe to the levels measured at the location.

This example describes a simple scenario but demonstrates the utility of receptor modeling. One can calculate the contribution of several sources to the atmospheric concentrations at a given location with knowledge of only source and receptor compositions. No information regarding meteorology, topography, location, and magnitude of sources is necessary.

A general mathematical framework can be developed for solution of problems similar to the example above. Suppose that for a given area there are m sources and n species.

If a_{ij} is the fraction of chemical species i in the particulate emissions from source j, then the composition of sources can be described by a matrix $\mathbf{A}$. For the conditions of the previous example

$$\mathbf{A} = \begin{pmatrix} a_{\text{Si,soil}} & a_{\text{Si,power}} \\ a_{\text{Fe,soil}} & a_{\text{Fe,power}} \end{pmatrix} = \begin{pmatrix} 0.2 & 0.01 \\ 0.032 & 0.15 \end{pmatrix}$$

Let c_i be the concentration (in $\mu g\,m^{-3}$) of element $i(i = 1,2,\ldots,n)$ at a specific site, and let f_{ij} be a fraction representing any modification to the source composition a_{ij} due to atmospheric processes (e.g., gravitational settling) that occurs between the source and the receptor points. Then $f_{ij}\,a_{ij}$ will be the fraction of species i in the particulate concentrations from source j at the receptor. If s_j is the total contribution (in $\mu g\,m^{-3}$) of the particles from source j to the particulate concentration at the receptor site, we can express the concentration of element i at the site as

$$c_i = \sum_{j=1}^{m} f_{ij}\,a_{ij}\,s_j, \qquad i = 1,2,\ldots,n \tag{26.7}$$

The corresponding equations for the example are

$$\begin{aligned} c_{\text{Fe}} &= f_{\text{Fe,soil}}\,a_{\text{Fe,soil}}\,s_{\text{soil}} + f_{\text{Fe,power}}\,a_{\text{Fe,power}}\,s_{\text{power}} \\ c_{\text{Si}} &= f_{\text{Si,soil}}\,a_{\text{Si,soil}}\,s_{\text{soil}} + f_{\text{Si,power}}\,a_{\text{Si,power}}\,s_{\text{power}} \end{aligned} \tag{26.8}$$

Usually f_{ij} is assumed equal to unity, thus assuming that the source signature a_{ij} is not modified by processes (reactions, removal, etc.) occurring during atmospheric transport between source and receptor. In this case we simply have

$$c_i = \sum_{j=1}^{m} a_{ij}s_j, \qquad i = 1,2,\ldots,n \tag{26.9}$$

Thus the concentration of each chemical element at a receptor site becomes a linear combination of the contributions of each source to the particulate matter at that site. Given the chemical composition of the ambient sample c_i and the source emission signature a_{ij}, (26.9) can then be solved to provide the source contributions s_j.

If there are k ambient aerosol samples, then let c_{ik} be the concentration of element i in the sample k. The source contributions will in general be different from sampling period to sampling period depending on wind direction, emission strength, and so on. Equation (26.9) can then be written in a more general form for the k samples as

$$c_{ik} = \sum_{j=1}^{m} a_{ij}s_{jk}, \qquad i = 1,2,\ldots,n \tag{26.10}$$

where s_{jk} is the concentration (in $\mu g\,m^{-3}$) of material from source j collected in the sample k. A number of approaches based on (26.10) have been used to develop our understanding of source–receptor relationships for nonreactive species in an airshed. These methods include the *chemical mass balance* (CMB) used for source apportionment, the *principal-component analysis* (PCA) used for source identification, and the *empirical orthogonal function* (EOF) method used for identification of the location and strengths of emission sources. A detailed review of all the variations of these basic methods is outside the scope of this book. For more information, the reader is referred to treatments by Watson (1984), Henry et al. (1984), Cooper and Watson (1980), Watson et al. (1981), Macias and Hopke (1981), Dattner and Hopke (1982), Pace (1986), Watson et al. (1989), Gordon (1980, 1988), Stevens and Pace (1984), Hopke (1985, 1991), and Javitz et al. (1988).

26.1.1 Chemical Mass Balance (CMB)

The CMB model combines the chemical and physical characteristics of particles or gases measured at the sources and the receptors to quantify the source contributions to the receptor (Winchester and Nifong 1971; Miller et al. 1972). CMB is a method for the solution of the set of equations (26.9) to determine the unknown s_j. The source profiles a_{ij}, that is, the fractional amount of the species in the emissions from each source type, and the receptor concentrations, with appropriate uncertainty estimates, serve as input to the CMB model. We start by analyzing the case where one particulate sample is available. The first assumption of CMB is that *all* sources contributing to the measured concentrations c_i in the receptor have been identified. Each measured concentration c_i can then be expressed as the sum of the true value $\tilde{c}_i$, and a random error e_i:

$$c_i = \tilde{c}_i + e_i, \qquad i = 1, 2, \ldots, n \qquad (26.11)$$

It is assumed in CMB that the measurement errors e_i are random, uncorrelated, and normally distributed about a mean value of zero. These errors can be characterized statistically by the standard deviation σ_i of their normal distributions.

For an initial guess of source contributions s_j, the predicted concentrations p_i for all elements are given by

$$p_i = \sum_{j=1}^{m} a_{ij} s_j \qquad (26.12)$$

If c_i are the corresponding measured elemental concentrations, we would like to minimize the "distance" between the measurements c_i and the predictions p_i. This distance can be expressed by the sum

$$\sum_{i=1}^{n} (c_i - p_i)^2$$

Because of the measurement uncertainty, no choice of s_j values will result in perfect agreement between predictions and observations. The measurement uncertainty depends on the element, and to account for these different degrees of uncertainties, $1/\sigma_i^2$ are used as weighting factors. Summarizing, one needs to minimize

$$\xi^2 = \sum_{i=1}^{n} \left(\frac{1}{\sigma_i^2} (c_i - p_i)^2 \right) \qquad (26.13)$$

by choosing appropriate values of the contributions s_j. Note that by using the weighting factors $1/\sigma_i^2$, elements with large uncertainties contribute less to the ξ^2 function compared to elements with smaller uncertainties. Combining (26.12) and (26.13), we obtain

$$\xi^2 = \sum_{i=1}^{n} \left[\frac{1}{\sigma_i^2} \left(c_i - \sum_{j=1}^{m} a_{ij} s_j \right)^2 \right] \qquad (26.14)$$

where n is the number of species and m is the number of sources. The solution approach is to minimize the value of ξ^2 with respect to each of the m coefficients s_j, yielding a set of m simultaneous equations with m unknowns $(s_1, s_2, \ldots, s_m)$. This is the common multiple regression analysis problem. The solution is the vector $\mathbf{s}$ of source contributions given by

$$\mathbf{s} = [\mathbf{A}^T \mathbf{W} \mathbf{A}]^{-1} \mathbf{A}^T \mathbf{W} \mathbf{c} \qquad (26.15)$$

where $\mathbf{A}$ is the $n \times m$ source matrix with the source compositions a_{ij}, $\mathbf{W}$ is the $n \times n$ diagonal matrix with elements of the weighting factors, $w_{ii} = 1/\sigma_i^2$, $\mathbf{A}^T$ is the $m \times n$ transpose[1] of $\mathbf{A}$, $\mathbf{c}$ is the vector with the measurements of the n elements, and $\mathbf{s}$ is the vector with the m source contributions. Note that $[\mathbf{A}^T \mathbf{W} \mathbf{A}]$ is an $m \times m$ square matrix so it can be inverted.

The solution of the receptor problem using (26.15) considers uncertainties in the measurements c_i but neglects the inherent uncertainty in the source contributions a_{ij}. Let us denote by $\sigma_{a_{ij}}$ the standard deviation of a determination of the fraction a_{ij} of element i in the emissions of source j. The solution can then be calculated by an expression analogous to (26.15) (Watson 1979; Hopke 1985):

$$\mathbf{s} = [\mathbf{A}^T \mathbf{V} \mathbf{A}]^{-1} \mathbf{A}^T \mathbf{V} \mathbf{c} \qquad (26.16)$$

Here $\mathbf{V}$ is the diagonal matrix with elements

$$v_{ii}^{-1} = \sigma_i^2 + \sum_{j=1}^{m} \sigma_{a_{ij}}^2 s_j^2 \qquad (26.17)$$

The unknown source contributions s_j are included in the elements of the $\mathbf{V}$ matrix, and therefore an iterative solution of (26.16) is necessary. The first step is to assume that $\sigma_{a_{ij}} = 0$,

---

[1]The transpose of an $n \times m$ matrix $\mathbf{A}$ denoted by $\mathbf{A}^T$ is simply the $m \times n$ matrix obtained by interchanging all the rows and columns.

solve (26.16) directly, and calculate the first approximation of s_j. Then v_{ii} can be calculated from (26.17) and a second approximation is found. If this approach converges, the solution is found. This approach using (26.16) and (26.17) is known as the *effective variance* method.

The major assumptions used by the CMB model are

1. Compositions of source emissions are constant.
2. Species included are not reactive.
3. All sources contributing significantly to the receptor have been included in the calculations.
4. There is no relationship among the source uncertainties.
5. The number of sources is less than or equal to the number of species.
6. Measurement uncertainties are random, uncorrelated, and normally distributed.

These assumptions are fairly restrictive and may be difficult to satisfy for most CMB applications. When they are not satisfied, the CMB predictions may be unrealistic (e.g., negative contributions) or may include significant uncertainties.

The application of CMB to an area poses a number of difficulties in addition to the assumptions of the method. Let us assume that a particulate sample has been collected in the area of interest and its elemental composition has been determined. The first issue that one needs to address is which sources should be included in the model. If an emission inventory exists for the region, it can be used to determine the major sources. The second issue is which source profiles should be used. Profiles used by studies in other areas may be applicable to only that specific source. For example, the emission fingerprint of a power plant in Ohio may not be representative of a power plant in Texas. Local sources of road and soil dust are usually different from location to location. To complicate things even further, emission profiles often change with time. For example, motor vehicle emission composition has changed dramatically in the last 40 years with the introduction of new fuels (unleaded gasoline), new engines, and control technologies. Uncertainties or errors in the CMB results can be reduced noticeably by obtaining source profile measurements that correspond to the period of the ambient measurements (Glover et al. 1991). It is clearly essential for the CMB application to know the area that is to be modeled (Hopke 1985).

When multiple samples are available CMB should be applied to each sample separately and then the results can be averaged (Hopke 1985). This approach, even if more time-consuming, is more accurate than the CMB application to the averaged measurements. Information is generally lost during averaging of sample composition data and cannot be recovered later by CMB.

CMB Application to Central California PM   Chow et al. (1992) apportioned source contributions to aerosol concentrations in the San Joaquin Valley of California. The source profiles used for CMB application are shown in Table 26.1. The standard deviations $\sigma_{a_{ij}}$ of the profiles (three or more samples were taken) are also included. To account for secondary aerosol components in the CMB calculations, ammonium sulfate, ammonium nitrate, sodium nitrate, and organic carbon were expressed as secondary source profiles using the stoichiometry of each compound. The average elemental concentrations observed at one of the receptors—Fresno, California, in 1988–1989—are shown in Table 26.2. The ambient concentrations of some species (e.g., Ga, As, Y, Mo, Ag) included in the source profiles were below the detection limits. These species

TABLE 26.1 Source Profiles (Percent of Mass Emitted) for Central California

Chemical Species	Paved Road Dust	Vegetative Burning	Primary Crude Oil	Motor Vehicle	Limestone
NO_3^-	0 ± 0.47	0.462 ± 0.123	0 ± 0.002	0 ± 0.001	0 ± 0.001
SO_4^{2-}	0.547 ± 1.17	1.423 ± 0.423	20.32 ± 4.24	3.11 ± 3.55	3.06 ± 0.3
NH_4^+	0 ± 0.008	0.0852 ± 0.057	0.0076 ± 0.005	0 ± 0.001	0 ± 0.001
Na^+	0.181 ± 0.055	0.143 ± 0.052	0.762 ± 0.399	0 ± 0.001	0 ± 0.001
EC	2.69 ± 1.44	15.89 ± 5.80	0 ± 0.072	54.15 ± 19.78	0 ± 0.001
OC	19.5 ± 4.67	44.60 ± 7.94	0.0894 ± 0.118	49.81 ± 24.15	0 ± 0.001
Al	9.34 ± 1.11	0.0019 ± 0.027	0 ± 0.009	0.077 ± 0.051	2.11 ± 0.21
Si	23.2 ± 2.62	0 ± 0.015	0.011 ± 0.016	0.957 ± 1.39	6.5 ± 0.65
P	0.304 ± 0.05	0 ± 0.022	0 ± 0.17	0.057 ± 0.02	0 ± 0.001
S	0.520 ± 0.17	0.521 ± 0.176	5.45 ± 0.39	1.037 ± 1.182	1.02 ± 0.1
Cl	0.163 ± 0.031	1.908 ± 0.64	0.024 ± 0.021	0.029 ± 0.02	0.46 ± 0.05
K	1.95 ± 0.28	3.993 ± 1.24	0.044 ± 0.054	0.008 ± 0.008	0.16 ± 0.04
Ca	2.98 ± 0.43	0.0659 ± 0.056	0.062 ± 0.005	0.072 ± 0.079	29.52 ± 2.95
Ti	0.499 ± 0.067	0.0009 ± 0.016	0.012 ± 0.002	0.001 ± 0.003	0.08 ± 0.04
V	0.0311 ± 0.008	0.0005 ± 0.007	0.823 ± 0.058	0.001 ± 0.002	0 ± 0.1
Cr	0.0299 ± 0.003	0 ± 0.0016	0.007 ± 0.025	0 ± 0.002	0 ± 0.01
Mn	0.106 ± 0.016	0.0007 ± 0.001	0.0056 ± 0.001	0.028 ± 0.024	0.05 ± 0.03
Fe	5.41 ± 0.88	0.0006 ± 0.001	0.2134 ± 0.022	0.001 ± 0.005	1.04 ± 0.1
Co	0.0059 ± 0.076	0.0001 ± 0.001	0.0185 ± 0.002	0 ± 0.001	0 ± 0.001
Ni	0.0111 ± 0.001	0.0001 ± 0.001	0.789 ± 0.093	0 ± 0.002	0 ± 0.1
Cu	0.02 ± 0.002	0.0001 ± 0.001	0.0009 ± 0.003	0.005 ± 0.003	0.02 ± 0.01
Zn	0.172 ± 0.026	0.0866 ± 0.036	·0.260 ± 0.034	0.053 ± 0.028	0.1 ± 0.01
Ga	0.0003 ± 0.006	0 ± 0.0021	0.0132 ± 0.002	0.002 ± 0.002	0 ± 0.001
As	0.0014 ± 0.042	0.0002 ± 0.002	0.0006 ± 0.001	0.004 ± 0.012	0 ± 0.001
Se	0.0001 ± 0.002	0.0004 ± 0.001	0.0114 ± 0.002	0 ± 0.002	0 ± 0.001
Br	0.0095 ± 0.001	0.0096 ± 0.002	0.0003 ± 0.0002	0.264 ± 0.152	0.03 ± 0.01
Sr	0.0794 ± 0.006	0.0007 ± 0.001	0.0015 ± 0.0003	0 ± 0.003	0 ± 0.001
Y	0.0025 ± 0.004	0.0001 ± 0.001	0.0008 ± 0.0003	0 ± 0.004	0 ± 0.001
Zr	0.0091 ± 0.002	0 ± 0.0019	0.0006 ± 0.0004	0 ± 0.019	0 ± 0.001

Species	Marine	$(NH_4)_2SO_4$	NH_4NO_3	Secondary OC	$NaNO_3$
Mo	0.0004 ± 0.006	0 ± 0.0033	0.0168 ± 0.002	0 ± 0.012	0 ± 0.001
Ag	0 ± 0.016	0.0003 ± 0.007	0.0002 ± 0.002	0 ± 0.016	0 ± 0.001
Cd	0.0015 ± 0.017	0.0007 ± 0.008	0.0006 ± 0.002	0 ± 0.02	0 ± 0.001
In	0.0030 ± 0.02	0.0001 ± 0.009	0.0009 ± 0.002	0 ± 0.026	0 ± 0.001
Sn	0.0037 ± 0.027	0 ± 0.012	0.0007 ± 0.003	0 ± 0.031	0 ± 0.001
Sb	0.0054 ± 0.03	0.0022 ± 0.014	0.0006 ± 0.003	0 ± 0.069	0 ± 0.001
Ba	0.064 ± 0.103	0.0095 ± 0.05	0.0013 ± 0.011	0 ± 0.129	0 ± 0.001
La	0.0142 ± 0.117	0.0016 ± 0.056	0.0041 ± 0.013	0 ± 0.236	0 ± 0.001
Hg	0.0015 ± 0.008	0 ± 0.0037	0 ± 0.0009	0 ± 0.002	0 ± 0.001
Pb	0.265 ± 0.032	0.004 ± 0.003	0 ± 0.0013	0.373 ± 0.207	0.27 ± 0.03
NO_3^-	0 ± 0.001	0	77.5	0	72.9
SO_4^{2-}	10.0 ± 4.0	72.7	0	0	0
NH_4^+	0 ± 0.001	27.3	22.5	0	0
Na^+	40.0 ± 4.0	0	0	0	27.1
EC	0 ± 0.001	0	0	0	0
OC	0 ± 0.001	0	0	100	0
S	3.3 ± 1.3	0	0	0	0
Cl	40.0 ± 10.0	0	0	0	0
K	1.4 ± 0.2	0	0	0	0
Ca	1.4 ± 0.2	0	0	0	0
Br	0.2 ± 0.05	0	0	0	0

Source: Chow et al. (1992).

TABLE 26.2   Annual (1988–1989) Aerosol Composition in Fresno, California

Chemical Species	$PM_{2.5}$, $\mu g\,m^{-3}$	PM_{10}, $\mu g\,m^{-3}$
NO_3^-	9.43 ± 11.43	10.26 ± 10.52
SO_4^{2-}	2.75 ± 1.32	3.20 ± 1.51
NH_4^+	4.04 ± 3.89	4.06 ± 3.85
EC	6.27 ± 5.68	6.73 ± 5.68
OC	8.05 ± 5.31	12.89 ± 7.66
Al	0.15 ± 0.18	2.94 ± 2.63
Si	0.38 ± 0.46	7.49 ± 6.31
P	0.013 ± 0.012	0.072 ± 0.067
S	1.12 ± 0.54	1.27 ± 0.76
Cl	0.17 ± 0.22	0.34 ± 0.37
K	0.28 ± 0.18	0.85 ± 0.53
Ca	0.072 ± 0.068	0.85 ± 0.64
Ti	0.016 ± 0.015	0.14 ± 0.12
V	0.0034 ± 0.0019	0.01 ± 0.008
Cr	0.0016 ± 0.002	0.0081 ± 0.0061
Mn	0.0073 ± 0.0058	0.035 ± 0.029
Fe	0.17 ± 0.16	1.48 ± 1.22
Ni	0.0023 ± 0.0024	0.0051 ± 0.0029
Cu	0.069 ± 0.064	0.0077 ± 0.095
Zn	0.069 ± 0.052	0.087 ± 0.066
Se	0.0016 ± 0.003	0.0019 ± 0.004
Br	0.017 ± 0.011	0.017 ± 0.008
Sr	0.0007 ± 0.0006	0.0043 ± 0.0035
Zr	0.0011 ± 0.0028	0.0031 ± 0.0029
Ba	0.013 ± 0.014	0.044 ± 0.034
Pb	0.051 ± 0.034	0.067 ± 0.034

Source: Chow et al. (1993).

cannot be used for source apportionment in this specific case. Results of the source apportionment using the CMB method (using CMB for each sample and then averaging the results) are shown in Table 26.3. The major contributors to the annual average PM_{10} concentrations that exceeded $50\,\mu g\,m^{-3}$ were primary geologic material and ammonium nitrate. For the $PM_{2.5}$, secondary NH_4NO_3 and $(NH_4)_2SO_4$, together with primary motor vehicle emissions and vegetative burning, were the major contributors.

CMB Evaluation   A method often used to evaluate the CMB method is use of only selected measurement elements for estimation of source contributions and then use of the remainder of the measurement elements and predictions as a test of the analysis. For example, Kowalczyk et al. (1982) used the CMB and nine elements (Na, V, Pb, Zn, Ca, Al, Fe, Mn, As) to calculate contributions of seven sources to the Washington, DC, aerosol. Each of the selected elements was characteristic of a source: Na for seasalt, V for fuel oil, Pb for motor vehicles, Zn for refuse incineration, Ca for limestone, Al and Fe for coal and soil, and As for coal. The authors used 130 samples from a network of 10 stations. Cr, Ni, Cu, and Se were significantly underestimated, but the concentrations of the remaining elements were successfully reproduced by CMB. Kowalczyk et al. (1982) repeated the

TABLE 26.3   Estimated Annual Average Source Contributions ($\mu g\, m^{-3}$) to PM$_{10}$ and PM$_{2.5}$ in Fresno, California, Based on CMB

Source	PM$_{10}$	PM$_{2.5}$
Geologic	31.78	2.26
Motor vehicle	6.80	9.24
Vegetative burning	5.10	5.92
Primary crude oil	0.29	0.25
$(NH_4)_2SO_4$—secondary	3.58	3.48
NH_4NO_3—secondary	10.39	12.35
Seasalt ($NaCl$–$NaNO_3$)	0.96	0.45
OC—secondary	0.07	0.36
Calculated mass	58.97	34.34
Measured mass	71.49	49.30

Source: Chow et al. (1992).

exercise using 9–30 marker elements and found little difference in the results as the key elements (Pb, Na, and V) were included. They also observed that including some elements (namely, Br and Ba) as markers gave erroneous results for several other elements.

The absolute accuracy of the CMB cannot be tested easily, because the true results are unknown. However, artificial data sets can be created by assuming a realistic distribution of sources, source strengths, and meteorology, simulating the scenario with a deterministic transport model (see Chapter 25), and using CMB to apportion the source contributions to the modeled concentrations. Gerlach et al. (1983) reported the results of such a test using a typical city plan and 13 known sources. The results of the CMB application indicated that the contributions of nine of the sources were accurately predicted (errors less than 20%) while errors as much as a factor of 4 were found for the remaining four sources. The contributions of the six most important sources were accurately predicted by CMB and the errors were associated with sources of secondary importance.

CMB Resolution   A final issue that may complicate the application of the CMB on ambient data sets is existence of two sources with similar fingerprints or, more generally, a source whose profile is a linear combination of other source profiles. This is called the *collinearity problem*. If this is the case then the matrix $[\mathbf{A}^T \mathbf{W} \mathbf{A}]$ used in (26.15) has two columns that are almost similar, or a linear combination of several others. This matrix from a mathematical point of view is close to singular and the result of its inversion is extremely sensitive to small errors. Often, if this is the case, the results of CMB are large positive and negative source contributions. The simplest solution to this problem is identification of the "offending" sources and elimination of one of them. Physically, because the sources are too similar, it is difficult for CMB to quantify the contribution of each. Thus there are limits to how far source contributions can be resolved even with almost perfect information; only significantly different sources can be treated by CMB. Similar sources have to be combined into a lumped source. Henry (1983) and Hopke (1985) have proposed algorithms that can be used for the a priori identification of estimable sources and the estimable source combinations that can be determined for a given source matrix.

We should note, once more, that during the derivation of (26.15) and (26.16) we have assumed that the atmospheric transformation terms f_{ij} are equal to unity [see also (26.17)].

Therefore these equations should not be applied to species that are produced or consumed (e.g., sulfate) during transport from source to receptor. Gravitational settling is often assumed not to modify a_{ij} (the elemental fractions of the source emissions) to a first approximation, even if it changes the net concentrations of these elements. This assumption is equivalent to assuming that all elements have the same size distribution. Application of (26.15) to gaseous pollutants that react in the atmosphere is generally not appropriate.

26.1.2 Factor Analysis

If the nature of the major sources influencing a particular receptor is unknown, statistical factor analysis methods can be combined with ambient measurements to estimate the source composition. Assuming that for a particular location several ambient particulate samples are collected and analyzed for several elements, the resulting data will probably include information about the fingerprints of the sources affecting the location. Principal-component analysis (PCA) is one of the factor analysis methods used to unravel the hidden source information from a rich ambient measurement data set. Factor analysis models are mathematically complex, and their results are often difficult to interpret.

Let us consider first a simple example given by Hopke (1985). A specific location, without us knowing it, is heavily influenced by two sources—automobiles and a coal-fired power plant. All samples are analyzed for aluminum (Al), lead (Pb), and bromine (Br). We assume that Al is emitted only by the power plant, while lead and bromine (mass ratio $3:1$) are emitted only by automobiles. Our samples, depending on the prevailing wind direction, traffic intensity, and so on, will have different Al, Pb, and Br concentrations. A three-dimensional plot of these concentrations (Figure 26.1) reveals little information about the underlying sources. However, all the data points are actually located on the same plane defined by the Al axis and the line $Br = \frac{1}{3} Pb$ (Figure 26.2). If the z-Al plane is rotated, all the measurements can be plotted on a two-dimensional graph, with axes defined by Al and z (Figure 26.3). Note that the three-dimensional data set has collapsed to

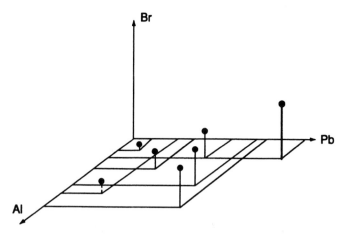

FIGURE 26.1 Measured aerosol composition of three elements in seven samples for a site influenced by automobiles (emitting Pb and Br) and a coal-fired power plant (emitting Al).

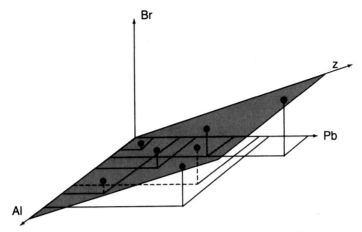

FIGURE 26.2 Plane passing through the data points of Figure 26.1. The z axis is defined by $Al = 0$ and $Br = \frac{1}{3} Pb$.

a two-dimensional data set and the two axes correspond to the composition of the sources. The Al and Br $= \frac{1}{3}$ Pb axes are the principal factors influencing the aerosol concentration at the receptor.

If there are more than three aerosol species, then we need to work with higher than three-dimensional spaces, and locating hyperplanes passing through (or close to) all the data points becomes a complicated exercise. The first step in the procedure is, of course, the collection of the data set, say, k samples of n aerosol species. These species measurements are then analyzed for the calculation of the *correlation coefficients*. If Al and Pb were two of the species measured, one would have available the values of $(c_{Al,1}, c_{Al,2}, \ldots, c_{Al,k})$, and $(c_{Pb,1}, c_{Pb,2}, \ldots, c_{Pb,k})$, where $c_{Al,i}$ and $c_{Pb,i}$ are the Al and Pb concentrations in the ith sample. The mean Al and Pb values will be

$$\bar{c}_{Al} = \frac{1}{k} \sum_{i=1}^{k} c_{Al,i} \qquad \bar{c}_{Pb} = \frac{1}{k} \sum_{i=1}^{k} c_{Pb,i} \qquad (26.18)$$

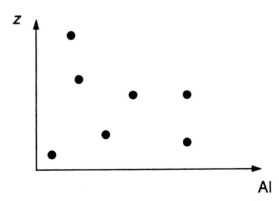

FIGURE 26.3 Two-dimensional depiction of the data shown in Figures 26.1 and 26.2.

The correlation coefficient around the mean between lead and aluminum $r_{Pb,Al}$ is then defined as

$$r_{Pb,Al} = \frac{\sum_{i=1}^{k}(c_{Al,i} - \bar{c}_{Al})(c_{Pb,i} - \bar{c}_{Pb})}{\left[\sum_{i=1}^{k}(c_{Al,i} - \bar{c}_{Al})^2\right]^{1/2}\left[\sum_{i=1}^{k}(c_{Pb,i} - \bar{c}_{Pb})^2\right]^{1/2}} \qquad (26.19)$$

and is a measure of the interrelationship between Pb and Al concentrations. If σ_{Al} and σ_{Pb} are the standard deviations of the corresponding samples, the correlation coefficient is given by

$$r_{Pb,Al} = \frac{1}{k-1}\sum_{i=1}^{k}\frac{(c_{Al,i} - \bar{c}_{Al})}{\sigma_{Al}}\frac{(c_{Pb,i} - \bar{c}_{Pb})}{\sigma_{Pb}} \qquad (26.20)$$

If the two are completely unrelated, $r_{Pb,Al} = 0$. If the two variables are strongly related to each other (positively or negatively), they have a high correlation coefficient (positive or negative). It is important to stress at this point that high correlation coefficients do not necessarily imply a cause-and-effect relationship. Variables can be related to each other indirectly through a common cause. For example, let us assume that for a given receptor Pb and Al concentrations are highly correlated; high Al values are always accompanied by high Pb values and vice versa. One would be tempted to conclude that they have a common source. However, the same correlation can also be a result of the fact that the lead source (a major traffic artery) is next to the Al source (a coal-fired power plant), and depending on the wind direction, their concentrations in the receptor vary proportionally to each other.

Correlation coefficients can be calculated for each pair of elements and a correlation matrix $(n \times n)$ can be constructed. Note that because

$$r_{Pb,Al} = r_{Al,Pb} \qquad r_{Pb,Pb} = r_{Al,Al} = 1 \qquad (26.21)$$

the matrix will be symmetric, and the elements in the diagonal will be equal to unity. A correlation matrix for nine elements measured at Whiteface Mountain, New York, is shown in Table 26.4. The correlation matrix $\mathbf{C}$ is the basis of PCA. Let $\lambda_1, \lambda_2, \ldots, \lambda_n$ be its

TABLE 26.4   Correlation Matrix for Elements Measured in Whiteface Mountain, New York

	Na	K	Sc	Mn	Fe	Zn	As	Br	Sb
Na	1	0.48	0.03	0.22	0.21	0.14	0.13	0.04	0.08
K	0.48	1	0.46	0.57	0.61	0.03	0.30	0.28	0.15
Sc	0.03	0.46	1	0.82	0.72	−0.26	0.64	−0.06	0.07
Mn	0.22	0.57	0.82	1	0.88	−0.08	0.65	−0.03	0.13
Fe	0.21	0.61	0.72	0.88	1	−0.03	0.46	0.08	0.09
Zn	0.14	0.03	−0.26	−0.08	−0.03	1	−0.12	0.04	0.62
As	0.13	0.30	0.64	0.65	0.46	−0.12	1	−0.18	0.07
Br	0.04	0.28	−0.06	0.03	0.08	0.04	−0.18	1	0.27
Sb	0.08	0.15	0.07	0.13	0.09	0.62	0.07	0.27	1

Source: Parekh and Husain (1981).

TABLE 26.5   Normalized Eigenvectors for Whiteface Mountain, New York

	Eigenvector			
Element	1	2	3	4
Na	0.174	0.262	0.403	0.696
K	0.381	0.222	0.397	0.083
Sc	0.455	−0.182	−0.148	−0.193
Mn	0.499	0.044	−0.104	−0.035
Fe	0.468	0.014	0.027	−0.095
Zn	−0.05	0.591	−0.394	0.188
As	0.375	−0.158	−0.309	0.096
Br	0.024	0.357	0.504	−0.617
Sb	0.081	0.588	−0.377	−0.191

Source: Parekh and Husain (1981).

eigenvalues and $\mathbf{e}_1, \mathbf{e}_2, \ldots, \mathbf{e}_n$ the corresponding eigenvectors. Each eigenvector corresponds to a particular aerosol composition influencing the receptor. Each aerosol sample collected at the receptor can be expressed as a linear combination of these eigenvectors. The corresponding eigenvalues can be viewed as a measure of the importance of each eigenvector for the receptor. For example, for the correlation matrix for Whiteface Mountain, the eigenvalues are 3.6, 1.83, 1.16, 1.01, 0.54, 0.32, 0.31, 0.16, and 0.078. Eigenvectors corresponding to eigenvalues close to zero are neglected because they are usually artifacts due to either the numerical or the sampling and analysis procedures. The decision of which eigenvector should be retained is not straightforward. In practice, eigenvectors corresponding to eigenvalues larger than one are usually retained. Following this rule of thumb, the four corresponding eigenvectors are shown in Table 26.5.

At this stage one is left with a set of vectors corresponding to the specific elemental combinations influencing the receptor (Table 26.5). These factors often have little physical significance. For example, the factors in Table 26.5 are difficult to interpret and so a further transformation is performed. The remaining eigenvectors are rotated in space in such a way as to maximize the number of values that are close to unity (Hopke 1985). This rotation, known as "varimax rotation" (Kaiser 1959), is a rather controversial step because it is an arbitrary procedure. In any case, the final eigenvectors are usually interpreted as fingerprints of emission sources based on the chemical species with which they are highly correlated. The rotated eigenvectors are shown in Table 26.6, and their physical significance can be discussed even if there are only a few elements available. For example, factor 1 containing elements coming from crustal sources is probably soil and other crustal materials. Factor 2, characterized by Zn and Sb, apparently results from refuse incineration. Factor 3, containing Na and K, could be marine aerosol with road salt. Bromine on factor 4 indicates a motor vehicle component. One should note that the above solution is not unique. Hopke (1985) reanalyzed the same data using a different rotation and concluded that there were five factors affecting the location with composition different from the one in Table 26.6.

The major assumptions of PCA modeling are

1. The composition of emission sources is constant.
2. Chemical species used in PCA do not interact with each other, and their concentrations are linearly additive.

TABLE 26.6   Rotated Factors for Whiteface Mountain, New York

Element	Factor			
	1	2	3	4
Na	0.047	0.086	0.943	−0.071
K	0.478	0.020	0.628	0.436
Sc	0.901	−0.128	−0.015	0.104
Mn	0.913	0.006	0.210	0.103
Fe	0.815	−0.006	0.251	0.251
Zn	−0.152	0.897	0.144	−0.068
As	0.803	0.021	0.038	−0.274
Br	0.134	0.126	0.141	0.890
Sb	0.756	0.887	−0.039	0.227

Source: Parekh and Husain (1981).

3. Measurement errors are random and uncorrelated.

4. The variability of the concentrations from sample to sample is dominated by changes in source contributions and not by the measurement uncertainty or changes in the source composition.

5. The effect of processes that affect all sources equally (e.g., atmospheric dispersion) is much smaller than the effect of processes that influence individual sources (e.g., wind direction, emission rate changes).

6. There are many more samples than source types for a statistically meaningful calculation.

7. Eigenvector rotations (if used) are physically meaningful.

Examples of PCA application can be found in Henry and Hidy (1979, 1981), Wolff and Korsog (1985), Cheng et al. (1988), Henry and Kim (1989), Koutrakis and Spengler (1987), and Zeng and Hopke (1989). PCA provides a rather qualitative description of source fingerprints, which can be used later as input to a CMB model or a similar source apportionment tool. For more information about other factor analysis approaches the reader is referred to Hopke (1985).

26.1.3 Empirical Orthogonal Function Receptor Models

Receptor models can also be used together with spatial distribution of measurements to estimate the spatial distribution of emission fluxes. The empirical orthogonal function (EOF) method is one of the most popular models for this. Henry et al. (1991) improved the EOF method by using wind direction in addition to spatially distributed concentration measurements as input. We describe this approach below.

Let us assume that during a given period the ground-level concentration field of an atmospheric species can be written as

$$C(x, y, t) = \sum_{i=1}^{N} a_i(t)\phi_i(x, y) \qquad (26.22)$$

where $\phi_i(x, y)$ are N orthogonal functions and $a_i(t)$ are time weighting functions. The functions $\phi_i(x, y)$ include the information about the sources of the species while the time weighting functions represent its atmospheric transport. Neglecting vertical concentration variations and dispersion, the atmospheric diffusion equation for this species is

$$\frac{\partial C}{\partial t} + u \frac{\partial C}{\partial x} + v \frac{\partial C}{\partial y} = Q(x, y, t) \tag{26.23}$$

where $u(x, y, t)$ and $v(x, y, t)$ are the wind components and $Q(x, y, t)$ is the net source term for the species including emissions and removal processes. Using (26.22)

$$\frac{\partial C}{\partial x} = \sum_{i=1}^{N} a_i(t) \frac{\partial \phi_i}{\partial x} \tag{26.24}$$

and

$$\frac{\partial C}{\partial y} = \sum_{i=1}^{N} a_i(t) \frac{\partial \phi_i}{\partial y} \tag{26.25}$$

Substituting (26.24) and (26.25) into (26.23) and integrating from time zero to the end of the sampling period T, one gets

$$\overline{Q}(x, y) = \frac{C(x, y, T) - C(x, y, 0)}{T} + \sum_{i=1}^{N} u_i(x, y) \frac{\partial \phi_i}{\partial x} + \sum_{i=1}^{N} v_i(x, y) \frac{\partial \phi_i}{\partial y} \tag{26.26}$$

where $\overline{Q}$ is the average source strength spatial distribution and

$$u_i(x, y) = \frac{1}{T} \int_0^T a_i(t) u(x, y, t) dt \tag{26.27}$$

$$v_i(x, y) = \frac{1}{T} \int_0^T a_i(t) v(x, y, t) dt \tag{26.28}$$

Equation (26.26) suggests that the spatial distribution of the source strength of the species can be found using concentration and wind data if the EOFs $\phi_i(x, y)$ and the time weighting functions $a_i(t)$ can be calculated (Henry et al. 1991).

The EOFs can be found with the following procedure. Assume that a given species is measured simultaneously at s sites during n sampling periods. The measurements can be used to construct the $n \times s$ concentration matrix $\mathbf{C}$. The first column of $\mathbf{C}$ contains all the measurements at the first site, the second column the measurements at the second site, and so on. Equation (26.22) can be viewed as the continuous form of the singular value decomposition of matrix $\mathbf{C}$

$$\mathbf{C} = \mathbf{UBV}^T \tag{26.29}$$

where $\mathbf{U}$ is an $n \times n$ orthogonal matrix whose columns are the eigenvectors of $\mathbf{CC}^T$, $\mathbf{V}$ is the matrix of the eigenvectors of $\mathbf{C}^T\mathbf{C}$, and $\mathbf{B}$ is a diagonal matrix of singular values (square roots of the eigenvalues). In this case there are s nonzero singular values. If a singular value is zero or very small, the corresponding eigenvectors can be removed from the matrices $\mathbf{U}$ and $\mathbf{V}$, leaving us with N significant eigenvectors. This step is similar to selection of the principal components during PCA. Let $\mathbf{U}^*$ be the $n \times N$ matrix with the significant eigenvectors, $\mathbf{B}^*$ the $N \times N$ diagonal matrix with the remaining singular values, and $\mathbf{V}^*$ the corresponding $s \times N$ matrix. Then let

$$\Phi = \mathbf{B}^* \mathbf{V}^{*T} \tag{26.30}$$

The columns of Φ are then the discrete EOFs and the columns of $\mathbf{U}^*$ are the discrete-time functions. These discrete EOFs can be interpolated in space to obtain the continuous EOFs using, for example, $1/r^2$ interpolation. Then (26.26) can be applied to obtain the spatial distribution of the source strength.

Note that while PCA is applied to many samples from the same site taken over a number of sampling periods, the EOF operates on many samples from many sites taken over the same period.

Assumptions implicit in the use of the EOF are

1. The fluxes of spatially distributed species are linearly additive.
2. Species are homogeneously distributed in the mixed layer.
3. Measurement errors are random and uncorrelated.
4. The number of sampling sites exceeds the number of sources.
5. Measurements are located in areas where there are significant spatial concentration gradients.

The last two assumptions are rarely met in practice. The spatial resolution of the EOF is limited by the number of observation sites and the distance between them. Sudden changes of windspeed and direction during a sampling period often result in problems.

Applications of the EOF have been presented by Gebhart et al. (1990), Ashbaugh et al. (1984), Wolff et al. (1985), and Henry et al. (1991). Henry et al. (1991) compared simulated two-dimensional data generated by a simple dispersion model and the above-described version of the EOF using simple wind fields. One of the comparisons is shown in Table 26.7. For this comparison a sampling site was located in each square and the model was able to reproduce the location of the two sources. However, the source strength is underpredicted as a result of numerical diffusion to the neighboring cells.

TABLE 26.7 Predicted Emission Location and Strenth[a]

−0.5	−0.5	−0.8	−1.5	−1.6	−1.3	0.8	2.9	3.7
−0.3	−0.2	−1.0	−0.8	−0.5	−0.1	2.4	3.0	4.2
−0.6	−0.5	−1.9	2.2	9.4	9.7	5.2	2.9	3.3
−1.6	−2.0	−2.6	7.3	28.5	31.3	10.8	0.3	−0.9
−0.6	−1.2	−0.6	8.1	25.5	28.4	11.4	1.2	−1.5
2.0	−4.6	5.7	5.5	6.2	7.0	4.1	0.4	−0.2
3.9	14.3	15.5	5.5	−1.0	−1.4	−1.3	−1.5	−0.4
4.0	13.3	14.2	5.4	0.1	−0.4	0.4	0.6	0.4
2.1	3.1	3.6	2.2	1.2	0.7	0.7	0.8	0.7

[a]Actual emissions in the shaded cells are 25 and 50 emission units. For the rest the actual emissions are zero.
Source: Henry et al. (1991).

26.2 PROBABILITY DISTRIBUTIONS FOR AIR POLLUTANT CONCENTRATIONS

Air pollutant concentrations are inherently random variables because of their dependence on the fluctuations of meteorological and emission variables. We already have seen from Chapter 18 that the concentration predicted by atmospheric diffusion theories is the mean concentration $\langle c \rangle$. There are important instances in analyzing air pollution where the ability simply to predict the theoretical mean concentration $\langle c \rangle$ is not enough. Perhaps the most important situation in this regard is in ascertaining compliance with ambient air quality standards. Air quality standards are frequently stated in terms of the number of times per year that a particular concentration level can be exceeded. In order to estimate whether such an exceedance will occur, or how many times it will occur, it is necessary to consider statistical properties of the concentration. One object of this chapter is to develop the tools needed to analyze the statistical character of air quality data.

Hourly average concentrations are the most common way in which urban air pollutant data are reported. These hourly average concentrations may be obtained from an instrument that actually requires a 1-h sample in order to produce a data point or by averaging data taken by an instrument having a sampling time shorter than 1 h. If we deal with 1 h average concentrations, those concentrations would be denoted by $c_\tau(t_i)$, where $\tau = 1\,h$. For convenience we will omit the subscript τ henceforth; however, it should be kept in mind that concentrations are usually based on a fixed averaging time. There are 8760 h in a year, so that if we are interested in the statistical distribution of the 1 h average concentrations measured at a particular location in a region, we will deal with a sample of 8760 values of the random variable c.

The random variable is characterized by a probability density function $p(c)$, such that $p(c)\,dc$ is the probability that the concentration c of a particular species at a particular location will lie between c and $c + dc$. Our first task will be to identify probability density functions (pdf's) that are appropriate for representing air pollutant concentrations. Once we have determined a form for $p(c)$, we can proceed to calculate the desired statistical properties of c.

If we plot the frequency of occurrence of a concentration versus concentration, we would expect to obtain a histogram like that sketched in Figure 26.4a. As the number of data points increases, the histogram should tend to a smooth curve such as that in Figure 26.4b. Note that very low and very high concentrations occur only rarely. We recall that aerosol size distributions exhibited a similar overall behavior; there are no particles of

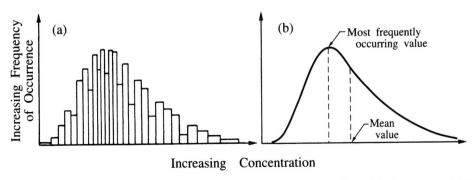

FIGURE 26.4 Hypothetical distributions of atmospheric concentrations: (a) histogram and (b) continuous distribution.

zero size and no particles of infinite size. Thus a probability distribution that is zero for $c = 0$ and as $c \to \infty$ is also desired for atmospheric concentrations.

Although there has been speculation as to what probability distribution is optimum in representing atmospheric concentrations from a physical perspective (Bencala and Seinfeld 1976), it is largely agreed that there is no a priori reason to expect that atmospheric concentrations should adhere to a specific probability distribution. Moreover, it has been demonstrated that a number of pdf's are useful in representing atmospheric data. Each of these distributions has the general features of the curve shown in Figure 26.4b; they represent the distribution of a nonnegative random variable c that has probabilities of occurrence approaching zero as $c \to 0$ and as $c \to \infty$. Georgopoulos and Seinfeld (1982) summarize the functional forms of many of the pdf's that have been proposed for air pollutant concentrations [see also Tsukatami and Shigemitsu (1980)]. In addition, Holland and Fitz-Simons (1982) have developed a computer program for fitting statistical distributions to air quality data.

Which probability distribution actually best fits a particular set of data depends on the characteristics of the set. The two distributions that have been used most widely for representing atmospheric concentrations are the lognormal and the Weibull, and we will focus on these two distributions here.

26.2.1 The Lognormal Distribution

We have already dealt extensively with the lognormal distribution in Chapter 8 for representing aerosol size distribution data. If a concentration c is lognormally distributed, its pdf is given by,

$$p_L(c) = \frac{1}{(2\pi)^{1/2} c \ln \sigma_g} \exp\left(-\frac{(\ln c - \ln \mu_g)^2}{2 \ln^2 \sigma_g} \right) \tag{26.31}$$

where μ_g and σ_g are the geometric mean and standard deviation of c.

Figure 26.5 shows sample points of the distribution of 1 h average SO_2 concentrations equal to or in excess of the stated values for Washington, DC, for the 7-year period December 1, 1961–December 1, 1968. A lognormal distribution has been fitted to the high-concentration region of these data.

We recall that the geometric mean, or median, is the concentration at which the straight-line plot crosses the 50th percentile. The slope of the line is related to the standard geometric deviation, which can be calculated from the plot by dividing the 16th percentile concentration (which is one standard deviation from the mean) by the 50th percentile concentration (the geometric mean). (This is the 16th percentile of the *complementary* distribution function $\overline{F}(c)$; equivalently, it is the 84th percentile of the cumulative distribution function $F(c)$.)$\overline{F}(x) = \text{Prob}\,\{c > x\} = 1 - F(x)$. For the distribution of Figure 26.5, $\mu_g = 0.042$ ppm and $\sigma_g = 1.96$. Plots such as Figure 26.5 are widely used because it is important to know the probability that concentrations will equal or exceed certain values.

26.2.2 The Weibull Distribution

The pdf of a Weibull distribution is

$$p_W(c) = (\lambda/\sigma)(c/\sigma)^{\lambda-1} \exp[-(c/\sigma)^\lambda] \qquad \sigma, \lambda > 0 \tag{26.32}$$

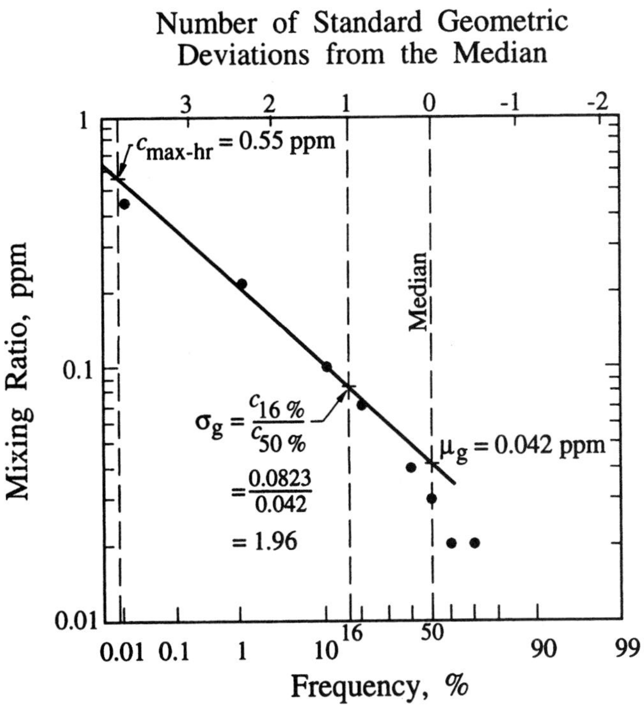

FIGURE 26.5 Frequency of 1-h average SO_2 concentrations equal to, or in excess of, stated values at Washington, DC, December 1, 1961 to December 1, 1968 (Larsen 1971).

As was done with the lognormal distribution, it is desired to devise a set of coordinates for a graph on which a set of data that conforms to a Weibull distribution will plot as a straight line. To do so we can work with the complementary distribution function, $\overline{F}_W(c)$. The complementary distribution function of the Weibull distribution is

$$\overline{F}_W(c) = \exp[-(c/\sigma)^\lambda] \qquad (26.33)$$

Taking the natural logarithm of (26.33) and changing sign, we obtain

$$\ln\left(\frac{1}{\overline{F}_W(c)}\right) = \left(\frac{c}{\sigma}\right)^\lambda \qquad (26.34)$$

Now take the logarithm (base 10) of both sides of (26.34):

$$\log\left[\ln\left(\frac{1}{\overline{F}_W(c)}\right)\right] = \lambda(\log c - \log \sigma) \qquad (26.35)$$

Therefore, if we plot $\log[\ln(1/\overline{F}_W(c))]$ versus $\log c$, data from a Weibull distribution will plot as a straight line. The values $\log[\ln(1/\overline{F}_W(c))]$ are the values on the ordinate scale of Figure 26.6, so-called extreme value probability scale. The scale is constructed so that a computation of $\log[\ln(1/\overline{F}_W(c))]$ of each value of $\overline{F}_W(c)$ yields a linear scale.

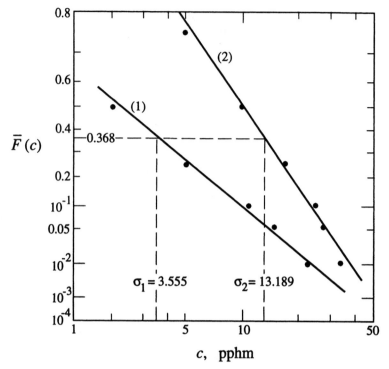

FIGURE 26.6 Weibull distribution fits of the 1971 hourly average (1) and daily maximum hourly average (2) ozone mixing ratio at Pasadena, California (Georgopoulos and Seinfeld 1982).

By setting $\log c = \log \sigma$ in (26.35), we obtain $\overline{F}_W(c) = e^{-1} = 0.368$. This point is shown in Figure 26.6. The parameter λ, which is the slope of the straight line, can be found by using any other point on the line and solving (26.35) for λ:

$$\lambda = \frac{\log\left[\ln\left(\frac{1}{\overline{F}_W(c)}\right)\right]}{\log c - \log \sigma} \tag{26.36}$$

Suppose that we use, for a second point, that point on the line that crosses $\overline{F}_W(c) = 0.01$. Then

$$\lambda = \frac{0.663}{\log c_{0.01} - \log \sigma} \tag{26.37}$$

26.3 ESTIMATION OF PARAMETERS IN THE DISTRIBUTIONS

Each of the two distributions we have presented is characterized by two parameters. The fitting of a set of data to a distribution involves determining the values of the parameters of the distribution so that the distribution fits the data in an optimal manner. Ideally, this fitting is best carried out using a systematic optimization routine that estimates the parameters for several distributions from the given set of data and then selects that distribution that best fits the data, using a criterion of "goodness of fit" (Holland and Fitz-Simons 1982). We present two methods here that do not require a computer optimization routine. The first method has, in effect, already been described. It is formally called the *method of quantiles*. The second is the method of moments, which requires the computation of the first (as many as the parameters of the distribution) sample moments of the "raw data." The procedure suggested should start with the construction of a plot of the available data on the appropriate axes (that gives a straight line for the theoretical distribution), in order to get a preliminary notion of the goodness of fit. Then one would apply one of the following methods to evaluate the parameters of the "best" distribution.

26.3.1 Method of Quantiles

We have seen that the parameters μ_g and σ_g of the lognormal distribution can be estimated by using the 50th percentile and 16th percentile values. These percentiles are called *quantiles*. Also, (26.36) and (26.37) indicate how this approach can be used to determine the two parameters of the Weibull distribution.

For the lognormal distribution, for example, we have already illustrated how μ_g and σ_g are estimated from the concentrations at the 50% and 84% quantiles:

$$\ln c_{0.50} - \ln \mu_g = 0$$

$$\ln c_{0.84} - \ln \mu_g = \ln \sigma_g$$

Choosing as a further illustration the 95% and 99% quantiles, we obtain

$$\ln c_{0.95} - \ln \mu_g = 1.645 \ln \sigma_g$$

$$\ln c_{0.99} - \ln \mu_g = 2.326 \ln \sigma_g$$

For the Weibull distribution, the quantile concentration is given by

$$c_{q_i} = \sigma \left[\ln \left(\frac{1}{1 - q_i} \right) \right]^{1/\lambda}$$

Using the 0.80 and 0.98 quantiles, for example, we obtain the estimates for the parameters λ and σ as

$$\lambda = \frac{0.88817}{\ln c_{0.98} - \ln c_{0.80}}$$

$$\sigma = \exp(1.53580 \ln c_{0.80} - 0.53580 \ln c_{0.98})$$

26.3.2 Method of Moments

To estimate the parameters of a distribution by the method of moments, the moments of the distribution are expressed in terms of the parameters. Estimates for the values of the moments are obtained from the data, and the equations for the moments are solved for the parameters. For a two-parameter distribution, values for the first two moments are needed.

The rth noncentral moment of a random variable c with a pdf $p(c)$ is defined by

$$\mu'_r = \int_0^\infty c^r p(c)\,dc \tag{26.38}$$

and the rth central moment (moment about the mean) is

$$\mu_r = \int_r^\infty (c - \mu'_1)^r p(c)\,dc \tag{26.39}$$

The mean value of c is μ'_1, and the variance is μ_2, which is commonly denoted by σ^2, or sometimes Var $\{c\}$.

Consider first the estimation of μ_g and σ_g for the lognormal distribution. Let $\mu = \ln \mu_g$ and $\sigma = \ln \sigma_g$. The first and second noncentral moments of the lognormal distribution are

$$\mu'_1 = \exp\left(\mu + \frac{\sigma^2}{2}\right) \tag{26.40}$$

$$\mu'_2 = \exp(2\mu + 2\sigma^2) \tag{26.41}$$

After solving (26.40) and (26.41) for μ and σ^2, we have

$$\mu = 2 \ln \mu'_1 - \frac{1}{2} \ln \mu'_2 \tag{26.42}$$

$$\sigma^2 = \ln \mu'_2 - 2 \ln \mu'_1 \tag{26.43}$$

μ'_1, μ'_2, and μ_2 are related through $\mu_2 = \mu'_2 - {\mu'_1}^2$ and are estimated from the data by

$$M'_1 = \frac{1}{n} \sum_{i=1}^n c_i \tag{26.44}$$

and

$$M'_2 = \frac{1}{n} \sum_{i=1}^n c_i^2 \tag{26.45}$$

$$M_2 = \frac{1}{n-1} \sum_{i=1}^n (c_i - M'_1)^2 \tag{26.46}$$

where n is the number of data points. Thus the moment estimates of the parameters of the lognormal distribution are given by

$$\mu = 2 \ln M'_1 - \frac{1}{2} \ln M'_2 \tag{26.47}$$

$$\sigma^2 = \ln M'_2 - 2 \ln M'_1 \tag{26.48}$$

For the Weibull distribution, the mean and variance are given by

$$\mu'_1 = \sigma\Gamma(1 + 1/\lambda) \tag{26.49}$$

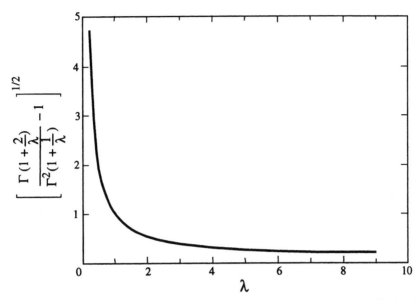

FIGURE 26.7 Calculation of parameter λ in fitting a set of data to a Weibull distribution by the method of moments (Georgopoulos and Seinfeld 1982).

where Γ is the gamma function and

$$\mu^2 = \sigma^2[\Gamma(1 + 2/\lambda) - \Gamma^2(1 + 1/\lambda)] \tag{26.50}$$

In solving these equations for σ and λ, we can conveniently use the coefficient of variation given by $(\mu_2)^{1/2}/\mu_1'$. Then the moment estimators of the sample correspond to λ so that

$$\left(\frac{\Gamma(1 + 2/\lambda)}{\Gamma^2(1 + 1/\lambda)} - 1\right)^{1/2} = \frac{(M_2)^{1/2}}{M_1'} \tag{26.51}$$

$$\sigma = \frac{M_1'}{\Gamma(1 + 1/\lambda)} \tag{26.52}$$

To aid in the determination of λ in fitting a set of data to the Weibull distribution, one can use Figure 26.7, in which the left-hand side of (26.51) is shown as a function of λ for $0 < \lambda < 9$.

As an example, we consider the fitting of a Weibull distribution to 1971 hourly average ozone data from Pasadena, California. The data consist of 8303 hourly values (there are 8760 h in a year). The maximum hourly value reported was 53 parts-per-hundred million (pphm). The arithmetic mean and standard deviation of the data are $M_1' = 4.0$ pphm and $M_2^{1/2} = 5.0$ pphm, and the geometric mean and geometric standard deviation are 2.4 and 2.6 pphm, respectively. If one assumes that the hourly average ozone mixing ratios fit a Weibull distribution, the parameters of the distribution can be estimated from (26.51) and (26.52) to give $\lambda = 0.808$ and $\sigma = 3.555$ pphm.

It is interesting also to fit only the daily maximum hourly average ozone values to a Weibull distribution. For these data there exist 365 data points, the maximum value of which is, as already noted, 53 pphm. The arithmetic mean and standard deviation of the data are $M_1' = 12.0$ pphm and $M_2^{1/2} = 8.6$ pphm; the geometric mean and geometric standard deviation are 9.1 and 2.2 pphm, respectively. Table 26.8 gives a comparison of the

TABLE 26.8   1971 Hourly Average Pasadena, California, Ozone Data Fitted to a Weibull Distribution

	Mixing Ratio (pphm) Equaled or Exceeded by the Stated Percent of Observations								
	1%	2%	3%	4%	5%	10%	25%	50%	75%
Data (hourly average)	24	20	18	16	15	11	5	2	1
Weibull distribution	23.5	19.2	16.8	15.1	13.8	10	5.3	2.3	0.76
Data (daily maximum)	34	33	32	30	28	25	17	10	5
Weilbull distribuion	38.8	34.6	32	30.1	28.6	24.8	16.6	10.2	5.5

data and the Weibull distribution concentration frequencies in the two cases. Both fits are very good, the fit to the daily maximum values being slightly better.

26.4   ORDER STATISTICS OF AIR QUALITY DATA

One of the major uses of statistical distributions of atmospheric concentrations is to assess the degree of compliance of a region with ambient air quality standards. These standards define acceptable upper limits of pollutant concentrations and acceptable frequencies with which such concentrations can be exceeded. The probability that a particular concentration level, x, will be exceeded in a single observation is given by the complementary distribution function $\overline{F}(x) = \text{Prob}\{c > x\} = 1 - F(x)$.

When treating sets of air quality data, available as successive observations, we may be interested in certain random variables, as, for example:

The highest (or, in general, the rth highest) concentration in a finite sample of size m

The number of exceedances of a given concentration level in a number of measurements or in a given time period

The number of observations (waiting time or "return period") between exceedances of a given concentration level

The distributions and pdf's, as well as certain statistical properties of these random variables, can be determined by applying the methods and results of *order statistics* (or statistics of extremes) (Gumbel 1958; Sarhan and Greenberg 1962; David 1981).

26.4.1   Basic Notions and Terminology of Order Statistics

Consider the m random unordered variates $c(t_1), c(t_2), \ldots, c(t_m)$, which are members of the stochastic process $\{c(t_i)\}$ that generates the time series of available air quality data. If we arrange the data points by order of magnitude, then a "new" random sequence of ordered variates $c_{1;m} \geq c_{2;m} \geq \cdots \geq c_{m;m}$ is formed. We call $c_{i;m}$ the ith *highest-order statistic* or ith *extreme statistic* of this random sequence of size m.

In the exposition that follows we assume in general that

- The concentration levels measured in successive nonoverlapping periods—and hence the unordered random variates $c(t_i)$—are independent of one another.
- The random variables $c(t_i)$ are identically distributed.

26.4.2 Extreme Values

Assume that the distribution function $F(x)$, as well as the pdf $p(x)$, corresponding to the total number of available measurements, are known. They are called the *parent* (or initial) distribution and pdf, respectively. The probability density function $p_{r,m}(x)$ and the distribution function $F_{r,m}(x)$ of the rth highest concentration out of samples of size m are evaluated directly from the parent pdf $p(x)$ and the parent distribution function $F(x)$ as follows.

The probability that $c_{r,m} = x$ equals the probability of $m - r$ trials producing concentration levels lower than x, times the probability of $r - 1$ trials producing concentrations above x, times the probability density of attaining a concentration equal to x, multiplied by the total number of combinations of arranging these events (assuming complete independence of the data). In other words, the pdf of the rth highest concentration has the trinomial form

$$p_{r;m}(x) = \frac{m!}{(r-1)!(m-r)!} [F(x)]^{m-r} [1 - F(x)]^{r-1} p(x)$$

$$= \frac{1}{B(r, m-r+1)} [F(x)]^{m-r} [1 - F(x)]^{r-1} p(x) \tag{26.53}$$

where B is the beta function (Pearson 1934). In particular, for the highest and second highest concentration values ($r = 1, 2$) we have

$$p_{1;m}(x) = m[F(x)]^{m-1} p(x) \tag{26.54}$$

and

$$p_{2;m}(x) = m(m-1)[F(x)]^{m-2}[1 - F(x)]p(x) \tag{26.55}$$

The probability $F_{r;m}(x)$ that $c_{r;m} \leq x$ is identical to the probability that no more than $r - 1$ measurements out of m result in $c_{r;m} > x$. Every observation is considered as a *Bernoulli trial* with probabilities of "success" and "failure" $F(x)$ and $1 - F(x)$, respectively. Thus

$$F_{r;m}(x) = \sum_{k=0}^{r-1} \binom{m}{k} [1 - F(x)]^k [F(x)]^{m-k} \tag{26.56}$$

For the particular cases of the highest and the second highest values ($r = 1, 2$), (26.56) becomes

$$F_{1;m}(x) = [F(x)]^m \tag{26.57}$$

$$F_{2;m}(x) = m[F(x)]^{m-1} - (m-1)[F(x)]^m \tag{26.58}$$

It is worthwhile to note the dependence of the probability of the largest values on the sample size. From (26.57), we obtain

$$F_{1;n}(x) = [F_{1;m}(x)]^{n/m} \tag{26.59}$$

Thus if the distribution of the extreme value is known for one sample size, it is known for all sample sizes.

Let the pdf of the rth highest concentration out of a sample of m values be denoted $p_{r;m}(c)$. Once this pdf is known, all the statistical properties of the random variable $c_{r;m}$ can be determined. However, the integrals involved in the expressions for the expectation and higher-order moments are not always easily evaluated, and thus there arises the need for techniques of approximation. The most important result concerns the evaluation of the expected value of $c_{r;m}$. In fact, for sufficiently large m, an approximation for $E\{c_{r;m}\}$ is provided by the value of x satisfying (David 1981)

$$F(x) = \frac{m - r + 1}{m + 1} \tag{26.60}$$

In terms of the inverse function of $F(x)$, $F^{-1}(x)$ [i.e., $F^{-1}[F(x)] = x$], we have the asymptotic relation

$$E\{c_{r;m}\} \simeq F^{-1}\left(\frac{m - r + 1}{m + 1}\right), \qquad \text{as } m \to \infty \tag{26.61}$$

26.5 EXCEEDANCES OF CRITICAL LEVELS

The number of exceedances (episodes) $N_x(m)$ of a given concentration level x in a set of m successive observations $c(t_i)$ is itself a random function. Similarly, the number of averaging periods (or observations) between exceedances of the concentration level x, another random function called the *waiting time*, *passage time*, or *return period*, is of principal interest in the study of pollution episodes.

In the case of independent, identically distributed variates, each one of the observations is a Bernoulli trial; therefore the probability density function of $N_x(m)$ is, in terms of the parent distribution $F(x)$

$$p_N(N_x; m, x) = \binom{m}{N_x} [1 - F(x)]^{N_x} F(x)^{m - N_x} \tag{26.62}$$

Thus the expected number of exceedances $\bar{N}_x(m)$ of the level x in a sample of m measurements is

$$\bar{N}_x(m) = m(1 - F(x)) = m\bar{F}(x) \tag{26.63}$$

The expected percentage of exceedances of a given concentration level x is just $100\bar{F}(x)$.

26.6 ALTERNATIVE FORMS OF AIR QUALITY STANDARDS

In the evaluation of whether ambient air quality standards are satisfied in a region, aerometric data are used to estimate expected concentrations and their frequency of occurrence. If it is assumed that a certain probability distribution can be used to represent

TABLE 26.9   Alternative Statistical Forms of the Ozone Air Quality Standard[a]

Number	Form
1	0.12 ppm hourly average with expected number of exceedances per year less than or equal to 1
2	0.12 ppm hourly average not to be exceeded on the average by more than 0.01% of the hours in 1 year
3	0.12 ppm annual expected maximum hourly average
4	0.12 ppm annual expected second highest hourly average

[a]For most practical purposes forms 1 and 3 can be considered equivalent.

the air quality data, the distribution is fit to the current years' data by estimating the parameters of the distribution. It is then assumed that the probability distribution will hold for data in future years; only the parameters of the distribution will change as the source emissions change from year to year. If the parameters of the distribution can be estimated for future years, then the expected number of exceedances of given concentration levels, such as the ambient air quality standard, can be assessed.

Table 26.9 gives four possible forms for an ambient air quality standard. We have used ozone as the example compound in the table. The ambient air quality standards involve a concentration level and a frequency of occurrence of that level. In this section we want to examine the implications of the form of the standard on the degree of compliance of a region. The choice of one form of the standard over another can be based on the impact that each form implies for the concentration distribution as a whole (Curran and Hunt 1975; Mage 1980).

The first step in the evaluation of an air quality standard is to select the statistical distribution that supposedly best fits the data. We will assume that the frequency distribution that best fits hourly averaged ozone concentration data is the Weibull distribution. Since the standards are expressed in terms of expected events during a 1-year period of 1-h average concentrations, we will always use the number of trials m equal to the number of hours in a year, 8760. We would use $m < 8760$ only to evaluate the parameters of the distribution if some of the 8760 hourly values are missing from the data set.

Let us now analyze each of the four forms of the ozone air quality standard given in Table 26.9 from the point of view that ozone concentrations can be represented by a Weibull distribution.

1. Expected Number of Exceedances of 0.12 ppm Hourly Average Mixing Ratio

The expected number of exceedances $\overline{N}_x(m)$ of a given concentration level in m measurements is given by (26.63), which, in the case of the Weibull distribution, becomes

$$\overline{N}_x = m \exp[-(x/\sigma)^\lambda] \tag{26.64}$$

If we desire the expected exceedance to be once out of m hours, that is, $\overline{N}_{x_1} = 1$, the concentration corresponding to that choice is

$$x_1 = \sigma(\ln m)^{1/\lambda} \tag{26.65}$$

For $m = 8760$, (26.65) becomes

$$x_1 = \sigma(9.08)^{1/\lambda} \tag{26.66}$$

2. The Hourly Average Mixing Ratio Not to Be Exceeded on the Average by More than 0.01% of the Hours in 1 Year

The expected percentage of exceedance of a given concentration x is given by $100\,\overline{F}(x)$, which, for the Weibull distribution, is

$$\overline{\Pi}(x) = 100 \exp[-(x/\sigma)^\lambda] \tag{26.67}$$

Equation (26.67) can be rearranged to determine the concentration level that is expected to be exceeded $\overline{\Pi}(x)$ percent of the time:

$$x = \sigma[\ln(100/\overline{\Pi}(x))]^{1/\lambda} \tag{26.68}$$

Therefore we can calculate the mixing ratio that is expected to be exceeded 0.01% of the hours in 1 year as

$$x_{0.01} = \sigma(9.21)^{1/\lambda} \tag{26.69}$$

3,4. The Annual Expected Maximum Hourly Average Mixing Ratio and Annual Expected Second Highest Hourly Average Mixing Ratio

The expected value of the rth highest concentration for a Weibull distribution is

$$E\{c_{r;m}\} = \frac{m!}{(r-1)!(m-r)!} \int_0^\infty (x/\sigma)^\lambda \\ \times \{1 - \exp[-(x/\sigma)^\lambda]\}^{m-r}\{\exp[-(x/\sigma)^\lambda]\}^r dx \tag{26.70}$$

We wish to evaluate this equation for $r = 1$ and $r = 2$, corresponding to standards 3 and 4, respectively, in Table 26.9, to obtain $E\{c_{1;m}\}$ and $E\{c_{2;m},\}$ the expected highest and second highest hourly concentrations, respectively, in the year, with $m = 8760$. Unfortunately, the integral in (26.70) cannot be evaluated easily. Even numerical techniques fail to give consistent results, because of the singularity at $x = 0$. Thus the asymptotic relation for large m, (26.61), must be used in this case. For the Weibull distribution we have

$$1 - \exp\left[-\left(\frac{E\{c_{r;m}\}}{\sigma}\right)^\lambda\right] = \frac{m-r+1}{m+1} \tag{26.71}$$

For $m = 8760$ and $r = 1,2$ we must solve, respectively, the equations

$$1 - \exp\left[-\left(\frac{E\{c_{1;m}\}}{\sigma}\right)^\lambda\right] = \frac{8760}{8761} \tag{26.72}$$

and

$$1 - \exp\left[-\left(\frac{E\{c_{2;m}\}}{\sigma}\right)^{\lambda}\right] = \frac{8759}{8761} \tag{26.73}$$

to obtain

$$E\{c_{1;m}\} = \sigma(9.08)^{1/\lambda} \tag{26.74}$$

and

$$E\{c_{2;m}\} = \sigma(8.38)^{1/\lambda} \tag{26.75}$$

Evaluation of Alternative Forms of the Ozone Air Quality Standard with 1971 Pasadena, California, Data   Earlier, 1971 hourly average and maximum daily hourly average ozone mixing ratios at Pasadena, California, were fit to Weibull distributions. We now wish to evaluate the four forms of the ozone air quality standard with these data. For convenience all mixing ratios values will be given as pphm rather than ppm.

1. Expected Number of Exceedances of 12 pphm Hourly Average Mixing Ratio

The expected number of exceedances of 12 pphm, based on the Weibull fit of the 1971 Pasadena, California, hourly average data, is, from (26.64)

$$N_{12} = 8760 \exp\left[-\left(\frac{12}{3.555}\right)^{0.808}\right] = 605.2$$

The hourly average ozone mixing ratio that is exceeded at most once per year is from (26.65)

$$x_1 = 3.555(\ln 8760)^{1/0.808}$$
$$= 54.41 \, \text{pphm}$$

which agrees well with the actual measured value of 52 pphm.

If, instead of the complete hourly average Weibull distribution, we use the distribution of daily maximum hourly average values, the expected number of exceedances of a daily maximum of 12 pphm is

$$\overline{N}_{12} = 365 \exp\left[-\left(\frac{12}{13.189}\right)^{1.416}\right] = 152.2$$

and the daily maximum 1 h mixing ratio that is exceeded once per year, at most, is

$$x_1 = 13.189 \, (\ln 365)^{1/1.416}$$
$$= 46.2 \, \text{pphm}$$

It is interesting to note that this value is underpredicted if we use the distribution of daily maxima instead of the distribution based on the complete set of data.

2. The Hourly Average Mixing Ratio Not to Be Exceeded on the Average by More than 0.01% of the Hours in 1 Year

The expected percentage of exceedances of 12 pphm is, from (26.67)

$$\Pi(12) = 100 \exp\left[-\left(\frac{12}{3.555}\right)^{0.808}\right] = 6.91\%$$

The mixing ratio that is expected to be exceeded 0.01% of the hours in the year is, from (26.68)

$$x_{0.01} = 3.555\left(\ln\frac{100}{0.01}\right)^{1/0.808} = 55.5\,\text{pphm}$$

This form of the standard cannot be evaluated from the distribution of daily maxima since it is stated on the basis of a percentage of all the hours of the year.

3,4. The Annual Expected Maximum Hourly Average Mixing Ratio and Annual Expected Second Highest Average Mixing Ratio

The annual expected maximum hourly average mixing ratio is obtained from (26.74) for $E\{c_{1;m}\}$, and for $\sigma = 3.555$, $\lambda = 0.808$. We have

$$E\{c_1\} = 54.41\,\text{pphm}$$

(whereas the observed (sample) maximum hourly average value was 53 pphm). Similarly, for the annual expected second highest hourly average mixing ratio we have, from (26.75)

$$E\{c_2\} = 49.40\,\text{pphm}$$

An interesting question of interpretation arises when an ambient air quality standard involves an averaging period of several hours. For example, the 8 h National Ambient Air Quality Standard for carbon monoxide is 9 ppm, not to be exceeded more than once per year. Two principal interpretations of the 8 h standard have been proposed (McMullin 1975). One approach is to examine all possible 8 h intervals by calculating a moving 8 h average (twenty-four 8 h averages each day). The other approach is to examine three consecutive nonoverlapping 8 h intervals per day, usually beginning, for sake of convenience, at midnight, 8:00 a.m., and 4 p.m. The principal appeal of the moving average is that it approximates the body's integrating response to cumulative CO exposure. A disadvantage is that the moving 8 h average affords no reduction in the number of data points to be examined compared with the input of 1 h values. The consecutive interval approach offers the convenience of reducing a year's 8760 hourly values to a set of 1095 consecutive 8 h periods. McMullin (1975) examined 1972 data for three sites (Newark, New Jersey; Camden, New Jersey; and Spokane, Washington) and found that the maximum and second highest concentration values derived from moving averages can be at least 20% higher than corresponding values detected by the consecutive 8-h intervals. He found that the natural fluctuation in the time of day when the maximum occurs and the variability and episode length make it doubtful that any framework of consecutive 8 h intervals can adequately portray the essential characteristics of CO exposure. He therefore recommended the moving 8 h average as more sensitive to actual maximum levels and to short episodes.

26.7   RELATING CURRENT AND FUTURE AIR POLLUTANT STATISTICAL DISTRIBUTIONS

The reduction R in current emission source strength to meet an air quality goal can be calculated by the so-called rollback equation

$$R = \frac{E\{c\} - E\{c\}_s}{E\{c\} - c_b} \tag{26.76}$$

where $E\{c\}$ is the current annual mean of the pollutant concentration, $E\{c\}_s$ is the annual mean corresponding to the air quality standard c_s, and c_b is the background concentration assumed to be constant. Since, as we have seen, the air quality standard c_s is usually stated in terms of an extreme statistic, such as the concentration level that may be exceeded only once per year, it is necessary to have a probability distribution to relate the extreme concentration c_s to the annual mean $E\{c\}_s$. We assume that if, in the future, the emission level is halved, the annual mean concentration will also be halved. The key question is: "If the emission level is halved, what happens to the predicted extreme concentration in the future year; is it correspondingly halved or does it change by more or less than that amount?"

To address this question, assume that the concentration in question can be represented by a lognormal distribution (under present as well as future conditions). If a current emission rate changes by a factor $\kappa (\kappa > 0)$, while the source distribution remains the same (if meteorological conditions are unchanged, and if background concentrations are negligible), the expected total quantity of inert pollutants having an impact on a given site over the same time period should also change by the factor κ. The expected concentration level for the future period is therefore given, for a lognormally distributed variable, by (recall $\mu = \ln \mu_g$ and $\sigma = \ln \sigma_g$)

$$E\{c'\} = e^{\mu' + \sigma'^2/2} = \kappa e^{\mu + \sigma^2/2} \tag{26.77}$$

where the primed quantities of c', μ', and σ' apply to the future period, and the unprimed quantities apply to the present. On an intuitive basis, if meteorological conditions remain unchanged, the standard geometric deviation of the lognormal pollutant distribution should remain unchanged; that is, $e^{\sigma'} = e^{\sigma}$. Thus $e^{\mu'} = \kappa \, e^{\mu}$, or

$$\mu' = \mu + \ln \kappa \tag{26.78}$$

The probability that future concentration level c' will exceed a level x is

$$
\begin{aligned}
\overline{F}_{c'}(x) &= 1 - \frac{1}{\sqrt{2\pi}} \int_{-\infty}^{(\ln x - \mu')/\sigma'} e^{-\eta^2} d\eta \\
&= \overline{F}_c(x/\kappa)
\end{aligned}
\tag{26.79}
$$

using (26.78). Similarly

$$
\begin{aligned}
\overline{F}_{c'}(\kappa x) &= 1 - \frac{1}{\sqrt{2\pi}} \int_{-\infty}^{(\ln x - \mu')/\sigma'} e^{-\eta^2} d\eta \\
&= \overline{F}_c(x)
\end{aligned}
\tag{26.80}
$$

Thus the probability that the future level κx will be exceeded just equals the probability that, with current emission sources, the level x will be exceeded. Therefore, with equal σ, all frequency points of the distribution shift according to the factor κ. This results in a parallel translation of the graph of $F(x)$ or $\overline{F}(x)$, when plotted against $\ln x$.

Figure 26.8 shows two lognormal distributions with the same standard geometric deviation but with different geometric mean values. The geometric mean concentrations of the two distributions are 0.05 and 0.10 ppm, and the standard geometric deviations are both 1.4. The mean concentrations can be calculated with the aid of $\ln E\{c\} = \ln \mu_g + \frac{1}{2}\ln^2 \sigma_g$ and the material presented above. We find that $E_1\{c\} = 0.053$ ppm, and

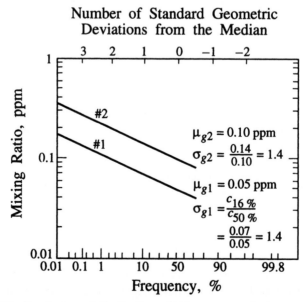

FIGURE 26.8   Two lognormal distributions with the same standard geometric deviation.

$E_2\{c\} = 0.106$ ppm for the two distributions, respectively. The variances can likewise be calculated with the aid of $\mathrm{Var}\{c\} = [\exp(2\mu + \sigma^2)][e^{\sigma^2} - 1]$ to obtain $\mathrm{Var}_1\{c\} = 0.00034\,\mathrm{ppm}^2$ and $\mathrm{Var}_2\{c\} = 0.00134\,\mathrm{ppm}^2$.

Suppose that distribution 1 represents current conditions, and therefore that the current probability of exceeding a concentration of 0.13 ppm is about 0.0027 (which corresponds to about 1 day per year if the distribution is of 24 h averages). If the emission rate were doubled, the new distribution function would be given by distribution 2. The new distribution has a median value twice that of the old one, since total loadings attributable to emissions have doubled. In the new case, a concentration of 0.13 ppm will be exceeded 22% of the time, or about 80 days a year, and the concentration that is exceeded only 1 day per year rises to 0.26 ppm.

PROBLEMS

26.1$_A$ Show how to construct the axes of an extreme-value probability graph on which a Weibull distribution will plot as a straight line.

26.2$_A$ Figure 26.P.1 shows the frequency distributions of SO_2 at two locations in the eastern United States from August 1977 to July 1978. Using the data points on the figure, fit lognormal distributions to the SO_2 concentrations at Duncan Falls, Ohio and Montague, Massachusetts. If the data points do not fall exactly on the best-fit lognormal lines, comment on possible reasons for deviations of the measured concentrations from log normality.

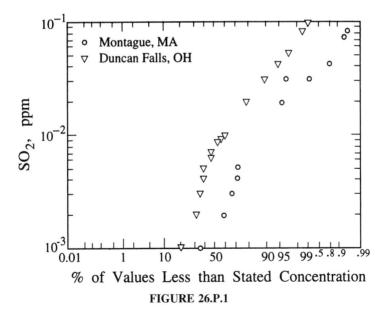

FIGURE 26.P.1

26.3$_A$ Figure 26.P.2 shows the frequency distributions of CO concentrations measured inside and outside an automobile traveling a Los Angeles commuter route. Fit lognormal distributions to the exterior 1 min and 30 min averaging time data.

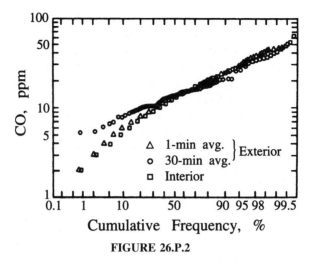

FIGURE 26.P.2

If the data points do not fall exactly on the best-fit lognormal lines, comment on possible reasons for deviations of the measured concentrations from lognormality. What can be said about the relationship of the best-fit parameters for the lognormal distributions at 1 min and 30 min averaging times.

26.4$_B$ Table 26.P.1 gives CO concentrations at Pasadena, California, for 1982—in particular, the weekly maximum 1 h average concentrations and the monthly mean of the daily 1 h average maximum concentrations.

a. Plot the weekly 1 h average maximum values on log–probability paper and extreme-value probability paper.

b. Determine the two parameters of both the lognormal and Weibull distributions by the method of moments and the method of quantiles. On the basis of your

TABLE 26.P.1 Carbon Monoxide Concentrations at Pasadena, California, for 1982: Some Statistics of 1 Hour Average Data, ppm

Month	Weekly 1 Hour Average Maxima	Highest 1 Hour Average Value	Mean of Daily 1 Hour Average Maxima
January	17, 12, 8, 10	17	7.2
February	10, 8, 9, 7	10	5.3
March	5, 7, 6, 6	7	4.0
April	3, 5, 4, 5 ,2	5	2.9
May	3, 4, 3, 3	4	3.2
June	5, 4, 3, 4	5	2.5
July	2, 3, 3, 4, 3	4	2.4
August	6, 3, 5, 4	6	2.8
September	7, 6, 4, 8	8	3.8
October[a]	8, 12, 11, —, 11	(12)	(7.2)
November	13, 10, 10, 11	13	6.9
December	15, 14, 20, 8, 13	20	8.9

[a]No data were available for the fourth week of October.

results, select one of the two distributions as representing the best fit to the data.

c. What was the expected number of exceedances of the weekly 1 h average maximum CO concentration at Pasadena of the National Ambient Air Quality Standard for CO of 35 ppm?

d. What is the expected number of weekly 1 h average maximum observations between successive exceedances of 35 ppm?

e. What is the variance of the number of weekly 1 h average maximum observations between successive exceedances of 35 ppm?

f. For the expected exceedance of 35 ppm to be once in 52 weeks, how must the parameters of the distribution change?

26.5$_B$ Consider a single elevated continuous point source at a height h of strength q. Assume that the wind blows with a speed that is lognormally distributed with parameters μ_{ug} and σ_{ug} and with a direction that is uniformly distributed over 360°. No inversion layer exists.

a. Determine the form of the statistical distribution of $\langle c(x, y, 0) \rangle$ in terms of the known quantities of the problem.

b. Assuming that the source strength changes from q to κq, determine the form of the new distribution of $\langle c(x, y, 0) \rangle$.

c. Assuming that the air quality standard is related to the value that is exceeded only once a year, derive an expression for the change in this value as a function of location resulting from the source strength change.

26.6$_B$ Given that the CO emission level from the entire motor vehicle population is reduced to one-half its value at the time the data in Figure 26.P.2 were obtained, calculate the expected changes in the three CO frequency distributions. (Note that in order to calculate this change one needs the results of Problem 26.3) How does the frequency at which a level of 50 ppm occurs change because of a halving of the emission rate? How does the expected return period and its variance change for the 50 ppm level from the old to the new emission level?

26.7$_D$ Calculate the contributions of sources (vegetative burning, primary crude oil, motor vehicles, limestone, ammonium sulfate, ammonium nitrate, and secondary OC) to the PM$_{2.5}$ concentrations observed in Fresno, CA during 1988–89 (Table 26.2). Use the CMB for nitrate, sulfate, ammonium, EC, OC, Al, Si, S, K, and V, with the source profiles given in Table 26.1. Use only the measurement uncertainties and compare your results with the estimates of Chow et al. (1992) in Table 26.3.

26.8$_D$ Using the results of Problem 26.7 as an initial guess, apply the effective variance method to the same problem. Discuss your results.

26.9$_D$ Verify the calculations of Section 26.1.2 applying the principal-component analysis method to the correlation matrix for elements measured in Whiteface Mountain, New York (Table 26.4).

REFERENCES

Ashbaugh, L. L., Myrup, L. O., and Flocchini, R. G. (1984) A principal component analysis of sulfur concentrations in the western United States, *Atmos. Environ.* **18**, 783–791.

Bencala, K., and Seinfeld, J. H. (1976) On frequency distributions of air pollutant concentrations, *Atmos. Environ.* **10**, 941–950.

Cheng, M. D., Lioy, P. J., and Opperman, A. J. (1988) Resolving PM_{10} data collected in New Jersey by various multivariate analysis techniques, in *PM_{10}: Implementation of Standards*, C. V. Mathai and D. H. Stonefield, eds., Air Pollution Control Association, Pittsburgh, PA, pp. 472–483.

Chow, J. C., Watson, J. G., Lowenthal, D. H., Solomon, P. A., Magliano, K. L., Ziman, S. D., and Richards, L. W. (1992) PM_{10} source apportionment in California's San Joaquin Valley, *Atmos. Environ.* **26A**, 3335–3354.

Chow, J. C., Watson, J. G., Lowenthal, D. H., Solomon, P. A., Magliano, K. L., Ziman, S. D., and Richards, L. W. (1993) PM_{10} and $PM_{2.5}$ compositions in California's San Joaquin Valley, *Aerosol Sci. Technol.* **18**, 105–128.

Cooper, J. A., and Watson, J. G. (1980) Receptor oriented methods for air particulate source apportionment, *J. Air Pollut. Control Assoc.* **30**, 1116–1125.

Curran, T. C., and Hunt, W. F. Jr. (1975) Interpretation of air quality data with respect to the national ambient air quality standards, *J. Air Pollut. Control Assoc.* **25**, 711–714.

Dattner, S. L., and Hopke, P. K. (1982) Receptor models applied to contemporary pollution problems: An introduction, in *Receptor Models Applied to Contemporary Pollution Problems*, E. R. Frederick, ed., APCA SP-48, Air Pollution Control Association, Pittsburgh, PA.

David, H. A. (1981) *Order Statistics*, 2nd ed., Wiley, New York.

Gebhart, K. A., Lattimer, D. A., and Sisler, J. F. (1990) Empirical orthogonal function analysis of the particulate sulfate concentrations measured during WHITEX, in *Visibility and Fine Particles*, C. V. Mathai, ed., AWMA TR-17, Air & Waste Management Association, Pittsburgh, PA, pp. 860–871.

Georgopoulos, P. G., and Seinfeld, J. H. (1982) Statistical distributions of air pollutant concentrations, *Environ. Sci. Technol.* **16**, 401A–416A.

Gerlach, R. W., Currie, L. A., and Lewis, C. W. (1983) Review of the Quail Roost II receptor model simulation exercise, in *Receptor Models Applied to Contemporary Pollution Problems*, S. L. Dattner and P. K. Hopke, eds., Air Pollution Control Association, Pittsburgh, PA, pp. 96–109.

Glover, D. M., Hopke, P. K., Vermette, S. J., Landsberger, S., and D'Auben, D. R. (1991) Source apportionment with site specific source profiles, *J. Air Waste Manag. Assoc.* **41**, 294–305.

Gordon, G. E. (1980) Receptor models, *Environ. Sci. Technol.* **14**, 792–800.

Gordon, G. E. (1988) Receptor models, *Environ. Sci. Technol.* **22**, 1132–1142.

Gumbel, E. J. (1958) *Statistics of Extremes*, Columbia Univ. Press, New York.

Henry, R. C. (1983) Stability analysis of receptor models that use least-squares fitting, in *Receptor Models Applied to Contemporary Pollution Problems*, S. L. Dattner and P. K. Hopke, eds., Air Pollution Control Association, Pittsburgh, PA, pp. 141–157.

Henry, R. C., and Hidy, G. M. (1979) Multivariate analysis of particulate sulfate and other air quality variables by principal components, I. Annual data from Los Angeles and New York, *Atmos. Environ.* **13**, 1581–1596.

Henry, R. C., and Hidy, G. M. (1981) Multivariate analysis of particulate sulfate and other air quality variables by principal components, II. Salt Lake City, Utah and St. Louis, Missouri, *Atmos. Environ.* **16**, 929–943.

Henry, R. C., and Kim, B. M. (1989) A factor analysis receptor model with explicit physical constraints, in *Receptor Models in Air Resources Management*, J. G. Watson, ed., APCA Transactions 14, Air Pollution Control Association, Pittsburgh, PA, pp. 214–225.

Henry, R. C., Lewis, C. W., Hopke, P. K., and Williamson, H. I. (1984) Review of receptor model fundamentals, *Atmos. Environ.* **18**, 1507–1515.

Henry, R. C., Wang, Y. J., and Gebhart, K. A. (1991) The relationship between empirical orthogonal functions and sources of air pollution, *Atmos. Environ.* **24A**, 503–509.

Holland, D. M., and Fitz-Simons, T. (1982) Fitting statistical distributions to air quality data by the maximum likelihood method, *Atmos. Environ.* **16**, 1071–1076.

Hopke, P. K. (1985) *Receptor Modeling in Environmental Chemistry*, Wiley, New York.

Hopke, P. K. (1991) *Receptor Modeling for Air Quality Management*, Elsevier, New York.

Javitz, H. S., Watson, J. G., Guertin, J. P., and Mueller, P. K. (1988) Results of a receptor modeling feasibility study, *J. Air Pollut. Control Assoc.* **38**, 661–667.

Kaiser, H. F. (1959) Computer program for varimax rotation in factor analysis, *Educ. Psychol. Meas.* **33**, 99–102.

Koutrakis, P., and Spengler, J. D. (1987) Source apportionment of ambient particles in Steubenville, OH using specific rotation factor analysis, *Atmos. Environ.* **21**, 1511–1519.

Kowalczyk, G. S., Gordon, G. E., and Rheingrover, S. W. (1982) Identification of atmospheric particulate sources in Washington, D.C. using chemical element balances, *Environ. Sci. Technol.* **16**, 79–90.

Larsen, R. I. (1971) *Mathematical Model for Relating Air Quality Measurements to Air Quality Standards*, EPA Publication A-89, U.S. Environmental Protection Agency, Research Triangle Park, NC.

Macias, E. S., and Hopke, P. K. (1981) *Atmospheric Aerosol: Source/Air Quality Relationships*, ACS Symposium Series, Vol. 167, American Chemical Society, Washington, DC.

Mage, D. T. (1980) The statistical form for the ambient particulate standard annual arithmetic mean versus geometric mean, *J. Air Pollut. Control Assoc.* **30**, 796–798.

McMullin, T. B. (1975) Interpreting the eight-hour national ambient air quality standard for carbon monoxide, *J. Air Pollut. Control Assoc.* **25**, 1009–1014.

Miller, M. S., Friedlander, S. K., and Hidy, G. M. (1972) A chemical element balance for the Pasadena aerosol, *J. Colloid Interface Sci.* **39**, 165–176.

Pace, T. G. (1986) *Receptor Methods for Source Apportionment: Real World Issues and Applications*, APCA Transactions TR-5, Air Pollution Control Association, Pittsburgh, PA.

Parekh, P. P., and Husain, L. (1981) Trace element concentrations in summer aerosols at rural sites in New York State and their possible sources, *Atmos. Environ.* **15**, 1717–1725.

Pearson, K. (1934) *Tables of the Incomplete Beta Function*, Biometrika Office, London.

Sarhan, A. E., and Greenburg, B. G. (1962) *Contributions to Order Statistics*, Wiley, New York.

Stevens, R. K., and Pace, T. G. (1984) Overview of the mathematical and empirical receptor models workshop (Quail Roost II), *Atmos. Environ.* **18**, 1499–1506.

Tsukatami, T., and Shigemitsu, K. (1980) Simplified Pearson distribution applied to an air pollutant concentration, *Atmos. Environ.* **14**, 245–253.

Watson, J. G. (1979) *Chemical Element Balance Receptor Model Methodology for Assessing the Source of Fine and Total Suspended Particulate Matter in Portland, Oregon*, Ph.D. thesis, Oregon Graduate Center, Beaverton.

Watson, J. G. (1984) Overview of receptor model principles, *J. Air Pollut. Control Assoc.* **34**, 619–623.

Watson, J. G., Chow, J. C., and Mathai, C. V. (1989) Receptor models in air resources management: A summary of the APCA international specialty conference, *J. Air Pollut. Control Assoc.* **39**, 419–426.

Watson, J. G., Henry, R. C., Cooper, J. A., and Macias, E. S. (1981) The state of the art of receptor models relating ambient suspended particulate matter to sources, in *Atmospheric Aerosol: Source/*

Air Quality Relationships, E. S. Macias and P. K. Hopke, eds., ACS Symposium Series, Vol. 167. American Chemical Society, Washington, DC, pp. 89–106.

Winchester, J. W., and Nifong, G. D. (1971) Water pollution in Lake Michigan by trace elements from pollution aerosol fallout, *Water Air Soil Pollut.* **1**, 50–64.

Wolff, G. T., and Korsog, P. E. (1985) Estimates of the contributions of sources to inhalable particulate concentrations in Detroit, *Atmos. Environ.* **19**, 1399–1409.

Wolff, G. T., Korsog, P. E., Stroup, D. P., Ruthkosky, M. S., and Morrissey, M. L. (1985) The influence of local and regional sources on the concentration of inhalable particulate matter in south-eastern Michigan, *Atmos. Environ.* **19**, 305–313.

Zeng, Y., and Hopke, P. K. (1989) Three-mode factor analysis: A new multivariate method for analyzing spatial and temporal composition variation, in *Receptor Models in Air Resources Management*, J. G. Watson, ed., APCA Transactions 14, Air Pollution Control Association, Pittsburgh, PA, pp. 173–189.

APPENDIX A
Units and Physical Constants

The units used in this text are more or less those of the International System of Units (SI). We have chosen not to adhere strictly to the SI system because in several areas of the book the use of SI units would lead to cumbersome and unfamiliar magnitudes of quantities. We have attempted to use, as much as possible, a consistent set of units throughout the book while attempting not to deviate markedly from the units commonly used in the particular area. For an excellent discussion of units in atmospheric chemistry we refer the reader to Schwartz and Warneck (1995).

A.1 SI BASE UNITS

Table A.1 gives the seven base quantities, assumed to be mutually independent, on which the SI is founded, and the names and symbols of their respective units, called "SI base units."

TABLE A.1   SI Base Units

Base Quantity	SI Base Unit Name	Symbol
Length	meter	m
Mass	kilogram	kg
Time	second	s
Electric current	ampere	A
Thermodynamic temperature	kelvin	K
Amount of substance	mole	mol
Luminous intensity	candela	cd

A.2 SI DERIVED UNITS

Derived units are expressed algebraically in terms of base units or other derived units (including the radian and steradian, which are the two supplementary units). For example, the derived unit for the derived quantity molar mass (mass divided by amount of substance) is the kilogram per mole, symbol $kg\,mol^{-1}$. Additional examples of derived units expressed in terms of SI base units are given in Table A.2.

Atmospheric Chemistry and Physics: From Air Pollution to Climate Change, Second Edition, by John H. Seinfeld and Spyros N. Pandis.   Copyright © 2006 John Wiley & Sons, Inc.

TABLE A.2   Examples of SI Derived Units Expressed in Terms of SI Base Units

	SI Derived Unit	
Derived Quantity	Name	Symbol
Area	square meter	m^2
Volume	cubic meter	m^3
Speed, velocity	meter per second	$m\,s^{-1}$
Acceleration	meter per second squared	$m\,s^{-2}$
Wavenumber	reciprocal meter	m^{-1}
Mass density (density)	kilogram per cubic meter	$kg\,m^{-3}$
Specific volume	cubic meter per kilogram	$m^3\,kg^{-1}$
Current density	ampere per square meter	$A\,m^{-2}$
Magnetic field strength	ampere per meter	$A\,m^{-1}$
Amount-of-substance concentration (concentration)	mole per cubic meter	$mol\,m^{-3}$
Luminance	candela per square meter	$cd\,m^{-2}$

Certain SI derived units have special names and symbols; these are given in Table A.3. Table A.4 presents examples of SI derived units expressed with the aid of SI derived units having special names and symbols. Table A.5 presents standard prefixes.

Concentration units are used in connection with chemical reaction rates and optical extinction. Traditional concentration units are $mol\,L^{-1}$ (the SI unit of the liter is the cubic decimeter, dm^3), $mol\,cm^{-3}$, or $mol\,m^{-3}$. Concentrations are also expressed as number

TABLE A.3   SI Derived Units with Special Names and Symbols, Including the Radian and Steradian

	SI Derived Unit			
Derived Quantity	Special Name	Special Symbol	Expression in Terms of Other SI Units	Expression in Terms of SI Base Units
Plane angle	radian	rad		$m\,m^{-1} = 1$
Solid angle	steradian	sr		$m^2\,m^{-2} = 1$
Frequency	hertz	Hz		s^{-1}
Force	newton	N		$m\,kg\,s^{-2}$
Pressure, stress	pascal	Pa	$N\,m^{-2}$	$m^{-1}\,kg\,s^{-2}$
Energy, work, quantity of heat	joule	J	$N\,m$	$m^2\,kg\,s^{-2}$
Power, radiant flux	watt	W	$J\,s^{-1}$	$m^2\,kg\,s^{-3}$
Electric charge, quantity of electricity	coulomb	C		$s\,A$
Electric potential, potential difference, electromotive force	volt	V	$W\,A^{-1}$	$m^2\,kg\,s^{-3}\,A^{-1}$
Capacitance	farad	F	$C\,V^{-1}$	$m^{-2}\,kg^{-1}\,s^4\,A^2$
Electric resistance	ohm	Ω	$V\,A^{-1}$	$m^2\,kg\,s^{-3}\,A^{-2}$
Electric conductance	siemens	S	$A\,V^{-1}$	$m^{-2}\,kg^{-1}\,s^3\,A^2$
Magnetic flux	weber	Wb	$V\,s$	$m^2\,kg\,s^{-2}\,A^{-1}$
Magnetic flux density	tesla	T	$Wb\,m^{-2}$	$kg\,s^{-2}\,A^{-1}$
Inductance	henry	H	$Wb\,A^{-1}$	$m^2\,kg\,s^{-2}\,A^{-2}$
Celsius temperature	degree Celsius	°C		K
Luminous flux	lumen	lm	$cd\,sr$	$cd\,sr$
Illuminance	lux	lx	$lm\,m^{-2}$	$m^{-2}\,cd\,sr$

TABLE A.4   Examples of SI Derived Units Expressed with the Aid of SI Derived Units Having Special Names and Symbols

Derived Quantity	SI Derived Unit		
	Name	Symbol	Expression in Terms of SI Base Units
Angular velocity	radian per second	$rad\,s^{-1}$	$m\,m^{-1}\,s^{-1} = s^{-1}$
Angular acceleration	radian per second squared	$rad\,s^{-2}$	$m\,m^{-1}\,s^{-2} = s^{-2}$
Dynamic viscosity	pascal second	$Pa\,s$	$m^{-1}\,kg\,s^{-1}$
Moment of force	newton meter	$N\,m$	$m^2\,kg\,s^{-2}$
Surface tension	newton per meter	$N\,m^{-1}$	$kg\,s^{-2}$
Heat flux density, irradiance	watt per square meter	$W\,m^{-2}$	$kg\,s^{-3}$
Radiant intensity	watt per steradian	$W\,sr^{-1}$	$m^2\,kg\,s^{-3}\,sr^{-1}$
Radiance	watt per square meter steradian	$W\,(m^2\,sr)^{-1}$	$kg\,s^{-3}\,sr^{-1}$
Heat capacity, entropy	joule per kelvin	$J\,K^{-1}$	$m^2\,kg\,s^{-2}\,K^{-1}$
Specific-heat capacity, specific entropy	joule per kilogram kelvin	$J\,(kg\,K)^{-1}$	$m^2\,s^{-2}\,K^{-1}$
Specific energy	joule per kilogram	$J\,kg^{-1}$	$m^2\,s^{-2}$
Thermal conductivity	watt per meter kelvin	$W\,(m\,K)^{-1}$	$m\,kg\,s^{-3}\,K^{-1}$
Energy density	joule per cubic meter	$J\,m^{-3}$	$m^{-1}\,kg\,s^{-2}$
Electric field strength	volt per meter	$V\,m^{-1}$	$m\,kg\,s^{-3}\,A^{-1}$
Electric charge density	coulomb per cubic meter	$C\,m^{-3}$	$m^{-3}\,s\,A$
Electric flux density	coulomb per square meter	$C\,m^{-2}$	$m^{-2}\,s\,A$
Permittivity	farad per meter	$F\,m^{-1}$	$m^{-3}\,kg^{-1}\,s^4\,A^2$
Permeability	henry per meter	$H\,m^{-1}$	$m\,kg\,s^{-2}\,A^{-2}$
Molar energy	joule per mole	$J\,mol^{-1}$	$m^2\,kg\,s^{-2}\,mol^{-1}$
Molar entropy, molar heat capacity	joule per mole kelvin	$J\,(mol\,K)^{-1}$	$m^2\,kg\,s^{-2}\,K^{-1}\,mol^{-1}$

TABLE A.5   Standard Prefixes

Factor	Prefix	Symbol
10^{-18}	atto	a
10^{-15}	femto	f
10^{-12}	pico	p
10^{-9}	nano	n
10^{-6}	micro	μ
10^{-3}	milli	m
10^{-2}	centi	c
10^{-1}	deci	d
10^{1}	deca	da
10^{2}	hecto	h
10^{3}	kilo	k
10^{6}	mega	M
10^{9}	giga	G
10^{12}	tera	T
10^{15}	peta	P
10^{18}	exa	E

concentration, molecules cm^{-3}. (In expressing number concentration, centimeter-based units are more widely used than meter-based units.)

A.3   FUNDAMENTAL PHYSICAL CONSTANTS

Fundamental physical constants used in this book are given in Table A.6.

TABLE A.6   Fundamental Physical Constants

Speed of light in vacuum	c	2.9979×10^8	$m\,s^{-1}$
Planck constant	h	6.626×10^{-34}	$J\,s$
Elementary charge	e	1.602×10^{-19}	C
Electron mass	m_e	9.109×10^{-31}	kg
Avogadro constant	N_A	6.022×10^{23}	mol^{-1}
Faraday constant	F	96485	$C\,mol^{-1}$
Molar gas constant	R	8.314	$J\,mol^{-1}\,K^{-1}$
Boltzmann constant	k	1.381×10^{-23}	$J\,K^{-1}$
Molar volume (ideal gas), RT/p	v_m	22.414×10^{-3}	$m^3\,mol^{-1}$
(at 273.15 K, 101.325×10^3 Pa)			
Stefan–Boltzmann constant	σ	5.671×10^{-8}	$W\,m^{-2}\,K^{-4}$
Standard acceleration of gravity	g	9.807	$m\,s^{-2}$

Source: Cohen and Taylor (1995).

A.4   PROPERTIES OF THE ATMOSPHERE AND WATER

Tables A.7–A.9 give properties of the atmosphere and water.

TABLE A.7   Properties of the Atmosphere[a]

Standard temperature	$T_0 = 0°C = 273.15\,K$
Standard pressure	$p_0 = 760\,mm\,Hg$
	$= 1013.25\,millibar\,(mbar)$
Standard gravity	$g_0 = 9.807\,m\,s^{-2}$
Air density	$\rho_0 = 1.29\,kg\,m^{-3}$
Molecular weight	$M_0 = 28.97\,g\,mol^{-1}$
Mean molecular mass	$= 4.81 \times 10^{-23}\,g$
Molecular root-mean-square velocity $(3RT_0/M_0)^{1/2}$	$= 4.85 \times 10^4\,cm\,s^{-1}$
Speed of sound $(\gamma RT_0/M_0)^{1/2}$	$= 3.31 \times 10^4\,cm\,s^{-1}$
Specific heats[b]	$\hat{c}_p = 1005\,J\,K^{-1}\,kg^{-1}$
	$\hat{c}_v = 717\,J\,K^{-1}\,kg^{-1}$
Ratio	$\hat{c}_p/\hat{c}_v = \gamma = 1.401$
Air molecules per cm^3	$N = 2.688 \times 10^{19}$ (at 273 K)
	$= 2.463 \times 10^{19}$ (at 298 K)
Air molecular diameter	$\sigma = 3.46 \times 10^{-8}\,cm$
Air mean free path $(\sqrt{2}\pi\,N\sigma^2)^{-1}$	$\lambda_a = 6.98 \times 10^{-6}\,cm$
Viscosity	$\mu = 1.72 \times 10^{-4}\,g\,cm^{-1}\,s^{-1}$
Thermal conductivity	$k = 2.40 \times 10^{-2}\,J\,m^{-1}\,s^{-1}\,K^{-1}$
Refractive index (real part)	$(n-1) \times 10^8 = 6.43 \times 10^3$
	$+\dfrac{2.95 \times 10^6}{146-\lambda^{-2}} + \dfrac{2.55 \times 10^4}{41-\lambda^{-2}}$ (λ in μm)

[a] Dry air at $T = 273\,K$ and 1 atm.

TABLE A.8   U.S. Standard Atmosphere, 1976

Height, km	Pressure, hPa	Temperature, K	Density, $g\,m^{-3}$	Water Vapor Moist Stratosphere, $g\,m^{-3}$	Water Vapor Dry Stratosphere, $g\,m^{-3}$
0	1.013×10^3	288	1.225×10^3	5.9	5.9
1	8.98×10^2	282	1.112×10^3	4.2	4.2
2	7.950×10^2	275	1.007×10^3	2.9	2.9
3	7.012×10^2	269	9.093×10^2	1.8	1.8
4	6.166×10^2	262	8.194×10^2	1.1	1.1
5	5.405×10^2	256	7.364×10^2	6.4×10^{-1}	6.4×10^{-1}
6	4.722×10^2	249	6.601×10^2	3.8×10^{-1}	3.8×10^{-1}
7	4.111×10^2	243	5.900×10^2	2.1×10^{-1}	2.1×10^{-1}
8	3.565×10^2	236	5.258×10^2	1.2×10^{-1}	1.2×10^{-1}
9	3.080×10^2	230	4.671×10^2	4.6×10^{-2}	4.6×10^{-2}
10	2.650×10^2	223	4.135×10^2	1.8×10^{-2}	1.8×10^{-2}
11	2.270×10^2	217	3.648×10^2	8.2×10^{-3}	8.2×10^{-3}
12	1.940×10^2	217	3.119×10^2	3.7×10^{-3}	3.7×10^{-3}
13	1.658×10^2	217	2.666×10^2	1.8×10^{-3}	1.8×10^{-3}
14	1.417×10^2	217	2.279×10^2	8.4×10^{-4}	8.4×10^{-4}
15	1.211×10^2	217	1.948×10^2	7.2×10^{-4}	7.2×10^{-4}
16	1.035×10^2	217	1.665×10^2	5.5×10^{-4}	3.3×10^{-4}
17	8.850×10^1	217	1.423×10^2	4.7×10^{-4}	2.8×10^{-4}
18	7.565×10^1	217	1.217×10^2	4.0×10^{-4}	2.4×10^{-4}
19	6.467×10^1	217	1.040×10^2	4.1×10^{-4}	2.1×10^{-4}
20	5.529×10^1	217	8.891×10^1	4.0×10^{-4}	1.8×10^{-4}
21	4.729×10^1	218	7.572×10^1	4.4×10^{-4}	1.5×10^{-4}
22	4.048×10^1	219	6.450×10^1	4.6×10^{-4}	1.3×10^{-4}
23	3.467×10^1	220	5.501×10^1	5.2×10^{-4}	1.1×10^{-4}
24	2.972×10^1	221	4.694×10^1	5.4×10^{-4}	9.4×10^{-5}
25	2.549×10^1	222	4.008×10^1	6.1×10^{-4}	8.0×10^{-5}
30	1.197×10^1	227	1.841×10^1	3.2×10^{-4}	3.7×10^{-5}
35	5.746	237	8.463	1.3×10^{-4}	1.7×10^{-5}
40	2.871	253	3.996	4.8×10^{-5}	7.9×10^{-6}
45	1.491	264	1.966	2.2×10^{-5}	3.9×10^{-6}
50	7.978×10^{-1}	271	1.027	7.8×10^{-6}	2.1×10^{-6}
70	5.220×10^{-2}	220	8.283×10^{-2}	1.2×10^{-7}	1.8×10^{-7}
100	3.008×10^{-4}	210	4.990×10^{-4}	3.0×10^{-4}	1.0×10^{-9}

TABLE A.9   Properties of Water

	Mass Basis	Molar Basis
Specific heat of water vapor at constant pressure, $\hat{c}_{pw}$	$1952\,J\,kg^{-1}K^{-1}$	$35.14\,J\,mol^{-1}\,K^{-1}$
Specific heat of water vapor at constant volume, $\hat{c}_{vw}$	$1463\,J\,kg^{-1}\,K^{-1}$	$26.33\,J\,mol^{-1}\,K^{-1}$
Specific heat of liquid H_2O at 273 K	$4218\,J\,kg^{-1}\,K^{-1}$	$75.92\,J\,mol^{-1}\,K^{-1}$
Latent heat of vaporization		
At 273 K	$2.5 \times 10^6\,J\,kg^{-1}$	$4.5 \times 10^4\,J\,mol^{-1}$
At 373 K	$2.25 \times 10^6\,J\,kg^{-1}$	$4.05 \times 10^4\,J\,mol^{-1}$
Latent heat of fusion, 273 K	$3.3 \times 10^5\,J\,kg^{-1}$	$5.94 \times 10^3\,J\,mol^{-1}$
Surface tension (water vs. air)	$0.073\,J\,m^{-2}$	

TABLE A.10   Typical Ranges of Values in Atmospheric Chemistry

System	Molecule, cm^3 Units	Mole, m^3 Units
Concentration	10^4–10^{19}	10^{-14}–10
unit	molecule cm^{-3}	mol m^{-3}
Bimolecular rate constant	10^{-18}–10^{-10}	1–10^8
unit	cm^3 molecule^{-1} s^{-1}	m^3 mol^{-1} s^{-1}
Termolecular rate constant	10^{-36}–10^{-29}	1–10^7
unit	cm^6 molecule^{-2} s^{-1}	m^6 mol^{-2} s^{-1}

Source: Schwartz and Warneck (1995).

A.5   UNITS FOR REPRESENTING CHEMICAL REACTIONS

The rate of a bimolecular reaction between substances A and B may be written as

$$\frac{dc_A}{dt} = -kc_A c_B$$

where c_A and c_B are the concentrations of A and B, respectively. The second-order rate constant k has units concentration^{-1} time^{-1}. Possible units for k are

$$cm^3 \text{ molecule}^{-1} s^{-1}$$
$$m^3 \text{ mol}^{-1} s^{-1}$$

Table A.10 compares ranges of concentrations and bimolecular and termolecular rate constants pertinent to atmospheric chemistry for, these two units. For example, the hydroxyl (OH) radical has an average tropospheric concentration of about 8×10^5 molecule cm^{-3}. This is equivalent to 1.3×10^{-12} mol m^{-3} (1.3 pmol m^{-3}).

A.6   CONCENTRATIONS IN THE AQUEOUS PHASE

Concentrations of substances dissolved in water droplets and present in particulate matter are of great importance in atmospheric chemistry. A commonly used unit of concentration in the chemical thermodynamics of solutions is *molality*, mole of solute per kilogram of solvent (mol kg^{-1}). One advantage of the use of molality is that the value is unaffected by changes in the density of solution as temperature changes.

For aqueous-phase chemical reactions the commonly used concentration unit is mol L^{-1}. Aqueous solutions in cloud and raindrops are characterized by concentrations in the range of μmol L^{-1}. For a dilute aqueous solution molality (mol kg^{-1}) is approximately equal to molarity (mol L^{-1}). The conversion between molar concentration c and molality m is

$$m = \frac{c}{\rho - Mc}$$

where ρ is the density of solution, kg m^{-3}; M is the molecular weight of solute, kg mol^{-1}; and c is concentration, mol m^{-3}.

A.7 SYMBOLS FOR CONCENTRATION

The following symbols for concentration are used in this book:

n_A Gas-phase number concentration of species A, molecule cm^{-3}

c_A or [A]   Gas-phase molar concentration of species A, $mol\, m^{-3}$

[A]   Aqueous-phase concentration of solute A, $mol\, L^{-1}$

m_A Molality of solute A, $mol\, kg^{-1}$

REFERENCES

Cohen, E. R., and Taylor, B. N. (1995) The fundamental physical constants, *Phys. Today* Aug., BG9-BG16.

Schwartz, S. E., and Warneck, P. (1995) Units for use in atmospheric chemistry, *Pure Appl. Chem.* **67**, 1377–1406.

APPENDIX B
Rate Constants of Atmospheric Chemical Reactions

See Tables B.1–B.10.

TABLE B.1 Second-Order Reactions: $k(T) = A \exp(-E/RT)$

Reaction	A, cm^3 molecule^{-1} s^{-1}	E/R, K	$k(298\text{ K})$, cm^3 molecule^{-1} s^{-1}
O_x and $O(^1D)$ reactions			
$O + O_3 \rightarrow O_2 + O_2$	8.0×10^{-12}	2060	8.0×10^{-15}
$O(^1D) + O_2 \rightarrow O + O_2$	3.2×10^{-11}	-70	4.0×10^{-11}
$O(^1D) + H_2O \rightarrow OH + OH$	2.2×10^{-10}	0	2.2×10^{-10}
$O(^1D) + N_2 \rightarrow O + N_2$	1.8×10^{-11}	-110	2.6×10^{-11}
$O(^1D) + N_2O \rightarrow N_2 + O_2$	4.9×10^{-11}	0	4.9×10^{-11}
$\rightarrow NO + NO$	6.7×10^{-11}	0	6.7×10^{-11}
$O(^1D) + CH_4 \rightarrow OH + CH_3$	1.5×10^{-10}	0	1.5×10^{-10}
$HO_x - O_x$ reactions			
$O + OH \rightarrow O_2 + H$	2.2×10^{-11}	-120	3.3×10^{-11}
$O + HO_2 \rightarrow OH + O_2$	3.0×10^{-11}	-200	5.9×10^{-11}
$OH + O_3 \rightarrow HO_2 + O_2$	1.7×10^{-12}	940	7.3×10^{-14}
$OH + HO_2 \rightarrow H_2O + O_2$	4.8×10^{-11}	-250	1.1×10^{-10}
$HO_2 + O_3 \rightarrow OH + 2\,O_2$	1.0×10^{-14}	490	1.9×10^{-15}
$HO_2 + HO_2 \rightarrow H_2O_2 + O_2$	2.3×10^{-13}	-600	1.7×10^{-12}
$\xrightarrow{M} H_2O_2 + O_2$	1.7×10^{-33} [M]	-1000	4.9×10^{-32} [M]
NO_x reactions			
$O + NO_2 \rightarrow NO + O_2$	5.6×10^{-12}	-180	1.0×10^{-11}
$O + NO_3 \rightarrow O_2 + NO_2$	1.0×10^{-11}	0	1.0×10^{-11}
$OH + NH_3 \rightarrow H_2O + NH_2$	1.7×10^{-12}	710	1.6×10^{-13}
$HO_2 + NO \rightarrow NO_2 + OH$	3.5×10^{-12}	-250	8.1×10^{-12}
$NO + O_3 \rightarrow NO_2 + O_2$	3.0×10^{-12}	1500	1.9×10^{-14}
$NO + NO_3 \rightarrow 2\,NO_2$	1.5×10^{-11}	-170	2.6×10^{-11}
$NO_2 + O_3 \rightarrow NO_3 + O_2$	1.2×10^{-13}	2450	3.2×10^{-17}
$OH + HNO_3 \rightarrow H_2O + NO_3$	See footnote[a]		

(*continued*)

Atmospheric Chemistry and Physics: From Air Pollution to Climate Change, Second Edition, by John H. Seinfeld and Spyros N. Pandis.   Copyright © 2006 John Wiley & Sons, Inc.

TABLE B.1   (*Continued*)

Reaction	A, cm^3 molecule^{-1} s^{-1}	E/R, K	k(298 K), cm^3 molecule^{-1} s^{-1}
HO$_x$ reactions			
OH + CO → CO$_2$ + H	1.5×10^{-13} $(1 + 0.6 p_{atm})$	0	1.5×10^{-13} $(1 + 0.6 p_{atm})$
OH + CH$_4$ → CH$_3$ + H$_2$O	2.45×10^{-12}	1775	6.3×10^{-15}
OH + HCN → products	1.2×10^{-13}	400	3.1×10^{-14}
OH + CH$_3$CN → products	7.8×10^{-13}	1050	2.3×10^{-14}
HO$_2$ + CH$_3$O$_2$ → CH$_3$OOH + O$_2$	4.1×10^{-13}	−750	5.2×10^{-12}
HCO + O$_2$ → CO + HO$_2$	5.2×10^{-12}		5.2×10^{-12}
CH$_2$OH + O$_2$ → HCHO + HO$_2$	9.1×10^{-12}	0	9.1×10^{-12}
CH$_3$O + O$_2$ → HCHO + HO$_2$	3.9×10^{-14}	900	1.9×10^{-15}
CH$_3$O$_2$ + CH$_3$O$_2$ → products	9.5×10^{-14}	−390	3.5×10^{-13}
CH$_3$O$_2$ + NO → CH$_3$O + NO$_2$	2.8×10^{-12}	−300	7.7×10^{-12}
C$_2$H$_5$O + O$_2$ → CH$_3$CHO + HO$_2$	6.3×10^{-14}	550	1.0×10^{-14}
C$_2$H$_5$O$_2$ + NO → products	8.7×10^{-12}	0	8.7×10^{-12}
CH$_3$C(O)O$_2$ + NO → CH$_3$C(O)O + NO$_2$	8.1×10^{-12}	−270	2.0×10^{-11}
FO$_x$ reactions			
OH + CH$_3$F → CH$_2$F + H$_2$O	2.5×10^{-12}	1430	2.1×10^{-14}
OH + CH$_2$F$_2$ → CHF$_2$ + H$_2$O	1.7×10^{-12}	1500	1.1×10^{-14}
OH + CHF$_3$ → CF$_3$ + H$_2$O	6.3×10^{-13}	2300	2.8×10^{-16}
OH + CH$_3$CH$_2$F → products	2.5×10^{-12}	730	2.2×10^{-13}
OH + CH$_3$CHF$_2$ → products	9.4×10^{-13}	990	3.4×10^{-14}
OH + CH$_2$FCH$_2$F → CHFCH$_2$F + H$_2$O	1.1×10^{-12}	730	9.7×10^{-14}
OH + CH$_3$CF$_3$ → CH$_2$CF$_3$ + H$_2$O	1.1×10^{-12}	2010	1.3×10^{-15}
OH + CH$_2$FCHF$_2$ → products	3.9×10^{-12}	1620	1.7×10^{-14}
OH + CH$_2$FCF$_3$ → CHFCF$_3$ + H$_2$O	1.05×10^{-12}	1630	4.4×10^{-15}
OH + CHF$_2$CHF$_2$ → CF$_2$CHF$_2$ + H$_2$O	1.6×10^{-12}	1660	6.1×10^{-15}
OH + CHF$_2$CF$_3$ → CF$_2$CF$_3$ + H$_2$O	6.0×10^{-13}	1700	2.0×10^{-15}
OH + CH$_2$FCF$_2$CHF$_2$ → products	2.1×10^{-12}	1620	9.2×10^{-15}
OH + CF$_3$CF$_2$CH$_2$F → CF$_3$CF$_2$CHF + H$_2$O	1.3×10^{-12}	1700	4.4×10^{-15}
OH + CF$_3$CHFCHF$_2$ → products	9.4×10^{-13}	1550	5.2×10^{-15}
OH + CF$_3$CH$_2$CF$_3$ → CF$_3$CHCF$_3$ + H$_2$O	1.45×10^{-12}	2500	3.3×10^{-16}
OH + CF$_3$CHFCF$_3$ → CF$_3$CFCF$_3$ + H$_2$O	4.3×10^{-13}	1650	1.7×10^{-15}
ClO$_x$ reactions			
O + ClO → Cl + O$_2$	3.0×10^{-11}	−70	3.8×10^{-11}
O + OClO → ClO + O$_2$	2.4×10^{-12}	960	1.0×10^{-13}
OH + HCl → H$_2$O + Cl	2.6×10^{-12}	350	8.0×10^{-13}
OH + HOCl → H$_2$O + ClO	3.0×10^{-12}	500	5.0×10^{-13}
OH + ClNO$_2$ → HOCl + NO$_2$	2.4×10^{-12}	1250	3.6×10^{-14}
OH + CH$_3$Cl → CH$_2$Cl + H$_2$O	2.4×10^{-12}	1250	3.6×10^{-14}
OH + CH$_2$Cl$_2$ → CHCl$_2$ + H$_2$O	1.9×10^{-12}	870	1.0×10^{-13}
OH + CHCl$_3$ → CCl$_3$ + H$_2$O	2.2×10^{-12}	920	1.0×10^{-13}
OH + CH$_2$ClF → CHClF + H$_2$O	2.4×10^{-12}	1210	4.1×10^{-14}
OH + CHFCl$_2$ → CFCl$_2$ + H$_2$O	1.2×10^{-12}	1100	3.0×10^{-14}
OH + CHF$_2$Cl → CF$_2$Cl + H$_2$O	1.05×10^{-12}	1600	4.8×10^{-15}
OH + CH$_3$CCl$_3$ → CH$_2$CCl$_3$ + H$_2$O	1.6×10^{-12}	1520	1.0×10^{-14}
OH + C$_2$HCl$_3$ → products	8.0×10^{-13}	−300	2.2×10^{-12}
OH + CH$_3$CFCl$_2$ → CH$_2$CFCl$_2$ + H$_2$O	1.25×10^{-12}	1600	5.8×10^{-15}

(*continued*)

TABLE B.1   (*Continued*)

Reaction	A, cm^3 molecule^{-1} s^{-1}	E/R, K	k(298 K), cm^3 molecule^{-1} s^{-1}
OH + CH$_3$CF$_2$Cl → CH$_2$CF$_2$Cl + H$_2$O	1.3×10^{-12}	1770	3.4×10^{-15}
OH + CH$_2$ClCF$_2$Cl → CHClCF$_2$Cl + H$_2$O	3.6×10^{-12}	1600	1.7×10^{-14}
OH + CH$_2$ClCF$_3$ → CHClCF$_3$ + H$_2$O	5.6×10^{-13}	1100	1.4×10^{-14}
OH + CHCl$_2$CF$_3$ → CCl$_2$CF$_3$ + H$_2$O	6.3×10^{-13}	850	3.6×10^{-14}
OH + CHFClCF$_3$ → CFClCF$_3$ + H$_2$O	7.1×10^{-13}	1300	9.0×10^{-15}
OH + CH$_3$CF$_2$CFCl$_2$ → products	7.7×10^{-13}	1720	2.4×10^{-15}
OH + CF$_3$CF$_2$CHCl$_2$ → products	6.3×10^{-13}	960	2.5×10^{-14}
OH + CF$_2$ClCF$_2$CHFCl → products	5.5×10^{-13}	1230	8.9×10^{-15}
HO$_2$ + ClO → HOCl + O$_2$	2.7×10^{-12}	− 220	5.6×10^{-12}
Cl + O$_3$ → ClO + O$_2$	2.3×10^{-11}	200	1.2×10^{-11}
Cl + CH$_4$ → HCl + CH$_3$	9.6×10^{-12}	1360	1.0×10^{-13}
Cl + OClO → ClO + ClO	3.4×10^{-11}	− 160	5.8×10^{-11}
Cl + ClOO → Cl$_2$ + O$_2$	2.3×10^{-10}	0	2.3×10^{-10}
→ ClO + ClO	1.2×10^{-11}	0	1.2×10^{-11}
ClO + NO → NO$_2$ + Cl	6.4×10^{-12}	− 290	1.7×10^{-11}
ClO + ClO → Cl$_2$ + O$_2$	1.0×10^{-12}	1590	4.8×10^{-15}
→ ClOO + Cl	3.0×10^{-11}	2450	8.0×10^{-15}
→ OClO + Cl	3.5×10^{-13}	1370	3.5×10^{-15}
BrO$_x$ reactions			
O + BrO → Br + O$_2$	1.9×10^{-11}	− 230	4.1×10^{-11}
Br + O$_3$ → BrO + O$_2$	1.7×10^{-11}	800	1.2×10^{-12}
BrO + ClO → Br + OClO	9.5×10^{-13}	− 550	6.0×10^{-12}
→ Br + ClOO	2.3×10^{-12}	− 260	5.5×10^{-12}
→ BrCl + O$_2$	4.1×10^{-13}	− 290	1.1×10^{-12}
BrO + BrO → 2 Br + O$_2$	1.5×10^{-12}	− 230	3.2×10^{-12}
S reactions			
OH + H$_2$S → SH + H$_2$O	6.0×10^{-12}	75	4.7×10^{-12}
O + OCS → CO + SO	2.1×10^{-11}	2200	1.3×10^{-14}
OH + OCS → CO$_2$ + HS	1.1×10^{-13}	1200	1.9×10^{-15}
OH + CH$_3$SH → products	9.9×10^{-12}	− 360	3.3×10^{-11}
OH + CH$_3$SCH$_3$ → CH$_2$SCH$_3$ + H$_2$O^b	1.2×10^{-11}	260	5.0×10^{-12}
NO$_3$ + CH$_3$SCH$_3$ → CH$_3$SCH$_2$ + HNO$_3$	1.9×10^{-13}	− 500	1.0×10^{-12}
HOSO$_2$ + O$_2$ → HO$_2$ + SO$_3$	1.3×10^{-12}	330	4.4×10^{-13}

^{a}The following apply:

$$k(T) = k_0 \frac{k_3[M]}{1 + k_3[M]/k_2}$$

$$k_0 = 7.2 \times 10^{-15} \exp(785/T)$$

$$k_2 = 4.1 \times 10^{-16} \exp(1440/T)$$

$$k_3 = 1.9 \times 10^{-33} \exp(725/T)$$

^{b}The rate constant given is for the DSM–OH abstraction path.

Source: Sander et al. (2003).

TABLE B.2   Association Reactions

Reaction	Low Pressure Limit[a] $k_0(T) = k_0^{300}(T/300)^{-n}$, cm^6 molecule^{-2} s^{-1}		High Pressure Limit[a] $k_\infty(T) = k_\infty^{300}(T/300)^{-m}$, cm^3 molecule^{-1} s^{-1}	
	k_0^{300}	n	k_∞^{300}	m
$O + O_2 + M \rightarrow O_3 + M$	6.0×10^{-34}	2.3	—	—
$O + NO + M \rightarrow NO_2 + M$	9.0×10^{-31}	1.5	3.0×10^{-11}	0
$O + NO_2 + M \rightarrow NO_3 + M$	2.5×10^{-31}	1.8	2.2×10^{-11}	0.7
$OH + NO + M \rightarrow HONO + M$	7.0×10^{-31}	2.6	3.6×10^{-11}	0.1
$OH + NO_2 + M \rightarrow HNO_3 + M$	2.0×10^{-30}	3.0	2.5×10^{-11}	0
$NO_2 + NO_3 + M \rightarrow N_2O_5 + M$	2.0×10^{-30}	4.4	1.4×10^{-12}	0.7
$CH_3 + O_2 + M \rightarrow CH_3O_2 + M$	4.5×10^{-31}	3.0	1.8×10^{-12}	1.7
$CH_3O + NO + M \rightarrow CH_3ONO + M$	1.4×10^{-29}	3.8	3.6×10^{-11}	0.6
$CH_3O + NO_2 + M \rightarrow CH_3ONO_2 + M$	5.3×10^{-29}	4.4	1.9×10^{-11}	1.8
$CH_3O_2 + NO_2 + M \rightarrow CH_3O_2NO_2 + M$	1.5×10^{-30}	4.0	6.5×10^{-12}	2.0
$CH_3C(O)O_2 + NO_2 + M \rightarrow$ $\quad CH_3C(O)O_2NO_2 + M$	9.7×10^{-29}	5.6	9.3×10^{-12}	1.5
$ClO + NO_2 + M \rightarrow ClONO_2 + M$	1.8×10^{-31}	3.4	1.5×10^{-11}	1.9
$ClO + ClO + M \rightarrow Cl_2O_2 + M$	1.6×10^{-32}	4.5	2.0×10^{-12}	2.4
$BrO + NO_2 + M \rightarrow BrONO_2 + M$	5.2×10^{-31}	3.2	6.9×10^{-12}	2.9
$OH + SO_2 + M \rightarrow HOSO_2 + M$	3.0×10^{-31}	3.3	1.5×10^{-12}	0

[a]Rate constants for association reactions of the type

$$A + B \rightleftharpoons [AB]^* \xrightarrow{M} AB$$

are given in the form

$$k_0(T) = k_0^{300} \left(\frac{T}{300}\right)^{-n} \text{ cm}^6 \text{ molecule}^{-2} \text{ s}^{-1}$$

where k_0^{300} accounts for air as the third body. Where pressure falloff corrections are necessary, the limiting high-pressure rate constant is given by

$$k_\infty(T) = k_\infty^{300} \left(\frac{T}{300}\right)^{-m} \text{ cm}^3 \text{ molecule}^{-1} \text{ s}^{-1}$$

To obtain the effective second-order rate constant at a given temperature and pressure (altitude z) the following formula is used:

$$k(T, z) = \left\{ \frac{k_0(T)[M]}{1 + (k_0(T)[M]/k_\infty(T))} \right\} 0.6^{\{1 + [\log_{10}(k_0(T)[M]/k_\infty(T))]^2\}^{-1}}$$

Source: Sander et al. (2003).

TABLE B.3   Alkane–OH Rate Constants: $k(T) = CT^2 \exp(-E/RT)$

Alkane	$k(298\ \mathrm{K}) \times 10^{12}$, cm^3 molecule^{-1} s^{-1}	$C \times 10^{18}$, cm^3 molecule^{-1} s^{-1}	E/R, K
Ethane	0.254	15.2	498
Propane	1.12	15.5	61
n-Butane	2.44	16.9	-145
n-Pentane	4.00	24.4	-183
2-Methylbutane	3.7		
n-Hexane	5.45	15.3	-414
2-Methylpentane	5.3		
2,3-Dimethylbutane	5.78	12.4	-494
n-Heptane	7.02	15.9	-478
2,2-Dimethylpentane	3.4		
2,4-Dimethylpentane	5.0		
2,2,3-Trimethylbutane	4.24	8.46	-516
Methylcyclohexane	10.0		

Source: Atkinson (1997).

TABLE B.4   Alkene–OH Rate Constants at 760 torr Total Pressure of Air[a]**:** $k(T) = A \exp(-E/RT)$

Alkene	$k(298\ \mathrm{K}) \times 10^{12}$, cm^3 molecule^{-1} s^{-1}	$A \times 10^{12}$, cm^3 molecule^{-1} s^{-1}	E/R, K
Ethene[b]	8.52	1.96	-438
Propene[c]	26.3	4.85	-504
1-Butene	31.4	6.55	-467
cis-2-Butene	56.4	11.0	-487
trans-2-Butene	64.0	10.1	-550
2-Methylpropene	51.4	9.47	-504
1-Pentene	31.4		
3-Methyl-1-butene	31.8	5.32	-533
2-Methyl-1-butene	61		
2-Methyl-2-butene	86.9		
1-Hexene	37		
2,3-Dimethyl-2-butene	110		
1,3-Butadiene	66.6	14.8	-448
Cyclohexene	67.7		

[a]Except for ethene and propene, these are the high pressure rate constants k_∞.
[b]$k_\infty = 9.0 \times 10^{-12}\ (T/298)^{-1.1}$
[c]$k_\infty = 2.8 \times 10^{-11}\ (T/298)^{-1.3}$

Source: Atkinson (1997).

TABLE B.5 Alkene–O_3 Rate Constants: $k(T) = A \exp(-E/RT)$

Alkene	$k(298\ \text{K}) \times 10^{18}$, cm^3 molecule^{-1} s^{-1}	$A \times 10^{15}$, cm^3 molecule^{-1} s^{-1}	E/R, K
Ethene	1.59	9.14	2580
Propene	10.1	5.51	1878
1-Butene	9.64	3.36	1744
cis-2-Butene	125	3.22	968
trans-2-Butene	190	6.64	1059
1-Pentene	10.0		
2-Methyl-2-butene	403	6.51	829
1-Hexene	11.0		
2,3-Dimethyl-2-butene	1130	3.03	294
1,3-Butadiene	6.3	13.4	2283
Cyclohexene	81.4	2.88	1063

Source: Atkinson (1997).

TABLE B.6 Alkene–NO_3 Rate Constants: $k(T) = A \exp(-E/RT)$

Alkene	$k(298\ \text{K})$, cm^3 molecule^{-1} s^{-1}	A, cm^{-3} molecule^{-1}s^{-1}	E/R, K
Ethene[a]	2.05×10^{-16}		
Propene	9.49×10^{-15}	4.59×10^{-13}	1156
1-Butene	1.35×10^{-14}	3.14×10^{-13}	938
cis-2-Butene	3.50×10^{-13}		
trans-2-Butene[b]	3.90×10^{-13}		
2-Methyl-2-butene	9.37×10^{-12}		
2,3-Dimethyl-2-butene	5.72×10^{-11}		
1,3-Butadiene	1.0×10^{-13}		
Cyclohexene	5.9×10^{-13}	1.05×10^{-12}	174

[a]$k = 4.88 \times 10^{-18} T^2 \exp(-2282/T)$ over the range 290–523 K.
[b]$k = 1.22 \times 10^{-18} T^2 \exp(382/T)$ over the range 204–378 K.
Source: Atkinson (1997).

TABLE B.7 Aromatic–OH Rate Constants at 298 K: $k_{ab} =$ Abstraction Path and $k_{ad} =$ Addition Path

Aromatic	$k(298\ \text{K}) \times 10^{12}$ (cm^3 molecule^{-1} s^{-1})	$\dfrac{k_{ab}}{k_{ab} + k_{ad}}$
Benzene	1.22[a]	0.05
Toluene	5.63[b]	0.12
Ethylbenzene	7.0	
o-Xylene	13.6	0.10
m-Xylene	23.1	0.04
p-Xylene	14.3	0.08
o-Ethyltoluene	11.9	
m-Ethyltoluene	18.6	
p-Ethyltoluene	11.8	
1,2,3-Trimethylbenzene	32.7	0.06

(*continued*)

TABLE B.7 (*Continued*)

Aromatic	$k(298\,\mathrm{K}) \times 10^{12}$ $(\mathrm{cm}^3\ \mathrm{molecule}^{-1}\,\mathrm{s}^{-1})$	$\dfrac{k_{ab}}{k_{ab} + k_{ad}}$
1,2,4-Trimethylbenzene	32.5	0.06
1,3,5-Trimethylbenzene	56.7	0.03
Styrene	58	
Benzaldehyde	12.9	
Phenol	27	
o-Cresol	41	
m-Cresol	68	
p-Cresol	50	

$^a k(T) = 2.33 \times 10^{-12} \exp(193/T)$.
$^b k(T) = 1.81 \times 10^{-12} \exp(338/T)$.
Source: Calvert et al. (2002).

TABLE B.8 **Oxygenated Organic–OH Rate Constants:** $k(T) = CT^n \exp(-E/RT)$

Organic	$k(298\,\mathrm{K}) \times 10^{12}$, $\mathrm{cm}^3\ \mathrm{molecule}^{-1}\,\mathrm{s}^{-1}$	C, $\mathrm{cm}^3\ \mathrm{molecule}^{-1}\,\mathrm{s}^{-1}$	n	E/R, K
Aldehydes				
HCHO	9.37	1.20×10^{-14}	1	-287
CH_3CHO	15.8	5.55×10^{-12}	0	-311
CH_3CH_2CHO	19.6			
$HOCH_2CHO$	9.9			
Ketones				
$CH_3C(O)CH_3$	0.219	5.34×10^{-18}	2	230
$CH_3C(O)CH_2CH_3$	1.15	3.24×10^{-18}	2	-414
α-Dicarbonyls				
$(CHO)_2$	11.4			
$CH_3C(O)CHO$	17.2			
$CH_3C(O)C(O)CH_3$	0.238	1.40×10^{-18}	2	-194
Alcohols				
CH_3OH	0.944	6.01×10^{-18}	2	-170
CH_3CH_2OH	3.27	6.18×10^{-18}	2	-532
Ethers				
CH_3OCH_3	2.98	1.04×10^{-11}	0	372
$C_2H_5OC_2H_5$	13.1	8.91×10^{-18}	2	-837
Carboxylic acids				
HCOOH	0.45	4.5×10^{-13}	0	0
CH_3COOH	0.8			
Hydroperoxides				
CH_3OOH	5.54	2.93×10^{-12}	0	-190
Unsaturated carbonyls				
$CH_2{=}CHCHO$	19.9			
$CH_2{=}C(CH_3)CHO$	33.5	1.86×10^{-11}	0	-175
$CH_3CH{=}CHCHO$	36			
$CH_2{=}CHC(O)CH_3$	18.8	4.13×10^{-12}	0	-452

Source: Atkinson (1997).

TABLE B.9   Biogenic Hydrocarbon–OH Rate Constants: $k(T) = A \exp(-E/RT)$

Biogenic	$k(298 \text{ K}) \times 10^{12}$, cm^3 molecule^{-1} s^{-1}	$A \times 10^{12}$, cm^3 molecule^{-1} s^{-1}	E/R, K
Isoprene (2-methyl-1,3-butadiene)	101	25.4	− 410
α-Pinene	53.7	12.1	− 444
β-Pinene	78.9	23.8	− 357
Myrcene	215		
Ocimene (*cis-* and *trans-*)	252		
Camphene	53		
2-Carene	80		
3-Carene	88		
Limonene	171		
α-Phellandrene	313		
β-Phellandrene	168		
Sabinene	117		
α-Terpinene	363		
γ-Terpinene	177		
Terpinolene	225		

Source: Atkinson (1997).

TABLE B.10   Biogenic Hydrocarbon–O$_3$ and –NO$_3$ Rate Constants: $k(T) = A \exp(-E/RT)$

Biogenic	$k_{O_3}(298 \text{ K}) \times 10^{18}$, cm^3 molecule^{-1} s^{-1}	$k_{NO_3}(298 \text{ K})$, cm^3 molecule^{-1} s^{-1}
Isoprene (2-methyl-1,3-butadiene)	12.8[a]	6.78×10^{-13} [c]
α-Pinene	86.6[b]	6.16×10^{-12} [d]
β-Pinene	15	2.51×10^{-12}
Myrcene	470	1.1×10^{-11}
Ocimene (*cis-* and *trans-*)	540	2.2×10^{-11}
Camphene	0.90	6.6×10^{-13}
2-Carene	230	1.9×10^{-11}
3-Carene	37	9.1×10^{-12}
Limonene	200	1.22×10^{-11}
α-Phellandrene	2980	8.5×10^{-11}
β-Phellandrene	47	8.0×10^{-12}
Sabinene	86	1.0×10^{-11}
α-Terpinene	21100	1.4×10^{-10}
γ-Terpinene	140	2.9×10^{-11}
Terpinolene	1880	9.7×10^{-11}

[a] $k(T) = 7.86 \times 10^{-15} \exp(-1913/T)$.
[b] $k(T) = 1.01 \times 10^{-15} \exp(-732/T)$.
[c] $k(T) = 3.03 \times 10^{-12} \exp(-446/T)$.
[d] $k(T) = 1.19 \times 10^{-12} \exp(490/T)$.

Source: Atkinson (1997).

REFERENCES

Atkinson, R. (1994) Gas-phase tropospheric chemistry of organic compounds, *J. Phys. Chem. Ref. Data* (monograph 2), 1–216.

Atkinson, R. (1997) Gas-phase tropospheric chemistry of volatile organic compounds: 1. Alkanes and alkenes, *J. Phys. Chem. Ref. Data* **26** (2), 215–290.

Calvert, J. G., Atkinson, R., Becker, K. H., Kamens, R. M., Seinfeld, J. H., Wallington, T. J., and Yarwood, G. (2002) *The Mechanisms of Atmospheric Oxidation of Aromatic Hydrocarbons*, Oxford Univ. Press, Oxford, UK.

Sander, S. P., Golden, D. M., Kurylo, M. J., Huie, R. E., Orkin, V. L., Moortgat, G. K., Ravishankara, A. R., Kolb, C. E., Molina, M. J., and Finlayson-Pitts, B. J. (2003), *Chemical Kinetics and Photochemical Data for Use in Atmospheric Studies*, Evaluation 14, Jet Propulsion Laboratory, Pasadena, CA (available at http://jpldataeval.jpl.nasa.gov/).

INDEX